D1072753

PRINTED CIRCUITS HANDBOOK

Other McGraw-Hill Reference Books of Interest

Handbooks

Avallone and Baumeister · STANDARD HANDBOOK FOR MECHANICAL ENGINEERS
Coombs · BASIC ELECTRONIC INSTRUMENT HANDBOOK
Croft and Summers · AMERICAN ELECTRICIANS' HANDBOOK
Di Giacomo · VLSI HANDBOOK
Fink and Beaty · STANDARD HANDBOOK FOR ELECTRICAL ENGINEERS
Fink and Christiansen · ELECTRONICS ENGINEERS' HANDBOOK
Harper · HANDBOOK OF COMPONENTS FOR ELECTRONICS
Harper · HANDBOOK OF ELECTRONIC SYSTEMS DESIGN
Harper · HANDBOOK OF THICK FILM HYBRID MICROELECTRONICS
Harper · HANDBOOK OF WIRING, CABLING, AND INTERCONNECTING FOR ELECTRONICS
Hecht · THE LASER GUIDEBOOK
Hicks · STANDARD HANDBOOK OF ENGINEERING CALCULATIONS
Inglis · ELECTRONIC COMMUNICATIONS HANDBOOK
Johnson and Jasik · ANTENNA ENGINEERING HANDBOOK
Juran · QUALITY CONTROL HANDBOOK
Kaufman and Seidman · HANDBOOK OF ELECTRONICS CALCULATIONS
Kaufman and Seidman · HANDBOOK FOR ELECTRONICS ENGINEERING TECHNICIANS
Kurtz · HANDBOOK OF ENGINEERING ECONOMICS
Perry · ENGINEERING MANUAL
Stout · HANDBOOK OF MICROPROCESSOR DESIGN AND APPLICATIONS
Stout and Kaufman · HANDBOOK OF MICROCIRCUIT DESIGN AND APPLICATION
Stout and Kaufman · HANDBOOK OF OPERATIONAL AMPLIFIER CIRCUIT DESIGN
Tuma · ENGINEERING MATHEMATICS HANDBOOK
Williams · DESIGNER'S HANDBOOK OF INTEGRATED CIRCUITS
Williams and Taylor · ELECTRONIC FILTER DESIGN HANDBOOK

Encyclopedias

CONCISE ENCYCLOPEDIA OF SCIENCE AND TECHNOLOGY
ENCYCLOPEDIA OF ELECTRONICS AND COMPUTERS
ENCYCLOPEDIA OF ENGINEERING

Dictionaries

DICTIONARY OF SCIENTIFIC AND TECHNICAL TERMS
DICTIONARY OF COMPUTERS
DICTIONARY OF ELECTRICAL AND ELECTRONIC ENGINEERING
DICTIONARY OF ENGINEERING

PRINTED CIRCUITS HANDBOOK

Clyde F. Coombs, Jr.
Editor-in-Chief
Hewlett-Packard Company
Palo Alto, California

Third Edition

McGRAW-HILL BOOK COMPANY

New York St. Louis San Francisco Auckland
Bogotá Hamburg London Madrid Mexico
Milan Montreal New Delhi Panama
Paris São Paulo Singapore
Sydney Tokyo Toronto

Library of Congress Cataloging-in-Publication Data

Printed circuits handbook / Clyde F. Coombs, Jr., editor-in-chief.—
3rd ed.

p. cm.
Includes index.
ISBN 0-07-012609-7 : $59.50
1. Printed circuits—Handbooks, manuals, etc. I. Coombs, Clyde
F.
TK7868.P7P76 1988
621.381'74—dc19 87-20985
CIP

1234567890 DOC/DOC 893210987

ISBN 0-07-012609-7

The editors for this book were Daniel A. Gonneau and Laura Givner, the designer was Mark E. Safran, and the production supervisor was Thomas G. Kowalczyk. It was set in Times Roman by University Graphics, Inc.

Printed and bound by R. R. Donnelley & Sons Company.

To Ann

CONTENTS

PART 2 ENGINEERING

CONTRIBUTORS

Andrade, A. D. *Sandia Corporation* (CHAP. 29, ACCEPTABILITY OF FABRICATED CIRCUITS)

Block, J. P. *Laminate Corporation of America* (CHAP. 10, DRILLING AND MACHINING)

Brooks, C. T. *Westinghouse Corporation* (CHAP. 7, SURFACE MOUNT TECHNOLOGY MATERIALS)

Cole, H. *Consultant* (CHAP. 23, DESIGNING AND PRESOLDERING CONSIDERATIONS FOR SOLDERING, *contributing author: J. Langan, consultant;* CHAP. 24, SOLDER MATERIALS AND PROCESSES; CHAP. 25, CLEANING OF SOLDERED BOARDS; CHAP. 26, QUALITY CONTROL IN SOLDERING)

Duffek, E. F. *Adion Engineering Co.* (CHAP. 12, PLATING; CHAP. 14, ETCHING)

Gause, S. A. *Westinghouse Corporation* (CHAP. 6, BASE MATERIALS)

Ginsberg, G. *Component Data Associates* (CHAP. 4, ENGINEERING PACKAGING INTERCONNECTION SYSTEM; CHAP. 8, CIRCUIT COMPONENTS AND HARDWARE; CHAP. 9, ACQUISITION AND SPECIFICATIONS)

Gurley, S. *Semflex* (CHAP. 35, FLEXIBLE CIRCUIT DESIGN, MATERIALS, AND FABRICATION)

Henningsen, C. G. *Insulectro* (CHAP. 6, BASE MATERIALS; CHAP. 10, DRILLING AND MACHINING)

Hinch, S. W. *Hewlett-Packard Company* (CHAP. 3, SURFACE MOUNT TECHNOLOGY)

Hohl, J. S. *Consultant* (CHAP. 19, COMPONENT ASSEMBLY)

Hroundas, G. *Trace Corp.* (CHAP. 17, BARE BOARD TESTING)

Hymes, L. C. *General Electric Co.* (CHAP. 27, SURFACE MOUNT SOLDER PROCESS)

Joslin, W. L. *Consultant* (CHAP. 18, MANUAL ASSEMBLY)

Konsowski, S. G. *Westinghouse Corporation* (CHAP. 7, SURFACE MOUNT TECHNOLOGY MATERIALS)

Langley, F. J. *GenRad* (CHAP. 21, TESTING IN ASSEMBLY)

Marx, P. C. *Fortin Industries* (CHAP. 32, MULTILAYER MATERIALS)

Mikschl, J. J. *Unisys* (CHAP. 31, DESIGNING FOR MULTILAYER)

Moleux, P. G. *Baker Brothers/Systems* (CHAP. 15, POLLUTION CONTROL AND RECOVERY SYSTEMS)

Nakahara, H. *PCK Technology* (CHAP. 2, TYPES OF PRINTED WIRING BOARDS; CHAP. 13, ADDITIVE PLATING)

Osborne, A. G. *Osborne Technical Associates* (CHAP. 33, FABRICATING MULTILAYER CIRCUITS; CHAP. 34, LAMINATING MULTILAYER CIRCUITS)

Schnorr, D. P. *General Electric Co.* (CHAP. 22, ASSEMBLY REPAIR)

Smith, G. A. *Trace Laboratories* (CHAP. 28, QUALITY ASSURANCE)

Wallig, L. R. *E. I. du Pont de Nemours & Company* (CHAP. 11, IMAGE TRANSFER; CHAP. 16, SOLDER RESIST)

Waryold, J. *Humiseal Division, Columbia Chase Corporation* (CHAP. 20, CONFORMAL COATINGS)

Williamson, M. A. *Westinghouse Corporation* (CHAP. 7, SURFACE MOUNT TECHNOLOGY MATERIALS)

Yager, T. *Hewlett-Packard Company* (CHAP. 30, RELIABILITY)

PREFACE

The "printed circuit" (or printed wiring) process continues to be the basic interconnection technique for electronic devices. Virtually every packaging system is based upon this process and undoubtedly will continue to be in the foreseeable future. However, this technology also continues to grow and change, just as the industries which it serves continue to grow and change. Of the several changes in the printed wiring field that have occurred since the publication of the second edition of this book, undoubtedly the greatest impact has come from the development of surface mount technology (SMT). SMT has emerged as a major, and pervasive, interconnection and packaging technology, introducing almost revolutionary change in all aspects of the electronic packaging field.

At the same time, strong evolutionary factors in the continuing growth of the more traditional printed wiring technology have pushed this process further into the classification of "high technology."

All of these changes, some revolutionary and some evolutionary, have combined to increase requirements for denser geometries, stricter process control, greater reliability, faster design cycles, lower total cost, and alternative materials and components. For continued success in electronic packaging and interconnection, these changes in all elements of the total printed circuit design, fabrication, and assembly process must be understood and effectively integrated into the total system.

This book continues the approach taken by the first two editions and addresses the need for detailed process information on all aspects of printed wiring technology. We have, therefore, gathered in this one volume the information needed to design, manufacture, test, and repair printed wiring boards and assemblies. We have added extensive new material while revising and updating most of the existing material. The result is an emphasis on those elements which have the greatest total impact—such as surface mount, multilayer, automation, process control, reliability, pollution control, and new materials—as well as a description of how all those elements interact. Thus we have an essentially new book, not just a revision of the previous editions.

The term "printed circuits" has been controversial since it was originally coined. Many have found it less definitive than "printed wiring" or "etched wiring," and over the years these more descriptive terms have generally been included in more and more official documentation. However, "printed circuits" has passed into the world's language as the term that is most often recognized and used with regard to what is described in this book. Therefore, we continue to title the book *Printed Circuits*

Handbook while using the term "printed wiring" as the industry-preferred term in the text itself.

Special appreciation is due for the help of the IPC (Institute for Interconnecting and Packaging Electronic Circuits); its executive director, Ray Pritchard; and its technical director, Dieter Bergman. Their contribution to the printed wiring industry in particular and to the worldwide electronics industry in general cannot be overstated. In addition, I would like to acknowledge the continuing support of John Fischer, general manager of the Printed Circuit Board Division of Hewlett-Packard. And, finally, I want to thank John Doyle, executive vice president of Hewlett-Packard, for his support and friendship over the years.

Clyde F. Coombs, Jr.

PRINTED CIRCUITS HANDBOOK

P · A · R · T · 1

INTRODUCTION TO PRINTED WIRING

CHAPTER 1
INTRODUCTION TO CONNECTIVITY

Clyde F. Coombs, Jr.
Hewlett-Packard Company, Palo Alto, California

1.1 *CONNECTIVITY TRADE-OFFS*

The printed wiring board (PWB) is an essential part of a total electronic circuit packaging system. This system starts with the components required to achieve product functionality, continues with the component package that provides the connection to the internal component contacts, and ends with the PWB interconnecting the package leads into a total circuit. With the introduction of semiconductor components, the technologies of packaging and interconnection have been strained to provide the capability to use the smaller, faster, cheaper integrated circuits available. As more functions are integrated on a chip, more connections off the chip are required, and more circuit traces are needed to interconnect them. This has resulted in a great deal of effort being put into increasing "connectivity," or the ability of a printed wiring process to provide electric connection to all the nodes in a circuit. The result was a concurrent reduction in line (trace) width (see Fig. 1.1) and an increase in conductor layers (see Fig. 1.2) and, therefore, a dramatic increase in capability for circuit density and complexity.

The question for the PWB designer has been: "How can board layout be optimized for economic and reliable manufacture?" For example, is it better to use more layers at a given line width, or fewer layers but a finer line? Each alternative has strong cost implications. With the rise of surface mount technology (SMT), where there is no longer a need for component lead holes that penetrate the entire board, this question has expanded to include the size of the hole, the annular ring around the hole, and even the need to have a hole between all layers. These parameters have made the design trade-offs even more complex.

This chapter gives structure to developing a design methodology that defines these trade-offs. It also, however, assumes that the designer has a basic understanding of the technical capability and cost structure of the facility which will produce the board, as a way of quantifying the impact of design option.

1.2 *COMPONENT LEAD DENSITY AND SURFACE MOUNT TECHNOLOGY*

While the dual-in-line package (DIP) was the standard integrated circuit component package, the requirement of 0.100-in lead spacing provided a consistent set

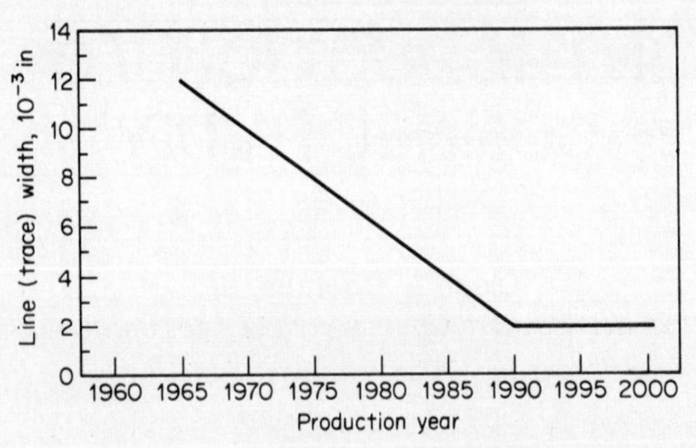

FIG. 1.1 Line (trace) width trends over time. The minimum available line width has decreased steadily with time.

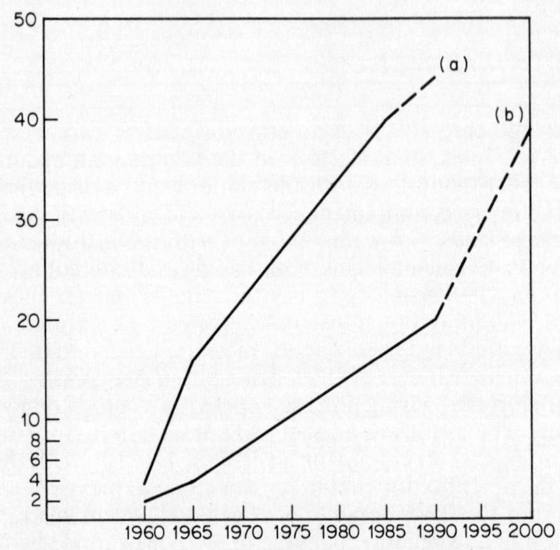

FIG. 1.2 Increase in total layer count available for special and commercial multilayer board designs. The top line (a) indicates the maximum number of layers available for special non-cost-sensitive purposes, originally in the aerospace and defense industries but now in large mainframe computer companies as well. The bottom line (b) indicates the number of layers available for commercial market-oriented products. The space between these lines indicates the range of increasing complexity and cost to achieve more layers.

of parameters for PWB designs. For example, the hole size was fixed to be able to receive a standard lead, and the hole went all the way through the board connecting all layers, whether they were needed or not. With the advent of SMT, several things changed:

1. No component leads penetrate the board, so holes can now be whatever size is achievable in production.

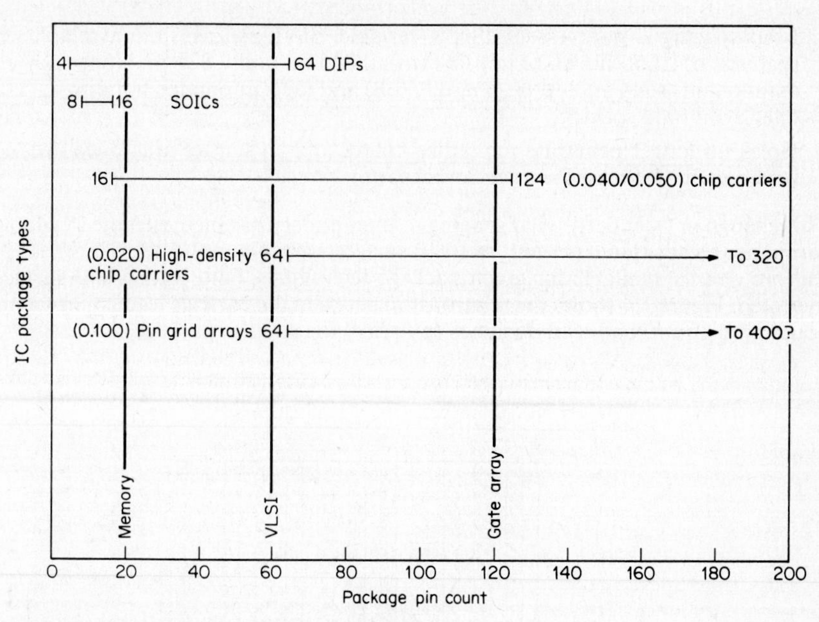

FIG. 1.3 Integrated circuit package capabilities.

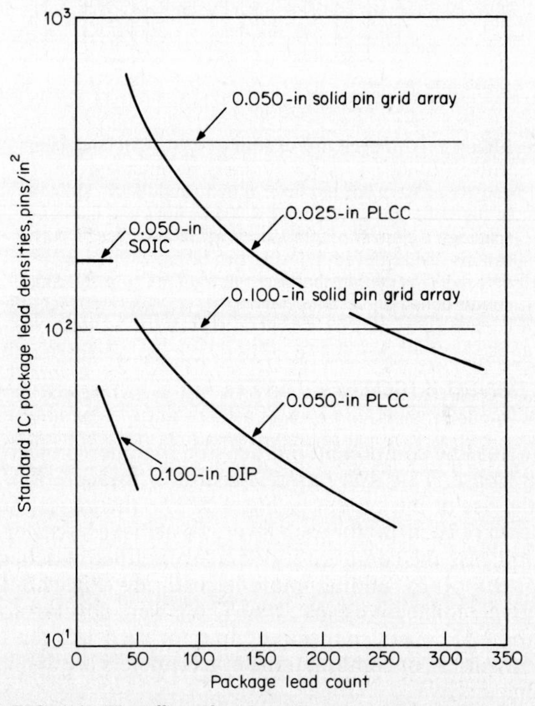

FIG. 1.4 The effect of package format on lead densities.

2. Lead spacing is now, essentially, a variable. SMT packages are available on spacings of 0.050 in, 0.032 in (0.8 mm), 0.025 in, and 0.020 in. Nonpackaged components such as chip-on-board (COB) and tape automated bonding (TAB) offer even finer spacing.
3. Holes no longer penetrate the entire board, and so surface area once needed for holes can be used for trace continuation.

This change in "standard" lead spacing to finer pitches has increased the available component leads (pins) per unit of PWB surface area. Figure 1.3 shows some typical pin counts available for given package techniques, both leaded and surface mounted. Figure 1.4 shows the relationship between the package lead count (component pin count) and lead densities (pins/in^2) for popular packages.

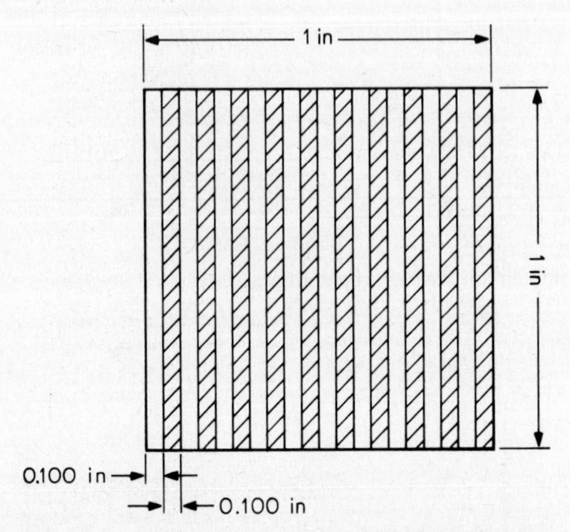

FIG. 1.5 Connectivity example. With 0.010-in lines with 0.010-in spacing (or 10 lines), an interconnection density of 10 in/in^2 per layer is achieved. With 20 lines in the same space (requiring 0.005-in lines and 0.005-in spacing), a density of 20 in/in^2 is achieved. This density can also be achieved by using two layers with 10 in/in^2 density. This defines the essential trade-off in increasing circuit density.

1.3 CIRCUIT DENSITY (CONNECTIVITY)

To respond to increasing component pin density, the interconnecting circuit density must also increase. This first requires a unit of measure for "connectivity" and has led to the use of the system of Fig. 1.5. The theoretical limit of the connection lines shown is 10 in/in^2 for each layer. To achieve a greater density, therefore, one must increase the connectivity on a layer either by reducing line width and spacing (pitch) or by adding more layers. See Fig. 1.6 for alternative approaches for given connective capacities. These were developed using a widely accepted factor of 2.25 in of "connective" line for each I/O pin (node) used. It can be seen that a variety of combinations of parameters are available to achieve the required connectivity.

Maximum pin density, I/O per in² →	14	20	30	50	100	200	
Required connective capacity, in/in² →	32	45	68	113	226	450	← 2.25 × maximum pin density
Printed wiring grid = 0.100 in*	4	5	7	12	23	45	Number of printed wiring signal layers required
0.05 in †	2	3	4	6	12	23	
0.025 in‡	1	2	2	3	6	12	

*Based on connectivity capacity of 10 in/in² per layer.
†Based on connectivity capacity of 20 in/in² per layer.
‡Based on connectivity capacity of 40 in/in² per layer.

FIG. 1.6 Effects of package formats and grid spaces on printed wiring board signal layer requirements.

1.4 DESIGN TRADE-OFFS

Figure 1.7 shows a cross section of a PWB with the main trade-off parameters defined. In Fig. 1.7 these parameters are given alternative dimensions that combine to achieve desired connectivity density. The table cannot be expected to list all the possible combinations, but the basic approach is clear, and the reader can develop a similar special table based on specific needs and the economic impact of each alternative. For example, if the question is whether it is cheaper to add another two layers or to reduce the trace geometry to 0.005-in lines and spaces, such a table can indicate how many layers will be required for a given line width. A fairly simple calculation then defines the total cost of more layers versus the cost of finer lines.

PWB hole grid, in	PTH pad diameter, in	PTH diameter, in	Line width, in	Line spaces, in	Available channel spaces, in	Interconnect lines/channel	Connectivity, in/in² per level
0.100	0.065	0.040	0.008	0.008	0.035	1	10
0.100	0.030	0.015	0.008	0.008	0.070	3	30
0.100	0.065	0.040	0.007	0.007	0.035	2	20
0.100	0.030	0.015	0.007	0.007	0.070	4	40
0.100	0.065	0.040	0.005	0.005	0.035	3	30
0.100	0.030	0.015	0.005	0.005	0.070	6	60
0.050	0.030	0.015	0.008	0.008	0.020	1	20
0.050	0.030	0.015	0.007	0.007	0.020	1	20
0.050	0.030	0.015	0.005	0.005	0.020	1	20
0.050	0.025	0.015	0.005	0.005	0.025	2	40

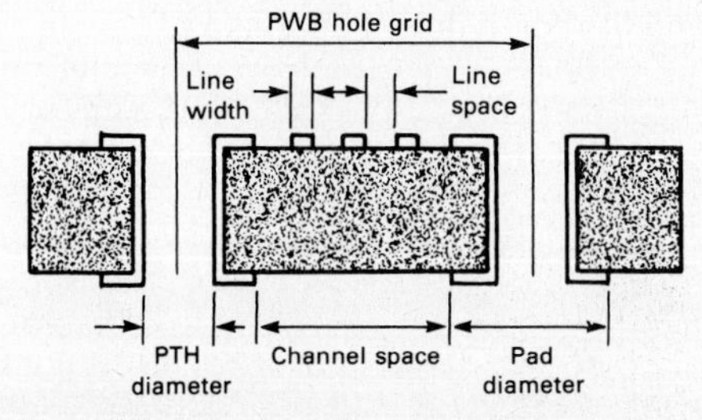

FIG. 1.7 Typical lines, pads, and connectivities.

Another way of looking at this line width versus layer count is shown in Fig. 1.8. An example of this chart use is as follows: Assume that there is a density of 100 pins/in^2, as required by a PLCC with 0.050-in lead spacing. If the fabrication process can achieve connectivity of 20 in/in^2, 11 or 12 signal layers will be required. If connectivity of 40 in/in^2 is available, then only five signal layers would be necessary. Conversely, if a maximum layer count is given, then the connectivity can be determined and the surface geometries of line, space, annular ring, hole size, etc., can be defined. Here, once the connectivity has been defined, the number of layers required can be estimated, or, knowing a maximum layer count possible, the needed connectivity (line width, hole size, annular ring, and grid spacing) can be defined.

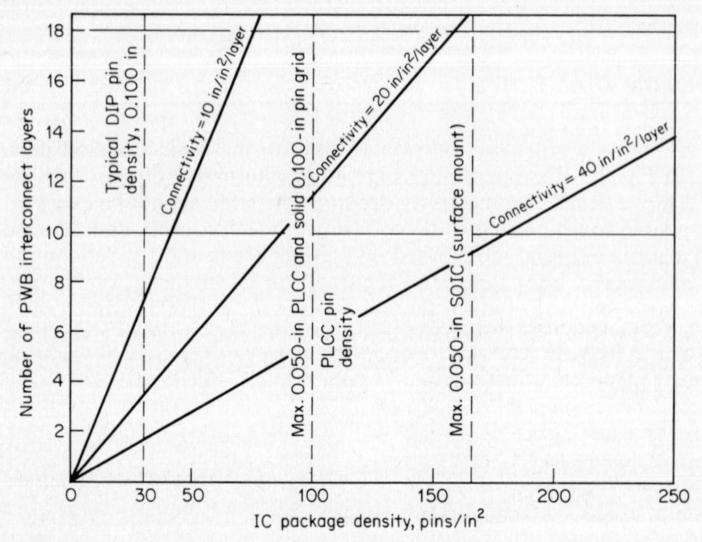

FIG. 1.8 The effects of package formats and connectivities on layer requirements.

1.5 SUMMARY

Surface mounted components have increased lead count per unit area dramatically. This, in turn, has led to an increase in circuit density and has pushed PWB fabrication processes to new extremes. The designer has several new options to consider in developing the optimal layout. This requires an understanding of the components themselves and the board design options and how to achieve an economic set of trade-offs.

CHAPTER 2
TYPES OF PRINTED WIRING BOARDS

Dr. Hayao Nakahara
Director of Far Eastern Operations,
PCK Technology Division,
Kollmorgen Corporation

2.1 INTRODUCTION

More than half a century has passed since Dr. Paul Eisner invented printed wiring technology in 1936. Since that time, several methods and processes have been developed for manufacturing printed wiring boards (PWBs) of various types. The purpose of this chapter is to give a general overview of the most common types of PWBs and to describe their various manufacturing methods and processes, with special emphasis on "through-hole" technologies.

For simplicity, the terms "printed wiring board," "PWB," and "board" will be used synonymously throughout this chapter.

2.2 CLASSIFICATION OF PRINTED WIRING BOARDS

Printed wiring boards may be classified according to their various attributes. This section defines the most basic ones first, then describes the derivative branches (see Fig. 2.1).

2.2.1 Graphical and Discrete-Wire Boards

Printed wiring boards may be classified into two basic categories based on the way they are manufactured: *graphical* and *discrete-wire* interconnection boards. "Graphical interconnection board" is another term for the standard PWB, in which the image of the master circuit patterns is formed photographically on a photosensitive material such as glass plate or film. The image is then transferred to the circuit board by screening or photoprinting the artworks generated from the master. Discrete-wire interconnection does not involve an imaging process for the formation of signal connections. Rather, conductors are formed directly onto the wiring board with insulated copper wire. Wire-Wrap, UNILAYER-II, and Multiwire (Fig. 2.2) are some of the best-known discrete-wire interconnection

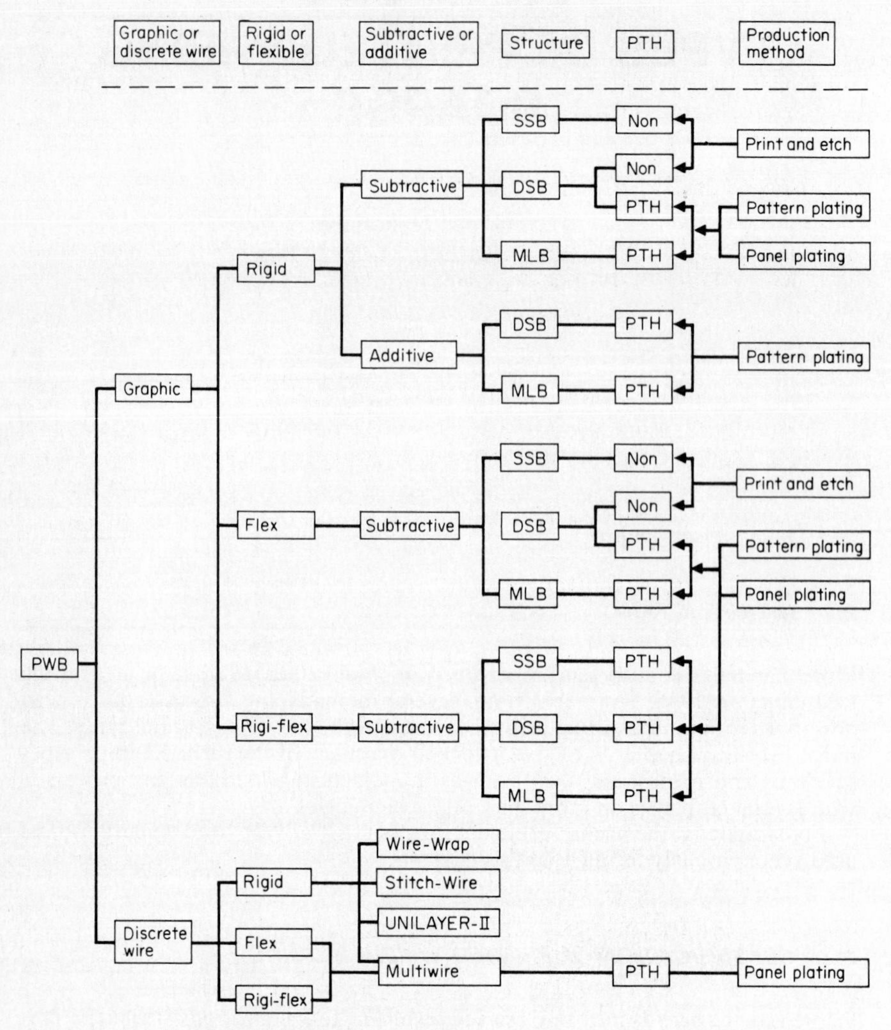

FIG. 2.1 Classification of printed wiring boards.

technologies.* Circuit designs in these technologies are easier than the circuit design in graphical interconnection technology, and their production period is shorter. However, they are sequential, and so their productivity suffers somewhat. In Fig. 2.1, the structures, methods of manufacture, and other characteristics of these PWB classifications are illustrated.

2.2.2 Rigid and Flexible Boards

Another class of boards is made up of the *rigid* and *flexible* PWBs. Whereas rigid boards are made of a variety of materials, flexible boards generally are made of polyester and polyimide bases. "Rigi-flex boards," a combination of rigid and

*Wire-Wrap, UNILAYER-II, and Multiwire are registered trademarks of Gardner-Denver Co., AUGUT Inc., and Kollmorgen Corporation, respectively.

flexible boards usually bonded together, are gaining widespread use in electronic packaging (Fig. 2.3). Most rigi-flex boards are three-dimensional structures that have flexible parts connecting the rigid boards, which usually support components; this packaging is thus volumetrically efficient.

2.3 GRAPHICALLY PRODUCED BOARDS

Since the majority of PWBs are graphically produced rigid boards, this chapter is concerned mainly with this type of board.

2.3.1 Single-Sided, Double-Sided, and Multilayer Boards

Over the years, PWBs have been produced most often by laminating copper to one or more surfaces of a sheet of plastic reinforced by paper or glass fiber. They can also be produced by adding copper chemically during the production process.

2.3.2 Single-Sided Boards

The single-sided board (SSB) has circuits on only one side of the board and is frequently referred to as the "print-and-etch" board because the resist protecting

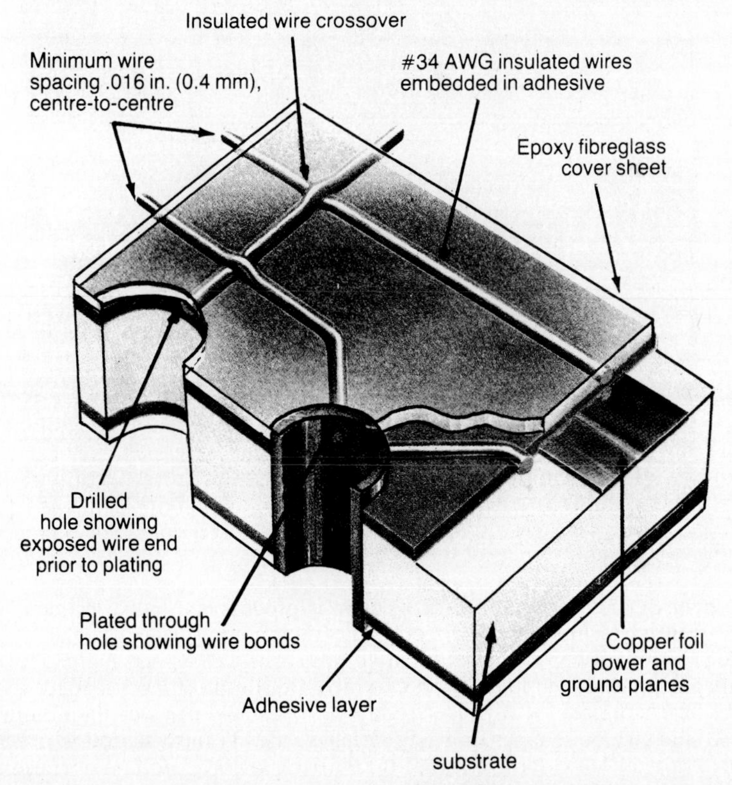

FIG. 2.2 Structure of multiwire board showing plated through-hole interconnection to wires. *(PCK Technology Division, Kollmorgen Corporation.)*

FIG. 2.3 Rigi-flex printed wiring board.

the copper during the circuit-forming etching process is usually "printed" on by screen-printing techniques. Its major applications are in packaging consumer electronic products. When the production volume is high, SSBs are manufactured by so-called print-and-etch lines, most of which are completely automated.

The base laminate for SSBs has a copper foil, usually 1 or 2 oz thick, laminated on one side (1 oz of copper in a 1-ft^2 sheet is 0.00137 in thick and is referred to as "1-oz copper"—by its size rather than its thickness). The first operation is to cut a large laminate sheet into panels of appropriate sizes. When the laminate is a glass-reinforced epoxy material, the panel is then drilled, in contrast to paper-reinforced laminate, which is often punched. Most print-and-etch lines adopt one of two types of registration systems. One depends on the right-angle edge of the panel, and the other on pilot holes (normally two holes drilled at one edge of the panel). The hole registration system is superior because of its accuracy. The flatness of laminates is of absolute importance in such automated print-and-etch lines, particularly when circuit patterns are not distributed evenly over the panel. The panel tends to warp after copper is removed during etching, allowing stresses in the material to be relieved. An excessively warped panel may not register properly for subsequent operations.

Most etch-resist, solder mask, and legend inks used for the manufacture of SSBs are curable by ultraviolet light. These inks make the print-and-etch line shorter, easier to maintain, and more economical in terms of power consumption.

When the laminate is punchable, holes are punched mechanically in the last operation. When the laminate is a glass-enforced epoxy material, holes are drilled normally as the first operation after the laminate is cut into panels.

With more interconnections made on the chip level and the advent of surface mounting technology (SMT), the use of SSBs is not decreasing as expected. In fact, SSBs are still considered an effective interconnection packaging method.

A deviation of SSBs occurs when a portion of the components counted on them are simple wires. These actually provide connection through the holes in the boards, interconnecting to the other side (the component side) of the board, and, in fact, provide wiring on the second side. Such a board may no longer be called an SSB.

2.3.3 Double-Sided Boards

Double-sided boards (DSBs) have circuits formed on both sides of the boards. They can be categorized into two classes—one without through-hole metallization and one with through-hole metallization. There are two types of metallized through-holes: *plated-through holes* (PTH) and *silver-through holes* (STH), in which the hole is completely filled with conductive silver ink.

Double-sided boards with conductive ink are usually made on inexpensive phenolic-paper laminate of thicknesses ranging from 0.0032 to 0.045 in. After copper patterns are etched onto both sides, the panel is screened once or twice, depending on its thickness, with conductive silver ink; a solder mask pattern is used to fill the holes. Since STHs have a relatively high resistance compared with PTHs,the application of STH boards is limited, but their use provides a tremendous economic advantage in some applications, such as mass-produced consumer products (see Fig. 2.4).

A variation of DSB is built sequentially with two layers of conductors that are placed on one side. A panel having copper foil on one side is first screened with conductor patterns and etched. Next, an insulating layer is screened, usually with appropriate solder mask ink, exposing pads to be interconnected to the conductors to be formed later. Then, conductors are screened using carbon- or silver-filled conductive ink, depending on the resistance requirements. These circuit boards are also popular in packaging consumer electronic products (Fig. 2.5). For this market, the fully additive technology may also be used, which is discussed later in this chapter.

Double-sided boards with PTH circuitry on both sides of an insulating plastic are connected by metallizing the walls of a hole in the plastic that intersects the circuitry on both sides. This technology, which is the basis for most printed circuits produced, is described below.

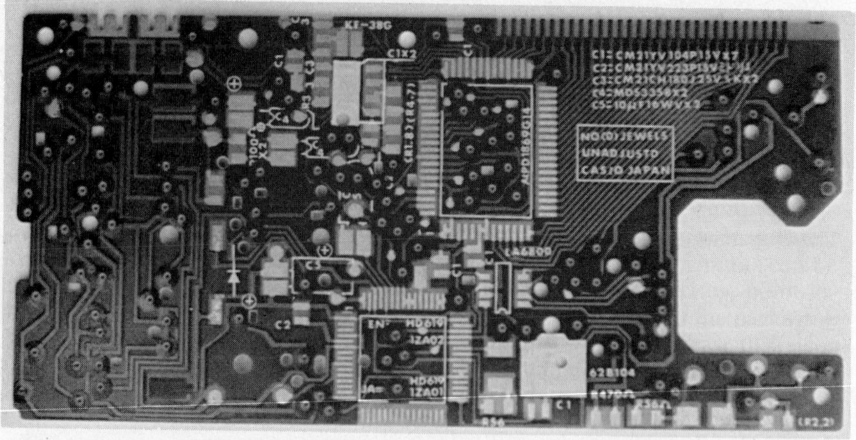

FIG. 2.4 Through-hole connections made by filling holes with silver conductive ink. *(Matsushita Electronic Component Co., Ltd.)*

2.3.4 Multilayer Boards

By definition, multilayer boards (MLBs) have 3 or more circuit layers; some boards have as many as 60 layers. The most widely practiced method of making MLB is by bonding, or laminating, layers of patterned, pre-etched, undrilled copper-clad laminate together. Interconnections between different layers are made through PTHs. After lamination, the subsequent manufacturing processes for MLBs are more or less the same as those used for DSBs made with the PTH process (see Fig. 2.6).

Because of the advent of the mass lamination technology, four-layer boards, and in some cases, six-layer boards, can be made with almost the same ease as DSBs. Therefore, four-layer boards are by far the most popular MLBs.

To increase the interconnection density, *buried via* holes or *blind via* holes are used frequently in high-level MLBs (Fig. 2.6). To make vias, inner layer pairs are fabricated exactly like double-sided PTH boards and are then assembled into an

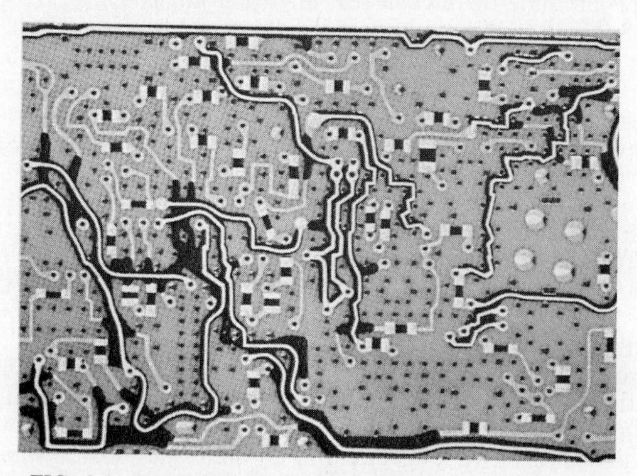

FIG. 2.5 Single-sided board with screened conductive paste. This offers a conductor structure of two layers, or a double-sided board without the plated-through process.

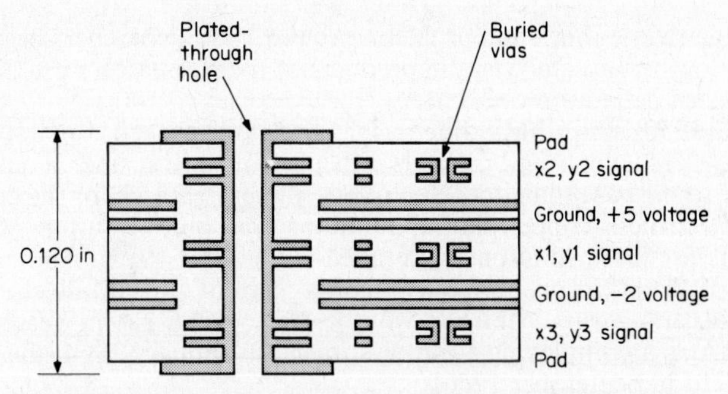

FIG. 2.6 Cross-section multilayer board with buried via holes. Buried vias are built into each of the double-sided boards that make up the final multilayer structure.

MLB, using standard techniques. In most MLBs used for mainframe computers, the use of three conductor tracks between through-hole lands on 0.100-in centers or two tracks between surface mount lands on 0.050-in centers is quite common. The width of such tracks is invariably 0.004 in or less, and tracks are formed either by etching the copper after an initial electroplating process or by a fully additive buildup of copper on a bare laminate.

Since the cost and reliability of the MLB process have improved, the use of MLBs is no longer limited to computers, telecommunication, and military equipment. It has spread everywhere, even into the toy industry.

2.4 PLATED-THROUGH-HOLE TECHNOLOGIES

In 1953, the Motorola Corporation developed a PTH process called the "Placir" method,[1] in which the entire surface and hole walls of an unclad panel are sensitized with $SnCl_2$ and metallized by spraying on silver with a two-gun spray. Next, the panel is screened with a reverse conductor pattern, using a plating resist ink, leaving metallized conductor traces uncovered. The panel is then plated with copper by an electroplating method. Finally, the resist ink is stripped and the base silver removed to complete the PTH board. One problem associated with the use of silver is the migration caused by silver traces underneath the copper conductors.

The Placir method was the forerunner of the semiadditive process, which is discussed in a later chapter.

In 1955, Fred Pearlstein[2] published a process involving electroless nickel plating for metallizing nonconductive materials. This catalyzer consists of two steps. First the panel is sensitized in $SnCl_2$ solution, and then it is activated in $PdCl_2$ solution. This process presented no problem for metallizing nonconductive materials.

At the same time that Pearlstein's paper was published, copper-clad laminates were starting to become popular. Manufacturers of PWBs applied this two-step catalyzing process to making PTHs using copper-clad laminates. This process, however, turned out to be incompatible with the copper surface. A myriad of black palladium particles called "smads" were generated between copper foil and electrolessly deposited copper, resulting in poor adhesion between the electroless copper and the copper foil. These smads and electroless copper had to be brushed off with strong abrasive action before the secondary electroplating process could begin. To overcome this smad problem, around 1960 researchers began attempting to develop better catalysts; the products of their research were the predecessors of modern palladium catalysts.[3]

The mid-1950s was a busy time in the area of electroless copper-plating solutions. Electrolessly deposited nickel is difficult to etch. But since it adheres somewhat better to the base than does electroless copper, research for the development of stable electroless copper-plating solutions was quite natural. Many patent applications for these solutions were filled in the mid-1950s. Among the applicants were P. B. Atkinson, Sam Wein, and a team of General Electric engineers, Luke, Cahill, and Agens. Atkinson won the case, and a patent[4] teaching the use of Cu-EDTA as a complexing agent was issued in January 1964 (the application had been filed in September 1956).

2.4.1 Subtractive and Additive Processes

Photocircuits Corporation was another company engaged throughout the 1950s in the development of chemicals for PTH processes. Copper-clad laminates were expensive, and a major portion of expensive copper foil had to be etched ("subtracted") to form the desired conductor pattern. The engineers at Photocircuits, therefore, concerned themselves with plating ("adding") copper conductors wherever necessary on unclad materials for the sake of economy. Their efforts paid off. They were successful in developing not only the essential chemicals for reliable PTH processes but also the fully additive PWB manufacturing technology known as the CC-4* process. This process is discussed in detail in Chapter 13.

With the use of $SnCl_2$–$PdCl_2$ catalysts and EDTA-base electroless copper-plating solutions, the modern PTH processes became firmly established in the 1960s. The process of metallizing hole walls with these chemicals for the subsequent formation of PTHs is commonly called the "copper reduction process." In the subtractive method, which begins with copper-clad laminates, pattern plating and panel plating are the two most widely practiced methods of making PTH boards. These methods are discussed in the following subsections.

2.4.2 Pattern Plating

In the pattern-plating method, after the copper reduction process, plating resist layers of the reverse conductor image are formed on both sides of the panel by screening resist inks. In most fine-line boards, photosensitive dry film is used instead. There are some minor variations in the pattern-plating method (see Fig. 2.7):

1. Catalyzing (preparing the nonconductive surface to cause copper to come out of solution onto that surface)
2. Thin electroless copper (0.00001 in) followed by primary copper electroplating; thick electroless copper (0.0001 in)
3. Imaging (application of a plating resist in the negative of the desired finished circuit)
4. Final electroplating copper
5. Solder plating (as etching resist) 0.0002 or 0.0006 in

*CC-4 is a registered trademark of Kollmorgen Corporation.

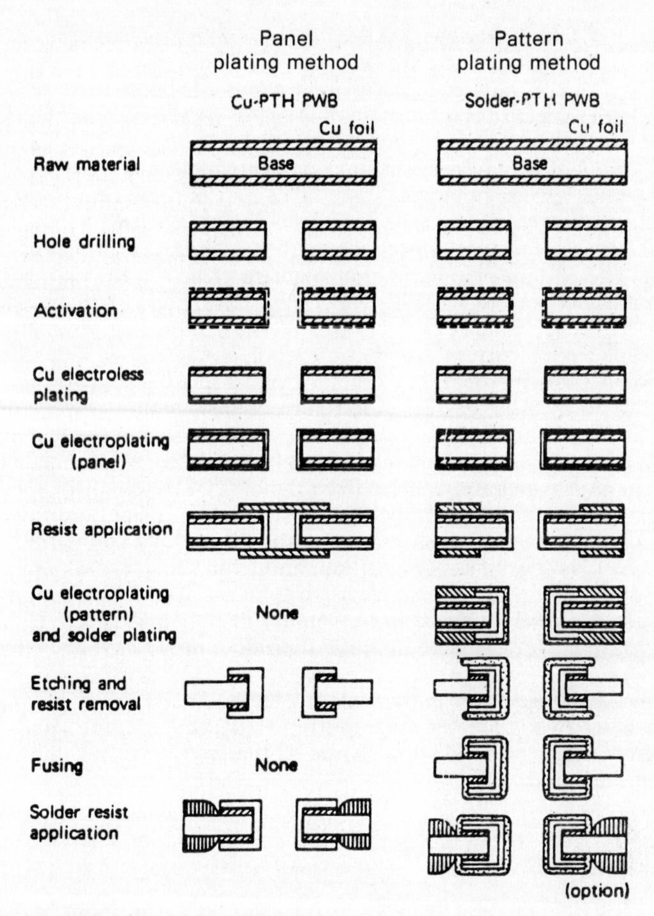

FIG. 2.7 Key manufacturing steps in panel-plating and pattern-plating methods.

6. Stripping plating resist
7. Etching of base copper
8. Solder etching (0.0002-in case); solder reflow (0.0006-in case)
9. Solder mask followed by hot-air solder coater leveler if solder etching is used
10. Final fabrication and inspection

Most manufacturers of DSBs with relatively wide conductors employ thick electroless copper plating. However, thin electroless copper followed by primary electroplating is preferred for boards having fine-line conductors, because a considerable amount of surface is brushed off for better adhesion of dry film. This provides a higher reliability for PTHs. Solder reflow boards had been preferred by many customers, particularly in military and telecommunications applications, until the emergence of hot-air solder coater levelers. Although the solder-over-copper conductors protect the copper from oxidization, solder reflow boards have some limitations. Solder mask is hard to apply over reflowed solder, and it tends to wrinkle and peel off in some areas when the boards go through component soldering. A more serious problem is the solder bridging that occurs when the conductor width and clearance become very small.

In step 9, the entire surface of the board except for the pads is covered by solder mask, and then the board is immersed into the hot-air solder coater leveler, resulting in a thin coating of solder over the pads and the hole walls. The operation sounds simple, but it requires constant fine tuning and maintenance of the hot-air solder coater leveler; otherwise some holes may become heavily clogged with solder and are then useless for component insertion.

One advantage of the pattern-plating method over the panel-plating method is in etching. The pattern-plating method needs to etch only the base copper. The use of ultrathin copper foil (UTC), which is usually $\frac{1}{8}$ or $\frac{1}{4}$ oz thick, offers a real advantage. However, as long as electroplating is used, the pattern-plating method cannot escape from a current distribution problem, regardless of the thickness of the base foils. The panel-plating method by electroplating suffers from the same problem but to a lesser degree. Good current distribution is very difficult to achieve when the boards are not of the same size or type, and particularly if some have large ground planes on the outer faces being plated. When the board has a few holes in an isolated area remote from the bulk of the circuitry, these tend to become overplated, making component lead insertion difficult during assembly. To minimize this current distribution problem, various countermeasures are practiced, such as special anode position, anode masking, agitation, and plating thieves. But none of these offers a decisive solution to the distribution problem, and they are extremely difficult to implement flexibly and effectively in a large plating operation, where a large number of product mixes have to be handled all the time.

Another advantage of the pattern-plating method is its ability to form padless micro-via holes of a diameter ranging from 0.012 to 0.016 in. Micro-via holes enable better usage of conductor channels, thereby increasing the connective capacity of the board.

2.4.3 Panel Plating

In the panel-plating method, there are two variations for finishing the board after the panel is plated with electrolytic copper to the desired thickness. In the "hole-plugging" method, the holes are filled with alkaline-etchable ink to protect the hole walls from being etched; this is used in conjunction with screened etch resist. In the other method, called "tent-and-etch" or simply "tenting," the copper in the hole is protected from etching by covering the hole or tenting with dry film, which is also used as an etch resist for conductors on the panel surface. The simplified sequence of the panel-plating method is as follows (see Fig. 2.7):

1. Catalyzing
2. Thin electroless copper deposition (0.0001 in)
3. Electroplating copper (0.001 to 0.0012 in)
4. Hole plugging with alkaline-resolvable ink; tenting (dry film lamination)
5. Screen-print etching resist (conductor pattern); photoexpose the panel for conductor pattern
6. Etching copper
7. Stripping etching resist
8. Solder mask

9. Solder coater leveler (optional)
10. Final fabrication and inspection

The panel-plating method is ideal for bare copper board. However, it is a difficult way to make padless via holes, which are becoming more popular. Generally, the conductor width of 0.004 in is considered to be the minimum realizable by this method for mass production.

Although the use of the panel-plating method in the United States and western Europe is limited, nearly 60 percent of the PTH boards in Japan are manufactured by this method.

2.4.4 Additive Plating

Plated-through holes can be formed by additive (electroless) copper deposition, of which there are three basic methods: fully additive, semiadditive, and partially additive. Of these, semiadditive involves pattern electroplating for PTHs with very thin surface copper, but the other two form PTHs solely by electroless copper deposition. The additive process has various advantages over the subtractive process in forming fine-line conductors and PTHs of high aspect ratio. A detailed account of the additive process is given in a later chapter.

2.5 SUMMARY

Modern electronic packaging has become very complex. Interconnections are pushed more into lower levels of packaging. The choice of which packaging technology to use is governed by many factors: cost, electrical requirements, thermal requirements, density requirements, and so on. Material also plays a very important role. All things considered, PWBs still play important roles in electronic packaging.

REFERENCES

1. Robert L. Swiggett, *Introduction to Printed Circuits,* John F. Rider Publisher, Inc., New York, 1956.
2. Private communications with John McCormack, PCK Technology Division of Kollmorgen Corporation.
3. C. R. Shipley, Jr., U.S. Patent 3,011,920, Dec. 5, 1961.
4. R. J. Zebliski, U.S. Patent 3,672,938, June 27, 1972.

CHAPTER 3
SURFACE MOUNT TECHNOLOGY

Stephen W. Hinch
Hewlett-Packard Company, Corporate Manufacturing Engineering

3.1 INTRODUCTION

Through-hole printed wiring technology is no longer adequate to meet the needs of high-performance electronic assemblies. Manufacturers faced with constraints on size, cost, and electrical performance of through-hole components are finding that surface mount technology (SMT) is a promising alternative. In fact, SMT may be the most significant development in printed wiring technology in many years.

Surface mount technology offers many benefits over conventional through-hole technology. Whereas through-hole components are mounted to the printed wiring board (PWB) by way of leads inserted through the board (Fig. 3.1*a*), surface mount components (SMCs) are soldered directly to the copper conductors on its surface (Fig. 3.1*b*). To withstand the rigors of the insertion process, through-hole component leads must be relatively large. The typical diameter is 0.032 in, with 0.015 in being a practical minimum. Surface mount components do not require these large leads and can therefore be much smaller than their through-hole equivalents. Some SMCs lack leads entirely and depend instead on metallized terminations on the component body for mechanical and electrical connection to the board.

Strictly speaking, SMT is not new. The integrated circuit flat pack (Fig. 3.2) was one of the first integrated circuit (IC) packages developed. It predates the now common dual-in-line (DIP) package by several years. And SMCs have been a mainstay of the ceramic hybrid industry since the mid-1960s. But only since the early 1980s has SMT been economically competitive when used on PWBs. Although high-speed automated equipment for through-hole insertion became available before 1970, surface mount assembly was done by hand for another 10 years. Now that automated assembly equipment has brought SMT assembly costs to approximate parity with through-hole assembly, increased interest has been shown by the manufacturing community. The elimination of cost penalties between SMT and through-hole components has further increased usage of the technology.

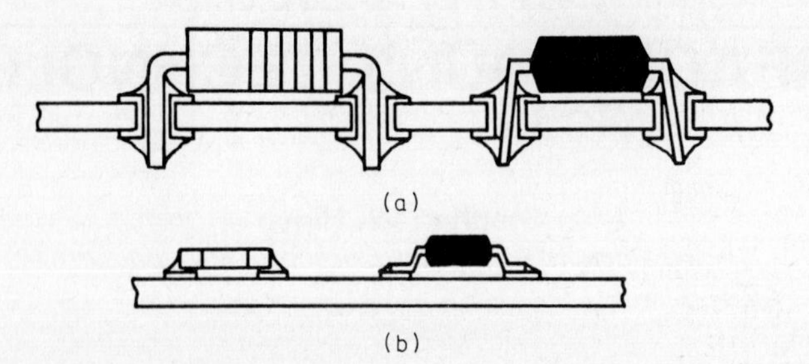

FIG. 3.1 Printed wiring board technologies. (*a*) Through-hole; (*b*) surface mount.

3.2 BENEFITS OF SURFACE MOUNT TECHNOLOGY

Printed wiring technology historically has been driven by factors relating to the cost, size, and performance of the end product. Surface mount technology provides benefits over through-hole technology in each area.

3.2.1 Size

Compared to an equivalent through-hole PWB, a typical SMT board can be made 30 to 50 percent smaller. Most of this savings can be attributed directly to the smaller physical size of surface mount components. Table 3.1 shows a comparison of PWB area required for selected surface mount and through-hole component types.

A related benefit is that of smaller drilled-hole diameters on the PWB. Surface mount via holes need be just large enough to ensure reliable electrical connection between layers. The area required on the PWB for such vias can be reduced by a factor of 3 or more over vias for through-hole components.

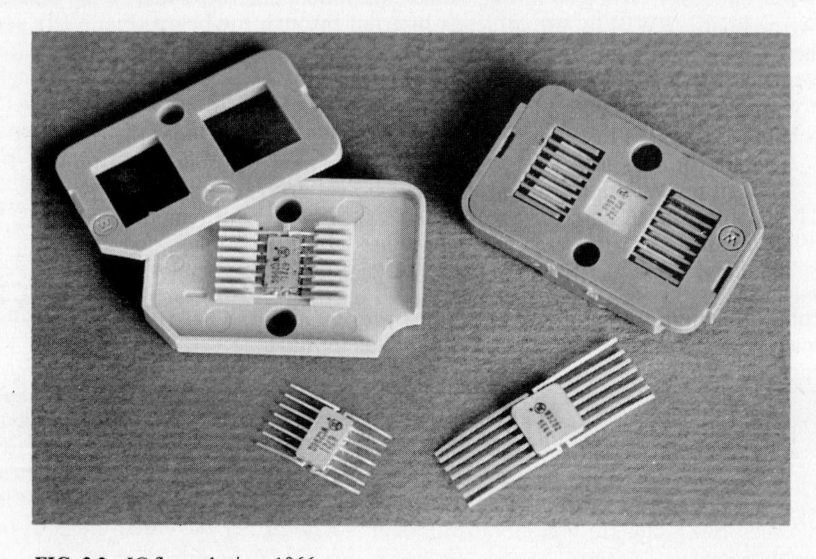

FIG. 3.2 IC flatpack circa 1966.

TABLE 3.1 Comparison of Physical Sizes of Selected Surface Mount and Through-Hole Components

Component type	Required land pattern area, in^2	
	Surface mount	Through-hole
¼-W resistor	0.009	0.0475
0.1-μF ceramic capacitor (X7R dielectric)	0.014	0.025
1000-μH inductor	0.025	0.046
Small-signal transistor	0.012	0.026
16-lead digital IC	0.096	0.248
68-lead digital IC	0.990	2.972*

*64-lead DIP package.

Smaller components also permit reduced overall thickness. Boards can be packaged more closely, and since there are no leads to penetrate the board, components can be placed on both sides to effectively double the available area.

3.2.2 Interconnectivity

Closely related to the issue of size is that of interconnectivity. This term refers to the number of electrical connections that can be realized per unit area on the PWB. It is an especially important characteristic when dealing with large numbers of integrated circuits. When the industry standard DIP package was developed, there was little need for packages with more than 16 leads. Today, ICs with 100 or more leads are common. At these lead counts the interconnectivity of a DIP package is unacceptable. An IC might require a package 15 or more times larger

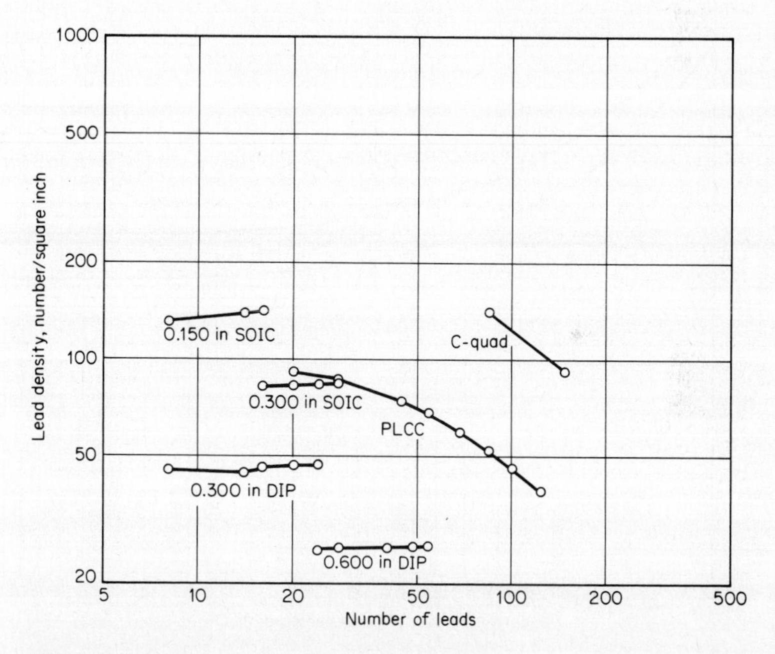

FIG. 3.3 Interconnectivities of common IC packages.

than the silicon chip simply to accommodate its leads. For example, a DIP-packaged integrated circuit with 16 leads on 0.100-in centers has an interconnectivity of less than 50 leads per square inch. The same IC in a small-outline integrated circuit (SOIC) package with leads on 0.050-in centers has an interconnectivity of more than 140 leads per square inch. The benefits become even more pronounced with higher-lead-count devices, where surface mount packages utilize tightly spaced leads on all four sides. The interconnectivities of many common package types are shown in Fig. 3.3.

3.2.3 Cost

The smaller physical size of SMT assemblies reduces the cost of the raw laminate. Drilling costs are also less. Every lead on a through-hole component needs its own drilled hole, but with SMT a hole is necessary only when electrical contact is required between layers. The net savings, however, is less than might be expected initially. Leads for through-hole components are often also used as via holes. For SMT boards these connections do not occur naturally and must be specifically added. Even so, a 30 to 50 percent reduction in the number of drilled holes can often result. Drilling costs are further complicated by the fact that although smaller drilled holes can be used, they are more expensive to drill than large holes.

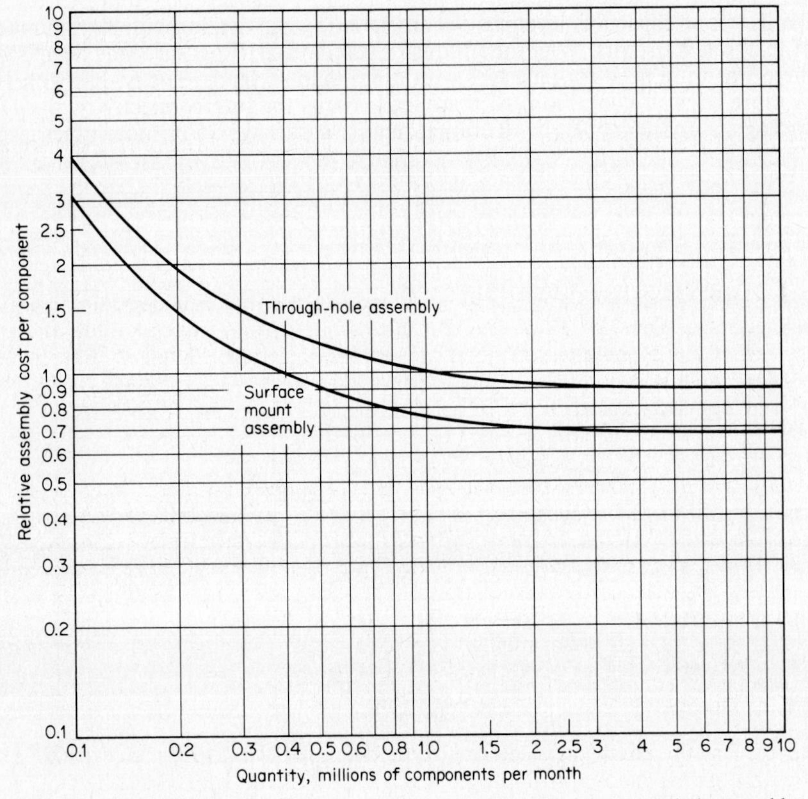

FIG. 3.4 Assembly cost as a function of factory size (relative to through-hole assembly of 1 million components per month).

Perhaps the most significant savings comes from combining on a single PWB circuitry that otherwise would require multiple boards. Since SMT can cut in half the PWB area required for a given circuit, this is frequently possible. By doing so, fewer system-level interconnects are necessary. Those that remain are less complex, reducing both material costs and assembly labor.

These advantages must be balanced against potential cost penalties. Some types of SMCs are more expensive than through-hole versions. Printed wiring board fabrication costs can also increase if finer lines and smaller drilled hole sizes are employed.

Assembly costs for an automated SMT line are roughly comparable to an automated through-hole assembly line. Such costs are usually measured on a cost-per-component basis as defined by

$$C = \frac{E}{N} \tag{3.1}$$

where E = factory operating costs per unit time and N = number of components placed per unit time.

Factory operating costs include those of direct labor; indirect labor, such as supervision and maintenance; depreciation of capital equipment; operating expenses such as supplies, electricity, and land and building costs; and overhead, including engineering and purchasing costs directly associated with the operation of the factory. Purchase prices of the components themselves are not included. Figure 3.4 shows an estimate of how relative assembly cost per component varies as a function of production rate for surface mount facilities and through-hole facilities of equivalent capacities.

Because of the range of variables, comparisons of total cost of SMT and through-hole technology vary widely. Cost savings of 50 percent or more have been reported, but poorly designed SMT circuits have also been shown to cost more than conventional circuits.

3.2.4 Performance

The shorter leads of SMT components reduce their parasitic inductances and capacitances. For this reason, they are preferred for high-frequency analog applications. Through-hole circuitry is rarely used above 500 MHz, but SMT circuits can be employed successfully in applications operating to 3 GHz and above. Digital circuitry also benefits from SMT packaging. Propagation delays are not only shorter but more uniform from lead to lead (see Fig. 3.5). This is especially important for digital circuits with clock rates above 10 MHz.

Another performance-related benefit is improved shock and vibration performance of SMT assemblies. This results from the smaller size and mass of components and from the smaller number of PWBs that often result. Finally, SMT assemblies are easier to shield against electromagnetic interference. Sensitive circuits can often be combined onto a single board to reduce the complexity of the shielding structure.

3.2.5 Quality

Because of the small physical size and close spacing of SMCs, assembling production boards manually is impractical. As a result, automation is required and SMT benefits from its higher inherent quality. The benefit is especially pronounced when an automated SMT facility is compared with a through-hole facil-

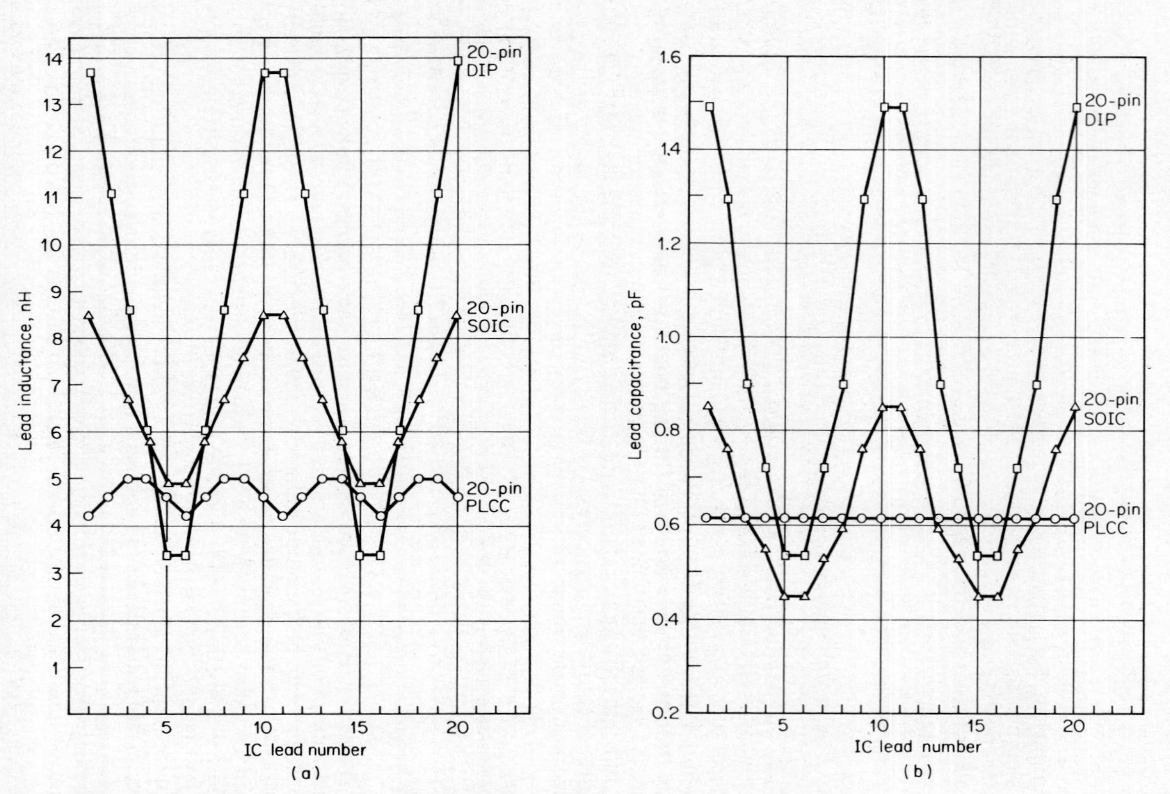

FIG. 3.5 IC parasitic reactances. (*a*) Lead inductances; (*b*) lead capacitances.

ity with a high percentage of manually inserted parts. Even when an automated SMT facility is compared with an automated through-hole facility, some improvement can be expected because surface mounting is a physically simpler process than inserting and crimping component leads. However, because SMT processes are less mature than through-hole processes, additional engineering effort may be necessary to achieve the high quality possible with the technology.

3.3 LIMITATIONS OF SURFACE MOUNT TECHNOLOGY

Surface mount technology is not without limitations. Designers contemplating the use of SMT should be aware of the following constraints.

3.3.1 Market-Entry Costs

The initial capital outlay for an SMT assembly operation may exceed that of a through-hole operation. Facilities with a large existing base of through-hole equipment need to make considerable additional investments to obtain SMT capability. For smaller production quantities, retaining the services of a contract assembly house that specializes in SMT assembly is sometimes more practical than establishing an internal facility.

3.3.2 Thermal Management

Component density is often higher with SMT, resulting in greater power dissipation per square inch of PWB. In addition, closely spaced components make forced-air cooling less efficient. Since these effects can cause SMCs to run hotter than through-hole components, reliability may be degraded unless more attention is paid to thermal management.

3.3.3 Standards

Standards in all areas of the technology are in a continuing state of review.[1] The many variations of SMT assembly process flows make standardization more difficult. Two areas that have been extensively addressed are standards for SMT components and standards for PWB design. Readers are encouraged to consult EIA/JEDEC Publication 95 for current component standards and IPC document SM-782 for PWB design guidelines.

3.3.4 Test

Assembled SMT boards must be carefully designed to be electrically testable. Standard in-circuit (bed-of-nails) test systems require access to all electrical nodes from the underside of the board. This may require adding additional via holes on the board to maintain this accessibility. An alternative approach is to use a double-sided test fixture to probe both sides of the board. Such fixtures are costly, and their additional complexity decreases their reliability.

Bed-of-nails probing is generally limited to probe spacings of 0.050 in or greater. Spacings of 0.100 in are preferred because the probes can be more ruggedly designed. Probes which can be placed on 0.040-in centers have been designed, but they are fragile and must be replaced often in production environments. Many types of SMCs use lead pitches of less than 0.050 in. High-lead-count ICs, for instance, need lead spacings of 0.032 in or even 0.025 in to maintain efficient use of PWB area. When such components are used, the designer must add access points on the board to ensure accessibility of all nodes.

3.4 SURFACE MOUNT COMPONENTS

Almost any type of component designed for through-hole assembly can be found in surface mount form (Fig. 3.6). Resistors, capacitors, inductors, transistors, diodes, ICs, connectors, and electromechanical devices can all be obtained from numerous manufacturers. The largest range of product offerings emphasizes low-power applications—less than 1 or 2 W of power dissipation. Above this range of power, offerings are more restricted. This is partly because the large physical size required by such devices makes surface mounting less practical. Cooling also becomes more of a problem. High-power SMCs must typically be mechanically mounted to the PWB for structural stability and to improve thermal dissipation.

FIG. 3.6 Typical surface mount components.

The availability of certain components that are physically large is also more restricted. Here again, the mechanical requirements of components such as power transformers, switches, and filter capacitors make surface mounting a less effective approach.

Standards have now been developed for many types of components. A brief review of some of the more commonly used components follows.

3.4.1 Resistors

3.4.1.1 Rectangular Chip. The rectangular chip resistor as shown in Fig. 3.7 has proven to be very popular. It consists of a ceramic substrate on which a thick-film resistor paste has been screened and fired. Electrical contact to the PWB is made through metallized terminations at the ends of the body. Often these terminations wrap around from the top (resistor) side to the underside of the body. Such a configuration allows the resistor to be mounted with the resistive material either face up or face down. This is important when using automated equipment that cannot distinguish orientation. If the metallized terminations do not wrap around to the underside, some means must be employed to ensure that the resistor is always mounted face down.

Although rectangular chip resistors have been manufactured in many sizes, the 1206 body style is the one most commonly used. This nomenclature refers to a product with a nominal length of 0.125 in and a width of 0.063 in. It is sometimes also referred to as a 3216 size, in reference to its dimensions in millimeters. This chip is typically rated at $\frac{1}{4}$ W power dissipation at 70°C. A similar but smaller chip designated 0805 has been developed for lower power applications in which small physical size is important.

Resistors manufactured using the thick-film process can be manufactured economically with tolerance ranges of ±1 to ±20 percent. Tighter tolerance devices often are manufactured using thin-film techniques at considerably higher cost.

3.4.1.2 Metal Electrode Face. The *m*etal *el*ectrode *f*ace (MELF) resistor shown in Fig. 3.8 is nothing more than a conventional leaded resistor without leads attached. Instead, the ends are metallized for surface mounting. These resistors

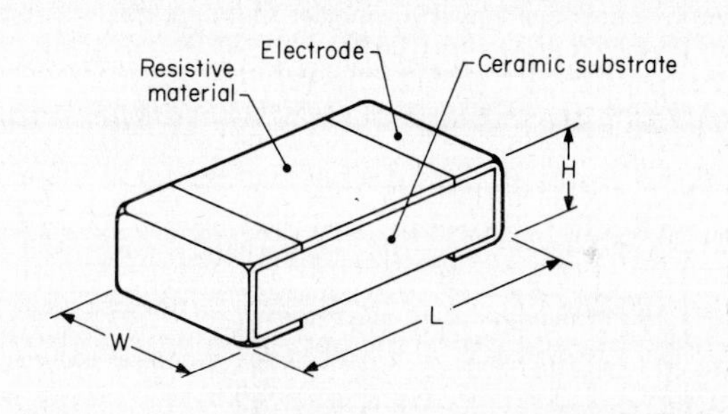

Size code	Power rating (70°C)	Nominal dimensions in inches (millimeters) L	W	H
1206 (3216)	0.125 W	0.125 (3.2)	0.063 (1.6)	0.023 (0.6)
0805 (2012)	0.100 W	0.080 (2.0)	0.050 (1.2)	0.023 (0.6)

FIG. 3.7 Rectangular chip resistors.

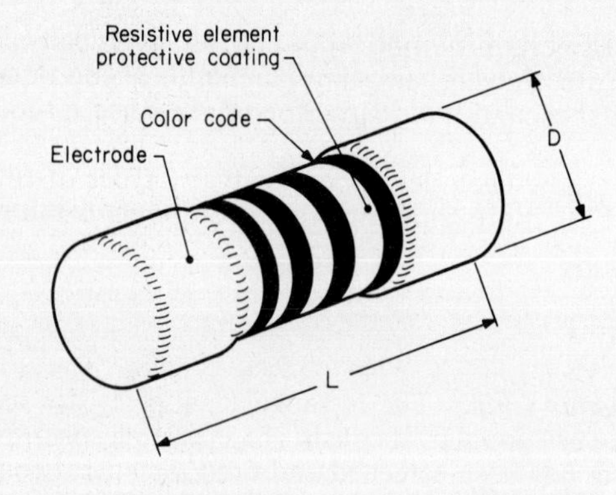

Power rating (70°C)	Nominal dimensions in inches (millimeters) L	D
0.125 W	0.138 (3.50)	0.055 (1.40)
0.250 W	0.230 (5.90)	0.087 (2.20)

FIG. 3.8 Metal electrode face resistor.

can be produced with the same equipment used to manufacture leaded resistors. They thus take maximum advantage of established manufacturing technology. Since the process is somewhat less complex than for rectangular chips, MELF resistors enjoy a slight cost advantage.

MELFs are most suitable for applications in which they are attached adhe-

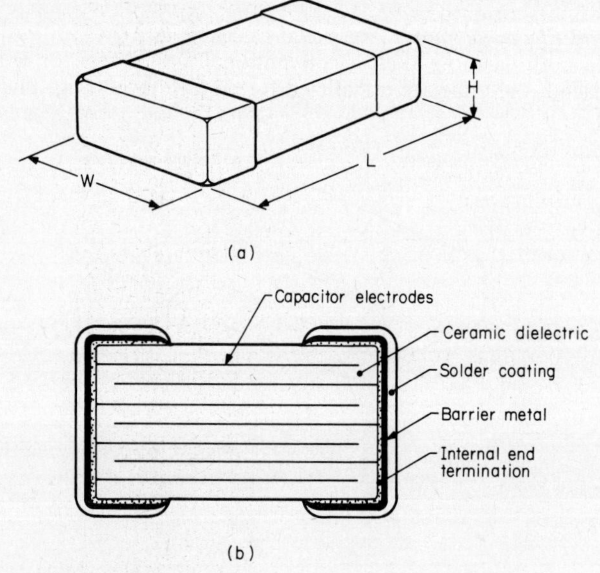

FIG. 3.9 Multilayer ceramic chip capacitor. (*a*) Physical appearance; (*b*) cross section.

TABLE 3.2 EIA Standard Ceramic Capacitor Sizes

Size code	Nominal dimensions, in (mm)		
	Length	Width	Height
0805	0.080 (2.0)	0.050 (1.2)	0.050 (1.2)
1206	0.125 (3.2)	0.063 (1.6)	0.060 (1.5)
1210	0.125 (3.2)	0.100 (2.5)	0.065 (1.7)
1812	0.180 (4.6)	0.125 (3.2)	0.065 (1.7)
1825	0.180 (4.6)	0.250 (6.4)	0.065 (1.7)

sively to the PWB. Since in-line reflow soldering equipment often employs inclined inlet and outlet conveyors, MELF packages tend to roll off the board if not secured in place. For this reason, the rectangular ceramic chip remains popular in the United States. In the Far East, where wave soldering is popular, both MELF and rectangular chips are used extensively.

3.4.2 Capacitors

3.4.2.1 ***Multilayer Ceramic.*** The most commonly used surface mount capacitor is the multilayer ceramic chip, as shown in Fig. 3.9*a*. It consists of a series of parallel precious metal electrodes separated by layers of ceramic dielectric (Fig. 3.9*b*). The five standard sizes shown in Table 3.2 have been adopted by the Electronic Industries Association (EIA) to cover the range of capacitances from about 10 pF to 2 μF. Commonly used dielectric materials include temperature-stable (COG), general purpose (X7R), and high-capacitance (Z5U).

3.4.2.2 ***Tantalum.*** The requirement for higher capacitance values is addressed by surface mount tantalum capacitors. Standards for these capacitors have not evolved as rapidly as for ceramics, and several competing body styles are presently available. The resin-molded body styles shown in Fig. 3.10 are widely used in many general-purpose applications. Capacitance values up to about 30 μF at an operating voltage of 10 V can be obtained in the largest package style. The flat molded body of this capacitor is especially suited for automatic assembly operations.

3.4.2.3 ***Other Capacitor Types.*** At present, various difficulties have prevented commercial manufacture of such capacitor types as polystyrene and aluminum electrolytic. Problems associated with designing a product that will withstand the rigors of reflow soldering and solvent cleaning have typically been difficult to overcome.

3.4.3 Inductors

3.4.3.1 ***Wire-wound.*** Wire-wound inductors for surface mounting are readily available. They typically consist of fine wire wrapped around a ceramic or ferrite

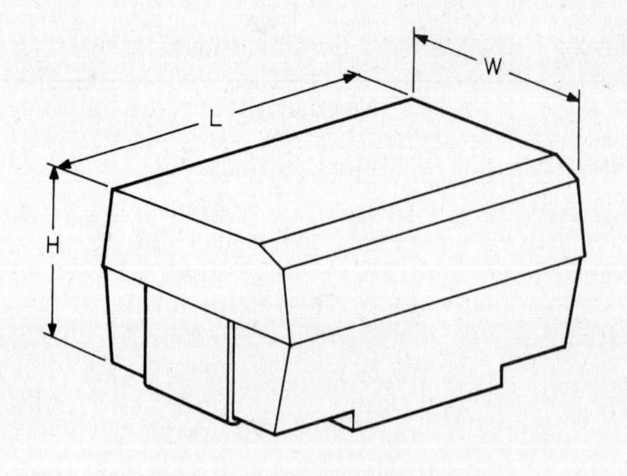

Size code	Nominal dimensions in inches (millimeters) L	W	H
3216	0.125 (3.2)	0.063 (1.6)	0.063 (1.6)
3528	0.138 (3.5)	0.110 (2.8)	0.075 (1.9)
6032	0.236 (6.0)	0.126 (3.2)	0.098 (2.5)
7343	0.287 (7.3)	0.169 (4.3)	0.110 (2.8)

FIG. 3.10 Resin-molded tantalum capacitors.

core material. The windings may be either vertical or horizontal (Fig. 3.11) and are protected in one of two ways. Conformal coating with an epoxy resin is the least expensive process. Its main drawback is that the geometry of the surface is difficult to control. This can lead to handling problems during automatic assembly. Somewhat more expensive plastic molded bodies are often preferred for automatic assembly because of their repeatable body dimensions.

3.4.3.2 Multilayer. Multilayer inductors are similar in construction to multilayer capacitors. They consist of alternate layers of ferrite paste and conductor paste which have been screen-printed and fired. Inductance values up to several hundred microhenries are commercially available in physical sizes compatible with ceramic capacitors.

3.4.4 Discrete Semiconductors

Surface mount semiconductors differ from through-hole semiconductors only in packaging format; the dies themselves are identical. As such, the concerns are those relating to this packaging difference.

3.4.4.1 Package Thermal Resistance. For many semiconductor types, the primary path for heat removal is via conduction through the leads. Convection and radiation play relatively minor parts. Since surface mount packages employ smaller leads than through-hole packages, their thermal resistances are higher.

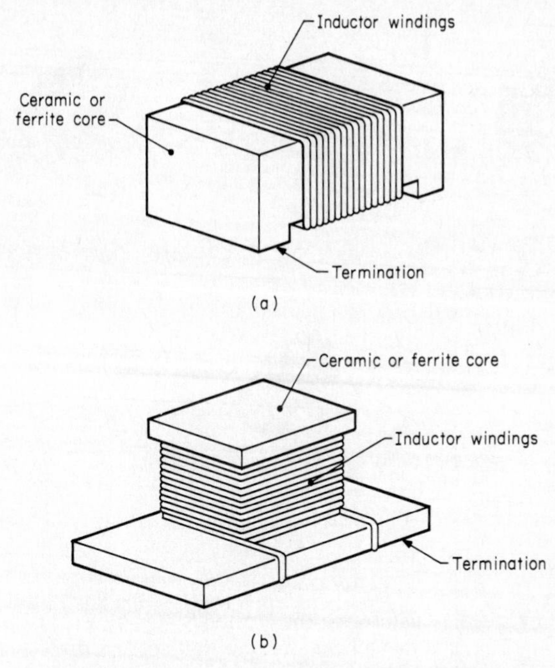

FIG. 3.11 Wire-wound inductors. (*a*) Vertical windings; (*b*) horizontal windings.

Design engineers must either restrict the power dissipations of the devices or employ more efficient cooling mechanisms when using these smaller packages.

3.4.4.2 Reliability. Most surface mount semiconductors are encapsulated in postmolded plastic packages. The amount of plastic surrounding the die is much less than for through-hole devices. Since moisture can penetrate the smaller package more easily, reliability failures due to corrosion are more of a concern. However, improvements in die passivation techniques have largely offset this concern. Multiple layers of silicon nitride, polyimide, or both, as used on many devices, provide a high degree of moisture protection for the die compared with older silicon oxide techniques.

3.4.5 Transistors

The most popular surface mount packages for transistors are the small outline transistor (SOT) configurations. Transistors with power dissipation not exceeding about 200 mW are packaged in the SOT-23 package (Fig. 3.12). Unfortunately, the version adopted in the United States by EIA/JEDEC (TO-236 package) varies somewhat from similar packages produced in the Far East. European manufacturers generally use a body outline similar to that used in the United States. Although the differences in body size and tolerances appear minor, they can pose problems for automatic equipment expected to accommodate either style interchangeably.

Larger transistors with power dissipations up to about 500 mW are packaged

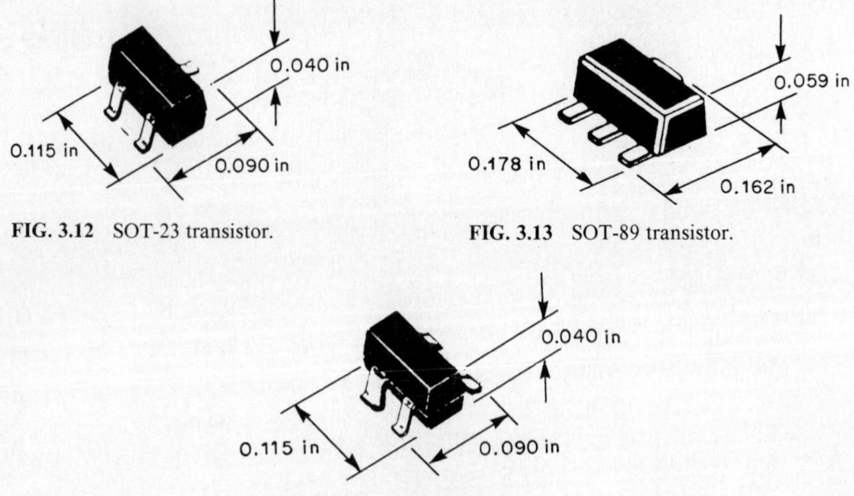

FIG. 3.12 SOT-23 transistor.

FIG. 3.13 SOT-89 transistor.

FIG. 3.14 SOT-143 transistor.

in the SOT-89 package, registered in the United States by EIA/JEDEC as the TO-243 package (Fig. 3.13). It has a heat sink tab that solders directly to the PWB to improve thermal conductivity.

Four-terminal devices with power dissipations under 200 mW are packaged in the SOT-143 package (Fig. 3.14). This package is an SOT-23 with four leads and has similar performance capability.

Above 500 mW, package offerings are more limited. The DPAK package (Fig. 3.15) has been developed for applications up to about 1.3 W, but exact mechanical dimensions vary among manufacturers. For higher power ranges, the conventional TO-220 package is sometimes used by preforming its leads for surface mounting. Thermal dissipation data for several SMT transistor packages is shown in Table 3.3.

3.4.6 Diodes

The same packages used for transistors are often used for surface mount diodes. The SOT-23 and SOT-143 are used for low-power single and dual diodes, respec-

TABLE 3.3 Approximate Thermal Resistances for Discrete Semiconductors

	Thermal resistance* (θja), °C/W	
Device type	Ceramic substrate	Epoxy-glass PWB
SOT-23	330	400
SOT-89	125	300
SOT-143	350	420

*Measured in still air.

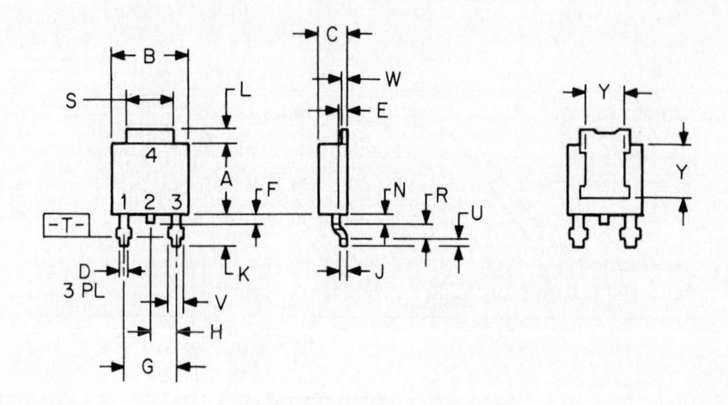

Dimensions	Millimeters		Inches	
	Min.	Max.	Min	Max.
A	5.97	6.19	0.235	0.245
B	6.35	6.73	0.250	0.265
C	2.18	2.38	0.086	0.094
D	0.69	0.88	0.027	0.035
E	0.89	1.14	0.035	0.045
F	0.64	0.88	0.025	0.035
G	4.57	BSC	0.180	BSC
H	2.87	BSC	0.090	BSC
J	6.46	0.58	0.018	0.023
K	2.59	2.89	0.102	0.114
L	0.89	1.27	0.035	0.050
N	0.81	1.11	0.032	0.044
R	1.07	1.37	0.042	0.054
S	5.21	5.46	0.205	0.215
U	0.51	—	0.020	—
V	0.76	1.14	0.030	0.045
W	0.48	0.55	0.019	0.022
Y	4.32	—	0.170	—

(a)

FIG. 3.15 DPAK transistor. (*a*) Dimensional outline drawing for surface mount DPAK; (*b*) DPAK internal construction. (*Courtesy of Motorola Inc.*)

tively. A two-terminal version of the SOT-89 has been designed for medium-power diodes. The DPAK is also being used for higher-power devices. In addition, two cylindrical MELF packages similar in design to MELF resistors are frequently employed. MELF-packaged diodes can be manufactured on the same type of equipment used for leaded diodes. For this reason they are less expensive packages than the SOT-23 or the SOT-89. MELFs can also be made hermetic and are

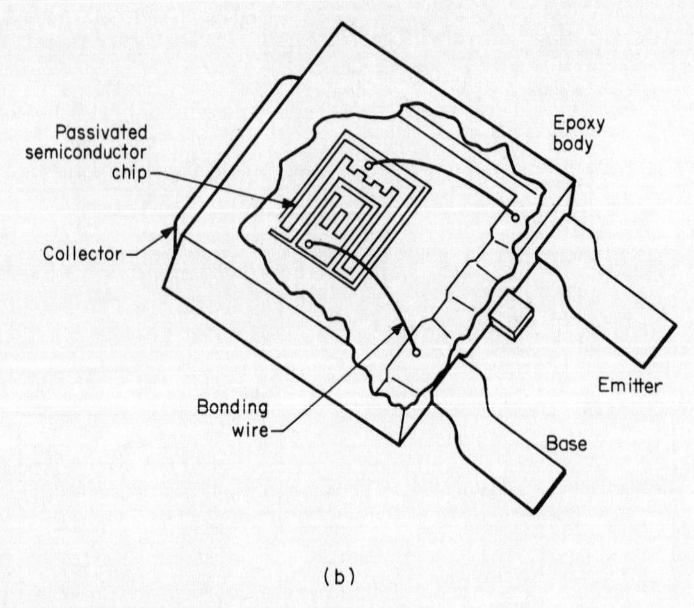

Fig. 3.15 (*Continued*)

preferred for high-reliability applications. But because MELF diodes are subject to the same assembly concerns as MELF resistors, SOT devices remain popular for many general-purpose applications.

3.4.7 Integrated Circuits

As with discrete semiconductors, surface mount ICs use the same silicon die and differ from through-hole versions only in package style. Again, the issues of thermal performance and package reliability are the primary concerns. A number of surface mount IC packages are available ranging from inexpensive plastic packages for consumer products to hermetic ceramic packages for military and high-reliability use. The characteristics of many common packages are described below.

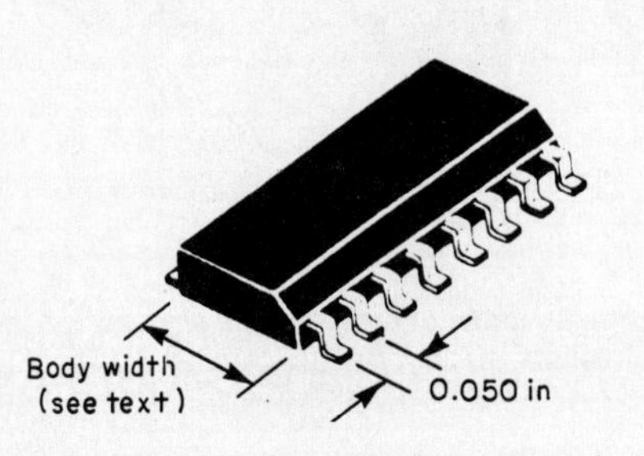

FIG. 3.16 Small-outline integrated circuit package.

3.4.8 Small-Outline Integrated Circuits

The small-outline integrated circuit, or SOIC, (Fig. 3.16) was originally designed for use in wristwatches. Similar to the DIP but smaller, it has leads on two sides with a lead pitch of 0.050 in. The leads are formed in a "gull-wing" pattern to sit flat on the PWB. Several body widths have been designed to accommodate various die sizes. In the United States, the EIA has adopted two body widths: 0.150 in (3.81 mm) for packages

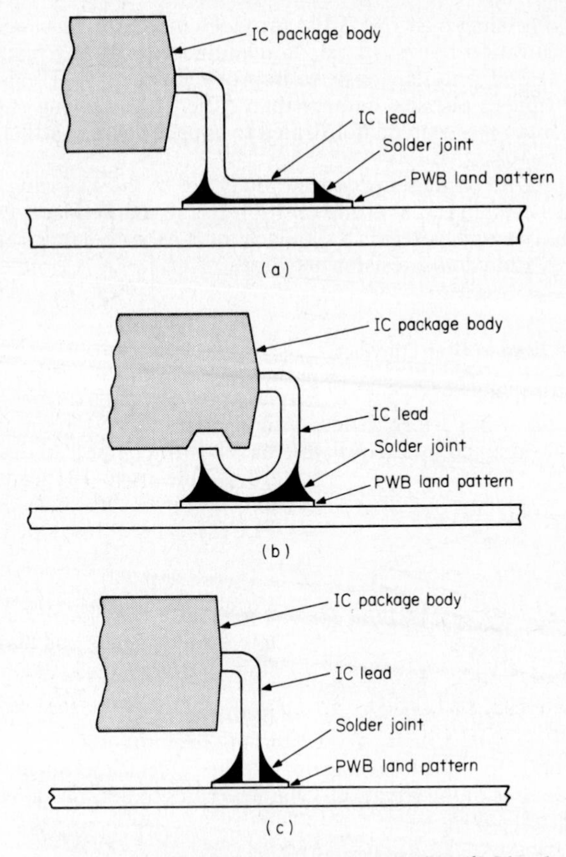

FIG. 3.17 IC lead configurations. (*a*) Gull-wing; (*b*) J-bend; (*c*) I-lead.

with 8 through 16 leads and 0.300 in (7.62 mm) for devices with 16 through 28 leads. In the Far East, packages with body widths of 4.4 mm (0.173 in), 5.4 mm (0.213 in), 7.5 mm (0.295 in), and 9.4 mm (0.370 in) are common.

3.4.9 Lead Configuration

The gull-wing lead design (Fig. 3.17*a*) promotes easy visual inspection of the soldered joint, but the small package size in relation to the total PWB area occupied restricts the size of the IC die that can be housed. The leads are also susceptible to damage in handling. If a lead is bent out of the seating plan by as little as 0.005 in, it may not solder properly during assembly. Other lead configurations have been developed to address these concerns. The "J-form lead" in Fig. 3.17*b* protects the leads against handling-induced damage and improves the ratio of die size to PWB area. In addition, the clearance between IC package and PWB is larger, making it easier to clean solder flux from under the package. Visual inspection of the solder joint, however, is much more difficult. Lead-forming of the J-form lead has proven more difficult than originally anticipated, adding cost and reducing yields for the IC manufacturer.

The I-lead joint shown in Fig. 3.17*c* provides many of the benefits of the J-form lead configuration but is simpler to manufacture. During manufacture, all leads may be sheared simultaneously to improve coplanarity. The I-lead joint is also less susceptible to physical damage than either the gull-wing or J-form lead. Mechanical strength has been demonstrated to equal or exceed either of the other approaches.

Virtually all SOIC packages produced today are of the gull-wing variety, and this package has established a strong position in the marketplace. However, the J-lead SOIC has been proposed as a standard for 1-Mbit dynamic random access memory devices and certain resistor networks.

3.4.10 Plastic Leaded Chip Carriers

For devices with higher pin counts, the plastic leaded chip carrier (PLCC) (Fig. 3.18) is preferred. It has J-form leads, with a pitch of 0.050 in on all four sides. Square PLCCs as defined by EIA/JEDEC have symmetrical lead designs ranging from 20 leads to 124 leads. For memory products, an 18-lead rectangular PLCC has been developed.

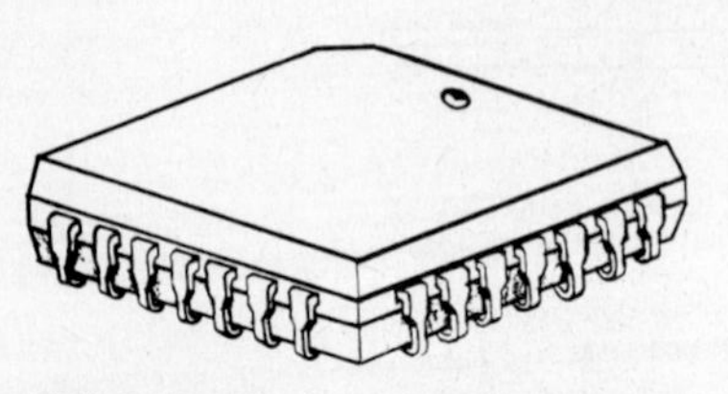

FIG. 3.18 Plastic leaded chip carrier.

3.4.11 Comparison of Small-Outline Integrated Circuits and Plastic Leaded Chip Carriers

Both the SOIC and PLCC are widely used, sometimes in competing applications. The choice of which package to use must be made on the basis of the specific requirements of the situation. Characteristics to be considered include the following.

3.4.11.1 Package Efficiency. The smallest practical PLCCs are the 20-pin square package for digital and linear circuits and the 18-pin rectangular package for memory products. The SOIC is more space-efficient for devices with lead counts below this limit. For devices with 28 pins or more, the PLCC is more space-efficient than the SOIC. In the range of 20 to 24 pins, the two packages are about equal in efficiency. For all pin counts, the SOIC has a lower profile than the PLCC, so in applications in which vertical clearance is of concern, the SOIC is preferred.

3.4.11.2 Assembly Processing. The gull-wing lead configuration of the SOIC is especially suited for wave soldering. Small SOICs can be mounted to the underside of the board and soldered in this manner. The J-form leads of the PLCC are not suited for wave soldering; these devices must be reflow-soldered.

3.4.11.3 Thermal Resistance. Neither the SOIC nor the PLCC is as efficient at removing heat from a device as is a DIP package. However, with copper alloy leads they both compare favorably to a DIP with alloy-42 leads. Relative thermal performance of the packages varies with lead count and die size. Table 3.4 compares the thermal performance of the SOIC and PLCC for average-size die, and

TABLE 3.4 Approximate IC Package Thermal Resistances

	Thermal resistance* (θja), °C/W	
	Package type†	
Lead count	SOIC	PLCC
8	160	—
14	120	—
16	110‡	—
	90§	
20	85	70
24	75	—
28	70	60

*Measured in still air.
†Copper lead frames.
‡0.150-in body.
§0.300-in body.

Fig. 3.19 shows how die size affects junction-to-ambient thermal resistance (θ_{ja}) for the 28-lead SOIC.

3.4.11.4 PWB Layout. Generally, it is easier to lay out a board using SOICs than using PLCCs. For the SOIC, traces can be routed under the package and out either end. The PLCC, with leads on all four sides, presents more of a challenge. Traces must either be routed between leads of the device or else brought down to inner layers of a multilayer board.

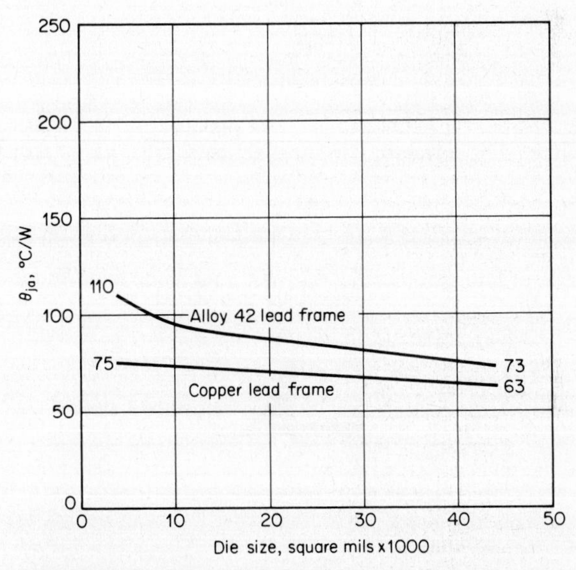

FIG. 3.19 Thermal resistance of 28-lead small-outline integrated circuit package.

3.4.12 Quadpacks

One of the earliest surface mount plastic IC packages is the quadpack (Fig. 3.20). Developed in Japan, it has gull-wing leads on all four sides. Lead pitches available include 1.25, 1.0, 0.8 and 0.5 mm. The fine lead pitches make this package especially suitable for devices with 84 leads and above. However, to achieve these pitches the lead frame must be manufactured from thin metal stock, making the leads extremely fragile. The handling difficulties associated with quadpacks have prevented their widespread use to date.

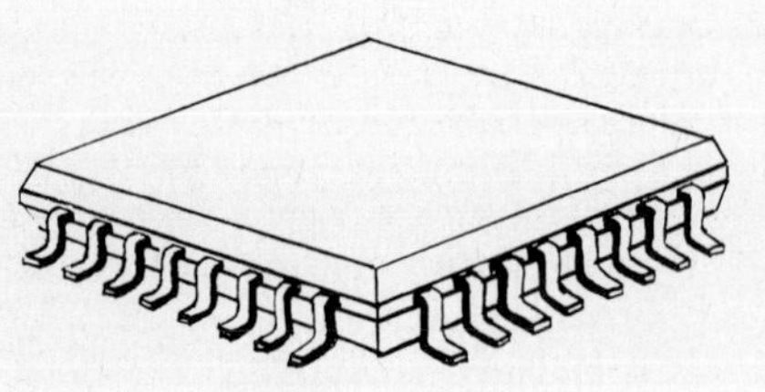

FIG. 3.20 Quadpack.

3.4.13 Leadless Ceramic Chip Carriers

The leadless ceramic chip carrier (LCCC) was designed for high-reliability and military applications. The ceramic design permits a lid to be soldered to form a

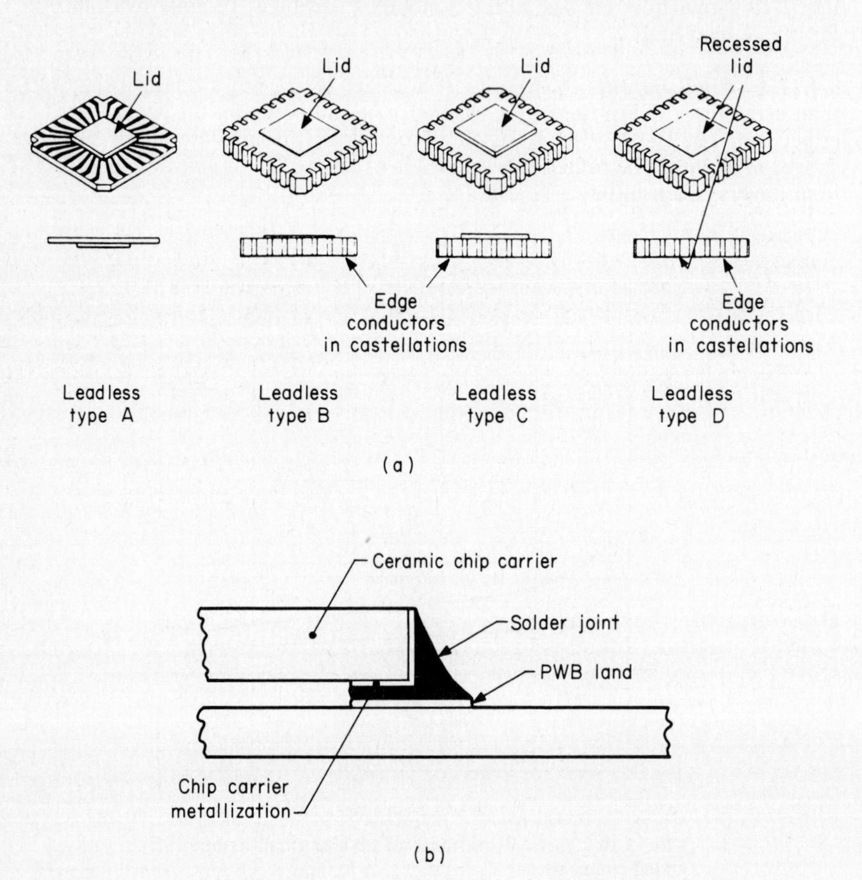

FIG. 3.21 Leadless ceramic chip carrier. (*a*) JEDEC versions; (*b*) solder joint cross section.

hermetic seal around the die. Four JEDEC-registered LCCC designs are shown in Fig. 3.21*a*. Types A and B were designed primarily for socket mounting with the lid either up (type A) or down (type B). Types C and D were designed for direct PWB mounting with the lid up (type C). or down (type D).

To permit an LCCC to be soldered to a PWB, the edge and underside of the ceramic material are metallized (Fig. 3.21*b*). Because the coefficient of thermal expansion (CTE) for ceramic is considerably less than that for an epoxy-glass PWB, the solder joint is subject to severe stresses over the full operating temperature range of the equipment. This can in turn lead to cracking and subsequent failure of the joint. A number of approaches have been proposed to solve this problem. One technique involves artificially constraining the expansion of the PWB. Typically this involves laminating the board to a base (such as a copper-Invar-copper composite) to approximate the expansion coefficient of ceramic. Other approaches include adding a compliant layer to the top of the PWB and adding tiny ball bearings between the LCCC and the PWB to aid compliancy.

Ceramic leaded chip carriers (CLCCs) were designed to avoid the problem of mismatched expansion coefficients. The leads act as compliant members between the chip carrier and the PWB. The two JEDEC-registered versions, the leaded types A and B, are shown in Fig. 3.22. Type A has integral leads, and type B is a leadless chip carrier to which clip leads have been attached. Because of the extra process steps involved with attaching leads, CLCCs are more expensive to manufacture than their leadless counterparts. The additional lead length can also degrade high-frequency electrical performance.

3.4.14 Other Component Types

Other SMCs that are commercially available include potentiometers, variable capacitors, resistor arrays, switches, connectors, quartz crystals, transient suppressors, and relays. Often the designs are simply leaded devices which have had their lead formed for surface mounting (potentiometers, switches, connectors). Entirely new package styles also have been developed (crystals, transient suppressors). In certain instances, manufacturers have been able to adopt an existing package such as the SOIC or PLCC (resistor networks). A high degree of standardization does not yet exist from one manufacturer to the next.

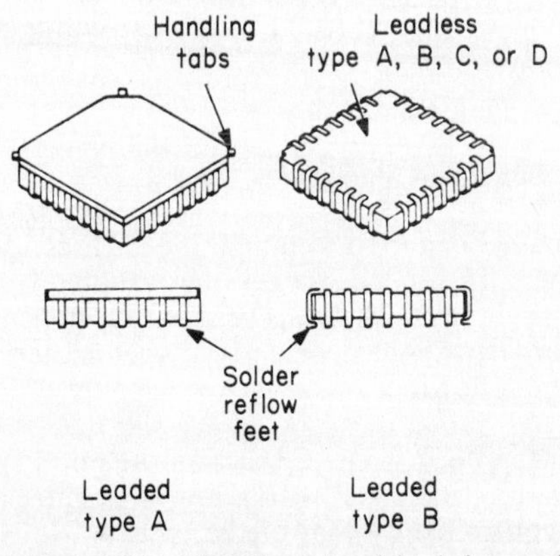

FIG. 3.22 JEDEC-registered leaded ceramic chip carriers.

TABLE 3.5 Coefficients of Thermal Expansion for Commonly Used Materials

Material	CTE, ppm/°C
Alumina	5–7
Epoxy-glass	12–16
Polyimide-glass	11–14
Copper-Invar-copper	5–6
Copper-clad molybdenum	5–6
Epoxy-Kevlar	6–7
Polyimide-Kevlar	5–7

3.5 DESIGN

Although many aspects of PWB design are independent of whether the components are surface mount or through-hole, several areas can be affected. Issues related to surface mount design are discussed in the following sections.

3.5.1 Solder Joint Reliability

The solder joints on SMCs serve both as electrical contacts and as the means for mechanically attaching the component to the board. As such, solder joint integrity is crucial to the reliability of the assembly. But while the mechanical strength of a through-hole solder joint is often an order of magnitude higher than the minimum required, the margin for SMCs can be as little as 20 percent.[2] The problem is compounded for leadless devices such as resistors, capacitors, and LCCCs, where differences in the CTE between device and substrate can place severe stresses on the joint.

Epoxy-glass and epoxy-paper PWB materials have CTEs ranging from 11 to 16 ppm/°C. The CTEs for ceramics range between 5 and 7 ppm/°C (see Table 3.5). The strain caused by this difference must be absorbed by the solder joint. Researchers have shown experimentally that when LCCCs are soldered directly to epoxy-glass PWBs, the solder joints may fail after less than 25 thermal cycles from −55 to +125°C.[3] Solutions to this problem run along two general lines: matching substrate and component CTEs, and providing a compliant layer to absorb the strain.

3.5.2 Matched CTE Substrate Materials

Several approaches have been used to match the CTE of PWB substrates to that of ceramic chip carriers. In general, these either involve constraining the expansion of the board through physical means or selecting substrates with CTEs lower than those of standard epoxy-glass or epoxy-paper materials. Commonly used techniques include the following.

3.5.2.1 Metal Core Approach. The CTE of an organic PWB may be artificially constrained by laminating the PWB with a low-expansion metal core. The CTE of the composite is controlled by that of the core. Various metal systems have been used for the core, including copper-plated alloy 42, copper-Invar-copper, Kovar, and copper-clad molybdenum. A typical approach using copper-Invar-

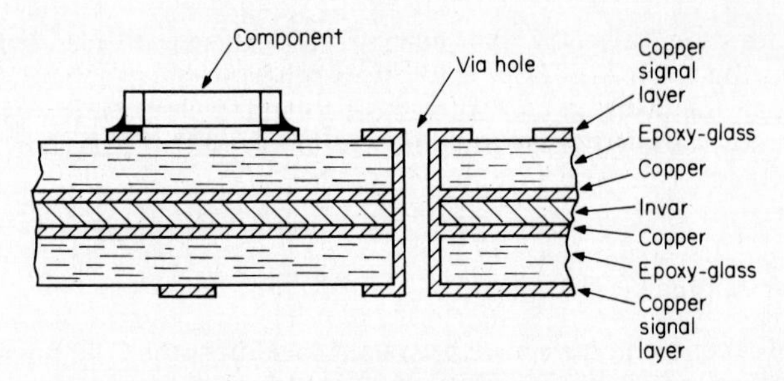

FIG. 3.23 Copper-Invar-copper board design.

copper is illustrated in Fig. 3.23. The ratio of copper to Invar is controlled to provide a CTE similar to that of ceramic. With this system, boards have been subjected to 1500 thermal cycles with no failures of solder joints.[4] In addition, the metal core can act as a heat sink for high-power components. Disadvantages of this approach include a considerable increase in weight of the assembly and a substantial cost increase over standard boards.

3.5.2.2 Aramid Fiber Approach. A commercially available aramid fiber, Kevlar, has been used as a replacement for glass-fiber mats in conventional PWB substrates. Kevlar has a negative CTE of −2 ppm/°C, which makes it possible to construct an epoxy-Kevlar board with a CTE similar to that of ceramic. Thermal cycling tests have shown that such assemblies show no evidence of solder joint failures after 200 cycles from −55 to +150°C. Kevlar presents a number of problems, however. Epoxy-Kevlar boards are two to three times more expensive than epoxy-glass. Kevlar also complicates the board fabrication process because it is difficult to cut and it readily absorbs moisture. Moisture absorption can lead to poor bonding between the Kevlar and epoxy and to reduced volume resistivity of the resulting laminate.

3.5.3 Compliant Layers

By adding a compliant layer to the PWB and making solder connections to the top of this layer, solder-joint stresses can be reduced. This approach is shown conceptually in Fig. 3.24. Various materials have been evaluated for the flexible layer, including Teflon film, silicone or polyurethane resins, and rubber nitrile blends. Circuit traces and pads are often applied to the top of this layer by way of an additive process.

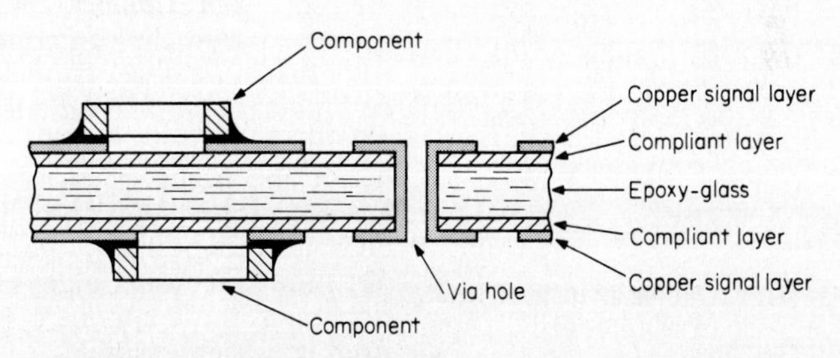

FIG. 3.24 Compliant layer board.

Boards designed this way can survive 1000 thermal cycles from −55 to +125°C, but the compliant layer introduces other problems. Via holes must be plated through the layer and so are subject to added stresses. In addition, after components are mounted, the compliant layer may resonate at certain frequencies and greatly increase the stress on the joints.

3.5.4 Other Approaches

A number of other techniques have been used to address the CTE mismatch problem. One approach is to use spacers such as solder-coated ball bearings or solder-coated copper wires to lift the component off the board. This is conceptually similar to adding leads to the device. The main drawbacks are the slow, labor-intensive nature of the process, making it a costly solution, and the high defect levels associated with manual assembly.

Another approach is to simply mount the devices directly to an ordinary glass-epoxy board and restrict the maximum permissible temperature excursion. Over a 0 to +70°C operating temperature range, this simplified approach may suffice for devices with 44 leads or less.

3.5.5 Component Land Patterns

Optimum assembly process yields are achieved with properly designed component land patterns. Poorly designed land patterns can contribute to the following assembly defects.

3.5.5.1 Component Misalignment. Component misalignment, illustrated in Fig. 3.25*a*, occurs when a properly placed component rotates unacceptably during the soldering process. Although electrical contact may be retained, the mechanical strength of the joint may be reduced to an unacceptable level. With integrated circuits, misalignment may cause component leads to shift such that some make contact with adjacent PWB lands, causing an obvious functional defect. Misalignment is caused by a lateral imbalance of forces during reflow. Surface tension of the molten solder pulls the component to a position of minimum solder surface energy. This is a result of a combination of poor land design and failure of all contacts to wet simultaneously.

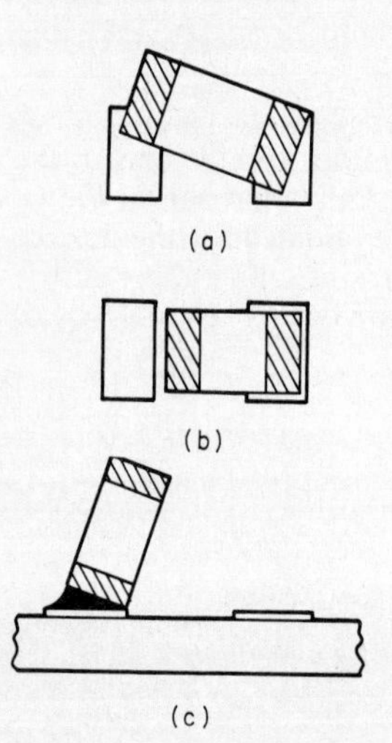

FIG. 3.25 Defects resulting from poor land pattern designs. (*a*) Misalignment; (*b*) open contact; (*c*) tombstoning.

3.5.5.2 Open Contact. When open contacts occur, the component moves such that no contact is made between component and land on one side (Fig. 3.25*b*). This is a definite functional defect and occurs when land patterns are not properly designed. If contacts on both sides of the component do not wet simultaneously, solder surface tension pulls the component back away from its design location.

3.5.5.3 Tombstoning (Drawbridging). Tombstoning (Fig. 3.25*c*) occurs when a component flips up to a vertical or near-vertical position. It is a problem primarily with small two-leaded chip components and never occurs with semiconductors. Although tombstoning has been attributed to a number of causes, land pattern design is a primary contributor. If both component terminations do not wet simultaneously, solder surface tension can pull on the end of the component, raising it to minimize total surface energy.

Each of the above defects can be reduced (although not entirely eliminated) by proper land pattern design. Industry associations, particularly IPC, have been instrumental in identifying optimum land patterns for many types of SMCs. The IPC document IPC-SM-782 describes a series of land patterns that have been determined experimentally to minimize defects. Although it is beyond the scope of this chapter to discuss individual land patterns in detail, certain general criteria apply.[5]

A generalized component land pattern is shown in Fig. 3.26. Key characteristics include the pad length and width, separation between pads, and the overlap and extension as defined in the figure. It is important that these dimensions be optimized to balance the forces at the joint during reflow. Excessive separation and extension, for example, contribute to opens. Tombstoning shows a strong correlation to extension, while misalignment is a function of pad width. Unfortunately, the wide dimensional tolerances of many chip components make it difficult to design a single land pattern to adequately accommodate all chips within the dimensional tolerances. For example, the tolerance on width for some small chip capacitors approaches ±20 percent. A land pattern designed for optimum dimensions may show a considerably higher incidence of defects when used with components at either end of the tolerance. Refer to IPC-SM-782 for a complete current treatment of recommended land pattern dimensions.

3.5.6 Other Design Considerations

In addition to land pattern design, certain other design practices can improve assembly yields.

1. Protect all circuit traces with a solder mask. This aids in preventing component movement by confining solder flow to the land itself. It also reduces the possibility of solder bridging to an adjacent trace.

2. Design lands such that they have similar thermal characteristics. This ensures that solder reflow will occur on all lands simultaneously, minimizing force imbalances. This is particularly important when one land is connected to a ground plane. When this occurs, the land should be thermally isolated from the ground plane through a short circuit trace.

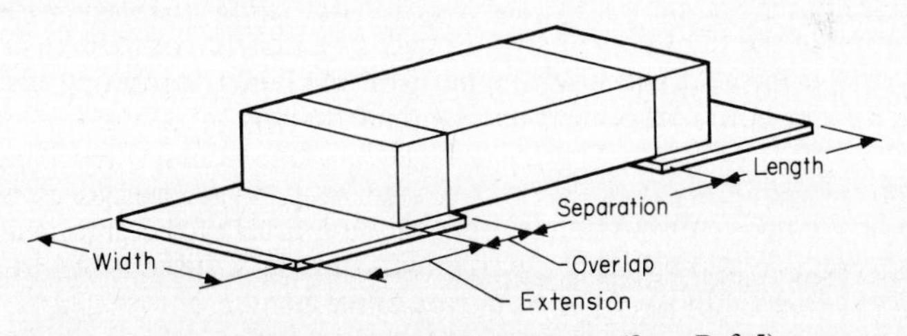

FIG. 3.26 Generalized component land pattern (from Ref. 5).

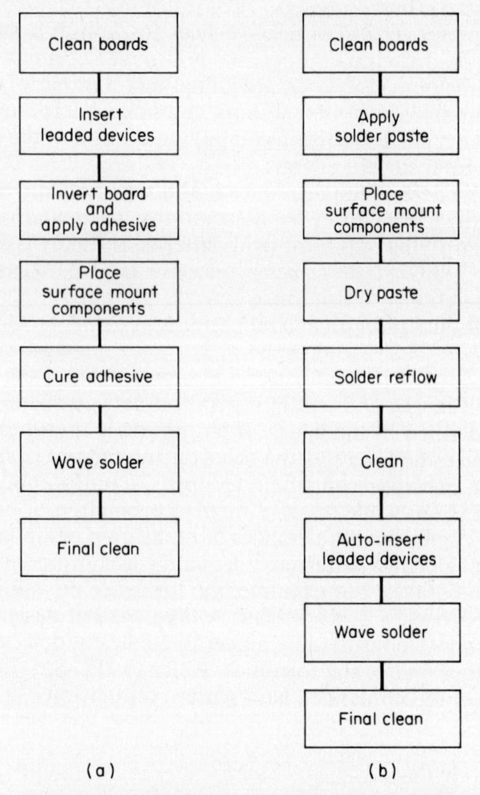

FIG. 3.27 Surface mount process flows. (*a*) Adhesive attach–wave solder; (*b*) reflow soldering (from Ref. 8).

3. Select components and PWB materials with adequate solderability. Poor solderability can cause component leads to wet at different times, resulting in unbalanced forces. Generally, solder-coated terminations are preferred.[6] Solder alloys near the eutectic composition exhibit the highest degree of solderability, although higher-tin-content finishes have also been reported to work well.[7] Alloys with less than 50 percent tin should be avoided because of poor solderability. Little difference has been found between plated finishes and those that have been applied through dipping the leads in molten solder. However, the shelf life for plated finishes that have not been fused is less than that for solder-dipped parts.

4. Minimize the thickness of solder mask underneath components. If the solder mask is too thick, it will hold the parts off the lands, inhibiting the ability of the solder to wet both component and land adequately.

Much of the above discussion is especially important for components that are reflow-soldered. Wave-soldered components are less susceptible to soldering defects because the adhesive prevents part movement.

3.6 MANUFACTURING PROCESSES

The two principal surface mount process technologies in use (Fig. 3.27) are "adhesive attach–wave soldering" and "reflow soldering."[8] The former, first widely used in Japan, involves attaching SMCs to the underside of the PWB with special adhesives. (Through-hole components are inserted on top of the board before the SMCs are attached.) The assembly is then sent through a wave- or drag-solder machine to complete the connections. In this approach the SMCs are actually immersed in the molten solder during the process. The reflow soldering method derives from traditional hybrid circuit techniques which place SMCs on the top of the board. The first step is to dispense a solder paste onto the raw board using either a silkscreen or a stencil. (For small production runs the paste may be applied manually.) SMCs are placed into the viscous paste, which is then dried to drive off solvents. No additional adhesive is necessary, since the paste holds the components in place. Solder reflow is accomplished by one of several common methods, including vapor phase, infrared (IR), or thermal conduction. The boards then go through a cleaning step to remove residual flux and any loose solder balls that may remain on the board. Finally, if through-hole components are needed, they are inserted, and the board is sent through a wave-solder operation. Normally, the top of the board does not get hot enough to reflow the solder holding the top side SMCs.

Surface mount assemblies are sometimes classified on the basis of how they combine surface mount and through-hole components. This scheme is defined as follows.

Type I—total surface mounting: Assemblies in this category consist entirely of SMCs and may include components on one or both sides of the board.

Type II—mixed technology: These assemblies combine all types of SMCs with through-hole devices on the same board. Components may be mounted on one or both sides of the board.

Type III—underside attachment: These assemblies are actually a subset of type II. They include only small SMCs such as capacitors, resistors, and SOT devices that are adhesively attached and wave-soldered to the underside of conventional through-hole assemblies.

This classification scheme, although useful in certain situations, says nothing about the processes used to mount the components. This chapter will classify processes as either adhesive attach–wave solder or reflow solder and will not make use of this alternative classification scheme.

3.6.1 Adhesive Attach–Wave Solder

The adhesive attach–wave solder process was developed in part to increase the utilization of through-hole process equipment. In theory, by using existing wave-soldering equipment, a PWB containing both through-hole components and SMCs could be assembled with minimum impact on existing process flows. In practice, wave soldering of surface mount assemblies is different from through-hole soldering in two respects. First, the components must be mechanically attached to the board before soldering. Second, the wave dynamics that have worked for through-hole assembly must be modified when soldering SMCs. Each of these differences is explained in detail below.

3.6.2 Adhesives

Components to be wave-soldered must be adhesively attached to the board so that they remain in place during the wave-soldering process. Once soldering is complete, the adhesive is no longer needed. It provides no long-term benefit and can be a liability if it exhibits mechanical or electrical characteristics that degrade circuit performance. The adhesive used must meet the following basic requirements.

1. It must hold components in the correct orientation upon placement, maintaining that orientation through the cure and wave-soldering operations. Although ideally the adhesive would cure without moving the part, movement of up to 0.010 in has been observed in practice.
2. It must have adequate adhesion to different surfaces and be unaffected by exposure to the adverse environment of solder flux and wave-soldering.
3. For reliability, it must be chemically inert throughout the life of the assembly.

Two types of adhesives are commonly used for SMT assembly: epoxies and acrylics. Other types are used on a more limited basis.

3.6.2.1 Epoxies. Epoxy resins used in SMT assembly are typically single-component heat-curing systems. Cure cycles from 10 min at 100°C to 2 min at 150°C are typical. Epoxies provide good insulation resistance and bond strength. They are available in a range of formulations for special requirements. A disadvantage of epoxies is that defective components are difficult to remove during repair.

3.6.2.2 Acrylics. Acrylic adhesives used for surface mounting are usually cured by a combination of UV and IR energy. The UV cure cycle requires that a portion of the adhesive extend outside the component body. This exposed adhesive is cured with a 15- to 30-s exposure. The subsequent IR cycle ranges from 90 s to 2 min at 150°C and cures the adhesive in areas that were not exposed to UV. Acrylic adhesives offer faster curing cycles than epoxies but are most suitable in low-density applications in which adequate adhesive can be exposed to the UV energy.

3.6.2.3 Other Adhesive Types. Cyanoacrylate adhesives cure rapidly, but their viscosity and poor working life make them unsuitable for general-purpose use. They have, however, been used successfully to hold quadpack ICs during the solder operation.[10]

Two-part epoxies can be used where faster cure cycles are required, but their working life is considerably shorter than that of single-component types. They must be mixed carefully to control their properties and prevent the introduction of air bubbles.

3.6.3 Adhesive Application

Adhesive can be applied through one of several techniques. The most common are syringe dispensing, pin transfer, and screen printing.

3.6.3.1 Syringe Dispensing. Adhesive is often dispensed by a nozzle on the pick-and-place machine immediately prior to placing the component. Adhesive is stored in an air-driven syringe mounted to the placement head or on a separate head preceding it. Some machines apply the adhesive to the PWB, whereas others apply it to the component. Since adhesive is dispensed to only one component location at a time, syringe dispensing is most suitable for low- to moderate-speed assembly operations using sequential placement equipment.

3.6.3.2 ***Pin Transfer.*** This technique makes use of a matrix of pins held in a fixture. The pins are arranged so that when the fixture is placed over the PWB, they touch the board in the locations where adhesive is to be dispensed. In operation, the pins are first dipped into a thin layer of adhesive. Surface tension causes a quantity of adhesive to remain on the pin after withdrawal. The fixture is located over the PWB and lowered to within close proximity of the board (the pins do not typically touch the board). The adhesive makes contact with the PWB, and a portion is transferred from the pins to the board. Adhesive quantity is controlled by carefully controlling the viscosity of the adhesive and the proximity of the pins to the board. The pin transfer method is suitable for high-volume applications. The adhesive used must have a long working life at room temperature, typically greater than 24 h.

3.6.3.3 ***Screen Printing.*** Adhesive may be screen-printed onto a PWB in a manner similar to that used in the graphic arts. Screens consist of a wire mesh held taut in a frame and covered with a photosensitive emulsion. Before initial use, the emulsion is exposed in the desired pattern and processed such that windows are opened in it where adhesive is to be deposited. In use, the screen is registered to the board and held in contact with it. Adhesive is spread across the top of the screen, and a squeegee is drawn across it. This forces the adhesive through the windows on the emulsion and onto the board. The amount of adhesive dispensed is controlled by the thickness of the wire mesh and the size of the windows in the emulsion. Screen printing processes an entire board at once. Complex patterns of adhesive can be laid down by controlling the shape of the openings in the emulsion. The adhesive used in screen printing should be thixotropic: its viscosity should drop during the application process. This ensures that it will flow onto the board properly. It must also have a long working life (more than 24 h) and not be adversely affected by repeated screening operations. Since screen printing requires a flat board surface, it cannot be used on boards already containing through-hole components.

3.6.4 Component Placement

Surface mount components are placed on the board by one of three methods: automatic pick-and-place, robotic placement, or manual placement.

3.6.4.1 ***Automatic Pick-and-Place.*** Most surface mount components are assembled to the board by an automatic pick-and-place machine. This equipment is typically designed to handle a specific part size or range of sizes. Parts are delivered to the machine in bulk, in tubes, or in a tape-and-reel format. The computer-controlled pick-and-place machine removes the component from its packaging and places it on the board in the correct location. Centering and squaring jaws are often used to ensure accuracy of part location to ± 0.004 in or less.

Pick-and-place machines are designed for either sequential or simultaneous placement. Sequential machines place one component at a time at speeds from 1000 to 20,000 components per hour. Many types of sequential machines are designed to handle a wide range of components from small ceramic capacitors to large ICs without operator intervention. Component capacities range from about 20 component types on line for a small-volume machine to 200 component types on line for a high-volume machine.

A variation of sequential placement is in-line placement. In this approach, a series of placement heads is used to place all the components on a board sequentially. Each placement head places only one component, so the number of heads is dictated by the number of components on the board. In-line systems are used for high-volume production. Each head moves in only one preset pattern, so it can achieve a much faster cycle time than a head that must place a multitude of component types. Since boards move down the line sequentially, the effective cycle time for an entire board is just the time it takes to place a single component. Because of the large number of dedicated heads necessary, in-line factories require a high capital investment. They are also difficult to reconfigure for different board types.

Simultaneous placement machines are designed to mount all components on a board at once. Consequently they are capable of much higher placement rates than sequential machines and approach that of in-line equipment. Components are fed from underneath to an array of placement heads arranged to mount them directly onto the appropriate locations on the board. Cycle times of 5 to 10 s per board are typical. These machines are configured for a specific board, and like in-line equipment, they lack the flexibility found on sequential placement machines. For high-volume assembly of a single board type, however, they are a very economical approach.

3.6.4.2 Robotic Placement. Robotic equipment is used where extreme precision or versatility is required. The articulated arm of a general-purpose robot is capable of the complex motions necessary to place many forms of odd-shape components. It is also capable of more precise positioning, making it the machine of choice for high-accuracy requirements such as placing high-pin-count ICs. Robots are much slower than pick-and-place machines, with an average placement cycle time of 3 to 10 s.

3.6.4.3 Manual Placement. Manual placement is limited to low-volume prototype or production assembly. Experienced operators can place low-pin-count

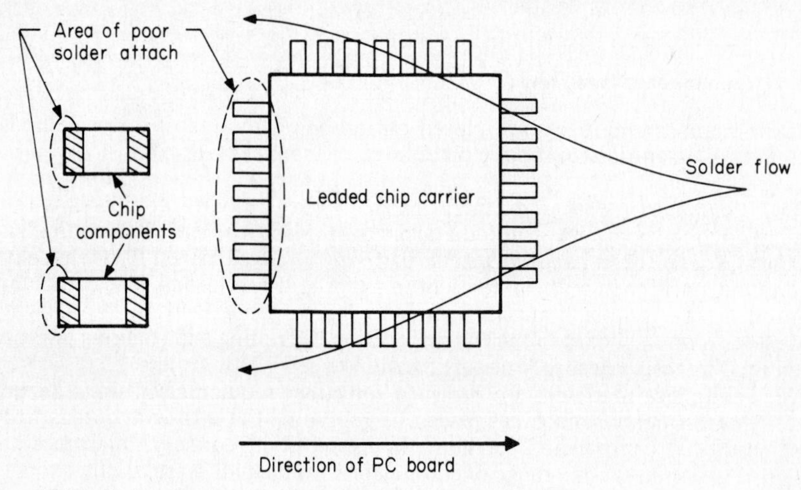

FIG. 3.28 Shadowing of solder joints during wave soldering.

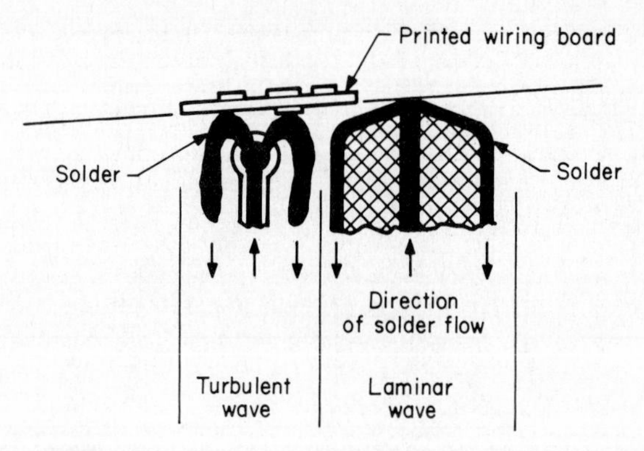

FIG. 3.29 Dual-wave soldering system.

devices at a rate of 50 to 150 devices per hour. Components with lead pitches finer than 0.050 in are very difficult to place manually.

Although manual placement can be done with tweezers, vacuum-tip tooling is preferred. Picking up larger components is difficult with tweezers. Component damage from electrostatic discharge is also more likely. Magnification from 3× to 10× is recommended. Experience has shown that defect levels for manual placement run 10 to 100 times higher than for automatic placement.

3.6.5 Wave Soldering

Once the components are attached, they are ready to be wave-soldered. The laminar waves used in through-hole soldering are not optimum when used for SMCs. Surface mount component bodies interfere with solder flow, causing a phenomenon called "shadowing" (Fig. 3.28). Because of this effect, leads on the trailing edge of a component frequently have insufficient solder or no solder at all. Equipment manufacturers have developed various modifications to improve soldering of SMCs. Typically, a second, turbulent wave is added to the machine preceding the laminar wave (Fig. 3.29). This turbulence ensures adequate wetting of all joints. The subsequent laminar wave removes excess solder and eliminates solder bridges. Preheating is necessary to minimize thermal shock that might otherwise damage components. A thermal profile typical of that employed in dual-wave soldering equipment is shown in Fig. 3.30.

With these modifications, wave soldering is effective for chip components and small-pin-count ICs. Larger chip carriers may still have problems with bridging or incomplete wetting because of their large body heights and close lead spacings.

Using the wave-soldering process, SMCs can be readily combined with through-hole components on the same board. When automated assembly equipment is used, through-hole components are first inserted and clinched (but not soldered). The board is then inverted and the SMCs mounted adhesively. Finally, both leaded components and SMCs are wave-soldered. This sequence ensures that the SMCs are not damaged by the cut-and-clinch tool of the autoinserter.

3.6.6 Reflow Soldering

The reflow soldering approach is potentially more versatile than the adhesive attach–wave solder process. Components such as variable resistors and some

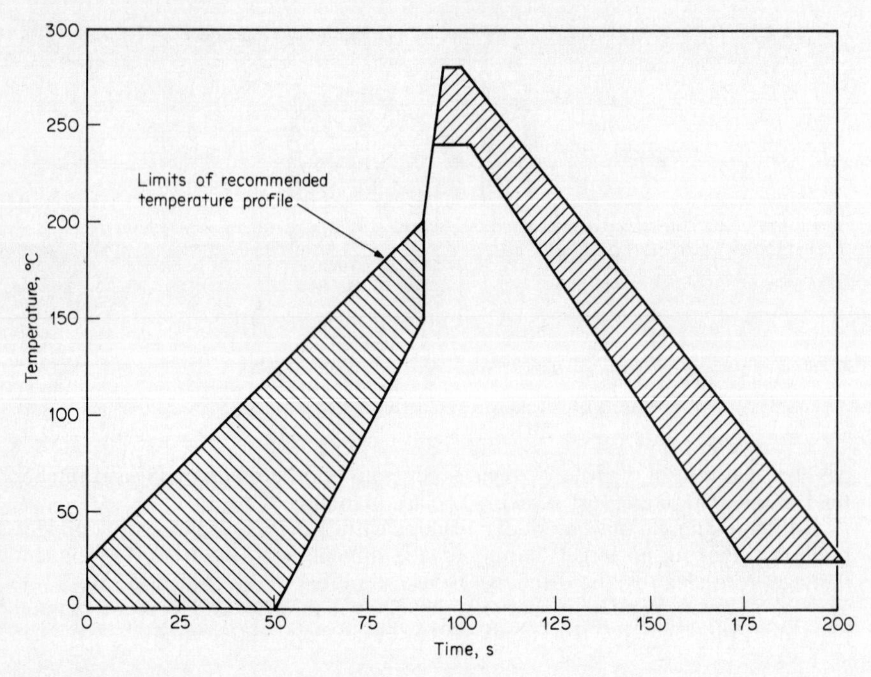

FIG. 3.30 Typical dual-wave soldering temperature profile.

wire-wound inductors cannot survive immersion in a solder wave. Others, such as high-pin-count chip carriers have not been wave-soldered with acceptable yield.

3.6.7 Solder Paste

The most common method of reflow soldering makes use of solder paste (sometimes called solder cream). Solder paste consists of a paste flux impregnated with small nodules of solder. Solvents are added to improve flow characteristics during the application process and to control the working life of the paste. Tin-lead or tin-lead-silver solders with RMA-grade fluxes are commonly specified. Pastes for surface mount applications must meet the following criteria.

1. The viscosity and particle size must be suitable for the intended method of application. For screen printing, solder particle sizes from 15 to 55 μM in diameter are generally used.[11] The shape of the particles is an important factor. Normally, spherical particles are preferred. For fine-line soldering, pastes with a percentage of nonspherical solder particles may work better than those consisting entirely of solder spheres.[12]
2. The paste must be formulated so as to resist the tendency to form loose solder particles, called "solder balls," after reflow.
3. The solder must reflow at a temperature compatible with the materials to be joined.
4. The flux residue must be removable in standard solvent or by aqueous cleaning processes.

Solder balling is the most widely discussed characteristic of solder pastes. Isolated spheres of solder can be left on the board if the solder does not completely flow to the connection during reflow. These balls can cause reliability problems if not subsequently removed.

Several factors contribute to this phenomenon. If the paste is reflowed before the solvents have fully evaporated, they can boil off and eject particles of solder. The solder purity is also important. Pastes with high oxide content have more problems than those with low oxide levels. Finally, the basic solderability of the materials to be joined can play a part. Tin-lead–coated component terminations are often specified because of their high inherent solderability.

3.6.8 Solder Paste Application

Screen printing is a widely used method of applying solder paste to PWBs. The process is identical to that described above for applying adhesive. The volume of solder paste dispensed is more critical than for adhesives. The amount must be adequate to ensure proper soldering of all joints but not so much as to promote bridging between adjacent joints. Screen printing of solder paste is sometimes performed in temperature- and humidity-controlled rooms to maintain adequate control of the process.

A variation of screen printing that is often used for solder paste involves the use of stencils. In this approach, a solid metal sheet replaces the wire mesh. Openings are chemically etched into the sheet in a manner similar to that used for screens. A stencil has the following advantages over a screen.

1. It has a considerably longer useful lifetime. The solid sheet is less likely to undergo plastic deformation than is a screen.
2. Thicker layers of adhesive may be deposited. Paste depth is controlled by the thickness of the metal sheet rather than by an organic emulsion.
3. The unobstructed openings in a stencil are more tolerant of material and process variations than is the wire mesh of a screen.

For these reasons, stencils are preferred for high-volume production. Disadvantages of stencils include high initial fabrication cost and less flexibility in possible pattern shapes. (For example, a doughnut-shaped opening is easy to achieve with a screen but more difficult with a stencil.)

For small-volume production, syringe dispensing of paste is frequently employed. Although air-driven syringes can be used, better control of solder volume is obtained by using a positive-displacement pump to drive the syringe. Such systems use a screw-driven piston to deliver a controlled amount of paste to the board.

3.6.9 Solder Paste Selection

When selecting a solder paste, one should observe the following guidelines.

1. Select a suitable solder alloy. For components that are tin- or tin-lead–coated, Sn63 or Sn60 alloys are recommended (Table 3.6). When soldering silver-bearing terminations, the Sn62 alloy is recommended. The silver content in the solder prevents scavenging of silver in the termination.
2. Select a paste with a high metals content. Metals weight percentages from

TABLE 3.6 Solder Alloys for Surface Mounting

	Nominal metal content, wt %			
Designation	Tin	Lead	Silver	Other metals
Sn60	60	40	—	0.08
Sn62	62	36	2	0.08
Sn63	63	37	—	0.08

88 to 92 percent are generally preferred to prevent slumping of the paste and subsequent bridging of closely spaced lines.

3. Select the lowest activity flux consistent with the application. Although the highly active RA* fluxes improve solderability, they can be difficult to remove. For many applications, mildly activated RMA* fluxes are sufficient. Such fluxes are less likely to cause long-term reliability problems if left on the PWB.

4. Specify the viscosity and desired working life of the paste. Typical viscosity for a screen-printing application would be 400,000 to 500,000 cP. Working life is controlled by the types of solvents added to the paste. It could range from a few hours to 24 or more, depending on how soon after screen printing the components are to be applied.

5. Dry the paste before reflow to prevent solder balls. The drying step is essential to remove solvents that could otherwise boil off during reflow and scatter solder particles across the board. The drying process must be tailored to the specific paste. Typical parameters range from 5 to 30 min at temperatures from 80 to 125°C.

3.6.10 Other Solder Application Methods

Because of these problems with solder pastes, and because screen printing is inherently a "batch mode" process, other techniques for applying solder to boards have been tested with varying success. One technique that has shown promise is to apply a heavy layer of solder (0.007 to 0.010 in thick) directly to the lands on the bare PWB as part of the board fabrication process. This has been called a "solder bump" process in reference to the appearance of the lands after application. Two variations of this technique have been used. In the first, the solder is plated onto the board as tin and lead from solution.[13] In the second, solder paste is screen-printed and immediately reflowed on the bare board. When the board is ready for component placement, it is first coated with a paste flux which serves to hold the components in place. Subsequent processing is identical to that of screen-printed solder paste. These approaches have the following advantages:

1. While solder paste must be reflowed within a few hours of application, pretinned boards can be stored for long periods with the solder in place.
2. Solder balls are easily eliminated. Cleaning solder balls off a bare board is much easier than cleaning a board with components in place.
3. Pretinned boards can be economically assembled in small lots. In contrast, screen-printing setup times must be amortized over large batches to be economical.

*"RA" stands for "rosin—activated." "RMA" stands for "rosin—mildly activated."

Because of the additional process steps required, fabrication costs for solder bumped boards are higher than for standard boards. Also the process is more sensitive to variations than when solder paste is used. To ensure proper joint formation, intimate contact is necessary between the component leads and solder on the pads. Since leads rest on top of the solder rather than being buried in the paste, lead coplanarity is more critical. These problems have kept the solder bumping process from enjoying widespread popularity, but it is being used successfully in certain applications.

3.6.11 Reflow

Three techniques are commonly used for reflowing the screened solder paste onto an SMT assembly: thermal conduction, IR, and vapor phase. Other methods have also been used in certain circumstances.

3.6.11.1 Thermal Conduction. The simplest method is reflow by thermal conduction, historically used in the hybrid circuit industry. In this approach, boards are placed on a belt which travels over two heater stages. The first serves to preheat the board, minimizing thermal shock and driving off solvents. The second is the reflow zone where actual soldering is accomplished. The thermal profile is controlled by adjusting heater stage temperatures and by varying the belt speed. Heat is transferred through the board, and so the technique works best for flat substrates.

3.6.11.2 Infrared. Infrared reflow is another method commonly used in the hybrid industry. Because it does not depend on heat conduction through the substrate material, it is not confined to soldering flat surfacees. Infrared systems employ one of three types of emitters: tungsten filament lamps, Nichrome alloy quartz lamps, or area-source emitters.[14]

3.6.11.2.1. Tungsten Filament Lamps. Tungsten filament emitters are only partially suited for solder reflow processes. The peak wavelength of the commonly used T-3 tungsten lamp operated at rated voltage is 1.15 μm, corresponding to a filament temperature of 2246°C. Such lamps do not operate efficiently at temperatures below about 1100°C. Since lamp temperatures are far above the reflow temperature of tin-lead solder, PWB assemblies can be damaged if the equipment is not properly adjusted. Another problem is that of color selectivity (differential heating of materials of different colors). At the near-infrared wavelengths generated by tungsten lamps, metal conductors reflect much of the energy, while the black epoxy used to encapsulate ICs absorbs most of it. By the time the solder joints have absorbed enough energy to reflow, the IC may well have been damaged.

Finally, since tungsten lamps depend on direct illumination for heat transfer, they are known as focused systems. Large components can shadow IR energy from the solder joints, resulting in uneven temperature distribution and incomplete reflow across the board.

3.6.11.2.2. Nichrome Alloy Quartz Lamps. An improvement over the tungsten lamp is the nichrome alloy quartz lamp. This type of source operates efficiently over the temperature range of 500 to 1100°C. Nichrome alloy sources exhibit less color selectivity than tungsten lamps, both because of their lower operating temperature and because a portion of the heat transfer (about 10 percent) occurs via thermal convection rather than through direct radiation. Heating irregularities caused by component shadowing are also less severe.

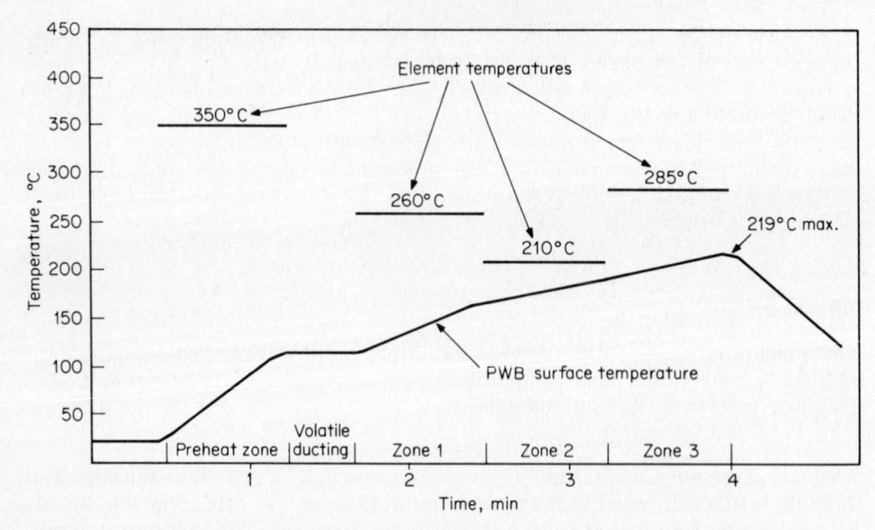

FIG. 3.31 Typical area-source infrared reflow temperature profile (from Ref. 14).

3.6.11.2.3. Area-Source Emitters. Area-source emitters exhibit the least amount of color selectivity and shadowing of any IR source. Much of this improvement can be attributed to the fact that thermal convection accounts for up to 40 percent of the heat transfer. Emitters consist of resistive elements embedded in a ceramic base backed by a refractory insulation material. A high-emissivity material is attached to the front of the assembly to diffuse the resulting IR emission. Area-source emitters are designed for a maximum peak wavelength of about 2.7 μm (800°C) but are typically operated at temperatures of 200 to 400°C. At these temperatures, they can heat air very efficiently to promote reflow by thermal convection. The combination of convection with direct IR heating reduces color sensitivity and shadowing. Such systems are known as nonfocused systems. One limitation of area-source systems is the length of time required to change temperatures. Because of the high percentage of heat transfer through thermal convection, it may take 20 min for the system to stabilize after adjustments have been made.

Most SMT reflow is now done using either nichrome alloy or area-source emitters. Typical systems use four to seven independent zones to define the thermal profile. Such a profile is shown in Fig. 3.31. For any given combination of emitter temperature and belt speed, the temperature profile of a board depends on its thermal mass. Each different type of board must have reflow parameters optimized specifically for it. Often this can be accomplished merely by changing belt speed.

3.6.11.3 Vapor Phase. Vapor phase reflow was developed by Western Electric Co. to try to overcome the problems associated with IR and thermal conduction techniques. It is accomplished by heating a long-chain fluorinated hydrocarbon liquid to its boiling point, approximately 215°C. This creates above the liquid a vapor zone at the boiling point. The cool PWB is immersed into the hot vapors, which transfer heat to the board. Since the liquid boils at a temperature higher than the melting point of tin-lead solder, this causes the solder to reflow. Temperature is precisely controlled by the liquid's boiling point.

The fluid used in vapor phase reflow is extremely expensive (more than $500 per gallon), and steps must be taken to minimize loss through evaporation. One widely used technique is to add a secondary fluid that boils at a lower temperature. The vapors of the secondary fluid form a blanket over the primary vapors, reducing loss through evaporation. Systems that make use of a secondary vapor are called "dual-vapor systems."

Vapor phase systems operate in either a batch mode or a continuous ("in-line") mode, as described below.

3.6.11.3.1. Batch Mode. Batch-mode systems derive from traditional vapor degreasing systems used for solvent cleaning processes. Typical construction is illustrated in Fig. 3.32. Boards to be soldered are placed in a basket which is lowered automatically into the reflow tank. When their rate of descent through the secondary vapor is controlled, some control over the temperature profile seen by the board can be maintained. This is especially important when soldering components that are sensitive to thermal shock.

Because of the large open area at its entry, a batch-mode system is susceptible to loss of the primary fluid through evaporation and drag-out. For this reason, production systems are invariably dual-vapor designs. To further minimize vapor loss, the basket can be programmed to dwell in the secondary vapor for an extended period during withdrawal.

Batch systems are best suited for relatively low production rates. Except when soldering very small boards, only a single board may be reflowed at one time. The long cycle time, especially when minimizing drag-out, further reduces throughput.

3.6.11.3.2. Continuous Mode. Continuous systems are designed to support higher production rates than batch systems. As shown in Fig. 3.33, boards are carried on a conveyor belt through the reflow zone and out the other side. Some systems control the temperature profile by adding a preheat zone at the inlet of the machine. The outlet may be designed to directly feed an in-line cleaning system. Continuous systems may be either single-vapor or dual-vapor. Single-vapor systems must be carefully designed to minimize evaporation loss of the primary fluid.

Continuous systems have two drawbacks compared to batch systems. First is the reduced ability to control the temperature profile seen by the board. Whereas a batch system can be programmed for independent dwell times in the primary and secondary vapors, such control is not possible with a continuous system. Increased control is obtained through the addition of a separate preheater. To gain

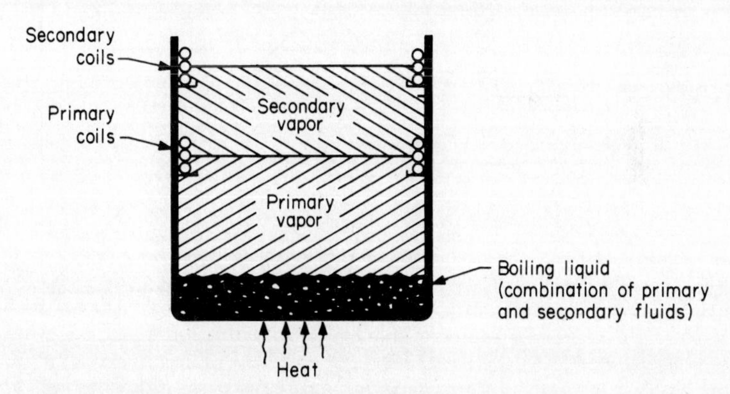

FIG. 3.32 Batch-mode vapor phase system.

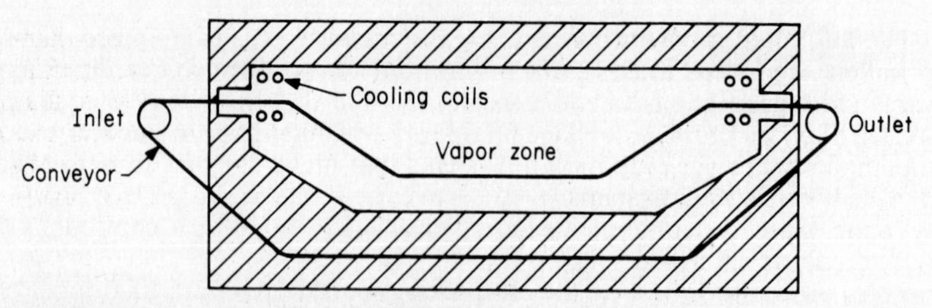

FIG. 3.33 Continuous-mode vapor phase system.

full control of the thermal profile, a postheater is also necessary. An idealized thermal profile for a continuous vapor phase system with preheating is shown in Fig. 3.34.

The second drawback comes from the inclined entry and exit throats often used on continuous systems. Inclination angles from different manufacturers vary from 5 to 10° and may be sufficient to allow some components to slide or roll away from their desired locations.

3.6.11.4 Other Reflow Methods. Several other methods of reflow have been used in certain applications. For various reasons they have not found widespread application in general production. These methods include the following.

3.6.11.4.1. Soldering Irons. Soldering irons have been popular for assembly of leaded components for many years. Standard soldering irons are impractical for SMT assembly, but various special modifications have been developed. Typically these involve using special tips that contact all joints of a device simultaneously. To avoid damaging the PWB material, these tools often incorporate temperature control of the tip. A certain amount of operator skill is necessary for

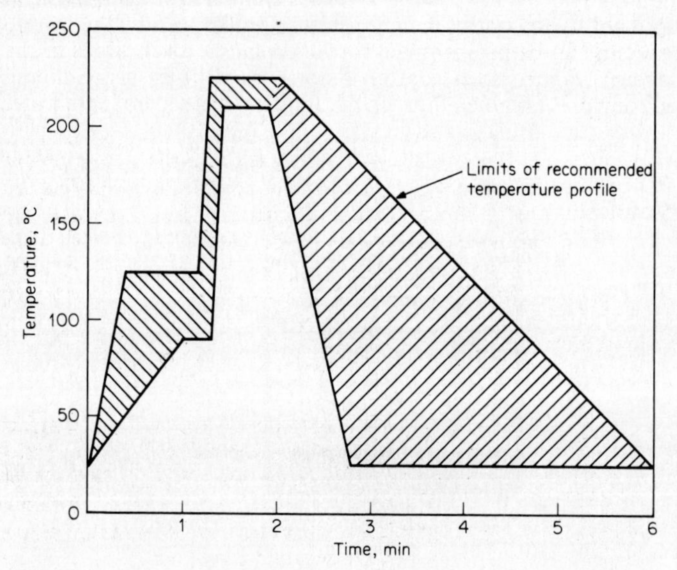

FIG. 3.34 Typical vapor phase temperature profile (continuous mode).

success. Since is is possible to solder only one component at a time, and since an assortment of different tips must be maintained, soldering irons are most suited for SMT repair.

3.6.11.4.2. Hot Gas. This process involves heating an inert gas to reflow temperature and directing it via a suitable nozzle onto the solder joint. Nitrogen is most commonly used as the thermal transfer medium. Where high reflow temperatures are required, helium is sometimes used because of its higher heat capacity. Nozzles typically are shaped to match the component, although for small devices a circular nozzle that heats a spot about 0.1 in in diameter can be used. Hot-gas systems generally have elaborate mechanisms for positioning the board under the nozzle and for protecting the operator from the high-temperature gas.

This approach is somewhat easier to use than thermal conduction but suffers from similar limitations. Temperature control is imprecise, so damage to the PWB or component is a real possibility. Since each component must be soldered individually, the process is slow. And there is risk of partially reflowing joints on adjacent components that have already been soldered. Hot-gas reflow is most suitable as a repair process.

3.6.11.4.3. Laser. By directing the output of a high-power Nd:YAG or CO_2 laser onto a solder joint, the joint can be made to reflow.[15] Since each joint is reflowed independently, the laser is usually mounted on a computer-controlled head programmed to reflow all joints in quick succession.

Laser soldering has the advantage of minimizing disturbance to surrounding components. Individual SMT devices may be added to a board after all other components (SMT and through-hole) have been mounted. Since each joint is at soldering temperature for as little as 50 ms, laser soldering also minimizes formation of intermetallic compounds in the joint. Although not efficient as a mass soldering or repair process, it can be used effectively for reflowing solder joints on quadpacks and similar devices.

3.6.12 Comparison of Reflow Techniques

Thermal conduction is the least expensive reflow alternative. Since it depends on intimate contact between circuit and heater plate, it is best suited for small ceramic substrates that lie flat on the belt and provide good thermal conductivity. It is less suitable as a general process for larger PWBs and for boards with components on both sides. However, with carefully designed fixturing, thermal conduction can be made to work adequately for small production runs.

The IR process can be successfully used for reflow on many types of substrates. Although more expensive to operate than thermal conduction, it is more economical than vapor phase. Its main limitation is that emitter temperatures and belt speeds must be set differently for boards of different thermal mass. While this is not a problem for production of large quantities of a single board type, it can be a serious obstacle for facilities that must routinely handle various sizes of boards.

Vapor phase reflow is the most versatile of the mass soldering processes. It provides precise temperature control and can accommodate the widest range of components. However, for two reasons it is not the ideal process.

1. Vapor phase is an expensive process. The Fluorinert compound costs $500 to $700 per gallon, and special precautions must be taken to minimize vapor loss. Even the best systems may lose liquid at the rate of several dollars per hour. Vapor phase systems also consume more power than either conduction or IR systems, further increasing operating costs.

2. The process is subject to environmental restrictions. Some communities

TABLE 3.7 Comparison of Solder Reflow Methods

Method	Advantages	Disadvantages
Thermal conduction	Least expensive capital cost. Least expensive operating cost.	Requires flat substrate with good thermal conductivity. May damage epoxy-glass PWBs.
Infrared	Compatible with all substrate materials. Lower capital cost than vapor phase.	Temperature profile must be adjusted for boards of different thermal mass. Color-selective heat absorption may damage certain components.
Vapor phase	Compatible with all substrate materials. Does not require adjustment for boards of different thermal mass.	High capital cost. Highest operating cost. Temperature profile cannot be controlled unless a separate preheater is added.
Laser	Ability to precisely reflow localized areas. Reduced possibility of intermetallic formation.	Only reflows one joint at a time. High capital cost.

regulate the operation of systems that emit fluorocarbons, and others may do so in the future. One potentially serious concern is that if the liquid is heated to the point of decomposition (as may happen in certain types of system failures) highly toxic gases are produced.

Because of these concerns, both thermal conduction and IR reflow systems are preferred where practical. Advantages and disadvantages of the various reflow processes are summarized in Table 3.7.

3.6.13 Mixed Assembly

When combined with through-hole assembly, the reflow approach is more complicated than wave soldering. Surface mount components must be mounted and reflowed to the top of the board before the leaded components are inserted. Adequate clearance must be maintained to ensure that autoinsertion tooling does not interfere with previously mounted SMCs.

3.6.14 Double-Sided Process Flows

A significant advantage of SMT over through-hole technology is its ability to mount components on both sides of a board. Double-sided SMT may be obtained through several alternative processes.

3.6.14.1 Double-Sided Reflow. In double-sided reflow, components are mounted to the top side of the board, using the reflow process described above. The board is then inverted and components are mounted to the underside, again using the reflow process. During this second reflow step, the previously reflowed top side components are on the underside of the board. Experience has shown that for many component types, surface tension of the molten solder will hold them in place. Large components such as certain chip carriers may need to be attached adhesively to be prevented from falling off.

Double-sided reflow is best suited for boards consisting entirely of SMCs. If solder paste is screen-printed, through-hole components cannot be mounted until after the SMCs are mounted on both sides. Otherwise, their leads interfere with the second screen-printing operation. Because of this limitation, through-hole components cannot be wave-soldered. Instead, they are typically soldered individually by hand.

3.6.14.2 Mixed Technology. It is sometimes possible to reflow SMCs on the top side of the board and wave-solder them on the underside. Such a process is particularly useful when surface mount and through-hole technologies must be mixed on a single board. The first step is to mount SMCs to the top side of the board, using the reflow process. Next, the through-hole components are inserted. The board is then inverted, and the underside SMCs are mounted adhesively to the board. Wave soldering of both through-hole components and underside SMCs is the final step.

The major limitation of this mixed-technology approach is that the underside SMCs must be restricted to those that are compatible with wave soldering. This excludes many types of high-pin-count ICs and a number of more sensitive passive devices.

3.6.14.3 Other Process Flows. Other approaches to double-sided SMT have been applied in special situations. These include laser reflow of small numbers of devices, localized wave soldering of SMT and through-hole devices, and modifying the screen-printing process so that both SMCs and through-hole components may be reflow-soldered. Such alternatives may be suitable in special circumstances, but the lack of published data makes it impossible to determine their feasibility as a more general process.

3.7 TEST

Testing of assembled surface mount boards is an important step in the overall manufacturing process. The cost to repair defects has been estimated to increase by an order of magnitude at each succeeding level of electronic assembly. Thus, repairing defective components found after assembly onto boards is 10 times more costly than repairing those found prior to assembly. And repairing a defective PWB after installation is 10 times more costly than repairing the final product before installation. Therefore, it is imperative that all defective PWB assemblies be identified and repaired before final assembly.

3.7.1 In-Circuit Testing

The most common test method for assembled PWBs is in-circuit ("bed-of-nails") testing. It makes use of a matrix of probes which contact circuit nodes on the underside of the board; the tester applies a signal to stimulate an individual node and compares the measured response to the expected one. Surface mount boards present several problems for in-circuit testers:

1. Standard bed-of-nails fixtures probe only the underside of the board. Frequently, not all nodes on a surface mount board are accessible from the underside.

2. When SMCs are mounted on the underside of a board, they may interfere with the ability of the probes to make contact with test points on the board.
3. The 0.100-in probe spacing on standard bed-of-nails fixtures is incompatible with SMCs, many of whose leads are on centers 0.050 in and smaller.

Bed-of-nails fixtures designed specifically for surface mounting often employ smaller probes designed to be used on 0.040- or 0.050-in centers. Often, such fixtures also probe both sides of the board. Although they address many of the technical concerns, they have certain limitations. The smaller probes are less robust than those designed for 0.100-in spacings and must be replaced more often. This also leads to a higher level of "testing defects"—good boards diagnosed improperly because of poor probe contact with the board. In addition, fixtures for double-sided probing are much more expensive than those for single-sided probing (by as much as a factor of 5) and so are used only when there is no other alternative.

3.7.2. Design for Testability

Many problems with testing can be avoided by careful design of the original board. Key features that can minimize the complexity and expense of in-circuit testing are as follows:[16]

1. Avoid direct probing of SMT device leads or solder joints. Such probing may force temporary contact between lead and pad on an otherwise defective joint.
2. Provide test pads which are sufficiently large to ensure consistent and reliable contact. For probes designed for 0.50-in spacings, a pad diameter of 0.040 in is recommended. The pad should be solder coated to provide a compliant surface for probing.
3. Use standard 0.100-in center probes whenever possible. When 0.050-in center probes are necessary, use them only in the locations where they are actually needed.
4. When probing on the component side of a board, select probes with sufficient probe travel to prevent damage to components or the probe.
5. Avoid using double-sided fixtures. Add test pads where necessary so that all nodes are accessible from one side.

3.8 REPAIR

Once a board has been diagnosed as being defective, it must be repaired. Issues to be considered in the repair process include (1) design for repair, (2) heating method, and (3) component removal and replacement.[17]

3.8.1 Design for Repair

The repair process is considerably simplified if the board has been designed for compatibility with the repair process. Although details will vary depending on the specific process employed, certain general guidelines apply:

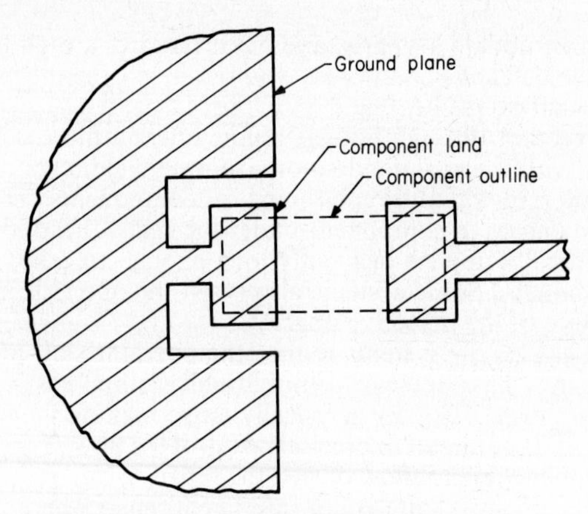

FIG. 3.35 Thermal isolation of ground planes.

1. Provide enough clearance around components to allow proper operation of repair tooling. In many instances, the requirements for repair will be the limiting factor on how closely components may be spaced. Tools employing thermal conduction as a reflow method require clear access to the periphery of the part in order to reflow the solder and remove the component. Hot-gas reflow, although usually less restrictive, also requires a minimum clear area around the component. The necessary clearance depends on the specific equipment selected for the process.

2. Design ground planes to minimize thermal heat sinking effects. Component leads should not be connected directly to a large ground plane. Instead, they should be thermally isolated by using short conductors to ground (see Fig. 3.35). This minimizes the time at temperature necessary to remove the component and reduces the likelihood of damaging the PWB or adjacent components.

3. All outer-layer PWB traces should be protected by solder mask. This prevents accidental bridging of traces when a new component is soldered in place.

4. If components are assembled via the adhesive attach–wave solder process, select a relatively low strength adhesive. Acrylic adhesives are preferred over epoxies for this reason.

5. If heat sinks are mounted to components, design them so that they do not interfere with repair. Generally this implies that heat sinks should not extend beyond the periphery of the device and should be thermally isolated from the component leads. Alternatively, heat sinks may be designed to be easily removable to allow access for repair.

3.8.2 Heating Method

Either thermal conduction or thermal convection (hot-gas) techniques may be used to remove defective SMCs. The most common thermal conduction technique uses soldering irons with specially shaped tips. This approach has two advantages: (1) tooling is inexpensive, making it an easily accessible process; and (2) soldering irons may easily be transported to remote locations for on-site repair. The disadvantages of thermal conduction are: (1) it requires a wide range of tooling to accommodate all parts, and (2) it requires a high level of operator skill to remove the defective component without damaging the board or adjacent components through excess heating.

Thermal convection through hot-gas reflow offers a more controlled process than is available with thermal conduction. Commercial hot-gas systems provide for semiautomatic removal and replacement of components. The operator need only position the defective component under the reflow head; once in place the machine automatically applies heat only long enough to reflow the solder and remove the component. Some systems also assist the operator in locating a new component in place of the defective one.

Although hot-gas reflow systems reduce the operator skill level necessary to perform repairs, they have their own limitations. Capital cost can be 100 times that of a soldering iron. They are physically large, making them impractical for on-site repair. And it is difficult to precisely control the flow of gas, so solder joints on nearby components may reflow during the repair cycle.

Soldering irons are most suitable for infrequent repair where the operator skill level can be controlled. They can be used safely in laboratory environments and by field service technicians who routinely travel to customer sites to perform repairs. Hot-gas systems are better suited for use by personnel in production line environments where a reliable and repeatable repair process regardless of operator skill is of primary importance.

3.8.3 Component Removal and Replacement

It is of utmost importance that the repair process not adversely affect the reliability of the PWB assembly. Not only must the new solder joints be reliable, but the replacement component, the board, and the surrounding components cannot be damaged in the process. Successful repair requires careful attention to the following:

3.8.3.1 Preparation. The PWB should first be cleaned to remove contamination. This is especially important for boards that have been in the field for extended periods, as they are more likely to have accumulated dirt and grime over time. Next, any adjacent components that may interfere with the repair process should be carefully removed and saved. Nearby components that are sensitive to heat must be protected through the use of shields or insulation. Finally, flux should be applied to the joints to promote reflow.

3.8.3.2 Preheating. Boards must be preheated to minimize the deleterious effects of thermal shock. It is recommended that the preheat temperature be within 100°C of the reflow temperature and that the temperature gradient from room temperature to the preheat temperature not exceed 5°C/s. Increasing the temperature at too high a rate may damage certain types of components. Ceramic capacitors, particularly those which employ X7R or Z5U dielectric types, are especially sensitive.

3.8.3.3 Component Removal. The choice of heating technique depends on the specific requirements of the situation. Regardless of which method is employed, steps must be taken to minimize the amount of time that the PWB is at reflow temperature. Excessive time at soldering temperature can damage both the PWB itself and adjacent components. Times of greater than 10 s at temperatures in excess of 240°C should be avoided.

3.8.3.4 Component Replacement. Once the defective device is removed, the new component must be precisely positioned and soldered. It is best to remove as much of the original solder as possible by using a vacuum solder sucker or a similar device. New solder can then be applied either by dispensing paste from a syringe or by pretinning the pads using solder wire and a soldering iron. The new component should be aligned as closely as possible to the land pattern on the board. Generally no more than half of any component lead may extend beyond the periphery of the pad. When reflowing the new solder, it is again essential to minimize time at temperature.

REFERENCES

1. Stephen W. Hinch, "SMT Component Standards Needed Now," *Circuits Manufacturing,* March 1986, pp. 45–56.
2. Howard H. Manko, "Surface Mounting and Fine Line Boards—Part 2," *Electri·onics,* October 1984, pp. 15–19.
3. Robert W. Korb and David O. Ross, "Direct Attachment of Leadless Chip Carriers to Organic Matrix Printed Wiring Boards," *IEEE Transactions on Components, Hybrids, and Manufacturing Technology,* vol. CHMT-6, no. 3, September 1983, pp. 227–231.
4. Howard W. Markstein, "Surface-Mount Substrates: The Key in Going Leadless," *Electronic Packaging and Production,* June 1983, pp. 50–55.
5. Elizabeth A. Kress and Kent Wicker, "Solder Pad Geometry Studies for Surface Mount of Chip Capacitors," *Proceedings of the 35th Electronic Components Conference,* 1985.
6. John K. Hagge, "Critical Component Requirements for Vapor Phase Soldering Leadless Components on Circuit Board Assemblies for Military Electronics," *IEEE Transactions on Components, Hybrids, and Manufacturing Technology,* vol. CHMT-6, no. 4, December 1983, pp. 443–454.
7. Anup Patel, Vijay Sajja, and Jag Belani, "Integrity of Solder Joints as a Function of Reflow Temperature," presented at Northcon/85, October 22–24, 1985.
8. Stephen W. Hinch and Yeng P. Wong, "Setting Up Production of Surface Mount Assemblies," *Electronic Packaging and Production,* January 1984, pp. 66–71.
9. W. James Hall, "Soldering Techniques for Manufacturing Surface Mounted Circuits," HTC Corporation, 1983.
10. John M. Altendorf, "Precision Surface Mount Assembly Using Robots," presented at Northcon/84, October 2–4, 1984.
11. Barbara Roos-Kozel, "Designing Solder Paste Materials to Attach Surface Mounted Devices," *Solid State Technology,* October 1983, pp. 173–178.
12. Dr. Wallace Rubin, "Smoothing the Way for Solder Creams," *Circuits Manufacturing,* October 1983, pp. 74–80.
13. Hendrik B. Hendriks and Bruce E. Inpyn, "Fluxless SMD Soldering," *Circuits Manufacturing,* October 1984, pp. 40–44.
14. Stephen J. Dow, "Use of Radiant Infrared in Soldering Surface Mounted Devices to Printed Circuit Boards," *Solid State Technology,* November 1984, pp. 191–195.
15. Carl Miller, "Soldering with Light," *Photonics Spectra,* May 1983, pp. 83–86.
16. Mike Horton, "Dual-Sided Fixtures and SMD Probing," Hewlett-Packard Co., Palo Alto, Calif., 1984.
17. John B. Holdaway, "Guide for Removal, Replacement of Surface Mounted Devices—Parts 1, 2, and 3," *Electri·onics,* December 1984, January 1985, February 1985.

P · A · R · T · 2

ENGINEERING

CHAPTER 4
ENGINEERING PACKAGING INTERCONNECTION SYSTEM

Gerald L. Ginsberg
President, Component Data Associates, Inc., Lafayette Hill, Pennsylvania

4.1 INTRODUCTION

Advances in interconnection systems have occurred in response to the evolution of electronic technology, particularly the growth in circuit component sophistication, complexity, and the number of terminals to be connected. Therefore, it should come as no surprise to find that printed wiring remains the most popular and universal method of interconnection.

Printed wiring was developed initially to adapt an interconnection method to mass-production and assembly schemes and to economize on weight and space in military equipment. It was soon apparent, however, that other advantages would accrue from its use. The close control over and reproducibility of electrical parameters was soon recognized as an important and valuable attribute. Wiring bulk was greatly reduced. Wiring errors—except those in the initial design—were all but eliminated.

Printed wiring was soon found to have the design flexibility to provide characteristics (electrical and mechanical) compatible with all types of electronic equipment. The adaptability of printed wiring to automation in both fabrication and use not only resulted in cost savings but, more important, facilitated controls for enhanced reliability of the total electronic system.

No finished product is ever better than its original design or the materials from which it is made. At best, the manufacturing process can merely reproduce the design. That is as true of printed wiring assemblies as of any other product. The entire design cycle must be repeated with each new board, and the design (board) layout is usually done by a drafter or another nonengineer. The need for formalizing layout and design methods and procedures can, therefore, assume critical proportions. The purpose of this chapter is to provide information that leads to thoughtful decisions and helps to ensure that all pertinent design and layout variables have been considered.

4.2 GENERAL CONSIDERATIONS

"Design and layout" includes the perspective of total system hardware, which includes not only the printed wiring but each and every component in its final orientation. Design and layout considerations must also encompass the relations between, and interactions of, the components and assemblies throughout the system. Therefore, some major considerations will be the following:

1. User system objectives
2. Product specifications
3. Life expectancy
4. Electronic circuit gain, impedance, voltage, etc.
5. Maintenance concepts at the component level, in service (dynamic), out of service (bench repair), and one-shot with no adjustment or maintenance (throw-away)
6. Environment for storage, assembly, transportation, use, and repair
7. Compatibility of manufacturing method with organization plan, size of production lot, and degree and type of mechanization
8. Material and component sources, performance data availability, cost, and verification screening for specific relevant design data

These functional and performance requirements, including any special considerations, must be properly balanced. As in all kinds of design, many trade-offs must be made in the course of developing the optimum final solution. It is not

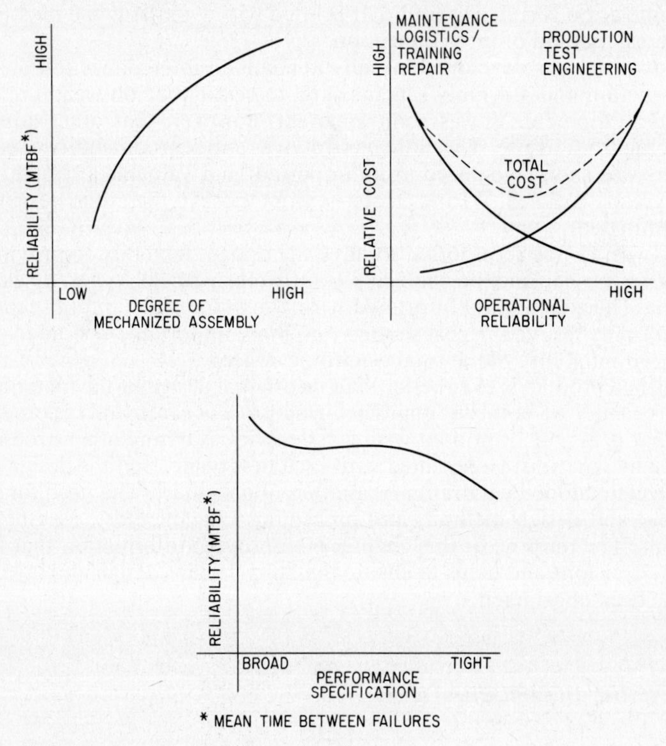

FIG. 4.1 Typical reliability trade-off curves.

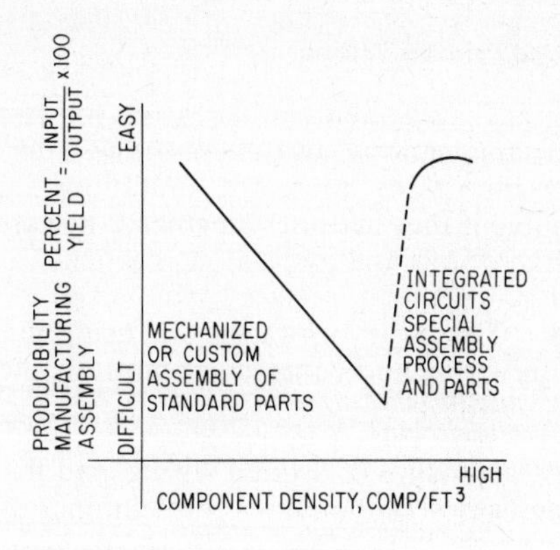

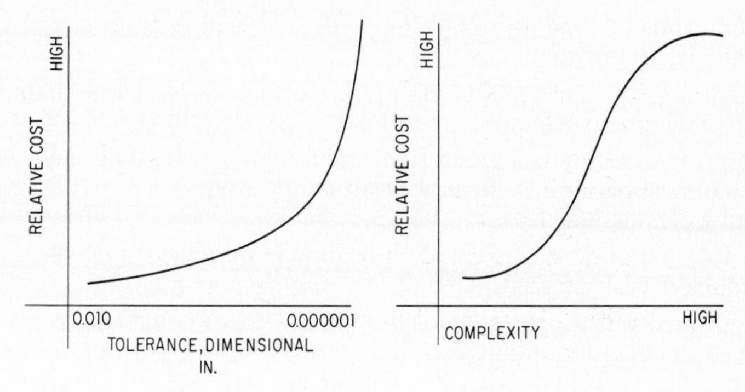

FIG. 4.2 Typical cost trade-off curves.

possible to fulfill all aims completely and also give top priority to every trade-off. Therefore, the fewest possible constraints should be imposed upon the design and the associated artwork.

The trade-off curves of Figs. 4.1 and 4.2 demonstrate a few of the contradictions that often must be resolved. Accurate and meaningful analysis can be best achieved by using the combined judgments of people with diverse design and manufacturing experience.

4.2.1 Printed Wiring Selection

The basic function of printed wiring is to provide support for circuit components and to interconnect the components electrically. To achieve these results, numerous printed wiring types have been developed; they vary in base dielectric material, conductor type, number of conductor planes, rigidity, etc. As previously stated, printed wiring designers should be familiar with the variations and their effect on cost, component placement, wiring density, delivery cycles, and functional performance. Otherwise, they will be unable to select the printed wiring structure with the optimum combination of features for the particular electronic apparatus or system requirements.

4.2.2 Decision to Use Printed Wiring

Although there are many good and definite reasons for using printed wiring, there are also potential misapplications; awareness of the latter will preclude ultimate failure. As with any good thing, there are limitations and associated problems which can cause trouble if they are not recognized. By careful review of the technology, most such limitations and associated problems can be avoided.

4.2.2.1 ***Advantages.*** The basic advantages of using printed wiring instead of other interconnection wiring and component-mounting techniques are found in the following areas:

1. The physical characteristics of printed wiring lend themselves to greater versatility in packaging design than does conventional wiring.
2. Wiring is permanently attached to the dielectric base, which also provides a mounting surface for the circuit components.
3. When printed wiring is properly applied, miswiring or short-circuited wiring usually is not possible.
4. A high level of repeatability affords uniformity of electrical characteristics from assembly to assembly.
5. The printed wiring technique significantly reduces the bulk and weight of interconnection wiring. Planar construction provides a very neat means of routing conductors.
6. The location of parts is fixed, identification is simplified, and color coding of conductors is eliminated.
7. The printed wiring process lends itself to the use of visual aids to speed accurate component mounting. Assembly errors are thereby minimized, and the complexity, time, and cost of the testing and checking process are reduced.
8. Printed wiring personnel require minimal technical skill and training.
9. Mass-manufacturing processes and automated techniques can be used.
10. Maintenance of electronic equipment and systems is simplified.

4.2.2.2 ***Disadvantages.*** Some of the apparent disadvantages of using printed wiring instead of alternative techniques are as follows:

1. The design of printed wiring requires special skills in layout of components and interconnections because of the regimentation imposed by the essentially planar structure.
2. Utilization of equipment space is limited to planar partitioning.
3. Lead time from initial design stage to delivery of the end product may be lengthy.
4. The design is difficult to change after it has been documented and tooled.
5. Low-quantity tooling is relatively expensive.
6. Repair of printed wiring is sometimes difficult and in some applications is not permitted.

4.2.3 Printed Wiring Types

There are three basic types of printed wiring structures. They are listed below in ascending order of interconnection wiring and component density (either flexible or rigid):

1. Single-sided, with conductors on only one surface of a dielectric base
2. Double-sided, with conductors on both sides of a dielectric base that are usually interconnected by plated-through or reinforced holes
3. Multilayer, with three or more conductor layers separated by dielectric material and usually interconnected by plated-through holes

4.2.4 Packaging Density

Packaging density is one of the most difficult items about which to generalize. That is due mainly to the differences in printed wiring component sizes and shapes, the number of component leads, and the variable complexity of interconnection. Also, the component lead pattern places a limit on the amount of circuitry that can be effectively packaged on a printed wiring structure. As a measure of the packaging density of a printed wiring structure, the number of component mounting holes per square inch of usable surface can be used. The ratio, although not perfect in application, can be used to estimate the amount of circuitry that can be efficiently interconnected on different printed wiring board types. Based on this design parameter, the following generalized relationships can be stated:

Printed wiring structure	Holes/in^2
Single-sided	3–10
Double-sided	10–20
Multilayer	20+

Multilayer boards are often used with integrated circuits when the component densities possible with double-sided boards must be exceeded. For instance, on two-sided boards (all of the interconnections must be distributed on two surfaces) a usual maximum is 2.0 TO cans per square inch. With multilayer boards, that packaging density can be increased to more than 3.0 if secondary board space allocations are taken into account. A similar condition exists for flat packs. Two-sided boards can interconnect a maximum of approximately 2.5 flat packs per square inch; multilayer boards are capable of interconnecting over 4.0, assuming a minimum of four layers. At least two of the four are for interconnection wiring and separate layers for ground and voltage distribution. In some designs, it is possible to place components on both sides of the multilayer board and thus double the component density without appreciably increasing the volume of interconnections.

Dual-in-line integrated circuit package (DIP) densities per square inch of overall multilayer board surface are, on average, up to 2.0, which is twice that normally achieved with two-sided boards.

4.2.5 Relative Costs

Increased interconnection and component densities relate directly to increased costs, manufacturing process complexity, and longer delivery cycles. Relative costs of various printed wiring structures also are difficult to generalize about, since the quantity of each structure produced has a significant effect on the amortization of tooling and setup charges. In addition, the quantity to be fabricated determines the sophistication of the tooling to be used, and that, in turn, affects fabrication costs.

The difference in material costs must also be taken into consideration when relative cost comparisons are made. A convenient means of comparing printed wiring structure costs is provided in the following generalized relationships:

1. Single-sided printed wiring, rigid, paper-base, phenolic substrate with punched holes; relative cost (RC) equals unity (RC = 1).
2. Double-sided printed wiring, rigid, glass-base, epoxy substrate with drilled and plated-through holes; RC = 7.
3. Multilayer printed wiring composite, rigid, glass-base, epoxy dielectric substrate with drilled holes and plated-through holes that are etched back or chemically cleaned, seven layers; RC = 40.

After the technical problems have been resolved, in the design of any product, an item of concern is cost. More often than not, potential cost factors change the course of a particular project, equipment, or product. Since initial costs are usually estimated, it is important that all factors be considered in proper perspective to avoid canceling the most cost-effective approach. Some of the key elements of multilayer board costs are (1) board layout, (2) choice of specification, (3) selection of base and B-stage material, (4) selection of copper weight, (5) choice of multilayer board thickness and overall size, (6) choice of terminal area pad size and hole sizes, and (7) provisions for maintenance, testing, and repair.

4.3 TECHNOLOGY DESCRIPTION

Printed wiring, since its inception during World War II, has been predominantly of the rigid type, but the later development of flexible wiring planes necessitated classification as rigid printed circuit board or flexible printed wiring. An important subclassification is by the number of planes, or layers of wiring, which constitute the total wiring assembly or structure.

The art of printed wiring has been particularly susceptible to inventions and processes that are of almost infinite variety and novelty; but in spite of the apparent diversity, a relatively small number of processes are still employed to any extent. The most common processes are described in this section. They are classified under the general headings of chemical and mechanical methods, but they are broken down into single-sided, double-sided, and multilayer. Boards of the latter type are presently being produced predominantly by chemical processes.

Generally, the first—and currently the most widely used—form of printed wiring, the rigid printed board, was early recognized as providing not only a conductive wiring path but also support for and protection of the components it connected and a heat sink to aid in the thermal management of the total package. The chemical methods used in producing rigid printed circuit boards are either (1) subtractive, or etched-foil, or (2) additive, or plated-up. The mechanical methods include stamped wiring, metal-sprayed wiring, embossed wiring, and molded wiring.

4.3.1 Single-Sided Printed Boards

"Single-sided" means that wiring is found on only one side of the insulating substrate. This arrangement represents a large volume of printed boards currently produced. Single-sided boards are used for relatively unsophisticated and simple circuitry, and they are applicable when circuit types and speeds do not demand unusual wiring electrical characteristics.

4.3.1.1 Subtractive, or Etched-Foil, Type. As illustrated by Fig. 4.3, the etched-foil process begins with a base laminate composed of a variety of insulators clad with copper or some other metallic foil. Following suitable cleaning and other preparation, a pattern of the desired circuit configuration is printed by using a suitable negative-resist pattern for photoresist or ink resist or a positive-resist pattern if plating is to be used as an etchant resist. In the latter, gold or solder plating is applied to the nonphotoresist areas. In etching, the next step, all copper not protected by the resist material is removed by the etchant. Following etching, the printed board is stripped of resist and otherwise cleaned to ensure that no etchant remains. The board is then ready to fabricate by drilling, trimming, or some other means.

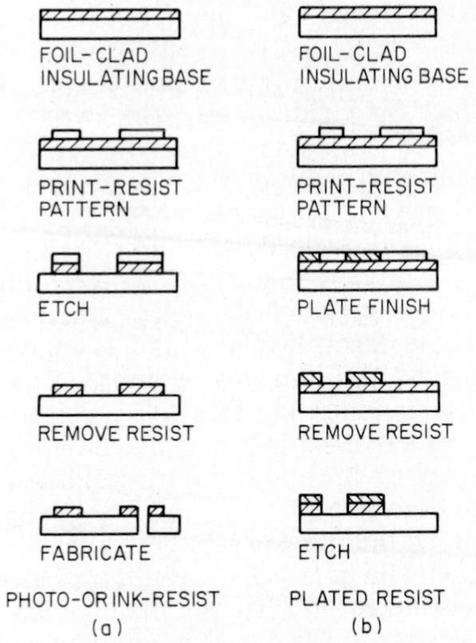

FOIL-CLAD INSULATING BASE | FOIL-CLAD INSULATING BASE
PRINT-RESIST PATTERN | PRINT-RESIST PATTERN
ETCH | PLATE FINISH
REMOVE RESIST | REMOVE RESIST
FABRICATE | ETCH
PHOTO-OR INK-RESIST (a) | PLATED RESIST (b)

FIG. 4.3 Etched-foil process.

4.3.1.2 Stamped, or Die-Cut, Type. This process requires a die that carries an image of the wiring pattern desired on the final product. As illustrated in Fig. 4.4, the die may be either photoengraved by conventional photoresist techniques or machine-engraved. In either case, the die is heated, and when it forces the adhesive-coated foil into the base material, it effects a bond as well as cutting the circuit pattern and embedding it in the base at the conductor edges. Blanking or piercing of the board may be done at the same time.

4.3.2 Double-Sided Printed Boards

When more than one layer of wiring is needed, circuit patterns are placed on both sides of a printed board, and the necessity that arises is for a means of interconnecting the two wiring layers. The usual connection goes through the board rather than around the edge; hence the name "through connection." In many cases the hole through the board serves a dual purpose: it accommodates a component lead, and it provides a location for some method of interconnecting the two sides. The many processes for effecting interconnection between circuit layers may be generalized as including a plating or a mechanical technique.

4.3.2.1 Plated-Through-Hole Type. There are two variations of the plated-through process; each utilizes a conductor plated through the hole to make the connection. For comparison, the subtractive process shown in Fig. 4.5*a* is an extension of the board fabrication process for single-sided boards described above. The additive process shown in Fig. 4.5*b* is described below.

The subtractive plated-through-hole (PTH) process starts with a double-clad laminate with a series of holes corresponding to the locations where a through connection is needed. The holes are drilled and deburred, and an electroless coating of copper is applied over the entire board surface, including the holes. Next, copper is electrodeposited on the exposed copper foil and sensitized walls of the hole, usually to a thickness of 0.001 in. A negative-, or plating-resist, pattern is then applied and registered to both sides of the material. Resist covers all areas of foil where base copper conductor is not required, and the surplus conductor will subsequently be etched off. The next plating step is to electrodeposit a thin layer of a suitable etch-resist plating, usually solder or gold. The original resist—screen or photoresist—is removed, and the circuit pattern is defined by etching away the exposed copper in a suitable etchant. The choice of etchant depends on the type of plating resist used.

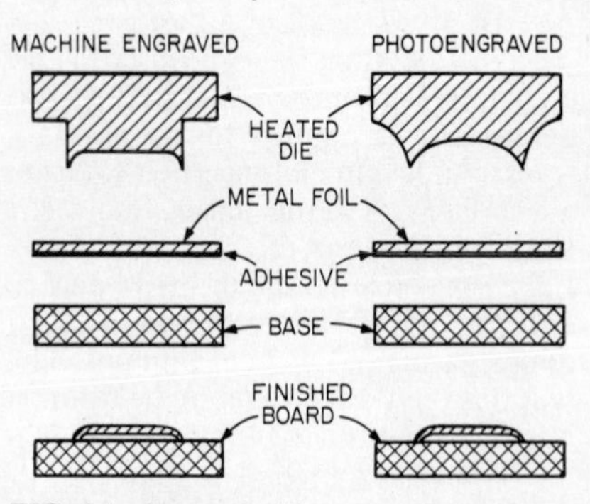

FIG. 4.4 Stamped wiring process.

The additive process, unlike the subtractive process, requires no etching, and the circuit pattern is defined at the same time the through connection is made. Holes are again drilled at desired connection points, and a thin layer of a suitable adhesive is applied as illustrated in Fig. 4.5*b*. Electroless deposition of copper occurs on the entire surface and holes. No electroplating is used in the pure additive process.

The semiadditive process starts the same as the additive process, but after initial sensitizing of the material, the circuit pattern and hole walls are built up using standard electroplating processes. After the registered printing of a plating-resist pattern to both sides of the board, copper is electrodeposited to the desired thick-

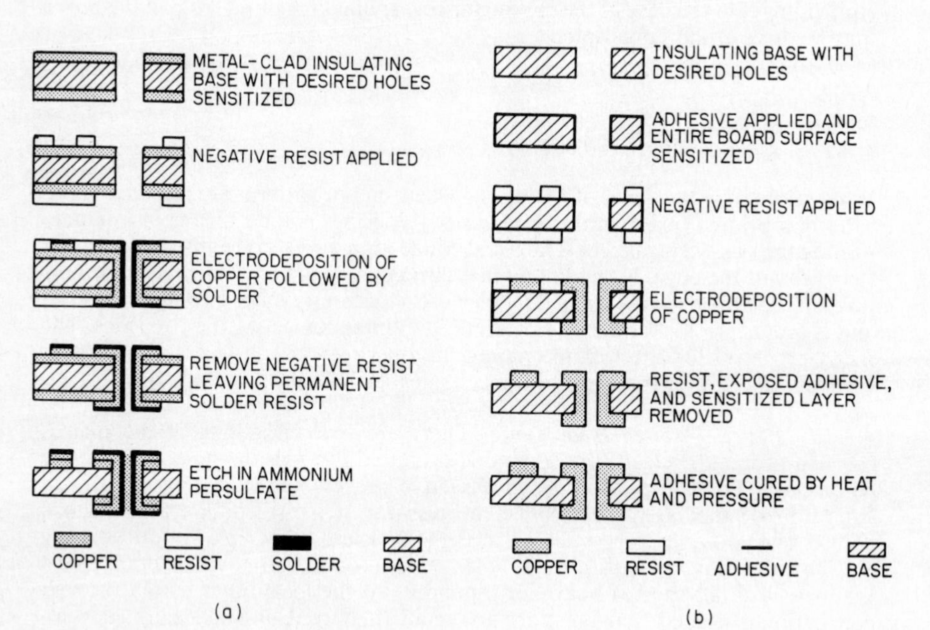

FIG. 4.5 Plated-through-hole process. (*a*) Subtractive; (*b*) additive.

ness on exposed areas. Resist is removed, unwanted electroless copper is "flash-etched" off, and excess adhesive is removed with appropriate solvents. The last step is the curing of the adhesive by subjecting it to a heat-and-pressure cycle.

4.3.2.2 Mechanical Through-Hole Type. The nonplating techniques illustrated in Fig. 4.6 preceded through-hole-plating processes for double-sided boards, and they continued to be used exclusively until the reliability of the latter was proved and the economic import of mass-interconnection techniques was realized. Although not to be construed as an integral part of the board fabrication procedure, as represented by the plated-through hole, mechanical interconnections—actually board assembly techniques—serve the purpose of connecting the two sides of a double-sided board.

A simple and easily made interconnection is represented by the clinched jumper wire, as illustrated in Fig. 4.6*a*. A formed, uninsulated, solid lead wire is placed through the hole and clinched and soldered to the conductor pad on each side of the board. The lead wires of parts are not usually considered to be interfacial connections.

Three types of eyelets also are used for double-sided board interconnection; these are shown in Fig. 4.6*b* to *d*. Funnel-flanged eyelets are soldered to the terminal areas on the component side of the board before insertion of component leads. The other connection is made at assembly when the boards are dip-soldered. The funnel-flange, by definition, has an included angle between 55 and 120°. Split-funnel-flanged eyelets differ only in the split and in the fact that they need not be soldered to the terminal area on the part side of the board before insertion of part leads.

4.3.3 Multilayer Printed Boards

Multilayer printed wiring emerged as a solution to the problem of interconnecting miniature electronic components into complex systems. The volume taken up by conventional wiring was disproportionately greater than the volume and weight of the components interconnected. By virtue of the multiplanar conductor struc-

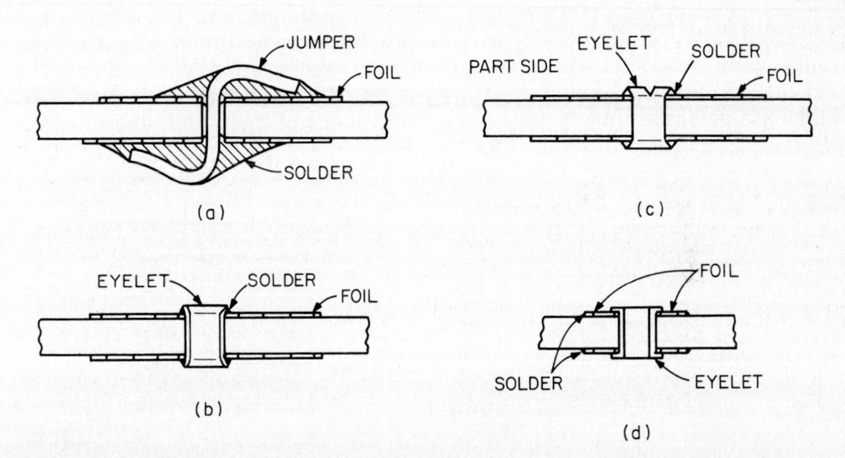

FIG. 4.6 Mechanical interconnections. (*a*) Clinched jumper wire; (*b*) funnel-flanged eyelet; (*c*) split funnel-flanged eyelet; (*d*) fused-in-place eyelet.

ture, multilayer printed wiring made possible a reduction in weight and volume of the interconnections commensurate with the size and weight of the components it interconnected.

Multilayer circuits normally consist of two or more layers of separate circuit patterns that have been laminated together under heat and pressure to produce a strong unit. Figure 4.7 illustrates the lay-up details of two multilayer boards, one composed of single-sided boards (Fig. 4.7*a*) and the other of double-sided boards (Fig. 4.7*b*).

The area of application that shows multilayer printed wiring to its greatest advantage is the interconnection of integrated circuits. Integrated circuits can be interconnected by other wiring techniques such as welded matrices, point-to-point wiring, or two-sided printed boards, but with these methods the packaging is done at the expense of interconnection density and with an increase in weight and volume. By a combination of integrated circuits and multilayers, drastic reductions in the overall size and weight of a complete system can be achieved.

Another distinct application of multilayer boards is to accommodate the problems of heat distribution and heat removal in systems utilizing integrated circuits. With multilayers, all interconnections can be placed on internal layers, and a heat sink of thick, solid copper or another material can be placed on the outer surfaces. Components can then be mounted directly on the metallic surface.

A third feature of multilayer printed wiring is the possibility of incorporating ground or shield planes. Solid copper planes, with clearance holes etched out in areas where no contact is desired, can be placed anywhere within the structure of the multilayer board. The planes either serve as electrical decoupling to minimize the noise interference or cross talk between various critical circuits or are used for shielding some circuits within the board, or the entire board, from any internal or external interference. A number of such planes can be incorporated within one board and can also be used for power distribution. This unique feature of multilayer printed boards offers some interesting possibilities, and often it provides the only solution to some interconnection problems. The ground plane feature of multilayer boards can be very important in high-frequency circuitry.

The expected growth of multilayer printed wiring has been realized, and it continues to expand with the advent of more sophisticated semiconductor devices (LSI). System design engineers should become thoroughly familiar with its potentials and limitations. The new generation of electronic systems requires ever-increasing complexity and density of interconnections, and the definite trend toward higher operating frequencies demands extremely careful design of even a single conductor. The width, length, location, and relations of conductors, the spacing, the dielectric material, etc., are interrelated, and they often exert critical influence on signal propagation, noise ratio, shape of the pulses, and other elec-

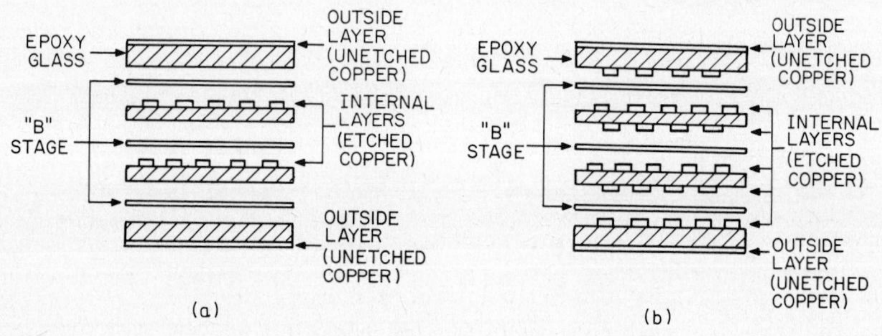

FIG. 4.7 Lay-up details. (*a*) Four-layer board; (*b*) eight-layer board.

trical characteristics of the system. Therefore, the involvement of system design engineers as well as of electrical design engineers in electronic packaging has become much more intensive.

The conditions that will necessitate the use of multilayer printed wiring generally are the following:

1. When weight and volume savings in interconnections are of paramount importance, as in military, airborne, and missile and space applications, or when integrated circuits are utilized
2. When the complexity of interconnection in subsystems will require very complicated and costly wiring or harnessing
3. When coupling or shielding of a large number of interconnections is required
4. When frequency requirements call for careful control and uniformity of conductor impedances with minimal distortion in signal propagation and when uniformity of those characteristics from board to board is of high importance
5. When the spacing of terminal points does not allow a sufficient density of interconnection on two-sided circuit boards

4.3.3.1 Advantages versus Disadvantages. To make an assessment as to whether to use multilayer boards, a comparison should be made of the different printed wiring forms that could be used in the end-product equipment. A determination should be made as to how many double-sided boards of various sizes would be required to mount and interconnect all of the circuitry associated with a given piece of equipment. A similar determination should be made for multilayer boards. The costs of the candidate technologies can be obtained and a cost-effective decision made. The comparative costs are often not limited to dollars and cents; they can include the cost to the user in volume required, amount of back-panel interconnection wiring, design-through-production time cycle, logistics support, etc. An equivalent analysis can be performed only after appropriate weighting factors for the intended application have been applied.

Several factors can be noted as favoring multilayer circuitry as compared with wiring utilizing conventional round hookup cable with conductors individually insulated. One of the most obvious advantages is that in practically all multilayer circuits up to about five layers, the presence of breaks, lesions, etc., in circuitry can generally be examined because of the translucency of the film or glass-reinforced materials. Another obvious advantage is in the tools of miniaturization which the multilayer circuits offer. In many cases, lightweight adhesives can replace the heavy hardware and awkward nuts, bolts, clamps, and brackets used as fastening devices in a conventional harness. An etched T splice is obviously a fraction of the cost and weight of such a junction in regular wiring.

The ability of the multilayer circuit to incorporate ground planes which (1) can be etched according to a predetermined pattern, (2) may consist of a wire mesh screen, (3) can be fabricated from a solid layer of foil, or (4) can even be grounded, by means of a tap-off connection to supporting metal frames or cases, permits the ready utilization of structural hardware for shielding purposes.

A further advantage of incorporating the ground plane into the three-dimensional circuit lies in high-speed applications, insofar as that capability makes possible the production of constant-impedance conductors of much lower cost for strip-line application. The ability to combine several circuits into a small capsule obviously reduces the chance for errors and the cost of testing of back panel wiring. The potential advantages and disadvantages of multilayer wiring are as follows:

Advantages	Disadvantages
1. Controlled assembly process through mechanization	1. Realistic simulation during breadboard testing difficult; may require sophisticated computer programs
2. Space saving through use of thin films and high terminal density	2. Special tools and skill needed for repairs
3. Wiring errors eliminated	3. High cost of small quantities
4. Uniform electrical impedance and coupling (especially with strip-line)	4. Extended design time
5. Possible cost savings with high quantity and proper tooling	5. Long lead time for fabrication
6. Assembly time reduction by simplification	6. Thermal sensitivity (warp)
7. Capable of high inherent reliability (dependent upon process control and type of interconnection)	7. Inspection of end product difficult, usually requiring microsectioning
8. Can combine structural and electrical functions	8. Difficult and costly to change completed board

4.4 GENERAL DESIGN CONSIDERATIONS

Probably the most important considerations in the design of electronic equipment are reliability, satisfactory performance, and maintainability. Those factors are not inherent in printed wiring per se; but by adequate design and proper selection of materials and manufacturing techniques, they can be introduced into the system.

When the printed wiring assemblies for an electronic equipment are being developed, there is an optimal limit to the size and cost of any discrete unit or functional subassembly. It is uneconomical or inefficient to increase size or complexity beyond that limit, because the unit then becomes too specialized and therefore is not easily interchangeable or is too costly to be thrown away. Investigations into various approaches to achieve a desired component density indicate that multilayer concepts lend themselves to a great many packaging requirements.

4.4.1 Design Factors

The following is a list of the factors that should be taken into account in any multilayer board design:

4.4.1.1 Design Elements

1. Board size (maximum and minimum)
2. Number of layers
3. Hole sizes
4. Composite thickness
   a. Thickness of internal layers
5. Conductor widths (internal and exposed layers)
6. Terminal area (internal and exposed layers)
7. Spacings
   a. Center-to-center mounting holes

b. Terminal area to terminal area
c. Terminal area to conductor
d. Conductor to conductor
e. Terminal area to board edge
f. Conductor to board edge
g. Conductor to mounting hole

8. Size of grid
9. Tolerances (design), both vertical and horizontal
10. Plating (type and thickness)
11. Drawing requirements
12. Terms and definitions

4.4.1.2 Performance and Properties

1. Tensile strength
2. Flexure strength
3. Vibration
4. Mechanical shock
5. Temperature cycling
6. Thermal shock
7. Moisture resistance
8. Fungus resistance
9. Salt spray (corrosion resistance)
10. Warp or twist
11. Solderability and resolderability
12. Insulation resistance (bulk and surface)
13. Dielectric withstanding voltage
14. Machinability
15. High- and low-temperature storage
16. Conductor temperature rise
17. Interlayer connection resistance (ΔR)
18. Conductor resistance
19. Composite texture
20. Flame resistance
21. High altitude
22. Continuity

4.4.2 Initial Design Development

No generally approved method exists for the development of a design for a printed board, but the following program sequence is useful:

1. Start with a careful design of the electrical circuit and prepare a schematic or logic diagram.

2. Build a breadboard; then analyze the circuit operation by using worst-case component values or operating conditions. Use qualified components that have been tested for reliability. Revise the schematic as necessary.

3. Furnish the final schematic or logic diagram and component list to the designer.

4. Select the shape and size for the board to accommodate all components and to fit the available space within the enclosure or the available area.

5. Select appropriate input/output connectors to accommodate all necessary input/output signals and be compatible with space and environmental requirements.

6. Supply specific information on widths of conductors, spacing of conductors and terminal areas, most suitable or critical routing of conductors, hole sizes and locations, type of electrical interconnections, shape and bulk of components, distance between components, method of component lead termination, electroplating requirements, etc., to the board designer. Use information from appropriate specifications.

7. Locate all holes on a modular grid system. When holes for a component having rigid terminals cannot fall on the grid intersection, locate at least one hole on a grid intersection and dimension other hole locations that are not on a grid intersection.

The major factor in poor manufacturing yields of printed boards is the attempt to produce boards that realistically cannot be made to meet their requirements. That situation arises from one or more of the following conditions:

1. The design of the board does not take into account sufficient manufacturing variation.
2. The manufacturing equipment does not produce the required tolerances.
3. Either the process is not under rigid control, or manufacturing personnel are not following the prescribed processing standards.

4.4.3 Overall Cycle

Printed wiring designers must be aware of the total time required to produce printed wiring. Although printed wiring is usually designed at the end of the equipment design schedule, the printed wiring itself is usually required early in equipment assembly. An inadequate understanding of printed wiring design, tooling, and fabrication times can lead to serious problems. The prime variables affecting those times are the type of structure involved, the degree of tooling required, the manufacturing equipment available, and the quantity involved.

4.4.4 Design Checklist

The following checklist covers the univeral areas of concern in the design cycle. Other items should be added for specific applications.

4.4.4.1 General

1. Has the circuit been analyzed and divided into basic areas for a smooth signal flow?
2. Does the organization of item 1 permit short or isolated critical leads?
3. Has shielding been effectively used where necessary?
4. Has a basic grid pattern been fully utilized?
5. Is the board size optimum?
6. Are preferred conductor widths and spacings used wherever possible?
7. Are preferred pad and hole sizes used?
8. Are the photomaster and schematic compatible?
9. Is jumper use kept to a minimum? Do the jumpers clear components and hardware?
10. Is lettering visible after assembly, and is it of the proper type and size?

11. Are the larger areas of copper broken up to prevent blistering?
12. Have tool-locating holes been provided?

4.4.4.2 Electrical

1. Have conductor resistance, capacitance, and inductance effects been analyzed, especially the critical voltage drops and grounds?
2. Are conductor and hardware spacings and shape compatible with insulation requirements?
3. Is insulation resistance controlled and specified where it is critical?
4. Are polarities adequately identified?
5. Do conductor spacings split the geometric mean with respect to leakage resistance, voltage, and noise?
6. Have the dielectric changes related to surface coating been evaluated?

4.4.4.3 Physical

1. Are terminals and control locations compatible with the total assembly?
2. Will the mounted board meet shock and vibration requirements?
3. Are specified standard component lead spacings used?
4. Are unstable or heavy parts adequately retained?

4.5 MECHANICAL DESIGN FACTORS

Although the printed board lends mechanical support to the components, it should not be used as a structural member. The equipment should be supported at peripheral intervals of at least 5 in at the board edge. An analysis of the board for any equipment shock and vibration requirements should reveal whether additional support is required. Board warpage should be considered in the design of the mechanical mounting arrangements. When the edge-board connector is to be used for the printed board, a careful design is required to assure that the overall board thickness remains within the tolerance capability of the connector. The design of the board mounting means that slides, handles, and other mechanical design features should be such that close tolerances are not required on the board dimensions.

Factors that must be considered in the selection and design of a printed board include the following:

1. Configuration of the board size and form factor(s)
2. Need for mechanical attachment or connector type
3. Circuit compatibility with other circuits and environments
4. Horizontal or vertical board mounting as a consequence of other factors such as heat and dust
5. Environmental factors requiring special attention, such as heat dissipation and ventilation, shock and vibration, humidity, salt spray, dust, and radiation
6. Degree of support
7. Retention and fastening
8. Ease of removal

4.5.1 Board Mounting

Printed boards should be supported within 1 in of the board edge on at least three sides. As a good practice, boards between 0.031 and 0.062 in thick should be supported at intervals of at least 4 in and boards thicker than 0.093 in at intervals of at least 5 in. This practice increases the rigidity of the board and breaks up possible plate resonances.

The choice of a particular board-mounting technique is made after consideration of such design factors as the following:

1. Board size and shape
2. Input/output terminations
3. Available equipment space
4. Desired accessibility
5. Type of mounting hardware
6. Heat dissipation required
7. Shielding required
8. Type of circuit and its relation to other circuits

4.5.2 Board Guiding

A major advantage of using plug-in printed wiring assemblies rather than other circuit-packaging techniques is the suitability of printed boards for use with board guides for ease of maintenance. The exact board guide technique to be used depends on the shape of the board, the degree of accuracy needed to assure proper mating alignment, and the degree of sophistication desired.

4.5.3 Board Retaining

Quite often, shock, vibration, and normal handling requirements necessitate that the printed board be retained in the equipment by mechanical devices. The selection of a proper board retainer is important, since the retaining devices will reduce the amount of board area available for component mounting and interconnection and will add significantly to the cost of the electronic equipment.

4.5.4 Board Extracting

A number of unique principles have been applied to solve the various problems of printed board extraction, and the result has been a constant increase in the types of extracting tools. Many of these tools use a minimum of board space, thereby maximizing board area for circuitry and component mounting. They also protect both the board and the associated input/output connectors from damage during extraction. When selecting among the many different types of printed wiring board extraction tools, one should consider the following:

1. Board area required to be free of circuit components
2. The extractor's effect on board-to-board mounting pitch
3. The need for special provisions in the printed board design, such as mounting holes and notches

4. The size of the extractor, especially if the extractor is to be stored in the equipment with which it is used
5. The need for an extraction device that is permanently attached to the board assembly, usually by riveting
6. The need for special design considerations in the printed board mounting chassis, such as load-bearing flanges
7. The suitability of the extractor to be used with a variety of sizes, shapes, and thicknesses of printed board assemblies
8. The cost of using the extractor, both in piece price and added design
9. The degree of access required inside the equipment to engage and use the extraction tool

4.6 BOARD SIZE AND SHAPE

The final choice of printed board size and shape will probably be a compromise. Some factors will indicate a large board, others a small one. Herewith is a review of some of the major factors which affect the selection of board size and shape.

4.6.1 Size

To reduce manufacturing costs, it is common practice to process a multiplicity of smaller boards on a larger panel and then separate them in the last steps of the manufacturing cycle. This practice can also be used in the manufacture of multilayer boards. If the tolerances are wider, larger panels with more small individual circuits can be processed through manufacturing without any difficulty. When closer terminal area spacing and tight tolerances are demanded, the size of such panels must be reduced to ensure accuracy of registration and hole locations. Therefore, a smaller number of multiple circuits on the panel can be processed simultaneously. That is reflected in the manufacturing costs, and the combined increase due to tighter tolerances and smaller panels can be as much as 80 percent more in the cost of individual circuits.

The final finished board dimensions, the length and width as well as the thickness, affect board cost. Most copper-clad base material is available in 24- × 36-in sheets or in multiples of 1-ft increments. A 1-in margin must be allowed around the final size of the board to accommodate tooling holes. Then the final size of the board must be limited so that a good yield is obtained from each standard-size sheet. The purchaser always pays for the scrap, and the amount of scrap can be considerable, as Fig. 4.8 shows.

Another useful purpose of the margin around the finished multilayer board is for sampling. Extra holes and lines should be made in the margin so that the sample can be tested and sectioned through for examination after processing to determine the condition of the finished board. Such samples, known as "coupons," will provide an excellent indication of the integrity of the entire board.

The final board thickness will depend on the number of conductor layers and on the electrical layer-to-layer spacing requirements of the design. In multilayer printed wiring, the increase in cost is not directly proportional to the increase in the number of conductor layers. For example, doubling the number of layers from 5 to 10 will probably increase costs by only 30 percent. In many cases, additional layers may so simplify the design, reduce conductor density, and increase con-

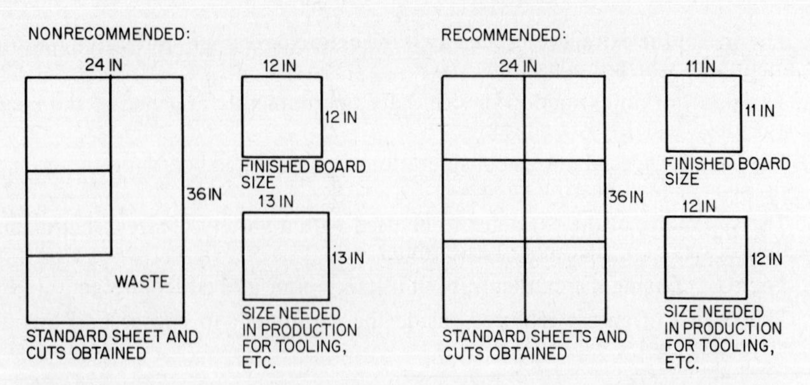

FIG. 4.8 Adjusting finished board size to the standard sizes of clad sheet material will reduce the waste for which the purchaser pays.

ductor size that costs are actually lowered. However, if the number of conductor layers exceeds 10, the extra-layer costs may increase at a rapid rate.

There is no standard rule for the optimum thickness of printed wiring nor for the number of multilayer conductive layers. The difference in base laminate costs due to variations in thickness has a small effect on the finished printed wiring cost.

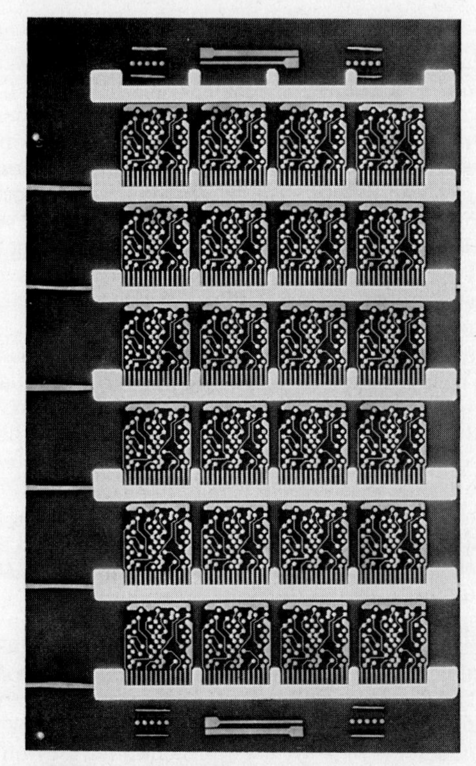

FIG. 4.9 Multiple-image board fabrication panel.

Occasionally, the limiting factor for printed wiring thickness is the diameter of the smallest hole, especially when the holes are plated through.

4.6.2 Thickness

Printed board thicknesses can vary from 0.020 to 0.250 in, but most rigid boards have a thickness of 0.0625 in ($\frac{1}{16}$ in). That thickness (0.0625 in) is also common to plug-in printed wiring. In addition, the structure thickness must be sufficient to mount components and meet the expected environmental requirements.

4.6.3 Cost of Manufacturing

Board manufacturing costs vary with board size, tolerances and clearances, finish, quantity, and delivery requirements. Assuming that any size of board can be made (i.e., that the limits mentioned above have been removed), the first point to consider is the material cost. It will vary as the number of boards that can be cut from a sheet of raw laminate varies. The costs of drilling and of the chemicals used vary directly with the area of the board.

Small boards would generally be grouped together on one production blank (see Fig. 4.9), so labor costs can be assumed to follow a step function similar to that for raw material. There will be some constant charges to be added, and an allowance for a fall in yield with larger boards must be made.

4.6.4 Board Strength

Some of the problems in choosing the size of board depend on the circuits used; other problems are mechanical. Since the mechanical problems can usually be solved in several ways to fit the best electrical or circuit requirements for the equipment under consideration, they can be considered first. Matching the length of the board to the board connector is obvious, and since the length of the connector will be determined by the number of contacts, length can be regarded as an electrical or circuit requirement. The amount of force needed to insert or withdraw a board must be kept in mind. When connectors that have many contacts are used, the boards can be very difficult to move without the use of some form of tool or cam mechanism. The other purely mechanical considerations are the strength and resistance to vibration of the board, any manufacturing limitations, and the ease or difficulty of dissipating the heat generated in use.

Board strength is unlikely to limit board size; bowing, warping, and the effects of vibration are far more likely to be restrictive. Large boards can be stiffened with supports riveted or bonded in place, or spacers can be mounted on the boards to assure minimum clearance between boards.

4.6.5 Partitioning

Minimizing the total number of interconnections can help in determining board size. Partitioning equipment in different ways can result in very different numbers of interconnection points being required. A skeleton block diagram of the whole equipment should be made that shows all the necessary block-to-block connections. The diagram can be examined to determine the points at which it can be divided so as to break the smallest number of connections; then similar diagrams of the major logic functions within each block can be prepared and split down into smaller areas.

Generally, as board size is increased, the number of interconnections per board increases. The increase is fairly rapid at first; then comes a point at which doubling the board size adds very few interconnections; and then a further increase in board size can add considerably more.

4.6.6 Testing Considerations

The time necessary to test each board and locate and repair faults has been cited as a strong argument in favor of the use of very small boards. The reason for making the boards at all is to make a complete set of equipment or, more probably, a set of equipment plus spares, and the only cost worthy of consideration is that of testing the boards for the whole equipment. With any form of automated tester, the handling time will be greater than the testing time, so the larger the boards, the less the total testing cost.

Another point on board size which must not be overlooked is that the larger the boards are made, the more money will be tied up while any faulty components found during testing are replaced. Also, the more components put on a board, the higher the probability that the board will have to be repaired.

The amount of attention that should be paid to this point in selecting a board size depends largely on what facilities can be provided for rapid repair of faults found upon testing.

4.6.7 Fault Location

It is desirable to minimize the time necessary to locate and replace a faulty board. That clearly indicates the use of the largest board possible, because if the whole equipment can be made on one board, the time taken to decide which board to replace will be zero. However, with most equipment, fault diagnosis time is not so easy to determine, and attempts to squeeze too much onto each board can make fault location more, rather than less, difficult. The entire equipment should first be considered and then partitioned in such a way that a faulty block can be found in the shortest possible time.

4.6.8 Size Definition

With so many factors to be taken into account, board size must inevitably be a compromise, unless the board must be fitted into a frame already designed. In that case, however, the size is not really in question, and only the packing density and type of connector to be used need be settled.

4.7 MATERIAL SELECTION

Printed wiring substrate materials are chosen for their mechanical and electrical characteristics and, of course, for their relative costs to buy and fabricate. A comparison of the various substrate material properties will help the printed wiring designer select the optimum substrate for the application.

4.7.1 Mechanical Considerations

Mechanical properties of substrate materials that often have an important bearing on the printed wiring assembly are water absorption, coefficient of thermal expansion, thermal rating, flexural strength, impact strength, tensile strength, shear strength, and hardness. All these properties affect both the functioning and producibility of the structure.

The most popular material for use as printed board with plated-through holes is epoxy-fiberglass. Its dimensional stability is adequate for its use with high-density wiring and minimizes the incidence of cracks in plated-through holes. Its relatively good dimensional stability and its availability in a semicured prepreg stage also make it the most desirable type of material for rigid multilayer construction.

One drawback of epoxy-glass laminate is the difficulty of punching it in the thickness range usually associated with printed boards. For that reason, the holes are usually drilled, and routing operations are used to form the board outline.

Although warp and twist in an etched printed board can be attributed to the fabrication process or the conductor pattern configuration, they are more frequently induced during the manufacture of copper-clad laminate. The designer can help minimize the effects in two ways: First, the direction of conductors should be such that bending stresses released in the etched copper foil on one side of the board are opposed by a similar stress pattern on the other side. This implies that there is also a balance between large conductor areas on both sides of the board, when they exist. Second, large conductor areas used as shields and grounds should be broken by etched slots.

4.7.2 Electrical Considerations

The most important electrical characteristics of the insulating substrate in dc and low-frequency ac applications are insulation resistance, arc tracking resistance, and flashover strength. In high-frequency and microwave applications, the important considerations are capacitance, dielectric constant, and dissipation factor. In all applications, the current-carrying capacity of the printed conductor is important. The electrical characteristics of two commonly used printed board materials are shown in Table 4.1.

TABLE 4.1 Electrical Characteristics of Two Board Materials

Parameter	G-10	G-11
Dielectric strength, perpendicular to laminate, typical, V/in $\times 10^{-3}$	475–525	475–525
Dielectric strength, parallel to laminate, typical, kV/in	45–89	45–80
Insulation resistance,* MΩ	0.2–3 $\times 10^6$	0.1–3 $\times 10^6$
Surface resistance, MΩ	0.1–2 $\times 10^6$	0.01–1 $\times 10^6$
Volume resistivity,* MΩ · cm	0.6–3 $\times 10^6$	0.1–1 $\times 10^6$
Dissipation factor, as received, at 1 MHz	0.019–0.030	0.013–0.025
Dissipation factor, 24 h in H_2O, at 1 MHz	0.025–0.045	0.015–0.045
Dielectric constant, as received, at 1 MHz	4.6–5.2	4.4–5.2
Dielectric constant, 24 h in H_2O, at 1 MHz	5.0–5.8	4.5–5.8
Dielectric constant at 10 MHz	5.0–5.6	
Dielectric constant at 25 MHz	4.7–5.2	

*Temperature- and humidity-conditioned 96 h at 35°C, 90% RH.

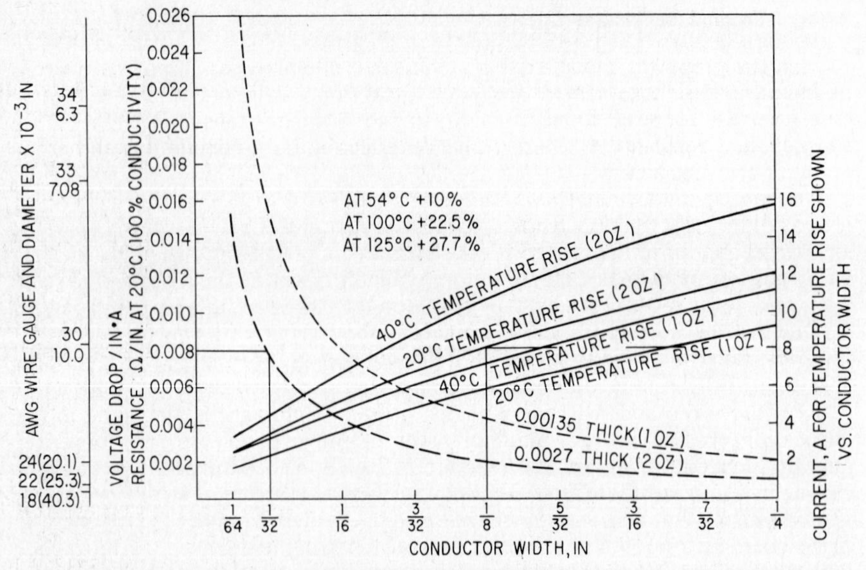

FIG. 4.10 Conductor width. Copper conductor resistance voltage drop and temperature rise conversion chart, rectangular printed wiring vs. round wire. IACS resistance is 0.67879 $\mu\Omega/\text{in}^3$ and 0.67879/0.00135–503 $\mu\Omega/\text{in}^2$ (1 oz).

4.8 ELECTRICAL DESIGN FACTORS

This section concerns a few of the important electrical circuit factors that must be accounted for in proper printed board design. The information provided here is by no means all-inclusive. The values and equations given should be used only as guides for estimating approximate values of a given characteristic. All electrical parameters of a board are highly interdependent, and empirically measured values will, therefore, rarely coincide exactly with calculated values. This is especially true of high-frequency regions, in which it is impossible to make a complete analytic prediction of board characteristics. Data given in this section are simplified for purposes of compactness, and many contributing factors have been eliminated.

4.8.1 Resistance

Current-carrying capacity of conductors is not usually a problem, but the ohmic resistance may be a problem when conductive paths are unusually long or when voltage regulation is critical. An approximate calculation of conductor resistance and temperature rise can be made from Fig. 4.10. The resistance and temperature rise may be calculated from the expression

$$R = 0.000227W$$

where R is the resistance per linear inch, in ohms, and W is the width of the line, in inches, based on 99.5 percent minimum-purity copper 0.0027 in thick (or 2-oz copper).

4.8.2 Capacitance

Capacitance may be of considerable importance, especially at high frequencies. The distributed capacitance between conductors located one above another must be accounted for when high-frequency circuits are involved; it is of the order of 1 pF/ft. The basic capacitance equation serves as a useful, if approximate, guide:

$$\text{Capacitance (pF/in)} = \frac{\text{conductor width } (10^{-3}\text{ in}) \times \text{dielectric constant}}{\text{dielectric separation } (10^{-3}\text{ in}) \times 4.45}$$

When the conductor width is at least 10 times greater than the dielectric separation, the equation is generally in close agreement with empirical values obtained but may be on the low side. The capacitance coupling between conductors can be minimized by limiting the lengths of conductors running in the same vertical plane. Capacitance between adjacent conductors is a function of conductor width, thickness, and spacing, as well as of the board material itself. It can be calculated from the expression

$$C = 0.31\,\frac{a}{b} + 0.23\,(1 + k)\log 10\left(1 + \frac{2b}{d} + 2b + \frac{b^2}{d^2}\right)$$

where k = material dielectric constant
a = conductor thickness, in
b = width of conductors, in
d = distance between conductors, in

Special attention must be given to circuits located over a shield or ground plane, since the length of the conductor is capacitance-coupled to the plane and consequently to similar conductors of similar relations.

For critical high-frequency circuitry, the electrical characteristics of single-sided circuits made of $\frac{1}{16}$-in epoxy-glass or epoxy-paper are not adequate, and associated ground plane micro-strip-line construction is a must. The electrical characteristics can also be used as a process control; they give a good indication of process variables.

4.8.3 Characteristic Impedance

The characteristic impedance of parallel conductors is determined as follows:

$$Z_0 = \frac{R + jwL}{G + kwC}$$

where Z_0 = the apparent impedance of an infinitely long line, Ω
R = resistance per unit length of line, Ω
L = inductance per unit length of line, H
G = conductance per unit length of line, mhos
C = capacitance per unit length of line, F

The method of achieving transmission-line capability with double-sided printed wiring is called "microstrip" (see Fig. 4.11). A specific formula microstrip is the following:

$$Z_0 = \frac{h}{W}\,\frac{377}{\epsilon_r}$$

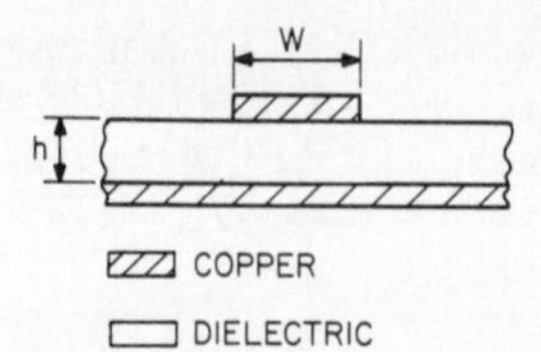

FIG. 4.11 Microstrip transmission line.

where h and W are given in the diagram, and ϵ_r is the effective dielectric constant of the material (considering the effect of air). This analytical method disregards fringing effects and leakage flux and gives validity to the notion that more reliable values may be obtained by measurement. To account for the fringing effects of microstrip transmission line, and if the analytical method must be used, the following formula is recommended:

$$Z_0 = \frac{h}{W} \frac{377}{\sqrt{\epsilon_r}\{1 + (2h/\pi W)[1 + \ln(\pi W/h)]\}}$$

4.9 ENVIRONMENTAL FACTORS

Printed wiring can usually be expected to exhibit the same properties as the base materials, i.e., the laminate and cladding from which it is constructed. Environmental requirements, therefore, relate back to those elements. An exception is the case of monolithic assemblies of printed boards, which are, by nature of the lamination and interconnection procedure, additionally affected by the environment.

In addition to serving as an electrical connection medium, printed wiring planes provide mechanical support for the active or passive components they are interconnecting. In that way they become an integral part of the package or assembly and must therefore be able to withstand the environmental stresses associated with the entire structure.

4.9.1 Shock and Vibration

Vibration, flexing, and bowing can become a problem on larger boards. When the board is normally mounted, the board connector will keep one edge of it held firmly, and the two sides will be supported to some extent by the board guides. That will leave only one edge free to bow either because of stresses from manufacturing or under the influence of vibration. A handle along the free edge will help to eliminate those effects and will leave the center of the board as the area most likely to be affected. The problems of minimizing the effects of vibration and warping are exactly the same as those met in any other form of engineering, and similar solutions can be used. A stiffening web, which can be a piece of metal angle, a plastic extrusion, a "top hat" section, or a solid bar, can be fitted across the middle of a large board. Alternatively, smaller pieces of suitable material can be riveted or bonded to the board to change the major resonant frequency.

One of the greatest dangers of bowing or vibration is that components with electrically live cases will be brought into contact with the soldered joints on the back of an adjacent board. Stiffening or the application of damping strips may not be enough to prevent natural warpings of a large board from introducing a dangerous bow. If any risk of bowing exists, the board can be fitted with spacers higher than any of the packages on it—as many as may be required being fitted at suitable places.

4.9.2. Thermal

In facing the problem of heat removal, the printed wiring designer should take advantage of the available conductive cooling, employ high-temperature components where possible, and ensure thermal isolation of temperature-sensitive components from high-heat-emitting sources.

All three principal modes of heat transfer—conduction, convection, and radiation—find uses in cooling electronic packages.

4.9.2.1 Conduction. Enhancement of heat transfer by conduction may be brought about by use of (1) materials with high thermal conductivities, (2) direct paths to large heat sinks, and (3) good thermal bonding or coupling between parts involved in the conduction path.

The use of direct large-cross-sectional paths to heat sinks is particularly important in conductive cooling. The shorter the path to the heat sink and the fewer the number of thermal bonds or joints present, the better. The printed conductor itself must be considered in the thermal path and should be as large as possible to help effect heat transfer. The contacting surfaces should be large and not subject to oxidation.

4.9.2.2 Convection. Convective cooling can be enhanced by (1) increasing the velocity of the mobile substance over the surfaces to be cooled, (2) increasing the surface area available for heat transfer, and (3) replacing laminar flow with turbulent flow, which increases the heat-transfer coefficient and ensures good scrubbing or wiping action around the parts to be cooled.

4.9.2.3 Radiation. Radiative heat transfer can be enhanced by (1) using materials with high emissivity and absorptivity, (2) raising the temperature of the radiating body or lowering the temperature of the absorbing body, and (3) arranging the geometry to minimize back reflection to the radiating body itself.

Power transistors and high-wattage resistors are among the components that must be given special consideration to eliminate local hot spots that can damage adjacent components or the board itself. In general, hot components should be mounted close to the frame, which serves as the heat sink, in order to shorten the thermal path. Temperature-sensitive components should be placed in the hottest part of the package—either in the center or closest to the hot components.

In estimating the heat dissipation in an electronic package, one must analyze the circuit to obtain the maximum power dissipation for each component and then determine the maximum operating surface temperature to be expected. With that information known, the design must then be arranged, by using the techniques already discussed, to maintain the components below their maximum operating temperature under the worst conditions.

The maximum temperature allowable in the package must be considered in light of the associated insulation present as well as the components themselves.

Some idea of the amount of heat that can be effectively removed from a sealed unit and from an external surface by various types of cooling is given in Table 4.2, for sealed units and external surfaces.

TABLE 4.2 Effective Heat Dissipation as a Function of Type of Cooling

Type of cooling	Heat removed, W/in³
Sealed unit	
Free convection, air	0.15
Metallic conduction	2–5
Forced convection, air	To 7
Direct liquid cooling	To 10
Vaporization cooling	To 20
External surface, based on a 40°C temperature rise	
Free convection, air	0.25–0.50
Free convection plus radiation	To 2
Direct liquid cooling	To 4
Vaporization cooling	To 7

4.9.3 Protective Coating

It is desirable, and often mandatory, to provide protection for printed wiring assemblies against dust, dirt, contamination, humidity, salt spray, and mechanical abuse. There are many insulating (conformal coating) compounds that can be applied; predominant among them are epoxies, polyurethanes, silicones, acrylics, polystyrenes, and varnishes. Those compounds are similar to encapsulating materials of the same chemical type, but frequently they are used in solution form. The number of variations and combinations of the materials is very large. For very thin conformal coatings (0.0006 in thick and less) vacuum-deposited paraxylene is often used. The most widely used coatings for military applications are in accordance with MIL-I-46058.

Some of the more important technical considerations involved in selecting a printed wiring assembly coating are as follows:

1. Its ability to prevent the occurrence and migration of circuit, component, and current-carrying corrosion products
2. The protection it affords to the printed wiring insulation
3. The effect of humidity and temperature on insulation resistance
4. The effect of its thickness and the temperature and humidity on important electrical properties such as dissipation factor, dielectric constant, and Q
5. Its flexibility
6. Its resistance to cracking during thermal shock
7. Ease of its application and processing
8. Its ability to be modified in order to control its thickness
9. Its optical clarity to allow for the viewing of the board and component markings
10. The ease by which it is removed before the printed wiring assembly is repaired

CHAPTER 5
LAYOUT AND STANDARDS*

5.1 INTRODUCTION

The creation of the master artwork that is subsequently used to provide the phototool for the "image transfer" process is a critical step in the printed wiring fabrication process. The function of the finished printed wiring board (PWB) is determined at this stage. In addition, this is the point at which the manufacturability of the board, its basic reliability, and other elements of the board as a component in the finished electronic product are determined.

This creation of the master artwork is usually called the "layout" of the board. It is the definition of the conductive traces, hole location, land size and shape, and number of layers.

Although computer-aided design and layout of the PWB are sophisticated and evolving, the basic rules are fairly constant. Whether the actual layout is done by a computer-aided system or by a manual process, the engineers and layout technician responsible need to understand these basic rules in order to create the most effective final design. This chapter defines these fundamental rules and provides the background to use them.

5.2 CONDUCTOR PATTERNS

The simplest process for forming conductor patterns is to etch the patterns from the foil laminated to the printed wiring base. The process requires a minimum number of steps and has been widely used in mass production. Foil can be etched on either side or on both sides of the laminate, but etching does not provide for interconnections between conductor planes except by hole plating or by mechanical means. Similar types of printed wiring patterns can be produced by proprietary additive processes or special types of resist.

5.2.1 Routing and Location

The printed conductor should travel by the shortest path between components within the limits of the wiring rules imposed. The latter may restrict parallel runs of conductors owing to coupling between them. Good design dictates that the

*Developed from material originally written by Dieter W. Bergman and G. L. Ginsberg for *Printed Circuits Handbook,* 2d ed., Clyde F. Coombs, Jr. (ed.), McGraw-Hill, New York, 1979.

minimum number of layers of wiring be used and also that the maximum line width and terminal area size be commensurate with the density of packaging required. Since rounded corners and smooth fillets avoid possible electrical as well as mechanical problems, sharp angles and bends in conductors should be avoided.

Translation of a circuit schematic to its proper physical realization generates many side effects, and crosstalk is one of them. Crosstalk is often a ground plane problem. There are two aspects to the problem: the crosstalk generated in individual boards and crosstalk through the system ground. If careful routing of individual sections of each circuit provides an exclusive conductor to a single ground at essentially a single point, the problem need hardly exist to the point of being harmful. This indicates that a major concern in determining internal circuit layout will be the grounding circuitry. Accordingly, it will partially determine physical subsection size and workable unit grouping.

Conductor routing is the very heart of printed board design, yet many engineers appear to feel it is best left to the layout team to struggle with as best they can. Somewhat belatedly, a realization of the importance of package placement and routing in getting the best possible circuit performance and also the realization that, with more and more complex boards, layout takes an excessive time and causes serious bottlenecks in production have led to the use of computer programs for both routing and layout. So far, however, it seems the best boards can result from manual routing when a well-considered system is used. A computer could do the job if it were fed with sufficient information, but it would take a long time to establish a good layout for a large, complex board and a full routing program. It appears that the best solution is computer-aided design (CAD), in which the computer gets on with the tedious "donkey work," and calls for assistance every time it meets a problem. A moderately experienced layout drafter will spot several possible ways through an apparently blocked board area in the time it would take the computer to try the first couple of dozen logical approaches. The operator might suggest moving several lines or even changing the package allocation to clear the trouble, whereas the computer might give up the problem as insoluble. If the computer displays its interim results on a cathode-ray-tube display, the operator can watch the board layout "grow" and can possibly even prevent the computer from doing something which might block subsequent runs.

One great advantage of CAD for routing is that the final output can be in the form of control tapes for a coordinate-plotting machine and for a numerically controlled drill.

5.2.2 Conductor Shapes

Sharp corners and acute angle bends in conductors should be avoided when possible because of the additional stresses imposed on both the conductor itself and the adhesive bond and also because of the electrical problems arising from local field intensification. Therefore, although more costly from a drafting standpoint, rounded corners at conductor bends and smooth fillets at the junction of conductors and terminal areas, or lands, are desirable. Such rounded contours will not only minimize conductor cracking, foil lifting, and electrical breakdown but also facilitate solder distribution. The generally preferred and nonpreferred printed wiring conductor shapes are illustrated in Fig. 5.1.

5.2.3 Width and Thickness

The width of any conductor is a function of the current carried and the maximum allowable heat rise due to resistance. The conductor width should be as generous

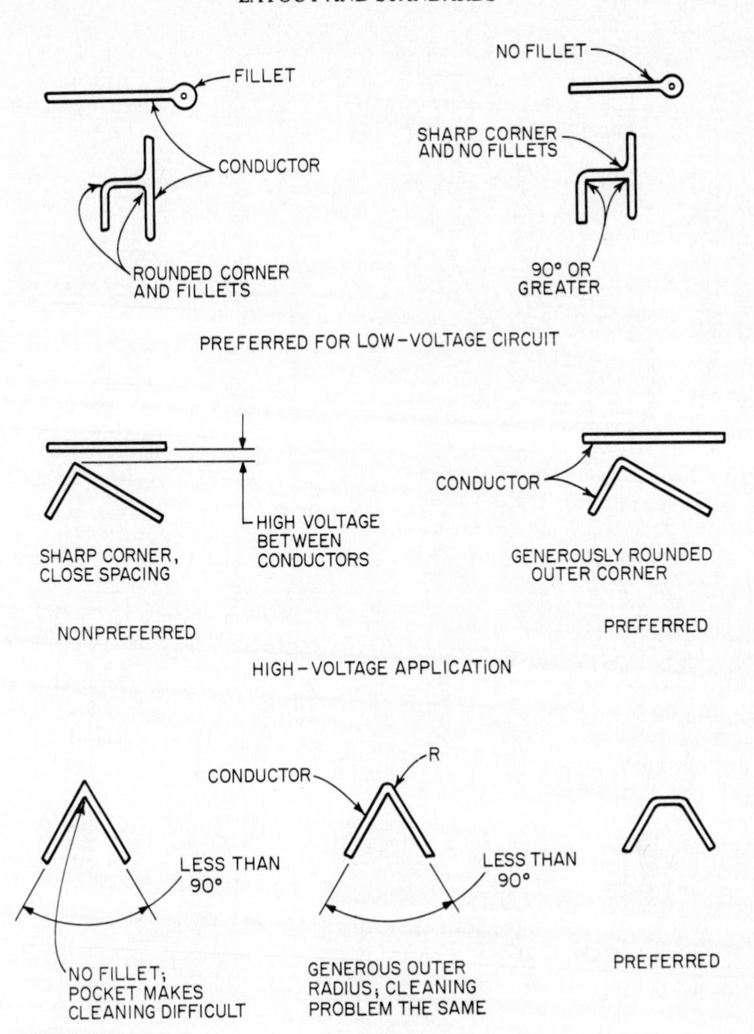

FIG. 5.1 Conductor shapes.

as possible to tolerate the normal amount of undercutting caused by etching and the manufacturing process, as well as nicks and scratches in the artwork caused by careless handling. The amount of undercutting or line reduction is equal to twice the thickness of the conductor.

The current-carrying capacity of etched-copper conductors for rigid boards is given in Figs. 5.2 and 5.3. For 1- and 2-oz conductors, allow a nominal 10 percent derating (on a current basis) to provide for normal variations in etching methods, copper thickness, and thermal differences. Other common derating factors are 15 percent for conformally coated boards (for base material under 0.032 in and copper over 3 oz) and 30 percent for dip-soldered boards.

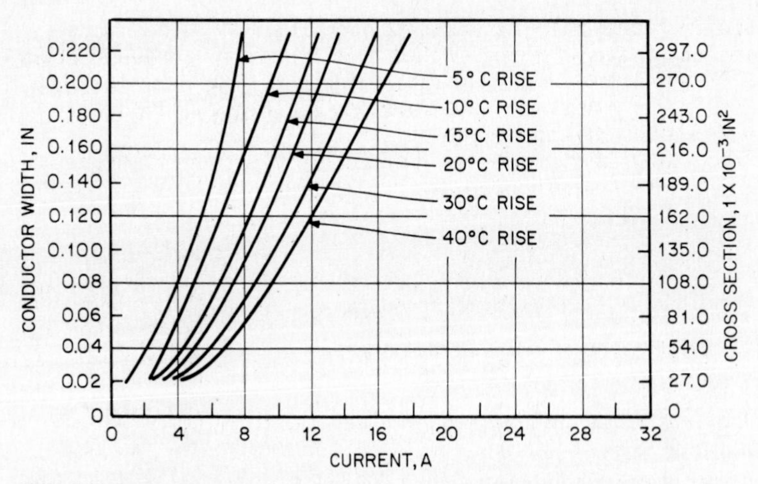

FIG. 5.2 Temperature rise versus current for 1-oz copper.

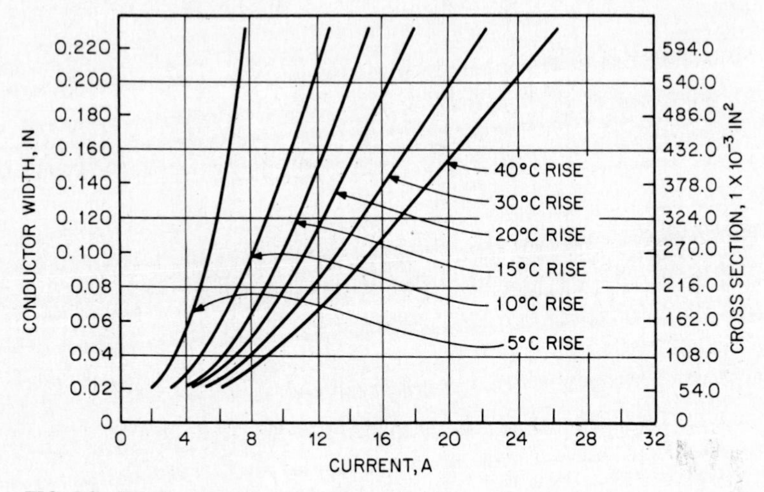

FIG. 5.3 Temperature rise versus current for 2-oz copper.

5.2.4 Spacing

Minimum conductor spacing must be determined to preclude voltage breakdown or flashover between adjacent conductors. The spacing is variable and depends on several factors:

1. Peak voltage difference between adjacent conductors
2. Atmospheric pressure (maximum service altitude)
3. Use of a coating
4. Capacitive coupling parameters

The logician, or circuit designer, may place on board layout certain restrictions which must be heeded. For instance, critical-impedance or high-frequency components are usually placed very close together to reduce critical stage delay. Transformers and inductive elements should be isolated to prevent coupling, and inductive signal paths should cross at nominally right angles. Components which may produce any electrical noise from movement of magnetic fields should be isolated or rigidly mounted to prevent excessive vibration.

Minimum conductor spacings for various voltages and altitudes for both coated and uncoated boards are shown in Table 5.1.

5.2.5 Conductor Pattern Checklist

The following checklist pertaining to conductive patterns should be used to help assure that the printed wiring design is of sufficient quality. (An affirmative answer to each question is desirable.)

1. Are conductors as short and direct as possible without compromise of function?
2. Are conductor width restrictions adhered to?
3. Has necessary minimum spacing been maintained between conductors, between conductors and mounting holes, between conductors and terminal areas, etc.?
4. Is close paralleling of all conductors, including component leads, avoided?
5. Are sharp angles, 90° or less, avoided in the conductor pattern?
6. Have all large conductive areas been properly relieved?
7. Is the circuit pattern geometrically arranged to avoid solder bridging?
8. Are component derating practices followed?
9. Has a worst-case analysis been run on the design?
10. Has proper spacing been maintained between conductive patterns and terminal areas and component and board mounting holes?

5.2.6 Hole Fabrication

Tolerance capability for hole formation is dependent upon the way in which the holes are made. Although holes may be punched in some rigid materials

TABLE 5.1 Spacing of Conductors for Various Voltages

Uncoated boards sea level to 10,000 ft		Uncoated boards over 10,000 ft and coated boards		
			Minimum space, in	
Voltage between conductors, dc or ac peak, V	Minimum space, in	Voltage between conductors, dc or ac peak, V	Above 10,000 ft uncoated	Conformally coated, all altitudes
0–150	0.025	0–50	0.025	0.010
151–300	0.050	51–100	0.060	0.015
301–500	0.100	101–170	0.125	0.020
Above 500	0.0002 (in/V)	171–250	0.250	0.030
		251–500	0.500	0.060
		Above 500	0.0001 (in/V)	0.0012 (in/V)

(especially paper-phenolic punching grades) and most flexible materials with a positional accuracy of ±0.003 in, the technique is limited to single-sided wiring only. For double-sided and multilayer boards, holes are drilled in a variety of ways. Tolerances for various methods of drilling holes are as follows:

Drilling by eye	±0.010 in
Jig-board drill plate	±0.005 in
Drilling with optical aid	±0.005 in
Tape-controlled drill, eight spindles	±0.003 in
Tape-controlled drill, one spindle	±0.001 in

5.3 LAYOUT

The printed wiring layout is defined as being a sketch that depicts the printed wiring substrate, the physical size and location of all electronic and mechanical components, and the routing of conductors that serve to interconnect the electronic parts. The layout is usually prepared in sufficient detail to permit the generation of documentation and artwork.

The engineer should furnish the drafter a well-drawn schematic with as few crossovers and loops as possible. He or she should bring to light the areas of coupling incompatibility and those where short circuit paths and isolation of one circuit from another are required.

The position and quantity of the off-board connections to be used to interface with the interconnection system wiring play a very important part in determining where the inputs, outputs, and grounds will be. The board contacts will greatly affect the total order of the final layout. They should be designated alphabetically on the schematic to help guarantee that they will be clearly understood and any necessary fixed contact assignments should be noted. Shields and grounds also must be clearly defined for the layout personnel, especially if they are used for reducing undesired noise or interference from couplings.

5.3.1 Layout Procedure

Layouts are usually preferred to be as viewed from the component side. Two-sided designs may be represented on a single sheet by a coding system or by another sheet representing the individual sides of the design. The latter method can cause errors and is not the one preferred. Multilayer conductor layouts also are laid out singularly in this manner, as are ground and voltage planes.

If a standard component matrix can be achieved, the preprinted layout format can be made in advance. That is true of the voltage and ground layers as well when those layers are required for the design.

Proper fabrication indexing hole locations are necessary for each layout; two or three indexing holes are desirable. It is preferred that the holes be located in the *x* or *y* datum plane, or in both.

5.3.1.1 Input Data. The normal input data for manual layout design is a rough logic symbol schematic. That is a document depicting the electronic circuit in symbolic form by utilizing both discrete and logic formats.

Schematic circuit flow is indicated going from left to right on the diagram. Terminals, components, and connector contacts are labeled, as are all input-output terminal designations. Associated with the data input is a list of components and specified critical circuit design holes.

The critical circuit design rules encompass such electrical areas as capacitance coupling, feedback, current, and clock signal grouping. They are the circuit design engineer's responsibility and should be clearly defined in his circuit design rules before the layout designer prepares the design layout.

An alternative input may be a line diagram combined with proposition and logic listings associated with the electrical circuitry design.

5.3.1.2 Development. The sequence of events in the development of a printed board design layout may vary, but the following are the general steps to be taken in the layout development after the input data have been received.

1. An initial evaluation is made of the schematic input, parts list, and special circuit rules. It includes comparison of the data with the physical limits of the board. Physical limitation is concerned with the specific usable board area available.

2. Components are reviewed as to their circuit relation, the necessity of implementing short conductor interconnections, heat sinks, ground and voltage connections, and special conductor width requirements.

3. Signal input and output connector interconnections are then reviewed to assimilate the logic flow organization as it is related to the individual design interface.

4. Specific company standards concerning automatic component insertion requirements, component matrix location parameters, mechanical hardware, and connector usage are then considered as they have impact on the design.

Since most of the above process is mental, the actual "hands on" procedure for the manual design layout continues to be carried out by the designer.

5. An underlay of printed-grid polyester base material, utilized as the background for the design layout, is attached to the designer's working surface.

6. Standard board sizes permit the use of a preprinted polyester base material (Mylar, Cronaflex) of the master printed board configuration. All specific rules and standards become a part of the master outline.

5.3.1.3 Methodology. It is very convenient initially to make a layout by using templates (paper dolls) of the components or the components themselves and making repeated freehand point-to-point wiring diagrams until the interrelations of all components and wiring are compatible with good design practice with respect to mutual coupling and inductive and capacitive effects. The process, with quick rearrangements of components and conductors between steps, will evolve more and more refined designs. It is recommended that at least three of the steps be taken quickly, and three or more additional steps of refinement are perhaps worthwhile for the most complicated layout. Each should be considered to be only a stepping-stone to the next one and not in any way a complete entity.

Electrical parameters of the printed wiring must be literally designed into the more critical circuits. In particular, the resistivity and dielectric properties of the insulating base require consideration for the higher frequencies and higher-impedance circuits, especially when stability in environmental extremes is an important design criterion.

Whenever possible, the component layout should permit automatic assembly unless a short production run is anticipated. Similar units should be placed together in rows or groups—bearing in mind the spacing requirements arising from the manufacturer's recommendations, use of modular grid locations, heat dissipation, electrical coupling, and repair considerations. The component placement should, if possible, allow the electrical values and code numbers to be read easily from one direction, and preference should be given to the electrical values. That will not generally be feasible in the case of automatic assembly. Components should be mounted on only one side of the board and be so spaced and located that each component can be removed without removing any other part, unless, of course, the package is to be expendable.

Heavy components should be located over or near supported areas and heat-emitting components near the board sides or heat sink areas. More is said about this in connection with designing for thermal shock and vibration stresses.

The insulator bodies of tube sockets, connectors, and similar parts subjected to insertion and withdrawal forces must be rigidly anchored without relying on the soldered connections for mechanical support. If tube sockets are not used on subminiature tubes and the flexible tube leads are soldered directly to the printed wiring, the tube envelope must be rigidly supported.

5.3.1.4 Orientations. Assembly and inspection of boards are much easier if all packages are mounted in the same way. Cases do arise when it appears that to reverse the orientation on alternate rows of packages will simplify the interconnections or the power distribution, but in most cases a number of minor snags crop up later in the interconnections and the final gain on the finished board is negligible. If package orientation is standard, automatic or semiautomatic placing and reflow soldering of flat packs, or insertion of dual-in-line packages (DIPs), is possible without having to reverse the board or reset the machine as would be necessary if alternate rows of packages were inverted. Also, fault finding on large boards is easier if the orientation is standard. The operator knows, without checking, which is pin 1, which pin is ground, and which is power. Finally, mistakes in routing are far more likely if there are any inversions in orientation.

5.3.1.5 Grid Systems. All component hole locations, mounting holes, and even the outside board configuration should be laid out on a modular grid system. The basic modular units of location are based on a 0.100-, 0.050-, or 0.025-in system applied along both the x and y axes. Unless dictated by a particular fixed design parameter, there are few acceptable reasons for a distance between mounting holes or the board length or width to have a dimension of, say, 2.093 in when 2.100 in would suffice or 4.120 in instead of 4.125 in. Manipulation of tolerances can normally adjust any dimension to the nearest 0.025 in and satisfy the grid requirements. Should a high production run be anticipated, numerically controlled program drilling of printed boards can now become a reality based on the 0.025-in rectangular coordinate system.

5.3.1.6 Conductor Routing. The same principles that make a good layout for point-to-point conventional wiring apply to printed wiring. For instance, to guard against undesirable feedback requires either adequate separation or shielding between input and output. Though the principles are the same, however, the dimensions are different. There is virtually only one plane, and it requires more intercircuit planning because normal hookup wire is insulated and may be routed on intersecting paths without concern for contact. Printed circuits, on the other hand, establish a connection at each junction. To overcome that disadvantage, the routing of each conductor must be thoughtfully planned. As described above, that sometimes means some leads may be longer than desired, but with extra thought a jumper or crossover can be used to great advantage.

Conductor routes should best utilize the space available. That, as an example, is where the problem illustrated in Fig. 4.4 exists. It should be noted that the route picked takes precisely the geometric mean with respect to such hazards as

1. Insulation resistance (leakage)
2. High voltage
3. High noise
4. Undesirable feedback or coupling due to electrostatic or electromagnetic coupling
5. Signal losses due to capacitance effects

If all things are considered equal, the conductor route should split the difference at every turn and have equal spacing over parallel runs as shown in Fig. 5.5.

5.3.1.7 Marking and Identification. Almost all printed circuits require the use of nomenclature on the boards to identify components, circuits, test points, and the board itself. The nomenclature may be applied after the board is etched, or it may be etched along with the circuit. If it is applied later, it is usually screen process printed or stamped. In either case, the marking material must be carefully chosen for its electrical and chemical properties. For that reason and for reason of the extra processing steps involved, designers usually include nomenclature in their artwork and etch it along with the circuit. Although such nomenclature is conductive, it does not normally present a problem. The nomenclature should be as small as it can be to survive etching and still be legible. Thickness of the copper and the location of the nomenclature relative to conductors will determine how much deterioration the nomenclature will receive during etching. Very small nomenclature should not be located in open areas when 2- or 3-oz copper is being used.

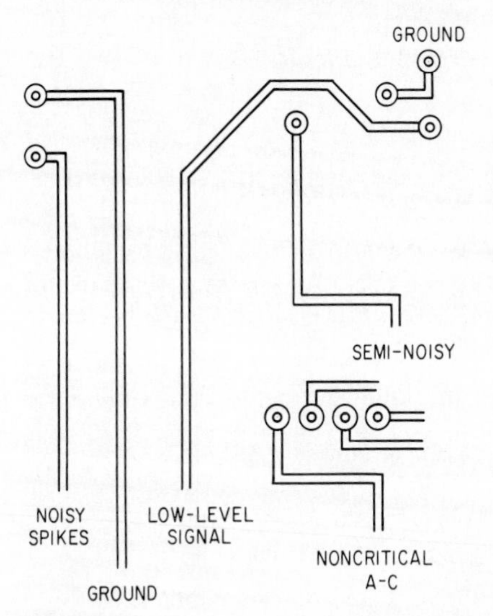

FIG. 5.4 Conductor routes utilize space availability.

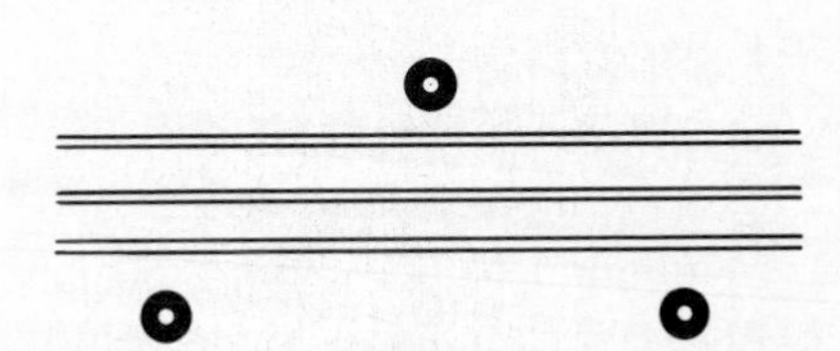

FIG. 5.5 Parallel-run conductor configuration.

To facilitate testing and servicing a printed board assembly—and to link the assembly visual aid to documentation—component reference designation marking should be considered. In addition to reference designators, polarities of capacitors and diodes should be indicated as close to their component mounting holes as possible. Reference designators and polarities should be completely visible with the components in place. On a dual-in-line integrated circuit, pin 1 should always be indicated, as should be the tab for a 6-, 8-, 10-, or 12-lead TO-5 type of integrated circuit.

Printed boards should carry two distinct identifications. Since the fabricator machines and plates the printed board before the electrical components are installed, he or she needs documentation and a method of identifying the board. The identification is the printed board fabrication drawing number (master drawing). The printed board assembly number should be located on the component side where assembly of electrical components takes place, from which orders for replacement of spare parts are made, and where servicing and testing are normally done. The number should always be preceded by "ASSY NO." In addition, the serial number of the assembly is necessary for tracing purposes. If space permits, display your company logo on the board. All those features are easily applied by the etched-foil method.

5.3.1.8 Summary. When printed wiring assemblies are being designed, the following basic concepts should be considered:

1. Specific contractual requirements.
2. Selection of standard parts, materials, and processes according to MIL-STD or commercial standards, in that order of preference.
3. Design that will accommodate the use of standard tools.
4. Safety factors for the equipment.
5. Accessibility of parts.
6. Minimum need for adjustment in operating the equipment.
7. Provision for simple maintenance.
8. Selection of parts resistant to damage by the heat of soldering. Parts to be mounted must not only meet the storage and operating conditions specific for the equipment but must also maintain their integrity after having been subjected (during assembly) to heat soldering.

Proper mounting design should fulfill the following basic requirements for mounting parts.

1. Ease of maintenance
2. Easily removable fasteners
3. Minimum loose hardware
4. Adequate clearance between printed conductor and mounting brackets
5. Board supported at proper interval
6. Positive board retention

5.3.2 Automated Techniques

The application of automated techniques to printed wiring design has become more prevalent. A relatively complete automated system for printed wiring design includes the following set of functions: logic assignment (component, gate, and pin allocation), logic partitioning (grouping logic functions by physical board), component placement, conductor routing (interconnection layout), design checking, and artwork generation. The desirable final step is the generation of control tapes for numerically controlled production equipment. Building the system requires a hierarchy of optimization steps, with intermediate design data being passed to successive higher levels of assembly.

When fully implemented, an automated design system will replace nearly all of the manual operations typically required. Key among the manual operations eliminated are component placement, interconnection layout design, and taping or digitizing from a rough layout of the board. A number of the widely used board design conventions or "standards" exist simply because they are convenient to a manual design environment and because they impose no restrictions on the human designer. By way of example, the 0.0625-in pad that is in common use provides no particular benefit and seems to derive solely from the 0.250-in "round number" at 4:1 tape-up scale. Another example is the frequent "odd number" 0.156-in spacing of connector pads or fingers. Although these and other conventions may represent acceptable or convenient practice for manual design, they are often awkward or limiting in an automated design environment. A comprehensive CAD system will optimize the pad (terminal area) size to complement the complete system.

5.3.2.1 ***Grid Pattern.*** One of the more fundamental considerations in an automated design environment is the use of a grid pattern to define the relative positions of component pins, obstacles, connectors, conductors, etc., on the board. Typical grid sizes are 0.100 in, 0.050 in, and 0.025 in, where the value defines the dimension from grid line to grid line. The physical dimensions of integrated circuit components generally support the latter two sizes, although it is important to note that the components influence rather than control the grid size used.

The minimum practical grid size is typically determined by the desired width of printed circuit conductors and of the air gap clearance between conductors. For example, a 0.020-in conductor and a 0.030-in conductor clearance will result in a 0.050-in grid size. A 0.012-in conductor and a 0.013-in clearance requirement will support a 0.025-in grid size. Selection of the grid size to be used is influenced by interconnection complexity, pad sizes, and conductor-pad clearances.

It is significant that the use of automated design techniques will often permit the use of a smaller grid size than nonautomated practices will permit. That results from the high precision with which the finished artwork can be produced and the hole drilling process controlled; each provides improved response to manufacturing tolerance requirements. Accordingly, organizations using automated design facilities for the first time should carefully inspect the basis for existing spacing practices. If an approximate 0.100-in grid size is in use for manual work, a 0.050-in size should be strongly considered. If an approximate 0.050-in grid size is in use, a 0.025-in grid size should be considered for the most complex designs.

5.3.2.2 ***Component Density.*** Perhaps the single most difficult parameter to generalize upon is the component density (number of components per square inch of board area) that can be supported by automated design techniques. The reason is simple enough: Component density is a dependent parameter affected in some degree by all the other factors under review.

Because there are so many variables which bear on component density, any quantitative statement requires a number of qualifying remarks. For example, nominal maximum values of component density are respectively 1.2 and 1.8 for dual-in-line and flat packs in two-sided boards and 3.0 and 4.0 for multilayers. The values are stated for a component count of about 25 integrated circuits (two-sided board values) and a grid size of 0.050 in. Assuming that preferred design conventions are present (some already reviewed, others to be reviewed later), the values are further sensitive to board size and proportions, grid size, and pad or tab sizes.

Board size and proportions can be expected to have a direct effect on conductor density—the concentration of conductors in certain portions of the board, which in turn affects component density. The concentration of conductors in the center of a board usually increases as the board size increases. Regarding board proportions, a 5- × 6-in board will support a higher component density than a 3- × 10-in board owing to an improved balance of conductor channels for side-to-side communication.

Grid size has a direct bearing on the gross number of points available for conductor routing. A smaller grid size will therefore support a higher component density with its increased pin count. Although a 0.025-in grid size provides 4 times the number of points as a 0.050-in grid size, increases in component density in the same or nearly the same proportion should never be assumed. Grid size changes affect only the gross point count and not the effective (available) point count.

5.3.2.3 Power and Ground Bus Patterns. In general, power and ground patterns utilized for manual designs are suitable for automated design techniques also. The prior emphasis on the need for close control of factors which tend to reduce the effective conductor routing area suggests, however, that the width of power and ground buses should be carefully examined and held to the smallest possible dimension. Excessively wide buses alone can reduce the number of conductor channels by 10 percent.

There may be a preferred orientation in which the narrower of the buses, if there is one, would best be placed on the board side which provides the greatest number of unobstructed channels flowing toward the connector strip and be oriented parallel with those channels. Board proportions also can influence the preferred orientation.

Designs which utilize power and ground plane(s) construction provide an advantage in that they do not consume conductor routing area. The manufacturing basis for the board may introduce restrictions, however, if the ground and power plane(s) must be prefabricated with a standard feed-through clearance hole pattern.

5.3.2.4 Summary. Not all printed wiring designs will permit all the recommendations to apply all the time. The important consideration is that the recommendations be reviewed for every new product design to maximize the number of the preferred conventions which can be used. The absence of any one, or a few, of the conventions does not void the opportunity to utilize automated design techniques.

5.3.3 Layout Checklist

The following checklist pertaining to board layout may be used as a guide to help assure that the printed wiring design is of sufficient quality. An affirmative answer to each question is desirable.

1. Are all noncritical component mounting hole locations at the grid intersections?
2. Are all component mounting hole spacings correct?
3. Are indexing holes adequately spaced?
4. Do component mounting hole terminal areas contain the proper size hole?
5. Is the diametral clearance of the component mounting hole within acceptable limits?
6. Are all board-mounting holes properly indicated on the drawing list and dimensionally located and specified on the board drawing?
7. Has the number of different hole sizes been kept to a minimum?

8. Have all tolerances been considered to establish the proper relation of terminal area to hole?
9. Are terminal area sizes compatible with the projection system on the numerically controlled drafting machine?
10. Are terminal areas elongated along the direction of lead clinch?
11. Is the space required for the printed board optimal?
12. Are the drawing list and schematic compatible?
13. Are the printed side and the component side properly indicated on the drawing list?
14. Has a suitable tolerance been allowed between plug-in and mating assemblies?
15. Are all tolerances complete and reasonable?
16. Are all board dimensions scaled to the basic grid?
17. Do all corners of the board have a specified radius?
18. Have all required test patterns been provided?

5.3.4 Artwork Factors

The problem is to obtain a high production yield in a product such that a reject is not detectable until the investment is high. To solve the problem, most vendors have instituted a rigid inspection of the incoming artwork, or film, before the actual fabrication is even begun. That, coupled with rigid inspection of individual layers, is their guarantee of good registration. The drilled hole, after all, must be aimed at the center of the termination pad. Accuracy is absolutely essential.

Several techniques commonly have been used to accomplish accurate layer-to-layer registration. "Accurate" is, of course, a word that means different things to different people. Some might consider a master with a misregistration of 0.010 in to be accurate, whereas others might quarrel with anything over 0.005 in. The several techniques described next are presented in order of ascending difficulty; the more accurate techniques are presented last.

The more basic, but least accurate, technique would be the completely taped or completely inked artwork. Inking is seldom used because of the physical difficulty of producing complicated artwork with it. Tape, on the other hand, is more difficult to photograph and commonly requires the preparation of a contact autopositive as the photographic master. Cut-and-peel films produce high-quality photographic masters, but their preparation is more complex and expensive than tape.

In today's complex printed wiring market, many companies are using automated equipment for the preparation of the artwork master from the printed wiring layout information.

5.4 ARTWORK GENERATION

Artwork is basically a manufacturing tool, although it is often prepared by the user (the designer). It is the most important manufacturing tool used in fabricating printed wiring because it uniquely defines the pattern to be placed on the board. Since it is one of the first steps in the overall process, the final product can only approach its quality and accuracy. Owing to the nature of the production processes used, the artwork must provide for certain process allowances; and thus close coordination with the fabricator is necessary.

5.4.1 Methods

The method required to produce artwork to the established sophistication is highly dependent on the size (area) and accuracy required. Those two factors are established by the board design and are affected by the method of manufacture. A guide to the tolerance capability of the two most common methods for producing artwork (i.e., manual and automatic) is given in Table 5.2. Final scale tolerance is then the tolerance at working scale divided by the reduction factor with a practical limit of ±0.001 in for location and ±0.0002 in for size. Except for transmission line applications, accuracy of artwork is directed primarily toward location and size of terminal areas (pads).

TABLE 5.2 Artwork Tolerance Capabilities at Working Scale

Line and pad	Tolerance, in, for application by	
	Manual taping*	Automatic NC machine†
Size tolerance	±0.005	±0.0005
Location tolerance	±0.012	±0.001

* Application on film material in normal room environments.
† Application on glass material by NC machines. Controlled environment assumed.

Most printed circuit layout drafters can position terminal areas and wiring on the masters to within 0.005 or 0.010 in with the proper equipment. Therefore, 10:1 layouts are used on artwork requiring very high accuracy, and 4:1 or 2:1 layouts are used on all other artwork. Drafters doing the layout must remember that the scale is applied to all dimensions (i.e., target diameters, line widths, etc.) and not just to those relating to placement. Sometimes it is necessary to choose a scale which is compatible with available artwork material.

The automatic artwork generator utilizes a precision *xy* plotter with an accuracy of more than ±0.001 in over a usable surface of 4 × 5 ft. The accuracy of the precision plotter is such that artwork is usually prepared at a 1:1 scale, which eliminates photo reduction. The manual technique requires the artwork to be produced at several times working scale and relies on the camera reduction to provide the accuracy required.

Input to the precision plotter can be either paper or magnetic tape. The data may be prepared by either a digitizer or a computer program. The digitizer method is presently more prevalent, but preparation of computer programs for conductor routing is making considerable progress. Output of the digitizer is either paper tape or punched cards; the punched cards are then converted to magnetic tape with standard computer programs. The output of the computer program is usually magnetic tape, but paper tape also can be obtained.

5.4.2 Materials

Materials used in artwork are either glass or polyester film. Glass plates have considerably better dimensional stability than polyester film when there are variations in temperature and humidity. Thus, when accurate artwork (NC machine) is required, glass is preferred.

5.4.3 Considerations

The use of multiple patterns enables the fabricator to optimize the many batch-type processes in printed wiring and thus increase production and reduce costs without reducing the need for artwork quality and accuracy.

In addition to the obvious gain in accuracy by generating artwork automatically, by-products such as drill tapes and test tapes for automatically testing the board are developed concurrently with the information required to generate the artwork. As CAD techniques are perfected and utilized, outputs of the programs can be used directly in plotting artwork.

Keeping artwork masters extremely clean is as important to the photographic process as clean metal is to photoresist coating. It is vital if photographers are expected to produce quality photo masks at a reasonable cost. If dirty or stained artwork is submitted for photographing, it should be rejected as unusable until it has been cleaned. Otherwise, photo masks made from it will have to be hand-opaqued to eliminate unwanted pinholes, lines, and ragged image edges. Opaquing takes time, is very expensive, and results in low-quality photo masks which produce low-quality printed circuits.

5.4.4 Camera Techniques

The general principles outlined in the following paragraphs are not intended to tell all about printed circuit photography. Obviously, the printed circuit photographer must be one of the best. Although expertise in studio photography or continuous-tone work is not necessary, the photographer should understand all photographic principles very well. In particular, he or she must thoroughly understand the use and maintenance of a process camera. The photographer should have complete facilities for making and checking film reductions of artwork, touch-up tables, and a place to make screens for screen-printing circuit boards.

1. Proper selection and use of photographic materials is a major consideration in photography for printed circuit work. That is particularly true for film bases and photographic emulsions.

2. Camera scales are notoriously inaccurate for making photo masks to the accuracies frequently required for printed circuits. Each photo mask should be checked for both image size and distortion before it is approved for such use. Such measurements may be made by using a micro rule or an optical comparator if close tolerances are specified.

3. All photo images should be inspected for pinholes, scratches, and other faults which are touched out with photo opaque between stages in the photo-reduction procedure. Magnifiers such as those used by photoengravers and lithographers are widely employed for this inspection and touch-up step.

5.5 DOCUMENTATION

Whether a board is to be a one-of-a-kind prototype or a high-volume production article, it should have some documentation that describes the means to an end. Exactly what should be documented and how the documentation should be prepared must be given serious consideration if a quality product is to be produced within a given budget and schedule. A printed wiring drawing package should include at least the following drawings.

1. A *schematic diagram* indicates with graphic symbols the electrical connections and functions of a circuit and thereby enables the user to test, evaluate, and troubleshoot.
2. The *master drawing* contains all the information necessary for the manufacture of the printed board.
3. An *assembly drawing* is a pictorial representation of the finished, assembled board and lists all electromechanical items contained on the assembly.
4. *Artwork* is an accurately scaled configuration of the printed circuit from which the master pattern is made photographically.
5. *Miscellaneous drawings* are prepared to support either the master or the assembly drawings.

5.5.1 How Much Documentation?

Too little printed board documentation results in excessive scrappage owing to misinterpretation, productivity decrease that is due to efforts to fill information gaps, and loss of uniform configuration that is due to word-of-mouth manufacturing. Information becomes too dependent on individuals rather than documentation.

Too much documentation results in increased drafting hours and labor costs and decreased manufacturing productivity owing to a time-consuming interpretation of overemphasized and confusing drawings.

Adequate documentation conveys to the user the basic electromechanical design concept, the type and quantity of parts and materials required, special manufacturing instructions, and up-to-date revisions. The use of adequate documentation offers increased profits by enabling schedule and budget commitments to be met, and the result is satisfied customers who receive quality products on time and at an equitable price.

5.5.2 Minimal Drawing Package

In a minimal drawing package, the goal is to prepare the information at the lowest possible drafting cost without compromising the design and integrity of the board. Since that type of documentation is used primarily by model shops for quick-reaction prototype development, the drawings should be neat, legible, and informative.

Master drawings may be prepared by a variety of methods. Miscellaneous drawings, such as mechanical details and modifications to purchased or existing parts, should be made in sketch form at no compromise to design integrity.

5.5.3 Formal Documentation

Formal documentation should begin only when a management decision has been made to manufacture and market a product for which only minimal documentation exists. The drawing package must be more comprehensive and presentable because formal documentation contains information pertinent to the function of several manufacturing processes. The drawing package need not be expensive. In fact, the cost of formal documentation is only moderately greater than that of the minimal drawing package, provided the integrity of the minimal drawing package is maintained.

5.6 SPECIFICATIONS AND STANDARDS

Specifications are often written to set forth exact needs or help qualify a product or service. They help the contracting agency or buyer monitor or judge the acceptability of the material or services supplied, and they put the buyer and supplier on common ground. For products which are obtained on a continual basis or are grouped in a general class, such as printed wiring, specifications and standards which may be supplemented by other documentation or written orders have been developed.

Specifications and guides pertaining to printed wiring may be divided into three general categories: those specifying design requirements, those concerned mostly with materials and acceptability criteria, and those setting forth the performance requirements for printed wiring and associated items.

Although it is not possible to include all of the specifications and standards applicable to printed wiring boards, the most common ones are covered briefly in the following paragraphs.

5.6.1 Institute for Interconnecting and Packaging Electronic Circuits (IPC)

Since 1957 the IPC has prepared and released many specifications and standards which are now widely used throughout the printed wiring industry. The initial standard of the IPC, published in 1960, covered the recommended standard tolerances for printed circuit boards. That standard (identified as IPC-D-300) has been revised on several occasions to reflect the increasing sophistication of the printed wiring state of the art.

5.6.2 Department of Defense (DOD)

Many aspects of printed wiring design, material selection, and process control are covered by military specifications and standards. It is important that the printed wiring designer have a complete understanding of the military documents that directly affect the design of printed wiring structures. The designer should also be familiar with the documents which affect design support activities.

5.6.3 American National Standards Institute (ANSI)

Nearly 4000 standards have been approved to date by the American National Standards Institute (ANSI), formerly the USA Standards Institute. An American National Standard implies a consensus of those substantially concerned with its scope and provisions; it is intended as a guide to the manufacturer, consumer, and general public.

5.6.4 International Electrotechnical Commission (IEC)

The "international" specifications covered by this chapter have been prepared by the International Electro-technical Commission (IEC), which is affiliated with the International Organization for Standardization (ISO). A foreword explains the intent of these documents:

> The informal decisions or agreements of the I.E.C. on technical matters, prepared by Technical Committees on which all the National Committees having a special inter-

est therein are represented, express, as nearly as possible, an international consensus of opinion on the subject dealt with.

5.6.5 Major Specifications and Standards

This section contains abstracts of the major documents that are applicable to the task of printed wiring design. It is intended that an abstract will help the printed wiring designer decide whether to obtain a copy of the latest revision of the document before proceeding with the design.

1. *IPC-D-300 (Printed Wiring Board Dimensions and Tolerances, Single and Two Sided Rigid Boards):* This specification covers dimensioning and tolerancing limits based on industry capabilities and fabrication costs. It provides five classes for dimensional features to reflect progressive increases in sophistication and cost of tooling, material, and processing.

2. *IPC-D-310 (Suggested Guidelines for Artwork Generation and Measurement Techniques):* This document has been written to provide a general reference to aid in the making of master patterns required to produce printed circuits. The guide is composed of information submitted by members of IPC committees who represent various companies and government agencies, and it includes techniques suggested in military and commercial practice.

3. *IPC-D-350 (End Product Description in Numeric Form):* This standard describes record formats for the transmission of end-product description data, in digital form on punched cards and magnetic tape, that are necessary and sufficient for the preparation of tooling or the manufacture and testing of printed boards. The record formats may be used in the transmittal of information and data between the printed wiring design phase and the manufacturing facility when the design has been formed by computer-aided processes or when the tooling for manufacture of the printed boards is to be performed by numerically controlled machines.

The data contained in each record are general in nature, are not in any particular machine language, and are in a form usable for manual as well as digital interpretation and readout.

4. *IPC-D-390 (Guidelines for Design Layout and Artwork Generation on Computer Automated Equipment for Printed Wiring):* This guide illustrates accepted techniques used to make printed-wiring artwork by semi- and fully-automated methods. The purpose of this document is to help improve design quality and accuracy of the master pattern, and to help reduce documentation costs.

5. *IPC-A-600 (Acceptability of Printed Wiring Boards):* This publication is a compilation of visual quality acceptability guidelines for printed boards prepared by the Repairability and Acceptability Committees of the Institute of Printed Circuits. The illustrations in the guide were made to portray certain specific points noted in the title of each. The compendium has been assembled to standardize the interpretations of specifications on printed wiring boards.

6. *IPC-CM-770 (Guidelines for Printed Circuit Board Component Mounting):* This document has been written to provide a general reference to aid in assembling components to printed boards. The guide is composed of selections from mounting specifications submitted by various companies and government agencies and includes techniques suggested for space flight, military, and commercial applications.

7. *MIL-STD-100 (Engineering Drawing Practices):* This military standard prescribes procedures and format authorized for the preparation of Form I engineering drawings and associated lists prepared by or for the departments and agencies of the Department of Defense as prescribed by military specification MIL-D-1000.

8. *MIL-STD-275 (Military Standard, Printed Wiring for Electronic Equipment):* This standard establishes design principles governing the fabrication of rigid single- or double-sided printed boards and the mounting of parts (including integrated circuits) and assemblies thereon for use in electronic equipment. The requirements do not apply to parts, such as resistors, inductors, capacitors, or transmission lines, fabricated by using the techniques.

9. *MIL-STD-429 (Printed Wiring and Printed Circuits Terms and Definitions):* This standard establishes terms and definitions which have a specific meaning when applied to printed wiring and printed circuit nomenclature.

10. *MIL-D-1000 (Drawings, Engineering, and Associated Lists):* This specification prescribes general requirements for the preparation of engineering drawings and associated lists and for the application of intended-use categories to acquisition of the drawings. It reflects Department of Defense policy to buy only the engineering drawings that are needed and to encourage procurement of commercial drawings when they are adequate for the purpose. "Engineering drawings," as used in this specification, include associated lists.

11. *MIL-I-46058 (Insulating Compound, Electrical, for Coating Printed Circuit Assemblies):* This specification covers conformal coatings which are suitable for application to printed circuit assemblies by dipping, brushing, spraying, or vacuum deposition.

12. *MIL-P-55110 (Printed Wiring Boards):* This specification covers printed boards consisting of a conductor pattern on the surface of one or two sides of an insulating base and associated interfacial connections and standoff terminals.

13. *USASI YI4.5-1966 (Dimensioning and Tolerancing for Engineering Drawings):* This standard establishes the rules, principles, and methods of dimensioning and tolerancing used to specify design requirements on engineering drawings. It also establishes uniform practices for stating and interpreting the requirements.

14. *ANSI Y14.15 (Electrical and Electronics Diagrams):* This standard contains definitions and general information applicable to most of the commonly used electrical and electronic diagrams. It also includes detailed recommendations on preferred practices for use in the preparation of such diagrams. The recommended practices are intended to eliminate divergent electrical and electronics diagram drafting techniques.

15. *ANSI Y32.16 (Reference Designations for Electrical and Electronics Parts and Equipments):* This standard covers the information and application of reference designations for electrical and electronic parts and equipment. The reference designations are intended for uniquely identifying and locating discrete items on diagrams and in a set and for correlating items in a set, graphic symbols on diagrams, and items in parts lists, circuit descriptions, and instructions.

16. *IEC 97 (Grid System for Printed Circuits):* The recommended grid system for printed circuits will ensure compatibility between the printed circuits and the parts to be mounted on them.

17. *IEC 321 (Guidance for the Design and Use of Components Intended for Mounting on Boards with Printed Wiring and Printed Circuits):* This document gives guidance to the designer, manufacturer, and user of components on matters relating to the specification, design, production, supply, and application of components particularly suited for use with printed circuits. In its present form, the report is intended to be applied to components which are to be soldered to printed circuits.

18. *IEC 326 (General Requirements and Measuring Methods for Printed Wiring Boards):* This recommendation relates to printed wiring boards, however manufactured, when they are ready for mounting of the components. It is intended as a basis for agreement between purchaser and vendor.

CHAPTER 6
BASE MATERIALS

Charles G. Henningsen
Insulectro, Mountain View, California

Smith A. Gause
Westinghouse Electric Corporation, Hampton, South Carolina

6.1 OVERVIEW

It is important to understand the types of copper-clad laminates that are available, how they are made, where they are used, and the advantages and disadvantages of each before selecting the material most suitable for the intended application.

Many types of copper-clad materials are available (see Fig. 6.1). The copper-clad laminates most widely used in the manufacture of printed circuit boards, however, are FR-2, CEM-1, CEM-3, FR-4, FR-5, and GI. These are the materials that are primarily discussed in this chapter.

The FR-2 laminates are composed of multiple plies of paper that have been impregnated with a flame-retardant phenolic resin. The major advantages of FR-2 are their relative low cost and their good electrical and punching qualities. FR-2 is typically used in applications where tight dimensional stability is not required, such as in radios, calculators, toys, and television games.

FR-3, the other all-paper base laminate, is also made of multiple plies of paper that have been impregnated with an epoxy-resin binder. The FR-3 laminate has higher electrical and physical properties than the FR-2 but lower than epoxy laminates that have woven glass cloth as a reinforcement. FR-3 is used to manufacture printed circuits used in consumer products, computers, television sets, and communication equipment.

CEM-1 is a composite material having a paper core impregnated with epoxy resin. Woven glass cloth impregnated with the same resin covers the two surfaces. This construction allows the material to have punching properties similar to those of FR-2 and FR-3, with electrical and physical properties approaching those of FR-4. CEM-1 is used in smoke detectors, television sets, calculators, and automobiles as well as in industrial electronics.

CEM-3, a composite of dissimilar core material, uses an epoxy-resin-impregnated nonwoven fiberglass core with epoxy-resin-impregnated woven glass cloth surface sheets. It is higher in cost than CEM-1, but it is more suitable for plated-through holes. CEM-3 is used in applications such as home computers, automobiles, and home entertainment products.

FR-4 (military type GF) laminates are constructed on multiple plies of epoxy-

Laminate designations and materials

Grade	Resin			Reinforcement				
	Epoxy	Polyester	Phenolic	Cotton paper	Woven glass	Mat glass	Glass veil	Flame retardant
XXXPC			•	•				
FR-2			•	•				•
FR-3	•			•				•
FR-4	•				•			•
FR-5	•				•			•
FR-6		•				•		•
G-10	•				•			
CEM-1	•			•	•			•
CEM-2	•			•	•			
CEM-3	•				•	•		•
CEM-4	•				•	•		
CRM-5		•			•	•		•
CRM-6		•			•	•		
CRM-7		•				•	•	•
CRM-8		•				•	•	

FIG. 6.1 Laminate designations and materials. (*Reprinted from Electronic Packaging and Production magazine.*)

resin-impregnated woven glass cloth. It is the most widely used material in the printed circuit board industry because its properties satisfy the electrical and mechanical needs of most applications. Its excellent electrical, physical, and thermal properties make it an excellent material for high-technology applications. It is used in aerospace, communications, computers and peripherals, industrial controls, and automotive applications.

FR-5 types (military type GH) are laminated using multiple plies of woven glass cloth impregnated with mostly polyfunctional epoxy resin. The glass transition temperature T_g is typically 150 to 160°C, as compared with an FR-4 with a glass transition temperature of 125 to 135°C. FR-5 is used where higher heat resistance is needed than is attainable with FR-4 but not where the very high thermal properties of GI type materials are needed.

GI type materials are composed of multiple plies of a woven glass cloth impregnated with a polyimide resin. The materials have a glass transition temperature in excess of 200°C, which virtually eliminates "drill smear" caused by heat during the drilling process. It also exhibits excellent mechanical properties and *z* axis dimensional stability at high temperatures. GI materials have a lower interlaminar bond strength than the epoxy systems; therefore, care should be taken when drilling and routing.

6.2 GLASS TRANSITION TEMPERATURE

The T_g has become a measure of how well a laminate resin system resists softening from heat. When the T_g temperature is reached, the resin changes from its "glassy" state and causes changes in the laminate's properties. The T_g is not a measure of the resin's melting point, but rather a point at which molecular bonds begin to weaken enough to cause a change in physical properties (dimensional stability, flexural strength, etc.), and its value is determined by the intersection of the two slopes of the temperature-property–change curve (Fig. 6.2*a*). FR-4 epoxy exhibits a T_g of 115 to 125°C and polyimide 260 to 300°C.

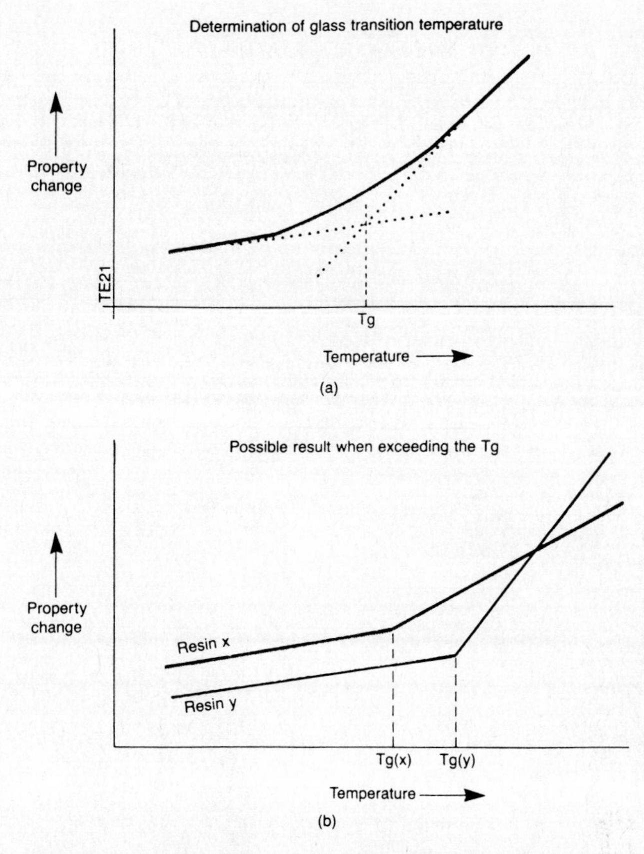

FIG. 6.2 Laminate glass transition temperature is the point where properties exhibit significant change (*a*). The higher the T_g, the more dimensionally stable the laminate. (However, when the T_g is exceeded, it becomes meaningless.) (*Reprinted from Electronic Packaging and Production magazine.*)

Although the T_g value is a measure of the toughness of a material under heating, it is not meant for comparison purposes once the T_g is exceeded. For example, materials with a higher T_g are desirable because they maintain their stability over a wider temperature range up to the T_g. However, once the T_g is exceeded, the material properties of a high-T_g resin could change much more rapidly than a material with a low T_g. At a higher temperature, the lower-T_g material may often exhibit superior properties (Fig. 6.2*b*).

6.3 LAMINATE MANUFACTURING PROCESS

6.3.1 Treating the Base Material

The base raw material—paper, glass matte, woven glass cloth, quartz, or Kevlar—is impregnated or coated with resin. The resin is then polymerized to a point suitable for storage and final pressing. A machine called a "treater" or "coater," as shown in Fig. 6.3, is used for treating the material. First the material passes

FIG. 6.3 Treater or coater. *(Westinghouse Electric Corp.)*

through a dip pan of resin, where it is impregnated, and then through a set of metering rollers (squeeze rollers) and a drying oven. The oven is air-circulating or infrared and can be up to 120 ft long. Most of the volatiles such as solvents in the resin are driven off in the oven, and the resin is polymerized to what is called a "B-stage." This semicured material is also known as "prepreg." The prepreg is dry and nontacky. The treater illustrated in Fig. 6.4 is a vertical treater that runs principally woven glass cloth; the horizontal treater in Fig. 6.5 impregnates mainly paper and glass matte.

Rigid process control is applied during treating so that the ratio of resin to base material, the final thickness of the prepreg, and the degree of resin polymerization

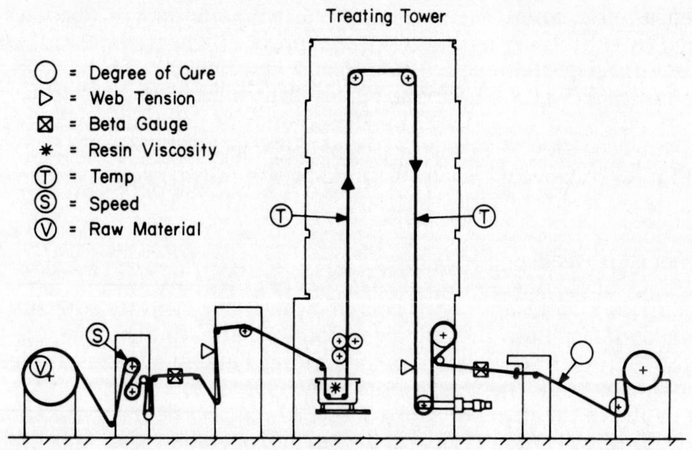

FIG. 6.4 Location and type of process controls on the vertical tower processing system. *(Westinghouse Electric Corp.)*

can be monitored. Beta-ray gauges may compare the raw material with the final semicured product and automatically adjust the metering rollers above the resin dip pan so that the proper ratio of resin to base material is maintained. The degree of polymerization of the resin is controlled by the treater air temperature, air velocity, and speed at which the material passes through the treater.

The prepreg material is usually stored in an area where the temperature is controlled below 70°F and below 35 percent humidity until the time of the pressing operation. Each roll or stack of prepreg is tagged with the processing date and the test results for resin content, gel time, resin flow, cured thickness, and volatile content.

6.3.2 Copper Inspection

Besides the base material and resin, the other principal component of copper-clad laminates is copper foil. Today, almost all copper is electrodeposited rather than

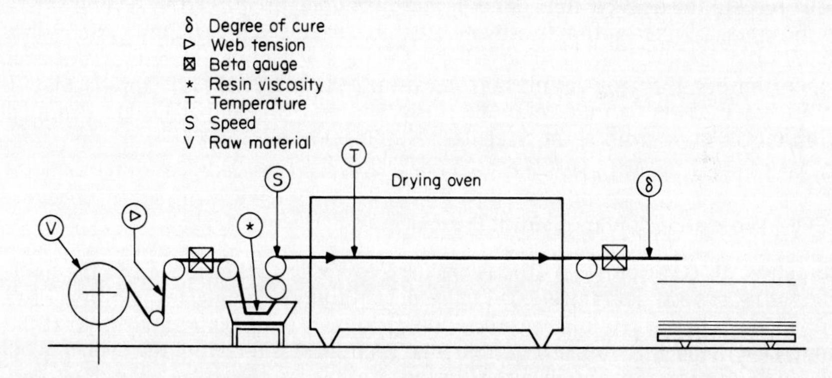

FIG. 6.5 Horizontal treater. *(Westinghouse Electric Corp.)*

rolled. Each roll of copper is inspected by the laminator for visual surface quality and pinholes, and a sample is taken for trial pressing. The trial pressing sample is tested for copper peel strength, solder blister resistance, copper oxidation after heat exposure, and general surface quality. The side of the foil to be pressed against the prepreg is treated with an alloy to improve adhesion of the copper to the laminate. The alloy is a proprietary coating, usually of zinc or brass, in a controlled ratio to enhance the chemical bond between the copper and the resin.

6.3.3 Laminate Buildup

Most laminators build up their sheets in clean room facilities with filtered air-conditioned systems to control the temperature and humidity as well as to keep dust particles from the copper and prepreg during buildup. Electrostatic attraction of dust particles to the treated material and copper before pressing is a source of contamination in the laminate and a source of pits and dents on the copper surface of the finished laminate. During the buildup operation, the copper foil is first laid against a large polished stainless steel press plate. Then a number of sheets of prepreg are laid on top of the copper. The number of layers depends on the desired thickness of the laminate and the characteristics of the prepreg material. Some of the art of laminating comes in balancing all of these variables to produce a final dense material to relatively close tolerances. The final sheet of copper foil is placed on top of the prepreg if the material is to have copper on both sides. If copper is desired on only one side of the laminate, a release film such as Tedlar replaces one of the sheets of copper.

6.3.4 Laminate Pressing

The press plates, with the material, are removed from the buildup room and stacked in a large, multiopening press, as illustrated in Fig. 6.6. Several sheets are pressed into each press opening, with typical presses being capable of molding 80 sheets 36 × 48 in to 250 sheets 48 × 144 in, $\frac{1}{16}$ in thick. The presses, which are hydraulic, are capable of developing pressure in excess of 1000 psi. Steam is a typical heat source. It is released into the press platens until the platens reach the uniform laminating temperature. The packs or books of material are then loaded into the press, and the desired pressure is applied so that the material is cured into a final homogeneous sheet. To ensure that all of the sheets in the pack or book receive the desired state of cure, thermocouples are placed in several sheets in the press. A timer automatically records time against a preset cure cycle. When the desired stage of cure is achieved, the steam is automatically cut off and cold water pumped through the press platens until the material is at approximately 80°F. The material is then removed from the press, and the edges are trimmed from the sheet to remove the irregular resin flow areas.

6.3.5 Laminating Quality Control System

Statistical process control methods are used to verify that each step in the manufacturing process is controlled. Figure 6.7 outlines laminate traceability. Each press load is tested according to the appropriate sampling plan to ensure that the material will meet customer requirements. Each sheet is then identified as to manufacturer, appropriate specification, and load or lot number. Most manufacturers retain samples of material from each load for at least 1 year to enable them to

FIG. 6.6 Multiopening press. *(Westinghouse Electric Corp.)*

effectively check any processing problems or questions which may result from the use of that lot in the field. Periodically, laminators are required to run a complete set of physical and electrical tests as set forth by NEMA or MIL-P-13949.

6.4 LAMINATE EVALUATION, SPECIFICATION, AND QUALITY CONTROL

Various types of manufacturers of copper-clad laminate must be evaluated before the criteria necessary for design and fabrication process control can be estab-

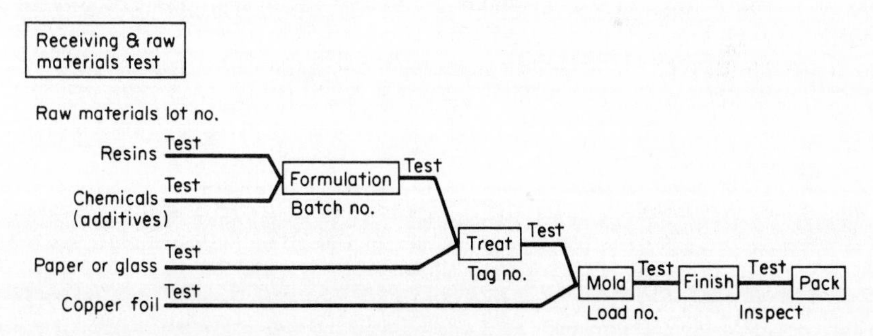

FIG. 6.7 Laminate flowchart for laminate quality control.

lished. Too often the evaluation is based upon one sheet or sample submitted by each manufacturer. Generally, electrical and mechanical design and the particular fabrication process will demand that the final material have certain features and controls. It is, therefore, imperative that evaluation tests and specifications be determined by the demands on the finished product. Standards often are set unrealistically high in many areas that have no relationship to the fabrication process requirements of the final product. Some of the tests most commonly used today to determine the pertinent areas for test and evaluation are described in the following sections.

6.4.1 Surface and Appearance

Probably the most difficult test to define adequately has been that of surface and appearance standards. Perhaps more laminate has been rejected for pits and dents in the copper than for any other reason. This is particularly frustrating to the laminator, who has used a great deal of technical skill to build a consistent insulating material only to find it rejected by the user for cosmetic reasons. Most laminators believe that surface and appearance standards must be applied in light of the finished boards. Some purchasers of laminate require copper surface standards only on critical areas, such as the areas for tips inserted into edge connectors. In those cases, the user usually provides the laminators with an overlay of the critical copper areas to be inspected on each sheet of material before it is shipped. Thus the surface standard is applied only to areas that are pertinent to the finished board. With more than 90 percent of the copper being ultimately removed, the chance of a pit or dent affecting a critical area is small indeed. Nevertheless, many customers will pay premium prices by requiring the laminator to use special techniques or special selection to provide pit-free copper-clad material.

6.4.2 Copper Surface

Copper pits and dents are best defined in the surface standards established in MIL-P-13949. The specification defines the longest permissible dimension of a pit or dent and supplies point values for rating all pits and dents.

1. *Grade A:* The total point count shall be less than 30 for any 12- × 12-in area.

2. *Grade B:* The total point count shall be less than 30 for any 12- × 12-in area.

There shall be no pits or dents with the longest dimension greater than 0.015 in. Pits with the longest dimension greater than 0.005 in shall not exceed three in any square foot.

3. *Grade C:* The total point count shall be less than 100 for any 12- × 12-in area. The point system is as follows:

Longest dimension, in	Point value
0.005 to 0.010, inclusive	1
0.011 to 0.020, inclusive	2
0.021 to 0.030, inclusive	4
0.031 to 0.040, inclusive	7
Over 0.040	30

TABLE 6.1 Copper Foil Thickness and Tolerance

Nominal weight, oz/ft²	Tolerance, wt % Class 1	Class 2	Nominal thickness, in*	Tolerance, in*
1/8	±10	±5	0.00020	—
1/4	±10	±5	0.00036	—
3/8	±10	±5	0.00052	—
1/2	±10	±5	0.0007	±0.0001
3/4	±10	±5	0.0010	±0.0002
1	±10	±5	0.0014	±0.0002
2	±10	±5	0.0028	±0.0003
3	±10	±5	0.0042	±0.0004
4	±10	±5	0.0056	±0.0006
5	±10	±5	0.0070	±0.0007
6	±10	±5	0.0084	±0.008
7	±10	±5	0.0098	±0.0010
10	±10	±5	0.0140	±0.0014
14	±10	±5	0.0196	±0.0020

*Derives by weight test method 2.2.12 of IPC-CF-150.

Scratches are permitted that have a depth less than 140 μin, or a maximum of 20 percent of the foil thickness, when measured with a Johannson surface finish indicator N533. Foil thickness is specified in MIL-F-55561 and is as shown in Table 6.1. The scratches to be tested should be located visually using 20/20 vision.

6.4.3 Color

Color variance from lot to lot on any particular grade of laminate is usually caused by variation in color of batches of resin, types of paper used, or variation of alloy coating on the copper. All the raw materials are supplied to the laminator, and only by careful inspection of incoming raw materials can a laminator be in complete control of their final effect on color. If coloration must be specified in copper-clad laminate, it should be done by working closely with the laminator. A set of samples that will illustrate the acceptable color extremes must be established.

6.4.4 Punchability and Machinability

The test for punchability and machinability has been subject to much debate among laminators and users. As of this writing, no universally accepted test method exists for either characteristic, and research has found little correlation with values such as Rockwell or Barcol hardness.

6.4.4.1 Drilling. Each laminator should be consulted on recommended drilling speeds and feeds prior to any evaluation. Sometimes, poor drilling from one material to another occurs because each laminator was not given the benefit of recommendations for the material. Sectioning the board will help in evaluating the type of hole being obtained and is particularly important on plated-through applications. Often sectioning will show that the material has been heated by drilling to such an extent that the surface is smooth and not platable, is smeared with resin, or is so roughened that glass fibers protrude and will inhibit continuous plated-through holes.

6.4.4.2 ***Punching.*** Punchability can be measured in a die that simulates the conditions used in the fabrication process. Various hole sizes, spacings, and configurations should be incorporated in the test die. Many paper-base grades will tend to vary in punchability from lot to lot, so care must be taken to measure a wide range of sample panels. Careful physical inspection and sectioning of the holes will tell what type of punching is being obtained. The material must be carefully inspected to ensure that no cracking occurs and that there is no lifting of copper around the hole. Each laminator should be consulted as to the recommended die tolerance and punching temperatures.

6.4.5 Peel Strength

6.4.5.1 ***Before the Soldering Operation.*** The basic test pattern for testing peel strength or copper-bond strength specified in MIL-P-13949 and by NEMA standards is illustrated in Fig. 6.8. The pattern should be processed by the same fabrication techniques as in the user's final process, with the exception of exposure to various plating solutions or solder which will also be tested. When being tested for peel strength, the specimen should be mounted on a flat, horizontal surface. The wide copper end of each trace should be peeled back approximately 1 in so that the line of peel is perpendicular to the edge of the specimen. The end of the peeled strip should then be gripped by a clamp which is attached to a force indicator or tensile tester adjusted to compensate for the weight of the clamp and connecting chain. The copper foil is then pulled from the material at a rate of 2 in per minimum, and the minimum load of the force indicator is recorded.

Peel strength of a $\frac{1}{8}$-in trace is converted to pounds per inch width of peel by dividing the indicated force by the measured width of the strip. The peel strength test pulls the adhesion area directly under the radius formed between the copper being peeled at right angles and the material. Since 2-oz copper forms a larger radius than 1-oz copper, a greater area of adhesion is being pulled. As a conse-

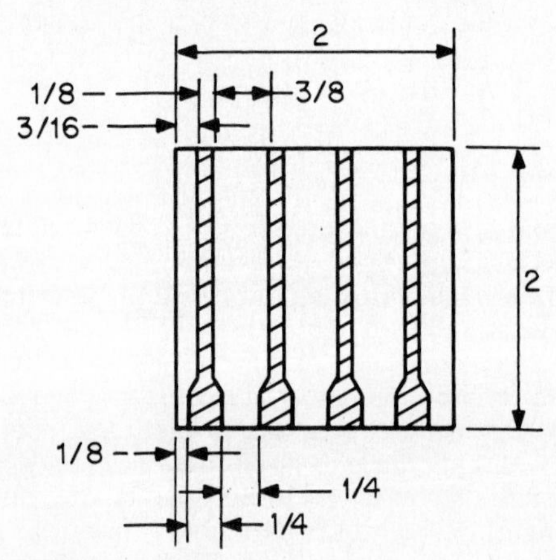

FIG. 6.8 Copper peel strength pattern. Dimensions are in inches.

quence, peel strength for 2-oz copper is increased. For that reason, it is always important to maintain a peeling force that is 90° to the copper surface so that the radius of the peel is constant. Generally, a 5° deviation from perpendicular is acceptable; if a large enough distance is allowed between the specimen and force indicator, the 5° variance will not be exceeded during the source of the test. However, many users and laminators have found it desirable to use a testing machine which moves the force indicator along the test specimen as the copper is being pulled, thus keeping the angle of pull constant.

6.4.5.2 ***During Soldering.*** As circuit traces and pads have become smaller, the problem of bond strength retention at dip-soldering or solder-touchup temperatures has become increasingly important. At an elevated temperature, the NEMA test uses the same test pattern. The test sample, if G-10 or FR-4, is immersed in silicone oil for 6 min at 125°C and if G-11 or FR-5 at 150°C. Peeling is done, as above, while the sample is submerged in oil, and the average peel retention is noted.

The long immersion is required to bring both the specimen and the test jig to a constant temperature; shorter periods of immersion can result in extremely inconsistent test data. Specimens can also be placed in an air-circulating oven for 60 ± 6 min. The peel test can then be run in the oven at the appropriate elevated temperature. Underwriters Laboratories has also developed a test to determine long-term effects of heat aging on bond strength. The test is the normal peel test after the sample has been aged at 125°C for 1344 h.

6.4.5.3 ***After Soldering.*** It is usually important to test for peel strength after the dip-soldering cycle. The specimen should be floated on the solder pot at 550°F for 5 to 20 s, depending on the material being used. No flux should be used on the copper, and all excess solder must be removed. Any solder on the specimen will result in extremely irregular peeling. Sometimes, petroleum jelly may be applied to the copper before soldering to prevent any solder wetting. After excess solder is removed, the specimens should be visually examined for evidence of blistering and delamination of the metal foil.

6.4.5.4 ***After Plating.*** Plating solutions, particularly cyanide gold, can affect adhesion on some copper-clad laminates. Therefore, it is recommended that a peel test be considered after exposure to a normal plating operation. MIL-P-13949 recommends that the pattern illustrated in Fig. 6.9 be used if peel tests are to be run after plating. The procedure recommended is as follows:

1. Immerse in methylene chloride for 75 ± 15 s at 20 to 25°C (68 to 77°F).
2. Dry specimens 15 ± 5 min at 105 to 148.9°C.
3. Immerse in a solution of 10 g/L sodium hydroxide at 85 to 95°C for 5 ± 1 min.
4. Rinse in hot water at 50 to 55°C for 5 ± 1 min.
5. Immerse for 30 ± 5 min in a solution of 10 g/L sulfuric acid (sp gr 1.836) and 30 g/L boric acid solution at 55 to 65°C.
6. Rinse in hot water at 50 to 55°C for 5 ± 1 min.
7. Dry for 30 ± 5 min at 105 to 148.9°C.
8. Immerse in hot oil at 215 to 225°C for 40 ± 5 s.
9. Immerse in 1,1,1-trichloroethane at 20 to 25°C for 75 ± 15 s to remove hot oil.

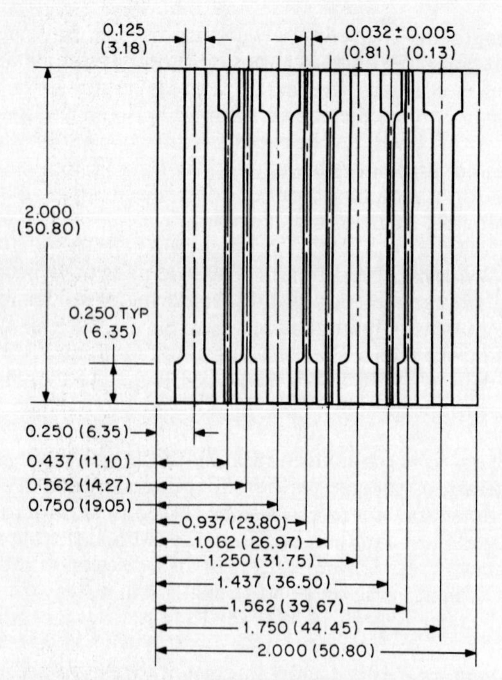

FIG. 6.9 MIL-P-13949 peel strength test pattern. Dimensions are in inches; numbers in parentheses are metric equivalents in millimeters.

10. Air-dry and inspect specimens for such defects as delamination, wrinkles, measling, warping, twisting, blisters, and cracks.

The $\frac{1}{32}$-in lines shall be used. The peel strength shall be calculated using the actual width of each line tested.

6.4.6 Bow and Twist

Bow and twist in manufactured sheets with either or both dimensions 18 in or over is determined by suspending the sheet so that the horizontal level plane touches both corners. No pressure shall be applied to the straight edge of the sheet. The bow or twist is calculated as follows:

$$C = \frac{36D}{L^2} \times 100\%$$

where C = bow and twist for 36-in dimension
D = maximum deviation from horizontal straight edge
L = length in inches along the horizontal straight edge

Cut-to-size panels should be placed unrestrained on a flat surface, with the convex surface of the panel upward and measuring the maximum vertical displacement.

The deviation is expressed in inches per inch rather than in terms of a 36-in dimension.

6.4.6.1 Bow and Twist Variation—Cut Panels.

Bow and twist of cut panels with either dimension less than 18 in are per MIL-P-13949. See Table 6.2. For intermediate thickness, the next greater thickness value applies.

6.4.7 Solder Resistance

Solder resistance should be measured and evaluated according to the NEMA grade for the laminate. When the laminate is to be used for military applications, MIL-P-13949 should be followed. Solder resistance is measured by horizontally floating specimens on a solder bath for a time dependent upon the laminate grade and specification temperature given. The specimens are then visually evaluated for blistering, measling, delamination, and weave exposure.

NEMA specification LI-1-1983 requires both clad and etched specimens cut 2 × 2 in to be floated at 500°C for 5 to 10 s for paper-base laminates, epoxy, and paper phenolic, respectively, and 20 s for all other laminate grades. MIL-P-13949 does not recognize any paper-reinforced laminates or glass with paper composites. MIL-P-13949 requires clad and etched specimens 2 × 2 in to be floated at 550°F for 10 s. Etched specimens are tested with and without solder flux of type R or RMA.

Some fabricators specify other test requirements to further guarantee the integrity of laminate for its specific application.

For many laminates, the NEMA and MIL-spec requirements are only minimal, and the actual solder resistance is greater.

6.4.8 Autoclave

The autoclave, often referred to as the "pressure cooker" test, is used to select laminates of high moisture resistance which offer improved yields of printed wiring boards through infrared fusing and wave soldering.

Specimens are etched and sheared to size (1 × 5 in to 4 × 6 in), set vertically

TABLE 6.2 Bow or Twist of Cut-to-Size Panels

		Total variation, maximum, %			
		Laminate—Class C			
		All weights of foil, one side		All weights of foil, two sides	
Thickness,* in	Panel size (maximum dimension), in	All other types	Types GP, GR, GT, GX, GY	All other types	Types GF, GR, GT, GX, GY
0.02 and over	8 or less	2.0	—	1.0	—
	8 to 12	2.0	—	1.5	—
	Greater than 12	2.5	—	1.5	—
0.030 or 0.031	8 or less	1.5	3.0	0.5	3.0
	8 to 12	1.5	3.0	1.0	3.0
	Greater than 12	2.0	3.0	1.0	3.0
0.060 and over	12 or less	1.0	1.5	0.5	1.5
	Greater than 12	1.5	1.5	0.5	1.5

*For nominal thicknesses not shown in this table, the bow and twist for the next lower thickness shown shall apply.

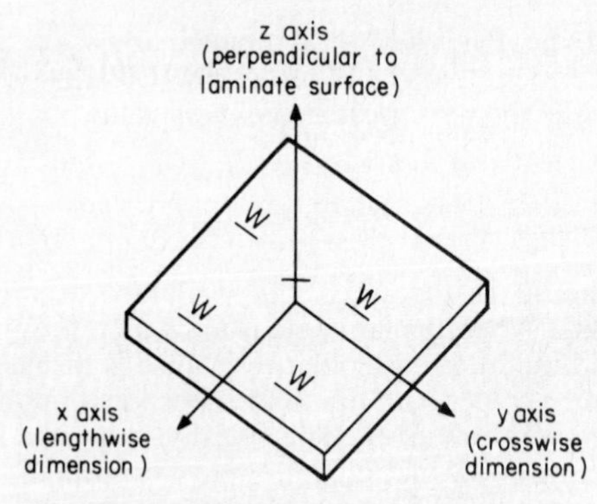

FIG. 6.10 Typical laminate axis definition. Note that *x* (lengthwise) direction is defined by a specific part of the manufacturer's watermark.

into a metal rack, and placed into a pressure cooker or steam autoclave and held at 15 psi for 30 min. The samples are removed to cool for 5 min and are then dipped vertically into a solder bath at 500°F for 20 s. The specimens are then inspected for the presence of blisters, measling, crazing, or weave texture, which are reasons to suspect irregularity in the laminate. In some instances, more strenuous autoclave conditions, such as 550°F solder and autoclave exposure up to 75 min, are used to further screen materials for critical applications.

6.4.9 Dimensional Stability

The need for circuit board dimensional stability has increased with increased circuit density. In general, the dimensional stability in the lengthwise or crosswise dimension *x, y* illustrated in Fig. 6.10 is a function of the laminate reinforcement (glass or paper). The thickness expansion *z* is generally a function of the resin system or resin matrix. It is important to recognize which is the lengthwise direction of the treated material going into laminate construction. The laminator usually identifies the lengthwise dimension by the vertical direction of the watermark, as shown.

To measure dimensional stability, one specimen should be taken from the lengthwise edge and one from the center of each sample sheet. Specimens should conform to Fig. 6.11, which is taken from MIL-P-13949F military specification.

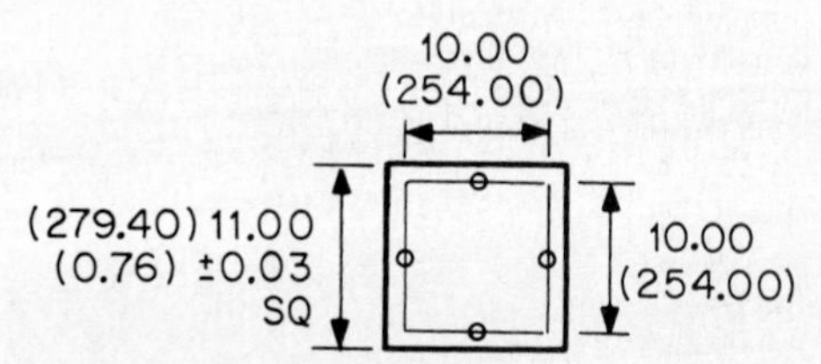

FIG. 6.11 Dimensional stability specimen. Dimensions are in inches. Metric equivalents (to the nearest 0.01 mm) are given for general information only and are based on 1 in = 25.4 mm. Millimeters are in parentheses.

6.5 DEGREE OF CURE

The curing of epoxy-resin-based printed circuit boards is directly related to the quality and performance of the boards. Methods for monitoring the curing process and of defining the extent of cure are needed to ensure that the circuits will meet design requirements. One method of defining cure is by measuring the glass transition temperature of the epoxy resin. Such measurements have been made in several ways and related to curing parameters via automatic dielectrometry. Based on these results, incoming inspection routines for materials have been defined, board pressing cycles established, and in-process controls fixed.

The extent of cure of epoxy resins can be measured by determining the glass transition temperature T_g, which is the temperature at which a polymer changes from a rigid, brittle, glasslike material to a softer, rubberlike product. At high temperatures, polymer molecules can move slightly, relative to one another. As temperature is decreased, a point T_g is reached where the thermal energy supplied is no longer enough to encourage this movement. Since the relative motion of the molecules in an epoxy resin also is dependent on the amount of cross-linking (cure), T_g and cross-linking are related.

Certain properties of epoxy resins have been found to be different above and below T_g. These include the heat content, the volume expansion coefficient, dielectric losses, refractive index, stiffness, and hardness. Of these, the heat capacity, the expansion, and the electrical losses are easily measured and are excellent tools with which to measure T_g and, hence, the degree of cure.

Limiting our discussion to epoxy-glass laminates and the methods suitable for those not having extensive laboratory facilities, the following techniques are suggested.

6.5.1 Thermal Analysis

Use a differential scanning calorimeter to determine the residual caloric value of the cured sample. The instrument measures heat from the exothermic reaction, caused by the resin curing, as a caloric value. When a laminate is fully cured, some residual caloric value always remains. Measurement of that value in a differential scanning calorimeter usually gives low numbers compared to the very high exothermic reaction measured when uncured samples are tested. Therefore, the measured amount of residual caloric values on acceptable standard laminates will provide a working reference for residual values and for the high exotherm of uncured samples.

6.5.2 Spectral Analysis

In some instances, spectral analysis is applicable, but it has some inherent drawbacks. Sample extraction and determination of weight loss during the extraction plus the spectral analysis of the extractables can sometimes provide a measure of the degree of cure.

6.5.3 Dip-Soldering Techniques

Many users determine relative degree of cure by inserting a 6- × 6-in sample with a 1-in strip of copper etched on it into a 500°F solder pot at a 45° inclination. Measurement of deflection as well as warp and twist will often give a measure of the relative degree of cure if enough data have been gathered to determine the basic working range of a particular resin system.

6.5.4 Dielectric Analysis

Dielectric analysis[1] consists of the dynamic measurement of dielectric properties of a material during its cure and postcure. Properties measured are dissipation factor and capacitance. They represent the response of polar groups in the material to a changing ac electric field. The response depends principally on the temperature and frequency of measurement, the number of polar groups per unit volume, and the mobility of the polar groups. It is the latter property that is of greatest concern in dielectric analysis, because it directly reflects the curing process.

The equipment used for dielectric analysis consists of a capacitance bridge with a frequency range of 0.1 to 1.0 kHz. Support equipment consists of a pneumatic press test cell with temperature control capability of ± 1°C and a three-channel recorder to record sample temperature, capacitance, and dissipation factor. This equipment has been used in production presses by attaching its sample leads to aluminum foil electrodes placed on either side of a prepreg sheet. Usually one electrode is separated from the sample by an inert film. Generally 0.001-in-thick polyimide film is used.

6.5.5 Microdielectrometry

An improved dielectric measuring technique known as "microdielectrometry" is available to monitor the cure of epoxy resins.[2] Integrated circuit technology is used to develop a miniaturized probe that combines a small size with built-in amplification to measure dielectric properties of polymers at frequencies as low at 1 Hz. The integrated circuit device consists of a planar interdigitated electrode structure with a pair of matched field-effect transistors. The electrode geometry does not change during cure and is reproducible from device to device.

The system, in essence, takes the relative gain and phase of the sensor output compared with sensor input (imposed sinusoidal voltage under command from a programmed computer) and, using an internally stored calibration, converts the data into permittivity ε', loss factor ε'', and their ratio, the loss tangent or dissipation factor $\tan \delta = \varepsilon''/\varepsilon'$. Thus ε' and ε'' can be measured for any material that is on the surface of the integrated circuit chip or sensor. The sensor consists of a 2- $\times$ 4-mm integrated circuit mounted in a flat cable package. Both electrodes used in the dielectric measurement are placed on the same surface to form an interdigitated capacitor, and on-chip amplification produces high signal-to-noise ratios.

6.5.6 Dynamic Dielectric Analysis

Dynamic dielectric measurements made over a wide range of frequency provide a sensitive and convenient technique for monitoring the cure process, and onset of flow, the onset of the reaction, changes in resin age, and changes in resin composition. This technique has been used to study the imidization and addition reactions of a polyimide dielectrically.[3]

6.6 INSULATION RESISTANCE

The insulation resistance between two conductors or holes is the ratio of the voltage to the total current between the conductors. Insulation resistance is composed of both the volume and surface resistance in a copper-clad laminate. Results of insulation resistance tests can vary widely if careful control of environmental conditions and process techniques of the test samples is not exercised. Although the actual value of insulation resistance may be important initially, the change in resistance under a specified environmental condition is usually more significant. The insulation resistance test is of greatest value when the test specimen is subjected to the same environment as will be required in the final application.

6.6.1 Parameters and Test Conditions

The insulation resistance of copper-clad laminates decreases both with increasing temperature and with increasing humidity. Volume resistance is particularly sensitive to temperature changes, and surface resistance changes widely and very rapidly with changes in humidity. Since extended periods of conditioning are required to determine the effects of humidity on surface resistivity, it is recommended, for example, that 96 h at 35°C in a 90 percent relative humidity environment be used. Test data also show that some materials will recover from humidity conditioning much more rapidly than others. Therefore, it may be desirable to cycle humidity and temperature in accordance with Method 106 of MIL-STD-202 (except steps 7a and 7b shall be omitted). Measurement shall be made at high humidity.

For consistent results to be obtained, extreme care must be taken in processing and handling the test boards. Etching of the pattern must leave well-defined lines, with no cracked or ragged areas. Rubber gloves—preferably surgical—must be used at all times while handling the test boards. Fingerprints can reduce the value of surface resistivity by as much as three decades. The following procedure for cleaning test specimens is recommended before any testing proceeds.

1. Place specimens from etch bath immediately in racks in running water at 60°F for 5 min.
2. Place specimens in 10 percent oxalic acid with agitation for 10 min.
3. Scrub with fine pumice.
4. Place specimens in running water at 60 ± 5°F for 30 min.
5. Scrub with demineralized water (1 MΩ minimum). Rinse with fresh demineralized water. Remove with gloves.
6. Stand specimens on rack in oven at 80°C for 1 h (use double clip).
7. Remove with rubber gloves (washed with alcohol), and put into desiccator used as dust-free container.
   a. Wash gloves in alcohol and store in same dust-free container.
   b. Touch only sample edges or clip leads; wear rubber gloves.

6.6.2 Specimen Preparation

Either the metal foil of the specimen may be completely removed by etching, or metal foil outlines may be left on the specimen to form the edges of the electrodes.

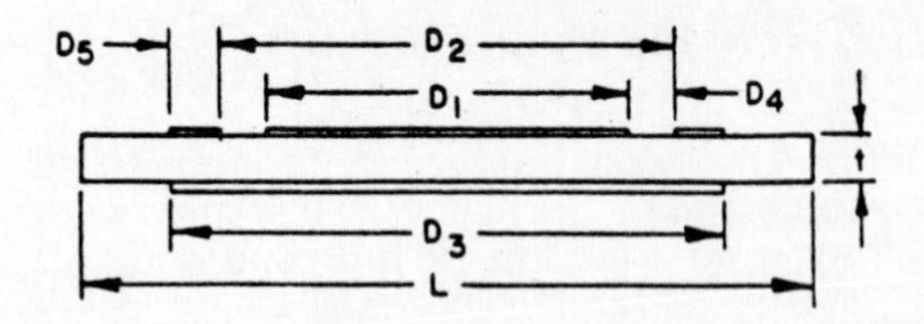

Thickness (t)	D_1 dia	D_2 dia	D_3 dia	D_4	D_5	Length of one side of specimen (L)
0.031 or less (.79)	1.000 (25.40) ±0.005 (.13)	1.020 (25.91) ±0.005 (.13)	1.375 (34.93) ±0.005 (.03)	0.010 (.25) ±0.001 (.03)	0.177 (4.50) ±0.005 (.13)	2.000 (50.80) ±0.015 (.38)
.137 (3.48) or less	2.000 (50.80)	2.500 (63.50)	3.000 (76.20)	.250 (6.35)	.250 (6.35)	4.000 (101.60)
.138 (3.51) to 0.250	3.500 (88.90)	4.500 (114.30)	5.500 (139.70)	.500 (12.70)	.500 (12.70)	6.500 (165.10)

(Table header: Dimensions)

FIG. 6.12 Electrode configuration of volume resistivity and surface resistivity test. Dimensions are in inches; numbers in parentheses are metric equivalents in millimeters.

The electrodes are generally completed with an application of a porous, conductive silver paint to both sides of the specimen. However, for comparative purposes, copper electrodes may be used rather than the conductive paint.

6.6.2.1 Surface Resistance. The surface resistance between two points on the surface on any insulation material is the ratio of the dc potential applied between the two points to the total current between them. One of the most commonly used surface resistance patterns is the ASTM pattern illustrated in Fig. 6.12. This is a circular pattern with a three-electrode arrangement for measuring the surface resistance parallel to the laminate. The third electrode is a guard conductor which intercepts stray currents that might otherwise cause error. For measurement of surface resistance, the resistance of the surface gap between electrodes 1 and 2 is measured. The measured current flows between electrodes 1 and 2, while stray current flows between electrodes 1 and 3, as shown in Fig. 6.13.

Measurements should be made after 500 V dc has been applied to the specimens for 60 s + 5 − 0 by means of a megohm bridge having an accuracy of ±6 percent at 10^{11} Ω. Measurements are to be made at the end of the prescribed conditioning time and condition.

Surface resistivity is calculated as follows:

$$r's = \frac{R'P}{D}$$

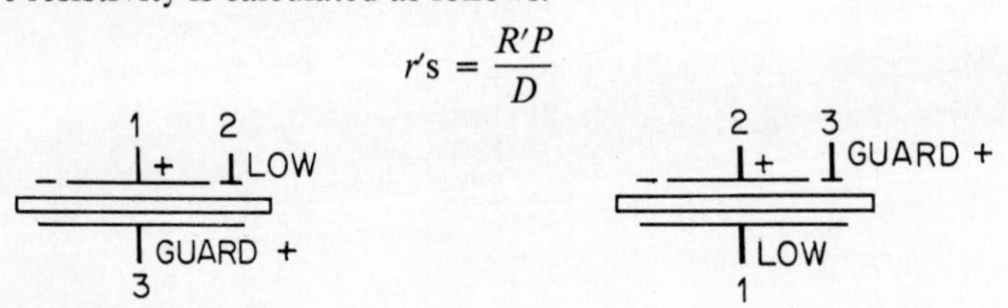

FIG. 6.13 Guarding circuit for surface resistance parallel to the laminate.

FIG. 6.14 Guarding circuit for volume resistance.

where VRw = surface resistivity, mΩ
R' = measured surface resistance, MΩ
P = measured perimeter of the guarded electrode, cm
D = distance between inner circle and outer guard ring, cm

6.6.2.2 Volume Resistance. Volume resistance is the ratio of the dc potential applied to electrodes embedded in a material to the current between them. It is usually expressed in ohm-centimeters. The ASTM test pattern shown in Fig. 6.12 can also be used for volume resistance. For the measurement of volume resistance, the measured current flows between electrodes 1 and 3 while stray current flows between electrodes 2 and 3, as shown in Fig. 6.14.

6.6.2.3 Steps

6.6.2.3.1 Comb Patterns. Comb patterns such as the one illustrated in Fig. 6.15 have also been commonly used in surface-resistance tests. The patterns may simulate the final configuration of most copper-clad laminate boards, since the lines on the pattern approximate trace spacing in the board. Because of the narrow, close pattern, care must be taken to achieve clean, well-defined lines with no ragged areas.

6.6.2.3.2 Specimen Preparation. Tinned terminals should be soldered to the land areas on the pattern using a 25- to 40-W soldering iron. The solder or resin should not spread beyond the land areas.

6.6.2.3.3 Removal of Fingerprints and Flux. Thoroughly clean and dry the specimens as follows and, until completion of treating, handle them by the edges only. Brush with a bristle brush under running tap water that is between 15.5 and 25.5°C (60 and 80°F). The hardness of the tap water should not exceed 175 ppm (expressed as calcium carbonate). Deionized water may be employed. Dry with oil-free compressed air. Brush while submerged in isopropyl alcohol, removing all excess resin. Dip into fresh isopropyl alcohol and dry with oil-free compressed air. Dry in an oven for 2 h between 49 and 60°C (120 and 140°F), remove from the oven, and then condition for 24 h before testing at 23°C (73.4°F) and 50 percent RH.

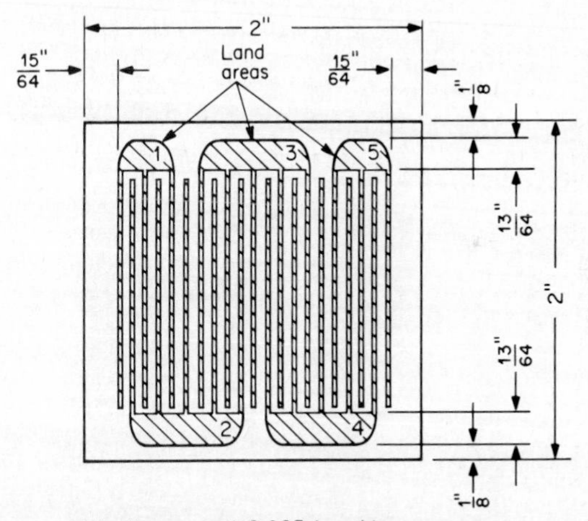

FIG. 6.15 Insulation resistance test pattern.

TABLE 6.3 Standard Materials

NEMA grade	Military designation MIL-P-13949F	Resin system	Base	Color	Description
XXXPC	None	Phenolic	Paper	Opaque brown	Phenolic paper with punchability at or above room temperature
FR-2	None	Phenolic	Paper	Opaque brown	Phenolic paper, punchable, with flame-resistant (self-extinguishing) resin system
FR-3	PX	Epoxy	Paper	Opaque cream	Epoxy resin, paper base with flame-resistant resin system, cold punching, and high insulation resistance
CEM-1	None	Epoxy	Paper-glass composite	Opaque tan	Epoxy resin paper core with glass on the laminate surface, self-extinguishing, economic fabrication of paper base, mechanical characteristics of glass
CEM-3	None	Epoxy	Glass matte	Translucent	Epoxy resin nonwoven glass core with woven glass surfaces, self-extinguishing, punchable with properties similar to FR-4
FR-6	None	Polyester	Glass-matte	Opaque white	Polyester, random glass fiber, flame-resistant, designed for low-capacitance or high-impact applications
G-10	GE	Epoxy	Glass	Translucent	Epoxy-glass, general purpose
FR-4	GF	Epoxy	Glass	Translucent	Epoxy-glass with self-extinguishing resin system
G-11	GP	Epoxy	Glass	Translucent	High-temperature epoxy-glass with strength and electrical retention at elevated temperatures
FR-5	GH	Epoxy	Glass	Translucent	High-temperature epoxy-glass with flame-resistant resin system with strength and electrical retention at elevated temperatures
None	GI	Polyimide	Glass	Translucent dark brown	Polyimide resin, glass laminate with high continuous operating temperature and high property retention at temperature, low-*z* dimensional expansion

TABLE 6.4 Materials for High-Frequency Application

NEMA grade	Military designation MIL-P-13949	Resin system	Base	Color	Description
GT	GT	TFE	Glass	Opaque brown	Glass fabric base, PTFE (Teflon) resin, controlled dielectric constant
GX	GX	TFE	Glass	Opaque brown	Glass fabric base, PTFE (Teflon) resin dielectric constant with closer controlled limits than GT
		Polystyrene	Glass	Opaque white	Polystyrene cast-resin base for low-dissipation-factor applications
		Cross-linked polyethylene	Glass	Opaque white	Polyethylene cast base, radiation cross-linked for low dissipation factor

Note that the above cleaning procedure does not replace the rinse after etching but supplements it.

6.6.2.3.4 Specimen Conditioning. Condition the specimen in the chamber for 96 h at 35°C in a 90 percent RH environment. Afterwards, apply a potential of 100 to 500 V dc measurement to be made at above temperature and humidity. Take measurements between the following terminals: 1 and 2; 2 and 3; 3 and 4; 4 and 5.

Take the readings after an electrification time of 60 s + 5 − 0 by means of a megohm bridge having an accuracy of ±6 percent at 10^{11} Ω.

6.6.2.4 Properties of Copper-Clad Laminates. To establish the design parameters necessary for any copper-clad printed board application, one must know the principal laminate properties. Listed in this section are the characteristics that are most readily required in both electrical and mechanical design applications. Some of the tests involving those properties are described in the section on laminate evaluation.

6.7 GRADES AND SPECIFICATIONS

See Tables 6.3 to 6.5.

6.7.1 Material Designations

The most common method of designating copper-clad materials is described in MIL-P-13949 and illustrated in Fig. 6.16. As an example, GFN-0620-CN/CI-A-2-A means no coloring in flame-retardant glass-epoxy laminate, 0.062 in thick, $\frac{1}{2}$-oz/ft² copper, drum side out, on one side, and 1-oz/ff² copper, drum side out, on the other side, grade A pits and dents, class 2 thickness tolerance, and class A warp and twist.

TABLE 6.5 Materials for Additive Circuit Processing

NEMA grade	Military designation MIL-P-13949	Resin system	Base	Color	Description
			Adhesives		
XXXPC	None	Phenolic	Paper	Opaque brown	These laminates are designed for use with additive processes using adhesive bonding techniques.
FR-2	None	Phenolic	Paper	Opaque brown	
FR-3	None	Phenolic	Paper	Opaque brown	
FR-4	None	Epoxy	Glass	Translucent	
			Seeded and coated		
XXXPC	None	Phenolic	Paper	Opaque brown	These laminates are used in the patented seeded and coated processes. They are seeded with a small percentage of the catalytic seeding agent dispersed throughout the resin system and are coated with a catalyzed adhesive.
FR-2	None	Phenolic	Paper	Opaque brown	
FR-3	None	Phenolic	Paper	Opaque brown	
FR-4	None	Epoxy	Glass	Translucent	
CEM-1	None	Epoxy	Paper-glass composite	Opaque tan, blue, white	
CEM-3	None	Epozy	Glass-matte composite	Translucent	
			Sacrificial aluminum-clad		
FR-2	None	Phenolic	Paper		These laminates are the patented sacrificial additive process. They are clad on two sides with a specially anodized aluminum foil. The sacrificial aluminum cladding makes the laminate surface acceptable for the additive process.
FR-3	None	Phenolic	Paper		
FR-4	None	Epoxy	Glass, glass-paper composite		
			Swell and etch		
FR-4	None	Epoxy	Glass	Translucent	This laminate is used in the commercially available swell-and-etch additive process. The epoxy-glass laminate has a 0.0015-in-thick "resin-rich" surface and a specially designed surface for swell-and-etch chemicals.

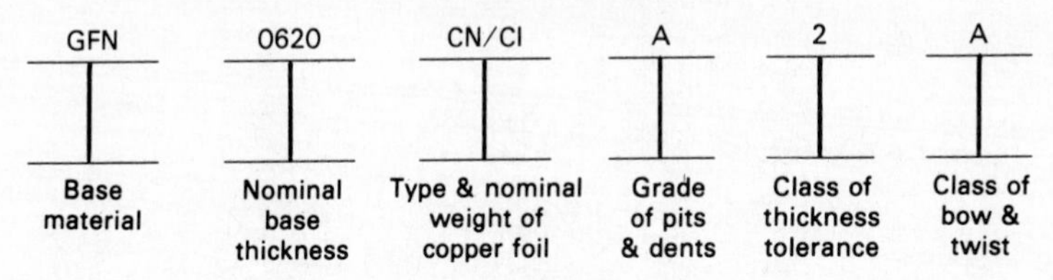

FIG. 6.16 Designation of copper-clad laminates.

6.7.2 Conditioning Designations

The conditioning designations used to describe the environment in which tests were run are as follows:

Condition A: As received; no special conditioning
Condition C: Humidity conditioning
Condition D: Immersion conditioning in distilled water
Condition E: Temperature conditioning
Condition T: Test condition

Conditioning procedures are designated in accordance with the following:

1. First, a capital letter indicates the general condition of the specimen to be tested, i.e., as received or conditioned to humidity, immersion, or temperature.
2. A number indicating the duration of the conditioning, in hours, follows.

6.8 MECHANICAL PROPERTIES OF LAMINATES

The principal characteristics usually required in the mechanical design of printed boards are outlined here. If a specific test is required, the test is listed under a particular heading. Whenever they are available, the minimum standards set forth in MIL-P-13949 or NEMA standards for industrial laminates are used.

6.8.1 Flexural Strength

Method: ASTM D 790. Unit of value: psi. This test is measure of load that a beam will stand without fracture when supported at the ends and loaded in the center, as shown in Fig. 6.17. See Table 6.6.

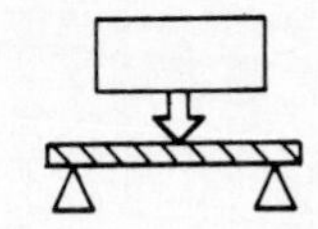

FIG. 6.17 Flexural strength test.

6.8.2 Weight of the Base Material

Unit of value: ounce/square foot (add 1 or 2 oz/ft^2 per side for the weight of copper). See Table 6.7.

TABLE 6.6 Flexural Strength—Condition A, Minimum Average psi

Material	Lengthwise	Crosswise
XXXPC	12,000	10,500
FR-2	12,000	10,500
FR-3	20,000	16,000
FR-4	60,000	50,000
FR-5	60,000	50,000
FR-6	15,000	15,000
G-10	60,000	50,000
G-11	60,000	50,000
CEM-1	35,000	28,000
CEM-3	40,000	32,000
GT	15,000	10,000
GX	15,000	10,000
GI	50,000	40,000

TABLE 6.7 Base Material Weight, oz/ft^2

Material	$\frac{1}{32}$ in	$\frac{1}{16}$ in	$\frac{3}{32}$ in	$\frac{1}{8}$ in
FR-2	3.3	6.7	10	13.4
FR-3	3.6	7.2	11	14.5
CEM-1	4.0	7.6	11.6	14.7
CEM-3	—	9.2	—	—
G-10	4.5	9.9	14.7	19.5
G-11	4.5	9.8	13.4	19.6
FR-4	5.0	10.0	15.0	20.0
FR-5	5.2	10.2	15.4	20.4

TABLE 6.8 Nominal Thickness and Tolerances for Laminates, per MIL-P-13949F

Thickness, in × 10^{-3}	Class 1 PX paper base only	Class 1 Glass reinforced	Class 2 glass reinforced	Class 3 glass reinforced	Class 4 for microwave application
0010 to 0045	—	±0010	±00075	±0005	—
0046 to 0065	—	±0015	±0010	±00075	—
0066 to 0120	—	±0020	±0015	±0010	—
0121 to 0199	—	±0025	±0020	±0015	—
0200 to 0309	—	±0030	±0025	±0020	—
0310 to 0409	±0045	±0065	±0040	±0030	±002
0410 to 0659	±0060	±0075	±0050	±0030	±002
0660 to 1009	±0075	±0090	±0070	±0040	±002
1010 to 1409	±0090	±0120	±0090	±0050	±0035
1410 to 2400	±0120	±0220	±0120	±0060	±0040

*These tighter tolerances are available only through product selection.

6.8.3 Thickness Tolerance

Nominal thicknesses and tolerances should be specified below per MIL-P-13949. At least 90 percent of the area of a sheet should be within the tolerance given, and at no point should the thickness vary from the nominal by a value greater than 125 percent of the specified tolerance. Cut sheets less than 18 × 18 in should meet the applicable thickness tolerance in 100 percent of the area of the sheet. Class of tolerance is as specified in the type designation. For nominal thicknesses not shown in this table, the tolerance for the next-greater thickness shown applies. The nominal thicknesses include the metal foil except microwave materials, which are without foil. See Table 6.8.

6.8.4 Bow and Twist Percent Variation

Bow and twist values are per MIL-P-13949. The values apply only to sheet sizes as manufactured and to cut pieces having either dimension not less than 18 in. For intermediate thickness, the next-greater thickness value applies. See Tables 6.2 and 6.9

6.8.5 Maximum Continuous Operating Temperature

See Table 6.10.

6.8.6 Peel Strength

Method: See Sec. 6.4.6. Unit of value: pound/inch of width. See Table 6.11.

6.8.7 Coefficient of Thermal Expansion

Method: ASTM D 696 (at 130°F). Unit of value: in/°C. The coefficient of thermal expansion, Table 6.12, is the change in length per unit of length per degree change in temperature. The coefficient may vary in different temperature ranges; so the temperature range must be specified.

6.8.8 Water Absorption

Unit of value: Percent water absorption is the ratio of weight of water absorbed by the material to the weight of the dry material. See Table 6.13.

6.8.9 Copper Bond Strength Retention

See Table 6.14.

6.8.10 Flammability

According to Underwriters Laboratories, materials tested for flammability are classified 94V-0, 94V-1, 94V-2. For material flame ratings, see Table 6.15. Defi-

TABLE 6.9 Bow and Twist, per MIL-P-13949

Thickness, in†	Total variation, maximum, % (on basis of 36-in dimension)*					
	Class A			Class B		
	All types, all weights, metal (one side)	All types, all weights, metal (two sides)		All types, all weights, metal (one side)	All types, all weights, metal (two sides)	
		Glass	Paper		Glass	Paper
0.20 and over	—	5	—	—	2	—
0.030 or 0.031	12	5	6	10	2	5
0.060 or 0.062	10	5	6	5	1	2.5
0.090 or 0.093	8	3	3	5	1	2.5
0.120 or 0.125	8	3	3	5	1	2.5
0.240 or 0.250	5	1.5	1.5	5	1	1.5

*These values apply only to sheet sizes as manufactured and to cut pieces having either dimension not less than 18 in.

† For nominal thicknesses not shown in this table, the bow or twist for the next lower thickness shown shall apply.

TABLE 6.10 Operating Temperature

Materials	Temp., °C, for Electrical factors	Temp., °C, for Mechanical factors
Ordinary applications		
XXXP	125	125
XXXPC	125	125
FR-2	105	105
FR-3	105	105
CEM-1	130	140
CEM-3	130	140
FR-6	105	105
G-10	130	140
FR-4	130	140
G-11	170	180
FR-5	170	180
GI	260	260
High-frequency applications		
GT	220	220
GX	220	220
Polystyrene	110	110
Cross-linked polystyrene	100	100

TABLE 6.11 Peel Strength

Ordinary applications				
	Condition A, oz		Condition E-1/150, oz	
Materials	1	2	1	2
XXXP	6	7		
XXXPC	6	7		
FR-2	6	7		
FR-3	8	9	5	6
CEM-1	8	10	5	6
CEM-3	8	10	5	6
FR-6	7	8		
G-10	8	10	5	6
FR-4	8	10	5	6
G-11	8	10	5	6
FR-5	8	10	5	6
GI	9	10	8	9
High-frequency applications				
GT	8	10	2	3
GX	8	10	2	3
Polystyrene	6.0	7.0		
Cross-linked polystyrene	6.0	7.0		

Additive (typically processed)			
Adhesive-coated	Condition A, 1 oz	Sacrificial aluminum	Condition A, 1 oz
XXXPC	9	FR-2	9
FR-2	9	FR-3	9
FR-3	9	CEM-1	9
FR-4	9	FR-4	9
CEM-1	9		
CEM-3	9		
Seeded and coated	Condition A, 1 oz	Swell and etch	Condition A, 1 oz
XXXPC	11	FR-4	9
FR-2	11		
FR-3	11		
FR-4	12		
CEM-1	12		
CEM-3	12		

TABLE 6.12 Coefficient of Thermal Expansion

Material	Coefficient, × 10^{-5}	
	Lengthwise	Crosswise
Ordinary applications		
XXXP	1.2	1.7
XXPC	1.2	1.7
FR-2	1.2	2.5
FR-3	1.3	2.5
CEM-1	1.1	1.7
CEM-3	1.0	1.5
FR-6	1.0	1.0
G-10	1.0	1.5
FR-4	1.0	1.5
G-11	1.0	1.5
FR-5	1.0	1.5
GI	1.0	1.2
High-frequency applications		
GT	1.0	2.5
GX	1.0	2.5
Polystyrene	7.0	7.0
Cross-linked polystyrene	5.7	5.7

TABLE 6.13 Water Absorption, Condition D 24/23

Material	$\frac{1}{32}$ in	$\frac{1}{16}$ in	$\frac{3}{32}$ in
Ordinary applications			
XXXP	1.3	1.0	0.85
XXXPC	1.3	0.75	0.65
FR-2	1.3	0.75	0.65
FR-3	1.0	0.65	0.60
CEM-1	0.50	0.30	0.25
CEM-3	0.50	0.25	0.20
FR-6		0.40	
G-10	0.50	0.25	0.20
FR-4	0.50	0.25	0.20
G-11	0.50	0.25	0.20
FR-5	0.50	0.25	0.20
GI		1.0	
High-frequency applications			
GT	0.20	0.10	0.09
GX	0.20	0.10	0.09
Polystyrene		0.05	
Cross-linked polystyrene		0.01	

TABLE 6.14 Copper Bond Strength Retention

Material	%	Material	%
XXXP	50	G-11	50
FR-2	50	FR-4	50
FR-3	50	FR-5	50
G-10	50	CEM-1	50
		CEM-3	50

TABLE 6.15 Flammability Classifications

Grade	UL classification	Grade	UL classification
XXXPC	94HB	G-10	94HB
FR-2	94V-1	FR-4	94V-0
FR-3	94V-0	G-11	94HB
CEM-1	94V-0	FR-5	94V-0
CEM-3	94V-0	FR-6	94V-0

TABLE 6.16 Fungus Resistance Tests

Test organism	Culture no.	Military specification
Aspergillus niger	WADC 215-4247	MIL-F-8261
Aspergillus favus	WADC 26	MIL-F-8261
Trichoderma T-1	WADC T-1	MIL-F-8261
Chaetomium globosum	USDA 1042.4	MIL-E-4970
Aspergillus niger	USDA tc 215-4247	MIL-E-4970
Penicillium luteum	USDA 1336.1	MIL-E-4970
Aspergillus flavus	WADC 26	MIL-E-4970
Memnoniella echinata	WADC 37	MIL-E-4970
Myro-thecium verrucaria	ATCC 9095	MIL-E-5272
Aspergillus terreus	ATCC 10690	MIL-E-5272
Penicillium	ATCC 9849	MIL-E-5272

TABLE 6.17 Fungus Resistance Test Results

Material	Construction	Military specifications		
		MIL-F-8261	MIL-E-4970	MIL-E-5272
XXP	Paper-phenolic	Not resistant	Not resistant	Not resistant
XXXPC	Paper-phenolic	Not resistant	Not resistant	Not resistant
FR-3	Paper-epoxy	Resistant	Resistant	Resistant
FR-4	Glass-epoxy	Resistant	Resistant	Resistant
FR-5	Glass-epoxy	Resistant	Resistant	Resistant

nitions of those classifications, as tested by the Underwriters Laboratories flammability procedure, are outlined below.

94V-0: Specimens must extinguish within 10 s after each flame application and a total combustion of less than 50 s after 10 flame applications. No samples are to drip flaming particles or have glowing combustion lasting beyond 30 s after the second flame test.

94V-1: Specimens must extinguish within 30 s after each flame application and a total combustion of less than 250 s after 10 flame applications. No samples are to drip flaming particles or have glowing combustion lasting beyond 60 s after the second flame test.

94V-2: Specimens must extinguish within 30 s after each flame application and a total combustion of less than 250 s after 10 flame applications. Samples may drip flame particles, burning briefly, and no specimen will have glowing combustion beyond 60 s after the second flame test.

94HB: Specimens are to be horizontal and have a burning rate less than 1.5 in/min over a 3.0-in span. Sample must cease to burn before the flame reaches the 4-in mark.

6.8.11 Fungus Resistance

See Tables 6.16 and 6.17.

6.9 ELECTRICAL PROPERTIES OF LAMINATES

6.9.1 Relative Humidity on Surface Resistivity of Glass-Epoxy

The effect of humidity on surface resistance of glass-epoxy was measured by using the ASTM three-electrode circular pattern, starting with 97.5 percent RH at 40°C and decreasing the humidity to 64 percent. The results, shown in Fig. 6.18, indicate that the surface resistivity decreases logarithmically with an increase in humidity at approximately the rate of one decade per 20 percent humidity change.

6.9.2 Thermal Conductivity of Base Materials Without Copper

See Table 6.18.

6.9.3 Dielectric Strength

(Perpendicular to the laminations at 23°C.) Method: ASTM D 149. Unit of value: V/mil. Dielectric strength is the ability of an insulation material to resist the passage of a disruptive discharge produced by an electric stress. A disruptive discharge is measured by applying 60-Hz voltage through the thickness of the laminate, as shown in Fig. 6.19. All the tests are run under oil. In the short-time test, the applied voltage is increased at a uniform rate of 0.5 kV/s. In the step-by-step test, the initial voltage is 50 percent of the short-time breakdown voltage; then the voltage is increased in increments according to a predetermined schedule at 1-min intervals. The test values for dielectric strength vary with the thickness of material, the form and size of electrodes, the time of application of the voltage, the temperature, the frequency and wave shape of the voltage, and the surround-

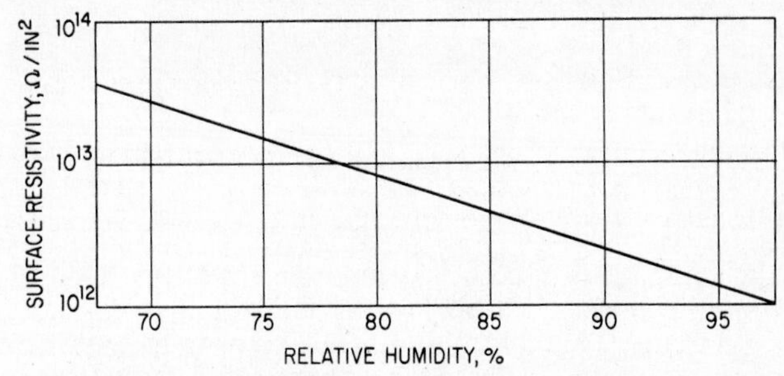

FIG. 6.18 Surface resistivity vs. relative humidity.

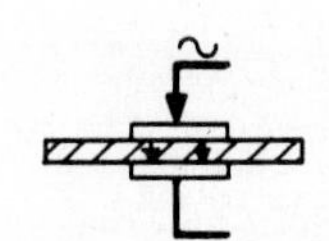

FIG. 6.19 Dielectric strength test.

FIG. 6.20 Dielectric constant test.

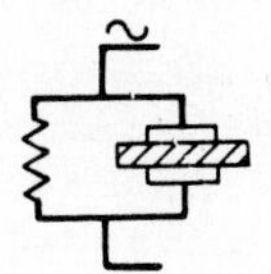

FIG. 6.21 Dissipation factor test.

TABLE 6.18 Thermal Conductivity of Base Materials

Material	Conductivity	Material	Conductivity
XXXP	1.7	G-11	1.8
FR-2	1.8	FR-4	1.8
FR-3	1.6	FR-5	1.8
G-10	1.8	CEM-1	1.8
		CEM-3	1.8

TABLE 6.19 Dielectric Strength Data

Material	V/mil	Material	V/mil
XXXP	740	G-10	510
FR-2	740	G-11	600
FR-3	550	FR-4	500
CEM-1	500	FR-5	500
CEM-3	500	GI	750

ing medium. Step-by-step data for $\frac{1}{16}$-in-thick material are as shown in Table 6.19.

6.9.4 Dielectric Breakdown

(Parallel to the laminations at 23°C.) Method: ASTM D 149. Unit of value: kV. Condition D 48/50. Dielectric breakdown is the disruptive discharge measured between two electrodes (Pratt and Whitney No. 3 taper pins) inserted in the laminate on 1-in centers perpendicular to the laminations. All tests are run under oil. The short-time and step-by-step tests are performed as in the test for dielectric strength perpendicular to laminations. Step-by-step data on $\frac{1}{16}$-in-thick material are given in Table 6.20.

6.9.5 Permittivity

Method: ASTM D 150. Unit of value: dimensionless. Dielectric constant is the ratio of the capacitance of a capacitor with a given dielectric to the capacitance of the same capacitor with air as a dielectric, as illustrated in Fig. 6.20. The dielectric constant is a measure of the ability of an insulating material to store electrostatic energy. It is calculated from the capacitance as read on a capacitance bridge, the thickness of the specimen, and the area of the electrodes. It varies with temperature, humidity, and frequency. See Table 6.21.

6.9.6 Dissipation Factor

(Average at 1 MHz.) Method: ASTM D 150. See Fig. 6.21. Unit of value: dimensionless. In an insulating material, the dissipation factor is the ratio of the total power loss, in watts, in the material to the product of the voltage and current in a capacitor in which the material is a dielectric. It varies over a frequency range. See Tables 6.22 and 6.23.

6.9.7 Current-Carrying Capacity as a Function of Ambient Temperature

Unit of value: A. The design chart, Fig. 6.22, has been prepared as an aid in estimating temperature rises (above ambient) versus current for various cross-sectional areas of etched copper conductors. It is assumed that normal design conditions prevail when the conductor surface area is relatively small compared with the adjacent free panel area. The curves as presented include a nominal 10 percent

TABLE 6.20 Dielectric Breakdown Data

Material	kV	Material	kV
XXXPC	15	G-10	40
FR-2	15	FR-4	40
FR-3	30	G-11	40
FR-6	30	FR-5	40
CEM-1	40	GT	20
CEM-3	40	GX	20

TABLE 6.21 Permittivity

Materials	Permittivity at 1 MHz condition D 24/23
Ordinary applications	
XXXPC	4.1
FR-2	4.5
FR-3	4.3
CEM-1	4.4
CEM-3	4.6
FR-6	4.1
G-10	4.6
FR-4	4.6
G-11	4.5
FR-5	4.3
GI	4.8
High-frequency applications	
GT	2.8
GX	2.8
Polystyrene	2.5
Cross-linked polystyrene	2.6

TABLE 6.22 Dissipation Factors

	Dissipation factor at 1 MHz	
Materials	Condition A	Condition D 24/23
Ordinary applications		
XXXP	0.028	0.03
XXXPC	0.028	0.03
FR-2	0.024	0.026
FR-3	0.024	0.026
CEM-1	0.027	0.028
CEM-3	0.020	0.022
FR-6	0.020	0.028
G-10	0.018	0.019
FR-4	0.018	0.020
G-11	0.019	0.020
FR-5	0.019	0.028
GI	0.020	0.030
High-frequency applications		
GT	0.005	0.006
GX	0.002	0.002
Polystyrene	0.00012–0.00025*	0.00012–0.00066*
Cross-linked polystyrene	0.0004–0.0005†	0.0005–0.0005

*Condition A, 10 MHz.
†Condition A, 10 GHz.

TABLE 6.23 Permittivity and Dissipation Factor of FR-4, Condition D 24/23

Frequency	Dielectric constant	Dissipation factor
100 Hz	4.80	0.009
1,000	4.75	0.012
10,000	4.70	0.015
100,000	4.65	0.018
1 MHz	4.60	0.020
10	4.55	0.022
100	4.50	0.024
1,000	4.45	0.025
10,000	4.40	0.025

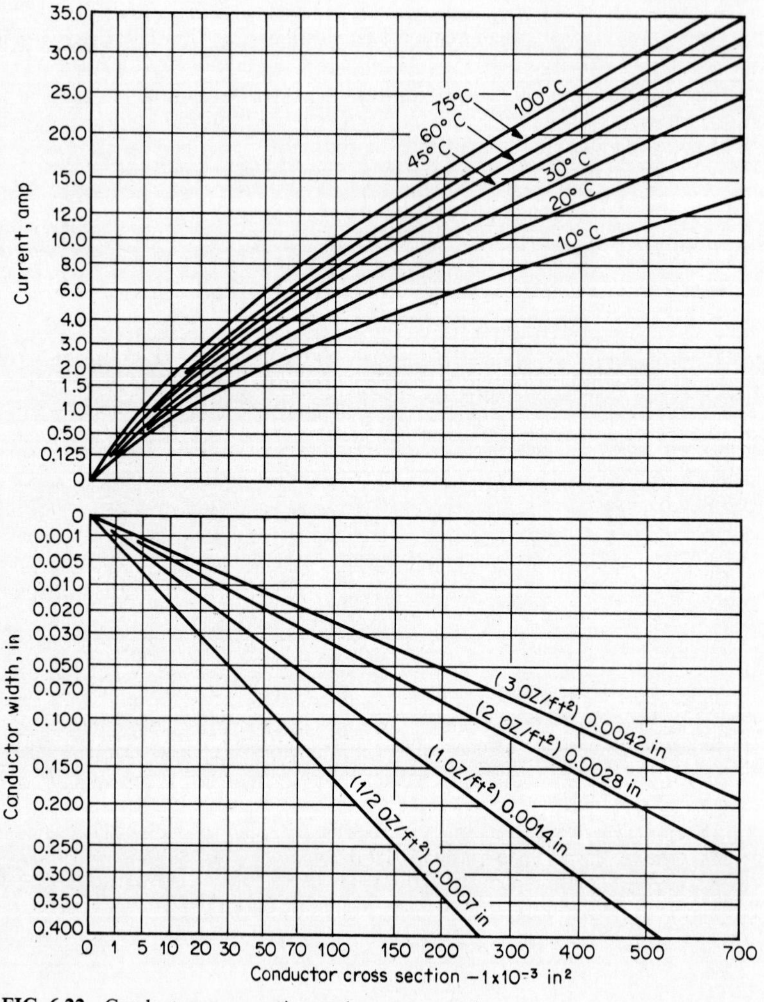

FIG. 6.22 Conductor cross section and current capacity.

derating (on a current basis) to allow for normal variations in etching techniques, copper thickness, conductor width estimates, and cross-sectional area.

Additional derating of 15 percent is suggested under the following conditions:

1. For panel thickness of $\frac{1}{32}$ in or less
2. For conductor thickness of 0.0042 in (2 oz/ft^2) or greater

For general use, the permissible temperature rise is defined as the difference between the maximum safe operating temperature of the laminate and the maximum ambient temperature at which the panel will be used. For single-conductor applications, the chart may be used for determining conductor widths, conductor thickness, cross-sectional areas, and current-carrying capacity for various temperature rises. Remember to calculate cross-sectional area with the final plating thickness.

For groups of similar parallel conductors, if closely spaced, the temperature rise may be found by using an equivalent cross section and an equivalent current. The equivalent cross section is equal to the sum of the cross sections of the parallel conductors, and the equivalent current is the sum of the currents in the conductors.

The effect of heating due to attachment of power-dissipating parts is not included. The conductor thicknesses in the design chart do not include conductor overplating with metals other than copper.

6.10 LAMINATE PROBLEMS

6.10.1 Traceability

It is impossible to build circuit boards in any quantity without having some difficulties that will be blamed principally on the copper-clad laminate material. Too often, the base material appears to be the cause of trouble, when actually the fabrication process is out of control. Even a carefully written and executed laminate specification will fail to specify the tests necessary to help identify the laminate as a cause of or a contribution to a process problem. Listed in this section are some of the most common laminate problems and how to recognize them.

Once a laminate problem is encountered, it should be considered for addition to the material specification. Often the addition, if not made, allows continuing variations and subsequent rejections. Usually any material problem traced to laminate variations will occur in discrete raw material or press load batches manufactured by the laminator. Few users keep records extensive enough to permit identification, in any processing area, of a particular press load or batch of material. So usually the boards continue to be manufactured and loaded; warpage in the solder pot, for example, may continue, and a large amount of labor and expensive components will be lost. If the load number is immediately known, the laminator can check resin batches, copper lots, cure cycles, etc. If the user does not provide continuity with the laminator's quality control system, the user is penalized in the long run. The common problems associated with base material in the fabrication process are discussed in the following sections.

6.10.2 Measling and Blistering

Cause	Corrective action
Entrapped moisture	Check drill hole quality. Check for delamination during drilling.
Excessive exposure to heat during fusing and hot-air leveling.	Monitor equipment for proper voltage regulation and temperature.
Laminate weave exposure	Ensure that you have sufficient butter coat. Be sure all wet processes are checked when materials are changed.
Measles related to stress on boards with heavy ground planes	Postbake panels prior to drilling. Be sure laminate has balanced construction. Use optimum preheat conditioning before IR fusing or hot-air leveling.
Handling when laminate temperature exceeds its glass transition temperature	Allow boards to cool to ambient temperature before handling.
Measling when large components or terminals are tight enough to cause excessive stress when heated	Check tooling for undersized holes. Loosen tight terminals.

6.10.3 Dimensional Stability

Cause	Corrective action
Undercured laminate	Check laminate vendor for glass transition temperature.
Distorted glass fabric (FR-4)	Examine etched panels for yarns parallel to warp and fill direction.
Dense hole patterns with fine lines and spaces	Prebake laminate in panel form before drilling.
Dimensional change parallel to grain differs from that of cross grain	Have laminator identify grain direction.

6.10.4 Warp and Twist

Cause	Corrective action
Improper packing	Assure that packing skids are flat and have sufficient support.
Storage	Stack material horizontally rather than vertically.
Distorted glass fabric (FR-4)	Examine etched panels for yarns parallel to warp and fill direction.
Excessive exposure to heat	Maintain proper temperature and exposure times in all heat-related fabricating processes (IR fusing, hot-air leveling, and solder mask applications).
Improper handling after exposure to heat applications	Cure solder mask horizontally. Use proper cooling techniques after IR fusing and hot-air leveling.
Unbalanced laminate construction	Work with laminator to get balanced construction.

6.10.5 Copper Foil

Cause	Corrective action
Fingerprints due to improper handling	Handle copper cladding with gloves.
Oils from punching, blanking, or drilling	Degrease with proper solvent.
Poor solderability on print-and-etch boards	Degrease to remove contaminates. Use highly activated flux. Check laminator for procedure to remove excessive antistaining compound.

6.10.6 Copper Bond Strength

Cause	Corrective action
Pad or trace lifting during wave soldering	Check for undercut due to overexposure during etch. Check with laminator to be sure solvents used in the circuit manufacturing process are compatible with the laminate.
Pad or trace lifting during hand soldering	The hand-soldering device is too hot or wattage is too high for the application. The device operator is applying heat too long to the soldered pad.
Pad or trace lifting during processing	Check laminate supplier for bond strength properties of the lot of material in the process. Check for undercut during etch. If pads lift after solder leveling, check fuser voltage or hot-air leveling equipment temperature.

REFERENCES

1. Z. N. Sanjana and R. N. Sampson, "Measuring the Degree of Cure of Multilayer Circuit Boards," *Insulation/Circuits,* April 1981, pp. 89–92.
2. S. D. Senturia, N. F. Sheppard, H. L. Lee, and S. B. Marshall, "Cure Monitoring and Control with Combined Dielectric Temperature Probes," *SAMPE,* vol. 19, no. 4, 1983, pp. 22–26.
3. D. E. Kranbuehl, S. E. Belos, and P. K. Jue, "Dynamic Dielectric Characterization of the Cure Process: LARC-160," *SAMPE,* vol. 19, no. 4, 1984, pp. 18–21.

CHAPTER 7
SURFACE MOUNT TECHNOLOGY MATERIALS

Stephen G. Konsowski
Advisory Engineer, Westinghouse Defense Center, Baltimore, Maryland

Carl T. Brooks
Senior Manufacturing Engineer, Westinghouse Defense Center, Baltimore, Maryland

Michael A. Williamson
Materials Engineer, Westinghouse Defense Center, Baltimore, Maryland

7.1 INTRODUCTION

The use of surface mount technology (SMT) in printed wiring introduces a set of new needs in substrate materials. Surface mount technology often creates a greater need for very fine line geometries, tighter dimensional stability, and coefficients of thermal expansion (CTE) that are compatible with different component packaging. This may require different considerations in base materials other than the most widely used material, FR-4. In addition, the process elements in SMT are somewhat different from more classical printed wiring, thus stressing the material in new ways. This chapter discusses these two areas of the interaction of surface mount technology and materials.

7.2 MATERIALS

7.2.1 Resin Materials

Surface mount technology often requires that resins used for printed wiring boards (PWBs) have increased thermal and chemical resistance properties over standard epoxy-glass PWBs because of the processing that will occur during manufacturing. The top and bottom PWB surfaces or resin "buttercoat" will be required to survive several harsh environments during manufacturing. For example, the resin may be heated to dry it before processing, as well as being heated at least twice during the solder reflow and parts attachment. If through-hole components are also used, the board will be heated during wave soldering and potentially during repair with a soldering iron. Also, the resin will naturally be subjected to the usual variety of process chemicals, such as etching solutions for removing copper if the subtractive process is used, solutions for plating on copper and tin-lead, and a variety of cleaning solvents used before and after soldering to remove contaminants. In addition to thermal and chemical properties, low electrical loss properties may be needed if signal transmission speeds are of concern. These properties are highly dependent upon the resin use. The materials used for SMT can be either thermosets or thermoplastics.

7.2.1.1 Thermosets. Traditionally the thermoset materials have been used for PWBs because of the ability to B-stage the resin, which improves the handling, the glass transition temperature T_g, and the chemical resistance. The thermosets are epoxies, modified epoxies, polyimides, modified polyimides, and bis-maleimide triazines (BT).

7.2.1.2 Thermoplastics. Thermoplastic materials are used because of their excellent electrical properties, their tailorable CTE, and their potential ease of processing. The thermoplastic materials most commonly used are polytetrafluoroethylenes (PTFE) and polysulfones. Although some work has been done with polysulfone, the most familiar thermoplastic is PTFE.

Modifications of the epoxies, accomplished by adding smaller amounts of tetrafunctional epoxies, polyimides, or BT are usually made to improve the T_g, which may also reduce the electrical losses. Modifications of the polyimides, accomplished by adding smaller amounts of epoxy, are often done to reduce the temperature-time cycle required to manufacture PWBs, to reduce the moisture absorption of the PWB and to improve the adhesive properties.

7.2.2 Thermal Coefficient of Expansion

Many surface mount devices (SMDs) are leadless and as such are meant to be mounted and soldered directly to the PWB. Capacitors and resistors in chip form as well as integrated circuits in leadless chip carriers (LCCs) constitute the bulk of such SMDs. As SMDs become larger, the strain resulting from the different coefficients of expansion of the SMDs and the epoxy-glass PWB becomes too great for the solder joints to withstand. Consequently, solder joint cracking is frequently encountered in assemblies where large chip carriers are used. The cracking may not be obvious initially, but repeated thermal cycling either under test conditions or in field equipment will eventually cause joint cracking. As a general rule, it is not advisable to solder large SMDs—that is, LCCs with more than 12 input/output pads on any side of the component.

To solve this thermal expansion mismatch, two approaches have been taken. One is to choose PWB materials which match the expansion of the components; the other is to restrain the PWB expansion with a core material. Activity to match the CTE of the PWB with that of the chip component, which is approximately 6 to 8 ppm/°C, has been extensive.

Many materials have been formulated, with variations in both resins and reinforcements. Several of these are listed in Table 7.1 along with significant mechanical properties. The data are presented for measured expansion coefficients in the plane of the PWBs as "warp" and "fill" according to the orientation of the reinforcement weave, and along the board thickness axis or z direction. Also listed is the T_g of each material. A high T_g is desirable because the expansion coefficient is

TABLE 7.1 Physical, Mechanical, and Thermal Data on Advanced Substrates

Material type	Resin content		CTE × 10^{-6}/°C*		CTE × 10^{-6}/°C†	T_g, °C
	Wt %	Vol %	Warp	Fill	z direction	
Aramid, mod. polyimide	58	61	6.5	7.2	66	167
Quartz, polyimide	37	46	8.5	9.8	34.6	188
Aramid, mod. epoxy	59	61	5.5	5.6	100.4	137
Aramid, mod. epoxy	62	63	7.4	7.2	114.3	106
Glass, polyimide	38	55	12.6	13.0	41	249
Quartz, FR-4 epoxy	50	64	11.0	13.6	62.6	125
Quartz, FR-4 epoxy with phenolic	51	65	10.2	12.4	39.6	172
Quartz, epoxy/polyimide (80/20)	50	54	12.2	14.0	60.3	109
Aramid, low-T_g epoxy	66	68	7.7	9.1	107.4	82
Aramid, high-T_g epoxy	58	60	5.8	6.2	118	157

*Data in these columns were derived using a horizontal quartz dilatometer.
†Data in this column were derived using a thermal mechanical analysis technique.

constant up to that temperature. A material with a high T_g is usually structurally stable in many thermal environments, is resistant to chemicals used for etching, cleaning, and assembly, is processible using state-of-the-art treater and laminating equipment, and can withstand component unsoldering and resoldering several times without significantly reducing the copper to PWB adhesion.

One of the consequences of using a low expansion reinforcement to restrain expansion of the resin in the plane of the PWB is that the expansion takes place in the *z* direction, because the reinforcements within individual layers are not tied together except through the resin itself. Large expansions in the *z* direction accelerate cracking of vertical platings such as the barrels of plated-through holes and the buttons of plated vias. Another consequence is that in aramid reinforcements the resin surface or buttercoat exhibits microcracking. This has caused researchers to concentrate on modifying the resins to lower the overall coefficient of expansion of the PWB materials. The five material types in Table 7.2 represent this latter group.

The reinforcements can also be selected on the basis of the electrical properties

TABLE 7.2 Properties of Emerging PWB Materials

Material type	Resin content, wt %	CTE × 10^{-6}/°C			T_g, °C
		Warp	Fill	z direction	
Aramid (matte/fabric), FR-4 epoxy	60	5.5	6.0	66	125
Glass, epoxy with compliant layer	50	16.0	12.7	59	115
Aramid, BT epoxy	52	7.7	6.9	114.3	185
Quartz, polytetrafluorethylene	67	7.5	9.4	88	19
Aramid (matte/fabric), mod. polyimide	61	10	10.4	113	177
Hybrid (aramid/quartz), polyimide	41	8.3	8.4	60	249

TABLE 7.3 Solid Core Heat Sink Mechanical Parameters

Material	CTE × 10^{-6}/°C	Thermal conductivity, W/°C in	Density, lb/in^3	Specific thermal conductivity, W in^2/lb °C	Relative cost	Young's modulus "E" psi × 10^6
Aluminum	23.6	5.49	0.098	56.0	1	10
Copper	17.6	9.92	0.323	30.7	1.2	17
Alloy 42	5.3	0.396	0.28	1.41	3.3	21
Molybdenum	4.9	3.71	0.369	10.1	5.0	47
Invar	1.3	0.263	0.30	0.88	3.3	21
Copper-Invar 20-60-20	6.5	4.12*	0.31	13.3	2	18.5
Copper molybdenum 13-74-13	6.5	5.32†	0.36	14.8	4	31.2
Boron aluminum MMC‡ 20%	12.7	4.65§	0.095	48.9	100	27
Graphite aluminum MMC 0–90° crossply	5.8	4.08	0.087	46.9	400	21
SiC aluminum MMC	16.2	3.51	0.103	34.1	200	15

*Conductivity in normal direction is 0.430 W/°C in.

†Conductivity in normal direction is 4.43 W/°C in.

‡MMC stands for metal matrix composites, a class of exotic metal matrices reinforced with ceramics such as boron, silicon carbide, or graphite.

§Conductivity in normal direction is 1.93 W/°C in.

needed by the user and can take the form of microspheres, powdered fillers, paper, and continuous or noncontinuous fibers such as aramid, quartz, glass, and graphite. Microspheres and powdered fillers are often selected to reduce the flexural modulus of the resin and to alter the electrical properties. The CTE of resins reinforced with these materials is isotropic, since the fillers are noncontinuous.

The use of a core within the PWB to control expansion has been limited. In that approach, a metal plate consisting either of claddings on a core or of a metal matrix whose combination of thermal conductivity, density, and CTE make it a desirable design choice is bonded on one or both sides to a multilayer PWB (Table 7.3). Although the core functions to conduct heat from the PWB as well as to control the thermal expansion in the plane of the board, it also adds considerable weight and cost.

7.2.3 Copper Cladding

The CTE of the copper foil and of the PWB to which the copper is adhered are different, and this necessitates a high copper peel strength. In addition, the line widths used in SMT are smaller than those used with through-hole PWBs, and the dissimilar CTEs could result in a break of the conductor path. The peel strength of copper at room temperature for epoxy and modified epoxy resins is usually adequate, but it is not adequate for polyimide and modified polyimide resins. To improve the copper to polyimide peel strength, it is necessary to perform an oxide treatment. This is described in greater detail in the section on processing special PWB materials. Because the conductor line widths will be narrow, the copper cladding must be chosen so that the surface to which the resin will adhere will be rough and have large hills and valleys. Such cladding is called "high-profile" material. To reduce etching undercutting of copper-clad epoxies or polyimides, a copper foil having a thickness of 0.0007 in ($\frac{1}{2}$-oz copper foil) or less should be used. Without these precautions, undercutting will lower the peel strength and increase the amount of breakage that will occur during handling.

7.2.4 Electrical Performance

Improved electrical performance is another feature which certain SMT designs require. In particular, as circuit computing speeds continue to increase, the PWB becomes a transmission medium. Its dielectric constant and dissipation factor can bear heavily on the speed and attenuation of signals carried within it. Both properties must be as low as possible for minimum degradation of signal integrity. Table 7.4 lists those properties for a variety of resin-reinforcement combinations. It is important to note that the dielectric constant varies, sometimes significantly, with operating frequency. The designer should be aware of this fact and should base material selection on the dielectric properties if high computing speed is a consideration.

7.3 PROCESSING SPECIAL PRINTED WIRING BOARD MATERIALS FOR SURFACE MOUNT TECHNOLOGY

7.3.1 Lamination

The lamination of circuit layers built of low-expansion materials presents a special set of problems to the PWB fabricator. Among those are prelamination con-

TABLE 7.4 Dielectric Properties of Advanced PWB Materials

Material	Dissipation factor, I / Dielectric constant, II	Frequency, MHz			
		1	10	100	1000
Aramid, modified epoxy	I	0.0024	0.0268	0.029	0.12
	II	4.0	4.5	4.2	8.6
Aramid (paper/fabric) modified polyimide	I	0.0048	0.0051	0.0036	0.0039
	II	3.49	3.64	3.63	4.27
Quartz, polytetrafluoroethylene	I	0.00042	0.0024	0.0001	0.00032
	II	2.40	2.50	2.50	2.69
Aramid/quartz, polyimide	I	0.0059	0.0080	0.0027	0.0028
	II	3.38	2.51	3.51	4.00
Aramid (paper/fabric) FR-4	I	0.0097	0.0125	0.0049	0.0055
	II	3.76	3.84	3.81	3.99
Glass, FR-4 (with compliant layer)	I	0.0077	0.0104	0.0096	0.0099
	II	4.54	4.69	4.64	5.11
Aramid, BT epoxy	I	0.0064	0.0116	0.0036	0.0041
	II	3.66	3.80	3.80	4.07

ditioning of the copper, storage of the prepreg, press parameters, and postbake treatment. The most important of these is the prelamination conditioning of the internal layer copper surfaces to improve bonding between that surface and the prepreg. The most widely used process utilizes the deposition of the cupric oxide onto the copper surface to provide a surface with a topography that is "rough" and lends itself to wetting by the B-stage resin system. It is of importance to note that most materials used in surface mount applications will require high-temperature lamination cycles. Some versions of cupric oxide break down and become brittle above 350°F. For this reason, high-temperature oxides must be used, especially with exotic materials such as quartz and aramid. The choice of oxides must be made in light of other process variables that precede that operation, such as etching and resist stripping. Some precautions should also be taken to reduce the amount of time the treated inner layer spends waiting before the actual lamination cycle. One of the first concerns is the handling and storage of the prepreg materials. Most notable is the effect of moisture absorption on new resin-fiber combinations such as polyimide-aramid. A radical process change that will most likely be required is the baking or drying of any prepreg made with aramid, which is a strong candidate for SMT. A forced-air oven operating at 250°F for 1 to 2 h is needed. This will assure that during lamination no entrapped moisture will volatilize and cause the laminate to explode at elevated temperatures. Certainly the inner layers themselves must undergo the same drying cycle with increased time (2 to 4 h).

Determination of the press cycle will depend on the type of press to be used, the resin system, the reinforcing fiber, the resin-to-fiber ratio, and board thickness requirements. As with any thermal bonding process, the amount of resin or bonding material left in the completed part is important to the overall performance of the part. The press cycle chosen must allow pressure application at the most strategic point in the viscosity range. High pressure (greater than 350 psi) will most likely produce best results along with high laminating temperatures (greater than 400°F but less than 500°F). This is true because those fibers and resin systems used for surface mount applications may resist interlaminar wetting as well as wetting to the copper surfaces. In some cases, depending on the resin system chosen, a "kiss" cycle may be most advantageous. For instance, if a modified polyimide-aramid or modified polyimide-quartz or a regular polyimide system is used, lower pressure may need to be applied while the laminate is being heated to about 300 to 325°F. At this point, which may take 10 to 20 min, high pressure (350 to 500 psi) should be applied and held for 1.5 to 2.5 h. The laminating temperature should be maintained (whatever is chosen based on product test runs) as consistently as feasible. Thermocouples should be used to monitor the activity of the press during the cycle. For any aramid-based laminates, the control of the temperature rise profile is critical because of dimensional shifts and because resin wettability of the fibers is poor. One departure to be taken from normal processing techniques when any polyimide resin system is used is the application of full pressure during the cool-down cycle to promote interlaminar adhesion, since those bonds are inherently weak. Postbaking is also a necessity to provide some assurance of complete cure within the laminate. That postcure should be performed at the laminating temperature and maintained for 4 to 6 h, especially in cases where polyimide aramid is used. The laminated panels should be separated in the postcuring oven so that all surfaces are exposed. If care is taken through the lamination operation, good bonds between B stage and copper may be routinely achieved. A solder float check should be used once the board has been drilled, plated, and reflowed to verify laminate integrity.

7.3.2 Drilling

The ultimate aim of surface mount technology is interconnecting the increased number of I/Os available on the device package. This will require plated-through holes of very small diameters (less than 0.010 in) located in very close proximity to each other (0.020-in centers). Some holes may be as small as 0.004 in. It is easy to see that major changes will be required to standardize approaches for drilling holes. Figure 7.1 shows the very discernible difference between a standard drill and the one required for producing small holes for very high I/O count packages.

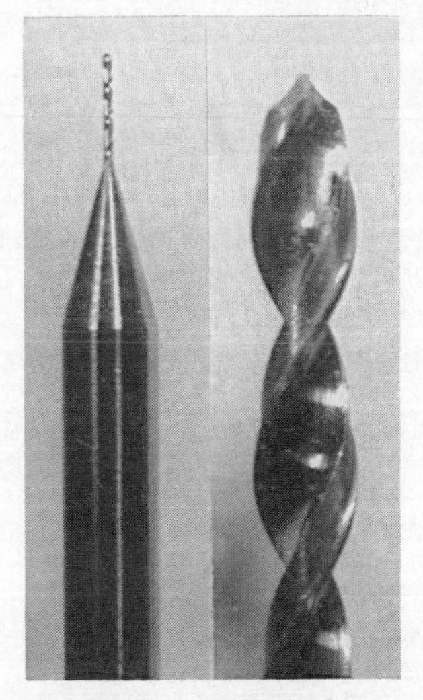

FIG. 7.1 Comparison of small-hole drill with conventional drill.

Some changes in bit point and flute geometry have been made to increase concentricity and overall stability. Once the tool has been manufactured and delivered, subsequent handling by the user other than ring setting and measuring must be minimized. One necessary handling operation, however, is 100 percent drill point inspection.

To use these micro drills successfully, some baseline criteria must be established regarding materials and

TABLE 7.5 Effects of Drilling Conditions

	Effects				
Drilling Condition	Hole smear	Drill wear	Delamination	Drill wander	Location accuracy
Entry material	X	X			
Backup material	X	X			
Drill bit r/min	X	X			
Drill configuration	X	X	X	X	X
PWB materials	X	X	X	X	X
Feed variation				X	

equipment. The entrenched practice of measuring the drill tips with hand-held micrometers must end if any drills are to survive the trip to the spindle. As Table 7.5 shows, there are many variables affecting the drilling of holes, especially small ones in aramid- and quartz-reinforced materials.

1. *Spindle/drill speeds:* These speeds must be increased to produce the maximum drill speed measured in surface feet per minute (sfm). For most spindles, which are capable of reaching only 80,000 r/min, drill speeds will be under 150 sfm for diameters under 0.010 in. This is much lower than the 600 to 700 sfm said to be optimum for DIP plated-through-hole technology (0.037-in-diameter holes in epoxy-glass material). Conversely, the in-feed rate must be slowed significantly to produce much lower chip loads. A drill speed of 10 in/min is not unreasonable when a 0.006-in drill is used.

2. *Spindle runout:* This must be held to a bare minimum (less than 0.0001 in). Any discontinuities in concentricity will be fatal to the very fragile drill. Indicators should be used both on the spindle sleeve and on the removable spindle collet before each use.

3. *Entry/material:* Although some success has been achieved drilling small holes with aluminum entry plates of varying thicknesses from 0.003 to 0.010 in, the length of the flute of the micro drill is greatly reduced. Therefore, the less material to be drilled, the better. Since surface mount boards will employ thin copper, the likelihood of burrs being generated from drilling without entry is very low. Backup material may be used strictly to elevate the board from the drill table.

4. *Pressure foot:* Two things are important here: (*a*) The foot must apply a minimum of 60 psi to the part, and (*b*) the drill should be recessed at least 0.1 in from the bottom of the pressure foot.

5. *Spindle retract rates:* These rates must be lowered to compensate for the inherent dynamic instabilities in drilling machines that use 1000 in/min as a benchmark. This number may need to be lowered to as little as 10 to 50 in/min, depending on the overall condition of the equipment.

One of the factors that will determine the overall hole quality is the bit's usable life. How many holes may be safely drilled before they deteriorate to the point of no return? Figures 7.2 to 7.5 show the effect of the number of holes drilled in different materials. Quartz-based materials are the most taxing to the drill because of their severe abrasiveness. When all machine and process parameters are optimized, holes as small as 0.004 in may be drilled in materials of varying thicknesses (Figure 7.6). This represents a giant step forward in the reduction of space required for interconnecting high-I/O surface mounted devices.

FIG. 7.2 Unused tungsten carbide drill tip.

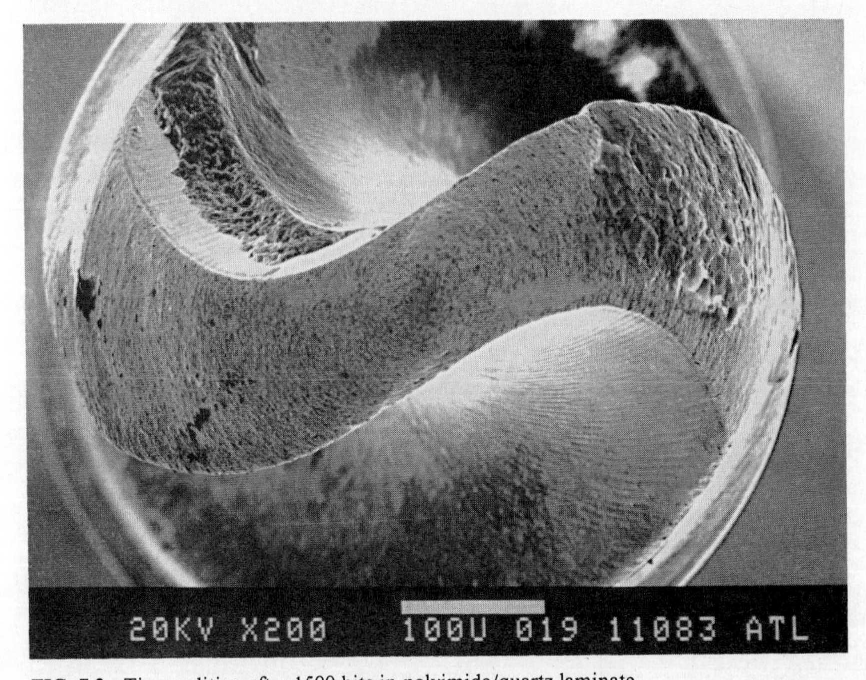

FIG. 7.3 Tip condition after 1500 hits in polyimide/quartz laminate.

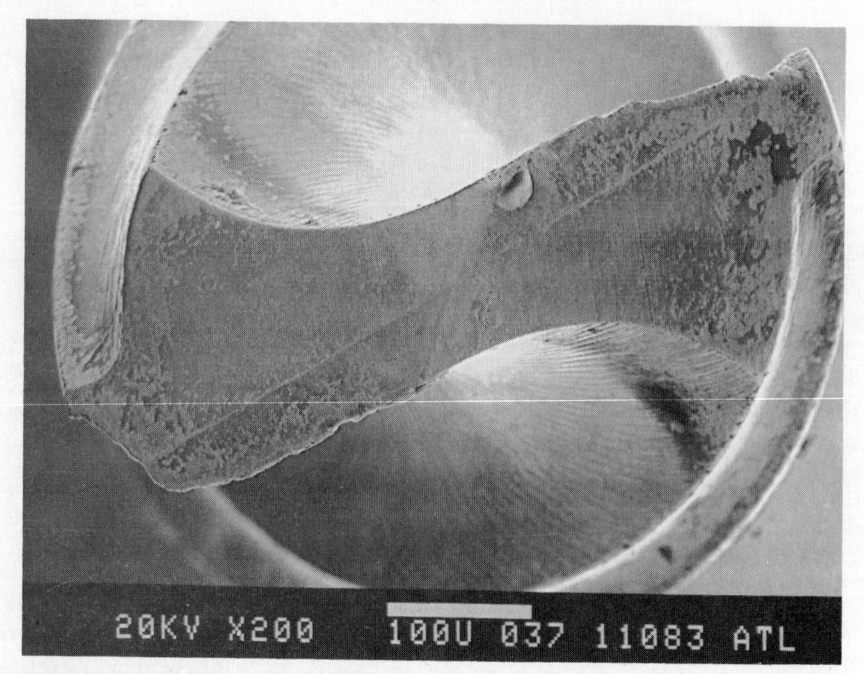

FIG. 7.4 Tip condition after 1500 hits in epoxy-aramid laminate.

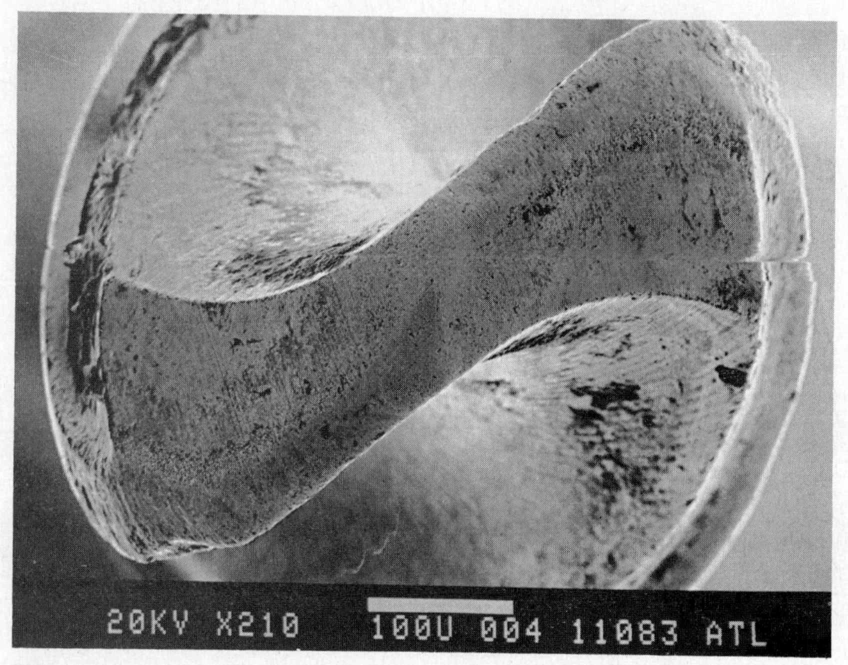

FIG. 7.5 Tip condition after 1500 hits in epoxy-quartz laminate.

7.3.3 Solder Pastes

In order to attach surface mounted devices to PWBs, relatively thick solder (0.003 to 0.004 in) on the pad surface is required. This thickness allows the leads of flat pack devices to seat properly and provide a solder joint with some integrity through the vapor phase reflow operation. One method of increasing this solder thickness is the application of a solder paste. It is exactly what the name implies: a screenable paste applied to selected areas with the use of a mask or screen. This allows the pad areas to be covered adequately. The concept is plausible if certain precautions are taken to assure proper and consistent coverage by the paste. The condition of the screen used for application must be checked for imperfections, as with most screening operations. A somewhat coarse mesh screen, preferably made of stainless steel, should be used. Eighty mesh has proven to work well with pad sizes of 0.04 to 0.05 in on a side, but the edge definition will tend to deteriorate with decreasing pad sizes and line widths. However, some drawbacks to paste use do exist. Most solder pastes may be ordered to specific requirements, depending on particular applications. Some of the most important properties of the paste are type of flux used, particle size, and lead-to-tin ratio. Particle size is important when the application is for SMT. Some pastes consist of solder ball clusters that may range from a few microns to several mils in diameter. One of the problems with pastes containing larger particle sizes is line definition. Other problems occur when via holes are located near the pad sites to be pasted. Some solder ball clusters may be come entrapped within those holes and because of their high surface energy become difficult to remove. In vibration environments, they can become dislodged. Care must be exercised when specifying the type of board cleaning to be performed prior to the application of the paste. Board cleanliness will have great bearing on the adhesion as well as on the coverage consistency during the vapor phase cycle. Understanding the process limitations of solder paste application will lead to a more productive soldering operation.

FIG. 7.6 A 0.004-in drilled hole plated through to finished diameter 0.0025 in for 0.008-in-thick epoxy-glass.

7.3.4 Plated Solder Pads

One alternative to paste application is the electroplating of those pad areas where surface mounted devices are to be attached. Although simple in theory, from a practical point of view many elements of plating, masking, cleaning, and reflow must come together to form an acceptable product. One of the first concerns involves the use of masking to selectively plate pad areas. Since 0.003 to 0.004 in of solder will be required, a plating resist of that thickness must be developed. Once this obstacle has been overcome (some military fabrication houses have developed it to the point of production), the concern becomes control of the solder plating bath so that the following occur: (1) The composition of the plated tin-lead is maintained at the 63:37 percent ratio that corresponds to the composition of the tinned leads of the package. A mismatch here can produce higher solder reflow temperature and poor wetting of the joints. (2) The grain structure of the deposition must be controlled to within limits that will allow such heavy plating without sacrificing panel distribution consistency. Since the industry is accustomed to plating only 0.0003 to 0.0005 in of solder, this heavy plate represents a formidable challenge and requires a similar advance in electroplating rectifier controls. Much more sophisticated and repeatable controls on current distribution, anode configuration, and chemical analysis are required for uniformity of composition and thickness.

7.3.5 Reflow and Interconnection of SMDs

Once the proper amount of solder has been deposited on the appropriate pads of the PWB, it is reflowed. The process involves heating the entire board in a controlled fashion after soldering flux has been applied. If solder paste or cream was used as the solder material, no additional flux should be required. Electroplated solder should, however, be coated with flux. The method of reflowing that solder can be a matter of preference, the selection being made from infrared, hot-air, or vapor phase techniques. Infrared utilizes powerful heating lamps, hot air uses a directed stream of air heated to a temperature above the melting point of the solder, and vapor phase uses a fluid heated to its vaporization temperature (slightly above 200°C). The board is immersed in the vapor in a large tank containing both the vapor and the condensed fluid. If vapor phase reflow is the method selected, a high-T_g PWB material will perform best. At the very least, a BT epoxy should be used. The vapor does not escape the tank in very large quantities because of cooling coils which line the walls of the tank and cause the vapor to condense and fall back into the sump. As the board is immersed in the vapor, the latent heat of vaporization causes the board and its solder to rise in temperature until the solder begins to flow. Since the vapor is always at the same temperature, the board cannot be overheated and scorched. The vapor also aids in preventing oxidation of the solder and carbonization of the flux. After the solder has melted, the board is retracted from the vapor and cooled. The flux is removed, and the board is inspected and made ready for component placement. It should be noted that some manufacturers have found it convenient not to reflow prior to component attachment, simply soldering the SMDs during the initial reflow step. That alternative may be cost-effective if the components are not complicated and there are only a few SMDs on a board. If there are more than 1000 solder joints to be made on one board, it is advisable to reflow the solder and to inspect the solder mounts for adequacy of volume and for random solder spheres on the surface of the board. This is especially critical if solder paste or cream is used. Failure to reflow before component attachment could result in solder spheres migrating under the edges of components, where they could cause shorting of exposed conductors, or even an electrical breakdown later because of reduced conductor separation.

After the SMDs and any other components have been assembled to the board, it is again fluxed and placed in the vapor, where the solder flows to the SMD terminations. This second reflow should result in the same joint characteristics as one is used to seeing with wave-soldered components; that is, the fillet should be smooth and the surface should be shiny. Earlier work has led to the conclusion that it was not possible to achieve shiny joints using vapor phase techniques, but refinement of the reflow processes and fluxing techniques has proved otherwise.

CHAPTER 8
CIRCUIT COMPONENTS AND HARDWARE

Gerald L. Ginsberg
President, Component Data Associates, Inc., Lafayette Hill, Pennsylvania

8.1 INTRODUCTION

The electrical circuit components and mechanical hardware used in printed wiring assemblies vary in both type and shape (Fig. 8.1). Often the selection of such components is limited by specification, availability, cost, or all three. However, many commonly available components, such as resistors and capacitors, can be obtained in several configurations, which allows the printed wiring designer some freedom in selecting or specifying the component size, shape, and configuration.

The following text discusses the basic features of printed wiring components and hardware and provides guidelines for their selection.

8.2 THROUGH-HOLE MOUNT COMPONENTS

8.2.1 Selection Criteria

To design an efficient and economical assembly, the printed wiring designer should take into account the following component selection criteria.

1. *Component size and shape:* Profile (the height above the printed wiring mounting surface) and printed wiring mounting area are initial considerations to be applied to all circuit components. The parameters govern the placement and density of components on the board and the mounting relation among boards installed side-by-side within equipment.

2. *Lead size and spacing:* Lead size and shape are key parameters affecting printed wiring board layout and assembly. Lead size is a prime factor in determining hole size (both drilled and plated) and terminal area size. These in turn relate with lead spacing to determine allowable conductor-routing path locations.

3. *Mechanical tolerances:* Proper tolerances on size and shape (for both the component body and its leads) are important to assure that the board and components can be assembled cost effectively and with a minimum of difficulty. For

FIG. 8.1 Printed wiring components and hardware. *(Winchester Electronics, Oakville, Conn.)*

components with flexible leads, tolerances on lead size and location are not normally troublesome, although assembly time may be more than that associated with rigid leads. The relative magnitude of this consideration also depends on the assembly method used, because compensations that can be made during manual assembly are not possible with high-speed automatic assembly equipment.

4. *Component mounting:* Whether the component is self-supporting, has built-in mounting provisions (i.e., threaded inserts or studs), or requires special mounting hardware (such as clamps or clips) is an important consideration. A general rule of thumb is that components weighing $\frac{1}{4}$ oz or more per lead should have a mechanical means of support to ensure that the lead-termination joint is not relied upon for component support.

5. *Thermal:* The heat-dissipation properties and provisions of a component are sometimes of significant interest. When small quantities of heat are involved, radiation of heat from the component body or leads and conduction of heat through leads and mounting brackets, lugs, etc., are usually sufficient to maintain correct operating temperatures. When larger amounts of heat are to be dissipated, heat sinks can be used with most components to remove heat more efficiently.

8.2.2 Circuit Components[1]

8.2.2.1 Axial-Lead Components. Perhaps the most common type of printed wiring component is the axial-lead component (Fig. 8.2). It usually is cylindrical and has a lead exiting from each of its ends along its neutral axis. The lead is usually round in cross section and the body is most often formed by molding or dipping; axial-lead components are the types most suited for automatic insertion. Many resistors, capacitors, and diodes are supplied in this configuration.

FIG. 8.2 Axial-lead component.

8.2.2.2 ***Radial-Lead Components.*** Radial-lead components are used in many printed wiring assemblies. This type of component has all of its leads exiting from a common side of the component. The actual body shape is variable; two common types of radial lead components are dipped capacitors and transistor TO cans.

8.2.2.3 ***Multiple-Lead Components.*** One of the major advantages of using a printed wiring structure for component mounting and interconnection is the suitability of the structure for use with multiple-lead components, usually integrated circuits (ICs). Multiple-lead components are packaged in several sizes and shapes, each of which has distinct advantages and disadvantages.

8.2.3 Multiple-Lead Components: Circuit Design Considerations

8.2.3.1 Package Types

1. An outgrowth of the radial-lead transistor can is the multiple-lead can. This type of component consists of a hermetically sealed can with up to 12 round leads exiting from the bottom of the device (usually in a circular pattern).

2. The dual-in-line (DIP) multiple-lead component (Fig. 8.3) closely resembles a large flat pack with its leads formed at a right angle after they exit from the component body. DIP leads exit from the body in a ribbon form but are usually shaped into a V or reduced in size prior to entering the printed wiring mounting hole. The DIP body can be plastic or ceramic. DIP devices are available with up to 50 leads on 0.100-in centers.

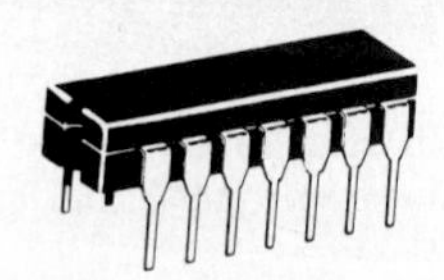

FIG. 8.3 Dual-in-line component.

8.2.3.2 ***Package Selection.*** Some major considerations for the selection of a multiple-lead-component configuration are the following.

1. Circuit speed has a dominant influence on the size, spacing, and tolerances associated with multiple-lead packages. Circuit conductor lengths and layout relations to other components become critical at speeds of 10 ns and faster.

2. Reliability, serviceability, and environment dictate the need for device hermeticity and ruggedness, and they affect lead configuration and suitability for soldering, welding, or bonding as well. The component mounting method also is dictated.

3. Quantity usage, especially production, determines the degree of assembly tooling and mechanization to be used. Some devices are better suited for high-volume or mechanized applications than others.

4. Printed board area usage affects the number of ICs per board assembly and quite often the number of assemblies. Close multiple-lead-component spacing affects terminal area size, conductor size, conductor spacing, number of conductor layers, and overall fabrication tolerances. Lead spacing also affects those selection parameters.

5. Heat-dissipation characteristics of the package and package suitability with heat sinks must be considered. High-component densities readily create hot spots, even though heat dissipation per component may appear to be minimal.

8.2.3.3 Board Layout Considerations. The following general considerations should be taken into account when printed wiring assemblies with multiple-lead components are designed:

1. The physical dimensions of the multiple-lead component
2. The limitations placed on the board layout area by mechanical or electrical requirements
3. Fabrication restraints or requirements affecting hole size and spacing
4. Terminal area size, lead forming, and lead clinching
5. Distribution buses for ground and voltage(s)
6. Artwork size, registration, and tolerances
7. Photography, glass master, or screen registration tolerances
8. Registration tolerances for board screening, etching, plating, drilling, or punching
9. Automatic, semiautomatic, and manual component insertion tolerances and restraints
10. Component dimensions and tolerances

8.2.4 Grid Arrays[2]

The higher the I/O lead count becomes, the lower the percentage of the total package area die cavity of any given size will occupy. (For example, the 96-lead, 40-mil-center chip carrier is more than 1 in square.)

For optimum packaging density, this percentage should be as high as possible. Therefore, 0.025- and 0.020-in chip carriers (as previously mentioned) and "grid arrays" are used (Fig. 8.4).

Grid arrays are pinned or leadless carriers with I/O contacts that populate one surface of the package on a 0.100-in (eventually 0.050-in) grid. The use of a solid grid necessitates placing the die cavity on the side opposite the I/O contacts. A double or multiple concentric row grid (Fig. 8.4) permits having die cavity and I/O contacts on the same side, with an optional heat sink on the opposite side.

FIG. 8.4 149-pin array package.

Figure 8.5 shows the I/O density achievable with the various packages. As can be seen, for lead-count requirements in excess of 100, a 0.100-in centerline solid grid provides greater I/O density than either 0.050- or 0.040-in centerline chip carriers.

The 0.024-in centerline chip carrier (CC) is clearly more I/O efficient than is a 0.0001-in grid array. However, a double-row 0.050-in grid array is equivalent to a 0.025-in centerline CC. The 0.050-in solid grid provides the greatest I/O density.

With pins brazed on 0.100-in centers, pin grid arrays are a logical carrier for high lead counts intended for through-hole (not surface mount) board assembly that is compatible with the use of DIPs. This should serve to extend the useful life of present-day technology for some high-lead-count applications.

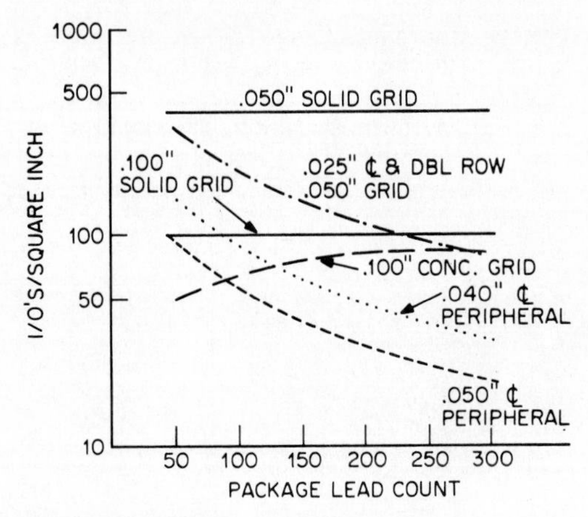

FIG. 8.5 I/O density versus lead count.

8.2.5 Adjustable Components

Adjustable components (usually resistors) are common to many printed wiring assemblies. These devices meet the description provided for radial-lead components but have the additional provision of an adjustment screw. Access to the adjustment feature can be via an exposed edge of the printed board, when a right-angle adjustable component is used. Access to the adjustment feature not on the exposed edge of the printed wiring assembly is achieved with the straight-through type of component.

8.2.6 Sockets

Some printed wiring applications call for the frequent insertion and removal of some or all circuit components. Printed board component sockets have been developed for that reason. There are appropriate sockets for almost all printed wiring circuit components, including those for DIPs (Fig. 8.6).

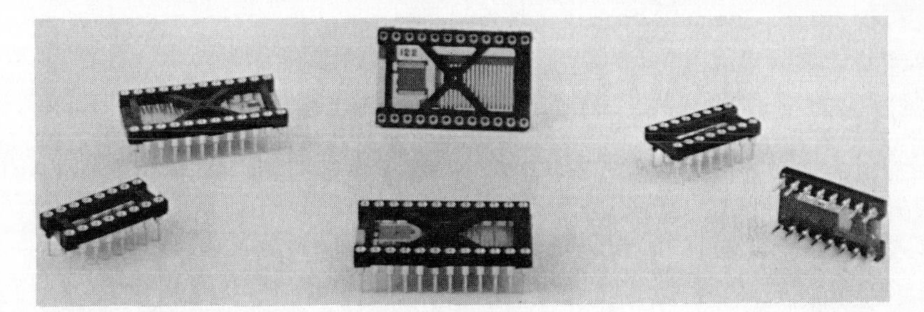

FIG. 8.6 Multiple-lead sockets.

8.2.7 Bus Bars

Board bus bars are constructed with two or three copper conductors (usually tin- or tin-lead–plated) laminated together with a thin dielectric separator. The exterior is generally covered with a plastic barrier material to prevent accidental shorting. The dimensions of the bars depend on their positions on the board. Bus bars that mount under DIPs are slightly less in width than the pin spacing. Bars that stand vertically have widths corresponding to the maximum height of mounted components. Bar length is determined by circuit board dimensions. A typical bus bar configuration for mounting under DIPs is shown in Fig. 8.7.

FIG. 8.7 Typical bus bar configuration. *(General Atronic Company.)*

8.2.8 Miscellaneous Component Types

There are many other types of printed wiring components which have both electrical and mechanical functions; among them are standoff terminals, relay cans, and switches.

8.2.9 Interconnection Devices

One of the major advantages of using printed board structures, as opposed to other types of component mounting and interconnecting methods, is ease of maintainability through plug-in usage. Board connectors consist of the following types for almost all applications:

1. One-part (card edge)
2. Two-part (plug-and-receptacle assemblies)
3. Discrete-contact (plug, receptacle, or both)

One-part board connectors, the most common type, use one edge of the printed board as the plug dielectric and printed and plated conductors as the contacts. The other half of the connector is usually an assembly of mating contacts in a chassis-mounted receptacle assembly (Fig. 8.8).

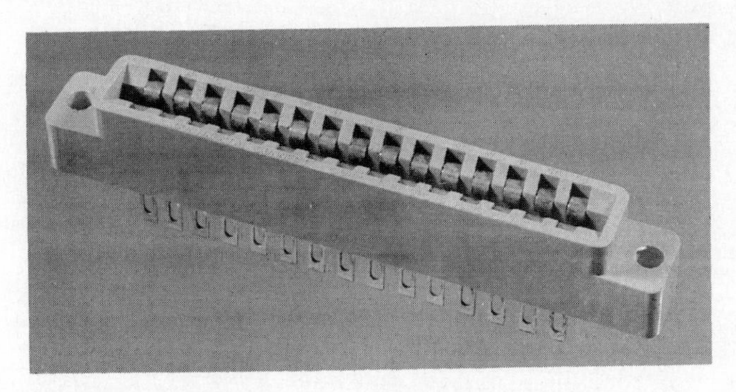

FIG. 8.8 Chassis-mounted receptacle assembly.

Two-part board connectors consist of self-contained multiple-contact plug-and-receptacle assemblies. Usually, but not always, the plug (male) contact assembly half of the connector mounts to the plug-in assembly (Fig. 8.9*a*), and the receptacle (female) half of the connector mounts to an interconnection wiring panel (mother board) or plate (Fig. 8.9*b*).

The three basic disadvantages of using one-part and two-part board connectors are as follows:

1. Connector location is usually limited to one edge of the board assembly.
2. The direction of insertion (plug-in) of the board assembly is usually limited.
3. The plug or receptacle halves are usually limited to a specific number of contacts based on the chosen connector.

When those disadvantages cannot be overcome by acceptable restraints on the design, the use of discrete (plug, receptacle, or both) contact connectors is recommended.

8.2.9.1 Contact Types. In addition to the three major distinctions among printed wiring connectors just described, a secondary description of a connector relates to the contact types. The most common printed board connector contact types are

1. One-part
   a. Bellows contact (Fig. 8.10*a*)
   b. Tuning fork contact (Fig. 8.10*b*)
   c. Cantilever contact (Fig. 8.10*c*)
2. Two-part discrete contact
   a. Pin-and-socket contact (Fig. 8.11*a*)
   b. Blade-and-fork contact
   c. Hermaphroditic contact (Fig. 8.11*b*)

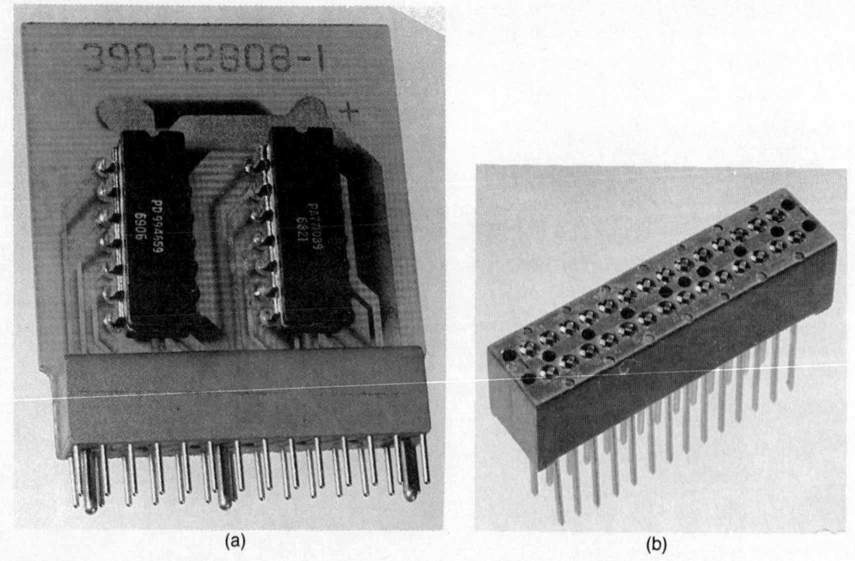

FIG. 8.9 (*a*) Two-part connector, printed board assembly (with plug half); (*b*) two-part connector (receptacle half).

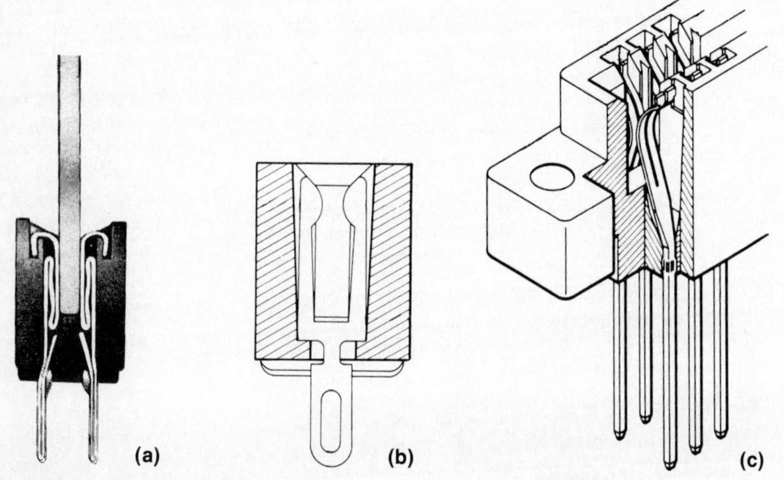

FIG. 8.10 One-part connectors. (*a*) Bellows contact; (*b*) tuning fork contact; (*c*) cantilever contact.

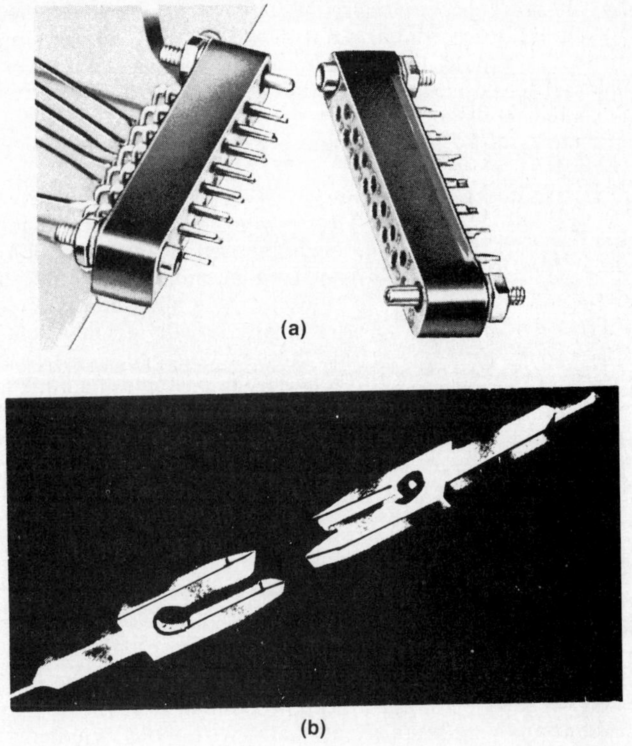

FIG. 8.11 Two-part connectors. (*a*) Pin-and-socket contact; (*b*) hermaphroditic contact.

8.2.9.2 Termination Types. The one-part connector and the non-plug-in half of the two-part and discrete-contact connectors are available in several termination types.

1. *Solder terminations:* For solder termination interconnection wiring, the connector can be soldered directly to a (mother) board by hand, wave, or dip soldering. When discrete hookup wiring is to be soldered to the contact terminal, various eyelets, tabs, and tongues are used. An example is shown in Fig. 8.12.

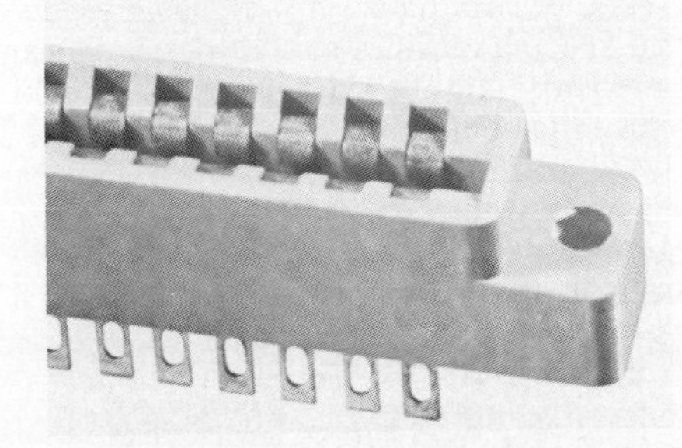

FIG. 8.12 Solder eyelet contact terminal.

2. *Solderless (crimp) terminations:* Many discrete wire hookup terminations are made to printed board connectors with the crimp method. Crimp terminations are made by crimping the bared portion of the hookup wire either to the contact portion of the connector or to a tab or pin which mates with the contact terminal. In the former instance, the wire becomes a part of a connector assembly; in the latter, the crimped tabs become a part of the interconnection wiring harness.

3. *Solderless (mechanized) terminations:* Increasing in popularity are solderless printed board connector terminations made by numerically controlled automatic and semiautomatic machines and handguns. In these applications, the connector terminal is usually a solid-metal rectangular-cross-section post. The connection between the post and the hookup wire can be made by tightly wrapping the bared portion of the wire around the post (solderless wrap), as in Fig. 8.13, or by restraining the wire against the post, using a suitable clip.

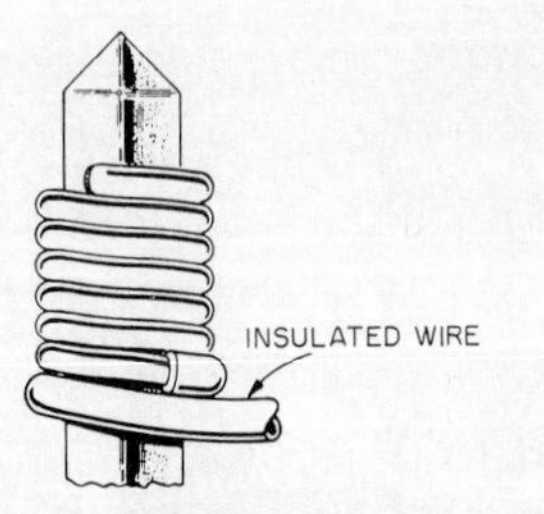

FIG. 8.13 Solderless wrap termination.

8.3 SURFACE MOUNT COMPONENTS

8.3.1 Flat Packs

One of the smallest of the multiple-lead component types is the flat pack (Fig. 8.14). The body of such a component can be as small as $\frac{1}{8}$ in wide, $\frac{1}{4}$ in long, and $\frac{1}{32}$ in thick. The component leads are normally flat ribbons mounted on 0.050-in centers. Flat packs are available with up to 50 leads.

FIG. 8.14 Flat-pack multiple-lead component.

8.3.2 Chip Carriers[3]

Recognizing that no single outline can satisfy all semiconductor packaging requirements, the Joint Electronic Device Engineering Council (JEDEC) has established a standard for CCs that allows for multiple design approaches, manufacturing techniques, and attachment means, thus enabling designers to choose and tailor the packages to their applications.

JEDEC has standardized two basic package styles (one with 0.050-in-center terminal spacing and another with 0.040-in) and has provided interchangeability within similar terminal-spacing outlines so that designers can interchange types to suit changing requirements without redesigning the interconnecting assembly.

The 0.050-in center family contains seven variations (Fig. 8.15). Four leadless types—A (Fig. 8.16), B, C, and D—mount in different orientations depending on the type, mounting substrate, and preferred thermal orientation. These packages are ceramic with hermetically sealed metal or ceramic lids.

The leaded type A category comprises two standards. The first, a one-of-a-kind 24-pin-only outline, is strictly a premolded plastic that is designated MS006. The second leaded type A package, MS007, has two versions—a substrate clip on a leadless type A package (Fig. 8.17) and a molded-plastic version. The second leaded type A unit comes in a wide assortment of pinouts.

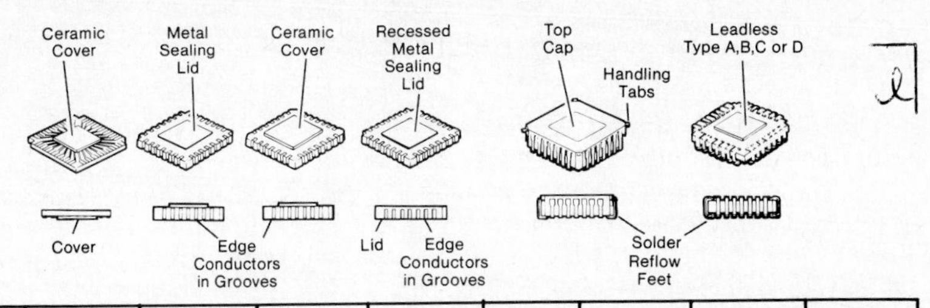

Type	Leadless Type A	Leadless Type B	Leadless Type C	Leadless Type D	Leaded Type A	Leaded Type A	Leaded Type B	0.040″ Spacing (Mil)
Designation	MS002	MS003	MS004	MS005	MS006	MS007	MS008	MS009
No. of Terminals Available	28, 44, 52, 68, 84, 100, 124, 156	28, 44, 52, 68, 84, 100, 124, 156	16, 20, 28, 44, 52, 68, 84	28, 44, 52, 68, 84, 100, 124, 156	24	28, 44, 52, 68, 84, 100, 124,156	28, 44, 52, 68, 84, 100, 124, 156	16, 20, 24, 32, 40, 48, 64, 84, 96

NOTE: 0.050-inch (1.27-mm) spacing unless otherwise specified.

FIG. 8.15 JEDEC outlines for chip carriers.

FIG. 8.16 Leadless type A chip carrier without cover.

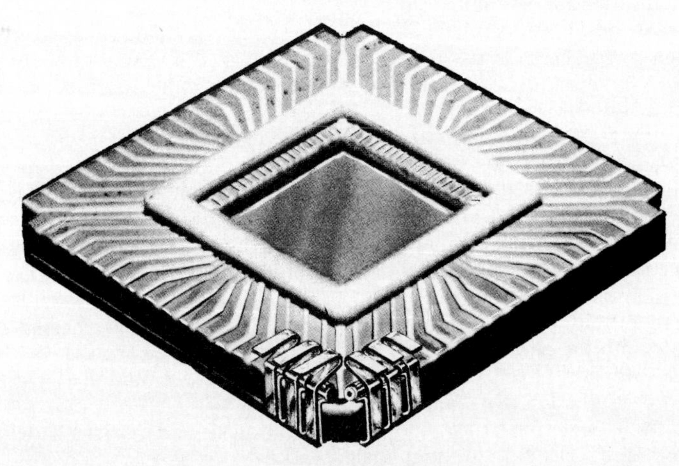

FIG. 8.17 Leaded type A chip carrier with clips.

The differences among leadless-type CCs center on internal features and mounting orientations with respect to substrate and thermal performance. The latter primarily concerns designers who can choose a CC based on its cavity-up or cavity-down thermal orientation. These terms define the cavity position and the back side of the die, which serves as the package's primary heat-transfer or heat-removal surface.

The cavity-up configuration places the heat-radiating surface down, next to the mounting surface. This arrangement limits the heat-transfer effectiveness of the forced-air-cooled systems and the attachment of heat sinks, but it does allow for a larger cavity size, and thus a larger die.

The cavity-down orientation allows heat sinks to mount on the primary heat-radiating surface but, as noted, results in a smaller cavity size. Designers must make the trade-off—thermal performance and die orientation versus cavity size—for each application.

Although most of the industry is concentrating heavily on the 0.050-in-center CCs, the 0.040-in-center devices were the family's forerunners and are part of the JEDEC standards. The 0.040-in-center CCs have been and continue to be applied mainly in military programs, where the need for the highest possible packaging density rates high on the requirements list. Thus, specifically for military-oriented users, the 0.040-in-center spacing device was defined and registered and carries JEDEC's standardization number MS009.

Turning again to the 0.050-in-center family, note that all of these devices mount on a common pattern. It is therefore possible to attach them directly to the mounting substrate or put them in a socket.

The 0.050-in leadless type A package, when used in the cavity-down configuration, must be socketed on both ceramic and printed-wiring substrates because of its package lid. The other leadless types, however, offer the choice of socketing or soldering on ceramic substrates. The leaded type A, on the other hand, is unusual in that it can be soldered or socketed directly onto either mounting substrate, as opposed to the leaded type B, which requires soldering because its lead configuration does not permit socketing.

Although the question of soldering or socketing might present some trade-off considerations, at least there is little concern about the number of pinouts available. There is a variety of sizes in the 0.050-in-center family, including 28-, 44-, 52-, 68-, 84-, 100-, 124-, and 156-I/O terminal sizes (Table 8.1). The family also includes smaller terminal counts, and the 0.040-in-center family furnishes smaller sizes and additional terminal counts.

Note that the CC concept can apply to multichip hybrid circuitry as well as to monolithic dice. The CC provides an alternate approach to high-density designs that hybrids have traditionally serviced. A CC hybrid can be assembled using conventional reflow-soldering techniques, and because the CCs attach to the substrate by this method, the rework operation basically consists of a heat, remove, replace, and reflow operation. Thus, the temperature excursion experienced by the CC hybrid is limited to the reflow temperature of the solder, which minimizes temperature-induced damage to the circuit.

Several semiconductor companies have submitted outlines for a package that is identified as leadless type E. This rectangular configuration is specifically targeted for memory applications.

The two pinout versions proposed for the type E carrier are designed to be compatible with the popular leadless type C package. The 28-pin 0.350- × 0.550-in outline and the 32-pin 0.550- × 0.450-in outline are intended for ROMs, PROMs, EPROMs, DRAMs, and static RAMs.

While there are numerous types of CC packages and packaged devices, it should be noted that there are two basic categories—plastic and ceramic. These basic material categories have substantially different properties.

TABLE 8.1 JEDEC Chip Carrier Sizes

Type[a]	Spacing, in	Number of leads	Mounting
Leadless type A[b]	0.050	28, 44, 52, 68, 84, 100, 124, 156	Socket only/PC
Leadless type B[b]	0.050	28, 44, 52, 68, 84, 100, 124, 156	Socket or solder/PC
Leadless type C[b]	0.050	16, 20, 28, 44, 52, 68, 84	Socket or solder/PC
Leadless type D[b]	0.050	28, 44, 52, 68, 84, 100, 124, 156	Socket or solder/PC
Leaded type A[c]	0.050	24[d]	Socket or solder/PC
Leaded type A[c]	0.050	28, 44, 52, 68, 84, 100, 124, 156	Socket or solder/PC
Leaded type B[e]	0.050	28, 44, 52, 68, 84, 100, 124, 156	Socket or solder/PC
Leadless	0.040[f]	40, 48, 64, 84, 96	Solder/ceramic

[a]All the above are square—memory chip carriers are to be standardized in a separate family.
[b]*X-Y* dimensions for given lead count are compatible.
[c]Primarily plastic units.
[d]Special package manufactured by AMP (only).
[e]*X-Y* dimensions as *b*. Leads may be crimped on after testing.
[f]The 0.040-in package approved in one outline only—*X-Y* dimension not compatible with other carriers.

In addition, there are several critical electrical properties of CC packages based both on materials and on the configurations of CC packages, as compared with DIPs, for instance. These include electrical loss properties and electromigration. Thermal expansion is the major mechanical property to be considered.

8.3.3 Small Outline Packages[3,4]

With all the recognition being given to CCs, another entry in the microminiature packaging arena, the small outline (SO) package, has found its applications. Resembling miniature versions of molded-plastic DIPs, small outline transistors (SOTs) and small outline integrated circuits (SOICs) are quickly gaining acceptance at several major semiconductor companies.

The primary advantage of the SO package is its small size. The SO package occupies an area some 30 to 50 percent less than an equivalent DIP, with a typical thickness that is 70 percent less. Consequently, assemblies that incorporate SO devices can serve applications where space and weight are at a premium.

8.3.4 Discrete Chip Components[5]

A discrete chip component is nothing more than a miniature axial-leaded component minus the leads. They are available in both cylindrical and flat rectangular shapes (Fig. 8.18).

Some of the factors promoting interest in chip components are as follows:

1. Manufacturers are successfully using CCs in high quantities.
2. Insertion equipment manufacturers are claiming increasingly higher insertion rates.
3. Increased component density and mounting techniques are compatible with CC technology.

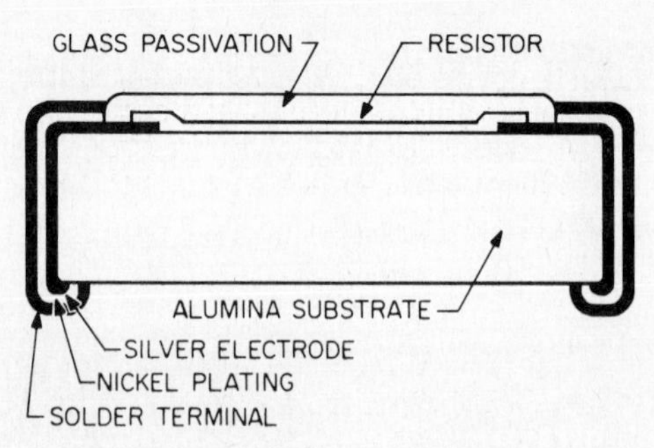

FIG. 8.18 Flat rectangular chip resistor.

4. Chips often sell in a price range that is comparable to or better than that for axial-leaded components.
5. Since there are no leads, basic circuit reliability is improved.
6. Chip components are ideal for high-frequency applications, since there are no leads.

8.4 HARDWARE

For obvious reasons, circuit components and interconnection devices perform a vital function. Often the function of support hardware (e.g., board mounting, component mounting) is not as obvious, but it is almost as vital in determining the effectiveness of the application of the printed board assembly.

8.4.1 Board-Mounting Hardware

A wide variety of hardware exists for mounting the printed board assembly in the end-product equipment; the range is from complete packaging systems to indi-

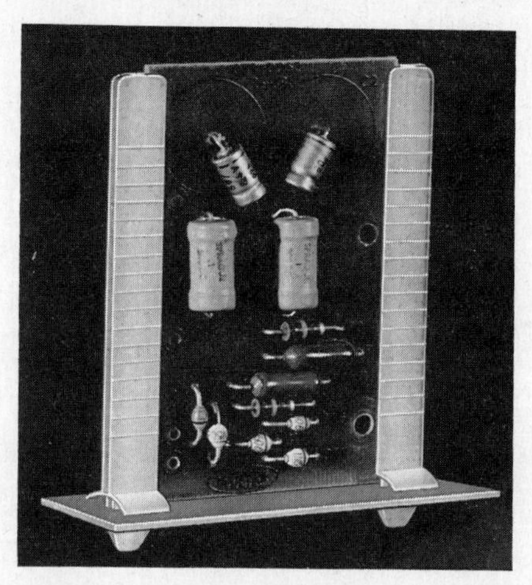

FIG. 8.19 Example of individual board hardware. *(Richo Plastics, Chicago, Ill.)*

vidual board hardware (Fig. 8.19). Between the extremes, there is hardware for locally mounting a group of assemblies, such as the stacking and spacing hardware shown in Fig. 8.20. The exact hardware to be used for a specific application depends on factors too numerous to mention here, but the broad selection that is available assures success in finding the proper board-mounting hardware for nearly any possible requirement.

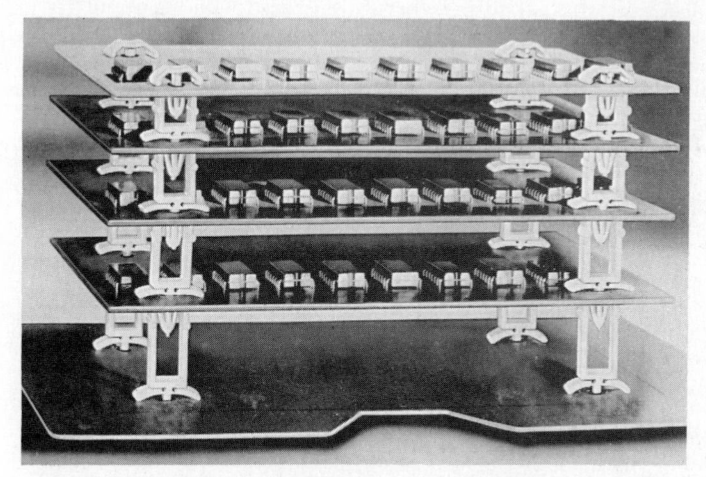

FIG. 8.20 Stacking and spacing hardware. *(Richo Plastics, Chicago, Ill.)*

8.4.2 Component Mounting

The shock and vibration to which printed wiring mounted components are subjected during normal handling and environmental testing can damage the lead terminations and lead-to-component body seals. For that reason, many printed wiring mounted components, especially those weighing more than $\frac{1}{4}$ oz per lead, should be mechanically secured to the mounting base prior to lead termination and during assembly. The more commonly used component-securing devices are (1) clips, clamps, and brackets; (2) wires and elastic straps; and (3) adhesives.

8.4.2.1 Clips, Clamps, and Brackets. The following are the basic requirements which should be adhered to when components are mechanically secured by clips, clamps, or brackets.

1. All clips, clamps, or brackets should be secured to prevent their rotation, as by using two fasteners or one fastener and a nonturn device. Holders designed for single-hole mounting should be capable of withstanding a 14-in·oz torque without rotating.
2. Clamps and brackets which require removal for component replacement should be secured with a threaded or other nonpermanent fastener unless the subassembly in which they are used is considered to be disposable or nonrepairable.
3. Spring clips which need not be removed during component replacement may be secured with permanent fasteners such as rivets or eyelets.

4. Spring clips should require their positive displacement for the component to be removed.
5. The use of twist-type lugs, tabs, or ears and the clipping of glass envelope components should be avoided.

*8.4.2.2 **Straps.*** When an elastic strap is used for mechanical securing, the strap is wrapped over the component body and passed through holes in the mounting base. When wire is used, it is clinched and soldered in the same manner as component leads to terminal areas. When wire is used with heat-sensitive or fragile components, the part of the wire that touches the component should be covered with a suitable sleeving.

Smaller holes are used with elastic straps than are used with wire. The elastic strap is secured by being stretched to reduce its cross section below that of the hole, and it is returned to its larger-than-hole size by relieving the tension after it has been passed through the hole. The resiliency of the strap holds the component in place.

*8.4.2.3 **Adhesives.*** Whenever possible, components should be secured by conventional means; when that is not possible, as in the case of oddly shaped components or when special support is required or when there are special design requirements (limited space, heat transfer, limited access, etc.), a suitable adhesive may be used.

8.4.3 Heat Sinks[6]

The increasing density of components on printed board assemblies and the use of higher-power components often necessitates the use of board-mounted heat sinks (Fig. 8.21). A heat sink must be considered as a complete thermal system. It consists of four elements: mounting hardware, interfacing materials to bring the mounting surface of the component into intimate contact with the surface of the sink, a surface coating of the metal heat dissipater, and the copper or aluminum

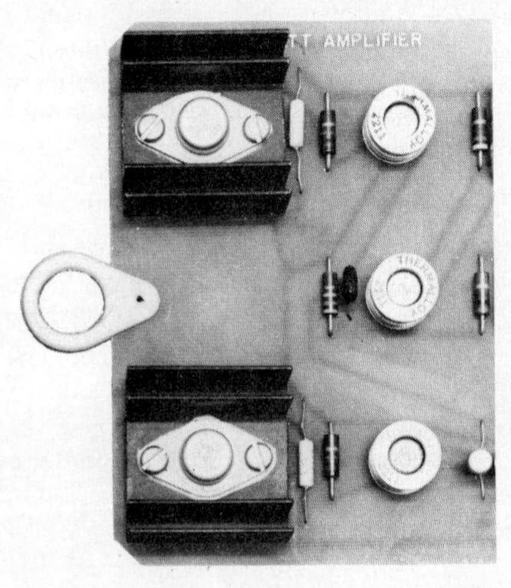

FIG. 8.21 Board-mounted heat sink. *(Thermalloy Inc., Dallas, Tex.)*

thermal dissipater itself. Each element affects overall thermal efficiency, cost, and the frequency of maintenance of the heat sink systems.

Heat sinks range in complexity from a simple, flat metal plate to which a heat-producing component is bolted to a system using thermal feedback sensors to control liquid coolant flowing through tubes surrounding the part. The difference between such thermal dissipaters and the machined metal plate with fingers reaching into and cooled by the ambient atmosphere is that the latter is a passive system that protects a component from thermal destruction at the lowest economically feasible temperature.

For practical purposes, heat sink design must be a compromise. The ideal heat dissipater would encase the device to be protected in a spherical metal mass consisting entirely of black, needlelike protrusions radiating directly from the thermal point source. That would provide the most efficient radiating area for a given volume. Although the ideal shape could not be mounted to any flat surface, practical coolers borrow the radial-fin structure from it. When heat sinks of equal volume are compared, the radial-fin coolers are more efficient radiators than are coolers with a series of parallel fins machined or soldered perpendicularly to the surface of the sink. The trade-offs in thermal efficiency, based on shape, are cost, ease of assembly, and installation.

8.4.4 Test Points

A printed board component which, unlike most others, does not perform an electrical circuit function is the test point. Test points do serve an important maintainability function. This type of printed wiring component can be of the right-angle board-edge type, the straight-through type for local access, or the combination type.

REFERENCES

1. *Printed Wiring Design Guide,* Institute for Interconnecting and Packaging Electronic Circuits, Evanston, Ill.
2. B. M. Hargis and D. J. Westervelt, "High Lead Count Packaging and Interconnection Alternatives," *Electronic Packaging and Production,* September 1981, pp. 82–89.
3. J. Tsantes, "Technology Update: Leadless Chip Carriers Revolutionize IC Packaging," *EDN,* May 27, 1981, pp. 49–74.
4. J. Resendes, "Production Techniques with S. O. Packages," *Electronic Packaging and Production,* March 1982, pp. 115–128.
5. B. Winkelmann, "The New Chip Resistors," *Appliance,* June 1981.
6. David Hegarty, "Treat the Heat Sink as a System," *Electronic Products,* United Technical Publications, Garden City, N.Y., Jan. 21, 1974.

CHAPTER 9
ACQUISITION AND SPECIFICATIONS[1]

Gerald L. Ginsberg
President, Component Data Associates, Inc., Lafayette Hill, Pennsylvania

9.1 INTRODUCTION

The procurement of printed wiring (printed circuit) boards requires that three basic items be understood and accomplished prior to the start of purchasing activity:

1. Knowledge of the environment to which the boards will be subjected.
2. Proper evaluation of the capabilities of the board fabricator.
3. Appropriate documentation which clearly defines the requirements of the product. When items 1 and 2 have been properly assimilated, so that the documentation is complete and does not overspecify or underspecify the printed wiring product, the vendors who have been deemed capable of fabricating the particular board may be contacted to obtain competitive pricing.

Printed board products run the gamut from the very simple single-sided variety to sophisticated multilayered versions. Within each category, there are many variations. A board whose end-product use is to provide circuitry for a home entertainment product does not require the critical handling, testing, or protection that may be necessary for a board of similar complexity that will see service on a military ocean-going vessel. End-product service considerations make the difference in the cost of the printed board. Too many times, vendor-user relations are disappointing to both parties, either because specifications for a board are too stringent and the cost is higher than expected, or because the specifications are too loose and the product does not perform as required. Either failure has a detrimental effect on vendor-user relations.

Just as many types of boards are required by the industry, so there are many types of fabricators. Not every fabricator is capable of producing satisfactory results. That is not meant to imply that the capability scale goes in only one direction and the high-quality house is capable of making all of the various types of printed boards. The opposite is usually true. In a fabrication facility that is geared to producing a high-reliability, high-quality product, it is sometimes difficult to compete in making a board that is less complex.

With all those considerations to be taken into account, a company that requires outside services for the fabrication of printed wiring products must have knowledgeable personnel assigned to the procurement of the products.

9.1.1 Procurement Practices[2]

Historically, the printed circuit has been most at home in the high-volume production field. There the economy and reproducibility permitted by the automatic and semiautomatic methods are dominant factors. That leaves a great many military and industrial projects on the fringes. Although the crossover point varies with board design and specifications, there are printed circuit vendors who have developed some unique approaches to serve the user who needs only half a dozen prototype boards, or perhaps even 500 units for a small project.

9.1.2 Prototype Kits

Starting with a single unit for a prototype or preprototype operation, several firms offer kits that can be used to make circuit boards from specially prepared materials. In addition to the actual material required, the kit normally includes several full-scale layout sheets to aid in designing the board.

As might be expected, the kits reflect several approaches. For example, one company offers its kit in several sizes and a choice of laminate. Conductor runs are applied in the form of tape. Mounting holes are drilled where required. A different approach is one in which the boards contain holes preformed on a 0.100-in grid. The holes, and in fact the whole board, are plated. The desired circuit is either drawn with etch resist or covered with tape and board-etched.

These do-it-yourself boards can be very helpful, but most prototype and limited-production circuits do not lend themselves to the approach. To meet the limited-volume need, most manufacturers have established separate short-run and prototype departments or ordering procedures. Nearly all vendors acknowledge that some change in procedure must be made for the limited-volume order. Whether a separate department or simply expedited handling and simplified paperwork is the best choice is something that the buyer must determine by contact with individual suppliers.

9.1.3 Limited Production Quantities

An innovation in the limited-production area is a semistandard program to reduce the cost. The program is aimed at the 100- to 1000-board market. The quantity range offers some serious challenges in that it is too high for the production techniques often used for short runs but does not justify the cost of fixed tooling. The approach is to limit the selection of mechanical variations while offering full freedom of conductor and hole pattern design. Since, in most cases, that provides the designer with all the freedom needed, the required circuitry can be purchased in limited quantities at reduced price and with quick delivery.

9.2 PURCHASING GUIDELINES

Printed wiring poses some special problems because it is unlike the usual electronic component. Although its function is primarily electronic, testing must be largely mechanical.

9.2.1 Evaluation of Supplier

In view of the complications, buying short runs by competitive bid may be overly expensive. More practical may be the procedure of selecting a vendor or vendors (depending on volume) to handle all orders for a fixed period. That will permit both a regular review of performance and good buying without adding excessive purchasing costs to each small order. It should be noted that the boards should be of a similar type.

Since even with a rigorous test program a vendor's quality performance is hard to ensure (thorough tests are destructive), vendors should be selected with great care. Printed circuits would seem to be an area in which the purchasing evaluation trip is particularly important. A plant visit permits the buyer and other members of the buying team to inspect the facilities and meet the people who will be responsible for the quality.

Buying printed circuits does not require a radically different approach but does demand good procurement practices. The best buying tip for users is to purchase only what is required in design tolerances and parameters. Demands for tight tolerances, much beyond operating requirements, are unnecessary and only add to costs.

Many designers and buyers seem to forget that printed circuits are primarily electrical components rather than mechanical parts and that the reliability of their electrical function is more important than extremely close mechanical tolerances. It is much harder to specify the processing and cleaning procedures necessary to produce printed circuits of high reliability. Since some of the conditions that may lead to unreliability may not show up until long after the circuit is in use, it is advisable to consider only manufacturers with a long record and a reputation for the production of high-quality boards. Buying on price alone can be extremely expensive in this field.

9.2.2 The Make-or-Buy Question

The make-or-buy question has been a hard one since printed wiring became an industrial reality. It is natural that many equipment manufacturers favor making, since the printed board replaces hand wiring done in their plant and so is unlike the traditionally purchased component. Many major firms currently supply their own requirements of printed boards, at least in part. Quite a number of others, however, either have never made their own boards or have stopped making them. The economics of board manufacturing are complex, and there is no single best solution for all users.

To evaluate this problem better, several printed circuit manufacturers were asked to relate the major factors in the make-or-buy decision. Manufacturers are obviously biased in favor of a buy decision, but the comments are pertinent to making a sound evaluation.

The major make-or-buy considerations are price, delivery, and quality. Unless an internal facility is set up to serve the needs of many divisions of a large electronic complex, there will be serious problems in meeting fluctuating requirements. If the internal facility is planned to meet the average yearly requirements anticipated, there will be serious delivery delays during periods of peak requirements and underutilization of equipment and personnel during slack periods.

In a make-or-buy study, shop-loading figures for six captive facilities were compared with those of an independent fabricator for a similar period. The ratio of minimum requirements to maximum requirements for the captives averaged 1:5. The independent fabricator's ratio was 1:12. That verifies the suspicion that many captive facilities find it necessary to solicit outside business during slack seasons to remain economically healthy, only to find that they can't compete with the independent printed wiring manufacturer in quality or service.

The technical requirements brought about by advanced design and construction place a severe strain on the capital equipment budget of the new entrant into the field. The trend toward more complex boards and more specialized techniques makes it less attractive for companies to consider making their own circuitry. The days when a silk screen and an etching tank were enough to start a printed circuit department have long since passed.

Capital equipment such as drilling machinery and plating controls is often obsolete in two or three years and must be amortized in a relatively short period of time. Those outside the industry are often unaware of that situation and plan on a payback period that either cannot be met or is met only by continuing to use outdated, inefficient equipment at increasingly uncompetitive costs. The printed circuit manufacturer with a larger volume and a stable shop load can afford the shorter amortization period and still realize a profit.

Board users faced with the make-or-buy decision must be able to determine their own costs accurately, because they are often deluded by understated overhead costs. Then unrealistic cost figures are arrived at, and these become the basis for their decisions.

Quality is a major reason why many companies go into competition with their suppliers. In printed circuits, however, most firms are hard pressed to match the independent shop. The basic reasons are closely related to the causes of the price and delivery problem. The captive is usually smaller and less efficient than the specialist. Because of those factors, neither equipment nor skills are the best available. Research on printed circuitry is the prerogative of the specialists and, of course, the large captive operations.

A more subtle attack on quality can affect even the largest captive shop unless it serves a very diversified company, and that is technical isolation. Within an operation, there usually is a tendency to employ as few basic packaging techniques as possible. Although that is often good economy, it can be very dangerous unless there is constant "cross-pollination" with other operations. The independent circuit board producer provides an information exchange link. In the procurement of printed circuits, as of other components, state-of-the-art information is one of the most important services your vendors can provide.

9.2.3 The Purchase Order and Legal Ramifications[3]

It is neglect of duty of the worst sort to issue a purchase order for a printed board without (1) carefully investigating and evaluating suppliers or (2) having meetings with the prospective vendor to discuss artwork requirements, delivery, tooling, tolerances, and performance standards.

When it becomes necessary for a court to construe the meaning of a word employed in the purchase order, the court refers first to a standard dictionary. Words used in connection with a particular trade, however, are given the meaning attached to them by experts in that trade. A trade usage or custom that is known to the parties and prevails in the community where the contract is to be executed and performed is incorporated into the agreement by force of law.

If there is a variance between the common standard English meaning and the legal significance of a word, the legal significance is adopted. When the words or terms used in the purchase order have a definite legal meaning, the parties will be presumed to have intended such words or terms to be given the effect that has been established by the courts, unless a contrary intention appears in the instrument.

The law even permits an implication as to price, if a price is not stated in the contract. If, through inadvertence, no price is established in the contract, the courts infer that a reasonable value was the price understood by the parties.

The implications contained in a contract, even though not expressed in the writing, include an obligation on the part of both parties to avoid any act which will make it impossible for the other party to perform.

If a question of interpretation of a purchase order is given to a court, the court will consider the whole instrument, if necessary, to reach a proper interpretation. A contract should be interpreted as a whole, and a meaning should be gathered from the entire context and not from scattered or isolated words, phrases, or clauses. It is the spirit rather than the letter of the contract which must control the interpretation. It is also the purpose rather than the name given to it by the parties which gives the contract its real meaning. Calling a purchase order a contract for sale on "trial or approval" does not make it so if it is, in fact, an absolute contract for sale. Calling the agreement a lease does not make it so if the legal effect of the words employed is to make it an outright sale.

After the general purposes are ascertained, the language employed will be construed to serve those general purposes and not destroy them.

9.3 VENDOR-USER RELATIONS

Buying boards is similar, but not the same, when the boards are single-sided, double-sided, or multilayered. Admittedly, the standard advice in any complex buy situation is to evaluate the source. In the case of multilayer boards, that advice cannot be ignored. Thorough vendor-user communications are imperative in order to have a quality buy.

Multilayer boards can be of great value, but one should not evaluate the state of the art by buying the cheapest board available and then deciding that it is the best that industry can provide. The small manufacturer of single- and double-sided boards cannot convert to producing multilayer boards without establishing an elaborate in-process inspection and control procedure. Microsectioning and inspection are called for after several of the many manufacturing steps in the multilayer fabrication process. Multilayer boards can cost between 4 and 8 times more than conventional two-sided boards having the same number of holes. Still, multilayer boards pay for themselves when they are the answer to complex interconnection problems.

As has been stated previously and will be reiterated throughout this chapter, a positive vendor-user relation is the most important item in purchasing satisfactory printed wiring products.

9.3.1 Multilayer Specification Items

In preparing specifications for the fabrication of printed boards, many items should be considered. They relate to the end-product requirements and the manner in which the end product is derived, i.e., who is responsible for what. The list in Sec. 9.3.2 indicates the items that are usually reviewed before going into a multilayer fabrication cycle. Some of the items are also pertinent to the fabrication of single- and double-sided boards. The conclusions that are reached in reviewing this list have a direct impact on the cost of the individual printed board.

9.3.2 Design Trade-Offs

Design trade-offs should be considered early in the cycle to optimize the items that have the most significant cost impact. The list should be reviewed with the particular board fabricator to ascertain the items that are the most cost-sensitive in the manufacturing processes:

1. Conductor thickness per layer
2. Conductor width (specify minimum when required)
3. Spacings (specify minimum when required)
4. Tolerances plus or minus (state all that are not standard)
5. Dielectric material to be used
6. Thickness of dielectric between layers
7. Overall laminate thickness with tolerances
8. Number of holes
9. Hole size
10. Hole tolerance
11. Hole location preferable on a grid using a 0.025-in base
12. Actual overall dimensions
13. Number of layers
14. Supplier of artwork
15. Quantity of boards required
16. Thickness of plating in hole (state minimum)
17. Other plating
18. Applicable military or industry specifications
19. Other applicable specifications
20. Type of testing
    - *a.* Bond (copper to laminate)
    - *b.* Cleavage (laminate to itself)
    - *c.* Dip solder
    - *d.* Hot oil
    - *e.* Dielectric constant and dissipation factor
    - *f.* Dielectric strength
    - *g.* Capacitance
    - *h.* Impedance
    - *i.* Vibration
    - *j.* Temperature cycling
    - *k.* Altitude
    - *l.* Humidity cycling
    - *m.* Insulation resistance
    - *n.* Flammability

9.4 STEPS IN BUYING SATISFACTORY BOARDS[4]

The quick guide to what can go wrong and the avoidance of such problems, as listed later in this section, is the compilation of inputs from a number of suppliers and users. Careful consideration of this information should save much anxiety and anguish. The points covered take on added meaning when viewed in the light of a total program necessary to use single-sided, double-sided, or multilayer boards successfully. Such a program would include the following.

9.4.1 The Total Program

1. Establish specifications consistent with design standards. Do not assume that normal printed wiring specifications are adequate or suitable. Question each vendor as to what is considered typical, normal, or within reason. Do not over-specify unless you are prepared to relax other requirements.

2. Establish basic board design; use all possible means to reduce the quantity of expensive tooling required.

3. Establish a performance specification to allow complete evaluation of thermal shock, dielectric breakdown, temperature cycling, solder resistance, warpage, etc.

4. Establish or adopt a standard test pattern such as that currently in use by the IPC for multilayer boards.

5. Evaluate the vendor's samples. Agree with the vendor on realistic reliability-testing procedures.

6. Make microscopic examinations of encapsulated microsections of sample holes from test strips.

7. Produce artwork masters at the precision required for the product. For multilayer boards, attempt to achieve as near to perfection as possible.

8. Consult with knowledgeable vendors to avoid built-in trouble areas. Include a test pattern on the master to allow future testing of each board produced.

9. Give vendors an accurate picture of present and future requirements and indicate possible areas of change. Cooperation in this area will decrease the number of tooling changes and allow better tooling to be produced and thereby greatly affect the price per card.

10. Establish on-the-line training in the handling, soldering, and assembly of boards to avoid needless damage.

11. Force the development of good specifications and design guides. "Reproducibility" and "application" are the key words.

12. Ascertain vendor qualifications to produce to applicable specifications.

13. Develop good contacts between vendor and user. In-house contacts must be technically aware of all facets of the board.

14. Make sure that all vendor clarification calls get "quick reaction disposition."

15. Establish firm tooling control.

16. Do not allow engineers or any others except those responsible for technical vendor contact (designated by buyer) to communicate with the vendor. Also, encourage vendors to establish contacts in their houses.

17. Allow adequate time in the procurement cycle for a good job. Although delivery can be forced, the product cannot be made reliably under pressure.

9.4.2 What Can Go Wrong?

Things that most commonly go wrong are as follows:

1. Layer-to-layer connection discontinuity
2. Opening between conductor and through-hole connection
3. Delamination
4. Measling

Other problems are:

1. Inadequate customer-furnished artwork (registration and tolerances cannot be held)
2. Inadequate plating in hole (high resistance)
3. Poor drilling that creates high resistance in some areas
4. Weak design and inefficient layout (excessive layers or fine-line circuitry)
5. Cramming of too much circuitry into too little space
6. Unrealistic tolerances

9.4.3 How to Avoid Problems

1. Evaluate the vendor. Pick an experienced vendor.

2. Follow recommendations. To prevent delamination or measling (blistering) during soldering, follow the manufacturer's recommendation on temperature and soldering time. Boards stored before soldering ought to be dried 2 h at 250°F.

3. Use quality control. To avoid getting boards with discontinuities due to poor plated-through holes, make sure the manufacturer has stringent process controls. Inspections should be made at many steps. Also, severe heat shocks should be avoided.

4. Test. Do test preproduction samples and test strips. Make microscopic examination of encapsuled microsections. (Use IPC test patterns or coupons.)

5. Establish vendor-user understanding. Make sure that there is complete communication between vendor and user and that specifications are thoroughly understood. Do not assume that normal board specifications are adequate or suitable. Performance specifications should allow complete evaluation of thermal shock, dielectrical breakdown, temperature cycling, solder resistance, warpage, etc.

6. Design correctly. Question several vendors to make sure that design and circuit layout take full advantage of capabilities, without going too far. Ask what is typical, normal, or within reason. Have only experienced people—perhaps the vendor—do artwork. Masters should be the proper size for the tolerance capability [1:1 for precision numerically controlled (NC) artwork generations, and 4:1 or more for manually prepared artwork].

9.5 COST FACTORS

When ordering printed boards, conventional or multilayer, remember that the quoted price is based heavily on the number of boards ordered, since it is high-quantity processing that gives good economy. That means the manufacturer will process all boards at the same time. One should not expect deliveries spread out over a period of time. For example, do not order 1000 pieces and then expect the vendor to ship them at a rate of 250 every three months just to keep in step with internal equipment production schedules.

Tight tolerances mean high prices, too, because the manufacturer cannot avoid extra problems in meeting the specs. Price depends heavily on yield at a particular tolerance level. With tight tolerances, yields must decrease. Price must therefore go up. Watch the platings. Using gold or other noble platings where solder would do and using solder where no plating at all would do can greatly increase the price. Solder plate not only is cheaper than gold but also leads to a greater processing yield.

All these points can be summed up with one statement that applies to all buying: Don't overspecify in terms of tolerances, materials, or configurations, and try to standardize board formats.

Do not forget the cost of testing. Try to ascertain the vendor's capability to produce and deliver acceptable boards. The costs of inspection, return procedures, and production delays often can outweigh the money saved by choosing the lowest bidder.

9.5.1 Pricing of One- and Two-Sided Boards[5]

Simple one- and two-sided boards are fairly well understood by users, except for one factor: price. Since each job is special, it has often been difficult to compare boards parameter by parameter. With standard price lists, however, one is able to compare price changes caused by changing single parameters. By using standard price indexes, one can make the following comparisons based on a board with these characteristics:

Board size: 4 × 6 in

Base material: $\frac{1}{16}$-in one-sided epoxy-paper

Board finish: Solder plating

Number of holes: 80

Number of different hole sizes: 4

Board outline: Rectangular with rounded corners

Initial quantity required: 10

Legend: Printed

The standard price index gives a unit price of x each for 10 boards with the above characteristics. If the initial order is increased to 100, the unit price drops to 25 percent of the price for 10; if the order is increased to 500 pieces, the price is 15 percent of the price for 10. This illustrates the tremendous economy of ordering in quantity. The prices are only theoretical and are subject to many changes due to inflation and the labor-machine mixture in the various companies, but the relations between a small and a quantity purchase serve the intended example.

Suppose one tries to reduce the board size in an attempt to reduce the price. A 3 × 4.5-in board would reduce the material cost per board and allow the manufacturer to process more circuits on one panel. The prices would then drop to 92 percent of unit price *x* each for 10 boards, to 24.5 percent for 100, and to 12.3 percent for 500. With an order of 100 boards, the total price would thus drop by only 0.5 percent of unit price *x*, which would not be worth the extra effort, but the 500-piece order would price out at savings of 2.7 percent of unit price *x*. That would be an economy.

If G-10 instead of epoxy-paper is used, the prices increase, but the offsetting gains are the superior mechanical and electrical characteristics of G-10. If XXXP, or FR-2, is used, the prices go down.

The original specification called for solder plating, apparently because a longer shelf life was needed. If there is a switch to simple etched copper with a protective water-dip lacquer, the prices can be decreased. The savings with 500 boards can be 22 percent.

If nickel-gold is used for longer shelf life, corrosion resistance, and low contact resistance, then the prices go up. The premium here is thus about 42 percent for 500 boards.

Now suppose that the number of different hole sizes is reduced from four to two and the number of drilling steps is thereby reduced. The prices drop.

Finally, if the rounded edges are considered not really necessary, but are included only for aesthetic reasons, the price for the 500 quantity can be reduced to save about 3.6 percent. Putting some of the savings together will revise the characteristics as follows:

Board size: 6 × 8 in (only half as many are needed)

Base material: XXXP

Board finish: Etched copper

Number of different hole sizes: 2

Board outline: Square corners

Legend: Enough space available—etched

With the new characteristics, the board prices at the 500 level drop to a point that results in a total saving of 53 percent.

9.5.2 Pricing of Multilayer Boards

To get an estimate of the relative costs of multilayer versus two-sided boards, several manufacturers were asked to give the cost factor multiplier for a multilayer board with twice, 4 times, and 8 times the conductor density of the two-sided boards for a typical quantity of 100 boards. One must remember, however, when making such comparisons that the costs saved in connectors, assembly, and handling will have a major effect on the overall cost and thus make the multilayer approach attractive. The increased reliability also should be considered.

The most common response was that if the conductor density was twice as great, the cost of the multilayer board would be three times as great because of the unavoidable extra work in setting up the multilayer boards with the necessary tolerances to assure acceptable registration. With 4 times the conductor density, the cost goes up to 5 times as great for the same reasons as in the first case. With 8 times the conductor density, the cost goes up to 8 times as much for the multilayer compared with the two-sided board. At that point, the extra multilayer costs are beginning to be absorbed, and the cost per unit density is decreasing.

One respondent gave a rough rule of thumb: Take the cost of a two-sided board and multiply it by the costs of the layers and add a 20 percent factor for bonding and checkout. Thus for a two-sided board of unit cost y, for four layers we get $4y$ and $0175y$ extra for bonding, giving $4.75y$ per multilayer board. One factor that can be attached to the economy of multilayer is the cost of connecting each integrated circuit package. A rough way to calculate costs for quantities of about 100 boards of a type is to take between $1z$ and $1.5z$ per simple integrated circuit package, where z is the base price. For large-scale integration, the price would be about $2z$ to $3z$ per IC. Again, those dollar estimates are general and are based on the standard price index established above.

Another factor that affects the price of multilayer boards is the number of layers. Many vendors can fabricate boards in excess of 20 layers, but most vendors and users prefer to stay around eight layers per board. A limit in the costing is the ratio of the final board thickness to the hole diameter. When the ratio exceeds 4:1, the plating in the hole begins to suffer, and special plating techniques which increase cost are required.

One manufacturer concluded that the density of terminal points or holes has a much greater effect on cost than the overall size, number of layers, or complexity of the interconnections has. Costs go up sharply when hole spacings drop below 0.100-in centers. The concept is that the entire tolerance structure for a board with closely spaced holes must be much more sophisticated. The price for the added precision is, of course, added cost.

9.5.3 Delivery Schedules

How long does it take to get delivery on single-sided, double-sided, and multilayer boards in large quantities? The answer to that basic buying question depends mainly on whether the buyer supplies the manufacturer with usable artwork. Minimum delivery time from the date artwork is finally approved is often 1 to 2 weeks for single-sided boards, $1\frac{1}{2}$ to 3 weeks for double-sided boards, and 4 to 6 weeks for multilayer boards. That is the time needed for the actual production of the boards. Artwork could take from 1 to 3 more weeks for production and checkout, which would result in a complete delivery cycle of 10 weeks or more for a sophisticated multilayer board.

Some users are not as skilled in laying out multilayer boards as they are in laying out single- and double-sided boards. In the cases of the latter, the manufacturer will prepare the artwork from sketches and wiring tables. The increased use of NC artwork generators is making that characteristic of the industry disappear.

Changes in wiring can have a profound effect on the circuit layout and artwork and can push delivery dates further back. Be sure that the circuit is complete before starting into the fabrication cycle.

9.6 PRODUCT DEFINITION[6]

Basic documentation requirements for printed boards are not unrealistic for even the smallest manufacturer to adopt as a basis for in-house standards. Definition of the printed board product in documented form is still one of the most misunderstood aspects of the industry. The misunderstanding comes about because so many companies are concerned about the high cost of preparing documentation that they attempt to take shortcuts in defining the items that assure them an acceptable product.

9.6.1 Documentation

Documentation may be very variable as to the quality of the actual recorded information. There are also many techniques for preparing the document that defines the end-product board. One document, however, is the nucleus of all the others, and that is the master drawing. In the IPC-T-50 Terms and Definitions, the master drawing is defined as follows:

> A document that shows the dimensional limits or grid locations applicable to any or all parts of a printed wiring (circuit) board (rigid or flexible), including the arrangement of conductive and nonconductive patterns or elements; size, type, and location of holes, and any other information necessary to describe the product to be fabricated.

Figure 9.1 shows a simplified flowchart of how the master drawing fits into the total product development cycle.

Printed wiring documentation can be divided into three basic groups:

1. *Minimum documentation:* Used for prototype and small-quantity runs.
2. *Formal documentation:* Used for a standard product line and boards built in production quantities.
3. *Military documentation (per MIL-D-1000):* Complies with government contracts specifying procurement drawings for the manufacture of identical items by anyone other than the original manufacturer.

To save time on the master drawing in the minimum documentation phase, a sepia print of the noncomponent (circuit) side artwork can be used. It should indicate board material, finish plating, outline dimensions (if no outline drawing is referenced in the parts list), and drilling data. Notes can be lettered freehand; capital single-stroke characters a minimum of $\frac{1}{8}$ in in size are generally acceptable. The contact sepia print should be processed with a light box or similar apparatus. Taped artwork should never be processed through a machine that administers heat, fumes, or a rolling action, because serious damage to the equipment and artwork can result.

Formal documentation should be generated only upon a management decision to manufacture and market a product for which only minimum documentation exists. It can make good use of the technique of "form drawings," which, when coupled with the fullest use of the in-house photo facility, make an excellent documentation tool.

Any drafting facility can realize cost savings through the reduction of drafting hours by using form drawings to which information is added that is required by each type of document. For example, a master drawing form might contain the location of the hole chart and an outline of the printed board indicating board dimensions, notes, and title block. An assembly drawing form might contain the board outline, notes, an IC package outline indicating pin identification, and a title block. In either case, the only information remaining to be added is that which is peculiar to the particular printed board, such as the drawing title, the number, and drilling information.

The photographic facility has been used primarily to transform artworks into master patterns for making printed boards. However, by capitalizing on its extended capabilities, reduced drafting costs can be realized by making screen-tint photopositive transparencies for the drawing package. Here photography is used not only to reduce the photopositive transparency to a scale convenient for use in the form drawing but also to produce a screen time (halftone) of the circuit

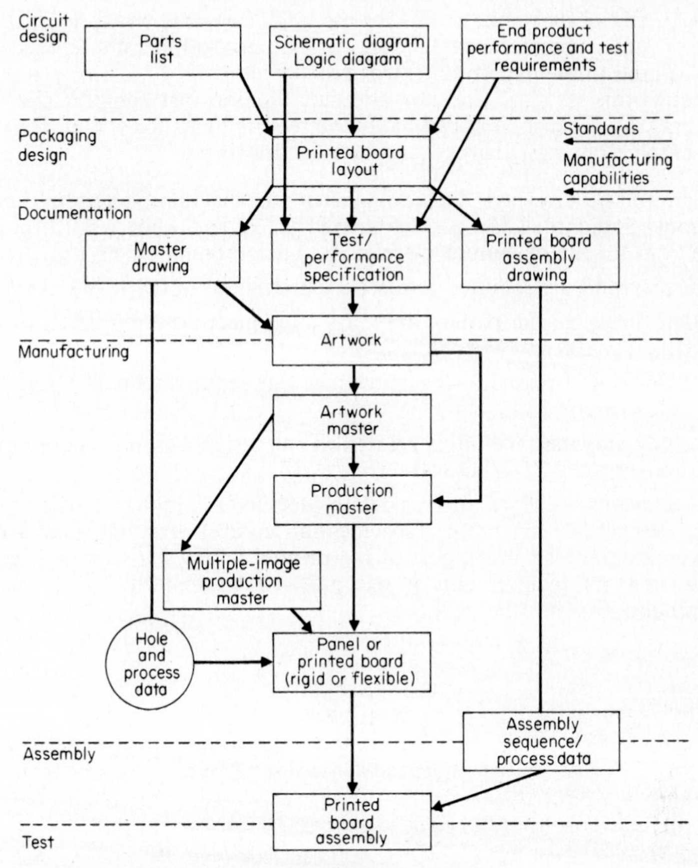

FIG. 9.1 Flowchart of master drawing.

pattern, which is then superimposed over a precision grid. The practice assures positive registration between terminal-area centers and their corresponding grid intersections, which is of primary importance when master drawings are made in compliance with military requirements.

9.6.2 Military Standards

Military documentation indicates that design establishments are obligated to prepare engineering drawings and associated lists in accordance with MIL-D-1000.

MIL-D-1000 specifies general requirements for the preparation of engineering drawings and associated lists and for application of intended-use categories for their acquisition. The specification reflects the policy of the Department of Defense to buy only the engineering drawings that are needed and to encourage procurement of commercial drawings when they are adequate for the purpose.

MIL-STD-100 primarily prescribes the procedures and format authorized for the preparation of form 1 engineering drawings and associated lists prepared for the Defense Department. The standard includes a paragraph stating in part that cost-reduction techniques in engineering drawing preparation may be used when the techniques do not impair the reproducibility quality of MIL-M-9868 or clarity and design disclosure requirements for the kind, category, and form of the engineering document being prepared. Specifically, photographic drafting techniques should be used to the maximum extent practicable.

Although this standard does not stipulate the drawings comprising a printed wiring drawing package, it does specify the requirements for schematics, master and assembly drawings, artwork, and support drawings:

1. Diagrammatic drawings, including schematic diagrams, refers to American National Standards (ANSI), ANSI/ANSI Y32.2 and ANSI Y32.16 providing directions for use of symbology relative to diagrammatic drawings.
2. Logic diagrams are required to meet the provisions of USAS Y32.14.
3. Printed wiring master pattern drawings must meet the requirements specified in MIL-STD-100 and MIL-STD-275.
4. Printed wiring master drawings must meet the requirements of MIL-STD-100 and MIL-STD-275.
5. Assembly drawings, including printed wiring assemblies, must conform to the requirements of MIL-STD-100.
6. Miscellaneous drawings, such as details, detailed assemblies, tabulated assemblies, inseparable assemblies, specification control drawings, and wire lists, which are prepared in support of the printed wiring drawing package, must conform to the requirements of MIL-STD-100 as applicable to the particular document.

REFERENCES

1. Dieter W. Bergman, "Specifications and Purchasing," *Printed Circuits Handbook,* 2d ed., McGraw-Hill, New York, 1979.
2. "Buyers Report on Printed Circuits," *Electronic Procurement,* August 1963.
3. Orland H. Ellis, "Construing the Purchase Order," *Electronic Procurement,* August 1963.
4. "Evaluation Guide: Multilayer Printed Wiring Boards," *Electronic Procurement,* April 1965.
5. "Specifying Guide: PC Boards," *IEEE,* April 1965.
6. George R. Jacobs, Jr., "Documenting Printed-Wiring Packages," *Machine Design,* May 1969.

P · A · R · T · 3

FABRICATION

CHAPTER 10
DRILLING AND MACHINING

James P. Block
President, Laminating Company of America, Garden Grove, California

Charles G. Henningsen
Vice-President, Insulectro, Mountain View, California

Leland E. Tull
President, Contouring Technology, Mountain View, California

10.1 INTRODUCTION

Laminate machining consists of the mechanical processes by which circuit boards are prepared for the vital chemical processes of image transfer, plating, and etching. Such processes as cutting to size, drilling holes, and shaping have major effects on the final quality of the printed board. This chapter will discuss the basic mechanical processes which are essential to producing the finished board.

10.2 PUNCHING HOLES (PIERCING)

10.2.1 Design of the Die

It is possible to pierce holes down to one-half the thickness of XXXPC and FR-2 laminates and one-third that of FR-3 (Fig. 10.1). Many die designers lose sight of the fact that the force required to withdraw piercing punches is of the same magnitude as that required to push the punches through the material. For that reason, the question of how much stripper-spring pressure to design into a die is answered by most toolmakers: "as much as possible." When space on the dies cannot accommodate enough mechanical springs to do the job, a hydraulic mechanism can be used. Springs should be so located that the part is stripped evenly. If the board is ejected from the die unevenly, cracks around holes are almost certain to occur. Best-quality holes are produced when the stripper compresses the board an instant before the perforators start to penetrate. If the stripper pressure can be made to approach the compressive strength of the material, less force will be required and the holes will be cleaner.

If excessive breakage of small punches occurs, determine whether the punch breaks on the perforating stroke or on withdrawal. If the retainer lock is breaking, the cause is almost certain to be withdrawal strain. The remedy is to grind a small taper on the punch, no more than $1\frac{1}{2}''$ and to a distance no greater than the thickness of the material being punched. If the grinding is kept within those limits, it will have no measurable effect on hole quality or size. The other two causes of punch breakage are poor alignment, which is easily detected by close examination of the tool, and poor design, which usually means that the punch is too small to do the job required.

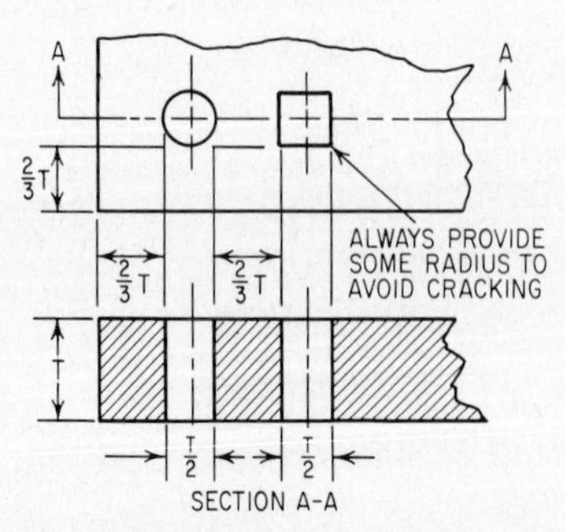

FIG 10.1 Illustration of the proper sizing and locating of pierced holes with respect to one another and to the edge of paper laminates. Minimum dimensions are given as multiples of laminate thickness *t*.

10.2.2 Shrinkage of Paper-Base Laminates

When paper-base laminates are to be punched, it must be remembered that the materials are resilient and that their tendency to spring back will result in a hole slightly smaller than the punch which produced the hole. The difference in size will depend on the thickness of the material. Table 10.1 shows the amount by which the punch should exceed the print size in order to make the holes fall within tolerance. The values listed should not be used for the design of tools for glass-epoxy laminates, the shrinkage of which is only about one-third that of paper-base materials.

TABLE 10.1 Shrinkage in Punched Hole Diameters, Paper-Base Laminates

Material thickness	Material at room temp.	Material at 90°F or above
$\frac{1}{64}$	0.001	0.002
$\frac{1}{32}$	0.002	0.003
$\frac{3}{64}$	0.003	0.005
$\frac{1}{16}$	0.004	0.007
$\frac{3}{32}$	0.006	0.010
$\frac{1}{8}$	0.010	0.013

10.2.3 Tolerance of Punched Holes

If precise hole size tolerance is required, the clearance between punch and die should be very close; the die hole should be only 0.002 to 0.004 in larger than the punch for paper-base materials (Fig. 10.2 and Table 10.2). Glass-base laminates generally require about one-half that tolerance. Dies have, however, been constructed with as much as 0.010 all-around clearance between punch and die. They are for use where inspection standards permit rough-quality holes.

A die with sloppy clearances is less expensive than one built for precision work, and wide clearance between punch and die causes correspondingly more break and less shear than a tight die will cause. The result is a hole with a slight funnel shape that makes insertion of components easier. Always pierce with the copper side up. Do not use piercing on designs with circuitry on both sides of the board, because lifting of pads would probably occur.

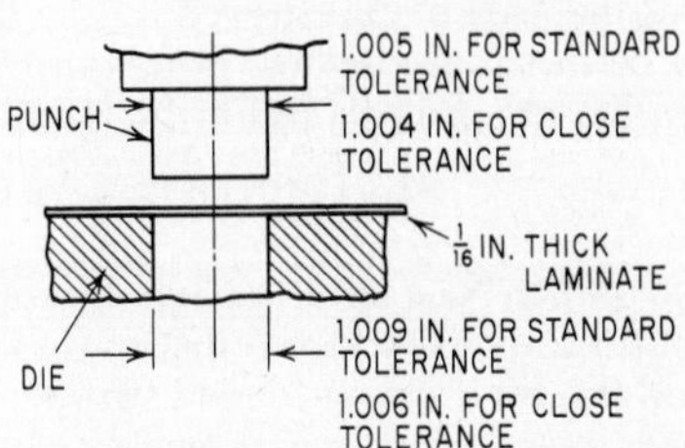

FIG. 10.2 Example of proper tolerance of a punch and die.

10.2.4 Hole Location and Size

Designs having holes whose distance from the edge of the board or from other holes approaches the thickness of the material are apt to be troublesome. Such designs should be avoided; but when distances between holes must be small, build the best die possible. Use tight clearance between punch and die and punch and stripper, and have the stripper apply plenty of pressure to the work before the punch starts to enter. If the distance between holes is too small, cracks between holes may result even with the best of tools. If cracks between holes prove troublesome, plan the process so that the piercing is done before any copper is etched away. The reinforcing effect of the copper foil will help eliminate cracks. Most glass-epoxy laminates may be pierced, but the finish on the inside of the holes is sometimes not suitable for through-hole plating.

10.2.5 Warming Paper-Base Material

The process of punching paper-base laminates will often be much more trouble-free if the parts are warmed to 90 or 100°F. That is true even of the so-called cold-punch or PC grades. Do not overheat the material to the point at which it crumbles and the residue is not ejected as a discrete slug. Overheated material will often plug the holes in the die and cause rejects. Opening the taper on the take-away holes will reduce plugging, but the most direct approach is to pierce at a lower temperature. Glass-epoxy is never heated for piercing or blanking.

10.2.6 Press Size

The size of the press is determined by the amount of work the press must do on each stroke. The supplier of copper-clad sheets can specify a value for the shear

TABLE 10.2 Tolerances for Punching or Blanking Paper-Base Laminate

Material thickness	Base material	Tolerance on hole size, in	Tolerances, in, on distance between holes and slots, 90°F: Up to 2 in	2 to 3 in	3 to 4 in	4 to 5 in	Tolerances for blanked parts, overall dimension, in
To and including $\frac{1}{16}$ in	Paper	0.0015	0.003	0.004	0.005	0.006	0.003
Over $\frac{1}{16}$ in to and including $\frac{3}{32}$ in	Paper	0.003	0.005	0.006	0.007	0.008	0.005
Over $\frac{3}{32}$ in to and including $\frac{1}{8}$ in	Paper	0.005	0.006	0.007	0.008	0.009	0.008

strength of the material being used. Typically, the value will be about 12,000 psi for paper-base laminate and 20,000 psi for glass-epoxy laminate. The total circumference of the parts being punched out multiplied by the thickness of the sheet gives the area being sheared by the die. If all dimensions are in inches, the value will be in square inches. For example, a die piercing 50 round holes, each 0.100 in in diameter, in 0.062-in-thick laminate will be shearing, in square inches:

$$50 \times 0.100 \text{ in} \times 3.1416 \times 0.062 \text{ in} = 0.974 \text{ in}^2$$

If the paper-base laminate has 12,000-psi shear strength, 11,688 lb of pressure, or about 6 tons, is required just to drive the punches through the laminate. Bear in mind that, if a spring-loaded stripper is used, the press will also have to overcome the spring pressure, which ought to be at least as great as the shear strength. Therefore, a 12-ton press would be the minimum which could be considered. A 15- or 20-ton press would be considerably safer.

10.3 BLANKING, SHEARING, AND CUTTING OF COPPER-CLAD LAMINATES

10.3.1 Blanking Paper-Base Laminates

When parts are designed to have shapes other than rectangular and the volume is great enough to justify the expense of building a die, the parts are frequently punched from sheets by using a blanking die. A blanking operation is well adapted to paper-base materials and is sometimes used on glass-base ones.

In the design of a blanking die for paper-base laminates, the resilience, or yield, of the material previously discussed under Piercing applies. The blanked part will be slightly larger than the die which produced it, and dies are therefore made just a little under print size depending upon the material thickness. Sometimes a combination pierce and blank die is used. The die pierces holes and also blanks out the finished part.

When the configuration is very complex, the designer may recommend a multiple-stage die: The strip of material progresses from one stage to the next with each stroke of the die. Usually in the first one or two stages holes are pierced, and in the final stage the completed part is blanked out.

The quality of a part produced from paper-base laminates by shearing, piercing, or blanking can be improved by performing the operation on material which has been warmed. Caution should be exercised in heating over 100°F because the coefficient of thermal expansion may be high enough to cause the part to shrink out of tolerance on cooling. Paper-base laminates are particularly anisotropic with respect to thermal expansion; that is, they expand differently in the x and y dimensions. The manufacturer's data on coefficient of expansion should be consulted before a die for close-tolerance parts is designed. Keep in mind that the precision of the manufacturer's data is probably no better than ± 25 percent.

10.3.2 Blanking Glass-Base Laminates

Odd shapes that cannot be feasibly produced by shearing or sawing are either blanked or routed. Glass blanking is always done at room temperature. Assuming a close fit between punch and die, the part will be about 0.001 in larger than the die which produced it. The tools are always so constructed that a part is removed from the die as it is made. It cannot be pushed out by a following part as is often true when the material has a paper base. If material thicker than 0.062 in is blanked, the parts may have a rough edge.

The life of a punch, pierce, or blank die should be evaluated with reference to the various copper-clad materials that may be used. One way to evaluate die wear caused by various materials is to weigh the perforators, or punches, very accurately, punch 5000 pieces, and then reweigh the punches. Approximately 5000 hits are necessary for evaluation, because the initial break-in period of the die will show a higher rate of wear. Also, of course, the quality of the holes at the beginning and end of each test must be evaluated. Greatly enlarged microphotos of the perforator can be used for visual evaluation of changes in the die.

10.3.3 Shearing

When copper-clad laminates are to be sheared, the shear should be set with only 0.001 to 0.002 in clearance between the square-ground blades (Fig. 10.3). The

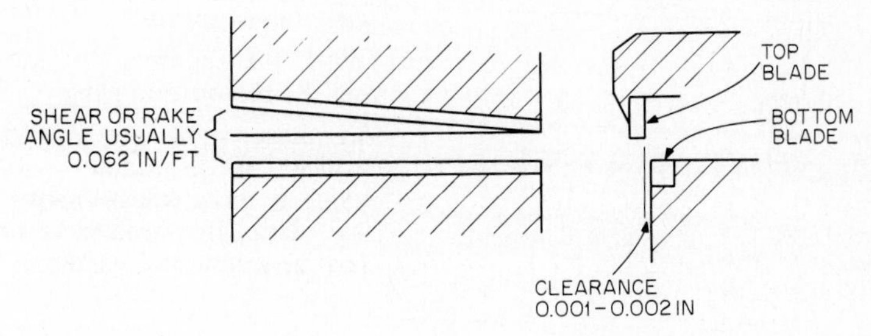

FIG. 10.3 Typical adjustable shear blades for copper-clad laminates.

thicker the material to be cut, the greater the rake or scissor angle between the top and bottom shear blade. The converse also is true: The thinner the material, the smaller the rake angle and the closer the blades. Hence, as in many metal shears, the rake angle and the blade gap are fixed; the cutoff piece can be twisted or curled. Paper-base material can also exhibit feathered cracks along the edge that are due to too wide a gap or too high a shear angle. That can be minimized by supporting both piece and cutoff piece during the shear operation and decreasing the rake angle. Epoxy-glass laminate, because of its flexural strength, does not usually crack, but the material can be deformed if the blade clearance is too great or the shear angle is too large. As in blanking, the quality of a part produced from paper-base laminates by shearing can be improved by warming the material before performing the operation.

10.3.4 Sawing Paper-Base Laminates

Paper-base laminates are much harder on sawing tools than are the hardest woods, and therefore a few special precautions are necessary for good saw life. Sawing paper-base laminates is best accomplished with a circular saw with 10 to 12 teeth per inch of diameter at 7500 or 10,000 ft/min. Hollow-ground saws give a smoother cut; and because of the abrasive nature of laminated materials, carbide teeth are an excellent investment. (See Fig. 10.4 for tooth shape.) When a saw does not last long enough between sharpenings, use the following checklist.

(These steps could have a cumulative effect and change saw life by a factor of 4 to 5.)

1. Check the bearings for tightness. There should be no perceptible play in them.
2. Check the blade for runout. As much as 0.005 in can be significant.
3. When carbide teeth are used, inspect them with a magnifying glass to make sure a diamond tool no coarser than 180 grit was used in sharpening them.
4. If the saw has a thin blade, use a stiffening collar to reduce vibration.
5. Use heavy pulleys with more than one V belt. Rotating parts of the system should have sufficient momentum to carry the sawtooth through the work smoothly and without variation of speed.
6. Check the alignment of the arbor and the motor mounting.

All these steps are intended to reduce or eliminate vibration, which is the greatest enemy of the saw blade. If vibration is noticed, find the source and correct it.

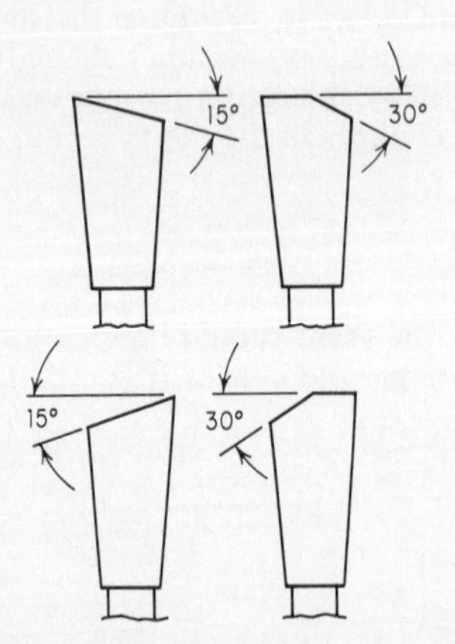

FIG. 10.4 Commonly used saw-tooth designs for paper and cloth laminates. At left, two successive teeth on a 15° alternate-bevel saw. At right, two successive teeth on a 30-ft alternate-corner-relieved (AC-30) saw.

10.3.5 Sawing Glass-Base Laminates

When glass-base laminates are to be sawn, carbide-tooth circular saws can be used; but unless the volume of work is quite low, the added investment required for diamond-steel-bonded saws will be paid for in future savings. The manufacturer's recommendation of saw speed should be followed; usually it will be for a speed in the neighborhood of 15,000 ft/min at the periphery of the saw blade. When economics dictate the use of carbide-tooth circular saws for cutting glass, use the instruction previously given for paper-base laminates (see Fig. 10.4 for tooth shape) and remember that each caution regarding runout, vibration, and alignment becomes more important when glass-reinforced laminates are sawn.

10.4 ROUTING LAMINATES

Routing offers the fabricator the advantages of superior edge finishes and closer tolerances than can be obtained from shearing or sawing. It can eliminate the expense of blanking dies and the long lead times associated with hard tooling. Now, when coupled with the use of multiple-spindle machines, the labor costs of routing can be competitive with those of die blanking. When a design has traces close to the board edge, routing may be the only blanking method that will produce acceptable boards.

Routing is basically a machining operation similar to milling, but it is done at much higher cutter speeds and feed rates. Three routing systems are available to the fabricator: pin routing, stylus or tracer routing, and NC routing.

10.4.1 Pin Routing

Pin routing requires a machined template, usually of aluminum or steel. The template, which is made to the board outline, includes registration pins for positioning the boards, which are usually stacked three or four high. The package is then routed by tracking the template against a pilot pin that protrudes from the table. The pin height should be a few thousandths below the top of the template; alternatively, a spacer can be used between the template and the adjacent board. The cutter is the same diameter as the pilot pin. Usually two or more passes are required to assure proper tracking of the template, because the cutter tends to force the work away from itself and the pilot pin. Work should be fed against the rotation of the cutter to prevent grabbing. Dwell can cause the cutters to load up with chips and overheat. Pin routing, although labor-intensive not only in cost but also in operator skill, may prove satisfactory for irregular shapes when volumes are low or edge finishes are critical.

10.4.2 Stylus or Tracer Routing

In tracer routing a stylus traces the board outline on a template similar to that used for pin routing. The stylus may control the movement of the spindles over a fixed table, or it may control a movable table under fixed spindles. The latter approach is most commonly used with multiple-spindle machines, and the following discussion will concentrate on that machine type.

Templates may be machined to the board outline, with the stylus tracing the external edge, but a closed track 0.005 to 0.010 in wider than the stylus is recommended for greater control and accuracy. In practice, the operator makes his first cut by tracking the outside edge of the template track. (For an outside cut, this assumes tracing clockwise around the template with clockwise cutter rotation.) A second, cleanup pass also is made in a clockwise direction, but with the stylus tracking the inside of the template track. This light cut will relieve most of the cutter load and give the operator better control for final sizing of the board.

Stylus routing provides better operator control, and accuracy is greater than by pin routing. With multiple-spindle machines, the system can be competitive with die blanking because up to 20 boards are routed at one time. Sometimes the limiting factor is the ability of the operator to provide the force needed to feed the work past the cutters, but power-assisted table movements can help with some of the load. Tolerances of ± 0.010 in can be achieved at production rates with moderate operator skill and by manipulating stylus sizes; closer tolerances are possible with more skilled operators. Cutter diameters of $\frac{1}{8}$ in can be used successfully on these machines, but cutter breakage is directly related to operator skill and control.

10.4.3 Numerical Control Routing

When production volume, flexibility in board design, or short lead time becomes paramount, NC routing may be the best approach to final blanking. Through numerical control of table or spindle movement, the routing operation is reduced to loading and unloading the machine. Complex shapes can be cut to close tolerances in very high volume. Programs can be written quickly, usually under two hours, and design changes can be made in minutes by simple program modifications.

NC router programs consist of a series of simple commands that direct the machine to follow the desired cutter path. Coordinate information is taken from the board blank; added commands must also be entered to direct such functions as feed rate and head up–head down.

Labor costs go down as equipment costs and accuracies are increased by use of the more sophisticated systems. The use of solid carbide cutter bits for routing any of the commonly used board materials is mandatory. Spindle speeds may vary from 12,000 to above 24,000 rpm, and spindles must have sufficient power to make a cut without an appreciable drop in r/min.

Tooling or registration holes inside the board perimeter are preferred. External tooling holes may be used by leaving tabs at the holes, but the tabs must be removed in a secondary operation.

Although square outside corners can be achieved by routing, all inside corners will have a radius equivalent to the cutter radius. In most designs, that will be acceptable. When square inside corners are required for part clearance, an undercut of 0.015 to 0.030 in along either axis, or a diagonal cut at 45°, will remove material enough that the mating part will have sufficient clearance. The inside corner may also be squared by a secondary operation.

The following subsections discuss in detail NC routing procedures.

10.4.4 Cutter Offset

Since the cutter must follow a path described by its centerline, it must be offset from the desired board edge by an amount equal to its effective radius. That is the basic cutter radius, and it will vary with the cutter tooth form. Since the cutters deflect during the routing operation, it is necessary to determine the amount of deflection to be added to the basic cutter radius before expending large amounts of time on programming parts (Fig. 10.5).

Variables which affect deflection are thickness, type of material, direction of cut, feed rate, and spindle speed. To reduce those variables, the manufacturer should:

1. Standardize on cutter bit manufacturer, diameter, tooth form, and end cut.
2. Fix spindle speed (24,000 r/min recommended for epoxy-glass laminates, 12,000 r/min for Teflon-glass laminates).
3. Maintain a stack height of 0.250 ± 0.050 in.
4. Rout in clockwise direction on outside cuts, counterclockwise on inside cuts.
5. Standardize on single or double pass.
6. Fix feed rates for given materials. (Note that higher rates will increase part size and slower feed rates will decrease part size.)

10.4.5 Direction of Cut

A counterclockwise direction of feed (climb cut) will leave outside corners with slight projections and inside corners with small radii. A clockwise direction of

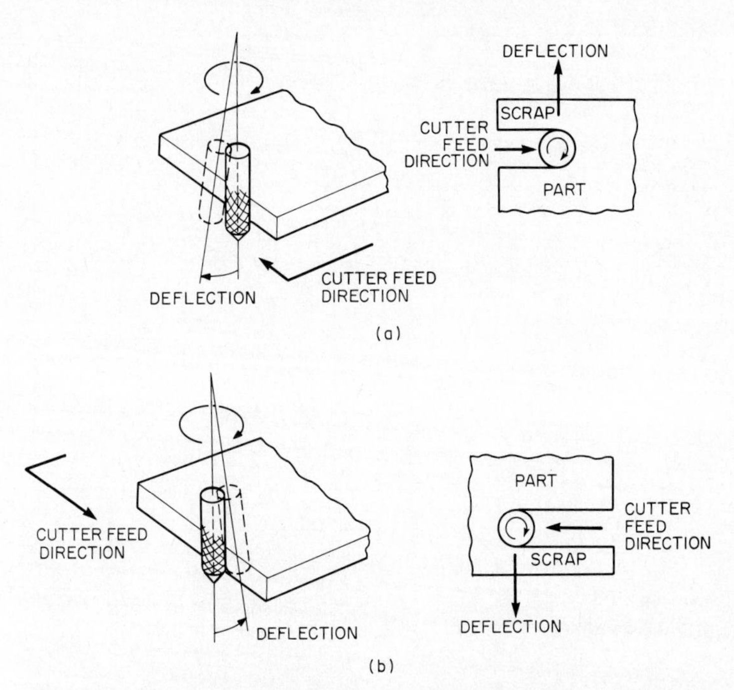

FIG. 10.5 Effect of cutter deflection on part size and geometry. (*a*) Clockwise cutting (recommended for outside cuts) deflects cutter away from part. That leaves outside dimensions large on first pass unless compensated for in programming. (*b*) Counterclockwise cutting (recommended for inside cuts and pockets) deflects cutter into scrap. Therefore, inside dimensions of holes or cutouts will measure small unless compensated for in programming.

feed (rake cut) will give outside corners a slight radius, and perhaps give inside corners a slight indentation. These irregularities may be eliminated by reducing the feed rate or cutting the part twice.

10.4.6 Cutter Speed and Feed Rate

The variables affecting cutter speed are usually limited to the type of laminate being cut and the linear feed rate of the cutter. A cutter rotation of 24,000 r/min and feed rates up to 150 in/min may be used effectively on most laminates. Teflon-glass and similar materials, the laminate binder of which flows at relatively low temperatures, require slower spindle speeds (12,000 r/min) and high feed rates (200 in/min) to minimize heat generation. The graph in Fig. 10.6 shows recommended feed rates and cutter offsets for most standard laminates at various stack heights. The cutter used is a standard $\frac{1}{8}$-in-diameter burr type.

10.4.7 Cutter Bits

Because of the precise control of table movement in NC routing, cutter bits are not subjected to the shock encountered in pin routing and stylus routing, and

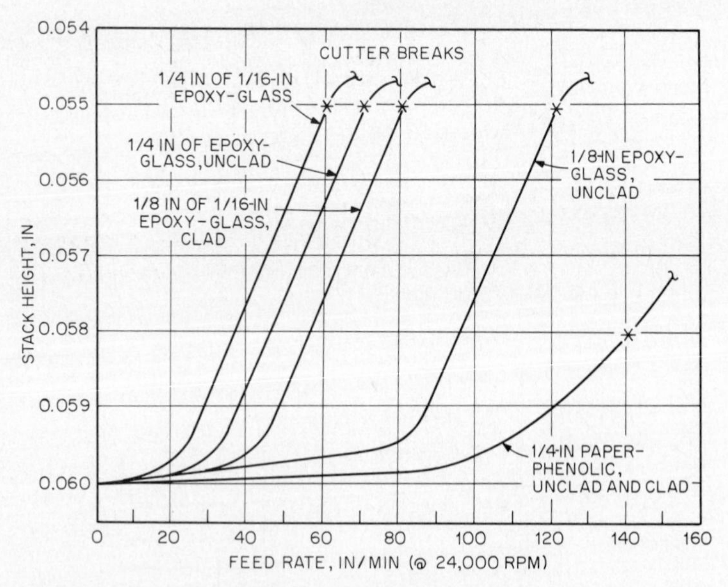

FIG. 10.6 Recommended feed rate, using $\frac{1}{8}$-in-diameter burr cutter at 24,000 r/min, for varying stack heights of specific thickness of material.

therefore small-diameter cutters may be used successfully. However, the fabricator would do well to standardize on $\frac{1}{8}$-in-diameter cutters because they are suitable for most production work and are readily available from a number of manufacturers in a variety of types. The resulting 0.062-in radius on all inside corners is usually acceptable if the board designer is aware of it.

Cutter tooth form is more important in NC than in other routing. Because of the faster feed rates possible, it is important that a cutter have an open tooth form that will release the chip easily and prevent packing. Many standard diamond burrs available on the market will load with chips and fail rapidly. The carbide cutting bit will normally cut in excess of 15,000 linear inches of epoxy-glass laminate before erosion of the teeth renders the cutter ineffective or too small. Cutter life can be prolonged by periodic cleaning to remove epoxy buildup in the reliefs of the teeth. Cleaning can be done by brushing or with chemical cleaners used on drills.

If extremely smooth edges are required, a fluted cutter may be used. Single- or two-fluted cutters with straight flutes should be used when cutting into the foil if minimum burring is desired. It should be noted that such cutters will be more fragile than a standard serrated cutter, and feed speeds should be adjusted accordingly. When a slightly larger burr can be tolerated, two- and three-flute left-hand spiral cutter bits should be used because of their greater strength. The left-hand spiral will force the workpiece down rather than lift it, assuming a right-hand-turning spindle.

10.4.8 Tooling

To simplify tooling and expedite loading and unloading operations, effective hold-down and chip-removal systems should be provided as part of the machine design. Various methods may then be devised to mount the boards to the machine table while properly registering them to facilitate routing the outline. Some machine designs will have shuttle tables available so that loading and unloading may be accomplished while the machine is cutting. Others will utilize quick-change secondary tooling pallets or subplates that allow rapid exchange of bench-loaded pallets with only a few seconds between loads.

10.4.8.1 ***Tooling Plates.*** Tooling plates utilize bushings and a slot on the centerline of active pattern under each spindle. They are doweled to the machine table (Fig. 10.7). The plates may be made by normal machine shop practice, or the router may be used to register and drill its own tooling plate. Mounting pins in the tooling plate should be a light slip fit.

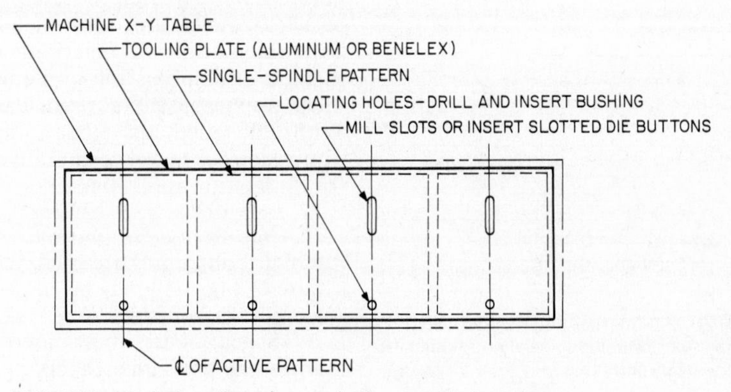

FIG. 10.7 Typical tooling of numerically controlled routing.

10.4.8.2 ***Subplates.*** Subplates should be made of Benelex, linen phenolic, or other similar material. Subplates should have the pattern to be routed cut into their surfaces. The patterns act as labyrinths and aid in chip removal. Part-holding pins should be an interference fit in subplates and a tight to loose fit in the part, depending on cutting technique used (Fig. 10.8).

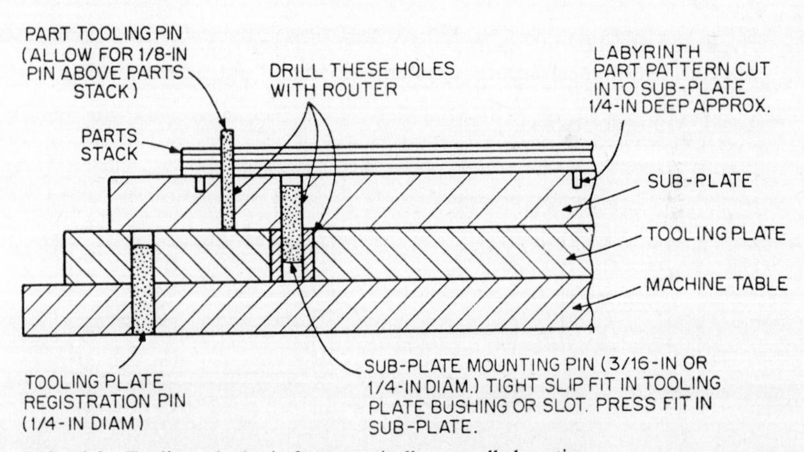

FIG. 10.8 Tooling schematic for numerically controlled routing.

It is recommended that the programmer position the tooling and hold-down pinholes in addition to the routing program. That will provide absolute registration between the tooling holes and the routing program. Each subplate, full- or part-size, may be used on both sides for a multiplicity of setups and is indexed to the table or master tooling plate with dowel pins to allow multiple tool changes without retooling.

10.4.9 Cutting and Holding Techniques

Since the precision required for cutting board outlines, as well as the placement of tooling holes for registering boards, will vary, a number of different cutting and holding methods may be used. Three basic methods are illustrated here. Experimentation will determine which method or combination of methods is most applicable to a particular job. With all methods, the minimum dimension for board separation with a 0.125-in cutter is 0.150 in.

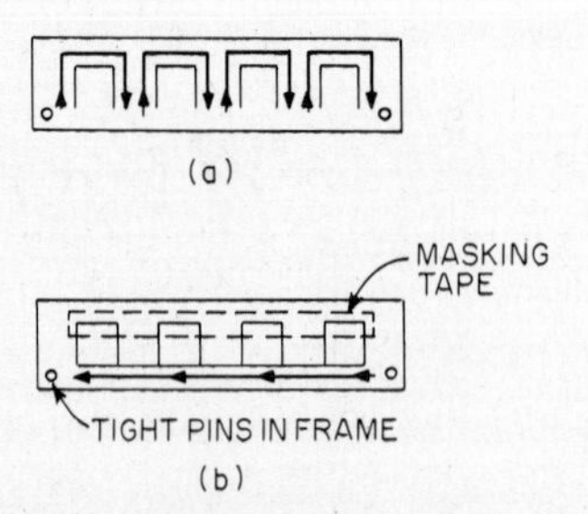

FIG. 10.9 No-internal-pin method. Step 1: Cut three sides (*a*). Step 2: Apply masking tape (*b*). Step 3: Cut parts free.

1. *No-internal-pin method:* If no internal tooling pins are used, the procedure of Fig. 10.9 may be employed.
Accuracy: ±0.005 in
Speed: Fast (best used with many small parts on a panel)
Load: One panel high—each station

2. *Single-pin method:* The single-pin method is illustrated in Fig. 10.10.
Accuracy: ±0.005 in
Speed: Fast (quick load and unload)
Load: Multiple stacks

3. *Double-pin method:* In the two-pin method, there is a double pass of cutter offset; see Fig. 10.11. Make two complete passes around each board, the first pass at a recommended feed rate and the second at 200 in/min. Remove scrap after first pass.
Accuracy: ±0.002 in
Speed: Fast (highest-accuracy system—load and unload slower than single-pin method due to tight pins)
Load: Multiple stacks.

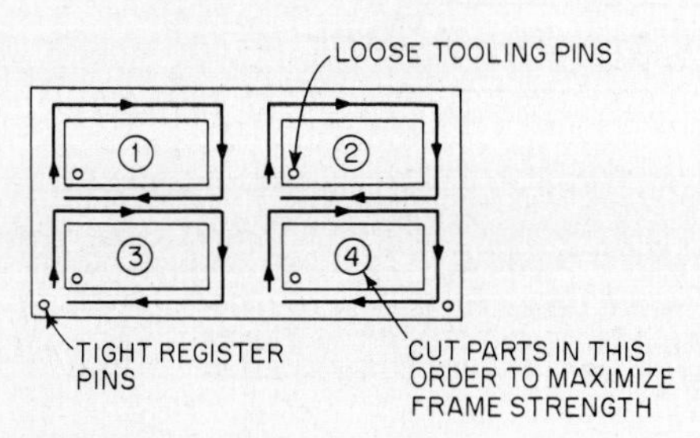

FIG. 10.10 Single-pin method.

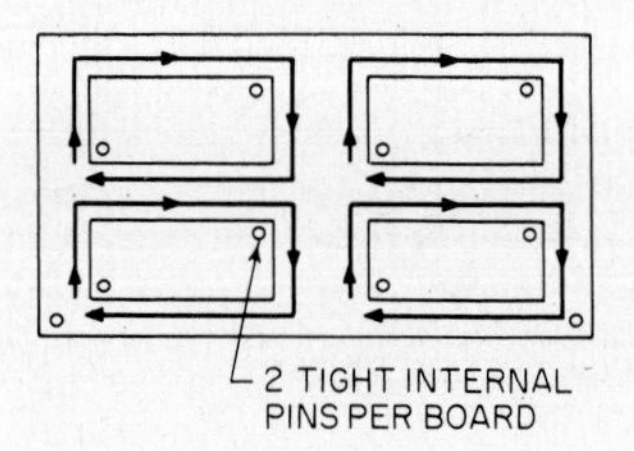

FIG. 10.11 Double-pin method.

10.5 DRILLING

The purpose of drilling printed circuit boards is twofold: (1) to produce an opening through the board which will permit a subsequent process to form an electrical connection between top, bottom, and sometimes, intermediate, conductor pathways; and (2) to permit through-the-board component mounting with structural integrity and precision of location.

The quality of a drilled hole through a printed circuit board is measured by its interface with the following processes: plating, soldering, and forming a high-reliability nondegrading electrical and mechanical connection.

It is possible to drill holes meeting the previously described requirements with high productivity, consistency, and yield.

The elements of this process are materials, such as laminates and drills; processes including machine parameters, techniques, and operating personnel; and control evaluation of hole quality, drills, and process machinery.

When all these elements are properly developed and implemented, high-quality printed circuit board holes are a natural result. Such holes can be plated directly, eliminating remedial processes of deburring, desmearing, and etchback, and resulting in process simplification, higher yields, and lower costs.

The interaction and practice of elements shown in Fig. 10.12 can optimize the drilling process for the fabrication of printed circuit boards. While all of the ele-

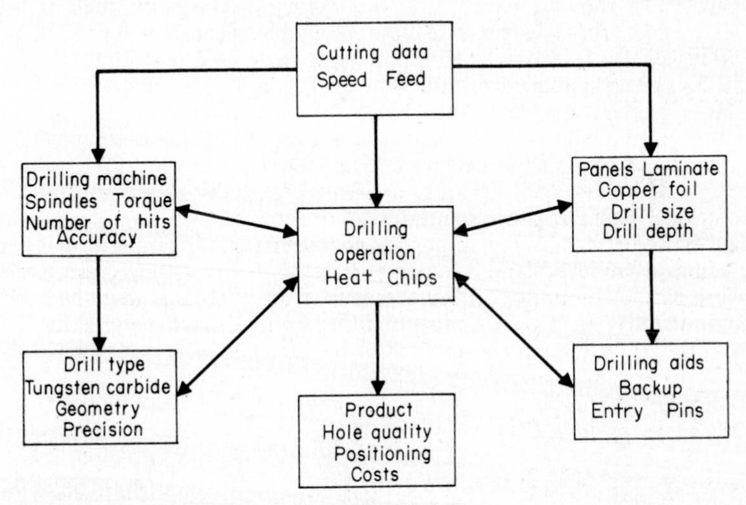

FIG. 10.12 Essential elements in good PWB drilling.

ments shown are interdependent, it is hole quality and its location accuracy that ultimately steer decisions on feeds and speeds, material choices, and productivity.

10.5.1 Drills

Drills for making holes in printed circuit boards are usually made of tungsten carbide. This is due to interrelated needs of cost, wear properties, machinability, and handling properties. No other material has proved to be as suitable.

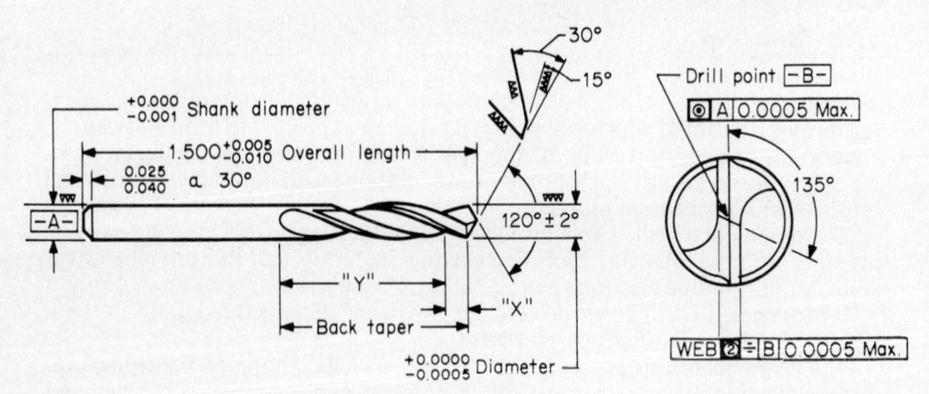

FIG. 10.13 Straight-shank drill.

Drills are of two types: common shank and straight shank. Figures 10.13 and 10.14 illustrate these types of drills.

The design of a drill is just as important as the materials used. The design and the wear of the drill affect its drilling temperature, ability to remove chips, tendency to create entry and exit burrs, and smoothness of the hole wall (all directly related to hole quality).

Figure 10.15 shows a typical drill bit geometry. The point angle is usually between 90 and 110° for paper-base materials and between 115 and 130° for glass-base materials. By far the most common point angle used in drilling is 130°.

The flute or helix angle determines the drill's ability to remove chips from the hole. Helix angles vary from 20 to 50°. A 20° angle provides fast chip removal with poor cutting efficiency. A high helix angle (50°) creates a smaller material plastic zone but yields slower chip removal, as shown in Fig. 10.16, angle A. A helix angle of 30° is a good compromise between a small plastic zone and quick chip removal. This compromise minimizes drilling temperature.

Figure 10.15 also shows an important characteristic of high-performance drills described as a "relieved land." Increases in temperature during drilling can also be caused by the amount of drill surface area in direct contact with the hole wall. To minimize this surface area, most manufacturers remove material just behind

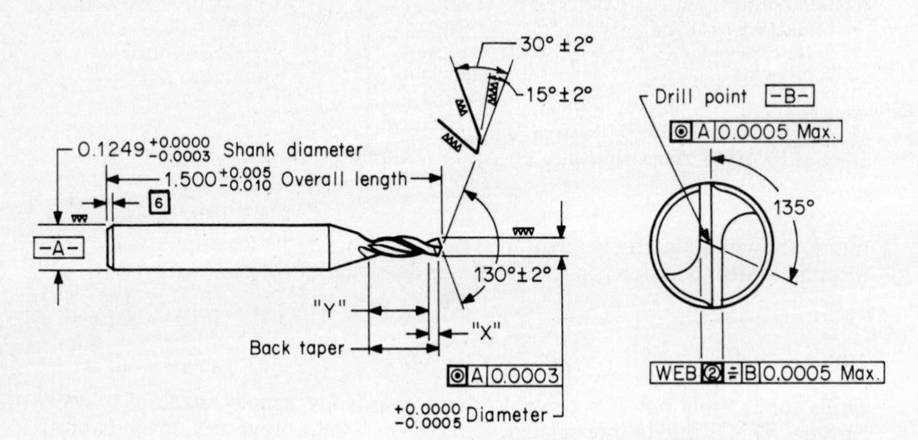

FIG. 10.14 Common-shank drill.

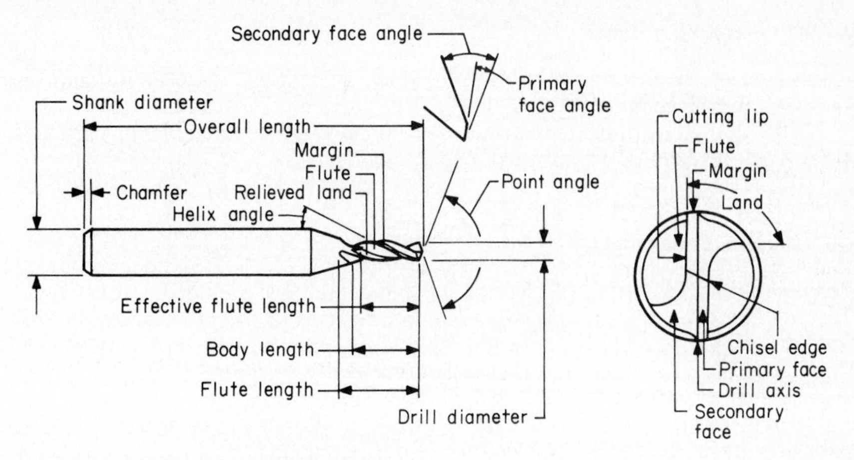

FIG. 10.15 Typical drill bit geometry.

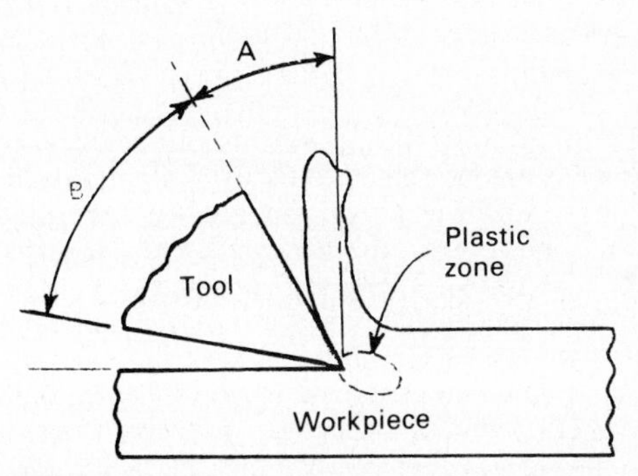

FIG. 10.16 Drill bit, section view of the cutting process.

the margin or cutting area to reduce friction and thus lower the temperature of drilling. Other geometries being equal, the narrower the margin, the cooler the drilling temperature.

The surface finish of the drill is important; that is, the smoother the surface, the cooler the drilling. Surface finishes lower than 4 μin should be used.

The volume of empty space in the drill flutes is another important design consideration. The greater the open volume, the higher the capacity of the drill to remove chips efficiently. Conversely, greater volume (thinner center web) implies a weaker drill which is more prone to breakage.

In general, the best drill designs are those that drill with the lowest drill temperature. Drills which drill cooler have a good surface finish, thin webs, no geometrical defects or chips, sharp cutting edges, and durable carbide. In addition, the best operating parameters for drills are those which also minimize drill temperatures. Finally, as drills wear, they drill hotter and hole quality decreases. Run length must be carefully determined.

It should not be assumed that all manufactured drills are of equal quality. Differences in design, manufacturing processes, surface finish, raw materials, and consistency vary widely and should be evaluated prior to the purchase of production quantities of drills.

10.5.2 Handling of Drills

Cutting edges are all important. Drill-handling procedures which do not damage these cutting edges must be used.

Cutting edges or points should never be pressed against metallic or hard surfaces. Drill bits should never touch one another.

10.5.3 Drill Bit Inspection

Drills should be cleaned upon receipt. Approximately $\frac{1}{8}$ of each drill's flute should be worked into an orange stick (wood commonly used to clean instruments). This applies a slight pressure to the drill's tip and loosens and removes inadequately bonded carbide and other substances from the drill's surface.

The cutting surfaces of each drill should be immersed and suspended in an ultrasonic solution (1 percent TSP—trisodium phosphate—in water) for 30 s. This process loosens and removes surface oils and debris.

Geometric tolerances in drill dimensions (length and diameter) should be measured by a contactless method. Micrometer techniques can easily damage the point or cutting edges and should never be used. Laser and other optical measuring devices are available that permit higher accuracy and contactless measurement.

A good inspection program is important for both new and repointed drill bits. Micrograin carbide drills are fabricated to industry standards, but variations from those standards produce drill bits of varying quality. Drill inspectors should use a 40 to 140× microscope to examine the quality of the drills. Drills should be held in a fixture like that shown in Fig. 10.17.

Each drill should be inspected for flaws with a stereooptical microscope. Rejection parameters for chips are based on the observation of chips within the cutting edges of a drill at the magnification power listed in Table 10.3.

Other flaws (layback, overlap, offset, gap negative, flair, and hook) can be

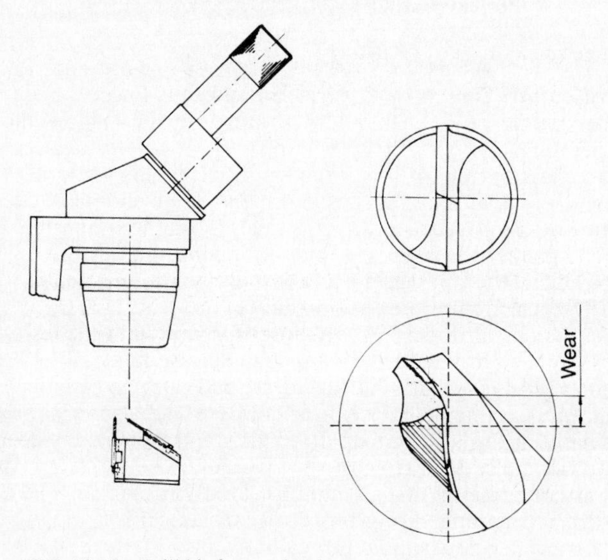

FIG. 10.17 Drill bit fixture, in place on microscope stage.

TABLE 10.3 Rejection Parameters for Drills with Chips

Drill size, in	Magnification
0.0006–0.0135	50×
0.0135–0.0625	40×
0.0625–0.125	30×
0.125–0.25	20×

observed by comparison to the examples shown in Fig. 10.18. An inspection magnification of 140× is required for drill diameters 0.020 in and smaller.

Drill wear should be determined by measuring the land wear and corner rounding of the cutting edge (as shown in Fig. 10.19). Total wear is the sum of corner rounding and land wear. Total wear for each process can be plotted as a function of the number of holes drilled.

Drills which have geometric defects, chips, poor surface finish, excessive wear, and poor design *do not* give good hole quality. In addition, rapid drill wear may be a sign of poor, fast-wearing carbide.

10.5.4 Repointing of Drills

Prior to repointing, drills should be cleaned to remove all foreign substances.

When drills of good quality are used, three repointing operations are often possible. The guide to repointing is for the total drill length removed by the repointing operations never to exceed 15 percent of the drill diameter.

Following the repointing operation, all drills should be inspected as previously described and stored in a container that will prevent any contact with another drill or hard surface.

If definite signs of wear and contamination occur quickly during a run, or if drill bits break frequently, use higher-quality drill bits or better repointing techniques.

Since drill bit quality is only one of the factors essential to good drilling, other factors may be involved when contaminated or quickly dulled drill bits are found. These factors may be improper feeds and speeds, improper entry and backup materials, or poor laminates.

10.5.5 Drill Entry

Entry materials are flat, thin sheets placed on the drill entry side of the laminate during the drilling operation. They should prevent damage to the top copper laminate surface from the pressure foot, reduce entry burrs, and minimize drill wander. They also should not cause drill damage or excessive wear and should not contaminate holes or increase drilling temperatures.

Smooth, flat surfaces are important attributes for entry materials. They should not be warped or twisted or become warped during drilling.

It is important for entry materials to have hard, smooth surfaces to center the drill and prevent burr formation on the entry side of the top laminate.

It is also important that entry materials do not contain contaminants like organic resins with oils that could be deposited or smeared on the hole wall.

Entry materials which contribute to rapid drill wear cause poor hole quality

Point geometry

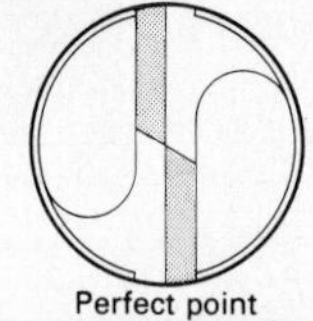

Drill point defects		Drill point defects	
Chips (unacceptable defect)	None visible at 40× on sizes up to ⅛ in. in diameter None visible at 20× on sizes over ⅛ in. in diameter	Layback (unacceptable defect)	None
Overlap (major defect)	1% maximum of drill diameter	Offset (major defect)	1% maximum of drill diameter
Gap (minor defect)	1% maximum of drill diameter	Negative (minor defect)	1° maximum
Flair (minor defect)	2° maximum	Hook (minor defect)	1% maximum of drill diameter

Diameter and length tolerances	
Drill diameter	0.0002 in maximum
Shank diameter	0.0002 in maximum
Overall length	+0.004 — 0.008 in
Concentricity, TIR (total indicator runout)	
Drill diameter to shank	Better than 0.0002 in
Drill point to shank	Better than 0.0005 in
Lip to lip	Better than 0.0005 in

FIG. 10.18 Drill point flaws.

and extensive drill usage. Entry materials which contribute to higher drilling temperatures cause greater amounts of smear.

Table 10.4 lists the available entry materials in order of performance. The best entry material provides hard surfaces for accurate drill entry, burr reduction, and laminate protection. It also is very easy to drill, exerts low torque on the drill, decreases drill temperature, and is void of contaminants affecting the plating processes.

The worst entry materials have combinations of problems caused by soft surfaces, abrasives, hole contaminants, or hard-to-drill substances that heat the drill and cause resin smear.

The contaminants often can be observed on the drill following a long drill run on the drill point or margin relief areas. Phenolic entry can build resin on the drill

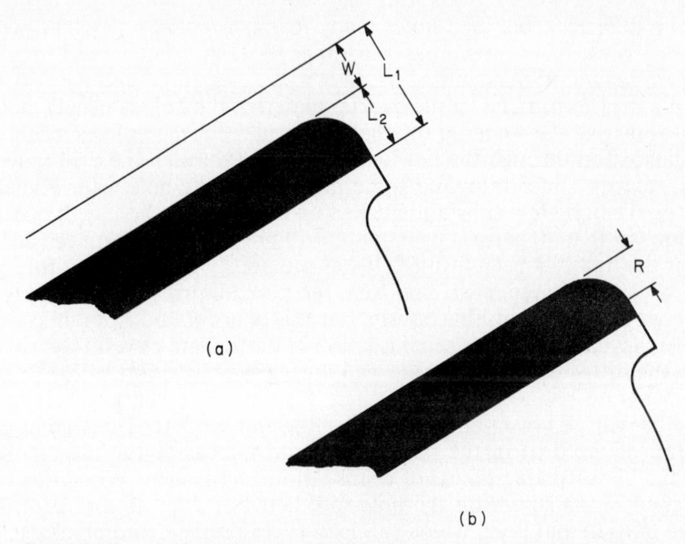

FIG. 10.19 (*a*) Measurement of drill land wear. L_1 = new drill land, L_2 = used drill land, W = land wear, $L_1 - L_2$. (*b*) Measurement of corner rounding. R = corner rounding.

because of excessive temperature. Worse yet, they can directly increase the level of smear on the hole wall. Aluminum entry can be problematic if it is thick and soft, which builds drill residues, decreases cutting efficiency, and increases temperatures.

Entry materials should always be used. The choice of material is dependent on the cost-performance trade-off desired. It is always good practice to choose materials that have the lowest probabilities of causing downstream failures.

10.5.6 Drill Backup

"Backup" is a material placed under the bottom laminate to terminate the drill at the bottom of its stroke and to prevent an exit burr from forming.

Although simple in concept, the backup has surprising and profound effects on the quality of the hole. Given good drilling practice in all other areas, the backup

TABLE 10.4 Commercially Available Entry Materials

Designation	Description
Aluminum-clad	Two thin Al alloy skins bonded to a noncontaminating core
Solid aluminum	5–30 mils thick, various alloys
Phenolic, resin-based	12–22 mils thick 60% + resin, balance paper
Phenolic, paper-based	15–22 mils thick 60% + paper, balance resin
Paper	5–10 mils thick (Kraft paper or the like)

can be the difference between holes ready for plating and holes that need additional processing to accept plating or perhaps those that can never be plated. The reasons are easy to understand.

When a drill terminates in the backup material, the hole is usually cut cleanly and smoothly. All the material that has been cut from the volume of the backup is then carried up through the newly formed holes through the drill flutes, where it bangs, scrapes, slides into, and sometimes gouges the hole wall. The choice of this material can have a substantial effect on the resultant hole-wall quality.

It is known that 90 percent or more of drill breakage occurs on the drill retraction after the hole has been formed. It was previously assumed that this occurred because of table movement. If that were the case, all drills on a multiple-spindle machine would break simultaneously, but this is not found in practice. The real answer lies in the binding or seizing action of particulate or resin debris wedging between the drill sides and the hole wall during the retraction.

Two separate phenomena occur. The first involves solid particulate debris from the backup or lower laminates collecting and binding. The second is caused by friction of the drill in the backup material increasing the drill temperature beyond the T_g (softening point) of resins within the system. When this happens, the softened resins smear on the hole wall between the wall and drill, building excessive friction and drag, which can lead to drill seizure and breakage.

It is clear that the drilling process need not deteriorate to that level before substantial damage is done to the hole walls by the chips or materials removed from the backup.

Aside from the direct frictional effects of the drilling process, the temperature problems are compounded even further by the very poor thermal conductivity of tungsten carbide. Heat from the drilling friction tends to remain at the drill's cutting edges and at the point of the drill. Because the heat cannot be conducted away, temperatures rise rapidly, softening and melting the organic resins. This greatly increases drill bit contamination problems and reduces hole quality through smearing.

Because of these factors, the proper backup material must, at the least, be devoid of low-temperature organic resins, cut into easily transportable and soft chips that will not damage the hole wall, free of hole-wall contaminants, and soft enough to reduce friction on the drill tip during the dwell time at the bottom of the stroke.

In addition, the backup must have a surface hard enough to prevent burrs from forming as the drill bit exits the copper-clad laminate.

Meeting these requirements is not an easy task. Correct backups are almost restricted to laminated composite structures and necessitate an exacting choice of raw materials to permit the drilling of quality holes and reliable interconnections. Nevertheless, a broad range of backup materials is currently available. They are shown in Table 10.5 in order of performance.

The best drilling backup structures meet the overall requirements outlined

TABLE 10.5 Commercially Available Backup Materials

Designation	Description
Vented aluminum	Aluminum structure permitting the drill to terminate in air
Aluminum-clad	Two thin aluminum skins bonded to a noncontaminating core
Paper- or phenolic-clad	Two thin phenolic or paper skins bonded to a wood core
Phenolic	60%+ phenolic resin, balance paper
Hardboard	Pressed wood fibers with resins and oils

above with the best trade-offs. Evaluation of various backups must be performed under rigorously controlled conditions and use hole quality as the dependent variable. Hole defects such as smearing, gouging, torn glass fibers, hole debris, delamination, nailheading, and burr height must be carefully monitored to understand differences in high-performance backups.

Easily observable performance failures of low-end backups are higher drill breakage, obvious hole-wall smearing, large exit burrs, burned and contaminated drill bits, and burned blind holes in the backup. Down the process line, poor backups usually increase the percentage of blowholes after soldering, which can be directly monitored with statistical quality control (SQC) process control.

Differences in backups have been observed to cause a factor of 10 difference in the number of blowholes on a multilayer printed circuit board.

Other obvious differences are that poor backups are not flat and do not offer enough support to the laminate, which causes burrs. Good backups of any type should be dimensionally stable and uniform in thickness and density. They should not be warped, twisted, or bowed.

Drilling depth into the backup should be set to provide a drill lip penetration depth into the backup of 0.005 to 0.015 in. Figure 10.20 shows a drill terminating in the backup after drilling through the laminate and entry material.

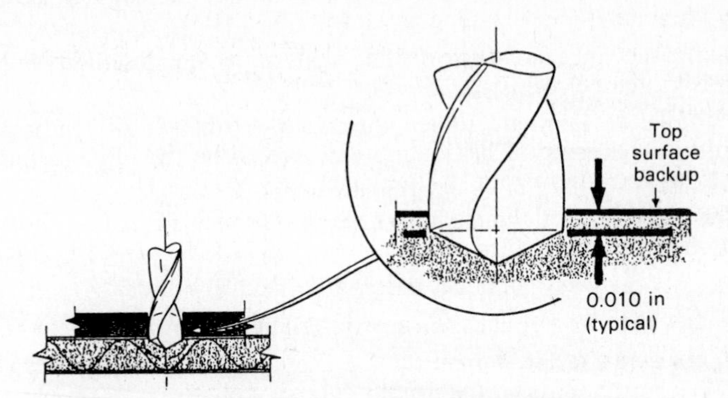

FIG. 10.20 Setting drill termination depth in the backup.

10.5.7 Laminates

All laminates are manufactured to specifications which take into account electrical and physical properties. Ease of drilling ordinarily is not considered by the laminate engineers. It is left up to the printed wiring board fabricator to develop the correct parameters to drill laminates for the best hole quality.

The various types of laminates are formed from several resins and supporting fibers. These vary from common to exotic types which differ in the ease with which they may be drilled. From easiest to most difficult to drill, some commonly available laminates are G-10 epoxy-glass, FR-4 epoxy-glass, polyimide-glass, phenolic-paper, Teflon-glass, polyimide-quartz, and exotics.

The type of weave and the fiber thickness affect drill wander. The finer the fiber, the less drill wander. In addition, laminate is produced with various thicknesses of copper, either in double-sided laminates or in multilayers. The ratio of the copper thickness to total thickness of the laminate changes the optimum drilling parameters so that feeds and speeds should be adjusted based on the ratio of copper to substrate.

Dimensional stability, warping, bow, and twist specifications are important to the drilling operations. For example, laminates which are not flat drill with high burrs.

The thickness of the laminate to be drilled is controlled by the smallest drill diameter to be used. The controlling factor is 10 times the drill diameter. For example, the maximum material thickness for a 0.0312-in-diameter drill is 0.312 in.

As drill depth increases, the drill's deviation from true center increases. For minimum-sized drills, the deviation increases by 0.001 in for each 0.0625 in of laminate thickness. For example, a 0.0312-in-diameter drill drilling through 0.312-in laminate provides deviations of 0.005 in. If this is unacceptable, laminate thickness must be reduced.

Good multilayers and double-sided laminates should not require prebaking before drilling to fully cure the laminate. Prebaking is often practiced to stress-relieve the laminate. Laminate edges should be free of burrs so that stacking can occur with good interlaminate contact.

To produce higher-quality holes for each series or type of laminate, the following generalities can be stated:

1. The higher the T_g of the laminate, the better.
2. The laminate should be flat, uniform in thickness, and smooth, and it should exhibit high copper peel strength.
3. Prebaking the laminate before drilling helps dimensional stability but does not "correct" uncured resin.
4. The storage of laminate under controlled conditions of temperature and humidity is necessary. These conditions should be the same as the ambient conditions surrounding the drilling machine.
5. The more glass layers and the higher the glass to resin ratio, the more drill wear is obtained per hole drilled.
6. The thicker the glass diameter, the greater the drill wander.
7. Laminates that are not flat will cause interlaminate entry and exit burring.
8. The more abrasive the supporting fibers (quartz, for example), the greater the drill wear and the shorter the drilling run per drill.

Stack Perparation. Laminate panels are normally stacked three high between entry and backup materials for 0.062-in-thick double-sided laminates and one or two high for multilayers. They are placed in a pinning machine which drills a minimum of two holes and inserts tooling pins to hold the stack firmly.

Foreign materials and burrs on the laminate must be removed. This is true for entry and backup materials as well. The stack must be tight. Loose pinning causes burrs and poor registration. Poor pin alignment can cause bowed stacks. Pins and bushings should be checked for wear, and drilling machine operators should be instructed on the importance of cleanliness when handling laminates and entry and backup materials.

10.5.8 Drilling Machines

Drilling machine types vary over a wide range, from single-spindle, manually operated, bottom-drilling types, using templates for locating holes, to multiple-drill-station, CNC, automatic machines that accept stacks of laminate, eject laminate, and change tools automatically.

Machines are designed to accept laminate panels of various sizes, up to 24 ×

24 in. Costs range from $10,000 at the low end to $500,000 for the largest and most automated machines. The type of machine needed depends on production capacity and the type of process design.

Whatever the type, it is important to prepare and design the machine's environment. Machines should be located in temperature- and humidity-controlled, dust-free environments. Floors should be adequately designed to carry the machine weight. Isolated electrical power and grounding are necessary.

Plan on hiring machine operators who are experienced machinists. Machinists with jig-boring experience often make the best PWB drilling operators.

Rigorous and well-planned maintenance programs are important. The items in the following list are most important, but the list is not inclusive.

1. The actual revolutions per minute and feed rate of the machine should be determined by an independent method. Do not depend on the machine readout.
2. Vacuum systems should be maintained at high flow rates and full efficiency. It is better to overdesign this function. Filters should be replaced regularly.
3. Spindle cooling systems, coolant, and heat exhangers should be checked and kept clean so that they work efficiently.
4. Spindles and collets should be kept clean at all times, using noncorrosive cleaners that keep thin protective films on the metallic parts.*
5. Spindle runout should be kept to less than 0.005 in for heart of the range drilling and to less than 0.002 in for small hole drilling.
6. Drill bars, springs, and air-pressure seals should be carefully checked and maintained.
7. Pressure foot pressures should remain as high as the laminate will allow.
8. The *z*-axis alignment should be checked often.

Machine manufacturers' instructions regarding warmup before use should be followed carefully. Machines should be kept scrupulously clean. Dirt, chips, and debris left on machine tables should be removed with a vacuum cleaner. Never blow surfaces with compressed air to remove dirt.

There are two aspects to drill wander: precision and accuracy. Only accuracy is affected by machine performance. The machine accuracy can be checked by drilling a square matrix of holes. The lines of holes can be defined in the x and y directions by comparing the mean lines to reference holes via regression analysis. These best-fit lines are compared with the desired lines which the machine was instructed to place. From the slopes, the orthogonality of the machine can also be determined.

Precision of drilling is dependent on many factors. One of these is spindle runout. The less the spindle runout, the greater the precision of drilling. With good spindles and a well-maintained machine, precision is 6 times better than accuracy for 0.043-in-diameter drills and about equal for 0.0135-in-diameter drills. These figures are based on drilling 0.100 in of laminate.

Precision is also affected by chip load, roughness of laminate surface, entry material roughness, entry material construction, drill bit diameter, glass weave, glass thickness, drill design, and drill concentricity.

---

*Often a dirty or worn collet or a dirty collet seat will increase the total indicated runout (TIR) or cause the runout to be excessive. Before replacing a spindle due to excessive runout, it is wise to first clean or replace the collet, then remeasure the TIR.

10.5.9 Feeds and Speeds

Feed rate in inches per minute is converted to chip load (CL) by using this formula:

$$CL = \frac{\text{feed rate, in/min}}{\text{r/min}}$$

Table 10.6 gives the best surface speeds and chip loads for FR-4 laminate with 1-oz copper on both sides.

This table is an example of drilling parameters for double-sided 1-oz copper laminate. If the ratio of copper to substrate changes, the drilling parameters must change.

TABLE 10.6 Optimum Feeds and Speeds as a Function of Drill Diameter

Drill diameter, in	Chip load, mils/revolution	Surface speed, sfm	Spindle speed, r/min
0.075	1.8	450	25,000
0.070	2.0	500	27,280
0.065	2.2	525	30,850
0.060	2.5	550	35,010
0.055	2.8	575	39,930
0.050	3.0	600	45,840
0.045	3.5	600	50,930
0.040	4.0	600	57,300
0.035	4.5	600	65,480
0.030	4.0	600	76,390
0.025	3.0	524	80,000[a]
0.020	1.0	419	80,000[a]
0.015	0.5	314	80,000[a]
0.010	0.15	209	80,000[a]
0.005	0.05[b]	105	80,000[a]

Notes for Table 10.6: (*a*) It is assumed that 80,000 is the maximum r/min of most drilling machines. If your machine can exceed this r/min, use higher values, but limit the surface feet per minute (sfm) to 200 for drill diameters less than 0.015. (*b*) Chip loads are kept low to prevent drill breakage.

For drilling multilayers, sfm and chip loads are reduced to about 450 and 0.003 in per revolution, respectively, for 0.040-in-diameter drills because of the increase in the ratio of copper to substrate. Corresponding reductions in chip load and sfm are used for all drill sizes.

Exotics are drilled with different parameters, depending on type. These must be determined by each driller on the basis of hole quality and drill durability. As an example, optimum parameters for two types of Teflon laminates and two drill sizes are shown in Table 10.7.

10.5.10 Inspection and Grading Holes

It is extremely helpful to follow a drilling run by plotting hole quality against the number of hits or holes drilled. This can be done to optimize drilling parameters

TABLE 10.7 Optimum Parameters for Drilling Teflon

Drill size	Type I*	Type II†
57	Surface speed = 260 sfm CL = 2.0 mils/revolution	Surface speed = 425 sfm CL = 2.5 mils/revolution
80	Surface speed = 225 sfm CL = 0.88 mils/revolution	Surface speed = 237.5 sfm CL = 1.31 mils/revolution

*Type I = 10 high 0.012 Teflon with ½-oz Cu both sides.
†Type II = 2 high 0.062 Teflon with ½-oz Cu both sides.

for a new product or laminate, or to decide which drills, entry and backup system, or feeds and speeds are best. This type of plot is also important to determine how long a drill can be run.

In order to do this correctly, methodology is all important, since the methods used and the grading scale must be adequate in defining hole quality. The best method available is called the hole-quality standard.

The method works as follows: The bottom laminate is always used for analysis of hole quality. Either the bottom laminate is sacrificed, or coupons are designed into the laminates so that two holes are drilled in the x direction and two holes in the y direction for each hit region of interest. For example, for each drill size, at the beginning of the drilling run, hit 1000, hit 2000, and so on to the end of the run, two holes are drilled in each coupon after each hit region has been completed. The coupons are removed from the bottom laminate by routing or by using a cutting wheel to cut away the coupons, one for the x direction and one for the y direction. These coupons are shown in Fig. 10.21.

If, instead, the bottom laminate is sacrificed, each region of interest is sectioned. Hardened steel wheels, carbide, or diamond wheels are all adequate. Heart

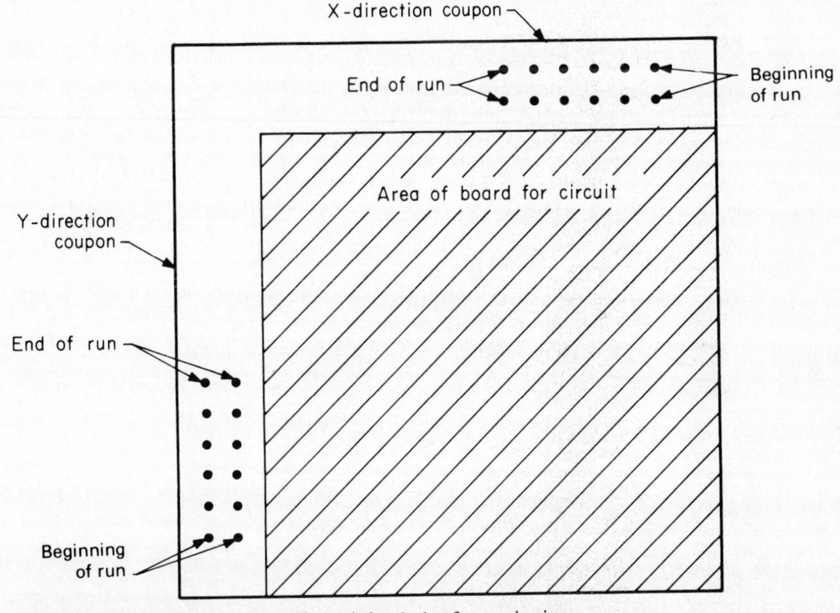

FIG. 10.21 Coupon method of obtaining holes for sectioning.

TABLE 10.8 Defects in Holes: Qualitative and Quantitative Factors*

Defects	Definition	Weighting factor (a)	Extent factor (b)
Copper defects:			
Delamination	Minimum magnification needed to clearly see the defect:	50.0	
	140×		0.01
	100×		0.08
	60×		0.30
	20×		1.20
Nailheading	Nailhead width:	1.5	
	0.00012 in		0.01
	0.00032 in		0.08
	0.00062 in		0.30
	0.00102 in		1.20
Smear	% of copper area covered with smear:	1.5	
	1%		0.01
	11%		0.08
	26%		0.30
	36%		1.20
Burr	Burr height:	1.0	
	0.00006 in		0.01
	0.00016 in		0.08
	0.00031 in		0.30
	0.00051 in		1.20
Debris	% of copper area covered with debris:	0.3	
	1%		0.01
	11%		0.08
	26%		0.30
	36%		1.20
Roughness	Minimum magnification needed to clearly see the defect:	0.2	
	140×		0.01
	100×		0.08
	60×		0.30
	20×		1.20
Substrate defects:			
Delamination	Minimum magnification needed to clearly see the defect:	20.0	
	140×		0.01
	100×		0.08
	60×		0.30
	20×		1.20
Voids	Minimum magnification needed to clearly see the defect:	0.8	
	140×		0.01
	100×		0.08
	60×		0.30
	20×		1.20

TABLE 10.8 Defects in Holes: Qualitative and Quantitative Factors (*Continued*)

Defects	Definition	Weighting factor (a)	Extent factor (b)
Debris pack	% of substrate area covered with debris pack:	0.8	
	1%		0.01
	11%		0.08
	26%		0.30
	36%		1.20
Loose fibers	% of substrate area covered with loose fibers:	0.3	
	1%		0.01
	11%		0.08
	26%		0.30
	36%		1.20
Smear	% of substrate area covered with smear:	0.3	
	1%		0.01
	11%		0.08
	26%		0.30
	36%		1.20
Plowing	Minimum magnification needed to clearly see the defect:	0.2	
	140×		0.01
	100×		0.08
	60×		0.30
	20×		1.20
Rifling	Minimum magnification needed to clearly see the defect:	0.2	
	140×		0.01
	100×		0.08
	60×		0.30
	20×		1.20

**Qualitative*: weighting factors relating defects to plating problems. *Quantitative*: values of the defects measured.

of the range holes are cut in half using the cutting wheel. Small holes are not cut in half; the cut is made near the edge. The section containing the five small holes is then sanded down until the holes are opened. In this manner, not much of the hole is wasted, and most of the wall area can be examined.

The coupon method is handled in the same way, except here, two holes for each coupon and hit region are sectioned.

The preparation for examination is to blow out each section with clean, dry air so that loose dust and debris from the cutting and sanding operations are removed.

The burr height is measured before sectioning with a profilometer or is examined after sectioning with a microscope.

Ordinary light microscope examination is then performed on the sections. Four or five hole sections are used for each hit region to avoid statistical anom-

HOLE QUALITY = 8.6

NOTE

The dark material on the copper is loose drilling debris.

DEFECTS	EXTENT	$a_i\ b_i$
Copper		
Smear	2%	.026
Burr	.00009 in.	.031
Debris	5%	.011
Substrate		
Debris Pack	2%	.014
Loose Fibers	5%	.011
		$\Sigma a_i\ b_i = .093$

$$\text{Hole Quality} = 10\,(.2^{(\Sigma a_i\, b_i)}) = 10\,(.2^{(.093)}) = 8.6$$

FIG. 10.22 High-quality hole.

HOLE QUALITY = 1.3

NOTE

The light gray area on the top copper is resin smear. The nail heading or internal burring is on the surface between the bottom copper and the substrate.

DEFECTS	EXTENT	$a_i\ b_i$
Copper		
Nail Heading	.00035 in.	.153
Smear	15%	.208
Burr	.00037 in.	.570
Debris	10%	.022
Roughness	100X	.016
Substrate		
Debris Pack	10%	.058
Loose Fibers	15%	.042
Smear	30%	.198
Plowing	100X	.016
		$\Sigma a_i\ b_i = 1.283$

$$\text{Hole Quality} = 10\,(.2^{(\Sigma a_i\ b_i)}) = 10\,(.2^{(1.283)}) = 1.3$$

FIG. 10.23 Poor-quality hole.

0-10 Hole quality rating

			Test		Date	
			Row		O.P.	
			Hole			

H.Q.	
S.E.M.	

Copper defects

	Q	0.01	0.08	0.30	1.20		E	a_ib_i
1.	50	0.500	4.0	15.0	60.0	Delamination		
2.	1.5	0.015	0.120	0.450	1.80	Nailheading		
3.	1.5	0.015	0.120	0.450	1.80	Smear		
4.	1.0	0.010	0.080	0.300	1.20	Burr		
5.	0.3	0.003	0.024	0.090	0.36	Debris		
6.	0.2	0.002	0.016	0.060	0.24	Roughness		

Copper Σa_ib_i

Substrate defects

	Q	0.01	0.08	0.30	1.20		E	a_ib_i
1.	20	0.200	1.60	6.00	24.0	Delamination		
2.	0.8	0.008	0.064	0.240	0.96	Voids		
3.	0.8	0.008	0.064	0.240	0.96	Debris pack		
4.	0.3	0.003	0.024	0.090	0.36	Loose fibers		
5.	0.3	0.003	0.024	0.090	0.36	Smear		
6.	0.2	0.002	0.016	0.060	0.24	Plowing		
7.	0.2	0.002	0.016	0.060	0.24	Rifling		

Substrate Σa_ib_i

Copper Σa_ib_i

Σa_ib_i

	Smear, loose fibers, debris	Burrs	Nail-head-ing	Plowing, roughness, voids, rifling, delamination
	Area, %	Mil	Mil	Visible
None				
0.01	1	0.06	0.12	140×
0.08	11	0.16	0.32	100×
0.30	26	0.31	0.62	60×
1.20	35	0.51	1.02	20×

$$H.Q. = 10\,(0.2^{\Sigma a_ib_i})$$

FIG. 10.24 Worksheet used to determine hole quality.

alies. If SEM and EDX are available, these can help to make a determination, but they are not absolutely necessary.

Hole quality is defined by separating defects of drilling into copper defects and substrate defects. Each defect is given quantitative and qualitative factors. Table 10.8 lists the defects, the qualitative factors (*a*), and the quantitative factors (*b*).

This method takes its perspective from the plating operation. How badly does each defect interfere with plating? If the defect is severe, then the qualitative and quantitative factors are larger.

This method asks these questions: Which defects are present? How much of each defect is present? How badly does each defect interfere with plating?

Using the qualitative and quantitative (*a* and *b*) factors, an empirical equation defines hole quality:

$$\text{Hole quality} = 10(0.2^{\Sigma a_i b_i})$$

This equation gives results between 0 and 10, where 10 is a perfect hole and 0 is an atrocious hole. The *a* and *b* factors are arranged so that a hole quality of 7 or higher can go to plating without deburring or desmearing.

Figures 10.22 and 10.23 show two examples of this rating process. Figure 10.24 shows a worksheet that can be used with the microscopic examination, and Fig. 10.25 shows a plot of hole quality versus the number of holes drilled.

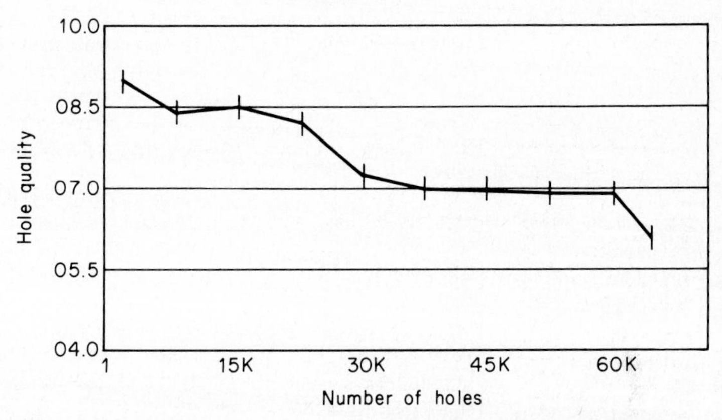

FIG. 10.25 Plotting hole quality versus the number of holes drilled.

Figure 10.25 shows that the dependent variable, "hole quality," can be followed as a function of chip load, surface feet per minute, drill bit, number of hits, or entry and backup material. Thus, rational decisions can be made about drilling. Factors can be introduced, logically, to achieve optimum drilling runs reproducibly.

10.5.11 Accuracy and Precision

It is important to understand how to produce not only high hole quality but also accurate, precisely drilled holes. "Accuracy" can be defined as how well the hole location agrees with the correct or target value. "Precision" is defined as how reproducible the hole location is.

Accuracy problems are usually due to drilling machine problems. Machine problems are due to mechanical wear, loss of computer data, and electromechanical error.

Precision problems are due to poor entry material, laminate thicknesses that are too great for the drill diameter, excessive chip loads, drill resonance, and excessive spindle runout.

Computer programs can separate and define the accuracy and precision elements in drill wander. It is important to do this to check out new machines, materials, and processes. The method uses a 100-hole square drilling program, reference holes, and use of a comparator to measure location.

10.5.12 Drilling Problems

Common drilling problems, probable causes, and solutions are as follows:

Copper defects	Causes	Solutions
Delamination of copper	Poor laminate	Replace laminate.
	CL too high	Check in-feed.
	Drill slippage	Check drill shank, diameter and collet tension.
Nailheading	Voids in laminate	Replace laminate.
	Improper drill bits	Check and replace drill bit.
	sfm too high	Check and modify sfm.
	Stack height too high	Reduce stack height.
	Laminate too thick	Reduce sfm.
	Poor entry and backup system	Replace entry and backup with cooler drilling materials.
Smear covers more than 20% of copper surfaces	Poor backup or entry	Change to higher-quality backup or entry.
	Uncured laminate	Replace laminate.
	Worn drill bit	Check and replace drill bit.
	sfm too high	Check r/min and drill size.
Entry burrs greater than 0.003 in	Poor entry	Change to higher-quality entry.
	Loose stacking	Tighten stack.
	Loose pins	Replace pinning drill.
	Pinning hole burrs	Deburr pinning holes.
	Dirt between surfaces	Clean laminate surfaces.
	Nonperpendicular pins	Repair worn fixture.
	Uncured laminate	Obtain better laminate.
	Low pressure foot pressure	Check springs, seals, and line pressure.
	Chip load too high	Reduce in-feed rate.
	Chipped drill	Check and replace drill.
Exit burrs greater than 0.003 in	Poor backup	Change to higher-quality backup.
	Loose stacking	Tighten stack.
	Loose pins	Replace pinning drill.
	Pinning hole burrs	Deburr pinning holes.
	Dirt between surfaces	Clean laminate surfaces.
	Nonperpendicular pins	Repair worn fixture.

Copper defects	Causes	Solutions
Debris on copper	Low pressure foot pressure	Check springs, seals, and line pressure.
	Chip load too high	Check for drill slippage.
	Retraction rate too high	Check and reduce retraction rate.
	Vacuum system weak or inoperative	Check and improve vacuum system.
	Drill's helix angle improper	Check drill against specs.
	Pressure foot channels plugged	Replace pressure foot.
Roughness of copper	Chipped drill	Check and replace drill.
	Worn OD margin	Check and replace drill.
	Excessive spindle runout	Check and clean spindle, collet, and collet seat.

Substrate defects	Causes	Solutions
Delamination of substrate	Poor laminate	Replace laminate.
	CL too high	Check in-feed rate.
	Drill slippage	Check drill shank diameter and collet tension.
Voids	Poor laminate	Replace laminate.
	Chipped or flawed drills	Check and replace drills.
Debris on substrate	Retraction rate too high	Check and reduce retraction rate.
	Vacuum system weak or inoperative	Check and improve vacuum system.
	Drill's helix angle improper	Check drill against specs.
	Pressure foot channels plugged	Replace pressure foot.
Loose fibers	Worn drill	Check and replace drill bit.
	Uncured laminate	Replace laminate.
	sfm too high	Check r/min and drill size.
Smear covers more than 20% of substrate surfaces	Poor backup or entry	Change to higher-quality backup or entry.
	Uncured laminate	Replace laminate.
	Worn drill bit	Check and replace drill bit.
	sfm too high	Check r/min and drill size.
Plowing	Worn drill	Check and replace.
	Uncured laminate	Replace laminate.
	sfm too high	Check r/min and drill size.
Rifling	Chipped drill	Check and replace.
	Excessive spindle runout	Check and clean spindle, collet, and collet seat.

Drill wander	Causes	Solutions
Hole locations off target	Poor machine	Check spindle runout and machine accuracy.
	Precision poor	Reduce CL, replace entry material, check drill concentricity, reduce stack height.
	Drill resonance	Drill at different r/min.
	Dirty or worn collets	Clean or replace spindle collets.

10.5.13 Drilling Summary

The drilling room should be run as a small and efficient business within the structures of the printed circuit board manufacturing plant. The success of the operation depends on many factors—each having importance—from a need for a management decision about desired hole quality to the drilling machine operator's need to handle materials with care.

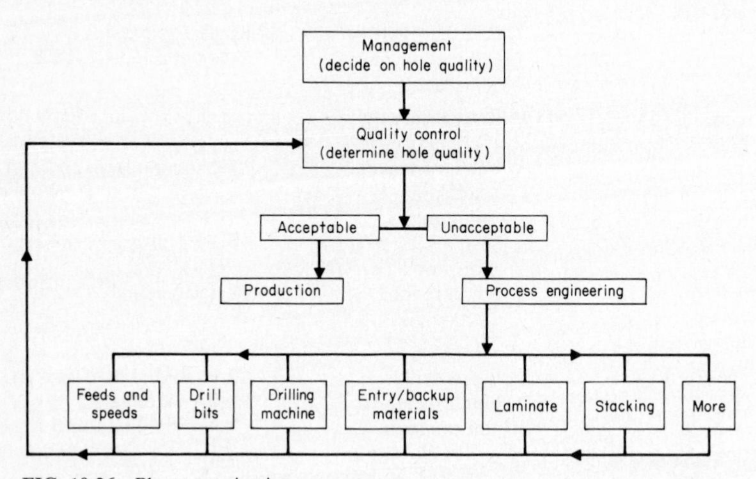

FIG. 10.26 Plant organization.

Figure 10.26 shows the interaction between the various departments involved in drilling.

The specific answer to drilling problems is to access the drilling process and optimize its individual elements. Drilling problems can be solved with proper understanding.

REFERENCES

1. A. J. Berlin and James Miller, *L.C.O.A. Drilled Hole Standard,* published privately by Laminating Company of America, Garden Grove, Calif., 1983.
2. Karl Hellwich, "The Holistic Approach to PCB Drilling," *Circuits Manufacturing,* August 1984.
3. Alvin J. Berlin, "A Computer Program to Calculate Accuracy and Precision in Drilling Printed Circuit Boards," *Printed Circuit Fabrication,* vol. 6, no. 4, 1983.
4. A. J. Berlin, "Drilling PCBs Using an Unusual Backup Material," *Printed Circuit Fabrication, Expo/85,* May 6–8, 1985, San Jose, Calif.
5. Jerry Murray, "Challenges in Drilling," *Circuits Manufacturing,* October 1984.
6. A. J. Berlin, "Excessive Drilling Heat Leads to Board Failures," *Electronic Packaging and Production,* January 1985.
7. James P. Block, "The Role of Backup and Entry Materials in PC Drilling Operations," *Electronic Packaging and Production,* November 1979.

CHAPTER 11
IMAGE TRANSFER

Lyle R. Wallig
E. I. du Pont de Nemours & Company, Wilmington, Delaware

11.1 INTRODUCTION

The image-transfer process as used in printed wiring board (PWB) manufacturing applies a resist material (screen ink or photoresist) to a copper-clad substrate. The resist materials define the circuit features which are electroplated or etched into the copper foil.

The two basic image-transfer processes are screen printing and photoprinting, which includes liquid resists and dry-film resists.

There are other PWB process techniques to form the circuit image, such as die stamping, milling, or printing the image without a scren (Cirtrak),[1] but they do not represent general industry use and will not be discussed here.

11.1.1 Image-Transfer Market Breakdown

Figure 11.1 shows that the growth of dry-film resist over the years has occurred primarily at the expense of liquid resist and some screen printing. Dry-film-resist image transfer approached parity with screen printing in the mid-1980s. Continuing developments, however, in liquid photoprinting systems such as W. R. Grace's Accutrace[2] and Ciba Geigy's Probimer[2] systems have specific applications and should be considered when selecting a system.

11.1.2 Technology Choices

The selection of the specific image-transfer technology for the fabrication of PWBs is among the most important and far-reaching decisions that a PWB fabricator will make. For most, the choices are 100 percent screen print, 100 percent dry film, or some combination of the two processes. See Tables 11.1 and 11.2 for image transfer process comparisons.

Key factors that affect this image-transfer technology decision include the following:

1. Typical or minimum line and space requirements
2. Requirement for a pattern-plating process

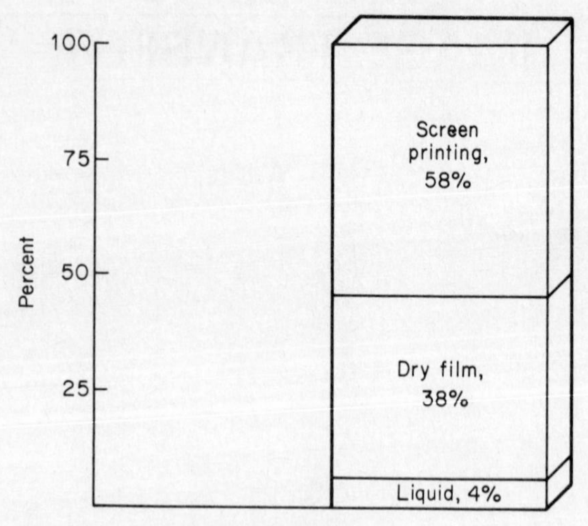

FIG. 11.1 Market breakdown (1985).

3. Nature of chemicals used in plating and etching
4. Yield experience
5. Annual volume and size of panels
6. Run length
7. Company marketing orientation, i.e., high-tech, military, consumer, multilayer
8. Capital equipment funding
9. Operator skill level

TABLE 11.1 Comparison of Image-Transfer Process Factors

Factor	Screen printing	Liquid resist	Dry film
Material cost (applied)	Low/moderate	Low/moderate	Moderate
Operator labor to process	Low/moderate	Moderate/high	Moderate
Ability to process 18 × 24-in panels	Moderate/low	Moderate/low	Moderate
Operator skill level required	Moderate	Moderate/high	Moderate
Setup and shutdown time	Moderate/high	Moderate	Low
Productivity—panels per operator hour	Moderate/high	Low/moderate	Moderate
Capital investment	Moderate	Moderate/low	High
Energy costs	Moderate low	Moderate	Low/moderate
Environmental	Low/moderate	Moderate	Moderate
Process latitude	Moderate	Low	Moderate/high
Turnaround time	Moderate	Long	Fast
Floor space	Moderate/high	Moderate	Moderate/low

TABLE 11.2 Comparison of Processing Steps for the Image-Transfer Processes

Liquid resist	Dry film	Screening
Prepare applicator.	Prepare laminator.	Prepare screen.*
Coat (spray, dip,* whirl).	Laminate.	Set up screen.*
Oven-dry.*	Expose.*	Screen image.*
Expose.*	Develop.	Oven- or infrared-dry or UV-cure.*
Develop.	Inspect.	Inspect; retouch.
Dye.	Etch or plate.	Etch or plate.
Wash and dry.	Strip.	Strip.
Inspect; retouch.		
Postbake.		
Etch or plate.		
Strip.		

*Coating or imaging and drying steps are repeated for double-sided boards except when dip-coated. Dry film is applied to both sides of double-sided boards simultaneously during the laminating step. Exposure is also carried out in double-side exposure equipment.

10. Shop floor space
11. Environmental and EPA issues
12. Screen-print solder mask or nomenclature required

A more detailed process description of the various image-transfer processes is given below.

11.2 SCREEN PRINTING PROCESS DESCRIPTION

Screen printing is a stencil-printing technique which uses a circuit pattern defined on the woven mesh of a screen fabric. A liquid resist material, usually called the "screen in," is forced through the open areas of the screen mesh not protected by the stencil onto the copper-clad substrate (positioned in register under the screen) by the pressure of a squeegee wiped across the top surface of the screen. The squeegee pressure deflects the screen downward in point contact with the substrate. As the squeegee passes a given point, screen fabric tension snaps the screen back, leaving the screen ink behind. See Fig. 11.2.

Screen printing is an excellent image-transfer process choice for many types of PWBs and fabrication process because of its relatively low materials cost and its high-volume throughput. As Fig. 11.1 shows, 58 percent of the PWBs in the mid-1980s were produced by screen printing. The majority of sreen-printed boards are produced for the consumer or for the electronic instrument and telecommunications markets.

Guidelines for favoring screen printing include the following:

1. Line and space resolution >0.3 mm (0.012 in)
2. Panel size <42 × 42 cm (18 × 18 in)
3. Registration requirements ±0.12 mm (0.005 in)
4. Run lengths >200 panels

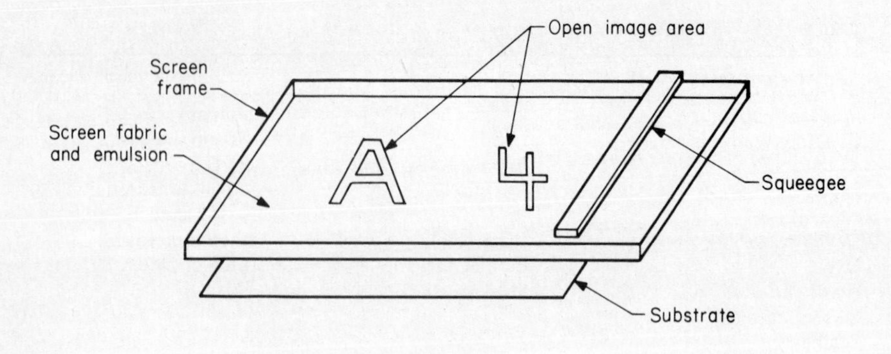

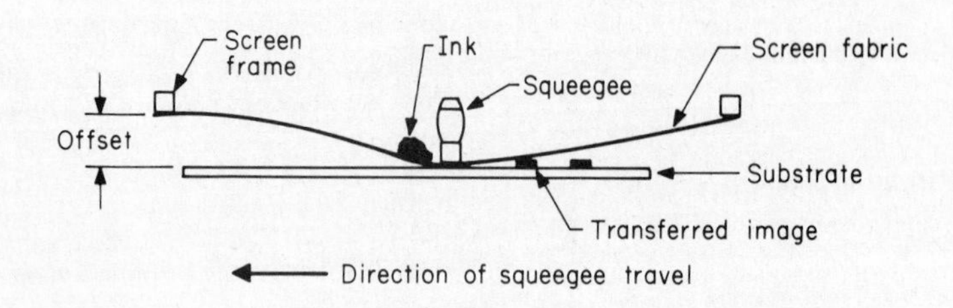

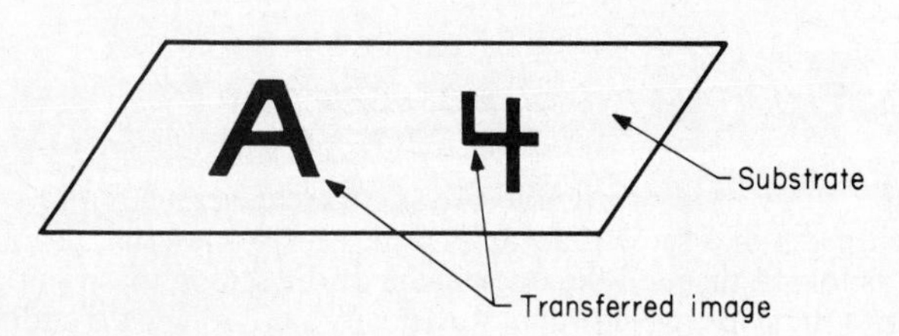

FIG. 11.2 Screen-printing process.

Some companies may profitably screen-print PWBs without following these guidelines, but photoprinting with dry film or liquid resist is usually considered a better approach.

11.3 SCREEN-PRINTING PROCESS

11.3.1 Screen Fabric

The woven mesh fabric used in screen printing may consist of stainless steel, silk, polyester, nylon, or a similar inert woven material. The screen fabric mesh is classified according to the number of threads or wires (openings) per linear inch. Thread or wire diameter, weave style, and monofilament or multifilament are factors to consider along with the mesh classification. The larger the mesh classification number, the more wires per linear inch and, therefore, the smaller the openings and the finer the resulting circuit image. It may also be much more difficult to push the screen ink through the small screen openings onto the substrate surface (see Table 11.3).

11.3.2 Screen Preparation

A screen is prepared by first analyzing the image-transfer line and space requirement, panel size, fabrication process, run length, and whether the job will be hand- or machine-screened. Only then can the fabric type, mesh size, frame, stencil, and ink compatibility be properly determined.

Once this analysis is accomplished, the fabric is cut and mounted into the rails of the frame and initially tensioned. The fabric strands should be parallel to the frame sides for strength and distortion reasons.

Tensioning should be done by uniformly tightening the fabric going from side to side on the frame and measuring tension with appropriate tension gauges. Operator feel is not adequate, especially with stainless steel. Stainless steel fabric should be tensioned slowly by bringing the fabric tension to its elastic limit. This may need to be measured and achieved over a 2- to 4-day period.

If fabric tensioning is not done correctly, the screen may subsequently stretch or tear in use. In between these extremes, there may be registration or image-distortion problems.

The screen frames themselves should be routinely inspected for squareness, twisted rails, broken corner joints, loose fittings, and similar defects that would interfere with making a properly tensioned screen. Aluminum frames are preferred because of their strength and long life; however, initial cost is high.

TABLE 11.3 Screen Fabric Selection Characteristics

Material	Characteristics
Stainless steel	High initial cost Long life Good chemical resistance When properly tensioned, maintains its tension 200 to 270 mesh typical Doesn't conform well to irregular surfaces More difficult to handle Easy to clean
Silk	Moderate cost Good solvent resistance High moisture absorption may cause loss of tension Low abrasion resistance 157 mesh (16XX) typical Difficult to clean off stencil
Polyester	Moderate to low cost Good solvent resistance Maintains tension Conforms well to uneven surfaces 196 mesh typical

11.3.3 Screen Stencil

The stencil image is applied to the screen fabric by techniques described as indirect, direct, direct-indirect, or capillary. Direct-indirect and capillary are generally the preferred stencil methods for PWB fabrication. Vendors[4,5] of these systems should be contacted for additional process information. Suffice it to say that proper screen cleaning, emulsion coating or application, exposure, development, and baking conditions all contribute to long stencil life and sharp resist images.

11.3.4 Squeegee

The squeegee is usually made from elastomeric materials such as a urethane. The durometer (hardness) of the elastomer is selected to be in the Shore 50 to 75 range. The squeegee blade should be checked frequently to ensure that it is straight and has sharp corners. The squeegee should be reground if the edges have become rounded through use; otherwise, shape image definition may be lost.

11.3.5 Environment

The actual screen-printing image-transfer operation should be done in an environment that is clean and well ventilated, has good illumination and wide aisles for carts, and is generally commensurate with the quality of work being produced. It would be difficult to screen-print fine line and space images with airborne particulate matter from a drilling operation falling into the screen-printing work area. Screen-printing ink and equipment manufacturers have long advocated placing the screen-printing operation in the same kind of environment that is routinely used for dry-film-resist processing.

11.3.6 Screen Ink Classification

Screen-printing inks are characterized by the types of chemicals used to strip them or by the curing method. The classifications are as follows: alkaline- or solvent-strippable and/or thermal-, UV-, or air-cured.

Chemically, the screen-printing inks are acrylic, epoxy, vinyl, asphaltic, or similar species which are resistant to PWB processing chemicals and conditions. They are low to moderate in cost and are easily obtained from many sources. Usually several screen-printing compositions are used in a shop so that the best ink and screen fabric can be selected for the job at hand.

The permanent screen inks used as solder masks and nomenclature resists are one- or two-part epoxy systems. The thermal curing inks are usually baked in a convection oven (30 to 60 min at 125°C) or cured by infrared (IR) energy in a conveyorized mode. The ultraviolet (UV) curing inks cure by exposure to high-

TABLE 11.4 Characteristics of UV Curing Inks

100 percent solids (no mixing on solvents)
High thixotropicity (increased film thickness)
UV cure (less energy expended)
Rapid cure (conveyorized)
High productivity (to print second-side image)

intensity mercury vapor lamps in a conveyorized mode. Both the IR and UV systems are single-sided; the second-side image, if required, is applied after the first side is cured.

As a class, the UV curing inks, especially the alkaline-strippable ones, have enjoyed great popularity because of the characteristics shown in Table 11.4. Typical problems with UV curing inks include the following:

Difficult to cure completely when wet-film thickness increases

May become brittle if overcured

Adhesion to copper more of a problem

Batch-to-batch consistency

11.3.7 Screen-Printing Equipment

Advances in screen-printing machine design and operation have accelerated the decline of the hand or manual screen-printing operation. Semiautomatic and fully automatic screen-printing machines[6] are available along with automatic feeders, rackers, and conveyorized curing units to make this operation very efficient. Standard panel sizes, 0.25- to 0.4-mm (0.0010- to 0.016-in) line and space, and long run lengths (i.e., >200 panels) are favored for machine printing. The highest resolution in screen printing is obtained by those skilled in hand screening. These artisans can control and adjust their techniques to changes in ink viscosity, humidity, screen tension, and condition of the substrate. But hand screening is generally too slow for most PWB operations.

11.3.8 Screen Process Troubleshooting

Table 11.5 shows typical problems, causes, and potential solutions for the screen image-transfer process.

11.4 LIQUID PHOTORESISTS

The commercial PWBs produced from the mid-1940s[7] to the late 1960s would have imaged by either screen printing or liquid photoresist material and processes. Screen printing was a favored process, since it could produce the required 1.0- to 10-mm (0.025- to 0.250-in) line and space circuit image at a fast rate (60 to 100 panel images per hour) and at low cost. Yields were high because of the relatively wide line widths, ferric chloride etching, and simple panel-plating processes. Liquid resists were later used by those companies seeking plated-through holes and high resolution.

Liquid photoresist materials and processes were taken from the chemical milling and printing plate industries and applied to the requirements of PWB fabrication. Kodak's[8] KPR and KMER quickly became the standards for making fine-line PWB images. These materials were followed by Shipley's[9] AZ positive-working resists and Kodak's KTFR. See Table 11.6 for general characteristics of liquid photoresists.

Liquid photoresists, however, have been difficult to work with. The processing was long and difficult to control, especially when dip coating was used to apply the resist. The coating thickness was uneven, and the plated-through holes (PTHs) had plugs of resist that were difficult to develop away. Pinholes (dirt) in the resist

TABLE 11.5 Typical Screen-Printing Troubleshooting Guide

Problem	Cause	Solution
Repetitive voids or spots	Material dried in screen	Clean screen.
	Old emulsion	Strip and remake stencil.
	Dirt	Move to clean area.
	Pinholes in stencil	Touch up screen.
	Emulsion breakdown	Remake stencil.
Bleed or smeared image	High solvent level	Reduce solvent level.
	Wrong type of ink	Contact vendor.
	Too much ink	Use finer-mesh screen.
	Viscosity of ink too low	Adjust.
	Too long a period between screening and curing	Cure immediately after screening.
	Excessive offset distance	Reduce distance.
Poor adhesion to copper (resist breakdown)	Inadequate copper cleaning	Clean and use water break test.
	Inadequate cure	Check time and temperature of curing.
	Chemical attack by high-pH cleaners or etchants	Reduce pH or switch to more stable ink.
	Wrong solvent in ink	Use fresh ink batch.
	Ink thickness too low	Use coarser-mesh screen or lower-viscosity ink.
	Stencil breakdown	Check stencil-making technique.
		Change to more compatible stencil.
	Ink viscosity incorrect	Adjust viscosity or change inks.
		Use finer-mesh screen.

image necessitated inspection and touch-up. Roller coating of the Shipley AZ positive-working liquid resists kept the resist out of the holes and was generally easier to work with than dip or spray coating the liquid negative photoresists. Resist thickness, however, remained a major shortcoming in the pattern-plating process for liquid photoresists.

TABLE 11.6 Characteristics of Liquid Photoresists

Solvent-based
Negative-working (except Shipley's AZ resist)
Applied by dipping, spray, whirl, or roller coating
Dried before exposure
Dried-film thickness 2.5 to 8.0 μm (0.0001 to 0.0003 in)
Relatively long exposures (2 to 5 min) carbon arc and UV sources
Developed in TCE vapor degreaser or similar chlorinated solvent
Image dyed after developed for inspection
1 to 2 h bake required to harden image for plating or etching
Difficult to strip
Pinholes a problem

11.4.1 Developments in Liquid Photoresists

In the early 1980s, W. R. Grace[2] introduced its Accutrace automated liquid photopolymer roll coating and exposure system for PWB fabrication. The proprietary system applies a 0.17- to 0.45-mm (0.0007- to 0.0018-in)-thick resist coating to one side of the panel by roll coating followed by an out-of-contact exposure. Development and stripping are done with low-cost aqueous chemistry. Productivity of 200 9.5- × 11.8-cm (24- × 30-in) single-sided or 100 double-sided panels per hour is claimed. Also claimed are image-transfer costs near those of screen printing but with the resist performance characteristics approaching dry-film resist. The primary application for the Accutrace system has been in high-volume print-and-etch operations, although the system can also be used for solder mask application and exposure.

In the late 1970s, Ciba Geigy[3] introduced a liquid epoxy-based photopolymer system for the solder masking market. The automated Probimer system coats one side of the panel in a curtain coating operation, followed by drying, exposure, solvent development, and curing. The capital equipment investment may be high, but the applied cost is claimed to be between that of screen printing and that of dry-film solder mask.[10,11] Performance of the solder mask resist is stated to be excellent. It is anticipated that the Probimer system could also be used for applying a photoresist that would be used for plating and etching.

Neither the Accutrace nor the Probimer system has demonstrated the capability to tent over and protect holes during plating or etching processes.

11.5 DRY-FILM PHOTORESISTS

Dry-film photoresist image-transfer technology was introduced to the electronics fabrication industry as an alternative to screen printing and liquid photoresists by Du Pont[12] in 1968, followed by others.[13,14] The introduction of dry-film technology revolutionized PWB fabrication. Specifications on high-density PWBs require 25 mm (0.001 in) copper minimum in the PTHs and 71 mm (0.0028 in) copper for the surface conductors. The increased thickness offered by the dry-film-resist structure (to contain the electroplating) over liquid resists[15,16] and screen printing was ideal to meet the specifications using a pattern-plating process.[17] PWB fabricators in the computer, military, and telecommunications industries converted to the dry-film-resist technology and to the pattern-plating process.

11.5.1 Dry-Film Selection

Factors in the selection of a dry-film-resist image-transfer system are outlined in Tables 11.7 and 11.8.

11.5.2 Dry-Film-Resist Structure

The dry-film-photoresist system is a unique method for producing the circuit image for plating or etching. The dry-film-photoresist structure consists of a photopolymer layer 17 to 75 μm (0.0007 to 0.003 in) thick sandwiched between polyester (Mylar) and polyolefin films (Fig. 11.3). The dry-film-resist system quickly replaced liquid-resist techniques as the image-transfer process for PWB fabrication. Increased thickness, uniformity, and ease of application were the strong points for the dry-film process.

TABLE 11.7 Selection of a Dry-Film Resist

Factor	Aqueous	Semiaqueous	Solvent
Material cost, ¢/ft^2 (ratio)	103	113	100
Developing cost:			
Materials	Low	Moderate	High
Labor	Low to moderate	Low to moderate	Low
Productivity	Moderate	Moderate	Moderate
Overall	Low	Moderate	Moderate
Stripping costs:			
Materials	Low	Moderate	High
Labor	Low to moderate	Low to moderate	Low
Productivity	Low	Low	High
Overall	Moderate	Moderate to high	Moderate
Imaging costs	Moderate	Moderate	Moderate
Capital equipment costs	Low to moderate	Low to moderate	High
Waste disposal or treatment cost	Low	Low	Low
Process latitude	Moderate	High	Moderate to high
Image resolution	Moderate to high	High	Moderate to high
Printout image	Yes	No	Yes

With the typical three-layer resist structure, the polyolefin separator sheet is automatically removed during the lamination step. The polyester film remains in place during exposure, providing both a protective physical surface and an oxygen barrier for the photopolymer layer. It is removed prior to development of the unexposed photopolymer areas. The polyester and polyolefin layers allow the dry-film resist to be wound, without sticking to itself, into rolls and packaged in lengths of 122 to 305 m (400 to 1,000 ft).

TABLE 11.8 Dry-Film Resist: Process Compatibility Comparison

Process	Aqueous	Semiaqueous	Solvent
Preplate cleaner:			
Hot alkaline	Poor	Good	Good
Warm alkaline	Fair to good	Good	Good
Acidic type	Good	Good	Good
Copper etchants:			
High-pH alkaline	Poor to fair	Good	Good
Low-pH alkaline	Fair to good	Good	Good
Acidic types	Good	Good	Good
Copper electroplating:			
Copper pyrophosphate plating	Poor	Good	Good
Acid copper sulfate	Good	Good	Good
Tin-lead electroplating	Good	Good	Good
Nickel-gold electroplating	Fair	Fair to good	Fair to good

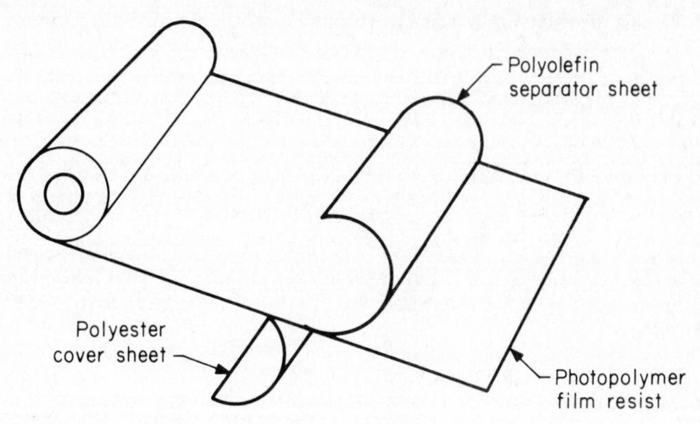

FIG. 11.3 Dry-film three-layer structure.

11.5.3 Resist Comparisons

Tables 11.9 and 11.10 define the main features and compare dry-film resists with other resist systems.

Figure 11.4 shows the magnified cross sections of screening and dry-film resist after pattern plating. Overplating or mushrooming of the plating is eliminated by containing the electroplating between the sidewalls of the resist.

The materials cost is 2.5 to 3.0 times higher for dry-film resist compared with screen printing or liquid resists. This material cost differential may be justified by increased productivity, ease of use, and yields achieved in the dry-film image-transfer process to actually reduce the *overall* PWB manufacturing cost[18,19] if the other features of fine-line requirements are needed.[18,19] An economic analysis of the process needs is important.

11.5.4 Classification of Dry-Film Resists

Dry-film resists can be classified in several ways.[20] The resists can be classified by how they respond to light energy; the terms used to describe this response are "negative working" and "positive working."

TABLE 11.9 Dry-Film-Resist Features

Feature	Benefit
Containment of electroplating between the resist sidewalls (Fig. 11.5)	Ability to pattern-plate fine lines
Tenting: Ability to cover over PTH holes and protect them from plating or etching	Eliminates secondary drilling operation for non-PTH
0.001 to 0.003 in thickness choices	Minimize effect of dirt and pinholes; pattern-plating latitude
Consistent thickness and chemical composition	Process reliability and performance
Predyed and printout image	Ease of inspection
Mylar cover sheet	Reduce handling and dirt damage
Variety of product offerings	Pick best product for job

TABLE 11.10 Comparison of Processing Steps for Image-Transfer Processes

Liquid resist	Dry film	Screening
Prepare applicator.	Prepare laminator.	Prepare screen.*
Coat (spray, dip,* whirl, roller).	Laminate.	Set up screen.*
Oven-dry.*	Expose.	Screen image.*
Expose.*	Develop.	Oven- or infrared-dry.*
Develop.	Inspect.	Inspect; retouch.
Dye.	Etch or plate.	Etch or plate.
Wash and dry.	Strip.	Strip.
Inspect; retouch.		
Postbake.		
Etch or plate.		
Strip.		

*Coating or imaging and drying steps are repeated for double-sided boards except when dip-coated. Dry film is applied to both sides of double-sided boards simultaneously during the laminating step. Exposure is also carried out in double-side exposure equipment.

In negative-working resists, the areas of the photopolymer exposed to UV energy become insoluble in the development chemistry employed.

In positive-working resists, the areas of the photopolymer exposed to UV energy become soluble and are washed away by the developing solution. In print-and-etch applications, positive-working systems appear less sensitive to opaque foreign bodies (dirt) during the exposure process. Also the positive-working resist can undergo multiple exposures and development. At this writing, there are no commercially available positive-working dry-film resists for PWB manufacture. Positive-working dry-film resists[21] have been tested successfully in microelectronics manufacture. Liquid positive resists applied by dip, spray, and roller coating have been used for many years in microelectronics as well as in a few PWB shops.

Another way to classify the dry-film resists is by the nature of the chemicals and solutions used to develop or strip the resist. Table 11.11 lists the developer and stripping chemistry for dry-film resists classified as (1) solvent-processible; (2) aqueous-processible, which is subdivided into totally aqueous and semi-aqueous; and (3) dry-processible.

11.5.5 Dry-Processible Photoresists

The dry-processible, or peel-apart, system[22–24] (Figs. 11.5 and 11.6) is designed to take advantage of differences in adhesion between the resist and the Mylar cover sheet and resist-to-copper surface in the exposed and unexposed areas. The actual "development" is due to a cohesive failure of the resist between the exposed and unexposed areas and an adhesive failure of the unexposed resist to the copper surface. So by simply peeling off the sheet of Mylar, the unexposed resist areas are moved by remaining attached to the Mylar. This dry development results in a tearing action (cohesive failure) which limits the line uniformity to approximately that achieved with screen printing. Most dry-processible films have been 25 mm (0.001 in) thick. There has been limited commercial interest in the dry-processing-resist technology because of its thickness limitations and the line quality after "developing." Additionally, aqueous resist development costs have also surpassed one of the major advantages offered by the peel-apart resists (i.e., low-cost development).

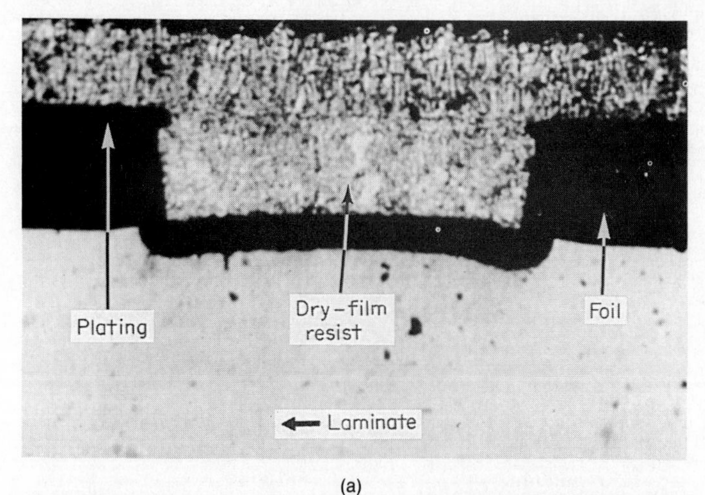

(a)

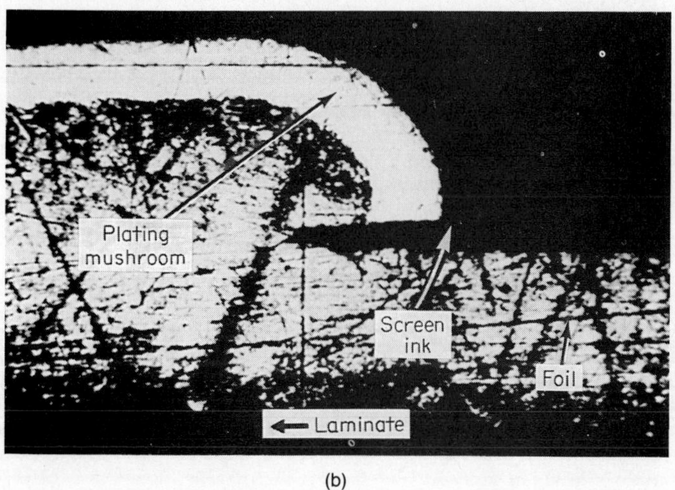

(b)

FIG. 11.4 Dry-film versus screening cross sections. (*a*) Dry film; (*b*) screening.

11.6 SELECTION OF A DRY-FILM-RESIST IMAGE-TRANSFER SYSTEM

The choice of a solvent- or aqueous-processible resist system as well as the selection of a specific dry-film resist is based on a number of important considerations (see Tables 11.7 and 11.8).

For some, the basic decision is to select either a solvent- or an aqueous-processible dry-film system. For others, the choice may be to use both solvent- and aqueous-processible resist systems along with screen printing. The aqueous resist systems would probably be a good choice for a dedicated print-and-etch line producing inner layers with acidic etchants such as cupric chloride. The aqueous

TABLE 11.11 Classification of Dry-Film Resists by Processing Chemistry

Resist classification or type	Developer chemistry and temperature	Resist removal/(stripping) chemistry and temperature	Equipment construction material
Solvent	1.1 trichloroethane, also known as methyl chloroform, 15–21°C (60–70°F)	Methylene chloride–alcohol mixtures, 15–21°C (60–70°F) Methylene chloride, 15–21°C (60–70°F)	Stainless steel
Aqueous			
Totally aqueous	1–2% sodium carbonate (by wt) at 40°C (105°F) 2% trisodium phosphate (by wt) at 29°C (85°F) 1% sodium carbonate (by wt) + 4% (by vol) butyl Carbitol* at 29–40°C (85–105°F) D-2000‡ at 29–40°C (85–105°F) *Note:* Dip in 5% H_2SO_4 may be required after development to harden image.	1–4% potassium hydroxide (by wt) at 40–60°C (105–150°F) 1–35 sodium hydroxide (by wt) at 40–50°C (105–130°F) S-1100X,‡ 32–54°C (90–120°F) 1–3% KOH or NaOH (by wt) + 4% (by vol) butyl Carbitol or butyl Cellosolve at 54–71°C (130–160°F)	PVC,† polypropylene, or stainless steel
Semi-aqueous	D-2000† at 29–40°C (85–105°F) 1% sodium carbonate (by wt) + 3–5% butyl Carbitol (by vol) at 29–40°C (85–105°F)	S-1100X,‡ 3–54°C (90–120°F) 1–4% KOH or NaOH (by wt) + 4% (by vol) butyl Carbitol or butyl Cellosolve at 54–71°C (130–160°F) Methylene chloride–alcohol mixtures at 15–21°C (60–70°F)	PVC, polypropylene, or stainless steel Polypropylene or stainless steel
Dry-processible (peel-apart resists)	None required	Methylene chloride–alcohol mixture 1–4% KOH or NaOH	PVC, polypropylene, or stainless steel

*Butyl Carbitol is Union Carbide's trademark for diethylene glycol monobutyl ether.
†PVC has 55°C (130°F) maximum service temperature.
‡Du Pont proprietary chemistry.

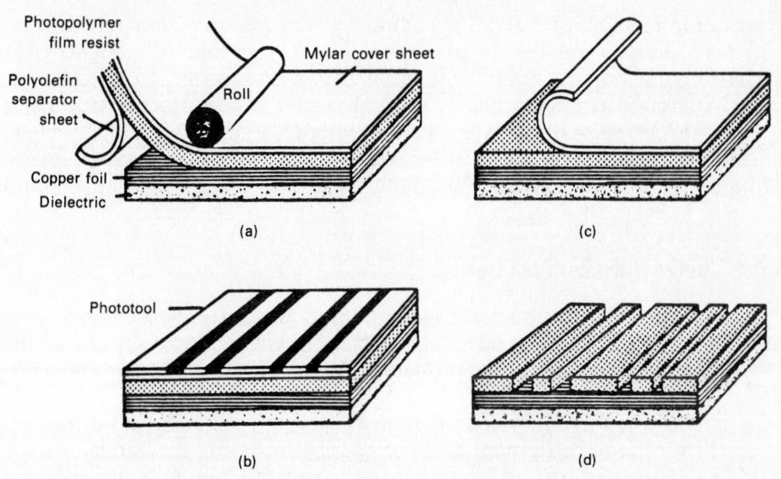

FIG. 11.5 How the dry-film process functions. (*a*) Remove polyolefin separator sheet and laminate resist to clean surface, using special laminator; (*b*) expose with UV source, using positive or negative phototool (positive to plate; negative for print-and-etch); (*c*) remove protective Mylar (readily removed by hand); (*d*) remove Mylar and develop using a special processor.

stripping chemistry would neither attack nor be absorbed into the laminate. These are very desirable characteristics for a resistant stripping process in multilayer PWB fabrication. A solvent resist system might be effectively used in a copper pyrophosphate pattern-plating process or a highly aklaline etching or a preplate cleaning environment which could be detrimental to aqueous film performance.

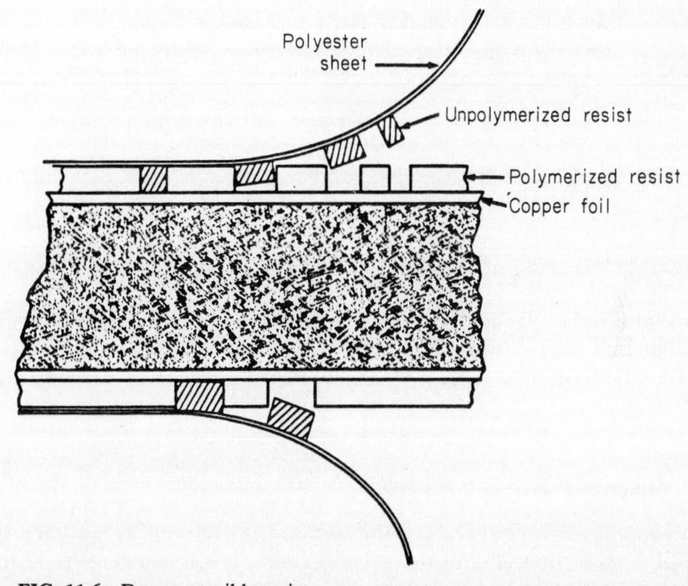

FIG. 11.6 Dry-processible resist.

In ammonia (NH_3)-based alkaline etching systems, the aqueous-processible dry-film resists perform reliably in pH environments less than 8.5. Higher-pH conditions (replenishment) may cause resist lifting or unsatisfactory resist performance. Solvent-processible resists generally offer broad process latitude along with closed-loop solvent recovery for those high-volume PCB fabricators willing to make the capital investment in the equipment. Or a solvent resist process and equipment system could be the means to meet local environmental regulations.

11.6.1 Solvent-Processible Resists

In general, the formulation technology behind the solvent-processible dry-film resist is mature. This resist chemistry has been used successfully since 1968 by most of the larger companies producing PWBs for the computer and telecommunications industries.

Criticisms of a solvent system include the necessity of using more costly stainless steel processing equipment and the possible airborne emissions of the chlorinated solvents used to develop and strip the resists. Counter arguments to these points are that the equipment is reliable and long-lasting, and a solvent airborne emission problem may be easier to deal with than waste treatment for an aqueous resist effluent discharged into the local sewer system. Both the chlorinated developer (1,1,1-trichloroethane) and the chlorinated stripper (methylene chloride-alcohol mixtures) are distillable for closed-loop operation. The resist solids are accumulated in the stills and disposed of by incineration or other locally acceptable solid waste disposal processes.

Both the solvent- and aqueous-processible dry-film resists are widely used throughout the world in pattern-plating and print-and-etch processes. The United States has a higher percentage of PWB manufacturers using aqueous resists than does the rest of the world, as shown in Table 11.12. A large portion of this usage is for print-and-etch inner-layer production.

TABLE 11.12 Estimate of Dry-Film Usage by Type

	Solvent	Aqueous
United States	31	69
Europe	65	35
Far East	90	10
Worldwide	47	53

A possible explanation for these differences in aqueous resist consumption may be the strict sewer dumping and water-quality regulations that are enforced in Europe and Japan. It is still remarkable, however, that the aqueous resists have captured the majority of the U.S. market in a relatively short time. Prime reasons are lower-cost equipment and process chemicals as well as easy sewer disposal of spent solutions.

11.6.2 Aqueous-Processible Resists

The aqueous-processible resist system was first introduced to the PWB industry circa 1971.[25] In comparison to the solvent resist, the aqueous resist formulation technology is comparatively young. The chemical performance tasks for the aqueous-processible resist are technically difficult, i.e., to develop and strip cleanly and quickly in low-temperature, dilute alkaline solutions, yet withstand hot alkaline cleaners, etching, and plating environments with good process latitude. Acceptance of the aqueous resists has accelerated as the industry has changed from hot alkaline preplate cleaners and the alkaline copper pyrophosphate plating process to room-temperature, acid preplate cleaners and the acid copper sulfate plating process.

The aqueous-processible resist system offers its users the potential for lower processing costs than those available in the solvent-processible system. The cost reduction is most evident in the chemicals used to develop and strip the resists. The aqueous system generally uses low concentrations of low-cost inorganic alkaline chemicals such as sodium carbonate, potassium or sodium hydroxide, and trisodium phosphate. Some aqueous resist systems use a semiaqueous process, i.e., 1 to 10 percent of an organic solvent such as butyl "Carbitol"* or butyl "Cellusolve"* in addition to the alkaline chemistry. The addition of the organic solvent increases the cost of aqueous developing and stripping but may also improve image quality or developing and stripping productivity.

The semiaqueous resists represent a hybrid resist technology in which the best characteristics of the solvent and aqueous films are brought together. These hybrid films have excellent process latitude and perform well in a wide range of plating and etching environments. For many users of dry-film resist, the semiaqueous products represent the performance and latitude of the solvent films with a processing cost which approaches that of the fully aqueous products. Each system (i.e., solvent, semiaqueous, or aqueous) has its champion (see Table 11.10). And the specific processes to which the resist is subjected, as well as how each user operates the process, supports the choice of a specific resist system.

11.7 DRY-FILM PHOTORESIST AND PWB FABRICATION PROCESSES

A typical process flow for the image-transfer process utilizing dry-film resists is shown in Fig. 11.7. The process flow is batch-oriented today with little automation and only fair equipment efficiency. Drilling of the holes, the electroplating, the image-transfer processes, and small lot sizes are major reasons why the PWB process remains batch-oriented. Some relief in these areas may be coming with improvements in high-speed plating technology[26,27] and an integrated image-transfer system such as Du Pont's "I" System.[28] (See Fig. 11.8.) The integrated image-transfer system is packaged in two modules. The first module automatically feeds panels vertically into the first module, which cleans the copper, applies an aqueous photoresist, trims the resist inside two of the panel edges, and removes the Mylar. In the second module, the panel is automatically registered and exposed.

The PWB manufacturing processes using an image-transfer step can be categorized into just two processes: subtractive and additive.

The subtractive manufacturing processes are as follows:

1. Print and etch
2. Sensitized and resist imaged and electrolessly plated (swell-and-etch process)

*Carbitol and Cellusolve are Union Carbide trademarks.

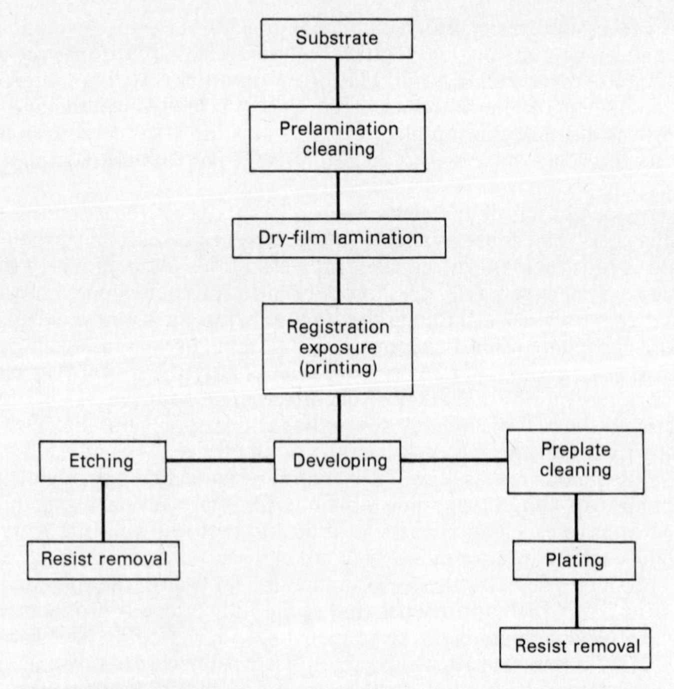

FIG. 11.7 Image-transfer process flowchart.

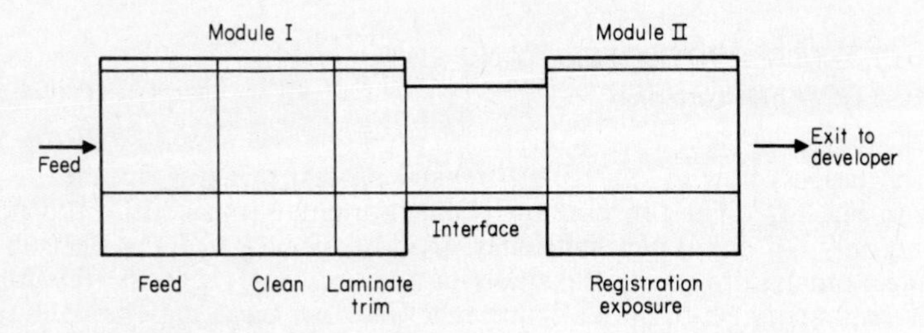

FIG. 11.8 Integrated resist system.

3. Print, pattern-plate multiple metals, and etch (includes semiadditive)
4. Print, tent, and etch

The additive manufacturing processes are as follows:*

1. Catalytic laminate and resist imaged and electrolessly plated (CC-4 process)[29]
2. Sensitized and resist imaged and electrolessly plated (swell-and-etch process)
3. Photoforming (TM)[29]
4. Photo adhesive and copper powder[30]

The basic subtractive fabrication processes are explained below.

*See Chapter 13 for complete process description.

11.7.1 Print and Etch

The print-and-etch process for PWB manufacture uses a positive resist image to protect the copper beneath the resist while the remaining unprotected copper is etched or subtracted away. The desired resist image is produced by exposing the photoresist (negative-working) with a negative image (clear circuit image) phototool (Fig. 11.9).

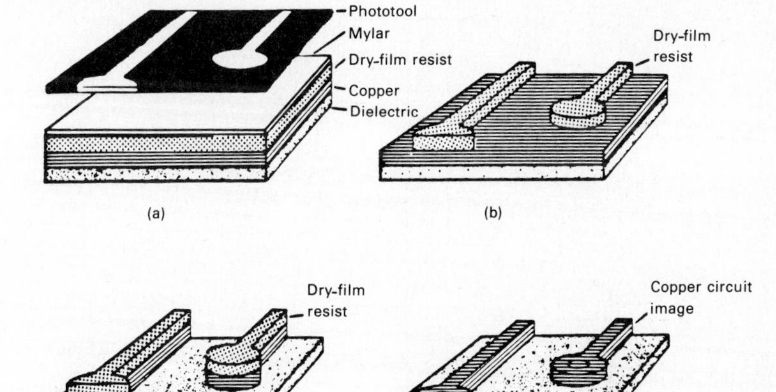

FIG. 11.9 Print-and-etch process. (*a*) Panel covered with dry-film resist is imaged using a negative phototool (clear lines with black background); (*b*) unexposed resist removed in development; (*c*) unwanted copper removed by etching; (*d*) remove resist with photoresist stripper (illustrated panel has been drilled after it was etched).

Generally, the thinner dry-film resists—17 to 30 μm (0.0007 to 0.0015 in)—are used in the print-and-etch process. If tenting of PTHs (i.e., buried vias) is a process requirement, thicker dry films may be preferred. Refer to the print, tent, and etch process in Fig. 11.14.

The thicker dry-film resists have shown more tolerance of dirt and handling and thus exhibit fewer image pinholes. Improved yields and less touch-up can easily justify the cost differential for a thicker dry-film-resist product. Obviously these benefits could be multiplied for a high-volume multilayer manufacturing operation producing a large number of inner layers. The use of an aqueous- or solvent-processible resist is a matter of choice. Both types of resist have good process characteristics in lamination, exposure, etching endurance, and strippability. Some resists may be favored because they exhibit a color change (printout image) in the exposed areas, which may aid in controlling the process. Others may offer a color contrast to the copper, which aids in the inspection process (see Fig. 11.9).

Aqueous-processible dry-film resists are preferred by many multilayer fabricators in the print-and-etch process used to produce the etched patterns for the internal layers. The main reason is that the inorganic stripping solutions used to remove the resist after etching do not attack the exposed substrate material (glass-epoxy or polymide) as severely as do the methylene chloride–based resist strippers used to strip the solvent-processible resists.

There is always anxiety on the part of the multilayer fabricator that the mul-

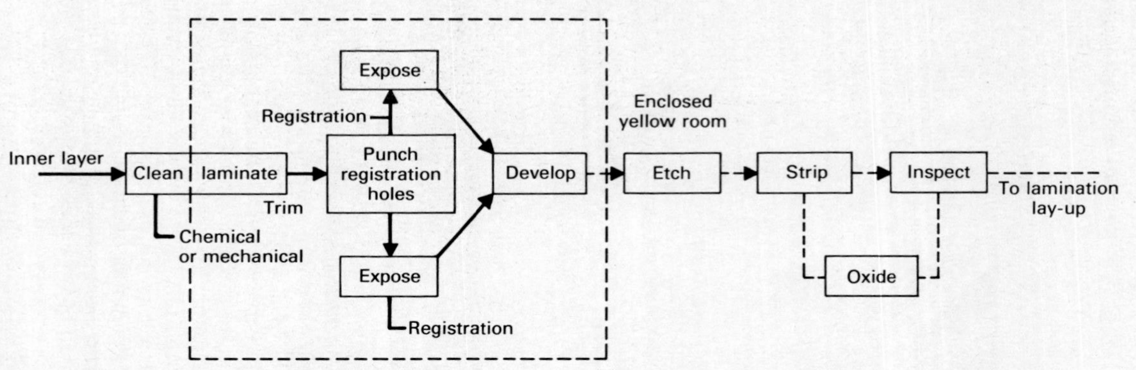

FIG. 11.10 In-line inner-layer production sequence.

tilayer boards (MLBs) may exhibit delamination at the time of wave soldering from rapid outgassing of trapped liquids, especially solvents used in resist stripping of the inner layer.

The dry-film process lends itself to an in-line all-aqueous process to produce etched inner layers, as shown in Fig. 11.10.

A separate dry-film production line (resist, laminator, and exposure and developing units) would be used to produce the images on the external layers as well as producing normal single- or double-sided panels. This approach allows for the optimization of the resists and equipment in each line, which results in high productivity, better work flow, and improved process control.

11.7.2 Pattern Plating

In this PWB manufacturing process, dry-film resist is used (after exposure and development) to define the image area which will be electroplated. The electroplating then serves as the etch resist after the dry film is stripped (Fig. 11.11). Typical electroplated metals which function as etch resists for PWBs include tin-lead, tin, tin-nickel, and nickel-gold.

When plated-through holes are required on the PWB, a panel-plating process followed by the print, plate, and etch process is used.

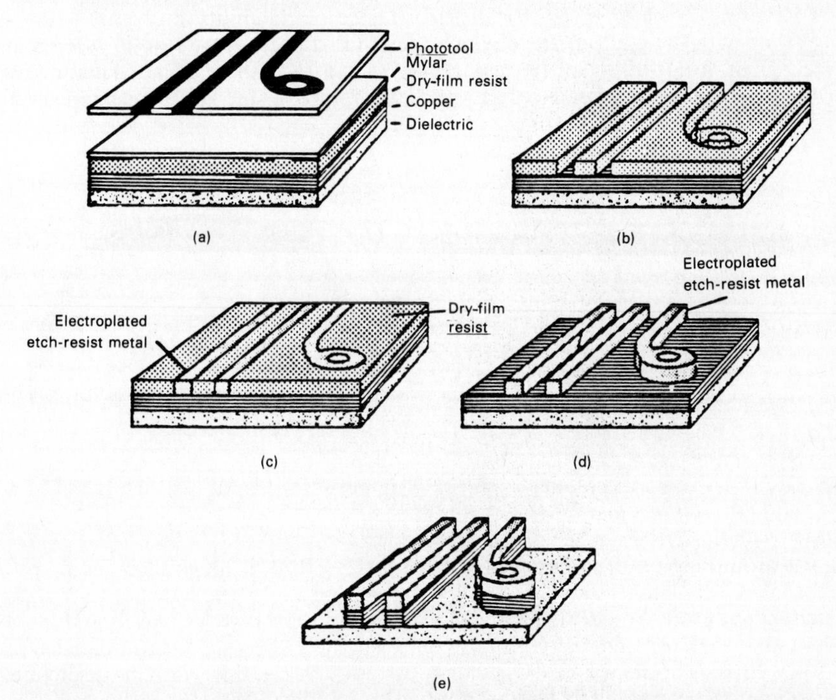

FIG. 11.11 Print, plate, and etch process. (*a*) Positive phototool used to image a panel laminated with dry-film resist; (*b*) unexposed resist, removed in development followed by preplate cleaning; (*c*) pattern plate resist metal (solder, tin, nickel-gold, etc.); (*d*) dry-film resist removed with a photoresist stripper; (*e*) unwanted copper areas removed by etching.

11.7.3 Panel Plating

1. Drill PTH.
2. Deburr.
3. Electroless copper deposition.
4. Electroplate copper.*
5. Laminate photoresist.
6. Expose (positive image phototool).
7. Develop resist image.
8. Electroplate etch-resist metal (tin-lead, etc).
9. Strip resist.
10. Etch copper.

The panel-plating process is best used on PWBs with circuit features greater than 0.4 mm (0.015 in). The reason for this is the undercut of the circuit features that results from etching the unwanted, relatively thick copper (foil thickness plus the copper electroplate). Even with 1:1 etching undercut, the line widths can be reduced 0.13 to 0.178 mm (0.005 to 0.007 in), due to the etching process. A potential problem associated with the etching undercut is the break-off of the etch-resist metal. This problem is called "slivering" and can result in electrical shorts between lines or pads.

Slivering can lead to electrical performance or reliability problems with the PWB or with the equipment in which the PWB is used. If tin-lead is used as the etch resist, a hot-oil or an IR fusing process can melt the tin-lead. This process (Fig. 11.12) effectively removes the overhang, improves the PWB appearance, and

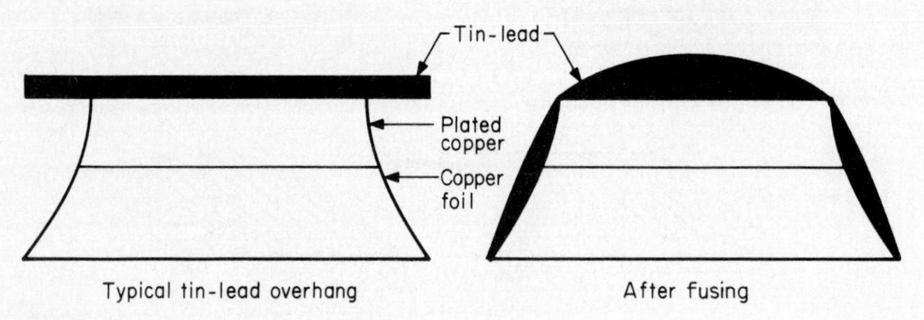

FIG. 11.12 Solder slivering and fusing.

gives an indication of the solderability of the PWB. But the pattern-plating process is preferred, since fine line widths can be produced with better control of the etching undercut. Solder fusing may still be specified, however, but more for solderability than for solder slivering. Solder fusing will show up contaminated areas as dewetted or nodular areas. Ordinarily, without fusing, this would be discovered only after the components had been mounted and the PWB wave-soldered.

*Note that no resist image is used, and the entire panel is plated.

11.7.4 Pattern Plating (Multiple Metal)

The pattern-plating process (see Fig. 11.13) is undoubtedly the most widely practiced manufacturing process to produce plated-through-hole PWBs. The process can produce PWBs with fine line definition (less etching undercut) and predictable costs.

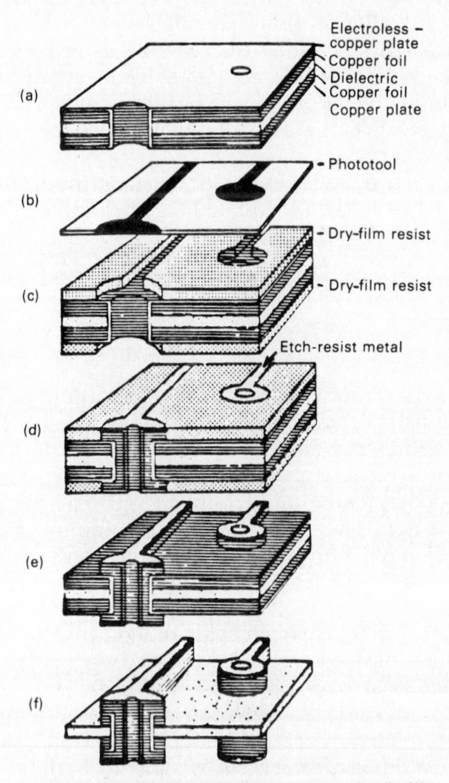

FIG. 11.13 Pattern-plating process. (*a*) Drill, deburr, and electrolessly plate copper through-holes; (*b*) a positive phototool used to image a panel laminated with dry-film resist; (*c*) develop and then preplate clean; (*d*) pattern-plate copper and then etch resist metals into the resist pattern and through-holes; (*e*) remove dry-film resist with chemical resist stripper; (*f*) spray-etch to remove unwanted copper areas.

The thickness property of dry-film resists makes them uniquely suited to the pattern-plating process. A 35-μm (0.0015-in)-thick aqueous- or solvent-processible resist product is a typical choice, although the resists are available in thicknesses from 18 to 100 μm (0.0007 to 0.004 in). Most PWBs have a plating specification of 25 μm (0.001 in) minimum copper in the hole. The 35-μm resist is thick enough to contain the copper electroplating and the etch resist metal plating (tin-lead) on the surface between the sidewalls of the resist. Conductor lines and PWB features can be produced which match the dimensions of the phototool. If the photoresist thickness is less than the thickness of the electroplating, the electroplate can grow or mushroom over the top of the resist (refer to Fig. 11.4).

This condition produces conductor line and PWB features of uncontrolled width and makes stripping of the photoresist more difficult. The mushroom effectively prevents the chemical stripping solution from diffusing into the resist. Aqueous-processible resists are particularly difficult to strip from overplated PWBs. The aqueous alkaline stripping chemicals tend to swell the resist, trapping the resist even more securely, rather than dissolving or breaking up the resist as methylene chloride–based strippers do with solvent-processible resists.

Incomplete stripping of the resist will result in copper areas that do not etch properly. These copper defects are unacceptable, since they could reduce the reliability of the PWB in service.

The pattern-plating process is also widely used on the external layers of MLBs. After the inner layers have been laminated together and drilled, the MLB can be treated as a double-sided PWB.

11.8 TENTING

Tenting (see Fig. 11.14) describes the ability of certain dry-film resists to cover, bridge, or span an unsupported area such as tooling holes or PTHs.[31] The tenting process is done with thick dry-film resists which protect the PTHs and conductor line pattern during the etching process. These dry-film resists are generally used to produce an all-copper, plated-through-hole PWB. The PWB is made by panel-plating the desired amount of copper in the holes and on the panel surface. The panel is etched using a dry film to protect the holes and conductor line pattern. Many PWBs are made by the tent-and-etch process followed by solder masking and hot-air leveling.

Hot-air leveling is a process which coats molten tin-lead on the copper areas in the PTH; the solder mask protects all the surface except the holes. This process is described as solder mask over bare copper (SMOBC).[32]

The tenting ability of some dry films can also be used to protect selected tooling holes in a pattern-plating process from receiving additional plating. This tenting ability reduces PWB fabrication costs by allowing both the circuit hole pattern and the required non-plated-through holes to be drilled at the same time. The alternative process to produce the non-plated-through holes is to return the PWBs after etching for drilling of the non-plated-through holes (secondary drilling operations). Drilling all the required holes at the same time is more efficient, more accurate, and less costly. The small amount of copper from the electroless heavy deposition (heavy dep) or flash electroplate is quickly etched out of the non-plated-through holes during the etching of the unwanted background copper areas.

In a typical copper–tin-lead pattern-plating process, the resist tent effectively keeps electroplating out of the non-plated-through holes up to 6.3 mm (0.250 in) in diameter.

11.9 THE COPPER SURFACE

The copper surface plays a major role in the success (yield) of all the image-transfer processes. The surface should be carefully inspected for pits, drilling burrs, and

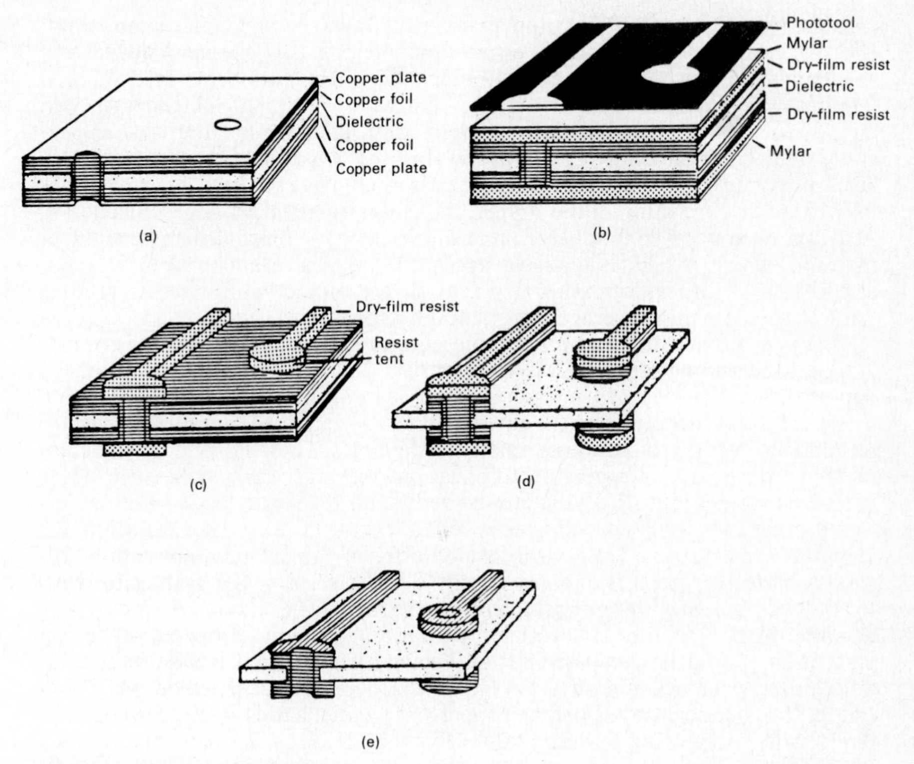

FIG. 11.14 Tenting. (*a*) Drill, deburr, and plate copper through holes to desired thickness; (*b*) laminate and image dry-film resist (use a negative phototool); (*c*) develop; (*d*) spray-etch; (*e*) remove dry-film resist with photoresist stripper.

irregularities such as electroless copper nodules or excessive chemical staining. If unacceptable defects are uncovered, no attempt should be made to process these panels through the dry-film-resist image-transfer process or any of the other image-transfer processes. The defective material should be returned to the vendor or errant workstations, where the problems can be identified and resolved.

11.9.1 Prelamination Cleaning

The physical and chemical nature of the copper surface are important parameters in the adhesion and ultimate chemical process performance of the dry-film resist. A properly cleaned surface provides good, trouble-free resist lamination. Good resist adhesion to the substrate opens a larger process latitude window in the subsequent plating and etching processes. An inadequate or poorly controlled prelamination cleaning process produces reduced process latitude and may result in resist breakdown and thus lower manufacturing yield. The more severe the chemical environment or process cycle, the more critical the prelamination cleaning process.

Cleaning, deburring processes, or material-handling practices which cause deep scratches in the copper surface should be avoided. Due to its high viscosity, dry-film resist has difficulty flowing into these deep scratches or defects during the resist lamination process. The proper choice of abrasives or chemicals can ensure effective surface cleaning without excessive damage to the copper surface.

Chemical cleaners (mild etchants or acid solutions)[33] are being used successfully by some PWB manufacturers for prelamination cleaning. Substrates seem to be particularly well suited to conveyorized chemical cleaning. The reasons generally offered are the difficulty of transporting and brush-cleaning the thin inner-layer substrate materials and concern that the abrasive rotation of the brushes may cause an elongation of the copper, thus impairing registration of this board. Also many multilayer producers are using treated or oxide-coated inner-layer material, which precludes abrasive brush or chemical cleaning prior to resist application. A vapor degreasing step may be the only cleaning used to remove loose particulate matter, process oils, and fingerprints.

The heavy dep electroless process presents an additional challenge to the prelamination cleaning process for dry-film resists. Many practitioners of the heavy dep electroless process do no mechanical cleaning prior to resist lamination. There are valid concerns that the abrasive brush action might remove the electroless copper in and around the perimeter of the holes. These fabricators rely completely on chemical cleaning (neutralization), several water rinses (including deionized water), and IR drying immediately after the electroless copper deposition process as sufficient cleaning prior to lamination of the resist. Cleaned in this way, the panels can be temporarily stored prior to the dry-film lamination process. A better approach is to apply the dry-film resist immediately after the electroless copper neutralization and then store the panels if necessary. Some aqueous dry-film resists, however, do exhibit a reduced hold time on copper, and the specific resist should be evaluated for any aging effects. Some fabricators use an antioxidant chemical dip after the electroless copper neutralization process to reduce the chemical activity of the copper surface. Such dips should also be evaluated with the specific dry-film resists for compatibility and performance. To avoid the cost and added process steps associated with copper panel (flash) plating, there is a definite industry trend to the heavy dep electroless copper process. Resist manufacturers are also developing aqueous resist products that have improved adhesion and aging characteristics on the electroless copper surface.

11.9.2 The Copper Surface Chemistry

Dry-film-resist adhesion is promoted by a high free copper state (Cu^0) and a low cupric ion (Cu^2) concentration. Scanning electron microscopy (SEM) and electron spectroscopy for chemical analysis (ESCA) have been used to characterize the copper surface conditions and to help in defining optimum cleaning methods.

Work by R. J. Adams and D. D. Keane[34] has shown that the aluminum oxide–reinforced brushes yield the cleanest copper surface with respect to residual abrasives. They have also shown that the copper surface is mainly Cu^0 immediately after cleaning. However, the deterioration rate of the Cu^0 to Cu^+ and Cu^{2+} is quite rapid in a typical manufacturing environment. The Cu^+ surface is not desirable for good resist adhesion. Dry-film-resist data sheets normally recommend that lamination be completed in as short a time as possible after cleaning and that the panels be recleaned if they are not resist-laminated within 4 h of the cleaning process. The copper surfaces are typically mechanically cleaned by abrasive brush-cleaning units employing aluminum oxide or silicon carbide abrasives. These abrasive materials are impregnated into a nylon or similar plastic matrix. The resulting brush construction can be either a compressed lamella or a filament type.* The silicon carbide 320-grit abrasive filament brush seems to be preferred

*Tynex A (DuPont).

over the compressed lamella brush for cleaning the copper surface prior to lamination of the dry-film resist, primary because of improved brush life. Coarser abrasive brushes may also be used in the deburring operation after drilling. Chemical cleaning using mold copper etchants (ammonium persulfate, hydrogen peroxide, etc.) and dilute acids may also be used alone or in conjunction with mechanical cleaning. Pumice cleaning using 4F-grit pumice in a water slurry with citric acid added is also a good cleaning system for copper. The equipment's mechanical components, however, may require considerable maintenance. Care must also be used in rinsing to completely remove all the pumice residue from the substrate. Chemical cleaning with mild etchants (ammonium persulfate, hydrogen peroxide, and acids) is very effective.

Determination of the cleanliness of the substrate or adequacy of the prelamination cleaning process is not straightforward. Often the report of resist breakdown in a plating operation will signal that the cleaning process is out of control. This is too late and after the fact. An in-process "water break" test can be used after the copper cleaning step to judge the adequacy of that process. The cleanliness of the copper surface can be described in terms of contact angle and wettability. A clean copper surface should have a very low contact angle ($180° > 0 > 0°$) and should hold a continuous film of clean water for 15 to 30 s. The water film should not break into droplets nor show areas of dewetting during this test. If a water break does occur, it indicates that the cleaning process is not being performed adequately and should be reevaluated. The water break test should be used at a minimum, at the beginning of every production run, to judge the adequacy of the cleaning process and the equipment setup. If the copper panels fail the water break test, check the equipment for brush setup and contamination on brushes (grease from bearings) or rollers. The copper surface roughness after abrasive brush cleaning, as measured by a profilometer, should have a root mean square (RMS) of 2.5 to 5.0 μm (10 to 20 μin) for optimum resist adhesion. Mechanical scrubbing under the recommended conditions will typically increase the copper surface area 3 to 4 times and produce the proper surface roughness.

11.10 RESIST LAMINATION

Dry-film resist is applied to the cleaned copper surface using heat and pressure. Roll lamination using heated rubber rolls and mechanical or pneumatic pressure is a standard practice. The HRL/LC 2400 System[35] (Fig. 11.15) is typical of the equipment used to apply the dry-film resist to one or both sides of the clean substrate.

Dry-film-resist products currently on the market are laminated to the copper-surface substrate using the hot roll lamination technique. Heating the resist in contact with the hot rolls at 100 to 120°C (212 to 248°F) just prior to the actual lamination process significantly lowers the resist viscosity. The roll pressure, 15 to 40 psi (100 to 275 kPa), then forces the softened resist into the microstructure of the copper surface to provide good mechanical keying. Most of the resist adhesion to the copper comes from this mechanical keying. A delay or hold time of 15 to 30 min after lamination allows the resist ingredients to equilibrate. The lower boiling point or more mobile ingredients have been driven by the heated rolls toward the cooler copper-resist interface. The hold time also allows time for the chemical adhesion promoting ingredients formulated into the resist to form stronger chemical bonds with the copper surface.

Preheating the substrates in a conveyorized IR system prior to entry into the resist laminator has been effective for improving the resist-to-copper adhesion for

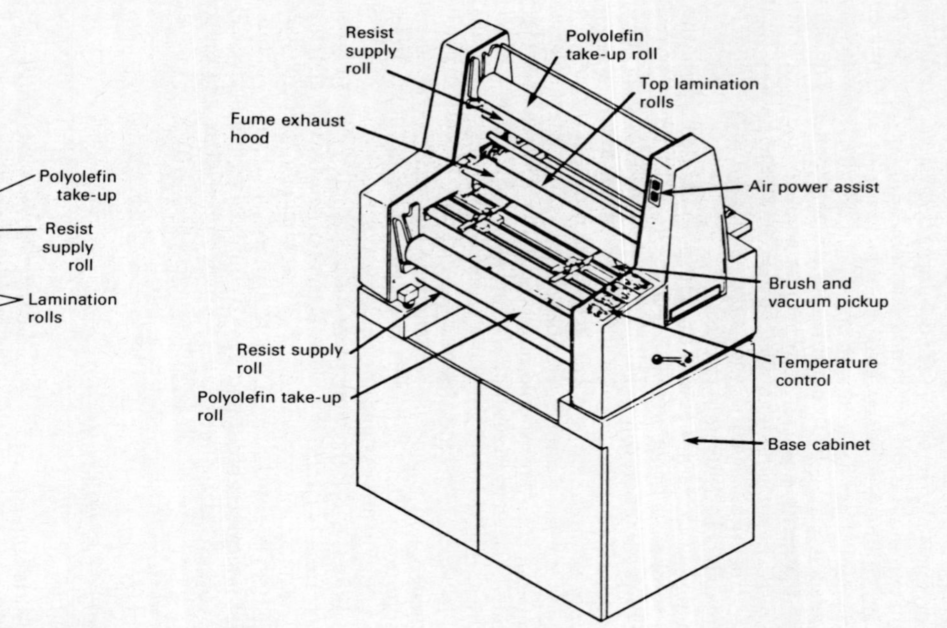

FIG. 11.15 HRL/LC 2400 resist laminator.

certain aqueous resist formulations; increased lamination speed is also a result of preheating.

Physical adhesion is favored by low-viscosity resists (better wetting) which flow into the copper microstructure with the application of pressure. But at room temperature, the viscosity of the resist must be high enough to allow it to be rolled up and shipped without the resist migrating out between the film laps, a deficiency called "edge fusion."

11.11 EXPOSURE (PHOTOPOLYMERIZATION)

The photopolymerization process comprises several chemical reactions followed by a development or readout step (see Fig. 11.16). Imaging of the negative-work-

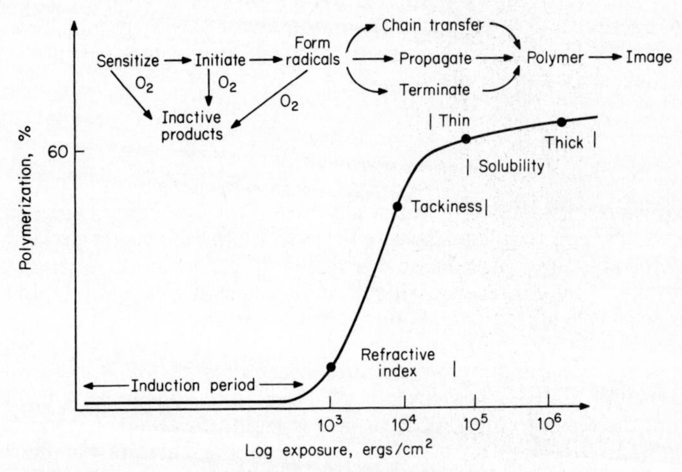

FIG. 11.16 Photopolymerization—A model.

ing dry-film resist is usually carried out with exposure units which produce UV light. The specific circuit image is placed on a high-contrast phototool (positive or negative) in contact with the resist surface. Most dry films, in practice, are exposed through a 25-μm (0.001-in) polyester cover sheet between the resist and the phototool. The dry-film resists have adequate phototool reproduction characteristics so that this out-of-contact separation is not a practical problem for the 0.125- to 0.25-mm (0.005- to 0.010-in) line and space PWBs that are produced today. This polyester sheet serves to prevent additonal oxygen from permeating the resist during the photopolymerization process. Oxygen inhibits the chain propagation reaction and results in the resist appearing to be underexposed. Inert atmospheres made by flooding the exposure chamber with nitrogen can be used if the cover sheet is removed to gain resolution. The resist can also appear underexposed if the polyester sheet is removed immediately (0 to 5 min) after exposure. Again, oxygen inhibits the chain propagation reaction which continues, though at a considerably reduced rate, for up to 30 min after the actual exposure has been made. For this reason, consistent hold times (with the polyester sheet in place) should be used when evaluating steps held in photoresists, changes in exposure unit operation, or other process parameters that relate to the photoresist image.

The polyester sheet protects the resist not only from oxygen but also from normal handling abrasion. It also provides a smooth, nontacky surface on which the phototool is placed during the exposure process. The polyester covering is best removed just prior to the development process.

A more recent development[36] is a dry-film-resist structure in which the polyester sheet has been replaced by a thin coating of polystyrene. This nontacky coating remains in place during exposure but then dissolves in the developing chemistry. Polyvinyl alcohol[26] solutions have been shown to work in a similar manner. Other developments include a new two-layer dry-film structure[28] which has no polyester sheet or flexible coating. This resist is less oxygen-sensitive, and its surface is not tacky so that the waxed phototool can be placed in direct contact with the resist without sticking during the registration-exposure process. Improved resolution is claimed along with in-line development, since there is no cover sheet to remove.

Dry-film resists are predyed photosensitive systems[37] containing sensitizers and initiators (ingredients which determine photospeed) which are primarily sensitive to the 365-nm actinic radiation emitted from mercury vapor lamp sources. Other UV energy sources such as carbon arcs, xenon, or lasers may be used, but a resist so exposed may exhibit slower photospeed and less desirable imaging results such as sloped sidewalls.

11.11.1 Exposure Control and Measurement

Proper exposure of the dry-film resist has been a misunderstood subject. Manufacturers' data sheets for the resist have listed recommended exposure equipment. But the exposure time required to reach a certain degree of polymerization is dependent on (1) light intensity, which can vary considerably with lamp type; (2) lamp age; (3) line voltage; and (4) the optical barriers between the lamp and the photoresist. It is readily apparent that using a fixed exposure time alone could lead to resist polymerization and performance problems. Ways to control resist exposure include time, step wedges (gray scales), and light integration. Physical adhesion of the resist to the copper surface is a limiting factor in obtaining good resolution when low-exposure energy values are used. Generally, the imaged resist remains soft and underpolymerized; in this condition, it is more easily washed away during the development process.

When higher-exposure energy values are used, resolution loss occurs primarily by blocking or plugging of the spaces between the lines. The resist between the lines (the opaque areas on a negative phototool which represent the spaces on the PWB) become susceptible to light undercutting (due to limited light collimation or to scatter from other sources) as the exposure time is increased. The resist in these areas becomes partially polymerized and does not wash out during the development process. A variable-density step tablet[38] based on a change in density has been used successfully to indicate the change in the level of resist polymerization with exposure time (see Fig. 11.17). Each step on the 17-step tablet represents a 0.05 change in the density (step 1 has a density of 0.50), so that the density at step 17 is 1.30.

The 17-step tablet is used by placing it in contact with a resist-coated panel and exposing the panel for a fixed amount of time. This process is repeated using several different exposure times. After the observation of consistent hold times for lamination and exposure as well as developing conditions, the panels are then developed. The step tablet image is read by determining the last step held after development. The last step held is usually defined as that step which covers at least 50 percent of the copper area beneath the step. Other interpretations can be used, but consistency is important. The resist covering that step may be wrinkled

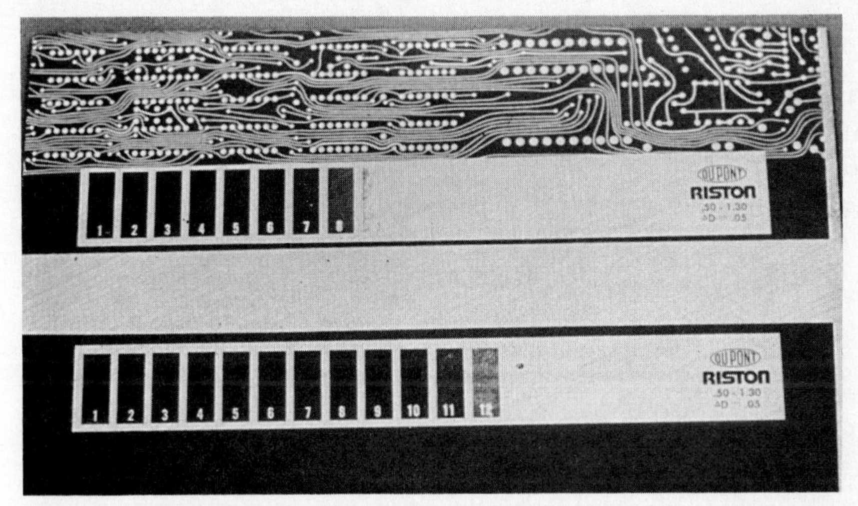

FIG. 11.17 Seventeen-step tablet placed both under and adjacent to diazo phototool.

or crinkled or partially removed by development. Tests show that this diazo phototool reduces the number of steps held from 16 to 7. If the resist required 10 steps, the exposure times would need to be increased if the diazo phototool were used.

Data sheets of most resist manufacturers now give recommended step-held ranges along with the approximate exposure times in the various commercial exposure units to achieve these steps held. The step-held range is a measure of the resist exposure latitude. This latitude is defined by such factors as artwork reproduction, resolution, sidewall geometry, and plating or etching performance.

Although step density tablets are useful in determining the initial exposure equipment setup, or as running spot checks on the image-transfer process, they are not suitable for use in controlling each and every exposure. Exposure energy is a more useful parameter to control the level of polymerization in a photoresist. Exposure units have been developed[39] which use integrating photometers (i.e., light integrators) to control the exposure level. The integrating photometers are set up using a step tablet to obtain the desired level of polymerization. This is accomplished by preparing five or six resist-coated and numbered substrates representative of the resist type, cleaning, phototools, and laminate that would be used in production. The exposure unit is cycled to bring the frame up to operating temperature. One prepared substrate is placed in the exposure frame along with the phototools and step wedges on both sides of the substrate. An arbitrary number is set on the integrator; the resist data sheet may be helpful in determining the initial integrator setting. Record the substrate number and the integrator setting. Complete the exposure. Remove the substrate and replace it with the second numbered substrate using the same phototools and step wedge. Select a different integrator setting and expose. Repeat this process for the remaining substrates. After at least a 15-min hold time, remove the cover sheet and develop using standard developing solutions, pressures, temperatures, and conveyor speeds. Examine the panels and read the step wedge values on both sides of the panel. Record these values next to the integrator settings. Select the integrator setting or step wedge value that best reproduces the phototool dimensions and has the resist characteristics that result in sharp line definition. Production work can now be exposed using the determined integrator settings; the integrator settings may be different for the top and bottom lamps. Once set, the integrating photometer is nearly independent of lamp intensity. If the exposure energy remains constant, the level of polymerization (i.e., step held) will be the same over a wide range of lamp intensities. The integrator automatically adjusts the exposure time as the lamp intensity changes to cause a consistent level of polymerization to occur in the dry-film resist.

The exposure energy E is the product of the lamp intensity I and the exposure time t:

$$E = I \times t$$

It is customary to express I in milliwatts per square centimeter, t in seconds, and E in millijoules per square centimeter.

As the PWB industry moves toward even finer lines and spaces (i.e., 0.1 to 0.2 mm, or 0.004 to 0,008 in), the ability to *measure* and *control* the key exposure parameters of intensity, time, and thus total exposure energy becomes extremely important. UV measuring devices,[40] such as integrating photometers on the exposure unit, are necessary to monitor and correct for the effects of aging lamps, spectrol shifts, line voltage changes, etc. For fine line imaging,[41] proximity and projection printing as well as laser direct imaging[42] are being evaluated by several companies.

11.12 RESIST DEVELOPMENT

The development process is the visual readout or proofing step of the photopolymerization process. With the typical negative-working photoresists, the unexposed or underpolymerized resist areas are removed by the developing solutions. Development of positive-working resists removes only the exposed areas. Positive resists, therefore, could be exposed a second time and developed again to define other required features on the panel.

The development process is the "bottom line" of a series of materials and processing events in the PWB manufacturing process. The results of the development process may indicate that the image-transfer process is well controlled, or it may reveal problems such as resist lifting or "swimming" conductor lines due to inadequate copper cleaning, poorly laminated resist, or overdevelopment. Underexposure of the resist may result in a soft and dull resist surface appearance after development. Faulty registration and exposure errors affecting line and space dimension will also be evident after development.

The development process used for most photoresists is a liquid spray technique, usually in conveyorized equipment. Such equipment allows for good control of development times (conveyor speeds), solution temperature, spray pressure, water rinsing, and an air knife for panel drying.

For horizontal development, the spray nozzle system should be designed to provide flooding of the panel (high volume, low pressure) in the entry section of the developing unit. Development rate is basically a diffusion-controlled process, so the developing solution must soak in or penetrate into the unexposed resist areas in order to begin the developing process. Once the diffusion process has begun, the softened resist can be dislodged by the use of nozzles spraying developing solution at higher pressures and lower volumes. The mechanical or abrading action of the spray impacting the resist seems to play a more significant role in the development of aqueous resist than in solvent-developed resists. To obtain front-to-back development uniformity, fan-type nozzles are often used on the top side and hollow-cone nozzles on the bottom.

Vertical processing equipment[43] has shown good performance with the aqueous dry-film resists, especially the thicker solder mask products.

Development times (conveyor speeds) are set to achieve complete development of the image in the first half or less of the developing chamber length. The remaining half of the chamber length is used to ensure that development is complete. Spray water rinsing and air knife drying are important and often neglected steps in the development process. The aqueous-processible resists, particularly, seem to require longer water rinsing times as well as better control (warmer) of the rinsewater temperature. Acid dips have been used after development to harden the resist and improve rinsing of developer residues.

11.12.1 Development Problems

Inadequate water rinsing of some resist formulations may leave resist or developer solution residues in the sidewall zone or in the developed channels, which can later result in process problems. These "scum" problems range from edge gapping to incomplete etching to electroplating metal peelers. Scum is defined as transparent resist residues (may be unpolymerized or polymerized) in the developed areas, which may cause metal-to-metal bond failure, skip plating, or incomplete etching. Edge gapping is a resist condition in which the electroplated metal does not completely fill the space within the resist channel. This is shown in Fig. 11.18.

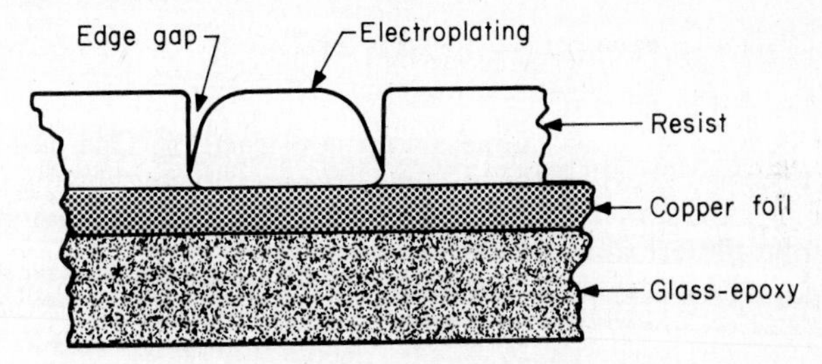

FIG. 11.18 Edge gapping.

This phenomenon results from certain resist ingredients, developer solution residues, or preplate hot-soak cleaning solutions which may migrate out from the resist sidewalls during the electroplating process and thus inhibit the electrodeposition of the metal. The copper-plating baths are particularly sensitive to showing edge gapping. Underexposure of the resist may aggravate the problem by making it more adsorptive or blotterlike. The development process does not remove all traces of the resist or the chemical adhesion promoters from the copper substrate surface. Prior to electroplating, these residues and adhesion promoters must be chemically etched away (10 to 20 μin) in the preplate cleaning process. Failure to do so can result in metal-to-metal bond failure of the electroplating. Fabricators using the heavy dep electroless process are particularly reluctant to use any preplate etching process. These fabricators may use hot-soak cleaners or reverse-current electrocleaning to remove the adhesion promoter. But the process must be carefully monitored and controlled.

Resist scum produced by incomplete development or inadequate rinsing is often transparent or invisible to the human eye. Long- and short-wavelength UV inspection lamps may have some value in finding scum. As a periodic process check, developed panels can be dipped in 2 percent potassium sulfide or cupric chloride to darken the copper. Scum areas will remain copper-colored. Alternatively, the panel may be flash-electroplated with tin-lead; scum areas will show up as unplated areas.

11.12.2 Solvent Development

Figure 11.19 shows the design of a solvent development system incorporating a solvent recovery still. The development solvent is a stabilized 1,1,1,-trichloroethane, and the equipment is of stainless steel construction. This design shows a countercurrent flow process for the solvent. Just prior to entering the water spray chamber, the panel is sprayed with solvent fresh from the still (chamber 3). This helps assure that no resist ingredients in solution are carried over into the water spray chamber, where they could precipitate out on the panel (scum) due to the nonmiscibility of the chlorinated solvent and water.

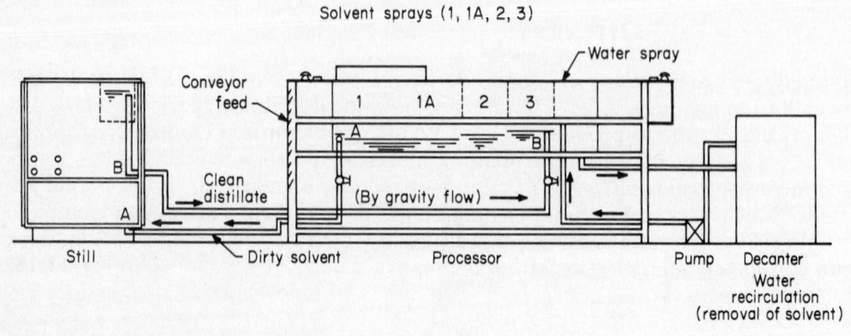

FIG. 11.19 Solvent spray developer.

11.12.3 Aqueous Development

In contrast to the continuous closed-loop operation of the solvent developing process, the aqueous development process is primarily a batch-oriented process. The developer solutions are usually one or more alkaline ingredients mixed with water (see Table 11.11). The developer solution is used until the development rate becomes unacceptably slow. At that point, the process is stopped, the solution discarded (usually into the local sewer system), and a fresh solution added to the developing unit. Potential sewer disposal problems include BOD, color, and pH. Dilution by the plant wastewater flow minimizes the problems. After the new solution is brought up to temperature, the process is ready to continue.

To minimize this equipment downtime, some users of aqueous resists are adopting a somewhat different approach to replenishing the developer solution.[44] Their premise is that the cost of the development chemicals is very low and that production equipment downtime is expensive. They have modified their equipment to use special pH probes which measure a decrease in solution pH as the amount of dissolved resist increases. When a certain pH level is reached, a measured volume of fresh developer solution is automatically added to the sump, and an equivalent volume is displaced out of the unit into the water-disposal system. This is described as a "bleed and feed" system. The discharged developer solution is approximately 50 percent spent; but due to its low cost, the fabricator is not concerned with this loss. What is gained is a continuous development process with a fairly steady-state development rate and minimum equipment downtime. Fewer sewer disposal problems result, since the discharge is less concentrated.

The aqueous development systems often add organic solvents (see Table 11.11) to improve the developing rate as well as to minimize scumming and edge-gapping problems.

Spraying alkaline solutions containing dissolved resist often results in excessive foaming of the solution inside the chamber. Foaming may reach a level which inhibits uniform development on the bottom side of the panel. Antifoam materials are usually added to the developer solution at makeup or as needed to control the foam. Antifoams should be carefully evaluated for causing edge gapping or plating metal-to-metal adhesion problems.

11.13 RESIST REMOVAL (STRIPPING)

The last step in the image-transfer process is the removal of the photoresist image from the PWB panel. The resist has served its purpose in the various plating and etching environments, and now its useful life has ended. Ideally, the resist will now be completely removed from the panel, quickly and easily, in a most cost-effective manner.

The stripping mechanism for dry-film resist is the same as the development mechanism; i.e., both are primarily diffusion-controlled processes.

11.13.1 Solvent Resist Stripping

The solvent-based stripping formulations (see Table 11.11) usually have chemicals (e.g., methylene chloride) which can rapidly diffuse into the bulk of the polymerized resist structure. This rapid diffusion causes the resist to swell. The swelling action causes strains or tensions to be built up in the resist which enhance the breakup and removal of the resist from the PWB panel without dissolution. Other stripper ingredients[45] (e.g., methyl alcohol) work to break the resist-to-copper bond and speed up the stripping process. Brushing or spraying the solution[46] also helps to physically dislodge the resist. Many of the dry-film-resist formulations have been optimized to support this diffusion and break-up process. Effective solvent spray equipment[47] has been developed to contain and recover the volatile chlorinated solvents for closed-loop operation and to concentrate the solid waste material. The stripper resist will dissolve with time in the solvent and does not cause any redeposition problems under normal use conditions.

11.13.2 Semiaqueous Resist Stripping

The aqueous and semiaqueous resist formulations and the stripper chemistry formulations present challenges to those designing aqueous stripping equipment. The totally aqueous stripping chemistry is usually 1 to 4 percent sodium or potassium hydroxide operated by 45 to 71°C (130 to 160°F). The intrinsic resist characteristics determine whether the diffusion process will lift the resist off in large sheets, large chunks, or small pieces. Without any organic solvents in the stripping solution, the resist pieces usually do not break up or dissolve. The stripped resist pieces need to be removed continuously from the solution, since they can absorb significant quantities of the stripping solution. The resist pieces can also plug spray nozzles as well as redeposit themselves on the panel surface. Removal of these resist "skins" is a challenge to the stripping-equipment designers.

11.13.3 Aqueous Resist Stripping

Stripping of aqueous resists that have been subjected to electroplating significantly thicker than the resist thickness (overplated) is difficult with aqueous stripping chemistries. The stripping solution diffuses into the resist and causes swelling of the resist to take place; the resist is now forced more tightly under the plating overhang. The overhang also impedes the stripping solution from diffusing into the resist under the overhang. Brushing, high-pressure sprays, or ultrasonic action[48] are often used separately or in combination to dislodge the resist trapped by the overhang and to improve stripping efficiency. But aqueous resist stripping is still 2 to 3 times slower than stripping solvent films in methylene chloride–alcohol solutions.

The use of organic[49] solvents (e.g., butyl "Carbitol" or butyl "Cellosolve") tends to speed up the diffusion process and to break up the larger resist skins into smaller pieces and make removal from the system easier. Increases in stripping productivity can be gained by the use of some organics (i.e., percent by volume) in the stripping solution. Many of the aqueous and semiaqueous resists can also be stripped in methylene chloride–alcohol solutions.

REFERENCES

1. Du Pont Cirtrak System, Wilmington, Del.
2. Accutrace System, W. R. Grace Co., U.S. Patent 4,436,806, Mar. 13, 1984.
3. Probimer System, Ciba Geigy, Ardsley, N.Y.
4. Advanced Process Supply, Chicago, Ill.
5. Tetko Inc., Elmsford, N.Y.
6. *Electronic Packaging and Production,* July 1983, pp. 203–204.
7. M. E. Pale-Baker, *Printed Circuit Fabrication,* December 1984, pp. 26–40.
8. Eastman Kodak, Rochester, N.Y.
9. Shipley Co. Inc., Newton, Mass.
10. Vacrel, Du Pont, Wilmington, Del.
11. J. J. Hickman, *Printing Wiring Design Aspects of Using Permanent Photopolymer Coatings,* First IPC World Conference, London, June 5–8, 1978.
12. U.S. Patents 3,469,982, Sept. 30, 1969, and 3,622,334, Jan. 23, 1974, and foreign patents.
13. Dynachem/Thiokol, Tustin, Calif., worldwide licensee, 1972.
14. A. G. Kalle/Hoechst, European licensee, 1980.
15. S. E. Scrupski, "Dry Photoresists Gain Ground," *Electronics,* vol. 43, Dec. 21, 1970, p. 95.
16. W. B. Tucker, *IPC Technical Review,* no. 125, October 1970, pp. 4–11, reviewed in *Circuits Manufacturing,* vol. 11, no. 2, February 1971, pp. 12–16.

17. MIL-STD-275B, *Printed Wiring for Electronic Equipment,* Dec. 18, 1964.
18. D. L. Harris and T. J. Wallace, "The Case for Dry Film Resist Systems," *Electronic Packaging and Production* vol. 13, no. 1, pp. 106–116.
19. G. Severin, *Dry Film or Screen Printing—A Quantitative Rating System for Choosing the Most Effective Imaging Process,* IPC-TP-225, 1978.
20. W. S. DeForest, *Photoresist: Materials and Processes,* McGraw-Hill, New York, 1975.
21. Minnesota Mining and Manufacturing Co., St. Paul, Minn., U.S. Patent 3,782,939, Jan. 1, 1974.
22. U.S. Patents 3,547,730, Dec. 15, 1970, and 3,770,438, Nov. 6, 1973.
23. U.S. Patents 3,060,023, Oct. 23, 1962, and 3,202,508, Aug. 24, 1965.
24. U.S. Patent 4,050,936, Sept. 27, 1977.
25. Laminar A, Dynachem/Thiokol Corp., Tustin, Calif.
26. Shipley/Micro Plate Div., St. Petersburg, Fla.
27. G. Peterson, *Printed Circuit Fabrication,* July 1984, pp. 50–58.
28. Integrated Resist System, Du Pont, Wilmington, Del.
29. Photocircuits Division, Kollmorgen Corp., Glen Cove, N.Y.
30. U.S. Patent 4,469,777, Sept. 4, 1977.
31. L. R. Wallig, "Tenting with Dry Film Resist," *Electronic Packing and Production,* July 1976, pp. 56–60.
32. Howard W. Markstein, "Solder Coating and Levelling Competes with Tin Lead Electroplating," *Electronic Packaging and Production,* December, 1982, pp. 30–35.
33. G. D. M. Menzies, "Treatment of Copper Surfaces in Printed Circuit Manufacture," *Circuit World,* vol. 7, no. 2, 1981, pp. 38–41.
34. "3M Building Services and Cleaning Products Division," *Circuits Manufacturing,* September 1971.
35. HRL/LC-2400, Du Pont, Wilmington, Del.
36. Asahi Chemical Co., Tokyo, Japan.
37. U.S. Patent 3,218,167, Nov. 16, 1985.
38. Data Sheet, "17-Step Table," Du Pont Riston Products Division, Wilmington, Del.
39. PC-30 Exposure Unit, Du Pont Riston Products Division, Wilmington, Del.
40. IL 700 Research Radiometer with Model XR 140B Detector, International Light Inc., Newburyport, Mass.
41. R. A. Carboni and R. H. Wopschall, *Imaging for Fine Line Circuitry,* IPC-TP-227, September 1978.
42. Excellon Automation, Torrance, Calif.
43. Circuit Services Inc., Minneapolis, Minn.
44. E. T. Hackett and R. H. Wopschall, *Automatic pH Controlled Replenishment of Developers for Aqueous Processable Dry Film Resists,* IPC-TP-399, April 1980.
45. U.S. Patent 3,813,309, IBM-Stripping Resists, May 28, 1984.
46. U.S. Patent 3,469,982, Celeste Solvent Swelling and Brushing, Sept. 30, 1969.
47. CS-24/SR-24 Solvent Stripping System, Du Pont, Wilmington, Del.
48. UAS-24 Ultrasonic Aqueous Stripping Unit, Du Pont, Wilmington, Del.
49. U.S. Patent 3,796,602, Mar. 12, 1974.

CHAPTER 12
PLATING

Edward F. Duffek, Ph.D.
Adion Engineering Company, Cupertino, California

12.1 INTRODUCTION

A major part of manufacturing printed circuit boards involves "wet process" chemistry. The plating aspects of wet chemistry include deposition of metals by electroless (metallization) and electrolytic (electroplating) processes. Topics to be described here are multilayer processing, electroless copper, electroplating of copper and resist metals, nickel and gold for edge connector (tips), and tin-lead fusing. Specific operating conditions, process controls, and problems in each area will be reviewed in detail. The effects of plating on image transfer, strip, and etching are also described in this chapter (Fig. 12.1).

Two driving forces have had a major influence on plating practices, namely, the precise technical requirements of electronic products and the demands of environmental and safety compliance. Recent technical achievements in plating are evident in the capability to produce complex, high-resolution multilayer boards. These boards show narrow lines (3 to 7 mils), small holes (13 mils), surface mount density, and high reliability. In plating, such precision has been made possible by the use of improved automatic, computer-controlled plating machines, instrumental techniques for analysis of organic and metallic additives, and the availability of controllable chemical processes. MIL-Spec quality boards are produced when the procedures given here are closely followed.

12.2 PROCESS DECISIONS

Process and equipment needs dictate the physical aspect of the facility and the character of the process, and vice versa. Some important items to consider are the following.

12.2.1 Facility Considerations

1. *Multilayer and two-sided product mix:* Need for lamination presses, inner layer processes, etc.

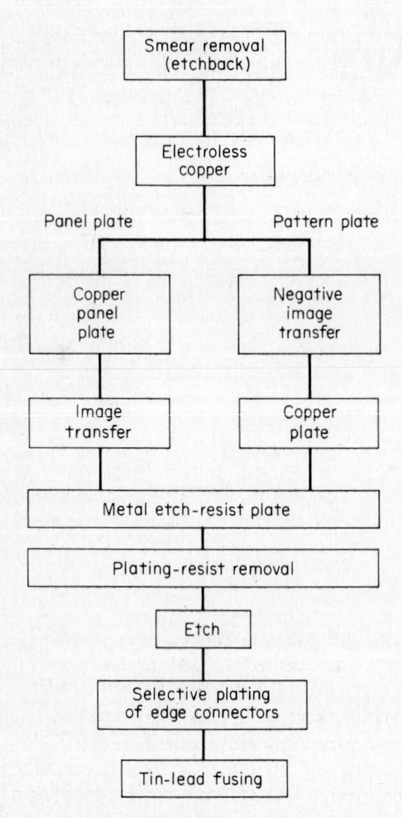

FIG. 12.1 Printed wiring board, plating flowchart.

2. *Circuit complexity:* Need for dry film, clean room, etc.
3. *Level of reliability (application of product):* Need for extra controls, testing, etc.
4. *Volume output:* Need for equipment sizing and building space
5. *Use of automatic versus manual line:* Need for productivity, consistency, workforce
6. *Wastewater treatment system:* Need for water and process control capability
7. *Environmental and personnel safety; compliance to laws*
8. *Costs*

12.2.2 Process Considerations

1. *Material:* The principal printed board material discussed will be NEMA grade FR-4 or G-10, i.e., epoxy-fiberglass clad with $\frac{1}{2}$-, 1-, or 2-oz copper. Other materials will be briefly mentioned because they can significantly alter plating and related processes.

2. *Standard:* Plated-through hole (PTH) is the current standard of the industry. The following purposes, objectives, and requirments apply to both multilayer and two-sided boards:

a. Purposes
Increased circuit density
Double-sided circuitry

b. Objectives
Side-to-side electrical connection
Ease of component attachment
High reliability

c. Requirements
Complete coverage
Even thickness
Hole-to-surface ratio
0.001-in minimum
No cracks
No voids, nodules, inclusions
No pullaway
No epoxy smear
Minor resin recession
Optimum metallurgical structure
M/L compatibility

3. *Image transfer:* Dry film or screening of plating resists will depend on board complexity, volume, and labor skills.

4. *Electroless copper:* The type chosen will depend on the method of image transfer as well as on the need for panel plating. These processes are readily automated.

5. *Electroplating processes:* Deposit requirements are as follows:
Electrical conductivity
Mechanical strength
Ductility, elongation
Solderability
Tarnish and corrosion resistance
Etchant resistance
Compliance to MIL-specification

Details emphasizing operation, control, and MIL-Spec plating practices are given in Sec. 12.8. Metal-plated structures of completed PC boards are as follows:
Copper/tin-lead alloy
Copper/tin (SMOBC)
Copper/tin-nickel (nickel)/tin-lead
Copper/nickel/tin
Nickel/silver

6. *Strip, etching, tin-lead fusing:* Methods required by these steps are determined by the above processes and by the need for possible automation.

12.3 PROCESS FEEDWATER

12.3.1 Water Supply

Printed circuit board fabrication and electronic processes require process feedwater with low levels of impurities. Large volumes of raw water must be readily available, either of suitable quality or else treatable at reasonable cost. New facilities must consider water at an early stage of the site selection and planning process.

12.3.2 Water Quality

Highly variable mineral content causes board rejects and equipment downtime, as well as reduced bath life, burdened waste treatment, and difficulty with rinse-water recovery. Many water supplies contain high levels of dissolved ionic minerals and possible colloidals that can cause rejects in board production. Some of these impurities are calcium, silica, magnesium, iron, and chloride. Typical problems caused by these impurities are copper oxidation, residues in the PTH, copper-copper peelers, staining, roughness, and ionic contamination. Problems in the equipment include water- and spray-line clogging, corrosion, and breakdown. The best plating practices suggest using good-quality water for high yields. The need for water low in total dissolved solids (TDS), calcium hardness, and conductivity is well known. Good water eliminates the concern that the water supply may be responsible for rejects. Although water quality is not well defined for plating and PC board manufacturing, for general usage, some guidelines can be assigned as below. Where "high-purity" water is required, see Sec. 12.3.3.

Typical quantities are as follows:

TDS	4–20 ppm
Conductivity	8–30 μS/cm
Specific resistance	0.12–0.03 MΩ
Carbonate hardness ($CaCO_3$)	3–15 ppm

Somewhat higher values are acceptable for less critical processes and rinses.

12.3.3 Water Purification

Two processes widely used for water purification are reverse osmosis and ion exchange. In the reverse osmosis technique, raw water under pressure (1.4 to 4.2 MPa or 200 to 600 psi) is forced through a semipermeable membrane. The membrane has a controlled porosity which allows rejection of dissolved salts, organic matter, and particulate matter, while allowing the passage of water through the membrane. When pure water and a saline solution are on opposite sides of a semipermeable membrane, pure water diffuses through the membrane and dilutes the saline water on the other side (osmosis). The effective driving forces of the saline solution is called its "osmotic pressure." In contrast, if pressure is exerted on the saline solution, the osmosis process can be reversed. This is called the "reverse osmosis" (RO) process and involves applying pressure to the saline solution in excess of its osmotic pressure. Fresh water permeates the membrane and collects on the opposite side, where it is drawn off as product. Reverse osmosis removes 90 to 98 percent of dissolved minerals and 100 percent of organics with molecular weights over 200, as shown in Table 12.1.

Other requirements are as follows:

pH	6.5–8.0
Total organic carbon	2.0 ppm
Turbidity	1.0 NTU
Chloride	2.0 ppm

TABLE 12.1 Purified Water Supply Values
Typical in/out RO values

TDS, ppm	SiO_2, ppm	Conductivity, μS	Hardness—$CaCO_3$, ppm
170/4	30/1	130/8	24/1
240/7	45/2	200/14	35/2
300/10	60/2	250/20	45/3

A small quantity of dissolved substances also facilitates deionized (DI) water production, wastewater treatment, and process rinsewater recovery, since it makes recycling less costly and more feasible. An RO system will result in lower costs for DI water preparation and for process water recycling. The setup for recycling requires additional equipment for polymer addition, filtration, and activated carbon treatment.

Deionized water purification is used when high-purity water is required, for example, in bath makeups, rinses before plating steps, and final rinses necessary to maintain low-ionic residues on board surfaces. MIL-Spec PC boards must pass the MIL-P-28809 test for ionic cleanliness. This is done by final rinsing in DI water.

Deionized water is made by the ion exchange technique. This involves passing water containing dissolved ionics through a bed of solid organic resins. These convert the ionic water contents to H^+ and OH^-. Deionized water systems are more practical when using feedwater low in ionic and organic content.

12.4 MULTILAYER PTH PREPROCESSING

Two-sided printed circuit boards are usually processed by first drilling and deburring, followed by the electroless copper process. Multilayers require treatment involving resin smear removal and etchback prior to the electroless copper.

12.4.1 Smear Removal

"Drill smear" refers to the epoxy resin that coats the inner-layer copper surface in the drilled hole and is due to heating during the drilling operation. Control of this smear is difficult due to variability in dielectric materials, inconsistency in curing stage, and poor drill quality. This smear must be removed before electroless copper to get full electrical continuity from the inner-layer copper to the PTH. The inner-layer connections will be flush with the drilled hole after smear removal (see Figs. 12.2 to 12.4).

12.4.2 Etchback

This term refers to the continuation of the smear removal process to expose 0.5 mil on the top and bottom surface of the inner-layer copper. Physically, the inner-layer copper will now protrude from the drilled hole three-point connection for copper bonding, which is required on some MIL-Spec boards (see Fig. 12.5).

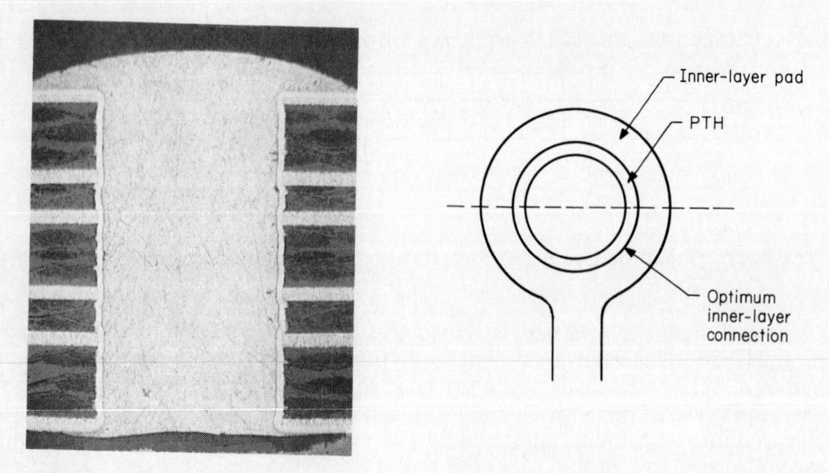

FIG. 12.2 PTH vertical and horizontal cross sections illustrating optimum inner-layer connection and smear removal.

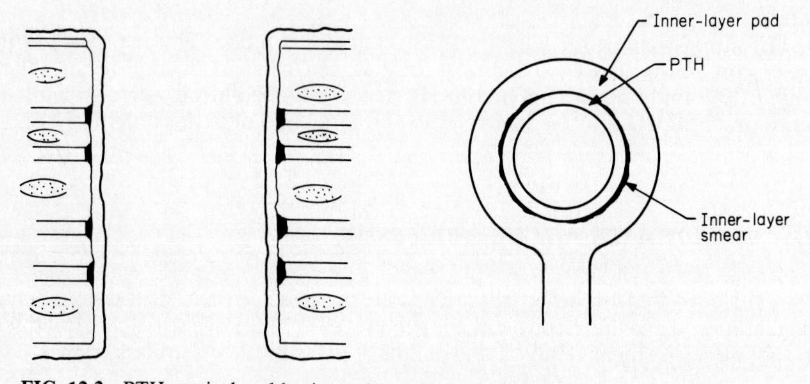

FIG. 12.3 PTH vertical and horizontal cross sections illustrating inner-layer smear.

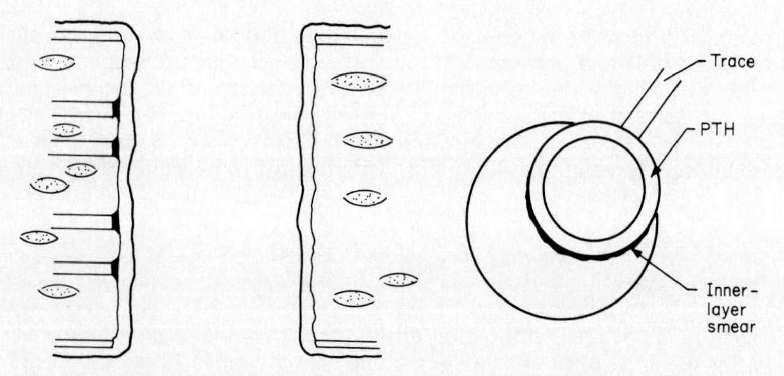

FIG. 12.4 PTH vertical and horizontal cross sections illustrating inner-layer smear and misregistration.

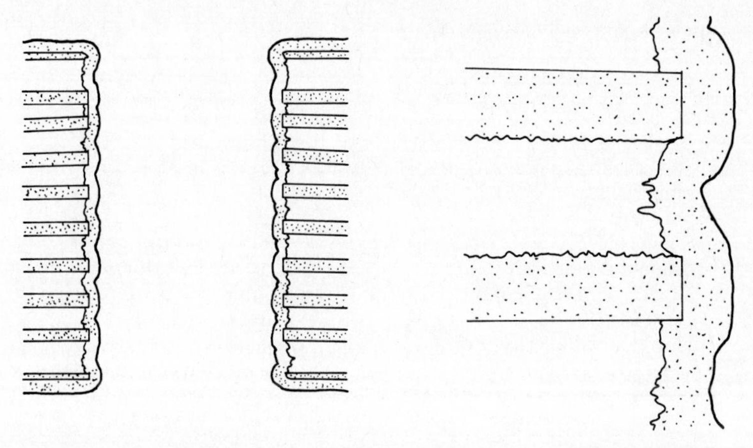

FIG. 12.5 PTH vertical cross section illustrating optimum inner-layer connection and etchback.

12.4.3 Smear Removal and Etchback Methods

The four methods that are commonly used utilize hole-wall epoxy or dielectric oxidation, neutralization-reduction, and glass etching:

12.4.3.1 Sulfuric Acid. This process has been used extensively for many years because of its ease of operation and reliability of results. A major disadvantage is a lack of control, which leads to hole-wall pullaway and rough holes. Operator safety is crucial, since concentrated sulfuric acid must be used.

12.4.3.2 Chromic Acid. This method provides more control and longer bath life. However, problems with copper voids due to Cr^{6+} poisoning, wastewater pollution, and contamination of plating processes must be considered. Etchback is possible by double processing.

12.4.3.3 Permanganate. This method is rapidly gaining acceptance due to improved copper adhesion to the hole wall (less pullaway), smoother PTHs, and better control.[1] Problems result from sludge by-production formation, dark copper holes, and from the operators' lack of experience. Permanganate is also used as a second step in conjunction with the other methods to enhance hole quality.[2]

12.4.3.4 Plasma. This is a dry-chemical method in which boards are exposed to oxygen and fluorocarbon gases. Persistent problems with the process are nonuniform treatment of holes—i.e., higher etch rates on the edges—and the high cost of equipment. This process has few steps and eliminates the use of large quantities of chemicals. Controls must be provided to prevent air pollution.

12.4.4 Process Outline: Smear Removal and Etchback

The four common methods for smear removal and etchback are given in the following table. Combinations of these methods are also in use because of added reliability to both process and product.

Smear Removal and Etchback Processes*

Sulfuric acid	Chromic acid
Rack panels	Rack panels
Sulfuric acid, 96%	Chromic acid
20-s dwell, 15-s drain room temperature	3 min, 140°F
Neutralizer	Reducer
8 min, 125°F	3 min, room temperature
Ammonium bifluoride	Ammonium bifluoride
3 min, room temperature	3 min, room temperature
Unrack panels	Unrack panels
High-pressure hole cleaning	High-pressure hole cleaning
Release to electroless copper	Release to electroless copper

Permanganate	Plasma etch
Solvent conditioner	Plasma
90°F, 5 min	Oxygen, CF_4, 30 min
Alkaline permanganate	Glass etch (optional)
170°F, 10 min	
Neutralizer—120°F, 5 min	High-pressure hole cleaning
Glass etch—4 min, room temp	Release to electroless copper
High-pressure hole cleaning (optional)	
Release to electroless copper	

*Water rinse after each step is not shown.

Polyimide and polyimide-acrylic systems are processed in chromic acid or plasma. Teflon* and R T Duroid† materials are treated (before the electroless copper operations) in sodium-naphthalene mixtures to yield void-free, high-bond-strength copper in the PTH.

12.5 ELECTROLESS COPPER[3–8]

12.5.1 Purpose

This second series of chemical steps (after smear removal) is used to make panel side-to-side and inner-layer connections by metallizing with copper. The process steps needed include racking, cleaning, copper microetching, hole and surface catalyzing with palladium, and electroless copper. Typical steps are as follows:

1. *Cleaner-conditioner:* Alkaline cleaning is used to remove soils and condition holes.
2. *Microetch:* This slow acid etching is used for removal of copper surface pretreatments, oxidation, and presentation of uniformly active copper. Persulfates and sulfuric acid–hydrogen peroxide solutions are commonly used.
3. *Sulfuric acid:* Used for removal of persulfate residues.
4. *Predip:* Used to maintain balance of the next step.

*Registered trademark of E. I. du Pont de Nemours & Company.
†Registered trademark of Rogers Corporation, Chandler, Arizona.

5. *Catalyst (activator):* Neutral or acid solutions of palladium and tin are used to deposit a thin layer of surface-active palladium in the holes and on the surface.
6. *Accelerator (postactivator):* Used for the removal of colloidal tin on board surfaces and holes.
7. *Electroless copper:* Alkaline chelated copper reducing solution that deposits thin copper in the holes (20 to 100 μin) in the holes and surfaces.
8. *Antitarnish:* This is a neutral solution that prevents oxidation of active copper surfaces by forming a copper conversion coating.

12.5.2 Mechanism

Equations 12.1 and 12.2 illustrate the process.

$$Pd^{2+} + Sn^{2+} \rightarrow Pd + Sn^{4+} \quad (12.1)$$

$$CuSO_4 + 2HCHO + 4NaOH \xrightarrow{Pd} Cu + 2HCO_2Na + H_2 + 2H_2O + Na_2SO_4 \quad (12.2)$$

12.5.3 Electroless Copper Processes

Selection from several types available depends on the type of image transfer desired. Operation and control of three bath types and the function of constituents are given in Tables 12.2 and 12.3.

The problems encountered with this system are as follows:

1. *Uncoppered holes:* This may appear as dark, hazy, or voided copper in the holes. To correct this, check the operation of the smear remover, cleaner, catalyst, accelerator, and the items listed in Table 12.2. Voids may also be due to copper plate precleaner attack.

TABLE 12.2 Electroless Copper
Operation and control

	Low deposition	Medium deposition	Heavy deposition
Copper	3 g/L	2.8 g/L	2.0 g/L
HCHO	6–9 g/L	3.5 g/L	3.3 g/L
NaOH	6–9 g/L	10–11 g/L	8 g/L
Temperature	65–85°F	115 ± 5°F	115 ± 5°F
Air agitation	Mild	Mild/moderate	Moderate
Filtration	Periodic	Continuous	Continuous
Tank design	Static	Overflow, separate sump	Overflow, separate sump
Heater	Teflon	Teflon	Teflon
Panel loading	0.25–1.5 ft²/gal	0.1–2.0 ft²/gal	0.1–2.0 ft²/gal
Replenish mode	Manual	Manual or continuous	Automatic
Idle time, control	70–85%	Turn off heat	Turn off heat
Deposition time	20 min	20 min	20–30 min
Thickness	20 μin	40–60 μin	60–100 μin

TABLE 12.3 Electroless Copper
Function of constituents

	Constituent	Function
Copper salt	$CuSO_4 \cdot 5H_2O$	Supplies copper
Reducing agent	HCHO	$Cu^{2+} + 2e \rightarrow Cu^0$
Complexer	EDTA, tartrates, Rochelle salts	Holds Cu^{2+} in solution at high pH; controls rate
pH controller	NaOH	Controls pH (rate) 11.5–12.5 optimum for HCHO reduction
Additives	NaCN, metals, S, N, CN organics	Stabilize, brighten, speed rate, strengthen

2. *Hole-wall pullaway:* This refers to copper pulling off the PTH and is observed either as large blisters or by cross sectioning. This may be due to spent sulfuric acid smear removal or to items given above. Pullaway is controlled by maintaining copper plating smooth and over 1 mil in thickness. This may also be observed on solder mask on bare copper (SMOBC) boards due to shock of hot-solder immersion.

3. *Bath decomposition:* This is rapid plating out of the copper. Common causes are bath imbalance, overloading, overheating, lack of use, tank wall initiation, or contamination.

4. *Electroless copper to copper-clad bond failure:* Review initial step in process and items in 1.

5. *Staining:* Copper oxidation is due to moisture or contamination on the copper surface. Corrective action involves dipping boards in antitarnish or hard rinsing in DI water.

12.5.4 Process Outline

This outline presents the typical electroless copper process steps:

Rack
Clean and condition
Water rinse
Surface copper etch (microetch)
Water rinse
Sulfuric acid (optional)
Water rinse
Preactivator
Activator/catalyst
Water rinse
Postactivator and accelerator
Water rinse
Electroless copper
Rinse
Sulfuric acid or antitarnish

Rinse
Scrub (optional)
Rinse
Copper flash plate (optional)
Dry
Release to image transfer

12.6 PANEL VERSUS PATTERN PLATING

In this process, the copper conductive coating over the insulating material (the epoxy-glass holes) bridges the plated-through hole with the copper cladding on both sides of the board. The two standard routes to get metal buildup into the holes and the circuit traces are referred to as "panel plating" and "pattern plating." Refer to the printed circuit plating flowchart (Fig. 12.1). *Panel plating* is the process wherein the entire surface area and the drilled holes are copper-plated, stopped-off with resist on the unwanted copper surfaces, and then plated with the etch-resist metal. *Pattern* (or selective) *plating* is the process where only the desired circuit pattern and holes receive copper buildup and etch-resist metal plate. A small amount of copper, 20 to 100 μin, and in some cases an additional 0.2 mil of electroplated copper, remains on the entire panel plus base copper (foil). During pattern plating, the circuit lines and pads increase in width on each side about as much as the surface thickness during plating. An increase of 0.001 in of the surface thickness results in pad and conductor line width increases of up to 0.002 in. Allowance for this must be made on the master artwork. Both processes have advantages and disadvantages. In choosing between them, it is important to consider the specific needs and the facilities available. Table 12.4 lists the differences.

12.6.1 Panel Plating—Typical Process

Receive boards directly from electroless copper
Rerack
Clean, using acid or alkaline solution

TABLE 12.4 Panel vs. Pattern Plating

	Panel	Pattern
Copper plating	Before imaging Fast plating (high CD) Thickness control Cathode fixturing	After imaging Fewer steps Less power, fewer materials Resist contamination
Image transfer	Thinner resists	
Metal etch resist	Optional Thicker deposits More steps	Thinner deposits
Etching	Double etching time More chemicals More pollution	Less etching Less undercut Greater circuit density

Water rinse, using spray or counterflow
Acid-dip, using sulfuric acid
Copper-plate, 1 mil in PTH
Drag-rinse
Water rinse, spray or counterflow
Dry

Dry film

Transfer image (dry film resist)
Rack
Clean, using acid solution
Water rinse, spray or counterflow
Microetch

Screen resist

Transfer image (screen resist)
Rack
Clean, using acid solution
Water rinse, spray or counterflow
Acid-dip (optional)

Water rinse, using spray or overflow
Acid-dip in sulfuric acid
Water rinse, using spray or overflow
Acid-dip in fluoboric acid
Tin-lead–plate
Drag-rinse
Proceed to strip and etch

12.6.2 Pattern Plating—Typical Process

Screen resist

Electroless copper (20 μin)
Scrub (optional)
Transfer image (screened resist)
Rack
Clean, using acid
Water rinse, spray or counterflow
Acid-dip (optional)

Dry film resist

Electroless copper (60 μin)
Scrub (optional)
Transfer image (dry film)
Rack
Clean, using acid
Water rinse, spray or counterflow
Microetch

Water rinse, spray or counterflow
Dip in sulfuric acid
Rinse in DI water (optional)
Copper-plate, 1 mil in PTH
Drag-rinse
Water rinse, spray or counterflow
Dip in fluoboric acid
Tin-lead–plate, 0.3–0.8 mil

Drag-rinse
Proceed to strip and etch

12.7 IMAGE TRANSFER AND EFFECTS ON PLATING

Photosensitive dry film and screened resist inks are the most commonly used resist materials. Dry film resists are selected for boards with narrow lines and spaces (3 to 7 mils), whereas screened resists are used on boards with wider lines and spaces ($\geq$8 mils). The processes of plating and image transfer depend on each other for success in the production of quality PC boards. Thus, in the preceding pattern-plating examples, dry film resist is needed to produce a narrow-line (6-mil) board but requires a thicker electroless copper (60 μin). This is due to the need for microetching prior to copper plating. In using alkaline-soluble screened resist for a wider-line (12-mil) board, a thin electroless copper (20 μin) is suitable, since preplate cleaning is done in an acid solution with minimal (if any) copper microetching. Problems in plating due to unsatisfactory image-transfer processing include copper-copper peeling, uneven plating, breakdown, and lifting. Strip and etching problems from plating are mainly overplating on the resists, over- and underetching, and slivers.

12.8 ELECTROPLATING: PATTERN BUILDER OF PRINTED CIRCUIT BOARDS

Electroplating is the production of adherent deposits of metals on conductive surfaces by the passage of an electric current through a conductive metal-bearing solution. The rate of plating depends on current and time, as expressed by Faraday's law:

$$W = \frac{ItA}{nF} \tag{12.3}$$

where W = metal, g
I = current, A
t = time, s
A = atomic weight of the metal
n = number of electrons involved in metal ion reduction
F = Faraday's Constant

Plating occurs at the cathode, the negative electrode. Accordingly, deposit thicknesses are determined by time and by the current impressed on the surface being plated; for example, 0.5 mil (0.0005 in) of tin-lead alloy is plated at 17 A/ft^2 for 15 min. Section 12.2.2 gives the features of electroplating. See references on electrochemistry and electroplating.[9–25]

Since most plating solutions are similar in nature, their use and the quality of the resulting deposits depend on the processing variables involved. Increasing the concentration, temperature, and agitation will enable faster plating rates and higher cathode efficiencies but will decrease the throwing power (surface-to-hole ratio) and the bath stability. Most plating solutions require proprietary additives that are either organic or metallic.

The rate of electroplating is limited by the supply of metal ions at the cathode

TABLE 12.5 Properties of Electrodeposits

	Cu	SnNi	Ni	Au	SnPb	Sn
Melting point, °F	1980	2200	2600	1945	361	450
Hardness, VHN	150	650	250	150	12	4
Coefficient of thermal expansion, 10^{-6}, in/°F	9.4	9.5–10.0	8.0	8.2	12.2	12.8
Conductivity, % IACS	101	32	25	73	11.9	15.6
Electrical resistivity, μΩ/cm	1.67	5.4	6.8	2.19	14.5	11.1
Thermal conductivity, CGS, °C	0.97	0.3	0.25	0.71	0.12	0.15

surface. This is of concern in high-speed plating and PTH uniformity and is expressed by the limiting current density given by the following equation:

$$I_L = \frac{nFDC}{\partial} \tag{12.4}$$

where D = metal ion diffusion constant
C = metal ion concentration in solution bulk
∂ = diffusion layer thickness

Smooth, bright electroplating cannot be achieved at I_L. Reduced diffusion layer thickness, as achieved by rapid solution agitation systems, allows increased plating speed. In addition, the electrodeposition of alloys presents other factors in addition to those in single-element plating.[21] Conditions must be controlled so that uniform composition is maintained over a wide range of operating variables. Table 12.5 gives selected properties of deposits.

For our purposes, multilayer and the two-sided, plated-through boards are similar in most processing respects. Highly reliable MIL-Spec–quality circuit boards can be produced following good plating practices and the standard procedures decribed later. The flowchart in Fig. 12.1 illustrates the subtractive processes. Pattern plating is the preferred method for manufacturing PC boards. Standard thicknesses are as follows: 1 mil copper, 0.5 mil tin-lead traces, pads and holes, and 0.2 mil nickel and 50 μin of gold on the connector tips. A 1-mil copper deposit in the PTH is specified in MIL-STD-275 and provides the following:

1. High electrical conductivity
2. Good solderability
3. High ductility and mechanical strength to withstand pulling of component terminal
4. Full copper coverage from surface into PTH
5. Repeated component replacements

12.9 COPPER ELECTROPLATING

Because of its high electrical conductivity, strength, ductility, and low cost, copper is the most commonly used metal for the structure of a printed circuit board. In addition, copper is readily plated from simple solutions and is easily etched. MIL-STD-275 states that electrodeposited copper shall be in accordance with MIL-C-104550 and shall have a minimum purity of 99.5 percent as determined by ASTM

E 53. The minimum thickness shall be 0.001 in (1 mil). Requirements for good soldering also indicate the need for 1 mil of copper and smooth holes.[26] Copper plating is generally regarded as the slow step in manufacturing PC boards. New methods which cut plating times by as much as 50 percent include high-speed additives, pulse plating, and rapid solution-impingement machines.[27–33]

12.9.1 Key Factors for Uniform Plating

To have day-to-day control and to achieve ductile, strong deposits and uniform copper thickness, the following controls are required:

1. Maintain equipment following best practices, such as uniform air agitation in the tank, equal anode-cathode distances, rectifier connection on both ends of the tank, low resistance between rack and cathode bar, etc.
2. Maintain narrow range control of all chemical constituents, including organic additives and contaminants.
3. Conduct batch carbon treatment regularly.
4. Control temperature at 70 to 85°F.
5. Eliminate contaminants in tank from preplate cleaners, microetchants, and impure chemicals.
6. Plate at one-eighth to one-half of the conventional cathode current density when using thick boards (0.100 in) with small holes (0.025 in) and fine lines (6 to 8 mils).

12.9.2 Acid Copper Sulfate

The preferred industrial process uses an acid copper sulfate solution containing copper sulfate, sulfuric acid, chloride ion, and organic additives. Using the proper additives, the resultant copper is fine-grained with tensile strengths of 50,000 psi (345 MPa), a minimum of 10 percent elongation, and 1.2 surface-to-hole thickness ratio. Table 12.6 gives acid copper properties.

12.9.2.1 Operation and Control. See Table 12.6.

12.9.2.2 Agitation. Air (vigorous) from oil-free source, at 70 to 80°F.

12.9.2.3 Filtration. Continuous through 3- to 10-μ filter to control solution clarity and deposit smoothness.

12.9.2.4 Carbon Treatment. New baths do not require activated carbon purification. Occasional circulation through a carbon-packed filter tube can be used to control organic contamination but is advised only after consultation with the supplier. The need for batch carbon treatment is indicated by corner cracking after reflow; dull, pink deposits; haze; "haloing"; or "comet trails" around the PTH. Carbon treat about every 1500 (A·h)/gal.

12.9.2.5 Procedure for Batch Carbon Treatment.

1. Pump to storage tank.
2. Clean out plating tank:
   a. Rinse and clean tank.
   b. Leach with 10% H_2SO_4.

TABLE 12.6 Acid Copper Sulfate
Operation and control

	Conventional	High-speed
Operating variables:		
Copper	2–3 oz/gal	3–4.5 oz/gal
Copper sulfate	8–12 oz/gal	12–18 oz/gal
Sulfuric acid	22–28 oz/gal	24–36 oz/gal
Chloride	40–80 ppm	40–80 ppm
Additives	As required	As required
Temperature	70–85°F	75–100°F
Cathode current density	20–40 A/ft^2*	40–150 A/ft^2
Anodes:†		
Type	Bars or baskets	
Composition	Phosphorized 0.04–0.06% P	
Bags	Closed-napped polypropylene	
Hooks	Titanium or Monel	
Length	Rack length minus 2 in	
Anode current density	10–20 A/ft^2 conventional; 25–50 A/ft^2 high speed	
Properties:		
Composition	99.8% (99.5% min, ASTM E 53)	
Elongation	10–25% (6% min, ASTM E 8 or E 345)	
Tensile strength	40–50 kpsi (36 kpsi min, ASTM E 8 or E 345)	

*A/ft^2 refers to amperes per square foot and is sometimes expressed as ASF.
†Operating anodes should have a thin, brown or black, easily removed film.

   c. Adjust agitators.
   d. Clean anodes.
3. Heat solution to 120°F.
4. Add 1 to 2 qt of hydrogen peroxide (35 percent per 100 gal of solution). Dilute with 2 pt water, using low stabilized peroxide.
5. Air-agitate or mix for 1 h.
6. Maintain heat at 120 to 140°F.
7. Add 3 to 5 lb powdered carbon per 100-gal solution. Use Supercarb,* Norit SGII,† or Darco.‡ Mix for 1 to 2 h.
8. Pump back to plating tank promptly or within 4 h.
9. Analyze and adjust.
10. Dummy-plate at 10 A/ft^2 for 6 h. Panels should be matte and dull. Replenish with additive.
11. Follow supplier instructions for electrolyzing and startup.

12.9.2.6 Contaminations. In general, acid copper tolerates both organic and metallic contaminants. Organic residues may come from cleaners, resists, and certain sulfur compounds. Dye systems usually are more resistant than dye-free sys-

*Product of M&T Chemicals, Rahway, New Jersey.
†Product of American Norit Company, Jacksonville, Florida.
‡Product of ICI Americas, Inc., Wilmington, Delaware.

tems with respect to certain cleaner constituents. Metals should be kept at these maximums: chromium, 25 ppm; iron, 500 ppm; tin, 300 ppm; antimony, 25 ppm. Nickel, lead, and arsenic may also cause roughness, etc.

12.9.2.7 Process Controls.

12.9.2.7.1 Bath Composition. Copper sulfate is the source of metal. Low copper will cause deposit "burning"; high copper will cause roughness and decreased hole-to-surface thickness ratios (wider lines). Sulfuric acid increases the solution conductivity, allowing the use of high currents at low voltages. However, excess sulfuric acid lowers the plating range, whereas low acid reduces hole-to-surface ratio (throwing power). It is important to control the chloride ion (Cl^-) at 60 to 80 ppm. Below 30 ppm, deposits will be dull, striated, coarse, and step-plated. Above 120 ppm, deposits will be coarse-grained and dull. The anodes will get polarized, causing plating to stop. Excess chloride is reduced by bath dilution or by electrolysis. Excess or insufficient additive will cause deposit "burning" and corner cracking. This condition can be judged by metallographic cross sectioning and etching. Optimum-quality plated metal shows no laminations or columnar patterns. The use of DI water and contamination-free materials such as low chloride and iron will give added control and improved deposit quality. Proper analysis and control of the additive components are critical for consistent product quality. Methods available use cyclic voltammetry stripping (CVS)[34–36] and liquid chromatography (LC).[37,38]

12.9.2.7.2 Temperature. Optimum throwing power and surface-to-hole ratios are obtained by operating at room temperature, i.e., 70 to 80°F. Lower temperatures cause brittleness, "burning," and thin plating. Higher temperatures cause haze in low-current density areas and reduced throwing power. Cooling coils may be necessary during a hot summer or under heavy operation.

12.9.2.7.3 Deposition Rate. A thickness of 0.001 in (1.0 mil) of copper deposits in 54 min at 20 A/ft^2, in 21 min at 50 A/ft^2.

12.9.2.7.4 Hull Cell. Operation at 2 A will show the presence of organic contamination, chloride concentration, and overall bath condition. However, an optimum Hull cell panel is only a small indication that the bath is in good operating condition, since test results are not always related to production problems. More reliable results are obtained by adjusting bath before Hull cell testing. See page 12.44 for procedures on Hull cell and cross sectioning.

12.9.2.7.5 Cross-Sectioning Results. Sectioning with etching provides information on the plated copper that explains PTH quality in terms of processing factors. Besides showing the overall quality, cross sectioning gives information on thickness and on possible problems such as drilling, cracking, blowholes, and multilayer smear. Copper deposits with columnar or laminar patterns indicate inferior copper properties. Cross sections of optimum copper deposits show very small nongranular particles ("structureless") upon etching.

12.9.2.7.6 Inferior Copper Deposits. These may be caused by any of the following:

- Either low or excess additives
- Chloride out of range, i.e., too high or too low
- Organic, metal, or sulfur (thiourea) contamination
- Excess dc rectifier ripple (greater than 10 percent)
- Low copper content with bath out of balance
- Roughness in drilling, voids in electroless steps, or other problems introduced in earlier processing

12.9.2.7.7 Cracking and Ductility. Resistance to cracking is tested by the following:

Solder reflow or wave soldering and cross sectioning.

Elongation: Two-mil copper foil should exceed 6 percent elongation. Acid copper elongation should range between 6 and 25 percent. Frequent testing gives more meaningful results.

Float solder test: This test includes prebaking and flux, using a 5- to 20-s float in a solder pot (60:40) at 550°F, followed by cross sectioning for evaluation.

Copper foil bulge test: This test measures tensile strength by puncturing copper at high pressure.

Board aging: Evidence indicates that ductility and surface properties change with the time elapsed after plating. Boards should not be tested immediately after plating.[39]

dc Ripple: High values of rectifier ripple (8 to 12 percent) may cause inferior copper deposits and poor distribution of thickness.

12.9.2.7.8 Visual Appearance. As plated, copper has a semibright appearance at all current densities. Unevenness, hazy or dull deposits, cracking and haloing around the PTH, and low-current-density areas indicate organic contamination. Carbon treatment is required if these conditions persist. "Burned," dull deposits at *high* current densities indicate low additive content, contamination, solution imbalance, or low bath temperatures. Dull, coarse deposits at *low* current densities mean that the chloride ion is not in balance. When chloride is high or bath temperature is too low, anodes may become heavily coated and polarized (current drops). Decreased throwing power (surface-to-hole ratio), reduced bath

TABLE 12.7 Printed Wiring Board Copper Plating Defects

Defect in copper process	Cause
Corner cracking	Excess additive, organic contamination in solution; also uncured laminate
Nodules	Particulate matter in solution; also drilling, deburring residues
Thickness distribution	See page 12.15
Dullness	Off-balance solution, organic contamination
Uneven thickness in PTH	Organic sulfur (thiourea) contamination
Pitting	Additive malfunction, also previous steps in electroless copper and preplate cleaning
Columnar deposits	Low additive, rectifier malfunction
Step plating, whiskers	Excess or defective additive
Defect, overall manufacturing process	**Cause**
Voids	Malfunction of electroless copper steps, also preplate cleaner etching
Inner layer smear	Drilling or malfunction in smear removal
Roughness	Drilling or drilling residues
Hole-wall pullaway	Malfunction of smear removal or electroless copper steps
Copper-copper peeling	Surface residues from image transfer; also off-balance copper plating process, i.e., low air agitation and high CD
Soldering blowholes	Drilling roughness, voids, and thin plating

conductivity, or poor-quality plating may also indicate contamination and is corrected by any of the following:

1. Maintaining solution balance and chloride content at 60 to 80 ppm.
2. Circulating solution through filter continuously, passing through a carbon canister periodically, or by batch carbon treatment.
3. Analyzing organic additives by CVS, LC, or Hull cell.
4. Checking metal contaminations every 3 months.
5. Controlling temperature between 70 and 85°F.
6. Checking anodes daily and replacing bags and filters (rinsed in hot water) every 3 to 4 weeks.

12.9.2.7.9 Problems. Table 12.7 lists problems that appear after copper plating. Two groups are listed, with the first group readily correlated to the copper-plating process. Thin, rough copper plating in the PTH may also be exhibited by outgassing and blowholes during wave soldering.[26,40–44] Figures 12.6 to 12.15 illustrate some of these effects.

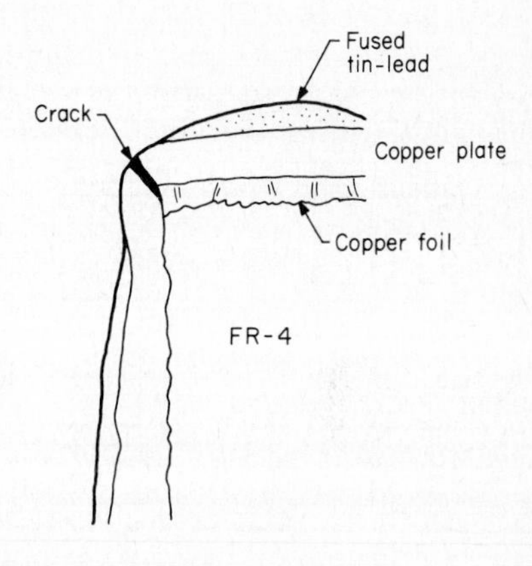

FIG. 12.6 PTH vertical cross section illustrating copper corner cracking.

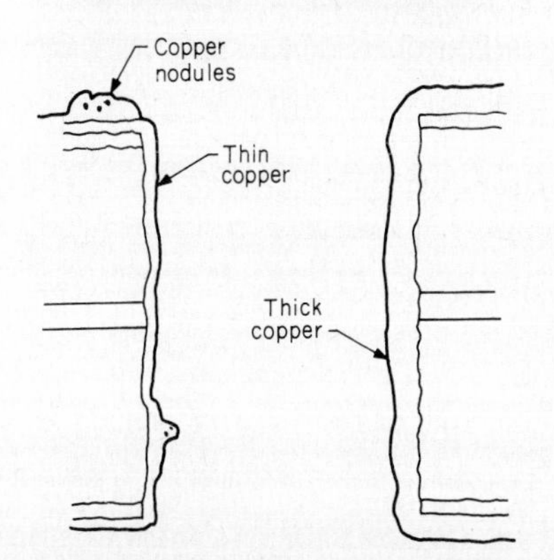

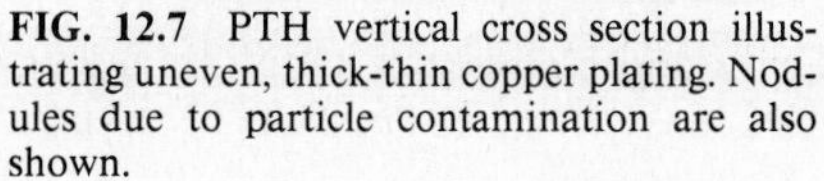

FIG. 12.7 PTH vertical cross section illustrating uneven, thick-thin copper plating. Nodules due to particle contamination are also shown.

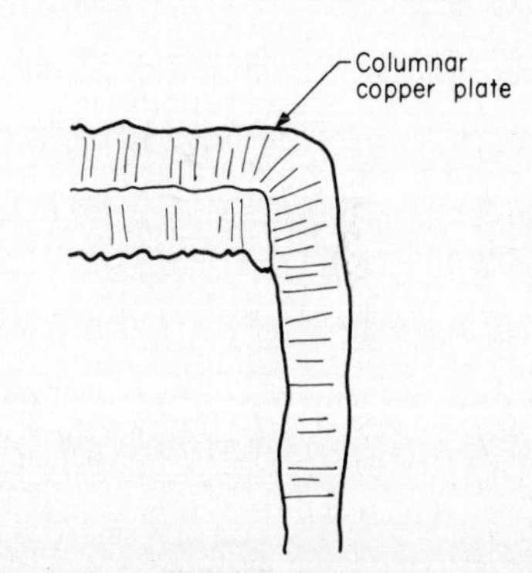

FIG. 12.8 PTH vertical cross section illustrating columnar copper deposit structure.

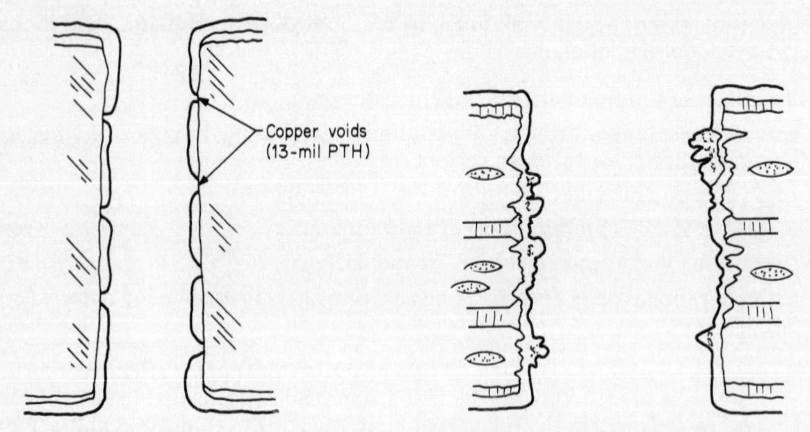

FIG. 12.9 PTH vertical cross section illustrating copper voids.

FIG. 12.10 PTH vertical cross section illustrating rough, nodular copper plating due to drilling roughness and residues. Nailheading is also shown.

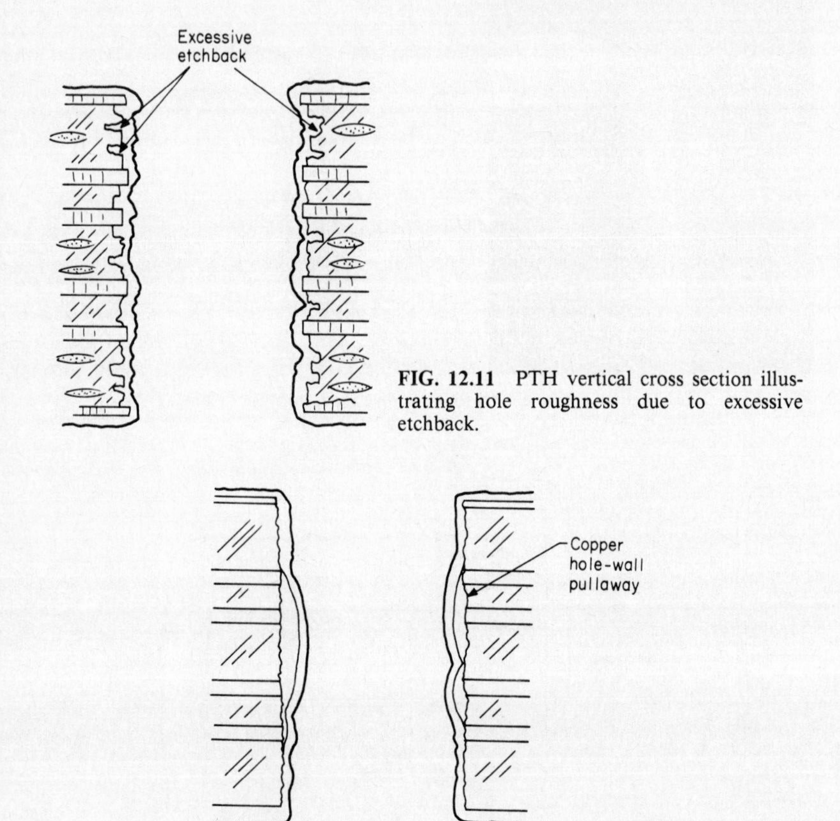

FIG. 12.11 PTH vertical cross section illustrating hole roughness due to excessive etchback.

FIG. 12.12 PTH vertical cross section illustrating copper hole-wall pullaway.

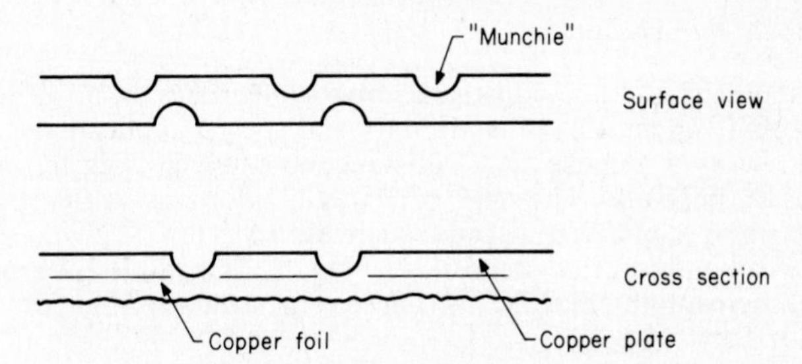

FIG. 12.13 Trace surface view and vertical cross section illustrating "munchies" and pitting.

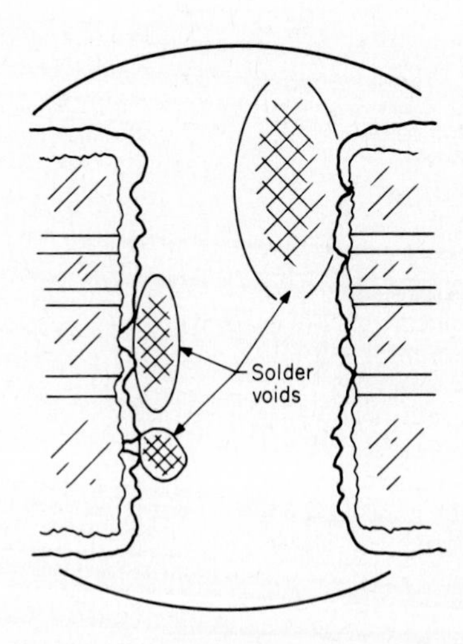

FIG. 12.14 PTH vertical cross section illustrating wave-soldering blowholes and thin, rough copper plating.

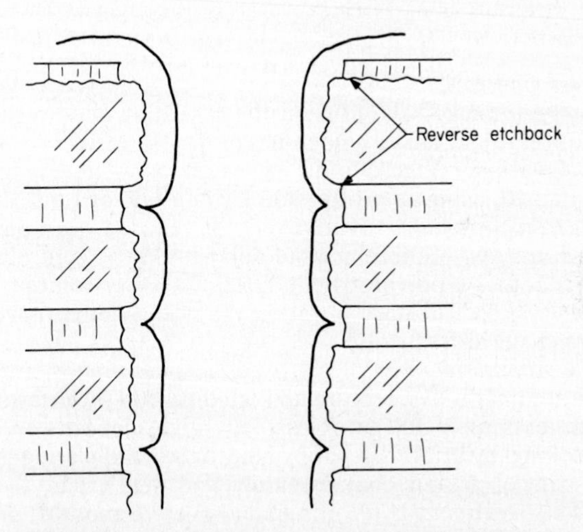

FIG. 12.15 PTH vertical cross section illustrating reverse etchback.

12.9.3 Copper Pyrophosphate

Once the standard of the industry,[16,45] pyrocopper has been almost entirely replaced by acid copper, except in military and special applications. Pyrocopper continues to be used because of its resistance to cracking, high throwing power, and purity of deposits. The use of organic additives with pyrocopper is optional,[46,47] but it is preferred because such additives provide wider tolerance of cracking control and improve deposit quality.[48–51] The additive PY 61-H,* identified as dimercaptothiadiazole, is effectively controlled in production by CVS.[52–54] See Table 12.8.

TABLE 12.8 Copper Pyrophosphate

Operation and control

Operating conditions:	
pH	8.1–8.5 copper
Copper	2.7–3.5 oz/gal
Pyrophosphate	19.4–26.3 oz/gal
Orthophosphate	8 oz/gal max
Ammonia (NH_3)	0.2–0.3 oz/gal
Ratio (pyro/Cu)	7.5–8.0/L
Temperature	115–125°F
Cathode CD	20–35 A/ft^2
Aqua ammonia	As needed; about 1–2 qt per day
PY 61-H*	0.25–0.75 mL/(A·h) Control with CVS
Anodes:†	
Type	Bars or baskets
Composition	OFHC copper
Bags	Optional, none required
Current density	20–30 A/ft^2
Hooks	Titanium
Length	Rack length minus 2 in

*Product of M&T Chemicals, Rahway, New Jersey.
†Operating anodes should have a thin golden or light tan film.

12.9.3.1 Process Controls.

12.9.3.1.1 Appearance. The film on the operating anodes should be easily removed, and appear light tan or golden in color.

12.9.3.1.2 Agitation. Vigorous air agitation between anode and cathode should be maintained, using an oil-free source, at a rate of 1.5 ft^3/min/ft^2.

12.9.3.1.3 Orthophosphate. Heat should be reduced when air is turned off.

12.9.3.1.4 Filtration. Smoothness of deposit may be controlled by the continuous use of a 3- to 5-μ polypropylene filter, with four solution turnovers per hour. A 1-μ (or less) second-stage filtration may be required if PY 61-H is not used. Filters are changed frequently.

12.9.3.1.5 Carbon Treatment. All pyrophosphate baths, regardless of age and with or without additives, require activated carbon purification. This is evidenced by either extremely dull or excessively bright deposits on a Hull cell, by PTH corner cracking, by brittleness, or by hazy or banded plating. A batch carbon treatment will control organic contamination and thus ensure the best deposits and ductility. This treatment is recommended every 6 months, or more often if

*Product of M&T Chemicals, Rahway, New Jersey.

needed. Clearly, shop facilities should be designed to conduct carbon treatment easily without causing downtime. Though continuous or occasional circulation through a carbon-packed filter tube is sometimes used, this method is not reliable because of poor maintenance and the ineffectiveness of carbon. The proper method for batch carbon treatment is described in Sec. 12.9.2.5.

12.9.3.1.6 Contaminations. Impurities harmful to pyrocopper are the following:

Organic: These come from additive breakdown, resists, cleaners, and oil. To remove haze and dullness, filter through a carbon pack prior to batch carbon treatment. If haziness persists, add 60 ml of hydrogen peroxide per 100 gal of solution.

Metals: The maximum allowable by atomic absorption are lead, 10 ppm; iron, 50 ppm; nickel, 50 ppm. To control this contamination, dummy (plate) on scrap panels or corrugated sheets at 5 A/ft^2 for 2 to 6 h once a week.

Nonmetals: Maximum levels allowable are chloride ion (Cl^-), 40 ppm; cyanide (CN^-), 0 ppm; and sulfur (as S^{2-}), 0 ppm.

12.9.3.2 Solution Controls. *Copper pyrophosphate* is the source of metal. Low copper (below 2.5 oz/gal) will cause deposit "burning." High copper (above 4.0 oz/gal) will cause roughness and decreased hole-to-surface ratio (decreased line spacing).

Potassium pyrophosphate ($K_4P_2O_7$) is the complexing agent for copper. A pyrophosphate-to-copper weight ratio of 7.5:1 to 8.0:1 is required for optimum plating control. A low ratio (<7.0:1) causes roughness, reduced throwing power, and banded deposits. A high ratio (>8.5:1) promotes orthophosphate (K_2HPO_4) formation and decreases bright plating range (dullness).[46,47]

The best *pH range* is 8.1 to 8.5, but pH remains satisfactory up to 8.8. Higher pH (>9.0) will cause CD range decrease (dullness), roughness, and poor anode corrosion. Low pH (<7.0) causes orthophosphate buildup and reduced throwing power.

Ammonium hydroxide is added to increase deposit luster and anode corrosion. Low ammonia (<0.1 oz/gal as NH_3) causes dullness, poor anode corrosion, and roughness. High ammonia (>0.4 oz/gal) may cause attack of resist and may reduce adhesion.

PY 61-H is needed for deposit leveling, copper ductility, and roughness (nodule) control and for providing some brightening. A typical addition per 100 gal is about 150 to 200 ml per day, or as required to get uniform plating. The amount added can vary somewhat and can be judged visually, by CVS analysis, by a Hull cell, or by metallographic cross sectioning and etching of the PTH. Excess PY 61-H causes step plating and corner cracks.[31]

Current density requires plating at 20 to 35 A/ft^2. One mil (0.001 in) will be deposited in 42 min at 25 A/ft^2.

The *Hull cell* test is widely used in evaluating bath condition and process problems. A Hull cell run at 2 A will indicate organic contaminants and overall bath condition. An optimum Hull cell will generally mean that the bath is in proper operating condition. See page 12.44 for Hull cell and cross-sectioning procedures.

Cross-sectioning results: Besides indicating overall quality, thickness, and other problems such as drilling, cracking, and blowholes, this method gives information on the plated copper and PTH that will tie the copper structure to improvements in processing. Copper deposits plated from balanced baths with the correct amounts of PY 61-H show laminar patterns. PY 61-H is present at suitable concentrations if the PTH corner radius is slightly less than the surface thickness or the hole copper thickness, as shown in Figs. 12.16 to 12.18.[55]

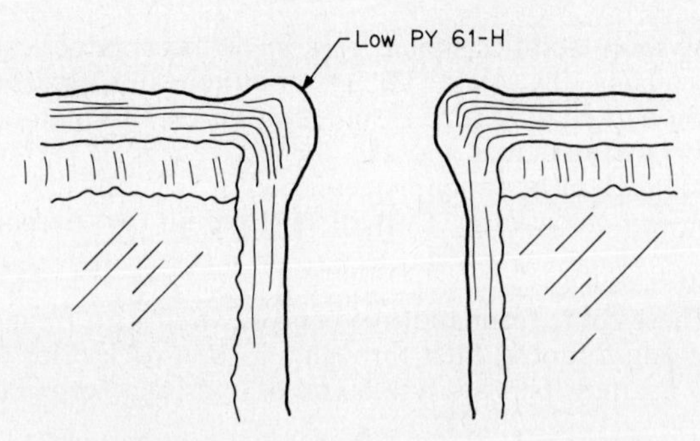

FIG. 12.16 PTH vertical cross section illustrating low additive (PY 61-H) concentration in copper pyrophosphate.[55]

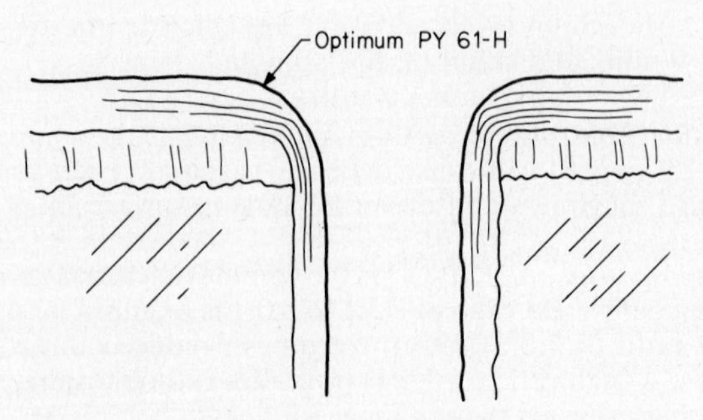

FIG. 12.17 PTH vertical cross section illustrating optimum additive concentration in copper pyrophosphate.[55]

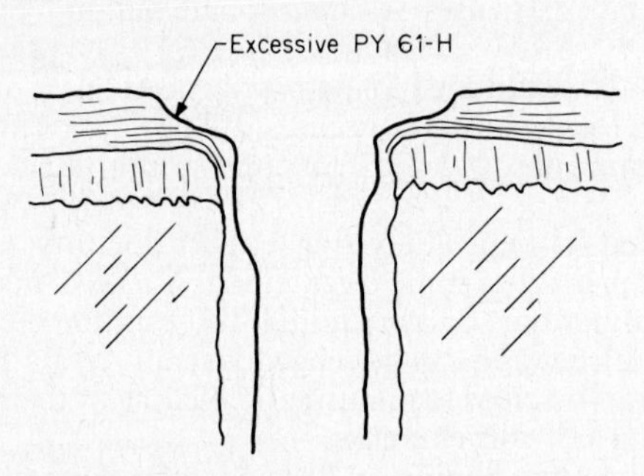

FIG. 12.18 PTH vertical cross section illustrating excessive additive concentration in copper pyrophosphate.[55]

Cracking and deposit ductility: Resistance to cracking is determined by the same tests as for acid copper. The physical properties of pyrophosphate copper deposits are as follows:

	Tensile strength, psi (avg.)	Elongation, % (avg.)
Organic-free	43,000	3–8
PY 61-H	87,000	5–18

Visual observation of plating: As plated, copper has a matte to satin appearance. When PY 61-H is used, the copper should appear semibright to satin at all current densities. Nodule formation (roughness) generally indicates particulate matter in the solution and the need for improved filtration and improved maintenance. Common causes of nodule formation are also related to anode control (low-anode CD), surface contamination, high copper, high pH, low ratio, and low PY 61-H. Some indications of organic contaminations are hazy, dull deposits; extremely dull or else bright zones on the Hull cell or the work; PTH corner cracking; brittleness; haloing around the PTH holes; and streaky deposits. This applies to baths with or without additives. Dull deposits may also be caused by low ammonia and foreign metals (lead >10 ppm). Reduced throwing power (increased surface-to-hole ratio) and poor-quality plating may indicate metal or chloride contamination, which may also reduce bath conductivity. In general, peelers are due to faulty precleaning steps. To correct these conditions, the following may be done:

1. Monitor pH, temperature, and anode filming daily.
2. Make daily additions of ammonia and additive.
3. Dummy-plate daily if baths are not used.
4. Analyze solution twice per week.
5. Circulate solution continuously through a well-maintained filtration system.
6. Analyze additive weekly by CVS or Hull cell.
7. Monitor metal contaminants monthly.
8. Batch-carbon-treat every 6 months.

12.10 SOLDER (TIN-LEAD) ELECTROPLATING

Solder plate (60 percent tin–40 percent lead) is widely used as a finish plate on printed circuit boards. This process features excellent etch resistance to alkaline ammonia, good solderability after storage, and good corrosion resistance. Tin-lead plating is used for several types of boards, including tin-lead/copper, tin-lead/tin-nickel/copper, SMOBC, and surface mount (SM). MIL-STD-275 states that tin-lead shall be 0.0003 in thick minimum, as plated on the surface. Fusing shall be required on all tin-lead-plated surfaces.

The preferred composition contains a minimum of 55 percent and a maximum of 70 percent tin. This alloy is near the tin-lead eutectic, which fuses at a temperature lower than the melting point of either tin or lead and thus makes it easy to reflow (fuse) and solder. (The composition of the eutectic is 63 percent tin–37 percent lead with a melting point of 361°F.) Fusing processes include infrared (IR), hot-oil, vapor-phase, and hot-air leveling for SMOBC.

Plating solutions currently available include the widely used high-concentration fluoboric acid-peptone system, as well as low-fluoboric, nonpeptone, and a nonfluoboric organic aryl sulfonic acid process. These processes are formulated to have high throwing power and give uniform alloy composition.[56] Sulfonic acid process has the advantage of using ball-shaped lead-tin anodes but is difficult to operate. Table 12.9 gives details of operation and control of two high-throw tin-lead (solder) baths.

12.10.1 Agitation

Solution is circulated by filter pump, without allowing air to be introduced.

TABLE 12.9 Tin-Lead Fluoborate

Operation and control

	High HBF_4/peptone	Low HBF_4/proprietary
Lead	1.07–1.88 oz/gal	1.4–2.0 oz/gal
Stannous tin, Sn^{2+}	1.61–2.68 oz/gal	2.8–4.0 oz/gal
Free fluoboric acid	47–67 oz/gal	15–25 oz/gal
Boric acid	Hang bag in tank	Same
Additive	Use as needed by Hull cell and A·h usage	
Temperature	60–80°F	70–85°F
Cathode current density	15–18 A/ft²	10–30 A/ft²
Agitation	Solution circulation	Mechanical and solution circulation
Anodes:		
Type	Bar	
Composition	60% tin–40% lead	
Purity	Federal specification QQ-S-571, or see Reference 26	
Bags	Polypro	
Hooks	Monel	
Length	Rack length minus 2 in	
Current density	10–20 A/ft²	

12.10.2 Filtration

A 3- to 10-μ polypro filter is need to control solution cloudiness and deposit roughness.

12.10.3 Carbon Treatment

Solution is batch-treated at room temperature every 4 to 12 months. If clear and colorless, additive is added as for new bath makeup. Do not use Supercarb or hydrogen peroxide.

12.10.4 Contaminations

Organic: Comes from peptone or additive breakdown, screen ink, and dry film. Periodic carbon treatment is needed.

Metallic: Copper is the most serious of these contaminations. It causes dark deposits at low current densities (in the plated-through hole) and may coat the anodes. The maximum levels of metallic contaminants allowable are copper, 15 ppm; iron, 400 ppm; and nickel, 100 ppm.

Nonmetallic: The maximum levels allowable are chloride, 2 ppm; and sulfate, 2 ppm.

Dummy: Copper is removed by "dummy" plating at 3 to 5 A/ft² several hours each week. ("Dummy" plating is plating at low current density, using corrugated metal sheets or scrap panels.) Other metals such as iron and nickel in high concentrations may contribute to dewetting and cannot be easily removed.

12.10.5 Solution Controls

Stannous and lead fluoborates are the source of metal. Their concentrations and ratio must be strictly maintained, as they will directly affect alloy composition.

Fluoboric acid increases the conductivity and throwing power of the solutions. Boric acid prevents the formation of lead fluoride. Additives promote smooth, fine-grained, "tree-free" deposits. Excess peptone (3 to 4 times too much) may cause pinholes ("volcanoes") in deposit when reflowed. Testing by Hull cell and periodic carbon treatments are indicated. The peptone-add rate is about 1 to 2 qt per week for a 400-gal tank. Only DI water and contamination-free chemicals should be used, for example, <10 ppm iron-free and <100 ppm sulfate-free fluoboric acid. A clear solution is maintained by constant filtration.

12.10.6 Deposition Rate

A layer of 0.5 mil tin-lead is deposited in 15 to 17 min at 15 to 17 A/ft^2. The best practice is to plate at 10 to 25 A/ft^2 ($\frac{2}{3}$ copper current). Higher currents lead to coarse deposits and more tin in the alloy. Excessive current causes "treeing" and sludge formation.

12.10.7 Deposit Composition

This is 60 percent tin–40 percent lead. Variation in composition is not a problem, since monitoring is maintained by solution analysis. However, the composition should be confirmed with a deposit assay. Alloy composition is determined by the ratio of tin and lead in the solution; thus

$$\text{Sn (oz/gal)} \div \text{total Sn (oz/gal)} + \text{Pb (oz/gal)} = \%\ \text{Sn}$$

12.10.8 Hull Cell

This test shows overall plating quality and the need for peptone, additives, or carbon treatment, as well as the presence of dissolved copper in the solution.

12.10.9 Visual Observation of Plating

As plated, solder has a uniform matte finish. The deposit should be smooth to the touch. A coarse, crystalline deposit usually indicates either the need for additives or peptone or too high a current density. Peeling from copper is generally related to procedure. The fluoboric acid predip should not be used as a holding tank (for more than 5 min) prior to solder plating. Acid strength and cleanliness are also important in this case. Load the tank with some residual current (5 to 10 percent). Rough, dark, thin, or smudged deposits may be due to organic contamination and may require carbon treatment. Dark deposits, especially in low-current-density areas and in the plated-through hole, are due to copper contamination or to thin plating. A Hull cell test will confirm deposit, and a dummy plate can then be used to remove copper.

12.10.10 Corrective Actions

1. Keep bath composition in balance.
2. Use contaminant-free chemicals.
3. Dummy-plate once a week at 3 to 5 A/ft^2.
4. Circulate solution through filter continuously.
5. Maintain additives by Hull cell and by analysis.

6. Analyze for copper, iron, and nickel every 3 months.
7. Carbon-treat on schedule.

12.11 COPPER/TIN-LEAD PROCESS STEPS

1. Receive panels from screening.
2. Rack panels.
3. Soak clean in acid cleaner for 5 min at 125°F.
4. Rinse in cold water (CWR) with spray and overflow.
5. Optional: Dip in 10 percent HCl for 2 min at room temperature.
6. Rinse in cold water (CWR) with spray and overflow.
7. Dip in 10 percent sulfuric acid for 5 min at room temperature.
8. Copper-plate using acid sulfate at 25 A/ft^2 for 1 h.
9. Drag out with CWR.
10. Rinse with cold water (CWR) with spray and overflow.
11. Dip in 15 percent fluoboric acid for 5 min at room temperature.
12. Tin-lead–plate, using high-throw fluoborate at 17 A/ft^2 for 15 min.
13. Drag out.
14. Rinse with cold water (CWR) with spray and overflow.
15. Unrack panels.
16. Release to etching.

12.12 TIN-LEAD FUSING

Fusing (or reflow) is the process of melting the plated tin-lead deposit after etching or edge-connector plating. The purposes of this step are to remove the tin-lead etching overhang (sliver), to cover exposed copper trace edges, to improve appearance and solderability (especially after storage), and to enhance corrosion resistance. Several methods are used.

1. *Infrared:* The principle of this technique is to melt the tin-lead with non-focused IR rays. This method is widely used, since it can be conveyorized and gives consistent quality with high productivity and safety. Machines operate by fluxing, preheating, and melting of tin-lead, followed by cleaning and drying. The melting point of eutectic solder (63 percent Sn–37 percent Pb) is 361°F (183°C). If either the tin or the lead varies widely from this composition, reflow will not take place in the IR. Equipment has been developed to fuse multilayers with large ground planes, which have given problems in the past.

2. *Hot oil:* This method was much used in earlier printed board production because of its ease of setup and its high tolerance for all types of boards and surface preparations. Hot oil is not currently preferred, due to safety and fire hazards and to slow productivity.

3. *Vapor phase:* This process uses the principle of condensation heating. It involves immersion of a board in a saturated vapor which condenses on the board, causing heating and subsequent tin-lead melting. The heat transfer liquids used are very costly.

4. *Hot-air leveling:* This process is used extensively to produce SMOBC boards. The principle involves dipping a board in pure molten solder and blowing off excess between air knives. Single-unit and high-production conveyorized machines are available. Surface activation with microetching and low copper ($<$ 0.30 percent) in solder must be controlled for solderability.

12.12.1 Problems in Tin-Lead Fusing

1. *Dewetting:* This refers to very uneven, thick-thin areas on surfaces of solder, on surfaces of pads, or on the traces after reflow. The most common cause of this problem is tin-lead contamination from etching residues. Solder must be cleaned after etching by chemical or mechanical means, followed by baking and fusing. Other causes of dewetting and nonwetting are underplate copper contamination, exhausted hot oil, and contaminated or inadequate circulation of the tin-lead plating solution.

2. *Outgassing during fusing:* This refers to pinholes, nodules, or small "volcano eruptions" on the tin-lead surface after fusing. This is due to codeposition of organic additives and is corrected by batch carbon treatment. Exhausted reflow oil may also cause this effect.

3. *Nonfusible tin-lead:* This refers to a condition where the tin-lead deposit will not fuse on heating. The cause may be thin plating (less than 0.25 mil), excess peptone or additives, or the need for carbon treatment of plating solution.

4. *Gritty or granular deposits:* These may be due to etching residues, embedded abrasives, soiled surfaces, or additive problems.

5. *Hole plugging:* This refers to closure of the PTH and is due to excessive tin-lead plating.

6. *Poor solderability:* The fused tin-lead deposit is not usually the cause of poor solderability of components. More likely, the causes are contamination and moisture buildup after storage. Correct by cleaning and baking boards at 250°F for 4 h and soldering immediately. Blowholes (gassing) during soldering may be due to thin, rough copper and/or voids in the PTH.[26,41,42]

12.13 TIN VERSUS TIN-LEAD ALLOY

12.13.1 Comparison of Usage

Although tin and tin-lead alloy are often equally regarded for finish-plate applications, there are many differences to be considered before a choice is made. For example, tin plate is not suitable for fusing and leaves overhang etch slivers; however, it is rigid under solder mask during wave soldering. Tin-lead is fused but will flow on traces under solder mask when wave-soldered. This causes distortion and flaking of the solder mask. Tin is free of water-polluting lead; tin-lead is not. Tin may form whiskers; fused tin-lead will not. Both tin and tin-lead can be used to produce SMOBC boards,[57,58] but tin may be a better choice, since both are stripped off after etching. Tin will not add to wastewater pollution.

12.14 TIN ELECTROPLATING

Tin is used extensively for plating electronic components and PC boards because of its solderability, corrosion resistance, and metal etch-resist properties. The current MIL-STD-275 does not include tin plating, although earlier versions stated a required thickness of 0.0003 in. Specifications covering tin plating are MIL-T-10727 and MIL-P-38510, which say that tin must be fused on component leads.

12.14.1 Acid Tin Sulfate

This is the most widely used system. A variety of processes are available, from bright deposits for appearance and corrosion resistance to matte deposits which can be fused and are solderable after long-term heating. Tin sulfate baths are somewhat difficult to control, especially after prolonged use.[59,60] Operation and control are given in Table 12.10.

12.14.1.1 Process Controls.

1. *Agitation:* Solution is circulated by filter pump without allowing air to be introduced. Cathode rod agitation is also useful for a wider range of plating current densities.
2. *Filtration:* A 3- to 10-μ polypropylene filter is needed to control excess cloudiness and sludge formation.
3. *Temperature:* The preferred temperature for deposit luster and visual appearance is 55 to 65°F. Use cooling coils. Baths can operate up to 85°F but may result in smoky, hazy deposits.
4. *Carbon treatment:* Batch treatment at room temperature removes organic contaminations. New baths are also made up if problems contine with deposit quality, solderability, thickness control, and cost effectiveness.

TABLE 12.10 Acid Tin Sulfate
Operation and control

Operating conditions:	
Tin	2 oz/gal
Sulfuric acid	10–12% by volume
Carrier, additives	Replenish by A·h usage and spectrophotometry
Temperature	55–65°F preferred 65–85°F hazy
Cathode current density	20–35 A/ft^2
Current efficiency	100%
Solution color	Milky, white
Plating rate	0.3 mil @ 25 A/ft^2 for 10 min
Anodes:	
Type	Bars
Composition	Pure tin
Bags	Polypropylene
Length	Rack length minus 2 in
Hooks	Monel or titanium
Current density	5–20 A/ft^2

5. *Contaminations:*
   a. *Organics:* These come from additive and breakdown of resist.
   b. *Metallic:* The effects of metallic contaminants on the plated deposits and the maximum levels allowable are as follows:

Copper	Stress	5–10 ppm
	Dullness	150 ppm
Cadmium	Dullness	50 ppm
Zinc	Dullness	50 ppm
Nickel	Streaks	50 ppm
Iron	Dullness	120 ppm

   c. *Nonmetallic:* The maximum level allowable is 5 ppm of chloride.
6. *Anodes:* Remove when bath is idle to maintain tin content.

12.14.1.2 Solution Controls.

1. *Bath constituents:* Stannous sulfate and sulfuric acid are maintained by analysis; additives by spectrophotometry, A·h usage, Hull cell, and the percentage of sulfuric acid additions. Electronic-grade chemicals must be used in order to control metallic contaminations of cadmium, zinc, iron, etc.
2. *Control additive:* Low levels of additives must be maintained.
3. *Hull cell:* This test is useful for control of additive levels and plating quality. See page 12.44 for procedure.
4. *Rinsing after plating:* Adequate rinsing after plating is important to control white or black spots on tin surfaces. See Sec. 12.15.
5. *Visual observation:* As plated, tin has a uniform lustrous finish. The deposit should be smooth to the touch.

12.14.2 Problems with Tin Electroplating

1. *Dull deposits:* These are due to out-of-balance condition of the main solution constituents, i.e., low acid (<10 percent), high tin, (>3 oz/gal), improper additive levels, contamination by metals or chlorides, or high temperatures (>65°F).
2. *Peeling tin:* This comes off due to low acid (<10 percent).
3. *Slivers:* These are caused by overetching. A review of etching practice is indicated, as well as the use of $\frac{1}{2}$-oz copper foil.
4. *Pitting:* If substrate is not the cause, check precleaning, solution balance and contaminants, current efficiency, and current densities. High current densities may cause pitting.
5. *Strip and etching residues:* Tin is attacked by strong alkaline solutions. To control residues and spotting, use mild room-temperature stripping and alkaline-ammonia etching.
6. *Poor solderability:* This may be caused by excess additives or contaminations in the bath, by poor rinsing, by bath age, or by excessive thicknesses (>0.3 mil).

12.15 SMOBC PROCESS STEPS

1. Receive panels from dry film.
2. Rack panels.
3. Soak clean with acid cleaner for 5 min at 125°F.
4. Rinse with cold water (CWR), using spray and overflow.
5. *Optional:* Microetch using sulfuric acid–hydrogen peroxide for 1 min at 100°F.
6. Rinse with cold water (CWR), using spray and overflow.
7. Acid-dip in 20 percent sulfuric acid for 5 min at room temperature.
8. Copper-plate, using acid sulfate at 25 A/ft^2 for 1 h.
9. Drag out and rinse in cold water (CWR).
10. Rinse with cold water (CWR), with spray and overflow.
11. Acid-dip in 10 percent sulfuric acid for 5 min at room temperature.
12. Rinse with cold water (CWR), using spray and overflow.
13. Tin-plate with acid sulfate for 7 min at 20 A/ft^2.
14. Warm water with overflow.
15. Unrack.
16. Etch copper.
17. Strip tin.
18. Release to hot-air leveling.

12.16 COPPER/NICKEL/TIN PROCESS STEPS

1. Receive panels from dry film.
2. Rack panels.
3. Soak clean with acid cleaner for 5 min at 125°F.
4. Rinse with cold water (CWR), using spray and overflow.
5. *Optional:* Microetch, using sulfuric acid–hydrogen peroxide for 1 min at 100°F.
6. Rinse with cold water (CWR), using spray and overflow.
7. Acid-dip in 20 percent sulfuric acid for 5 min at room temperature.
8. Copper-plate using acid sulfate at 25 A/ft^2 for 1 h.
9. Drag out and rinse in cold water (CWR).
10. Rinse with cold water (CWR), with spray and overflow.
11. Acid-dip in 10 percent sulfuric acid for 5 min at room temperature.
12. Rinse with cold water (CWR), using spray and overflow.
13. Nickel-plate with sulfamate for 10 min at 25 A/ft^2.
14. Drag out.
15. Rinse with cold water (CWR), using spray and overflow.
16. Dip in 10 percent sulfuric acid for 5 min at room temperature.
17. Rinse with cold water (CWR), using spray and overflow.

18. Tin-plate with acid sulfate for 10 min at 20 to 30 A/ft^2.
19. Rinse twice with warm water, using overflow.
20. Unrack panels and release to etching.

12.17 NICKEL ELECTROPLATING

Nickel plating is used as an undercoat for precious and nonprecious metals. For surfaces such as contacts or tips that normally receive heavy wear, the uses of nickel under a gold or rhodium plate will greatly increase wear resistance. When used as a barrier layer, nickel is effective in preventing diffusion between copper and other plated metals. Nickel-gold combinations are frequently used as metal etch resists. Nickel alone will function as an etch resist against the ammoniacal etchants.[61] MIL-STD-275 calls for a low-stress nickel with a minimum thickness of 0.0002 in. Low-stress nickel deposits are generally obtained using nickel sulfamate baths in conjunction with wetting (antipit) agents. Additives are also used to reduce stress and to improve surface appearance.

12.17.1 Nickel Sulfamate

Nickel sulfamate is commonly used both as undercoat for through-hole plating and on tips. Conditions given in Table 12.11 are applicable for through-hole and full board plating.

12.17.1.1 Process Controls.

1. *pH:* With normal operation, pH increases. Lower with sulfamic acid (not sulfuric acid). A decrease in pH signals a problem, and anodes should be checked.
2. *Temperature:* The preferred temperature is 125°F. Low temperature causes

TABLE 12.11 Nickel Sulfamate
Operation and control

Operating conditions:	
pH	3.5–4.5 (3.8)
Nickel	10–12 oz/gal
As nickel sulfamate	43 oz/gal
Nickel chloride	4 oz/gal
Boric acid	4–6 oz/gal
Additives	As required
Antipit	As required
Temperature	130 ± 5°F
Cathode current density	20–40 A/ft^2
Anodes:	
Type	Bars or chunks
Composition	Nickel
Purity	Rolled depolarized, cast or electrolytic; SD chips in titanium basket
Hooks, baskets	Titanium
Bags	Polypro, Dynel, or cotton
Length	Rack length minus 2 in

stress and high CD "burning"; higher temperatures increase softness of deposit nickel.

3. *Agitation:* Solution circulation between panels is done by filter pump and/or cathode rod agitation.
4. *Filtration:* Filtration is done continuously through 5- to 10-μ polypro filter, which is changed weekly.
5. *Deposition rate:* For 0.5 mil, plate 25 to 30 min at 25 A/ft^2. For 0.2 to 0.3 mil, plate 15 min at 25 A/ft^2.
6. *Contaminations:*
   - ***a.*** *Metals:* Maximum allowable by atomic absorption: iron, 250 ppm; copper, 10 ppm; chromium, 20 ppm; aluminum, 60 ppm, lead, 3 ppm; zinc, 10 ppm; tin, 10 ppm; calcium, 0 ppm. These metal contaminants lower the deposition rates and cause nonuniform plating. Copper and lead cause dark, brittle deposits at low CD. Dummy at 3 to 5 A/ft^2. Iron, tin, lead, and calcium cause deposit roughness and stress.
   - ***b.*** *Organics:* These cause pitting, brittleness, step plating, and decreased ductility.
   - ***c.*** *Sulfates:* These cause solution breakdown and should not be added.
7. *Carbon treatment:* Circulate through carbon canister for 24 h, or batch-treat to remove organics. The basic method to batch-carbon-treat is as follows:
   - ***a.*** Heat to 140°F.
   - ***b.*** Transfer to treatment tank. Do not adjust pH.
   - ***c.*** Add 3 to 5 lb carbon per 100 gal of solution. Premix outdoors.
   - ***d.*** Stir 4 h at temperature.
   - ***e.*** Let settle 1 to 2 h.
   - ***f.*** Filter back to cleaned tank.

12.17.1.2 Common Problems. Pitting, stress (cracking), and burned deposits are common problems.

1. *Pitting:* This is caused by low antipit, poor agitation or circulation, boric acid imbalance, or the presence of organics. In testing for antipit, the solution should hold bubble for 5 s in a 3-in wire ring and for less than 5 s in a 5-in ring. In the case of severe pitting, cool to room temperature, add 1 pt/200 gal hydrogen peroxide (35 percent), bubble air, and reheat to 140°F. Carbon-treat as above. Filter quickly.

2. *Stress:* This refers to the cause of deposit cracking. While low nickel chloride causes poor anode corrosion (rapid decrease in nickel content), high chloride causes excess stress. To prevent this, maintain pH, current density, and boric acid for control below 10 kpsi.

3. *Burns:* A low level of nickel sulfamate causes high current density "burns."

4. *Hull cell:* Plate at 2 A, 5 min with gentle agitation. This test is useful for bath condition and contaminants.

5. *Visual observation:* The plated metal has a matte, dull finish. "Burned" deposits are caused by low temperature, high current density, bath imbalance, and poor agitation. Rough deposits are due to poor filtration, pH out of spec, contamination, or high current density. Pitting is due to low antipit, poor agitation, bath imbalance, or organic contamination. Low plating rates are due to low pH, low current density, or impurities. Gassing at the cathode is a sign of low plating rates.

12.17.2 Nickel Sulfate

This is typically plated with an automatic edge connector (tip) plating machine. Table 12.12 gives the operating conditions that apply to these systems. For additional instruction, follow details in Sec. 12.17.1.

TABLE 12.12 Nickel Sulfate
Operating conditions

Operating Conditions:	
pH	1.5–4.5
Nickel	15–17 oz/gal
Nickel chloride	2–4 oz/gal (with soluble anodes)
Boric acid	3–4 oz/gal
Stress reducer	As required
Antipit	As required
Temperature	130 ± 5°F
Cathode current density	100–600 A/ft^2
Current efficiency	65%
Anodes:	
Composition	Nickel or platinized titanium

Nickel anodes are preferred in tip machines because the pH and metal content remain stable. The pH will decrease rapidly when insoluble anodes are used. The pH should be maintained at 1.5 or higher with additions of nickel carbonate. Stress values are higher than in sulfamate baths with values of about 20 kpsi.

12.18 TIN-NICKEL ALLOY

Printed circuit boards produced with tin-lead over tin-nickel[62] and solder masking over tin-nickel with tin-lead–plated pads and holes show major improvements over the standard methods of solder masking over tin-lead or copper.[63] When compared with the standard solder mask on copper/tin-lead board, improvements are noted in solder mask stability, reduced entrapment of fluxes, fine-line processing, no tip line, and reduced need for touch-up after wave soldering.[63] When compared with SMOBC, improvements are noted in resistance to hole-wall pullaway and corrosion, and ease of component replacement. Surface mount boards also benefit with tin-nickel (or nickel) due to improvements in solder pad thickness control and flatness, terminal strength, repeated component replacement, and reduced touch-up. Terminal pull tests using component lead wires soldered into a series of PTHs and pulled with a tensile tester gave the following results:

	Pull, lb/in^2	
Board thickness	Copper/tin-lead	Copper/tin-nickel/tin-lead
0.060 in	65	87
0.093 in	71	93
MIL-P-55110	50	50

Tin-nickel alloy also finds uses for "burn-in" boards and high-temperature applications due to its ability to provide a scratch-, thermal-, and corrosion-resistant barrier on the copper traces. Tin-nickel alloy is commonly plated from a fluoride-chloride electrolyte with no additives. This solution has exceptional throwing power and deposits a 1:1 atomic tin-nickel alloy of 65 percent tin and 35 percent nickel. Operating conditions are listed in Table 12.13.

TABLE 12.13 Tin-Nickel Alloy

Operation and control

Operating conditions:	
pH	2.0–2.5 (pH paper)
Nickel	10–12 oz/gal
Stannous tin	3.5–4.2 oz/gal
Fluoride (total)	4.5–6.6 oz/gal
Additives	None
Temperature	160 ± 5°F
Cathode current density	5–17 A/ft^2
Anodes:	
Type	Bars
Composition	Nickel
Purity	Rolled depolarized or cast
Bags	Nylon
Hooks	Monel
Length	Rack length minus 2 in
Deposit composition	65% Sn, 35% Ni (by weight)

12.18.1 Process Controls

1. *Agitation:* Solution is slowly circulated by filter pump, without allowing air to be introduced. The agitation is turned off during plating.
2. *Filtration:* A 5- to 10-μ polypro filter is used to remove solids, keeping the outlet under solution.
3. *Carbon treatment:* Circulate the solution through the carbon canister periodically. This is best done when the bath is not in use. Batch treatment is generally not required, unless there is known organic contamination, i.e., ink leaching, or hazy, pitted plating. A poor Hull cell will also indicate need for carbon treatment.
4. *Contaminations:*
   - ***a.*** *Organic:* These come from the screen ink and are carbon-treated.
   - ***b.*** *Metals:* Maximum allowable by atomic absorption: copper, 25 ppm; iron, 50 ppm; lead, 10 ppm.
   - ***c.*** *Dummy:* Plate at 5 A/ft^2, several hours each week to remove metal contaminants.
5. *Solution controls:* Stannous and nickel chloride provide metal ions. Total fluoride, added as ammonium bifluoride, must be equal to or higher (oz/gal) than the sum of the stannous (Sn^{2+}) and the stannic$^{(4+)}$ content. The bath pH is very critical and should be maintained at 2.0 to 2.5 (by pH paper). Low pH reduces the plating current density range and causes stressed, cracked plating under solder after reflow. General tendency is for pH to decrease. Add aqua

ammonia (ammonium hydroxide) to adjust. Low fluoride causes dull, black plating and should be maintained at 6.5 oz/gal minimum. Use only DI water and contaminant-free chemicals.

6. *Temperature:* Temperature control is important for optimum deposits. Below 150°F, the plating range will be reduced, causing cracked white deposits at higher current densities.
7. *Deposition rate:* Rate at 5 to 17 A/ft^2. A thickness of 0.0002 in (0.2 mil) is plated in 15 min at 15 A/ft^2. Excessive current density may cause cracked deposits after solder reflow.
8. *Deposition composition:* This is 65 percent tin–35 percent nickel. Composition remains constant over a wide range of operating conditions.
9. *Hull cell:* When run at 1 A for 5 min at 160°F, this useful test will give indications of pH control, copper and iron contamination, organic contamination, and overall bath conditions. Hull cell testing has proven to be a valuable aid in solution and process adjustment.
10. *Visual observation of plating:* As plated, metal is bright over a wide current density range. Cracked, stressed deposits seen after solder IR reflow are due to low temperature, low pH operation, or high current density. Dull, hazy plating is due to bath imbalance or to organic or metal contaminants. Black plating indicates a need for fluoride. Dark plating in low current density areas is caused by copper contamination. White, cracked deposits in high current density areas indicate iron contamination or bath imbalance. The corrections needed are as follows:
    - ***a.*** Check and maintain pH and temperature daily. Stir well.
    - ***b.*** Maintain fluoride content. Add stannochlor twice weekly.
    - ***c.*** Test solution with Hull cell once a week. This is important for tin-nickel control.
    - ***d.*** Analyze for lead, copper, and iron contamination every month. Check Sn^{4+} every 3 months.
    - ***e.*** Dummy-plate once a week.
    - ***f.*** Check deposit integrity by immersion in tin-lead stripper.

Newer processes are available which use amine additives and operate at lower temperature and higher pH.[64]

12.19 COPPER/TIN-NICKEL/TIN-LEAD PROCESS STEPS

1. Receive panels from image transfer.
2. Rack panels.
3. Soak clean with acid cleaner for 5 min at 125°F.
4. Rinse with cold water (CWR) with spray and overflow.
5. *Optional:* Dip in 10 percent sulfuric acid for 5 min at room temperature.
6. Rinse in cold water, using spray and overflow.
7. Dip in 10 percent sulfuric acid for 5 min at room temperature.
8. Copper-plate, using acid sulfate at 25 A/ft^2 for 1 h.
9. Drag out and rinse in cold water (CWR).

10. Rinse in cold water (CWR) with spray and overflow.
11. Acid-dip, using 10 percent HCl for 5 min at room temperature.
12. Rinse in cold water (CWR), using spray and overflow.
13. Tin-nickel–plate, using chloride-fluoride, pH 2.2, at 160°F and 15 A/ft^2 for 15 min.
14. Drag out and rinse in cold water.
15. Rinse in cold water with spray and overflow.
16. Dip in 15 percent fluoboric acid for 5 min at room temperature.
17. Tin-lead–plate at 17 A/ft^2 for 15 min.
18. Drag out and rinse in cold water.
19. Rinse in cold water with spray and overflow.
20. Unrack panels.
21. Release to etching.

12.20 GOLD ELECTROPLATING[23]

Early printed circuit board technology used gold extensively. In addition to being an excellent resist for etching, gold has good electrical conductivity, tarnish resistance, and solderability after storage. Gold can produce contact surfaces with low electrical resistance. In spite of its continued advantages, the high cost of gold has restricted its major application to edge connectors (tips) and selected areas, with occasional plating on pads, holes, and traces (body gold). Both hard alloy and soft, pure gold are currently used. Plating solutions are acid (pH 3.5 to 5.0) and neutral (pH 6 to 8.5). Both automatic plating machines for edge connectors and manual lines are in use.

12.20.1 Edge Connectors

12.20.1.1 Acid Hard Gold. To a large extent, acid golds are used for compliance to MIL-STD-275, which states that gold shall be in accordance with MIL-G-45204, Type II, Class 1. The minimum thickness shall be 0.000050 in (50 μin); the maximum shall be 0.000100 in in areas that are to be soldered. A low-stress nickel shall be used between gold overplating and copper. Nonmilitary applications require 25 to 50 μin. Type II refers to 99.0 percent minimum gold purity, alloy gold composition. Type II hard gold is not suited for wire bonding. These systems use potassium gold cyanide in an organic acid electrolyte. Deposit hardness and wear resistance are made possible by adding complexes of cobalt, nickel, or iron to the bath makeup. Automatic plating machines are being used increasingly because of the enhanced thickness (distribution) control, efficient gold usage, productivity, and quality. A comparison of automatic versus manual plating methods for edge connectors is given in Table 12.14.

12.20.1.2 Process Controls.

1. *Analysis:* Gold content, pH, and density should be maintained at optimum values. Operation at too low a gold content causes early bath breakdown with loss of properties, less current efficiency, and lowered cost savings. The pH is raised by using potassium hydroxide and is lowered with acid salts. Solution

TABLE 12.14 Acid Gold-Cobalt Alloy

Operation and control

	Manual	Automatic
Gold content, troy oz/gal	0.9–1.1	1–3
pH	4.2–4.6	4.5–5.0
Cobalt content	800–1000 ppm	800–1200 ppm
Temperature	90–110°F	100–125°F
Solution density	8–15 Be	12–18 Be
Replenishment per troy oz gold	8 A·h	6.5 A·h
Current efficiency	50%	60%
Agitation	5 gal/min	50 gal/min
Anode to cathode distance	2–3 in	$\frac{1}{4}$ in
Anodes, composition	Platinized titanium	Platinized titanium
Cathode current density	1–10 A/ft^2	50–100 A/ft^2
Thickness	40 $\pm$ 10 μin	40 $\pm$ 2 μin
Deposition rate for 40 μin	3–6 min	0.3–0.6 min
Deposition composition	99.8% gold, 0.2% cobalt	99.8% gold, 0.2% cobalt
Hardness	150 Knoop	150 Knoop

Gold solution contaminants

Metal	Maximum ppm	Metal	Maximum ppm
Lead	10	Iron	100
Silver	5	Tin	300
Chromium	5	Nickel	300–3000
Copper	50		

Organics: Tape residues, mold growth, and cyanide breakdowns

conductivity is controlled by density, which is adjusted with conductivity salts. Hull cell is not recommended for this purpose.

2. *Anodes:* Platinized titanium should be replaced when operating voltages are excessive or when thick coatings develop on the anode surfaces.
3. *Recovery:* Plating solutions should be replaced when contaminated, when plating rates decrease, or after about 10 total gold content turnovers.

12.20.1.3 Problems. Difficulties can be controlled by proper gold bath and equipment maintenance. Some typical situations are the following:

1. *Discolored deposit:* This may be due to low brightener; to metal contaminants such as lead, low pH, low density, or organic contaminations; or to leaking from tape. Gold plate is stripped to evaluate nickel. For these problems, first try to lower pH, raise density, or increase brightener before replacing bath or using decreased current densities.
2. *Gold peeling from nickel:* This is generally due to inadequate solder stripping, leaky tape, or to poor activation. Methods to increase nickel activation are to increase acid strength after nickel plating and use fluoride activators, cathodic acid, or gold strike. A gold strike is available which is compatible with gold plate and has low metal content and low pH. Its main purpose is to maintain adhesion to the nickel substrate.
3. *Wide thickness range:* To narrow thickness range, improve solution movement between panels; clean, adjust, or replace anodes; and adjust or replace solution.

4. *Low deposition rate:* This is characterized by excessive gassing and low efficiencies. To correct this condition, adjust solution parameters and check for contaminants such as chromium.
5. *Pitting:* Strip gold and evaluate nickel and copper by cross sectioning.
6. *Resist breakdown:* Low current efficiencies cause this condition. Use solvent soluble-screened or dry film for best results.

12.20.2 Microelectronics

12.20.2.1 Pure 24-Karat Gold. High-purity, 99.99 percent gold processes are used for boards designed for semiconductor chip (die) attachment, wire bonding, and plating solder (leaded) glass devices, for their solderability and weldability. These qualities comply with Types I and III of MIL-G-45204. The processes are neutral (pH 6 to 8.5) or acid (pH 3 to 6). Pulse plating is frequently used. Table 12.15 gives typical conditions for a neutral bath.

TABLE 12.15 Neutral Pure Gold

Operation and control

Gold content	0.9–1.5 troy oz/gal
pH	6.0–7.0
Temperature	150°F
Agitation	Vigorous
Solution density	12–15 Be
Replenishment	4 (A·h)/troy oz
Current efficiency	90–95%
Cathode current density	1–10 A/ft^2
Deposition rate for 100 μin, 5 A/ft^2	8 min
Deposit composition	99.99% + gold
Hardness	60–90 Knoop

12.20.2.2 Alkaline, Noncyanide Gold. Various processes for alloy and for pure gold deposits are available. Solutions are based on sulfite-gold complexes and arsenic additives and operate at a pH of 8.5 to 10.0. A decision to use this process is based primarily on the need for uniformity (leveling), hardness (180 Knoop), purity, reflectivity, and ductility. PC board use is limited to body plating, since wear characteristics of the sulfite-gold are not suitable for edge connector applications. The microelectronics industry uses these processes for reasons of safety and gold purity. Semiconductor chip attachment, wire bonding, and gold plating on semiconductors are possible applications and are enhanced by using pulse plating and without metallic additives.

12.20.2.3 Gold Plate Tests. Several routine in-process and final quality-control tests are performed on gold.

1. *Thickness:* Techniques are based on beta-ray backscattering and x-ray fluorescence. Thickness and area sensitivity are as low as 1 μin with 5-mil pads.
2. *Adhesion:* Standard testing involves a tape pull test.
3. *Porosity:* Tests involve nitric acid vapor and electrographics.
4. *Purity:* Lead is a common impurity that must be controlled to <0.1 percent.

Other tests include discoloration by heating, electrical contact, and wear resistance.

12.21 PLATINUM METALS

Interest in using these systems usually soars in a climate of high gold prices, even though plating results are not always as dependable as those obtained when using gold.

12.21.1 Rhodium

Deposits from the sulfate or phosphate bath are hard (900 to 1000 Knoop), highly reflective, extremely corrosion-resistant, and highly conductive (resistivity is 4.51 $\mu\Omega$/cm). Rhodium plate is used where a low-resistance, long-wearing, oxide-free contact is required. In addition, rhodium plate minimizes the noise level for moving contacts. The use of rhodium as a deposit on nickel for edge connectors has been replaced by gold. This is due to difficulty in bath control, problems with organic and metallic contaminants, and cost. Table 12.16 shows details of rhodium plating.

TABLE 12.16 Rhodium Sulfate

Operation and control

Rhodium	4–10 g/L
Sulfuric acid	25–35 mL/L
Temperature	110–130°F
Agitation	Cathode rod
Anodes	Platinized titanium
Anode-to-cathode ratio	2:1
Cathode current density	10–30 A/ft^2
Plating rate	At 20 A/ft^2, 10 μin will deposit in 1.4 min, based on 70% cathode current efficiency

12.21.2 Palladium and Palladium-Nickel Alloys

Deposits of 100 percent palladium, 80 percent palladium–20 percent nickel, and 50 percent palladium–50 percent nickel find use as suitable deposits for edge connectors. Deposits are hard (200 to 300 Knoop), ductile, and corrosion-resistant. A palladium-nickel undercoat for gold shows good wear and electrical properties.[65]

12.21.3 Ruthenium

Deposits of ruthenium are similar to rhodium but are plated with easier control, greater bath stability, and high current efficiency at lower cost. Deposits are usually stressed.

12.22 SILVER ELECTROPLATING

Silver is not widely used in the PC industry, although it finds applications in optical devices and switch contacts. Thicknesses of 0.0001 to 0.0002 in (0.1 to 0.2

mil) in conjunction with a thin overlay of precious metal are specified. Silver plating should not be used when boards are to meet military specification. The reason is that under certain conditions of electrical potential and humidity, silver will migrate along the surface of the deposit and through the body of insulation to produce low-resistance leakage paths. Tarnishing of silver-gold in moist sulfide atmospheres also produces electrical problems on contact surfaces due to diffusion of the silver to the surface. Another reason for the lack of acceptance of silver is that it is plated from an alkaline cyanide bath, which is highly toxic. Bright plating solutions produce deposits with improved tarnish and corrosion resistance, relative freedom from porosity, and greater hardness. Plating troubles are usually related to "black anodes" and are due chiefly to solution imbalance, impurities in anodes, or solution roughness and pitting. Most metals to be plated, particularly the less noble metals, require a silver strike prior to silver plating to ensure deposit adhesion.

12.23 IMMERSION AND ELECTROLESS PLATING

12.23.1 Tin Immersion

Other than electroless copper, immersion tin is the only other process in widespread use. Immersion tin is used to clean boards after etching, to cover copper trace edges with tin, and to act as a soldering aid. Deposits are thin, about 30 μin, and must be processed immediately for effective results. The copper content of this process must be monitored to control tin-lead fusing problems.

12.23.2 Electroless Gold[66]

Processes are available for depositing up to 100 μin of pure gold on copper and gold-isolated circuitry by true electroless (autocatalytic) means. These systems contain organic amine boranes and borohydride reducing agents and cyanides and operate at a high pH and temperature. Deposits meet the requirements of MIL-G-45204, Type III, Grade A with 99.99 percent purity and hardnesses of 90 KHN. They are suitable for edge connectors, semiconductor wire bonding, and die attachment.

Other processes of this type are electroless nickel, electroless palladium,[67] and immersion gold.

12.24 LABORATORY PROCESS CONTROL

12.24.1 Conventional "Wet" Chemical Analysis

The traditional wet chemical methods for metals and nonmetal plating solution constituents are available from suppliers and in the literature.[68,69] These methods also make use of pH meters, ion electrodes, spectrophotometers, and atomic adsorption. The composition of liquid concentrates for plating solutions is shown in Table 12.17.

12.24.2 Advanced Instrumental Techniques

New techniques have been developed for the control of organic additives in copper plating. Continued development is in progress in the area of measurement of

TABLE 12.17 Composition of Liquid Concentrates

Chemical	Formula	Weight, lb/gal	Percent	Metal, oz/gal
Acids:				
Sulfuric	H_2SO_4	15.0	96	—
Hydrochloric	HCl	9.8	36	—
Fluoboric	HBF_4	11.2	49	—
Alkaline:				
Sodium hydroxide	$NaOH$	12.8	50	—
Ammonium hydroxide	NH_4OH	7.5	28	—
Metals:				
Copper sulfate	$CuSO_4 \cdot 5H_2O$	9.7	27	9
Copper fluoborate	$Cu(BF_4)_2$	12.9	46	25.4
Stannous fluoborate	$Sn(BF_4)_2$	13.3	51	44.3
Lead fluoborate	$Pb(BF_4)_2$	14.4	51	65.0
Nickel sulfate	$NiSO_4 \cdot 6H_2O$	11.0	44	17.8
Nickel sulfamate	$Ni(NH_2SO_3)_2$	12.9	50	24
Nickel sulfamate	$Ni(NH_2SO_3)_2$	12.3	43	20
Nickel chloride	$NiCl_2 \cdot 5H_2O$	11.2	54	23.7

such additives in nickel, gold, and tin solutions. New methods include liquid chromatography–ultraviolet/visible (UV/VIS) spectrophotometry, cyclic voltammetry stripping (CVS), ion chromatography, UV-persulfate oxidation, and polarography. These techniques can detect contaminations in various processes, and they show the need for and effectiveness of carbon treatment. Table 12.18 lists and references these techniques, which are having a major influence on plating process capabilities.

12.24.3 Metallographic Cross Sectioning[73,74]

A method for cross sectioning PC boards is as follows:

1. *Bulk cutting:* Removal of manageable-size piece of board or assembly by shearing or abrasive cutting.
2. *Precision cutting:* Low-speed sawing with a diamond wafer blade to produce vertical sections about $1 \times \frac{1}{2}$ in, and horizontal sectioning cut next to PTH pads.
3. *Mounting:* Encapsulate vertical and horizontal sections in epoxy resin.
4. *Fine grinding:* Hand grinding using 240-, 320-, 400-, and 600-grit silicon carbide papers. Rinse sample between grits.

TABLE 12.18 Advanced Instrumental Analysis Techniques

Technique	Constituent	References
Cyclic voltammetry stripping	Organics and inorganics	34–36, 52–54
Liquid chromatography with UV/VIS	Organics and inorganics	37, 38
Ion chromatography	Ionic species	70
Polarography	Organics and inorganics	11, 25, 69
Ion-selective electrode	Ionic metals, nonmetals	71, 72
Atomic absorption	Metals, nonmetals	—
UV oxidation	Total carbon	38

5. *Polishing:* Diamond (6-μ) on nylon cloth and 0.3-μ alumina on nap cloth polishing on rotating wheel. To polish sample, place it on the rotating wheel polisher and move it slowly in the opposite direction. Polish for 4 min if using a 6-μ diamond on nylon, and 1 min on 0.3-μ alumina on nap cloth. Clean and dry between polishing compounds.
6. *Etching:* Apply cotton swab for 2 to 5 s soaked in a solution of equal parts of ammonium hydroxide and 3 percent hydrogen peroxide. Rinse in water and dry carefully.
7. *Documenting:* Observe and photograph sample with microscope at 30× to 1500× magnification.

12.24.4 Hull Cell

Although the advanced techniques discussed above provide precise control of plating solutions, Hull cell testing is still widely used in the industry. Its advantages are low cost, simplicity of operation, and its actual correlation with plating production. Its main disadvantage is that defects in copper plating frequently are not shown by this testing. For example, Hull cell testing will not help in detecting dull plating, roughness, or pitting. The procedures starts with brass panel preparation, in the following order: Remove the plastic film, treat with cathodic alkaline cleaner, soak in 10 percent sulfuric acid, and rinse. Repeat these steps until panel is water-break-free. Proceed with the Hull cell as follows: Rinse with test solution, fill to mark, adjust temperature and agitation, attach panel to negative terminal, and plate. Agitation should be similar to tank operation, i.e., vigorous air bubbling for copper and gentle stirring for tin and tin-lead, none for tin-nickel. Plate copper and nickel at 2 A and other metals at 1 A. The effects of bath adjustment, carbon treatment, dummy plating, etc., are readily translated from the Hull cell to actual tank operations. See previous sections on metal plating for Hull cell results, and consult supplier for test equipment.

REFERENCES

1. C. A. Deckert, E. C. Couble, and W. F. Bonetti, "Improved Post-Desmear Process for Multilayer Boards," *IPC Technical Review,* January 1985, pp. 12–19.
2. G. Batchelder, R. Letize, and Frank Durso, "Advances in Multilayer Hole Processing," MacDermid Company.
3. G. C. Kourtures, "Trouble Shooting an Electroless Deposition Line for Printed Circuits," *Proceedings of the Electronics Industry Symposium,* American Electroplaters' Society, Palo Alto, Calif., 1971, pp. 69–79.
4. F. E. Stone, "The Maintenance of Electroless Copper Plating Solutions," *Plating in the Electronics Industry Symposium,* American Electroplaters' Society, Indianapolis, Ind., 1973, pp. 3–12.
5. J. Murray, "Plating, Part 1: Electroless Copper," *Circuits Manufacturing,* vol. 25, no. 2, February 1985, pp. 116–124.
6. F. Polakovic, "Contaminants and Their Effect on the Electroless Copper Process," *IPC Technical Review,* October 1984, pp. 12–16.
7. K. F. Blurton, "High Quality Copper Deposited from Electroless Copper Baths," *Plating and Surface Finishing,* vol. 73, no. 1, 1986, pp. 52–55.
8. C. Lea, "The Importance of High Quality Electroless Copper Deposition in the Pro-

duction of Plated-Through-Hole PCB's," *Circuit World,* vol. 12, no. 2, 1986, pp. 16–21.

9. S. Glasstone, *Introduction to Electrochemistry,* D. Van Nostrand, New York, 1942.
10. C. Prutton and S. H. Maron, *Fundamental Principles of Physical Chemistry,* rev. ed., Macmillan, New York, 1951.
11. E. C. Potter, *Electrochemistry,* Cleaver-Hume Press, Ltd., London, 1961.
12. D. A. MacInnes, *The Principles of Electrochemistry,* Reinhold Publishing Corp., New York, 1939.
13. S. E. Glasstone, *The Fundamentals of Electrochemistry and Electrodeposition,* American Electroplaters' Society, 1943.
14. E. Raub and K. Muller, *Fundamentals of Metal Deposition,* Elsevier, Amsterdam, 1967.
15. A. T. Vagramyan and Z. A. Solov'eva, *Technology of Electrodeposition,* R. Draper Ltd., Teddington, 1961.
16. F. A. Lowenheim, *Modern Electroplating,* 3d ed., J. Wiley and Sons, New York, 1974.
17. F. A. Lowenheim, *Electroplating,* McGraw-Hill, New York, 1978.
18. W. H. Safranek, *The Properties of Electrodeposited Metals and Alloys,* 2d ed., American Electroplaters and Surface Finishers Society, Florida, 1986.
19. R. Pinner, *Copper and Copper Alloy Plating,* pub. 62, Copper Development Association, London, 1962.
20. R. Brugger, *Nickel Plating,* R. Draper Ltd., Teddington, England, 1970.
21. A. Brenner, *Electrodeposition of Alloys,* vols. I and II, Academic Press, New York, 1963.
22. J. W. Price, *Tin and Tin Alloy Plating,* Electrochemical Publications Ltd., AYR, Scotland, 1983.
23. Frank H. Reid and William Goldie, *Gold Plating Technology,* Electrochemical Publications Ltd., AYR, Scotland, 1974.
24. J. Fischer and D. E. Weimer, *Precious Metal Plating,* R. Draper Ltd., Teddington, England, 1964.
25. A. J. Bard and L. R. Faulkner, *Electrochemical Methods, Fundamentals and Applications,* J. Wiley and Sons, New York, 1980.
26. E. F. Duffek, "Soldering Process vs. the PC Board," *Printed Circuit Fabrication,* vol. 6, no. 10, 1983, pp. 62–70.
27. G. Peterson, "Precision Plated-Through-Holes," *Printed Circuit Fabrication,* vol. 7, no. 7, 1984, pp. 50–58.
28. W. Mastie, "High Speed Copper," *Circuits Manufacturing,* vol. 23, no. 6, pp. 25–32.
29. N. M. Osero, "Overview of Pulse Plating," *Plating and Surface Finishing,* vol. 73, no. 3, 1986, pp. 20–23.
30. P. P. Pellegrino, "Apparatus for Electroplating, Deplating or Etching," U.S. Patent 4,174,261, 1979.
31. D. A. Luke, "Electroplating Copper for Printed Circuit Manufacture," *Circuit World,* vol. 13, no. 1, 1986, pp. 18–23.
32. M. Carano, "High Speed Copper Plating for Printed Wiring Boards," *Printed Circuit Fabrication,* vol. 6, no. 7, 1983.
33. B. Sullivan, "Electroplating Theory and the High Speed Copper Debate," *Printed Circuit Fabrication,* vol. 8, no. 8, 1985, pp. 35–54.
34. R. Haak, C. Ogden, and D. Tench, "Cyclic Voltametric Stripping Analysis of Acid Copper Sulfate Plating Baths, Part I: Polyether-Sulfide-Based Additives," *Plating and Surface Finishing,* vol. 68, no. 4, 1981, p. 52; "Part II: Sulfoniumalkanesulfonate-Based Additives," *Plating and Surface Finishing,* vol. 69, no. 3, 1982, p. 62.
35. M. Carano and T. Barringer, "Optimization of an Acid Copper Plating Bath for Through-hole Plating," *Proceedings of the 12th AES Plating in the Electronics Industry Symposium,* Orlando, Fla., January 1985.

36. P. Bratin, "New Developments in the Use of Cyclic Voltametric Stripping for Analysis of Plating Solutions," *Proceedings of AES Analytical Methods Symposium,* Chicago, Ill., March 1985.

37. T. R. Mattoon, P. McSwiggen, and S. A. George, "Printed Circuit Plating Bath Process Control," *Metal Finishing,* vol. 83, Parts I, II, and III, 1985.

38. K. Heikkila, "Selection and Control of Plating Chemistry for Multilayer Printed Wiring Boards," *ElectriOnics,* August–September 1985.

39. P. J. Darikh, "Electro-deposited Copper for Hi-Rel PC's," *Electronic Packaging and Production,* vol. 17, no. 3, 1977, pp. 61–65.

40. M. W. Jawitz, "Trouble Shooting Manual for Printed Circuit Production," *Insulation/Circuits,* vol. 22, no. 4, 1976, pp. P-5–P-36.

41. C. Lea, F. H. Howie, and M. P. Seah, "Blowholing in the PTH Solder Filets," *Circuit World,* vol. 12, no. 4, 1986, Parts 1–3, pp. 14–25.

42. C. Lea, F. H. Howie, and M. P. Seah, "Blowholing in the PTH Solder Filets," *Circuit World,* vol. 13, no. 3, 1986, Part 8, pp. 11–20.

43. J. Wynschenk, "Copper Plating," *Insulation/Circuits,* vol. 23, no. 3, 1977.

44. B. F. Rothschild and R. P. McCluskey, "Plated-Through-Hole Cracking: Causes and Cures," *Electronic Packaging and Production,* vol. 10, no. 5, 1970, pp. 114–124.

45. G. R. Strickland, "Pyrophosphate Copper Plating in Printed Circuit Manufacture," *Product Finishing,* no. 4, 1972, pp. 20–24.

46. B. F. Rothschild, "The Effect of Ortho-phosphate in Copper Pyrophosphate Plating Solutions and Deposits," *Metal Finishing,* vol. 84, no. 1, 1978, pp. 49–51.

47. J. W. Dini, H. R. Johnson, and J. R. Helms, "Effect of Some Variables on the Throwing Power and Efficiency of Copper Pyrophosphate Solutions," *Plating,* vol. 54, no. 12, 1967, p. 1337.

48. B. F. Rothschild, "Copper Electroplating Systems: An Evaluation," *Electronic Packaging and Production,* vol. 15, no. 8, 1975, pp. 102–107.

49. C. J. Owen, H. Jackson, and E. R. York, "Copper Pyrophosphate Plating without Additives," *Plating,* vol. 54, 1967, pp. 821–825.

50. L. E. Hayes, "Organic Additives for Pyrophosphate Copper: Panacea or Poison?" *Electronic Packaging and Production,* no. 17, 1977, pp. 102–104.

51. D. E. Sherlin and L. K. Bjelland, "Improve Electrodeposited Copper with Organic Additives and Baking," *Insulation/Circuits,* vol. 24, no. 9, 1978, pp. 27–32.

52. D. Tench and C. Ogden, "A New Voltametric Stripping Method Applied to the Determination of the Brightener Concentration in Copper Pyrophosphate Plating Baths," *J. Electrochem. Soc.,* vol. 125, 1978, p. 194.

53. C. Ogden and D. Tench, "New Methods for Understanding and Controlling Plating Systems Applied to Circuit Board Plating from Copper Pyrophosphate Baths," *Proceedings of 7th AES Symposium on Plating in Electronics Industry,* San Francisco, January 1979.

54. C. Ogden and D. Tench, "On the Mechanism of Electrodeposition in the Dimercaptothiadiazole/Copper Pyrophosphate System," *J. Electrochem. Soc.,* vol. 128, 1981, p. 539.

55. B. F. Rothschild, J. G. Semar, and H. K. Omata, "Carbon Treatment of Pyrophosphate Copper Baths for Improved Printed Wiring Board Production," *Proceedings of the Printed Wiring and Hybrid Circuits Symposium,* Ft. Worth, Tex., November 1975.

56. B. F. Rothschild, "Solder Plating of Printed Wiring Systems," *Proceedings of the Printed Circuit Plating Symposium,* California Circuits Association, Nov. 5–6, 1969, pp. 10–21, and Nov. 12–13, 1968, pp. 61–65.

57. B. Bean, "SMOBC: A Fabrication Point of View," *Circuit Manufacturing,* vol. 25, no. 3, 1985, pp. 26–33.

58. J. P. Langan, "Air-Leveled Printed Circuit Boards," *Plating and Surface Finishing,* vol. 73, no. 1, 1986, pp. 26–28.

59. G. F. Jacky, "Soldering Experience with Electroplated Bright Acid Tin and Copper," *Circuit Technology Today,* California Circuits Association, October 1974, pp. 49–74.

60. P. E. Davis and E. F. Duffek, "The Proper Use of Tin and Tin Alloys in Electronics," *Electronic Packaging and Production,* vol. 15, no. 7, 1975.

61. R. G. Kilbury, "Producing Buried Via Multilayers: Two Approaches," *Circuits Manufacturing,* vol. 25, no. 4, 1985, pp. 30–49.

62. E. Armstrong and E. F. Duffek, "Tin-Nickel Increases PC Reliability," *Electronic Packaging and Production,* vol. 14, October 1974, pp. 125–130; *Electronic Production Methods and Equipment* (G. B.), vol. 4, no. 2, 1975, pp. 41–42.

63. E. F. Duffek, "P. C. Processing Using Solder Mask Over Tin-Nickel," *Electronic Packaging and Production,* vol. 19, no. 6, 1979, pp. 71–74.

64. M. Carano, "Tin-Nickel Plating: An Alternative," *Printed Circuit Fabrication,* vol. 7, no. 7, 1984.

65. R. G. Baker and R. Duva, "Electrodeposition and Market Effect of Palladium and Its Alloys," *Plating and Surface Finishing,* vol. 73, no. 6, 1986, pp. 40–46.

66. H. O. Ali and R. A. Christie, "A Review of Electroless Gold Deposition Processes," *Circuit World,* vol. 11, no. 4, 1985, pp. 10–16.

67. W. J. Hawk, "Electroless Palladium for Electronics," *Metal Finishing,* vol. 84, no. 3, 1986, pp. 11–14.

68. D. G. Foulke, *Electroplaters' Process Control Handbook,* rev. ed., R. E. Krieger Publishing Company, Huntington, N.Y., 1975.

69. K. E. Langford and J. E. Parker, *Analysis of Electroplating and Related Solutions,* 4th ed., R. Draper Ltd., Teddington, England, 1971.

70. K. Haak, "Ion Chromatography in the Electroplating Industry," *Plating and Surface Finishing,* vol. 70, no. 9, 1983.

71. M. S. Frant, "Application of Specific Ion Electrodes to Electroplating Analyses," *Plating,* vol. 58, no. 7, 1971.

72. W. C. McDonnell, "Ion Selective Electrodes for Analysis in Metal Finishing," *Plating and Surface Finishing,* vol. 73, no. 11, 1986, pp. 32–35.

73. J. A. Nelson, "Basic Steps for Cross Sectioning," *Insulation/Circuits,* 1977.

74. P. Wellner and J. Nelson, *High Volume Cross Sectional Evaluation of Printed Circuit Boards,* IPC WC-4B-1, Evanston, Ill., 1981.

CHAPTER 13
ADDITIVE PLATING

Dr. Hayao Nakahara
Director of Far Eastern Operations, PCK Technology Division, Kollmorgen Corporation

13.1 INTRODUCTION

Additive methods for the manufacture of printed wiring boards (PWBs) may be classified into three basic processes: fully additive, semiadditive, and partially additive (see Fig. 13.1). All these processes involve electroless copper plating to form conductors or plated-through holes (PTHs) at one point in their production steps. The most widely used additive process is the CC-4* process, in which conductors and PTHs are formed simultaneously. It offers many other advantages[1] over the conventional subtractive process, such as lower plating thickness variation, ability to plate holes of high aspect ratio (Fig. 13.2), ability to form fine-line conductors, etc. Since the first commercial production of PWBs by the CC-4 process by Photocircuits Corporation in 1964, a countless number of fully additive boards have been manufactured for a variety of applications. In spite of the aforementioned advantages and production record, the adoption of additive processes by the PWB industry has been slower than one might expect. This has been partly because of the relative difficulty of the processes and partly because of the lower physical properties of electrolessly deposited copper compared to those of electrolytically deposited copper. As a result, the applications of fully additive PWBs have been confined mainly to packaging consumer electronic products.

This situation has changed, however. The quality of electrolessly deposited copper has been continuously improved (see Table 13.1). This has been because of improved understanding of the electroless copper deposition mechanism, improved process control, the introduction of the automatic bath controller, and better materials. All of these have greatly enhanced the advantages of the additive processes.

The purpose of this chapter is to examine the additive processes from various viewpoints such as electroless copper deposition, mechanical fabrications and their ability to form fine-line conductors, and high-aspect-ratio holes.

*CC-4 is a registered trademark of Kollmorgen Corporation.

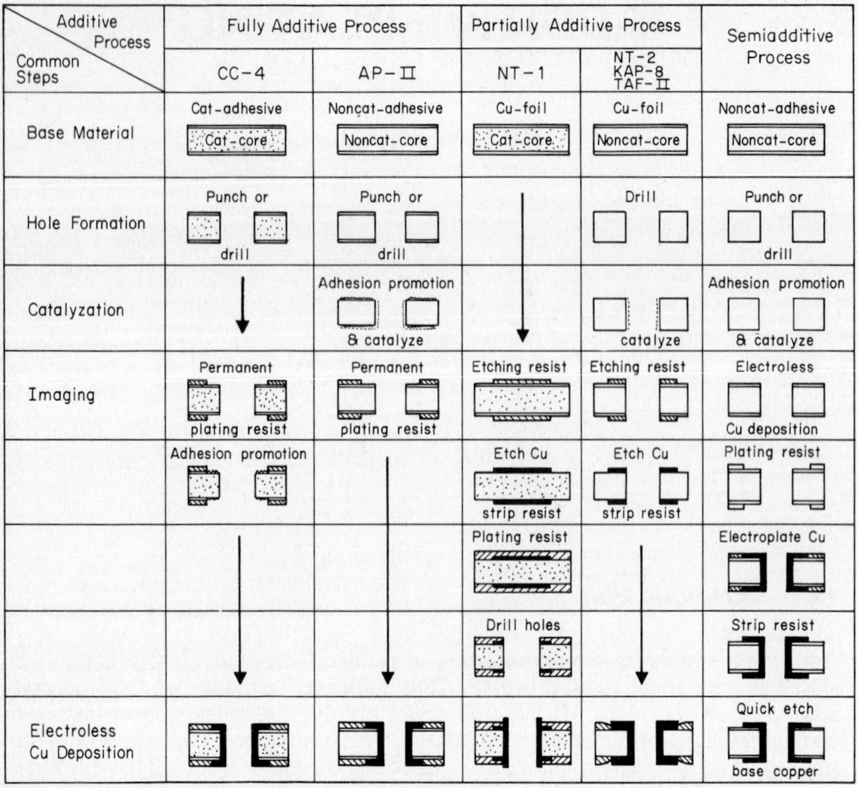

Additive Process / Common Steps	Fully Additive Process: CC-4	Fully Additive Process: AP-II	Partially Additive Process: NT-1	Partially Additive Process: NT-2 KAP-8 TAF-II	Semiadditive Process
Base Material	Cat-adhesive Cat-core	Noncat-adhesive Noncat-core	Cu-foil Cat-core	Cu-foil Noncat-core	Noncat-adhesive Noncat-core
Hole Formation	Punch or drill	Punch or drill		Drill	Punch or drill
Catalyzation		Adhesion promotion & catalyze		catalyze	Adhesion promotion & catalyze
Imaging	Permanent plating resist	Permanent plating resist	Etching resist	Etching resist	Electroless Cu deposition
	Adhesion promotion		Etch Cu strip resist	Etch Cu strip resist	Plating resist
			Plating resist		Electroplate Cu
			Drill holes		Strip resist
Electroless Cu Deposition					Quick etch base copper

FIG. 13.1 Simplified illustration of additive process.

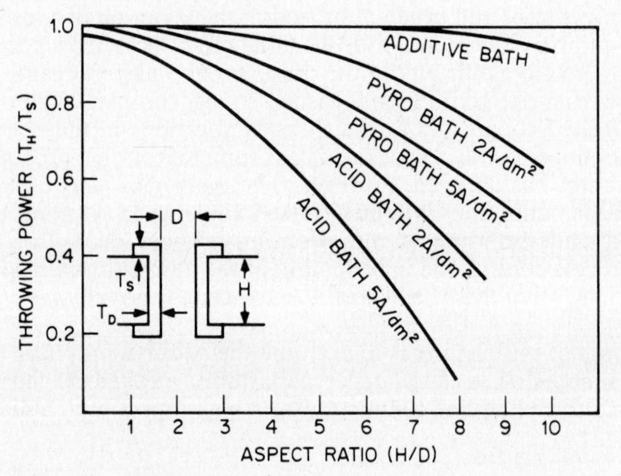

FIG. 13.2 Ability of electroless bath to plate hole of high aspect ratio without reducing throwing power.

TABLE 13.1 Various Electroless Copper Baths for the Additive Process

	Kollmorgen,[2] CC-4	Hitachi Chem.,[3] L-59	Kanto Kasei,[4] KAP-8	Hitachi Ltd.,[5] AP-II	Hitachi Ltd.,[6] SK-4	Matsushita,[7] DUSTON
$CuSO_4$	7 g/L	8 g/L	0.06 mol/L	10 g/L	12 g/L	0.03 mol/L
HCHO, 37%	3.5 mL/L	3 mL/L	0.5 mol/L	3 mL/L	6 mL/L	0.10 mol/L
NaOH	—	—	—	—	12 g/L	0.23 mol/L
EDTA	As 4Na 27 g/L	As 2Na 60 g/L	0.12 mol/L	As 2Na 30 g/L	As 2Na 35 g/L	As 2Na 0.04 mol/L
pH	12.0	12.1–12.4	12.5	12.2	12.2	12–12.5
Additive	NaCN, 15–20 mg/L	A: 1 g/L B: 20 mg/L	2,2′-dypiridil 10 mg/L	2,2′-dypiridil 10 mg/L	2,2′-dypiridil	2,2′-dypiridil NaCN, 10 mg/L
	2-MBT, 0.1 mg/L	C: 1 mg/L		PEG, 80 mg/L	Ag_2S	10–40 mg/L
Surfactant	Polyethelene oxide 1.0–8.0 mg/L	—	REG-1000 500 mg/L	PEG-600	REG-SA 50 mg/L	—
Temperature, °C	72	70	70	70–72	70	80–90
Speed, μm/h	2–3	2–3	7–10	2.5–3.0	3.5	7
Tensile strength, kg/mm^2	30–35	30–35	—	30–40	45	—
Elongation, %	6–12	7–12	—	6–12	3–5	—

13.2 CLASSIFICATION OF ADDITIVE PROCESSES

Additive processes are classified as fully additive, semiadditive, and partially additive. These are discussed in detail below.

13.2.1 Fully Additive Process

There are two basic variations of the fully additive process. One uses catalytic base laminates coated with catalytic adhesive, and one uses noncatalytic base laminates with noncatalytic adhesive. The former is known as the CC-4 process and the latter as the seeded process. Simplified process sequences of the fully additive processes are shown in Fig. 13.1.

13.2.1.1 CC-4 Process.[8] The essential manufacturing steps of the CC-4 process are as follows (see Fig. 13.1):

1. Catalytic base laminate with catalytic adhesive coated on both sides of the laminate
2. Formation of holes by drilling (FR-4) or punching (XXXPC, FR-2, FR-3, CEM-1, CEM-3)
3. Mechanical abrasion of the adhesive surfaces for better adhesion of plating resist
4. Application of plating resist (screening ink or laminating dry film)
5. Formation of microporous structure of exposed adhesive surfaces by chemical etching in acid solution such as dipping the panels into CrO_3/H_2SO_4 or $CrO_3/$ HBF (adhesion promotion)
6. Electroless copper deposition on the conductor tracks and hall walls
7. Baking of the panel
8. Application of solder mask and legends
9. Final fabrication and test

In catalytic laminate and adhesive, palladium-based catalyst is dispersed homogeneously in their respective resins.[9–12] The catalytic sites are very fine particles, and the distance between neighboring sites is large enough not to cause degradation in insulation characteristics. An example of such adhesive consists of a thermosetting resin such as disphenol A-type epoxy resin, a rubber, and, normally, a filler. When the adhesive surface is treated with an oxidizing agent such as chromic–sulfuric acid solution, rubber particles are etched, yielding a microporous surface structure (Fig. 13.3). It is this microporous structure that increases the contact surface area between the copper film subsequently formed by electroless copper deposition and the base and enhances the adhesion of the film to the base. The coating of the adhesive onto the base laminate is done by dip coating, curtain coating, or transfer coating.

In the CC-4 process, plating resist applied on the surfaces of the panel in the form of a reverse conductor pattern must withstand a highly acidic solution (step 5) and also a highly alkaline solution (step 6). Screening ink stays on permanently. This "permanent plating resist" yields two advantages to additive PWBs. Since the board surface is relatively flat (see Fig. 13.4), the application of solder mask is easier, and the damage to the conductor due to scratching is minimized. Another advantage is in surface mounting multiterminal flat-pack ICs. In the pro-

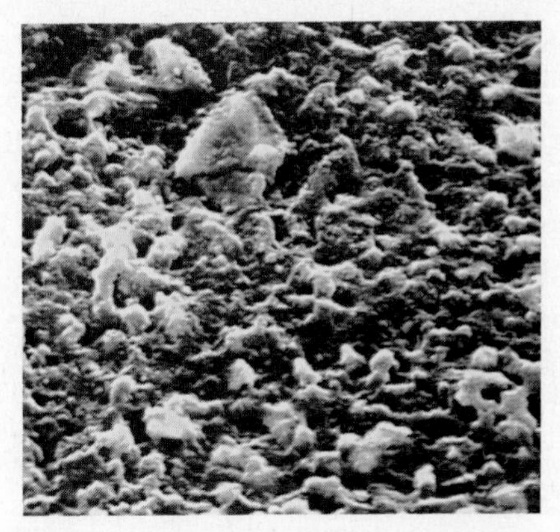

FIG. 13.3 Scanning electron microscope (2000 × magnification, 30° tilt) surface view of adhesive after etching in chromic acid bath. Microporosity of "activated" surface layer ensures adhesion of plated metals. *(Additive Products Division, Kollmorgen Corp.)*

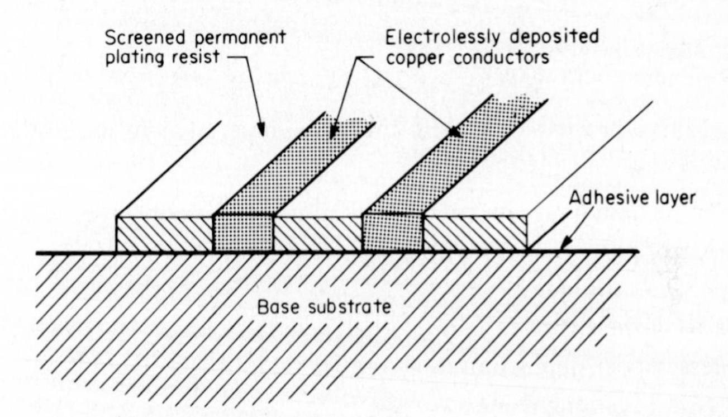

FIG. 13.4 Flush surface due to permanent plating resist.

cess of reflow soldering IC leads, solder tends to be repelled by the solder mask and is attracted to the pads, thus providing self-alignment of component leads to the proper pads. This effect is more pronounced in fully additive boards because of their flush surface.

Dry-film plating resist comes in two categories: strippable and permanent. Manufacturers of additive PWBs lean toward the use of permanent dry film because of the aforementioned advantages.

As for electroless copper deposition (step 6), a detailed account will be given later. The rest of the process steps are basically the same as for the conventional subtractive process.

Electrolessly deposited copper solders very well, so additive PWBs are normally shipped to customers as bare copper boards on which the exposed conductors and PTHs are covered with organic lacquer. The conductors and PTHs may be covered with a thin layer of solder using a hot-air solder coater leveler for extra protection from oxidation.

13.2.1.2 Seeded Process. There are many variations of the seeded process. AP-II process[13] is one variation, the production sequence of which is as follows:

1. Noncatalytic base laminate coated with noncatalytic adhesive
2. Formation of holes
3. Brushing of adhesive surfaces
4. Adhesion promotion
5. Catalyzing of the surfaces and holes
6. Application of plating resist
7. Electroless copper deposition

The rest of the steps are essentially the same as in the CC-4 process.

This process suffers from somewhat inferior surface resistivity because the catalytic sites in step 5 are not as well dispersed as in catalytic adhesive and remain underneath the plating resist. The variation of this process is catalyzing the panel after plating resist is applied and brushing off excess catalysts from the surface of the resist. This tends to brush off the catalytic sites from the conductor paths as well. To avoid this problem, plating resist which contains inhibitor[14] for catalysts was developed, but it has not led to commercial success with conventional palladium catalyst.

13.2.2 Semiadditive Process[15,16]

The semiadditive process starts with the same material as in the seeded process (see Fig. 13.1):

1. Noncatalytic base laminate coated with noncatalytic adhesive
2. Formation of holes
3. Adhesion promotion
4. Catalyzing of the entire surface and hall walls
5. Electroless copper deposition to about 0.0002 in thick
6. Application of plating resist
7. Electrolytic copper deposition to the desired thickness
8. Strip plating resist
9. Quick etch of the base electroless copper

The rest of the steps are the same as before. This process requires more production steps than does the fully additive processes. One advantage of this process is the use of electrolytic plating, which is quick and gives reliable copper. Another advantage is that existing PWB facilities with PTH capability can be used for this process with only minor modifications. However, the process suffers from the inherent problem of electrolytic plating: uneven plating thickness. Unless the plating field is carefully controlled, the holes on the outer periphery of the panel can be overplated and cause severe problems at the time of component insertion. The semiadditive method cannot offer surfaces as level as those attained by the fully additive method.

In this process, electroless nickel may be used instead of electroless copper. Nickel tends to adhere better to the adhesive surface but is harder to etch cleanly than copper. Furthermore, the presence of a copper-nickel interface can increase corrosion due to the galvanic effect.

13.2.3 Partially Additive Process

13.2.3.1 NT-1 Process.[17–19] This process is completely different from the previous processes in that it uses conventional copper-clad base laminate (see Fig. 13.1). Its steps are as follows:

1. Copper-clad base laminate with 1-oz copper foil on both sides
2. Screen printing or photoprinting of etching resist
3. Copper etched to form conductor pattern
4. Application of hydrophobic catalyst repellent insulating mask over entire board surface
5. Formation of holes by drilling
6. Selective catalyzing of the hole walls to the reception of electroless metal
7. Electroless deposition of copper in the hall walls
8. Removal of minor deposits of copper on the mask by quick etching

One basic problem with the PWBs made by this process is the difficulty of forming bonding pads for surface-mounting components.

13.2.3.2 NT-2 Process.[20,21] Variations of this process are KAP-8[22,23] and TAF-II.[24] This process also starts with copper-clad material, as in the NT-1 process.

1. Copper-clad base material
2. Formation of holes
3. Catalyzation of hole walls and board surface and brushing off of catalysts from the surface
4. Screen printing or photoprinting of etching resist on both sides of the board
5. Formation of conductor patterns by etching and stripping of the resist
6. Screen plating of resist over the entire surface, leaving pads and holes uncovered
7. Electroless copper deposition onto the pads and hall walls to the desired thickness

This process offers advantages over other additive processes in its (1) use of conventional copper-clad material, (2) formation of surface bonding pads, and (3) formation of fine-line conductors by screening method down to 0.006 to 0.007 in because etching is for 1-oz copper. In addition, the use of catalytic copper-clad laminate eliminates the necessity of catalyzing hole walls (step 3).

13.3 ELECTROLESS COPPER DEPOSITION[25,26]

One of the key elements in the additive processes is electroless or chemical copper deposition, which is done without the assistance of an external supply of electrons. There have been many variations of electroless copper deposition solutions, as shown in Table 13.1, but a fundamental principle is that a material to be plated with copper is immersed in an aqueous solution which is composed of a water-soluble copper compound, a copper complexing or chelating agent, a reducing agent for the copper ion to be deposited, an agent capable of adjusting pH, agents to improve the physical property of deposited copper, and stabilizing agents. As shown in Table 13.1, the most popular main agents are

NaOH	pH adjustment
HCHO	Reducing agent
$CuSO_4$	Source of copper ion
EDTA*	Chelating agent

The chemical reaction of copper deposition may be represented by

$$Cu\,(EDTA)^{2-} + 2HCHO + 4OH^- \rightarrow$$
$$Cu^\circ + H_2 + 2H_2O + 2CHOO^- + EDTA^{4-} \tag{13.1}$$

Under the alkaline pH condition, the cupric ions would normally combine with OH^- to produce cupric hydroxide [$Cu(OH)_2$], a useless precipitate. When a chelating agent such as EDTA is added, it prevents $Cu(OH)_2$ formation by maintaining the CU^{2+} in solution. Once the deposition of metallic copper starts, the reaction in Eq. 13.1 continues because of the autocatalytic nature of the reaction.

13.3.1 Side Reactions

While the main reaction represented by Eq. 13.1 takes place, other side reactions proceed in competition with this reaction. The conditions for the deposition reaction must be chosen to minimize these side reactions as much as possible.

One of the major difficulties is the lowering of formaldehyde concentration in the solution due to its disproportionate consumption in alkaline solution. This is known as the Cannizzaro reaction, which may be characterized by

$$2HCHO + OH^- \rightleftarrows CH_3OH + HCOO^- \tag{13.2}$$

This reaction continues independently. When no part to be plated is in the solution, the solution will deplete quickly, particularly at high pH and temperature because of the above reaction. Fortunately, methanol (CH_3OH), one of the by-products of the Cannizzaro reaction, tends to shift the equilibrium of the reaction in Eq. 13.2 to the left[27] and thus prevent the decrease of formaldehyde concentration by an unproductive reaction. By controlling the plating conditions properly, the wasteful consumption of formaldehyde by the Cannizzaro reaction can be retained within 10 percent of the consumption in the reaction in Eq. 13.1.

In addition to the Cannizzaro reaction, the following side reaction also competes for formaldehyde:

$$2Cu^{2+} + HCHO + 5OH^- \rightarrow CU_2O + HCOO^- + 3H_2O \tag{13.3}$$

Under conditions which would favor the Fehlings-type reaction (Eq. 13.3), spontaneous decomposition of uncatalyzed plating solutions produces precipitation of finely divided copper with attended vigorous evolution of hydrogen gas. The finely divided copper produced is due to the disproportionate quantity of Cu_2O under alkaline conditions:

$$Cu_2O + H_2O \rightarrow Cu^0 + Cu^{2+} + 2OH^- \tag{13.4}$$

---

*Ethylene-diamine-tetracetete-acid. Other chelating agents are Rochelle salt, Quadrol, and tartrates.

Furthermore, formaldehyde may act as a reductant for the cuprous oxide to produce metallic copper:

$$Cu_2O + 2HCHO + 20H^- \rightarrow 2\,Cu^0 + H_2 + 2HCOO^- + H_2O \quad (13.5)$$

The Cu^0 nuclei thus produced according to Eqs. 13.4 and 13.5 are not relegated to deposition on substrate but are produced randomly throughout the solution and become the catalytic sites for further undesirable copper deposition.

13.3.2 Use of Stabilizers[28,29]

Since reaction 13.3 indicates the greatest degree of instability of the plating bath, various measures are taken to counteract this reaction. Alkaline cyanide has been a popular stabilizer used for CC-4 bath, since cyanide forms strong complexes with Cu^+, $Cu^+ + 2CN^- \rightarrow Cu\,(CN)^{2-}$, but produces relatively unstable complexes of Cu^{2+}. However, since cyanide also reacts with HCHO, it is difficult to control. Note that 2,2′-dypiridil also chelates Cu^+ and does not react with HCHO. Therefore, it is a more favored stabilizer today (see Table 13.1). Aeration[8] of the plating solution is known to stabilize the solution and is widely practiced today.

13.3.3 Use of Wetting Agents

Because of the large microporous surface character produced on the surface portion of the adhesive by activation of CrO_3/H_2SO_4 and the need for solution to be in intimate contact with the catalytic nuclei, a reduction in surface tension is required to permit thorough surface solution interaction. For this reason, a wetting agent or surfactant is included in the bath formulation (see Table 13.1). The wetting agent also prevents the accumulation of gaseous material (air or H_2 bubbles) on the surface of the board, particularly around the top of the holes, which would also tend to lower the surface-solution interaction and thus seriously impair proper metal deposition.

13.3.4 Electrodialysis[30–32]

In addition to formate ions generated as a result of reactions 13.1, 13.2, 13.3, and 13.5, other anions such as sulfate ions and carbonate ions are generated in the solution. When accumulated in the solution (these anions will capture divalent copper ions), they will result in deterioration of the stability of the plating solution, decomposition of the plating solution, and lowering of physical properties of deposited copper film.

To remove such harmful anions from the solution, the use of electrodialysis has been suggested and put to practical use.[22]

A cathode compartment that contains sodium hydroxide is separated by an anionic membrane from the plating solution (Fig. 13.5). This is also separated from the anode compartment containing dilute sulfuric acid. The plating solution is circulated through the center compartment. When the electromembrane cell is energized, hydroxyl ions are transferred from the cathode compartment into the center compartment. Hydrogen is discharged from the cathode. The anode compartment receives sulfate and formate ions and a small amount of hydroxide ions from the center compartment. The formate is oxidized to carbon dioxide at the

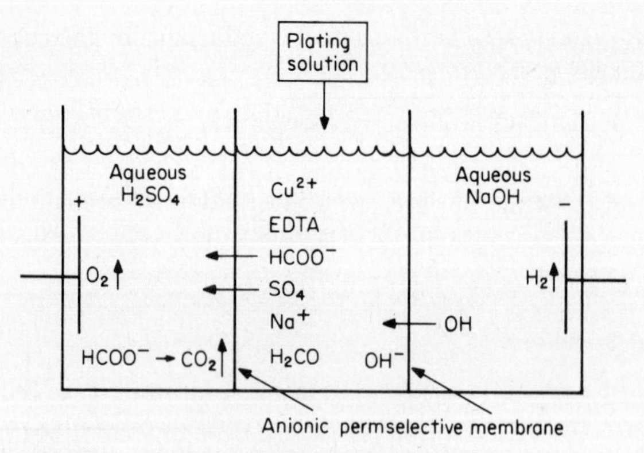

FIG. 13.5 Principles of electrodialysis.

anode and passes off as a gas. The only discharged material is dilute sulfuric acid, which can be used elsewhere. Essential ingredients of the plating solution are restored, particularly copper in chelated form, and put back into the main plating tank. This operation is very much like kidney dialysis. The effectiveness of the electrodialysis method depends on the electrode efficiency and economy.

13.3.5 Chemical Replenishing

Since the essential components of the plating solution are consumed continuously, they must be replenished to maintain desirable plating conditions. In so doing, a considerable solution overflows because of the addition of water. The overflow also contains the reaction products mentioned previously; that is, they are continuously removed from the main solution as a part of the overflow.

After a certain amount of overflow solution is accumulated, sodium hydroxide and formaldehyde may be added to the overflow solution and heated to promote spontaneous decomposition of the plating solution to precipitate copper metal. The remaining solution can be treated with H_2SO_4 to recover over 95 percent of the EDTA in the solution. Plating baths using other complexing agents are difficult to recover and can be a pollution problem. Ninety-eight percent of the solution is recovered as permeate and is used in such processes as rinsing. The concentrate, the other 2 percent, is high in salt content (Na_2SO_4 and HCOONa) and may be evaporated to produce a small amount of solid waste, or discharged into the ocean if the plant is located near the ocean (Fig. 13.6).

This is another proven economical way to waste-treat the rinsewaters. The construction of a plating tank and the methods of solution circulation and replenishment also play important roles in the stable operation of the plating solution and deposition of copper film of superior physical properties.

13.3.6 Photochemical Imaging

There are methods to catalyze the insulating surface selectively by means of photoreduction of metallic salt as a basis for the selective deposition of electroless copper. A principal one is illustrated in Fig. 13.7. All of the methods to be described use the same basic principle.

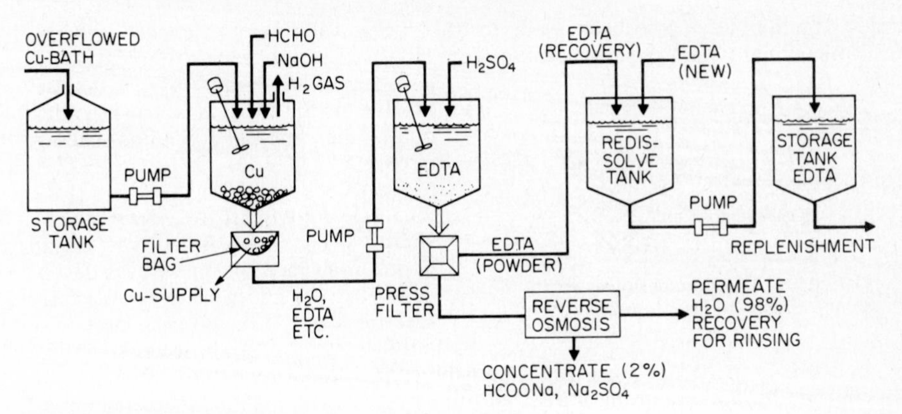

FIG. 13.6 Chemical recovery system of copper and EDTA.

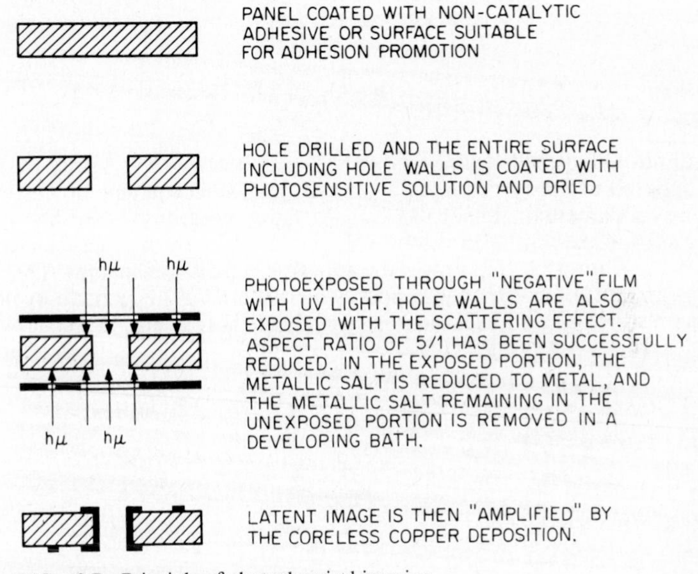

FIG. 13.7 Principle of photochemical imaging.

13.3.6.1 ***Photochemical Circuits.***[33] The photochemical circuits (PCC) process was developed by Nippon Telephone and Telegraph Company (NTT). PCC makes use of the light reduction reaction by glutamine acid salt involving silver. When exposed through UV light of appropriate wavelength, the silver salt is reduced to silver metal (granules) forming a metallic latent image in the exposed areas, which then act as catalysts in subsequent electroless copper deposition to the desired thickness.

$$\mathrm{RCOOAg} \xrightarrow{h\mu} \mathrm{RCOO^-} + \mathrm{Ag^+} \xrightarrow{h\mu} \mathrm{RCOO^\circ} + \mathrm{Ag^\circ} \tag{13.6}$$

Attempts have also been made to use an argon ion laser to expose a three-dimensional PWB[34] as shown in Fig. 13.8. However, silver has one inherent problem: migration. Because of this, the use of silver as the catalytic site is not adequate for fine-line or fine-spacing circuits.

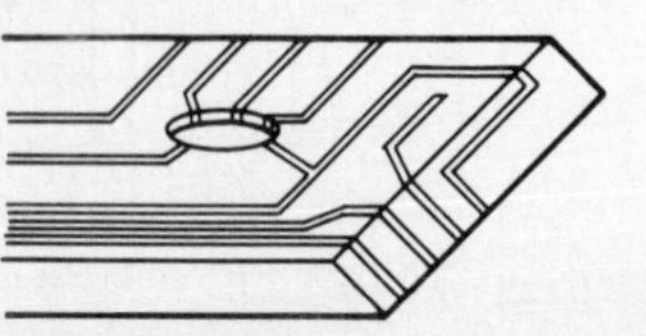

FIG. 13.8 Three-dimensional circuit with photoreduction by laser beam.

13.3.6.2 ***Photosensitive Metal Deposition.***[35] The photosensitive metal deposition (PSMD) process was developed by AT&T (formerly Western Electric Co.). The process uses the photo-oxidation of absorbed stannous ions (Sn^{2+}) to stannic ions (Sn^{4+}) and the conventional reduction of palladium chloride by the remaining unreacted (unexposed) stannous ions to metallic palladium. In the absence of light, stannous ions are absorbed onto the surface in the form of stannous oxide, and the hydroxyl ion takes part in the reaction.

$$SnO + Pd^{2+} + 2OH^- \rightarrow SnO_2 + H_2O + Pd^\circ \tag{13.7}$$

In the presence of UV light and oxygen (air), stannous oxide is oxidized to stannic oxide

$$SnO \xrightarrow{h\mu,\ O_2} SnO_2 \tag{13.8}$$

which cannot reduce palladium ions. PSMD may be viewed as a "negative" process as opposed to "positive" imaging of PCC and other imaging processes. This poses one fundamental difficulty. Because of this negative nature of exposure, catalytic sites of palladium develop in the pocket area of micropores created in the adhesion promotion step unless the adhesive is light-transparent. This results in a large amount of extraneous copper, which is difficult to remove by conventional brushing operation because it is strongly anchored (Fig. 13.9).

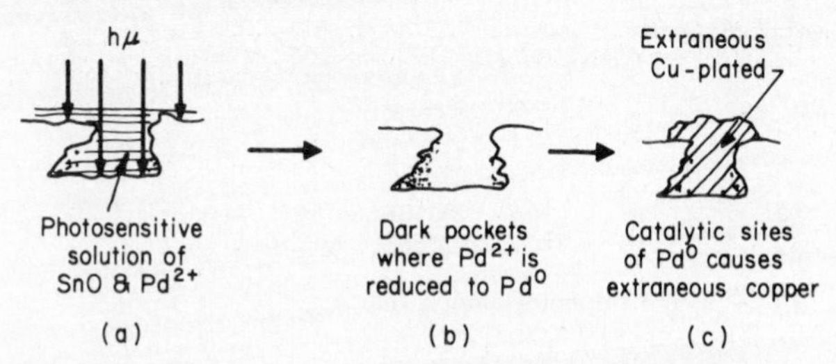

FIG. 13.9 Shadow effect of PSMD ("positive imaging").

AT&T also developed another photochemical imaging method called photoselective copper reduction (PSCR). This method makes use of the photoreduction reaction of copper salt. As a sensitizer, ferrous oxalate $Fe_2(C_2C_4)_3$ is used. Since this method is a negative imaging method and is the same in principle as the photoformation method, it will not be treated here.

13.3.6.3 ***Photoformation.***[37,38] The photoformation (P/F) process was developed by the PCK Technology Division of Kollmorgen Corporation. It uses cupric acetate as the reducible metallic salt. When the panel is exposed with UV light through "negative" film, the cupric acetate is reduced to metallic copper in the presence of a radiation-sensitive reducing agent such as aromatic diazo compound. The panel is then washed with developer to remove the unreacted radiation-sensitive composition remaining, leaving metallic copper nuclei in a positive form in the desired circuit pattern and the inner wall of PTH, followed by electroless copper deposition.

This process has been used successfully on a commercial scale for plating on plastic and metal-core PWBs.[39]

13.3.6.4 Physical Development-Reduction.[40] Physical development-reduction (PD-R) was developed by Philips in the Netherlands. It is based on the light sensitivity of TiO_2. The adhesive consists of usual thermosetting and flexible components in which fine solid particles of TiO_2 are dispersed homogeneously. A very thin aqueous layer of $PdCl_2$ is applied over the adhesive, and the panel is exposed with actinic light. As a result of this exposure, electrons are liberated from TiO_2. These electrons accomplish the reduction of Pd^{2+} to Pd° nuclei, creating a latent palladium image. Since the adhesion of these palladium nuclei to the adhesive is strong and that of Pd^{2+} is weak, rinsing in water or complexing agent for Pd^{2+} can remove the Pd^{2+} in the nonexposed areas. Since PD-R depends on the reduction reaction by TiO_2, it suffers from the necessity of using a special core material with dispersed TiO_2 particles when PTH must be formed.

The motivation for the development of the aforementioned photochemical imaging methods is the cost reduction in imaging, particularly for fine-line applications. In the case of photosensitizers using inexpensive nonprecious metal ions such as in the P/F or PSCR process, the cost reduction objective has been satisfied to some degree. However, there is one problem common to all these processes: growth of conductor line width (see Fig. 13.10). Since the latent image functions as the anchor for the adhesion of subsequent electrolessly deposited copper conductor to the base as well as the latent image, the latent conductor image cannot be made wide enough to render sufficient peel strength when a very fine conductor with reasonable conductivity must be built. For example, if the finished conductor width is desired to be 0.003 in with a thickness of 0.001 in, the latent image can only be 0.001 in wide, not enough for strong peel strength. The only way to build such a conductor is to use a dry film to retain the copper growth within the walls of plating resist.

Nevertheless, photochemical imaging will find its applications in metallizing various plastic and metallic bases, particularly in three-dimensional configurations (Fig. 13.11).

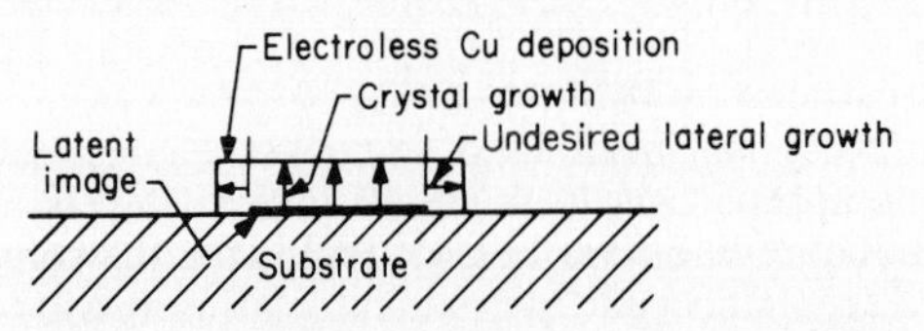

FIG. 13.10 Undesired lateral growth of electroless copper deposition in photochemical imaging method.

13.4 APPLICATION OF THE ADDITIVE PROCESS TO OTHER TYPES OF PLATED-THROUGH-HOLE BOARDS

The discussions so far on the additive process have centered on the manufacture of double-sided PTH boards. In this section, the application of the additive pro-

FIG. 13.11 Three-dimensional circuitization by photoformation process features selective additive plating on multiplanes. Bosses in foreground illustrate capabilities for plating up sidewalls, around edges or corners on nonplanar surfaces. *(AT&T Technologies/Bell Laboratories.)*

cess to the manufacture of multilayer boards and boards made of different types of materials is discussed.

13.4.1 Multilayer Boards

With the advent of VLSI technologies, more and more functions and interconnections are being incorporated on the chip level to meet the demand for faster and more powerful computers. At the board level, conductors are becoming finer, and the number of layers is growing larger to minimize the delay in signal propagation. Internal signal layers of some multilayer boards today carry conductors as narrow as 0.002 to 0.003 in, and some multilayer boards have holes whose aspect ratio is frequently in excess of 12 or 13. Forming conductors of 0.003-in width with straight edges by a conventional etching process in a mass-production environment is a difficult task. Forming PTHs with aspect ratios in excess of 10 by electroplating is also difficult, if not impossible, particularly when the hole diameter becomes smaller. The additive process provides an answer to these problems, as demonstrated by IBM.[41,42]

The additive process has been used to make reliable, low-cost multilayer boards as well. One approach is to make all internal layers by a conventional etching process, including blind via holes.[43] Following the conventional lamination process, holes are drilled and permanent plating resist is screened with a reverse conductor pattern using a solder mask. Then the panel is catalyzed with a copper colloidal solution. Copper colloidal particles tend to be strongly absorbed on the hydrophilic surface of the hole walls prepared by an alkaline cleaner but weakly absorbed on the hydrophobic surface of the solder mask, thus allowing easy removal of copper catalyst sites from the surface of the solder mask by brushing it with a sponge brush. The panel is then subjected to fully electroless copper deposition on the surface conductor pattern as well as hole walls. Low-cost multilayer boards up to six layers have been successfully made in this manner (Fig. 13.12).

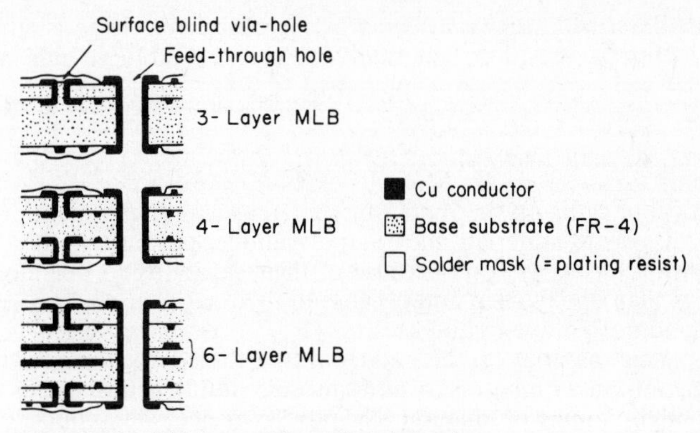

FIG. 13.12 Additive multilayer board structures.

Another alternative is a sequential method which renders more economical multilayer boards, particularly four-layer boards. It starts with the construction of ground and power layers, either by a conventional etching process using ordinary double-sided copper-clad material or by a fully additive process using catalytic material. The fully additive process is preferred because of the flush surface which results from the use of permanent plating resist. This flush surface provides for easier application of insulating layers and adhesive in subsequent operations. After the formation of power and ground planes, the panel is processed through usual black oxide treatment and coated with a thin layer of insulating resin followed by a coating of adhesive. The resultant panel is then treated in the same manner as the adhesive-coated, catalytic base material used to make double-sided PTH boards by the additive process. In this approach, economy is realized by eliminating the lamination process. Paper materials have been used also to make four-layer multilayer boards by this sequential method, realizing further reduction in cost because of the use of inexpensive material and holes being punched en masse.

13.4.2 Metal-Core Substrates

Metal-core substrates have been used for many years as an alternative material to conventional phenolic paper or epoxy-glass laminates. Metal-core substrates offer mechanical strength and can be formed into three-dimensional shapes after circuits are formed. They are also used as an effective heat sink. The most popular metals are iron, aluminum, copper, copper-invar alloy, etc.

Holes are formed by punching or drilling, and the entire surface of the substrate, including hole walls, is covered by insulating epoxy resins. Fluidized bed coating and electrophoresis are the most popular coating methods. When high density with small holes is required, oversized holes are formed first, again by punching or drilling, and holes are filled with insulating resin. At the same time, surfaces are coated with a thin layer of insulating material. Then, smaller holes are drilled through the filled holes.

In this case, the insulating material acts as an adhesive as well. The surface and hole walls are swelled by a solvent and etched in CrO_3/H_2SO_4 solution to render a microporous structure for strong adhesion of the subsequently deposited electroless copper. The adhesion-promoted base can then be screened with catalytic ink (positive conductor pattern) or given the latent image by photochemical imaging methods described previously. Finally, the conductor image is amplified by electroless copper deposition. Alternatively, the semiadditive process can be employed.

13.4.3 Thermoplastic Materials

Photochemical imaging on thermoplastic materials was briefly mentioned previously. Thermoplastic materials such as polysulfone, polyetherimide, etc., have drawn attention in recent years because of their capability to operate at higher temperatures than epoxy resin and because of their superior electrical characteristics at Rf and microwave frequencies.

Thermoplastic material can be molded into a three-dimensional structure or planar configuration with necessary holes formed simultaneously, thus providing economy in high-volume production. The process of electroless copper deposition on plastic materials starts with the swell-and-etch operation for adhesion promotion. The rest of the operations are nearly the same as in circuitizing metal-core substrate.

13.4.4 Ceramic Materials

Ceramic circuit boards are usually made by thick-film and thin-film technologies. Multilayer ceramic boards have also been made by both thick- and thin-film technologies. However, these technologies are not without shortcomings. Conductive inks used in thick-film technology have higher resistivity by a magnitude compared with pure copper. The finest line width achievable by a screening method in a mass-production environment is about 0.003 in. Thin-film technology is expensive, and there is a limitation to the thickness of metal deposited by this technology. Direct copper deposition on ceramic substrates can fill the gap between thick- and thin-film technologies.

The Ceraclad* process starts with forming holes on a ceramic substrate such as 96 or 99 percent alumina by laser drilling the substrate or punching it in green state. The substrate is then adhesion-promoted by etching the surface in alkaline solution, and the entire surface including hole walls is catalyzed. Following the catalyzation, the substrate is deposited with a thin or relatively thick layer of copper by the fully additive method. In the case of thin electroless copper deposition, the subsequent process is exactly the same as the ordinary semiadditive process. The purpose of depositing thick electroless copper is to use the tent-and-etch process commonly used for making bare copper boards. Conductors having a width of 0.0008 in or less with a thickness of 0.0006 in or more can be formed with relative ease by this technology. Nickel-gold can be plated over the copper surface either electrolessly or electrolytically for direct chip bonding.

13.5 SUMMARY

With the improved physical properties of electrolessly deposited copper, the additive process is finding its way into the manufacture of low-cost PTH boards using paper-based substrates for consumer electronic products as well as highly sophisticated multilayer boards for supercomputers. It is also opening its horizon of applications to plating copper directly onto ceramic, metal, and thermoplastic

*Ceraclad is a trademark of Kollmorgen Corporation.

substrates to satisfy various packaging needs. As it is made easier in the future with improved materials and more scientific control of plating baths, the additive process will be used more widely.

REFERENCES

1. G. Messner and H. Nakahara, *Electronic Packaging and Production,* December 1984, pp. 54–61.
2. Japanese Patent 48-34975.
3. T. Okamura et al., *Hitachi Chemical Technical Report,* Jpn., vol 11, no. 3, January 1984, pp. 11–16.
4. K. Saito and H. Honma, *Metal Surface Finishing,* Jpn., vol. 29, no. 4, 1978, pp. 190–195.
5. M. Wajima et al., *Electronic Technology,* Jpn., vol. 24, no. 4, 1982, pp. 92–95.
6. Japanese Laid-Open Patents 17333 (1977), 77833 (1977), 51143 (1978).
7. M. Hirohata et al., *Metal Surface Finishing,* Jpn., vol. 29, no. 10, 1978, pp. 485–489.
8. U.S. Patent 2,938,805, 1960.
9. U.S. Patent 3,546,009, December 1970.
10. U.S. Patent 3,560,257, February 1971.
11. U.S. Patent 3,600,330, August 1971.
12. U.S. Patent 3,625,758, December 1971.
13. Japanese Laid-Open Patent 232488, December 1984.
14. U.S. Patent 3,443,988.
15. U.S. Patent 3,625,758.
16. B. F. Barclay, *Circuit Manufacturing,* May 1985, pp. 34–42.
17. U.S. Patent 3,628,999, December 1971.
18. U.S. Patent 3,799,802, March 1974.
19. U.S. Patent 3,799,816, March 1974.
20. U.S. Patent 3,600,330, August 1971.
21. Japanese Patent 43-16929, July 1968.
22. N. Ohtake, *Electronics Technology,* Jpn., vol. 27, no. 7, June 1985, pp. 55–59.
23. Japanese Patent 58-6319, February 1983.
24. Advertising brochure by Hitachi, Ltd., Tokyo.
25. R. M. Lukes, *Plating,* November 1964, pp. 1066–1068.
26. W. Goldie, *Plating,* November 1964, pp. 1069–1074.
27. U.S. Patent 3,595,684, 1971.
28. U.S. Patent 3,607,317, September 1971.
29. U.S. Patent 3,095,309, May 1960.
30. U.S. Patent 4,289,597, September 1981.
31. U.S. Patent 4,324,629, April 1982.
32. R. E. Horn, *Plating and Surface Finishing,* October 1981, pp. 50–52.
33. A. Iwaki et al., *Photographic Science Engineering,* vol. 20, no. 6, 1976, p. 246.
34. H. Tabei, *Electronic Material,* Jpn., June 1979, pp. 53–57.
35. J. F. D'Amico and M. A. DeAngelo, *J. Electrochem. Soc.,* vol. 120, no. 11, 1973, p. 1469.

36. D. Dinella, *Proceedings of the PC World Convention II, Session 2b,* June 1981, pp. 6–17.
37. U.S. Patent 3,772,078, November 1973.
38. G. Messner, *Technical Proceedings of NEPCON,* 1976, pp. 65–71.
39. D. F. Frisch, *Electronic Packaging and Production,* February 1985, pp. 194–203.
40. J. F. Mansfeld and J. M. Jans, *Plating and Surface Finishing,* vol. 66, no. 1, 1979, p. 14.
41. J. R. Bupp et al., *IBM J. Res. Development,* vol. 26, no. 3, May 1982.
42. W. A. Alpaugh and J. M. McCreary, IPC technical paper, IPC-TP-108, fall meeting, 1976.
43. K. Kobayashi et al., IPC technical paper, *Proceedings of the PC World Convention IV,* May 1984.

CHAPTER 14
ETCHING

Edward F. Duffek, Ph.D.
Adion Engineering Company, Cupertino, California

14.1 INTRODUCTION

One of the major steps in the chemical processing of subtractive printed boards is etching, or removal of copper, to achieve the desired circuit patterns. Etching is also used for surface preparation with minimal metal removal (microetching) during inner-layer oxide coating and electroless or electrolytic plating. Technical, economic, and environmental needs for practical process control have brought about major improvements in etching techniques. Batch-type operations, with their variable etching rates and long downtimes, have been replaced with continuous, constant-etch-rate processes. In addition, the need for continuous processing has led to extensive automation along with complete, integrated systems.

The most common etching systems are based on alkaline ammonia, hydrogen peroxide–sulfuric acid, and cupric chloride. Other systems include persulfates, ferric chloride, and chromic-sulfuric acids. Process steps include resist stripping, precleaning, etching, neutralizing, water rinsing, and drying. This chapter describes the technology for etching high-quality, fine-line (0.004 to 0.006 in) circuits in high volume at a lower cost, as well as closed-loop continuous processing, constant-etch rates, control at high dissolved-copper capacities, regeneration-recovery, less pollution, and increased safety. Problems of waste disposal and pollution control have been minimized by adapting these principles.

Typical procedures are given for etching organic (i.e., dry film) and metal-resist boards, and for inner layers. Strippers and procedures for resist removal are described, based on resist selection, cost, and pollution problems. The properties of available etchants are also described in terms of finish plate compatibility, control methods, ease of control, and equipment maintenance. Other considerations include chemical and etchant effects on dielectric laminates, etching of thin-clad copper and semiadditive boards, solder mask on bare copper (SMOBC), equipment selection techniques, production capabilities, quality attained, and facilities.

14.2 GENERAL ETCHING CONSIDERATIONS AND PROCEDURES

Good etching results depend on proper image transfer in both inner-layer print-and-etch and plated-metal etch resists. Etch personnel must be familiar with screened, photosensitive, and plated resists commonly used. The etching of printed boards must begin with suitable cleaning, inspection, and pre-etch steps to ensure acceptable products. Plated boards also require a careful and complete resist removal. The steps after etching are important because they are necessary to remove surface contaminations and yield sound surfaces. This discussion considers the various types of resists and outlines typical procedures used to etch printed boards by the use of organic and plated resist patterns.

14.2.1 Screened Resists

Screen printing is a common method for producing standard copper-printed circuitry on metal-clad dielectric and other substrates. The etch-resist material is printed with a positive pattern (circuitry only) for copper etch-only boards or with a negative image (field only) when plated-through holes and metal resist are present.

The type of resist material used must meet the requirements for proper image transfer demanded by the printer. From the metal etcher's point of view, the material needs to provide good adhesion and etch-solution resistance; be free of pinholes, oil, or resin bleed-out; and be readily removable without damage to substrate or circuitry. A description of screen printing is given in Chapter 11.

Alkali-soluble and vinyl-based screen-printed resists are quite suitable for printed circuit manufacture. They are available commercially in consistent form in grades suitable for screen and offset printing. Chemical resistance of the applied coatings is high enough to protect against the action of acidic etchants, including sulfuric acid–hydrogen peroxide, cupric chloride, ferric chloride, sulfuric-chromic, and ammonium persulfate, as well as the low-pH and mildly alkaline solutions (pH 8 to 9) commonly used in printed-circuit plating. Alkaline etchants containing high concentrations of ammonium hydroxide are not recommended for use with vinyl or alkali-soluble resists.

When cured without excessive baking, etching resists withstand abrasion and yet are adherent and pliable. Mechanical cleaning without damaging the resist can be accomplished by using a soft-bristle brush and fine, soft abrasives. Pumice and other hard substances can easily cause damage. Commercial cleaners can be used as soak cleaners or as electrocleaners at temperatures up to 160°F.

When precleaning is excessive, or when plating current densities are too high, plating may occur on top of the resist. The overplating, plus the metal nodules which grow from pinholes in the resist, are usually removed along with the resist.

A process for the print-and-etch production of copper-printed boards and inner layers is as follows:

1. Clean (soak or electroclean), at 150°F, neutral or acidic process.
2. Water rinse.
3. HCl-dip, 20 percent by volume (optional).
4. Water rinse.
5. Microsurface-etch, i.e., sodium persulfate, 30 to 60 s.
6. Water rinse.
7. Sulfuric acid–dip, 10 percent by volume.
8. Water rinse.
9. Examine exposed copper for water breaks. If present, preceding steps should be repeated.

10. Etch using appropriate conditions.
11. Water rinse.
12. Neutralize etchant.
13. Water rinse.
14. Blow-dry; inspect.
15. Strip resist (see Sec. 14.3).
16. Rinse in clean stripper to ensure complete removal of resist.
17. Scrub panel with a soft brush or with a light abrasive wheel in machine scrubber.
18. Water rinse and blow-dry.

Adequate cleaning before etching is necessary for consistent quality. Typical problems are excessive undercut, slivers, unetched areas, and inner-layer shorts in multilayer boards. In addition, conductor line lifting may occur when the copper-to-laminate peel strength is below specification.

14.2.2 Hole Plugging

Plugged-hole, copper-only boards use alkaline-soluble screen resists in a unique manner. The technique called "hole plugging" makes the SMOBC board possible. The main steps are as follows:

1. Prefab and drill laminate.
2. Electroless copper–plate.
3. Copper panel–plate.
4. Coat plated-through holes with alkali-soluble resist (hole plugging).
5. Cure resist.
6. Brush surface and clean.
7. Screen-print positive pattern with alkali-soluble resist.
8. Cure resist.
9. Etch with acid etchant.
10. Strip resist. Use 2 percent sodium hydroxide or proprietary.
11. Apply solder mask.
12. Coat holes and traces with tin-lead solder or tin.

14.2.3 UV-Cured Screen Resists

Ultraviolet-cured solventless systems are available for print-and-etch and plating applications. These products are resistant to commonly used acidic plating and etching solutions. Stripping and applications with plating may result in rejected work.

14.2.4 Photoresists

Dry film and liquid photoresist materials are capable of yielding fine-line (0.004 to 0.006 in) circuits needed for production of surface mount circuit boards. Like screened resists, photosensitive resists can be used to print either negative or positive patterns on the metal-clad laminate. Although dry film and liquid materials differ in both physical and chemical properties, they will be considered together for our purpose. See Chapter 11.

In general, both positive- and negative-acting resists offer better protection in acidic rather than alkaline solutions; however, negative-acting types are more tolerant of alkaline solutions. Negative resists, once exposed and developed, are no longer light-sensitive and can be processed and stored in normal white light. The positive resists remain light-sensitive even after developing and must therefore be protected from white light. Liquid photoresists, although less durable, are capable of finer line definition and resolution.

Increasing the temperatures of processing solutions increases the chemical and swelling action on photoresists. However, those resists which can be baked onto the copper at high temperatures are affected the least. Thicker coatings of these materials offer greater resistance, but care must be taken not to sacrifice fine-line definition or acuteness.

The process steps for etching copper with photoresist films are similar to those given for screened resists, with the following cautions. Scrub cleaning must be done carefully. Photoresist films, even in a baked condition, will not hold up to abrasive cleaners. Here, as with chemical resistance, greater thicknesses help by providing a thicker blanket for protection. Electrocleaning is not favored, since it can lift the resist.

A light etch followed by a thorough water spray rinse, a 10 percent sulfuric acid dip, and another thorough water rinsing prior to either etching or plating is successful and does not appear to affect any of the common photoresists. With positive-acting resists, it is good practice to give the boards an acid dip just prior to plating or etching, since the resist developer is alkaline and may cause basic copper salts to appear on the surface. Scrubbing prior to plating improves adhesion.

Electroplating conditions should be reviewed for improved board quality. Excessive time in etch must be avoided, and any reason for excess time must be carefully analyzed. For example, spent etch solutions, low temperatures, excessive copper, improper cleaning, etc., can be readily corrected. Excessive time invariably results in overetching and undercutting and may lead to penetration of the etch-resist coating.

Positive-acting resists are subject to the same problems as are negative-acting resists, although they are easier to remove cleanly, after exposure, from areas to be etched or plated. Resist-breakdown problems are caused by undesired light exposure, excessive temperature, and alkalinity during processing. Alkaline etchants will destroy positive-acting resists. Excessive current densities result in gassing and also may lift the resist. Problems related to plating photoresist-coated boards are reviewed in Chapter 12.

14.2.5 Plated Etch Resists

At present, the most extensive use of metal-plated resists is found in the production of double-sided and multilayer plated-through-hole circuit boards. The most commonly used resists are solder plate (60 percent Sn, 40 percent Pb), tin, nickel, tin-nickel, and gold. Silver is used to some extent for light-emitting and liquid crystal applications. Details concerning the deposition of these metals are given in Chapter 12. The use of these resists is described in the following paragraphs.

14.2.5.1 ***Solder Plate.*** Tin-lead solder (0.0003 to 0.001 in thick) is the most commonly used plated etch resist. The 60 Sn–40 Pb alloy offers good etchant resistant with few problems. Increased reliability is achieved by the use of solder plate over tin-nickel.[1] Thin solder deposits (0.0002 in) can be used for the SMOBC process. The most suitable etchants are alkaline ammonia, sulfuric acid–hydrogen peroxide, chromic-sulfuric acid, and ammonium persulfate–phosphoric acid. Ferric and cupric chloride acid etchants cannot be used because of solder plate attack. Post-etch neutralization rinses are needed, especially with alkaline systems, to rinse away etchant residues and to maintain optimum surface properties.

A typical procedure for solder-plated or SMOBC board follows:

1. Resist stripping, bath 1.
2. Resist stripping, bath 2.
3. Water rinsing.
4. Examination of board after stripping. Make sure all resist has been removed. Return to stripping if not complete.
5. Examination and removal of solder or other materials in the copper area. Care must be taken not to damage circuit lines.
6. Application of touch-up ink and oven drying: 10 min at 120°F.
7. Alkali cleaning (optional). For boards that are heavily stained or oxidized, 3-min soaking at 160°F.
8. Water rinsing.
9. Sulfuric acid dip, 10 percent (by volume), 30 s (optional).
10. Water rinsing.
11. Blow drying.
12. Inspection.
13. Pre-etch (optional).
14. Inspection for resist residues.
15. Loading of boards on conveyor or etching rack.
16. Etching of boards (alkaline etchant).
17. Water rinsing.
18. Neutralization.
19. Water rinsing.
20. Blow drying.
21. For tin-lead board: Proceed to gold-edge connector plating or solder reflow.
22. For SMOBC: Stripping of tin-lead, solder mask, and tin-lead coating.

14.2.5.2 ***Tin Plating.*** Tin plating (directly over barrier layers of nickel or tin-nickel) is used because of its optimum solderability. Other etchants, such as sulfuric acid–hydrogen peroxide, chromic-sulfuric acid, alkaline ammonia, and ammonium persulfate–phosphoric acid, have been especially formulated for bright tin. Thin tin deposits (0.0002 in) are used to make SMOBC boards. Follow the procedure given above for solder plate.

14.2.5.3 ***Tin-Nickel and Nickel.*** Tin-nickel alloy (65 percent Sn, 35 percent Ni) and nickel-plate, used as is or overplated with gold, solder, or tin, are also preferred metal resists for etching copper in alkaline ammonia, sulfuric acid–hydrogen peroxide, and persulfates.

14.2.5.4 ***Gold.*** Gold with an underplate of nickel or tin-nickel provides excellent resistance to all the common copper etchants. Some etchants have a slight dissolving effect on gold.

14.2.5.5 ***Precious Metals and Alloys.*** Rhodium has been described as being a suitable resist for edge connectors on boards; however, the plating process is difficult to control. When plated over nickel, rhodium tends to be thin and porous and to lift during etching. Because of varying surface properties, 18K-alloy gold and nickel-palladium must be evaluated carefully when used as substitutes for pure-gold systems.

14.2.5.6 ***Silver.*** Although silver is not used on most printed boards (MIL-STD-275 states that it shall not be used), it has found some application for camera, light-emitting, and liquid crystal devices. Copper etching using silver as resist can be done with alkaline ammonia solutions. Silver loss is about 0.0001 in/min.

14.2.5.7 ***Etching Procedure: Recommendations for Plated-Metal Resists.*** Etching of the metal-resist-plated boards begins with removal of the resist using commercial solvents and strippers. Gold, solder, and tin resists must be handled carefully because they scratch very easily. Tin-nickel alloy and nickel plate, however, are very hard and resistant to abrasion.

The oily residues left by certain resist strippers can be removed by high-pressure hole flushing, by the use of alkali cleaners (heated to 180°F and used as soak), or by electrocleaning. After cleaning and rinsing, the technician must look at the edges of circuit lines, at the plated-through holes, and at the centers of letters and numbers to detect organic resist residues. The most difficult materials to remove are the negative-acting photoresists and the UV-screened resists. These materials do not dissolve in their stripper; instead, they soften and form gel-like films which lift off the metal surface. Solder boards may be cleaned by high-pressure cleaning and soft-bristle brushing. This must be done carefully without the use of abrasives, since it is possible to scratch through the plating. Once the organic resists have been removed, surface oxides and other film on the exposed copper can be removed by rinsing the boards in a mild acid solution. After thorough water rinsing, the boards are dried prior to etching. Microetching can also be used to detect areas that are not completely clean. These areas are obvious at this point, since they do not have the matte appearance of the rest of the surface. Repeat cleaning steps and proceed with the etching.

The procedure after etching includes thorough water rinsing and acid neutralizing to ensure removal of etchant residues on the board surface and under the traces. Alkaline etchants are followed by treatment with proprietary ammonium chloride acidic solutions, ferric and cupric chloride with solutions of hydrochloric or oxalic acid, and ammonium persulfate with sulfuric acid. Alkaline cleaning is used for tin-lead boards etched in chromic-sulfuric acid. Etchant residues not removed before drying or reflow result in lowered electrical resistance of the dielectric substrate and in poor electrical contact and soldering on the conductive surfaces.[2]

Problems which occur in etching when a plated metal is used as the etch resist are due to artwork error, to a thin or porous organic resist, or to plating which is thin, nodular, or porous. In the etching process itself, lack of surface cleanliness, excessive etching time or temperatures, chemically unbalanced etchant, machine problems, excessive spray pressures, inadequate rinsing and neutralizing, or rough handling may all result in rejected boards.

A problem common to etching printed boards is not having the entire area etch clean at the same time. This occurs when etch action is more rapid at the edges of printed areas than in a broad expanse of copper. If very fine patterns and lines are required, the result can be loss of the pattern due to undercut, especially when the board is left in the etcher until all the field copper has etched away. Fine-line boards in high-volume production require special fine-line etchants, dry-film photoresist, thin-clad laminate, controlled plating distribution, spray etching, and fully additive or semiadditive processing.

14.3 RESIST REMOVAL

The method used for resist stripping is important when a resist is selected. The effect on board materials, cost and production requirements, and compliance with safety and pollution standards must be taken into account. Both aqueous and solvent stripping systems are widely used.

14.3.1 Screen Resist Removal

Alkali-soluble resist inks are generally preferred. See Sec. 14.2 for the cycle. Stripping in the case of thermal and UV-curable resists is accomplished in 2 percent sodium hydroxide or in proprietary solutions. The resist is loosened and rinsed off with a water spray. Advantages are low cost and compliance with disposal requirements. Adequate safety precautions must be taken, since caustics are harmful.

Conveyorized resist-stripping and etching machines use high-pressure pumping systems which spray hot alkaline solutions on both sides of the boards. Single-sided boards and certain laminate materials such as the polyimides may be attacked by alkaline strippers. Measling, staining, or other degradation is noted when strippers attack epoxy or other substrates.

Screened vinyl-based resists are removed by a dissolving action in solutions of chlorinated, petroleum, or glycol ether solvents. Methylene chloride and toluene are used extensively in "cold" stripper formulations.[3] Commercial cold strippers are classified according to their pH in a 10 percent stripper-water mixture. The most common strippers are acidic formulations which contain copper brighteners and swelling, dissolution, and water-rinsing agents. The usual procedure for static tank stripping involves soaking the coated boards in at least two tanks of stripper. Excessive time in strippers is to be avoided because of attack on the "butter" (top epoxy) coat, especially on print-and-etch or single-sided boards. Water is a contaminant in most cold strippers.

Solvent stripping machines are commercially available. High-volume users save costs by using conveyorized systems equipped with reclamation and pollution-control facilities. Distillable cold strippers generally contain methylene chloride or trichloroethylene (TCE). Some heating may be used, but only in closed systems because of health and fire hazards. In all cases, proper safety, ventilation, pollution control, and certified waste disposal must be provided. Methylene chloride, toluene, and TCE are priority pollutants (TTO) regulated by the EPA. Alternatives to methylene chloride are available as glycol ethers.

14.3.2 Photoresist Removal

14.3.2.1 ***Dry Film.*** Dry-film resists have been formulated for ease of removal in both aqueous-alkaline and solvent solutions. Strippers of each type are available for both static tank and conveyorized systems. Cold stripper solvent-type formulations are similar to those used on vinyl screen inks in which the primary solvent is methylene chloride.[4] Aqueous-alkaline stripping results in undissolved residues of softened resist films. These residues can be captured in a filter system and disposed of in accordance with waste-disposal requirements.

14.3.2.2 ***Negative-Acting Liquid Photoresist Removal.*** Negative-acting, liquid-applied photoresist can be readily removed from printed boards that have not been baked excessively. Baking is critical to removal because it relates to the degree of polymerization. Since overbaking is also damaging to the insulating substrates, processes should stress minimal baking—only enough to withstand the operations involved.

The negative-acting resists are removed by using solvents and commercial strippers. In this case, the resist does not dissolve; instead, it softens and swells, breaking the adhesive bond to the substrate. Once that has taken place over the entire coated area, a water spray is used to flush away the film. With resists that are more difficult to remove (baked, for example), a timed soak (depending on substrate resistance to the stripper) followed by soft scrubbing in the stripper has been used effectively. Thin coatings, when properly baked, are generally removed more readily than thick coatings. Cleaning after resist stripping of plated-metal-resist boards is essential for good, complete etching. Photoresist left at the edge of circuit traces or in the letters will result in incomplete etching.

14.3.2.3 ***Positive-Acting Photoresist Removal.*** Positive-acting photoresists are removed by dissolving in acetone, ketone, cellusolve acetate, or other organic solvents. Commercial organic and inorganic strippers are suitable if baking has not been excessive. Removal by exposure to UV light and subsequent dipping in sodium hydroxide, TSP, or other strong alkaline solutions is also effective. Overexposure is required to ensure against the shielding of metal overplate at trace edges on pattern-plated boards. Overbaking also makes removal difficult.

Machine stripping is done in a solution of 0.5*N* sodium hydroxide, nonionic surfactants, and defoamers.

14.4 ETCHING SOLUTIONS

This section is a survey of the technology and chemistry of the copper etching systems in common use. Changes from batch-type operation to continuous constant-rate systems with increased process automation represent major innovations in etching practices. Factors that account for these changes are as follows:

1. Fine-line board design
2. Thin-inner-layer handling
3. Compatibility with resist
4. Control of etch rate
5. Etch rate (speed)
6. Dissolved copper capacity

7. Ease of process control
8. Ease of equipment maintenance
9. Costs and economics
10. High yield
11. Regeneration and replenishment
12. By-products
13. Operator and environmental protection
14. Disposal and pollution control

All the above factors serve to evaluate copper etchants to be used. Introduction, chemistry, properties, and problems are given in this section, along with suggestions for selection and control.

Problems encountered in the control of equipment and etchant solutions are frequently difficult to separate. Often, the utmost in performance and life of the etchant is not attained. A current understanding is needed in the areas of fine-line production, regeneration and recycling of materials, and pollution control.

14.4.1 Alkaline Ammonia

Alkaline etching with ammonium hydroxide complexing is increasingly used because of its continuous operation, compatibility with most metallic and organic resists, minimum undercut, high capacity for dissolved copper, and fast etch rates. Both batch and continuous (closed-loop) spray machine systems are in use. Continuous operation provides constant etch rates, high work output, ease of control and replenishment, and improved pollution control. However, costs are relatively high, neutralization after etching is critical, and the ammonium ion introduced to the rinses presents a difficult waste-treatment problem. Complete regeneration with chemical recycling is not routinely practiced.

14.4.1.1 Chemistry. The main chemical constituents function as follows:

1. Ammonium hydroxide, NH_4OH, acts as a complexing agent and holds copper in solution.
2. Ammonium chloride, NH_4Cl, increases etch rate, copper-holding capacity, and solution stability.
3. Copper ion, Cu^{2+}, is an oxidizing agent that reacts with and dissolves metallic copper.
4. Sodium chlorite, $NaClO_2$, is also an oxidizing agent that reacts with and dissolves metallic copper.
5. Ammonium bicarbonate, NH_4CO_3, is a buffer and as such retains clean solder holes and surface.
6. Ammonium phosphate, $(NH_4)_3PO_4$, retains clean solder and plated-through holes.
7. Ammonium nitrate, NH_4NO_3, increases etch rate and retains clean solder.

Some batch formulations require mixing two solutions just before etching, since oxidizers and complexers are kept separate for stability. Continuous-process operations consist of single-solution makeup buffered to a pH of 7.5 to 9.5.

Alkaline etching solutions dissolve exposed field copper on printed boards by a chemical process of oxidation, solubilizing, and complexing. Ammonium

TABLE 14.1 Composition of Alkaline Etchants

Component	1[5]	2[6]	3[7]
NH_4OH	3.0 mol/L	6.0 mol/L	2–6 mol/L
NH_4Cl	1.5–0	5.0	1–4.0
Cu (as metal)		2.0*	0.1–0.6
$NaClO_2$	10.375		
NH_4HCO_3	0–1.5		
$(NH_4)_3PO_4$		0.01	0.05–0.5
NH_4NO_3	0–1.5		

*Starter solution only.

hydroxide and ammonium salts combine with copper ions to form cupric ammonium complex ions [$Cu(NH_3)_4^{2+}$], which hold the etched and dissolved copper in solution at 18 to 30 oz/gal.

Typical oxidation reactions for closed-loop systems are shown by the reaction of cupric ion on copper, and air (O_2) oxidation of the cuprous complex ion:

$$Cu + Cu(NH_3)_4^{2+} \rightarrow 2Cu(NH_3)_4^{+} \tag{14.1}$$

$$4Cu(NH_3)_2^{+} + 8NH_3 + O_2 + 2H_2O \rightarrow 4Cu(NH_4)_4^{2+} + 4OH^- \tag{14.2}$$

and for batch-operating systems by chlorite ion on copper:

$$2Cu + 8NH_3 + ClO_2^- + 2H_2O \rightarrow 2Cu(NH_3)_4^{2+} + Cl^- + 4OH^- \tag{14.3}$$

The calculated copper capacity achieved in batch etching with chlorite is 6 oz/gal of dissolved copper (Table 14.1, no. 1). Air oxidation (O_2) or a temperature increase is needed for higher copper capacity.

Closed-loop etching can be continued with the formation of $Cu(NH_3)_4^{2+}$ oxidizer from air during spray etching and as long as the copper-holding capacity is not exceeded.

14.4.1.2 Properties and Control. Early versions of alkaline etchants were batch-operated. They had a low copper capacity, and the etch rates dropped off rapidly as copper content increased.[5,8,9] It was found necessary to add controlled amounts of dissolved oxidizing agents to speed up the rate and increase copper capacity at a constant temperature. Batch operation is still practical for low-volume production.

Etching solutions are operated at 120 to 130°F and are well suited to spray etching. Efficient exhaust systems are required because ammonia fumes are released during operation.[10] Etching machines must have a slight negative pressure and moderate exhausting to retain the ammonia necessary for holding dissolved copper in solution. Currently available solutions offer constant etching of 1-oz copper in 1 min or less, with a dissolved copper content of 18 to 24 oz/gal.

14.4.1.3 Closed-Loop Systems. A practical method of maintaining a constant etch rate with minimal pollution uses automatic feeding controlled by specific gravity or density.[9] The process, which is generally referred to as "bleed and feed," is illustrated in Fig. 14.1. As the printed boards are etched, copper is dissolved, and the density of the etching solution increases. The density of the etchant in the etcher sump is sensed to determine the amount of copper in solution. When the density sensor records an upper limit, a switch activates a pump which automatically feeds replenishing solution to the etcher and simultaneously removes etchant until a lower density is reached.

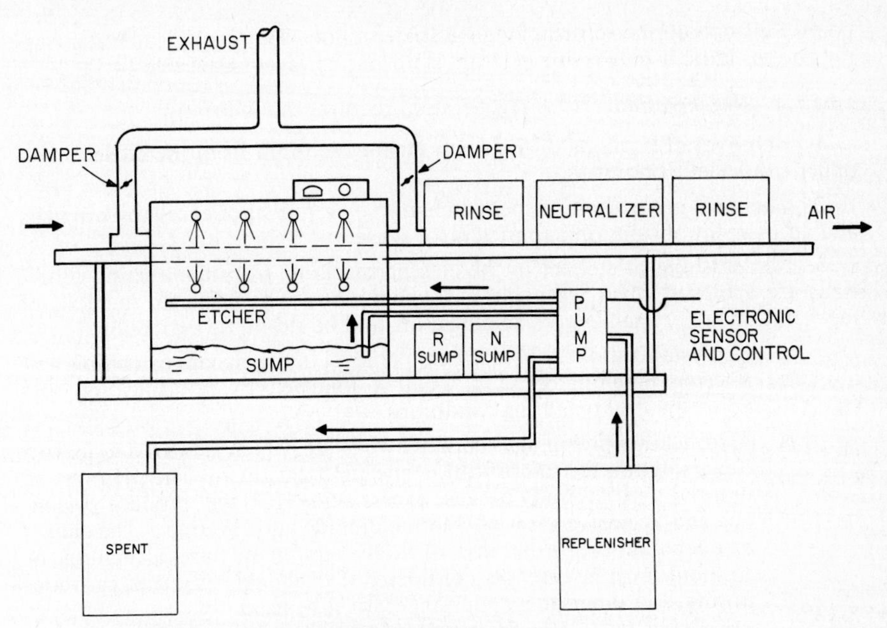

FIG. 14.1 Automated flow-through alkaline etching system.

A typical operating condition is as follows:*

Temperature	120–130°F (49–50°C)
pH	8.0–8.8
Specific gravity at 120°F (49°C)	1.207–1.227
Baumé, Be°	25–27
Copper concentration, oz/gal	20–22
Etch rate, 0.001 in/min	1.4–2.0
Chloride level	4.9–5.7 mol/L

A study of the etching rate of copper versus dissolved copper content shows the following effects:

0–11 oz/gal	Long etching times
11–16 oz/gal	Lower etch rates but solution control difficult
18–22 oz/gal	Etch rates high and solution stable
22–30 oz/gal	Solution unstable and tends toward sludging

All work must be thoroughly rinsed with water immediately upon leaving the etching chamber. Do not allow the boards to dry before rinsing. Etched circuitry with plated tin-lead solder resist also requires an acid neutralization and application of solder brightener. The main purposes are to remove etchant from under circuit edges and to clean and brighten the circuit surfaces and plated-through

*P. A. Hunt Chemical Corporation, West Paterson, New Jersey.

holes. This operation is critical for good solder reflow. Thorough clean water rinsing and air knife drying result in clean, stain-free surfaces (see Table 14.1).

14.4.1.4 Regeneration. True regeneration requires the following:

1. Removal of portions of the spent etching solution from the etcher sump under controlled conditions.

2. Chemical restoration of spent etchant, i.e., removal of excess by-products and adjustment of solution parameters for reuse.

3. Replenishing of etchant in the etching machine. Constant etching conditions are achieved when regeneration is continuous. The principal methods of regeneration are crystallization, acidification, and liquid-liquid extraction.

- ***a.*** Recrystallization reduces the copper level in the etchant by chilling and filtering precipitated salts. That is followed by refortification and adjustment of operating conditions.
- ***b.*** Acidification precipitates copper hydroxide, which is filtered to leave a clean solution for refortification and pH adjustment. The pH must be reduced very carefully because excess acid (HCl) will produce poisonous and explosive gases of chlorine dioxide and chlorine.[10] The danger of overacidification because of faulty equipment, miscalculation, or accident must be carefully considered. Dependable exhausts, gas monitoring, and other safety equipment must be used.
- ***c.*** Liquid-liquid extraction[11,12] is gaining acceptance because of its continuous and generally safe nature. The process involves mixing spent etchant with an organic solvent (i.e., hydroxy oximes) capable of extracting copper. The organic layer containing copper is subsequently mixed with aqueous sulfuric acid, which extracts copper to form copper sulfate. The copper-free etchant is restored, and the copper sulfate is available for electroless copper, acid copper plating, or copper recovery. Closed-loop regeneration systems reduce chemical costs, sewer contamination, and production downtime. However, regeneration by these methods is expensive and is thus limited to large printed-circuit facilities.

14.4.1.5 Special Problems Encountered during Etching[13]

1. *Low etch rate with low pH, <8.0:* This is caused by excessive ventilation, heating, downtime, and spraying when the solution is hot, under conditions of adequate replenishment or low ammonia. The pH must be raised with anhydrous ammonia. Automatic replenishment equipment must be checked.
2. *Low etch rate with high pH, >8.8:* This is caused by high copper content, water in etchant, or underventilation.
3. *Low etch rate with optimum pH:* This is due to error of copper thickness, oxygen starvation in etcher, or contamination of etchant.
4. *Solder attack:* This is caused by having excess chloride in etchant, or by the tin-lead deposit not being in compliance.
5. *Residues on solder plate, holes, and traces:* These may be caused by etchant solution imbalance or by spent finisher.
6. *Under- or overetching:* This may be due to improper pH.
7. *Sludging of etch chamber with low pH, <8.0:* See special problem 1, above. Sludge will be gritty and dark blue. This may be corrected by the addition of anhydrous ammonia.

8. *Sludging of etch chamber with high pH, >8.8:* See special problem 2, above. Sludge will be light blue and fluffy. This may be due to the copper concentration exceeding the capacity of the chloride concentration. It may be corrected by adding ammonium chloride.
9. *Presence of ammonia fumes:* The cause of this is leaks in the etcher. Ventilate for operator safety.
10. *Pollution:* This results when water coming from the etchers contains dissolved copper. If so, it must be chemically treated and separated from ammonia-bearing rinses. Thin-clad copper laminates present a further problem because the faster movement through the etcher increases the transport of etchant into the rinses.

14.4.2 Sulfuric Acid–Hydrogen Peroxide

Sulfuric-peroxide systems are used extensively for copper surface preparation (microetching), i.e., for oxide coating of inner layers, and for electroless and electrolytic copper plating.[14–16] The reasons for this wide acceptance are the ease of replenishment, simple waste treatment needed, closed-loop copper recovery, and optimum surface texture of the copper. In addition to these advantages, the compatibility of these systems with most organic and metallic resists, their steady etch rates, and the optimum undercut they provide make them especially suited to be used as final etch step.[15–21] Both tank immersion and etching systems are commercially available. Continuous-processing equipment is also available for the electroplating-through-etching operations.

14.4.2.1 Chemistry. Typical constituents of both immersion and spray etchants and their functions are as follows:

1. Hydrogen peroxide is an oxidizing agent that reacts with and dissolves metallic copper.
2. Sulfuric acid makes copper soluble and holds copper sulfate in solution.
3. Copper sulfate helps to stabilize etch and recovery rates.
4. Molybdenum ion is an oxidizing agent and rate exaltant.[22]
5. Aryl sulfonic acids are peroxide stabilizers.[23]
6. Thiosulfates are rate exaltants and chloride-ion controllers which permit lower peroxide content.[24]
7. Phosphoric acid retains clean solder traces and plated-through holes.[16,25]

The etching reaction is as follows:

$$Cu + H_2O_2 + H_2SO_4 \rightarrow CuSO_4 + 2H_2O \tag{14.4}$$

14.4.2.2 Properties and Control. The earlier technical problems of slow etch rates, peroxide decomposition, and foaming in spray systems have been solved, but critical concerns still remain. Among these are process overheating, etchant composition balance with by-product recovery, etchant contamination, and the dangers in handling concentrated peroxide solutions. Table 14.2 compares typical composition and operating conditions of immersion and spray-etching systems.

14.4.2.3 Closed-Loop Systems. Production facilities require continuous recirculating of the etchant through the etching tank or machine and the copper sulfate

TABLE 14.2 Sulfuric-Peroxide Final Etchants

	Immersion*	Spray†
Makeup:		
Sulfuric acid, 96%	10% by volume	17–20% by volume
Hydrogen peroxide, 50% (200 vol)	6–8% by volume	10–12% by volume
Copper sulfate	16–20 oz/gal	16–20 oz/gal
Stabilizers	As needed	6–8%
Other additives	As needed	6–8%
Operating Conditions:		
Copper	4–5 oz/gal	4–5 oz/gal
Etch rate	6–8 min	1.5 min
Temperature:		
Etching	115°F	130°F
Recovery	50–70°F	50–70°F

*Shipley Co., Newton, Mass.
†Electrochemicals, Youngstown, Ohio.

recovery operation. Etchant replenishment is controlled by chemical analysis and by additions of concentrates.

Copper sulfate recovery is based on lowering the solubility of $CuSO_4 \cdot 5H_2O$ by decreasing the etchant temperature to 50 to 70°F. See Fig. 14.2 and Table 14.2 for method and operating conditions.

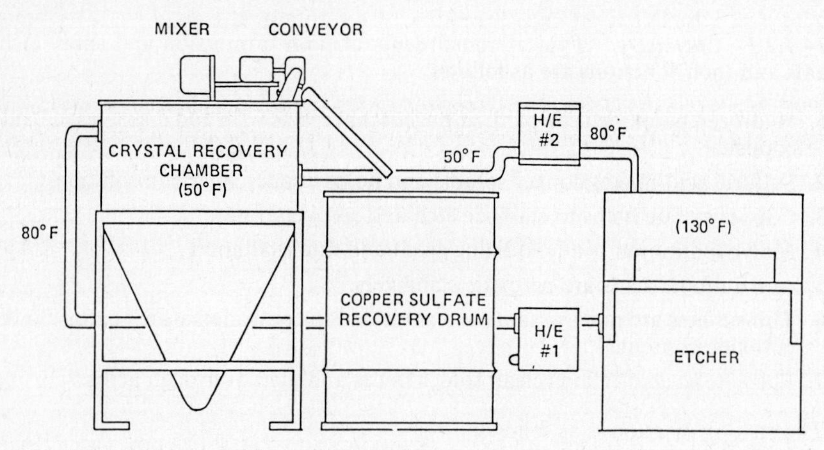

FIG. 14.2 Sulfuric-peroxide continuous etching and recovery system.[17]

14.4.2.4 Problems Encountered with Peroxide Systems

1. *Reduced etch rates:* This problem can be caused by operating conditions, solution imbalance, or chloride contamination.
2. *Under- and overetching:* A review of etching conditions, solution control, and the resist-stripping process may show deviations from normal. In the case of immersion etching, the solution and panel agitation may need to be increased. When spray etching, a check of nozzles and line clogging is indicated.

3. *Temperature changes:* Recirculating water rates and thermostats need to be examined regularly. Overheating may be due to high copper content, contamination, and rapid peroxide decomposition.
4. *Solder attack:* Review tin-lead process for copper contamination, and etchant for excess chlorides.
5. *Copper sulfate recovery stoppage:* Examine solution balance, heat exchanger, and other recovery equipment.
6. *Pollution:* Examine etcher for leaks, and rinsewater for excess copper.
7. *High equipment costs:* The need for customized etchers, etching machines, heat exchangers, and recovery units adds considerably to equipment costs.

14.4.3 Cupric Chloride

Cupric chloride systems are typical of the innovations to achieve closed-loop regeneration, lower costs, and a constant, predictable etch rate. Steady-state etching with acidic cupric chloride permits high throughput, material recovery, and reduced pollution. Regeneration in this case is somewhat complex but is readily maintained. Dissolved copper capacity is high compared with that of batch operation. Cupric chloride solutions are used mainly for fine-line multilayer inner details and print-and-etch boards.[26] Resists are screened inks, dry film, gold, and tin-nickel. Solder and tin boards are not compatible with cupric chloride etchant.

14.4.3.1 Chemistry. The etching reaction is as follows:

$$Cu + CuCl_2 \rightarrow Cu_2Cl_2 \tag{14.5}$$

Chloride ions, when added in excess as hydrochloric acid, sodium chloride, or ammonium chloride, act to solubilize the relatively insoluble cuprous chloride and thereby maintain stable etch rates. The complex formation can be shown as

$$CuCl_2 + Cl^- \rightarrow CuCl_3^{2-} \tag{14.6}$$

The etchant can be regenerated by reoxidizing the cuprous chloride to cupric chloride as illustrated by the following equations:

1. *Air:*

$$2Cu_2Cl_2 + 4HCl + O_2 \rightleftharpoons 4CuCl_2 + 2H_2O \tag{14.7}$$

   This method of regeneration is not used because the oxygen reaction rate in acids is slow and the solubility of oxygen in hot solution is low (4 to 8 ppm). Spray etching induces air oxidation.

2. *Chlorination:*

$$Cu_2Cl_2 + Cl_2 \rightarrow 2CuCl_2 \tag{14.8}$$

3. *Electrolytic:*

$$Cu^+ + e^- \rightarrow Cu \text{ (Cathode)} \tag{14.9}$$

$$Cu^+ \rightarrow Cu^{2+} + e^- \text{ (Anode)} \tag{14.10}$$

Refer to Table 14.3 for typical formulations.

TABLE 14.3 Cupric Chloride Etching Solutions

Component	Solution 1[27]	Solution 2[28]	Solution 3[29]	Solution 4[30]
$CuCl_2 \cdot 2H_2O$	1.42 lb	2.2 *M*	2.2 *M*	0.5–2.5 *M*
HCl (20° Be)	0.6 gal	30 mL/gal	0.5 *N*	0.2–0.6 *M*
NaCl		4 *M*	3 *M*	
NH_4Cl				2.4–0.5 *M*
H_2O	*	*	*	*

*Add water to make up 1 gal.

14.4.3.2 Properties and Control. Early cupric chloride formations had slow etch rates and low copper capacity and were limited to batch operation.[27–35] Regenerable continuous operation using modified formulations has brought useful improvements. Etch rates of 50 to 55 s for 1-oz copper are obtained from cupric chloride–sodium chloride systems operated at 130°F with conventional spray-etching equipment. Copper capacities are maintained at 15 to 20 oz/gal.

14.4.3.3 Closed-Loop Etching and Regeneration. Systems in use include chlorination, chlorate, and electrolytic regeneration.

14.4.3.3.1 Chlorination. Direct chlorination is the preferred technique for regeneration of cupric etchant because of its low cost, high rate, efficiency in recovery of copper, and pollution control. The cupric chloride–sodium chloride system (Table 14.3, no. 3) is suitable. Figure 14.3 shows the process. Chlorine, hydrochloric acid, and sodium chloride solutions are automatically fed into the system as required. Sensing devices include redox and colorimeter (Cu oxidation state), density (Cu concentration), etch rate monitor, level sensor, and thermostats. Chlorination is reliable and controllable. Other factors are safety and solution control.

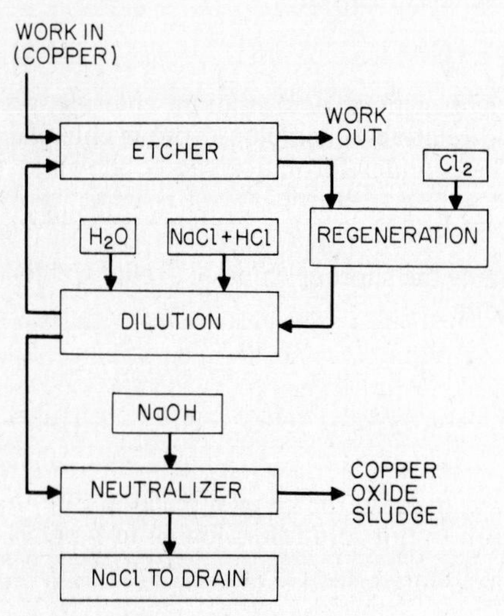

FIG. 14.3 Cupric chloride chlorination regeneration system.[29,36]

1. *Safety:* Use of chlorine gas requires adequate ventilation and leak-detection equipment.
2. *Solution control:* An increase in pH will cause the copper colorimeter to give erroneous readings caused by the turbidity in the solution. Excess NaCl at 18 to 20 oz/gal copper causes coprecipitation of salts when the solution is cooled.

14.4.3.3.2 Chlorate Regeneration. This method uses sodium chlorate, sodium chloride, and hydrochloric acid and is an alternate method similar to chlorination.

14.4.3.3.3 Electrolytic Regeneration. The electrochemical reversal of the etching of copper shown in Eq. 14.5 is claimed to be effective and economical. Descriptions of this system are given in the literature.[27] On a large scale, electrolytic regeneration requires a high investment in equipment and materials, as well as high power consumption.

The etchant is a solution of cupric chloride and hydrochloric acid (Table 14.3, no. 1). Etchant flows continuously between spray-etching machines and a plating tank. In the plating machine, two processes take place simultaneously: Copper is plated at the cathode, and regeneration of the spent etchant occurs at the anodes.

14.4.3.4 Batch Regeneration

14.4.3.4.1 Oxidizing Agent Treatment. As shown by Eq. 14.7, chemical regeneration is possible with oxygen. Similarly, it is possible with faster, more active oxidizing agents in batch processing, as shown by Clark[34] using hydrogen peroxide or sodium hypochlorite. Equations 14.11 and 14.12 show the oxidation steps:

$$Cu_2Cl_2 + 2HCl + H_2O_2 \rightarrow 2CuCl_2 + 2H_2O \qquad (14.11)$$

$$Cu_2Cl_2 + 2HCl + NaOCl \rightarrow 2CuCl_2 + NaCl + H_2O \qquad (14.12)$$

Ventilation must be adequate to draw the hydrochloric acid fumes. Each gram of dissolved copper requires 1.75 mL of 30 percent hydrogen peroxide (100 vol), or 14 mL of 5 percent available Cl_2 sodium hypochlorite solution. Common practice is to remove about half of the solution, add sufficient acid, and, while stirring vigorously, slowly add the amount of oxidizing agent required. The solution volume is then adjusted. This method is justified for small operations only.

14.4.3.4.2 Solution Replacement. Black and Cutler[35] have also shown that the reaction given in Eq. 14.7 can be used in principle to regenerate a cupric chloride etching solution. In this case, they used a solution at 100°F in a spray etch operation. After etching 1-oz copper boards, they replaced (by using a proportioning pump) three parts of the used etchant with two parts of hydrochloric acid (37 percent) and one part water. Solution replacement must be in keeping with the low copper capacity of the batch-type etchant, namely, below 4 oz/gal copper.

14.4.3.5 Etching Practice. Any of the common techniques, such as immersion, bubble, splash, or spray etching, can be used with cupric chloride. Spray, splash (paddle), and air-bubble etching are the most effective. Increasing temperatures and agitation speed up the etch action, although higher temperatures cause excessive fumes with HCl-type etchants and attack some resists. As with all other etchants, final rinsing must be thorough. A dip in 5 percent HCl (by volume) followed by thorough water rinsing is recommended. Spray equipment, tanks, and linings are usually made of PVC or other chemical-resistant plastic. Temperature controls and safety devices must be used to prevent plastic-lined tanks from overheating and burning. Quartz immersion heaters, titanium pump parts, heaters, racks, etc., are suitable. Stainless steel should not be used. Parts which do not come into direct contact with the etchant, such as door hinges, can be made of Monel.

14.4.3.6 Problems with Cupric Chloride Systems

1. *Slow etch rate:* This is frequently due to low temperatures, insufficient agitation, or lack of solution control. If temperature and agitation are under control, an increase of etching time and a dark green solution may result from a decrease in the cupric ion content. If etching time doubles, the solution should be renewed or regenerated. Acid must also be added to clarify cloudy solutions. In regenerative systems, the source of oxidation may be depleted.

2. *Sludging:* This occurs if acid is low or if water dilution occurs.

3. *Breakdown of photoresists.* This may occur with excess acid and elevated temperatures. Correction is readily made either by neutralizing with sodium hydroxide or by replacing part of the solution with water. If lifting occurs, a problem exists in cleaning prior to resist coating, in inadequate exposure, or in baking. Thick photoresists are more resistant to breakdown.

4. *Yellow or white residues on copper surface:* The yellow residue is usually cuprous hydroxide. It is water-insoluble and is left when boards are etched and alkali-cleaned. A white precipitate will probably be cuprous chloride, which can remain after etching in solutions that are low in chloride ion and acid. To eliminate both conditions, the solution in which the board is rinsed just before final water-spray rinsing should be 5 percent by volume hydrochloric acid.

5. *Waste disposal:* Spent etchant can be sold for its copper content. This is a reasonable option because of hazardous wastewater restrictions. Rinses must be chemically treated to precipitate copper in accordance with local wastewater ordinances. See Chapter 15 on pollution control.

14.4.4 Persulfates

Ammonium, sodium, and potassium persulfates modified by certain catalysts have been adopted for the etching of copper in PC manufacturing. Continuous regenerative systems and a batch system using ammonium persulfate are common. Wide use is made of persulfates as a microetch for inner-layer oxide coating and copper electroless and plating processes. Persulfate solutions allow all common types of resists on boards including solder, tin, tin-nickel, screened inks, and photosensitive films. Persulfate solutions are not suitable etchants for gold because of excess undercut and low etch factors.

Formulations of ammonium persulfate catalyzed with mercuric chloride have etch rates comparable to those of the chloride etchants and are preferred for solder, print-and-etch, and tin-nickel boards. Formulations with proprietary additives other than mercury catalysts are available and have been improved to give good etch factors.[37] Regenerative systems have made possible higher copper capacities and constant etch rates. In general, persulfate etchants are unstable and will exhibit decomposition, lower etch rates versus copper content, and lower useful copper capacity. The use of persulfate etching systems has declined recently because of high costs and other improvements in alkaline ammonia etchants.

14.4.4.1 ***Chemistry.*** Ammonium, potassium, and persulfates are stable salts of persulfuric acid ($H_2S_2O_8$). When these salts are dissolved in water, the persulfate ion ($S_2O_8^{2-}$) is formed. It is the most powerful oxidant of the commonly used peroxy compounds. During copper etching, persulfate oxidizes metallic copper to cupric ion as shown:

$$Cu + (NH_4)_2S_2O_8 \rightarrow CuSO_4 + (NH_4)_2SO_4 \qquad (14.13)$$

Persulfate solutions hydrolyze to form peroxy monosulfate ion (HSO_5^{1-}) and, subsequently, hydrogen peroxide and oxygen. This hydrolysis is acid-catalyzed and accounts for the instability of acidic persulfate etching solutions.

Ammonium persulfate solution, normally made up at 20 percent, is acidic. Hydrolysis reactions and etchant use cause a reduction of the pH from 4 to 2. The persulfate concentration is lowered, and hydrated cupric ammonium sulfate [$CuSO_4 \cdot (NH_4)_2SO_4 \cdot 6H_2O$] is formed. This precipitate may interfere with etching.

Solid persulfate compounds are stable and do not deteriorate if stored dry in closed containers. Solution composition is catalyzed by various agents, including organic matter and transition metals (Fe, Cr, Cu, Pb, Ag, etc.). Materials for storage must be chosen carefully. Persulfates should not be mixed with reducing agents or oxidizable organics. Avoid contact with solvents, oils, grease, nitric acid, HCl, halide salt solutions, strong caustics, or ammoniacal solutions. Consult supplier MSDS sheets for additional details.[38]

14.4.4.2 Properties and Control. Etchant compositions are given in Table 14.4. Solution 1 is used for organic and tin-nickel resists. Composition 2 is used for solder-plated circuits. Phosphoric acid (1 to ½ percent by volume) is added to eliminate the incomplete etching of copper next to the solder conductors ("runoff") and to minimize solder darkening.[25]

TABLE 14.4 Composition of Persulfate Etch Solutions

	Solution 1[37,39]	Solution 2[25,37]	Solution 3[25,37]
$(NH_4)_2S_2O_8$	2 lb/gal	2 lb/gal	
$Na_2S_2O_8$			3 lb/gal
$HgCl_2$	5 ppm	5 ppm	15 ppm
Additive	1 g/gal	1 g/gal	1 g/gal
H_3PO_4		57 mL/gal	57 mL/gal

Occasionally, incorrect plating conditions give rise to solder tin-lead deposits with the wrong composition. In those cases, the solder plate will darken during etching and may cause portions of the copper around the circuit to remain unetched. Processing temperatures range from 75 to 140°F, but the preferred temperature is about 115°F to balance etch rate, copper capacity, and solution decomposition. All common types of etching systems can be used, but spray etching in conveyorized equipment is preferred for final etching. Cooling coils are required to maintain a constant etching temperature and prevent runaway reactions.

The useful capacity of the etchant is about 7 oz/gal copper at 100 to 130°F. Above 5 oz of copper per gallon, it is necessary to keep the solutions at 130°F to prevent salt crystallization. The etch rate of a solution containing 7 oz/gal of dissolved copper is 0.00027 in/min at 118°F.

Copper concentration is readily controlled by specific gravity or colorimetric measurements. Table 14.5 shows the dependence of specific gravity on copper concentration in an etchant containing 2 lb/gal of ammonium persulfate.

14.4.4.3 Batch Operation. Composition 3 is used for batch-type spray etching.[37] Sodium persulfate is preferred because it has minimal disposal problems and somewhat higher copper capacity and etch rates. Etch rates vary throughout bath life and range from 0.0018 to 0.0006 in/min for copper content of 0 to 7 oz/gal. Prepared solutions must be aged for 16 to 72 h before etching when proprietary additives are used.

TABLE 14.5 Density of Ammonium Persulfate Etching Solutions

Copper concentration, oz/gal	Specific gravity at 120°F	Copper concentration, oz/gal	Specific gravity at 120°F
0	1.115	4	1.171
1	1.128	5	1.186
2	1.142	6	1.201
3	1.158		

14.4.4.4 Problems with Persulfates

1. *Low etch rates:* Since the solution may decompose, it will be necessary to replace the bath. If solution is new, add more catalyst, and check for iron contamination.

2. *Salt crystallization:* Salts crystallize on the board and cause streaks, damage the solder plate, and plug the spray nozzles or filters. When copper content is high, blue salts may precipitate.

White salts of ammonium persulfate also may precipitate when the solution becomes highly concentrated because of excess water evaporation. To prevent this, the etching temperature should be maintained above 95°F. Plastic tube filters installed at the sump inlet will protect the spray nozzles and the pump impellers.

3. *White films on solder surface:* This may occur normally, or when the lead content in the solder plate is too high. The films can be removed with a soft metal brush followed by rinsing with a solution of fluoboric acid-thiourea.

4. *Black film on solder:* This condition can result when the solder alloy is high in tin. If solder reflow or component soldering is to follow, activate by tin immersion or with solder brighteners. Adjust phosphoric acid content in etchant and solder-plating conditions.

5. *Spontaneous decomposition of etch solution:* This breakdown is due to contaminated, overheated, or idle solutions. Also, when ammonium persulfate etchants are placed in etching machines or tanks that were previously used for ferric chloride or other etchants, it is possible for the persulfate to decompose spontaneously. Thorough cleaning with acid solutions and rinsing with water are required. Ammonium persulfate etchants are unstable, especially at higher temperatures. At about 150°F, the solution decomposes quickly. Use it soon after mixing.

6. *Disposal:* The exhausted etchant consists mainly of ammonium or sodium and copper sulfate with a pH of about 2. As with other etchants, direct sewer discharge is not allowed. Two methods for disposal are suggested:

> ***a.*** Electrolytic deposition of the copper on the surface of passivated 300 series stainless steel. The spent etchant is acidified with sulfuric acid prior to electrolysis. Once the copper has been removed, the remaining solution can be diluted, neutralized, checked, and discarded. The copper can be removed from the cathode. Spent sodium persulfate can be treated with caustic soda.
>
> ***b.*** Addition of aluminum or iron machine turnings to a slightly acidified solution is another practical but possibly more difficult means of removing the dissolved copper. The reaction, especially in the presence of chloride ions, will be violent, and considerable heat will be given off if the solutions are not diluted. Because of problems in handling, the electrodeposition method and the use of a scavenger service are the best means of disposal.

14.4.5 Ferric Chloride

Ferric chloride solutions are used as etchants for copper, copper alloys, Ni-Fe alloys, and steel in PC applications, electronics, photoengraving arts, and metal finishing. Ferric chloride is used with screen inks, photoresist, and gold patterns, but it cannot be used on tin-lead or tin-plated boards. However, ferric chloride is an attractive spray etchant because of its low cost and its high holding capacity for copper.

TABLE 14.6 Composition of $FeCl_3$ Solutions*

	Low strength	Optimum		High strength
Percent by weight	28	34	38	42
Specific gravity	1.275	1.353	1.402	1.450
Baumé	31.5	38	42	45
lb/gal	3.07	3.9	4.45	5.11
g/L	365	452	530	608
Molarity	2.25	2.79	3.27	3.75

*Data taken at 68 to 77°F (20 to 25°C). Photoengraving $FeCl_3$ 42° Baumé has 0.2 to 0.4% free HCl. Proprietary etchants contain up to 5% HCl.

The composition of the etchant is mainly ferric chloride in water, with concentrations ranging from 28 to 42 percent by weight (see Table 14.6). Free acid is present because of the hydrolysis reaction.

$$FeCl_3 + 3H_2O \rightarrow Fe(OH)_3 + 3HCl \qquad (14.14)$$

This HCl is usually supplemented by additional amounts of HCl (up to 5 percent) to hold back the formation of insoluble precipitates of ferric hydroxide. Commercial formulations also contain wetting and antifoam agents. The effects of ferric chloride concentration, dissolved copper content, temperature, and agitation on the rate and quality of etching have been reported in the literature.[40–42] Commercial availability includes lump $FeCl_3 \cdot 6H_2O$ and aqueous solutions with and without additives. Ferric chloride with additives has the advantage of low foaming (reduced odor and fuming), fast and even etching (due partly to added strong oxidizers and surface-wetting properties), and reduced iron hydroxide precipitate formation, owing to the slight acidity and to the chelating additives. The useful life of ferric chloride etchants and uniformity of etching rates have been greatly improved by the manufacturers of proprietary solutions.

14.4.5.1 Chemistry. At the copper surface, ferric ion oxidizes copper to cuprous chloride with the formation of green ferrous chloride:

$$FeCl_3 + Cu \rightarrow FeCl_2 + CuCl \qquad (14.15)$$

In solution, cuprous chloride is further oxidized to cupric chloride:

$$FeCl_3 + CuCl \rightarrow FeCl_2 + CuCl_2 \qquad (14.16)$$

As cupric chloride builds up in the etching solution, a disproportionation reaction takes over:

$$CuCl_2 + Cu \rightarrow 2CuCl \qquad (14.17)$$

In practice, when a 42° Baumé solution contains 8 oz/gal (60 g/L) or more of dissolved copper, the etch time becomes longer than desired, and the etchant must be replaced. The etching time increases rapidly beyond 11 oz/gal of copper, but the solution has a large remaining capacity to dissolve copper. Many etching plants make use of partially depleted solutions by adding HCl at this point. See Sec. 14.4.5.4 for other corrective actions that may be taken to extend the useful life of the etchant.

14.4.5.2 Properties and Control. One of the most important considerations in the etching of copper is the criteria used to determine the exhaustion and usefulness of partially depleted ferric chloride. Because of the complex relations arising from changes in the ferric chloride–cupric chloride composition, methods other than chemical analysis (evaporation, dilution, material introduction, etc.) are not helpful in determining the useful properties of partially exhausted etching solutions. Thus, we will consider other procedures to evaluate the performance of ferric chloride as an etchant and the desirability of changing operating conditions.

The two criteria proposed are (1) etching time (etch rate) and (2) the degree of exhaustion of the etching solution.

Etching time is defined differently by each user, but it must include a measurement of the exact time required to dissolve a known weight of copper when all conditions are constant or specified (i.e., temperature, concentration, agitation, copper area exposed, solution volume). If the quality of etch is also to be judged, other factors must be known, such as procedures for cleaning, positions in etch machine, and size of board.

A 1-oz copper thickness should etch clean in 1 to $1\frac{1}{2}$ min with fresh 42° Baumé ferric chloride proprietary etchant at 100°F in a high-pressure oscillating spray etcher. If the etch time exceeds 7 min under those conditions, the solution is considered spent.

14.4.5.3 Effect of Variables

14.4.5.3.1 Concentration of Reagents. Figure 14.4 shows the concentration of ferric chloride (new solution) versus relative etch times at different temperatures. The best concentration for minimum etching time using fresh solutions varies from 30 percent ferric chloride at room temperature to 35 percent ferric chloride at 70°C (158°F). As expected, an increase in temperature causes an increase in etch rate. Spray etching is done at about 120°F.

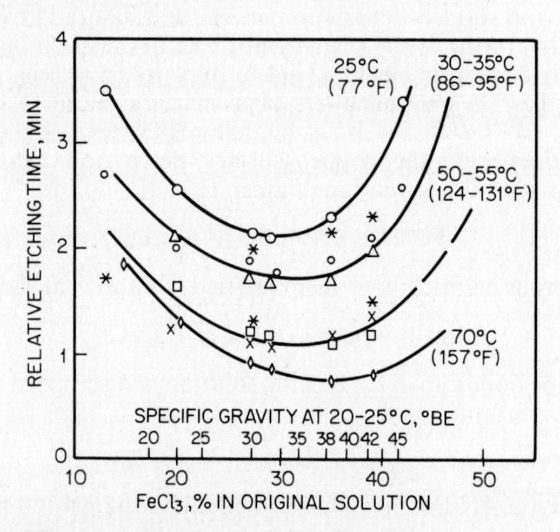

FIG. 14.4 Relative etching time versus $FeCl_3$ concentration at various times.[41]

14.4.5.3.2 Addition of HCl. Figure 14.5 shows how additional hydrochloric acid improves the relative etching time for ferric chloride solutions. Above 6 oz/gal of copper, hydrochloric acid depresses the formation of ferric hydroxide and aids in the etching reactions. In fact, the useful capacity is increased by 2 oz/gal of dissolved copper.

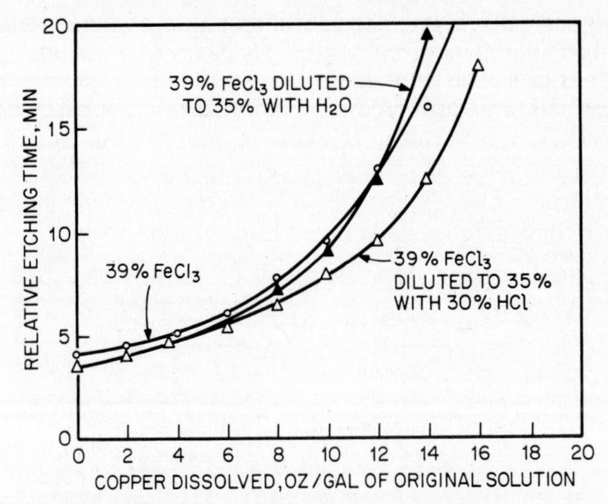

FIG. 14.5 Relative etching time versus dissolved copper for 42° Baumé $FeCl_3$ diluted with H_2O and HCl.[41]

14.4.5.3.3 Agitation. The etching time can be controlled by the type and extent of agitation used. The time required to etch in a still solution by immersion is high from the start and rapidly increases, as the dissolved copper increases. Etch action is increased considerably when air is used for agitation or when it is introduced with a spray or splash operation.

High etch factors are also obtained with air introduced into the spray or other etching processes. Additional studies on control of undercut of patterns are given in the literature.[43–47] Powderless etching methods are used to increase etch factors, and the ratio of depth of etch to lateral etch is increased by using additives such as thiourea, disulfide, and phosphoric acid esters, which form gelatinous films on the lateral shoulders and sidewalls of the traces.[43,48,49] The change to the newer technologies of thin-clad laminates and additive and semiadditive systems has reduced the need for using these compounds.

14.4.5.4 Regeneration. Ferric chloride is not easily regenerated. Regeneration systems are based on the need to separate copper chloride from ferric chloride.[34] The method shown in Figure 14.6 adds chlorine gas and recovers copper from the

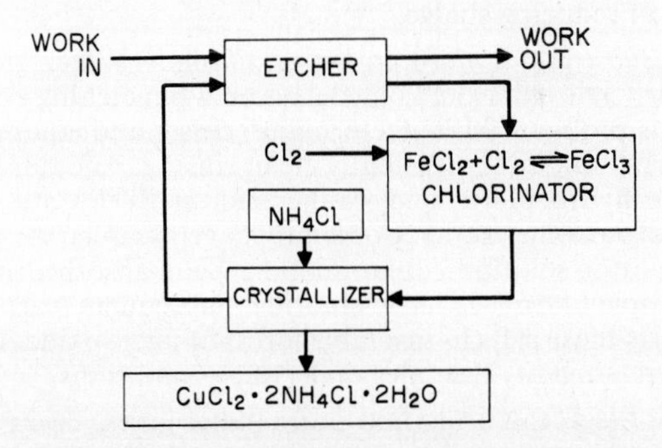

FIG. 14.6 Ferric chloride closed-loop regeneration and recovery process using recrystallization.[36]

ferric etchant by crystallization. The addition of ammonium chloride (2 *M*) and cooling to ambient temperature permit crystallization of a chloride double salt $CuCl_2 \cdot 2NH_4Cl \cdot H_2O$. Ferric chloride is maintained at 6 oz/gal of copper.

A second method for copper recovery involves passing spent etchant over iron plates that reduce dissolved copper to metal as shown in Eq. 14.18 and Fig. 14.7.

$$Cu^{2+} + Fe \rightleftharpoons Cu + Fe^{2+} \tag{14.18}$$

By-products are copper sludge and diluted iron chloride.

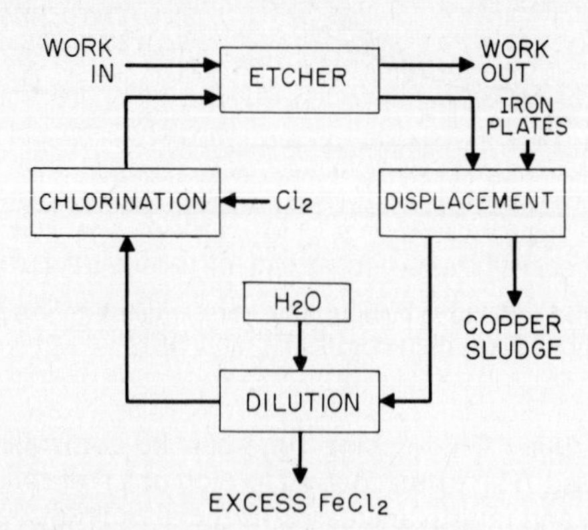

FIG. 14.7 Ferric chloride closed-loop regeneration and recovery by using iron displacement.[36]

14.4.5.5 Etching Problems

1. *Rapid decrease in the Baumé value and short bath life:* This is due to introduction of water into the etchant, which commonly occurs when automatic equipment includes a rinse cycle. Baumé gravity measurement is not recommended for etch-solution control.

2. *Excessive sludge formation:* This may be caused by excessive temperature that causes HCl evaporation, by the drag-in of alkali, or by the presence of materials that react with free HCl or hard water. A yellow precipitate sludge can be dissolved by adding 1.2 mL of concentrated HCl for each gallon of etching solution at 42° Baumé. Diluting the etchant to 38 to 39° Baumé with HCl has the added advantage of reducing sludge.

3. *Insulating substrate contamination (staining):* Water rinsing does not always remove the ferric chloride. Long delays between etching and final cleaning may aggravate this problem. Manufacturing processes and equipment must provide for a hard spray and for consistently clean rinsing. Another option is to use dilute solutions of hydrochloric acid, oxalic acid, or EDTA after the first rinse, followed by a vigorous water spray. Conductivity cells can be installed in critical-rinse tanks to monitor solution contamination. Salts absorbed on the substrate are present as complex forms of ferric chloride. Subsequent washing, hydrolysis, and drying converts these salts to insoluble forms of ferric oxide. Insulation resistance of the substrate readily degrades under such conditions.

4. *Corrosion:* Steel, cast iron, and other metal pipes, decking, racks, roofs, drains, etc., are quickly corroded by ferric chloride and its fumes. Use approved plastics for all construction.

14.4.6 Chromic-Sulfuric Acids

These etchants for solder- and tin-plated boards were preferred for many years. More recently, their use has been limited drastically because of the difficulty in regeneration, inconsistent etch rate, the low limit of dissolved copper (4 to 6 oz/gal), and especially the pollution concerns. Chromic acid etchant is suitable for use with solder, tin-nickel, gold, screened vinyl lacquer, and dry- or liquid-film photoresists. Although chromic acid etchants are strong oxidizing agents, they do not attack the solder, since insoluble lead sulfate is formed. Undercut is seen less with solder-plated patterns than with gold and organic resists. Proprietary etchants are made with chromic and sulfuric acids. The reaction between copper and chromate is:

$$3Cu + 2HCrO_4^- + 14H^+ \rightarrow 3Cu^{2+} + 2Cr^{3+} + 8H_2O \qquad (14.19)$$

Potassium chromate and potassium dichromate are also used as starting materials. Equations 14.20 and 14.21 show the reaction with sulfuric acid to produce the hexavalent Cr^{6+} oxidizing agent.

$$K_2CrO_4 + H_2SO_4 \rightarrow K_2SO_4 + H_2CrO_4 \qquad (14.20)$$

$$K_2Cr_2O_7 + H_2SO_4 + H_2O \rightarrow K_2SO_4 + 2H_2CrO_4 \qquad (14.21)$$

Compositions of chromic acid etchants are given in Table 14.7. Sulfuric acid is normally used as the source of hydrogen ions. Concentrations of 10 to 13 percent by volume are used to achieve optimum etchants[52] and maximum holding capacity for copper. The use of other acids has been successful. Nitric acid increases the etch rate but causes attack on the solder plate, phosphoric acid decreases the etch rate, acid fluorides attack titanium etcher parts and solder, and organic acids are decomposed by chromates. Hydrochloric acid must not be added to this etchant because of toxic chlorine gas[10] formation:

$$2HCrO_4^- + 6CL^- + 14H^+ \rightarrow 2Cr^{3+} + 3Cl_2 + 8H_2O \qquad (14.22)$$

Chromic-sulfuric mixtures etch copper slowly, and additives are needed to increase the etch rate. For example, sodium sulfate is used in formulation 1 of Table 14.7, and iodine compounds are used in formulation 2.[53]

TABLE 14.7 Composition of Chromic-Sulfuric Acid Etch Solutions*

	Formulation 1[50]	Formulation 2[51]
CrO_3	240 g/L	480 g/L
Na_2SO_4	40.5 g/L	
H_2SO_4 (96%)	180 g/L	31 mL/L
Copper		4.9 g/L

*Proprietary agents contain wetting, antifoaming, and chelating agents, and catalysts.

14.4.6.1 ***Properties and Control.*** Increases in temperature and agitation generally increase the etch rate. However, air agitation of tank solutions results in a much slower etch rate than other methods such as vibration, spray, or splash etching. In addition, air produces a mist or spray which is toxic. These fumes are very corrosive and contaminate plating solutions. Proprietary solutions contain antimisting agents.

A hydrometer is used to control the Baumé value of the solution. The density should be maintained at 30° Baumé at 68°F, and 28° Baumé at 80°F. Water is used to decrease high Baumé solutions. Colorimetric standards made up by adding known quantities of copper to fresh 30° Baumé etchant are used to estimate copper content.

14.4.6.2 ***Regeneration.*** Commercial systems for regeneration of chromic-sulfuric acids are not in common use because of the corrosive nature of the products, the handling hazards, and the pollution restrictions. Other methods in use include electrolytic deposition of copper[52] and electrodialysis.[54]

14.4.6.3 Problems with Chromic-Sulfuric Acid Systems

1. *Solder attack:* The protective value of solder depends on the formation of insoluble compounds on the surface. Solder is attacked if the sulfate content of the bath becomes very low or contains chloride or nitrates. The solder plate composition can also cause etchant attack. When the lead content becomes low, the sulfate film protection is lowered, and protection is lost.

2. *Slow or no etching of copper:* This can be caused by low chromic content, low temperature, insufficient acid, or high copper content. The solution is maintained as close as possible to 30° Baumé (pH about 0.1, temperature 80 to 90°F) and should be discarded when copper metal content exceeds 5.5 oz/gal.

3. *Staining of board materials:* The surfaces of dielectric substrates such as paper-based phenolics are attacked by chromic acid etchants. Removal is difficult, and the boards are generally rejected.

4. *Disposal:* Spent chromic acid etchants present a serious disposal problem. Disposal must comply with pollution standards and approved practice.

5. *Safety hazards:* Chromic acid is an extremely strong oxidizing agent. It will attack clothing, rubber, plastics, and many metals. Safety measures require adequate ventilation to keep fumes out of room air, synthetic rubber gloves, aprons, face and eye shields, and storage away from combustible materials. Dermatitis and nasal membrane damage are possible dangers.[10]

14.4.7 Nitric Acid

Etchant systems based on nitric acid have not found extensive application in PC manufacture. Copper etching is very exothermic, which may lead to violent runaway reactions. Problems with this system include solution control, attack on resists and substrates, and toxic gas fuming. However, nitric acid has certain advantages. These include rapid etching, high dissolved copper capacity, high solubility of nonsludging products, availability, and low cost.

14.4.7.1 ***Chemistry.*** Reaction in strong acids is shown by the following equation:

$$3Cu + 2NO_3^- + 4H^+ \rightarrow 3Cu^{2+} + 2NO_2 + 2H_2O \qquad (14.23)$$

Recent work shows that process improvements are possible.[55,56] Controlled etching has been attained in solutions containing 30 percent copper nitrate, water-soluble polymers, and surfactants. Dry-film resists work well in this etchant. An important finding was that straight wall trace edges were achieved using nitric. This could result in higher yields and density of fine-line boards.

14.5 OTHER MATERIALS FOR BOARD CONSTRUCTION

Printed board laminates are usually composed of copper bonded to organic dielectric materials, to ceramics, or to other metals.

1. *Organic dielectrics:* These are thermosetting or thermoplastic resins usually combined with a selected reinforcing filler. *Thermoset*-reinforced materials used for rigid and flexible boards provide overall stability, chemical resistance, and good dielectric properties. *Thermoplastic* materials are also used for flexible circuit applications. A factor in material selection is the effect of process solutions, etchants, and solvents on the material. In addition, the adhesives used in laminating metal to substrate can be softened, loosened, and attacked by some solutions.

2. *Thin-clad copper:* Etched printed boards with epoxy laminate of $\frac{1}{4}$-oz copper or less show minimal overhang and slivers.

3. *Semiadditive copper:* A copper thickness of 0.000050 to 0.000200 in with subsequent copper and resist metal plating shows no overhang or sliver formation.

14.6 METALS OTHER THAN COPPER

14.6.1 Aluminum

Aluminum-clad flexible circuits find use in microwave strip-line[57] and radiation-resistance applications. Aluminum and its alloys have good electrical conductivity, are lightweight, and can be plated, soldered, brazed, chem-milled, and anodized with good results. Laminate dielectrics include PPO,[57] polyimide,[58] epoxy-glass, and polyester.

Precleaning for resist application includes nonetch alkaline soak, water rinsing for 5 to 10 s in chromic-sulfuric acid, rinsing, and drying. Preferred etchants include ferric chloride (12 to 18° Baumé), sodium hydroxide (5 to 10 percent), inhibited hydrochloric acid, phosphoric acid mixtures, solutions of HCl and HF, and ferric chloride–hydrochloric acid mixture.

Screen-printed vinyl resists and dry-film photoresist are the most durable for deep etching or chem-milling. A dip in a 10 percent nitric or chromic acid solution will remove residues from the surface or edges of conductor lines which may be left on some alloys. Dilute chromic acid has also been used for this purpose. Spray-rinse thoroughly with deionized water after etching.

14.6.2 Nickel and Nickel-Based Alloys

Nickel is increasingly used as a metal cladding, electroplated deposit, or electroformed structure for printed wiring because of its welding properties. Nichrome- and nickel-based magnetic alloys are other examples of materials requiring special etching techniques.

The methods previously described are adaptable to image transfer and etching of nickel-base materials. Etching uses ferric chloride, 42° Baumé, at about 100°F. Other etchants include solutions made from one part nitric acid, one part hydrochloric, and three parts water, or one part nitric, four parts hydrochloric, and one part water.

14.6.3 Stainless Steel

Alloys of stainless steel are used for resistive elements or for materials with high tensile strength. Etching of the common 300 to 400 series can be done with the following solutions:

1. Ferric chloride (38 to 42° Baumé) with 3 percent HCl (optional).
2. One part HCl (37 percent) by volume, one part nitric acid (70%) by volume, one to three parts by volume water. Etch rate is about 0.003 in/min at 175°F, useful for high 300 to 400 series alloys.
3. Ferric chloride + nitric acid solutions.
4. One hundred parts HCl (37 percent) by weight, 6.5 parts nitric acid by weight, 100 parts water by weight.

14.6.4 Silver

Silver, the least expensive precious metal, has excellent properties, including superior electrical and thermal conductivity, ductility, visible-light reflectivity, high melting point, and adequate chemical resistance. As such, it is widely used throughout the electronics industry. Flexible circuit structures with silver are used in electronic cameras and LED products.

Standard image-transfer methods are suitable. Pre-etch cleaning should include a dip in dilute nitric acid. Mixtures of nitric and sulfuric acids are effective etchants. With silver on brass or copper substrates, a mixture of 1 part nitric acid (70 percent) and 19 parts sulfuric acid (96 percent) will dissolve the silver without adversely attacking the substrate. The solutions should be changed frequently to prevent water absorption and the formation of immersion silver on the copper.

Etching can be done with a solution containing 40 g chromic acid, 20 mL sulfuric acid (96 percent), and 2000 mL water.[59] This is followed by a rinse in 25 percent ammonium hydroxide. Thin films of silver are etched in 55 percent (by weight) ferric nitrate in water or ethylene glycol. Solutions of alkaline cyanide and hydrogen peroxide will also dissolve silver. Use extreme caution. Electrolytic etching is also possible with 15 percent nitric acid at 2 V and a stainless-steel cathode.

14.7 UNDERCUT, ETCH FACTOR, AND OVERHANG

During etching, as the depth of etch proceeds vertically, the sidewalls tend to etch sideways and produce an undercut action. The degree to which this occurs is known as the etch factor, as is defined as the ratio of depth to side attack (see Fig. 14.8). In practice, controlled spray etching vertical to the copper surfaces with

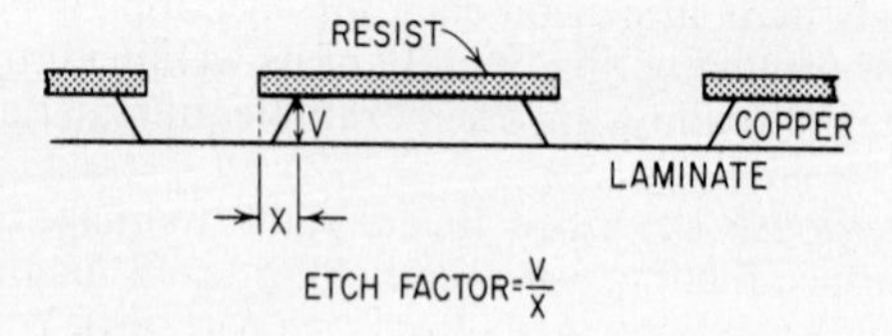

FIG. 14.8 Etch factor in printed board etching.

selected etchants leads to high etch factors.[47] Immersion etching generally results in low etch factors. Fine-line etching with a minimum of undercut is best achieved with copper foil of $\frac{1}{2}$ oz or less and is carried out by removing the board from the etching machine exactly at the time of completion, using fine-line etchants. Compensation for line width reduction and undercutting should be designed into the artwork, especially for panel-plated boards, thick metal cladding, and dense, fine-line inner- and outer-layer patterns. The etch factor can be minimized on metal parts by resist patterning and then etching both sides at once.

Undercut and overhang for resists are shown in Fig. 14.9. Excess overhang may fall loose as metallic slivers and cause electrical shorting, and thus they present reliability problems. After etching, removal requires soft-brass brushing, ultrasonic agitation and rinsing, or fusion in the case of solder-plated resists.

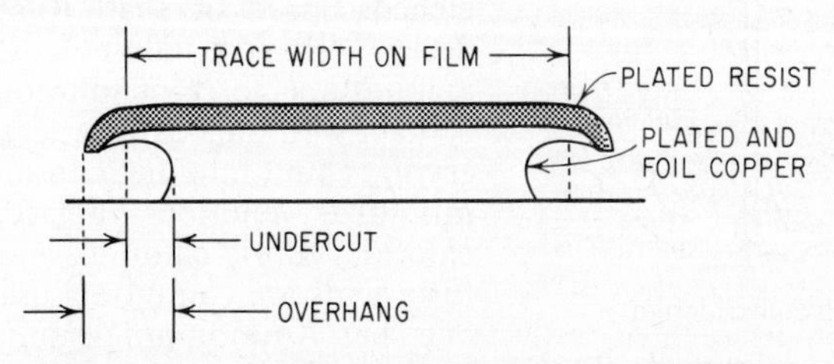

FIG. 14.9 Undercut and overhang for a plated-metal-resist pattern.

14.8 EQUIPMENT AND TECHNIQUES

Etching techniques and the equipment used today have evolved from four basic etching methods: immersion, bubble, splash (paddle), and spray etching. Spray etching is the most common method, since it is fast, well suited to high production, and capable of very fine line definition.

14.8.1 Immersion Etching

Deep-tank, or immersion, etching is the simplest method of etching. Pieces are immersed in the solution until etching is complete. Solution heating and agitation are required. This method is used for sulfuric acid–hydrogen peroxide final etching.

14.8.2 Bubble Etching

Bubble etching is a modified form of immersion etching. Air is bubbled through the solution past the work being etched. This air has two purposes: to ensure the availability of fresh etchant at the surface and rinse away dissolved metal, and to provide oxygen that increases the effectiveness of the etchant by providing additional oxidizing power and regenerating the etchant. Chromic-sulfuric acid etchant used on solder-plated copper boards is perhaps the widest application of bubble etching. Ammonium persulfate may also be used, but cooling coils are required.

Since boards are heavy, racks have to be rugged and built of materials not affected by the etching solutions. PVC has proved to be a suitable plastic because of its chemical resistance when used below 175°F and when not subjected to organic solvents.

14.8.3 Splash Etching

Paddle, or splash, etching offers advantages over bubble, or still, etching with respect to evenness of etch and minimization of undercut. However, it is limited to handling only a few pieces at one time (see Fig. 14.10).

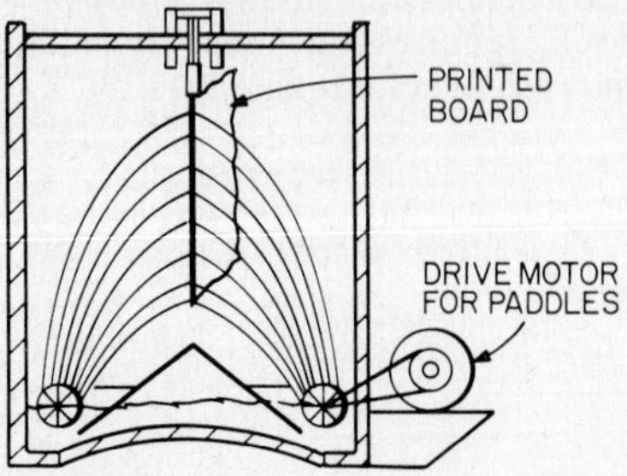

The operation involves picking up solution in a cup attached to a motor-driven shaft and throwing it toward boards being etched. Speed of shaft rotation and cup design (shape, volume) determine the amount and force of solution coming into contact with the board. A large-volume reservoir is built into the bottom of the tank to minimize solution replacement. Ferric chloride, cupric chloride, and chromic-sulfuric acids are commonly used in this type of etcher. Ammonium persulfate is not used because of the danger of overheating.

FIG. 14.10 Splash etcher design.

Slow etch rates have accounted for the obsolescence of this method in favor of faster, automatic spray-etching machines.

14.8.4 Spray Etching

Spray techniques include single- and double-sided etching with either horizontal or vertical positioning of the boards. These techniques yield high etch factors and short etching times, due in part to high solution controls and to the introduction of high quantities of air.[60–63] As in all etching procedures, however, the highest definition (fine-line patterns) or the use of thicker copper foils requires control of the undercutting by careful selection of equipment and etchants.

Spray-etching machines have evolved simultaneously with the availability of chemical-resistant metals and plastics essential to their construction (PVC and titanium alloys). Titanium is suitable for constant use in all common etchants except sulfuric acid–hydrogen peroxide, which requires stainless steel materials. Polycarbonate, polypropylene, and Hastelloy C alloys are also used.

An important caution is that PVC will distort when heated at temperatures above 130°F. PVC pipes used to transport hot solutions may expand and stretch to the point that moving parts in certain spray systems become locked and stop working.

A simple etching machine has a box-type chamber and a sump below. The solution is pumped from the sump and discharged from the spray nozzle onto the board surface. Thus, fresh solution is constantly impinged against the surface, which results in rapid etching. Evenness of etch depends on uniformity of spray pattern, force, drainage, and factors peculiar to the material being etched, as well as to pattern configuration.

Double-sided etching machines are common in production. To be effective, distances and pressures must be equal on both sides of the boards. A filter pump is needed to prevent metal slivers or crystallized salts from getting into the pump and plugging the nozzles. Spray patterns and pressure vary with nozzle and pump wear. Maintenance and repair are relatively easy, since machines are well built and designed for regular maintenance. Pressure gauges and regulators for each nozzle bank are available and are essential for even etching. Modifications to standard, off-the-shelf models can be made by using well-positioned, matching spray nozzles, with provisions for direct-temperature-readout thermostats which regulate precisely the current to heaters.

Ammoniated alkaline etchants, if allowed to crystallize excessively, may cause problems with plugged nozzles and cause damage and excessive wear of working parts of the machines.

14.8.4.1 Vertical Etching. Vertical etching is carried out by placing panels in a rack similar to a drawer frame that sits tightly when lowered into the spray-box area. Design of a vertical spray etcher is shown in Fig. 14.11.

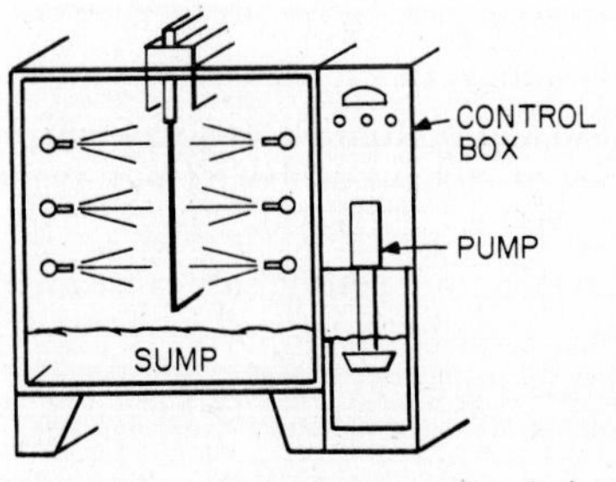

FIG. 14.11 Vertical spray etcher design.

More even etch is obtained if the nozzles are oscillated during the etch cycle. That is evidenced by the pattern of the spray breaking through the metal cladding (copper) sooner than at areas where the spray is less direct. A combination of nozzle movement and oscillation up and down or sideways, with a large number of nozzles with broad, solid patterns, provides optimum results. Some unevenness of etch may be produced by drawing of the etchant to the bottom of the board. This unevenness can be minimized by rotating the board during the etch cycle. Safety features must be provided when etchers are opened.[10] Machines of this type do not have built-in rinsing or neutralizing capability. Although shorter etching times are possible than with the bubble etchers, only a few boards or panels can be etched at one time. Most vertical etching machines are filled with etchant from the top.

14.8.4.2 Automatic Vertical Etching. This type of machine is designed for higher production rates. A mechanism carries a loaded rack through the etch chamber, where it is sprayed on one or both sides by oscillating banks of spray nozzles. The rack goes through water (spray-rinse) and neutralizing chambers. Cooling coils are available for sulfuric-peroxide etchants. Control of pressure to each bank of spray nozzles and on-off valves provide additional versatility.

The etchant sump has a larger volume of etchant than drawer-type vertical etchers, as well as a capability for continuous replenishment. Some potential problems of this machine are:

1. The rinsewater cycle may cause dilution of the etchant.
2. The mechanism used to feed the loaded racks suffers much wear and requires maintenance.
3. The feed mechanisms will be damaged if the loaded racks are not placed correctly into the machine.
4. Strong ventilation is required, since etchant fumes come out of the etcher when racks are entering or leaving the chambers.

Fine-line etching is attainable when spray nozzles, pressure, speed, and other variables are working optimally.

*14.8.4.3 **Horizontal Etching.*** The single-sided variety of horizontal etching was used for many years in the photoengraving arts prior to its application to PC manufacture. Its principal use was in fine-line and dot-printing plates. The principles of the upward vigorous spray and the use of additives to reduce undercut and increase etch factors were developed in the photoengraving trade and are now of use in printed board processing.

Double-sided horizontal etchers are generally preferred in PC manufacturing, since the majority of the boards are two-sided. The etcher is available with a drawer-type holding rack and also has a built-in sink at the left of the chamber. The etch operation proceeds and automatically pushes the rack into the rinse area after a timed cycle. Etching is done from independently controlled spray-nozzle banks at the top and bottom. Desirable features in horizontal etching machines include the following:

1. Double-sided etch capability.
2. Single-side (face-down) etching for very fine line work.
3. Built-in rinse area and automatic control; top and bottom water spray which flushes both sides at once and avoids drying of etchant on board and over-etching. Handling is eliminated.
4. Positive drawer seals from sink to prevent water from getting into rinse area during etching.
5. Sealed-in etch chamber, preferred because it requires no venting even when heated to 100°F.
6. Automatic cooling coils that prevent overheating and maintain constant temperature (for sulfuric acid–hydrogen peroxide and persulfate).
7. Independent gauges, pressure valves, and switching that allow compensation for differences between top and bottom etching.
8. Oscillating action and a large number of spray nozzles that result in even etching over the entire area.
9. Built-in transfer pump with intake going into the etchant sump and output into rinse tank, allowing for filling or emptying without spillage, damage, or hazard. The pump should be electrically reversible for both filling and emptying.
10. Minute- and second-timer switches for careful control of etch time.
11. Titanium heaters that withstand shock cracking.
12. Electrical controls well protected physically and electrically.
13. Accessibility to all parts of machine.
14. Simple yet effective mechanical linkages.
15. Filters at intake side of spray pumps that prevent solids from damaging or clogging pump and spray nozzles.
16. Screen insert tray for placing thin metal sheet prior to etching. Two screens are sometimes desired.
17. Structural soundness which allows moving or shipping, yet permits easy dismantling for repair.
18. Chemical and temperature stability of construction materials.

*14.8.4.4 **Automatic Horizontal Etching Machines.*** Made for high-volume production, these machines incorporate the features listed above for horizontal machines, plus the advantages of conveyorized loading and handling of boards up to 36 in wide and indefinite length, as well as built-in rinsing and neutralizing. In operation, machines are loaded by laying boards flat on an open horizontal conveyor belt which carries them progressively through the etch chamber and subsequent rinses. Rollers on conveyor belts are spaced so as to allow the bottom spray to reach the board. Automatic equipment is available for flow-through solution replenishment, which gives constant etch rates (see Fig. 14.12).

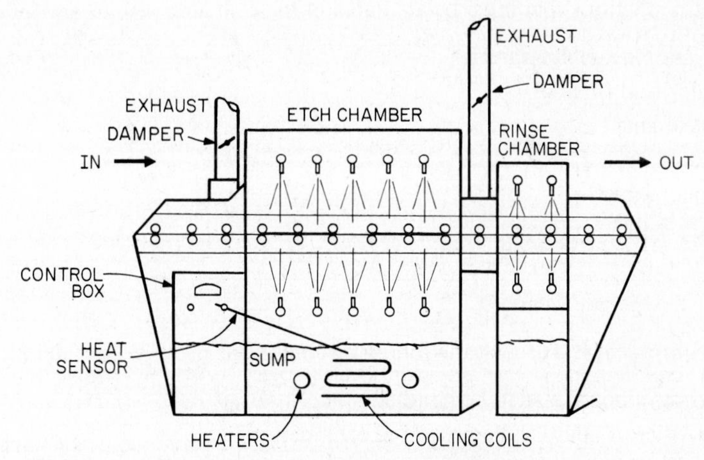

FIG. 14.12 Horizontal conveyorized spray etcher design.

Some limitations of this type of machine include the following:

1. The machines represent a considerable investment. Installation, plumbing, electrical work, and venting also are costly. These machines take a large amount of floor space, and to be economical they must not be idle.
2. Many moving parts, connections, gaskets and seals, hardware, etc., require competent maintenance.
3. Spray leaks reduce spray pressure. In addition, upper and lower sprays should hit the same area on opposite sides simultaneously. If not, small, lightweight boards can be tilted on the conveyor belt, and cause damage due to belt binding.
4. Vapor leaks when etching solutions are heated, so exhaust ducts must be provided for ammoniated alkaline etchants. Venting must be balanced with enough exhaust to keep fumes out of the etching room, but not to cause excessive loss of ammonia in the etchant.
5. Upper and lower pump pressures should have gauges; otherwise, adjustment to compensate for the differences between top and bottom etch action becomes a constant trial-and-error procedure.
6. Water drains must allow for proper drainage of rinsewaters. Water lines to spray units must have vacuum-breaking valves to prevent etchant from getting into water lines.
7. First water rinse coming out of etch chamber should be recirculated to help reduce waste treatment. In addition, several water rinse chambers are needed to reduce copper pollution.

14.8.4.5 Other Problems. Most mechanical problems are due to wear of moving parts. Sludging and crystallization also cause wear on mechanical parts. Process and equipment must be maintained in optimum condition.

14.9 ETCHING AREA REQUIREMENTS

The basic functions that must be available in the etch area include the following:

1. Pre- and post-etch cleaning
2. Etching and rinsing
3. Neutralizing
4. Resist stripping
5. Regeneration
6. Solder fusing (optional)

14.9.1 Equipment Selection

When equipment is selected, the following conditions must be considered:

1. Maximum board size to be handled
2. Quantities of board to be processed each day
3. Space available for etching
4. Type of etching to be done—for example, fine-line, wide-line, or print-and-etch only
5. Types of boards to be etched: metal-resist-plated, print-and-etch, etc.

The maximum board size to be handled will determine the size of tanks for etching, rinsing, cleaning, and neutralizing, as well as the size of the holding rack or conveyor. Extra room for easy loading and unloading should also be provided. Spray etchers require generous amounts of walk-around space.

The quantity of boards to be handled will also determine the type of etching equipment. It is essential that the equipment be able to complete the normal daily quota while allowing for solution adjusting time, machine maintenance, and actual hours of labor for etching. When large quantities of one type of board need to be processed, automatic equipment should be considered. Reduction of hand labor reduces costs, and final results will be more consistent. Using two machines offers the advantage of having two different etch solutions available at the same time.

Storage racks, bins, tables, and shelves are to be provided in each area where boards accumulate, as from plating, screening, photoresist coating, and etching. Boards must be dried prior to being placed in racks for storage.

Adequate electric power, exhaust systems, storage areas, and safety provisions are very important in an etching facility.

14.9.2 Facility

Floor construction should be chemical-resistant tile and mortar or acid-resistant epoxy-coated. An epoxy-fiberglass coating $\frac{1}{16}$ to $\frac{1}{8}$ in thick applied to cured and dry cement may be suitable. Subfloors of wood can be coated with similar plastic coatings. Tanks and pipes must be double-contained and kept off the floor to allow inspection and to keep them as dry as possible. Tanks should be supported well above the base floor and have proper inspection ports. Pipes, etc., run under the tanks, and aisles must be constructed of chemical-resistant materials to prevent corrosive attack. Spent solutions should never be put into a sewer. Spent chemicals and spills containing metal ions or other pollutants should be treated according to accepted pollution-control practice or disposed of off-site. Rinses must be waste-treated and discharged as acceptable effluents.

REFERENCES

1. E. Armstrong and E. F. Duffek, *Electronic Packaging and Production,* vol. 14, no. 10, October 1974, pp. 125–130.
2. S. W. Chaikin, C. E. McClelland, J. Janney, and S. Landsman, *Ind. Eng. Chem.,* vol. 51, 1959, pp. 305–308.
3. L. Fullwood, *Proceedings of the California Circuits Association Symposium,* San Francisco, May 1971.
4. W. S. Deforest, *Photoresist Materials and Processes,* McGraw-Hill Book Company, New York, 1975.
5. L. J. Slominski, U.S. Patent 3,466,208, 1969.
6. D. J. Sykes, E. Papaconstantinou, and K. Murski, U.S. Patent 3,868,485, February 1975.
7. J. G. Poor and G. F. Hsu, U.S. Patent 3,753,818, August 1973.
8. E. Laue, U.S. Patent 3,231,503, 1966.
9. E. King, U.S. Patent 3,705,061, 1972.
10. N. I. Sax, *Dangerous Properties of Industrial Materials,* rev. ed., Reinhold Publishing Corp., New York, 1957, p. 464.
11. R. W. Spinney, U.S. Patent 3,440,036, 1966.
12. *Solvent Extraction Technology,* Center for Professional Advancement, Somerville, N.J., 1975.
13. K. Murski and P. M. Wible, "Problem-Solving Processes for Resist Developing, Stripping, and Etching," *Insulation/Circuits,* February 1981.
14. H. Holden, *Proceedings of the California Circuits Association Symposium,* Irvine, October, 1975.
15. M. F. Good, "Surface Preparation of Circuit Boards for Plating—A New Approach," *Proceeding of the 1975 Spring Meeting of the Institute of Printed Circuits, Inc.,* Washington, D.C., 1975.
16. D. A. Luke, *Printed Circuit Fabrication,* vol. 8, no. 10, October 1985, pp. 63–76.
17. F. W. Kear, *Proceedings of the California Circuits Association,* Newport Beach, October 1979, pp. 20–26.
18. Shipley Company, *Printed Circuit Fabrication,* vol. 5, no. 2, February 1982, pp. 79–83.
19. A. G. Steger, *Printed Circuit Fabrication,* vol. 6, no. 10, January 1983, pp. 33–38.
20. W. Hastie, *Circuits Manufacturing,* vol. 23, no. 8, August 1983, pp. 25–28.
21. J. Tate, *Printed Circuit Fabrication,* vol. 7, no. 4, April 1984, pp. 22–30.
22. O. B. Dutkewych, C. Gaputis, and M. Gulla, U.S. Patent 4,130,454, 1978.
23. J. C. Solenberger, U.S. Patent 3,801,512, 1974.
24. M. L. Elias and M. F. Good, U.S. Patent 4,130,455, 1978.
25. B. J. Hogya and W. J. Tillis, U.S. Patent 3,476,624, 1969.
26. J. C. Swartzell, *Printed Circuit Fabrication,* vol. 5, no. 1, January 1982, pp. 42–47, 65.

27. G. Parikh, E. C. Gayer, and W. Willard, *Western Electric Engineer,* vol. XVI, no. 2, April 1972, pp. 2–8; *Metal Finishing,* March 1972, pp. 42, 43.
28. L. Missel and F. D. Murphy, *Metal Finishing,* December 1969, pp. 47–52, 58.
29. F. Gorman, "Regenerative Cupric Chloride Copper Etchant," *Proceedings of the California Circuits Association Meeting,* 1973; *Electronic Packaging and Production,* January 1974, pp. 43–46.
30. W. Thurmal, U.S. Patent 3,306,792, 1963.
31. P. D. Garn and L. H. Sharpe, U.S. Patent 2,964,453, 1960.
32. A. Jones and F. H. Haskins, U.S. Patent 3,083,129, 1963.
33. L. H. Sharpe and P. D. Garn, *Ind. Eng. Chem.,* vol. 51, 1959, pp. 293–298.
34. J. O. E. Clark, *Marconi Rev.,* vol. 24, no. 142, 1961, pp. 134–152.
35. O. D. Black and L. H. Cutler, *Ind. Eng. Chem.,* vol. 50, 1958, pp. 1539–1540.
36. Anonymous, *Circuits Manufacturing Magazine,* April 1972, pp. 55–57.
37. *Etching Metals with Ammonium Persulfate,* Food Machinery and Chemical Corporation, Princeton, N.J. See also *Tech. Bulletins III,* 52, 54, 55; *112,* 4, 29, 1, 35.
38. Ibid.; *Tech. Bulletin 110.*
39. P. A. Margulies and J. E. Kressbach, U.S. Patent 2,978,301, 1961.
40. E. B. Saubestre, *Ind. Eng. Chem.,* vol. 51, 1959, pp. 288–290.
41. W. F. Nekervis, *The Use of Ferric Chloride in the Etching of Copper,* Dow Chemical Co., Midland, Mich., 1962.
42. B. M. Schaffert, *Ferric Chloride Etching of Copper for Photoengravings,* Photo-Engravers Research, Inc., Columbus, Ohio, 1959.
43. P. M. Daugherty, U.S. Patent 3,144,368, August 1964.
44. J. R. Sayers and J. Smit, *Plating,* vol. 48, 1961, pp. 789–793.
45. E. C. Jubb, *Plating,* vol. 51, 1964, pp. 311–316.
46. H. S. Hoffman, "Methods of Producing Fine Lines in Printed Circuits," *Proceedings of the 1960 Fall Meeting of the Institute of Printed Circuits, Inc.,* Chicago, Ill., 1960.
47. R. Letize, *Printed Circuit Fabrication,* vol. 6, no. 2, February 1983, pp. 50–57.
48. R. D. Byers and J. G. Poor, U.S. Patent 3,351,555, 1967, and U.S. Patent 3,362,911, 1968.
49. P. M. Daugherty and H. C. Vaugh, U.S. Patent 3,033,725, 1962; J. Braday, L. Elston, and W. Burrows, U.S. Patent 3,033,793, 1962; L. Elston, U.S. Patent 3,148,100, 1964; P. Borth and J. McKeone, U.S. Patent 3,340,195, 1967.
50. T. D. Schlabach and B. A. Diggory, *Electrochem. Tech.,* vol. 2, 1964, pp. 118–121.
51. Anonymous, *Photoengravers Bull.,* vol. 36, no. 11, 1947, pp. 19–24.
52. S. Gowri, K. S. Indira, and B. T. Shemol, *Metal Finishing,* vol. 64, 1966, pp. 54–59.
53. L. J. Slominski, U.S. Patent 3,322,673, 1967.
54. F. Steward, *Proceedings of the California Circuits Association Symposium,* San Francisco, October 1974.
55. J. F. Battey, U.S. Patent 4,482,425, 1984.
56. N. J. Nelson, U.S. Patent 4,497,687, 1985.
57. F. T. Mansur and R. G. Autiello, *Insulation,* March 1968, pp. 58–61.
58. H. R. Johnson and J. W. Dini, *Insulation,* August 1975, p. 31.
59. P. F. Kury, *J. Electrochem. Soc.,* vol. 103, 1956, p. 257.
60. R. W. Lay, *Electronic Packaging and Production,* March 1982, pp. 65–83.
61. D. Ball and R. Markle, *Electronic Packaging and Production,* June 1983, pp. 74–76.
62. R. Keeler, *Electronic Packaging and Production,* November 1983, pp. 126–132.
63. H. Markstein, *Electronic Packaging and Production,* February 1985, pp. 168–171.

CHAPTER 15
POLLUTION CONTROL AND RECOVERY SYSTEMS

Peter G. Moleux, P.E.
Baker Brothers/Systems, Stoughton, Massachusetts

15.1 GENERAL INFORMATION

Both proposed and existing manufacturing facilities must meet specific pretreatment discharge requirements. Local municipalities may impose more stringent regulations than that issued by the United States Environmental Protection Agency (EPA). Due to the high cost of sludge disposal and the lack of available land disposal sites, every attempt must be made to recover metals and baths, recycle wastewaters and rinses, and reduce overall water consumption.

This chapter defines typical waste streams and proposes methods to recover both copper and process baths as part of the program to comply with the regulatory requirements.

The requirements for disposal of hazardous waste (i.e., sludge from a metal-finishing treatment system) will continue to be revised, and, therefore, the most recent publications should be consulted. The material in this section, however, is expected to represent the basic materials used in the printed wiring fabrication industry.

15.2 REGULATORY REQUIREMENTS IN THE UNITED STATES

15.2.1 Wastewater-Treatment Regulations

The "pretreatment" requirements for a company must be defined by the city (assuming that the pretreated wastes discharge into a municipal sanitary sewer) which receives that waste. Both the state and federal EPA will define specific regulations for companies with a direct discharge into a river, stream, or underground water system.

All regulatory agencies refer to the minimal EPA federal regulations, but they can impose more stringent regulations. The federal regulations are listed in the United States Code of The Federal Regulations, Chapter 40, Section 433 and Section 413 (for independent printed wiring board manufacturers and job shops).

The detailed regulations are available from the U.S. Government Book Stores and are defined in Table 15.1. All concentrations stated (so-called total metal) include the dissolved plus the suspended concentrations. The compliance date for the metal finishing regulations was February 15, 1986.

Most existing regulations use concentration as the basis to determine compliance. Some requirements are defined on a mass basis (e.g., pounds of copper per day).

For waterlike solutions, the mass of copper discharged per day is directly related to the discharge concentration (in milligrams per liter) and to the daily flow (in millions of gallons per day) by the following formula.

$$\text{lb/day} = \text{concentration} \times 8.34 \times \text{the daily flow}$$

If you know the expected flow produced from a facility in one day and the allowable concentration, you can determine the mass of a specific pollutant to be discharged.

A municipality may grant an industry a variance from the EPA pretreatment standards if this is allowed by the federal EPA. This would allow companies to meet less stringent requirements than those specified in the categorical pretreatment standards.

Regulations for subsurface disposal (i.e., after pretreatment the waste would discharge into a leaching field) are now being formulated and could be identical to the primary and secondary drinking water standards. These very stringent regulations cover more parameters than listed in the metal-finishing regulations. For example, sulfates, chlorides, and fluorides are parameters which could be regulated under certain circumstances.

The EPA has identified (see 40 CFR 433) certain organics in wastewaters classified as "total toxic organics" (TTO*). These are defined in the following list:

Acenaphthene
Acrolein
Acrylonitrile
Benzene
Benzidine
Carbon tetrachloride (tetrachloromethane)
Chlorobenzene
1,2,4-trichlorobenzene
Hexachlorobenzene
1,2-dichloroethane
1,1,1-trichloroethane
Hexachloroethane
1,1-dichloroethane
1,1,2-trichloroethane
1,1,2,2-tetrachloroethane
Chloroethane
Bis(2-chloroethyl)ether
2-chloroethyl vinyl ether (mixed)
2-chloronaphthalene
2,4,6-trichlorophenol
Parachlorometa cresol
Chloroform (trichloromethane)
2-chlorophenol
1,2-dichlorobenzene
1,3-dichlorobenzene
1,4-dichlorobenzene
3,3-dichlorobenzidine
1,1-dichloroethylene
1,2-trans-dichloroethylene
2,4-dichlorophenol
1,2-dichloropropane (1,3-dichloropropene)
2,4-dimethylphenol
2,4-dinitrotoluene
2,6-dinitrotoluene

*The term "TTO" stands for "total toxic organics," which is the summation of all quantifiable concentrations greater than 0.01 mg/L for the toxic organics in the list.

1,2-diphenylhydrazine
Ethylbenzene
Fluoranthene
4-chlorophenyl phenyl ether
4-bromophenyl phenyl ether
Bis(2-chloroisopropyl)ether
Bis(2-chloroethoxy)methane
Methylene chloride (dichloromethane)
Methyl chloride (chloromethane)
Methyl bromide (bromomethane)
Bromoform (tribromomethane)
Dichlorobromomethane
Chlorodibromomethane
Hexachlorobutadiene
Hexachlorocyclopentadiene
Isophorone
Naphthalene
Nitrobenzene
2-nitrophenol
4-nitrophenol
2,4-dinitrophenol
4,6-dinitro-o-cresol
N-nitrosodimethylamine
N-nitrosodiphenylamine
N-nitrosodi-*n*-propylamine
Pentachlorophenol
Phenol
Bis(2-ethylhexyl)phthalate
Butyl benzyl phthalate
Di-*n*-butyl phthalate
Di-*n*-octyl phthalate
Diethyl phthalate
Dimethyl phthalate
1,2-benzanthracene (benzo[*a*]anthracene)
Benzo[*a*]pyrene (3,4-benzopyrene)
3,4-benzofluoranthene (benzo[*b*]fluoranthene)
11,12-benzofluoranthene (benzo[*k*] fluoranthene)
Chrysene
Acenaphthylene
Anthracene
1,12-benzoperylene (benzo[*ghi*]perylene)
Fluorene
Phenanthrene
1,2,5,6-dibenzanthracene (dibenzo [*a.h*]anthracene)
Indeno (1,2,3-*cd*) pyrene (2,3-*O*-phenylene pyrene)
Pyrene
Telrachloroethylene
Toluene
Trichloroethylene
Vinyl chloride (chloroethylene)
Aldrin
Dieldrin
Chlordane (technical mixture and metabolites)
4,4-DDT
4,4-DDE (*p.p*-DDX)
4,4-DDD (*p.p*-TDE)
Alpha-endosulfan
Beta-endosulfan
Endosulfan sulfate
Endrin
Endrin aldehyde
Heptachlor
Heptachlor epoxide
(BHC-hexachlorocyclohexane)
Alpha-BHC
Beta-BHC
Gamma-BHC
Delta-BHC
(PCB-polychlorinated biphenyls)
PCB-1242 (Arochlor 1242)
PCB-1254 (Arochlor 1254)
PCB-1221 (Arochlor 1221)
PCB-1232 (Arochlor 1232)
PCB-1248 (Arochlor 1248)
PCB-1260 (Arochlor 1260)
PCB-1016 (Arochlor 1016)
Toxaphene
2,3,7,8-tetrachlorodibenzo-*p*-dioxin (TCDD)

The EPA requires that the sum of the concentrations for every item listed must be less than 2.13 mg/L. Compliance is determined by a statement from a company (remembering your liability) or by submitting chemical analysis.

TABLE 15.1 EPA Regulations Impacting PWB Manufacturers

1-28-81 Electroplating regulations			Metal-finishing regulations as of 7-15-83									
Independent PWB manufacturers,[a] pretreatment standards—dischargers to POTWs			Captive and independent[b] PWB manufacturers that are direct dischargers to a stream						Captive PWB manufacturers, pretreatment standards—dischargers to POTWs			
PSES[c]		Parameter	BPT[d]		BAT		NSPS		PSES		PSNS	
1-day max.	1-day avg.	Conc., mg/L	1-day max.	30-day avg.	1-day max.	1-day avg.	1-day max.	30-day avg.	1-day max.	4-day avg.	1-day max.	30-day avg.
1.2	0.7	Cadmium	0.69	0.26	0.69	0.26	0.11	0.07	0.69	0.26	0.11	0.07
7.0	4.0	Chrome[e]	2.77	1.71	2.77	1.71	2.77	1.71	2.77	1.71	2.77	1.71
4.5	2.7	Copper[e]	3.38	2.07	3.38	2.07	3.38	2.07	3.38	2.07	3.38	2.07
0.6	0.4	Lead	0.69	0.43	0.69	0.43	0.69	0.43	0.69	0.43	0.69	0.43
4.1	2.6	Nickel[e]	3.98	2.38	3.98	2.38	3.98	2.38	3.98	2.38	3.98	2.38
1.2	—	Silver	0.43	0.24	0.43	0.24	0.43	0.24	0.43	0.24	0.43	0.24
4.2	2.6	Zinc[e]	2.61	1.48	2.61	1.48	2.61	1.48	2.61	1.48	2.61	1.48
1.9	1.0	Cyanide (T)[f]	1.20	0.65	1.20	0.65	1.20	0.65	1.20	0.65	1.20	0.65
		Cyanide (A)[f]	0.86	0.32	0.86	0.32	0.86	0.32	0.86	0.32	0.86	0.32
10.5	6.6	Total regulated metals[e]										
2.13	—	TTO[a]	2.13	—	2.13	—	2.13	—	2.13	—	2.13	
		TSS	60.00	31.00			60.00	31.00				
		Oil and grease	52.00	26.00			52.00	26.00				
		pH	6–9	6–9			6–9	6–9				

Metals and cyanide nonintegrated facilities, 4-27-84 Metals and cyanide nonintegrated facilities 6-30-84 TTO standard, 7-15-86	Compliance dates	BPT requirements, ASAP	BAT requirements, 7-1-84	NSPS requirements upon commencement of discharge	Electroplaters non-wt. M&C, 4-27-84 Electroplaters integ. M&C, 6-30-84 Interim TTO Std. of 4.57 mg/L, 6-30-84 PSES metals and cyanide, 2-15-86 Final TTO std. of 2.13 mg/L, 2-15-86	PSNS requirements upon commencement of discharge

[a]An independent PWB manufacturing facility manufactures PWBs principally for sale to other companies.

[b]Differences between captive and independent limits: (1) captives will probably require daily monitoring; (2) cyanide (T) for captives is 0.65 mg/L 30-day average vs. electroplater's (independent) of 1.0 mg/L 4-day average; (3) chrome limit is 1.71 vs. 4.0 mg/L; (4) copper limit is 2.07 vs. 2.7 mg/L; (5) nickel limit is 2.38 vs. 2.6 mg/L; (6) silver limit is 0.43 (max.) vs. 1.2 (max.) mg/L; (7) zinc limit is 1.48 vs. 2.6 mg/L.

[c]Technology assumed for PSES—stream segregation for cyanide and hex chrome, cyanide destruct, chrome reduction, and final metal removal by precipitation/clarification plus solvent waste segregation.

[d]BPT = best practicable control technology, BAT = best available technology economically achievable, NSPS = new source performance standard (direct discharge), PSES = performance standard existing sources, PSNS = performance standard new sources, TTO = total toxic organics.

[e]Total regulated metals = chrome + copper + nickel + zinc.

[f]Cyanide—measured at end of pipe for electroplater's regulations and after cyanide treatment for all others. Cyanide (T)—total cyanide. Cyanide (A)—cyanide amenable to alkaline chlorination (to be negotiated with local POTWs).

[g]Additional monitoring requirements for TTO: (1) baseline monitoring for toxics reasonably expected to be present, (2) certification of compliance statement, (3) solvent management plan.

Source: Institute for Interconnecting and Packaging Electronic Circuits (IPC).

15.2.2 Air Pollution Requirements

Air pollution requirements vary from location to location. Circuit board manufacturers make provisions to exhaust the fumes from a plating room and selected process tanks and sometimes scrub all or a portion of the exhausted air to satisfy neighbors' fears of contamination as well as to meet specific requirements. The necessity to scrub the exhausted air from selected baths must be defined. The costs may be obtained from a fume scrubber supplier. Each manufacturer should contact the air pollution control district to determine the specific requirements.

15.2.3 Right-to-Know Laws

Every company must obtain material safety data sheets (MSDS) for every chemical purchased. From each chemical supplier, every company must obtain written verification, whether the items listed in Section 15.2.1 are present or not. These data sheets may state whether any complexing or chelating agents (such as amines) are present in the bath. The MSDS are required to satisfy the Right-to-Know Laws. These laws state that a company must identify what types of hazards exist with chemicals which are used, and each company must list procedures to be taken if misuse of the chemical occurs. Some states require each company to submit copies of each MSDS.

15.2.4 Hazardous Wastes: RCRA

Generally, the sludge produced from end of pipe line treatment systems is classified as hazardous wastes. The hazardous waste regulations are described in the rules and regulations (refer to the Code of the Federal Regulations, Chapter 40, Sections 260 through 280) of the Resource Conservation and Recovery Act (RCRA). These require circuit board manufacturers to do the following:

1. Obtain an EPA identification number. Request a permit to be a generator or a treatment, storage, and disposal facility of hazardous waste, as the case applies.

2. Obtain a permit if hazardous waste is treated or stored on site for more than 90 days or disposed of on site. Some recycled materials (e.g., spent ammonium etchant, spent cupric chloride etchant, gold solutions) that are returned to a chemical supplier will fall under this regulation.

3. Use appropriate containers for storage and disposal and approved manifests and labels for shipment.

4. Document via the manifest system that hazardous waste arrives at the designated and approved disposal facility.

5. Keep records for periodically reporting to the appropriate agencies.

This requirement applies to companies generating 100 kg (22 lb) per month of hazardous waste. Each circuit board manufacturer must be familiar with the requirements applicable in that state.

California, for example, enacted a law which specifies dates after which sludges containing toxic metals will not be allowed in landfills, assuming that adequate alternative technology exists for recovering these metals. With that in mind, the recovery of metals and process solutions must be more than just a goal.

The 1984 amendments to the U.S. Resource Conservation and Recovery Act imposed regulations for underground tanks (and their inlet and outlet pipes) containing regulated substances. Underground tanks have 10 percent or more of their volume underground. Companies with underground tanks (whether or not currently in use) are required to notify the designated state agency. These tanks are regulated by state and federal authorities to protect human health and the environment. As a minimum, monitoring wells, sampling and recording procedures, and leak-detection requirements must be followed. For proposed manufacturing facilities, we recommend that only aboveground storage tanks be used.

From November 1987 onward, no company may export hazardous waste out of the United States unless, among other items, the receiving country has agreed in writing to accept the waste.

All hazardous waste manifests must contain a generator certification that (1) the volume and/or quantity and toxicity of the waste has been reduced to the maximum degree economically practicable and that (2) the method used to manage the waste minimizes risk to the extent practicable.

This act also established new regulations for waste incinerators, land disposal facilities (requires a double liner and a leachate collection system), and surface impoundments.

15.3 PRELIMINARY INFORMATION REQUIRED BY DESIGNERS

To design either recovery or destruct-type systems, certain information must be obtained. This includes a floor plan of the manufacturing operations showing the location for each wet process, including laboratories, final inspection, cooling towers, water-cooled rectifiers, and fume scrubbers. The size, depth, and location of all building sewers, and the location of any monitoring facilities must be defined. Each source of liquid waste must be identified on the drawing. A numbering system should be used which identifies each bath, rinse, and source of liquid waste either continuously or periodically discharged. The flow rates from each rinse, the dumping frequencies of baths, and the operating volumes must be specified. The location of chemical storage facilities, receiving and shipping docks, and bulk containers for rubbish and hazardous wastes must be included on the drawing.

The clear ceiling height in both the wet process area and the potential area for waste treatment must be defined. Earthquake zone requirements, high groundwater table conditions, 100-year flood elevations, and special building requirements must also be addressed.

In smaller manufacturing operations, one rinse tank may be used to receive the drag-out from numerous process baths. If this is normal operating procedure, this must be identified. Where a rinse tank receives only the drag-out from one particular bath, this information must also be included on the table. Where the discharge from one rinse tank is the source of water to another rinse tank or operation, this fact must be noted.

Various process baths are dumped based upon the volume of boards processed through the line or when the concentration of copper exceeds a certain value. This should be noted on the table.

If the dumping frequency from each bath, its volume, and the concentration of copper (or other metals) in a bath just prior to dumping are known, the total pounds of copper produced from each bath (due to dumping) can be determined. If every rinse flow could be measured and analyzed for the concentration of copper, its contribution to the discharge can be determined. A chart should be prepared, by the process engineer, identifying all major sources of copper.

A work process flowchart showing the sequence in which parts travel through cleaners, rinses, activators, plating baths, treatment tanks, and conveyorized equipment should be provided.

An analysis taken at different times of the year from the various sources of intake water used should be provided. Occasionally, a municipality may provide water to its customers from one or more sources, depending upon weather conditions and economics. Municipalities are known to treat their water with various chemicals which may inhibit the rinsing technique proposed for your manufacturing facility. Hard water, for example, causes problems in rinse stations with fine spray nozzles due to the buildup of calcium carbonate. This will increase your maintenance costs unless a water softening process is included. Chemical analysis of the water is obtained from the city water department in each municipality. If water is extracted from a well, a local laboratory should be contacted to analyze the water.

All existing treatment processes and recovery systems must be well documented. This is important especially if the process engineer who purchased the equipment leaves the company. An itemized list (including quantities) of all major chemicals used, on an annual basis, should be provided. The facility water and sewer use rates must also be defined. The average and maximum daily flow of water must be defined.

Copies of all correspondence between the regulatory authorities and the company must be collated. Companies that are licensed to haul away hazardous waste and the hazardous waste-disposal facilities should be listed. Each hauler will define whether drums, bulk liquid, or large dumpsters are used to transport the wastes. This should definitely be determined in the design phase.

15.4 SELECTING AN EQUIPMENT SUPPLIER OR CONSULTANT

An "outsider" may be selected either to assist a company or to be in charge of the entire project. However, outside consultants should never be substituted for qualified in-house process or environmental engineers. Also, these in-house engineers should supervise the consultant or the equipment supplier. Without proper supervision and someone to review and implement the consultant's directives, the waste-treatment system may never operate properly.

Consultants can be either independents or specialists in the application of equipment to circuit board waste-treatment and recovery systems. Generally, independent consultants refer to one or more equipment suppliers with whom they are familiar for technical information prior to providing recommendations to a company. The equipment suppliers, on the other hand, are most familiar with their own equipment designs. When an equipment supplier is selected, that supplier must demonstrate adequate experience and knowledge for the entire manufacturing process, both process bath and metals recovery, and destruct-type waste-treatment technologies. Otherwise, numerous consultants (each perhaps requiring identical information but consuming more of the process engineer's time) may be required.

Do not assume that the consultant with the most experience is necessarily the ideal consultant for a particular client. Among the primary considerations for selecting a consultant are performance record, staff qualifications, ease in communicating ideas, and the degree of specialization. The consultant's ideas should be able to be challenged by the client.

Consultants must define who (in their organization) will be the project engineer. Consultants should list what laboratory capabilities are available at their facility to justify their recommendations. If a consultant must send samples outside the organization for analysis, the question of responsibility and liability for the waste-treatment system must be addressed. It is better to select consultants who have laboratories within their own facilities.

Novel and innovative techniques must be thoroughly tested. Consultants should review and approve all tests performed. The process engineer should evaluate the cost for this "breakthrough" and determine what provisions must be incorporated if the system fails.

The consultants must have a professional engineering license in the state or country in which your activity is to be conducted. (Occasionally both the corporation and the responsible engineer must be registered.) Usually, plans for a treatment system must be reviewed and approved by the appropriate regulatory agencies prior to ordering pollution-abatement equipment. This will require meetings between your consultant, your company, and the regulatory agencies to review the design.

A consultant may consume up to 6 months to design a complete treatment system, including all chemical feasibility studies. This period of time should be incorporated into your construction schedule.

The consultant or equipment supplier should, prior to receiving the equipment, supply a designated number of manuals including all electrical, mechanical, and process flow drawings; instruction and maintenance manuals; the location and phone number for all local equipment suppliers (for spare parts); and vendor literature.

The consultant should provide an equipment summary sheet. It should contain details about each individual piece of equipment—for example, its electrical and mechanical characteristics, its physical dimensions, the date that the supplier issued a purchase order to the manufacturer, the purchase order number, the scheduled date of receipt, and the actual date of receipt. Equipment warranties will depend upon the knowledge of this information.

The consultant or equipment supplier must specify how many days of training time are provided in the contract and when and for how long samples of treated water will be collected.

15.5 STAFFING YOUR COMPANY FOR WASTE MANAGEMENT AND RESOURCE CONTROL

15.5.1 General

There are three functions which can be classified as job descriptions to be performed by any manufacturing organization. These can be loosely defined as an environmental manager, a process engineer, and a waste-treatment operator.

15.5.2 Environmental Manager

Environmental managers must approve all new process equipment and all new process chemicals after determining their impact on the recovery and wastewater-treatment systems. They must coordinate the solution to any problems in the waste-treatment and recovery systems with the production manager and key operating personnel. They must keep up-to-date on all regulatory agency decisions and rules and regulations, be responsible for the periodic retraining of the waste-treatment operators, know the design bases used for the waste-treatment and recovery systems, be familiar with the chemistry and equipment in the PC shop, continually review methods (with the production people) to reduce the operating costs in the waste-treatment area, and review why changes occurred in the quality of the treated or untreated waste. For new facilities, managers must hire the consultant or equipment supplier and coordinate all activities between their company and the regulatory authorities.

15.5.3 Process Engineer

Process engineers must review existing process equipment to find ways to minimize the drag-out and reduce the rinse flow rate; ensure that all new or anticipated equipment or process chemicals are thoroughly reviewed for their impact on the existing recovery and wastewater-treatment systems; ensure that the existing process equipment is properly maintained to minimize floor spills; investigate and test all new recovery equipment available; propose the use of new recovery equipment to minimize the sludge haul-away cost; review, approve, and maintain engineering prints showing how the waste from each specific process equipment is plumbed to the recovery and wastewater-treatment systems; and order all spare parts for the waste-treatment and recovery systems. Process engineers should coordinate visits between their company and a hazardous-waste company to review methods for spill cleanup. For new facilities, they must decide how the waste will be collected and whether or not a basement will be provided. They must determine the factor of safety that the equipment supplier must use.

15.5.4 Waste-Treatment Operator

Waste-treatment operators should have a chemical and mechanical background and may have to be licensed. Operators should maintain all chemical supplies, coordinate with the production people when process baths may be dumped, prepare manifests to transport wastes, maintain the recovery and wastewater-treatment systems, periodically check the inventory of spare parts, monitor the chemical analysis of the waste coming into and leaving the wastewater-treatment and recovery systems, coordinate any floor spill cleanup problems, and be responsible for cleaning the floor and surrounding area associated with the waste-treatment and recovery systems.

15.6 INTRODUCTION TO WASTE TREATMENT

This section will define the chemical analysis and the treatment and recovery of typical wastes found in a printed circuit operation. It will identify each waste and recommend recovery or treatment techniques.

Waste can be either metal-bearing or non-metal-bearing. If the maximum concentration of copper or other ions in a particular waste is less than the existing pretreatment standard, this waste would be considered low- or non-metal-bearing. Likewise, for those rinse streams whose maximum concentration is equivalent to or greater than the pretreatment standard, this waste would be considered metal-bearing.

Wastewaters are divided into process baths, which are periodically dumped, and rinsewaters. Process bath dumps can be further categorized into metal-containing acid, metal-containing alkali, chrome-containing, non-metal-containing alkali, and chelated metal-containing dumps. Sometimes the latter category may have subdivisions due to the incompatibility of these wastes in a single treatment step. Typical rinsewater classifications include metal-containing rinses (where the minimum concentration of copper or other ions is higher than the concentration required in the discharge), low-metal-containing rinses, chrome rinses, and chelated metal-containing rinses.

15.7 MAJOR SOURCES OF COPPER

The major contributors of copper to the discharge from a circuit board manufacturer are deburrers, board scrubbers, fume scrubbers, the bath and the rinses following microetching, inner- and outer-layer etching, rack stripping, floor spills, the rinses following copper electroplating, the electroless-copper bath, its solution growth, and its rinses. Following these are the acidic cleaners.

The purposes for analyzing every bath and rinse are to (1) determine which wastes are metal-bearing and which are non-metal-bearing, (2) determine the major sources of copper and other metals, (3) size a recovery or destruct system properly, (4) control quality in a circuit board facility, and (5) determine the appropriate dilution ratio so that an untreated rinsewater can be simulated in a treatability study. Periodically, all baths and rinses should be analyzed even after recovery and treatment systems have become operational.

15.8 THE TREATABILITY STUDY

The purpose of the study is to simulate the treatment process or recovery system using "actual" samples. Treatability studies are the basis for a system performance warranty.

The performance warranty differentiates the independent consultant from the equipment (or systems) supplier. Independent consultants can only assume that the equipment they select will be purchased by the circuit board manufacturer. On the other hand, equipment suppliers can certify that their equipment will meet the guaranteed limits based upon the results of a treatability study.

For production facilities in the planning stage, all potential chemical suppliers must submit representative samples of used baths for the treatability study. Existing companies should use representative samples from their own operations for the study. Once an analysis is made for all the baths and rinses, the instructions for a treatability study are generated.

The maximum concentrations desired to be met in the laboratory should be less than one-half of the concentrations to be guaranteed.

To prepare a treatability study, one needs the drag-out ratio, the design rinse flow rates, and the bath dumping frequency. Wastewater samples are collected from each bath (when contaminant levels are highest) and from each rinse tank, just after the panels have exited the rinse tank.

The waste to be treated originates from every concentrated bath, whether it is a plating bath or a mild acid bath. The total waste load from a bath is the sum of the solution drag-out volume plus the volume of the bath which is dumped over the same period of time, where appropriate. The instructions for the treatability study define the volume from each concentrated bath to be added together in tap water in order to simulate the worst case (typically 50 to 200 ppm of copper) of incoming waste load to a treatment system. A typical calculation is as follows.

A rinse flow of 2 gal/min equals 4800 gal per week, assuming that it runs continuously for 8 h per day. If the average concentration of copper in the bath is 5000 mg/L and the average concentration of copper in the rinse is 10 mg/L, the dilution ratio is 500 to 1. The volume of drag-out to be treated is calculated by:

$$\frac{5000}{10} = \frac{4800}{X} \qquad \text{or} \qquad X = 9.6 \text{ gal/week of drag-out to be treated}$$

The contribution from all the baths would be analyzed in a similar fashion. A typical treatability study for the electroless-copper line of a company is shown in Table 15.2. Please note that if a waste is classified as a nonmetal, it is not added to this list. To determine the final discharge, the treated waste from the metals removal system is blended with the nonmetal waste, and the combination is analyzed to determine the anticipated effluent produced in the laboratory.

After a treatability study has been conducted, one can prepare a waste segregation drawing. This drawing identifies the destination of each process bath dump, each rinse, and the location for recovery systems. A typical schematic drawing is shown in Fig. 15.1.

Due to the autocatalytic nature of electroless copper and nickel, any samples, for the purpose of the treatability study, should be of the original components and not of the bath in use. These components should be combined just before the study so that the copper does not "plate" out of the solution. In other cases, ammonium ions, hydrogen peroxide, and other organics may volatilize from the sample.

All sampling techniques for purposes of chemical analysis are specified in the latest edition of *Standard Methods for the Examination of Water and Wastewater.** Care must be used in collecting samples.

*Prepared and published jointly by The American Public Health Association, The American Water Works Association, and The Water Pollution Control Federation, 1015 15th Street, N.W., Washington, D.C. 20005.

TABLE 15.2 Typical Treatability Study for the Electroless-Copper Line

Sample, electroless Cu/line	Rinse, gal/min	Average weekly dump, gal/week	Drag-out, gal/week	Sum to be treated, gal/week	%
MacDermid G-5B	3	100	22	122	19.3
	—				
H_2SO_4, 10%		100	14	114	18.0
	—				
MacDermid 9008		100	—	100	15.8
MacDermid 9070M	3	25	22	47	7.0
	—				
MacDermid 9074		50	22	72	11.0
H_2SO_4, 10%	3	100	14	114	18.0
HNO_3, 50%	3	50	14	64	10.9
Total	12	525	108	633	100%

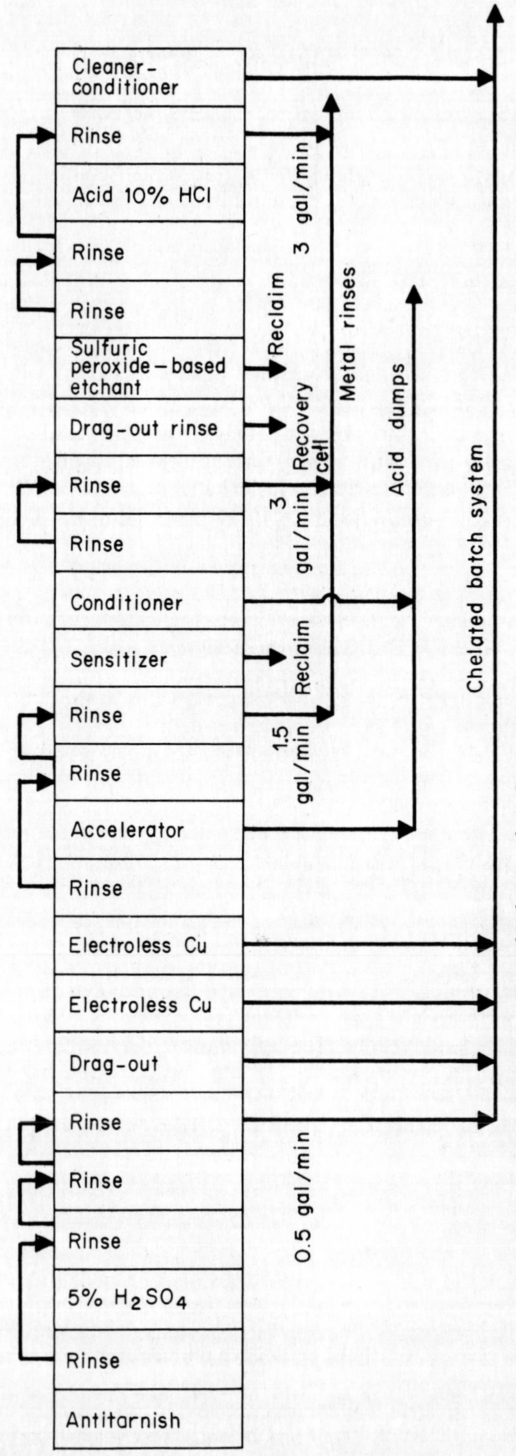

FIG. 15.1 Electroless-copper segregation. (*Courtesy of Baker Brothers/Systems.*)

15.9 RECOVERY AND WATER REUSE TECHNIQUES

Due to existing environmental regulations, the closing of hazardous waste landfills, increasing financial liability, and increased costs of disposing of sludge and liquid wastes, companies are forced to recover baths, metals, and rinsewaters. Several techniques for recovery are presented below.

15.9.1 Water Use Reduction Techniques

One of the most effective techniques to reduce the water consumed in any plant is to use *counterflow rinses* plus *flow restrictors*. In an automatic plating machine, a well-mixed single-stage rinse tank with air agitation uses water at a rate of 5 to 7 gal/min. A two-stage counterflow rinse uses approximately 3 to 5 gal/min. A three-stage counterflow rinse tank uses between ½ to 1 gal/min. Water reduction is obtained by substituting counterflow rinse tanks for single-station rinse tanks at the cost of additional floor space and, perhaps, productivity. However, where the installed treatment equipment cost can be over $2500 per gal/min, it pays to keep the flow rates low.

The best technique for maintaining a rinsewater flow rate is to use flow restrictors. Flow restrictors are available with flexible diaphragms and are rated for a specific flow rate in a specific pipe size. This assumes a minimum water pressure of about 20 psig. With flow restrictors, the designer can be sure that the flow rate will not be exceeded unless someone changes the flow restrictor. Without a flow restrictor, anyone could increase the flow rate of water in the rinse tank, possibly causing hydraulic problems in the treatment system.

Pulsating spray rinses are used to reduce the total water demand for a facility. The rinsewater pulsates on each side of a panel at different times so that penetration through holes can be achieved. This may prove difficult where very small holes are present. The effectiveness of spray rinsing depends upon the geometry of the board, the water quality, and the degree of rinsing required.

15.9.2 Deburrers and Board Scrubbers

There are four filtration systems which are used to remove particles from the wash water from deburrers and board scrubbers so it becomes suitable for up to 100 percent recycling, depending upon the application. These include a centrifuge, a paper filter, a sand filter, and gravity settling and filtration techniques, each of which has advantages and disadvantages. One or more process machines can be connected to one filtration system. Other filtration systems are available for pumice scrubbers.

15.9.3 Evaporative Recovery

Evaporative recovery is a technique that uses heat to concentrate contaminated rinses. Water molecules driven off as steam are cooled to produce contaminant-free water suitable for reuse. The remaining concentrate can be recovered or treated in a destruct system. High energy costs, however, have tended to limit its use for copper recovery in the printed circuit industry.

This technique is used to recover chromic acid in desmear operations from the rinsewater following that bath. The rinsewaters are recycled. Impurities (trivalent chrome) should be removed from the bath by using at least a cation exchanger. The rinses following concentrated sulfuric acid, alkaline permanganate, and phosphoric-nitric acid are usually classified as low-metal-containing, although the spent process baths have been reported to contain up to 300 mg/L of copper.

Metal recovery systems are sized by knowing the volume of drag-out and concentration of metals in that drag-out. Usually, systems are sized on the basis of the grams of metal to be recovered per hour. For proposed facilities, we use a suitable drag-out rate such as 5 to 7 mL per surface square foot of panel area.

Drag-out reduction is achieved by having the plater or a hoist hold the panels up in the air over the process tank for a period of time before proceeding to the adjacent rinse tank. The amount of time depends upon the process being used. Drag-out may be reduced, where appropriate, by the use of air knives on the hoist or on the process tank to blow off the drag-out from the rack and panels prior to proceeding to the next step. Air knives are practical in operations when the panels are stacked in one plane. In the electroless process where a number of panels are stacked side by side, the use of air knives may be impractical. The use of an air knife has been demonstrated* to reduce the drag-out by 50 percent from that before the use of an air knife. Typical drag-out rates are 5 to 7 mL of drag-out per surface square foot of panel. An 18 × 24-in panel contains 6 surface square feet. The use of an air blowoff reduces the drag-out rate to approximately 2.5 to 3.5 mL per surface square foot. The use of high-pressure water sprays plus the use of air knives has reduced the volume of rinsewater required from an average of 3 gal/min for a two-stage counterflow rinse tank in a line producing 100 panels per hour to less than 1 gal/min for the same application. The sprays would only run when the panels are in the rinse chamber or are entering or leaving the rinse chamber.

Drag-out rinse tanks are required for some recovery systems and where one desires to minimize the volume of chelated wastes to be treated. As a general rule, 90 percent of the drag-out from a bath is captured in the first drag-out rinse tank.

Figure 15.2 shows how a drag-out station is used in combination with a parallel recovery cell to maintain the concentration of copper in the drag-out station at 1 g/L. Without this setup, the two-station counterflow rinse would receive a solution containing a concentration of 22.5 g of copper per liter. With the recovery system, the counterflow rinse receives a solution containing 1 g of copper per liter (or a 95.5 percent reduction in the quantity of copper to be treated).

Figure 15.3 shows a slow-flowing drag-out rinse in combination with a recovery system. If the recovery system is directly connected to the drag-out rinse, the copper concentration in that rinse could be maintained at about 200 mg/L.

The drag-out bath adjacent to the process bath (Fig. 15.4) is periodically dumped to either a recovery or a destruct system. The frequency of dumping depends upon the concentration of copper in the flowing rinse. As long as the concentration of copper in the rinse is less than the pretreatment standard, the drag-out bath does not have to be dumped. Once the concentration of copper in the rinse approaches 1 to 2 mg/L, the first drag-out bath is dumped, and the second drag-out bath is transferred into the first tank. Fresh water is added to the second drag-out tank.

Rinses in the electroless plating line may be selectively reused without passing through a treatment system. In Fig. 15.1, observe that selected rinses are recycled within the plating line. Refer to your particular chemical supplier to ensure that adverse effects will not occur using this procedure with your water quality.

---

*S. Siniscalchi, "Improved Rinsing Techniques for Automatic Pattern Plating," *Printed Circuit Fabrication,* vol. 7, no. 7, July 1984.

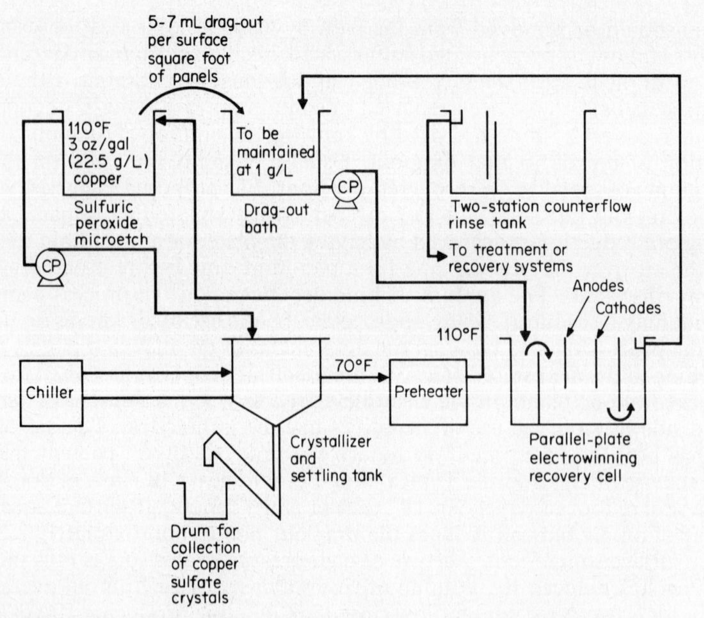

FIG. 15.2 Microetch recovery systems. CP = centrifugal pump. *(Courtesy of Baker Brothers/Systems.)*

15.9.4 Electrolytic Recovery

Electrolytic recovery systems are used in many applications. They are used to remove copper from the various microetching baths, their drag-out rinses, the drag-out rinse following acid copper, the rinse following electroless copper, and the various dilute sulfuric acid baths. Parallel-plate cells are used for reducing high (over 2 g/L) concentrations down to 0.5 g/L. Expanded mesh systems can be used to reduce the concentration to less than 1 mg/L in certain cases. Separate electrolytic modules are used to recover lead.

Drag-out rinses are recirculated to and from a single recovery cell, or they can be fed with a slow-flowing (e.g., $\frac{1}{2}$ gal/min) rinse. In the latter case, this rinse (see Fig. 15.3) and other slow-flowing drag-out rinses may flow into a collection tank. The contents of that tank would be recirculated to and from a plate cell, while a separate collection tank would receive fresh waste. When the copper concentration has been reduced to less than 1 mg/L, the contents of the collection tank are transferred directly to the final pH-adjustment system in this sludge-free approach.

In Fig. 15.2, a parallel-plate electrolytic cell is used to maintain the concentration of copper in the drag-out station (following a microetching bath) at 1 g/L. A high cathode current efficiency (over 95 percent) is obtained by using either copper foil or electropolished 316 stainless steel cathodes and iridium oxide on titanium (or, in some cases, 6 percent antimony in lead) anodes. Usually, a current density of 20 A/ft^2 is used where the copper concentration remains above 0.5 g/L in a parallel-plate cell. The advantage of using copper foil is apparent. The foil, along with the plated copper, can be removed from the plating cell without manually scraping the cathode.

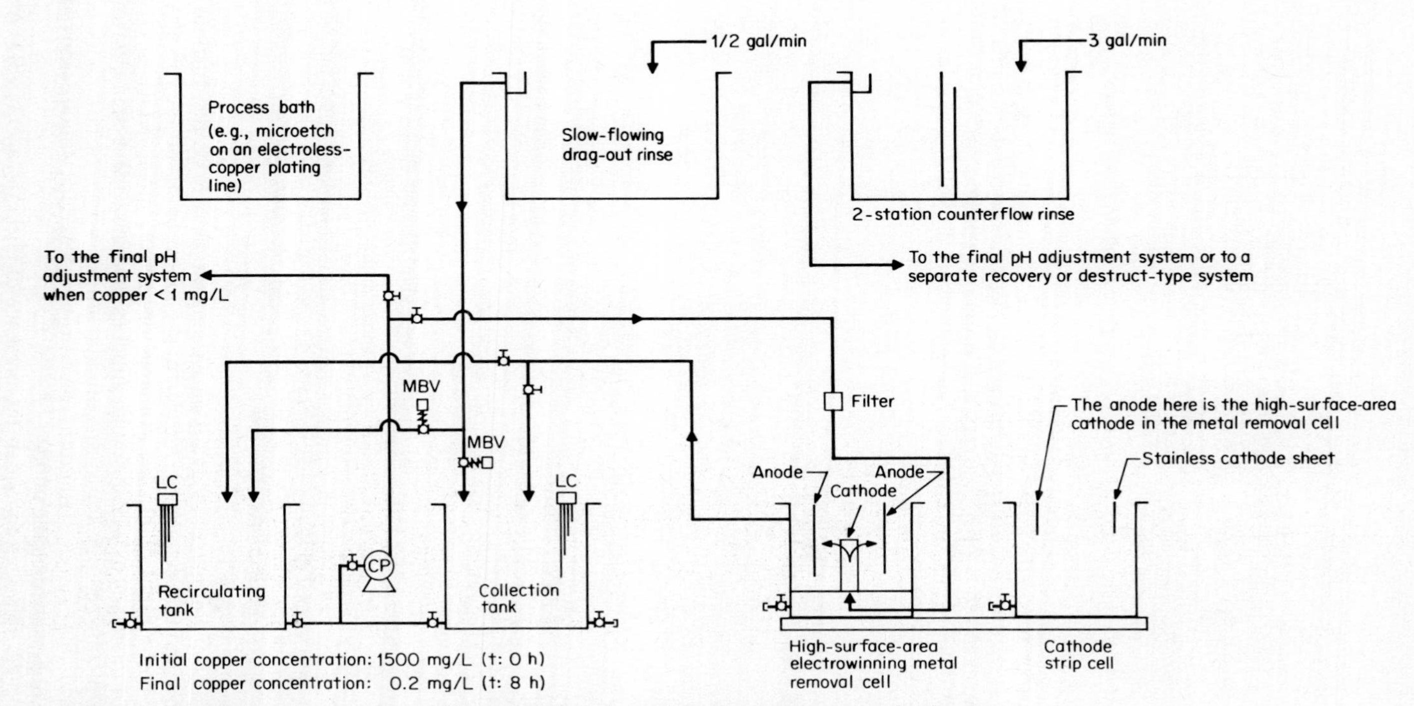

FIG. 15.3 Electrolytic recovery systems. This method is used to collect more than one slow-flowing drag-out rinse, including an electroless-copper rinse. MBV = motorized ball valve; LC = level control; CP = centrifugal pump. (*Courtesy of Baker Brothers/Systems.*)

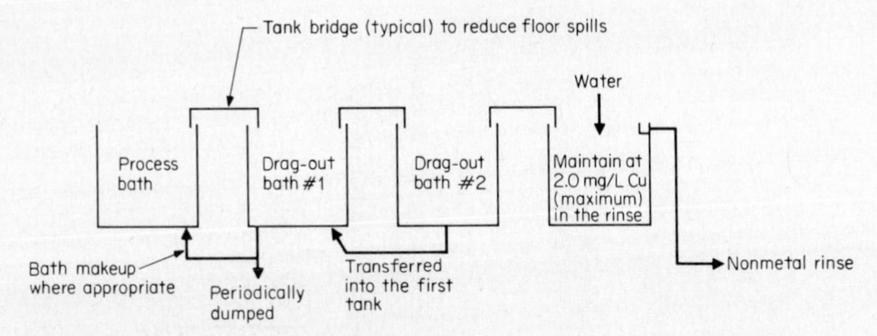

FIG. 15.4 Multiple drag-out stations. (*Courtesy of Baker Brothers/Systems.*)

Figure 15.5 shows the concentration of copper (remaining in solution) versus time. These results were obtained during a copper recovery study for a particular waste. The expected copper recovery rate (in grams per hour per square foot of cathode) is obtained by knowing the slope of the line, the volume of solution, and the area of the cathode.

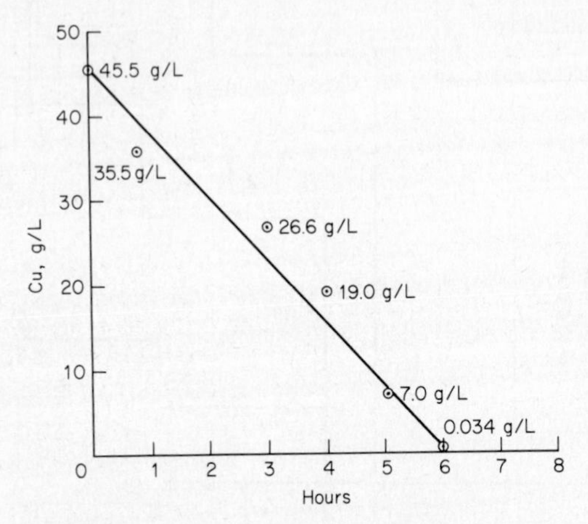

FIG. 15.5 Copper recovery rate: copper electrowinning from a 3 percent sulfuric acid solution. Volume, 300 mL; current, 2 A/1.8 V; anode, iridium on titanium, 2 × 2.5 × 0.032 in; cathode, copper sheet, 2 × 2.5 × 0.032 in; current density, 58 A/ft^2; current efficiency, 96% = 100 × {[(4 × 5.5 g/L − 0.034 g/L)/6 h] × 0.30 L}/(1.18 g Cu^{2+} /Ah × 2 A). (*Courtesy of Baker Brothers/Systems.*)

A number of commercially available systems can be used on the drag-out stations following microetch baths, the electroless-copper bath, and the copper-electroplating baths. Similar units are available for lead and other metals.

One commercially available system supplies the high-surface-area electrowinning. The plating cell portion of this module utilizes woven carbon fibers as the high-surface-area cathode. Titanium, coated with a proprietary material to reduce anodic gasing, is used as the anode.

This system removes copper in the presence of chelating agents (EDTA and ammonia) found in an electroless-copper bath. The price range starts at $18,000.

15.9.5 Copper Recovery from Sulfuric Peroxide Etchants

Typical microetch baths include sulfuric acid–hydrogen peroxide, sodium and potassium persulfate, or ammonium persulfate. Because of the trend to reduce chelating agents, and with recovery in mind, circuit board manufacturers are switching away from ammoniated microetching baths.

Some of the copper sulfate in the sulfuric acid–hydrogen peroxide baths can be crystallized by reducing the temperature (normally around 100°F) of the bath. This can be done on a batch basis for small shops or on a continuous basis for very high production shops. For batch applications, a three-tank system is utilized. The first tank (which is in the process line) contains the used etchant in the process line, which is high in temperature and high in copper concentration. The second (off-line) tank contains a fresh or recently recovered sulfuric-peroxide bath at ambient temperature. The third (off-line) tank is empty. At the end of the operating day, the first tank is dumped into the third tank. The second tank is pumped into the first tank. As the contents of third tank cool, copper sulfate crystals are formed and settle. The liquid portion is then decanted, and the solution is then transferred into the now empty second tank. The solution is analyzed for peroxide prior to its reuse. The copper sulfate crystals are then removed from the third tank.

In a continuous system (as shown in Fig. 15.2), the contents of the microetch or final etch bath are continuously recirculated from the microetch bath to a crystallizer, to a settling system, to a preheater, and then back to the etching tank.* The copper sulfate crystals are removed from the bottom of the clarifier through the use of an auger or another device. The solids are removed from the liquid by means of a filter bag or similar arrangement. By using the continuous system, the concentration of copper in the etching bath can be maintained, and a constant etching rate can be approached. The microetching baths in the electroless-copper line, the pattern-plate line (although the quantity of copper etched in this line is small compared with the other applications), and the black-oxide line may be combined in this system. Each circuit board manufacturer should consult the chemical supplier to determine if the sulfuric-peroxide final etching bath (if used) can be combined in one recovery system with the microetching baths.

15.9.6 Other Low-Sludge Metals Removal Techniques

Copper, from the electroless-copper bath, can be removed by combining the bath dump and solution growth with the spent catalyst solution in a tank. In this case, the copper "deposits" (is reduced) along the wall and bottom of the tank.

Proprietary consumable products for the electroless-copper bath or its solution growth are available. The solution is slowly metered through the container to remove the cooper.

Sodium borohydride is an effective but costly reducing agent manufactured by Mortin-Thiokol, Inc. (Ventron Division) and is available as a stabilized solution labeled VenMet. VenMet is used to reduce copper mostly in selected batch-treatment systems from both chelated wastes (where the presence of ammonium ions may be a problem) and nonchelated wastes.

The following reaction is typical of a metal-reduction process:

$$8MX + NaBH_4 + 2H_2O \rightarrow 8M + NaBO_2 + 8HX$$

where M is the cation with $+1$ valence and X is the anion.

*These and similar systems are commercially available.

VenMet contains 12 percent, by weight, sodium borohydride and 40 percent sodium hydroxide. A treatment procedure is to first react the waste with sodium bisulfite at a pH of 6.0 (with 15-min reaction time). The reaction is monitored using both pH and ORP (oxidation-reduction potential) controllers. Sodium bisulfite reduces divalent copper to its monovalent form. The VenMet solution is then added in a second stage (with 15-min residence time also at a pH of 6.0) to reduce the bisulfite to hydrosulfite and to reduce the monovalent copper to its particulate form. Following this reduction, copper will precipitate. It is important not to reoxidize (or aerate) the waste; otherwise, excess sodium borohydride will be required. Depending upon the solution to be treated, an anionic and a cationic polyelectrolyte may be required. Sodium borohydride should be used with care where boron is restricted in the discharge.

Selective ion-exchange systems are available. Cation exchange is a process where a resin is used to remove a positively charged cation (such as divalent copper) from a waste and to substitute, in its place, another positively charged cation (such as hydrogen or sodium). Such selective ion-exchange resins, which can extract various metal ions out of chelated waste streams, are readily available. Manufacturers of ion-exchange equipment can supply the equipment which can use these or other resins to extract copper from both chelated and nonchelated wastes. The regeneration step for the cation resin typically uses sulfuric acid and will produce a small volume of concentrated acidic solution containing copper. Copper in this solution can be recovered using electroplating techniques. Sometimes, untreated wastes are segregated and run through various ion-exchange modules using different resins, depending upon the waste to be treated.

Slow-flowing rinses containing copper may be pumped through an ion-exchange system to remove copper from the rinses. These cation columns are regenerated with sulfuric acid. Copper is removed from the regenerated waste by using a parallel-plate electrowinning cell. If EDTA is present, an anion exchanger must be used to remove the complex copper.

Liquid ion exchange (also known as "solvent extraction") is a generic name for a process used to extract copper from acidic chelated and nonchelated waste streams. The copper-containing aqueous wastes are carefully mixed with a stabilized organic liquid in one or more extraction vessels. The copper is quickly transferred from the aqueous waste into the organic phase. Following this, the liquid and organic phases are separated. The copper-free liquid phase is discharged to the final pH-adjustment module. The organic phase is pumped to a desorption column where copper is removed from the organic phase into a concentrated solution of sulfuric acid. The organic phase is recycled from the desorption vessels back to the extraction vessels. After passing through the desorption columns (or a mixer-settler), copper in the sulfuric acid is recovered by electroplating, and the sulfuric acid is then recycled from the electroplating system back to the desorption columns. This technique is used to remove copper from spent alkaline etchant. In this case, the process is modified to include a stage where ammonium sulfate is produced.

Each circuit board manufacturer may use one or more etching solutions for inner- and outer-layer applications. This generates the largest quantity of copper waste.

By proper equipment design, the rinse following the etching step can be classified as low-metal-containing. A schematic drawing showing a typical horizontal etcher is shown in Fig. 15.6.

A 3- or 4-stage low-flow (approximately 10 to 30 gal/h) rinse module can be used as a replenishing chamber following the alkaline etch chamber. Fresh etchant enters the multiple-stage replenisher module and flows back into the primary etch module. In some 4-stage replenisher modules, fresh etchant is also added directly

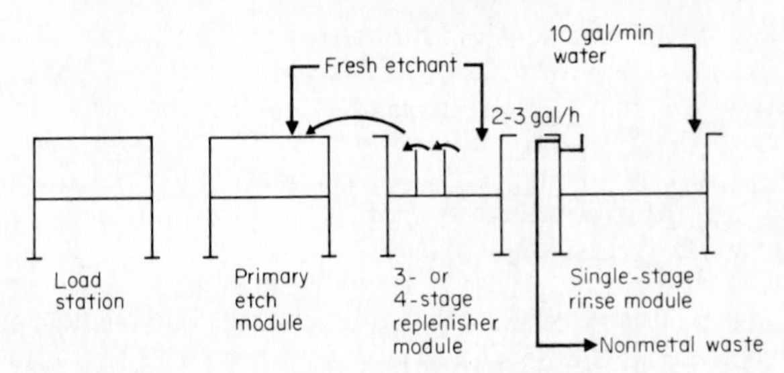

FIG. 15.6 Drag-out retention on an alkaline etcher. Rollers and air knives are not shown. (*Courtesy of Baker Brothers/Systems.*)

to the primary etch module. The fresh etchant in the 4-stage replenisher module theoretically removes 99.9 percent of the copper from the drag-out leaving the replenisher module. The rinse flow rate (following the replenisher module) is adjusted so that the maximum copper concentration is less than 2.0 mg/L. Due to its low copper concentration, the flowing rinse may be classified as a low-metal-containing waste.

For inner-layer etching, thorough rinsing of the copper foil must be accomplished prior to having the foil proceed from the etching step to the resist strip operation. Without complete rinsing, copper will enter the resist strip solution. If the resist solution contains amines, this waste must be treated for at least copper removal in the chelated treatment system.

Concerning the resist strip used in the outer layer, where solder plating is practiced, to minimize the lead concentration in the resist strip, the hydroxide concentration should be less than 5 percent. Some proprietary resist strips have a higher concentration of caustic.

15.10 CHEMICAL TREATMENT OF WASTES

The following techniques produce a sludge which must be hauled away to a suitable disposal site designed to accept these wastes. Typically, floor spills and nonrecoverable spent baths are treated using these systems.

Every attempt must be made to reduce the volume of sludge to be hauled away and reduce the volume of liquid waste to be treated. This includes installing recovery equipment, installing filtration systems for deburrers and board scrubbers, reducing the flow of those waste streams containing copper in concentrations exceeding the pretreatment standards, purchasing process equipment which minimizes both the drag-out volume and rinse flow rates, and selecting process baths without chelating or complexing chemicals.

In this text, the definition of a chelating (or a complexing) agent is any chemical by itself or in combination with other chemicals which will inhibit or prohibit the precipitation of copper (or other metallic ions) at an alkaline pH. Typical examples of chelating agents are EDTA, thiourea, formaldehyde, quadrol, tartrates, and other amines. Cyanide, unless destroyed, is considered a chelating agent.

15.10.1 The Collection System

Periodically, dumped baths are collected in tanks separate from rinses. The spent process baths are slowly metered into the appropriate rinsewater collection system. Rinsewater collection tanks are normally sized with a minimum residence time of 20 min. For a 10 gal/min flow, the tank volume is 200 gal. This should not include the minimum 12 in of freeboard in any tank. Tank volumes should be specified to be the operating volumes and not "filled-to-the-brim" capacities. If one or more rinse tanks are periodically dumped for cleaning purposes, this additional volume must be added to "normal" volume for specifying the actual operating capacity of a rinsewater collection tank.

A tank cover on the acid dump collection tank is desirable. The cover should be equipped with a suitable exhaust vent and a blastgate to regulate air flow. Typical exhaust rates are 200 to 300 SCFM (standard cubic feet per minute) of air per tank; however, ventilation experts should be consulted. Tank covers on other tanks such as the chelated batch-treatment tank, the chrome-reduction tank, the sodium bisulfite mixing tank, and other chemical mixing tanks may be required or desired.

Every attempt must be made to reduce the dumping frequency of process baths. Some chemical companies will add chemicals to their formulations to increase the useful life of a bath. Care must be observed, as sometimes "chelating agents" are added which can adversely effect the recovery or treatment chemistry. Process engineers should encourage the use of electroplating and other recovery techniques to prolong the use of a bath rather than dumping the bath to the waste-treatment system or adding chelating agents to the bath. Good quality control dictates that all baths should be periodically analyzed, and the dumping frequency can be scheduled on the basis of the analytical results. Automatic analyzers and chemical feed systems are available to accomplish this.

Chemical suppliers state that they have the right to change the chemical formulation of their bath as long as it increases the productivity of the bath. All circuit board manufacturers should stress to their chemical supplier that changes in any bath formulation must be clearly identified to the process engineer, the waste-treatment operator, and the environmental manager.

15.10.2 Nonchelated Metal Rinses and Dumps

Nonchelated metal rinses and dumps contain copper and other metals at a higher concentration than that allowed in the discharge. In a typical system, these wastes flow into a two-stage pH-adjustment system. As the pH in a waste is elevated, copper combines with hydroxides, available through the addition of caustic, to form insoluble copper hydroxide. This forms a sludge which is removed in a clarifier.

Figure 15.7 shows both 50 percent caustic soda and 66° Baumé sulfuric acid being added to the first-stage pH-adjustment tank for rough (on-off type) pH control and caustic being added in the second stage. Fifteen minutes of residence time is provided in each stage.

For adequate pH control and for proper diagnosis of system performance, pH recorders should be provided. Monitoring records should be stored for at least 3 years. Even with a programmable controller, a 30-day strip chart for every process variable is desired.

15.10.3 The Settling Process

Following the second-stage pH-adjustment tanks, the waste enters the flocculation chamber in the clarifier. Polyelectrolytes, other flocculating agents, and recycled sludge are added to the flocculation chamber and are rapidly mixed with a high-speed mixer. The waste then enters the second portion of the flocculation cham-

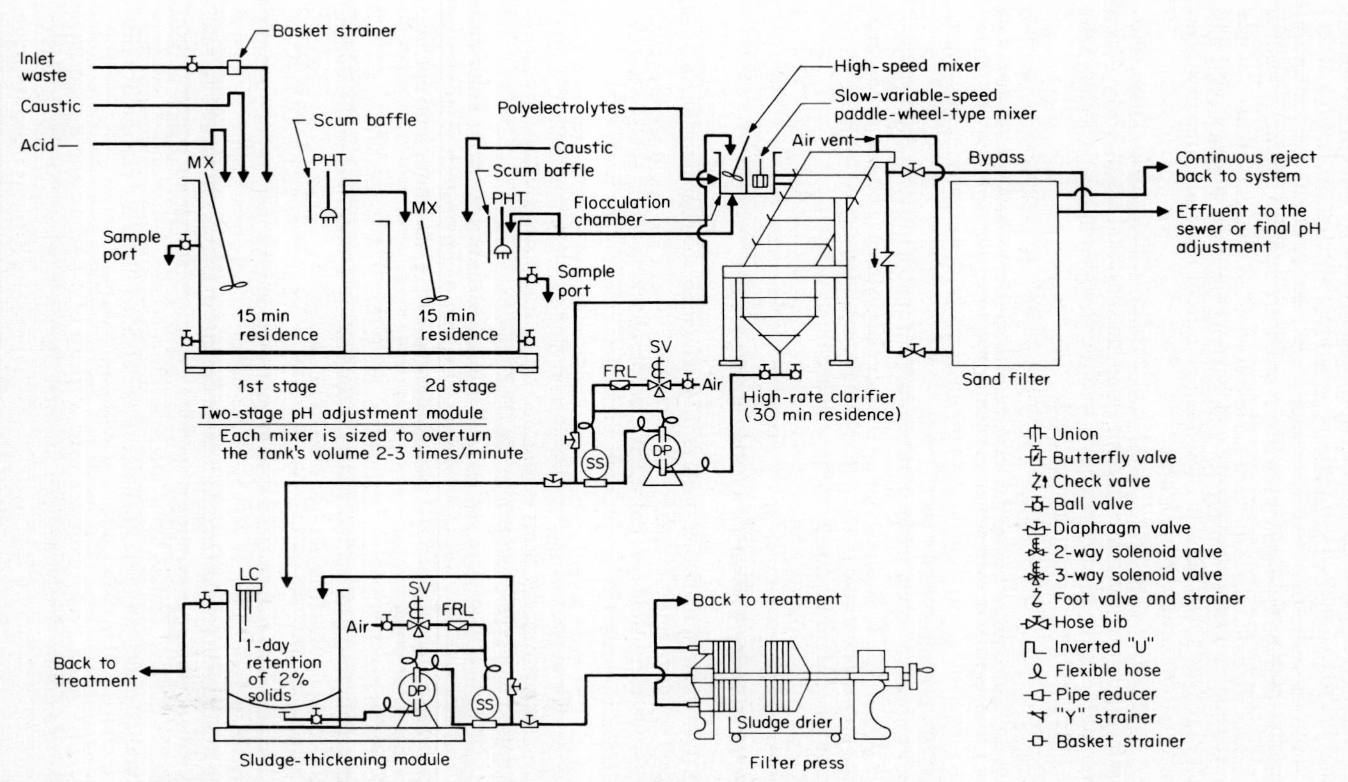

FIG. 15.7 Treatment of metal-bearing wastes. MX = mixer; PHT = pH transmitter; FRL = filer, regulator, lubricator; SV = solenoid valve; SS = surge suppressor; DP = diaphragm pump; LC = level control. (*Courtesy of Baker Brothers/Systems.*)

ber, where a slow-variable-speed paddle-wheel-type mixer is used to enlarge the oxide and hydroxide particles to make settling more effective. Following flocculation, the waste will flow down the upflow-type Lamella clarifier. Each settling tank supplier may recommend a minimum residence time or hydraulic loading (gallons per minute per square foot of settling area) for settling. A clarifier's ability to achieve good results (with discharge copper concentrations ranging from 1.0 to 5.0 mg/L) are best achieved with more than 30 min of residence time in a high-rate clarifier at maximum design flows.

Calcium salts are occasionally added in place of or in addition to caustic soda. These salts are added for more efficient (than caustic soda) solids settling (as a coprecipitation agent), solids dewatering, and for fluoride removal. Calcium salts include hydrated lime and calcium chloride.

Lead removal is enhanced by the addition of sodium carbonate (soda ash). Lead carbonate is more insoluble than lead hydroxide. Soda ash is purchased as a powder and is mixed with water to form a solution. The solution is metered into the second-stage pH-adjustment tank in direct proportion to the concentration of lead in the untreated waste stream.

The precipitated solids settle to the bottom of the clarifier, and the clarified water is discharged to a sand filter, a final pH-adjustment system, or the final discharge point. The dilute ($\frac{1}{2}$ to 2 percent total suspended solids) sludge from the bottom of the clarifier is transferred to a sludge-thickening tank.

15.10.4 The Sludge-Thickening and Dewatering Process

The function of the sludge-thickening tank is to increase the concentration of total suspended solids to approximately 2 to 3 percent.

The increase in solids concentration is accomplished by continuously decanting the water from the sludge-thickening tank and directing the water back to the metal rinse collection system. The solids in the sludge-thickening tank are dewatered by being pumped from the dish-bottom sludge-thickening tank to a filter press. The liquid from the press is returned to the metal rinse collection system. The operation of a press is shown in Fig. 15.8.

Filter press manufacturers include JWI, Inc.; Duriron; Industrial Filter and Pump Co.; Hoesch, Inc.; and The Shriver Division of Envirotech. Typically, a caulked gasketed recessed (CGR) plate is desired with a semiautomatic opening and closing of the press.

JWI, Inc., Baker Brothers, and others supply sludge dryers which increase the solids concentration (from 35 to about 70 percent) by adding heat to the sludge. This reduces the volume of sludge to be hauled away.

Sand filters (such as the upflow DynaSand filter, manufactured by the Parkson Corp.) are added to a treatment system following the clarifier to continuously remove the suspended material which occasionally discharges from the clarifier with the treated water. This condition exists because of changes in waste flow rate, process bath chemistry, incoming suspended solids concentrations, concentration of copper in the acid dumps being metered into a system, temperature, polyelectrolyte addition rate, and rate of sludge removal. Sand filters are especially useful where a 2.0 mg/L limit of copper is consistently required.

Cross-flow filtration systems are used in place of a clarifier and a sand filter when the maximum discharge limit for total copper is below 1.0 mg/L. Wastes are pretreated in a method similar to the precipitation system. In cross-flow treatment systems, the wastes are collected in a recirculation tank and pumped at turbulent flow and 10 to 35 psig through a series of capsules containing tubular filters. The majority of the recirculated wastewater plus all the suspended solids

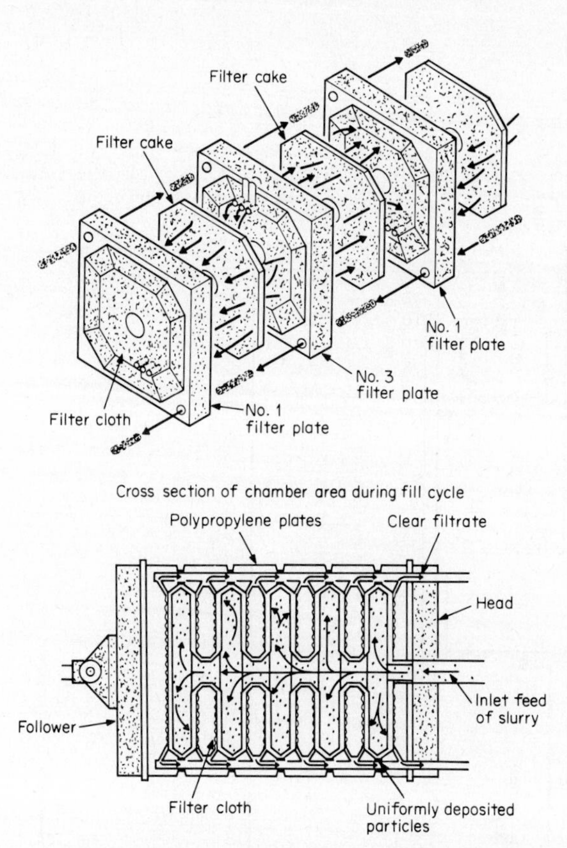

FIG. 15.8 Operation of a filter press. (*Courtesy of JWI, Inc.*)

returns to the recirculation tank. Each filter tube wall contains holes with diameters ranging from 0.1 to 1 μm. The suspended solids free water will pass through the sidewalls of the tubular filter. Like all membrane-type systems, the flow rate through the filters may be adversely effected by oil and the process chemistry used.

Ferrous sulfate, lime, sodium borohydride, calcium chloride, or diethyldithiocarbamate are chemicals sometimes used in this process. The objective is to produce an effluent with a low concentration of copper with the least amount of sludge at the lowest operating cost.

Figure 15.9 shows a cross-flow system with a sludge withdrawal line located near the discharge of the recirculation pump. The 3 percent (total suspended solids) sludge is pumped to a sludge-holding tank and then pumped to a filter press. Typical discharge concentrations of copper are 0.1 mg/L, or less, for each metal. Both Baker Brothers/Systems and Memtek supply these systems.

Chromic acid can be used in the smear-removal process. A typical chrome-reduction module is shown in Fig. 15.10.

Chrome reduction is completed in one step by using any one of a variety of reducing chemicals, such as sulfur dioxide, sodium hydrosulfite, or sodium bisulfite. For sulfur dioxide and sodium bisulfite the reduction takes place rapidly at a constant low pH (approximately 2.5 pH units).

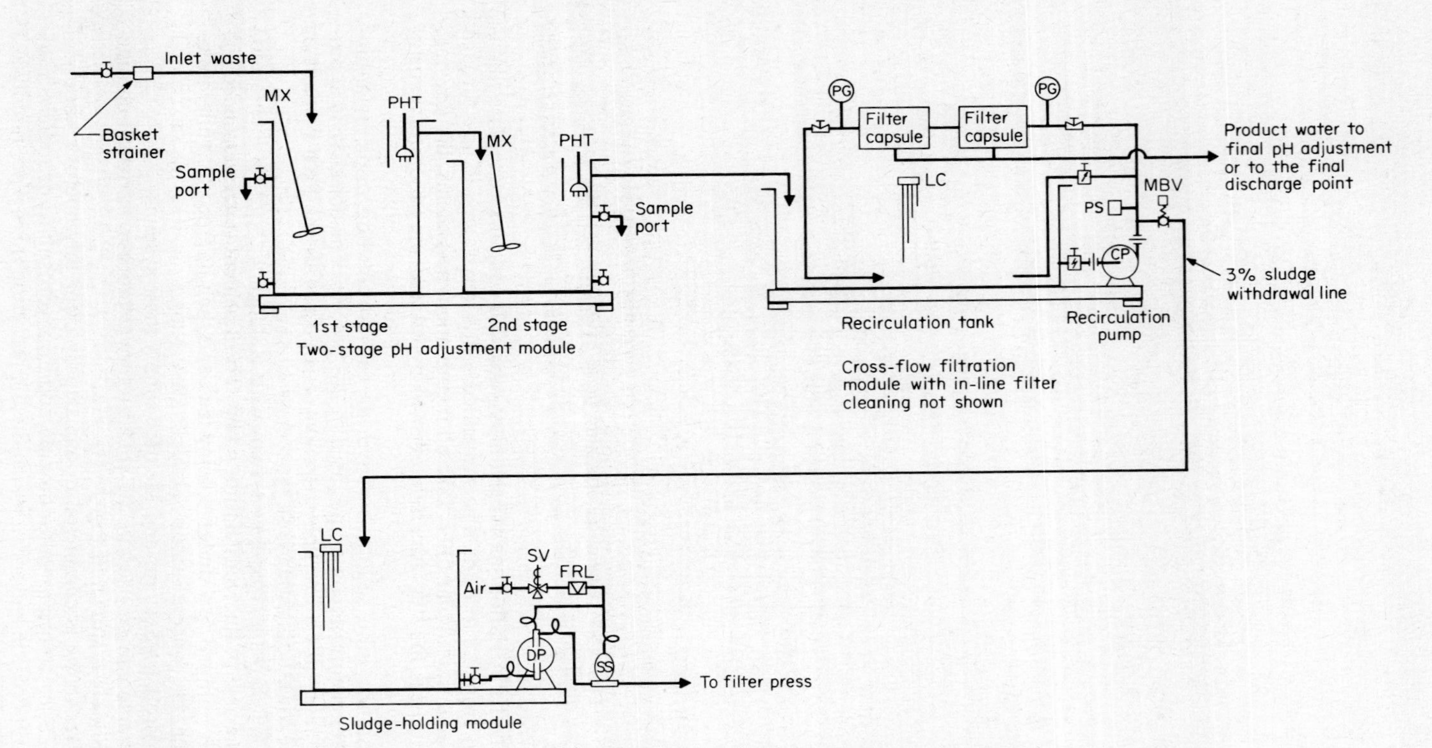

FIG. 15.9 Cross-flow filtration. MX = mixer; PHT = pH transmitter; PG = pressure gauge; LC = level control; PS = pressure switch; MBV = motorized ball valve; CP = centrifugal pump; SV = solenoid valve; FRL = filter, regulator, lubricator; DP = diaphragm pump; SS = surge suppressor. (*Courtesy of Baker Brothers/Systems.*)

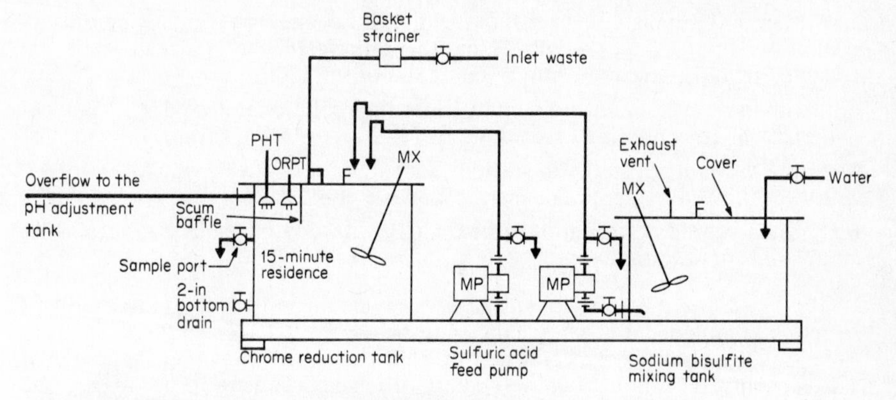

FIG. 15.10 Chromic acid smear-removal reduction module. PHT = pH transmitter, ORPT = ORP (oxidation-reduction potential) transmitter, MX = mixer, MP = metering pump. (*Courtesy of Baker Brothers/Systems.*)

Various techniques are available to reoxidize trivalent chrome to hexavalent chrome in the bath. This may prolong the life of the bath.

Nonmetal wastes should be segregated from the metal-containing wastes. These two categories can combine after the metal ions have been removed.

15.10.5 Nonmetal Rinses and Dumps

Nonmetal rinses and dumps originate from aqueous and semiaqueous photoresist stripping baths (containing a 3 to 5 percent sodium or potassium hydroxide solution, some butyl cellosolve or carbitol, and sometimes formulations which contain amines) and developing solutions (containing a 2 to 3 percent sodium or potassium carbonate solution).

Du Pont reports that "gooey" material (consisting of the photoresist) comes out of solution and will coat pipes, tanks, and valves at a pH less than 6. The photoresist redissolves at a pH above 11. At least for the Du Pont photoresist, no interference with metals removal (via pH adjustment) occurs. However, if the photoresists are present in the wastes entering a clarifier, some of the photoresist solids will sink, some will float, and some will remain suspended. This may interfere with the proper precipitation of the suspended solids and with solids dewatering.

Solvent-based developers and resist stripper baths should be recovered via distillation or through the use of activated carbon. The rinsewaters should be monitored for TTO concentration to ensure compliance with the regulations.

Chelated metal-containing wastes include electroless copper, ammonium persulfate, ammoniacal alkaline etching solutions and rinses, certain fluxes from reflow machines, resist strip baths with metals and amines, and certain other acid and alkali baths.

Although it is most desirable to collect a small volume and batch-treat these wastes, continuous-treatment systems are available for chelated wastes. A few sludge-producing continuous systems are outlined below:

1. Using the precipitation system and sand filtration with ferrous sulfate addition
2. Using ultrafiltration or cross-flow filtration as previously described

3. Adding sodium bisulfite to reduce the divalent copper to its monovalent state, followed by sodium borohydride (or other proprietary chemicals) addition, followed by ultrafiltration or by precipitation or sand filtration
4. Using sodium bisulfite and then hydrated lime solution followed by ultrafiltration or by precipitation or sand filtration
5. Using the Lancy International, Inc., sorption filter system, which can follow one of the above-mentioned clarification systems
6. Using a sulfide precipitation system along with diatomacious earth (to assist in solids dewatering)

Batch-treatment systems for chelated wastes are the most cost-effective method to reduce the concentration of copper to below 1 mg/L. Generally, it is more expensive (in the use of chemicals) to treat chelated wastes than nonchelated wastes. Thus, by minimizing the quantity of chelated wastes to be treated, one minimizes the operating cost. The containment of chelated wastes into a low enough volume to be treated in a batch system has been successfully demonstrated at numerous facilities by the proper purchase of production equipment. It is desirable, although not always possible, to have one collection tank and one batch-treatment tank for the chelated wastes. The treatability study, during the engineering phase, will determine if one or more batch-treatment systems are required for the various wastes. A typical semiautomated batch-treatment tank schematic for chelated wastes is shown in Fig. 15.11.

In this process, the pH is first reduced, via the addition of sulfuric acid, to a pH of 2. Ferrous sulfate is added, while the variable-speed mixer operates at high speed, at a ratio of approximately 10 parts of iron to 1 part of the sum of the metal ions. It is in batch systems where sodium borohydride and other proprietary chemicals are most useful. These will reduce the volume of sludge.

The polypropylene air diaphragm pump is continuously recycling solution from the batch-treatment tank, through the pH-monitoring tank, and back to the batch-treatment tank. In the pH-monitoring tank (or in a separate well within the batch-treatment tank) a pH probe is always immersed in a solution. After a 30-min mixing period, with the mixer working at a high speed, the pH is elevated by the addition of caustic soda or hydrated lime to approximately 10.0 pH units.

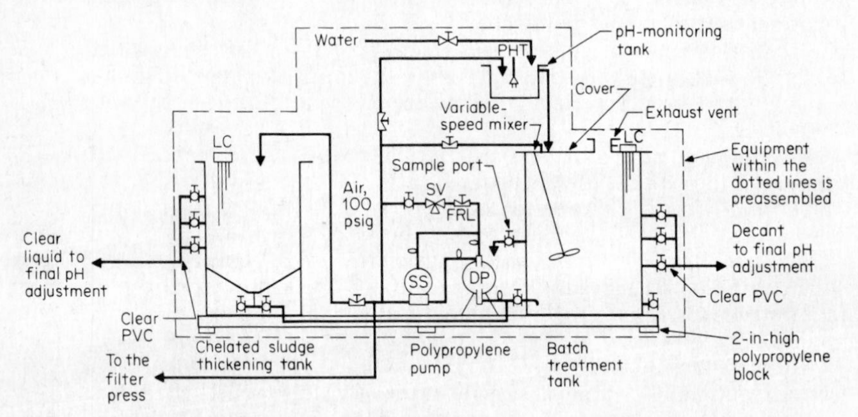

FIG. 15.11 Chelated batch-treatment module. LC = level control; PHT = pH transmitter; SV = solenoid valve; FRL = filter, regulator, lubricator; SS = surge suppressor; DP = diaphragm pump. (*Courtesy of Baker Brothers/Systems.*)

After mixing for 30 min, the mixer motor speed is reduced, thereby allowing large particles of insoluble copper and other hydroxides to form. A polyelectrolyte or other flocculant can be added to assist in the settling of solids in the dish-bottom tank. Following the formation of large particles, the mixer motor is stopped, and the solids settle to the bottom of the tank. Once a sample of the liquid has been analyzed to contain less than the desired concentration of copper, the liquid is decanted to the final pH-adjustment stage, and the solids are pumped to a chelated waste sludge-thickening tank.

Sludge containing a high concentration of chelating agents should be kept separate from other sludges not containing a high concentration of chelating agents. The liquid from the filter press containing the chelating agents should be kept separate from the nonchelated metals-removal process.

One difficulty which occurs is the resolubilization of copper in the chelated sludge-storage tank. The waste-treatment operator should analyze for the concentration of copper in the liquid above the sludge being dewatered. If the concentration of copper exceeds the pretreatment standard, these wastes must be retreated (i.e., the copper must be reduced and then precipitated). This can happen because of the effect of the chelating agents on the copper in the sludge.

As soon as possible, the chelated sludge should be pumped to a filter press for dewatering. The filter press should be thoroughly cleaned after its use. It is recommended that a separate filter press for chelated sludge be provided.

Floor spills should be segregated into similar segregated waste streams as exist for the waste-treatment system. The spills around an alkaline etcher or an electroless-copper plating tank should be kept separate, for ease in treatment, from other wastes.

The floor spill transfer pump should not automatically be actuated. The pump should be manually activated after the method for treatment of the waste is defined.

15.11 FACILITY CONSIDERATIONS IN BUILDING A TREATMENT PLANT

Prior to specifying the location of the waste-treatment plant, the following information must be considered. The presence of a high groundwater table will increase the construction costs. One must define the depth of the groundwater table during the rainy season. Additional information on the source (or sources) of potable and process water, its pressure, and its chemical constituents must be defined.

Air ventilation rates in the waste-treatment room should be specified. An exhaust rate of 6 to 10 air changes per hour, including both high- and low-level exhaust and supply air ducts, has been successfully used to minimize corrosion while maintaining a comfortable work area. The 10 air changes per hour should include the volume of air to be exhausted from the treatment tanks. The room temperature (or at least the temperature of 50 percent caustic soda) must be maintained above 65°F. Fifty percent caustic soda crystallizes at about 60°F. If 25 percent caustic is used, its crystallization temperature is about 20°F.

Each manufacturing company should verify that enough capacity exists in the existing or proposed municipal wastewater treatment facility to accept the maximum anticipated waste volume. One must review how the municipality will invoice each company for the water and sewer use. Water recycling systems can be justified on the basis of the cost for water as well as disposal of waste. It is anticipated that the sewer use tax will forever increase due to the increasing demand made upon municipalities.

Each manufacturing company should anticipate that a fiberglass or a suitable plastic catwalk will be required to permit access by the waste-treatment operator to the top of the treatment tanks. Stone-clad HT floor coatings or the equivalent may be required.

It is good practice (especially for the operator) to have a graphic display panel showing the motor status and alarm conditions. The color of each tank can be displayed on the panel to aid in the operator's understanding of the process flow. Small rooms are used to reduce noise, and this humanizes the area.

Prime consideration must be given to the seismic conditions. Construction costs can vary radically as seismic risk increases. Fiberglass-reinforced plastic tanks may have to be built using a wall thickness greater than that specified by the ASTM code (for contact-molded reinforced thermoset resin chemical resistant tanks, D-4097-82) and contain at least eight (5-in-wide) hold-down lugs (for tanks equal to or greater than 6 ft in diameter by 8 ft high) to meet seismic Zone IV requirements.

The use of a wet floor in the plating room area is discouraged and may be prohibited in the future. Tank bridges and drip pans should be installed to minimize or eliminate floor spills which could possibly leak through the floor and into the groundwater. Dikes should surround all wet or potentially wet areas.

Safety eyewashes and showers should be located where chemicals are mixed. A 30-min supply of air should be contained with an air mask device outside of the plating area so that, if toxic fumes should form, one or more operators can use the safety masks to enter the treatment area and rectify the problem.

If a treatment facility is to be located outdoors, it is recommended that a roof be provided to protect the equipment from rain and dust. Either sidewalls consisting of removable panels or a fence should be provided to prohibit unauthorized access into the treatment area. Appropriate diking for floor spills in the treatment area should be provided. Floor spills should be directed to one or more collection sumps.

Where restricted access must be controlled (e.g., around a filter press), a programmable safety light curtain may be desirable. If the light rays are disturbed, an alarm will sound. Scientific Technology Incorporated, among others, supplies units with transmitters, receivers, and a protected height from 6 in to 48 in. Mirrors are used to bend the invisible infrared light around the object to be protected.

15.12 EQUIPMENT AND INSTALLATION COSTS

Equipment and installation costs vary widely with the degree of preassembly, the degree of automation, the quality of the equipment used, and the amount of custom-designed (as opposed to off-the-shelf) systems. In most cases, only custom-designed equipment can adequately satisfy anyone's requirements.

As a rule of thumb, the installation cost can vary between 25 and 50 percent of the modularized capital cost. It is advisable to purchase equipment from a company which can provide installation supervision or a turnkey project. In a turnkey project, the equipment supplier would be responsible for rigging the equipment into place and for providing the plumbing and electrical works within the confines of the treatment system. It is usually the circuit board manufacturer's responsibility to provide the plumbing from the wet process equipment to the collection equipment for the wastewater-treatment system. Usually a general contractor is hired to construct dikes, install a floor coating, or modify any buildings.

Total equipment costs are the sum of the individual module costs, or unit operations, used to make up the treatment system. They include not only the equipment modules, but also the motor control centers, programmable controllers and software, and the amount of on-site start-up time.

Typically, a small (50 gal/min) circuit board manufacturer without recovery equipment and having a chelating batch-treatment system as well as a continuous system for the treatment of metal-bearing and non-metal-bearing wastes (including solids dewatering) can purchase preassembled equipment for approximately $200,000. Preassembled systems for 20 gal/min or less are available for about $150,000.

For systems generating 150 gal/min of total wastewater, the equipment cost can be $350,000, with installation costs of approximately $100,000.

Each manufacturing company should evaluate each equipment supplier on the basis of the quality of the equipment, the amount of time that the engineers have spent at the facility prior to quoting, the useful life of each piece of equipment, and the engineering support which that company can provide to the circuit board manufacturer. Although the distance from the circuit board manufacturer to the equipment supplier may be important, it is the proximity of spare parts inventories which should be a criterion in reviewing equipment suppliers.

The above costs are for systems producing an effluent which will meet U.S. EPA pretreatment standards. Occasionally, a manufacturing company may be forced to meet more stringent standards, or it may require its treatment system to produce a more consistent effluent with a lower concentration of copper. This may require an additional expense. The cost for this depends upon the flow rate as well as the type of treatment system utilized. Care must be exercised in evaluation of this requirement.

Operating costs also vary. Generally, for a destruct-type system, the operating cost will range between $0.10 and $0.60 per laminate square foot. Another calculation indicated a cost of about $2.00 per 1000 gal treated. This includes costs for chemicals purchased, sludge disposal (approximately $350 per ton), direct labor, depreciation, repair and maintenance, and tools. The purchase of chemicals in bulk and a preventative maintenance program will significantly lower the operating costs.

One should expect that the cost to dispose of sludge will continue to rise as more and more disposal sites are closed due to groundwater contamination or failure to comply with all the regulations issued in the 1985–1986 time period.

CHAPTER 16
SOLDER RESIST

Lyle R. Wallig
E. I. du Pont de Nemours & Company, Wilmington, Delaware

16.1 INTRODUCTION AND DEFINITION

IPC-T-50* b defines a solder resist (mask) as "a coating material used to mask or to protect selected areas of a printed wiring board (PWB) from the action of an etchant, solder, or plating."

A somewhat more useful working definition for a solder resist is as follows: a coating which masks off a printed wiring board surface and prevents those areas from accepting any solder during vapor phase or wave soldering processing (see Fig. 16.1).

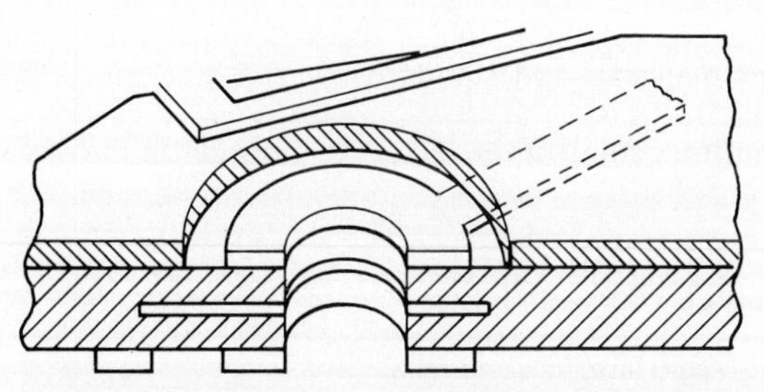

FIG. 16.1 Important factors of solder resist on a printed wiring board: (*a*) mask should be away from plated-through hole and its associated land or pad; (*b*) trace should be covered; (*c*) laminate area should be completely covered; (*d*) adjacent conductors shouldn't be exposed.

16.2 FUNCTIONS OF A SOLDER RESIST

The prime function of a solder resist is to restrict the molten solder pickup or flow to those areas of the PWB, holes, pads, and conductor lines that are not covered by the solder resist. PWB designers, however, often expect more functionality out of the solder resist than just a means to restrict the solder pickup. Table 16.1 lists the functions of a solder resist.

*Institute for Interconnecting and Packaging Electronic Circuits (IPC), Lincolnwood, Illinois.

TABLE 16.1 Functions of a Solder Resist

Functions
Reduce solder bridging and electrical shorts
Reduce the volume of solder pickup to obtain cost and weight savings
Reduce solder pot contamination (copper and gold)
Protect PWB circuitry from handling damage, i.e., dirt, fingerprints, etc.
Provide an environmental barrier
Fill space between conductor lines and pads with material of known dielectric characteristics
Provide an electromigration barrier for dendritic growth
Provide an insulation or dielectric barrier between electrical components and conductor lines or via interconnections when components are mounted directly on top of the conductor lines

16.3 DESIGN CONSIDERATIONS FOR SOLDER RESISTS

16.3.1 Design Goals

The design goals for the selection and application of a solder resist should be carefully considered. As with all design goals, one should try to achieve maximum design flexibility, reliability, and functionality at a cost consistent with the required level of system performance.

16.3.2 Design Factors

The system's performance and reliability requirements are the keys in determining the selection process for a solder resist. Critical life-support systems will require different materials and standards than a less critical system such as a VCR.

Table 16.2 outlines some of the factors to consider in the design process when selecting a solder resist.

TABLE 16.2 Design Factors

Design factors
Criticality of system's performance and reliability
Physical size of PWB
Metallization on PWB, i.e., SnPb, copper, etc.
Line and space (density) of the PWB
Average height of conductor line (amount and uniformity of the metallization)
Size and number of drilled plated-through holes (PTHs)
Annual ring tolerance for PTHs
Placement of components on one or both sides of PWB
Need to have components mounted directly on top of conductor line
Need to tent via holes in order to keep molten solder out of selected holes
Need to prevent flow of solder up via holes, which may have components sitting on top of them
Likelihood of field repair or replacement of components
Need for solder resist to be thick enough to contain the volume of solder needed to make good solder connections
Choice of specifications and performance class that will give the solder resist properties that are necessary to achieve the design goals

It is not very likely that a single solder resist material or application technique will satisfy all the design considerations that are viewed as necessary. It should also be noted that not all the design factors listed in Table 16.2 carry the same weight or value, so the designer needs to prioritize those design factors, analyze the necessary trade-offs, and then specify the solder resist material and process that gives the best balance of properties or characteristics.

16.4 ANSI/IPC-SM-840 SPECIFICATION

In an effort to aid PWB and systems designers who wanted to know what performance properties or characteristics they would receive when they asked for a solder resist (mask) an IPC industry group developed the first solder resist specification in 1979. This specification is IPC-SM-840, entitled *Qualification and Performance of Permanent Polymer Coating (Solder Mask) for Printed Wiring Boards.*

The IPC-SM-840 specification calls out three classes of performance which the designer can specify:

Class 1: Consumer—Noncritical industrial and consumer control devices and entertainment electronics

Class 2: General industrial—Computers, telecommunication equipment, business machines, instruments, and certain noncritical military applications

Class 3: High reliability—Equipment where continued performance is critical; military electronic equipment

In addition to calling out the performance classes, the specification assigns responsibility for the quality of the solder resist to the materials supplier, the PWB board fabricator, and, finally, the PWB user.

Responsibilities per IPC-SM-840

Role	Responsibility
Materials supplier	The solder material and the appropriate data on the chemical, electrical, mechanical, environmental, and biological testing. Data are gathered on the standard IPC-B-25 test PWB.
PWB fabricator	The entire PWB fabrication and production process. This includes the application and curing of the solder resist material. The fabricator is also responsible for determining the end use application and the conformance of the PWB and the solder resist to the specification class call-out on the PWB fabrication print.
PWB user	Monitoring of the acceptability and functionality of the completed PWBs.

16.5 SOLDER RESIST SELECTION

The solder resists available are broadly divided into two categories, i.e., permanent solder resists and temporary resists. The breakdown of the solder resist types is shown in Fig. 16.2.

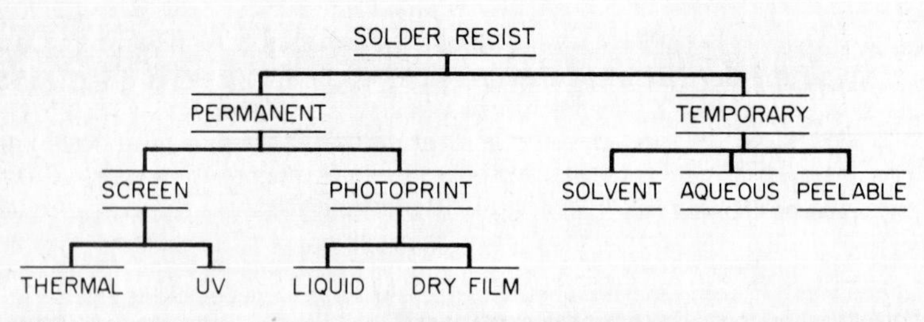

FIG. 16.2 Solder resist selection tree.

The permanent solder resist materials are classified by the means used to image the solder resist, i.e., screen printing or photoprint. In addition, the screen-printed resists are further classified by the curing technique, i.e., thermal or ultraviolet (UV) curing. The photoprint solder resists are distinguished from each other by whether they are in the form of liquid or dry film. The temporary resist materials are classified by chemistry or means of development.

16.5.1 Temporary Resists

A distinction is made between permanent and temporary solder resists. The temporary resists are usually applied to a selected or limited area of a PWB to protect certain holes or features such as connector fingers from accepting solder. The temporary resist keeps solder out of the selected holes and thus allows for certain process- or temperature-sensitive components to be added manually at a later time.

The temporary solder resists usually consist of a latex rubber material or any of a variety of adhesive tapes. These materials can be applied by an automatic or manual dispenser. Some of the temporary mask materials dissolve in the solvents or cleaning processes that are used to clean off the soldering flux residues. This is really a benefit, since it eliminates the need for a separate manual removal and/or cleaning step for the temporary resist.

16.5.2 Permanent Resists

Permanent solder resists are not removed and thus become an integral part of the PWB.

The demand for permanent solder resist coatings on PWBs has greatly increased as the trend toward surface mounting and higher circuit density has increased. When the conductor line density was low, there was little concern about solder bridging, but as the density increased, the number and complexity of the components increased. At the same time, the soldering defects, such as line and component shorts, greatly increased. Inspection, testing, and rework costs accelerated as effort went into locating and repairing the offending solder defects. The additional cost of the solder resist on one or both sides of the PWB was viewed as a cost-effective means to offset the higher inspection, testing, and rework costs. The addition of a solder resist also had the added value for the designer of providing an environmental barrier on the PWB. This feature was important and dictated that the materials considered for a permanent solder resist should have similar physical, thermal, electrical, and environmental performance properties that are in the laminate material. See Table 16.3 for a comparison of properties for permanent solder resist types.

TABLE 16.3 Permanent Solder Resist Selection Guide

Feature	Screen print		Dry film		Liquid photoresist solvent
	Thermal	UV	Aqueous	Solvent	
Soldering performance	1	1	1	1	1
Ease of application	1	1	2	2	2
Operator skill level	2	2	2	2	2
Turnaround time	2–3	2–3	2	2	2
Inspectability	2–3	2–3	3	3	3
Feature resolution	3	3	1	1	1
Adhesion to SnPb	1	3	1–2	1–2	1
Adhesion to laminate	1	1–2	1	1	1
Thickness over conduct or lines	3–4	3–4	1–2	1–2	2
Bleed or residues on pads	3–4	3	1	1	1
Tenting or plugging of selected holes	4	4	1	1	4
Handling of large panel size with good accuracy	3	3	1	1	1
Meeting of IPC-SM-840 Class 3 specification	3–4	3–4	1–2	1	1
Two-sided application	4	4	1	1	3–4
Capital equipment cost	4	4	1–2	1–2	1

Key: 1 = good or high; 2 = moderate; 3 = fair; 4 = poor or low.

16.5.3 Selection Factors

General considerations are as follows:

- Reliability and performance data on the solder resist material
- Cost effectiveness
- Past experience
- Vendor reliability and technical support
- Number of panels required
- Appearance and cosmetics of solder resist
- Number of sources for application and supply

Material considerations are as follows:

- IPC-SM-840 class designation call-out
- Cost and availability of materials
- Lot-to-lot consistency record
- Setup and cleanup times
- Working time and shelf life of solder resist
- Safety concern for release of noxious or toxic fumes during processing or curing steps
- Degree of workability

Process considerations are as follows:

- Operator skill level required
- Need for special applications or curing equipment
- Cleaning requirements for PWB before and after application
- Size of PWB
- Need for solder resist on one or both sides of PWB
- Number of panels to be processed
- Turnaround time required
- Machinability
- Need to tent selected holes
- Touch-up or rework limitations
- Inspectability and conformance to specification

Performance considerations are as follows:

- Testing to IPC-SM-840 specification requirements
- Adhesion after soldering and cleaning
- Bleed-out of resist onto pads or PTHs
- Solvent resistance to flux and flux cleaners
- Ionic contamination levels
- Integrity of resist after soldering and thermal cycling

16.6 SOLDER MASK OVER BARE COPPER (SMOBC)

A major solder resist application technology is called SMOBC. This name stands for the application of "solder mask over bare copper." A problem for conventional copper–tin-lead electroplated PWBs is the flow of tin-lead solder under the solder resist during the wave or vapor phase or infrared soldering. This flow of molten metal underneath the resist can prevent the resist from adhering to metal or laminate. If the resist fractures because of this hydraulic force, the surface integrity is lost and the effectiveness of the resist as an environmental or dielectric barrier can be severely impaired. In fact, such breaks in the resist can actually trap moisture, dirt, and soldering flux and serve as a conduit to direct liquids down to the resist-laminate interface. This solder resist situation could lead to serious reliability and/or performance concerns.

The SMOBC process addresses the tin-lead flow problem by eliminating the use of tin-lead electroplating on the conductor lines under the solder resist. An all-copper PTH printed wiring board is often produced by a "tent-and-etch" process. This process is one in which the PWB is drilled and plated with electroless copper, which is immediately followed by copper panel plating with the full-thickness copper required for the PTHs. A dry-film resist process is used with a negative phototool (clear conductor lines and pads) to polymerize the dry-film resist in only those clear areas of the phototool. This polymerized resist will now protect the lines and PTHs during a copper-etching process which will remove all the unwanted background copper. The photoresist is then stripped off, and the solder resist material is applied and processed through curing. Tin-lead is next added to the open component pads and PTHs by the hot-air leveling process.

An alternative process uses conventional procedures to create a pattern-plated board. After etching, however, the metal etch resist is removed chemically, leaving the underlying copper bare. Subsequent process steps are the same as SMOBC.

The primary function of tin-lead in the PTHs and on the component pads is to improve solderability and appearance. It is important to demonstrate the solderability of the holes and pads on the SMOBC panel. This is accomplished by a hot-air leveling process which places a thin coating of molten tin-lead on only those copper areas of the PTH that have not been covered by the solder resist. This hot-air leveling process proves the ability of the copper surface to be soldered and also improves the appearance and solderability after longer-term storage of the PTHs and pad surfaces.

Since there is no flowable metal under the solder resist during the hot-air leveling step or later during component soldering, the resist maintains its adhesion and integrity.

The lower metallization height of the conductor lines allows the use of a thinner dry-film resist and also makes the liquid and screen-printing application somewhat easier.

One variation on the basic SMOBC process is to make the PWB by the conventional pattern-plate copper and tin-lead process followed by etching of the background copper. Then another photoresist step is used to tent the holes and pads so that the tin-lead can be selectively stripped from the conductor lines. This would be followed by infrared or oil reflow, cleaning, and application of solder resist. A second process variation strips off the tin-lead plating completely and is followed by cleaning, solder resist application, and hot-air leveling. There are still other PWB fabricators who do not like either of the above processes and are opting to use a nonflowable, copper-etchant-resistant metal like tin-nickel under the solder resist. The major shortcoming to tin-nickel is that it is considered more difficult to solder with low-activity soldering fluxes.

16.7 CLEANING AND PWB PREPARATION PRIOR TO SOLDER RESIST APPLICATION

Optimum solder resist performance and effectiveness can only be obtained if the PWB surfaces are properly prepared prior to the application of the resist.

Surface preparation usually consists of a mechanical brush scrubbing for the non–tin-lead PWBs followed by an oven-drying step. The tin-lead PWBs should not be scrubbed and require less aggressive cleaning procedures. The cleaning options prior to solder resist application are shown in Table 16.4.

TABLE 16.4 Preparation for Solder Resist

	Panel metallization		
Operation	Copper	Tin-lead	Other
Mechanical brush	Yes	No	Yes
Pumice	Yes	No	Yes
Solvent degrease	Yes	Yes	Yes
Chemical cleaning	Yes	Yes	Yes
Oven drying	Yes	Yes	Yes

16.7.1 Surface Preparation

Dry-film and liquid-photopolymer resists applied to meet the IPC-SM-840 Class 3 requirements are particularly sensitive to the cleaning processes and baking step used to remove volatiles prior to the application of the resist.

Tin-lead-coated PWBs should not be mechanically brush-scrubbed because of the smearing potential of such a malleable metal as tin-lead. Scrubbing can cause a thin smear of the metal to be wiped across the substrate, leaving a potentially conductive path or, at a minimum, a decrease in the insulation resistance between the conductor lines. Pumice scrubbing is also unacceptable for tin-lead circuitry, since the pumice particles may become embedded in the soft metal, which can lead to poor soldering performance.

Solvent degreasing with Freon* or 1,1,1-trichloroethane cleaners is necessary with tin-lead PWBs in order to remove the light process oils from the solder reflow step, dirt, and fingerprints that are usually found on the PWB at the solder resist step. Solvent degreasing will not remove metal oxides or contaminates that are not soluble in the degreasing media.

The last step prior to the application of a solder resist to the PWB should be an oven-drying step in which absorbed surface moisture and low boiling volatiles are removed. This drying step should immediately precede the resist application step in order to minimize the reabsorption of moisture.

The most stringent performance specifications will in turn require the most stringent cleaning procedures prior to the resist application. Not all resists and performance specifications require the same degree of cleaning. In certain cases, where performance requirements are less stringent, some screen-printed resists may be used successfully without any particular cleaning or oven-drying steps.

16.8 SOLDER RESIST APPLICATIONS

Permanent solder resists may be applied to the PWB by any of several techniques or pieces of equipment. Screen printing of liquid solder resists (ink) is the most common; with regard to photoprint solder resists, the dry-film solder resists are applied to one or both sides of the PWB by a special vacuum laminator, and the liquid-photoprint solder resists are applied by curtain coating, roll coating, or blank-screen-printing techniques.

16.8.1 Screen Printing

Screen printing is typically carried out in manual or semiautomatic screen-printing machines using polyester or stainless steel mesh for the screen material. If solder resist is required on two sides, the first side is coated and cured or partially cured and then recycled to apply the solder resist on the second side using the screen pattern for that side, and then the entire PWB is fully cured.

16.8.2 Liquid Photoprint

For some liquid-photoprint solder resists, a screen-printing technique is used to apply the resist in a controlled manner to the surface of the PWB. The screen has

*Registered trademark of E. I. du Pont de Nemours & Company.

no image and only serves to control the thickness and waste of the liquid solder resist. There is no registration of the screen, since there is no image. The actual solder resist image will be obtained by exposing the coated PWB with ultraviolet light energy and the appropriate phototool image. The unexposed solder resist areas defined by the phototool are washed away during the development step.

Some liquid-photoprint solder resists require a highly mechanized process (roller or curtain coating) and therefore, because of the equipment costs and the setup, cleanup, and changeover costs, are best suited to high-volume production.

The liquid solder resists do not tent holes as effectively as the dry-film solder resist materials.

16.8.3 Dry Film

Dry-film solder resists are best applied using the vacuum laminators that have been designed for that purpose. The equipment removes the air from a chamber in which the PWB has been placed. The solder resist film is held out of contact from the PWB surface until atmospheric pressure is used to force the film onto one or both sides of the PWB.

The roll laminators that apply the dry-film resists for plating or etching are usually found to be unacceptable for solder resist application. The roll laminators were designed to apply a resist to a smooth, flat surface such as copper foil, and not to a three-dimensional surface such as an etched and plated PWB. Air is usually trapped adjacent to the conductor lines as the lamination roll crosses over a conductor line running parallel to the lamination roll. Entrapped air next to the conductor lines can cause wicking of liquids, which in turn causes reliability and/or performance concerns.

The dry-film resist thickness, as supplied, is usually 0.003 or 0.004 in and will meet the requirement of the Class 3 specification for a 0.001-in minimum thickness of solder resist on top of the conductor lines. The resist thickness chosen depends upon the expected thickness of the copper circuitry to be covered, allowing for filling of the spaces between circuit traces with resist.

16.9 CURING

Once the solder mask has been applied to the PWB, it must be cured according to the manufacturer's recommendations. Typically, curing processes are thermal curing by oven baking or infrared heating, UV curing, or a combination of the two processes. The general objective of the curing process is to remove any volatiles (if present) and to chemically cross-link and/or polymerize the solder resist. This curing toughens the resist to help ensure that it will maintain its integrity during the chemical, thermal, electrical, and physical exposure the PWB will see during its service life.

Undercuring, or an out-of-control curing process, is usually the prime cause for solder resist failure. The second leading cause for failure is inadequate cleaning prior to solder resist application.

Special Note—Inspection before Curing. All PWBs should be carefully inspected for defects prior to curing. Once a solder resist has been cured, it is usually impossible or impractical to strip the resist for rework without seriously damaging the PWB.

CHAPTER 17
BARE BOARD TESTING

George Hroundas
Trace Instruments, Canoga Park, California

17.1 INTRODUCTION

Justification for testing at the earliest stage in the manufacturing cycle results from increased complexity of printed wiring boards (PWBs), higher board failure rates, and greater costs for defects occurring at each stage of assembly. As assemblies become more complex, their circuitry requires an increasing number of through-holes and networks. A typical early PWB had an average of 400 through-holes—of which 25 percent were via holes—and an average of 200 networks (e.g., connections between through-holes and via holes). As the trend toward denser circuits continues, the typical board is likely to contain 2000 or more through-holes—40 percent of which are via holes—and to have an average in excess of 600 networks (see Fig. 17.1).

These highly populated boards have led to higher failure rates for PWBs and substrates; typically, failure rates have increased from approximately 3 to 5 percent to 5 to 20 percent. For this reason, the bare board itself is treated as a component, and testing requirements have been developed to prevent the propagation of bare board faults at the level of the assembled board.

17.2 REASONS FOR BARE BOARD TESTING

Certain intrinsic failure mechanisms have been identified which can lead to circuit board failure at advanced stages in the manufacturing and test process. These are especially apparent in high-density, multilayered boards which have trace separations of 0.01 in or less. One example of such a mechanism is the occurrence, under high humidity, of an electrochemical growth of a conductive copper compound along the internal glass filaments of the board. Another example is improperly applied dry-film solder mask, resulting in voids left between conductors which can trap contaminants and moisture. In some cases, although the insulation failure can be cured by baking the boards, the failure will reappear at a later stage. This type of board failure results in such faults as slivers, opens, cuts, leakage, and contamination. In almost all cases, these faults can be identified during

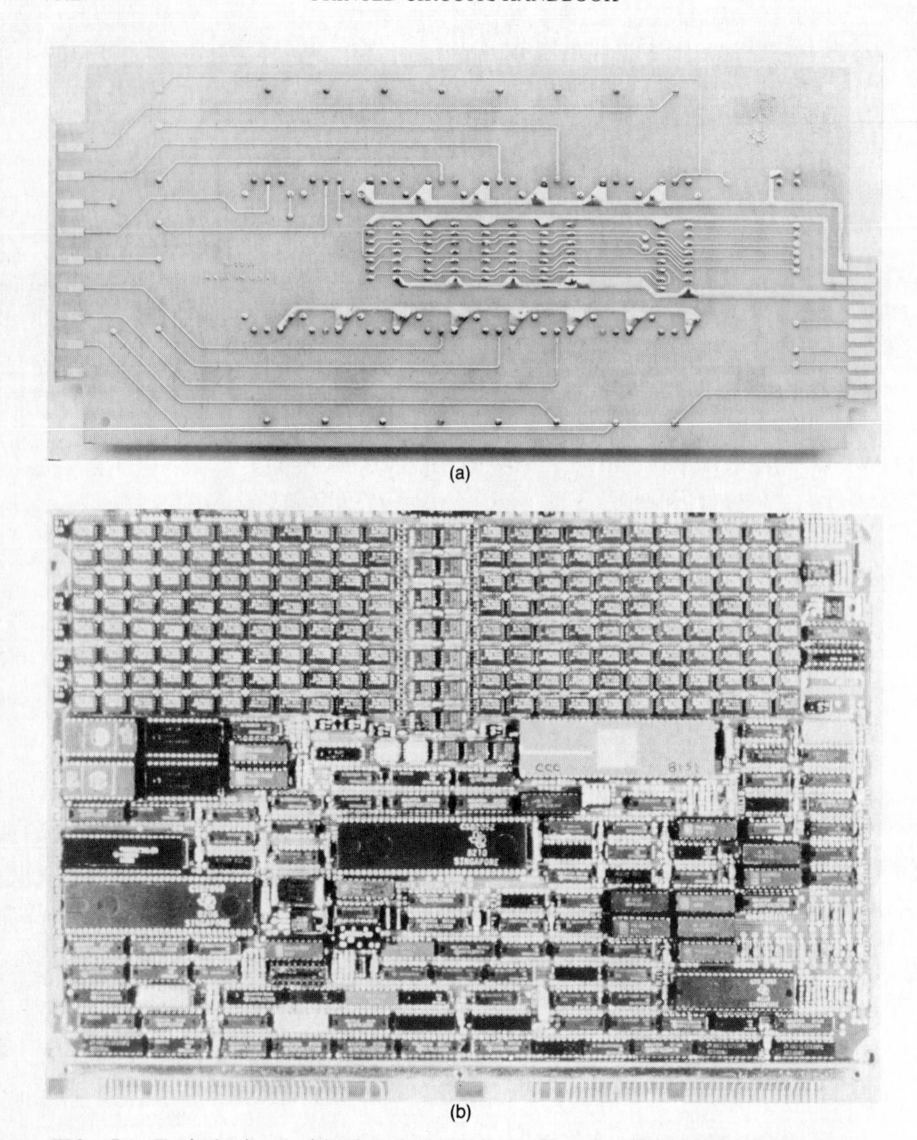

FIG. 17.1 Typical printed wiring boards. (*a*) Early configuration; (*b*) modern configuration.

the manufacturing process by isolation resistance threshold* testing at the standard level of 100 MΩ, with stimulus voltage of 100 V or higher, so that subsequent failures in the field can be avoided. The purpose of this chapter is to examine the various board test techniques and strategies that can prevent costly failures at the level of the assembled board.

*Isolation resistance is an electrical parameter that describes how well two points are electrically isolated from each other. The isolation threshold level is a high-resistance value (100 MΩ) that the industry accepts as defining that two points are in fact isolated.

17.2.1 Cost Comparison

Detecting and removing a board fault at the bare board level is almost always much less expensive than detecting and removing it later. For example, it often costs 50 cents per fault to detect it at the bare board level, as compared with $5 per fault at the loaded board level, $50 per fault at the product level, and at least $500 per fault for an in-the-field diagnosis. This is the so-called 10× rule.

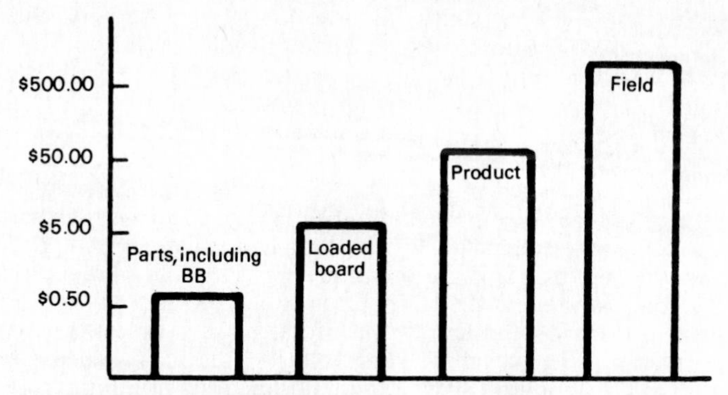

FIG. 17.2 Costs to find defects by level of complexity.

Depending on board complexity, failure rates of PWBs can range from 5 to 30 percent. Figure 17.2 illustrates the typical order of magnitude cost increase that occurs when faults are detected at higher levels in the manufacturing process.

17.2.2 Surface Mounted Devices and Substrates

The increasing complexity of printed wiring assemblies (PWAs), with high populations of LSI and VLSI integrated circuits, many of them surface mounted, has led to very high fault detection and removal costs at the assembled board and product levels. Reductions of more than 40 percent in PWB size have been achieved through advances in surface mounted device (SMD) technology. These increased densities require electrical testing at the bare board level, since the very complexity of the surface mount assembly makes faults costly to find and correct due to the degree of technical expertise required to diagnose them. Furthermore, unless the assembly was designed to be tested, it may be impossible to find the fault or to correct it, making it necessary to scrap the entire assembly. Visual or optical inspection of these assemblies is very slow and ineffective compared with electrical test techniques, which assure continuity and high isolation.

Ceramic substrates have been used successfully for some time in the manufacture of hybrid microelectronic devices. Ceramic circuit board assemblies have continued to increase in density and complexity. Substrates typically range from 2 in^2 to 10 in^2 in size, and even larger sizes are used for military, aerospace, and medical applications. Most problems that arise include contamination, shrinkage, bow (warp) and twist, and misregistration. Electrical connection to the substrate still remains the biggest general problem. Accessing only the outer connecting pads usually achieves inadequate diagnostic information. Those test methods that allow access to any point within the substrate are ideal for both in-process inspection and final electrical testing.

Unlike the photoresist and chemical etch processes used to make copper-clad

PWBs, in ceramic manufacture circuits are silk-screened onto the ceramic and fired in a high-temperature kiln. Also, unlike copper-clad PWBs, ceramic printed circuits have no tooling holes to use for reference during manufacture or testing. Therefore, all registrations are referred to the edges of the card. During one screening process, screening remains dimensionally stable to the outside edges of the ceramic circuit assembly. However, it is recommended that for each screened lot, test head alignment should be performed.

Various methods of locating and aligning the product are available. Spring contact probes (see Fig. 17.3) similar to those commonly used in bare board testers can also be used in ceramic circuit board probing applications. The probe spring pressure is a very important element in this procedure. Very high pressures can physically damage the pads. Very low pressures can cause a high resistance path, in which case faulty results will occur. Furthermore, probing ceramic substrates requires that the substrate under test be precisely aligned so that every electrical connection is well made. Otherwise, if any electrical connections are missed, false errors may be reported, resulting in high scrap rates. Or some errors may escape being reported altogether, resulting in a faulty finished product.

Experience in the manufacture of substrate products has proved that distribution of testing throughout the entire production process reduces overall costs. High-resistance electrical testing resolves problems such as leakage, contamination, and isolation. The stimulus-testing voltages used today range from 100 to 250 V, utilizing the standard 100 MΩ threshold, and are a minimum requirement.

17.2.3 Inner Layers

The inner layers in multilayer circuits consist of separate conducting circuit planes, each separated by insulation materials and bonded together into thin, uniform structures. Each circuit may have internal and external connections to other levels of circuitry, as needed. These are called "buried" or "blind" vias and are explained elsewhere in this book. Such circuits offer important weight and space savings, and in many cases allow easy replacement and simplified wiring or connection systems.

The reason for testing inner layers is to find open and short circuits caused by contamination, misregistration, and delamination. Electrical testing excels in determining circuit integrity, simply because circuit integrity easily translates into tests for opens in the conductor paths and shorts between adjacent conductors. Experience has shown that the electrical testing steps taken to assure circuit integrity also account for the future reliability of the board. This finding has been confirmed by printed circuit shops which have automated electrical testing facilities. Most manufacturers perform 100 percent electrical testing, using the standard isolation resistance of 100 V/100 MΩ for every layer.

17.3 FAILURE ENVIRONMENTS

Although the reliability of the printed circuit board is normally taken for granted unless a serious equipment failure occurs, several factors can affect the functioning of a circuit board. The PWB not only must perform well in its particular application but also must be able to withstand the environmental hazards it may be exposed to, depending upon the mechanical, electrical, and chemical nature of the product. To a large extent, this ability depends upon electrical parameters based on the quality of the copper-clad laminates used in the printed circuit man-

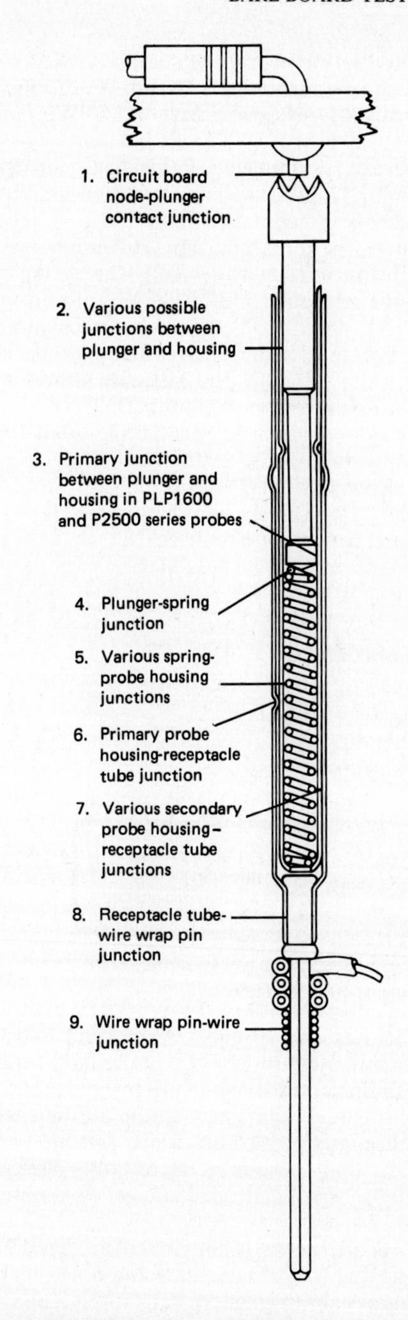

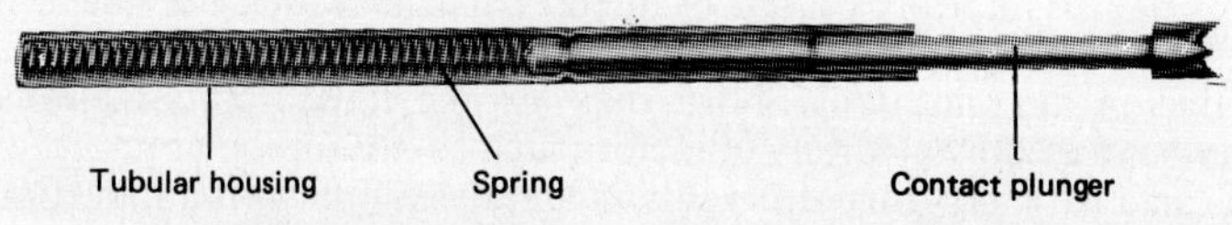

FIG. 17.3 Spring contact probes. *(Courtesy of QA Technology, Inc.)*

ufacturing process. Laminate quality, however, can be adversely affected by circuit manufacturing processes or assembly procedures. Harsh conditions such as heat, mechanical stress, and chemical exposure are all part of the PWB fabrication process.

Contamination on circuit boards can result in electrical leakage, corrosion, and failure of electronic components in high humidity. The equipment that tests contamination must detect and measure ionic contamination and also supply data that will allow isolation of the sources of contamination. Because contaminants are often electrically conductive substances, they will usually change the electrical characteristics of an uncontaminated substance. Unfortunately, at nominal resistance measurements (i.e., 2 MΩ or less), these problems can be initially undetected but can cause catastrophic board failure at a later stage. The performance of a high-isolation electrical test such as the standard 100 MΩ with stimulus voltage of 100 V or greater helps to detect most levels of contamination.

Some faults, such as those found on single or simple double-sided PWBs, are easy to remove at any level. For example, a sliver between two circuit paths caused by faulty artwork can be easily removed with a file or a knife. A short caused by a sliver in an inner layer of a multilayer board, however, is much more difficult to remove. Careful drilling through the circuit board may be required to clear the short; in some cases, the fault may be impossible to clear.

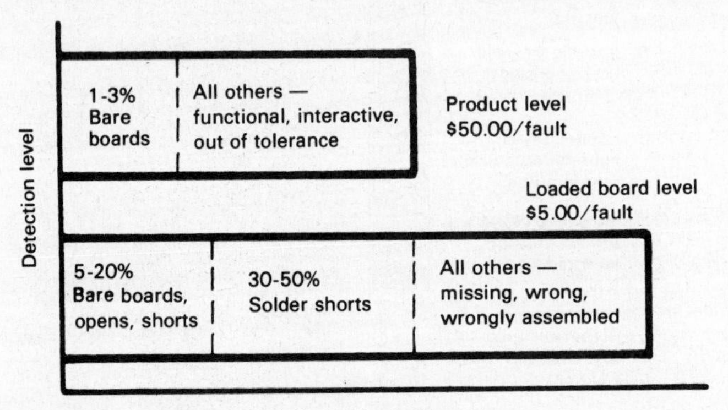

FIG. 17.4 Percentage of defects—loaded boards.

Figure 17.4 shows the typical fault distribution and cost per fault for detection and removal of various types of faults in the average electronics manufacturing process. Examination of these data shows that most faults are detected at the assembled board level and that those faults which do escape detection are expensive to find. Furthermore, it is clear that those bare board faults detected at the assembled board level constitute a significant percentage of the total fault spectrum.

A typical distribution of bare board faults is illustrated in Fig. 17.5. These include shorts due to slivers, opens due to overetch, cuts and holes, leakage, and contamination. Figure 17.6 illustrates examples of typical short circuits and open circuits. A short is usually caused by metal particles which may have been left behind from the etching process and which were not completely washed away, or by scratches in the phototool. Shorts may also be caused by contaminants. An open may be caused by a variety of factors such as dirt on the negative, overetching in the acid bath, clogging of the silk screen mask, improper application of acid resist, improper plating, and other errors in the PWB fabrication process.

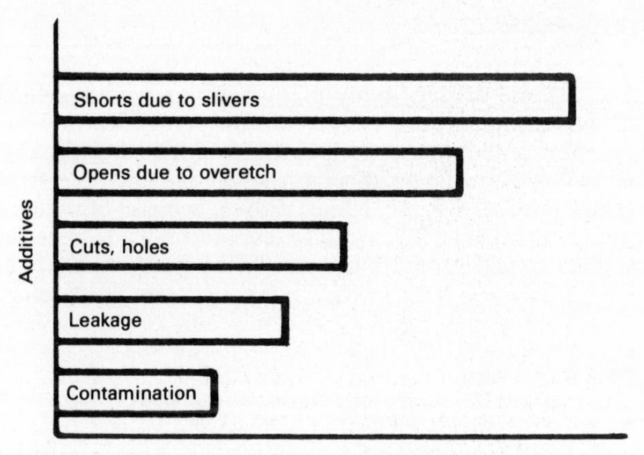

FIG. 17.5 Percentage of defects—bare boards.

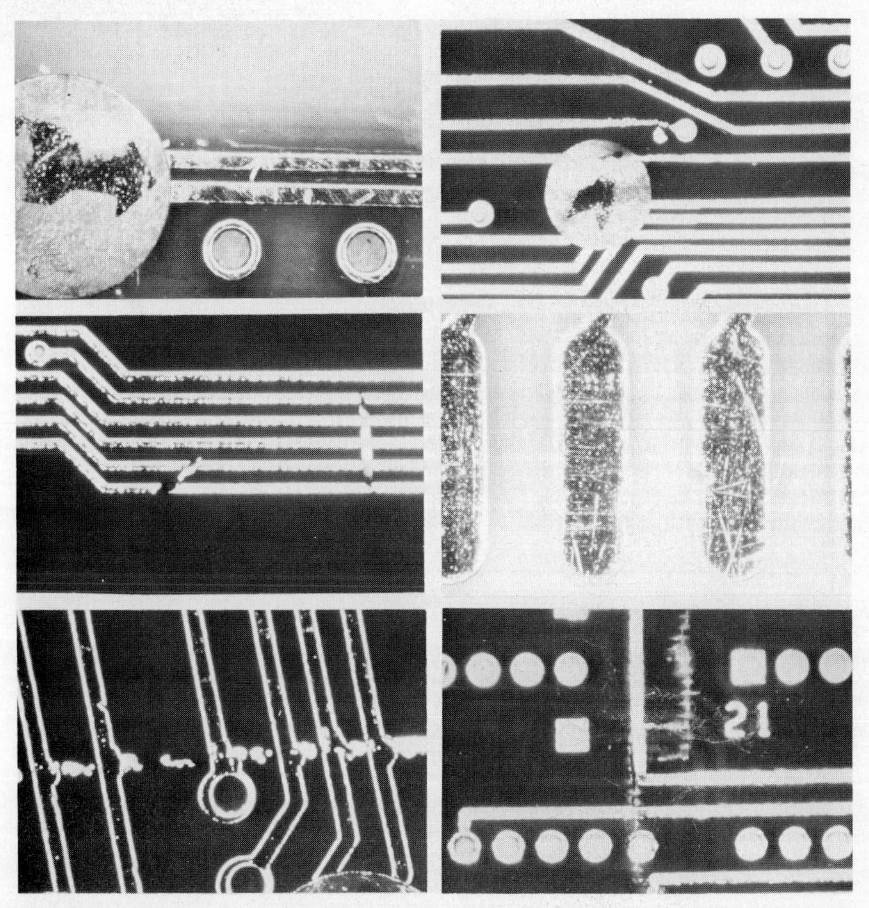

FIG. 17.6 Typical circuit board faults.

17.4 TESTING TECHNIQUES

As boards have become denser, many manufacturers have installed automatic test equipment for testing as many as 50,000 test points, with up to 10,000 networks. With so many potential tests to perform, board testing must be automatic, and individual switch elements must be very small to avoid excessively slow testing and bulky equipment. Also, as defined above, because of potential contamination, electrical tests must be high-voltage tests and must have isolation thresholds in the range of 10 to 100 MΩ in the case of isolation resistance testing.

17.4.1 Isolation Resistance

Isolation resistance tests are characterized by time and voltage *independence.* The exact test voltage must be ascertained accurately in order to calculate the resistance. The test should be made from each network to all other networks combined, rather than from network to network in turn. Additional tests can be performed, however, for rework purposes to identify a leakage path. Isolation resistance measurements are made with 100 to 250 V applied, and the failure threshold is typically 10 to 100 MΩ. In some cases it is necessary to limit the test current to avoid destroying the leakage path while leaving a potential for regrowth. Experience shows that the effective shunt leakage of the matrix should be less than 100 μA in order to be small compared with 1 μA, which is the most sensitive setting for isolation resistance tests (100 V divided by 1 μA equals 100 MΩ).

17.4.2 Breakdown

Unlike isolation resistance tests, breakdown tests are characterized by time and voltage *dependence.* The voltage required for isolation breakdown is often a function of the time the voltage is applied. Typically, a breakdown test must be applied for 10 ms or more to achieve consistent results. The discharge energy should be limited to avoid damage to the board under tests. Breakdown testing typically results in tens of milliamperes of fault current. This type of test may be unnecessary in cases in which the operating voltage applied to the board is low, as is the case with logic boards.

17.4.3 Continuity

Continuity testing is often required along with high-voltage testing. For a high-voltage test to be valid, a continuity test must be made between two probes attached to the same network. This assures that the probes are actually in place and that the network is connected to the probe. It is desirable that the same fixture and matrix be used to confirm the validity of the high-voltage tests. Continuity measurements are usually made in a 10 to 500 Ω range. Lower ranges are more accurate but are also slower to test, requiring a decision on these trade-offs.

For the continuity measurement to be accurate, the effective series resistance of the matrix should be less than 1 Ω to be small compared with 10 Ω, which is the lowest commonly used continuity measurement threshold.

17.5 CIRCUIT TECHNIQUES FOR PARAMETER CORRECTION

One technique for escaping the consequences of excessive series resistance in switches used to connect an unknown resistance to a measuring bridge is the kelvin, or 4-terminal, connection (see Fig. 17.7). With this connection, current passes through the resistance under measurement and through a pair of switches separate from those used to connect the potential measuring instrument. Thus the voltage drop across the current-carrying switches is not included in the voltage measurement, and the effective switch resistance is nearly zero.

A technique called "guarding" also exists for reducing the effects of the "off" leakage. For example, in a matrix of perhaps 40,000 points, there will be approximately 40,000 leakage paths around the measurement through the switches which are "off"; 4000 times 10 nA equals 40,000 nA, or 400 μA, which could certainly swamp the measurements. It is important that the tester have this guarding feature of source switches arranged so that groups of switches are isolated from the measurement bus by a single switch, and the leakage current of all the isolated switches is connected through a drain or guard wire.

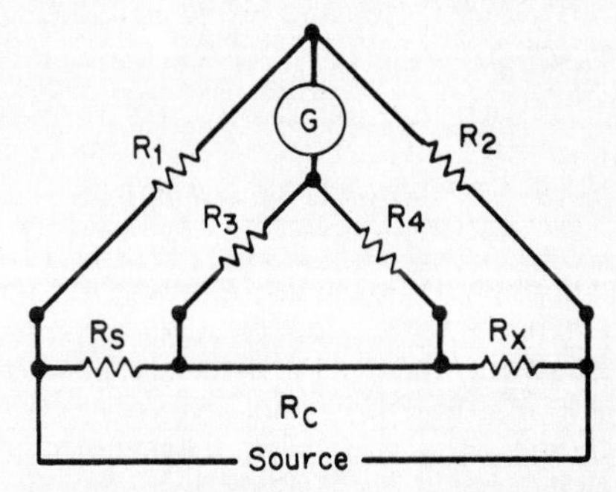

FIG. 17.7 Kelvin bridge—also called a double or Thomson bridge. A seven-arm bridge for comparing the resistances of two 4-terminal resistors or networks. Their adjacent potential terminals are spanned by a pair of auxiliary resistance arms of known ratio, and they are connected in series by a conductor joining their adjacent terminals.

17.5.1 Available Switch Elements

Devices considered suitable for switch elements include DMOS, FET, CMOS, MOS, BIPOLAR, SCR, and RELAY. DMOS, FET, and CMOS, although they have excellent "off" characteristics and no inherent offset voltage, cannot be used as high-voltage devices. The latching characteristics of the SCR make it awkward to control in a matrix. The relay has ideal electrical properties, such as high "off" resistance, low "on" resistance, and control isolation. However, the relay is unsuitable for low-voltage testing because of its size, cost, and slow rate of throughput, and it is only chosen for testing voltages of over 300 V. Mercury relays offer long life and low "on" resistance. These are used in the switch matrix of in-circuit and functional test systems. MOS and BIPOLAR transistors are preferred for a high-voltage switch; however, both devices have considerable "off" leakage and "on" resistance, which must be compensated for by guarding.

17.5.2 Elemental Matrix Cell

The above examples of source and sink switch guarding have been given with bipolar transistors for high-voltage elements and CMOS switches for control. Bipolar transistors can be used for source and sink switches because they are available at voltages over 300 V and can be photon-coupled for source switch use when the drive must be floating. Because of their excellent "off" characteristics, CMOS switches may be used for control.

Switches for kelvin connection must withstand high voltages and must have no offset (see Fig. 17.7). MOS devices are suitable for this application. CMOS switches may be used for control and to enable one MOS device to serve as either the source potential switch or the sink potential switch. These elemental matrix cells can be combined, using the series and parallel guarding techniques previously described, and using suitable address logic to form a very large switch matrix.

17.6 FIXTURING TECHNIQUES

Although bed-of-nails fixtures tend to look alike, there are a number of factors to consider so that reliable and cost-effective performance can be achieved. One important factor to consider is the maximum size of the boards to be tested. For this reason, most standard grids are wired for up to 50,000 test points, even though the individual test system may be purchased with fewer test probes and switch cards, each of which supports 128 test points. To minimize initial investment, the purchaser should have the option to postpone the additional test probe and switch card expense until additional test capacity is required. At that point, simply by adding the appropriate number of switch cards and inserting the corresponding test probes into the already wired empty sockets on the grid, the test area can be expanded as required.

17.6.1 Standard Grid

The decision as to the number of test points to support in any initial grid configuration is based on economic trade-offs. Etching several small circuits into a single large printed circuit board (PCB) can create reduced handling requirements and, consequently, significant cost savings. After etching, drilling, and plating—and possibly stuffing this large PWB with components—it is then cut into small, individual circuit boards. Testing these boards individually is always more expensive than testing them while joined together, and the resulting savings will most likely be more than offset by the cost of the additional switch cards and probes needed. Also, the testing of several small boards concurrently as part of a larger PWB may result in higher throughput so that the need to buy an additional tester is eliminated.

Another factor to be considered is whether to purchase a system that offers a universal or dedicated grid fixture (see Figs. 17.8 and 17.9). From an economic standpoint, the choice of a dedicated grid system usually results in a smaller initial capital outlay for the test equipment because of its typically smaller specific test probe array. However, this type of equipment requires separate test fixtures, each dedicated to a specific PWB type. Each test fixture has its own individual size, configuration probe array, and tooling pins and cannot be used to test other PWB designs. A dedicated fixture (see Fig. 17.10) can cost up to $4000* or more for each unique board tested. In addition, dedicated fixtures require large amounts of storage space and additional weight and complexity in handling, and they are

*1987 cost estimate.

FIG. 17.8 Example of universal grid test system. *(Courtesy of Trace Instruments.)*

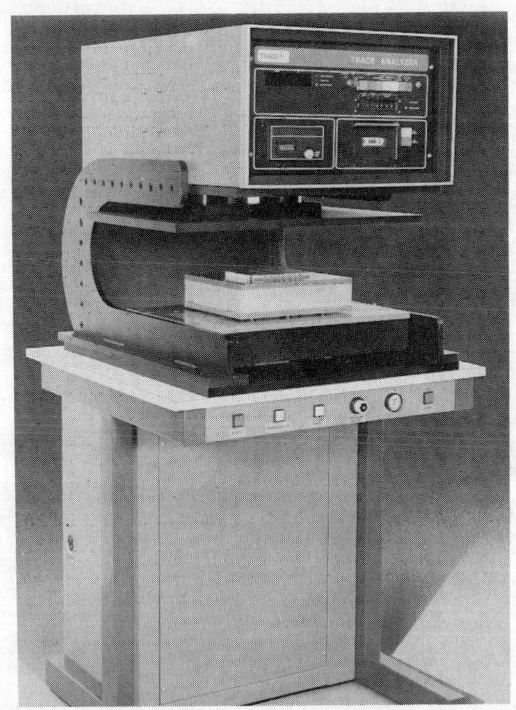

FIG. 17.9 Example of dedicated grid test system. *(Courtesy of Trace Instruments.)*

FIG. 17.10 Example of dedicated fixture. *(Courtesy of Trace Instruments.)*

usually time-consuming to fabricate. On the plus side, however, these fixtures can be used for any type of contact array, even mixed contact spacing.

The higher price of the universal test system can be offset by the economies inherent in its ability to test boards of various sizes and configurations. Even with the use of masks or translation modules, as described below, the universal system has proved to be more cost-effective for short to medium runs, particularly at the bare board level.

There are a number of test services that offer to test products for board manufacturers. It is important that the test service equipment meet the manufacturer's own quality criteria.

17.6.2 Soft and Firm Tools

Manufacturers of automatic test equipment are constantly developing new methods to lower system operating costs and to enhance the efficiency of their existing systems. One of these methods is the development of electronic masking. This unique and powerful feature makes it possible to eliminate expensive dedicated hard-tooled fixtures in favor of inexpensive soft- and firm-tooled fixtures. The advantage of soft-tooled fixtures is that they can be easily taken apart, and most of the parts can be used to build a new fixture. Soft-tooled fixtures cost about $300* and can be reused. These are best for long runs with very high variety. Firm-tooled fixtures cannot be disassembled after each test and cost about $500.*

*1987 cost estimate.

Fixture size should correspond to board size, which can range from 2 × 2 in all the way up to 22 × 24 in. Placing a board on a fixture which is too small could cause poor probe contact at the board perimeters. Conversely, placing a board on a fixture which is too large causes excessive probe compression. A good rule to follow in determining the relationship between board and fixture dimensions is that the fixture dimension should be about 30 percent greater than the board dimension. If the boards or ceramics that require testing are small or have very fine density and a small number of test points, then dedicated fixtures should be considered.

One problem faced by the user of the universal grid test fixture which is not shared by the user of the dedicated grid is that of accommodating off-grid test points. Because of nonstandard pinout locations on some components, even boards designed by computer-aided design (CAD) systems can contain off-grid pads. Off-grid test points can also result from individual packaging requirements that shift a component with leads on 0.050-in centers off-grid. With dedicated fixtures (see Fig. 17.10), it does not matter if these locations do not fall exactly on the 0.10-in centers. Users of the universal grid fixture, however, cannot directly access such off-grid locations. In this situation, a translation module is required.

17.6.3 Translation Modules

Translation modules can be thought of as fixtures themselves (see Fig. 17.11). They consist of two nonconductive pin support plates separated by spacers. The top plate is drilled to the board's actual test point locations, while the bottom

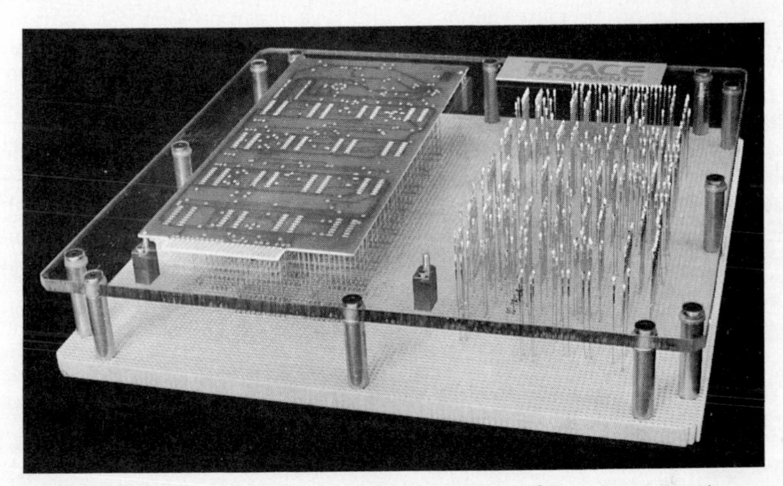

FIG. 17.11 Example of translation module. *(Courtesy of Trace Instruments.)*

plate is drilled to conform to the universal grid center-to-center spacing. A second set of test probes is loaded between each plate, connecting the corresponding holes. Although most of the test probes are straight-through connections, some must provide the needed offset between the universal grid and the off-grid test points. To accomplish this offset, two approaches are commonly used. These consist of the use of straight test probes, which are loaded at a slight angle, and the use of flexible probes. Straight probes, while relatively inexpensive, contact the board at a slight angle, and some engineers believe that this angle may negatively affect the contact integrity between the probe and the test point. In actual practice, however, the angle is so small (typically less than 4°) that its effects can be ignored. Flexible probes, however, do provide true perpendicular contact between the board and the test points. On the other hand, flexible probes may not last through as many operations as straight-angled probes, because under pressure they tend to become deformed. This deformation may prevent reuse of the probes in other translation modules; nevertheless, straight probes can provide a high level of serviceability combined with economy.

17.6.4 Mechanical versus Vacuum

Another factor to be considered is whether to choose a vacuum or pneumatic system. One of the criteria involved in this choice is whether universal or dedicated grid fixtures are preferred. Most dedicated fixtures are designed for use with a vacuum pull-down system. This type of equipment, which is less expensive than the pneumatic type, uses a vacuum to press the board against the test probes and is applicable to both bare and loaded boards. In the vacuum system, the test fixture sits on top of the test probes, and a compliant material mask is placed over the PWB to form a seal. Air is then drawn out of the fixture via a vacuum system, and atmospheric pressure forces the board firmly against the test probes for mechanical and electrical contact. However, at high board densities, the vacuum system has limitations not present in systems operated by pneumatics, even for dedicated fixtures. Many manufacturers prefer the pneumatic system exclusively because of its inherent reliability and testability. In addition, universal fixtures require much more pressure than a vacuum system can provide. For example, a universal grid with 49,000 test probes requires 3½ tons of air pressure to achieve only a 3-oz force on each probe. For this reason, most universal systems use pneumatic systems that utilize air cylinders to press the PWB against the test probes and thus provide higher pressures than vacuum-type systems. While the pneumatic system is usually slightly more expensive than the vacuum type, it allows faster operation, and thus higher throughput. Aside from the number of test probes utilized, both types of test systems are virtually identical. Both vacuum and pneumatic systems verify point-to-point continuity and check isolation between test point pairs or between each point and all other points on the board. The primary differences between them lie in the individual hardware options and unique software features provided with each system.

17.6.5 Probe Assembly Considerations

In selecting a spring probe fixture, it is always preferable to purchase it from the test system manufacturer. In this way, users can be sure that it was designed for their specific system and that it was thoroughly checked out to meet their particular requirements. The fixture should have the capability of testing a variety of products with different grid patterns and thus be capable of accepting different spring probe test heads. The fixture should be designed for access not only to all existing products but to future products as well. For this reason, consideration should be given to the maximum active probe area required, as well as to the ability of the fixture to perform accurately and consistently.

In universal probe assemblies, where center-to-center probe spacing is 0.075 in, probes are interchangeable; therefore, each probe should be mounted in its individual socket or sleeve. The plunger head of the probe should be designed to mate exactly with the specific products under test. To assure positive electrical contact with the product, the probes should exert 4 to 8 oz of force during normal engagement. In universal grid testing, the probes should have a total travel distance of 0.200 in to accommodate the use of through-hole masks and thus allow the testing of an entire family of PWBs. Internal resistance of the probes should be minimal, preferably less than 10 mΩ. In addition, the probes should be capable of at least 100 mA of continuous flow. The fixture's construction should also provide for positive isolation between each probe socket to prevent high-voltage breakdown due to dust, moisture, or contamination.

For universal probe assemblies with center-to-center probe spacing of 0.050 in, the smaller spring probe required has less force than larger probes. A probe with 7.25 oz of force is available and achieves satisfactory contact. In this size range, plunger travel is also limited to 0.130 in. For this reason, when through-hole masks are used, a Mylar material 0.007 to 0.010 in thick should be used. This material may be more difficult to use than the 0.031-in G-10 glass-epoxy, but it has the advantage of isolating those probes which are not required for a specific test.

17.7 PROGRAMMING TECHNIQUES

17.7.1 CAD/CAM

The use of networking can be effective in a computing environment where each user station on the network performs its own computing function. In the automated factory, a common database allows management to exercise controls that assure that each user works from the same design revision level. Such management control prevents errors and eliminates delays that inevitably result when engineers make changes that are not immediately passed on to the design group. In this type of network configuration, all users working on the same project receive the relevant updated information simultaneously.

In PWB applications, CAD systems must provide software that will allocate spaces proportionately for each component. This type of software requires algorithms that check each component placement with all other placements on the board, rather than commonly used pairwide interchange in which only the connections between two components are checked at any one time. The CAD system must have a flexible grid system to accommodate the fine lines required for SMT devices. These boards are designed to use lines of 0.005 in or smaller trace etch, and 0.005 in or smaller clearance between each feature. CAD systems must also accommodate components mounted on both the top and bottom of the board, with typically higher pin counts. The flexibility to create and store new electrical test parameters and, when necessary, to modify them is also a requirement of the CAD system, as is the capability to deal with different pin-to-pin spacings, routings, and grid sizes.

Many automatic board-testing systems offer downloading of data from the CAD/CAM system. When evaluating such a system, the buyer should make sure that all testing parameters can be transmitted, as well as such features as speed, default parameters, and final management reports. CAD/CAM technology is by far the most expensive way to communicate with an automatic board tester, but it has the advantage of eliminating all human error.

17.7.2 Self-Learn

All automatic board-testing systems can generate a program from a "known-good" board.* This is known as the "self-learn" method and is the least expensive; however, it requires electrical parameters to be manually set before the test (e.g., test voltage, continuity, isolation thresholds).

Older systems offer manual programming for all networks, test points, and electrical and default parameters. Such manual procedures usually are very costly in terms of setup time and expense and therefore are usually replaced by automated equipment whenever possible.

17.8 TEST STRATEGY DETERMINATIONS

17.8.1 Factors to Consider

There are several reliable methods for determining the optimum level at which various faults may be removed. The matrix shown in Table 17.1 illustrates the most common types of PWB faults: shorts, opens, cuts and holes, leakage, and contamination. With this matrix, once the fault has been identified, it is a straightforward task to select the appropriate test or tests.

17.8.2 Flowchart and Formula Approach

One reliable method for determining at which level and to what extent bare PWB testing should be implemented is to make a flowchart of the various strategies available during product manufacture. When this is done, a formula can be constructed similar to the formula shown with the flowchart of Fig. 17.12, and the actual costs for each strategy can be calculated. For an existing strategy, it is only necessary to flowchart the process, determine the existing total cost, and then

TABLE 17.1 Fault-Detection Methods at Bare Board Test

	Slivers	Overetch	Opens and cuts	Holes	High-resistance leakage	Contamination
Continuity test, Ω resolution			X			
Continuity test, multiohm resolution			X			
Isolation test, kΩ resolution	X					
Isolation test, MΩ resolution	X					
100-MΩ resolution	X				X	X

*Sometimes referred to as a "golden board."

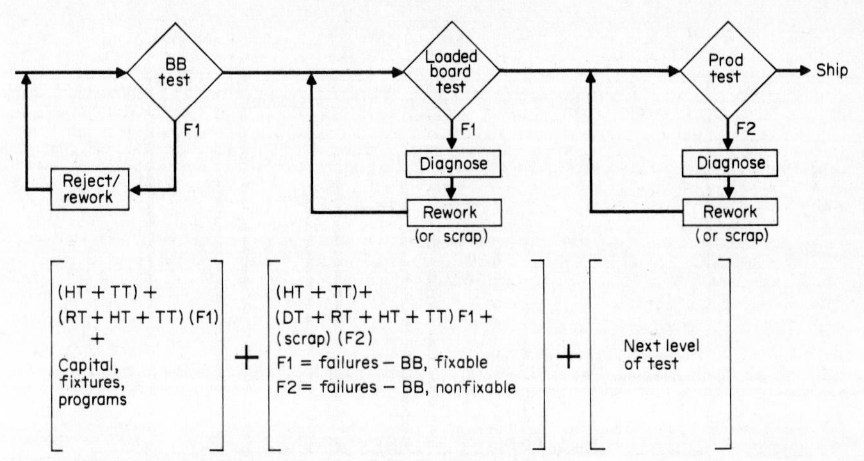

FIG. 17.12 The flowchart-formula method. F1 = first-time failure rate, F2 = second-time failure rate, HT = handling time, TT = test time, RT = rework time, DT = diagnostic time.

insert the bare board test step to see whether or not it saves more than it costs. If it does save more than it costs, then the test step should be implemented; otherwise, it should not.

To use this method, it is necessary to establish or estimate the costs for product handling time, test time, diagnostic time, rework time, and scrap, as well as the failure rates at each testing level. These figures are then entered into the formula to arrive at the recurring costs for each strategy for a given time period and for a given quantity of PWBs. It is important to note that of all the factors contained in the formula, the F1 (first-time failure rate) and DT (diagnostic time) have the greatest impact on the total recurring costs. Anything that can be done to lower F1 (e.g., to improve yield) at every level has a great impact on lowering the overall costs, and this impact increases at each increasing level of complexity. Similarly, everything that lowers diagnostic time, and thus costs, at each level has a great effect on the total. These factors are shown graphically in Figs. 17.13 and 17.14.

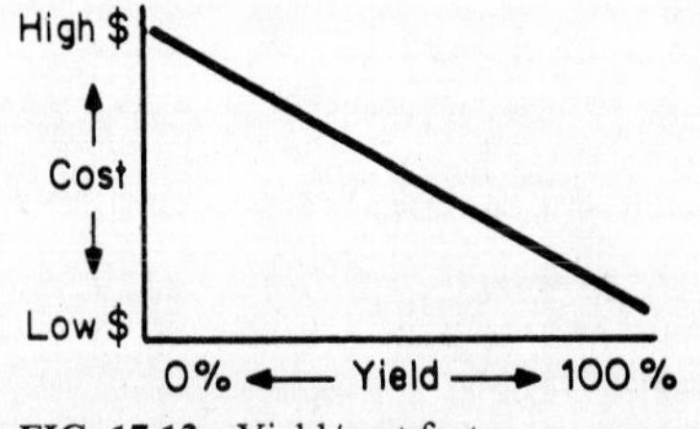

FIG. 17.13 Yield/cost factor.

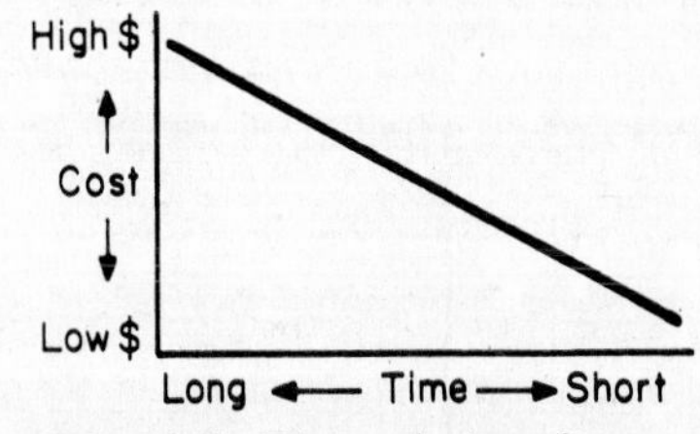

FIG. 17.14 Diagnostics/cost factor.

The recurring cost figure is only a portion, although usually the largest portion, of the total testing cost at any level. To determine the total cost of a given strategy, it is necessary to also consider capital equipment costs, fixturing costs, and programming costs. These are added to the recurring costs for each option, along with the costs associated with faults that "escape" a particular test and must therefore be found later (usually at a more expensive step in the process). Figures 17.15 through 17.17 show typical average costs for capital equipment, fixturing using various methods, and programming using various popular methods. In each case,

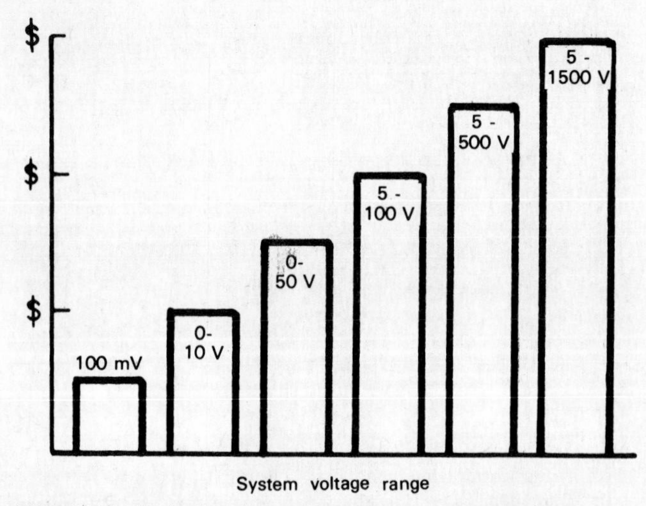

FIG. 17.15 Fault detection method characteristics—capital equipment costs. *(Courtesy of Trace Instruments.)*

there are trade-offs between the major nonrecurring costs and the major recurring costs. For example, as illustrated in Figs. 17.18 and 17.19, if test time is short because of the capabilities of an expensive piece of equipment, the cost of fault detecting is low. A definite "break-even" point can be calculated to show at which points it is better to spend less on the equipment, even though it will cost more to operate, and at which points it is better to spend more on the equipment in order to gain operation efficiencies. The same calculation of break-even points can be performed on the basis of the total number of tests to be performed (see Fig. 17.18) on a given set of PWBs, since test length contributes directly to test time in most instances.

In conjunction with the economic factors to be considered, there are several technical factors. Keeping in mind that it typically costs 10 times more money to find a fault at the next level of complexity, it is important that any previous test-

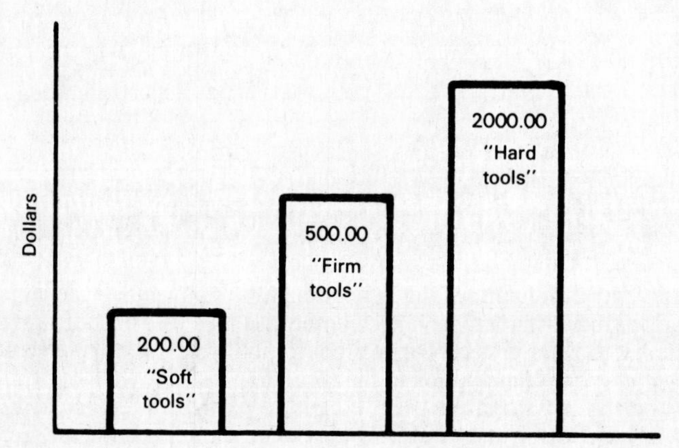

FIG. 17.16 Fault detection method characteristics—fixturing costs per 1000 points.

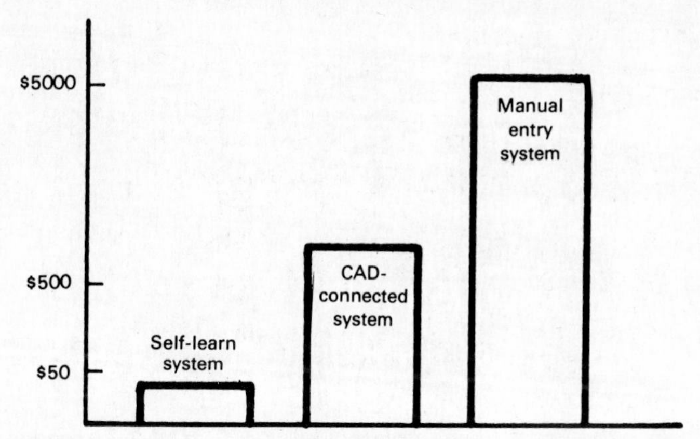

FIG. 17.17 Fault detection method characteristics—programming costs.

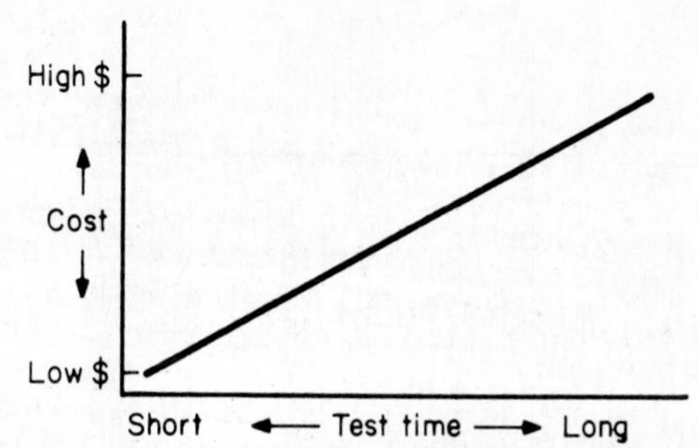

FIG. 17.18 Fault detection characteristics vs. operating costs.

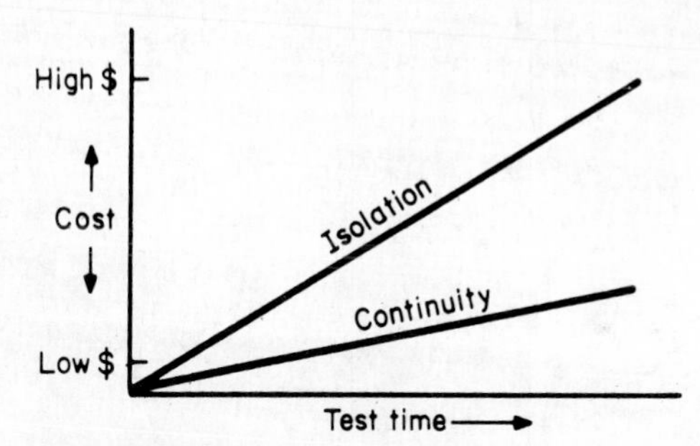

FIG. 17.19 Effects of test length on operating costs.

ing step find as large a percentage of the total number of faults as possible. This becomes one of the technical factors to be considered.

Another factor is test comprehensiveness, as illustrated in Fig. 17.20 and Table 17.2. Test comprehensiveness, and thus total test cost, is a function of both test stimulus voltage and resistance-measurement threshold. The relatively inexpensive and easy-to-achieve 1-MΩ threshold, depending on stimulus voltage level,

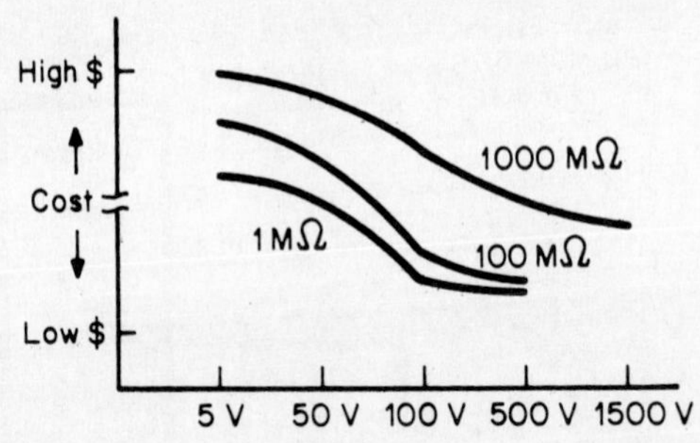

FIG. 17.20 Test cost vs. test voltage as a function of isolation resistance.

TABLE 17.2 Failure Rate Table Based on Test Comprehensiveness

Isolation resistance, MΩ	At PCB level, %	At product level, %
1	10	2
10	8	1.5
100	1	0.1
1000	0.1	0.01

can provide anywhere from 85 to 95 percent fault coverage at the bare PWB level. This means that, depending on capital equipment costs, test costs, and next assembly costs, a 1-MΩ threshold test can be quite effective economically—especially at high stimulus voltages (e.g., 200 to 500 V dc). The standard 100-MΩ threshold is more expensive to reach using low stimulus-voltage levels, so it becomes economical only with stimulus voltages of 100 V dc or more. In addition, the 1000-MΩ threshold is very expensive compared with the others, since the equipment and fixturing costs are usually very high to preserve electrical integrity at the 1000-MΩ level. In fact, if we look at the effect of stimulus voltage on capital equipment costs at the standard 100-MΩ testing threshold (see Figs. 17.21 and 17.22), we can see that it is difficult to achieve adequate resolution at

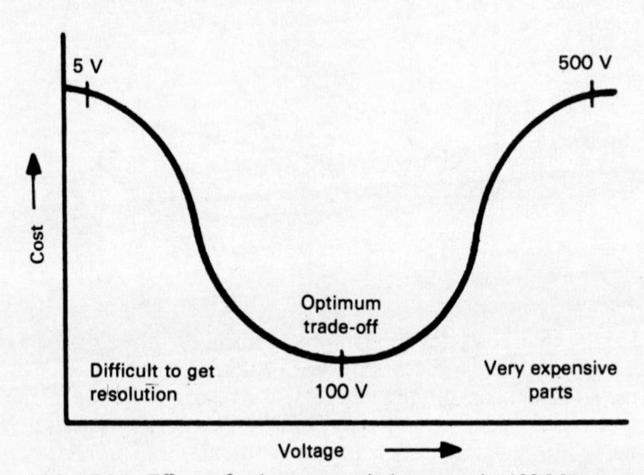

FIG. 17.21 Effects of voltage on capital costs at the 100-MΩ standard isolation resistance level.

5 V dc, and the resulting cost is high. At the opposite end of the voltage curve, where the standard 100-MΩ threshold resolution is easy to achieve, we find that the cost for 500-V-dc components is excessive. Thus, it turns out that the least expensive equipment capable of resolving the standard megohm threshold operates at about 100 V dc.

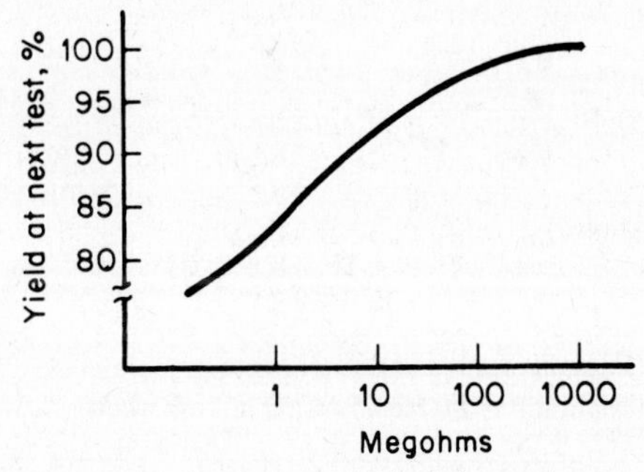

FIG. 17.22 Yield based on isolation resistance (fault coverage).

Another advantage to using 100 V dc for bare board testing at high isolation resistance levels is test speed. At 5 V dc, the test current for a standard 100-MΩ measurement is only 50 nA. This current is so small compared with the noise currents unavoidably injected into a test system (by 60-Hz power lines, radio transmitters, and other sources) that extensive filtering or integration is required to recover the signal with any degree of accuracy. In a 100-V system, the current signal-to-noise ratio is 20 times greater than in a 5-V system, and therefore the time constant of any filtering circuitry can be shorter by this same factor of 20. Thus, a 100-V system gives a 20:1 improvement in test speed as compared with a 5-V system. Above 100 V dc, however, the time needed to charge and discharge system stimulus and measurement paths and fixture capacitance erodes the gains made in the filter circuit time constant, extending test time again even though improving accuracy.

Tables 17.3 and 17.4 show the effect of measurement threshold on fault cov-

TABLE 17.3 Economic Example 1

10,000 boards, 10% defect rate, 10 V, 10-kΩ threshold

	With bare board test	Without bare board test
BB* capital equipment	$ 50K	$ —
BB fixtures	10K	—
BB programs	5K	—
BB operations	5K	—
LB* fault isolation	5K	50K
LB rework	5K	50K
LB scrap	1K	10K
FP* fault isolation	10K	20K
FP rework	1K	2K
FP scrap	1K	2K
	$ 93K	$134K

*BB = bare board; LB = loaded board; FP = final product.

TABLE 17.4 Economic Example 2
10,000 boards, 10% defect rate, 100 V, 100-MΩ threshold

	With bare board test	Without bare board test
BB* capital equipment	$ 60K	$ —
BB fixtures	10K	—
BB programs	5K	—
BB operations	5K	—
LB* fault isolation	1K	50K
LB rework	1K	50K
LB scrap	1K	10K
FP* fault isolation	1K	20K
FP rework	1K	2K
FP scrap	1K	2K
	$ 86K	$134K

*BB = bare board; LB = loaded board; FP = final product.

erage and the resulting failure rate at the loaded board and product level. It is clear that there is a significant increase in "next assembly" yield using the standard 100-MΩ threshold as compared with a 1-MΩ threshold. Yet the difference between the standard 100-MΩ threshold yield and the much more expensive-to-achieve 1000-MΩ threshold is minimal. Thus, in terms of return on capital equipment investment, there is much more leverage in using the standard 100-MΩ level and achieving virtually the same fault coverage. Such is obviously not the case with a 1-MΩ threshold, however, because even though the equipment may be cheaper, the higher next assembly test and fault isolation costs far outweigh the small savings in capital equipment.

17.8.3 Impact on Next Assembly Yield

To illustrate this point, Tables 17.3 and 17.4 show total test cost calculations using a 10-V-dc system with a 10-kΩ threshold (the low threshold is needed to achieve comparable speed with higher voltage system), and using a 100-V-dc system with the standard 100-MΩ threshold. While bare PWB testing is indicated in both cases, the total test costs with the 100-V-dc, 100-MΩ threshold system are

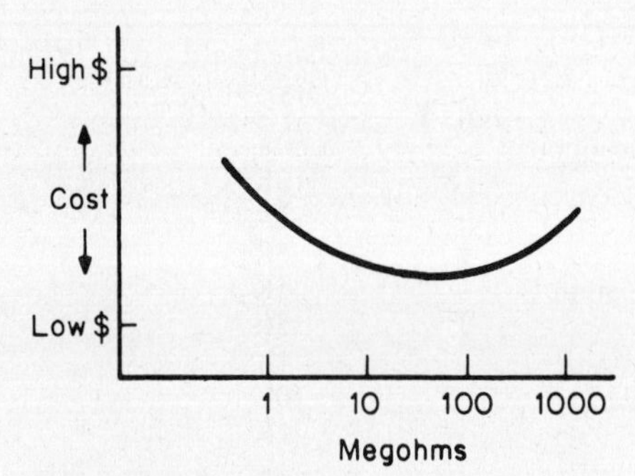

FIG. 17.23 Sensitivity analysis—overall test cost vs. isolation threshold.

certainly lower. The big difference is in increased yield at the loaded board and final product test levels. The cost to get from 10 to 100 V dc and from a low resistance threshold up to the standard 100-MΩ threshold is negligible. Figure 17.23 illustrates all the trade-offs. This chart combines capital equipment costs; programming, fixturing, and operating costs (including the effect of test speed and wrongly diagnosed problems); and next assembly costs all in one graph. It clearly shows that the standard 100-MΩ testing approach, which is technically the best, is also economically the best.

Another economy that is being implemented in production environments is automatic, or robotic, board handling. To determine whether these methods are justifiable in a given operation, enter the new HT (handling time) in the test flowchart formula in Fig. 17.12, and recalculate the total test cost.

17.9 ADDITIONAL TECHNICAL CONSIDERATIONS

Computer-aided design technology has been widely adopted in the production of circuit boards. CAD systems can be used for all steps of artwork generation, and in conjunction with plotters they can produce highly accurate artwork masters. With the sophisticated database and memory of these systems, a wide variety of board sizes and types can be produced. Advantages of using CAD techniques include not only ease of making design changes but also ease of generating complete documentation for the fabrication of circuit boards. This could include assembly drawings, parts lists, and even tapes for numerically controlled drilling and insertion equipment.

The original engineering schematic is always the first step in the PWB design and documentation process. When the schematic is entered into the system, it usually involves the placement, rearrangement, and interconnection of the schematic symbols as well as annotation and generation of the hardcopy input. Other steps in this process include generation of the list of test points, number of networks, components, component and signal names, pin names, device types, and connector information. Also included are the assignment of pins to gates and gates to multicomponent packages, placement of components on a matrix-grid pattern interconnected to other packages, editing, and checking of all electrical and mechanical elements.

The benefits of a sophisticated CAD system are most obviously realized at the final documentation step. At this point, the layout is output to a photoplotter or scribing plotter to generate the artwork for silk screens, soldering, masks, or drill templates. Tapes can be generated for numerically controlled drilling, board test fixtures, automatic insertion equipment, and automatic printed circuit board testers, or they can be generated for universal data exchanges to other systems. The schematic input is also used to generate schematic drawings and parts lists. The system can also supply information and reports relating to stock control, inventory, and other data needed in manufacturing.

17.9.1 Testability

In the fabrication of printed wiring boards, the role of testing is a vital one. Early electrical testing of bare boards, loaded boards, or components always leads to early identification of failures, thereby facilitating economies in the production process. After the product has been shipped, the cost of detecting and repairing the circuit fault increases dramatically. As previously noted, a defect identified and corrected on the bare board may cost 50 cents. After the board has been loaded, the cost will be $5; and if the defect appears after the product has been manufactured and shipped, the field service expense brings the costs up to $500 or more.

Printed wiring board faults can be categorized by defects caused by errors and deficiencies in board fabrication, component placement, soldering, or other processes. Some examples include opens and shorts due to broken or missing wiring, wrong component placement, missing or reversed-direction components, bent leads, defective solder joints, and substandard etching or plating. Component faults can be caused by defective substrates, diodes, transistors, or ICs. In this case, the fault is in the component itself and is not caused by the connection process that makes the component part of the circuit. Performance faults of loaded boards are characterized by test conditions, interaction of components, excessive noise, improper outputs, and distorted signals. In addition, the part or component may be voltage-sensitive or temperature-sensitive.

Bare board testing, the first step in the circuit testing process, will ensure that the PWB has no opens or shorts. Bare board testing programs can be automatically generated from a known-good board or input to the system as a point-to-point wire list. In this way, a defect can be located at the earliest possible moment. Testability itself can become a very expensive factor if it is not taken into consideration at a very early stage in the PWB fabrication process. Each board should be designed in a format so that its shape, size, and pad accessibility make it very easy to tool, probe, and test. Cost effectiveness of the finished product is heavily dependent upon board testability at each stage of the manufacturing process, including the field service stage.

17.9.2 Management Reports

Most electronics manufacturers consider improved productivity and quality a top objective. In addition to reducing production costs and increasing production control, automation offers the manufacturer a potential for collection and analysis of test data from many steps in the manufacturing process. This potential, however, is often not realized if each segment of the process remains isolated from every other segment instead of acting as an integral part of a unified whole. To form the foundation for an automated factory, a comprehensive network can be designed that can link these segments together. Such a networking approach requires an overall engineering plan plus applications software to support the network. By making use of nodes, or strategic points in the network, such software can better control manufacturing processes, perform more comprehensive tests, increase productivity, and give managers better data for making decisions.

There is a great deal of momentum for such networks on which test engineers can base an automated factory. In an ideal facility, all information will flow through the network for easy retrieval by management. Both equipment and personnel will share data as needed, and as a result, the automated factory will be a highly productive environment using company resources to the maximum extent. In such a competitive factory, materials, testing, and production planning systems will be able to use real-time information on orders, inventory, testability, and development of schedules. Test engineers will be able to provide managers with the latest test data to assist the manufacturing process. In addition, managers will be able to control and administer test time and yields and improve productivity with a minimum of paperwork.

In this kind of environment, all testing parameters could be downloaded from a host computer. Special programs could be written to provide detailed technical information; for example, the engineer could specify the number of test points and the number of networks and their formats, as well as the electrical testing parameters. These programs could yield the test results for each individual board tested, plus the overall results of the day's testing at the end of each shift. Such reports are vitally important to management and could instantly provide information on productivity, the number of products that were accepted, the ratio of accepted versus rejected products, and specific information on the types of failures, as well as the exact physical location of a fault on an individual board. This type of information could be very valuable to management, since it provides traceability that will allow correction of manufacturing problems at the point where the defect occurs.

Many large corporations which have plants located throughout the world maintain testing facilities at each plant which test products for various parameters. These facilities could be linked together to a central corporate headquarters. With a push of a button, management could obtain test results, yields, productivity information, and cost analyses for each plant. The combined information from all these plants could provide corporate officials with instant information to use for sales forecasts, delivery scheduling, market share data, profitability, and long- and short-term planning for growth.

17.9.3 Optical Inspection

The growth of automatic test equipment technology has pushed the development of automatic optical inspection techniques. As PWBs have become more complex, with smaller traces and spaces, visual inspection practices have become less effective and more expensive. Five to ten percent of all the workers involved with the fabrication of double-sided and multilayer boards are engaged in visual inspection, and many manufacturers would rate their inspection effectiveness as low as 10 percent. Even 20 percent effectiveness would be disastrous for those boards with 10 layers or more.

Shorts and opens can be easily found in finished boards by electrical testing, and many manufacturers now use these methods. However, visual inspection is still used in the manufacturing process. A cost-effective method is needed that will assure that only good boards go on to the final manufacturing stages. The best answer may be with automatic optical inspection and final electrical testing.

Printed circuit failure can sometimes be attributed to construction features and processing problems, and sometimes to handling. Environmental factors such as heat, mechanical stress, and chemical exposure can also play a part in the reliability or failure of the finished board. To assure ultimate quality and reliability, the product must be monitored and tested throughout the entire manufacturing process. Optical scanning of inner layers has proved to be an important step in the monitoring sequence. This type of inspection can detect numerous defects and thus increase throughput. It is obvious, however, that optical testing techniques cannot perform electrical testing. Inner layers and circuit boards, like electrical devices and components, must contact current through various traces and must provide high isolation throughout the board. Leakage, contamination, conductivity, and isolation cannot be detected by an optical tester. To assure the high reliability, performance, and high standards of the electrical parameters demanded from today's printed wiring boards, the electrical testing of every board has become an essential part of the fabrication process.

REFERENCES

1. Christopher Kalmus, "Types of Printed Circuits," *Electri-onics,* February 1983.
2. Thomas Neal, "Trends in Testing Ceramic Circuit Assemblies," *Evaluation Engineering,* March 1983.
3. Thomas N. Newsom, "Special Report: Networking," *Electronics Test,* June 1983.
4. James R. Robinson, "A View of the Technology," *Test & Measurement World,* December 1984.

P · A · R · T · 4

ASSEMBLY AND TEST

CHAPTER 18*
MANUAL ASSEMBLY

Wayne L. Joslin
Hewlett-Packard Company, Palo Alto, California

18.1 INTRODUCTION

Various types and configurations of components are attached to a printed board to make the board a functional electronic device. The manner in which the components are added has several parameters. The relative weight of each parameter in the consideration of the chosen method will depend upon the ultimate use of the parent electronic assembly. Printed boards used in airborne computers will certainly have different relative values placed on the constraints than will those used in low-cost transistorized radios. In between the extremes of concern for reliability and cost are the majority of printed circuit assembly requirements.

All factors should be reviewed when the assembly process is chosen. That will ensure that changing needs will not have been overlooked. It is not the intent of this chapter to supply a perfect pattern for use in all possible situations. It is intended that the data given will be useful in determining the course of action once the factors have been identified.

18.1.1 Reference Board

In order to indicate differences in approach, a basic reference must be chosen. A typical board has dual-in-line packages, transistors, carbon resistors, tubular diodes, and dipped mica capacitors and connectors; and it is a plug-in board. Although the board will usually have uniform hole spacing for the resistors and diodes, transistors and capacitors, and not have clinched leads, it can be shown what differences would occur if that were not the case. The board usually has plated-through holes. It is to be produced for commercial use in an industrial instrument. Shock and vibration reliability is no more than that encountered during shipment. Quality levels are those of good general workmanship. Specific assumptions will be made to match the process for such a board in the rest of this chapter.

---

*Adapted from Coombs, *Printed Circuits Handbook,* 2d ed., McGraw-Hill, New York, 1979, Chap. 10.

18.1.2 Assembly Process

Printed board assembly can be described as a process with a number of steps. A typical process consists of these steps:

1. Job put-up (the gathering of all parts required)
2. Component preparation (cutting off and bending of components before assembly)
3. Board cleaning (washing in a commercial cleaner or active flux and rinsing in water, etc., to prepare circuits for soldering)
4. Inspection (examining components and boards for proper dimensions and damage)
5. Kit components (placing of parts in trays in assembly sequence)
6. Board assembly
7. Inspection (inspecting sample for component placement and value)
8. Soldering (hand-solder, dip-solder, machine-solder)
9. Assembly cleaning (removal of all solder splashes and flux)
10. Addition of special components (components that might be damaged by normal cleaning or heat of soldering)
11. One hundred percent visual inspection (a check on solder, raised components, and aesthetic appearance of board)
12. Solder touch-up and correction of errors
13. Electrical testing (this will obviate the need to check components in step 10)
14. Defects repair and assembly retest
15. Humidity sealing (if applicable)
16. Stocking

All the above steps must be taken regardless of the details of the actual process, but some may be combined for small volume.

18.1.3 Economic Justification

No two assembly facilities are identical. Decisions as to the approach of assembly techniques must be made at the manufacturing location. The data submitted will aid in making those decisions. The figures in Table 18.1 are stated in hours. Multiplying by the applicable labor rate plus variable overhead will give the proper order of magnitude. The hour figures are meant for reference; their source is a predetermined time system verified by actual use. They may not be the same for any facility, but the order of magnitude is correct.

The equipment costs are close approximations made to determine if acquisition is feasible.

Rotary table with spare trays	$ 125–200
Lead former—axial components	1200–1500
Lead former—dual-in-line packages	2000–3000
Lead former—radial lead	200–250
Power unit for items 3 and 4	150–200
Push line (per station)	80–150

TABLE 18.1 Typical Assembly Times Per Component
(Figures in hours)

Component type	Hand form: Hand bend, insert, trim after solder	Machine form: Form	Machine form: Hand insert	Hydraulic form: Form	Hydraulic form: Hand insert
Axial lead (resistor, diode)	0.0037	0.0005	0.0010	0.0005	0.0007
Radial lead (not transistors)	0.0037	0.0010	0.0010		
Transistor	0.0032	0.0012	0.0015		

For each observation to check location from print or prototype, add 0.0010
For hand lead clinch (per lead) add 0.0006
Example: 100 axial lead, hand bent, clinched leads, 100 observations (100)(0.0037 + 0.0010 + 0.0012) = 0.590
100 axial lead, machine formed, four observations (100)(0.0005 + 0.0010) + (4)(0.0010) = 0.154
100 axial lead, die formed, four observations (100)(0.0005 + 0.0007) + (4)(0.0010) = 0.124

Note: Source of values is predetermined time verified by use.

Hydraulic press for lead forming	800–1000
Dies for press (axial lead only, per set)	150–200

18.2 BASIC CONSIDERATIONS

The constraints on facility organization have two major divisions, design and cost. Each of these has a number of subdivisions.

18.2.1 Product Design

The "engineered package" is the first subdivision, and it will be the most confining of all factors. Component type, variety, location on the board, and interface with the rest of the instrument will confine the assembly process. It is well to remember that review of a design before release by a skilled manufacturing engineer will usually remove unnecessary restrictions from the package. That will allow lower cost and normally better quality to be achieved.

Quality is the second subdivision of design. A board assembly must meet the required quality levels. An exact concept of quality is difficult to convey. Most firms identify their quality goals by means of definite phrases regarding workmanship, standard practices, aesthetic appearance of the end product, and the duration of time or number of repeated cycles the unit will operate, i.e., its effective lifetime.

Reliability is the third division of design, and it is often confused with quality. It is the degree of certainty that the device will function when required and will continue to operate for a specified period under given environmental conditions. Military specifications for reliability, if required, will probably alter the assembly process.

18.2.2 Cost

The other major constraint is cost. Regardless of all other requirements, the method chosen will need to achieve the considered ends at the lowest possible cost. It is understood that a board that requires special techniques will cost more than one that does not, but it should be manufactured with all available economies consistent with that need.

The subdivisions of cost are direct labor, volume, and equipment. The combination of the three will define the approach outlined by the design constraint.

18.3 SMALL-VOLUME PROCESS PLANNING

Generally, "small volume" is considered to mean lots of fewer than 100 in a period of a month. Equipment available should be as follows:

1. Lead former and trimmer for axial-lead components
2. Lead former and trimmer for radial-lead components
3. Lead former and trimmer for transistors
4. Die and press for cut and trim axial-lead components
5. Rotary table and bins with vertical rotary part carriers

18.3.1 Visual Aids

A sepia print is made of the board assembly. Each component is numbered to indicate assembly sequence. (Low-level components, such as diodes and resistors, are installed first; larger components are placed on the board later.) The next step is to list the component part numbers in assembly sequence and note hole spacing for preparation equipment. Copies of the board assembly and component list are made. The first copy is used as the assembly reference document. The second is the guide for arranging component trays, which are numbered according to the list at the staging area.

18.3.2 Learning Curve

The major problem in direct-labor assembly cost is the learning curve. One factor in line assembly is the operator's memorizing what to do. It is not possible to memorize large portions of an operation in a short period (e.g., the time required to build 100 boards), but it is possible to learn after one observation where a component should be placed. If an operator had only to place one component in each board, the learning curve would be almost eliminated. The problem, then, is to allow the boards to flow past the operator so that only one component is placed at a time; in that way learning time and the cycle for one board are minimized. The rotary table ("Lazy Susan") does exactly that. Any number of boards are placed on the table. As each component is placed, the table rotates and offers a new board for assembly. After one completed revolution, a new component is selected and the operation is repeated. If the boards are so small that 10 or more can be placed in front of the operator, the rotary table is not required.

18.3.3 Typical Right- and Left-Hand Analysis

Here a typical right- and left-hand analysis is given for prebent and cut $\frac{1}{4}$-W carbon resistors. Compare it with the second analysis for die-formed parts.

Left hand	*Right hand*
1. Get handful of parts from bin (once for each component type)	**1.** Hold pliers
2. Move parts to work area	**2.** Hold pliers
3. Move one part to fingertips	**3.** Advance table
4. Check for orientation	**4.** Hold pliers
5. Orient, if necessary	**5.** Hold pliers
6. Hold part	**6.** Grasp part with pliers at right-hand leads near body of part
7. Release part	**7.** Hold part
8. Move hand to table	**8.** Move part to board
9. Hold table	**9.** Insert component
10. Clear hand	**10.** Remove pliers
11. Repeat steps 1 to 10 for all boards on table	**11.** Repeat steps 1 to 10 for all boards on table
12. After all positions of same part are filled, replace remaining parts in bin	**12.** Hold pliers

Analysis for die-formed and cut parts on vertical parts carriers.

Left hand	*Right hand*
1. Index parts wheel (once per part value)	**1.** Hold pliers
2. Move to work area	**2.** Grasp part with pliers at right-hand leads near body of part
3. Hold table	**3.** Move part to table
4. Hold table	**4.** Insert part
5. Hold table	**5.** Remove pliers
6. Advance table	**6.** Reach to next part

The second analysis is obviously faster, but it has the limitation of requiring all boards made for different instruments to have the same component hole spacing. Otherwise, separate die costs become uneconomical.

18.3.4 Preforming Leads

The first analysis for formed leads allows for greater hole-spacing flexibility. However, if volume approaches the levels for machine insertion, the die-form technique is an excellent preliminary stage that prepares for eventual machine insertion.

FIG. 18.1 Preforming axial-lead component leads with die set. *(Hewlett-Packard Company.)*

Die forming is accomplished by applying double-faced pressure-sensitive tape to resistors and diodes before those components are removed from the commercial package. The strip is placed into a die of a hydraulic press that cuts and forms to proper dimensions (Fig. 18.1). (Suggested dimensions are 0.50-in hole spacing for $\frac{1}{4}$-W carbon resistors and diodes with 0.045 in sticking through the board. See Chap. 1.) The taped components are placed on a circular carrier that rotates vertically. The rows are numbered, as are the bins for rotary tables. Where applicable,

FIG. 18.2 Cutting and preforming transistor leads by machine. *(Hewlett-Packard Company.)*

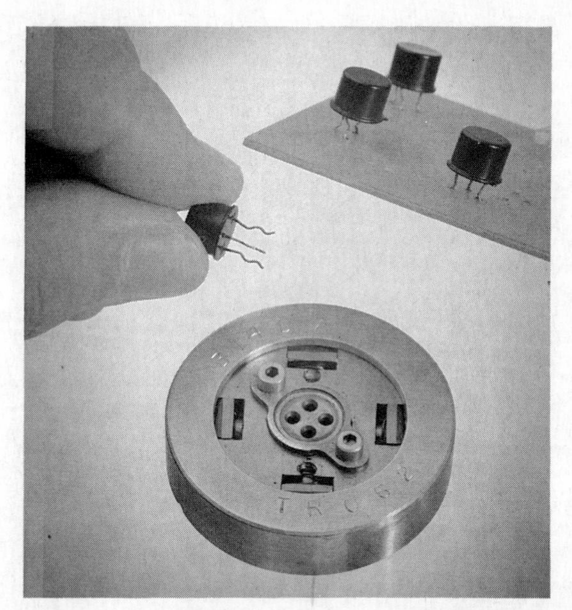

FIG. 18.3 Cutting and forming die for transistor leads, manual method. *(Hewlett-Packard Company.)*

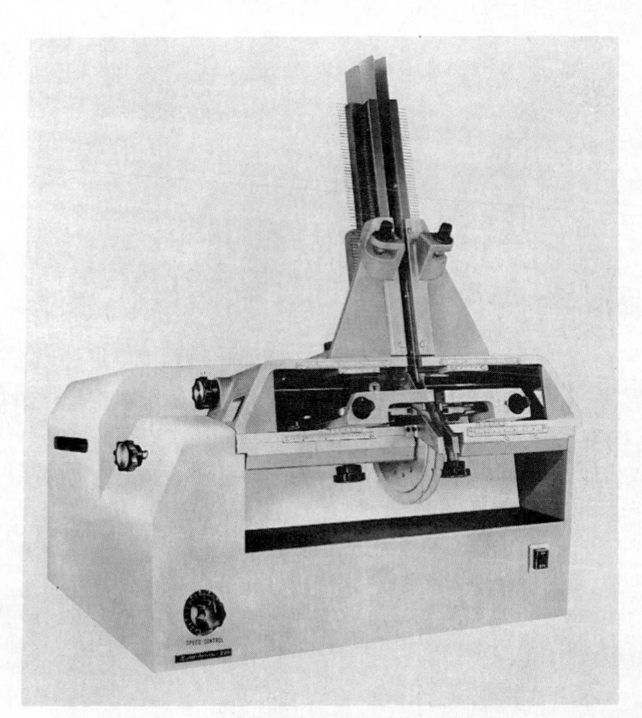

FIG. 18.4 Preforming axial-lead component leads with individual handling machine. *(Develop-Amatic Engineering.)*

combining vertical part carriers with rotary-table bins offers the least variable cost short of mechanical insertion. See Figs. 18.2, 18.3, and 18.4 for other preforming techniques.

18.3.5 Assembly-Station Layout

The assembly station is a 4-ft-diameter rotary table with one layer of bins and one vertical parts carrier. An assembly document has a numbered sequence of components which corresponds to the bin numbers or row of parts on the carrier. Boards are held in place on 1-in-thick Styrofoam pads or other appropriate

FIG. 18.5 Rotary workstation. *(Hewlett-Packard Company.)*

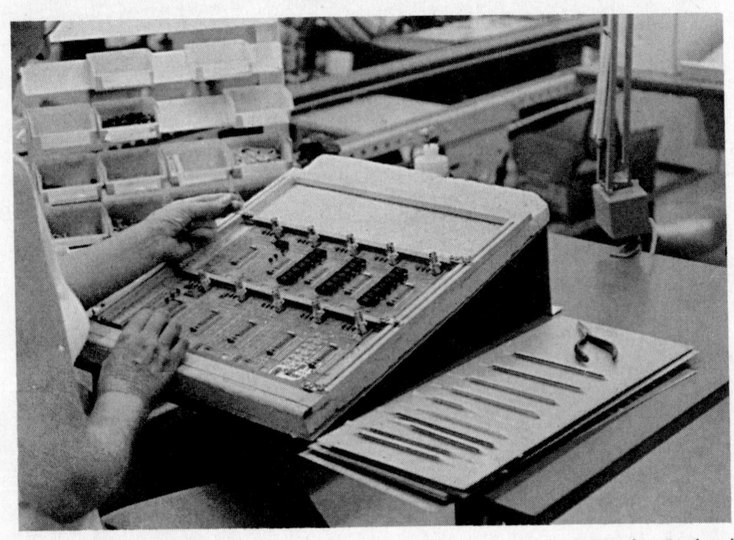

FIG. 18.6 Insertion operation with preformed parts. *(Hewlett-Packard Company.)*

device. The only hand tool required is a pair of long-nose pliers. The finished parts are transported on the Styrofoam pads (Figs. 18.5 and 18.6).

18.4 ASSEMBLY PROCESS

18.4.1 Lead Clinching

If the leads must be clinched to meet a design specification, the boards will be placed into a special fixture after assembly. This fixture consists of a frame nest for the board. A metal plate with $\frac{3}{4}$-in-thick foam rubber is placed on top. The fixture can now be inverted for clinching.

The operator uses a pair of flush-cutting diagonal wire cutters. The lead is grasped where it will be cut. The lead is bent against the circuitry and clipped in a single motion. The work begins at the lower left-hand corner and progresses across and upward.

Parts which have been precut will have to have excess stick-through length to allow for clinching. The normal dies mentioned earlier do not have this lead-clinching capability unless so specified.

18.4.2 Presoldering Inspection

One board of the assembled group per table load is 100 percent visually inspected for component placement and value. The purpose of the check is to find consistent errors. It is likely that if one component is misplaced on one board, it will be misplaced on other boards. Random errors are found in testing. Under normal operating conditions it is not economical to inspect further at this point.

Overlays made of Mylar* or similar material can be used to check for component placement. They are made by drawing or gluing the appropriate component outline and/or color coding on the transparent material. Another approach is to cut the component outline from heavier material. All components of a given type can be on the one overlay. The series of overlays will indicate which components are missing.

18.4.3 Soldering

The assembly is now ready for soldering. In normal commercial work water-soluble active fluxes can be used to get better and more consistent solder results. When rosin fluxes are required, the technique varies slightly. For military work refer to MIL-STD-275.

A dip pot is the most common form of soldering at the time of writing. The board has a rubber channel fitted to the plug-in fingers to serve as a solder mask. It is then placed into a tonglike device held in one hand. The board is passed over a foot-control spray device which applies the flux, and it is then placed into a skimmed pot of solder that is 490°F at the surface. One edge is placed first, and with a careful rolling motion of the wrist and forearm the solder is applied (Fig. 18.7). The board is set aside and allowed to cool.

The rubber channel is removed. The board is placed in a standard dishwasher

---

*Registered trademark of E. I. du Pont de Nemours & Company.

FIG. 18.7 Dip soldering. *(Hewlett-Packard Company.)*

and cleaned with a detergent. If the water is very hard, a softener may be desirable to avoid deposition of a film on the board.

If rosin flux is used, a bath in Freon* or similar solvent is substituted for the washing step. Extreme care must be used in choosing the solvent, because some solvents will remove the inks used for component identification. A number of manufacturers make equipment that will distill and recover the solvents. Equipment using ultrasonic transducers to clean is considered unacceptable because of a history of damage to semiconductor devices.

18.4.4 Special Components

If there are components that will not stand washing or the heat of the dip pot, they are installed now. Some examples are connectors, trim pots, and some tunnel diodes. A normal bench station is used, because the number of components is usually small.

18.4.5 Touch-up

This is a rework station. Its main purpose will be solder touch-up. The best iron for this purpose is not in excess of 50 W. The main danger here is the lifting of circuitry with excess heat. A suction device that operates with the iron is quite useful in removing excess solder. If it is necessary to remove a component, this device is quite effective.

Since rosin flux is used here, a solvent will be required to remove the flux at touch-up points. It is normally applied with a small brush or cotton swab. It is easily removed with a toothbrush.

---

*Registered trademark of E.I. du Pont de Nemours & Company.

18.4.6 Final Inspection

The boards are now 100 percent visually inspected for soldering, proper seating of components, and other visual defects. Each defect is noted and passed to the rework operator. It is essential that solder touch-up not be done by the inspector. The tendency is for the inspector to "dress" a board that does not require it. That results in higher touch-up costs and potential board damage. Identification of touch-up points with a dyed flux is preferable. The main requirement is adequate light. A 4× lighted magnifier is excellent for the purpose.

18.4.7 Test and Repair

This station is essential to minimize checkout time of the final instrument. It is obvious that fewer errors going into a system will mean fewer considerations necessary when malfunction of the assembled system occurs. Accessibility of the malfunction is an important factor too. The specific test equipment used will depend on the facility and its volume and the available personnel.

The station may also be used to install selected value components. Most small malfunctions are corrected here and the recheck made. If the facility is large, it may be better utilization to have the troubleshooting done separately. Again the decision is best reached at the facility.

18.4.8 Conformal Coatings

Humidity seals are applied by spraying or dipping. Spraying is preferable for production because of ease of application. If sealing is a requirement, the boards are usually dried in an oven prior to coating to ensure low initial moisture content.

Required equipment includes an oven, spray booth, and spray gun; or vented hood and drying area for dipping. A wire rack is usually sufficient for drying sprayed boards. Dipping requires a drip-off area. Use of clips on a traveling clothesline is very effective in this application.

18.4.9 Stocking

The assembled boards are now ready for the stockroom or final assembly area. Special containers for boards may seem to be an unnecessary expense, but replacement costs for damaged boards will pay for proper containers. Commercial containers are available.

18.5 PROTOTYPE PROCESS

The major difference here is that components will not be preformed and precut. All boards can be placed in front of the operator on 1-in-thick Styrofoam. If the boards are large and have many components, it will be worthwhile to build as though they were small-volume. A check of the justification sheet will confirm that statement.

If dip-solder facilities are not available, hand soldering will suffice. A low-wattage (not to exceed 50 W) iron is a must to avoid circuit damage.

18.6 LARGE-VOLUME PROCESS

The main difference between small volume and large volume is the use of the assembly line. Calculations will show that there is a slight theoretical advantage to the line assembly even when loss to line balance has been considered.

18.6.1 Planning

Additional planning is required for line assembly. All operators must have approximately equal portions of work if the total unit time is to be minimized. Unequal work content will allow wastage of time in the lighter-loaded stations when the heavier-loaded stations are working at a normal pace.

Line balancing must be done for various levels of production, since the number of operators on the line will affect the production rate per unit of time. Production requirements may vary widely over a period of time, or sufficient operators may not be available to operate at a given level.

A document indicating which components are to be assembled by each operator is necessary for consistent level of productivity.

18.6.2 Assembly-Station Layout

The assembly station will consist of a track on which the board slides, if push-type, or is moved by conveyor, if powered. A quantity of numbered trays will be evident, as in rotary-table assembly, except that they are stationary. The vertical parts carrier is also of advantage here.

18.6.3 Soldering

Soldering may be dip pot, but more likely a wave-soldering machine will be used. Either the boards are transferred from the assembly line to fixtures for soldering or the track extends so the soldering is an extension of the assembly stations. In either case the board is automatically fluxed and soldered (see Fig. 18.8 and Chap. 23). Two main advantages of such equipment are reproducibility of process and lower unit cost. Uniform quality is assured, and unit cost is lower because initial soldering is better and less rework is required.

The rest of the operations are similar to small-volume. They are often connected by conveyors to handle material flow.

18.7 IN-LINE VS. ROTARY WORKSTATION

When should the rotary table assembly be replaced by line assembly? That is a good question that is difficult to answer. However, as a rough rule of thumb, unless the calculated time values weigh *extremely* in favor of line assembly, the rotary table is preferable. Table 18.1 will aid in the necessary calculations. After the totals are compared, look at the comparative benefits.

Line assembly will provide operator pacing, reduce the time of the first completed piece, allow quick checks of completed-parts status, use less floor space per operator, and give the impression of mass-production savings. Also, it requires

FIG. 18.8 Soldering-machine console. *(Courtesy of Electrovert, Inc.)*

more service time for initial start-up and maintenance from manufacturing engineering, requires a skilled labor pool to offset absenteeism and production changes, and is held to the slowest operator. It does not allow job identification as does the rotary table.

Again this is an analysis best made at the facility, by a review of the facts.

18.8 SUMMARY

The small-volume printed board assembly process has been detailed with some alternatives. The prototype and large-volume processes were compared with the small-volume process. The constraints of design and cost were explored. Data were given to aid in reaching a decision as to a specific process. To develop a process:

1. Review the constraints, item by item.
2. Make the cost analysis.
3. Add good judgment.

This procedure will yield the optimum process for any facility.

CHAPTER 19
COMPONENT ASSEMBLY

John S. Hohl
Universal Instruments Corporation, Binghamton, New York

19.1 INTRODUCTION

The automatic assembly of electronic circuits, as a technology, started in the late 1950s. Early versions were used primarily to insert components with axial leads—resistors, diodes, and capacitors—into printed circuit boards (PCBs). Since there was no effective standardization at that time in the handling systems used to bring components into place, several schemes were tried: lead taping, body taping, and several forms of magazine feed. It did not take long for the lead-taping concept to become the standard handling system for all axial-lead devices (see Fig. 19.1).

The first axial inserters were simple, semiautomatic, bench-mounted units: an insertion head, a clinching means, usually a wiping anvil, and a reel of taped components (see Fig. 19.2). The wiping anvil provided a lead-bending function, not unlike that performed by an ordinary desk stapler. About the only difference between the wiping anvil and the stapler was that the wiping anvil could be rotated and locked, so that a clinched lead could be pointed in any direction.

The basic principles of early axial insertion heads are still used today: a component is fed into the cutting position; the leads are cut from the tapes and then formed into the shape of a "staple" within the grooved outside formers, or guides; finally, a downward push of the component inserts it into the prelocated holes of a PCB (see Fig. 19.3). For board-to-head registration, the PCB was either manually placed against fixed locators or it was moved into position by means of conveyor systems using pallets. This conveyor usually served a number of similar insertion heads.

In the early 1960s, "*X-Y* table" systems became available to move the PCBs under the insertion head, first with manually operated pantograph systems using a stylus and template to provide proper table positions. Later the *X* and *Y* table movements were controlled by numerically controlled (NC) equipment, and in the 1970s, these were replaced with computer controlled (CC or CNC) positioning systems. In most of these later machines the wiping clinch was replaced with a combination cut-and-clinch unit which gives better dimensional control for improved soldering.

In the early days of automatic electronic assembly, and for quite a few years to follow, the prime components used were of the axial variety. In fact, there were very few components available at that time which had the necessary common mechanical characteristics lending them to mechanization. Axial devices had one

FIG. 19.1 This early photograph shows both lead-taped (left) and body-taped axial-lead components on reels, with paper interliner (a layer-to-layer separator). Lead taping is the modern standard, and body-taped components are no longer available.

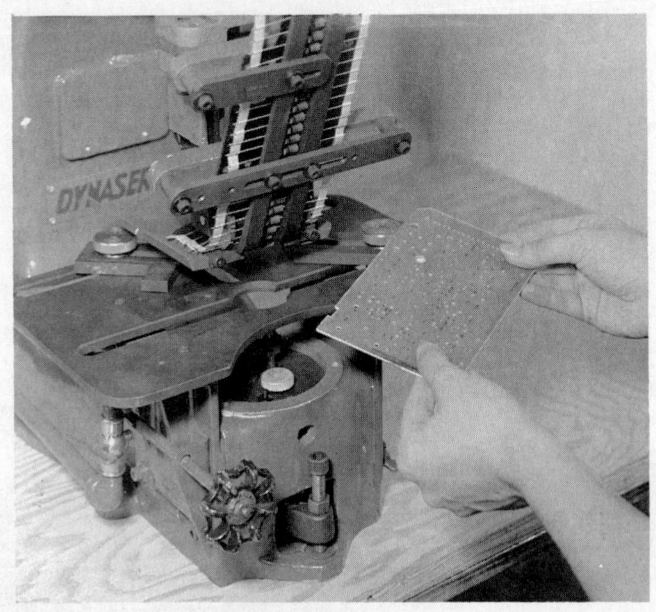

FIG. 19.2 Early single-station machine for inserting a specific component into manually positioned boards.

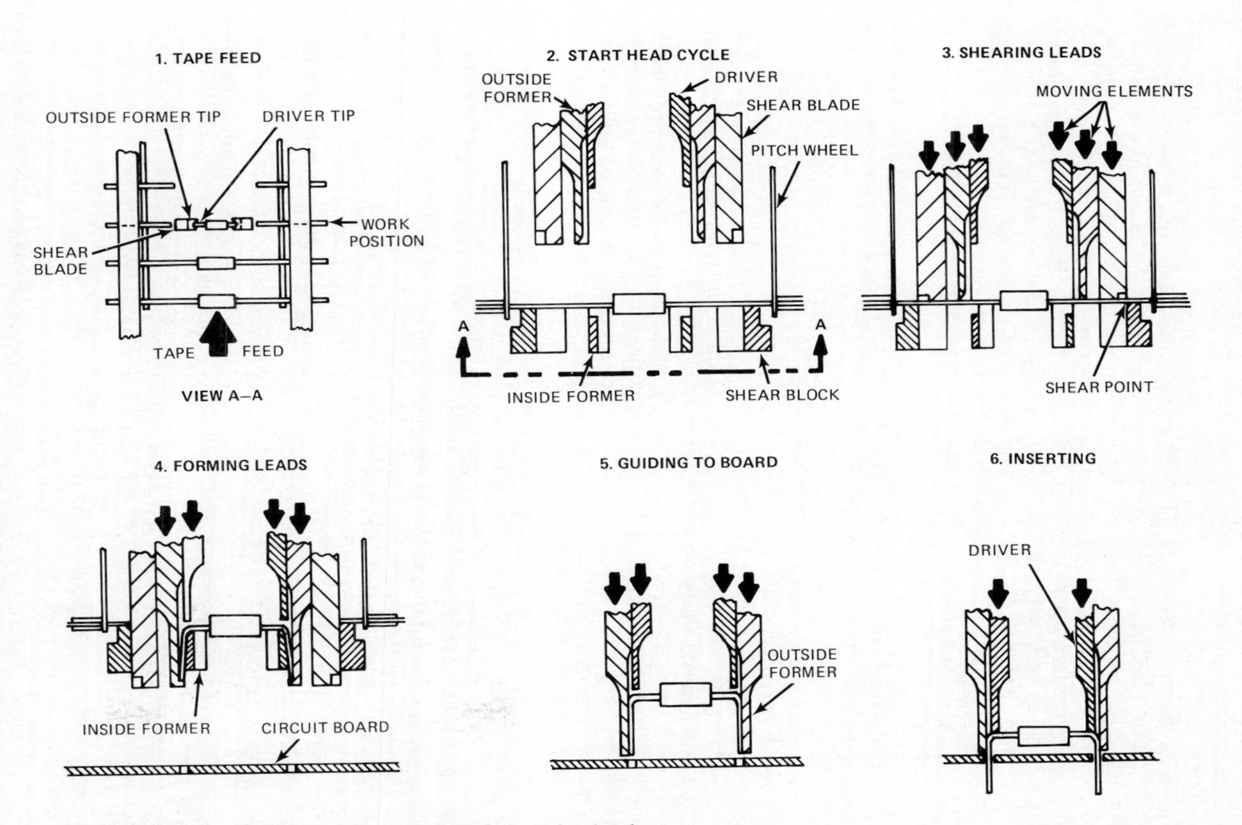

FIG. 19.3 The essential functions of an axial-lead insertion head.

important common parameter: all leads were of a reasonably similar size. Within related body-size ranges, the Electronic Industry Association (EIA) soon developed a taping standard, RS-296. This standard still exists, although it has been modified to fit worldwide needs and it covers all aspects of lead taping of axial components. It defines the body-diameter ranges allowed, shows the acceptable component pitches—for example, 0.200 and 0.400 in (5 and 10 mm)—defines how the tape should be applied, and provides reel size information and minimum reel-marking data. Today, one can buy lead-taped axial components on reels to the present RS-296 standard from virtually any supplier in the world.

Using simple pantograph-type inserters which provided a moving *X-Y* table, and using a 1:1 ratio precision-locating template, the early users quickly found ways to make board population easier. Because they had to work with "pure" reels of taped components, i.e., all components on a reel had the same value, the technique of using multiple break-away boards was an accepted practice. With four or eight identical printed circuit cards placed on the machine as one "board," even if there was only one component of a given type per card, up to eight insertions could be made for each board handling. This approach produced significant labor savings.

This shortcoming, or the inability to populate all of the insertable devices on a PCB with one pass through an *X-Y* type of inserter, prompted the development of the sequencer. This is a machine which takes a number of different input reels of lead-taped components, each containing all parts of the same value; then, under programmed control, it sequentially feeds and cuts these components from their reels, dispenses them onto a moving conveyor, then repitches and retapes them again in the order necessary to populate a specific circuit board. Sequencers generally put a "space-gap" (a missing part) in the retaped output to separate the repetitive sequences. Now, a reel of sequenced tape could be put on an *X-Y*-controlled inserter and all the insertable axial devices inserted with one pass through the machine.

With the sequencing problem solved, it became apparent that the various mechanical sizes of axial devices really should be inserted on different center spans. This brought about the variable center distance (VCD) insertion head machines, which have become the most widely used. Not only can the VCD head be programmed to handle different component spans (insertion centers), but it can also adapt to different body and wire diameters as well.

Meanwhile, other "leaded" devices with potentially high use were coming into the marketplace in the middle to late 1960s. Perhaps the most significant of these was the dual-in-line package (DIP) for integrated circuits (see Fig. 19.4). With a DIP, the designer was able to place an entire circuit function into a PCB with one stroke of an insertion machine. The development history of the DIP and the machines to insert them parallels that of the axial family. However, the DIP, with its greater number of leads, required a much more stringent set of physical dimensional specifications than did the simple two-leaded axial device. In the early years of the DIP, there was a high degree of integration between the component manufacturers and the makers of the machines which insert them. This cooperative effort covered areas such as lead dimensions, shape, tolerances, body thickness, variations, etc. Because of the poorly mechanized manufacturing process in the beginning, there were many growing pains with poor physical quality of the devices—leads bent well out of tolerance and two-piece ceramic bodies skewed. In time, these problems were cleared up as more mechanical handling came into being at the component manufacturers' plants. With time, the DIP component grew in size and number of leads, and into different lead-attachment forms, and so did the capability of the machines to insert them. Some of the limiting parameters of the early devices made them impossible to insert reliably, but as time

FIG. 19.4 Typical DIPs and DIP sockets.

went on, both the components and the machines were refined to the point where nearly all DIPs made could be automatically inserted. For instance, some of the early DIPs had molded "bumps" on the bottom surface which were intended to establish a seating plane. Since these "bumps" interfered with the escapement, spreading, and transfer portions of nearly all DIP inserters, they soon disappeared. The introduction of side-brazed leads is another example of change which the machine tooling had to be modified to accommodate.

Other families of through-hole devices are available, such as two- and three-lead radial devices, i.e., components with all leads extending from the same end. The effective automatic handling of these devices has been made possible by the establishment of a common and an internationally acceptable handling system promoted by the Japanese. This is reflected in EIA standard RS-468. This specification gives information similar to that contained in RS-296 for axial components, only this covers radial-lead components intended for automatic insertion. The lead-taping dimensions and pitch, with tolerances, as well as the critical lead-tape to component-body dimensions are stated and shown with figures and tables. Acceptable reels and other forms of packaging for shipment are also defined.

The list of through-hole insertable items is far from exhausted: we also have the single-in-line packages (SIPs), jumper wires, pins, connectors, sockets, wire-gripping devices, and terminals, plus an entire group which can be handled when one employs the manipulative capability of a robot.

So far we have only described that portion of automated electronic assembly where leaded devices are inserted through holes and mechanically attached to a PCB. In certain other specialized areas of electronic manufacturing, particularly where high density and high reliability are required, designers have called for ceramic substrates, instead of PCBs, with the components placed *on the surface* of the circuit, rather than being inserted through "holes." This approach has been used for many years with ceramic as the primary medium upon which the devices were placed. This method was generally referred to as "hybrid" assembly, as it could contain both active and passive components in various package forms—from bare chips to standard components with specially formed leads. More recently, PCBs have been used as the placement medium for these types of components, and we now refer to this as "surface mount technology" (SMT).

For the purpose of this handbook we shall concentrate primarily on the printed circuit board aspects of surface mounting, although the techniques used for ceramic or hybrid assembly do not differ greatly.

Because of the increasing need for more circuit complexity in smaller and smaller physical volumes, the design and manufacture of electronic equipment constantly needs to become more efficient. Surface mounted devices, because of their smaller size, become natural candidates to challenge the package designer.

19.2 FACTORS FOR SUCCESS

Excellent performance in any automation endeavor is a result of a blend of many ingredients: (1) understanding the total problem; (2) designing the product for automation; (3) learning the extent of existing industry standards, then taking full advantage of them; and (4) putting all of this together in such a way that the job is not only done right, but done right the first time. Let's take these ingredients one at a time.

19.2.1 Understanding the Problem

Here the problem is the methodology and the overall rationale of the automatic assembly process. All assembly machine suppliers offer either "design guidelines" or other "application information" for the equipment they offer. Generally, these documents will provide the user with specific information on how to design the printed circuit board to be most suitable for automatic assembly. Items covered would include:

1. Board considerations, material, size, and locating reference requirements
2. Component arrangement objectives, including those preferred and why
3. Maximum board sizes
4. Amount of board sag allowed, and ways to minimize this effect
5. Items for leaded components:
   - ***a.*** Hole diameter considerations, with relation to component lead, board, and machine tolerances
   - ***b.*** Tooling, or work-board holder clearance requirements for both stand-alone machines and those of an in-line arrangement
   - ***c.*** Specific detail information on the various types of machines supplied by the particular maker, including taping and component specifications for each type
6. Items for surface mounted devices:
   - ***a.*** Placement accuracy, including device centering and rotational capability
   - ***b.*** Input-handling systems and device capability

Any mechanized operation must work within a framework of simple rules, and it is up to the users to find out what these rules are for the particular process or machine they intend to use. Understanding these rules does not mean that the users are totally limited in their approach; it just means that to get closer to an optimum level of performance, they must be clever enough to apply the rules in a manner such that they are on their side. Another powerful tool is to know what the machines can do, how they operate, and where their potential strengths and weaknesses are. To list several examples: (1) Once a machine setup has been verified, it is rare to find a part in the wrong location; hence, on dense boards, error-free assembly is attainable; (2) machine insertion is much more consistent than hand assembly in both lead form and lead length (it is particularly useful for high-frequency circuits); (3) if boards have tight dimensional requirements—specifically, small lead spans for axial components, tight clearances from lead diameter to hole diameter, and worse than average accuracy between the reference holes and the insertion holes—the net yield will go down. On the other hand, by using good design practices for axials, DIPs, and radial devices, one should typically expect an insertion reliability of better than 99.9 percent (or a miss rate of less than 1 part per 1000). But by using the most generous clearances and the tightest tolerances affordable on the PCB design, even these values can be improved. With through-hole insertion, the whole basis for success is *tolerances* and *clearances.* These affect both the design of the PCB and the actual selection of the specific devices themselves.

19.2.2 Designing for Automation

Machines cannot always do the same thing hand assembly can do. For example, human hands can put a part with a lead diameter of 0.020 in into a hole with a 0.021-in diameter (a 0.500-mm wire into a 0.525-mm hole). No machine can do this, or at least do it reliably. Manually, one can also put parts into a PCB at any angle. Most machines have great difficulty in doing this. In hand assembly, parts with odd sizes or nonuniform shapes can easily be handled. Machine assembly, on the other hand, requires a much stricter set of dimensions. The machine needs to know where the leads are. It is clear that the requirements for automatic component assembly are substantially different from the "no-rules" world of hand assembly. Generally, the circuit layout for automatic assembly will have the parts arranged in one direction or two orthogonal directions; the latter approach is accomplished with either a rotary table or, in some cases, the insertion-head tooling. One must also consider clearances from lead diameter to hole diameter. Ideally these should be as large as possible for reliable insertion, but not larger than that required for good soldering. While detailed soldering considerations are covered elsewhere in this book, it might be helpful to point out that the maximum recommended lead-to-hole diameter clearance for good soldering, often given as approximately 0.015 in (0.4 mm), applies primarily to single-sided boards without plated-through holes. When plated-through holes are used in a PCB design, a much greater lead-to-hole diameter clearance can be used without affecting the quality of the solder joint. This, of course, will improve insertion yield in a direct manner. Therefore, for a good design, it is recommended that all machine and component parameters be taken at or near their most generous limit. Too many designs have suffered, in the long run, from the designer who has pushed each and every tolerance or clearance to its limit. This type of "limit design" should be avoided if one is truly striving for success. In most cases, the more generous approach rarely affects the overall circuit board size.

19.2.3 Industry Standards

To understand what industry standards are, one must recognize where and how they come about. Every major industrialized country has standards groups for nearly everything. In the United States there is the Electronic Industries Association (EIA), which is the prime nongovernmental standards group for electronic components. The military also maintains standards for many commonly used electronic devices. Therefore, the majority of standards that we see domestically come from these two sources. In arriving at what is called the "standards" for any given device, the user's Standards Committee usually finds out what the physical dimensions and tolerances are from all suppliers, pools them, and then publishes the nominal values with a totally encompassing tolerance. The point is that while industry standards are important, they need not be considered to be the final word, particularly where automatic assembly is concerned. All assembly machine makers will provide the recommended device outline specifications for their machines, and in some cases, these may be tighter than overall industry standards for optimum reliability. The reason for this seemingly unusual situation is simply that industry standards often are the result of compromise, and while the nominal values are generally all right, the tolerances can be somewhat exaggerated to permit all member suppliers to have their products included, as described above. Machines, being less adaptive than humans, may require a slightly tighter tolerance than the industry specification allows in order to meet the published insertion yields for that machine. This is rarely a problem to the user, as the majority of suppliers can meet these tighter requirements with no problem and, generally, with no increase in cost. This is because suppliers generally operate according to three sets of dimensional tolerance levels: (1) for manufacturing, (2) for checking, and (3) for selling. The tightest tolerance level is that for manufacturing. The purchase specification should reflect what the user's equipment requires for best performance. If these recommendations are not followed, there could be problems. In addition to the above, there are international standards on most electronic devices as well. These are issued by the International Standards Organization (ISO) and should be treated in a similar manner.

19.2.4 Putting It All Together

The net result of the above factors is potentially a successful automation project: one in which the production rates, yields, quantities, and cost either meet or are better than the original plan.

From the information obtained in the design guidelines for the assembly machines involved, plus similar information covering the remaining process functions, i.e., soldering, testing, etc., the user should be well on the way toward success.

The five most important ingredients to a good design implementation are as follows:

1. A solid electrical design; then sufficient electrical testing to assure that only good parts are used and, possibly, on-line verification in addition.
2. Selection of components with proper mechanical dimensional tolerances, with leads that are sufficiently stiff to give the necessary springback after forming, and with a minimum of lead-tin plating to prevent lead-tin buildup on the machine tooling.
3. Proper board design and manufacture. Good locating means and adequate lead-to-hole diameter clearances.
4. Good board rigidity, or equivalent support means.
5. Thorough training of all personnel involved: the machine operator, those who handle boards and supplies, etc. The latter will minimize any physical damage to either the boards or the components, which can affect yields.

One final thought: Once the process has been defined, the design completed, the parts selected and procured, and production is about to begin, recognize that there will be a learning-curve situation on the production floor (unless the user has been through this process many times before). Generally, optimum results will not be accomplished immediately; however, if the planning has been done as suggested, success will come.

19.3 IDENTIFYING COMPONENTS FOR ASSEMBLY

19.3.1 Hand Assembly

Depending on the production volume, the rules for hand assembly design and component selection range from basically "no rules" to those almost as stringent as the ones required for automatic insertion. Specifically, in low volumes, the only real requirement is that the part fit the PCB hole pattern and that it perform its electrical function.

As hand assembly volume increases, so do the "aids" provided for the assembler. When this happens, the part-handling and part-preparation, or "prepping," requirements generally encourage the procurement of components to nearly the same higher-level packaging and handling systems required for automatic assembly. This upgrading is necessary primarily for lead "prepping" and also for use with semiautomatic parts feeders and for fitting into light-indicating "stick" holders. "Prepping" means preparing the leads of the raw part into a form that allows the operator to insert it rapidly. Examples are shown in Fig. 19.5. For axial and radial devices this usually means cutting and forming the leads for either drop-in or snap-in assembly. There are many axial-lead formers designed to accept lead-taped input.

Semiautomatic, program-controlled "pointer systems" for hand assembly are often used for low to medium production volumes. These systems generally have the same component preparation requirements as the high-volume lines described above, and for the same reasons.

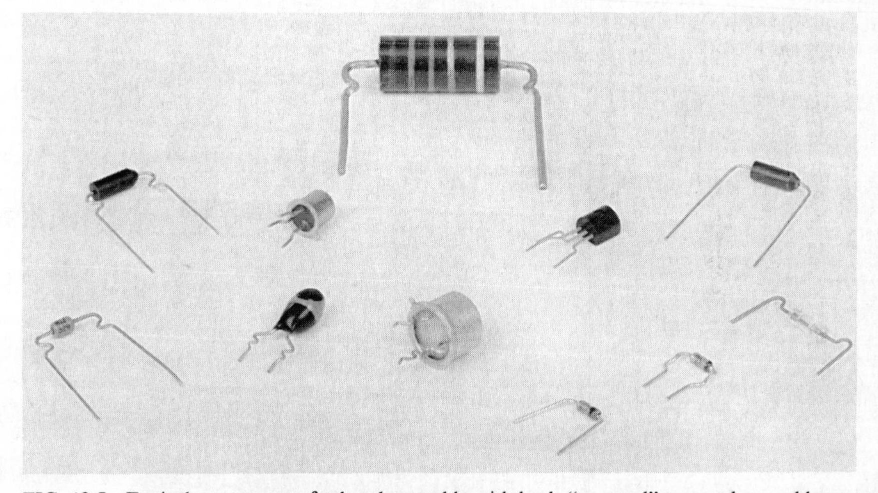

FIG. 19.5 Typical components for hand assembly with leads "prepped" to speed assembly.

19.3.2 Automatic Assembly

19.3.2.1 Through-Hole Insertion. At the component level there are some basic rules which aid in the definition of a "machine-insertable part." For machine insertion, it is best if all of the leads "are outside of the shadow of the body" when seen from above. If this condition is satisfied, conventional grooved hard tooling may be used for guiding the leads into the PCB holes. Consider both the two-leaded axial component and the multileaded DIP, Fig. 19.6, type I-A and type I-B. From a side view of the axial or an end view of the DIP, both devices look the same and similar techniques are used to insert both. This guiding approach works best with a lead material that has a modest amount of springback after forming. With most axial-lead components, even though the leads, after being cut from tape, are formed to 90°, they will not stay there, because the lead tips will spring back slightly toward their original shape. As a result, they will stay within the three-sided confines of the outside formers, or guides, in a stable manner. As these formers are lowered to close proximity to the PCB, the positions of the lead tips are very well controlled. For insertion, all that is necessary then is for the driver-assembly to push the component downward within these grooves, into the properly positioned holes immediately below. The same technique is used for DIPs. Here the leads are spread outward slightly, about 15°, then gathered into formers or guides with many grooves, and again finally pushed into the holes in the board. Leaded devices of this type with the leads outside of the body are relatively easy to mechanize, since the final lead tip positions can be totally controlled by hard tooling, which is a part of the insertion machine. A reasonable clearance value between the lead diameter (effective lead diameter for DIPs) and the hole diameter for either type I-A or type I-B insertion would be 0.015 in (0.4 mm), with the hole location tolerance on the board held to at least ±0.003 in (±0.075 mm) and with typical machine tolerances.

For components where the leads are inside the shadow of the body (i.e., a precision, multipin module; see type II on Fig. 19.6), the machine designer has several choices. If the lead-to-lead tip matrix dimensions and tolerances will mate with the hole matrix with a clearance such that interference will not occur, even under worst-worst tolerance conditions, then by repositioning the part body on the vacuum-tipped placement head using the leads as the reference, the part can be inserted simply by moving the head down to the board.

With a SIP (single-in-line package), if the body dimensions *and* the pin matrix are both of very good dimensional tolerances, it can be inserted with a high degree of reliability (see Fig. 19.7). For either a SIP or the multipin module, above, and

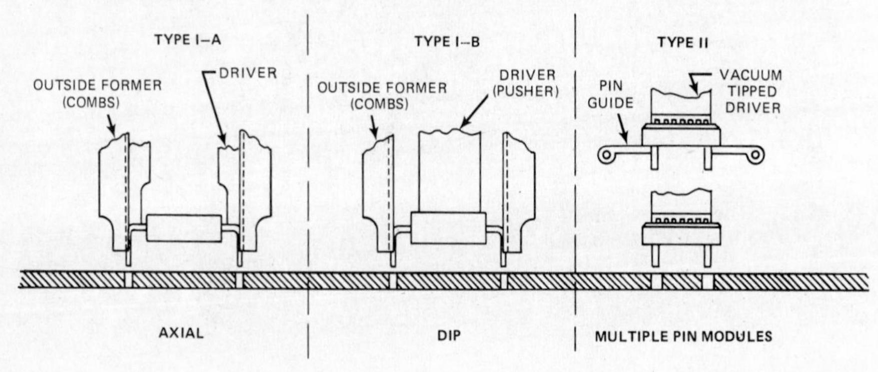

FIG. 19.6 Types of through-hole insertion.

using the same board and machine assumptions as given for the type I devices, to attain the same degree of insertion yield, one would have to increase the clearance between the pin diameter and the hole diameter to approximately 0.020 in (0.5 mm).

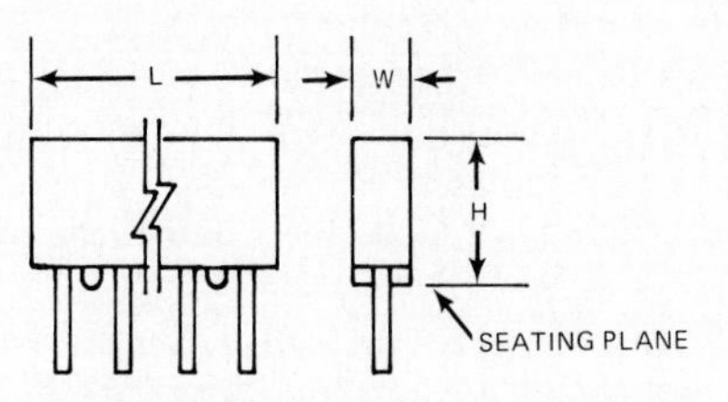

FIG. 19.7 A single-in-line package (SIP). Key dimensions shown.

Yet another approach is to use pivotable, grooved tooling which wraps around the lower extremity of the part to truly guide the leads. In this case, the tooling must spread apart after insertion to clear the body as the tooling withdraws. This guiding with hard tooling reclassifies this to a type I insertion, and it shares the same tighter relationship between lead diameter and hole diameter that exists for axials and DIPs. On the other hand, this larger "footprint" generally limits the side-to-side placement capability, sometimes to the point of being impractical.

The above suggests that for machine handling, components with the type I-A or I-B configuration, axials and DIPs, offer the highest order of insertion reliability and will provide the best results with respect to lead clearance, speed, cost, and yield.

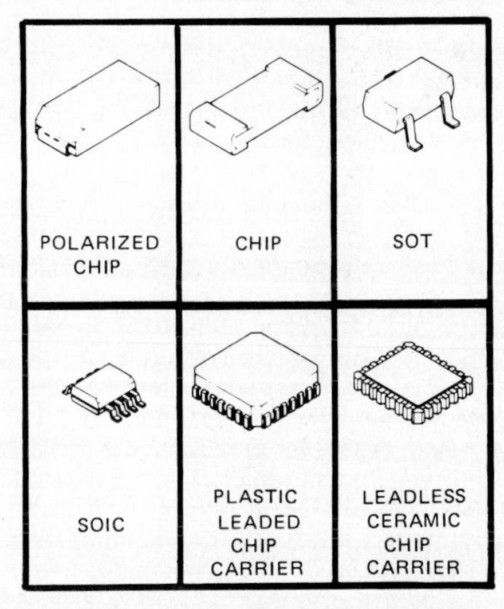

FIG. 19.8 Typical surface mounted devices.

19.3.2.2 Surface Mounted Devices. Components considered for automation in the category of surface mounted devices have two basic requirements: (1) a top surface which is generally flat so that they can be handled by a vacuum-tipped placement nozzle, and (2) availability in an industry standard handling system.

Surface mounted devices (see Fig. 19.8) generally fall into the following categories:

1. *Chip devices:* These are usually capacitors or resistors.

2. *Packaged semiconductor devices:* These are either transistors or complete integrated circuits in a packaged form:

   a. Small outline transistors (SOTs): These three-leaded plastic molded devices come in several sizes and are given size designations such as SOT-23, SOT-89, etc.

   b. Small-outline integrated circuits (SOICs): These devices are similar to the above except that the number of leads has increased substantially to match the function of the device. The lead spacing is much closer and there are many more leads, generally 8 to 16, although more are possible.

   c. Plastic IC packages or chip carriers (also called plastic leaded chip carriers—PLCC): The intent is to have compliant leads to allow for differential thermal expansion without device damage.

   d. Leadless ceramic chip carriers (LCCC): These devices utilize a small ceramic carrier with screened circuit traces on both top and bottom, with interconnections over the edge. The IC chip is then bonded to the top surface of this carrier and connected to the final termination pads on the bottom.

19.3.3 Handling Systems—Through-Hole and Surface Mounted Devices

Among designers of automation equipment there is an old saying: "Once you have control of a part, don't let it go." The part in this case is any device to be ultimately placed or assembled automatically. From a practical standpoint we must consider this part from the time it is being made at the component manufacturer's plant, where it must be under control at one or more steps in the process. "Don't let it go" then means to use a form of handling system that will maintain orientation and protect the mechanical integrity of the device from some point in its manufacture through the input to an assembly machine for final placement on, or in, a circuit.

A handling system can be defined as a widely accepted, standardized, intermediate means of packaging for any specific device family that (1) is cheap, (2) can carry a reasonable quantity of parts, (3) provides protection to the parts during shipment and storage, (4) can be labeled to describe the contents, (5) can be easily handled by one person, (6) allows simple replacement when exhausted, and (7) is easily adapted to automated equipment.

In the early days when the initial emphasis was on axial-lead components exclusively, the handling system that rapidly became the standard was lead taping (see Fig. 19.1). This does meet all of the above requirements. For surface mounted devices, most of the suppliers and industry standards groups worldwide have been working toward standardization of the systems to handle the wide variety of devices used. It is rather unfortunate that many of these very small devices adapt to more than one handling system, as this duplication tends to dilute the effort to achieve true standards.

To list the major handling systems currently in use for surface mounted devices, we have:

1. *Sprocket-driven tape:* 8 mm, 12 mm, 16 mm, 24 mm, 32 mm, 56 mm, and larger, both punched cardboard and vacuum-formed plastic. This is by far the preferred method and is currently covered by standards such as EIA RS-481.

2. *Matrix trays:* Made by several manufacturers and generally available in an overall size of 2 × 2 in (approximately 50 × 50 mm), and larger, with internal

waffle-type nests made to conform to the part size. This, unfortunately, results in quite a variety of trays available.

3. *Sticks:* Recently the semiconductor industry has introduced both flexible and rigid "sticks" for several sizes of SOICs. These have the advantage that a reasonable number of devices may be handled in each stick and also that, once loaded, parts will remain oriented and protected. Most machine makers can provide feeders to select and remove devices from these sticks.

4. *Bulk feeders:* These are usually small, linear, vibratory feeders, tooled to handle a given part size. By means of a "stop block" at a given point, these feeders can provide a ready source of components from which the placement machine pickup nozzle can find a part. These feeders require certain $L \times H \times W$ ratio relationships of the device to permit reliable mechanical orienting. If the devices are polarized, other handling systems would be recommended. Devices most suitable for bulk feeders include the ordinary, rectangular, metalized-end chip resistors and capacitors.

19.3.4 Robotic Assembly

Generally, the components targeted for robotic assembly are called odd-form or normally nonautomatically insertable devices. These include components such as relays, transformers, conformally coated devices, connectors, potentiometers, switches, etc. Although these devices may represent only 10 percent of the total number of components on a PCB, they often account for up to 80 percent of the total assembly labor content.

The components successfully used for automatic assembly have guidelines, specifications, standards, and handling systems, all of which have evolved, thus permitting a high level of acceptance. Odd-form devices which are candidates for robotics still need to be presented to the machine in a handling system, even if they do not meet the other remaining requirements for insertion or placement by a standard or even a special type of machine.

Two insertion types were discussed earlier, and both of these exhibited some common characteristics: either the leads were formed to a known dimensional matrix by hard tooling, or the device already had a lead or pin matrix where the lead-tip positions were stable and known to be suitable for reliable insertion without checking or straightening. By means of robotic assembly, we can adapt to even less stringent device parameters, but from a practical standpoint, even the robotic approach has its limits. A typical robotic assembly station is described in more detail later in this chapter.

Robotic assembly of odd-form components is not going to answer the total automation needs of every PCB manufacturer. There are, and always will be, a large percentage of PCBs in production designed to support the automatic assembly of only the more conventional insertable components, i.e., axial, radial, and DIP components, and for one or more very good reasons. Perhaps the main reason is the cost of additional equipment versus the cost of using hand assembly, or perhaps the length of the production run simply cannot justify this approach.

19.4 LEVELS OF AUTOMATION

This section describes different levels in the hierarchy of automation of electronic circuits, from the very lowest level which can qualify as automation to the very highest level being considered by the user.

19.4.1 Automated Hand Assembly

Under this classification are many different "operator assists," ranging from bench setups with lighted parts bins and parts-handling systems (such as carousels), to some rather sophisticated "pointer" and parts presentation systems, and finally to the conveyorized assembly system, sometimes called a "push-line," where the circuit to be assembled is moved in front of a number of operators' stations with each operator being responsible for placing a limited number of components. The common denominator for all of the above is that the part is put into or onto the circuit board *by hand.*

A detailed description of all of these methods is not within the scope of this chapter; however, we can say that as the level of hand assembly increases in sophistication, the capital equipment cost increases in a near equal manner. The net result is that partial or full automation would have produced the same, or greater, results for equal, or less, total investment.

19.4.2 Islands of Automation

This concept represents the first level of true automatic circuit assembly. It certainly was the earliest one used and even today offers high potential to the manufacturing operation where variety is the way of life. "Islands" refer to groupings of freestanding, semiautomatic assembly machines, each accomplishing one type of an insertion or placement function. In these machines, the actual insertion or placement of the part is by automatic means; however, the board or substrate handling is usually performed by an operator, who sometimes can handle more than one machine. The advantage of this approach is that each machine has the capability of handling many more components per unit of operator time than is possible with hand assembly. Also, the user reaps the rewards of the uniformity of machine assembly, a low error rate, high speed, and, in some cases, part value verification, as well.

The machines and machine functions included within the bounds of this approach are VCDs and sequencers for axial-lead components, DIP inserters, radial-lead inserters, pin and jumper-wire inserters, and terminal and socket inserters, as well as many special machines tailored to specific components and functions. The same island approach is also available for those who are placing surface mounted devices. Machines are available for the simple two- and three-terminal devices, and others for components requiring a much higher placement accuracy, such as SOICs and leaded or leadless carrier devices.

These islands may be an end in themselves or merely a stepping-stone to a higher level of automated manufacturing, as we will see next.

19.4.3 In-Line and Direct-Transfer Systems

This and the next higher category have certain aspects in common. In both categories automated board handling is required in one form or another. Both the in-line and the direct-transfer systems are best applied to situations where the product is quite stable and has a high volume. The in-line system, which was one of the earliest forms of high-volume automated electronic assembly and is still used, generally consists of a number of operating stations, each capable of inserting one part. The circuit is transported from station to station, and when accurately positioned, all heads operate simultaneously to insert or place one component at each station. The chief advantage of this approach is speed; each cycle of the machine produces a completed assembly, at least for those parts which can be handled automatically. Some of the disadvantages include a relatively high initial cost and a somewhat complex changeover routine.

The direct-transfer system generally is made up of standard *X-Y* types of insertion or placement machines enhanced with a fully automatic board-handling and transfer system. Often these systems will start with a raw-card feeder; then, after the first machine, the circuit boards are either placed in a magazine, which is then transported by a buffer-conveyor assembly to the next machine, or handled as single boards. The magazine-oriented buffer and transfer assembly approach takes full magazines from the output of one machine and puts them into a queue for the next assembly station, where they are called for as needed. This flow continues through the last machine. Here the circuits are ready for further process steps. The single-board system is similar except for the "batching" which the magazine provides. All of the assembly machines in this type of system are similar to their standard counterparts—VCD, DIP, pin inserter, etc.—and therefore each station is capable of putting many parts into the board. The number of actual machine stations needed is more dependent on the number of types of devices to be inserted than on the quantity of components of each type. "Balancing the line," on the other hand, may dictate that certain stations be repeated to even the machine-time per station. This system is best suited for relatively long runs of a dedicated product.

19.4.4 Random Routing or Flexible Manufacturing

This level of electronic circuit assembly characterizes the "automated factory" concept presently considered to be the optimum arrangement for high volume *and* high mix. Here the assembly machines that were used in the direct-transfer system, above, are essentially taken apart and made into a number of freestanding assembly stations. To this is added the means to bring to the machine input station a magazine of circuit boards ready for assembly of the parts which that machine is capable of handling. Once a magazine's worth, or batch of boards, has been assembled, an automatically guided vehicle (AGV), part of the material transport system, is dispatched to place a new magazine in the input station with another set of boards, move the empty input magazine to the output station, and, finally, pick up the full magazine. The material transport function may also be done manually.

With this approach, the large producer of electronic circuits can obtain a high degree of flexibility and regain the ability to handle small lot sizes as well. The random routing or flexible manufacturing system offers many additional advantages, including low work-in-process inventory and rapid turnaround. This system is commonly referred to as a computer integrated manufacturing (CIM) system.

19.5 SYSTEMS AND TECHNIQUES FOR THROUGH-HOLE INSERTION

Through-hole insertion is a relatively stable technology. The devices available and the handling systems used have reached a mature level. There are still components used in printed wiring assembly designs which have not yet qualified for automatic placement, and in these areas two things are happening. First, some of the physical parameters of devices are changing, which permits them to adapt to an acceptable handling system and be automatically inserted with relatively conventional means. Radial insertion with all devices supplied on the Japanese-generated taping standard is an excellent example. The second influencing factor is the entry of robotics into electronic assembly. As we have discussed, robots can manage to insert some of these complex parts into circuit boards, which has not been possible in the past.

Also covered here are typical areas where electronic hardware insertion is possible, what is available in this area, and how it can best be used.

19.5.1 Axial-Lead Insertion

The principle of axial insertion has already been described earlier in this chapter, and it is shown in detail in Fig. 19.3. The left-hand view of Fig. 19.9 shows a typical cut-and-clinch unit with the part just inserted. With the head "down," the clinch cutters move in to cut and clinch the leads, as shown on the right-hand view.

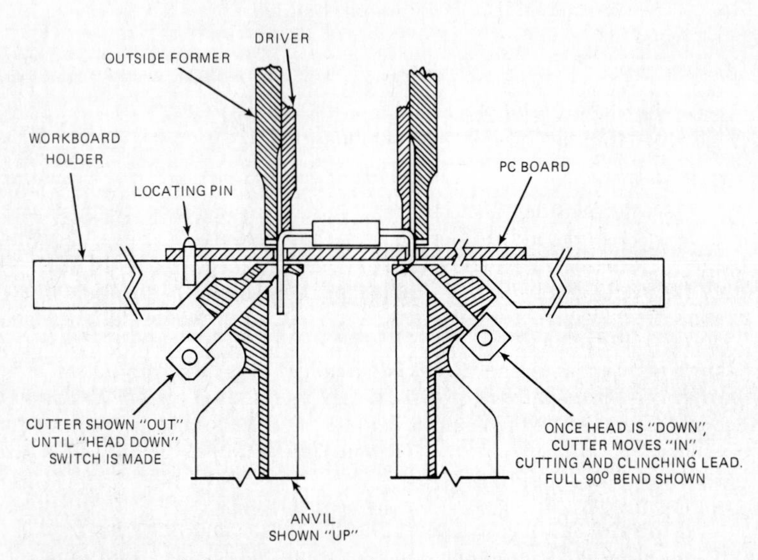

FIG. 19.9 A typical cut-and-clinch unit, shown in the "up" position, before and after cutting and clinching.

The majority of axial insertion machines sold offer a VCD capability and are program-controlled. There are inserters available with an on-board sequencing function (a sequencer-inserter); all others require a separate off-line sequencer. Sequencers are available with over 200 input stations, making them into a virtually complete "stock room."

Component verification is available in some axial sequencer-inserters and on most off-line sequencers. This testing capability allows a final check for proper part type and value at a point immediately prior to the actual insertion. Cost studies have shown that this is the best place to avoid expensive downstream test failures due to the placement of a component whose electrical characteristics are questionable.

19.5.2 Dual-in-Line Packages (DIPs), Including Sockets

DIP insertion uses the same principle of guiding gathered leads as the axial inserters. All DIP inserters are capable of on-board sequencing. Since the DIP is a higher-level part than those of the axial family, the need for a large variety of inputs is greatly reduced. DIP inserters, however, are available with up to 100 stick-type input stations for different types of devices. Some also offer multiple sticks at each input, allowing relatively long run times between the need to reload components. There are DIP insertion machines which can handle several sizes of DIPs. Within a given lead span, say 0.300 in, most machines can handle components ranging from a minimum of 6 leads up to 22 leads, which is usually the greatest number available. Figure 19.10 shows a close-up of the head tooling inserting a DIP. For the larger parts with more leads, those on a 0.400- or 0.600-in span, there are several approaches. Depending on the quantity to be inserted and the mix, machines are available with a single head, tooled for either of these larger devices, or with dual heads, tooled to handle any two of the three spans. A third approach uses variable-span tooling, which can handle any one of the three spans. Two-lead and four-lead DIPs present a totally different problem for the inserter. Since they tend to tumble in a stick and not retain their proper orientation, these devices need either a dedicated machine or a separate head and feed system on a dual-head machine.

Component verification is an available option on many DIP inserters. Again, this virtually eliminates the possibility of placing a DIP into a board which is reversed or which has questionable electrical characteristics.

FIG. 19.10 A 0.600-in DIP, still in the head tooling, being guided into a PCB.

Another capability available is the insertion of DIP sockets, as well as the DIP that goes into the socket. Socket specifications become more stringent than those for hand assembly, but selecting one suitable for automatic placement generally offers a substantial cost saving. Sockets are available which will allow DIPs to be inserted *before* the board is soldered. There are others which require that the DIP placement wait until after the soldering operation is complete. Figure 19.4 shows several varieties of DIPs and sockets.

19.5.3 Radial-Lead Insertion

Radial-lead insertion has been a long-time need in the industry. The prime candidate from the very beginning was the ceramic disc capacitor, the old "penny capacitor." Several versions of this component in its taped form are shown in Fig. 19.11. This inexpensive device was quickly designed into many circuits in great numbers, but since there was no real industry effort to develop a standardized handling system, all early attempts to mechanize the insertion of this part were doomed to failure. In the early 1970s, Japanese electronic manufacturers cooperated on a program to standardize the tape on which these types of devices were made. They also agreed to accept a common lead separation of 5 mm. The result is that the taping standards adopted in the United States for these components are identical to those originated in Japan. With the standard sprocket-fed tape as the basis, other components of the radial, or quasi-radial, variety fell into the system quite nicely, such as coils, chokes, and axial components formed into a "hairpin" shape. With components available on a common handling system, manufacturers of insertion equipment soon came up with fully automatic sequencing and insertion equipment for this family of devices. Because of the low cost of many radial components, they found rapid acceptance in the entertainment sector of the market. As more component suppliers began to use this standard, the

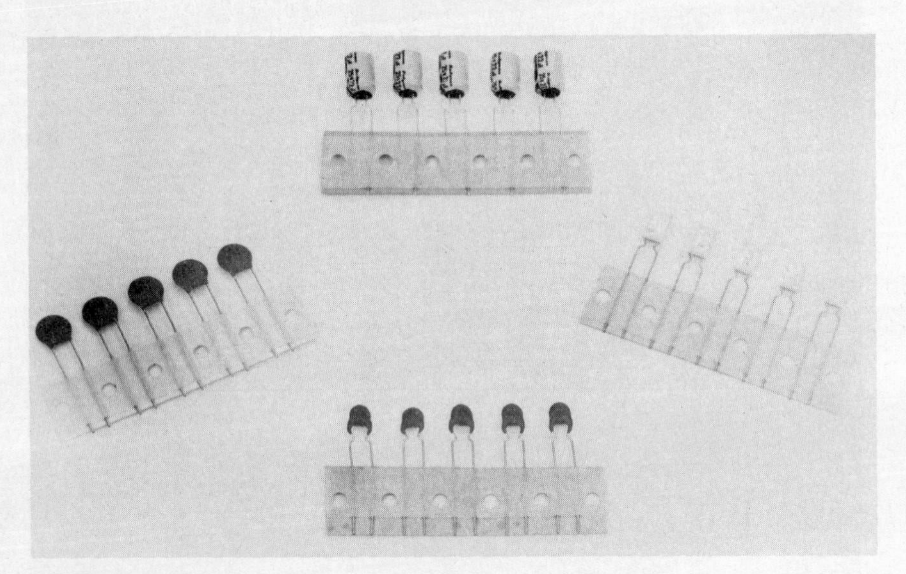

FIG. 19.11 Typical radial-lead components shown on standard sprocket-fed tape.

approach spread both into the United States and Europe. Today, radial-lead insertion is one of the major techniques available to the electronic equipment manufacturer.

Several insertion approaches are used for radial devices: one is similar to the type I approach used on axials and DIPs where the leads are spread somewhat prior to placing the part into the lead-gathering guide combs of the insertion tooling. This approach requires that the component body be held momentarily as the part is pushed through the tooling. In another approach the part is handled by the leads directly (see Fig. 19.12). Here, the part body is touched only by a downward-moving "pusher" when the leads have been accurately positioned over the holes in the PCB. After the lead tips have entered the holes in the PCB, the clamps loosen and the entire assembly swings away and upward, clearing both the part being inserted and the previously inserted parts. In either system, part-to-part clearances can be quite small. Component verifiers are available on several of the radial-lead sequencer-inserters.

19.5.4 Single-in-Line Packages (SIPs)

This is a special package used mostly for passive circuit networks—resistors, diodes, capacitors, and combinations (see Fig. 19.13). As a specialized component, its adaption to automatic insertion has been somewhat slow, but rewarding for those who have had the quantity requirements to justify its automation. SIPs are usually available with leads approximately 0.1 in long and spaced either 0.100 or 0.125 in (2.54 or 3.17 mm) apart. The number of leads varies from two to eight

FIG. 19.12 Close-up of the head tooling of a radial-lead insertion machine inserting a component into a PCB.

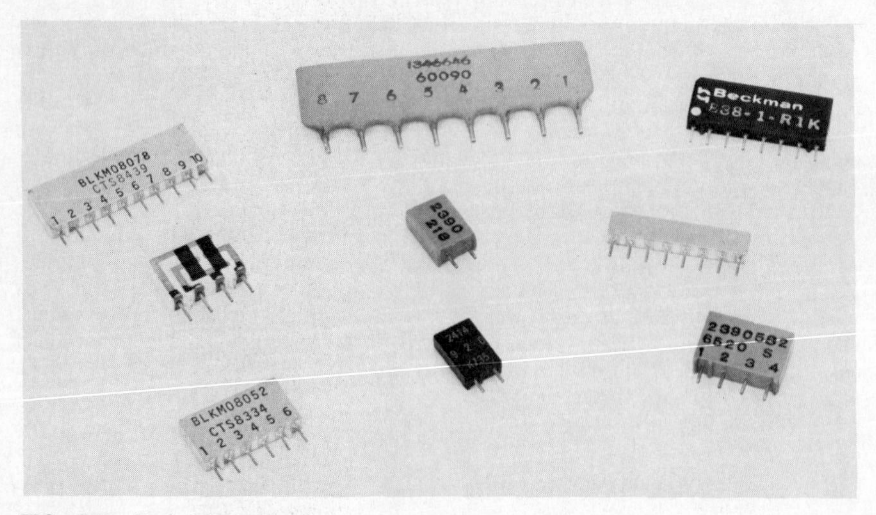

FIG. 19.13 Typical SIP packages.

or more. The body is either precision-molded or dipped into a conformal coating material. For automatic handling the conformally coated versions are of little use, except, possibly, through the use of robotics. The handling system most used is an extruded stick, similar to that used for DIPs, with the SIPs placed side by side in a way which fully protects the leads. SIP insertion is a modified form of type II. The devices are normally fed to a driver having a side pickup surface with a vacuum port large enough to cover the part with the fewest leads (see Fig. 19.14). Needless to say, the lead-to-hole clearance must be slightly greater than for axials

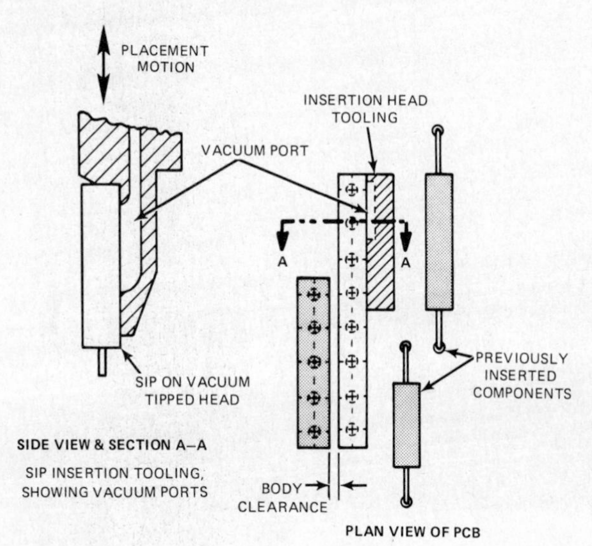

FIG. 19.14 Detail of SIP tooling, indicating size range of device and required clearances.

and DIPs to maintain the same insertion reliability. With this vacuum-holding method, SIPs may be placed side by side as close as body tolerances permit, a process sometimes called "brick-walled."

19.5.5 Nonstandard Components

19.5.5.1 Special Machines. If a high-volume design must be assembled and it contains nonstandard components, the most obvious route is to follow the automated assembly functions with a hand-assembly operation. Should this become too expensive or if there is an overwhelming need to further automate the product, then a special machine to handle one or more of these nonstandard items could be considered. Special machines mean just what the name implies: a machine built to do a specific job. These special machines vary from a "do-it-yourself" machine to a general-purpose assembly machine that is tooled to do the special job to a candidate for the new world of robotics. The final choice will depend largely on the return on investment (ROI).

19.5.5.2 Robotic or Flexible Machines. Robots can range from a very simple, two-position, pick-and-place mechanism to a very complex, multiaxis, controlled platform capable of being tooled to do one or more given tasks. Actually, the capabilities are almost unlimited, and the extent to which these techniques are generally utilized is a function of ROI. From a practical standpoint, robots used and being considered for electronic component assembly fall somewhere in the middle of the technology. They generally offer several controlled axes, stored programs, replaceable tooling, possibly some adaptive learning, and maybe some tactile feedback capability. The primary use is to place or insert components that cannot be handled by conventional hard-tooled machines. This typically means components used in relatively low volumes on each board and/or those whose physical shape requires special handling. Suppliers of robots can provide specifications outlining the capabilities of their specific models.

The essential ingredients of a practical robotic system are a moving set of "fingers" (often called end-effector tooling), which may or may not have tactile sensing capability, and usually a multiaxis positioning system to provide several degrees of freedom. The system also may have "vision capability" (optical recognition). In addition, there can be a gauging station where a part with poor body definition, but with a pin or lead-tip matrix which is known to be good (or which can easily be made good), can be relocated such that the lead tips are now at a known position with respect to the tooling. The part can then be picked up again by the robot's end-effector tooling, or fingers, and placed into the hole pattern of the prelocated PCB.

For a more graphic explanation of these system functions, a robotic placement sequence for an insertion task is shown in Fig. 19.15. View (*a*) shows an overall view of a robotic head with the fingers, or end-effector tooling, about ready to pick up a component from a magazine. View (*b*) shows the component gripped by the head fingers at the magazine pickup point. In view (*c*) the component is shown being placed into a gauging station. Here the fingers relax and the part is pushed into conical locating and straightening guides. Next, the floating fingers regrasp the component body with the leads now in a known position and then move to the insertion location. View (*d*) shows the component being placed through holes on a PCB. Precision parts can often be inserted without this gauging sequence. Other methods employ "vision" to inspect and register the pin pattern for insertion.

The robotic or flexible automation system offers one significant advantage: when the initial requirement has passed, the basic robot can be retooled or reprogrammed to handle a different component configuration or mix.

19.5.6 Electronic Hardware

These are nonelectronic items designed to be put into or onto a PCB in order for other components to be attached or for interconnection purposes. Interconnection includes pins or sockets for external cables for board-to-board connections, or for top-side to bottom-side connections. To make these hardware items suitable for automation, many of the same principles already outlined must apply. The best candidates for automatic assembly are those devices supplied on a continuous metallic strip, similar to the terminals shown in Fig. 19.16. These dual-spade receptacles are typical of the family of interconnected hardware items designed for automatic insertion.

Square wire, generally in 0.025- and 0.045-in (0.6- and 1.14-mm) square sizes, is commonly used to provide single pins, form rows of pins for use with external sockets, or make pin grid arrays used for wire wrapping. Inserters are available for pins of different types. These can use previously prepared and plated pins affixed to a carrier strip, or they can take raw, plated square wire material, pyramid the ends, and insert these in a nearly identical manner. The latter method is shown in more detail in the close-up of a continuous wire inserter, Fig. 19.17.

(a)

FIG. 19.15 Robotic placement sequence.

(b)

(c)

FIG. 19.15 *Continued*

(d)

FIG. 19.15 *Continued*

FIG. 19.16 Dual spade receptacles on a strip.

19.6 SYSTEMS AND TECHNIQUES FOR SURFACE MOUNTING

Surface mount technology (SMT) is the placement and attachment of active and passive electrical components directly to the prepared surface of a PCB. This technology is quite different from the process used with leaded devices, as there is no mechanical integrity of attachment at the assembly level. The common denominator in SMT is the need for a temporary attaching adhesive to hold the devices

FIG. 19.17 Close-up of square wire pins being inserted into a PCB. Note the pyramid-formed lead tips.

from the time that they are placed until permanent attachment takes place, usually by a soldering process. The temporary adhesive may take the form of a solder paste, usually only on the areas to be soldered, or a glue, which, after the devices are placed, is cured by one of several means.

19.6.1 Surface Mounted Devices and Handling Systems

There are several categories of body shapes and contact or lead arrangements for the most popular devices used with SMT. First, there are the two-terminal devices, most of which are rectangular in shape, including chip capacitors and chip resistors. Closely related to these are the two-terminal polarized capacitors, and these generally have deliberate shape factors to aid in orienting. There are also some cylindrical two-terminal devices used, mostly diodes.

The handling systems for all of the above devices generally fall into two categories: bulk, and sprocket-driven tape with covered pockets. Bulk parts are usually fed with a vibratory feeder which uses the dimensional parameters of the body to separate and/or orient and move them to a fixed location for pickup. The tape approach shown in Fig. 19.18 is the standard. Most of the more commonly used resistors and capacitors are available on 8-mm tape. Both laminated paper and formed plastic carrier strip materials are used. With both approaches, the component nest varies to accommodate the specific part size. Other tape widths are used for larger-size components.

Another grouping is the small outline (SO) devices, both the transistor and the IC. These are smaller versions of the now familiar plastic DIP, but with a modified lead form. Figure 19.19 shows a close-up view of a SOT (small outline transistor) on a vacuum placement spindle. SOTs have three leads asymmetrically spaced, while the SOICs (small-outline integrated circuits) have many leads equally spaced on opposite sides. Both may be purchased on sprocket-driven tape, again with various widths. The SOTs also can be fed from bulk using specially tooled vibratory feeders; however, this is not a preferred way. SOICs may also be obtained in rigid or flexible sticks which have their usual limitation as to quantity.

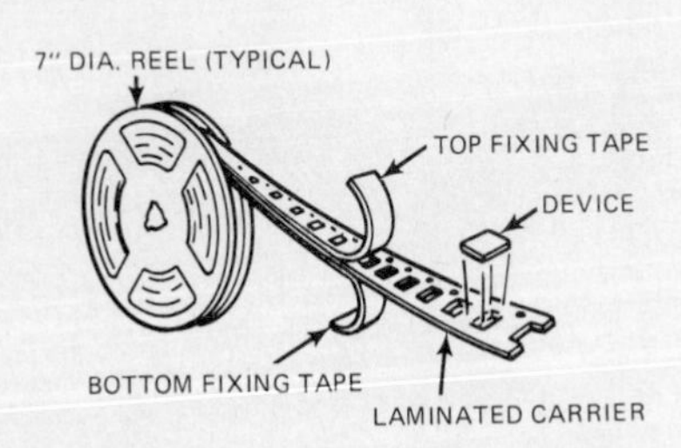

FIG. 19.18 Typical tape carrier used for surface mounted components. A laminated paper version is shown.

Other surface mounted devices are the leaded chip carriers, both plastic (PLCC) and ceramic (CLCC). They are becoming the most acceptable form for an IC in surface mounted design. Leaded packages help to mechanically isolate the IC chip from the circuit medium. This reduces the temperature differential stresses to a minimum. These packages are available on the larger widths of sprocket-driven tape.

Leaded chip carriers should not be confused with ceramic *leadless* carriers. The latter have noncompliant solder pads on the mounting surface and can generate a very large shear force in the solder joint from any significant temperature change, unless used on a circuit surface of similar thermal coefficient of expansion (TCE), specifically a ceramic substrate.

FIG. 19.19 A close-up of a vacuum placement spindle holding a SOT. This view also shows the open fingers of a centering device located just above the spindle tip position.

19.6.2 Automatic Placement Machines

Now that we have discussed the devices and their related handling systems, we need to categorize these same surface mounted devices in yet one more way: their required placement accuracy.

Placement accuracy is a major specification of any given machine, and it affects the machine's cost, speed, and complexity.

The placement accuracy required for each device is largely a function of the size, number, and spacing of the terminal pads or leads. Generally, the two- and three-lead devices do not require the same placement accuracy as a multileaded SOIC or a PLCC. These call for more sophisticated techniques in locating, centering, correcting rotational errors, etc. As a result, the latter group of devices requires a very different assembly machine than one designed exclusively for the more elemental group.

The above information on devices, handling systems, and machines may now be combined with the process information found elsewhere in this book. This will give a sound basis for detail planning, machine selection, and design planning.

19.6.3 Placement Techniques

The technique common to virtually all surface mount assembly machines is pick-and-place. In essence, there is a head and it carries a vacuum placement spindle which has certain common characteristics: it can move up and down or in and out. Often it can rotate after a device has been picked up. In the spindle's controlled up-and-down placement motion, the down pressure is often programmable into several force levels. This allows the device to be pressed into the PCB with a known force so that penetration into the solder paste or glue can be controlled.

Another feature often found on, or in support of, a placement head assembly is a centering device. In most cases the pickup positioning of devices to be placed—i.e., a tray, a linear feeder, the open nest in an 8-mm or larger tape—is only approximate. In these cases, the vacuum pickup nozzle has literally only a general idea where the part is with respect to its true reference center. Most pick-and-place systems have a vacuum sensor which indicates that the pickup has been made. If the part needs to be rotated prior to placement, this is generally done first. The centering mechanism next centers the part on the nozzle, or at least moves it to a consistent position, which becomes known as the placement center. Mechanical centering functions use the edge surfaces of the part; therefore, accuracy and repeatability are totally dependent on the part's edge-to-pad or lead-to-lead tolerances. Figure 19.19 also shows a close-up of a centering device. Other factors of the process must also be considered: the stretch and shrink of the circuit itself, the pad-to-pad tolerances of the part to be placed, and, finally, the remaining placement machine tolerances such as X and Y resolution, calibration, angular error, etc. There is a technique available called "vision" which can help with these very tight requirements. It uses optical recognition techniques to find and correct rotational and positional errors between the part and the surface pattern on which it will be placed.

From an overall machine standpoint, there are two basic variations: the *fixed workpiece and moving head machines* and the *fixed head and moving workpiece machines.* When many different component inputs are at various locations, the machine must move the pick-and-place mechanism, not only over the board to be populated but to and from a number of input locations. This is an example of the fixed workpiece and moving head variety. Conversely, a fixed head and moving workpiece machine will typically have the component pickup point fixed and part selection made using a carousel or some other moving feeder containing a large number of inputs.

Two other choices are automatic circuit board handling and manual board placement. Both approaches have their advantages and disadvantages. Figure 19.20 shows a large in-line surface mount placement machine with automatic circuit handling and a number of *X*-*Y*-controlled placement stations, each of which can place more than one variety of part.

FIG. 19.20 A large in-line surface mount assembly machine with automatic circuit board transfer.

19.6.4 Conclusion

We have discussed some of the advantages and pitfalls that can arise if devices are selected without a full understanding of the entire process. Many of the commercially available automatic assembly machines are handling-system-oriented. Others have a greater degree of input flexibility. The overall program planner, the electrical designer, and the circuit layout designer must be aware of all potential trade-offs.

19.7 MIXED ASSEMBLY, OR THE VARIOUS ROUTES TO SURFACE MOUNT TECHNOLOGY

There are generally three accepted paths for the application of surface mounted devices. Figure 19.21 shows these in flowchart form, and they range from pure application of surface mounted devices only to a very mixed arrangement of surface mounted and through-hole leaded components.

The process flow shown as method A in Fig. 19.21 uses surface mounted devices on only one side of the board. This is the method most used for ceramic substrate or hybrid designs, and it is, perhaps, the original classic version of SMT, having been in use the longest. In recent years there have been designs where the placement medium is a PCB with all active and passive devices surface mounted.

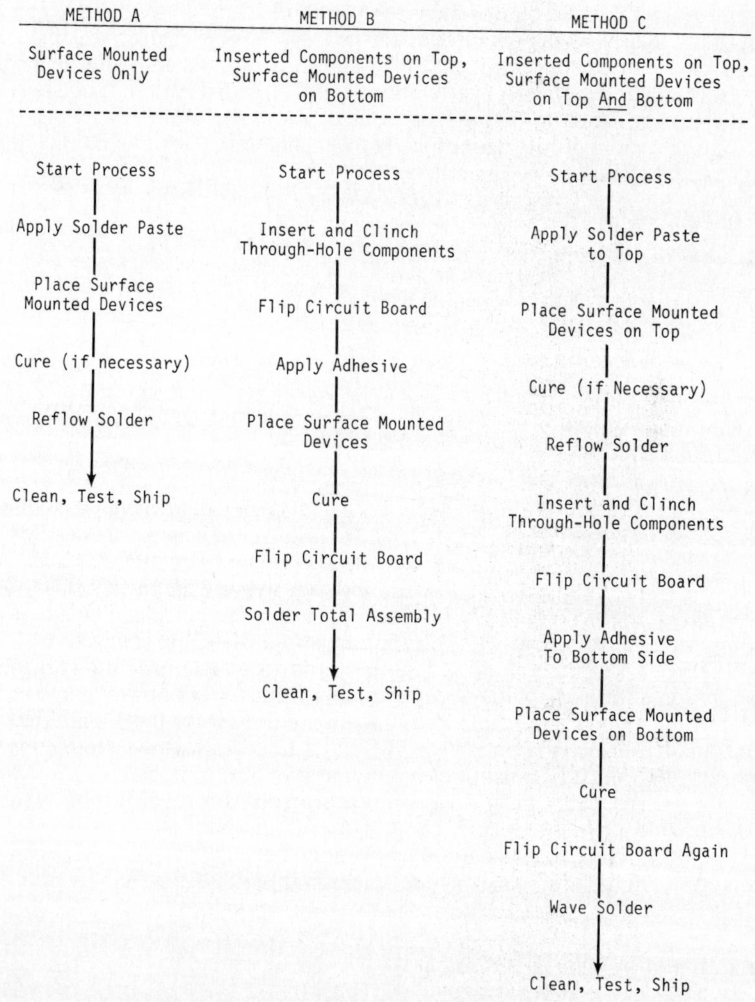

FIG. 19.21 The various routes to surface mount assembly.

Method B, on the same figure, shows the easiest form of mixed technology. It uses through-hole components on the top of the board and surface mounted devices on the bottom only. This is the approach used by many in making the transition from through-hole inserted assembly to the incorporation of SMT. It is, perhaps, the most tolerant approach to both design and process variables; therefore, a higher degree of success can be predicted.

Method C allows both surface mounted and through-hole components on a PCB, with surface mounted devices on both sides. It is the most complex of the three.

19.8 THE AUTOMATED FACTORY

19.8.1 The Commitment

The decision to take the route of the flexible manufacturing system factory should not be taken lightly, as the overall impact for any business enterprise considering this approach can be enormous. Flexible manufacturing is a computer-controlled configuration of independent work stations coupled with a material transport system, and it is designed to manufacture one or more part numbers, typically the high-volume with high-mix operation. The benefits of the flexible manufacturing system are as follows:

1. High equipment utilization
2. Reduced direct labor cost
3. Reduced work in process
4. Response to changing production requirements
5. Ability to maintain production through redundancy
6. High product quality
7. Operational flexibility
8. Accurate history database

With these advantages, the approach offers a high degree of attractiveness, but the ROI must be looked at very carefully. The commitment aspect requires involvement of the entire organization, from the top down. *Top management* must understand what is being done and why, and they must make the necessary funding available. *Engineering* must provide proper designs which can be adapted to an automated environment with a minimum of hard-tooling changeover. There must be proper design planning so that the automated machines can work within their operating limits. *Manufacturing* must be willing to retrain workers to be able to support higher levels of machine complexity. *Purchasing* must obtain the component quality and consistency needed for automatic assembly. *Production and material control* must be able to release material in optimum batch sizes. *Quality assurance* will probably notice a decrease in inspection requirements which is consistent with the higher quality resulting from automation.

This commitment must continue, not only through the system design and installation phases but also throughout the useful life of the system.

19.8.2 Board Handling and Tracking

Effective board handling starts with the initial design. Most automatic board-handling systems in factories prefer only one board size, as well as one with the same board locator references. For those operations where final board sizes must vary, multiple breakaway boards having a standardized overall outline should be considered. Another approach is to vary the length of the board only, keeping all other dimensions fixed.

Boards can be tracked either by individual board or by magazines. Both methods generally use bar-code labels that can be read by a person and by a machine. A major advantage of tracking the individual boards is to record known defects, e.g., where a part is known to have been misinserted, or missing. This information is useful in the board repair area.

19.8.3 Factory Floor Software

In Sec. 19.2 note that the subject of automation in printed circuit fabrication is covered from an overall systems standpoint. It is important to emphasize that nothing will happen in the factory of the future without the planning and integrating of the machines which do the actual productive work with the hardware and software of an information and data-flow system that is sound.

The earlier section takes basically a top-down view of the picture. By starting at the bottom, the machine level, and looking up, things may appear somewhat different. The software controlling the machine has to be able to run and monitor all the machine functions in real time, and since there will be no dedicated machine operator, it also must have the ability to communicate a number of status items upward. For example, the machine must be able to tell the next upward control level when it has completed a magazine, when it has stopped and for what reason, when it is unavailable, when there is an insertion error, etc. These status or event messages must be common throughout the entire machine family represented, and the next level system must be able to interpret them, handle some, and pass others on for human intervention.

Any automated factory system is a series of hierarchical levels of interrelated control systems. Starting at the machine level, several machine controllers are usually tied into one or more communication concentrators. These units perform the necessary two-way communication between a number of assembly machines and a host system, which can represent an entire factory section. Whether the material transport system (MTS) is connected into this host level depends on the final overall system network arrangement. In any event, the automatic assembly system host and the MTS must be wholly integrated.

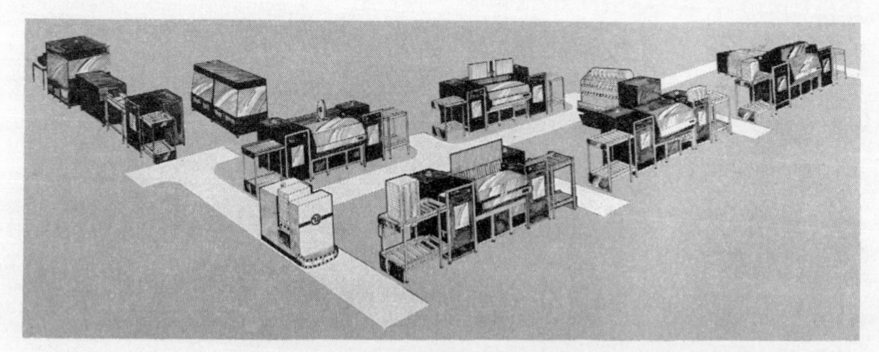

FIG. 19.22 A portion of a typical automatic factory for electronic assembly. Note the "islands" of assembly being serviced by an automatic guided vehicle (AGV).

19.8.4 Making It All Happen

Many of the higher-level functions of the automatic factory have been developed to a relatively mature stage. The most challenging portion will be to make a functioning and compatible integration between the actual assembly machines, the material transport function, inventory management, and the planned work schedule.

Figure 19.22 is an artist's concept of a portion of an automatic factory. It shows several high-level automatic assembly islands, complete with input and output magazine storage and handling units. Also shown is an automatic guided vehicle (AGV), part of the material transport system.

Once completed, the automatic factory with its flexible manufacturing capability will provide the user with the full array of benefits stated earlier. This flexibility, coupled with a substantial reduction of work-in-process inventory, should make the ROI acceptable.

CHAPTER 20*

CONFORMAL COATINGS

John Waryold
HumiSeal Division, Columbia Chase Corporation,
Woodside, New York

20.1 INTRODUCTION

Conformal coatings are systems of synthetic resins that are usually dissolved in balanced volatile solvent vehicles. The resins are selected with particular attention to their electrical and thermal properties and their ability to form, when properly cured, a thin but tough film capable of providing protection against various environmental stresses.

Of the vast number of synthetic resins now available, a selected few types of acrylics, polyurethanes, and epoxies have been found to have the best all-around properties and are currently being used as the basis for most general-purpose conformal coatings. For special applications, however, other resins are used occasionally. For instance, to meet high-temperature requirements, silicones and polyimides may be indicated; and when high dielectric properties are needed, polystyrenes are used.

Basic resins, plus accessory chemicals and solvent vehicles, are blended to achieve optimum desired operating characteristics in the cured film. Many variations within each resin group are made possible by the addition of chemicals. For instance, plasticizers may be added for increased flexibility, dyes for identification or inspection, and wetting agents to improve adhesion to the substrate.

20.1.1 Acrylic Coatings

Acrylics are excellent coating systems from a production standpoint because they are relatively easy to apply. Furthermore, application mistakes can be corrected readily, because the cured film can be removed by soaking the printed circuit assembly in a chlorinated solvent such as trichloroethane or methylene chloride. Spot removal of the coating from isolated areas to replace a component can also be accomplished easily by saturating a cloth with a chlorinated solvent and gently soaking the area until the cured film is dissolved.

Since most acrylic films are formed by solvent evaporation, their application

*Adapted from Coombs, *Printed Circuits Handbook,* 2d ed., McGraw-Hill, New York, 1979, Chap. 12.

is simple and is easily adaptable to manufacturing processes. Also, they reach optimum physical characteristics during cure in minutes instead of hours.

Acrylic films have desirable electrical and physical properties, and they are fungus-resistant. Further advantages include long potlife, which permits a wide choice of application procedures; low or no exotherm during cure, which avoids damage to heat-sensitive components; and no shrinkage during cure. The most obvious disadvantage of the acrylics is poor solvent resistance, especially to chlorinated solvents.

20.1.2 Polyurethane Coatings

Polyurethane coatings are available as either single- or two-component systems. They offer excellent humidity and chemical resistance and good dielectric properties for extended periods of time.

In some instances the chemical resistance property is a major drawback because rework becomes more costly and difficult. To repair or replace a component, a stripper compound must be used to remove effectively all traces of the film. Extreme caution must be exercised when the strippers are used, because any residue from the stripper may corrode metallic surfaces.

In addition to the rework problem, possible instability or reversion of the cured film to a liquid state under high humidity and temperature is another phenomenon which might be a consideration. However, polyurethane compounds are available to eliminate that problem.

Although polyurethane coatings systems can be soldered through, the end result usually involves a slightly brownish residue which could affect the aesthetics of the board. Care in surface preparation is most important, because a minute quantity of moisture on the substrate could produce severe blistering under humid conditions. Blisters, in turn, lead to electrical failures and make costly rework mandatory.

Single-component polyurethanes, although fairly easy to apply, require anywhere from 3 to 10 days at room temperature to reach optimum properties. Two-component polyurethanes, on the other hand, provide optimum cure at elevated temperatures within 1 to 3 h and usually give working potlifes of 30 min to 3 h.

20.1.3 Epoxy Resins

Epoxy systems are available, as two-component compounds only, for coating electronic systems. Epoxy coatings provide good humidity resistance and high abrasive and chemical resistance. They are virtually impossible to remove chemically for rework, because any stripper that will attack the coating will also attack epoxy-coating or potted components and the epoxy-glass printed board as well. That means that the only effective way to repair a board or replace a component is to burn through the epoxy coating with a knife or soldering iron.

When epoxy is applied, a buffer material must be used around fragile components to prevent fracturing from shrinkage during the polymerization process. Curing of epoxy systems can be accomplished either in 1 to 3 h at elevated temperature or in 4 to 7 days at room temperature. Since epoxies are two-component materials, a short potlife creates an additional limitation in their use.

The limitations with epoxies are short potlife, elevated temperature cure, poor repairability, and the need to use a buffer material around fragile components.

20.1.4 Silicone Resins

Silicone coatings are especially useful for high-temperature service (approximately 200°C). The coatings provide high humidity and corrosion resistance along with good thermal endurance, which makes them highly desirable for printed circuit assemblies that contain high heat-dissipating components such as power resistors.

Repairability, which is a prime prerequisite in conformal coating, is difficult with silicones. Because silicone resins are not soluble and do not vaporize with the heat of a soldering iron, mechanical removal is the only effective way to approach spot repair. That means the cured film must be cut away to remove or rework a component or assembly. In spite of some limitations, silicone coatings fill a real need because they are among the few coating systems capable of withstanding temperatures of 200°C.

20.1.5 Polyimide Coatings

Polyimide coating compounds provide high-temperature resistance and also excellent humidity and chemical resistance over extended periods of time. Their superior humidity resistance and thermal range qualities are offset by the need for high-temperature cure (from 1 to 3 h at 200 to 250°C). High cure temperatures limit the use of these coating systems on most printed circuit assemblies. Because the polyimides were designed for high-temperature and chemical resistance, chemical removal and burn-through soldering cannot be successful.

20.1.6 Diallyl Phthalate Coatings

Diallyl phthalate (DAP) varnishes also require high-temperature cure (approximately 150°C), which limits their use on printed circuit assemblies. Furthermore, their removal with solvents or with a soldering iron is difficult, owing to their excellent resistance to chemicals and high temperatures (350°F).

20.2 THE PURPOSE OF COATING

Printed boards and components are coated with a protective film primarily to avoid, or at least to minimize, degradation in electrical performance when subjected to environmental stresses encountered in specific operating conditions. There are no coatings that will totally resist the effects of all ambient stresses; most of the stresses are cumulative and will ultimately overcome the protection afforded by the coating. A coating is considered to have served its purpose if it has contained the performance degradation within acceptable levels for an acceptable period of time.

The most damaging, and usually the most prevalent, environmental stress is generally recognized to be humidity. Excessive humidity will drastically lower insulation resistance between conductors, accelerate high-voltage breakdown, lower Q, and corrode conductors.

Contaminants, of which any of a few hundred different types may be found on printed boards, are equally damaging. They can cause the same electrical degradation as humidity, corrode conductors, and even cause dead shorts. The contaminants most frequently found in electronic systems are various chemicals which may be residues of manufacturing processes. A few of them are fluxes, solvents, release agents, metal particles, and marking inks. An important group of contaminants are those inadvertently deposited by human handling, such as body greases, fingerprints, cosmetics, and food stains. Ambient operating conditions may also contribute a variety of contaminants such as salt spray, sand and dust, fuel, acid and other corrosive vapors, and fungi.

Although the list of possible contaminants is almost endless, it is a consolation to know that, in all but the most severe cases, the destructive action can be effectively eliminated with a good conformal coating.

Coatings are applied in a film thickness that rarely exceeds 0.005 in. Such a film will withstand the effects of mechanical shock and vibration as well as of thermal shock and operations at temperature extremes. However, it is a fallacy that such a light coating can be relied upon to provide mechanical strength or adequate thermal insulation for individual components mounted on the printed board. Components must be anchored by mechanical means and must have a suitable sealant of their own.

20.3 SELECTION OF THE COATING

Selection of a suitable coating is a major task, considering the vast array of available materials. The user without some working knowledge of coatings would be well advised to turn for assistance to one of several manufacturers specializing in formulating coatings specifically for dielectrical and electronic applications. The manufacturers' printed information is responsive to the needs of the electronics engineer and contains extensive data, including electrical performance characteristics, from which the engineer can determine whether a particular coating meets the requirements. Characteristics of a coating can usually be classified into two groups: application characteristics, which deal primarily with the physical and chemical characteristics of the liquid coating, and operating characteristics, which define the physical and electrical performance of the cured film.

In the selection of the proper type of conformal coating for a specific application, the basic characteristics of the five coating types described previously should be considered. Furthermore, considering the main environmental stresses anticipated during the life of the electronic assembly, some characteristics could be traded off against others to assure high performance reliability. Tables 20.1 and 20.2 give a comparison of the average characteristics of various resin groups.

20.3.1 Application Characteristics

20.3.1.1 Potlife. Potlife is the length of time during which a coating material can be used before curing sets in. It is an important consideration in planning the coating procedure, and it has an important impact on the economics of coating in large-scale production. Short potlife leads to a substantial waste of materials and to some lack of uniformity in results owing to rapid buildup in viscosity. An acceptable minimum potlife for most applications is about 30 min at room temperature.

Short potlife is common to most two-component coating systems. Single-component systems, on the other hand, have a very extended potlife that almost approaches their shelf life. Note that potlife most not be confused with shelf life. The latter is the length of time a material can be stored in unopened containers without degradation.

20.3.1.2 ***Viscosity.*** Viscosity, or resistance to flow, must be low enough (less than 200 cSt) to permit adequate flowing of the coating over and around components to assure complete coverage. It is particularly important when a high-density board with many components packed tightly together with low clearances is to be coated.

Viscosity will determine the thickness of coating per application. A relatively higher viscosity is desirable when the coating must cover components with sharp corners or protruding leads from which it tends to roll off. Coating viscosity must be established by trial and error to achieve optimum performance for a specific board configuration. The viscosity of any solvent-type coating system can be reduced by the addition of the proper thinner.

20.3.1.3 ***Solids Content.*** The solids content represents the portion of a coating which will cure into a film; it is opposed to the solvent content, which will evaporate. The value has a bearing on coating thickness per application, and consequently it has some impact on the economics of the coating operation.

Solvent-type coatings are usually supplied with a solids content ranging from 10 to 50 percent. Solventless systems are, of course, rated as "100 percent solids" coatings.

20.3.1.4 ***The Chemical Components.*** The chemical components of a coating, particularly the solvent vehicle, may adversely affect the materials used in some of the components such as styrenes, acrylics, some marking inks, and adhesives.

The chemical effects of the selected coating must be ascertained beforehand, and the components designed into a printed wiring circuit must be chemically compatible with the coating.

20.3.1.5 ***Ease of Application.*** If the number of boards to coat is substantial, it is naturally more advantageous to use a coating that does not require specialized

TABLE 20.1 Coating Selection Chart

	Acrylic	Urethane	Epoxy	Silicone	Polyimide	DAP
Application	A	B	C	C	C	C
Removal (chemically)	A	B		C		
Removal (burn through)	A	B	C			
Abrasion resistance	C	B	A	B	A	B
Mechanical strength	C	B	A	B	B	B
Temperature resistance	D	D	D	B	A	C
Humidity resistance	A	A	B	A	A	A
Humidity resistance (extended periods)	B	A	C	B	A	A
Potlife	A	B	D	D	C	C
Optimum cure	A	B	B	C	C	C
Room-temperature curing	A	B	B	C		
Elevated-temperature curing	A	B	B	C	C	C

Property ratings (A–D) are in descending order; A is optimum.

TABLE 20.2 Typical Characteristics of Various Coating Materials

Properties	Acrylic	Urethane	Epoxy	Silicone	Polyimide	DAP
Volume resistivity, Ω·cm (50%RH, 23°C)		2×10^{11}				
	10^{15}	11×10^{14}	10^{12}–10^{17}	2×10^{15}	10^{16}	1.8×10^{16}
Dielectric constant, 60 cycles	3–4	5.4–7.6	3.5–5.0	2.7–3.1	3.4	3.6
Dielectric constant, 10^3 cycles	2.5–3.5	5.5–7.6	3.5–4.5		3.4	3.6
Dielectric constant, 10^6 cycles	2.2–3.2	4.2–5.1	3.3–4.0	2.6–2.7	3.4	3.4
Dissipation (power) factor, 60 cycles	0.02–0.04	0.015–0.048	0.002–0.010	0.007–0.001		0.010
Dissipation (power) factor, 10^3 cycles	0.02–0.04	0.043–0.060	0.002–0.02		0.002	0.009
Dissipation (power) factor, 10^6 cycles	2.5–3.5	0.05–0.07	0.030–0.050	0.001–0.002	0.005	0.011
Thermal conductivity, 10^{-4} cal/(s) (cm²) (°C) (cm)	3–6	1.7–7.4	4–5	3.5–7.5		4–5
Thermal expansion, 10^{-5}/°C	5–9	10–20	4.5–6.5	6–9	4.0–5.0	
Resistance to heat (°F) continuous	250	250	250	400	500	350
Effect of weak acids	None	Slight to dissolve	None	Little or none	Resistant	None
Effect of weak alkalies	None	Slight to dissolve	None	Little or none	Slow attack	None
Effect of organic solvents	Attacked by ketones, aromatics and chlorinated hydrocarbons	Resists most	Generally resistant	Attacked by some	Very resistant	Resistant

equipment, skilled labor, or involved production controls for its application. In this respect, room-temperature cure is preferable to elevated-temperature cure, which requires an oven, and single-components systems are preferable to two-component coatings, since they do not need metering and mixing equipment and thus eliminate a frequent source of operator error.

To the greatest extent possible, the coating should be procured at a viscosity at or near the viscosity desired for application. Thus the operations required to thin or reduce the coating are minimized or even eliminated. Some consideration should also be given to toxicity and flammability of the coating, since those characteristics will often require that the material be given special handling.

20.3.1.6 Curing Temperature. Room-curing materials usually require up to 24 h to cure fully. However, over-cured materials (mostly two-component systems) may be cured in as little as ½ h. Generally, the higher the curing temperature the faster the cure. Although the shortest cure time is desirable to expedite the production process, the curing temperature is limited to the operating temperature rating of the board and the components mounted on it.

20.3.2 Operating Characteristics

20.3.2.1 Electrical Properties. One of the principal purposes of the coating applied to a printed board is to provide insulation to the otherwise bare printed wiring. The coating, therefore, must when cured provide insulation resistance and dielectric strength which satisfy design requirements over the entire temperature operating range and under adverse environmental conditions. In certain applications the dielectric constant and loss factor *(Q)* may also be significant values which should be considered when the coating is selected.

20.3.2.2 Thermal Properties. Operating temperature characteristics of plastics vary substantially. The coating selected must, naturally, have an operating temperature range at least equal to that of the system in which it is used. Within that range the coating must, in addition to meeting minimum electrical performance requirements, remain free from physical degradation such as embrittlement, cracking, or shrinking. A coating characterized by good flexibility will usually resist such physical deterioration.

Particular attention must be given to the hydrolytic stability of the coating; under conditions of high operating temperature in a high-humidity environment, certain coatings may, after a period of exposure, "revert" to a liquid state.

20.3.2.3 Humidity Resistance. Coatings should have low moisture absorption as well as low water vapor permeability. Most coatings marketed for printed board applications have more than adequate resistance to humidity. Yet, in special applications in which unusual and extended conditions of humidity are anticipated, the determination of minimum insulation resistance under humidity should be a guiding factor.

20.3.2.4 Resistance to Chemicals and Fungus. When operating conditions expose the coating to possible contact with chemicals, salt spray, or fungus, it is naturally necessary to determine the coating's ability to withstand the destructive effects of such external agents. Urethanes and epoxies, as a rule, have good chemical resistance. On the other hand, some of those coatings may contain nutrient inorganic additives which, unless neutralized by a fungicide, will promote fungus propagation. Acrylic coatings usually contain only nonnutrient organic components, but their resistance to solvents is generally low.

20.4 PREPARATION OF CIRCUIT ASSEMBLIES

20.4.1 Design

A major factor in the ability of a coated printed circuit assembly to function under severe adverse environmental conditions is the basic circuit design. Assemblies that have sharp corners, sharp pins, or variable components reduce the effectiveness of a smooth monolithic film, and component density and geometric positioning on the assembly also are major factors. The minimal vertical clearances and interface between dual-in-line packages, ICs, etc., create coating penetration problems.

The composition of various components and markings, as well as various insulations such as Teflon, creates problems of compatibility. When choosing materials and components, we must consider whether they possess enough chemical resistance to withstand a variety of cleaning solutions. Too often we choose a type of insulation (TFE, vinyl, etc.) without anticipating receptiveness for coatings.

20.4.2 Cleaning

The foremost consideration to remember is that coatings seal in as well as out. With that in mind, the cleaning procedure adopted should ensure an absolutely clean substrate. Listed below are a few simple steps which will produce a clean substrate and reduce the possibility of residual contamination (flux, fingerprints, grease, plating salts, moisture, release agents, etc.) being entrapped beneath a cured film. These steps should be followed after all the visual, physical, and electrical tests have been made:

1. Vapor-degrease (a suitable solvent would be trichloroethane or Freon) to remove residual greases and fluxes.

2. Rinse in deionized or distilled water, ethyl alcohol, or isopropyl alcohol to dissolve any salts which are not readily soluble in other solvents.

3. Oven-bake for 2 h minimum at 150°F to remove any residual traces of solvent and moisture which may still remain on the assembly. After cleaning and baking, the printed assembly should be handled with rubber or lintless gloves.

20.4.3 Storage

If the clean assemblies are not to be coated immediately, they should be stored in a desiccator cabinet or in sealed polyethylene bags. If any masking is required, a polyester tape, rather than the conventional pulp-type masking tape, is recommended. The polyester tape is less porous and less likely to be attacked by solvents during the coating procedure. Another tip which may help in masking is to use either RTV silicone or a thixotropic latex mask which can be applied to hardware such as the tuning screws of potentiometers. After the conformal coating has become tack-free, the tape or RTV can be pulled off.

20.5 APPLICATION OF COATINGS

There are four major methods for applying conformal coatings: spraying, dipping, brushing, and flow coating. Also, there are variations of each method. The order given is probably the order of preference, but it is not necessarily the order of efficiency.

20.5.1 Dipping

Dipping is by far the most efficient way to apply a conformal coating. It is the only effective way to ensure even deposits of coating and uniform coverage. For best results the temperature of the coating should be kept within the 70 to 90°F range. However, the most important factor influencing the results of the dip-coating application is the speed of the substrate's immersion; the coating should be allowed to seep into all the voids on the printed circuit assembly. Typical immersion speeds are between 2 and 12 in/min. Slow speed of immersion will allow the coating to displace the air surrounding the components. Too rapid immersion may result in trapping the air and thereby forming bubbles. The assembly should be left in the coating until all bubbling has ceased.

The immersion and withdrawal speeds depend, of course, on the size and complexity of the assembly. In most cases, they should not exceed 1 ft/min. Those considerations are especially important for the first dip. Subsequent dips may be made safely at higher speeds.

When conformal coatings are applied by dip, the evaporation of solvents may occur at such high rates that the viscosity of the bath will increase at a similar rate. In such cases, it is desirable to monitor the viscosity of the coating and, if necessary, add the proper thinner to restore the original viscosity.

20.5.2 Brushing

Brushing is the least effective application method because of the difficulty of getting uniform coverage, controlling bubbling, etc. It is neither a practical nor an efficient way to coat large quantities of printed circuit assemblies.

20.5.3 Flow Coating

Flow coating consists of pouring or flowing the coating as a curtain as the assembly passes through the coating. The method is especially effective if printed circuit assemblies have minimum flat packs, integrated circuits, or cordwood modules.

20.5.4 Spraying

Spraying is the most expeditious method for coating assemblies. With the proper combination of solvent reductions, spray pressures, nozzle pattern, and solids contents, excellent results are obtainable. Spray-coating is widely employed for protecting all types and sizes of assembly, and both hand and automatic equipment is used. Spraying is particularly adaptable for coating uneven surfaces. For spray application, coatings are used as supplied or thinned down to a spraying viscosity. If the viscosity is too high, difficulties are encountered in spraying a smooth coat. On the other hand, materials can be sprayed at extremely low viscosities if the solvent system is well formulated and the pressure and gun settings are adjusted properly.

Regardless of which technique is used, the coating assembly should be allowed to drain and air-dry. Turning the unit from time to time during the drying period prevents pooling of the coating and speeds up evaporation of solvents. If very sharp edges or asperities exist on one side of the board, especially where the leads protrude through the solder side of the board, it sometimes helps if, during the draining process, the board is held in a horizontal position with pins facing down. That allows "stalactites" to form on the tips of the sharp edges. It may be necessary to repeat the process two or three times in order to produce enough coating to cover the protrusions. Figure 20.1 further explains the need for multiple coats.

20.6 INSPECTION

After a printed circuit assembly has been coated, visual inspection is necessary. That becomes evident when one realizes that a void or bubble provides a path for moisture to penetrate to the substrate and defeat the original intent of the coating.

Inspection, usually under magnification, is simplified if the coating contains a fluorescent tracer. When a coated assembly is viewed under ultraviolet light, the areas not coated will produce a darkened or shadowed blot.

20.7 REPAIR OF COATINGS

An obvious prime consideration in the selection of a conformal coating is the degree of difficulty of removal if repairs should be necessary.

Acrylics: Their thermoplastic film-formers usually can be removed by most chlorinated solvents.

Urethanes: Most single- and two-component urethane materials may be removed by certain strippers. The strippers selectively dissolve the urethanes rather than destructively decompose them.

Epoxies: To date, chemical removal is not possible; obviously, anything that would dissolve the epoxy film would attack the board and components also.

Silicones: Some silicone coatings are soluble in a few select strippers.

In the use of any stripper, thorough rinsing is necessary to prevent any residual traces from becoming entrapped beneath the recoated area.

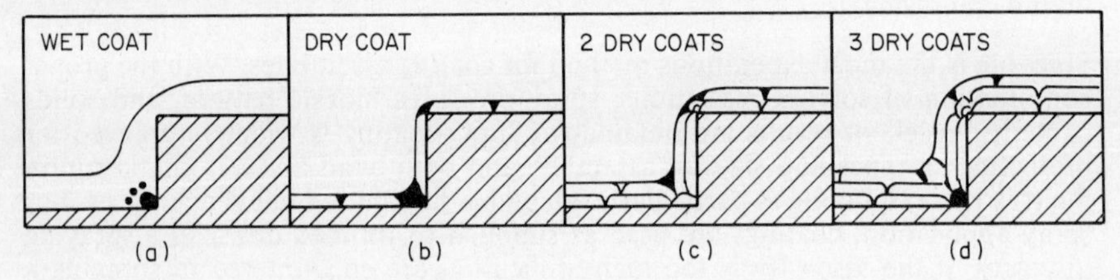

FIG. 20.1 An example of the need for multiple coats.

Most coating materials when applied in thicknesses of up to 0.004 in may be soldered through. However, the degree of difficulty of soldering through a coating is dependent upon the specific resin used, the film thickness, and the aesthetic quality of the reworked area.

20.8 STORAGE OF COATING MATERIALS

The flammability of liquid coatings is determined by the specific solvents used in the formulation and the flash points those chemicals have. The flash point is defined as "the temperature at which the vapors of that solution will ignite and burn." To overcome any potential hazard, we should eliminate all ignition sources. Liquid coatings should always be stored in their originally sealed container away from fire or open flame. The shelf life of a coating is the time a coating may be stored in the original unopened container at room temperature without noticeable deterioration of properties. Polyurethane coatings are sensitive to moisture. Whenever possible, they should be stored in the original container. If the container is partially filled, it should be purged with dry nitrogen.

20.9 SAFETY PRECAUTIONS

Whenever conformal coatings are applied, the following precautions should be observed:

1. Do not inhale vapors from the coating. Use a ventilated hood.
2. Provide adequate ventilation and exhausting.
3. Avoid contact, especially of lips and eyes, with liquid coating.
4. Use protective clothing, respirators, and goggles if exposure to vapors is constant.
5. Request a material safety data sheet to establish safe vapor threshold limits and handling procedures.

CHAPTER 21
TESTING IN ASSEMBLY

Frank J. Langley
GenRad Inc., Concord, Massachusetts

21.1 INTRODUCTION

The assembly of a fabricated printed wiring board (PWB) into a printed wiring assembly (PWA) requires 80 or more process steps when one includes fabrication of the board from the copper-clad glass-epoxy or alternative base material. Therefore, test and its close relative, inspection, are vital to the economical production of a quality product. That is, producing high-quality electronic modules at the lowest cost requires careful monitoring, maintenance, and control of the process to achieve high yields of good boards. It also requires the timely repair of faulty boards or assemblies before further assembly steps are added.

Although the PWB process is over 30 years old, some radical changes have occurred, both in the packaging of components and in the way boards are manufactured. The two major design driving trends are simply higher packing density, through surface mount technology, and increased automation. It can be readily appreciated that these two drivers support lower production cost through the reduction in the number of PWAs, manual labor, and the time required to build any given product. In addition, functional performance can be increased owing to the significant reduction in signal path lengths.

From a test viewpoint, the above changes require increased performance and greater integration of the test and inspection equipment with the process machinery as well as the board and assembly design and verification databases. The latter is important to utilize computer-based resources and thereby reduce the cost of the test.

In-circuit automatic test equipment is widely used to cope with the increasing complexity of PWBs. Quality management networks have also been introduced to eliminate paper test reports and make immediately available to the manager and repair technicians the results of automatic board tests, thus enabling better control of the process, increased yield, and facilitation of the repair of faulty boards. In addition, there is an increasing use of semicustom techniques involving design for test (DFT) and on-chip test (OCT). This simplifies the testing of one-of-a-kind components such as "application-specific integrated circuits" (ASICs).

It is the intent of this chapter to provide the tools which can help managers, engineers, and technicians alike to deal effectively with these changes and to capitalize on such innovations by obtaining information on the character and control of the process that produces the finished PWA.

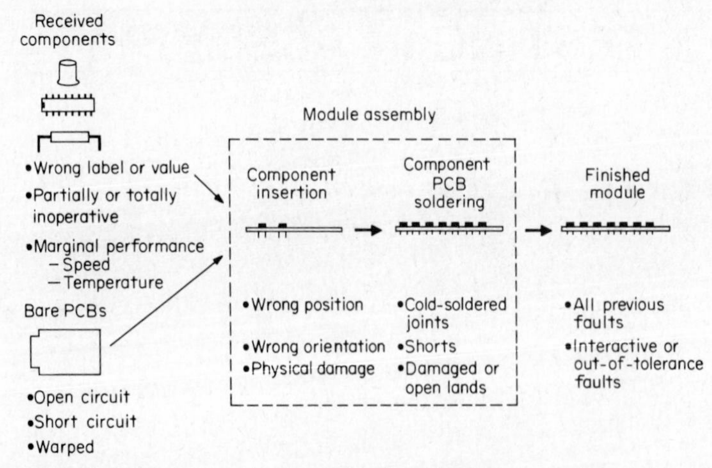

FIG. 21.1 Approximately 75 percent of module faults occur in assembly, 20 percent are component faults, and 5 percent PWB faults.

21.2 ASSEMBLY FAULTS

Human operators can contribute to poor quality through fatigue or lack of diligence. However, in terms of PWA production, faults usually have their origin in the manufacture of the components used or in the machines used in the assembly process (Fig. 21.1). This is true whether the faults are electrical components or the actual PWB. Therefore, in this section, both the faults and their origin will be discussed, since correcting the cause is as important as finding the fault.

There are two distinct forms of circuit packaging in use: surface mounted components and the traditional through-hole components. An introduction to these two packaging techniques is in order from the viewpoint of the faults they introduce to PWAs.

21.2.1 Through-Hole versus Surface Mount Technology PCB Faults

Figure 21.2 serves to illustrate the basic differences between the two methods of PWB manufacture from a defect point of view.

Although both packaging techniques commence with a bare PWB, the fault spectrum is different because of two fundamental distinctions arising from the components themselves and the soldering processes used.[1,2,3]

Whereas traditional through-hole components provide stable mechanical positioning and retention after component insertion and lead clinching, surface mounted components (SMC) use only the solder paste or, in some cases, glue or cement to hold them in place after they are placed on the surface of the board prior to solder reflow or wave soldering. Lateral and/or rotational misalignment can therefore occur in the case of SMCs either before or during the soldering operation. In addition, the reduction in lead spacing [from 0.100 in for dual-in-line packages (DIPs) to 0.050 in and less for SMCs] requires only 0.025 in of lateral displacement or 2.5° rotation from the true orthogonal position to cause bridging or shorting of leads in the case on an 18-lead plastic leaded chip carrier (PLCC). Moreover, in multilayer boards, the probability of inner-layer shorts increases, since a hole is drilled through the board for every component lead.

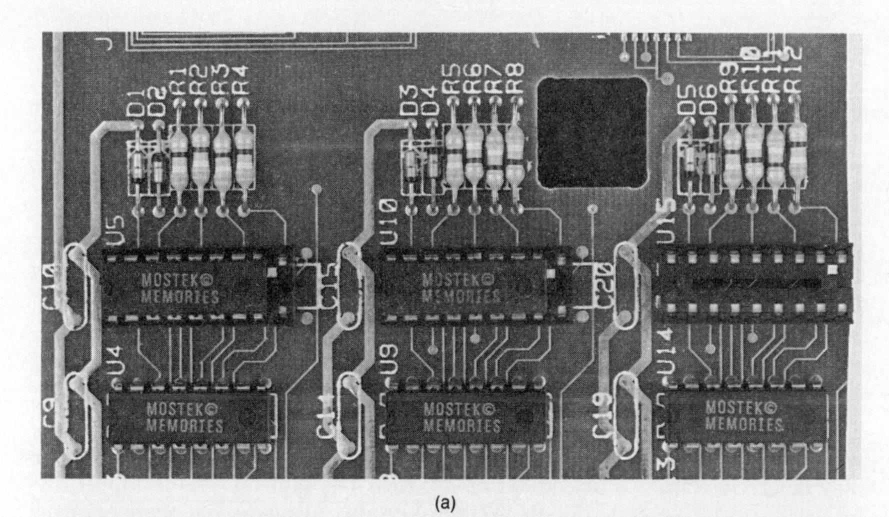

(a)

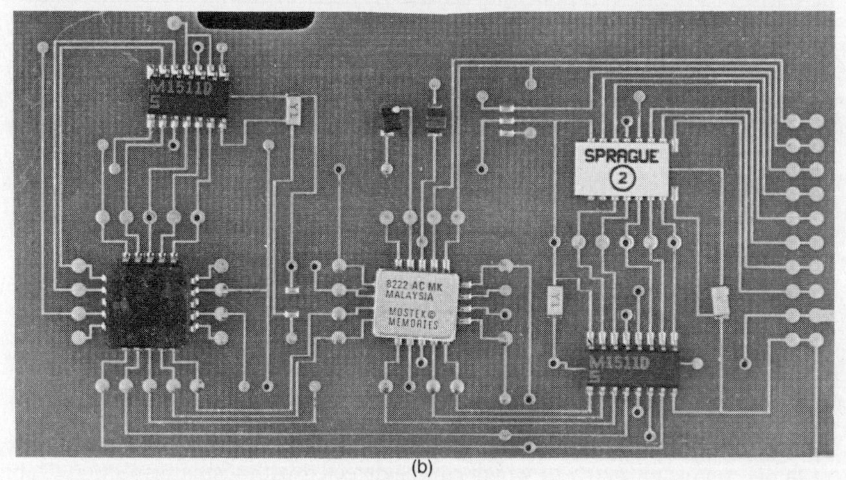

(b)

FIG. 21.2 Through-hole vs. surface-mount. (*a*) Example of through-hole printed wiring assembly (PWA); (*b*) example of surface mount PWA.

The soldering processes are radically different for wave-soldered through-hole components versus vapor-phase, or infrared (IR), reflow-soldered SMCs. Wave soldering bathes the whole surface of the PWB with solder, thereby enhancing the possibility of interpin shorts. Conversely, the silk screening of solder paste on carefully defined areas minimizes shorts due to the overall soldering operation. Figure 21.3 serves to illustrate the conceivable differences in the two fault spectra for two typical boards. This assumes, of course, that good PWB layout practices are adhered to in both cases, e.g., no conductor paths between contact pads.

21.2.2 Component Faults

At the component level, as a logical point of departure in the review of faults and their sources, the fault spectrum is shown in Table 21.1. Figure 21.4 shows three of the above faults.

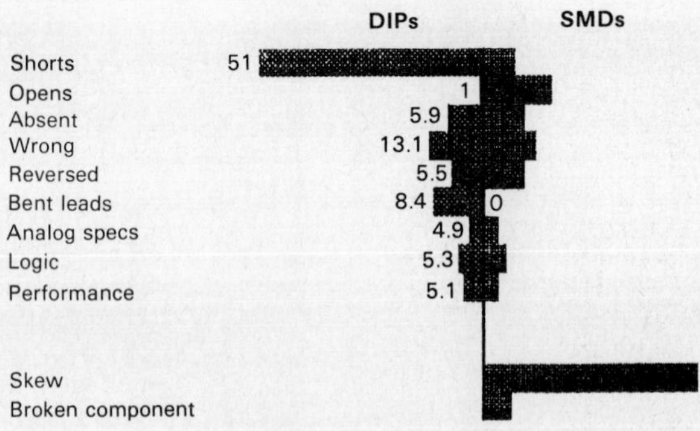

FIG. 21.3 DIP vs. SMD finished PWA faults.

Wrongly labeled or coded components, if similar in other physical respects, require some form of electrical test to detect such errors in manufacture, preferably before insertion or placement on a board. The same applies to inoperative or partly operative components. Wrong versions of any given semiconductor family, however, can usually be detected electrically or visually before assembly. Thereafter, only a functional electrical test can identify a wrong component when the performance of a board is found to be unacceptable.

Bent leads on through-hole components result in either their absence on a loaded board (uninsertable) or, in the case of one or two bent leads on a DIP, their noninsertion or deformation, e.g., bent under the component. On SMCs, bent leads are more likely to cause open circuits after solder reflow. In both cases, bent leads can be detected visually before or after PWB insertion or placement. A raised lead on a gull-wing-leaded SOIC package, however, can be particularly troublesome, since probing under electrical test makes a temporary contact.

Oxidized leads are detectable by careful visual inspection before insertion or placement but more easily after solder with an electrical test.

21.2.3 Bare Board Faults

With the bare board process involving more than 40 process steps, a wide variety of faults can occur, many of which can be detected with visual inspection. Table 21.2 lists some of the faults encountered with bare PWBs and their respective sources and causes.[4]

Faults arising from a predominantly photochemical etching process range

TABLE 21.1 Component Faults and Causes

Fault	Source(s)	Cause(s)
Wrongly labeled or coded	Part manufacturer	Sorting and marking equipment
Inoperative or partly operative	Part manufacturer	Wire-bond or die process
Wrong semiconductor family (e.g., bipolar vs. low-power Schottky)	Part manufacturer	Sorting and marking equipment
Bent lead(s)	Material handling and packaging	Tube end-stop, rough handling
Oxidized leads (poor solderability)	Storage	Humidity and polluted atmosphere
	Poor tinning	Tinning process

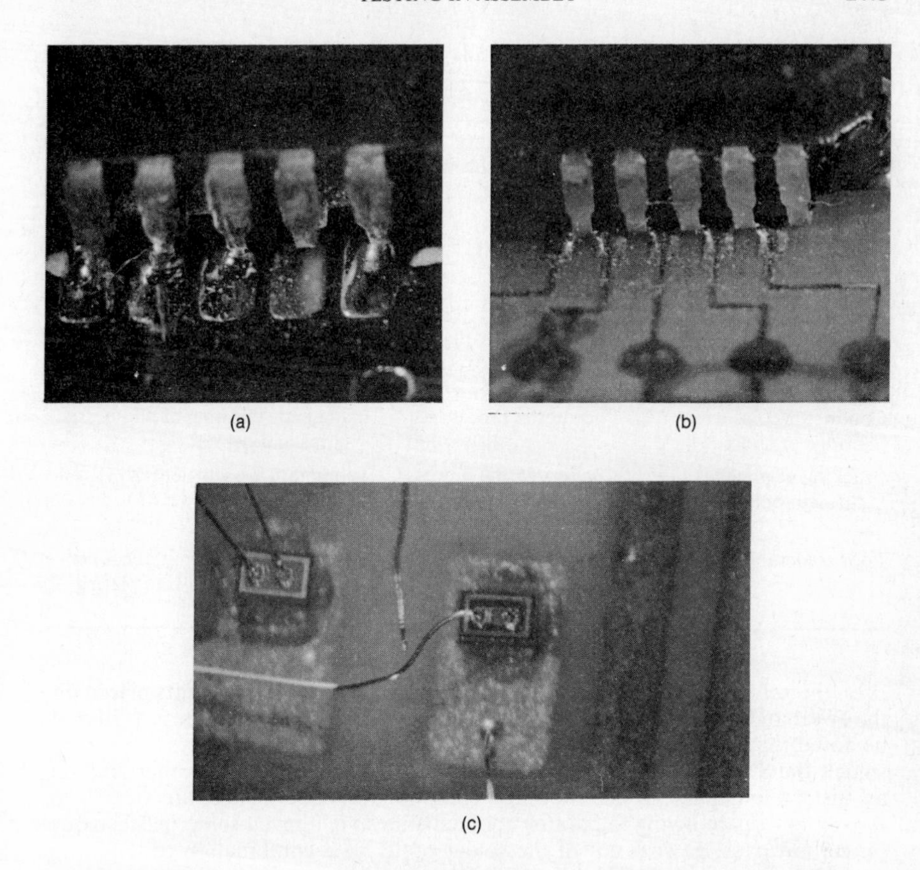

(a) (b) (c)

FIG. 21.4 Typical component faults. (*a*) Bent lead; (*b*) surface oxide; (*c*) broken wire bond.

from small cosmetic imperfections to electrically unacceptable flaws (Fig. 21.3). The cosmetic imperfections, however, often herald an oncoming bout of unacceptable faults. Table 21.2 is a simple example of faults produced in bare board manufacture and their relationship to process degradation.

Ambiguities in the sources of faults in the finished board indicate the need for additional "looking" points in the process (Fig. 21.5).

21.2.4 Loaded Board Faults (Presolder)

Compared with bare board fabrication, the selection and mounting of components for PWAs involve very few process steps. The three major faults encountered with through-hole components and their possible sources and causes are shown in Table 21.3.[5]

21.2.5 Loaded Board Faults (Postsolder)

In through-hole component mounting, the application of solder represents the tail end of PWA manufacture. For SMT, however, it can be considered a starting point, with the application of the solder paste. For the purposes of analysis, how-

TABLE 21.2 Some Bare Board Manufacturing Faults: Their Possible Sources and Causes

Fault or Imperfection	Source(s)	Cause(s)
Eccentric drill holes	Drilling machines Lamination process	Drilling data error Drill registration error Room temperature variation Quality of laminate
Poor plated-through hole	Drilling process	Drill quality, drill spindle speed and feed rate incorrect Inadequate epoxy smear removal
	Plating process	Inadequate analysis, replenishment, and maintenance of plating tanks
Conductor-conductor bridge or short	Photoresist process	Inadequate cleaning of photoresist chips and debris
Cracking of plated-through hole	Copper-plating process	Temperature coefficient of expansion of copper vs. laminate
Poor solderability	Cleaning process before tin-lead plating	Surface contamination of copper surface before tin-lead plating

ever, the solder processes will be assumed to start with all components placed on the PWB and end with them soldered and ready for final electrical test. It should be noted that whereas the former process effectively covers the entire board with solder, the SMT process applies solder in a manner as precise as the markings on an instrument dial. Surface mounted components, nevertheless, are subject to movement, again owing to poor or uncured adhesive, uneven solder paste, oxide on certain pads, gassing out of the solder paste, or a combination of the latter quality variations. Table 21.4 lists the types of faults discovered after the solder process and their possible sources and causes in through-hole components. Figure 21.6 illustrates these faults.

In SMCs, similar faults can occur, but the sources and causes are different

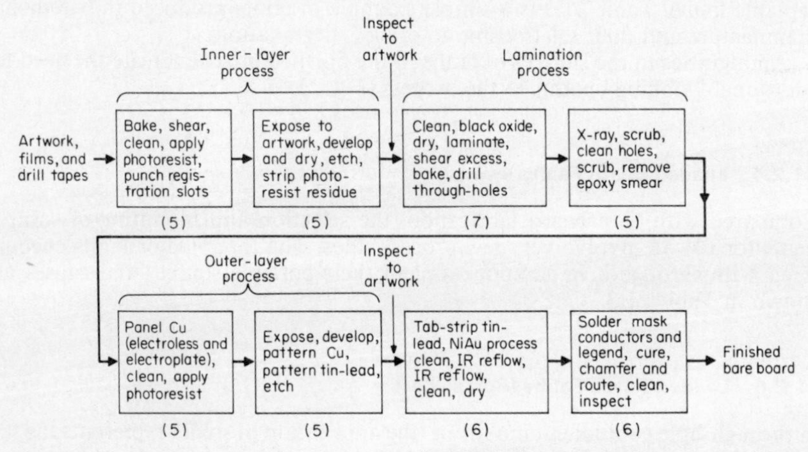

FIG. 21.5 Inspection points in a typical multilayer bare board fabrication process. Numbers in parentheses indicate the number of process steps.

TABLE 21.3 Some Loaded Board Manufacturing Faults: Their Possible Sources and Causes (Through-Hole Components)

Fault	Source(s)	Cause(s)
Missing component	Insertion machine Bare board drilling machine	Bent component leads Board positioning traverse malfunction Bare board drilling data error Bare board drill bit blunt Bare board drill positioning traverse malfunction
Wrong component	Component manufacturer PWB manufacturer	Tube packing data error Sequencer data error Sequencer selector malfunction
Wrong orientation of component	Component manufacturer PWB manufacturer	Sequencer selector malfunction

(Table 21.5). Figure 21.7 illustrates faults encountered with PLCCs after vapor-phase and IR reflow. Visual inspection of these soldered joints is more difficult than for through-hole components, which emphasizes electrical test.

21.3 TESTING TECHNIQUES

Given the fault spectrum associated with the PWA process previously outlined, there is a wide and seemingly confusing range of commercially available test and

TABLE 21.4 Soldered Board Process Faults: Their Possible Sources and Causes (Through-Hole Components)

Fault	Source(s)	Cause(s)
Solder shortage	Bare board Component lead Wave-soldering machine	Surface oxides (improper storage) Improper operation and maintenance of wave-soldering machine
Solder excess	Bare board Component lead Wave-soldering machine	Surface oxides (improper storage) Improper operation and maintenance of wave-soldering machine
Solder void	Wave-soldering machine	Trapped solder flux Wrong speed or temperature of wave-soldering machine
Blowhole	Wave-soldering machine	Trapped solder flux Wrong speed or temperature of wave-soldering machine
No solder	Bare board Component lead	Lead too small or hole too big Poor capillary action

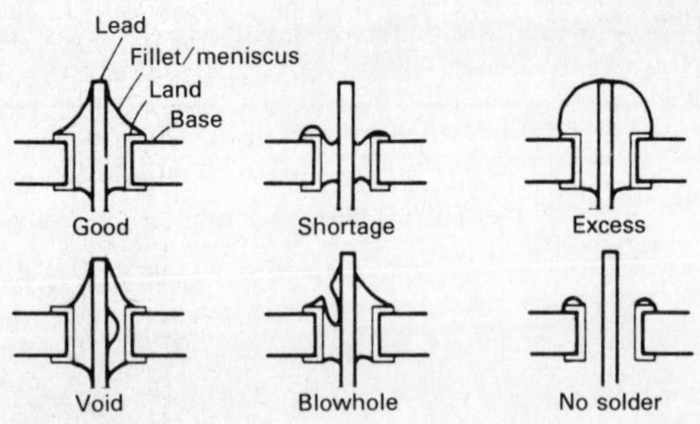

FIG 21.6 Through-hole component soldered joints and faults.

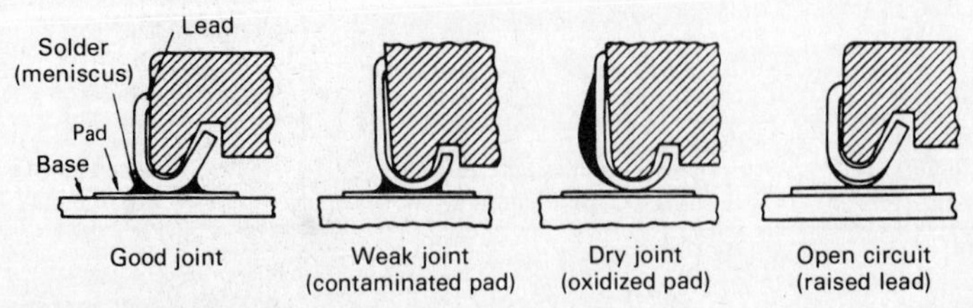

FIG. 21.7 SMT soldered joints and faults (PLCC).

inspection equipment to choose from. The best choice for a particular manufacturing situation becomes a question of the stages of the total assembly process involved, the type and complexity of the components and finished boards being used and assembled, and—last but most important—the production volume and mix of different board types manufactured. The latter factor in the decision process is dealt with in the extensive published literature. However, it becomes apparent that the fewer different assemblies produced on a line, the more practical it becomes to fine-tune the process and improve the "first-pass yield" to the point

TABLE 21.5 Soldered Board Process Faults; Their Possible Sources and Causes (Surface Mounted Components)

Fault	Source(s)	Cause(s)
Solder shortage	Bare board Component lead Solder paste applicator	Surface oxides Clogged screen Supply deficiency Paste consistency
Solder excess	Solder paste applicator	Gauge of screen Squeegee pressure Paste consistency
No solder	Bare board Component lead Solder paste applicator	Surface oxides Clogged solder paste screen Solder supply deficiency Squeegee pressure Paste consistency
Wicking	Bare board Solder paste applicator	Surface oxides Paste formula, silver deficiency

where test complexity at the finished board stage diminishes to a relatively simple problem. In this section, however, each major stage in printed wiring assembly will be considered, commencing with the basic circuit components, and the test equipment options will be presented and assessed.

21.3.1 Contacting versus Noncontacting Techniques

Contacting testers were once the only choice for determining whether PWAs in their various stages of manufacture met the expected quality standards. Various forms of "vision systems" have appeared on the market which have begun to replace human visual inspection of PWAs and, in some cases, outperform human inspectors.[6] High yields have been achieved through careful manual scrutiny of PWAs and their electrical components at every major stage in their assembly, followed by repairs as needed before passing the work on to a subsequent assembly stage, but while not replacing contacting electrical test, noncontacting vision systems can also catch and reduce "passed-on" faults which can accumulate to cause severe or excessive electrical test complexity. Table 21.6 lists the types of contacting and noncontacting testers available on the market for each major level of assembly. Almost all the testers listed are microprocessor-based, whether contacting or noncontacting, and these therefore provide considerable flexibility to cover many types of components or boards. Contacting testers achieve flexibility through appropriate test programs, component libraries, and the generation of binary test patterns for digital circuitry to verify their function according to known truth tables.

Simple discrete and integrated circuit component testers are relatively small (desktop) and low-cost. However, VLSI component testers are large (several racks) and can cost in excess of $1 million, due to the need for high speed and the

TABLE 21.6 Types of Contacting and Noncontacting Assembly Testers

Assembly level	Contacting	Noncontacting
Electrical components	RLC Linear ICs Digital ICs Memory ICs Elevated temperature	X-ray and ultrasonic: Wire-bond inspection
Bare boards	Two-probe: continuity and multiprobe insulation	Visual-band imaging: Shorts and opens (single layers) X-ray imaging: Inner-layer shorts and opens
Loaded boards	Shorts testers Manufacturing defects analyzers Functional testers In-circuit testers Combination testers	Visual-band imaging: Through-hole lead checking (bottom side) Component presence or absence Component orientation (in certain cases) Visual, x-ray, and IR-band imaging: Soldered-joint quality IR thermal imaging

complexity of the 1000-gate-plus devices being tested on these machines. Noncontacting component testers using x-ray and ultrasonic microscopy are typically limited to the detection of faulty wire bonds.

Bare board electrical testers range from simple point-to-point two-probe testers to multiprobe, bed-of-nails systems, taking a few seconds to test a complete board, and with prices ranging upwards of $200,000. Their noncontacting (visual) counterparts perform a dual role of artwork as well as actual bare board (inner-layer) inspection in the same price range, but much more slowly.

Loaded board electrical testers vary widely in speed and setup cost, covering and detecting the entire range of manufacturing faults, whereas noncontacting testers detect "visible" faults using carefully selected visible, infrared, and x-ray illumination of the boards and components.

From the foregoing, it can be seen that effective visual inspection (even though predominantly human visual inspection) at strategic points in an assembly process is an established way of controlling quality, particularly in a complex multistep process such as PWB manufacture. Consequently, with increasing automation, effective automatic vision systems are becoming an increasingly significant factor in overall test strategy.

21.3.2 Built-In Test, Built-In Self-Test

In some integrated circuits and loaded PWAs, built-in test (BIT) capabilities are included to facilitate testing, particularly in military and high-availability electronics.[7–20]

Such test facilities are also used when boards are installed in subsystems for future diagnostic purposes, i.e., system fault location to the board or module level. There are two major categories which come under the more loosely used heading of built-in test: namely, design for test (DFT), which makes provision for properly testing a very large scale integrated circuit (VLSIC) or PWA module, and built-in self-test (BIST), which actually incorporates self-testing circuits on the chip or board.

21.3.2.1 Design-for-Test Techniques. The evolution of DFT techniques for LSIC and VLSIC applications can be traced back to the use of built-in provisions for scanning the contents of storage elements using serial shifting techniques. This DFT method was later augmented with a race-free logic design technique in which the clocking of data was allowed to occur only when the logic levels were stable. The net result of these two techniques was a more reliable logic design and a means of testing it using external stimuli and test data analysis. Figure 21.8 illustrates the level-sensitive scan design (LSSD) for a single storage element. Note the addition of seven logic elements to provide the means of accessing the status of each five-element data latch in the circuit.

Serial-digital data transfer has distinct advantages in a system context in that it provides the potential for using a single, serial-data multiplex bus for access to all the LSI and VLSI circuits (Fig. 21.9).

For LSSD devices, two additional time-phased clock lines are needed to shift the test data out for analysis, i.e., correlation with expected results. As custom and semicustom VLSICs, or application-specific integrated circuits (ASICs), have become economically attractive for medium- to high-volume proprietary designs, the need to test such devices before paying the silicon broker or foundry has forced circuit designers to provide at least the means of isolating major functional blocks (Fig. 21.10). Such test provisions result in the addition of logic switching and associated control input decoding circuits. The external test interface arising

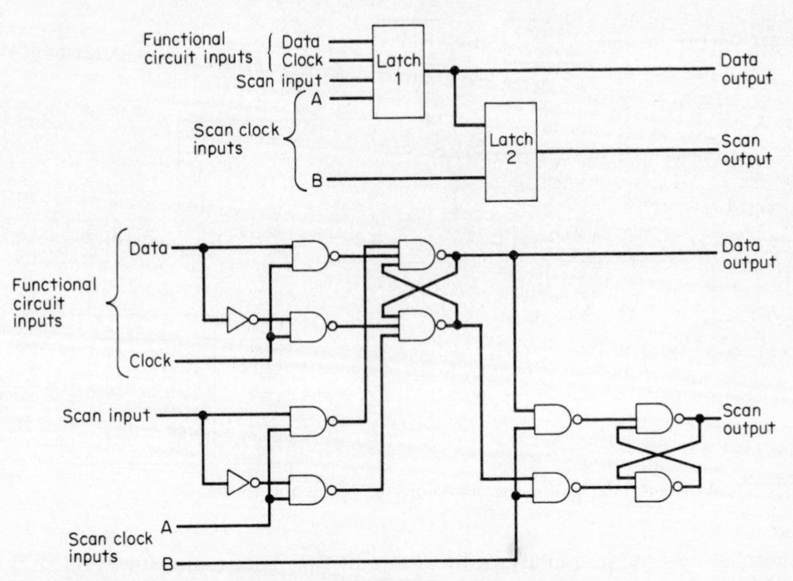

FIG. 21.8 Level-sensitive scan design, functional and test circuits.

from this DFT partitioning approach is the more traditional parallel form for test patterns and test data, similar to the case in which the subsystem is packaged in small-scale and medium-scale ICs on PWBs.

21.3.2.2 Built-In Test. Both of the DFT techniques reviewed above require off-chip or off-board test pattern generation and test data analysis. A more recent

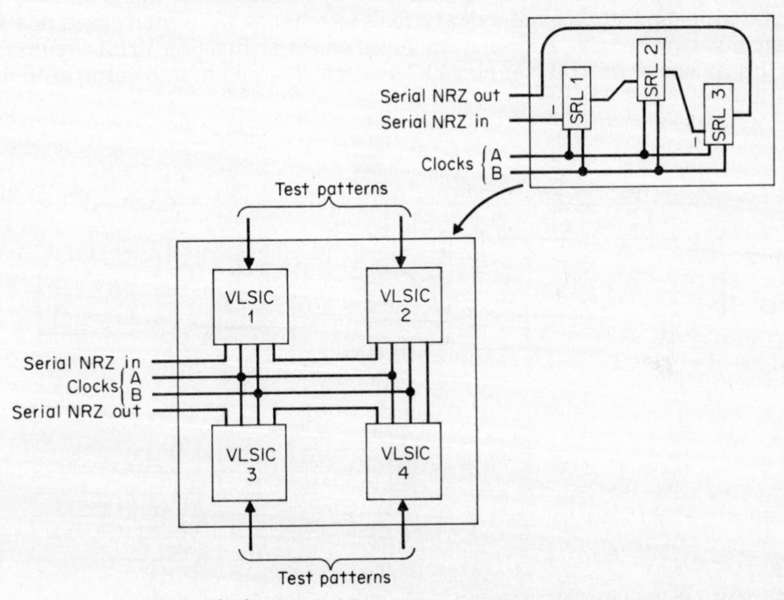

FIG. 21.9 LSSD circuits in system context.

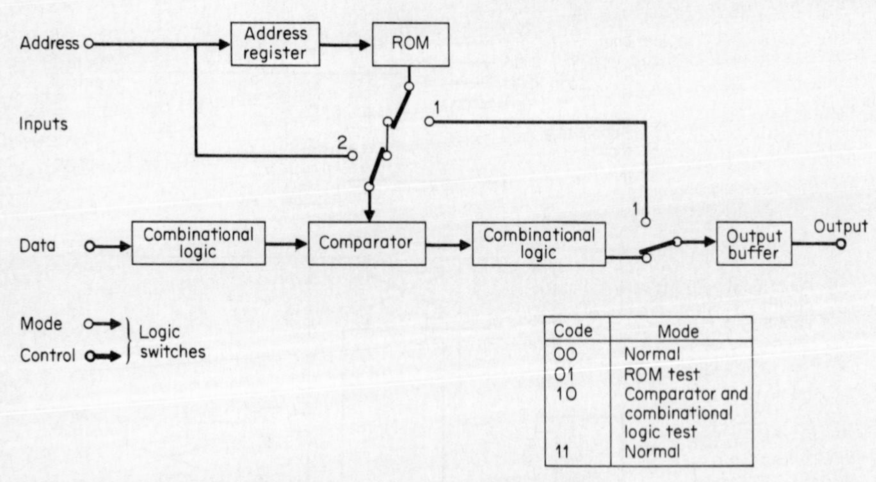

Code	Mode
00	Normal
01	ROM test
10	Comparator and combinational logic test
11	Normal

FIG. 21.10 Typical design-for-test provisions for ASICs and PWBs.

effort, using signature analysis, incorporates two *n*-stage multifunction shift registers on the chip to generate pseudorandom test patterns and to perform signature analysis, respectively (Fig. 21.11). In this arrangement, termed "built-in logic block observation" (BILBO), or on-chip test (OCT), each register is used to sandwich the functional circuits of the VHSI device (Fig. 21.12). Note, however, the convenient serial-NRZ interface from a system interface viewpoint. The additional logic required to add BILBOs to VHSI and VLSI circuits is on the order of 200 gates for the 8-bit registers shown. The advantage obtained with any OCT technique is high-speed testing without the drawbacks of test fixture wiring.

Nevertheless, none of the above DFT and OCT methods avoids the necessity of interrupting the functional circuit's normal operation. For high-reliability and high-speed applications, redundant circuit techniques, as formerly used in a subsystem-system context, are being employed on major-function VHSI circuits (Fig. 21.13). Whereas an 8-bit complex FFT butterfly (two-point transform) arithmetic

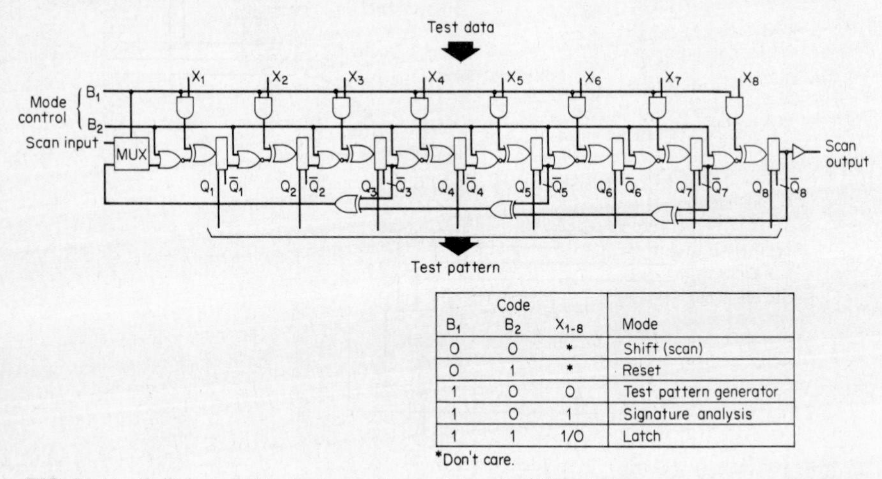

Code			Mode
B_1	B_2	X_{1-8}	
0	0	*	Shift (scan)
0	1	*	Reset
1	0	0	Test pattern generator
1	0	1	Signature analysis
1	1	1/0	Latch

*Don't care.

FIG. 21.11 Built-in logic block observation multimode shift register.

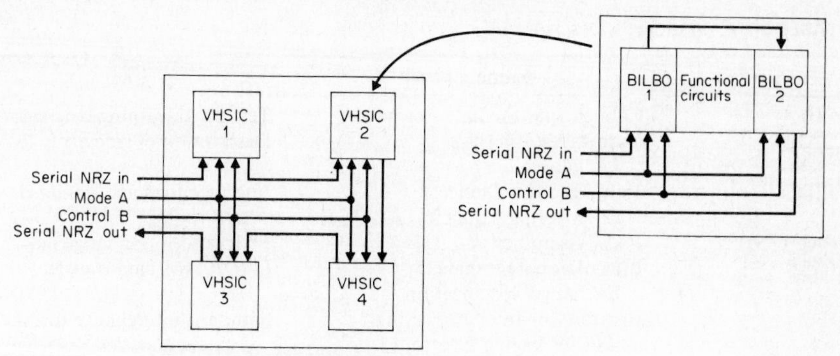

FIG. 21.12 BILBO circuits in system context.

unit would form the normal on-line functional circuit, a simple 2-bit mirror image can be used to form a duplex channel for "confidence level" test purposes. The silicon real estate penalty for the modulo-3 butterfly could be expected to be 25 to 30 percent additional logic. From an external test interface viewpoint, we have a minimal situation: one pass-fail status line.

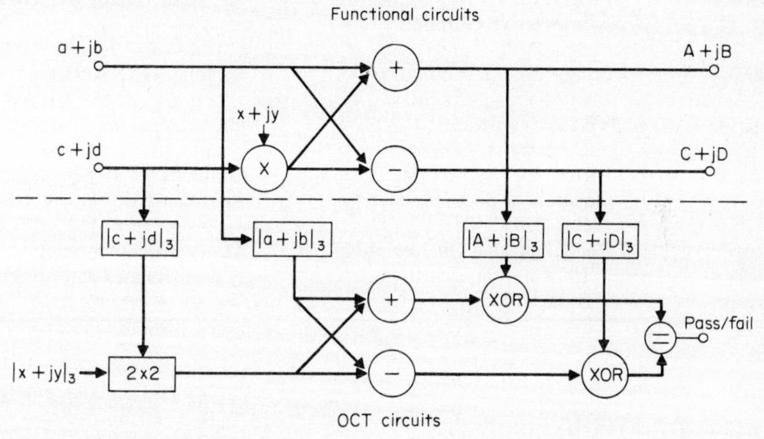

FIG. 21.13 Example of concurrent testing for VHSICs. FFT butterfly arithmetic unit using Modulo-3 residue arithmetic.

21.3.3 CAD and CAE Data Coupling

In cases where there is no built-in self-test or, quite often, no test provisions at all, the test engineer would be faced with the task of learning the circuit board and determining how best to test it given only the input and output pins. With the part types known and their respective interconnections, test input stimuli could be determined either by analysis for simple boards with simple parts or through automatic test-pattern generation by automatic test equipment.

With the evolution of computer-aided engineering (CAE) workstations as well as the well-established CAD stations, there is a need to couple such design data to automatic test equipment (ATE).[22,23] This integration, termed "feed forward," has not been simple due to the lack of common interface standards (Table 21.7).

TABLE 21.7 Standards Descriptions

Acronym	Name	Use
CIF	Caltech Intermediate Format	Test programming language
GAIL	Gate Array Interface Language	Description of gate array logic
JEDEC	Joint Electron Device Engineering Council	Interface formats, standard terminology
PALASM	PAL Assembler	PAL description language
IGES	International Graphic Exchange Specification	CAD/CAM data transfer
IPC	Institute for Interconnecting and Packaging Electronic Circuits	Standard interchange formats
ATLAS	Abbreviated Test Language for all Systems	Test programming language
CORE	Core	Graphic software interface
EDIF	Electronic Data Interchange Format	Hierarchical hardware description language
GKS	Graphics Kernel System	Graphics software interface
TIDB	Tester Independent Data Base	Neutral tester information
PHIGS	Programmer's Hierarchical Interactive Graphics Standard	High-end version of CORE
PMIGS	Programmer's Minimal Interface to Graphics	Minimal GKS subset
VHDL	VHSIC Hardware Description Language	Hierarchical hardware description language
VDM	Virtual Device Metafile	Graphics file storage
GPIB	General-Purpose Interface Bus (IEEE 488)	Bus standard protocol
ROLAIDS	Reconfigurable On-Line ATE Information Distribution System	Tester instrumentation protocol and data structure

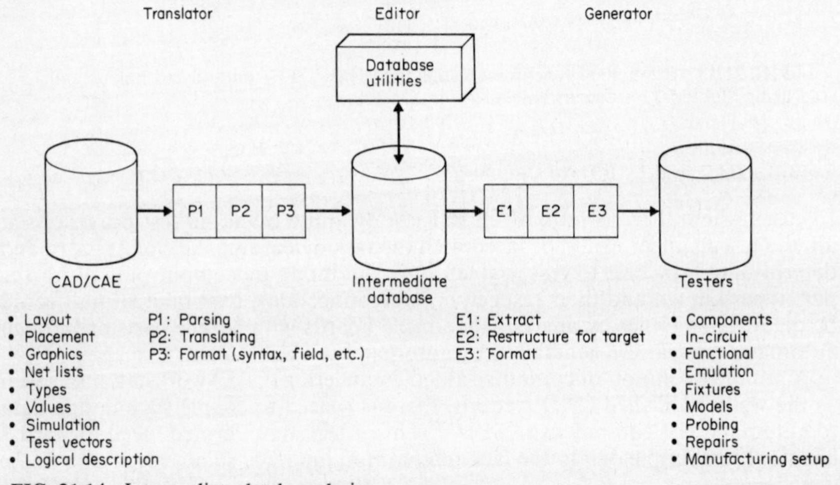

FIG. 21.14 Intermediate database design.

However, certain manufacturers of automatic test equipment have partly solved the problem by introducing circuit simulation and verification-test programs to CAE workstations, and the programs can, in turn, be used on the ATE when the hardware (ASIC or board or both) becomes available. In addition, by the creation of an intermediate database for a specific manufacturer's ATE product line, various sources of CAD and CAE data can be translated to the intermediate form, which can then be more easily used by the tester family (Fig. 21.14).

21.4 COMPONENT TESTING

If all components received for assembly on PCBs had been tested before shipment from the manufacturer, then, in the absence of shipping damage, there should be no need to perform incoming tests. This basically constitutes the "dock-to-stock" no-incoming-inspection philosophy. However, when this situation is impractical, there is an extensive range of testers to select from.

There are three basic classes of component testers on the market which enable devices to be characterized, failures to be analyzed, and vendors to be qualified. The three classes are as follows:

- Passive discrete component testers—RLC bridges.
- Low- or medium-performance integrated-circuit testers—linear, digital, and memory ICs.
- High-performance integrated-circuit testers—VLSICs and VHSICs.

The first two classes represent the benchtop units mentioned earlier, with the second of these classes being essentially high-performance "personal computer" class testers, with special-purpose hardware controllers, drivers, sensors, and fixtures, together with a library of test programs to cover the extensive range of standard components used in the electronics industry. The third class of component testers requires several racks of electronics and because of their cost, they are frequently available through a "test house" where such a resource can be shared among several users on a job-shop basis.

21.4.1 Simple Passive Discrete Component Testers

These benchtop testers range from $2000 to $40,000, depending on the degree of automation incorporated through software. Figure 21.15 shows a basic precision *RLC* tester which, nevertheless, allows remote programming to set values and limits for the acceptance of resistors (R), capacitors (C), and inductors (L).

Such remote control and monitoring interfaces vary from product to product (e.g., RS-232-C and IEEE 488 bus), but more important is the software required to support the networking of benchtop testers. For the price of one of these machines, a networking package can be purchased which provides the basic communication interface plus the means to process experienced data and thereby regulate the level of testing based on statistical results, which, of course, leads to more efficient quality control.

21.4.2 Linear Integrated-Circuit Testers

Active component testers need to cover a much wider range of device types than the relatively simple RLC bridge instruments. Linear IC testers and their digital

FIG. 21.15 Basic precision *RLC* tester.

counterparts constitute the "high end" of the benchtop component tester range (i.e., the upper $30,000 price bracket).

A typical range of component types which can be tested by such a machine is as follows:

- Operational amplifiers (including single rail, dual, quad, FET input types)
- Transconductance amplifiers
- Fixed-voltage regulators
- Resistor programmable regulators
- Adjustable regulators
- Comparators
- Current mirrors
- A/D converters (both straight and gated output types)
- D/A converters (both straight and latched input types)
- Line drivers
- Line receivers
- Transistors (small signal, and transistor arrays)
- Diodes (discrete, and diode arrays)
- Field-effect transistors (FETs)
- Analog multiplexers
- Opto couplers
- Hybrid devices
- Digital SSI and MSI devices

Additional features include automatic handler interface (testing which the components are on the handler), mass data storage (e.g., tape cartridge) for as many as 40 different device programs, and a family of plug-in boards to tailor the tester (hardware and software) to cope with the wide range of device types.

21.4.3 Digital Integrated-Circuit Testers

For digital small-, medium-, and large-scale IC testers, there are benchtop testers with test characteristics similar to those of large digital in-circuit testers, but without the sophistication necessary for testing ICs in situ on circuit boards. Such a component tester is capable of performing both static and functional tests for the following logic families:

Logic families	Tests
Bipolar	Continuity (open pins)
ECL	dc parametric
CMOS	4000 test vectors
NMOS	

21.4.4 Memory Testers

For this last major group of distinct IC types, equally distinct high-performance benchtop testers are available to test the various forms of semiconductor memories, i.e., static and dynamic RAMs, ROMs, PROMs, and EPROMs in ECL, TTL, NMOS, and CMOS. Memory tests require special test patterns and associated timing and address signals capable of detecting memory faults under normal operating conditions (e.g., dead cells, leaky cells, addressing or location faults).

21.4.5 VLSIC and VHSIC Testers

Where the testing of very large and/or very high speed integrated-circuit components with clock rates in the 120-MHz range is concerned, a radically different class of tester is necessary. Systems can accommodate devices with up to 288 pins, 80-MHz multiplexed data rate, and 32 independently programmable timing generations with a system accuracy of ± 750 ps. Other features are CAE-to-test interface links, networking, and rapid wafer to packaged VLIC test flexibility. As could be expected, these tester giants are considerably more expensive than their SSI, MSI, and LSI benchtop relatives.

21.5 LOADED BOARD INSPECTION AND TEST

There is an empirical rule of 10 associated with the cost to find and repair faults at each major stage in the assembly process, starting with approximately 5 cents per component before board assembly, 50 cents after insertion and placement, $5 after solder, etc. Hence, inspection and test before soldering the components to the PWB has a distinct economic advantage.

Electric prescreeners have been in existence for several years, but recently, machine vision systems have been introduced to check that component leads have been properly inserted and clinched by viewing the underside of the PWB.

21.5.1 Vision Systems

Topside inspection has proved to be both complex and unreliable with gray-scale imaging techniques, due to inconsistent component features (labeling, color, surface texture, and orientation marks) and board colors and textures. However, with the more reliable reflection characteristics of bright, tinned component leads, bare board contrast levels are obtainable and bottom-side lead-inspection machines have been successfully built and added to component insertion machines. Apart from checking lead integrity, such a through-hole lead-inspection system automatically verifies the presence or absence of a component.

21.5.2 Electric Prescreeners

The same machines which are often used for presolder tests (i.e., shorts testers and manufacturing defects analyzers) offer the means of detecting and locating many of the accumulated process faults except, of course, determining the functionality of the components or of the board as a whole. In certain situations where the fault spectrum, together with the devices used on a board, results in a low percentage of process and component faults but a high incidence of component or system interactive faults, prescreeners followed by functional tests could well suffice. However, whenever the process degrades or the components used change, the shift in fault spectrum could result in poor or inadequate diagnostic capability with the basic prescreener or functional approach.

The ability to run or upgrade the prescreener to a power-on in-circuit tester (ICT) would then become a great advantage by providing the necessary flexibility to solve such a problem without the need to purchase and program an ICT from scratch.

21.5.3 In-Circuit Testers

These machines, introduced in the late 1970s, provide a unique and effective solution to the shortcomings of earlier functional testers, which used only the edge-connector pins augmented with a single probe to determine whether a board functioned properly. As the complexity and number of components on PWBs increased, it became increasingly difficult and costly to isolate a fault to a specific component using only the edge connector and probe. "Access," or lack of it, constitutes the basic problem, and by probing the surface nodes on a PWB and driving the logic states of digital components to desired states, digital ICs can be checked against their respective truth tables in situ. Figure 21.16 illustrates a typical ICT, and Fig. 21.17 shows the basic internal architecture.

The two major components which make an in-circuit tester effective are (1) the test interface and (2) automatic test generation (ATG) software. The test interface consists essentially of a form of programmable cross-bar switching matrix through which composite driver-sensor amplifiers can be connected to an array of probes (nails) upon which a PWB is placed. Through the use of a vacuum system, the board is pulled down so that the probes contact specific nodes on the board. The probes used in the array are telescopic (spring-loaded) and aptly called a "bed of nails," with as many as 3000 such "nails" in large machines. The basic ICT is, in essence, a general-purpose tester which can be tailored to test a specific board by designing a fixture containing a unique array of nails with tooling pins to properly locate the board, contact nodes, and corresponding probes.[24–30]

Figure 21.18 shows the physical layout of a typical fixture which plugs into an in-circuit tester and its multiplexed driver-sensors.

The drilling of holes in the base plate to accommodate the test nail sockets is normally achieved on automatic drilling machines using the CAD data of the PWA. The wiring of such complex fixtures is also automatically performed on

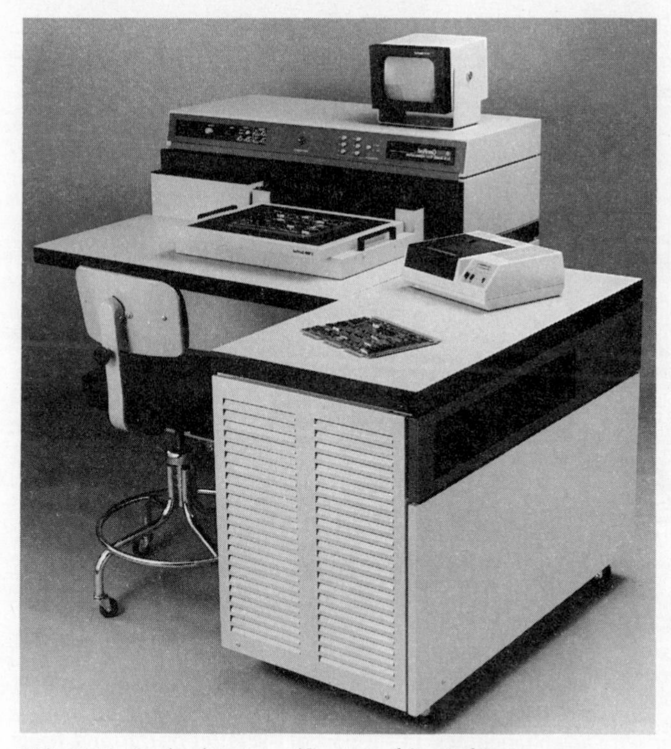

FIG. 21.16 In-circuit tester. *(Courtesy of GenRad Inc.)*

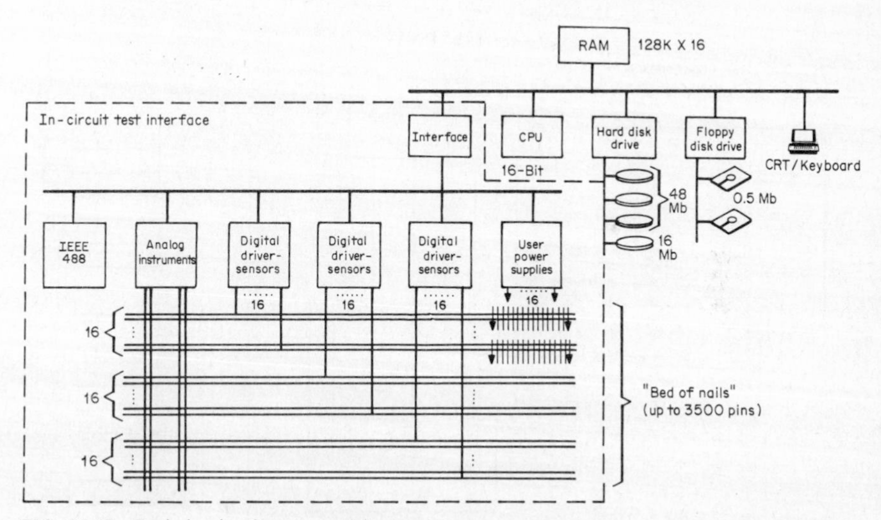

FIG. 21.17 Basic in-circuit tester architecture.

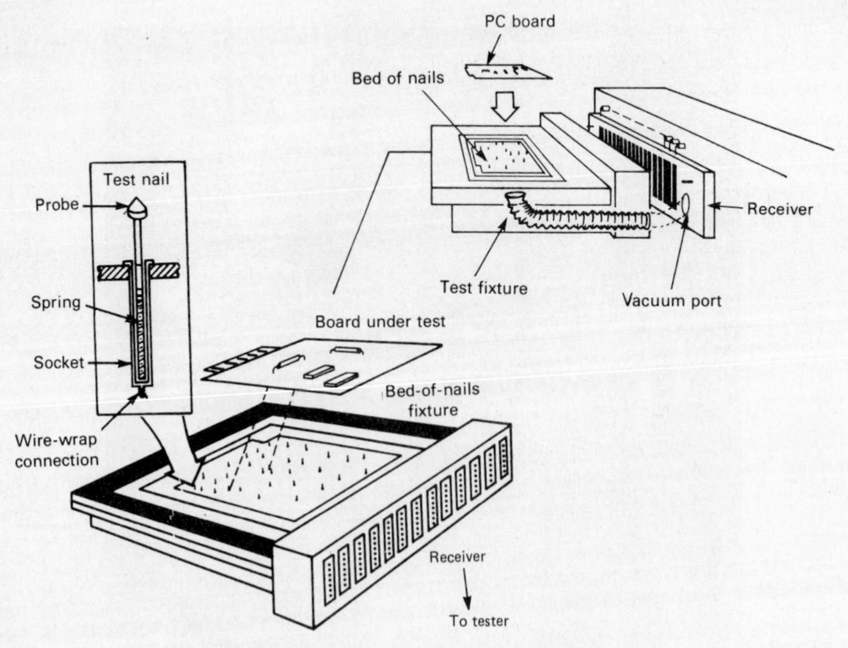

FIG. 21.18 Bed-of-nails fixture.

automatic wire-wrap machines. In support of the latter operations, tester software packages are available to automatically integrate the assignment of individual nails and driver-sensors with the test programs for PWA testing.

To increase the speed and performance of ICTs, the truth table test patterns used for digital IC testing are loaded into fast RAMs assigned to each pin and its associated driver-sensor. Figure 21.19 illustrates this high-speed interface and

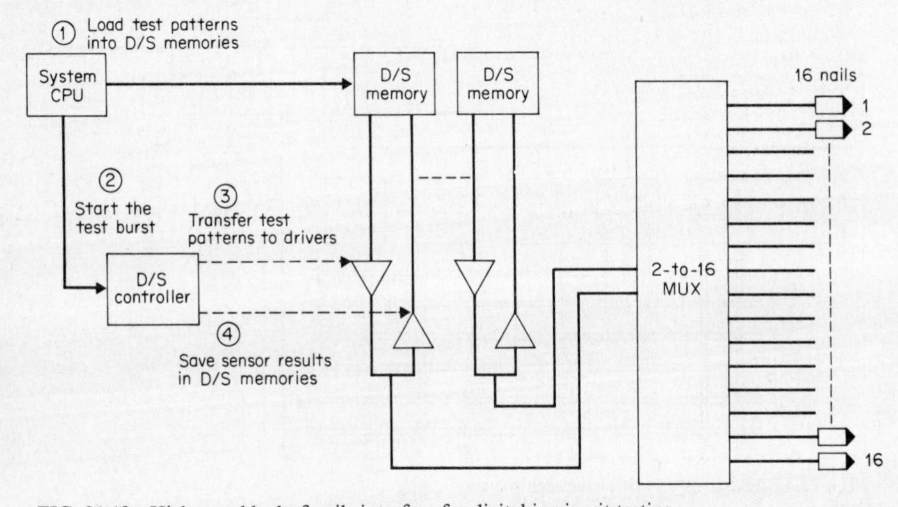

FIG. 21.19 High-speed bed-of-nails interface for digital in-circuit testing.

indicates the sequence of events where a "burst" of test patterns is applied to the test nodes and the resulting test data stored in the "pin RAMs."

In the figure shown, the multiplexer enables a pair of driver-sensors to access any of 16 nails, taking advantage of the small number of driver-sensors used for any given IC test versus the extensive number of nodes and corresponding pins used in the fixture.

Analog component testing is typically performed with built-in instruments or IEEE 488 bus-compatible instruments connected via a multiwire bus (Fig. 21.20).

In-circuit testing provides access to any desired node on a PWB, but the problem of isolating a component for test requires "guarding" for analog or passive components and "backdriving" in the case of digital ICs.

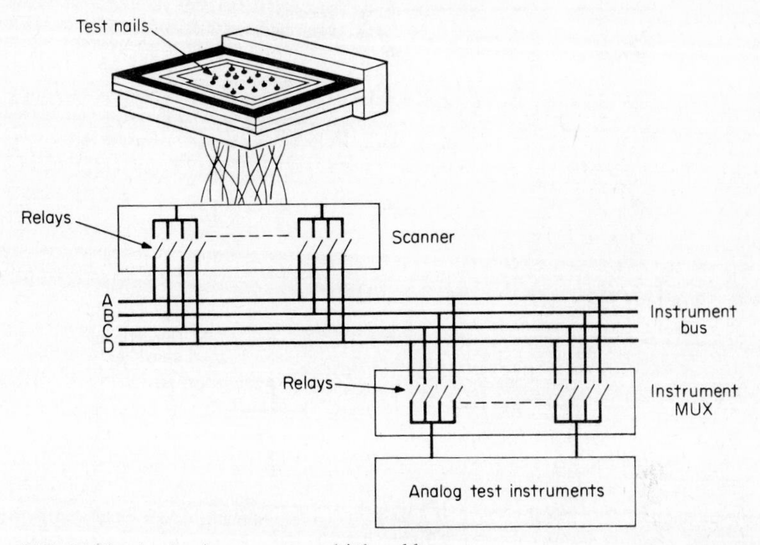

FIG. 21.20 Analog instrument multiplexed bus.

21.5.3.1 Guarding. A simple example will be used to describe this method of isolating analog components.[31] Figure 21.21 shows a resistor in isolation and in the PWB circuit context. There would very likely be some shunt paths around the resistor (R_x), which would divert some of the isolated resistor current from the ammeter. (Resistors R_1, R_2, R_3, and R_4 represent these shunt paths.)

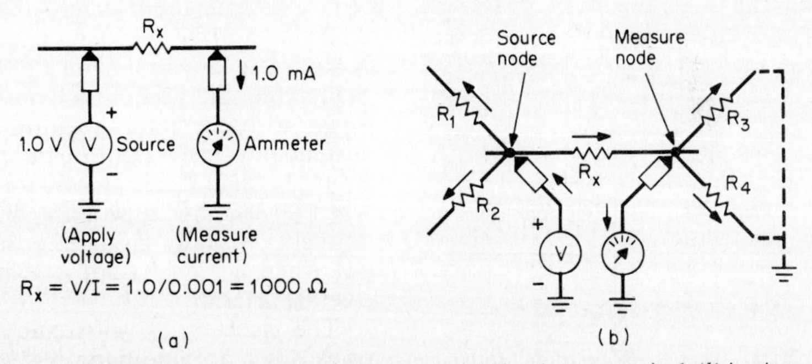

FIG. 21.21 Analog component measurements. (*a*) Simple two-terminal; (*b*) in-circuit configuration.

Instead of flowing directly to ground through the ammeter, the total current through R_x divides and flows through R_3 and R_4 as well as through the ammeter.

Depending on the resistance values of these shunt paths, the ammeter reading could be affected significantly, and therefore, the calculated value of R_x could also be significantly in error.

Guarding stops, or at least reduces significantly, the current flow through the shunt paths connected to the measure node by temporarily connecting all shunt paths around the component under test to ground (the guard voltage) and using an operational amplifier in the ammeter circuit (Fig. 21.22).

This arrangement, known as a four-terminal guard measurement, ensures that all the current through R_x flows through the ammeter.

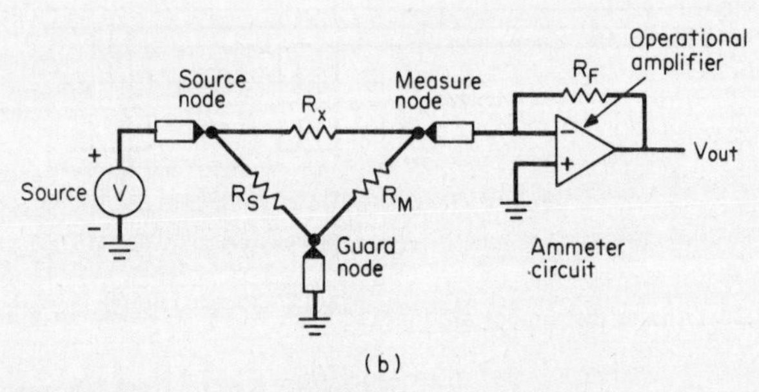

FIG. 21.22 Guarding (four-terminal). (*a*) Connecting all shunt paths to ground; (*b*) measurement via an operational amplifier.

21.5.3.2 Backdriving. In digital in-circuit testing, backdriving is used to force any logic input of a device under test (DUT) to a required logic level for the duration of the test.[32] Figure 21.23 shows a basic NAND gate and its truth table.

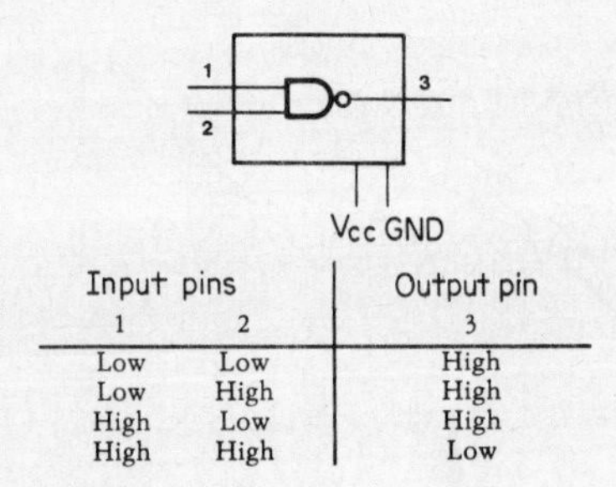

Input pins 1	Input pins 2	Output pin 3
Low	Low	High
Low	High	High
High	Low	High
High	High	Low

FIG. 21.23 Basic NAND gate and truth table.

Figure 21.24 shows the same NAND gate "in-circuit" on a PWB and the forcing states required to test it.

Backdriving simply requires a high current into the output of the gate to drive it "high," i.e., 3.5 V for TTL, when $Q1$ is a low impedance, or, alternatively, virtually grounding the output when a "low" state is required[33] (see Fig. 21.25).

The success of in-circuit testers owes much to automatic test genera-

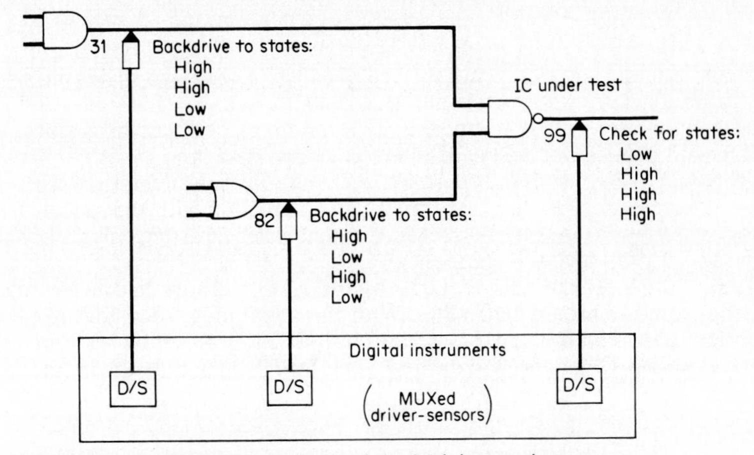

FIG. 21.24 Basic NAND gate in-circuit backdriving requirements.

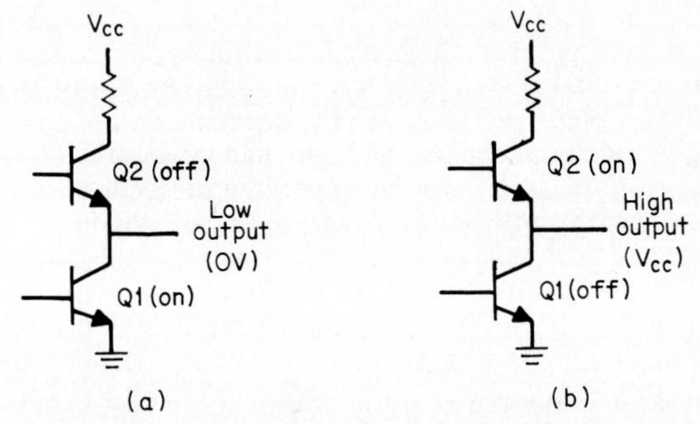

FIG. 21.25 Output circuit of TTL NAND gate. (*a*) Low output; (*b*) high output.

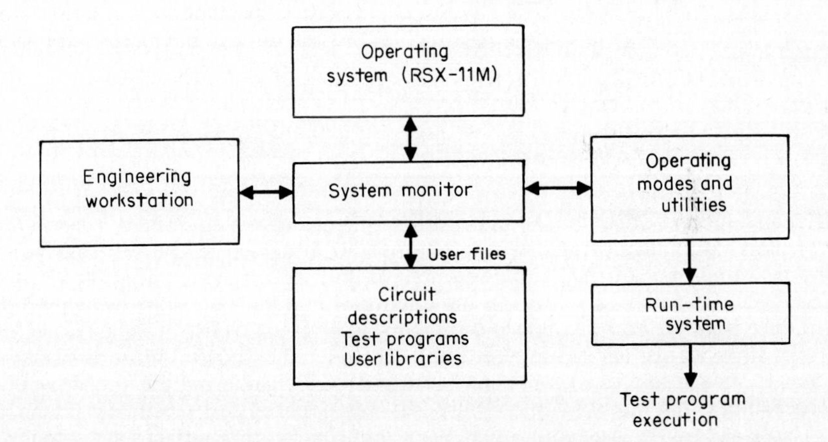

FIG. 21.26 Typical in-circuit tester ATG software configuration.

tion (ATG) software.[34] As the name implies, ATG automatically generates the sequence of the tests needed to check each component on any given board using the extensive nodal accessibility of the bed of nails. Figure 21.26 shows the typical organization of ATG software elements used for ICT.

Basically, a user enters component type numbers and resistor values, etc., together with printed wiring board interconnection data, and the ATG does the rest. The ATG achieves its purpose by referring to a comprehensive library of device characteristics and test data which, in conjunction with the specific board topology data, determines the correct component isolating arrangements and tests (analog and digital) for each component.

Initially, much of the data required by the ATG software had to be entered manually from the circuit schematic. With the continuing trend toward greater integration of workstation databases, links have been provided by ATE manufacturers to couple CAD PWB data and CAE ASIC data directly to the ATG software.

21.5.4 Functional Testers

As noted previously in the section on ICTs, functional testers became increasingly costly to program and less effective in fault isolation to the component level as board and component complexity increased. A functional tester, using only the edge-connector pins of a PWB to enter test conditions and sense the response, can determine whether a board works or not by means of a rapid go–no-go test. This form of testing could be adequate for high-yield product lines—typically one-board-type, high-volume situations. However, the problem arises, when the process degrades, of needing an ICT to salvage and repair the faulty boards. Hence, the disappearance of first-generation functional testers and the advent of "combination" testers described in the following section.

21.5.5 Combination Testers

The combination tester, i.e., one that combines in-circuit and functional PWB testing in a single product, provides the most comprehensive form of board test in a single workstation. In many ways it is the direct outgrowth of higher-density packaging in all its forms: SMT, VLSICs, and ASICs, together with 32-bit microprocessor technology and the fast, direct addressing of large databases with wide word formats it provides.

21.6 TEST DATA MANAGEMENT

The test data produced by the many different types of test equipment used in the assembly of quality printed wiring boards and discussed in the previous paragraphs represents an invaluable resource of information concerning the day-to-day, batch-to-batch variations in the quality of materials and processes used to produce both the components and the fully assembled PWBs.

In other words, test equipment can be considered as a form of process instrumentation whose effectiveness can be improved by gathering the test data and processing it for quality control purposes—as opposed to its primary intended purpose as fault-finding equipment. As a result, network interfaces are provided on component and board test equipment to serve the needs of two distinct users: (1) repair stations and (2) quality management.

The faults found by testers are translated into instructions for repair personnel at nearby repair station (CRT) terminals to facilitate the location and correction of faults on a particular PWB (usually identified by a bar-code wand) by a human operator. This arrangement eliminates paperwork and gives rise to the so-called paperless repair station.

The incidence and frequency of certain types of faults are also tracked and processed to provide the means of monitoring (and correcting) component and process quality trends to improve yields and reduce costs.

In summary, by adding a network interface, a processor, and appropriate software, the entire manufacturing process can be automatically monitored, and, more important, data that can be used directly by repair and quality personnel can be generated and accessed in a timely manner.[35–40]

REFERENCES

1. "Surface Mounting," *Circuits Manufacturing,* January and March 1984.
2. A. Rahn, "Wave Soldering Leadless Components," *Circuits Manufacturing,* May 1982.
3. *Introduction to Surface Mounting,* Report no. SMFA001, Texas Instruments, January 1984.
4. *Guidelines for Acceptability of Printed Boards,* IP-A-600C, Institute for Interconnecting and Packaging Electronic Circuits.
5. *Designing for Automation,* USM Corporation, 1978.
6. F. J. Langley, "Machine Vision Systems and Printed-Circuit Board Manufacture—Requirements vs. Capabilities for Improved Process Control," *Proceedings of Robots East,* Boston, October 1985.
7. M. J. Y. Williams and J. B. Angell, "Enhancing Testability of Large Scale Integrated Circuits via Test Points and Additional Logic," *IEEE Transactions on Computers,* C-22, 1973, pp. 46–60.
8. E. B. Eichelberger and T. W. Williams, "A Logic Design Structure for LSI Testability," *Proceedings of the IEEE Conference on Design Automation,* June 1977, pp. 462–468.
9. S. McMinn, "Semiconductor Manufacturing Considerations for VLSI Designers," *VLSI Design,* July/August 1982.
10. B. Elspas, "The Theory of Autonomous Linear Sequential Networks," *Transactions of the Institute of Radio Engineers,* CT-6, 45, 1959.
11. W. F. Rogers, "Character Cyclic Codes," *IBM Technical Disclosure Bulletin,* vol. 8, no. 4., September 1965.
12. B. Konemann, J. Mucha, and G. Zwiehoff, "Built-in Logic Block Observation Techniques," *Digest of Papers of IEEE Conference on Test,* October 1979, pp. 37–41.
13. R. Chandramouli, "Designing VLSI Chips for Testability," *Electronics Test,* November 1982.
14. D. Bahr, "Understanding Signature Analysis," *Electronics Test,* November 1982.
15. J. Clary, "Signal Processing Concurrent Test Considerations," IEEE VLSI Test Workshop, Atlantic City, March 1984.
16. J. B. Clary and R. A. Sacane, "Self-Testing Computers," *Computer,* IEEE, October 1979, pp. 49–59.
17. E. K. Nowicki, "Testing VLSI/Scan-Design with Today's Board Testers," *Proceedings of the ATE Silicon-Valley Conference,* April 1984, pp. IV-22–IV-26.

18. E. K. Nowicki, "VLSI Board Designs Demand Enhanced Testing Capabilities," *Electronics Test,* April 1984.
19. F. J. Langley, "VLSI/VHSIC and the Implications of On-Chip Test Techniques in Multi-Level Testing," *Proceedings of the Electronic Production Efficiency Exposition,* Birmingham, U.K., May 1985.
20. H. Flynn, "Augmenting VLSI Self-Tests with ATE Diagnostics," *Proceedings of ATE West,* Anaheim, January 1985.
21. Mark Swanson, "Preparing Tests for Boards Containing PALs," *Automated Testing for Electronics Manufacturing Conference Proceedings,* Oct. 4–6, 1983.
22. R. Roetzer, "Forging the Link between CAE and Program Preparation Software," *Proceedings of ATE Central,* Dallas, October 1984.
23. J. Sell, "Integrating Design and Test—A Feedforward Strategy for Reducing the Cost of Test Set Development," *Proceedings of ATE West,* Anaheim, January 1986.
24. H. Andrews, "Set-Up Costs—What It Takes to Make ATE Automatic," *Proceedings of Electronic Production Efficiency Exposition,* London, March 1986.
25. J. J. Faran, "The Implications of Ever-Higher Speed upon Wiring Practices for In-Circuit Test Fixtures," *Proceedings of the Electronic Production Efficiency Exposition,* Birmingham, U.K., May 1985.
26. Richard Barnes, "Fixturing for Surface Mounted Devices," *1983 International Test Conference Digest of Papers,* Philadelphia, Oct. 18–20, 1983.
27. Sam Meshulam, "In-Circuit Testing and the Manufacturing Process for Surface Mounted Devices on Circuit Boards," *Proceedings of ATE Central,* Dallas, Oct. 30–Nov. 1, 1984.
28. R. G. Bennets, "The Effect of Surface-Mounted Devices on Testing," Institute of Circuit Technology 7th Annual Northern Symposium, England, November 1984.
29. Sam Meshulam, "Testing Surface-Mounted Devices on Printed Circuit Boards," *Proceedings of ATE Northwest,* April 1985.
30. M. Eiger and M. Chabot, "Algorithms for High-Performance Fixture Wiring," *Proceedings of the International Test Conference,* Philadelphia, November 1985.
31. M. Khazam, "Predicting Test Accuracy for Analog In-Circuit Testing," *Proceedings of the International Test Conference,* Philadelphia, October 1983.
32. J. McPhee, "The Effects of Backdriving Integrated Circuits: An Accurate Electro-Thermal Model," *Proceedings of the International Test Conference,* Philadelphia, November 1985.
33. J. Congdon, "Driver-Sensor Design for High-Performance ATE," *Proceedings of the International Test Conference,* Philadelphia, November 1985.
34. M. Chabot, "In-Circuit Automatic Test Program Generation—Its Strengths, Limitations and Future Trends," *Proceedings of ATE East,* Boston, June 1984.
35. J. Hungerford and J. Lapidas, "Computer-Aided Test Models—What Do They Tell about Test Cost?" *Proceedings of ATE West,* Anaheim, January 1984.
36. P. Manikas and S. Eichenlaub, "Reducing the Cost of Quality through Test Data Management," *Proceedings of the International Test Conference,* Philadelphia, October 1983.
37. P. Manikas, "Putting a Value on Test Data Management," *Proceedings of the ATE Silicon-Valley Conference,* April 1984.
38. P. Harding and J. Howland, "Turning Test Data into Information," *Proceedings of the International Test Conference,* Philadelphia, October 1983.
39. J. Howland and P. Harding, "Estimating the Required Size of an Automated Test and Repair System from Subassembly Volume and Failure Information," *Proceedings of the International Test Conference,* Philadelphia, October 1983.
40. E. Schultz and T. Wray, "Transferring Data from a Multi-Vendor Test Facility into a Central Data Management System," *Proceedings of the ATE Silicon-Valley Conference,* April 1984.

CHAPTER 22
ASSEMBLY REPAIR

Donald P. Schnorr
RCA Corporation, Moorestown, New Jersey

22.1 INTRODUCTION

Assembly repair represents the type of repair normally acceptable for printed wiring assemblies. Bare boards, on the other hand, are often not repaired in the same manner because of the reliability risks associated with utilizing them later in assemblies and because of the comparatively limited investment in them. Normally, therefore, for high reliability and military applications, repairs or rework of bare boards are not allowed. For commercial applications, however, repairs may be performed on bare boards as part of in-process corrections, or rework, which are superficial in nature. When assembly repair and rework are acceptable, they are performed only when they are cost-effective; for example, when many expensive or hard-to-get components are used, assembly repair and rework are essential. The guidelines noted in this chapter for repair of assemblies are consistent with those defined for both commercial[1] and military[2] applications.

Printed wiring assemblies, driven by increasingly larger and more complex circuit elements such as VLSI, are demanding more responsive and reliable methods for repair and rework. New problems, dictated by the use of more surface mount technology and different and exotic materials and heat sinks, have required that increasingly more attention be given to reworking and repairing circuit assemblies.

The primary consideration in rework and repair of the board must be that any result of the repair will not degrade the electrical, mechanical, or physical properties of the printed wiring assembly. The repaired article, in all cases, must meet the intended original design requirements of the finished product. Repair techniques outlined in this chapter have been proven in actual application and use to be acceptable for normal applications. In all cases, finished product requirements should dictate reliability and quality standards for the repaired article.

22.2 GENERAL CONSIDERATIONS

The design of a printed wiring board should allow for change if a need for repair is anticipated. In all cases, the repair technique and function criteria for the operation should be subject to engineering approval. In many cases, it may be necessary to guarantee the reliability of the repair step by the use of a rudimentary performance test. The justification for repair should be flexible enough to consider the alternate solution to repair, namely, module or board replacement. Indeed, the economics of time, labor, material, and repair limitations, as well as customer preference, must be weighed against outright replacement.

22.2.1 Repair of Conventional Through-Hole Printed Wiring Board Assemblies

The repair of conventional through-hole printed wiring board assemblies represents by far the majority of boards produced. This board type and its assembly is characterized by through-hole mounting of components. As such, problems in component removal and replacement (with accompanying damage to the holes and pads) are primarily related to the application of heat or force. Other repair features include broken lines and traces, shorts, warped boards, and laminated defects, which will be treated in turn below.

22.2.2 Repair of Surface Mount Printed Wiring Board Assemblies

Unique problems exist with surface mounted assemblies when being considered for repair. The normal superficial feature repair (i.e., broken lines and traces, shorts, and laminate defects) are not unlike those found with conventional through-hole printed wiring boards, except that, in some cases, new and exotic materials may be used which may complicate repair. Surface mount boards may use exotic materials such as polyimide glass, polyimide Kevlar* or quartz reinforcements. In some cases, metal heat sinks or reinforcing layers are used to compensate for the differential thermal coefficient of expansion (TCE) normally experienced with surface mount of hermetic chip carriers (HCCs). Although much attention has been given to designing surface mounted devices and boards with compatible TCEs in repair operations, it must be remembered that identical expansion is only effective when all elements of the assembly are at or near the same temperature. Therefore, in repair, heat should be applied only to the terminal area to avoid overheating other areas of a finished assembly.

In many cases, double-sided mounting is incorporated to take advantage of an inherent density multiplier available with surface mounted assemblies. This complicates repair with regard to using localized heating so as not to damage or remove components on the wrong side of the board.

Another consideration in surface mount repair is the fact that many devices are large and have a large number of leads. When removal or replacement of leads is considered, it must be on a wholesale basis (not one lead at a time). Also, adhesive may be used for devices in situations where the coating needs to be removed, as in the case of conformal coating. The use of composite boards with heat sinks can complicate the removal process because of the inherent heat-sinking nature of the baseboard.

22.3 PREPARATION FOR REWORK

In addition to the assembling of suitable tools for repair and the designation of a definite repair workstation, the printed board must be prepared for the rework

*Kevlar is a trademark of E. I. du Pont de Nemours & Company, Wilmington, Delaware.

operations which are planned for it. One of the most important operations is the removal of conformal coating from the assembly. If the boards are unassembled, that is, of course, no problem, but for the majority of military contracts and for federal agency and commercial work, some type of moisture- and fungus-proof (MFP) coating is specified on the assemblies following the assembly soldering operation.[4,5] The need to remove all the conformal coating from the connections or area to be reworked exists because the coating creates a heat barrier which makes it difficult to melt solder joints and because the resoldered joint may become contaminated by any residual conformal coating. There must be no obstruction of any kind that would jeopardize the quality of the reworked assembly. In most cases, only spot removal of the MFP from the work area is necessary, but in some instances, it may be more efficient to remove all of the coating, by solvent means, on a production basis.

22.3.1 Conformal Coating Removal[1]

The type of problem faced by repair technicians is to determine which coating removal methods should be used for a specific coating. Obviously, within the factory system, generic and commercial identification of a specific coating is usually available, i.e., acrylic lacquer, varnish, silicone elastomers, polyurethane, epoxy, RTVs, etc. Consequently, the coating removal methods used in the factory can usually be specified and related to the known coatings being used.

However, when identification data is not available, a procedure of simple observation and testing will help identify the coating characteristics so that the technician can select the proper removal procedure. (*Note:* The generic or commercial identification of the coating material is not necesssary to accomplish coating removal.)

Hardness: Use a penetration test in a noncritical area to determine relative hardness. The harder the coating, the more suitable it is for removal by purely abrasive techniques. The softer and gummier the coating, the more suitable it is for removal by brushing.

Transparency: Obviously, transparent coatings are usually more suitable for removal than the opaque type. Removal methods used with opaque coatings must be far more controllable and less liable to damage the covered components and printed board surfaces, and they are usually slower.

Solubility: Test the coating for solubility characteristics in a noncritical area with trichloroethane, xylene, or other solvents with low toxicity and mild activity.

Thermal ability: Use a thermal parting device with controlled heating and without a cutting edge to determine whether the coating can be thermally parted at temperatures at least 50 percent below the melting point of solder. If the coating flows or gums up, the device is too hot or the coating is not suitable for thermal parting.

Strippability: Carefully slit the coating with a sharp blade in a noncritical area and try to peel back from the surface to determine if this method is feasible. Because of the adhesion required of coating materials, strippable techniques without chemical aids are usually very limited.

Thickness: Determine by visual means whether the coating is thick or thin. Thin coatings show sharp component outlines and no fillets, while thick coatings reduce sharp component outline and show generous fillets at points where components or leads intersect with the printed board. Thick coatings usually require two-step removal methods to prevent surface damage to the board. First, reduce the thick coating down to a thin one and then use pure abrasion methods to reach the surface of the board.

The specific coating to be removed may have one or more of these characteristics; consequently, the removal method selected should consider the composite characteristics.

22.3.2 Coating Removal Methods

There are three recommended coating removal methods: (1) solvent, (2) thermal parting, and (3) abrasion. The specific removal method to be employed is based on the generic type of coating used, the specific condition of the coating, the nature of the parts, and the nature of the board. These methods are covered briefly below.

22.3.2.1 Solvent Method. Mild solvents, such as xylene and trichloroethane or the equivalent, or harsh solvents, such as methylene chloride, may be used to remove specific soluble coatings on a spot basis by brushing or swabbing the local area a number of times with fresh solvent until the area is free of coating. If warranted, all the soluble-type coating can be removed by immersing and brushing the entire printed wiring assembly.

Caution: The primary note of caution in using the solvent method is the determination, on a module-by-module basis, of hazards to parts, etc., by short-term immersion in, or entrapment of, solvents. Solvent should be rapidly dried off through forced evaporation and, if required, neutralized.

If chloride-based or other harsh solvents are utilized, extreme care must be exercised to prevent damage to base material, component parts, plated-through holes, and solder joints. Sometimes coating removal methods can cause expansion (swelling) of the base material, which can degrade the printed board or printed board assembly. Under no circumstances should these solvents be used except in a closely controlled process (to prevent base material swelling). It is recommended that the printed board or printed board assembly be inspected to ensure that no damage has occurred.

The procedure for local or spot removal is as follows:

1. Dip the end of a cotton-tipped applicator in stripping solution, and apply a small amount to the coating around the soldered connections of the part to be replaced. Because various substances have been used as conformal coatings, the time required for a given coating to dissolve or soften will vary. (Repeat several times as the solvent evaporates very readily.)

2. Rub the treated surface carefully with a bristle brush or the wood end of the applicator to help dislodge the conformal coating. A wedge-shaped applicator tip, knife, or heated blade is very effective in removing some coatings, particularly polyurethanes.

3. Neutralize or clean the stripped area and dry.

4. Abrade the exposed solder connections and brush away the residue.

All coating may be removed in a single step by providing a continuous flow of fresh mild solvent which will carry off the contaminated solvent.

Alternatively, process the board in a series of tanks containing mild solvent until the solvent in the final tank remains clean. Start with a high-contamination tank and progress sequentially to a final, fresh solvent tank.

22.3.2.2 Thermal Parting Method. Thermal parting uses a controlled, low-temperature, localized heating method for removing thick coatings by overcuring or softening. Although military-approved coatings are now in the range from 0.001 to 0.003 in, the method is also useful for removing urethane resin strapping and adhesives holding down components.

Caution: Do not use soldering irons for coating removal since their high temperatures may char and delaminate the base substrate.

The procedure is as follows:

1. Select an appropriate thermal parting tip to suit the workpiece configuration. Set the nominal tip temperature by following the manufacturer's recommended procedure.

2. With a light pressure, apply the thermal parting tip to the coating. The coating material will either soften (epoxy) or granulate (polyurethane).

3. Gradually reduce the coating thickness around the area to be repaired without contacting the board surface. Low-pressure air or a bristle brush should be used to remove waste material.

4. Remove the remaining coating material by a combination of the thermal parting method and the abrasive method, below.

Another variation of the thermal parting method is to use the hot-air-jet method as follows:

1. Set up the hot jet per the manufacturer's instructions to provide the minimum-size jet. Adjust flow rate and temperature to suit specific coating removal application.

2. Apply jet to work area, and with the aid of a relatively soft-edged, nonmarring tool, use light pressure to remove the softened or overcured coating.

Caution: Never set the hot jet to heating levels that will cause scorching or charring of the coating material.

22.3.2.3 Abrasive Method. This method is often used in conjunction with the solvent and thermal parting methods after the coating has been softened. Rubberized abrasives of the proper grade and grit are ideally suited for removing thin hard coatings from flat surfaces, but not softer thick coatings, which would cause the abrasive to load with coating and become ineffective. A wide variety of tools may be employed, including twist drills, ball mills, and rotary brushes, to suit the various coating types and configurations.

The procedure is as follows:

1. Apply an appropriate rotating abrasive tool to the coating with various degrees of pressure to test the rates of coating removal. Use a coarse abrasive first, and change to a finer abrasive or bristle brush to clean up the area.

2. When all coating has been removed from the desired surface, the area should be cleaned with an appropriate solvent to remove any remaining contaminants.

Note: This method is primarily suited to the circuit, or solder, side of the board, which is readily accessible. On the component side of the board, use either the thermal parting or the solvent method.

22.3.2.4 ***Other Methods.*** Special coatings require other removal methods. Coatings such as parylene (MIL-I-46058, type XY)[5] are not removable by solvent methods. In this case, the abrasive-jet method or plasma etch may be used. The abrasive jet depends on a finely collimated spray of abrasive particles to cut through the tough conformal coating without harming the remainder of the circuit board. Many kinds of equipment are commercially available for this purpose. The abrasive is typically Airbrasive No. 4, suitable for parylene coating removal. A vacuumized dust cabinet with a glass see-through panel is required to house the removal operation.

In plasma cleaning, localized parylene removal cannot be used, since the entire board is placed in the vacuum chamber. This method is particularly effective in removing parylene coating.

22.4 CONDUCTORS

22.4.1 Damaged, Defective, or Missing Conductors

As indicated in Fig. 22.1, a fault in this category may consist of (1) a complete break in the conductor, (2) scratches, (3) nicks, or (4) pinholes, all of which may reduce the cross-sectional area of the conductor below that required by specification.

Note: The limitations are that rework should be limited to two repairs per conductor and not more than six repairs per board or as determined by the specific requirements of the customer. Of the repairs, no more than three per board should be made by the procedure described below. Conductor widths and spacings must not be reduced below the allowable tolerances.

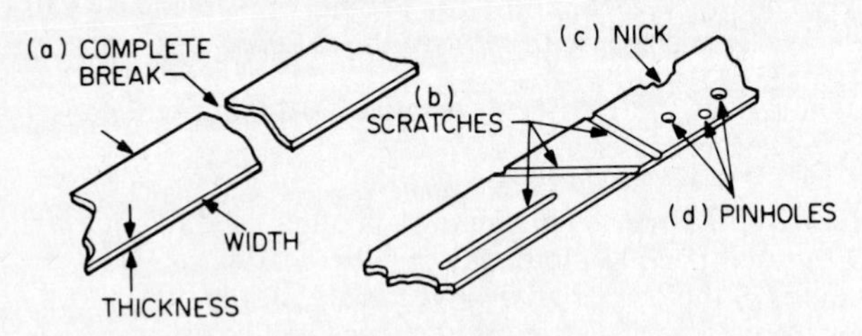

FIG. 22.1 Types of conductor damage.

22.4.2 Recommended Procedures for Repairing Damaged Conductors

22.4.2.1 ***Jumper Wire Method (Fig. 22.2)***[1]

1. Select component leads, unused through-holes, or terminals if available, and clean area to remove conformal coating if present. Clean with isopropyl alcohol. With soldering iron, clean excess solder off lead. Either side of board may be used.

2. Cut a piece of solid tinned copper wire of applicable gauge, and wrap each end around terminals or in the hole to be connected.

3. Flux connections and solder in place.

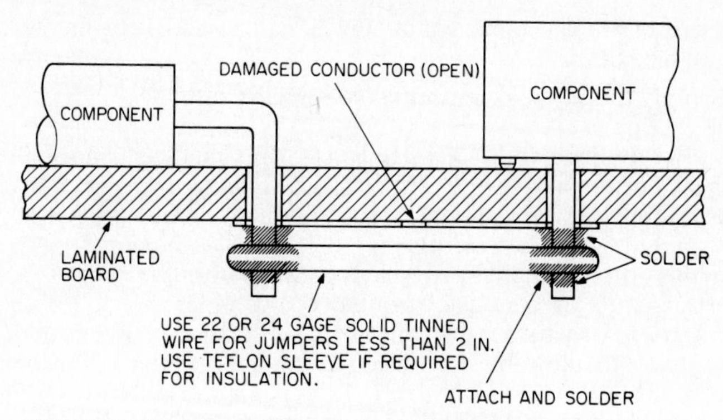

FIG. 22.2 Conductor repair, jumper wire method.

4. Clean flux residue from the joint with isopropyl alcohol.

5. Whenever possible, connect wires to a point where soldering an adjacent component will not cause the wire to become unsoldered.

6. Jumpers over 2 in long should be covered with Teflon* sleeving and be firmly secured to the board.

22.4.2.2 Foil Jumper Method[1]

1. Peel off any damaged or unwanted conductive pattern using a surgical or sharp knife. Heat can be applied to the conductor to add additional delamination where required.

2. Roughen the printed board surface under the removed circuitry using a suitable abrader.

3. Clean the roughed printed board surface using isopropyl alcohol and a lint-free industrial wipe.

4. Cut or etch a piece of copper sheet to the size and shape of the missing circuitry, allowing for at least a 3-mm ($\frac{1}{8}$-in) overlap to existing circuitry. (A piece of circuitry removed from a scrap board may be used or cut from commercially available conductor frames.)

5. Using a suitable abrader, remove the oxide coating from the ends (of the replacement pieces of circuitry) that are to be overlapped.

6. Clean the piece of replacement circuitry with isopropyl alcohol and a lint-free industrial wipe, and tin the conductor ends on the board with solder.

7. Place the replacement conductor in position, and flow-solder the conductor ends in place. Make sure the replacement conductor is in proper alignment with the printed board conductive pattern and that the solder-joint overlap requirements are obtained.

8. Remove the solder flux residues, and clean the area with a suitable solvent.

9. Rough-clean and fine-clean the area to remove any loose material and other contaminants.

---

*Registered trademark of E. I. du Pont de Nemours & Company.

10. Leave the replacement conductor as is, or bond it to the base material using a suitable adhesive.

11. Reapply the suitable conformal coating if required.

22.4.2.3 ***Welded Wire or Ribbon Method.***[1] This method is normally used for the repair of short breaks or opens in conductive patterns or inner layers of multilayer boards prior to lamination, or on the outer conductive patterns (surface layers) of printed boards or assemblies.

The method uses a parallel gap welder to weld a jumper wire across damaged conductors. A damaged conductor can be (1) a complete break, (2) scratches, (3) nicks, or (4) pinholes which reduce the cross-sectional area of the conductive pattern more than the allowable tolerances of the applicable specification (see Fig. 22.1).

The modified or repaired conductor pattern must not reduce the conductor widths, spacings, or current-carrying capacity below the specification tolerances.

The procedure for repair is as follows:

1. Rough-clean the area to remove any loose foreign material, flux, oils, grease, or other surface contaminants.

2. Select the wire or ribbon to be the same width as the conductive pattern being repaired ±0.05 mm (0.002 in).

3. Cut the wire or ribbon to be 3 mm (0.125 in) greater in length than the section being repaired.

4. Thoroughly clean the wire or ribbon, conductor, and base material surrounding the repair area with an abrader and then isopropyl alcohol.

5. Carefully place and center the wire or ribbon over the section to be repaired, using the microscope, leaving equal wire or ribbon end lengths on each side and parallel to the conductive pattern.

6. Carefully position the work under the electrodes, viewing through the microscope, so that as the electrodes are depressed to the area of the weld, the electrodes align with the one end of the wire or ribbon and away from the defective conductor area.

7. Hold the wire or ribbon in place with tweezers until the weld is completed using the weld schedule based on the accepted test sample.

8. Clean with isopropyl alcohol, and inspect the weld for quality, alignment, and/or all visual acceptance standards.

9. If the repair was performed on a tinned or soldered coated area, reflow the solder over the repair and clean with isopropyl alcohol if required.

10. Rough-clean and fine-clean the area to remove any loose material and other contaminants, and coat the repaired area with epoxy or other suitable conformal coating if required.

22.4.3 Surface Conductor Delamination Modification and Repair

The following procedures are all designated to repair conductors that have become delaminated from the base material.

22.4.3.1 ***Liquid Adhesive under Conductor***[1]

1. Clean the underside of the lifted conductor and surrounding area with isopropyl alcohol, and dry completely.

2. Remove all obstacles which prevent the lifted conductive pattern from making intimate contact with the surface of the base material.

Caution: Be careful while cleaning and removing all obstacles not to stretch the delaminated conductive pattern.

3. Carefully force the liquid adhesive under the entire length of the lifted conductive pattern, using a toothpick or small brush.

4. Press the lifted conductive pattern down into the liquid adhesive and into contact with the base material, place a release film over the area, and hold under pressure until the adhesive has set.

5. Allow the liquid adhesive to air-cure at an elevated temperature per the manufacturer's recommendations.

22.4.3.2 *Film Adhesive under Conductor (Attached)*[1]

1. Clean the underside of the lifted conductor and surrounding area with isopropyl alcohol, and dry completely.

2. Remove all obstacles which prevent the lifted conductive pattern from making intimate contact with the surface of the base material.

Caution: Be careful while cleaning and removing all obstacles not to stretch the delaminated conductive pattern.

3. Cut the film adhesive to approximate size.

4. Carefully place the film adhesive under the entire length of the lifted conductive pattern.

5. Press the lifted conductive pattern down against film adhesive and into contact with the base material.

6. Adhere the conductive pattern to the base material by applying heat (soldering iron recommended) until the film adhesive softens and bonding occurs.

7. Remove excess film adhesive from around the conductive pattern and the base material.

22.5 LANDS

Lands, or terminals, are important because they are points of attachment or interconnection between the component and the printed wiring pattern. For that reason, contact tabs are included in this category as input-output points. They share in functional importance with other board lands.

22.5.1 Land Modification and Repair

These modifications and repair procedures deal with various degrees of delaminated, lifted, and missing lands and associated conductive patterns. In some instances, through connections are an integral part of land modification, and repair procedures are included where appropriate.

Defects in this category consist of land areas which have become separated, loosened, or lifted or which, for some reason, have become unbonded from the base material. Included are land areas damaged by some means in excess of established limits (Figs. 22.3 and 22.4) and those that are missing or were inadvertently removed during the manufacturing process.

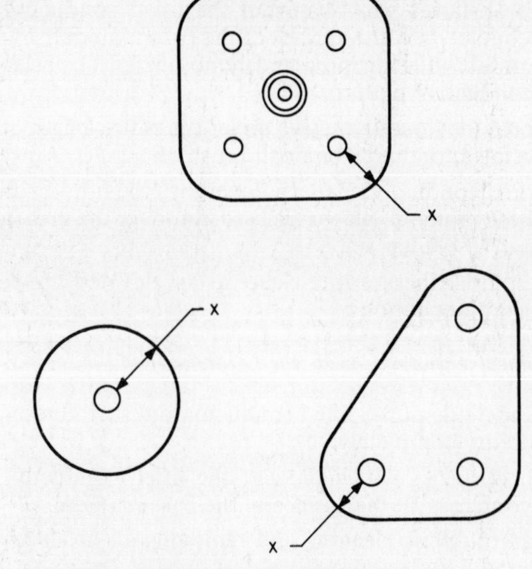

FIG. 22.3 Typical land patterns.

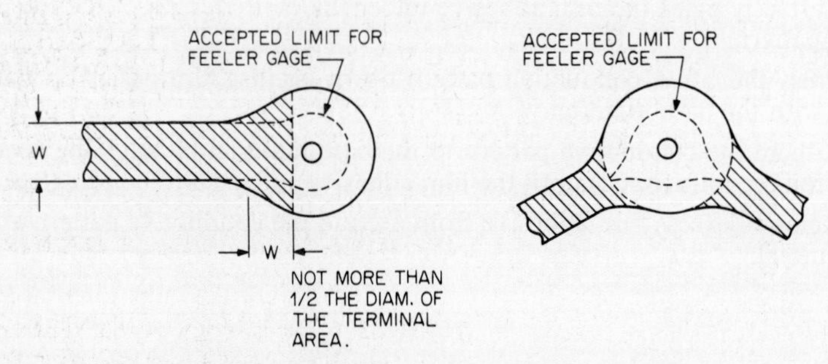

FIG. 22.4 Limits for land area repair. W = minimum width of connected circuitry within $\frac{1}{4}$ in of hole center.

Note: The limitations are as follows:

1. No repairs on bare (or unassembled) boards should be made without customer approval.

2. The spacing between conductors must not be reduced to less than the minimum specified in the original or governing specifications.

3. Repairs are made if the lifting or separation exceeds the established limits during inspection. The unshaded areas in Figs. 22.3 and 22.4 represent the land areas and/or land areas to be inspected. In those areas (other than noted), the 0.0015-in feeler gauge can penetrate a distance equal to or no more than one-half the distance from the interface of the hole to the nearest edge of the terminal area (annular ring) for not more than 180° of the periphery. When feeler gauge penetration does not exceed the limits described above, the land area is acceptable and is not considered as lifted. If, however, the feeler gauge penetration is more than described above, or if one defect can be visually determined, then the repair may be made in accordance with the procedures and conditions listed below.

The repair procedure for partially lifted land areas is as follows:

1. Free immediate area of solder by using a vacuum-type solder remover.

2. Gently but firmly clean the areas to be repaired by brushing with clean isopropyl alcohol. Air-dry for 5 min; then oven-dry for 30 min at 60° C.

3. Prepare adhesive. One recommended material is 10.0 g of Epon 828 mixed thoroughly with 15.0 g of Versamid 140. Deaerate by centrifuging for 2 min.

4. Apply adhesive immediately after removing the board from the oven; carefully flow it under and around the lifted pad with a camel's hair brush, a syringe, or some other suitable applicator. Press the land down with the applicator or clamp until adhesive is set.
Caution: The solder side of the terminal area must be free of contamination and/or epoxy.

5. Examine the board under ultraviolet light. The adhesive glows and should be seen to surround the reworked pad or trace.

6. Cure in 60°C oven for a minimum of 3 h. The surface buildup of adhesive should be smooth and neat in appearance.

The repair procedure for lifted land areas around bare holes is as follows:

1. Free immediate area of solder by using a vacuum-type solder remover.

2. Lift up the separated area without bending, and clean all dirt, flux, residue, and foreign matter from under and around the land area by brushing with isopropyl alcohol. Air-dry for 5 min; then oven-dry 30 min at 60°C.

3. Apply adhesive per steps 3 to 6 above for partially lifted land area.

- ***a.*** If the surface under the lifted land area is smooth and free of pits, adhesive film (Permacel P-18) may be used instead of liquid adhesive. Cut adhesive film to the exact circuit dimensions, and place in position.
- ***b.*** Adhere the loose area by applying the flat side of a clean soldering iron for approximately 5 s.

4. Clean the repaired areas with isopropyl alcohol or an approved solvent.

5. Install a fused or funnel eyelet of sufficient size to receive the component lead. Eyelet OD should be within 0.010 in of hole ID. The ID of the eyelet should not be a tight fit. For a lifted land area which has a flat side to facilitate circuitry spacing, the eyelet may be clipped by using a diagonal cutter to conform to the shape of the original pad. In no case shall the eyelet be clipped within 0.005 in of the OD of the eyelet. All other procedures outlined in this section apply here.

The repair procedure for lifted land areas which have been separated or otherwise broken away from the associated circuity (Fig. 22.5) is as follows:

1. Remove defective land area, if any; clean board surface with an applicator which has been saturated in isopropyl alcohol or other approved solvent; and allow the surface to dry.

2. Remove a similarly shaped section of good circuit from a scrap board or use a tinned copper foil of equal or greater thickness to replace the damaged or broken land area or pad. A minimum overlap of 0.125 in with the connecting circuit conductor should be allowed for the solder lap joint. The overlapping area

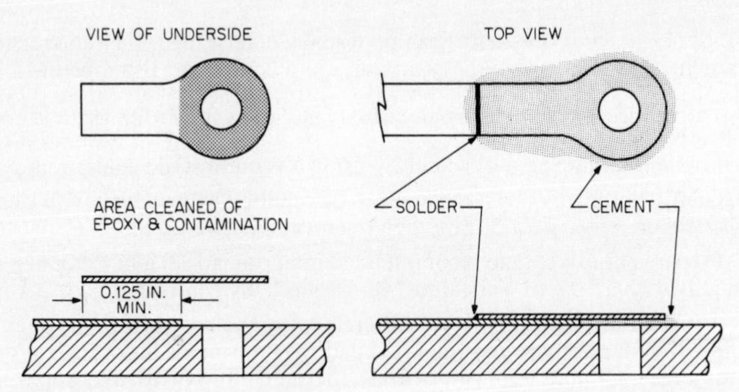

FIG. 22.5 Lifted and separated land area.

of land area segment and board circuitry should be completely cleaned of contamination by using an electric erasing machine or an ink eraser. Clean the area, after buffing it, with isopropyl alcohol or some other approved solvent.

3. Solder the newly prepared land area segment to the existing circuit. Take care to match the land area with the hole.

4. Cement the replaced land area segment to the board per steps 2 to 4 above for repairing lifted land areas around bare holes.

5. Install a fused or funnel eyelet (see step 5 above for repairing lifted land areas around bare holes).

The repair procedure for a fully lifted land area is as follows:

1. Free the immediate area of solder by using a vacuum-type solder remover.

2. Lift the separated area without bending, and clean all dirt, flux residue, and foreign matter from under and around the pad with isopropyl alcohol or some other approved solvent.

3. Insert Epon 828 or an approved epoxy resin under the copper with a camel's hair brush, a syringe, or some other suitable applicator. Make sure that the resin does not contaminate the solder surface of the terminal area.

Note: Adhesive film method per step 3 for repairing lifted land areas around bare holes may be used.

4. After adhering and curing, check repaired areas for adequate adhesion.

5. Clean the repaired area with alcohol or some other approved solvent.

6. Install a fused or funnel eyelet (see step 5 above for repairing lifted land areas around bare holes).

The repair procedure for lifted land areas which are not connected to circuitry (via holes) is as follows:

1. Remove defect land foil, if any, and clean the board surface around the hole with isopropyl alcohol or some other approved solvent.

2. Leave hole and board bare unless a mechanical solder joint is required.

3. If a mechanical solder joint is required, install a flat flange eyelet in the hole. (See step 5 above for repairing land areas around bare holes.)

22.5.2 Damaged or Defective Contact Tabs

Although this type of defect commonly involves pitted, peeled, or delaminated gold plating, only badly worn, missing, or damaged contact tabs will be discussed (see Fig. 22.6). Gold plating is normally repaired on unassembled boards by completely replating the board and on assembled boards by a selective, or brush-plating, procedure. Information on the latter method is best obtained from the manufacturer of the equipment.

Note: The following limitations apply. The number of repairs of this type per board should be determined by the specific requirements of the procuring agency and/or customer. Industry practice never exceeds 5 percent of the contacts per board. Repair should not be performed if there is serious damage to the underlying base laminate.

The procedure is as follows:[1]

1. Peel off the defective contact tab by using a razor knife, and clean and roughen the board surface by using a suitable abrasive.

2. Brush with 50 percent isopropyl alcohol in naphtha, and dry with a lint-free tissue.

3. Select a similar contact tab from a scrap board, remove it, and cut it to a length of at least 3 mm (0.0125 in) longer than the removed defective printed contact.

4. Prepare the new tab by abrading its back surface with a glass-fiber eraser or equivalent and tinning the end.

5. Tin the end of the etched conductor on the board.

6. Flux ends and reflow-solder in place with a minimum overlap of $\frac{1}{16}$ in.

7. Remove flux residue, and clean areas to be adhesive-bonded with 50 percent isopropyl alcohol in naphtha.

8. Use masking tape to protect the neighboring contact tabs as well as the tabs on the opposite side of the board.

9. Coat the back of the tab with a thin film of epoxy adhesive, and position the tab. Clamp the tab firmly in place until the epoxy is cured; use a slight, uniform, and constant pressure.

In some cases, worn or defective plating on contact tabs can be repaired by selective plating. Selective plating (sometimes known as spot plating, swab or brush plating, touch-up, or doctoring) is a repair procedure finding growing use on printed boards. While in theory it is pure electrodeposition, in practice it looks more like arc welding. A portable dc power pack has one flexible lead connecting to the board and a second to one of a series of working tools called styli. The anodes on these styli (in different sizes and shapes) are wrapped with absorbent cotton and then dipped into extra-high-speed proprietary plating solutions. When these saturated anodes are brushed across board conductors, lands, or contacts, metal electrodeposits are made wherever electrical contact is made. The thickness of the metal deposit can be determined either manually using current and time indicators or with an analog or digital ampere-hour meter.

22.5.3 Deleting Surface Conductive Patterns

One of the most common of all rework procedures consists of removing foil material. This is done in the case of shorts and spurs (see Fig. 22.7*a*) and as needed to

FIG. 22.6 Damaged (worn) contact tabs.[1]

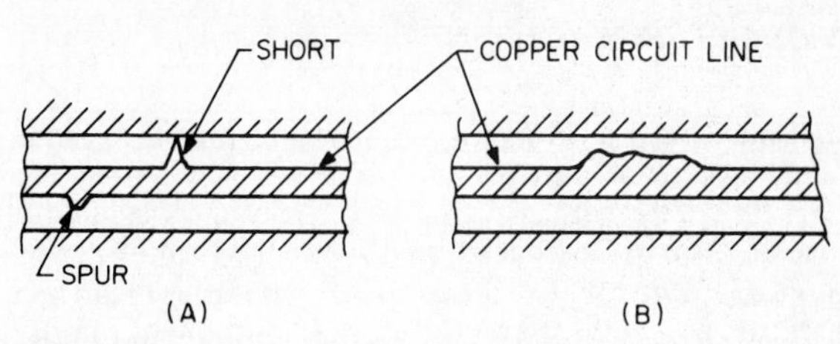

FIG. 22.7 Excess conductive materials. (*a*) Spur and short; (*b*) underetched area.

disconnect or interrupt a conductive pattern on printed boards or printed board assemblies. Rework of this type is especially important on internal layers of multilayer boards, where the presence of a short can have disastrous results after the board has been laminated and the processing has been completed.

The procedure is as follows:

1. By using a suitable pointed instrument such as a razor or knife, carefully cut the foil and separate the excess material from the body of the conductor. Take care not to cut into the base laminate.

2. By using a chisel-point knife, skive up the conductor from the base material.

3. Scrape the area under the removed conductor and inspect it to make certain that there are no conductor traces. Apply epoxy material or conformal coating if specified. (Not in the case of internal layers.)

22.5.4 Missing Holes or Terminals

Holes may be drilled in boards to introduce missing mounting holes or terminals, to interrupt internal layers, or to route added wires.

Note: The following limitations apply. Added terminals or holes drilled for any reason should not exceed eight per board. With the exception of drilling out holes in multilayer boards to break connections, no internal conductive layers should be pierced.

The procedure is as follows:

1. Drill the proper hole in the board according to the detail drawing. In the case of a missing terminal, install a terminal.

2. For interrupting internal conductive circuits, or for routing wires through holes, an oversized hole should be drilled (typically using an etch-cutting machine) and filled with epoxy as follows:

- ***a.*** Clean area to be filled with a stiff bristle brush dampened in perchloroethylene, and then rinse with isopropyl alcohol.
- ***b.*** Mask and/or plug adjacent holes to prevent their filling, and take care to keep the rework area localized.
- ***c.*** Select the proper bonding materials (resin and catalyst), and mix as recommended by the material supplier. Mix in Fiberglas No. 128, and stir until all the fibers are wetted.
- ***d.*** Fill hole and cure as recommended. Abrade to flush. Small plates coated with release agent may be clamped on the filled area to minimize the amount of abrading required to flush.
- ***e.*** Redrill the hole, if necessary, per the applicable drawing.

22.6 COMPONENT REPLACEMENT[6,7]

Printed wiring boards are used as interconnection media for increasingly more sophisticated and expensive electronic assemblies. A wide variety of circuit elements are currently used in an assortment of package styles, including axial-lead devices, flat packs, dual-in-line packages (DIPS), and a number of radial-lead or header-type devices. Additionally, the growing popularity of surface mounting for packaging has raised a need for new methods of component removal and replacement because of the particular constraints of surface mounting. When a device on most assemblies fails, it is usually cost-effective to remove the device and replace it, thereby salvaging the cost of the assembly. Of primary importance in removing and replacing failed devices is the avoidance of damage to the board and adjacent devices. It is of secondary importance to remove the device in good condition, since, often, much valuable information on the manufacture and application of the device may be obtained by examining the failure modes and mechanisms of an essentially intact "failed" device. It is, therefore, important to have procedures for removing devices in an expedient manner and without mechanical or electrical damage to the printed wiring board, the adjacent devices, and, if possible, the removed device itself.

22.6.1 Defective Components

In this category falls the removal of components which need to be replaced, including all axial-lead types, as well as device body styles incorporating multiple conductors with fixed leads and surface mounted devices.

Note: The number of repairs of this type per board is determined by the specific requirements of the procuring agency and/or customer. Extreme care is required to avoid damage to the circutry, base laminate, or other components. Reasons for damage could be excessive heat application, improper or sloppy use of tools, inadequate training of the operator, or rough handling of the printed board. Heat damage may occur when an iron is held on a terminal too long. A guideline for soldering iron application is this: No more than 3 seconds for a terminal connected to circuit traces; no more than 6 seconds for a terminal connected to a ground plane. The use of heater blocks for multiple-lead components in plated-through-hole joints is not recommended. It has not proved feasible to heat multiple leads uniformly; joints are either over- or underheated, and the result is either plated-through-hole damage or pad delamination and/or blistering.

22.6.2 Axial-, Radial-, and Multiple-Lead Components

22.6.2.1 Through-Hole Procedure

1. Identify component to be removed and note polarity.
2. Remove strapping material, if necessary, as follows:
   a. Use a thermal parting tool to remove large accumulations of adhesive. Take care not to cut into the board or discolor the board with heat.
   b. Thin applications of remaining material may be carefully removed by methods outlined previously under conformal coating removal methods.
3. Remove conformal coating, if applied, as outlined previously.

4. Remove component by applying one of the following techniques:

a. Apply heat and remove excess solder by one of the methods outlined below (solder joint removal methods).

b. If leads are clinched, apply heat, remove solder as above, and straighten and remove lead.

c. Cut component leads, apply heat, and remove solder as outlined below. Remove the lead.

5. After a component has been removed and the land area has cooled, melt all the excess solder on the land and remove it with an extractor as outlined below under solder joint removal methods.

6. Inspect the hole thoroughly for damage before attempting to replace any components.

7. Form the leads of the replacement component and trim them to size.

8. Mount the component on the board. Note the polarity.

9. Solder leads in such a fashion as to avoid any undesirable heat buildup, especially on multiple-lead conponents. Allow cooling time when necessary.

10. Clean flux residue from joint with cotton-tipped applicator and isopropyl alcohol when necessary.

11. Perform the electrical and/or functional tests.

12. Restrap the component by using a tie-down, an approved adhesive, or both if required.

13. Add conformal coating if necessary.

22.6.2.2 ***Surface-Soldered Components Procedure.*** This procedure is intended for use with *leaded* surface mounted components—a method for handling leadless surface mounted assemblies is covered later.

1. Identify component to be removed and note polarity.

2. Remove conformal coating, if applied, as noted previously.

3. Surface solder connections may be removed by one of the methods outlined below. If the board is double-sided, remove the solder joint on each side of the board as if the sides were two separate joints.

4. When all the leads are free, it may be necessary to debond the component body from the board by means of a razor knife or the thermal parting method already described.

5. Remove the solder remaining on the lead areas by means of the wicking or the vacuum extraction method (below).

6. Now resolder replacement components in place by using standard reflow soldering techniques.

22.6.3 Solder Joint Removal Methods[8]

High-density circuit boards characteristically incorporate tight lead-hole spacing so that the lead fits tightly into the plated-through hole. In addition, many times there are large ground planes located near or on the surface of these boards that aggravate component removal by adding thermal inertial properties that must be overcome during unsoldering and/or soldering.

There are many solder removal methods which have been used by the electronics industry; two of the most popular, which are recommended, are listed below:

22.6.3.1 Solder Sucker. The solder sucker usually employs a single-pulse vacuum-generating device utilizing either a squeeze bulb or a plunger mechanism with a Teflon orifice through which the vacuum is applied. The technique is to melt the solder joint with a soldering iron and insert the solder sucker tip with collapsed bulb or cocked plunger into the molten solder and then release the bulb or plunger (Fig. 22.8). The action creates a pulse of limited vacuum (level and flow) which removes the molten solder. Although the method offers a positive vacuum rather than the low capillary force of the wicking method, it has inherent limitations in its general application and is specifically used with plated-through-hole terminations.

Because of the resweat problem in plated-through-hole joint configurations, the solder sucker should be limited to use on single-sided or surface solder joints to which both the iron and the sucker tip can be applied simultaneously. This method is not recommended for plated-through-hole solder joints.

22.6.3.2 Solder Extractor. Most often used for production repair, the solder extraction method, like the solder sucker above, operates on a lead-by-lead basis, rather than as a mass removal method as used with surface mounted devices. It comprises a controlled combination of heat, pressure, vacuum, airflow, and manipulative qualities for the removal of any solder joint configuration. The solder extraction device is a coaxial, in-line instrument with the general configuration of a small soldering iron. It consists of a hollow tip, a heating element, a transfer tube, and a chamber located within the handle in which the waste solder and clipped leads are collected. The unit is operated by a power source that provides a controlled vacuum and electrical supply (Fig. 22.9).

The advantage of the solder extraction device, or continuous vacuum desoldering, is that it is easily manipulated and fully controllable. The power source

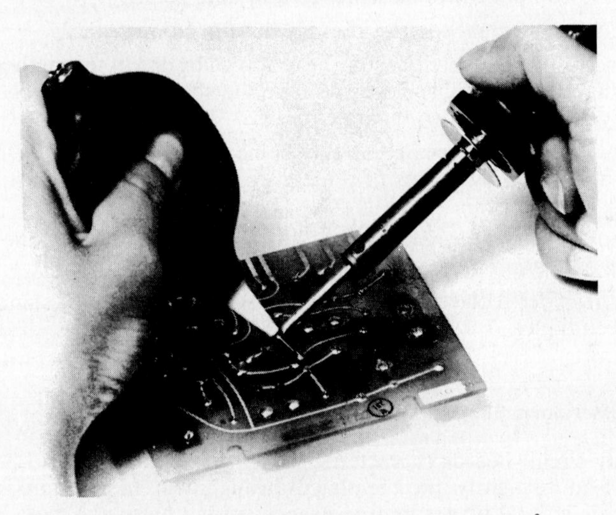

FIG. 22.8 Solder removal by the solder sucker technique.[9]

provides variable control over vacuum and flow rates, as well as temperature control over the heated tubular tip.

In the vacuum mode, the heated tip is applied to the solder joint; when a melt is noted, the vacuum is activated. That causes the solder to be withdrawn from the joint and deposited in the chamber. To overcome the critical problem of the lead resweating to the sidewalls of a plated-through-hole-type joint, the lead is oscillated with the tip while the vacuum is being applied. That permits cool air to flow into and around the lead and the hole sidewalls, which causes the lead and sidewalls to cool down and prevent resweat. Solder extraction with continuous vacuum is the only consistent method for overcoming the resweat problem for either dual- or multiple-lead devices terminating in through-hole solder joints. Therefore, it is the method recommended for removing through-hole solder joints. An illustration of the continuous vacuum desoldering method appears in Fig. 22.10*a* and *b*. Figure 22.10*a* illustrates the melting of the solder determined by movement of the lead or observation of melt. Figure 22.10*b* illustrates application of the vacuum to remove the solder.

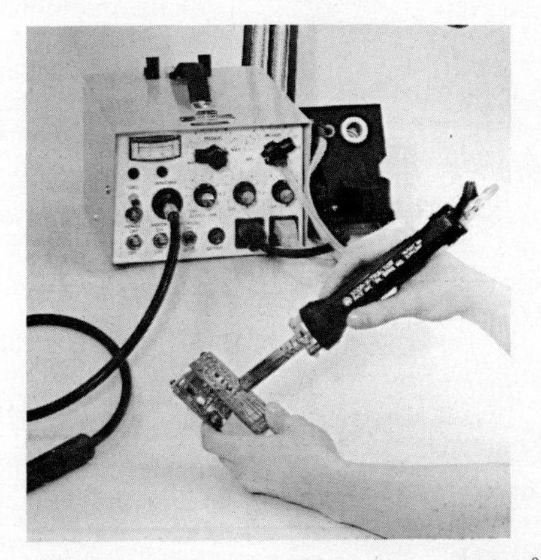

FIG. 22.9 Solder removal by solder extraction.[9]

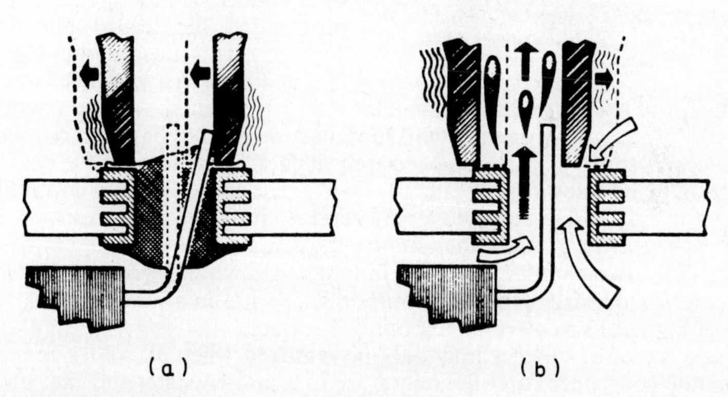

FIG. 22.10 Solder vacuum extraction method.

22.7 REPAIRING SURFACE MOUNT BOARDS[3,7]

Printed wiring boards incorporating surface mounted devices (SMD) require different repair techniques than those used for conventional, through-hole mounted packaging.

Surface mount boards are often so heavily populated and dense that the number of components alone impedes repair operations. Also, the high I/O count LSI and VLSI devices with which these boards will be stuffed (to bursting) present a very large number of individual pads to desolder.

The SMD board may carry components on both sides—sometimes a mixture of leaded and leadless surface mounted and leaded through-hole-mounting devices. Many operations have reported unsatisfactory results when reflowing entire boards a second time to reattach failed components.

Finally, without component mounting holes and plated-through holes for attaching jumper wires, the aesthetics of signal rerouting on an SMD board leave much to be desired.

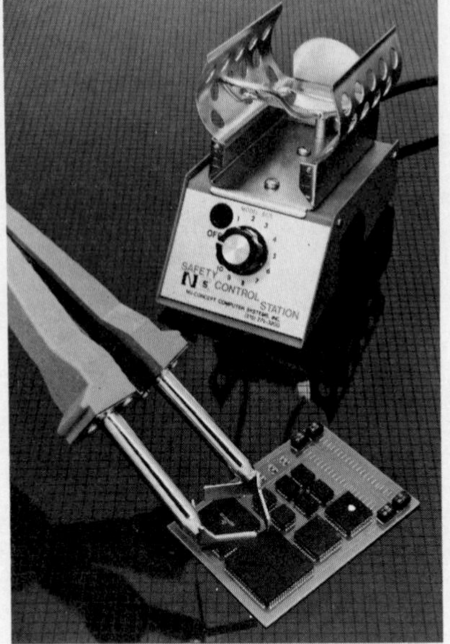

FIG. 22.11 SMD removal—conduction method. urtesy of Nu-Concepts Systems, Inc.)

Since traditional suction or extraction desoldering is too slow for production module repair of SMD boards, two techniques have been developed which are compatible with this packaging method and are covered below.

22.7.1 The Conduction Method

This method utilizes a direct contact method similar to a soldering iron, except that all the solder connections of the component are contacted simultaneously. Since this method depends upon good heat transfer, the solder joints should be examined for uniformity—if solder fillets are small or the junction is void of a fillet, solder should be added to the juncture. The tool recommended for leaded and leadless components is illustrated in Fig. 22.11. It is available in a variety of sizes from 0.185 to 1.3 in square and in rectangular sizes within these ranges. The tool has nickel tips independently heated and mounted on handles having a bidirectional hinge mechanism. Tips can then be exactly aligned to the sides of the chip carrier connection by pivoting slightly laterally. The tool is controlled by a controller unit as shown in Fig. 22.11. To remove and replace chip capacitors, resistors, single-in-line packages (SIPs), and other surface mounted devices, a different hand tool is used having two parallel tips 0.100 in long as shown in Fig. 22.12.

To remove a surface mounted component, secure in a holder similar to that shown in Fig. 22.13 or place on a grounded surface.

Surface mounted devices may only be removed when all solder connections between the component and the board are in a molten state, and the same condition must exist for attachment also. With some boards, especially when conductive heat sinks are included in the composite structure, a preheat cycle may

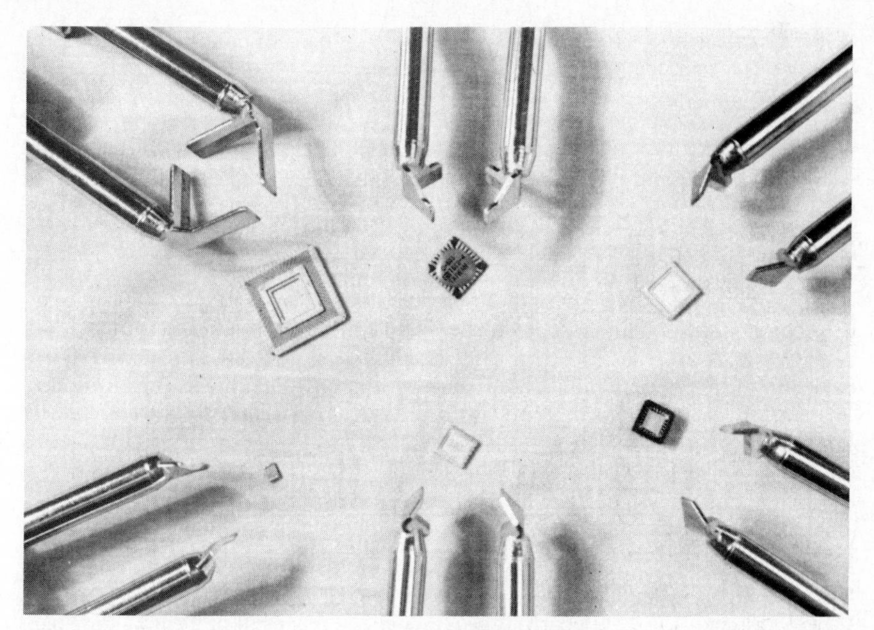

FIG. 22.12 SMD removal tools. (*Courtesy of Nu-Concept Systems, Inc.*)

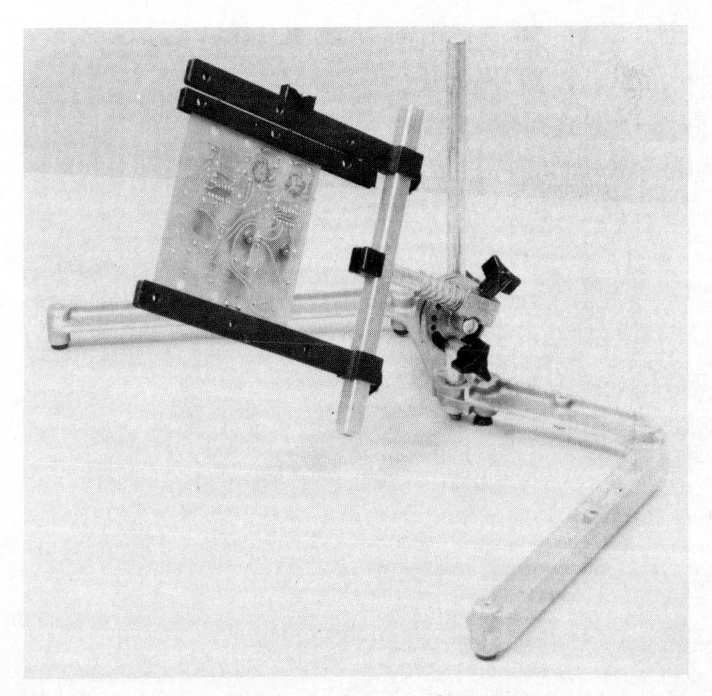

FIG. 22.13 Work handling unit. (*Courtesy of Pace, Inc.*)

be necessary prior to the removal and/or replacement of components. Ceramic base materials are also preheated to circumvent unequal thermal expansion and cracking of the board.

To remove components, the temperature range of the tool should be adjusted to meet board requirements. Flux should be applied to solder junction points and the tips of the tool placed on the outside periphery of the components. The component is grasped so that all the tips contact four sides of the component at the same time. When the solder is molten on all the component connections at the same time, the tips are raised carrying the component with it.

In resoldering components using the conduction method, pretinned components must be used. The footprint on the board surface should be examined for good and uniform solder coverage. The new component is placed into position and aligned. A probe is placed against the component to hold it in place. The tips of the hand tool are positioned against the component leads, and the bottoms of the tips are held flush with the surface of the board. When the solder is molten, the tips of the hand tool are removed from the joint and the probe is removed, allowing the component to settle in place on the designated footprint land areas.

22.7.2 The Convective Method[7]

The convective method depends upon delivery of hot air or gas to a predetermined limited area of the board. It also provides a means of preheating the board and components when needed. It utilizes low-pressure air (so as to leave small components undisturbed) directed to the component area to melt the solder connections. Equipment of this type is illustrated in Fig. 22.14 (Nu-Concept's Hot

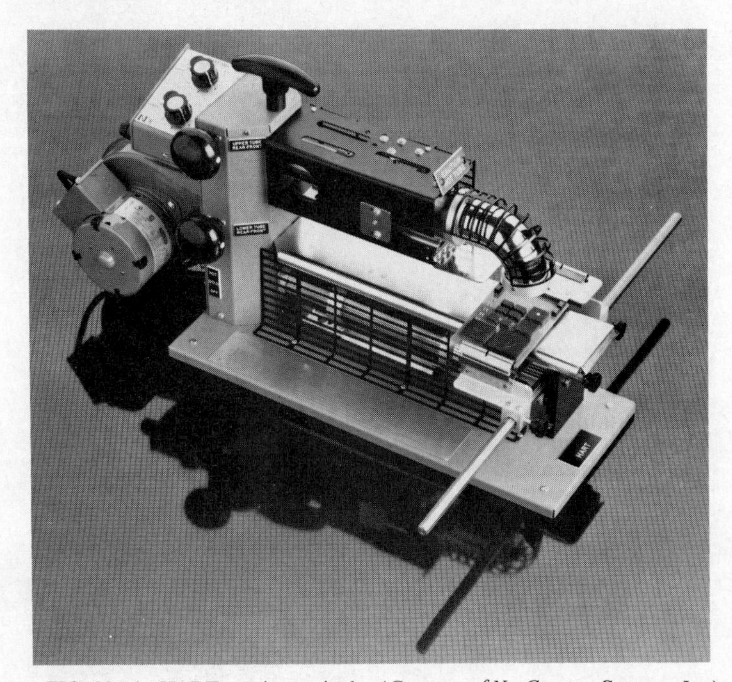

FIG. 22.14 HART repair terminal. (*Courtesy of Nu-Concept Systems, Inc.*)

Air Removal Technique, "HART"). The HART machine operates in a typical fashion and employs two hot-air tubes, one positioned above the board, the other below. First, the lower tube is used to raise the substrate to about 212°F (100°C). This temperature can be set by a chart provided, or a temperature-sensitive compound may be used which indicates when soldering temperatures have been reached, so as to minimize time at high temperatures and to prevent overheating of the component. The top tube is then positioned, reflowing the solder joints. The component is then removed with tweezers.

For reattachment of components using the convective method, the footprint solder amounts should be uniform as in the conductive method. The footprints should first be examined, and if insufficient solder remains, solder cream should be added. Next, the carrier is placed in position and the heat tubes are extended. Once the reflow temperature is reached, leadless chip carriers will "float" to the position intended automatically, relying on the surface tension of the solder. Leaded chip carriers can be positioned into place with a special probe.

22.8 PLATED-THROUGH HOLES

Plated-through holes, by the nature of their construction, are subject to damage or destruction for a number of reasons related to normal use, as well as normal misuse. If the plated "barrel" of copper is damaged, so that the electrical continuity it is designed to provide is interrupted, either it must be repaired or the assembly must be replaced. The repair of plated-through holes is typically restricted to double-sided interfacial connections only, as there has been no reliable method for repairing multilayer plated-through holes. If it is necessary to disconnect all interconnections on double-sided or multilayer printed boards or printed board assemblies, the common method is to drill out the plated-through hole, fill the hole with an adhesive material, and redrill a smaller hole in the center of the adhesive filler material. In some cases, a replacement pin or terminal may be placed in the hole for component termination, but connections to that point are hardwired from other parts of the board.

For repair of double-sided plated-through holes, two methods are recommended: the clinched C or Z jumper wire method or the fused eyelet method. Although both are covered in detail in IPC-R-700, the procedures are briefly covered below.[1]

22.8.1 Clinched Jumper Wire Method

The procedure is as follows:

1. Remove components and conformal coating as outlined previously if that is necessary to obtain access to the defective plated-through hole.

2. Remove solder by using a vacuum-type solder remover.

3. Cut a length of tinned copper wire and insert it into the defective plated-through hole. Clinch both sides so the lay of clinch is approximately $\frac{1}{16}$ in along the conductor.

4. Solder clinched leads on both sides by using rosin-core solder.

5. Replace components if removal was required for access to damaged lead. Solder.

6. Remove flux residue with cotton-tipped applicator moistened with solvent.

7. Test electrically, if applicable.

8. Add conformal coating if required.

22.8.2 Fused Eyelet Method

These repairs can be used with single-sided and double-sided printed wiring boards and with the outer layers of multilayer boards to establish a through connection between the respective lands or to reinforce a hole. As with the clinched jumper wire method above, the repair is restricted to assembled boards only, and the number of repairs per assembly should be determined by agreement between the supplier and the purchaser.

The eyelet repair method can be accomplished by either machine or manual setting. Although both fused-in-place eyelets and funnel-flange eyelets with hand set, upset end, and hand soldering are used, the former are considered more reliable, and only this type will be covered.

The procedure is as follows:

1. Rough-clean the area to remove any loose foreign material, flux, oils, grease, or other surface contaminants.

2. Remove conformal coating or polymeric coatings, if required.

3. Select a flat-flange tinned eyelet with an ID sufficient to receive the component lead, and one eyelet diameter longer than the board thickness.

4. Mill or drill out the hole 0.050 mm (0.002 in) larger in diameter than the eyelet OD. Clean the land surfaces where the eyelet is to be installed.

5. Apply an approved liquid flux to the land areas (on both sides of the printed board if double-sided).

6. Insert the flange eyelet into the hole, cold-set the eyelet according to the fusing-machine manufacturer's operating instructions, and remove from the machine to inspect for proper setting.

7. Replace the board with the cold-set eyelet into the machine, close the setting dies on the eyelet, and activate the power for the fusing operation. Fuse-set the eyelet.

8. Install the component lead and solder, if required. Clean the area and remove flux.

9. Replace polymeric coatings and conformal coatings, if required.

22.9 WARPED BOARDS

Warped boards, both in the unassembled and assembled state, are a very common and annoying defect. The cause is not well understood. The defect is particularly troublesome when the boards are mounted on close-tolerance centers or when the board is required to slide into guides or plug into a one-part (edge card) connector. There seems to be evidence that the anomaly is design-related and that unequal distribution of copper planes or pattern can result in warp or twist. Unequal stresses caused by poor distribution of copper planes or pattern can result in warp or twist. Unequal stresses caused by poor distribution of wired and adhesive layup in laminating can also cause this. It is usually accentuated following the heat of flow soldering, and this is the product which is most valuable and, therefore, a viable candidate for repair action.

22.9.1 Warped Boards without Components

The following limitations apply: (*a*) This technique is specifically for NEMA G-10 and FR-4 or MIL-P-13949 GE and GF; (*b*) this technique has limited application, since the boards may rewarp.

The procedure is as follows:

1. Place the board between two steel plates or in a suitable straightening fixture. Clamp with sufficient force to hold the board flat between the plates.

2. Oven-bake for 20 min at 240 to 300°F (115.6 to 148.9°C).

3. Remove from oven and cool to room temperature.

4. Remove from the straightening device and measure warp.

5. Repeat steps 1 to 4 if necessary.

Note: The number of times the procedure should be repeated is determined by results obtained from each cycle.

22.9.2 Warped Boards with Components

The following limitations apply: (*a*) Same as *a* and *b* in the preceding section; (*b*) components mounted on the board should be able to withstand the desired oven temperature without reduction of the service life of the components.

The procedure is as follows:

1. Place selected areas of the board between steel plates or in a suitable straightening fixture (usually edges).

2. Clamp with sufficient force to hold board flat between the plates. Arrange plates in such a way that only open areas of the board free of components are in contact with the clamping pieces.

3. Place in oven for 10 to 20 min at 240°F (115.6°C).
Caution: See limitations at beginning of this section.

4. Remove from oven and cool to room temperature.

5. Remove the straightening device and measure the warp.

Note: The number of times the procedure should be repeated is determined by results obtained after each cycle.

22.10 LAMINATE DEFECTS

Epoxy-glass laminate material is subject to a number of defects inherent in the nature of the material itself. Composed as it is of epoxy-impregnated glass cloth, it is prone to a large number of aesthetic and/or functional anomalies which may be inherent in the raw material or be aggravated by either the manufacture or repair of the printed board. Unfortunately, the defects are not usually evident until the final steps of manufacturing the board (i.e., outer-layer etch). Even worse, they may be introduced during some type of repair or component replacement step. An overriding consideration is the value of the board or assembly, which usually justifies the cost of the salvage operation.

22.10.1 Measling and Weave Exposure

"Measling" is defined as a condition existing in the base laminate in the form of discrete white spots or crosses below the surface of the base laminate reflecting a separation of fibers in the glass cloth at the weave intersection. "Weave exposure" is a surface condition in which the unbroken woven glass cloth is not uniformly covered by resin. Acceptance of either defect is subject to agreement between board vendor and user. The defect can result in a functional failure if the condition abridges terminals on the printed board because of a breakdown in insulation resistance between these points.

Note: The following limitation applies: The acceptable extent of the defect is subject to agreement between manufacturer and user, but not usually exceeding 5 percent of the board area.

The procedure is as followins:

1. Thoroughly clean area to be repaired by brushing with isopropyl alcohol.

2. Bake board at 100°C for at least 45 min.

3. Brush authorized conformal-coating material on the affected areas. The coating must extend a minimum of 0.062 in beyond the defect in all directions.

4. Cure conformal coating per manufacturer's recommendations.

22.10.2 Laminate Voids

Included in this category are blistering and, in the case of multilayer boards, surface-layer delamination. Frequently found after dip or wave soldering, the defect is usually the result of moisture entrapped in the board in combination with a basic material shortcoming. Gross cases of widespread internal delamination are not treated here; they should be cause for rejection of the multilayer assembly.

Note: The following limitations apply: (*a*) Number of repairs (or area repaired) per board should be determined by the specific requirements of the procuring agency or customer; (*b*) cost of the board or assembly should justify the rework; (*c*) voids caused by blisters should be accessible by the methods used herein; (*d*) only electrically functional boards should be repaired.

The procedure is as follows:

1. Clean the board thoroughly, and remove all surface contamination and loose foreign matter.

2. Puncture each blister with a dental pick in at least two small areas, opposite each other, around the perimeter of the blister.

3. Place the board in the oven, and use the time and temperature recommended to cure epoxy.

4. Remove from the oven, and add the epoxy material over one of the openings around the perimeter of each blister. Epoxy may be applied by using a hypodermic needle. The heat of the board will draw the epoxy into the void area and fill it completely.

5. (Optional.) If it is determined that evacuation is necessary to remove possible included air, it should be accomplished at this point.

6. Cure the epoxy by using the recommended time and temperature.

7. Perform electrical tests of all interconnections in and around reworked area.

22.10.3 Crazing

"Crazing" is defined as a condition existing in the base laminate in the form of connected white spots or "crosses" on or below the surface of the base laminate. It reflects the separation of fibers in the glass cloth and connecting weave intersections. It is serious because it can provide an entrance and/or dwelling place for moisture and, if terminals are abridged, a path for electrolysis and subsequent insulation resistance breakdown.

Note: The following limitations apply: (*a*) The number of repairs per board should be determined by the specific requirements of the customer; (*b*) the cost of the board or assembly should justify the rework; (*c*) only electrically functional boards should be repaired; (*d*) crazing in external layers should be a maximum length of 0.100 in.

The procedure is as follows:

1. Remove the affected laminate by means of the miniature machining abrasion method using a ball mill, or an S. S. White (or equivalent) air abrader with appropriate nozzles and abrasive material.

2. Clean areas to be filled by scrubbing with Freon TF.* Use a stiff nonmetallic brush, and take care not to scuff circuit lines, components, etc.

3. Preheat the board at 65°C for a minimum of 30 min.

4. While the board is hot, fill the holes by using a syringe which has been filled with an appropriate epoxy replacement material (such as Epon 828). The board must be hot, and it must be reheated if it cools to room temperature.

5. After filling the void with epoxy, allow the board to sit in a dirt-free atmosphere for 15 to 20 min; then cure it in a controlled oven for 4 h at 65°C.

22.10.4 Base Material Replacement

This method is used to repair mechanical or thermal damage to, or physical material defects in, laminated base materials. The method is used on inner base material laminate repairs (on single, double, or multilayer boards—where the internal layers are not affected) where the damage may exceed the outer two surface layers of glass cloth. For superficial base laminate repairs, a repair may be found in laminate voids, or crazing, above.

The procedure is as follows:

1. Rough-clean the area to remove any loose foreign material, flux, oils, grease, or other surface contaminants.

2. Remove the board material with a surgical knife and/or machining burr until the damaged area has been completely removed. Rough- and fine-clean the area. Hot-air-dry.

3. If a glass-filled epoxy is being used as the replacement base material, then fill the excavation with the filled epoxy (eliminate air bubbles) and allow the filled resin system to cure; then skip the next two steps.

4. If a filled epoxy is not being used, then partially fill the excavation with epoxy (eliminate bubbles).

5. Press a piece of glass cloth or cut fibers, approximately the same dimen-

*Registered trademark of E. I. du Pont de Nemours & Company.

sions as the excavated area, into the epoxy. The fibers should be prevented from breaking the epoxy surface, and the filler material should be about flush with the surface of the printed board.

6. Abrade the cured epoxy to be flush with the surface of the board.

7. Rough-clean and fine-clean the area to remove any loose material and other contaminants.

8. To seal the surface, a second coat of nonfilled epoxy may be applied to the filled epoxy layer and cured.

22.11 MULTILAYER BOARDS

In many ways, the multilayer board is similar to single- and double-sided boards when rework is involved. Nearly identical procedures are used for outer-layer rework, component replacement, repair of laminate defects, and conductor repair. Internal layers of multilayer boards are treated as single-sided boards prior to lamination. An added precaution when working on multilayer boards concerns susceptibility to damage from heat. Many of the repair techniques mentioned earlier in this chapter are applied in combination with some type of heat application (e.g., component replacement and conductor repair), and it is extremely important to minimize the amount of heat added to the multilayer assembly during any repair step. Among the problems which may be caused are damage to the plated-through hole, scorching of the board with subsequent possible loss of insulating properties, and hole contamination by flow of epoxy binder.

The only repair technique reserved exclusively for multilayer boards involves the defects which may be found in internal layers after assembly; in this case, they are internal shorts. Other repairs for this type of board may be found in detailed areas previously covered.

22.11.1 Internal Shorts

This defect consists of shorts between internal layers and/or the ground.

Note: The following limitations apply: (*a*) The number of repairs per board should be determined by the specific requirements of the customer; (*b*) shorts in the connector area are to be repaired in unassembled boards only.

The procedure is as follows:

1. Make a series of resistance readings and/or continuity checks, and record the data. Determine, if possible, the exact location of the short or shorted holes before attempting any drilling.

2. Drill out the barrel of the suspected hole or holes slightly larger than the barrel OD. Precision drilling is necessary.

3. Examine the hole carefully to be sure that the whole barrel has been completely removed.

4. Repeat resistance and/or continuity checks to verify removal of the short.

5. Clean the hole or holes with isopropyl alcohol, and mask the area around the hole.

6. Fill the hole with epoxy and cure.

7. Abrade flush to the surface of the board, and clean with isopropyl alcohol.

8. Hardwire, externally, all circuits removed by drilling, and solder with rosin-core solder.

Note: Care should be taken in routing external strappings to maintain wiring rules and circumvent cross talk.

9. Remove all flux residue with a cotton-tipped applicator and isopropyl alcohol.

10. Apply a conformal coat over the external wires.

11. Cement all external wiring to the board surface if required or if board does not have conformal coating.

Note: A good rule of thumb is to tack external wiring down with cement in increments of approximately 1 in. Each tack should be $\frac{1}{4}$ in in length.

REFERENCES

1. *Modification and Repair for Printed Wiring Boards and Assemblies,* IPC-R-700C, Institute for Interconnecting and Packaging Electronic Circuits, September 1985.
2. *Printed Wiring Assemblies,* MIL-P-28809A, Oct. 5, 1986.
3. John B. Holdway, "Guide for Removal, Replacement of Surface Mounted Devices," *Electri-Onics,* December 1984.
4. C. A. Harper, ed. *Handbook of Electronic Packaging,* McGraw-Hill Book Company, New York, 1969.
5. *Insulating Compound, Electrical (for Coating Printed Circuit Assemblies),* MIL-I-46058, October 1981.
6. *Maintainability Design Criteria Handbook for Designers of Shipboard Electronic Equipment,* Bureau of Ships, NAVSHIPS 94324, March 1965.
7. H. F. Vandermark, *The Removal and Reattachment of Surface Mounted Components,* Nu-Concept Systems, Inc., Colmar, Pa., 1968.
8. *Solder and Soldering,* Department of the Army Technical Bulletin TB-SIG222, Change No. 1 Unclassified.
9. William J. Siegel, *Support of Modern Electronic Equipment: Fact or Fantasy,* Pace, Inc., Silver Spring, Md., March 1979.

P · A · R · T · 5

SOLDERING

CHAPTER 23*

DESIGNING AND PRESOLDERING CONSIDERATIONS FOR SOLDERING

Hugh Cole

23.1 INTRODUCTION

This chapter deals with the specification of the materials system for the printed circuit and the design parameters which must be considered before a final circuit is laid out. The soldering operation must be considered from the inception of the board layout in order to ensure satisfactory performance. The rules are simple and straightforward; if they are followed, the operation should run smoothly and efficiently. If they are disregarded, however, the result is invariably recurring problems with bridges, icicles, and imperfectly formed fillets that will need handling by a large number of touch-up operators.

23.2 DESIGN CONSIDERATIONS

During the layout of the board, several soldering parameters should be carefully considered. They are (1) the wire-to-hole ratio, (2) the size and shape of the terminal areas, (3) the number and direction of extended parallel circuit runs, (4) the population density of the solder joints.

23.2.1 Wire-to-Hole Ratio

The wire-to-hole ratio represents a compromise between the ideal situation for assembly (large hole and small-diameter lead) and the ideal situation for soldering (smaller lead-to-wire ratio). The minimum hole size can be established by the rule of thumb that it should be no less than the lead diameter plus 0.004 in. The maximum hole diameter should be no more than 2.5 times the lead diameter. Of course, if the board is a plated-through-hole or a multilayer circuit board, the hole-to-wire ratio should be lower than 2.5 to encourage the capillary action of the flux and solder during the soldering operations.

---

*Adapted from Coombs, *Printed Circuits Handbook, 2d ed., McGraw-Hill, New York, 1979, Chap. 14.*

23.2.2 Size and Shape of the Land Area

The pad area around the solder joint is normally either circular to slightly elongated (teardrop). It should not be more than 3 times the diameter of the hole in the board. There is sometimes a tendency, particularly on low-density boards, to leave large irregular land areas around the holes. That should be avoided! Excessively large land areas expose too much copper to the solder pot, cause excessive quantities of solder to be used in joint formation, and promote bridging and webbing.

If the leads are to be clinched during assembly, the land should be so oriented that the clinched lead will be in the center of the elongated pad. Both the pad and the component should be so oriented that the clinched lead is parallel to the direction of solder flow in the solder wave and not perpendicular to it.

23.2.3 Number and Direction of Extended Parallel Lines

The use of automated printed board layout programs and the trend toward high-density circuit packaging have resulted in a tendency to group large numbers of circuit paths together and run them parallel to one another for long distances. If those paths are oriented perpendicularly to the direction of flow in the solder wave (i.e., at right angles to the direction of the conveyor), then they can contribute to bridging and webbing. Every effort should be made to maximize the spacing between lines which must be oriented perpendicular to the direction of the conveyor.

23.2.4 Population Distribution

An excessive number of joints in one area promotes bridging, icicling, and webbing. It may also cause a heat-sinking effect and interfere with the formation of a good solder joint.

23.3 MATERIAL SYSTEMS

In the soldering process there are two surfaces which must be considered before a solder flux is selected. They are the lead surface and the pad surface. The average printed circuit assembler has little control over the material systems used in component leads, since the selection is usually made by the component manufacturer. Furthermore, most components are mass-produced and supplied on large reels. It is not economically justifiable to treat each lead according to an individual assembly shop's specific requirement. Therefore, in selecting the components, care should be taken that the leads are solderable, and an incoming inspection should be established to ensure lead solderability.

The board itself, however, is a different story. Since each board is custom-manufactured, the assembly or soldering engineer can exercise a great deal of control over the material systems used on the board. Again, it is important, in order to keep defects to a minimum, that the board be made of a solderable material and that the solderability of the board be checked as a part of incoming material inspection. The next section will deal with some typical material systems encountered during the soldering process.

23.3.1 Common Metallic Surfaces

23.3.1.1 Bare Copper. Because of its low cost and ease of processing, one of the most common metallic surfaces encountered is bare copper. Chemically clean copper is the easiest material to solder; it can be soldered with even the mildest fluxes. But unless it is protected with a rosin-based protective coating, its solderability will rapidly degenerate because of oxides and tarnishes. As we shall see later in this chapter, however, the solderability of tarnished copper surfaces is easily restored with surface conditioners. If boards with bare copper surfaces are used, care should be taken to maintain solderability during handling and storage (storage time should be minimal), and the boards should not be stored in the presence of sulfur-containing material such as paper, cardboard, or newsprint. Sulfur produces a tenaciously adhering tarnish on copper which seriously impairs the solderability.

23.3.1.2 Gold. Gold is encountered most commonly on component leads and plug-in finger surfaces. It is a highly solderable material, but it is extremely expensive and it rapidly dissolves in the molten solder. Because it affects the properties of the solder joint, causing the joint to become dull and grainy, it is usually avoided or eliminated by pretinning the lead before soldering. Various studies have shown that all gold on a gold-plated lead can be dissolved in a solder pot of eutectic tin-lead solder within 2 *s* (plate thickness of about 50 μin). Therefore, pretinning is economical as well as easy.

23.3.1.3 Kovar. Many dual-in-line packages (DIPs) and related integrated circuitry are supplied with Kovar leads. Kovar is a very difficult metal to solder because it doesn't wet well. For that reason component manufacturers and/or assembly shops prefer to pretin Kovar. The pretinning is normally accomplished only with organic acid fluxes and certain proprietary acid cleaners.

23.3.1.4 Silver. Although silver was once very popular in the electronics industry, it is not used on terminal areas or component leads. The reason is the problem of silver migration, a phenomenon discovered in the late 1950s and extensively researched during the early 1960s. Silver should be avoided. If it must be used, it is an easily soldered material and should be treated similarly to the bare copper surface (i.e., avoid sulfur-bearing materials and minimize storage and handling).

23.3.1.5 Immersion Tin. Immersion tin coatings are electrolessly deposited coatings of tin metal on bare copper surfaces. When tin is initially deposited, the coating is extremely solderable. It does, however, deteriorate rapidly, and it becomes more difficult to solder than bare copper. Originally, immersion tin coatings were used to protect the solderability of bare copper surfaces and thereby extend the shelf life of the board. Experience has shown, however, that fused tin-lead plate is far superior for the purpose.

23.3.1.6 Tin-Lead. Tin-lead coatings are put on printed boards and component leads to preserve the solderability of the material. They can be applied by electroplating, hot dipping, or roller coating. The mechanics of the processes are discussed elsewhere in this chapter. A properly prepared tin-lead surface should exhibit excellent solderability and long shelf life (about nine months to one year). Tin-lead coatings can be soldered with most rosin-based fluxes, even the nonactivated types. However, optimum results are obtained with the activated rosin fluxes.

23.4 WETTING AND SOLDERABILITY

Soldering is defined as a metallurgical joining technique involving a molten filler metal which wets the surface of both metals to be joined and, upon solidification, forms the bond. From the definition it is apparent that the materials to be soldered do not become molten and therefore the bonding occurs at the interface of the two metals and is strongly dependent on the wettability or solderability of the base metal by the molten alloy. Although the base metal does not become molten, some alloying can take place if the base metal is soluble in the filler metal. The bond which is formed is strictly metallic in nature, and no chemical reaction which covalently or ionically bonds the metal to the surface occurs.

To understand the basic mechanism of soldering, it is necessary to understand the thermodynamics of wetting. Fortunately, however, in order to understand wetting it is not necessary to understand thermodynamics. Wettability or solderability of two materials is a measure of how well one material "likes" the other. The property can be easily visualized by using a water drop resting on the surface, such as the one shown in Fig. 23.1. When the water drop doesn't like the surface on which it rests, it pulls up into a ball and touches the surface, in the idealized case, at only one point. The angle between the drop and the surface at the point of contact is called the dihedral angle. If the drop likes the surface, it spreads out all over the surface and comes in intimate contact with it. Various degrees of wettability are therefore related to the ability of the drop to spread out or wet the surface. Figure 23.2 shows the relation between the dihedral angle and the various wetting states. Wettability or solderability is related to the surface energy of the material. Wetting is substantially improved if the surface is clean and active (i.e., if all dirt and grease are removed and no oxide layer exists on the metal surface).

Therefore, to form a solder bond efficiently, we must start with a material system which can be wet by the molten solder, and the cleanliness of the parts must be maintained.

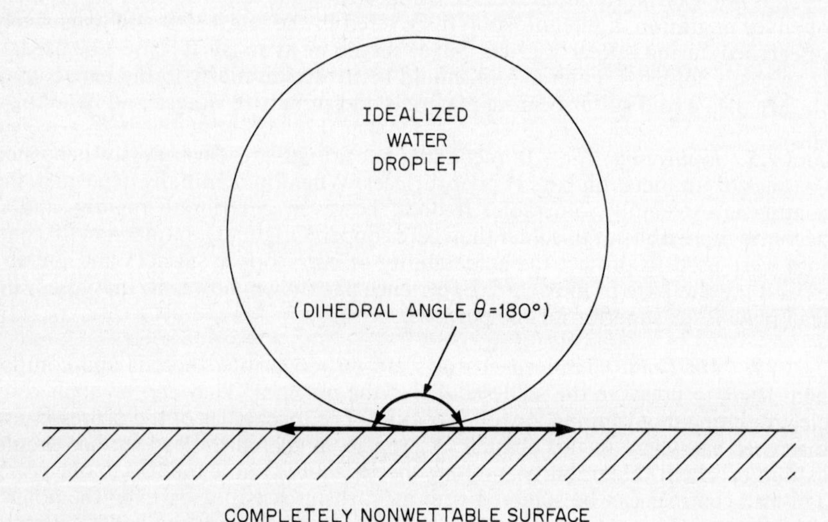

FIG. 23.1 Complete nonwetting of an idealized surface.

23.5 SOLDERABILITY TESTING

Solderability testing is an important quality control procedure in the electronics industry. It is a simple procedure but problems can occur if the fundamental test principles are not understood thoroughly. "Solderability" is a measure of the ease (or difficulty) with which molten solder will wet the surfaces of the metals being joined. When molten solder leaves a continuous permanent film on the metal surface, it is said to wet the surface. Wetting is a surface phenomenon which depends on cleanliness. Fluxing facilitates wetting by cleaning the surface, and the degree of surface cleanliness depends on the activity of the flux.

However, there is often a limitation on the activity of the flux in the electronics industry. Most electronics soldering operations require relatively weak rosin-based fluxes to avoid the possibility of current leakage caused by flux residues remaining on the part. To enhance the solderability of some surfaces and eliminate the need for active fluxes, electroplating often is employed to deposit a solderable coating over a base metal that tarnishes easily or is difficult to solder.

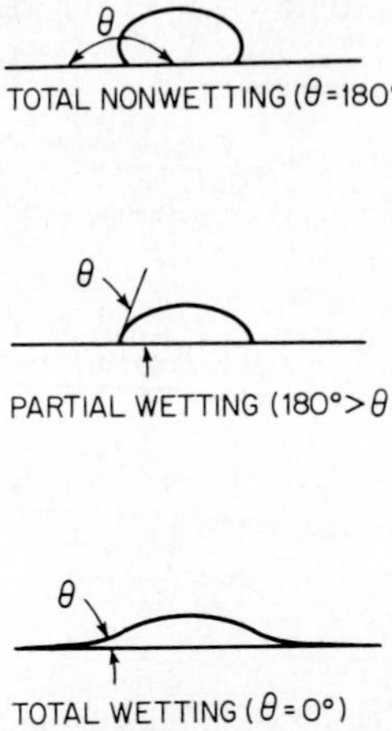

FIG. 23.2 Relation between dihedral angle θ and the degree of wetting.

23.5.1 Testing Procedures

Testing for solderabiltiy can be a simple procedure of inspecting production parts or of dipping an appropriately fluxed lead or portion of a printed board in a solder pot and observing the results. Good and bad wetting are then identified visually. The problem is to recognize borderline cases which simulate effective solderability but quickly deteriorate. In order to alleviate borderline solderability, the mildest possible flux should be used at the lowest soldering time-temperature relations which will give adequate results.

An effective solderability test involves the use of water-white rosin flux and a solder pot. The surfaces to be checked are fluxed and immersed for 3 to 4 s in the solder pot, which is maintained at approximately 500°F. The solder is then permitted to solidify, and the components are cleaned of flux residues prior to visual examination. The inspection is usually performed with either no or low magnification (5× to 10×). Most solderability tests will permit up to a 5 percent imperfection of the total surface, provided the entire imperfection is not concentrated in one area.

More elaborate solderability tests are described in governmental and industrial specifications. Testing of component leads is covered by Electronic Industries Association (EIA) Test Method RS17814, which is similar to the solderability method described in Military Specification 202, Method 208. The test incorporates a dip fixture which provides identical dip ratio and immersion times (Fig. 23.3).

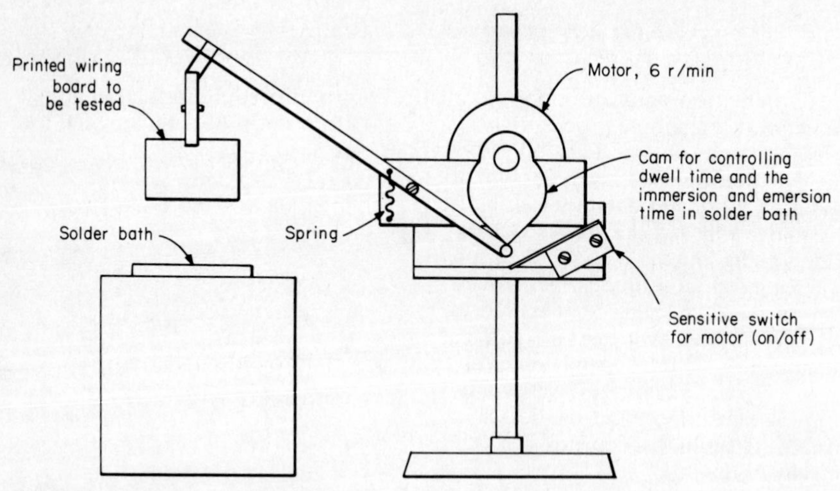

FIG. 23.3 Solderability dip tester.

23.5.1.1 Dip Test. For printed boards, the edge dip test, described in the EIA specification RS319 and by the IPC* in standard S801, often is employed. The edge of a printed board is dipped first in a mild flux and then in a solder pot for a predetermined time and temperature. After the flux residues are removed, the board is inspected visually for the quality of wetting. A similar test is employed to determine the solderability of solid wire leads, terminals, and conductive accessories of component parts normally joined by soft solder. Applicable test standards are EIA Standard RS178A, Solderability Test Standard, and MIL-STD-202C, Method 208 A, Solderability.

To perform the dip test, the operator places the item to be tested into a holding arm. The arm lowers the sample section into the solder pot. After the preset dwell time has elapsed, the arm automatically raises the sample. Then a visual determination of solderabiltiy is made. The dip test can also be performed as a manual operation, but that leaves too many variables to the discretion of the operator.

Since interpretation of results is based on a subjective judgment by the operator, it is essential that the operator be provided with examples of good, marginal, and poor solderability. It is imperative, too, that pot temperatures, cleaning and fluxing procedures, dwell times, and solder purity be controlled carefully in order to obtain meaningful results.

23.5.1.2 Globule Test. This test is one that is prevalent throughout Europe and is mandatory for European suppliers whenever specifications dictate. It is described in the International Electrochemical Commission Publication 68-2 Test T, Solderability. The globule test provides a numerical designation for the solderability of wires and component leads. It measures the ability of the solder to wet the lead.

A lead wire, coated with a nonactivated rosin flux, is placed in a holding fixture. It is gripped and straightened and then lowered into a globule of molten solder. The volume and temperature of the globule are controlled. As the lead

*The Institute for Interconnecting and Packaging Electronic Circuits is referred to as IPC.

wire bisects the globule, the timer actuates and measures the time between the moment the wire contacts the solder and the moment the solder flows around and covers the lead. At this second point, the timer stops, and the elapsed time is registered on a read-out. Elapsed time is indicative of the solderability of the lead: the shorter the time, the greater the solderability.

This globule test method is completely automatic and is designed for continuous operation. Time is measured to $\frac{1}{100}$ s. Wires that are 0.008 to 0.062 in in diameter can be tested, and special heat regulators can hold solder temperature to $\pm 2°F$.

When wire plated with a soluble or fusible coating is tested, it is advisable to perform a dip test to supplement and verify the globule test findings. The reason is that, under certain conditions, the plated coating might be totally reflowed or dissolved during the soldering operation, which would give misleading results.

23.6 PLATED COATINGS FOR PRESERVING SOLDERABILITY

The three commonly used types of plated coatings are generally referred to as fusible, soluble, and nonfusible and/or nonsoluble. Fusible electrodeposited coatings provide corrosion protection to a surface that has been activated for soldering. Whether the solder bond of a soluble electrodeposited coating is to the base metal or to the deposit depends on soldering conditions and coating thickness. Nonfusible and/or nonsoluble electrodeposited coatings are used frequently as barrier layers in electronic applications to prevent diffusion of solder and base metal.

23.6.1 Fusible Coatings

Tin and tin-lead electrodeposited coatings are commonly used in electronic applications to preserve solderability because they are fusible and do not contaminate the solder pot or fillet. Contamination can adversely affect tensile, creep, and shear strengths at a solder connection. Also, contamination of a part can reduce the flow and spread of the solder on the part.

If the coating operation is not closely controlled, an electrodeposited coating may be applied over a partially contaminated surface. If that happens, dewetting will occur in the contaminated areas because the electrodeposited coating is fused during soldering. Therefore, adequate cleaning prior to plating is essential to obtain good solderability. Figure 23.4 shows a surface which has dewet after a reflow operation.

Another point to remember is that plating thickness should be sufficient that porosity is virtually eliminated. Porosity and codeposited impurities will lower the protective value of the electrodeposited coating and will eventually cause poor solderability.

23.6.2 Soluble Coatings

Soluble coatings commonly used in electronic applications include gold, silver, cadmium, and copper. During soldering, these metal coatings are either completely or partially dissolved. The amount of dissolution depends on solubility of the coating metal, thickness of the deposit, and the soldering conditions. Silver

FIG. 23.4 Tin-lead-plated surface which exhibits dewetting after the reflow operation. (*Alpha Metals,Inc.*)

and copper tend to tarnish; and if a mildly activated flux is called for, they should be protected with a thin rosin coating. Cadmium offers sacrificial corrosion protection which often necessitates the use of highly activated fluxes to promote effective soldering.

Soluble gold coatings provide excellent corrosion and chemical resistance. However, because of the high cost, fairly thin coatings are used. Moreover, care should be taken because gold coatings of under 50-μin thickness tend to be porous and, as a result, lower the protective value of the metal. Corrosion of the base metal or barrier plate via pores causes soldering problems because the gold usually dissolves completely during soldering and it is difficult to wet the corroded base metal. Also, solderability decreases with the amounts of alloying elements that are used to increase hardness and are often codeposited with the gold. Thicker gold coatings, on the other hand, may cause brittleness in a solder connection because of formation of gold-tin intermetallic compounds.

23.6.3 Nonfusible and Nonsoluble Coatings

Nickel and tin-nickel electrodeposited coatings are considered to be nonfusible and/or nonsoluble because they provide an effective barrier to prevent alloying of solder to the base metal in electronic applications. They allow for effective soldering to such materials as aluminum and silicon. However, problems can occur from passivity, which is caused by codeposited impurities, or from some additional agents that are used in nickel plating to increase hardness and reduce internal stress. In such cases, nickel and tin-nickel electrodeposited coatings should be protected with tin or tin-lead to improve the shelf life of the soldered part.

Furthermore, nickel and tin-nickel have limited solubility in solder, and a flash coating is not an effective barrier. When electroless nickel is used as a barrier coating on aluminum, the required thickness will depend on soldering conditions. A thickness of 50 to 100 μin is sufficient for most operations.

23.6.4 Organic Coatings

Although the metallic coatings for preserving solderability are the most reliable and effective coatings, they are also the most expensive ones. For applications that do not require a long shelf life, considerable economies can be obtained by using an organic protective coating. There are several basic types of organic protective coatings—water-dip lacquers, rosin-based protective coating, and the benzotriazole-type coating. Organic protective coatings must be easily removable, and they must be compatible with the rosin-based fluxes normally used in the printed board industry.

Water-dip lacquers were once a very popular means of protecting the solderability of printed boards. However, they have fallen into disfavor because of their tendency to set up and polymerize with age. As they age, they become extremely difficult to remove. They also become insoluble in the flux solvent. If they are not properly removed after fluxing, they leave white residues which are not corrosive but are extremely unattractive.

Today, rosin-based protective coatings are much more prevalent than water-dip lacquers. They are applied by dip, spray, or roller coating, and, depending on the thickness, they will provide solderability protection for six weeks to four months. They are composed of the rosin solids material, and as such they are quite compatible with the rosin thinners and the normal cleaning method used to clean rosin fluxes. When those materials are used, it is important to ensure that adequate fluxing occurs and that the preheating time and temperature are sufficient to allow the protective coat to melt and be displaced by the flux. If the coat is not displaced, soldering will occur with a rosin nonactivated flux, no matter how active the solder flux really is.

A third alternative which some people have explored is the benzotriazole-type coating. Benzotriazole is an organic compound which is applied to the board surface during the final rinse operation of the plating line. It forms a thin nonporous film on the copper surface and prevents oxygen from reacting with the surface copper molecules. Benzotriazole films are very fragile, and they cannot be subjected to handling or scraping abuse.

23.7 TIN-LEAD FUSING

Fused coatings ordinarily are electrodeposited, low-melting metals or alloys that have been heated sufficiently above their melting points to become completely molten. In the molten state, alloying between the liquid and the basis metals is accelerated; on solidification, the deposit usually is dense and nonporous. The procedure is commonly known as "reflowing," and it usually connotes tin or tin-lead electrodeposits. However, it can also apply to the remelting of hot-dipped coatings.

Fused coatings are employed to guarantee that the cleaning procedure prior to plating is adequate, to produce a slightly denser deposit with less porosity, and to improve the appearance of the coating. Reflowing leaves a bright deposit which has a definite sales appeal. A typical conveyorized infrared reflow oven is shown in Fig. 23.5.

23.7.1 Thick Fused Coatings

Tin-lead electrodeposits that are reflowed are usually between 300 and 500 μin in thickness. That thickness provides adequate protection with a minimum of reflow

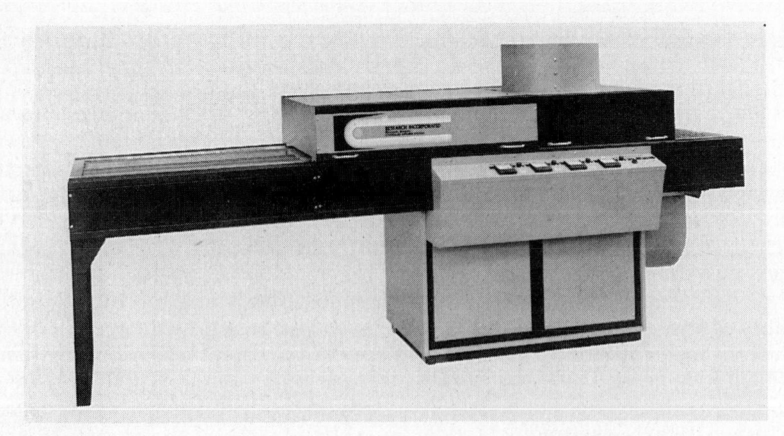

FIG. 23.5 Conveyorized infrared fusing machine. (*Research Incorporated.*)

problems. When printed boards are reflowed, the tin-lead deposit forms a meniscus on the conductor pad. For that reason the edges of the pad have a much thinner coating than the original plating, whereas the center is thicker. The average thickness is the same, but it is distributed differently. As the plating thickness increases over 500 μin, the surface forces are not always sufficient to hold the solder in the meniscus—especially on wide pads; the solder shifts upon solidification and the deposits appear uneven. On inspection, that may be mistaken for a dewetting condition, but it actually is a shifting of the molten solder before solidification.

When tin-lead with a thickness over 500 μin is reflowed, the boards must be in a horizontal position and be withdrawn in a smooth manner. Otherwise, shifting of the molten solder is inevitable. Because of that problem, there is a practical limit on the coating thickness. In recent years, some military specifications have been calling for 0.001 to 0.0015 in (1000 to 1500 μin) in reflow tin-lead plating. The main objective is to obtain maximum corrosion protection. Deposits of that thickness have been reflowed, but not without processing difficulties.

Improvements in infrared reflow equipment have increased the use of thick fused coatings in recent years. Reflowed tin-lead electrodeposits offer several advantages. One advantage is that they provide a 100 percent quality control check immediately after etching. By examining the boards after reflow, problems in hole drilling, cleaning procedures, plating, and etching can be detected and corrective action can be taken immediately. The procedure has had a dramatic effect in improving printed board reliability.

Another benefit is that solder slivers are eliminated. During etching there is undercutting where the tin-lead coating overhangs the conductor pad. Under certain vibration conditions, the overhang can fall off and cause a short circuit. Fusing eliminates the condition and adds to board reliability.

23.7.2 Thin Fused Coatings

Thin fused coatings level the solder during fusing when they are applied by procedures such as roller coating, spin coating, and hydro-squeegeeing. The leveling is accomplished by means of a hot liquid ejected from spray nozzles. The techniques usually result in coatings under 50 μin, hardly enough for adequate corrosion protection. Also, thin coatings can mask poor solderability when tin-lead is plated over unsolderable copper. For those reasons, use of leveling techniques has diminished.

23.7.3 Problems in Reflowing Plated Coatings

Codeposited impurities, especially copper in tin-lead, can cause dewetting in a reflowed deposit. Variations in alloy composition raise the melting point and cause reflow problems. That is particularly true when organic contamination results in poor solution throwing power and consequent high lead deposit in the plated-through holes. Heavy oxidation or tarnish films that result from chemical attack by etching solutions must be removed prior to reflowing. The films can act as insulating barriers and interfere with reflowing.

23.8 SOLDER RESISTS

Solder resists are polymer materials which are silk-screened or laminated onto the board to protect the circuitry selectively while the pads are being soldered. The solder resists minimize bridging and webbing and restrict the amount of solder dragged from the pot during the soldering operation. A solder resist also minimizes the amount of board area in contact with the solder pot and thereby restricts the metallic contamination buildup in the solder pot.

There are two basic types of solder resists, temporary and permanent. Permanent solder masks are meant to remain on the board, so they are usually made of durable scratch-resistant material such as epoxy. Permanent solder masks can be applied over copper or over solder plate. Resists which are applied over solder plate must be capable of containing the molten solder when it flows during the soldering operation.

Temporary solder masks are applied as protection during the soldering operation. For instance, a temporary mask would be applied to the gold contact fingers during soldering to prevent the fingers from being soldered and thereby ruined (see Chap. 16 for a detailed discussion).

23.9 SOLDERABILITY AND THE PLATING OPERATION

Blowholes in plated-through holes on printed boards normally are caused by solution entrapped in the hole, but they also can be caused by an excess of organics occluded in the electrodeposits. The heat involved in soldering causes moisture and entrapped chemicals in the laminate to build up in pressure and escape through voids or cracks in the plating. That results in blowholes. The problem can usually be alleviated by prebaking at 180 to 200°F.

In some cases blowholes can be caused by an excessive amount of occluded organics in the electrodeposit. The problem is more prevalent with bright plating deposits such as tin and tin-lead. When blowholes are caused by excessive organics occluded in the deposit, prebaking will not remedy the problem. The occluded material can be released only when the metal liquefies.

Figure 23.6 shows a cross section of a fused tin-lead surface that had an excessive amount of organic material. Note that the organic material tends to create

FIG. 23.6 Organic material from plating bath occluded in reflowed tin-lead deposit. (*Alpha Metals, Inc.*)

voids in the plate which rise to the surface. The result is a grainy, pitted appearance after solidification.

29.9.1 Effect of Organic Plating Additives

In order to solder successfully to electrodeposits that employ addition agents, it is extremely important that the additives be carefully controlled. Additives usually are essential to produce sound deposits of tin and tin-lead alloys. In electroplating, chemicals are added to the basic formulation of a plating solution to enhance the properties of the deposit. Some properties that can be improved by additives are throwing power, smoothness, hardness, leveling, brightness, and speed of deposition.

Normally, when foreign metals or organic materials are present in a plating solution, the properties of the deposit are adversely affected. In some rare cases, beneficial effects can be produced by codepositing small amounts of other metallic ions or occluding organic material. When that occurs, the materials are classified not as contaminates but as addition agents. When they produce a bright deposit, they are called brighteners. Strictly speaking, they are controlled impurities.

When organic additives are employed, some forms of the compounds are absorbed at the cathode surface during electrolysis. Often, the compound absorbed is a decomposition product of the original material. The amount that is absorbed is proportional to the nature and concentration of the compound and the time of electrolysis. Frequently, the organic decomposition products develop over a period of time and may affect the deposit adversely. Six months to a year may pass before critical concentrations are reached. If the breakdown products can be removed by an activated carbon treatment, the problem can be controlled. It always is a worthwhile practice to remove organic addition agents and their breakdown products by a carbon treatment at least two or three times a year. That assures continuous operation of the plating solution without unscheduled purification treatments during peak production.

23.9.2 Effect of the Plating Anode

When inorganic contaminants build up in a plating solution, increased concentrations of the additives are normally required to produce the desired effects. Poor-quality anodes constitute the major source of inorganic contaminants. In tin and tin-lead plating, at least a 99.9 percent purity anode is required. High-purity chemicals and efficient rinsing before plating are also essential to maintain a high-purity plating solution. Operation of tin-plating solutions at temperatures below 65°F can reduce the amount of organics occluded in the deposit.

When excess organics are occluded in a tin or tin-lead electrodeposit, bubbling is often observed during soldering. That often can be the cause of blowholes in soldering a printed board. However, it should be noted that the majority of blowholes are caused by entrapped plating or cleaning solutions. When postbaking does not alleviate the problem, occluded organics in the tin or tin-lead electrodeposits are a likely cause.

23.10 SOLDERABILITY CONSIDERATIONS FOR ADDITIVE CIRCUITRY

Additive printed circuitry is a process whereby conductive patterns are formed on a plastic laminate by means of electroless plating. In the process, the copper conductor pads are usually of extremely high purity and, as deposited, are usually very solderable. However, there can be a problem in preserving solderability. Although additive circuitry is not subjected to an etching solution that activates the copper and renders it susceptible to reoxidation, it will oxidize in time and thus can produce solderability problems. The key to soldering additive circuitry is protection of the copper during processing.

In subtractive circuitry, in which the board is laminated with copper and additional copper is deposited on both by electroplating, solderability is usually preserved by the deposition of tin-lead on the copper. The tin-lead serves the dual function of being an etch-resist metal and preserving solderability. The copper is rendered solderable in the cleaning process prior to the deposition of the tin-lead, and the coating usualy provides the necessary corrosion protection for preserving solderability during storage. That usually permits soldering with mildly activated fluxes. To obtain higher reliability, it is often specified that the tin-lead be reflowed. Reflowing is a process in which the coating is heated over the melting point and allowed to solidify in place. If nonsolderable conditions exist, they can be readily observed, because dewetting will occur or the tin-lead will not fuse. Proper fusing will result in a smooth, shiny surface. The process affords a 100 percent inspection immediately after etching and has had a marked effect on improving reliability of solder joints.

In additive circuitry, tin-lead cannot be plated because there is usually isolated circuitry that cannot be electroplated. Other means must therefore be employed to preserve solderability. One is to use rosin- or resin-type protective coats. The process is designed to preserve solderability of copper during in-plant processing, but it was never intended for long-term storage. If tarnishing of the copper occurs during storage, soldering can be a problem because, although the coating is a mild flux, it dilutes the activator content of the soldering flux. If soldering problems arise when this type of coating is present on a printed board, it is usually suggested that the coating be removed and precleaning be performed on the copper surface prior to the soldering process. In wave-soldering it is also important that adequate preheating be employed so that the activator in the soldering flux can be dispensed into the protecto-coat prior to soldering.

Another means of preserving solderability is to employ immersion tin, a chemical process by which a tin coating is applied over copper. The problem is that it is not a true electroless process and there are limitations on the thickness of the coating. Normally, 30 to 50 μin is the practical thickness that can be applied. The solution that deposits the coating acts as a cleaner and renders the copper solderable. However, a tin coating of under 50 μin is normally not sufficient to provide corrosion protection during long-term storage. Also, the tin alloys with the copper, and it can create an unsolderable condition during long-term storage.

There is often a misunderstanding of the electroless process; the process can be employed provided its limitations are thoroughly understood. If the coating could be applied with a thickness of 200 to 300 μin, the process would be an excellent one, but to date the technology has not developed sufficiently to get coatings that thick on a production basis.

A major limitation of the additive process is in preserving solderability. The future of additive circuitry is dependent to a large extent on the development of a process that can deposit a sufficient thickness of tin or tin-lead to provide the corrosion protection necessary for long-term storage. Another approach would be to improve organic protective coatings so they can provide long-term-solderability shelf life.

23.11 USE OF PRECLEANERS TO RESTORE SOLDERABILITY

In the electronics industry, a restriction is often placed on the activity of the flux that can be employed in soldering, because an assumption has been made that postsoldering cleaning may not always be 100 percent effective. If ionic residues are left on a printed board after cleaning, there is a possibility that voltage leaks could develop under high-humidity conditions. Because of that restriction, a situation in which effective soldering cannot be accomplished with the specified flux can arise. In that event, solderability must be restored or the parts must be scrapped.

23.11.1 Causes of Poor Solderability

It is important that the cause of poor solderability be understood. In some cases, oil, grease, or organic films may be responsible. A simple solvent or alkaline cleaning can remedy that situation. The most common cause of poor solderability, however, is heavy tarnish or oxidation on the surface of the metal being soldered. Precleaning in an acid cleaner usually will restore solderability. After acid cleaning, it is essential that the acid residues be thoroughly rinsed off. In critical applications, a neutralization step, followed by another rinse, is employed to ensure that all acid residues are removed. A quick, thorough drying is required after rinsing to prevent reoxidation.

It is also possible that the solderability problem may be a combination of the two situations. That would necessitate a solvent or alkaline cleaning to remove organic films and an acid cleaning to remove tarnish and oxidation. If the dual cleaning operation is impractical, then cleaning with an organic solvent containing acid should be considered. With that type of solution, effective cleaning can be performed in one operation followed by rinsing and thorough drying. That type of cleaning, which requires minimum space and equipment, is ideal when organic films on the surface are not extensive but do prevent 100 percent removal of oxides and tarnish.

A solvent containing acid cleaner is ideal for copper and brass, since it dissolves most organic films that could be on the surface and assures complete removal of tarnish and oxidation. Straight acid cleaners such as hydrochloric, sulfuric, and fluoroboric acids and sodium acid sulfate (sodium bisulfate) are completely effective only when organic films are removed in prior operations. In severe cases of copper oxidation, etching type cleaners such as ammonium persulfate sometimes are used. Although the solutions are very effective, it must be noted that they leave the metal in an active state so that it can easily reoxidize. Hence it is a good practice to follow this procedure with a mild acid dip and thorough rinsing and drying.

23.11.2 Cleaning Tin-Lead Surfaces

Leads and printed boards are often coated with tin or tin-lead to preserve solderability. When that type of coating is applied by electroplating, it is extremely important that the deposit be applied on a solderable surface. Adequate cleaning prior to electroplating is essential. Tarnishing of the tin or tin-lead coating during etching or storage can detract from solderability. Acid cleaners for removing tarnish from tin or tin-lead usually contain thiourea, fluoboric acid, wetting, and complexing agents. If spray equipment having titanium heating coils or rollers is used, then an acid cleaner containing fluoboric acid cannot be used. However, there are available equivalent cleaners that do not contain fluboric acid that will clean tin or tin-lead effectively in a spray operation. It should be noted that if the tin or tin-lead is plated over an unsolderable surface, then the only recourse is to strip the tin or tin-lead and replate. Solutions containing glacial acetic acid and hydrogen peroxide often are employed for the purpose. Stripping and replating can be a costly operation, and economic considerations may rule out the procedure in some cases.

23.12 STANDARD SOLDERABILITY TESTS

Although many solderability tests are used throughout industry, those below are agreed upon by the IPC as effective. If a printed board has a special surface treatment, a special test may be required. In general, however, these will give acceptable and repeatable results.

23.12.1 IPC Solderability Test Methods for Printed Wiring Boards*

1.0 SCOPE

1.1 SCOPE. This standard provides a recommended test method which may be used by both vendor and user to determine solderability of printed wiring boards, with or without surface coatings, which will be soldered.

1.2 PURPOSE. The solderability determination is made to verify that the printed wiring fabrication processes and storage have had no adverse effect on the solderability of the printed wiring board. This is determined by evaluating the ability of those portions of the printed wiring board normally soldered to be wetted by the new coat or solder. Determination is judged by nondestructive methods.

1.3 METHODS. This standard describes four methods by which both the surface conductors and plated-through holes may be evaluated for solderability. These are:

- Edge Dip Test (for surface conductors only)
- Rotary Dip Test (for plated-through holes and surface conductors)
- Wave Solder Test (for plated-through holes and surface conductors)
- Timed Solder Rise Test (for plated-through holes)

*IPC-S-804A, August 1986, Courtesy of IPC, Lincolnwood, Illinois.

Provisions are made for this determination to be performed at the time of manufacture, at the receipt of the boards by the user or just prior to assembly and soldering.

1.4 LIMITATION. This standard shall not be construed as a production procedure for preparing and soldering printed wiring boards.

2.0 APPLICABLE DOCUMENTS

The following documents of issue currently in effect form part of this specification to the extent specified herein.

IEC-68-2-20—Basic Environmental Testing Procedures
ASTM-B-32—Solder
QQ-S-571—Solder; Tin Alloy; Lead Tin Alloy; and Lead Alloy
TT-I-735—Isopropyl Alcohol
LLL-R-626— Rosin
IPC-T-50—Terms and Definitions
IPC-S-815—General Requirements for Soldering of Electrical Connections and Printed Board Assemblies

3.0 REQUIREMENTS

3.1 TERMS AND DEFINITIONS. The definition of terms shall be in accordance with IPC-T-50. (For convenience appropriate definitions from IPC-T-50 are shown in the following paragraphs.)

3.1.1 SOLDERABILITY. The property of metal to be wetted by solder.

3.1.2 WETTING. The formation of a relatively uniform, smooth unbroken and adherent film of solder to a base metal.

3.1.3 DEWETTING. A condition which results when the molten solder has coated the surface tested and then receded leaving irregularly shaped mounds of solder separated by areas covered with a thin solder film; base metal not exposed.

3.1.4 NON-WETTING. A condition whereby molten solder has contacted a surface, but has not adhered to all of the surface; base metal remains exposed.

3.2 MATERIAL.

3.2.1 SOLDER. The solder shall be composition 60B or 63B of ASTM-B32 or Sn60 or Sn63 of Federal Specification QQ-S-571. Other alloys may be used upon agreement between user and vendor.

3.2.2 FLUX. A non-activated rosin flux having a nominal composition of 25% by weight of water white gum rosin (LLL-R-626) in a solvent of isopropyl alcohol (99%) (TT-I-735) shall be used. The specific gravity of the flux shall be 0.843 ± 0.005 at 25°C (77°F). (See safety statement, paragraph 6.3.) A mildly active flux may be used for the wavesolder solderability test only.

3.2.3 FLUX REMOVER. The flux remover shall be either isopropyl alcohol or other suitable solvent. *Caution:* Do not use chlorinated solvents on silicone base materials as delamination and damage to finish may occur. (See safety statement, paragraph 6.3.)

3.3. EQUIPMENT. Equipment that is specific to the solderability test methods shall be as described in paragraph 4.1.1 (for edge dip test), 4.2.1 (for rotary dip test), 4.3.1 for wave solder test, and 4.4.1.1 for timed solder rise test. In addition, the following equipment applies to all methods.

3.3.1. SOLDER POT/BATH. A thermostatically controlled solder pot/bath of adequate dimensions to accommodate the specimens and containing no less than 2.3 kg (5 lbs.) of solder shall be used (see paragraph 3.5).

3.3.2 OPTICAL EQUIPMENT. Inspection is generally by the unaided eye (corrected vision glasses permitted) but on occasion either a direct or projection lens system with a maximum of 10× magnification may be used.

3.4 SAMPLE PREPARATION AND CONDITIONING FOR TEST. Sample preparation and test conditions shall be in accordance with the requirements of paragraphs 3.4.1 through 3.4.4.2.

3.4.1 AFTER MANUFACTURE. The manufacturer may use any of the pretreatments listed in paragraph 3.4.4. Solderability testing will be performed as soon as possible following the manufacture of the board. If pretreatment prior to solderability testing is used, all boards reflected by the test specimen shall receive the same pretreatment. After pretreatment, specimens shall be protected from contamination.

3.4.2 AT INCOMING. The printed boards shall be tested in the "as-received" condition from the vendor. Care shall be exercised to prevent contamination (by grease, perspiration, etc.) of the surface to be tested. Unless otherwise specified there shall be no pretreatment of the surface prior to testing. When pretreatment is agreed to between user and vendor the pretreatment in paragraph 3.4.4 may be used.

3.4.3 BEFORE ASSEMBLY. Printed boards should be tested before assembly without any pretreatment. If solderability criterion is not acceptable, pretreatment per paragraph 3.4.4 may be used, as necessary, to simulate production techniques, or to enhance solderability.

3.4.4 PRETREATMENTS. When agreed upon between user and vendor the specimen to be tested may undergo the following types of pretreatment in accordance with the normal manufacturing process:

Type

A Degreasing
B Aqueous Cleaning
C Copper and Solder Brightening
D Baking (see. 6.1)
E Other (as agreed upon between user and vendor)

3.4.4.1 ROSIN COATING. No rosin coating may be used as a pretreatment. If a rosin coating is used as a protection, it shall be removed using the flux removers indicated in paragraph 3.2.3.

3.4.4.2 BAKING. Multilayer boards shall be baked at 105°C (221°F) for 4 hours minimum prior to testing. Double-sided boards shall be baked as necessary to prevent outgassing. Test specimens shall be cooled to room temperature prior to testing (see paragraph 6.1).

3.5 SOLDER BATH REQUIREMENTS

3.5.1 The solder composition shall be analyzed periodically during use for conformance to Table 23.1. Table 23.1 provides the allowable contaminant levels for solder pots or wave solder equipment and inspection requirements. If the amount of any individual contaminant or the total of contaminants listed exceeds the percentage specified in Table 23.1, the solder shall be changed, or solder shall be added to bring the bath within allowable limits.

3.5.2 Solder oils may be intermixed with the molten solder and carried to the surface of the solder wave or applied to the surface of the solder wave or solder bath.

3.5.3 Solder in solder baths shall be chemically or spectrographically analyzed or renewed at the testing frequency levels shown in Table 23.1, column B. These intervals may be lengthened to the eight-hour operating days shown in column C when the results of analysis provide definite indications that such action will not adversely affect the purity of the solder bath. If contamination exceeds the limits of Table 23.1, intervals between analyses shall be shortened to those eight-hour operating days shown in column A or less until continued purity has been assured by analysis. Records containing the results of all analyses and solder bath usage shall be maintained.

3.6 TEST SELECTION

3.6.1 METHOD 1, EDGE DIP TEST. Surface conductors shall be edge dip tested prior to assembly under the procedure in paragraph 4.1. The conductors shall be considered solderable if they meet the requirements of paragraph 3.7.1.1 through 3.7.1.3.

3.6.2 METHOD 2, ROTARY DIP TEST. The solderability of printed wiring boards with surface conductors and plated-through holes shall be timed-solder-rise tested prior to assembly under the procedure in paragraph 4.2. The hole walls shall be considered

TABLE 23.1 Maximum Limits of Solder-Bath Contaminants

Contaminant[a]	Test soldering — Percent by weight[b] of contaminant	Testing frequency No. of operating shifts[c] [A	B	C][d]	Solder joint characteristic guidelines (if solder is contaminated)[e]
Copper	0.30	15	30	30	Sluggish solder flow, solder hard and brittle, gritty appearance.
Gold	0.20	15	30	30	Solder gritty and brittle, sluggish.
Cadmium	0.005	15	30	60	Porous, sluggish solder flow.
Zinc	0.005	15	30	60	Solder rough and grainy, frosty and porous. High dendritic structure, surface oxide on pot.
Aluminum	0.006	15	30	60	Solder sluggish, frosty and porous, surface oxide on pot.
Antimony[f]	0.50	15	60	120	Solder brittle, sluggish.
Iron	0.02	15	60	120	Compound on surface presents resoldering problems; gritty appearance.
Arsenic	0.03	15	60	120	Small blister-like spots, sluggish.
Bismuth	0.25	15	60	120	Lowering of melting point & surface oxidation.
Silver[g]	0.10	15	60	120	Dull appearance. May reduce fluidity.
Nickel	0.01	15	60	120	Blisters, formation of hard insoluble compounds.

[a]The tin content of the solder shall be maintained with ±1 percentage point of the nominal tin content of the alloy being used. Tin content shall be tested at the same frequency as testing for copper/gold contamination. The balance of the bath shall be lead and/or the items listed above.

[b]The total of copper, gold, cadmium, zinc, and aluminum contaminants shall not exceed 0.4% for assembly soldering.

[c]An operating shift constitutes any 8-hour period, or any portion thereof, during which the solder is liquified and used. The testing frequency may be extended as agreed to between user and vendor to compensate for solder bath additions.

[d]See 3.5.3 for application of these columns.

[e]Presented to assist in monitoring the solder operation and may be used to indicate increased frequency of testing. The given characteristics appear first at higher concentrations than given in this table and depend on several factors of which an important one is cooling.

[f]Minimum antimony requirements per QQ-S-571 are 0.20.

[g]Not applicable for Sn62 solder—limits to be 1.75–2.25 (both operations).

solderable if they meet the requirements in paragraphs 3.7.1.1, 3.7.1.3 and 3.7.2.1 through 3.7.2.6.

3.6.3 METHOD 3, WAVE SOLDER TEST. The solderability of surface conductors and holes on printed wiring boards may also be evaluated by wave soldering under the procedure in paragraph 4.3. The boards shall be considered solderable if they meet the requirements in paragraphs 3.7.1.1, 3.7.1.3 and 3.7.2.1 through 3.7.2.6.

3.6.4 METHOD 4, TIMED SOLDER RISE TEST. The solderability of plated-through holes in printed wiring boards shall be timed solder rise tested prior to assembly under the procedure in paragraph 4.4. The hole walls shall be considered solderable if they meet the requirements in paragraphs 3.7.1.1, 3.7.1.3 and 3.7.2.7.

3.7 EVALUATION

3.7.1 SURFACE CONDUCTOR SOLDERABILITY.

3.7.1.1 CRITERIA FOR ACCEPTABLE SOLDERABILITY. A minimum of 95% of the surface being tested shall exhibit good wetting. The balance of the surface may contain only small pin holes, dewetted areas, and rough spots provided such defects are not concentrated in one area. For less critical applications, a smaller percent coverage may be determined between vendor and user. There shall be no nonwetting or exposed base metal within the evaluated area.

3.7.1.2 An area of $\frac{1}{8}$ inch width from the bottom edge of each test specimen shall not be evaluated. Areas contacted by fixtures should not be evaluated.

3.7.1.3 EVALUATION AID. As an aid to evaluation of the test results, see Fig. 23.7. This aid is to be used primarily to illustrate types of defects rather than percentage of area covered.

3.7.2 PLATED-THROUGH HOLE SOLDERABILITY

3.7.2.1 INCOMING ACCEPTANCE. Solderability acceptance for plated-through holes in thick printed wiring boards (high aspect ratio board to hole) shall be agreed to between user and vendor.

3.7.2.2 Only plated holes that are at least 5 mm (0.20 in.) from any surface or fixturing structure supporting the specimen during the test will be evaluated.

3.7.2.3 Visual inspection with the unaided eye, corrected to 20/20, in a well-lighted room is adequate for inspection purposes. To aid the evaluation of borderline cases or when supplier and user agree that more critical viewing conditions are appropriate, the illumination and observation set up described in paragraph 3.3.2 is recommended.

3.7.2.4 The specimen has soldered successfully if solder has risen in all the plated holes. The solder must have wet out onto the land around the top of the hole (see Fig. 23.8). The solder need not have wet 100% of the land to be acceptable..

3.7.2.5 The solder shall have fully wetted the walls of the hole. There shall be no non-wetting or exposed base metal on any plated-through hole.

3.7.2.5.1 Partially filled holes are acceptable under the following condition: the solder in partially filled holes must exhibit a contact angle less than 90° relative to the hole wall (see Fig. 23.8, D and E).

3.7.2.5.2 All holes less than 0.060″ shall retain a full solder plug after solidification. Holes greater than 0.060″ shall not be rejected for failure to retain a full solder plug provided that the entire barrel of the hole and the surface of the top land have been wetted with solder.

3.7.2.6 Profile views of acceptable and unacceptable conditions are illustrated in Fig. 23.8.

3.7.2.7 METHOD 4. The criteria for acceptable solderability of plated-through holes for Method 4 shall be that all holes in the test sample shall fill within 5 seconds from the time the sample was placed on the bath of molten solder. There also shall be no evidence of gas evolution within this time, or after removal from the solder bath. After removal from the solder bath, all holes in the board shall have wetted completely both the barrel and 95% of the lands and the knee. There shall be no blow holes or voids in the solder-filled hole (see 6.2).

4.0 TEST METHOD PROCEDURES

4.1 METHOD 1, EDGE DIP TEST

4.1.1 EQUIPMENT

4.1.1.1 VERTICAL DIPPING DEVICE A dipping device as shown in Fig. 23.3 shall be used. A similar device may be used provided the following requirements are met:

4.1.1.2 Rate of immersion, dwell time and rate of withdrawal are within the test limits (paragraph 4.1.4).

4.1.1.3 Perpendicularity of board and solder surface are maintained.

STABILITY TEST SAMPLES

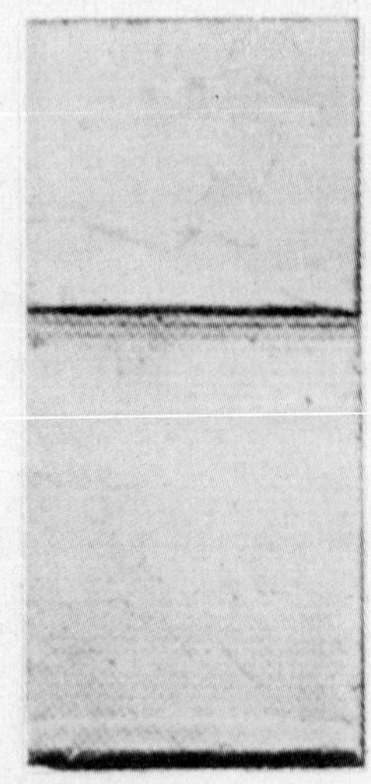
PREFERRED WETTING

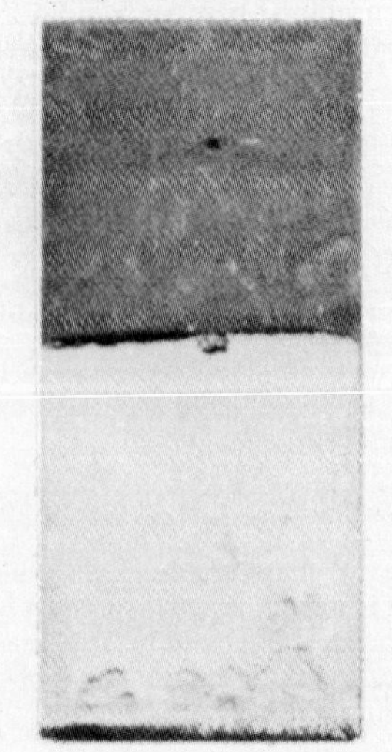
SMALL AMOUNT OF DEWETTING

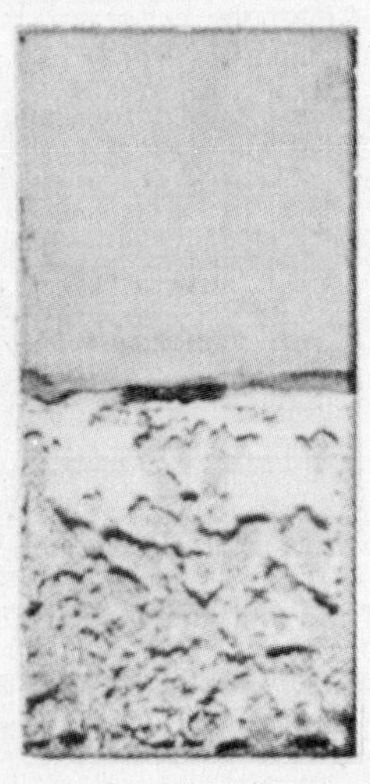
COMPLETE DEWETTING

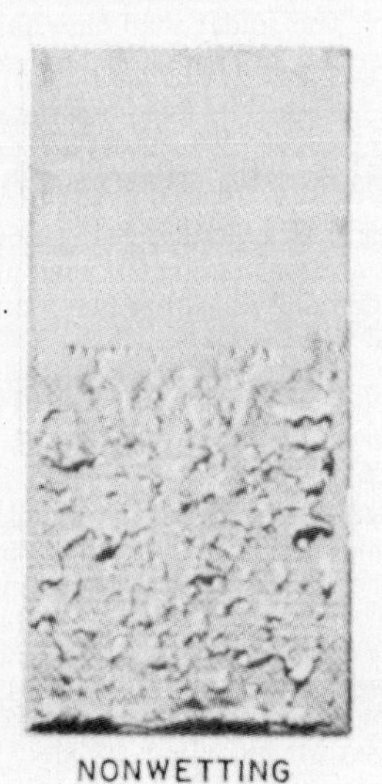
NONWETTING

FIG. 23.7 Aid to evaluation of circuit wetting. (*IPC-S-804A, August 1986.*)

Acceptable condition

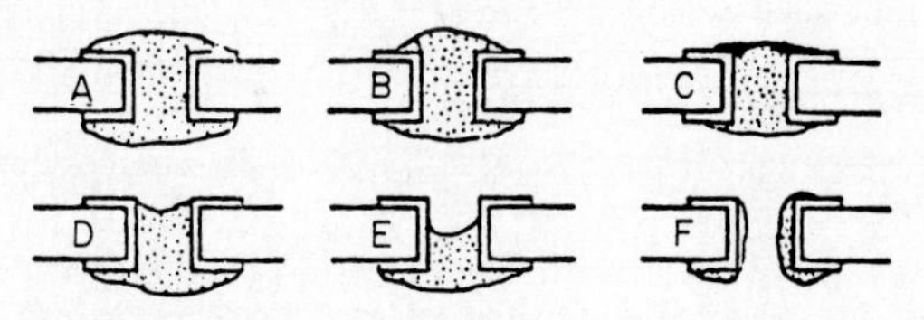

Not acceptable condition (solder has not wetted hole wall surfaces)

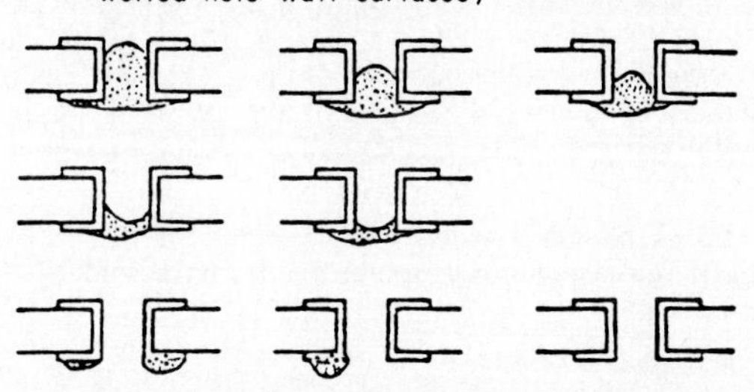

Note: Plated hole walls sometimes cause the formation of blowholes by the evolution of gaseous products during the heating cycle of soldering. Such evolution is often visible during soldering and also after soldering as voids left in the solidified metal. This outgassing of material is a separate issue from the ability of the hole wall to be wet by solder. It is the responsibility of the supplier and user to agree on the acceptability of this outgassing condition. See paragraph 6.1 (Prebake).

FIG. 23.8 Effectiveness of solder wetting of plated-through holes.

4.1.1.4 Wobble, vibration and other extraneous movements are elimianted.

4.1.2 TEST SPECIMEN REQUIREMENTS

4.1.2.1 The test specimen shall be a representative portion of the printed wiring board being tested such that an immersion depth of 25 cm (1 inch) is possible.

4.1.2.2 The test specimen shall be representative of the lot being tested. When this test is to be used as a criterion for material acceptance, the number of test specimens shall be defined by agreement between the user and vendor.

4.1.3 APPLICATION OF FLUX. The test specimens are to be dipped to the full depth to be soldered for 5–10 seconds in the flux of paragraph 3.2.2. After withdrawal from the flux, the specimen shall be allowed to vertically drain for 60 seconds. At the end of this time, the edge shall be blotted. The solder dip test shall then be performed not less than one minute and not more than five minutes after blotting.

4.1.4 SOLDER APPLICATION. The molten solder shall be initially stirred with a clean stainless steel paddle to assure that the solder is of a uniform composition and at a temperature of 245°C ±5° (474°F ±8°). Dross and burned flux shall be completely removed from the surface of the molten solder immediately prior to dipping. After fluxing and draining, the specimen shall be immersed into the molten solder edgewise to a depth of 25 ±6 mm (1.0 ±0.25 inches). The insertion and withdrawal rates shall be 25 ±6 mm (1.0 ±0.25 inches) per second. The dwell time in the molten solder shall be 3.0 ±0.5 seconds. After withdrawal, the solder shall be allowed to solidify by air cooling while the board is maintained in a vertical position.

4.2 METHOD 2, ROTARY DIPPING TEST

4.2.1 EQUIPMENT. A device to move the test specimen in a vertical circular path so that the flat surface of the specimen will contact the solder at a constant speed without stopping shall be used. The distance between the center of rotation and the center of the test specimen shall be 100 ±5 mm (4 ±0.20 in.). The specimen holder shall be of any design provided that it holds the specimen as described above and satisfies the following requirements (see Fig. 23.9).

4.2.1.1 The exposed length of specimen test face in the direction of travel shall be 25 ±1 mm (1 ±0.04 in.).

4.2.1.2 Those parts of the holder including the retaining spring (if fitted) which come into contact with the specimen and/or the solder should have low thermal capacity and conductivity.

4.2.1.3 The time of contact between any point of the test face of the specimen and the molten solder shall be determined by a timer activated by the electrical contact of the needle with the molten solder. The tip of the needle shall be located adjacent to the specimen and it shall be on the same axis and radius of rotation as the center of the test face of the specimen. The needle shall be kept clean. It shall be electrically insulated from the specimen holder which carries it.

4.2.1.4 A strip of 50 mm (2 in.) wide PTFE shall precede the test specimen in the test cycle in order to remove oxide or flux residue from the solder surface before the specimen is introduced.

4.2.2 TEST SPECIMEN REQUIREMENTS

4.2.2.1 The test specimen shall be rectangular, 25–31 mm (0.98–1.20 in.) on each side. Smaller specimens may be used if test results correlate to 25–31 mm (0.98–1.20 in.) specimens.

4.2.2.2 The minimum number of holes to be tested is 30; if there are not at least 30 holes in the test specimen, additional specimens shall be tested until at least 30 holes have been tested. The lot size applicable to a given test, the number of specimens to be tested and the instructions for selecting the test specimens shall be agreed to by user and vendor.

4.2.3 TEST PARAMETERS

4.2.3.1 The solder bath temperature shall be maintained at 245°C ±5°C (474°F ± 8°F). Other temperatures may be used when agreed to by user and vendor.

4.2.3.2 Adjust the test equipment to immerse the specimen 0.75–1.0 mm (0.03–0.04

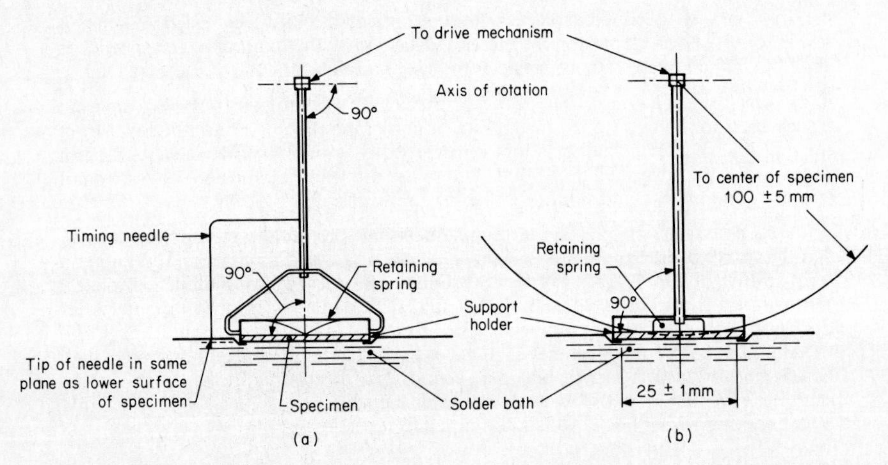

FIG. 23.9 Sketch of specimen holder and timing needle for performing the rotary dip test (time solder rise test). (*a*) Front view; (*b*) side view. (*From IEC 68-2-20, Basic Environmental Testing Procedure, Part 2: Tests—Test and Soldering.*)

inches) into the solder unless otherwise specified. Care must be taken to adjust the equipment so that solder does not flow over the upper face of the specimen.

4.2.3.3 Dwell time at the maximum depth shall be 3.0 ± 0.5 seconds.

4.2.3.4 Care shall be taken when handling the specimen to keep oxidation and contamination of the surfaces to be tested to a minimum.

4.2.4 APPLICATION OF FLUX. Immerse the test specimen in flux, taking care that no air bubble prevents the flux from wetting the total surface of the plated hole wall (slow immersion with the specimen inclined from the vertical may aid in eliminating air bubbles). Allow the specimen to drain in a nearly vertical position for 10–15 seconds to clear the holes of excess flux, then blot the lower edge of the specimen horizontally on fresh absorbent paper (not the surface to be soldered). The specimen should be soldered within 1–5 minutes after removal from the flux.

4.2.5 SOLDER APPLICATION

4.2.5.1 Mount the specimen to be tested in the test equipment specimen holder.

4.2.5.2 Activate the test equipment to expose the specimen to solder.

4.2.5.3 After the specimen has cleared the solder path, allow all the solder to solidify in the position which the machine stops before removing from the specimen holder.

4.3 METHOD 3, WAVE SOLDER TEST

4.3.1 EQUIPMENT. A commercially available wave soldering system adjusted to provide the parameters agreed to between printed wiring board manufacturer and user.

4.3.2 TEST SPECIMEN REQUIREMENTS

4.3.2.1 The test specimen shall be a production printed board or a representative of the lot being tested. When this test is to be used as a criterion for material acceptance, the number of test specimens shall be defined by agreement between the user and vendor.

4.3.3 TEST PARAMETERS

4.3.3.1 Specimens shall be fixtured, as much as is practical, so as to be representative of the production setup, without components inserted.

4.3.3.2 The application of solder shall meet the requirements of the applicable wave solder equipment specifically for depth of contact, angle of contact, and duration of contact. Solder temperature shall be 245°C ± 5° unless another temperature is agreed to by vendor and user.

4.3.3.3 The solder composition shall be analyzed periodically during use for conformance to paragraph 3.5.3.

4.3.4 APPLICATION OF FLUX. The test specimens are to be flux coated using flux meeting the requirements of paragraph 3.2.2 before proceeding with the solder application.

4.3.5 SOLDER APPLICATION

4.3.5.1 Wave solder the specimen according to the process agreed to between user and vendor. The following parameters must be established and noted: board fixturing (if required), conveyor, flux applicator, preheater, solder unit with or without oil intermix, machine process controls, incline, board preheat temperature, solder temperature.

4.3.5.2 CLEANING. After the solder has solidified and cooled flux shall be removed from the specimens to facilitate inspection. Flux removing material shall be as specified in paragraph 3.2.3. Flux removal may be done manually or by machine.

4.3.5.3 The time between soldering and cleaning shall not exceed one hour.

4.4 METHOD 4, TIMED SOLDER RISE TEST

4.4.1 EQUIPMENT

4.4.1.1 SOLDER POT. The solder pot shall meet the requirements of 3.3.1. In addition, the surface area of the pot shall be great enough to float the test specimen without it touching the sides of the pot.

4.4.1.2 SPECIMEN HANDLING TOOL. Stainless steel forceps, or other specially designed tools of stainless steel, shall be used to handle the specimen only by the edges.

4.4.1.3 TEST SPECIMEN. The test specimen shall be a portion of the printed board not greater than 50 × 50 mm (2 × 2 inches) or the complete board if it is smaller than this size. The specimen shall contain a representative sampling of all hole sizes and configuration as is possible.

4.4.2 PROCEDURE

4.4.2.1 FLUXING. The test specimen shall be immersed in the flux. After removal from the flux, it shall be drained by standing on edge for 1 minute.

4.4.2.2 SOLDER FLOAT. Slide the fluxed test specimen onto the surface of the molten solder at 245°C ±5°C (474°F ±8°F) after it has been skimmed to remove dross. Allow one second to float the test specimen. Begin timing as soon as the test specimen is fully floating on the solder bath. Stop timing as soon as all holes fill and wet out on the lands, or five seconds have elapsed. To remove the test specimen from the bath, slide from the solder bath surface. Allow the solder to solidify while holding the specimen still and in the horizontal plane above the pot surface. Remove all flux and flux residues per 3.2.3.
This is a test method, and the solder float and wetting time achieved shall not be construed as having a relationship with actual soldering times. Heavy internal grounds planes in multilayer boards may prevent full solder rise in some holes. The board design shall be taken into consideration during testing.

4.4.3 EXAMINATION. Examine all holes and top pads for fill and solder wetting.

5.0 PREPARATION FOR DELIVERY

This section is not applicable to this standard.

6.0 NOTES.

6.1 PREBAKING. The occurrence of outgassing, which may result in blowholes, measling, blisters or delamination, may be reduced by baking the printed board prior to soldering to eliminate moisture or solvents. Other factors, such as conveyor speed (for wave solder testing), solder temperature, contamination content, etc., may also be involved in producing defects and, therefore, should be analyzed if problems occur. Specimens should be baked in a suitable oven to remove any absorbed moisture. Time between bake and testing should not exceed 5 days depending on atmosphere humidity levels. Temperature and time of baking is to be determined on an individual basis.

Printed boards should not be prebaked if prebaking is not normally used as a production procedure. Baking should be kept to a minimum but at least equal to the production procedure to prevent excessive oxidation and intermetallic growth.

6.2 Heavy boards or multilayer boards with a heavy internal ground plane may require greater than 5 seconds for hole filling. If such a condition exists then the hole filling time should be agreed on between vendor and user.

6.3 SAFETY NOTE. Isopropyl alcohol used in paragraphs 3.2.2 and 3.2.3 is flammable. Care must be taken in both usage and storage to keep the isopropyl alcohol from sparks or flames. See MSDS for all solvents.

6.4 TEST EQUIPMENT SOURCES. The equipment sources described below represent those currently known to the industry. Users of this document are urged to submit additional source names as they become available, so that this list can be kept as current as possible.

6.4.1 EDGE DIP SOLDERABILITY TEST APPARATUS

Williams Machine, 2092 W. Main St., Norristown, PA 19403, (215) 539–1123

Multicore Solders, Cantiague Rock Road, Westbury, NY 11590, (516) 334–7997

Hybrid Machine Products, 1014–19 Morse Ave., Sunnyvale, CA 94086, (408) 734–4390

6.4.2 ROTARY DIP TEST APPARATUS

Pro-Tech, 11801–1 28th Street, St. Petersburg, FL 33702, (813) 577–5500

CHAPTER 24*

SOLDER MATERIALS AND PROCESSES

Hugh Cole

24.1 INTRODUCTION

The primary purpose of this chapter is to discuss the materials and machinery used in the soldering process. The emphasis is placed on automated soldering techniques, but most of the information is germane to other techniques. Because of the sequential nature of the soldering process and the need for a high degree of control of the processing parameters, soldering lends itself well to automated techniques. The large throughput required for most modern high-volume production installations makes automated soldering techniques mandatory.

24.2 THE SOLDER ALLOYS

Solder alloys are normally composed of tin and lead in ratios close to the eutectic point. The eutectic composition is a distinct alloy composition that has a single melting point, rather than a discrete melting range. The melting point of the eutectic is always less than that of parent metals. In the tin-lead phase diagram shown in Fig. 24.1, the eutectic point is marked as 63 percent tin and 37 percent lead. As shown in the figure, compositions different from the eutectic do not solidify immediately on cooling; instead, they have a temperature range in which they are partly liquid and partly solid. That region, called the "pasty range," becomes more pronounced the farther one departs from the eutectic. Large pasty ranges result in internal stresses and cold solder joints. Therefore, compositions at or near the eutectic are the most desirable for automated soldering.

Alloys other than tin-lead are available with melting temperatures covering the range from 200 to 600°F, but they are not often used in automated soldering processes. Table 24.1, from Federal Specification QQ-S-571, gives a rough idea of the number and types of alloys available. Solder alloy selection is made on the basis of soldering temperature (not melting temperature), mechanical strength, electrical characteristics, mechanical properties, and economy. Because tin is much

*Adapted from Coombs, *Printed Circuits Handbook,* 2d ed., McGraw-Hill, New York, 1979, Chap. 15.

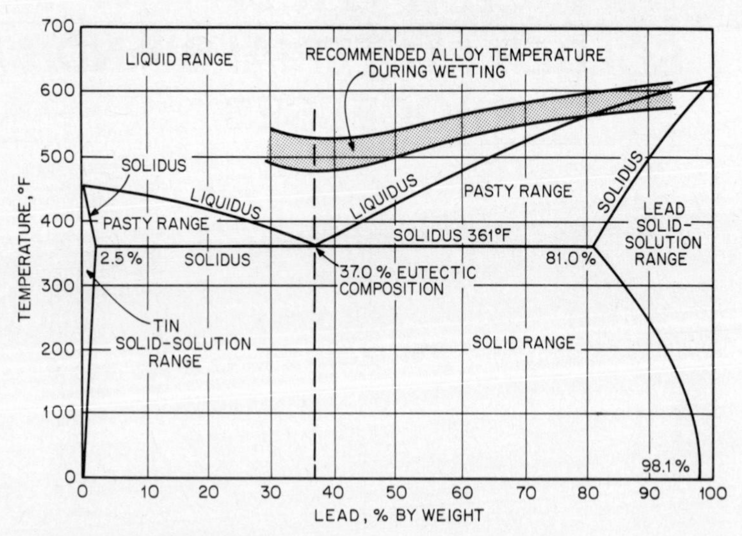

FIG 24.1 Binary phase diagram, tin-lead alloys. Note that the eutectic point occurs at 63 percent tin and 37 percent lead.

higher in cost than lead, solder alloy cost is basically a function of the amount of tin in the alloy. The higher the tin content, the higher the cost. The highest tin content used is the eutectic composition, 63 percent Sn and 37 percent Pb. It provides the optimum soldering properties.

24.3 PURITY OF SOLDER ALLOYS

The purity of solder, as it is received from the manufacturer, is subject to much confusion, which volumes of advertising literature, claims, and counterclaims of solder manufacturers have done little to dispel. The confusion exists basically because of a tangle of antiquated solder specifications and a lack of appreciation for the complexity of solder manufacture. Federal Specification QQ-S-571 and ASTM B-32, the most often quoted specifications, were written during World War II. Since tin supplies were greatly restricted and the refining and recycling techniques we have today were unavailable, solder impurity levels greater than those currently acceptable were tolerated for expediency's sake. Also at that time, modern high-speed production techniques were not available, and hand-soldered assemblies were most common. Contoured solder joints were rejected because they "didn't have enough solder on them to make a good joint," and the correlation between reliability and inspectibility wasn't fully recognized. In that environment, the type of difficulties caused by solder impurities simply were not recognized as problems.

Two basic types of impurities are of concern in tin-lead solder alloys; they are the metallic impurities, such as copper, gold, aluminum, and cadmium, and the nonmetallic impurities, such as the sulfides and oxides formed by the reaction of the solder alloy (tin-lead and the metallic impurities) with sulfur and oxygen. Impurities of both types are deleterious to the properties of the solder joint. Figure 24.2 shows two solder alloys. One was made in such a way as to minimize the metallic impurities; the other was further processed to remove the nonmetallics.

TABLE 24.1 Solder-Alloy Compositions

Composition	Tin, %	Lead, %	Antimony, %	Bismuth, max, %	Silver, %	Copper, max, %	Iron, max, %	Zinc, max, %	Aluminum, max, %	Arsenic, max, %	Cadmium, max, %	Total of all others, max, %	Approximate melting range, °C*	
													Solidus	Liquidus
Sn96	Remainder	0.10, max	—	—	3.6 to 4.4	0.20	—	0.005	—	0.05	0.005	—	221	221
Sn70	69.5 to 71.5	Remainder	0.20 to 0.50	0.25	—	0.08	0.02	0.005	0.005	0.03	—	0.08	183	193
Sn63	62.5 to 63.5	Remainder	0.20 to 0.50	0.25	—	0.08	0.02	0.005	0.005	0.03	—	0.08	183	183
Sn62	61.5 to 62.5	Remainder	0.20 to 0.50	0.25	1.75 to 2.25	0.08	0.02	0.005	0.005	0.03	—	0.08	179	179
Sn60	59.5 to 61.5	Remainder	0.20 to 0.50	0.25	—	0.08	0.02	0.005	0.005	0.03	—	0.08	183	191
Sn50	49.5 to 51.5	Remainder	0.20 to 0.50	0.25	—	0.08	0.02	0.005	0.005	0.025	—	0.08	183	216
Sn40	39.5 to 41.5	Remainder	0.20 to 0.50	0.25	—	0.08	0.02	0.005	0.005	0.02	—	0.08	183	238
Sn35	34.5 to 36.5	Remainder	1.6 to 2.0	0.25	—	0.08	0.02	0.005	0.005	0.02	—	0.08	185	243
Sn30	29.5 to 31.5	Remainder	1.4 to 1.8	0.25	—	0.08	0.02	0.005	0.005	0.02	—	0.08	185	250
Sn20	19.5 to 21.5	Remainder	0.80 to 1.2	0.25	—	0.08	0.02	0.005	0.005	0.02	—	0.08	184	270
Sn10	9.0 to 11.0	Remainder	0.20, max	0.03	1.7 to 2.4	0.08	—	0.005	0.005	0.02	—	0.10	268	290

TABLE 24.1 Solder-Alloy Compositions (*Continued*)

Composition	Tin, %	Lead, %	Antimony, %	Bismuth, max, %	Silver, %	Copper, max, %	Iron, max, %	Zinc, max, %	Aluminum, max, %	Arsenic, max, %	Cadmium, max, %	Total of all others, max, %	Approximate melting range, °C*	
													Solidus	Liquidus
Sn5	4.5 to 5.5	Remainder	0.50, max	0.25	—	0.08	0.02	0.005	0.005	0.02	—	0.08	308	312
Sb5	94.0, min	0.20, max	4.0 to 6.0	—	—	0.08	0.08	0.03	0.03	0.05	0.03	0.03	235	240
Pb80	Remainder	78.5 to 80.5	0.20 to 0.50	0.25	—	0.08	0.02	0.005	0.005	0.02	—	0.08	183	277
Pb70	Remainder	68.5 to 70.5	0.20 to 0.50	0.25	—	0.08	0.02	0.005	0.005	0.02	—	0.08	183	254
Pb65	Remainder	63.5 to 65.5	0.20 to 0.50	0.25	—	0.08	0.02	0.005	0.005	0.02	—	0.08	183	246
Ag1.5	0.75 to 1.25	Remainder	0.40, max	0.25	1.3 to 1.7	0.30	0.02	0.005	0.005	0.02	—	0.08	309	309
Ag2.5	0.25, max	Remainder	0.40, max	0.25	2.3 to 2.7	0.30	0.02	0.005	0.005	0.02	—	0.03	304	304
Ag5.5	0.25, max	Remainder	0.40, max	0.25	5.0 to 6.0	0.30	0.02	0.005	0.005	0.02	—	0.03	304	380

*For information only.

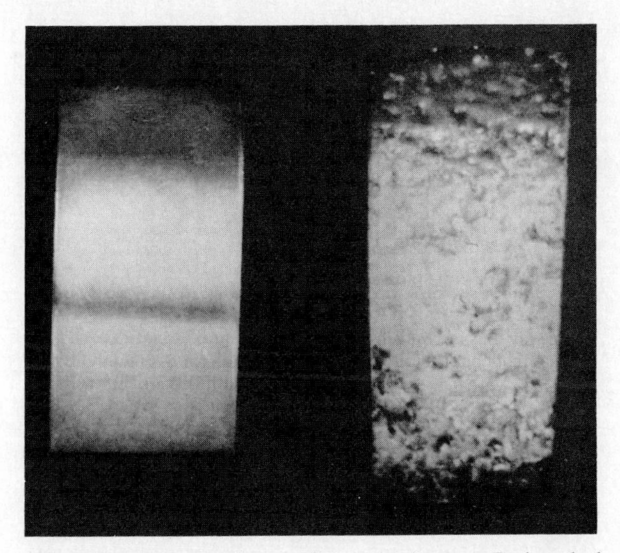

FIG. 24.2 Solder specimens after outgassing test. Pockmarked appearance is due to outgassing of non-metallic impurities. (*Alpha Metals, Inc.*)

Both specimens were then melted under vacuum. The gas pockets and blowholes in the specimen on the left result from the outgassing of the occluded nonmetallic impurities. Because the nonmetallic impurities affect the flow characteristics of the molten solder, in addition to the metallurgical properties of the joint, more joints can be made faster, and with less solder, if the nonmetallics are removed. Care should be taken, however, during the operation of the soldering process to minimize dross formation. Otherwise, the resultant buildup of impurities will negate the effects of initial impurity removal.

24.3.1 Metallic Impurities

Metallic impurities also affect the performance of solder. Common contaminants and their effects are discussed in the following subsections.

24.3.1.1 ***Copper.*** Copper forms two intermetallic compounds with tin: Cu_3Sn and Cu_6Sn_5. The compounds weaken the solder joint and cause the solder to become sluggish and gritty. Since copper is a very common material in the electronics industry, care must be taken to minimize contact of the copper with the molten solder.

Because the copper-tin intermetallics freeze at about 10 to 20°F above the melting point of the solder alloy, they can be removed by dropping the temperature to that level and scooping out the resultant copper-tin crystals with a special tool. The procedure will not remove all the copper, and it necessitates the readjustment of the alloy composition with tin-rich solder. It should be used only as a stopgap if the pot cannot be changed immediately.

24.3.1.2 ***Gold.*** Gold is readily soluble in molten solder. Although gold was once considered a highly solderable surface, it is now recognized that it rapidly embrittles the solder joint and forms a dull grainy fillet. Gold and copper can act synergistically to deteriorate the performance of a solder pot rapidly. In addition, the value of gold makes it well worthwhile to monitor the level of accumulation in the solder pot.

24.3.1.3 Iron. Iron impurities also can be detrimental. Iron forms two intermetallic compounds with tin: FeSn and $FeSn_2$. The formation of those compounds is more rapid at high temperatures such as 800°F. Iron surfaces in contact with the solder pot should be blued or heavily oxidized to prevent the dissolution of iron in the solder.

24.3.1.4 Zinc. Zinc is one of the most detrimental of solder contaminants. As little as 0.005 percent zinc will cause grittiness, lack of adhesion, and eventual failure of the joint. Other materials considered detrimental to the solder alloy are aluminum, magnesium, and cadmium. Table 24.2 lists the common contaminants and the levels at which they begin to affect the soldering operation.

Both types of impurities, metallic and nonmetallic, can be found in raw materials from primary sources such as tin mines and reclaimed sources such as solder scrap. With the chemical refining techniques available today, there is virtually no difference in purity between primary or virgin grade material and properly refined reclaimed material. Therefore, both constitute acceptable raw material sources for solder alloys. No matter which source is used, however, the solder alloy should be processed to remove both metallic and nonmetallic impurities.

24.4 SOLDER FLUX

The first step in most automated soldering operations is the fluxer or flux pot; but before the mechanics of the fluxing operation are described, it is essential to understand the requirements of the solder flux and the way in which various types of fluxes fill those requirements.

To form a reliable solder joint, both the board surface and the component lead must be free of oxidation and must remain so even at the elevated temperatures used in soldering. Also, the molten solder alloy must wet the surfaces of the metals

TABLE 24.2 Solder Contamination

Contaminant	QQ-S-571-E, %*	New solder, %†	Contamination limits, %‡
Aluminum	0.005	0.003	0.006
Antimony	0.2–0.5	0.3	—
Arsenic	0.03	0.02	0.03
Bismuth	0.25	0.006	0.25
Cadmium	—	0.001	0.005
Copper§	0.08	0.010	0.25
Gold§	—	0.001	0.08
Iron	0.02	0.001	0.02
Silver	—	0.002	0.01
Zinc	0.005	0.001	0.005
Others	0.08	0.01	0.08

*Limits established by federal specification QQ-S-571-E for acceptable contaminant levels for various metals.

†Levels of contaminants usually found in new solder delivered from the factory and before any use.

‡Contamination levels which indicate the solder must be replaced for any use.

§Copper and gold combined not more than 0.300.

to be joined. Without the solder flux, neither condition can be met reliably or economically. The requirements for a solder flux can be summarized as follows: The flux must react with and remove metal oxides from the surfaces to be joined and prevent the reoxidation of the cleaned surface even at elevated temperatures; it must promote wetting of surfaces to be joined by the molten solder alloy; it must act as a heat-transfer medium to ensure that the parts to be joined reach a temperature high enough to form a metallurgical bond; and finally, it must be either noncorrosive or easily removable.

24.4.1 Basic Categories of Flux

Solder fluxes can be divided into two basic categories; rosin-based and water-soluble. Rosin-based flux has three components: the solvent or vehicle, the rosin, and the activator. (For very mild fluxes, the activator is sometimes omitted.) The function of the solvent is simply to act as a vehicle for the flux. To be effective, the rosin and activator must be applied to the board in the liquid state. That is accomplished by dissolving them in the flux solvent. Solvent vehicles can be simple one-component systems such as isopropyl alcohol, or they can be complex, multicomponent systems such as those used in foaming fluxes. When the flux vehicle is a multicomponent system, maintaining it properly becomes difficult. Because different components evaporate from the flux at different rates, they must be replaced with flux thinners of carefully controlled composition. Most manufacturers supply thinners blended to maintain a constant composition of the flux vehicle. Maintenance of the flux and thinner ratio will be discussed in another section of this chapter.

24.4.2 Rosin Flux

Rosin is a naturally occurring substance extracted from pine trees. It is a mixture of many organic acids, and its purity and composition are difficult to control. The principle constituents, however, are abietic acid and pimeric acid, the structures of which are shown in Fig. 24.3. After rosin is extracted from the pine trees, it is given a superficial processing which removes undesirable impurities and neutralizes the acid residues remaining from the extraction operation. The purified material, called water-white rosin, is then used in the manufacture of rosin-based flux. Some manufacturers, in the hope of overcoming the difficulties associated with obtaining and processing natural rosin, have chemically synthesized substitute materials. Such materials are called "resins."

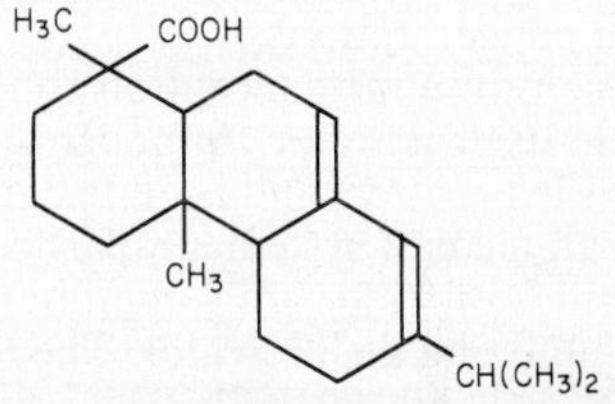

Rosin by itself is a very weak flux; its ability to remove surface oxides is so limited that only exceptionally clean surfaces of high solderability can be joined with it alone. When surfaces require a more active flux, chemical compounds called "activators" are added to the rosin. Activators are thermally reactive compounds, such as the amine hydrochlorides, which break down at elevated temperatures and release hydrochloric acid to dissolve the surface oxides and tarnishes. The precise nature of the activators used in a given manufacturer's flux is

FIG. 24.3 Typical organic acid (abietic acid) found in natural rosin, which is a mixture of many such acids in different concentrations.

normally a carefully guarded trade secret, but review of the literature on flux activators usually provides more than sufficient familiarity with the compounds available.

24.4.2.1 Design of a Flux. The design of a rosin-based flux is a careful balance of the ratio of rosin to activator. If there is too much activator, the flux will be too corrosive and difficult to remove; if there is too little, the fluxing action will be impaired. The amount of the rosin and activator mixture in the solvent vehicle, called the "solids content" or "density," is tailored for the specific application. The higher the solids content of the flux, the more rosin-activator mixture is deposited on the board and the more difficult the board will be to clean. To obtain adequate coverage, high-solids-content fluxes are used for boards with dense circuitry and small plated-through holes. Single-sided boards with less-dense circuitry can use fluxes with lower solids content. That is an important point: The rosin-to-activator ratio controls the activity of the flux, *not* the solids content. Solids content affects only the coverage. Thus a high-solids-content flux with a low ratio of activator to rosin may be the wrong choice for a surface with poor solderability. A more appropriate selection may be a lower-solids-content flux with a much higher activator-to-solids ratio.

24.4.2.2 Activity Levels. The activity levels of rosin-based fluxes have been divided into three categories: rosin (R), rosin mildly activated (RMA), and rosin-activated (RA). R-type fluxes contain no activators. RMA fluxes contain small quantities of activators that for all practical purposes are completely inert and noncorrosive after soldering. For that reason, most government contracts and some commercial supply contracts require the use of RMA fluxes. RA fluxes contain the greatest quantities of activator and hence are the most active.

24.4.3 Water-Soluble Fluxes

Water-soluble fluxes also are used to solder printed circuit assemblies. Water-soluble fluxes are much more active than rosin-based fluxes, and they will therefore rapidly and efficiently solder material of even very poor solderability. However, because they are so active, they are extremely corrosive and must be thoroughly and rapidly removed. Water-soluble fluxes are mixtures of compounds like glutamic acid hydrochloride and aniline hydrochloride in a solvent vehicle. In some cases, activators quite similar to those in rosin-based fluxes are used, but in much higher concentrations.

A common misconception concerning water-soluble fluxes is that the solvent vehicle itself is water. Although water can be used as the vehicle, it results in a flux which spatters and is quite difficult to use. More commonly, the solvent vehicle is an alcohol system similar to the systems used for rosin-based fluxes. Small quantities of water are sometimes added to multiple-component systems to aid in solubilizing the activator. The term "water-soluble flux" refers to the fact that the residues are generally water-soluble. Although that is true for very long rinse times and elevated rinse temperatures, it is better to neutralize the residues with ammonia water or a similar neutralizing agent. Neutralizing the residues greatly increases their solubility in water. That will be discussed in detail in a later chapter.

24.4.4 Application of Fluxes

Solder fluxes must be applied to the underside of the board in the most effective and economical manner possible. The resultant flux coat must be uniform and must thoroughly cover the areas to be joined. An excessive amount of flux not only presents a fire hazard if it drips on the hot preheater but also is wasteful and ineffective and greatly increases the work required to clean the printed circuit assembly. No matter how the flux is applied, an air knife or adjustable squeegee should follow the flux pot to remove excess solder flux.

For all practical purposes, most automated solder machines in commercial use apply the flux by using one of these techniques: wave, foam, brush, spray, or dip. Of those techniques, wave and foam fluxing are the most popular, and they will therefore be discussed in more detail.

24.4.4.1 Foam Fluxer. A foam fluxer consists of a sump and a long high nozzle which contains a porous stone. A typical arrangement is shown in Fig. 24.4. The nozzle, called a "chimney," shapes and constrains the foam head. The pump of

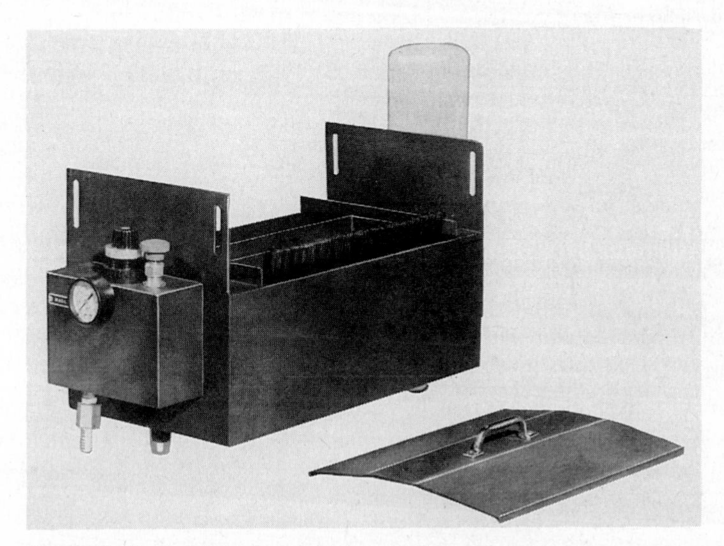

FIG. 24.4 Typical foam flux applicator. (*Electrovert Inc.*)

the fluxer is filled with flux until the porous stone is about 2 in below the surface. Air is pumped into the stone, and the foam head rises out of the chimney to meet the underside of the assembly. To prevent contamination of the foam flux with oils and water from the air lines, the inlet lines must be fitted with filters and traps. The pressure of the air line should be regulated to control the pressure and volume of air being passed through the stone. A nominal pressure of 3 to 5 lb is normally ample. The air line should also have a needle valve to adjust the air volume precisely and thereby control the quality of the foam head. The use of a foam fluxer allows a thin uniform layer of flux to be placed on the board surface. The foam fluxer is particularly effective when the assembly to be soldered has a lot of plated-through holes. Apparently the action of the bubbles bursting within the through-hole is one of the most effective ways to distribute flux on such structures.

24.4.4.2 Wave Fluxer. A wave fluxer consists of a sump and a nozzle similar to the one in the foam fluxer. In a wave fluxer, however, the flux is pumped through the nozzle to form a standing wave of flux. The parts to be fluxed are passed over the top of the wave. This method provides less control over the amount of flux deposited on a printed board, but it is quite adequate for high-speed board production.

24.4.4.3 ***Spray Fluxing.*** Spray fluxing is another technique used for depositing flux on a board surface. It has the advantage of very precise control over the quantity, uniformity, and location of solder flux, but it is a very messy technique and is subject to frequent maintenance. Fluxes used in this operation must have a very volatile solvent system. That makes them expensive and very difficult to control. Unless the precise control over flux deposition parameters afforded by the method is absolutely necessary, the foam fluxer may offer a more attractive alternative.

24.4.4.4 ***Brush and Dip Fluxing.*** Brush- and dip-fluxing techniques are self-explanatory. They are fairly gross techniques and are not readily adaptable to high-speed or high-volume work.

24.4.5 Fluxer Maintenance

The key to proper flux application and performance is regular maintenance. As mentioned previously, fluxes that incorporate volatile vehicles such as alcohol must be periodically thinned to make up for solvent evaporation. Thinning must be done several times during a normal production shift. The most convenient way of doing it is to monitor the specific gravity. The procedure for replenishing the flux pot is as follows:

1. Bring the flux in the pot back to the proper specific gravity by diluting it with the appropriate thinner. Specific gravity information is normally found on the manufacturer's data sheet.
2. Add new flux to the system to bring it back up to the proper level.

It is also important to completely empty the flux pot at periodic intervals and fill it with a fresh change of flux. That minimizes the effect of drag in contaminants and is fairly inexpensive preventive maintenance. Depending on production conditions, it should be done once every 30 to 40 h of operation. All parts used in the construction of a fluxer should be acid-resistant material. Stainless steel or polypropylene is often used for the purpose.

24.5 PREHEATERS

Before the solder flux can become effective, chemical reactions must take place within the flux to release the acidic activators. These chemical groups then react with the metal oxides to remove those oxides from the surface to be soldered. To initiate the reaction, the flux must be heated to what is called the "activation temperature"; for most rosin fluxes, it is about 190°F. If only the solder wave is relied upon to heat the flux to that temperature, the time in the wave is extended by the additional time required for the flux to clean the metal surfaces. A more palatable approach is to heat the flux to the activation temperature prior to the time it enters the wave.

Preheating of the assembly offers other benefits. In most fluxes, a large portion of volatile material will still be retained in the rosin when the assembly reaches the solder wave. Some organic acid fluxes contain water, which also is a relatively volatile component. If the boards are soldered in that state, the heat of the solder pot will rapidly volatilize the solvent and cause spattering and blowholes in the solder joint. By preheating the board, the excess volatiles are driven off, and that eliminates potential problems in the solder wave. If the flux contains water, the preheat temperature must be higher than it would have to be if it contained a more volatile solvent such as alcohol.

Preheating minimizes the thermal shock of the soldering operation by providing a more gradual increase in the temperature of the assembly. It thereby minimizes warpage of the board and moderates thermal stresses which could affect the mechanical integrity of the assembly. Preheating also minimizes the danger of damage to heat-sensitive components.

24.5.1 Types of Preheaters

Two basic types of preheaters are in widespread commercial use; they are the radiant preheaters and the volumetric preheaters. Radiant preheaters can be further divided into two subcategories: long electrically heated rods such as the Calrod type shown in Fig. 24.5 and flat-plate sources such as the one shown in Fig. 24.6.

FIG. 24.5 Calrod radiant preheater. (*Hollis Engineering.*)

24.5.1.1 Radiant Preheaters. Radiant preheaters transfer heat almost entirely by radiation; the temperature on the underside of the board is controlled by the heater temperature and the distance between the heater and the underside of the board. To increase their efficiency, heated-rod-type systems are provided with reflectors which focus the heat energy on the underside of the board. The reflectors are lined with aluminum foil, which can be changed regularly. Flux drippings, therefore, do not present a major maintenance problem. Plate-type heaters, on the other hand, are

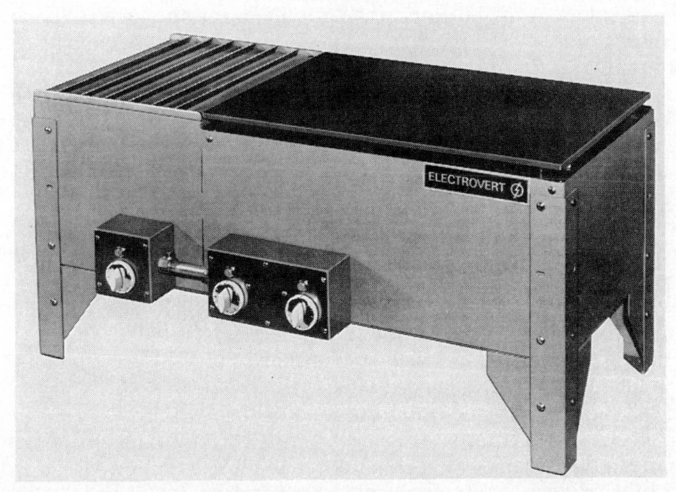

FIG. 24.6 Flat-plate radiant preheater with forced-air convection unit. (*Electrovert Inc.*)

situated under the conveyor in close proximity to the board surface. That greatly enhances the efficiency of the heat transfer between the plate and the board. Flux which drips onto plate-type preheaters must be regularly removed to eliminate the danger of fire and maintain the efficiency of the preheater.

24.5.1.2 Volumetric Preheaters. Volumetric preheaters pass a controlled volume of heated air across the underside of the board. Heat transfer is primarily by conduction and convection. The forced air currents aid in the rapid elimination of solvent vapors from the underside of the board. Volumetric preheaters are preferred when plated-through-hole boards are used, because the inside of the hole is often shielded from straight-line contact with radiant sources. That is particularly important in the case of multilayer boards. Many commercial applications use both techniques—a volumetric preheater to eliminate excess solvent followed by a flat-plate radiant heater to maintain the proper temperature.

24.6 SOLDER POT

The solder alloy is contained in a large reservoir called the "solder pot," and the solder pot is heated electrically to maintain the solder in a molten state. The pot should be well insulated and should have sufficient heating capacity to maintain the alloy temperature adequately. Since soldering is a time- and temperature-dependent phenomenon, accurate regulation of the heat input is essential. To control the temperature effectively, many solder pots have a powerful bulk heater to melt the metal rapidly and a smaller regulated heat supply to control the temperature thermostatically. The smaller heater also prevents thermal loads, such as the printed board, from affecting the pot temperature seriously. Of course, another approach is to maintain a very large volume of solder at the appropriate temperature.

Heater placement in the solder pot also is important. Heaters must be so placed that the solder heats uniformly and there are no hot spots. The temptation to heat only the bottom of the pot and not the sides is great, but it should be resisted. Careful consideration of the initial melting pattern of the solder is necessary. If the solder at the bottom of the pot melts and the surface remains intact, large stresses are built up as the molten solder expands. If the stresses are not relieved by melting the metal in the surface, they can build to such a level that an area on the surface will rupture, and molten metal will be propelled in a fountain several feet high. That is a very dangerous condition. To avoid it, the sides of the pot also should be heated. That applies equally to small dip pots and the larger pots used for automated soldering of assemblies, such as the one shown in Fig. 24.7.

When the solder pot is used in conjunction with a pump and a nozzle to make a solder wave, there are several other commonsense considerations. The pump should be provided with a safeguard such that it can't be operated until the metal is molten. The volume of the pot should be great enough that, when the wave is operational, the solder level will not be affected substantially. The pot should be easy to drain and refill with a new solder charge, and all parts should be easily accessible for regular maintenance.

All parts of the solder pot which come in contact with the molten metal must be made of metals which are not readily wetted by the solder and which will not dissolve in the solder and contaminate it. The most common alloy for that purpose is stainless steel. All tools used to maintain the solder pot, such as dross

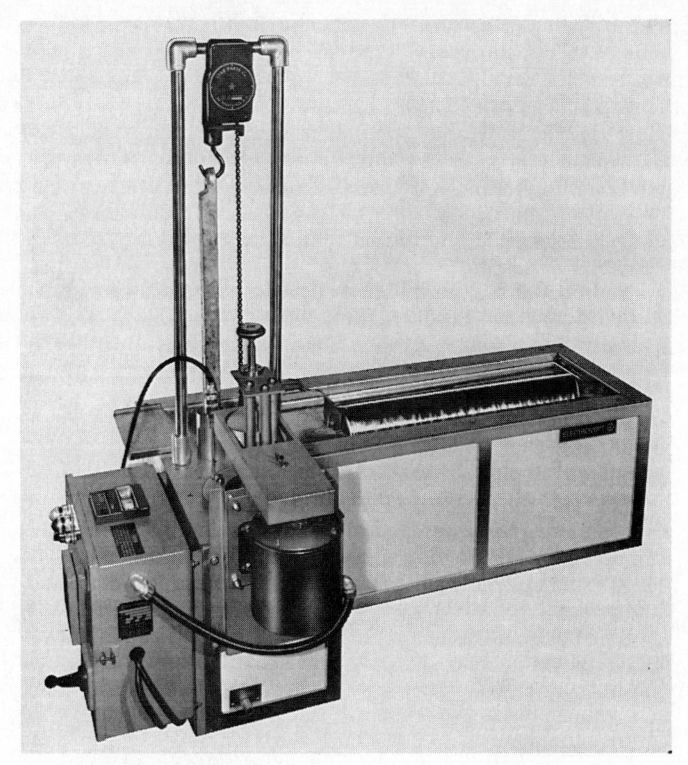

FIG. 24.7 Wave-solder pot with nozzle and automatic solder-level control. (*Electrovert Inc.*)

skimmers, sampling scoops, and perforated paddles, should also be made of stainless steel. Cleaning the inside of the solder pot with a wire brush prior to adding a new charge of solder should be avoided. The wire-brushing operation embeds contaminants in the wall of the solder pot and creates an iron-rich dust which is difficult to remove. Both the dust and the iron-rich particles imbedded in the wall will rapidly contaminate the new solder charge.

Wire brushing can cause damage to a cast-iron solder pot also. Cast-iron pots are normally heavily oxidized to prevent the iron from being dissolved by the tin in the molten solder. Wire brushing or similar abrasive cleaning and scraping can damage the integrity of the oxide layer and expose a potential source of iron contamination to the molten solder.

24.7 DROSS FORMATION AND CONTROL

"Dross" is the name given to the layer of metallic impurities on the surface of the solder pot. It is composed of the oxides of tin and lead formed either on initial melting or on contact of the molten metal with air. Dross formation is principally a function of solder pot design, i.e., exposed surface area on the pot, turbulence of the wave, etc. There is, however, a tendency for dross to be self-propagating.

The more dross in the solder pot, the more likely it is that dross will continue to form. Therefore, dross should be removed on a regular basis (at least once per shift), and new solder should be added to make up for the removed dross.

Dross is harmful to the soldering process and the solder wave. If the dross layer becomes thick enough to be pulled into the solder pump, it will rapidly abrade the impeller and increase maintenance costs. Dross in the wave results in a bumpy, uneven wave with excessive turbulence. If the dross particles are small, they may become occluded in the solder and produce a dull grainy joint. Larger particles of dross may adhere to the bottom of the board and contribute to webbing and bridging.

Dross formation can be minimized by using a blanket, such as rosin or oil, to reduce the amount of exposed solder surface area. When solder blankets are used, however, it is important to change them on a regular basis. Rosin-based blankets should be changed after 4 h of use; soldering oils can be used for 8 to 16 h before they need to be changed.

24.8 SOLDER WAVE DYNAMICS

One of the most critical but least understood variables in the automated soldering machine is the design of the solder wave. Historically, the solder wave was considered little more than a fountain of solder which supplied the molten alloy to the joint area. That was, of course, the logical extension of the drag-and-dip soldering operation. As the industry matured, it became apparent that the dynamics of the solder wave contributed greatly to the formation of smooth shiny joints that were free of webbing and icicling. To understand the importance of wave dynamics better, it is necessary to examine the constraints imposed by the use of automated soldering. Because of the high temperatures involved, the printed board should not be exposed to solder bath too long. However, to ensure formation of a good metallurgical bond, the board must remain in contact with the solder long enough to bring the joint areas up to the proper temperature. Molten solder moving past the underside of the board induces a scrubbing motion in the joint area which greatly aids the fluxing action. To aid in the scrubbing action, the solder wave should have maximum contact time with the board; but to minimize dross formation, the wave must have minimal contact with the air. To minimize icicling, the board should be moving at zero velocity relative to the solder when it leaves the wave. All these apparently conflicting requirements must be satisfied if optimum performance of an automated soldering process is to be attained.

In addition, the solder wave must be even and smooth enough to contact the entire board surface as the board passes over the solder wave. An uneven wave could cause molten solder to pass over the board, rather than under it, and thereby ruin the assembly. Dross formation must be minimized because dross accelerates wear of the solder-pumping system and causes an uneven wave. Excessive dross formation also promotes tin depletion and consumes inordinately large quantities of solder. From those conditions it is easy to see that the solder wave is much more than a fountain of molten metal.

24.8.1 Types of Solder Waves

Many solder wave systems, each purporting to optimize the above list of constraints, have been marketed. They can be divided into two classes: unidirectional (all solder flows in one direction) and bidirectional (solder flows in both directions). Some variations of the bidirectional system offer extenders to broaden and flatten the wave on the trailing side. We discuss one such nozzle in detail later on.

24.8.1.1 ***Undirectional.*** Variations of the unidirectional scheme have included a multiple-wave cascade such as the one shown in Fig. 24.8 and an arching stream of molten solder called a "solder jet." Neither technique is seen on newer machines.

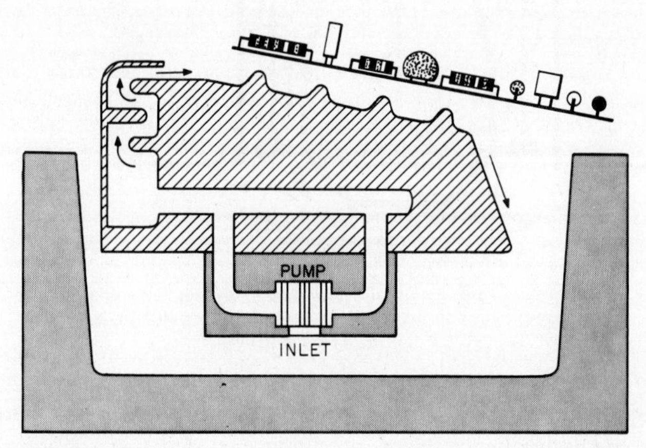

FIG. 24.8 Typical multiple-wave soldering scheme. (*RCA Corporation.*)

24.8.1.2 ***Bidirectional.*** The most commonly used system is the bidirectional wave. Because of the velocity profile on the surface of the wave, the bidirectional solder wave can be designed to minimize icicling. Since solder is flowing both forward and backward in the wave, it follows that there will be an area of zero velocity on the surface of the solder. The velocity profile of the flowing solder in the vicinity of the zero-velocity line is critical to the formation of icicle-free joints.

The bidirectional wave is formed by pumping the molten solder into a nozzle with a large plenum chamber. The resulting head of solder rises up through the lip of the nozzle to form the solder wave. The contour of the nozzle controls the shape of the solder wave and hence the wave dynamics (Fig. 24.9). Baffle screens can be placed in the nozzle to ensure laminar flow and a smooth wave. To minimize dross formation, the solder from the wave must be returned to the pot without excessive turbulence. Two schemes are normally used to accomplish the return. In the wave shown in Fig. 24.10, sluice gates are placed at the outside of the nozzle. The sluice gates return the solder to the pot below the surface level, and so the surface remains undisturbed. If sluice gates are not used, the other alternative is to return the solder to the pot via gradually sloping ramps. That minimizes surface turbulence, but it requires slightly more surface area than the sluice gates.

24.8.2 Surface Tension of Molten Solder after Bidirectional Waves

To understand how a bidirectional nozzle minimizes icicling, we must first understand the phenomenon of surface tension and its relation to wetting. Surface tension forces make liquids such as molten solder and water form into beads or drop-

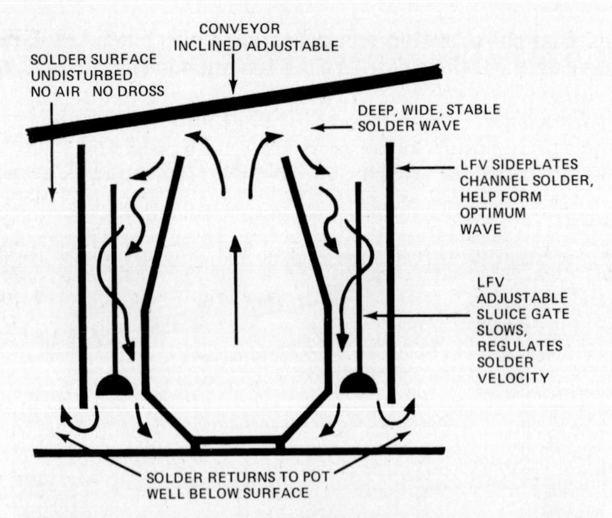

FIG. 24.9 Cross section of a nozzle on an automated solder wave. The large plenum on the nozzle is designed to provide smooth laminar flow through the nozzle. (*Hollis Engineering.*)

lets when they are incapable of wetting a surface. They control how well a surface "likes" a given liquid; hence they control how well solder wets the fluxed copper. In Fig. 24.11 we see a printed board passing over a bidirectional wave. The solder has wet the board surface and is being pulled from the solder wave in the form of a web. The size of the web is controlled by the surface tension of the solder, the velocity profile at the point where the wave contacts the web, and the weight of the molten solder in the web.

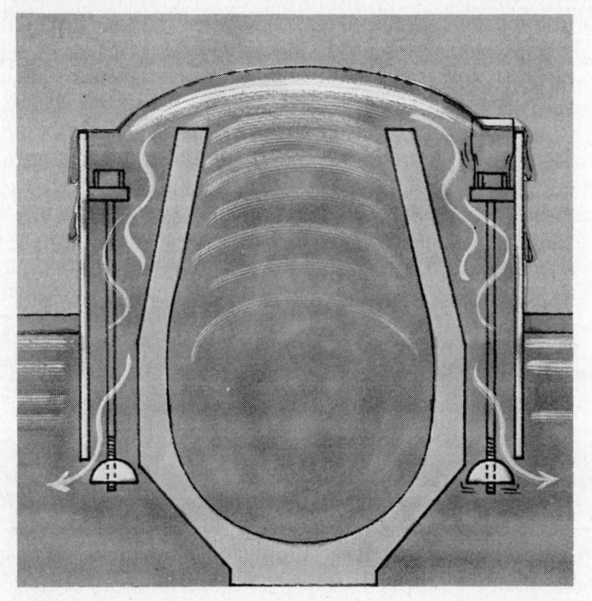

FIG. 24.10 Solder nozzle with sluice gates to minimize dross formation on the pot. (*Hollis Engineering.*)

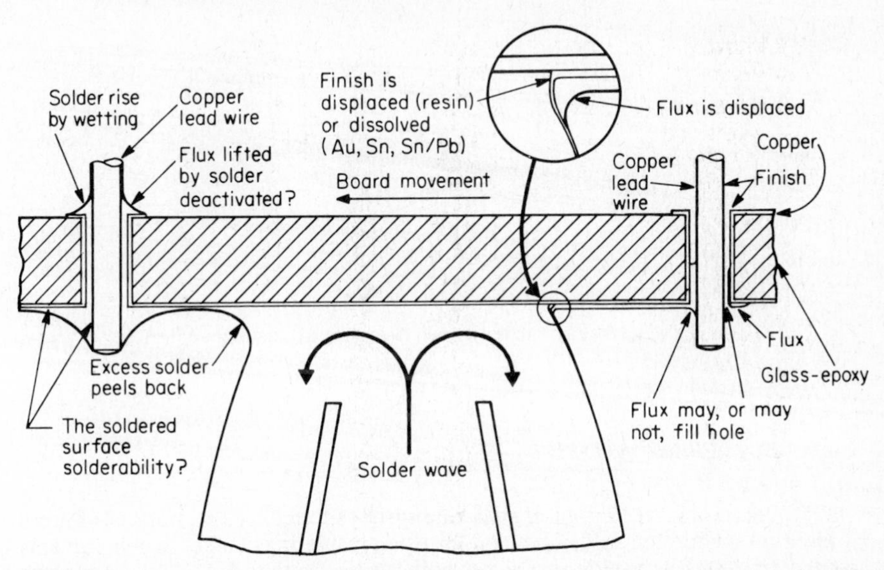

FIG. 24.11 Printed board passing over a typical solder wave. Note the tendency to pull a web of solder from the wave as the board passes over. (*Alpha Metals, Inc.*)

As the web becomes larger, it becomes more difficult for the surface tension forces in the molten wave to pull the excess solder back into the wave. At some critical web size, surface tension forces cause the web to separate; and if the excess solder has not been pulled back into the wave, an icicle is formed. From this very crude model we can see that our task is to minimize web formation by altering either the surface tension of the solder or the velocity profile of the wave at the point where the web occurs. That can be done in several ways. The surface tension of the solder is affected by the temperature of the solder. Higher temperatures will result in a lowering of the surface tension forces. The result, however, may be damage to heat-sensitive components, and the heating will not significantly improve the surface tension of the solder. Surface tension can be reduced by injecting oil into the solder wave. That will be discussed in detail later on.

The size of the web can also be decreased by inclining the conveyor. The friction coefficient of the solder is about 3°, so inclining a conveyor at 4 to 9° will aid in causing the solder to peel back more rapidly. Another technique which has been used is to shape the wave in such a way that it becomes quite wide. The wide wave, when used in conjunction with an inclined conveyor, allows the board to exit from the solder wave at or near the zero relative velocity point. That allows sufficient time for the surface tension forces to draw the web completely back into the solder wave. Figure 24.12 shows a nozzle which was designed by using that approach. Note also in this figure that the board is initially forced to contact the wave at a point of high velocity; thus the scrubbing action of the molten solder also is optimized. Several other features of this type of nozzle design are noteworthy. Since the shape, and hence the velocity profile, of the wave is controlled by positioning the baffles on the front of the nozzle, a region of zero velocity which covers a substantial portion of the front of the nozzle is created. Hence the board can be made to emit the wave at a zero relative velocity point for a wide range of inclination angles. After the board exits from the wave, its close proximity to the hot solder causes a postheating effect which aids in minimizing icicle formation. By using this type of nozzle, icicle-free soldering has been successfully accomplished at conveyor speeds in excess of 20 ft/min.

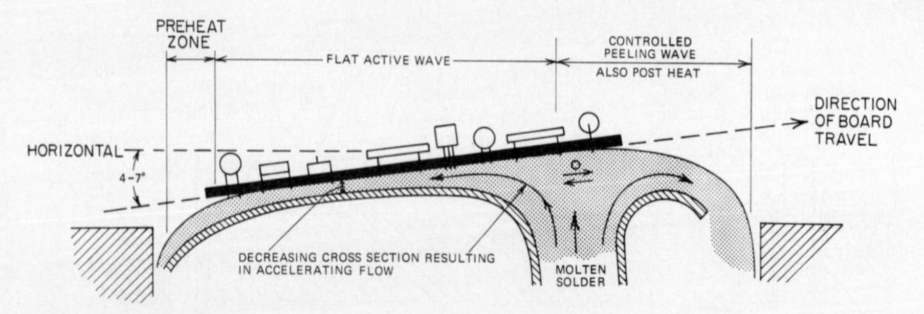

FIG. 24.12 Lambda wave: The system is designed to ensure exit conditions of zero velocity from the solder wave. (*Electrovert, Inc.*)

24.9 SOLDERING WITH OIL

Oil has commonly been used in soldering systems to reduce the amount of dross formed on the molten solder. The oil forms an insulating blanket which prevents contact of the molten solder in the pot with the air. Automated soldering systems which inject the oil into the solder wave at the pump impeller also have been perfected. The oil is dispersed uniformly through the wave and greatly reduces the surface tension of the molten solder. It also reacts with the tin and lead oxides which make up solder dross to form tin and lead soaps. That reduces the abrasive action of solder dross on the pump impeller and minimizes dross in the solder wave itself.

From the preceding discussion it should be apparent that the real reason for mixing oil with molten solder is to reduce the surface tension between the printed board and the molten solder at the point where the board departs from the wave. This effectively minimizes the solder web and eliminates icicles. Proponents of oil intermix systems have pointed out several other "processing type" benefits that can be obtained. The solder joint is coated with oil when it leaves the wave. The coating prevents oxidation and results in brighter and shinier joints which are easy to inspect. Since the tendency for icicling has been minimized, conveyor speeds can be accelerated and process throughput is increased. Lowered surface tension increases the ability of solder to wet copper pads on the circuit board, and the temperature of the solder wave can therefore be reduced substantially (20°F) without affecting joint formation. When oil is applied in proper amounts, it can assist in solubilizing rosin residues and thereby aid in the cleanup of the finished assembly.

According to a different, and somewhat more conservative, school of thought, soldering with oil results in inclusions of soldering oil in the solder joint. The oil in the inclusions could affect the mechanical and electrical properties of the joint and could be a source of corrosive acidic residues. The critics also point out that soldering with oil results in a messy, oil-coated assembly which is difficult to clean effectively. When soldering is with oil, cleaning is a necessity.

The key to effective use of soldering oil is process maintenance. The soldering oil must be changed on a regular schedule to prevent its degradation and carbonization. Carbonized oil in the solder wave negates most of the advantages that can be obtained by using oil. The soldering machine and the machine area must be kept spotless. Oil spills and the results of similar small handling accidents should be cleaned up before the cumulative effects turn the area into a pigpen. Finished assemblies should be cleaned as soon as they are processed, and the assemblies should be inspected to make sure all oil is removed.

24.10 TIN DRIFT

One other subject which is of interest to companies using automated soldering is tin drift, the slow depletion of tin in a solder pot as the pot is used. Tin drift can be very troublesome when compositions at or around the eutectic must be maintained. It occurs because tin is easier to oxidize than lead, and dross formation is therefore biased toward the formation of tin oxide rather than lead oxide. Basically, the reaction is as follows:

$$2Pb + O_2 \rightarrow 2PbO$$
$$2Sn + O_2 \rightarrow 2SnO$$
$$Pb + Sn + O_2 \rightarrow PbSnO_2$$
$$PbO + SnO \rightarrow Pb + SnO_2$$

When soldering oils are used, the reaction is somewhat different, but the results are generally similar. Tin drift can easily be corrected by replenishing the solder pot with solder that is slightly tin-rich. Another, and probably more palatable alternative, is to use tin ingots or tin bits to bring the solder pot back to specification.

24.11 CONVEYOR

Another parameter which should be considered when an automated soldering system is designed is the conveyor. It controls the speed at which the boards are passed through the process; hence it controls preheat time and temperature and dwell time in the solder wave. Conveyors are available in two basic types: the pallet type in which boards are held in carriers or holders, and the finger type in which boards are held in spring fingers. Finger-type conveyors can be adjusted to accommodate the width of the board; they can automatically discharge boards at the end of the process; and they are easier to load. Figure 24.13 shows a typical finger-type conveyor. Both pallet- and finger-type conveyors are available in either horizontal or inclinable models.

FIG. 24.13 Finger conveyor for holding printed boards in an automated soldering system without the need for pallets. Boards are held in place by spring tension of conveyor fingers. (*Electrovert, Inc.*)

24.12 AUTOMATED SOLDERING SYSTEMS

So far we have discussed each segment of the system in great detail; only post-soldering cleaning has been omitted because it is covered in another chapter. Most manufacturers sell complete systems rather than just component parts. Whenever possible, it is advisable to purchase a complete system and assure yourself that it is properly engineered and integrated into your entire process. Most automated soldering systems also contain provisions for adequate ventilation and proper employee protection. In today's legal and ecological environment, purchasing a complete system from a single manufacturer is an extremely effective way to shift a major portion of the burden of complying with federal and state regulations onto the machine manufacturer.

24.13 HAND SOLDERING

24.13.1 Soldering Irons

Although the great majority of solder joints are formed during the automated soldering operation, there are still many instances in which manual soldering techniques are essential. Individual joints may be faulty and therefore require repair and touch-up at a manual-soldering station. Heat-sensitive and solvent-sensitive components may need to be hand-inserted, or defective components may need to be replaced. Therefore, a knowledge of manual-soldering techniques is essential for the printed circuit processing engineer. Manual soldering is accomplished with a soldering iron and depending on the application, a length of wire or core solder or a solder preform. The function of the soldering iron is to transfer heat to the joint area rapidly and efficiently. Because of the wide variety of joint configurations and joining problems, there are many different sizes and types of soldering iron and tips. To help you evaluate your particular job and requirements, some of the important considerations in selecting a soldering iron are outlined below:

24.13.1.1 Wattage. This refers to the power consumption of the soldering iron. It is a rather coarse measure of the total amount of energy available for heating the soldering iron tip. All other things being equal, an iron with a higher wattage will have a higher soldering temperature.

24.13.1.2 Maximum Tip Temperature. This is the steady-state or equilibrium temperature of the soldering iron tip. A properly designed iron should maintain a constant tip temperature when operated at its rated voltage.

24.13.1.3 Heat Content of the Iron. Heat content is a measure of the thermal bulk of the soldering iron, the amount of thermal energy that can be stored by the metallic components of the tip. The heat content must be carefully matched to the job. If the iron is too small, the heat drain caused by making a single joint will lower its temperature below that necessary for making the next joint. Valuable production time will then be lost in waiting for the iron to reheat. If the heat content is too high, excessive heating of the components may cause damage or degradation of some of the electrical components.

24.13.1.4 Recovery Rate. The recovery rate is a measure of the recycle time of the soldering iron; it is the time necessary to reach the equilibrium tip temperature after a single joint has been made. Obviously, the recovery time is a function of wattage, heat content, and tip geometry.

24.13.1.5 Selection of Irons. In practice it is not feasible to maintain a diverse inventory of soldering irons and tips to match each production situation, nor is it anything but naïve to think that a single iron and tip will cover all requirements. Therefore, compromises must be made in the interest of economy and efficiency. The following general guidelines should be considered when soldering irons are selected for individual applications.

1. *The caloric requirements of the work:* This parameter depends on several other variables: the solder alloy and flux used, the thermal properties of the base metals, and the geometric configuration of the joint.

2. *The production environment in which the tool is to be used:* The number of joints to be produced per unit of time, the human factors involved in handling the tool, and the limitations imposed by the assembly to be soldered are important considerations.

24.13.2 Tip Selection

Careful tip selection is just as important as proper iron selection (see Fig. 24.14). As a matter of practice, the tip should be evaluated in conjunction with the soldering iron to be used. Soldering irons with equivalent wattage ratings but different efficiencies will yield different tip temperatures when used in conjunction with the same tip. Therefore, it is essential that the soldering iron and tip be evaluated as a system. The following guidelines will aid you in the selection of your soldering iron tip.

1. Select a configuration that affords maximum contact between the tinned areas on the work and the soldering iron tip. That maximizes the heat transfer between the iron and the work.

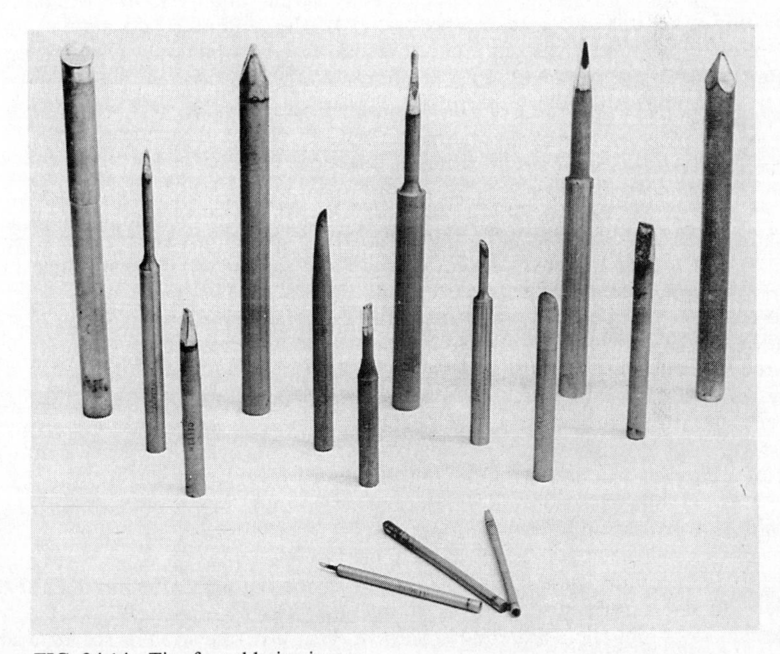

FIG. 24.14 Tips for soldering irons.

2. Select a tip that will facilitate the manual work of gaining access to the solder joint. For example, in some cases the use of a right-angle tip greatly simplifies the manual-handling problems.
3. The taper of the tip should be as short as possible to achieve optimal heat transfer to the tip.
4. Select the shortest reach available to minimize wobble and thermal loss due to convection heat transfer.
5. The diameter of the shank determines the heat transfer between the elements of the iron and the tip. The larger the diameter, the better the coupling between the iron and the tip.
6. The life of bare copper tips is short, and the cost of maintaining a copper tip's shape is relatively high. Iron-clad tips, on the other hand, have much longer lives, but they cannot be filed or shaped.

24.14 SOLDER PREFORMS AND SOLDER CREAMS

We have shown that solder joints can be made easily, reliably, and rapidly through the use of automated soldering equipment. However, there are applications and designs which simply are not compatible with wave- or dip-soldering techniques. They may be encountered when the circuit assembly is nonplanar or when the proximity of sensitive components makes the use of a hot soldering wave impossible. If, in addition, we broaden the definition of printed board to include the ceramic substrates used in microcircuit assembly, then the automated techniques previously described simply become inapplicable. For such applications, techniques based on either solder preforms or solder creams are often the most useful. In this section we will discuss both the nature and application of such materials. However, because of the inherent versatility and flexibility, we can provide only a few insights into the properties of the materials and a small cross section of typical applications. Your imagination and your own possible applications will soon provide you with an estimate of the design flexibility and cost-saving advantages of solder preforms and creams.

24.14.1 Stamped Preforms

Solder preforms are precision-manufactured solder parts made to the exact shape desired for an application. Typically, they are punched from a strip of solder alloy by using either a standard-shaped die such as a disk or washer or a die which has been custom-manufactured for some specific application. Obviously, the cost of each part is significantly higher if a custom-made die is necessary. Some typical stamped preforms are shown in Fig. 24.15.

Some solder manufacturers will also supply preforms stamped from laminated metal systems such as copper-solder or solder-copper-solder. Continuous strips of those material systems are made by soldering the molten solder alloy onto one side of a copper strip and then skiving the solder alloy to the desired thickness. Preforms made from such material systems offer some major advantages which are not so apparent on a first examination. For joining problems involving difficult surfaces such as heavily tarnished copper or nickel, the copper segment of the preform will provide a solderable surface in close proximity to the joint. Thus the solder will be retained in the joint area until the flux can activate the surfaces to be joined. Without the solderable surface to retain it, the molten solder might

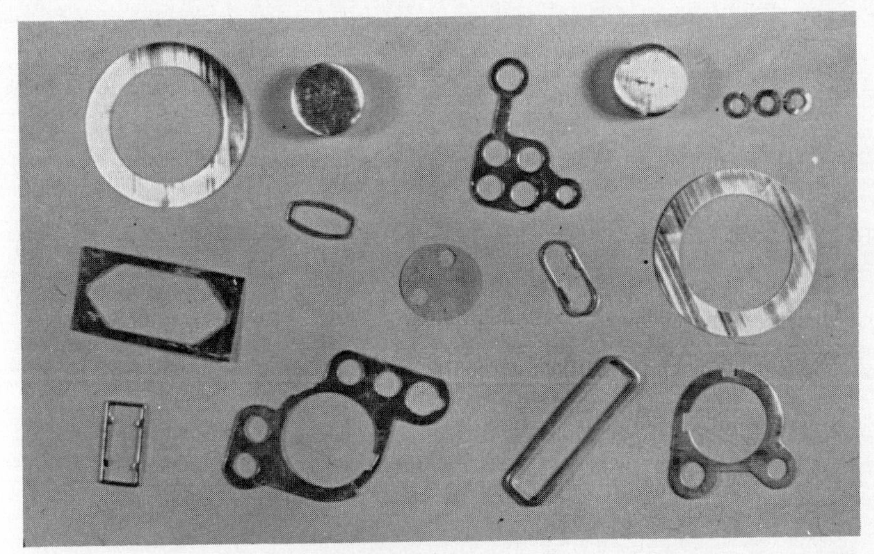

FIG. 24.15 Solder preforms can be stamped in a variety of shapes and sizes. Preforms simplify complex handling problems. (*Alpha Metals, Inc.*)

have dewetted and moved to another area of the assembly. In some instances the parts to be joined may have excessive tolerances that will produce a joint area that cannot be filled by the molten solder. In that case the copper core can act like a bridge or filler to maintain the integrity of the joint.

Preforms can be supplied with or without flux. The flux can be an integral part of the preform—that is, the preform can be flux-filled, or it can be coated on the outside of the preform in a subsequent manufacturing operation. In the case of the flux-filled preform, the strip from which the preforms are punched is made from a flattened piece of core solder. This results in a sandwich structure that has flux in the center and solder alloy on the outside. Preforms made from this type of material offer the advantage of very precise control over the final dimension of the part. The other side of the coin is that the flux is not released until after the metal is molten. In some applications, that could be a disadvantage. When precise dimensional control is not essential, it is a simple matter to coat the preform with flux. The flux coating results in a dull powdery appearing surface.

Solder preforms offer a number of advantages. They not only provide solder in a carefully controlled shape to aid in joint formation but also provide a controlled volume of solder alloy. The use of a solder preform also allows complete control over the placement of the solder alloy, which ensures that the joint is formed in areas specified by the designer and nowhere else. Considerations such as these lessen dependence on operator skill and judgment in the assembly of electronic equipment and put control in the hands of the designer.

24.14.2 Solder Spheres

Another type of solder preform which should prove quite useful is the solder sphere. Solder spheres are made by using one of two basic techniques. For large spheres, a wire of the desired alloy is chopped into small segments that contain the proper volume of solder. The segments are spherized by passing them through a bath of hot oil maintained at a temperature above the melting point of the solder. While the segments are passing through the heated section of the column, they melt and surface tension forces cause them to become spherical; and when they reach the cooler sections of the bath, they solidify as spheres. For smaller spheres, the molten solder is sprayed into an oil bath from a nozzle with a very small orifice. The technique can yield extremely small spheres, but the spheres must be sorted carefully in order to have them uniform in size. Some typical solder spheres are shown in Fig. 24.16.

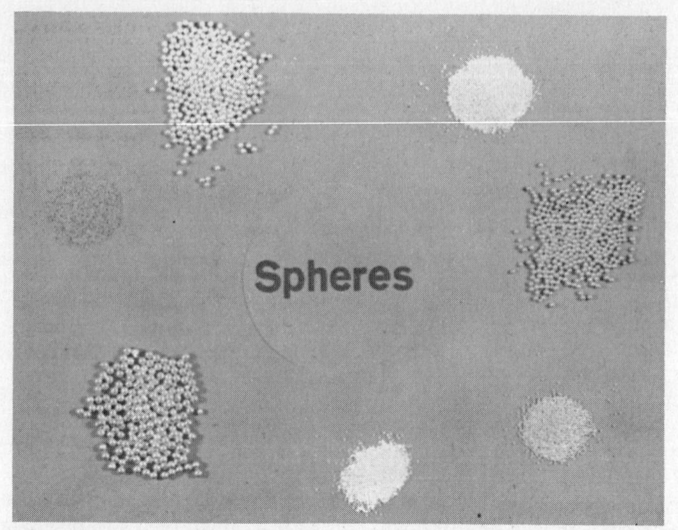

FIG. 24.16 Typical solder spheres. The spheres contain precisely controlled volumes of solder and are easily handled. (*Alpha Metals, Inc.*)

24.14.3 Handling of Solder Preforms

Solder preforms can be placed in position by using a variety of methods. For high-speed or high-volume applications, they are normally positioned on special tools or fixtures which are loaded by using a linear or rotary vibrator feeder. In this type of equipment, the preforms are loaded into vibrating bins or hoppers. The bins are structured in such a way as to accept the preform at the loading point only when it has vibrated to the proper altitude for dispensing. Preforms with a flux core or no flux at all are more suitable to these applications than are flux-coated preforms. The latter will lose some of their coating in the hopper, and the resultant dust will gum up the fixtures and adjacent machinery. Figure 24.17 shows some typical vibratory feed apparatus.

For applications in which the symmetry and spacing can be controlled appropriately, preforms can be manufactured in "daisy chains," that is, connected to one another in a symmetrical pattern. The operator can then simply take a precut length of daisy chain and position it on the assembly. When the preforms are heated, the connections will melt and wet into the appropriate joint. Examples of where such a technique would be applicable are pins from a plug-in connector or the wire wrap pins common to many forms of electrical assemblies.

Many types of specialized jigs and fixtures have been designed for manual placement of preforms. A common technique which is often used with disks and

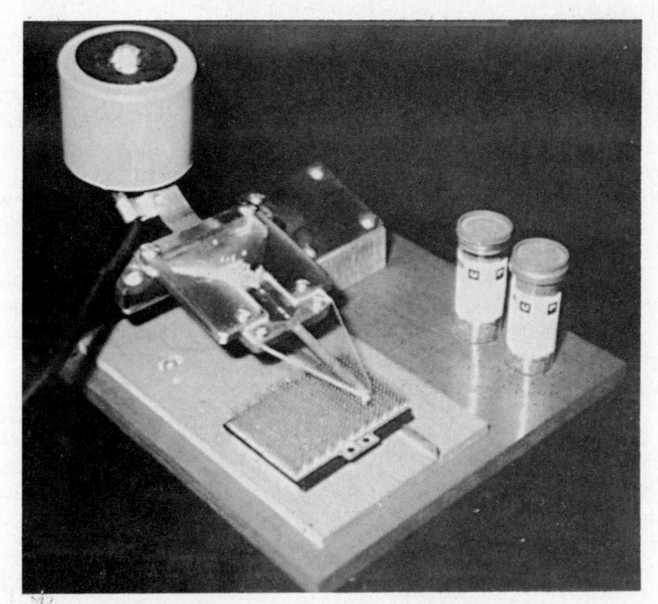

FIG. 24.17 Vibratory feeder for loading solder preforms. (*Alpha Metals, Inc.*)

washers is the "shuffle board." With the shuffle board, a large quantity of preforms is placed at one end of the fixture. The preforms are then pushed or brushed across the fixture plate in such a way that one preform drops into each hole in the plate. A retaining plate is then secured on the fixture and the loaded jig is moved to the assembly area.

A variation of the shuffle board technique is often used for loading solder spheres. In the vernacular, it is called the "seesaw," and the author has also heard it called a "pinball machine." Quite simply, it consists of two wells, one on either side of the fixture plate. The fixture has a pivot in the center so it can be tipped back and forth like a seesaw. The solder spheres are rolled across the fixture plate each time the fixture is tipped until the holes in the fixture are loaded.

Manual placement of the solder preform can be accomplished by using tweezers, hand tools, and small vacuum pickups. The methods normally require some operator skill and judgment, to say nothing of a steady hand.

24.14.4 Solder Creams

Solder creams are mixtures of finely powdered solder in a viscous vehicle. The vehicle contains the fluxing agent, and it controls the flow properties of the solder cream. Solder creams can be manufactured either with rosin-based or water-soluble fluxes, so they can be used in any application which requires a solder joint. Because of the many unique ways in which it has been used, solder cream is considered to be the most versatile method for soldering unusual assembly configurations. Solder cream, in fact, can be used as a fluid preform capable of being shaped and molded to fit the application. Solder creams can be cured in place and stored for several days before the joint is flowed to make the bond. They can be used as a kind of viscous putty to hold components in place until the reflow operation. Because they can be applied equally well to planar and vertical surfaces, they can be used to overcome some of the fixturing problems inherent in preforms.

The manufacture of a solder cream is a fairly sophisticated operation because of the tremendous number of variables which affect properties and performance. The viscosity of the final solder cream is a function not only of the viscosity of the base or vehicle but also of the particle size distribution, the amount of metal in the cream, and the shape and surface roughness of the individual powder particles. Variations in the properties of the cream caused by variations in the parameters may not affect some of the grosser applications, but they are critical to the reliability and reproducibility of many of the microelectric joining applications. Because an understanding of the importance of those parameters is critical to the proper use of solder creams, we will discuss each parameter in turn and try to show the interrelations among the factors.

24.14.4.1 ***Alloy Powder.*** Solder powder is made by flowing molten solder alloy through a small nozzle into a cooling liquid in much the same way solder spheres are made. Because of the manufacturing method used, a large variation in particle size distribution is unavoidable. The size variation results in even greater variations in the actual surface area of solder alloy for a given standard volume of solder powder. In particular, the presence of a large number of very fine particles, called "fines," greatly increases the actual surface area of the given standard volume.

To separate the powder, or, more appropriately, to grade it, two methods are normally used. In the first method, called an "air separator," the particles are introduced into a stream of air and are carried distances dependent on their masses. Bins placed along the stream collect particles of various sizes. The method is tedious and expensive, but it is highly effective. The second method is probably more commonly used. It involves the use of various grades of sieves to sort the particles. When the system is used, the powder is classified by the smallest screen mesh through which it has passed.

24.14.4.2 ***Flux and Vehicle.*** The properties of a solder cream are also to a very great extent dependent on the properties of the vehicle and flux mixture; it is those properties that make the solder cream so versatile. Depending on the application, solder creams that cure in place can be obtained; i.e., they dry to a nontacky finish and can be handled after deposition. To aid in deposition, the solder cream can be formulated to be thixotropic, i.e., shear thinning. That means they will run freely when a force is applied to them during deposition; but when the force is removed, they remain in place as thick viscous amalgams of solder and flux. Thus solder creams can be easily and rapidly deposited on the parts to be joined. Then, if desired, they can be cured and handled during subsequent operations and finally reflowed to form the final coating.

The viscosity of the solder cream is affected by the amount of metal in the cream. The more metal added, the more viscous the cream will be. However, as can be seen in Fig. 24.18, the relation between metal content and viscosity is nonlinear. A typical commercial solder cream will contain about 80 percent by weight of metal powder and have a viscosity in the 300,000- to 400,000-cP range.

24.14.4.3 ***Application of Solder Creams.*** One of the most popular methods for applying solder cream is by screening. In the technique shown in Fig. 24.19, a stainless-steel screen which conforms to the desired pattern of solder cream is made. The part is placed under the screen, and the solder cream is forced through the screen by using a squeegee. The method is similar to silk screening. Screening

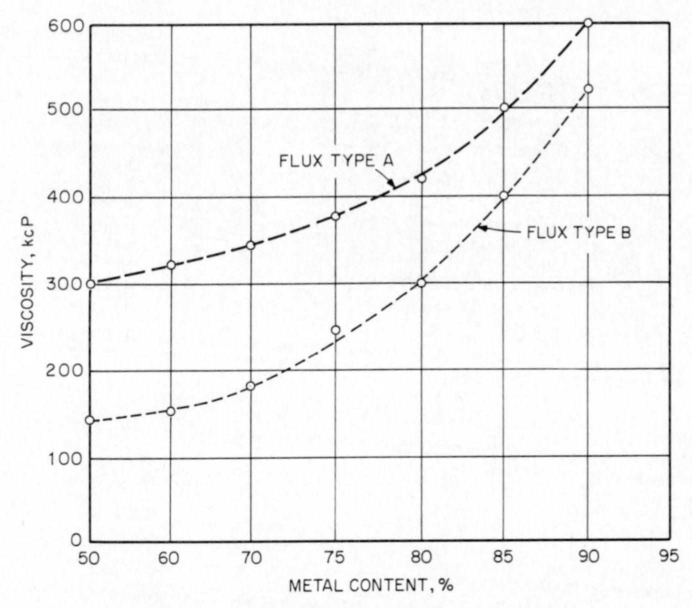

FIG. 24.18 Viscosity as a function of metal content in a solder cream.

can be used as a fairly gross technique, or it can be an accurate, precise way of depositing solder creams.

Solder cream can also be applied by using pneumatically operated dispensers which meter out precisely predetermined amounts of solder cream. A typical dispenser is shown in Fig. 24.20. Pneumatically controlled dispensers are capable of placing a 0.015-in dot of solder cream on a representative part, an attribute which has made them most useful in the fabrication of microcircuitry. For applications which do not require the precise metering control of a pneumatic dispenser, a manually operated syringe or even a plastic ketchup bottle can be used.

FIG. 24.19 Stainless-steel screen for automated application of solder cream. (*Alpha Metals, Inc.*)

An interesting application for solder creams takes advantage of the fact that the solder cream can be applied either by screening or by dispensing in a gross sort of way. Because the solder cream becomes molten solder when it is heated to the proper temperature, it will wet only surfaces which are properly fluxed and are wettable in the fluxed state. Therefore, a blob of solder cream can be applied to the general vicinity of a joint or a pad. When it is melted, the solder will retract onto the metal surface and form a bond. If a chip or large ceramic integrated circuit is placed on the solder cream before melting the cream,

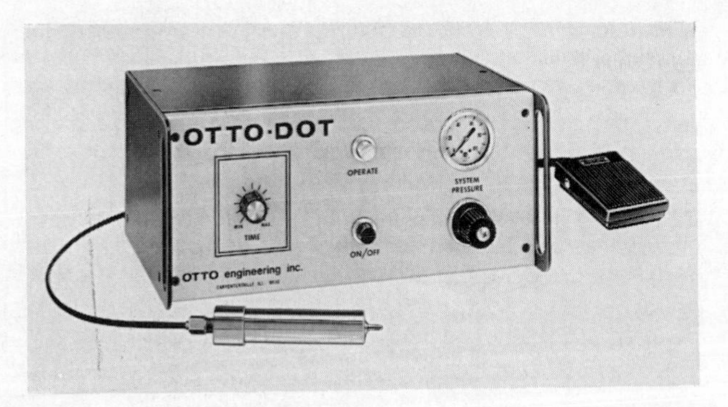

FIG. 24.20 Pneumatic solder cream dispenser. (*Otto Engineering, Inc.*)

it will be pulled into perfect alignment with the metal pads as the solder cream is melted.

24.14.5 Heating Techniques

Heating techniques are germane to the discussion of both solder creams and preforms. They can, therefore, be discussed without considering the form of the solder on the part. Just as solder creams and preforms challenge the imagination of the designer, so the techniques available for meeting the heating requirements of the assembly call for a great deal of versatility on the part of the assembly engineer. Some applications may simply require the touch of a hot soldering iron, but others will require the complex and intricate techniques inherent in induction and radiation heating. Do not be afraid to let your imagination wander a bit to take advantage of the flexibility of the available technologies.

24.14.5.1 Soldering Iron. The use of a soldering iron or a hot plate is a fairly straightforward technique. It is discussed elsewhere, so there is no need to dwell on it here.

24.14.5.2 Hot-Air Gun. A surprisingly versatile heating technique which is often overlooked is the electrically heated hot-air gun. Air from such a gun can be hot enough to flow even some of the high-temperature solders. The gun can be used in a gross sort of way to heat the entire assembly, or, through the use of nozzles, a stream of hot air can be directed into minute spaces of approximately the dimensions of a soldering iron tip. Hot-air guns can be manually or automatically operated. However, because of the critical relations between distance and temperature inherent in the hot-air technique, an automated setup generally enhances the reproducibility of the operation.

24.14.5.3 Infrared Heating. Infrared heating is a very common and very attractive technique. It is a form of radiation heating; therefore, like all radiation, it can be focused onto a small area. It can induce rapid temperature changes on metallic surfaces (surfaces which absorb infrared radiation) and, at the same time, cause only minor changes in the temperature of the epoxies generally used for printed board construction. Because infrared heating is done principally by radiation, some of the design parameters you must consider are the geometry of the parts to be joined, the emissivity and thermal conductivity of both the parts and the surrounding region, and the color of the parts and the surrounding area.

If it is desirable, shields and masks can be made to allow the parts to be heated in one region and not in another.

Infrared heating adapts itself well to automated processing and high-speed production. Most printed boards which are protected by reflowed tin-lead coatings are reflowed by using infrared techniques. Infrared heating is also a common technique for bonding silicon devices to substrates, bonding substrates to printed boards, and soldering and splicing flexible cables.

24.14.5.4 Induction Heating. Induction heating is a technique for heating electrically conductive materials to high temperatures. It is accomplished with high-frequency eddy currents induced in the parts by an external induction coil. Because of the current and voltage characteristics of the eddy currents, it is not a good idea to use the technique if sensitive active devices are present. However, the author has seen induction heating used quite successfully for soldering connectors on both flat and stranded interconnection cable. The large-scale use of induction heating is generally limited by the high power requirements and the bulky electrical equipment needed to generate the high frequencies that are required.

24.14.5.5 Hot Oil. Hot oil or hot glycol reflowing techniques are a common heating method. They are, however, quite messy, and they are somewhat difficult to control. They are not generally used except for reflowing solder plate onto printed boards or component leads. Do not confuse the hot-oil technique with the hydro squeegee. Reflowing with hot oil requires the parts to be immersed in a bath of hot oil or glycol; the hydro squeegee technique impinges jets of hot oil onto the surface of a printed board to reflow the plate under temperature and pressure. Hydro-squeegeed boards have been shown to have minimal thicknesses of reflowed tin-lead on their surfaces after processing, and they are therefore quite susceptible to solderability degradation during storage. Boards reflowed with hot glycol, on the other hand, retain all solder plate and have excellent shelf life if properly cleaned.

CHAPTER 25*
CLEANING OF SOLDERED BOARDS

Hugh Cole

25.1 INTRODUCTION

Cleaning has become one of the most important steps in printed board fabrication; it affects the ultimate reliability of the board. Ionic residues on a printed board lead to electrical leakage and dielectric breakdown, two phenomena which seriously impair the electrical performance of the product. Flux residues left on the board may become corrosive and thereby affect the mechanical and electrical properties of the board. The economics of processing the board are also significantly affected by the choice of a cleaning procedure. Particularly in today's competitive environment, it is important to understand the nature of the trade-offs between reliability and cost of various postcleaning systems.

25.2 TYPES OF CONTAMINATION

Dirt or contamination on a printed board can be divided into two categories: polar and nonpolar. Polar, or ionic, contamination ("polar" and "ionic" are used interchangeably in the electronics industry, a practice that is somewhat disconcerting to a theoretically oriented chemist) consists of chemical compounds that will dissociate in solution into positively and negatively charged species. It is undesirable both because it can carry a current in the presence of moisture and because it can enter into chemical reactions associated with corrosion. Ionic contamination typically comes from plating residues, finger salts, and flux activators. Nonpolar (or nonionic) contamination, although not as detrimental as polar contamination, will still interfere with circuit performance. Nonpolar dirt typically consists of rosin from solder fluxes and oils and greases associated with handling operations. Oils, greases, and rosin can form insulating films on plug-in contact surfaces, and the films can result in intermittent open circuits which are difficult and expensive to troubleshoot. Nonpolar soils, however, are readily removed by most chlorinated or fluorinated solvents. They do not usually present a problem

*Adapted from Coombs, *Printed Circuits Handbook*, 2nd ed., McGraw-Hill, New York, 1979, Chap. 16.

unless the mechanics of the cleaning process are improperly specified. For example, if pallet design causes entrapment of rosin residues or if dwell time in the cleaning solution is too short, problems are likely to occur.

25.3 MEASUREMENT OF CLEANLINESS

Before a level of desirable cleanliness is specified, it is necessary to establish a quantitative measure of the degree of dirtiness of the part. Cleanliness and dirtiness are very general terms, and, understandably, many tests have been devised to determine objectively the amount of dirt on a printed circuit assembly. As enforceable quality control standards, however, they all have one common disadvantage: They don't work! Most notable among these types are ultraviolet inspection and the Lieberman-Storch test. Both methods are based on the detection of rosin residues and the implicit assumption that, when all the nonpolar rosin residues are gone, all polar residues also will be gone. That simply isn't true. In addition, the Lieberman-Storch test is a carefully controlled laboratory procedure; it is tedious, cumbersome, and difficult to adapt to high-volume production.

To quantify the level of cleanliness, it is necessary to measure the amount of ionic soil on the board surface. Since ionic soil is the most dangerous, a control on the level of ionic dirt promises to be most effective. Several methods of control have been devised. Most notable among them are the solvent extract resistivity tests and the Ionograph. In both methods, the ionic soil on the electronic assembly is dissolved in a water and alcohol mixture, and the resistivity or conductivity of the solution is monitored. Variations use high-purity water as a solvent, and that, as we shall see in a subsequent part of this chapter, will result in error because water will not dissolve residual rosin. Therefore, ionics trapped in or under the rosin will not be measured.

When the water-alcohol solutions are monitored, most people prefer to use conductivity rather than resistivity. Conductivity is the reciprocal of resistivity. At the concentrations used to measure cleanliness of electronic assemblies, it is a linear function of the concentration of dissolved ionics; hence it is much easier to interpret. Because of the linear relation, the Ionograph can quantitatively measure the amount of ionic material present on the board surface.

The Ionograph, shown in Fig. 25.1, continuously recirculates the water-alcohol mixture through an ion-exchange column, and the result is a solution of very high purity. By monitoring the change of conductivity as the ionics are dissolved in the water-alcohol mixture, the Ionograph rapidly determines the amount of ionic material on the part. The use of a calibration solution of known ionic content permits quantitative determination of ionic levels on the electronic assembly. Thus the Ionograph can be used either for relative comparison of cleaned parts or for sensitive analytical determination of the ionic levels. Unlike the solvent resistivity test, which measures only the conductivity of a water-alcohol mixture used to wash the finished assembly, the Ionograph will determine the actual ionic level on the finished part.

25.4 SELECTING AMONG PROCESS ALTERNATIVES

When a printed circuit assembly operation is selected and installed, the process engineer is faced with three basic alternatives. Two of them involve a rosin-based flux, and the third involves a water-soluble flux. Figure 25.2 shows the alternatives and the implications of each one.

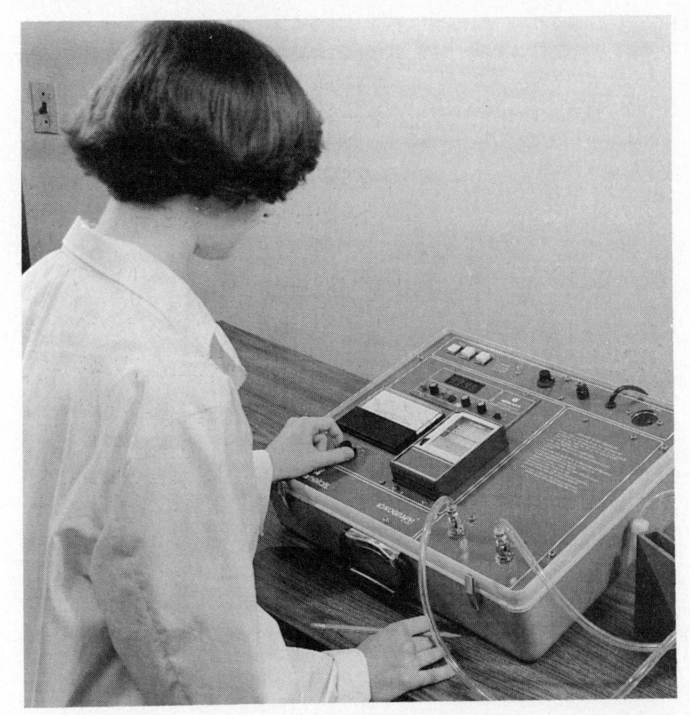

FIG. 25.1 The Alpha Ionograph, an instrument used to determine the amount of ionic contaminant on a circuit assembly. (*Alpha Metals, Inc.*)

If the most common path is chosen, the engineer will use a rosin-based flux and clean with an organic solvent. The mechanics of the cleaning process depend on the constraints of the particular application. The second alternative, and one which is growing in popularity, is to remove the rosin-based flux with a biodegradable (i.e., water-based) cleaner. A water-based cleaning system can offer some distinct advantages over solvent cleaning for certain types of assemblies. The third alternative is a water-based cleaning process for water-soluble fluxes. Details of the three cleaning alternatives will be covered in depth in this chapter.

25.5 USE OF SOLVENT-BASED CLEANING PROCESSES

Because soil on a printed board can be either ionic or nonionic, solvents used in the final cleaning process must be capable of dissolving both kinds of soil. A solvent capable of dissolving only nonpolar soils will remove oil and greases and rosins, but it will not remove plating residues and flux activators. Figure 25.3 shows the final result of an experiment which illustrates this point. In the experiment, a typical amine hydrochloride flux activator was dissolved in several solvents commonly used to clean electronic assemblies. The more activator dissolved in a given solvent, the greater its propensity to remove polar activators and residual plating salts. Nonpolar solvents such as perchloroethylene, 1,1,1,-trichloroethane, and trichlorotrifluoroethane are incapable of dissolving any of the flux activators. The other solvents dissolved various amounts of the activator.

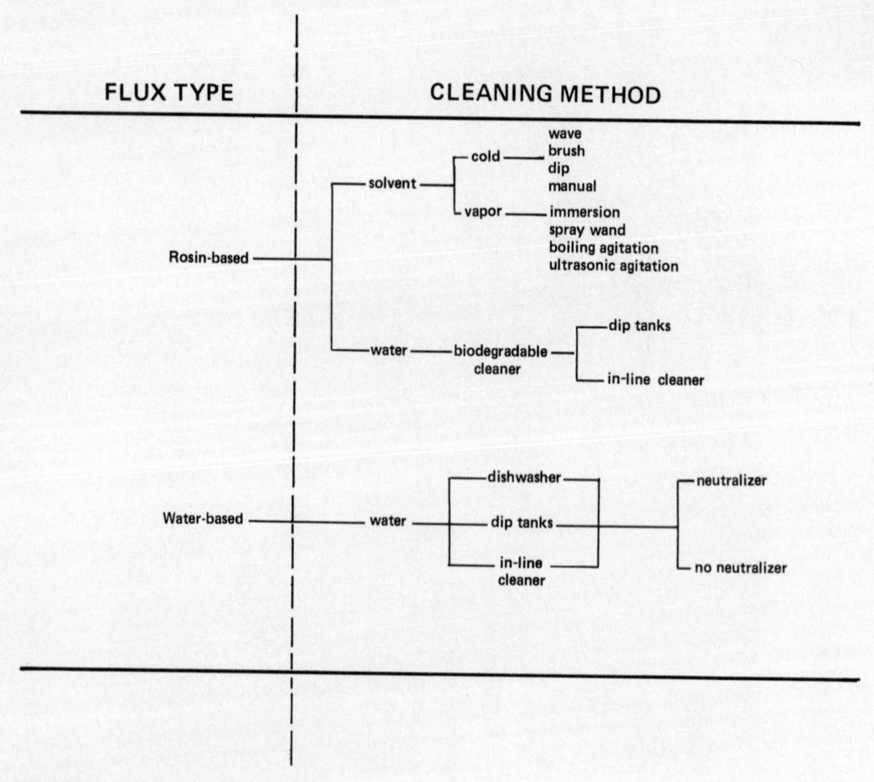

FIG. 25.2 Process alternatives; flux can be removed in a variety of ways. The proper sequence depends on the requirements of the printed board.

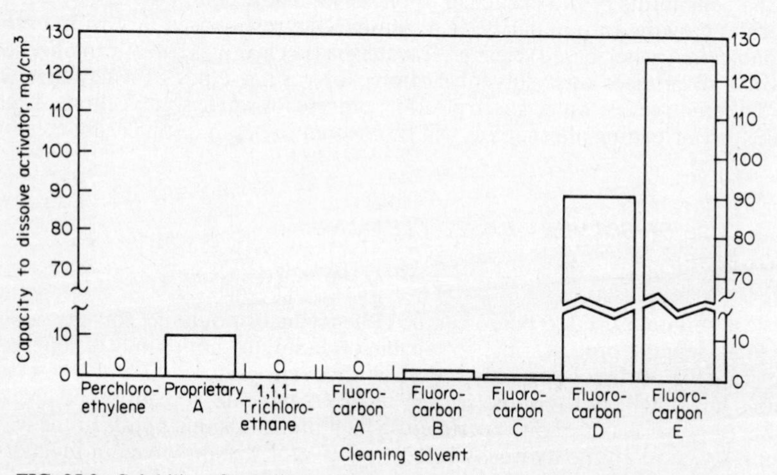

FIG. 25.3 Solubility of a typical ionic flux activator in commonly used flux solvents. Note that only a polar solvent will dissolve a polar salt. Single-component systems: perchlorethylene; 1,1,1-trichloroethane; fluorocarbon A. Designed blends: proprietary A and fluorocarbons B, C, D, and E.

The more polar the blend, the greater the amount of activator dissolved. So in choosing a solvent for postcleaning of electronic assemblies, it is important to choose one that has the capacity to remove both ionic and nonionic soil.

Another consideration is the compatibility of the solvent with the assembly to be cleaned. The solvent must be capable of rapidly removing rosin and handling soils and ionic residues without damaging marking inks or the materials used in the construction of the assembly. That property is related to strength as an organic solvent. If the solvent is too strong, it will dissolve more than the rosin. If, on the other hand, it is too weak, an excessively long time may be required for it to dissolve the rosin residues. The strength of an organic solvent is related to chemical structure and the effective temperature of the cleaning process. The role of temperature is straightforward: the higher the temperature the more active the solvent. The role of structure is more complex. We can generally relate the strength to the number of chlorine atoms on the molecule. When fluorine atoms are added to the solvent molecule, the molecule becomes less active. So, as a general rule of thumb, chlorocarbon solvents are stronger than fluorocarbon solvents. Chlorocarbon solvents have higher boiling temperatures and lower volatilities, and they are therefore adaptable to cold cleaning processes. Fluorocarbon solvents have lower boiling temperatures and higher volatilities, and they are much more adaptable than chlorocarbon solvents to vapor degreasing processes.

25.5.1 Redistillation

To be used economically in a cleaning process, the solvent must be easily redistilled. The individual components of any given solvent blend may boil over a very broad range, which will make recovery by distillation almost impossible. Unless solvents that boil within a very narrow range can be selected, redistillation will be economically infeasible. Solvent blends which boil at only one temperature are the most preferable alternatives. Those solvent compositions, called "azeotropes," are also ideal for vapor-cleaning operations. It is important to realize that azeotropes exist at one and only one temperature, the boiling point. An azeotropic mixture used at temperatures below boiling will rapidly change composition because of the unequal volatilization of the solvents. The ability to redistill the solvents also means that stabilizers and acid acceptors added to the original solvent must be present after distillation. They should come over during distillation in sufficient amounts to ensure the stability of the redistilled solvent.

25.5.2 Health Requirements

Another area of concern is safety and health requirements. In that respect the two most often quoted, and least understood, parameters are flash point and threshold limit value (TLV).

25.5.3 Flash Point

The flash point of a solvent can be measured in several ways: tag open cup test, tag closed cup test, Cleveland open cup test, and Cleveland closed cup test. For our purposes it is unimportant to know the mechanics of each test. It is enough to realize that the flash point of a given solvent measured by each of the tests could vary within about a 30°F range. Obviously, most manufacturers quote the most optimistic figure. So when you compare two solvent blends on the basis of flash point, make sure you are comparing apples and apples, not apples and oranges. OSHA requirements are standardized on the tag open cup test. Be sure you have that number before approving a solvent for use.

The flash point does not really measure flammability; it measures the temperature at which the solvent vapors will momentarily sustain a flame. In many cases, the solvent will flash and then extinguish itself. In general, most chlorocarbon solvents do not have flash points and are nonflammable materials; fluorocarbon solvents such as FC 113 and FC 112 also are nonflammable. Usually, flammable mixtures result from the addition of polar solvents such as the alcohols to increase the solvency power of ionic materials. In that case, it's strictly a question of the ratios. The more flammable the solvent in the vapors, the greater the propensity to flash. So when a solvent blend is designed, a delicate balance is normally struck between the ability to remove ionic soil (the principal objective of the solvent) and flammability.

25.5.4 Threshold Limit Value

The other parameter of concern is toxicity or TLV. It refers to the tolerance level of an "average" person for the solvent vapors. It is given as an index, running from 1 to 1000, that is derived from the maximum amount of solvent (given in parts per million) that a person can be exposed to during an 8-h working day. The lower the index, the more toxic the solvent. In general, chlorocarbons have lower TLVs than fluorocarbons. The index is somewhat subjective but is nonetheless an excellent tool for ranking the toxicity of solvents. However, TLV doesn't tell the whole story; the other half is evaporation rate. The more a solvent evaporates, the higher the concentration of solvent vapors in the ambient. Therefore, a solvent with a high TLV and a high evaporation rate may be just as toxic as, if not more toxic than, a solvent with a low TLV and a low evaporation rate. TLV is a factor inherent in the solvent itself; process modifications will not change it. Evaporation rate, on the other hand, can easily be minimized by proper equipment design and ventilation. Along those lines, don't forget the human element. The adverse effects of organic solvents have been given widespread publicity. Unless job assignments are made with a view toward the psychological implications, do not be surprised if delays, complaints, and union grievances result when sensitive individuals begin working around organic solvents.

25.6 WATER-BASED CLEANING PROCESSES

Water is one of the most effective solvents for rosin activators; but because it is a polar liquid, it will not dissolve such nonpolar materials as rosin. One way to overcome that difficulty is to react the rosin with other chemicals to form a water-soluble material called a "rosin soap." The reaction, called "saponification," is shown in Fig. 25.4. Like most chemical reactions, the saponification of rosin is highly dependent on time and temperature. If the temperature is too low or the saponification agents are not in contact with the rosin for a long enough time, the reaction will not go to completion. If that happens, the result is a very dirty board. Because saponification agents react with the rosin, they will be depleted according to the amount of rosin on the boards; therefore, they must be replenished on a carefully controlled basis. Unlike solvents, the exhausted material cannot be reclaimed and must be discarded. Once the rosin has been converted to a rosin soap, the soap must be completely rinsed from the printed board. That is nor-

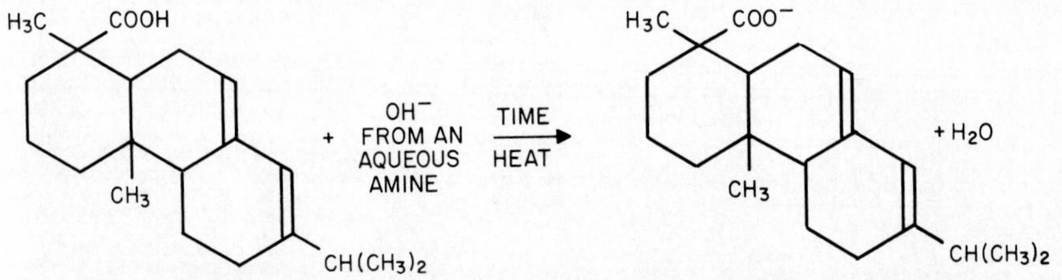

H_3C COOH ... + OH^- FROM AN AQUEOUS AMINE —TIME/HEAT→ H_3C COO^- ... + H_2O; CH_3; $CH(CH_3)_2$

FIG. 25.4 Saponification of rosin. When a biodegradable rosin cleaner is used, the rosin is changed chemically so that it is water-soluble.

mally done by spraying large quantities of water on the board surface. Several rinse steps are normally used to ensure that all the rosin has been completely cleaned off the board. The important point to remember here is that the parts must be thoroughly rinsed in water of as high a quality as possible.

If the circuit assembly has very high reliability requirements, a final rinse in deionized water is essential. Water can be one of the most deleterious contaminants a board will encounter. If the water is high in ionic salts, improper rinsing may leave the board more contaminated than if it had not been cleaned. If a board has capillary spaces, such as open relay coils, or if it has tight component spacing, such as the space between dual-incline packs and the board surface, then it may be impossible to rinse it thoroughly.

Once the parts have been thoroughly rinsed, they must be dried. Drying is accomplished by using air knives to blow off excess moisture and then baking the board at elevated temperatures to remove residual moisture. Since water has a lower volatility than fluorinated organic solvents, the boards must be baked for an extended period of time to remove the moisture. Alternatively, fluorocarbon drying systems which will replace the excess moisture with a higher-volatility organic solvent are available. The use of this type of procedure, however, negates the original objective of using the biodegradable rosin cleaner, namely, to avoid the use of organic solvents.

The process just described adapts itself quite well to automation, and several manufacturers have introduced conveyorized equipment to that end. An automated system such as the one shown in Fig. 25.5 can be readily adapted to a specific application. However, one fact to consider when such a conveyorized system is designed and installed is drying time. Because of the time and temperature involved in assuring complete drying of parts, a conveyorized system moving at any reasonable speed would be ludicrously long. Most manufacturers (rightly so) skimp on the length to allow the machine to operate in a reasonable manner. Be prepared, therefore, for either a supplemental baking operation or a reasonable dwell time at the end of process to allow all the moisture to evaporate.

25.6.1 Cleaning Water-Soluble Fluxes

Water is also used as the principal cleaning solvent for water-soluble fluxes, the residues from which are far more corrosive than those from an activated rosin flux. They must be thoroughly rinsed off to ensure trouble-free operation of the printed board. The cleaning of water-soluble flux residues, however, is quite a complex subject and one which seems to be very controversial. Some of the residues obtained after soldering with a water-soluble flux are "sparingly soluble" in water. That means that, in order to remove them with water alone, very high temperatures and very large volumes of water must be used. Even then, we are

FIG. 25.5 Typical conveyorized water cleaner. (*Hollis Engineering.*)

not assured of their complete removal. To hasten the removal of the residues and convert them into salts that are more soluble, many people use a neutralizing agent such as ammonia water, alkaline detergents, dilute ammonium hydroxide, or similar proprietary chemicals. Most of the agents have the disadvantage that they are probably more corrosive and more deleterious to board performance than the flux residue itself. Although they are for the most part essential to the effective cleaning of residues from water-soluble fluxes, the neutralizing agents must also be thoroughly rinsed with large volumes of clean deionized water to ensure their complete removal. Entrapment of any of those residues, or the flux residues themselves, in a capillary or beneath a component, virtually assures a failure of the printed board. Extreme care must be taken in the design of an assembly which is to be processed by these kind of techniques to avoid capillaries and potential sites of residue entrapment.

Also, to minimize the dangers of using a neutralizing agent, materials based on subliming ammonia salts should be selected. If they are trapped, they can be dissipated either in the heating stage of the cleaner or during storage at factory ambient.

The equipment for cleaning water-soluble flux residues can be as simple as a common dishwasher or as complex as the conveyorized systems previously described. For applications in which reliability of the finished product is important, however, it is highly recommended that the in-line conveyorized type of cleaner be used. It assures that parts are cleaned immediately, which provides minimal time for the formation of difficult-to-remove corrosion products. Also, the more automated the system is, the more uniform and controlled the cleaning and drying process will be.

Admittedly, the preceding discussion espouses a somewhat conservative viewpoint concerning water-based cleaning systems. That is because of the myriad of field problems encountered by companies unaccustomed to some of the nuances of water cleaning. The secret of an effective water cleaning system is design not just of the process, but of the assemblies to be cleaned. If either is incorrect, the disadvantages of water cleaning rapidly show up.

Unquestionably, water is a better solvent for flux activators than any organic solvent currently available, and it will obviate all of the legal and environmental requirements inherent in solvent cleaning. However, it is not a panacea. Not all assemblies are compatible with water cleaning; the probability of failure of an improperly water-cleaned unit is much higher than that of an improperly solvent-cleaned unit; and the cost advantage may be more imaginary than real. Weigh your alternatives carefully; the rewards for success are quite attractive, but the penalty for failure is unduly severe.

25.7 CLEANING EQUIPMENT—COLD CLEANING

The type of equipment chosen for a cleaning process will depend on the solvent used, the temperature at which the solvent is to be used, the level of cleanliness desired, and the process throughput. For applications requiring moderate cleanliness and a rapid in-line cleaning process at a minimum cost, an in-line wave or brush cleaner is the ideal choice. Cleaners of those types are generally used with chlorocarbon solvents, although some of the lower-volatility fluorocarbons have also been successfully used.

25.7.1 Wave Cleaning

A wave cleaner such as the one shown in Fig. 25.6 works in a similar manner to the solder wave discussed in the preceding chapter. The solvent is pumped through a nozzle to form a large standing wave of solvent. The lip of the nozzle is contoured to broaden and flatten the wave. That maximizes the contact time of the wave with the board to be cleaned. Because the board is still warm from the soldering operation, the solubility of the flux residues is greatly enhanced. The system should be connected directly to a still so that cleaning solvent can be con-

FIG. 25.6 Typical solvent wave cleaner with optimal brush installed. Solvent is pumped in a broad flat wave, and the boards are passed over the wave. (*Electrovert, Inc.*)

tinuously introduced into the system at the same time dirty solvent is being removed for reclamation. Otherwise, flux contamination will rapidly build up in the solvent and result in a redeposition of the flux residues on the printed board to be cleaned. Obviously, that will defeat the purpose of the cleaning operation.

25.7.2 Brush Cleaning

Brush cleaning is a variation of wave cleaning. In brush cleaning, a large rotating brush is added to the system in the center of the wave. The soft nylon bristles of the brush rotate in a direction counter to the forward motion of the printed board and add a scrubbing action to aid in cleaning. The height of the brush is adjustable so that the brush can remain just below the wave surface and still contact the board surface effectively. The brush must be changed periodically because flux residues have a tendency to make the bristles gummy and stiff. A typical brush cleaner is shown in Fig. 25.7.

Off line, the same type of cleaning operation can be done with a brush and a pan of solvent, but that poses many problems. If the operation is limited to hand dipping of parts, several immersions must be made, each in successively cleaner solvent, in order to minimize redeposition of the flux residues. The first tank must be changed often to minimize contamination buildup. After the contents of the first tank are dumped into an appropriate holding tank, the second tank becomes the first, etc., and the purest solvent is used in the last tank. If the procedure sounds tedious in theory, it is even more so in practice. Most often it degenerates into a single tank changed once a shift, if that often, and, under the guise of solvent economy, the dirty parts are sent to the next operation.

A variation of the procedure is brush cleaning after touch-up and repair. In that operation, the solvent is contained in individual dispensers at the work stations. Using a small brush, the operator dabs solvent on a localized glob of repair flux. Scrubbing vigorously, he or she redistributes the localized glob over the

FIG. 25.7 Typical brush cleaner. The brush can be mounted either in the cleaner or above it. (*Electrovert, Inc.*)

entire board; and when the solvent evaporates, he or she passes the uniformly contaminated board on to the next process step. Of course, not all the flux residue remains on the board; quite a bit remains on the brush to assist in contaminating the next part.

25.8 CLEANING EQUIPMENT—VAPOR CLEANING

For solvent-cleaning applications in which cleanliness is paramount, vapor degreasing is the most popular choice. Depending on the application, vapor cleaning can be done in a small off-line batch-cleaning unit or in an incline machine capable of high process throughput. Vapor degreasing is a technique long popular in the plating and metal-finishing industries, and it is ideally suited to cleaning of printed boards. It is based on the fact that the hot vapors of a boiling solvent will condense on any surface that is at a temperature lower than the boiling temperature of the solvent. Since the vapors are high-purity solvent, the part will be rinsed with solvent of very high purity until the temperature of the part equilibrates with the boiling temperature of the solvent.

25.8.1 Vapor Degreasers

A typical vapor degreasing unit is shown in Fig. 25.8. It consists of two sumps, one of which will continuously boil and contain the bulk of the flux residues. The other sump will contain continuously redistilled solvent held at any desired temperature. Any contaminant in the second sump will be continuously diluted as clean solvent drips into the tank and dirty solvent cascades into the boiling sump. The vapors from the boiling sump are condensed on coils maintained at low temperatures. The coils, situated above both sumps, prevent vapor from escaping

FIG. 25.8 Batch vapor cleaner. (*Electrovert, Inc.*)

from the degreaser and continuously supply the clean sump with redistilled solvent. Some degreasers also are equipped with a spray nozzle that draws solvent from either the cold sump or the distillate and allows it to be sprayed on the part to be cleaned. Large vapor degreasers can be designed with conveyor belts to adapt them for in-line use. The conveyor belts can be made to move through just about any desired sequence of spray, dip, and vapor cleaning.

25.8.2 Selection of Construction Materials

Stainless steel is the best material for the construction of a vapor degreaser. For economic reasons, some of the plumbing in the degreaser can be made from copper, but that does not include any plumbing which is likely to be carrying a water and organic solvent mixture. Active metals such as aluminum and zinc must be strictly avoided because they catalyze the breakdown of organic solvents and result in the formation of acidic groups that attack both the degreaser and the parts being cleaned. Materials containing iron, which could rust in the presence of moisture, should be avoided for the same reason. Most organic solvents used in vapor degreasing are fortified with stabilizers and acid acceptors to guard against acid breakdown. However, the use of active metals in the degreasing fixtures or in the material used in the construction of the degreaser rapidly negates the effectiveness of the stabilizers. The discussion has been confined to vapor cleaning, but similar considerations apply to the use of active metals in any solvent-cleaning operation.

25.8.3 Boiling and Ultrasonic Agitation

One other topic germane to our discussion of vapor cleaning operations is the role of boiling and/or ultrasonic agitation. There is a widespread uneasiness in the electronics industry concerning the use of ultrasonic agitation to remove soil. The uneasiness stems from the theoretical consideration that beam leads and fine wires used in the assembly of active microelectronic components might hit resonant frequencies and be destroyed. The possibility seems to delight many theoreticians who thrive on computer modeling of abstract concepts, but in practice the notion is farfetched. Properly designed and implemented ultrasonic agitation can provide an exceptionally effective scrubbing action to aid in the removal of some types of soil.

If ultrasonic agitation is too costly, or if the fears of the theorist are not easily quelled, then the other alternative is boiling agitation. Immersing the part to be cleaned in a sump of clean boiling solvent is an extremely effective procedure. Of course, all components of the assembly must be compatible with the boiling solvent and must remain intact after exposure to the boiling temperature of the solvent.

25.8.4 Nonazeotropic Vapor Cleaning

Some systems that have been marketed recently utilize nonazeotropic mixtures for vapor degreasing. A typical machine is shown in Fig. 25.9. The systems rely on the fractionation of mixtures of fluorocarbon 113 and alcohol during the distillation process. Each rinse sump will thus have a successively lower content of the lower-volatility alcohol. The vapor composition of these systems is fairly high

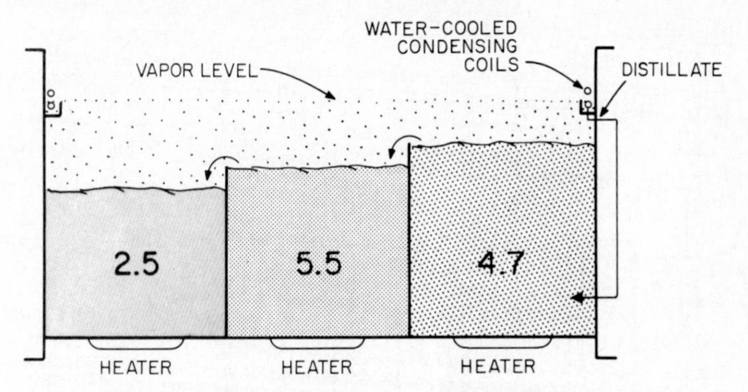

FIG. 25.9 Nonazeotropic vapor-cleaning processes offer the advantage of higher alcohol contents than azeotropic systems. Boldface numbers indicate alcohol concentration. (*Imperial Chemical Industries, Inc.*)

in alcohol and must be monitored carefully to ensure that the ratio of alcohol to fluorocarbon does not exceed the flammability point. The procedure is extremely effective for cleaning rosin-based fluxes, but it must be run on specially constructed equipment and it must be carefully maintained to ensure that the alcohol content does not reach the region of flammability.

25.9 EFFECTIVENESS OF A PROCESS ALTERNATIVE

It is always difficult to draw a comparison of one process with another. The effectiveness of a process depends on many factors which are quite independent from one specific application to the next. However, without offering a lot of weasel words and a long list of exceptions, we can examine some laboratory data obtained on boards of similar configurations with similar soldering procedures. The data should suffice at least to make generalized rules of thumb and broad generalizations.

The data shown in Fig. 25.10 were obtained by using the Ionograph to measure the cleanliness of boards fluxed with an activated rosin flux and cleaned with various solvents. The processes selected for evaluation were cold dip (or immersion) and two vapor-degreasing cycles. The length of the bar in the graphs represents the amount of ionic soil remaining after the cleaning operation. The results correlate well with the results obtained in Fig. 25.3 relating to the ability of the solvents to dissolve flux activators.

As can be seen from the data, vapor degreasing with a solvent designed to remove ionic material is the most effective process. Fluorocarbon solvents adapt best to the vapor-cleaning processes. Cold cleaning with nonpolar chlorinated solvents is essentially a futile attempt to clean the boards, whereas cold cleaning with a properly designed chlorocarbon cleaner can match vapor cleaning in effectiveness. When these data were presented, no data on water-based cleaning processes had been obtained. Since then, however, a wealth of data has been produced by several people. The data tend to show that a properly designed water-based cleaning process will be almost as effective as vapor cleaning with an effective fluorinated blend. However, the number of boards on which the cleaning was incomplete is alarming. Almost the same thing holds true for water-soluble flux–

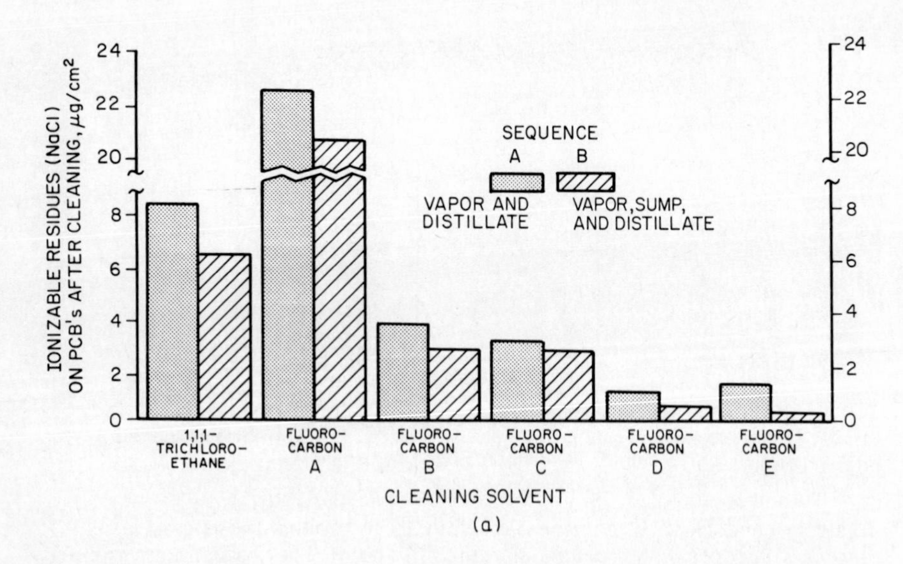

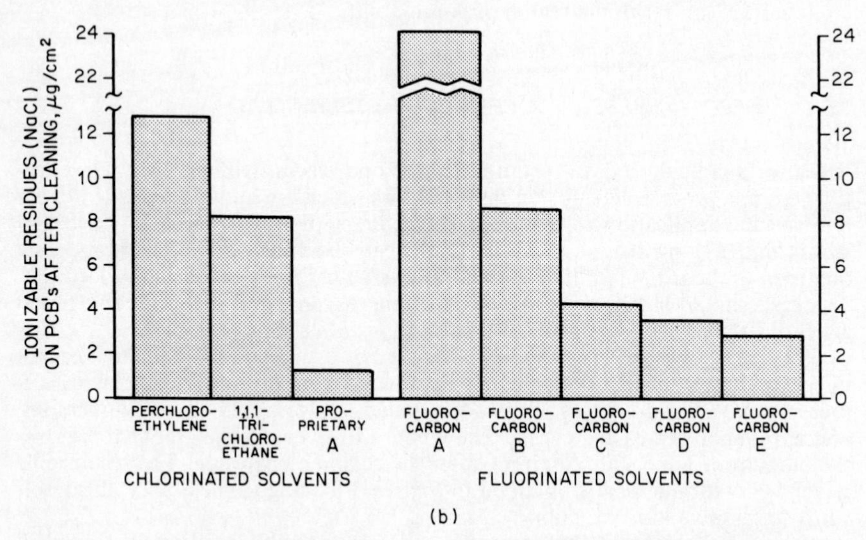

FIG. 25.10 Relative effectiveness of various cleaning solvents in (*a*) vapor cleaning and (*b*) cold cleaning. A solvent which is a blend of polar and nonpolar solvents should always be selected. (*Courtesy of Alpha Metals, Inc.*)

water-cleaning processes. Use of a properly designed process with an appropriate neutralizer and sufficient quantities of deionized rinse water results in an end product that is cleaner than any solvent-cleaned board measured. Again, however, the same qualification holds true. The number of highly contaminated boards seen in actual field measurement is exceptionally high.

25.10 DEFECTS RELATED TO CLEANING—WHITE RESIDUES

Two of the most common defects related to improper cleaning of the printed circuit assembly are white residues and mealing. A board exhibiting white residues is shown in Fig. 25.11. White residues have a number of causes, and not all problems identified in the field as white residues are the same thing. White residues can be caused by improper cure of the base laminate or by solvent attack on

FIG. 25.11 White residues on circuit board.

improperly cured solder resists. Those defects, however, are relatively rare. Most commonly, white residues are caused by the polymerization of rosin fluxes and/or water-dip lacquers during the soldering operation. When the rosin in those materials polymerizes, some of the rosin becomes a very long chain molecule which cannot be dissolved in the commonly used solvents. Therefore, the flux-cleaning solvents dissolve only the short-chain rosin segments and the original rosin and leave behind a tenaciously adhering white powder which is polymerized rosin. Once the polymerized rosin has formed, not even the best flux solvents will dissolve it. There is, however, a way to remove it. Very simply, like materials can dissolve one and another; so if the board is recoated with a rosin-based flux, the rosin in the flux dissolves the polymerized rosin. As long as there is sufficient rosin to maintain the polymerized rosin in solution, the entire baord can be readily recleaned by using normal cleaning methods (i.e., vapor cleaning).

25.11 DEFECTS RELATED TO CLEANING—MEALING

The other cosmetic defect directly related to improper cleaning procedures is mealing: white granular-appearing spots which form between the printed board and a conformal coating. The board shown in Fig. 25.12 exhibits mealing. Do not confuse mealing of the conformal coat with measling of the base laminate. Although the defects appear to be similar, measling is related to selective delam-

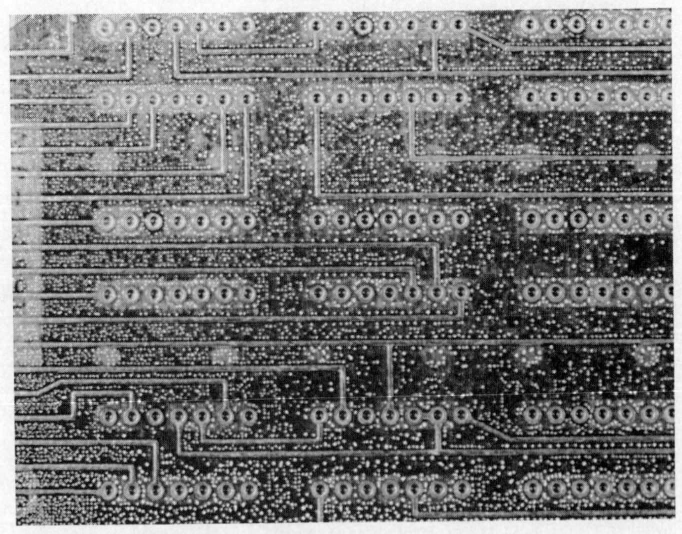

FIG. 25.12 Circuit board with mealing.

ination of multilayer printed boards. It is caused by resin-starved regions in the laminate, and it shows up long before the soldering operation.

There is a school of thought in the printed circuits industry that feels that mealing is not detrimental to the printed board. That is simply not true. Mealing is symptomatic of ionic contamination on a printed board. As such, it indicates inadequate cleaning of the assembly, and it casts serious doubt on the long-term reliability of the circuit.

Since conformal coatings are essentially polymer films, they will behave in a manner similar to that of thin polymer membranes. They are not impervious to water vapor; instead, they will experience a distinct osmotic pressure and, depending on the coating used, will be fairly porous to water vapor. When a void occurs in the conformal coat, either in the coating itself or between the board and the coating, the transmission of water vapor will essentially become localized to the void area. Hence the void induces a kind of localized pumping action for the water vapor. When ionic materials are near the void area, they dissolve in the water vapor and are carried to the void. Since the membrane transmits the moisture and not the salt, the salt concentrates in the area of the void to form a meal. Obviously, the effect is more than just cosmetic.

Mealing is sometimes cured by switching to a different conformal coat, one with a much lower propensity to transmit water vapor. That camouflages the problem, but it doesn't solve it. If mealing occurs, the problem is in the cleaning of the board, not the conformal coating. Cleaning procedures should be reexamined and modified to ensure that all ionic contaminants are removed.

25.12 CIRCUIT CLEANLINESS

In normal handling of circuit boards and final assemblies, a certain amount of contamination is inevitable. A cleaning process can be added, but it also adds expense and time to the total cycle. However, the complication of ionic contamination is the electrical effect of leakage paths at unknown resistance. To see if this is a potential problem, check the circuit for the resistance levels required and then test the board for contamination by means of one or more of the following tests. If unacceptable contamination is detected, both preventive measures in the basic process and a cleaning step should be installed to facilitate reaching the acceptance level.

25.12.1 MIL-P-28809 Excerpts on Cleanliness

3.7.1 Resistivity of solvent extract (see . . . 6.7). When uncoated printed-wiring assemblies are tested as specified in 4.8.3, the resistivity shall not be less than 2,000,000 ohm-centimeters. Equivalent test methods may be used in lieu of 4.8.3 only when specifically approved by the government procuring activity. Such approval will be determined on the basis that the alternate method is demonstrated to have equal or better sensitivity, and employs both polar and nonpolar solvents with the ability to dissolve flux residue as does the alcohol-water solution specified in 4.8.3.

4.8.3 Cleanliness and resistivity of solvent extract (see . . . 3.7.1). A convenient sized funnel shall be positioned over an electrolytic beaker. The printed-wiring assembly shall be suspended within the funnel. A wash solution of 75 percent by volume of ACS reagent grade isopropyl alcohol and 25 percent by volume of distilled water shall be prepared. This wash solution must have a resistivity equal to or greater than 6×10^6 ohm-centimeters (see 6.7). The wash solution shall be directed in a fine stream from a wash bottle onto both sides of the assembly, until 100 milliliters of the wash solution is collected for each 10 square inches of board surface (including both sides of the board, but not counting the surface area of the parts mounted thereon). The time required for the wash activity shall be a minimum of one minute. It is imperative that the initial washings be included in the sample to be measured for resistivity. The resistivity of the collected wash solution shall be measured with a conductivity bridge or other instrument of equivalent range and accuracy. Note: All laboratory ware must be scrupulously clean (see 6.7).

6.7 Resistivity of solvent extract (see 4.8.3). This test procedure, including solution preparation and a laboratory ware cleaning procedure, is documented in Materials Research Report No. 3-72, "Printed-wiring assemblies; detection of ionic contaminants on." Application for copies of this report should be addressed to the Commander, Naval Avionics Facility, Indianapolis, Indiana 46218.

25.12.2 Water Extract Conductivity Method for Determining Ionic Contamination

25.12.2.1 Outline of Method. The plated parts are placed in water of known conductivity and agitated for a specific time. The conductivity of the water extract is measured, and the increase in conductivity due to residual plating salts and other conducting impurities is calculated.

25.12.2.2 Applicability. This method is applicable to the detection of residual plating salts and other soluble, ionizable contaminants on parts that are completely plated.

25.12.2.3 Reagents and Apparatus

1. *Conductivity monitor:* A conductivity monitor with a primary scale of 0 to 2 μS/cm and additional scales for higher readings. An instrument such as the Balsbaugh Laboratories Model 910M-2C Monitor, obtainable from Balsbaugh

Laboratories Incorporated, 25 Industrial Park Road, South Hingham, MA 02043, or equivalent, may be used.

2. *Working conductivity cell:* A conductivity cell with an integral thermistor for automatic temperature compensation with a constant of 0.1 cm^{-1}. A conductivity cell such as the Balsbaugh Laboratories No. 910-.1TD Dip Conductivity Cell, or equivalent, may be used.

3. *Standard conductivity cell:* A duplicate of the working conductivity cell to be used only for checking the calibration of the working conductivity cell.

4. *Cell selector switch:* A cell selector switch for connecting either the working or standard conductivity cell to the conductivity monitor. A switch such as the Balsbaugh Laboratories No. 910SS-2 Selector Switch, or equivalent, may be used.

5. *Patch cord:* A shielded patch cord for cell selector switch to conductivity monitor connection. A patch cord such as the Balsbaugh Laboratories No. 910PC-10 Patch Cord, or equivalent, may be used.

6. *Jogger:* A jogger (agitator), 3600 cycles/min (60-H current), with adjustable amplitude. A jogger such as the Single Action Model DL-1A, obtainable from Syntron, a division of FMC Corporation, Homer City, PA 15748, or equivalent, may be used. Joggers with larger tables are also available.

7. *Water bottle:* A polyethylene bottle of about 1.9-L (1-gal) capacity with a polyethylene or polypropylene screw cap.

8. *Deionized water:* Deionized water with a conductivity of 0.5 μS/cm or less (resistivity 2 MΩ or more).

25.12.2.4 Preparation of Equilibrated Water and Calibration of Working Conductivity Cell. Put the working and standard conductivity cells into separate container of deionized water. Before removing them to make a conductivity measurement, *always* measure the water in which they are stored to be certain it is less than 1 μS/cm. That ensures the cells are clean.

CAUTION: *The standard conductivity cell is used only for checking the working conductivity cell in equilibrated water. It must not be put in any other liquid.*

Rinse the polyethylene water bottle thoroughly with deionized water; then half fill the bottle with deionized water. Shake the bottle vigorously for 2 min to equilibrate the water with the carbon dioxide in the air. (See Note.)

NOTE: Carbon dioxide is a normal and necessary component in air. It is soluble in water and forms carbonic acid, which ionizes and is at equilibrium at 0.8 μS/cm. In a closed polyethylene bottle, the equilibrated water will remain in the range of 0.8 to 1.0 μS/cm for at least a week.

Transfer some of the equilibrated water to a Pyrex or other borosilicate glass beaker of sufficient size to hold both the working and standard conductivity cells. Insert both cells and measure the conductivity with the standard cell first. Check to see that the level of equilibrated water is sufficient to cover the side holes in the cells. That is necessary to obtain a correct reading. The conductivity reading should be between 0.8 and 1.0 μS/cm; if not, see the following note.

NOTE: If the conductivity is higher, either the polyethylene bottle or the beaker was contaminated. Clean both by rinsing several times with deionized water and prepare new equilibrated water. If the conductivity of the equilibrated water is lower, continue shaking until water in the desired range is obtained.

Next, measure the conductivity with the working conductivity cell. Both conductivity cell readings should agree within 0.1 μS/cm; if they do not, see the following note.

NOTE: If the conductivity cells do not agree within 0.1 μS/cm, clean the cell surface by washing the working conductivity cell in reagent grade acetone, followed by rinses with equilibrated water.

Store the working and standard conductivity cells in equilibrated water in separate Pyrex or other borosilicate glass beakers.

25.12.2.5 Procedure for Determining Conductivity of Extract of Plated Parts

1. *Test facilities:* This test procedure should not be performed in a plating shop or in the presence of acid or alkaline mist or in a laboratory with noticeable acid fumes.

2. *Sample:* Calculate the surface area of the plated parts to be extracted. A sample of the plated parts having a total surface area of 32.2 cm^2 (5 in^2) should ordinarily be used and extracted in 100 cm^3 of equilibrated water. For other sample sizes, the volume of equilibrated water should be adjusted to keep the ratio of surface area to volume of water fixed at 0.32 cm^2 (0.05 in^2) to 1 mL of water.

3. *Extraction:* Measure the correct volume (see sample, 2 above) of equilibrated water into a clean, dry Pyrex or other borosilicate glass beaker. Measure and record the conductivity of the equilibrated water (C_w) with the working cell. The conductivity should be in the 0.8- to 1.0-μS/cm range. If the conductivity is greater than 1.0 μS/cm, see the following note.

NOTE: Either the equilibrated water or the beaker or both are contaminated. Check the conductivity of the equilibrated water and wash the beaker to correct the condition.

Add the plated parts to the beaker containing the equilibrated water and place the beaker on the jogger. Do not touch the parts with bare fingers, and be certain that the parts are completely covered by the water.

NOTE: To prevent the lateral movement of the beaker while the sample is being agitated, cover the top of the jogger with an approximately 1.27-cm-(0.5-in-) thick rubber pad with holes of suitable size to accommodate the beakers.

Turn on the jogger and adjust the amplitude of the jogger until a standing wave exists on the surface of the liquid.

CAUTION: *Do not increase the amplitude of the jogger to a point at which droplets of liquid will be expelled from the beaker.*

After 15 min of operation, turn off the jogger and remove the beaker.

Measure the conductivity of the water in which the working cell has been stored. If it is greater than 1.0 μS/cm, replace with equilibrated water.

Measure and record the conductivity of the sample extract (C_E) in the beaker with the working cell.

Calculate the increase in conductivity of the sample extract due to residual plating salts and other conducting impurities as follows:

$$\Delta C = C_E - C_W$$

where ΔC = increase in conductivity, μS/cm
C_E = conductivity of sample extract, μS/cm
C_W = conductivity of water, μS/cm

25.12.3 Lieberman-Storch and Halphen-Hicks Tests for Rosin

25.12.3.1 Scope. This method covers procedures for the qualitative detection of rosin in varnishes by the Lieberman-Storch and Halphen-Hicks tests. The rosin may be present as either free rosin (abietic acid), esterified rosin, or metal salts.

25.12.3.2 Apparatus

1. Test tubes, 2.5 × 15 cm
2. Filter funnel, 75 mm in diameter.
3. Porcelain spot plate.

25.12.3.3 Reagents

1. Acetic anhydride
2. Sulfuric acid reagent. Add 35.7 mL of concentrated sulfuric acid (H_2SO_4, sp. gr. 1.84) slowly to 34.7 mL of water, and cool to room temperature. Store in a glass-stoppered bottle.
3. Phenol reagent. Dissolve one part by volume of phenol in carbon tetrachloride.
4. Bromine reagent. Dissolve one part by volume of bromine in four parts by volume of carbon tetrachloride.

25.12.3.4 Procedure

1. *Lieberman-Storch test:* Place 0.1 to 0.2 g of the sample in a test tube and add 15 mL of acetic anhydride. Heat gently until the sample is dissolved or dispersed. Cool and filter, with an ashless rapid filter paper, into a clean test tube. Place a few drops of the clear solution in a depression of the spot plate and add one drop of sulfuric acid reagent, so that the acid will mix slowly with the filtrate. If rosin is present, a fugitive violet color develops immediately. A pink or brown coloration should be ignored. A control sample containing rosin should be run simultaneously.

2. *Halphen-Hicks test:* Dissolve a small quantity of the sample in 1 to 2 mL of the phenol reagent. Pour the solution into a cavity of the spot plate until it just fills the depression. A portion of the solution will spread out on the flat part of the plate a short distance beyond the rim of the cavity unless too much of the carbon tetrachloride has been lost by evaporation, when a drop or two more should be added to produce the spreading effect referred to. Immediately in an adjacent cavity, place 1 mL or more of the bromine reagent, so that the bromine vapors evolved will contact the surface of the solution in the other cavity. Sometimes it is necessary to blow a gentle current of air in the proper direction to accomplish this satisfactorily, or both cavities may be covered by a watch glass of suitable size. The development of a fugitive violet coloration, best observed upon a flat portion of the test plate, indicates the presence of rosin.

25.12.4 Chloride Determination with Silver Chromate Paper

25.12.4.1 Application. This test method is used to determine qualitatively if chlorides or bromides are present on solder joints or other materials or surfaces. The chloride or bromide must be on the surface or otherwise soluble in water.

25.12.4.2 Test Material

1. *Source:* Silver chromate reagent paper may be obtained from A. Eichhorn Co., 644 Salem Ave., Elizabeth, NJ 07209, (201)351-6975.

2. *Characteristics:* Silver chromate test paper is extremely sensitive. The paper should not be touched with the hands or allowed to come in contact with materials that have been contaminated by handling with bare hands. It should be handled with clean forceps; contact with clean scissors, glass rod, etc., is permissible. The use of disposable plastic gloves is advisable.

The paper is also light-sensitive, and its original brick-red color will fade if exposed to direct daylight for a prolonged time. It should always be kept in clean containers and shielded from daylight when not in use.

25.12.4.3 Procedure

1. Moisten a piece of silver chromate reagent paper of suitable size with distilled water. Use forceps or other suitable means to handle the paper. *Do not permit the paper to come in contact with the hands.*

Drain excess water by touching the edge of wet reagent paper with blotting paper. Place moist reagent paper in intimate contact with the surface area to be tested.

Use a clean glass rod or forceps to press the reagent paper down. Allow the paper to remain in position for 1 min; then remove it and examine it for color change. An off-white color indicates the presence of soluble chlorides or bromides.

2. As an alternate procedure, pipette a small quantity of distilled water into a capillary tube and transfer it to the area to be tested. One drop should be sufficient. Allow the water to remain in contact with the test area for 1 min. Remove water with the capillary tube and transfer it to the test paper. Observe for the color change to off-white or yellow-white, which indicates the presence of soluble chlorides or bromides.

NOTE: Certain acidic solutions may react with the reagent paper to provide a color change similar to that obtained with chlorides. When a color change is observed, it is advisable to check the acidity of the affected area by means of pH indicating paper. If pH values of 3 or less are obtained, the presence of chlorides should be verified by other analytical means.

25.12.5 Determination of Ionizable Surface Contamination

25.12.5.1 Scope

1. *Purpose:* This test method establishes a procedure for determining the amount of surface ionic soil on a circuit board. The soil must be soluble in water, alcohol, or some mixture of the two. The determination can be made on either a quantitative or a qualitative basis.

2. *Restrictions:* The Ionograph does not differentiate between specific ionic species. It determines their presence and ranks them according to their ionic mobilities. Salts with higher ionic mobilities are weighted heavier than salts with lower ionic mobilities.

25.12.5.2 Uses. This method has application as a quality control tool and also as a method for developing and evaluating cleaning process parameters. As a quality control tool, it can be used to inspect parts to determine if they conform to predetermined levels of cleanliness. In process development, the procedure can be used to evaluate solvent and process efficiency and also to set levels of acceptable cleanliness.

25.12.5.3 *Theory*

1. *Description of measurement technique:* The Ionograph utilizes a dynamic measurement of ionic conductivity of a rinse solution which extracts the ions from the surfaces being measured. The solvent is pumped through a recirculating loop which includes a plastic tank, conductivity cell, and ion-exchange column to remove all traces of ions from the solvent before entering the tank. The conductivity cell used is temperature-compensated to avoid reading variations caused by temperature changes. A metering pump pumps the solution through the loop at a constant rate.

Without a sample in the tank, a condition will be established in which the conductivity of the solvent, as measured by the conductivity cell, will attain a constant low value. With the introduction of a contaminated sample into the tank, the conductivity reading measured at the cell will rise rapidly. A recorder is used to follow the change of conductivity with time. The sample remains immersed in the solvent until the conductivity of the solvent returns to its original equilibrium level. At that point, no further ionic material can be removed from the sample.

2. *Theory of measurement technique:* The entire amount of ionic material removed from the sample can be related to the integral of the conductivity readings over the period of time required to dissolve the material and purge it through the system as follows: At any instant t the number of moles n_t of ionic material within the conductivity cell is $n_t = V_c \times C_t$, where C_t is the concentration of ions and V_c is the cell volume, which is constant. Over an infinite amount of time, the total number N of moles of ions passing through the cell N will be

$$N = \int_0^\infty nt\,dt = V_c \int_0^\infty C_t\,dt$$

Since we are dealing with very low concentrations (10^{-4} N), we can assume complete ionization; therefore,

$$\text{Conductivity} = L = kC$$

(assuming one salt to be present. Of course, different ionic salts with different ionic mobilities will give different conductivities for a given concentration).

$$N = kV_c \int_0^\infty L_t\,dt$$

If the monitor and recorder responses are linear with respect to L, then, according to this last equation, the area under the conductivity-time curve which is charted on the recorder is a linear function of N, the total amount of ions removed from the sample.

3. *Solvent systems:* The Ionograph can be used with either pure water or water-alcohol mixtures. Water is used when only water-soluble salts such as plating salts are to be measured. The use of pure water in the Ionograph results in a measuring fluid with an initial conductivity of about 0.1 μS (10 MΩ).

Water-alcohol systems are used when nonpolar soils might encapsulate or otherwise mask the water-soluble ionic soils. Various alcohols have been used successfully. The preferred systems use either *n*-propanol or isopropanol as the alcohol solvent. Because of the high dielectric constant of the alcohols, excessive alcohol in the mixture will generally degrade the sensitivity of the measurement. To obtain maximum sensitivity and to ensure sufficient alcohol to readily remove all nonpolar residue, the recommended mixtures are 40 percent (by volume) *n*-

propanol and 60 percent water or 50 percent (by volume) isopropanol and 50 percent water. Mixtures with as high as 75 percent by volume isopropanol have been successfully used.

25.12.5.4 Test Method

1. *Calibration:* Once the fluid in the system has established a stable level of conductivity for the pump rate, solvent mixture, and tank size selected for the test, a precision hypodermic needle is used to inject a known volume of a solution containing 1 $\mu g/\mu L$ of sodium chloride. A 30-μL injection is normally an adequate initial one. The output of the Ionograph chart recorder is then integrated to determine the total area under the curve corresponding to the initial injection of salt. Integration can be accomplished by using an electronic integrator or a planimeter or simply by weighing the cut-out area under the curve on a sensitive analytical balance.

Additional volumes of the calibration solution are then injected until a curve such as the one shown in Fig. 25.13 is obtained for the chosen set of test conditions.

2. *Test procedure:* Once the system has been calibrated in accordance with 1, above, the test sample is immersed in the sample tank. Care must be taken not to handle the sample or any of the appliances used to insert the sample into the tank. Finger dirt contains highly mobile ionic soils and may give spurious readings on the Ionograph.

During the course of the measurement, the conductivity of the solution will depart from the baseline of conductivity and then gradually return. When it has returned to the baseline, no additional soil can be removed and the measurement is complete.

The curve for the sample is then integrated as per 1, above, and the calibration curve is used to determine the amount of contamination on the part.

3. *Treatment of test data:* The number obtained from 2, above, will be the ionic contamination on the surface of the board in terms of equivalent micrograms of sodium chloride. (If the calibrating solution contained a salt other than

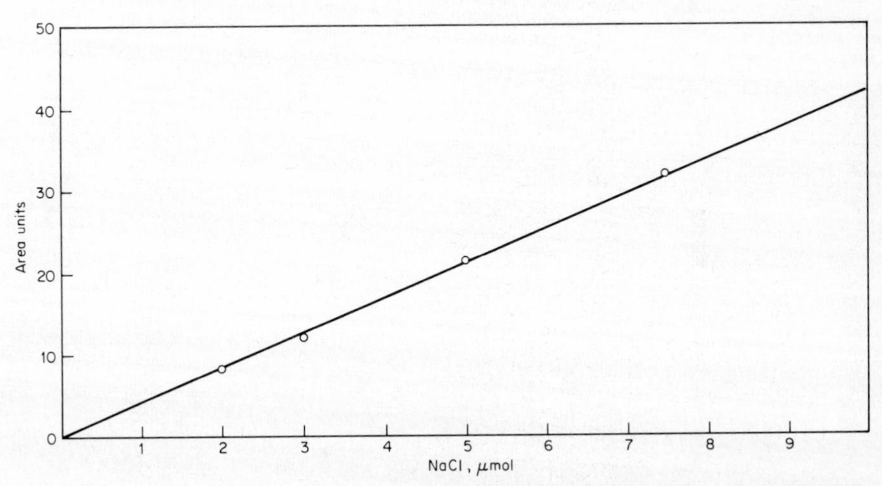

FIG. 25.13 Ionograph calibration: Area measured under recorded curve vs. micromoles of NaCl.

sodium chloride, the number will be in terms of equivalent micrograms of that salt.) It is common practice to divide that figure by the total area on both sides of the board and present the data in terms of equivalent micrograms of salt per unit of area.

By using a scale of measurement based on only one salt, i.e., sodium chloride, the ionic contaminants are being measured in terms of their ionic mobility. The more mobile or active an ion is, the more likely it is to cause trouble on a circuit assembly. Thus while the Ionograph method will not differentiate between specific ions, it is an effective way of quantifying the presence of many ions.

25.12.5.5 Miscellaneous: Other Uses of the Ionograph. Because the Ionograph measures the ionic activity of any part or solution which contains ionic material, it has been used for various other purposes. A partial list appears below:

1. Incoming inspection of reflowed tin-lead boards to determine if residues have been completely removed
2. Measurement of purity of incoming and redistilled solvents
3. Measurement of amount of activated rosin flux dissolved in the boiling sump of a vapor degreaser
4. Measurement of activity level of activated rosin fluxes

CHAPTER 26*

QUALITY CONTROL IN SOLDERING

Hugh Cole

26.1 INTRODUCTION

From the preceding chapters it should be apparent that an automated soldering operation can be a smooth-running, efficient way to assemble printed boards. When the processing parameters remain within their specified boundaries, the operation requires minimal maintenance and supervision. However, when one or two of the critical parameters drift away from their expected values, the result is mounting backlogs of quality control rejects and an inordinate amount of high-level supervisory and engineering time spent in hunting down the cause. That is why maintenance and quality control of the soldering process are so important. With a well-designed, well-documented quality control program, the loss of many hours of valuable time and much unwanted visibility can easily be avoided. This chapter will cover the quality control programs required for an automated soldering operation, and it will touch on various common defects and their causes.

26.2 PROCESS CONTROL

Once a suitable process has been designed (i.e., the proper flux, the proper method of application, the proper alloy, etc.), a few simple maintenance operations can easily be set up to ensure its smooth operation. This discussion covers a "typical soldering operation" and should serve as a skeleton on which to build your own program.

26.2.1 Solderability

Both boards and component leads are subject to a deterioration of the solderability of their surfaces as a result of sloppy manufacture or excessive storage environments. To prevent solderability problems from disrupting the smooth flow of

*Adapted from Coombs, *Printed Circuits Handbook,* 2d ed., McGraw-Hill, New York, 1979, Chap. 17.

work through a soldering process, solderability checks should be made on all incoming materials and on all materials entering the soldering process from storage. Boards and components with marginal and inferior solderability should either be rejected or have their solderability restored with an appropriate surface conditioner (see Chap. 23). Components placed in bins for hand insertion operations should also be checked periodically, since there is a great tendency to unknowingly subject some of those components to the factory environment for extended periods of time because of the last-in first-out scheme inherent in most bins.

The solderability of a surface is related to the ability of the molten solder to wet or spread out on the surface to be soldered. Several parameters can be examined to gain a good understanding of the solderability of a surface. However, the most common and the simplest indicators are based on the observations of either the total area covered by a controlled volume of solder or the angle which the solder makes with the surface to be tested. Since solderability is an empirical concept, rather than one which can be precisely formulated, solderability measurements are often surrounded with a mystical, semiquantitative set of procedures and numerical scales designed to give the tester a sense of making a precise analytical determination. That is not really the case, because the solderability of a surface is greatly dependent on your requirements and quite independent of any definition the author might formulate for solderability. In the final analysis, the solderability test should be an objective attempt to answer the question, "Will it work in my process?" rather than a subjective attempt to quantify the dihedral angle of a solder blob.

Solderability tests should be performed away from the production area, and they should be performed by using the flux and alloy to be used in the final assembly. They can be as simple as a hand-dipping operation on a statistically significant number of parts or as complex as a 100 percent inspection of all parts using the meniscograph. (For more information on solderability testing, see Chap. 23.) The type and extent of the solderability testing done for each process is a function of the activity of the flux (recall that RA* and water-soluble organic acid fluxes are much more forgiving of solderability defects than RMA* or R-type fluxes) and the final requirements of the assembly.

Do not forget that the solderability check is in reality a check both of your vendor's ability to produce solderable parts and your ability to store the parts in a suitable environment. Along those lines, work with your vendors to ensure that he or she knows your requirements and rapidly feed back any difficulties you encounter. Also, maintain a clean storage area with minimal handling of the parts and minimal dwell time.

26.2.2 Fluxes

The flux density should be checked at least twice during an 8-h shift. The simplest way is to use a hydrometer. The flux should be thinned with the appropriate thinner and makeup flux added.

Depending on the process throughput, the flux should be dumped after 30 to 40 h of operation and new flux added to the appropriate level. If a foam fluxer is being used, the stone should be cleaned periodically by soaking it in a tank of thinner and bubbling air through it for 10 to 15 min to assist in removing any flux residues which may be trapped. Since the foaming properties of most fluxes

*"RA" stands for "rosin—activated." "RMA" stands for "rosin—mildly activated."

are affected by moisture and oil, air lines feeding the stone must be fitted with traps to remove oil and moisture from the lines. The traps must be maintained periodically to ensure trouble-free operation. If the flux pot has brushes or squeegees to spread the flux on the board, they should be cleaned regularly (once a week) to prevent flux buildup.

26.2.3 Preheaters

The secret to the successful and safe use of preheaters lies in the regular and thorough cleaning of flux drippings from the preheater surface and/or the reflectors. Preheaters should be cleaned about once each shift to ensure proper operation. The preheater temperature should also be checked to ensure that temperature settings are accurate and all heating elements are operating properly. If forced-air preheaters are used, the air-intake filter and/or grating must be periodically checked to ensure unrestricted passage of air across the heating elements.

26.2.4 Solder Pot

The most important solder pot parameters to be controlled are temperature and composition. The temperature should be verified with an independent thermometer at least once per shift to ensure reliability of the solder process. The alloy composition should be tested at least once per month to guard against rapid buildup of metallic impurities. Also, dross should be removed periodically and replaced with new solder alloy to maintain the proper solder level. Depending on the process throughput, it is also a good idea to set up a schedule of regular additions of fixed amounts of solder chunks to the pumping system. If oil is being used, a regular cleaning and replacement schedule is essential. No soldering oil should be used for more than 16 h. Most oils should be changed every 8 h to be effective.

26.2.5 Cleaning Systems

If a solvent-based cleaning system is used, the level of acid acceptors in the solvent should be checked periodically to guard against acid degradation. Most manufacturers will supply procedures or simple test kits to perform the test. If the solvent is recovered by batch distillation, the test should be performed after each distillation. If vapor degreasing is used, it is usually sufficient to test the level of acid acceptance once every three or four weeks.

If a water-based cleaning system is used, the conductivity or resistivity of the rinse water should be continuously monitored to ensure adequate cleaning. In any case—whether solvent or water cleaning, a check of some fraction of the boards with a device such as the Ionograph should be made. That ensures that adequate levels of cleanliness are being maintained.

26.3 CAUSES FOR DEFECTIVE JOINTS

Under ideal soldering conditions, perfect wetting of the component leads and the printed board terminal area results in solder joints with smooth, bright, concave fillets that exhibit very low contact angles with both the lands and the component leads. Such a solder joint is shown in Fig. 26.1.

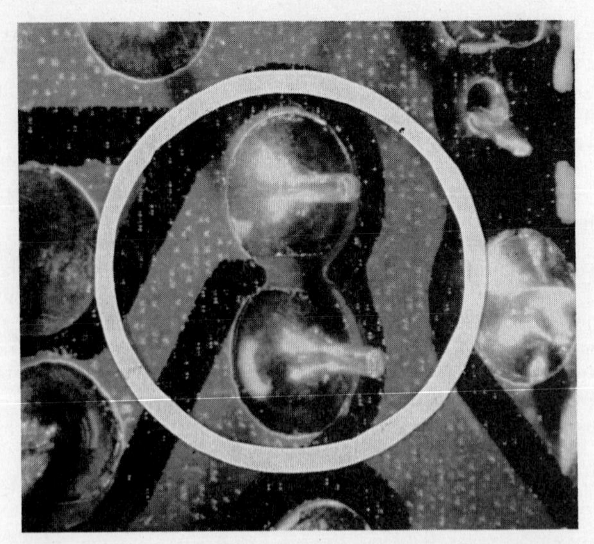

FIG. 26.1 An acceptable solder joint. Note that the contour of the lead is clearly visible.

If ideal conditions prove somewhat elusive, then the solder joint will be imperfect. The material in this section is designed to identify the types of defects that are indications of serious problems somewhere in the soldering process and that call for immediate corrective action.

Imperfect solder joints may be traced to one or more of the three following general sources:

1. Poor solderability of the printed board terminal area and/or the component lead
2. The soldering process itself (i.e., incorrect adjustment or control of the soldering conditions)
3. Incorrect design of the assembly to be soldered

26.4 DEFECTS ARISING FROM POOR SOLDERABILITY

When the basis material has marginal or poor solderability, incomplete wetting by the molten solder will result in a discontinuous film of solder on the termination or board pad. This will result in two basic types of defects: dewetting and nonwetting, which are sometimes difficult to separate. In the case of nonwetting, the color of the basis metal will be readily visible through the discontinuous film of solder. However, if the basis metal is similar in color to the solder, as would be true with a roller-tinned or palladium-coated surface, the nonwet surface may be extremely difficult to discern from one which has simply dewet. A typical dewetted surface is shown in Fig. 26.2. By contrast one which has not wet the terminal areas is shown in Fig. 26.3. Both dewetting and nonwetting are serious problems and indicate inadequate bonding of the solder to the circuit metal.

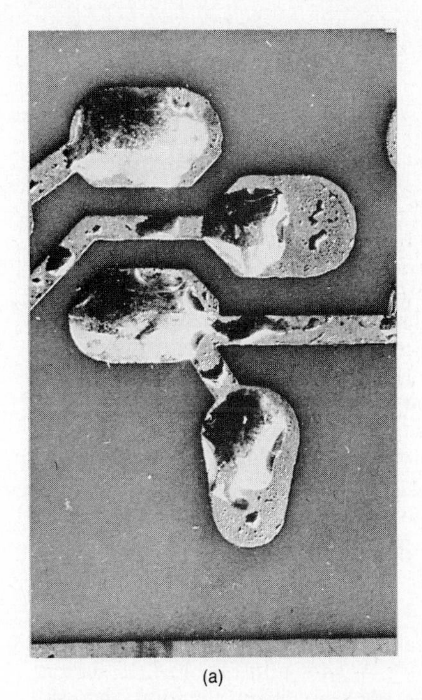

(a)

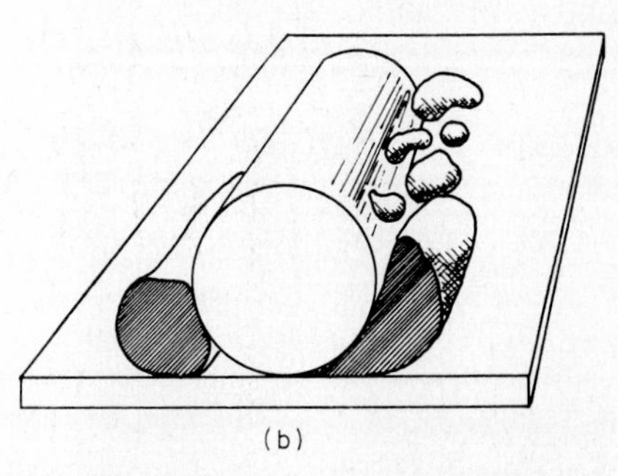

(b)

FIG. 26.2 A dewetted surface. The solder has initially wet the terminal areas (*a*) and then drawn back into discontinuous blobs (*b*).

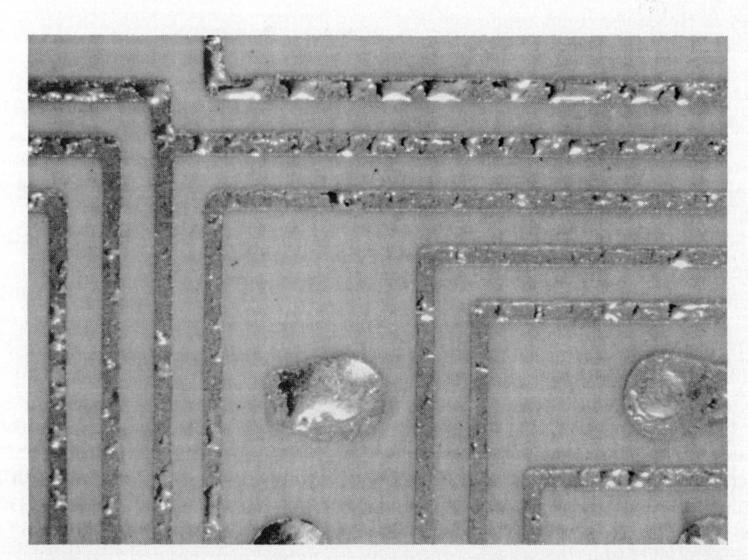

FIG. 26.3 A nonwetted surface. The solder has never wet the terminal areas. The metallic surface below is visible.

26.4.1 Nonwetting

Nonwetting can occur for a number of reasons, some of the most typical of which are the following: (1) inadequate solderability of the base metal in relation to the activity of the flux used; (2) oil or grease on the surfaces, which prevents the flux from coming in contact with the surface; (3) improperly controlled time-temperature cycle during the soldering process.

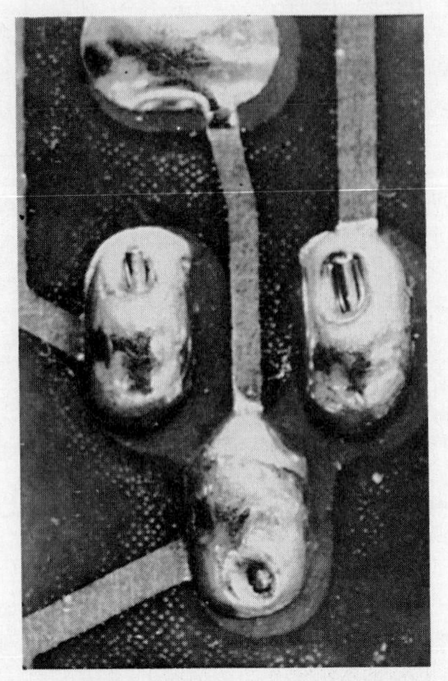

FIG. 26.4 Nonwetted component leads on a highly solderable terminal area. Note that the solder has completely pulled away from the lead.

Figure 26.4 shows a solder joint which exhibits good wetting of the wired paths but nonwetting of component lead. Note that partial wetting of the base of the lead results in a cup-shaped depression around the lead. Although the outline of the lead is quite visible, it is not because the solder has wet the lead and feathered away from it as it drained; it is because the complete lack of adherance of the solder to the lead has resulted in the lead itself being visible. Nonwetting can also be masked by excessive solder on the joint, as shown in Fig. 26.5. If there is too much solder on the joint, the lead becomes encapsulated in the solder and any visual indication of its condition is destroyed.

26.4.2 Dewetting

In contrast to nonwetting, a typical dewet condition results when the solder first wets the basis metal and then retracts, leaving a thin but continuous coating of solder over the basis metal and discrete globules of solder spaced discontinuously across the surface. The large globules of solder have very large contact angles where they meet the basis metal. The large contact angles are indicative of a poor wetting condition.

Dewetting of the basis metals can be traced to sources similar to the nonwet condition. In addition, however, dewetting can be caused by certain types of contaminants on the surface of the basis metal; embedded cleaning abrasives are an example. Metallic impurities in the solder bath can, in sufficient concentrations, also result in dewetting.

In extreme cases of poor solderability due to gross contamination of a surface, both nonwetting and dewetting may be seen to occur simultaneously on the same surface. In such cases, solder fillet formation will be poor or nonexistent. On the other hand, a very slight tendency for nonwetting or dewetting on a surface may merely show as pinholes in the solder coating on, for example, printed board conductors; the pinholes are, however, unlikely to have a significant adverse effect on joint formation.

FIG. 26.5 Variation in the amount of solder required to form the joint. Note that excess solder results in a rounded configuration, completely masking the lead contour.

26.5 PROCESS-RELATED DEFECTS

26.5.1 Excess Solder

The presence of an excessive amount of solder in a joint, enough to cause the fillet to have a convex surface, may mask the tendency for nonwetting on either the land or the component termination. If no termination is visible, it is not possible to decide whether the component termination wire was too short or whether, in fact, no component has been mounted on the board. For those reasons, excessive solder is undesirable. It may occur because a dip- or wave-soldering bath is operated at too low a temperature, but it more generally occurs in hand-soldering operations.

26.5.2 Bridging

Bridging can take the form of a web or film between adjacent vertical terminations, as shown in Fig. 26.6, or if it is more extensive in nature, it can appear as a web of solder joining the legs of adjacent conductors (Fig. 26.7).

The cause of bridging may be the use of such incorrect soldering conditions as low temperature, insufficient flux, or the presence of a tenacious oxide film on the surface of the molten solder. The oxide film may be due to the presence of an impurity such as zinc or aluminum. Soldering machine design factors, such as the form of the wave and the angle at which the printed board approaches and leaves the molten metal during wave soldering, may have a strong effect on the tendency for bridging to occur, and the design of the printed circuit assembly is also an important factor.

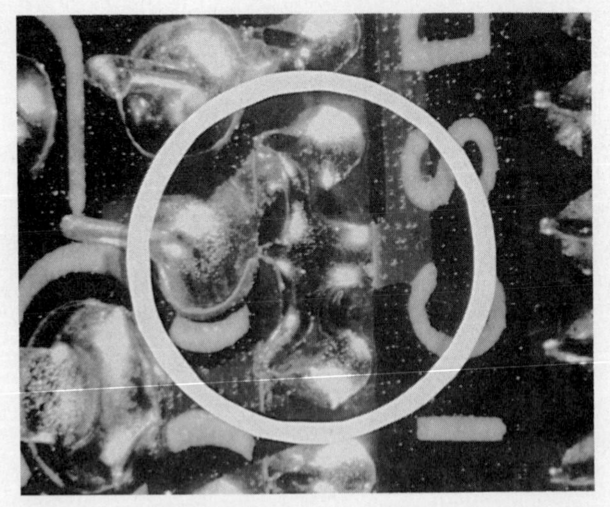

FIG. 26.6 Solder bridging: an unwanted short between two conductors.

FIG. 26.7 Solder webbing. Note that in webbing, nonmetallic surfaces can be involved and many conductors can be shorted together.

26.5.3 Icicle Formation

Stalactites or icicles of solder such as those shown in Fig. 26.8 arise mostly in wave-or-drag soldering operations when the draining of the molten solder from the soldered board is restricted. Icicle formation may often be alleviated by causing the printed circuit assembly to leave the solder bath surface at a small angle, for example, 7°. Solder temperature, conveyor speed, and activity of the flux also

FIG. 26.8 Icicling, a common soldering defect.

may influence the tendency for icicle formation. As in the case of bridging, icicle formation may be exaggerated by the presence of tenacious oxide films due to high levels of an impurity such as zinc or aluminum in the bath.

26.5.4 Blowholes

Blowholes, or small spherical cavities in the surface of a solder fillet, may arise from solidification of the solder around entrapped bubbles of air or flux vapor. A typical blowhole is shown in Fig. 26.9. Usually the interior of the cavity can be

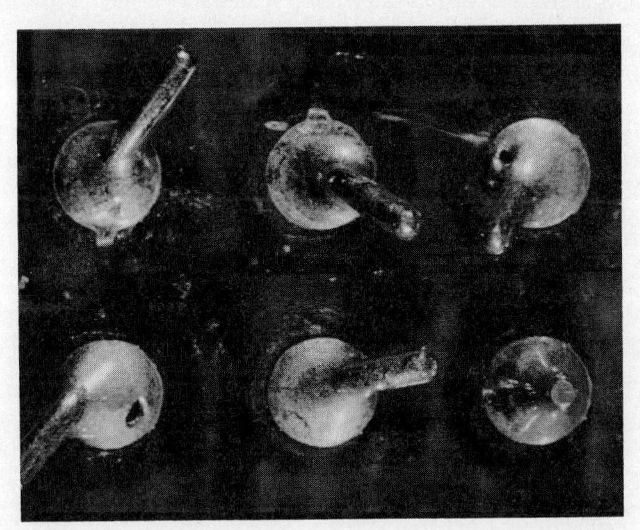

FIG. 26.9 Blowholes, a defect caused by the trapping of liquid or vapor inside the joint as the joint is forming.

seen to be bright and smooth. Blowholes may be quite small, but they should not be confused with the normal surface identations due to solder contraction upon solidification.

Excessive flux or insufficient evaporation of the flux solvent before soldering can give rise to blowholes and internal porosity in the joint. Blowholes may also result when moisture or organic plating residues become trapped in plated-through holes.

26.5.5 Cold or Fractured Joints

Cold joints are caused when the component lead moves as the solder joint is freezing. The movement of the lead results in uneven freezing of the solder joint, and ultimately the stress set up can cause a crack in the joint. Any indication of a cold or fractured joint casts serious doubt on the integrity of the joint, and the joint should be reworked. If cold joints are a continuing problem, the difficulty can easily be corrected by providing a smooth transfer of the printed board from the solder pot to the next station. A typical cold joint is shown in Fig. 26.10.

26.5.6 Dull or Grainy Solder Joints

As outlined in one of the preceding chapters, the principal cause of dull solder joints is the accumulation of metallic impurities in the solder pot. Metallic impurities, particularly gold and copper, will rapidly cause the solder joints to take on a dull gray appearance. The problem can easily be corrected by replacing the solder in the pot. Several stopgap measures, outlined in Chap. 24, will allow continued operation until a new solder charge can be obtained. To prevent significant buildup of metallic impurities, it is important to have the solder pot analyzed on a regular basis.

A dull appearance can also result from a solder alloy which is low in tin. The bright shiny appearance of the solder joint is due to the tin, not the lead. Therefore, low-tin alloys such as 50/50 will be substantially duller than high-tin alloys such as 63/37. If the tin becomes depleted from the solder alloy, then the result will also be duller solder joints.

Dull joints will also result when the residues of certain highly active fluxes are left on the solder joint for an extended period. In that case, the dull appearance is the result of chemical attack on the solder joints and the solder process should be modified to ensure rapid cleaning of the boards.

Dross particles or carbonized oil in the solder wave often gives rise to a solder

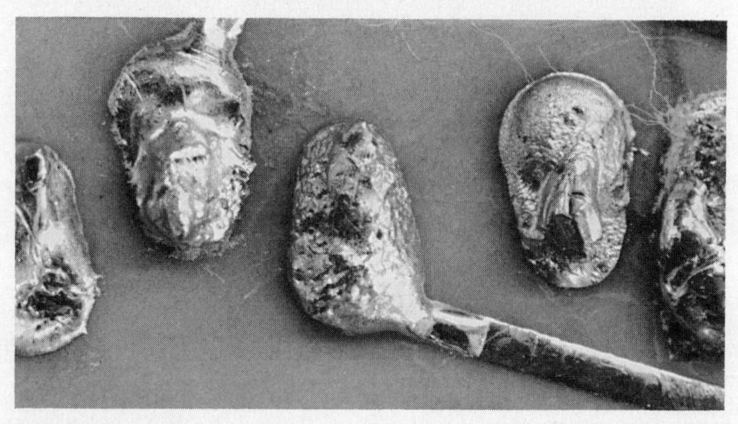

FIG. 26.10 A cold joint. Cold joints are caused by excessive movement of the joints during formation.

joint with a grainy, lumpy appearance. Both dross and carbonized oil result from improper maintenance of the solder pot. If dross has been sucked into the solder wave, the pot level is too low and additional solder must be added as makeup.

Once dross has entered the solder pump, it is advisable to remove the pump and clean it, since dross is a highly abrasive material and will not purge itself through the pump. Carbonized oil results from using inferior soldering oils or from running a good soldering oil at too high a temperature for too long a time. When that occurs, the oil should be changed immediately and new oil should be passed through the system to purge the carbonized material. In extreme cases, the solder pot must be dumped and recharged.

26.6 DESIGN-RELATED DEFECTS

An important feature of the design of printed circuit assemblies is the selection of the optimum component termination size for a given board hole diameter. Even with perfect solderability and soldering conditions, excessive clearance between the terminations and the hole in a terminal area, in the case of unplated holes, often leads to incomplete solder fillet formation. A marginally suitable clearance may become unsuitable when the termination is displaced from the axis of the hole, a condition which is extremely difficult to avoid in the real world of production line assembly.

Insufficient spacing between terminations can lead to bridging, and incorrect printed circuit conductor spacing, size, and layout may also result in bridging or webbing. A large area of copper, or a high density of terminal areas and terminations, tends to act as a heat sink which may, again, cause bridging, or, in extreme cases, localized nonwetting.

CHAPTER 27
SURFACE MOUNT SOLDER PROCESS

Leslie C. Hymes
General Electric Company, Medical Systems Group, Waukeshaw, Wisconsin

27.1 INTRODUCTION

Surface mount technology (SMT) uses surface mounted devices (SMDs—passive chip components, discrete components, and chip carriers) as a means of achieving higher-density electronic assemblies. Increasing acceptance of SMT has contributed to an evolution in the soldering technology employed in the manufacture of printed wiring assemblies (PWAs). This chapter deals with characteristics of the soldering process associated with implementation of SMT. Criteria are provided for evaluating the interrelationship of design, components, substrates, materials, and processes with acceptable-quality solder joints and PWA reliability.

27.2 DESIGN REQUIREMENTS AND IMPACTS

Much of the application of SMT results in PWAs with both through-hole mounted and surface mounted components. These mixed-technology assemblies require careful consideration of manufacturing technology from the onset of the design and component selection activity. Quality, reliability, and total cost of the end product can be significantly affected by the combination of layout, components, and substrates selected and the impact of this combination on PWA manufacturing process.

27.2.1 Layout and Land Patterns[1]

The land patterns used with chip carriers fall into two main categories—high-density and low-density—based on the local routing density associated with the surface mounted device. The selection, design, and position of the land geometry in relation to the chip carrier have a significant impact on the resultant solder joint configuration.

The attachment lands for surface-mounting leadless chip carriers should be the same width as the component termination maximum plus 0.005 in. whenever possible. The land length should extend between 0.015 and 0.040 in. beyond the maximum chip carrier outline on each of all four sides to create a horizontal solder fillet length equal to the vertical fillet rise. This design aids in creating a more reliable solder joint.

27.2.1.1 Low-Density Designs. A suitable land pattern for use in low-density application is shown in Fig. 27.1. The number of lands and the overall size of these patterns is governed by the particular chip carrier that is to be used.

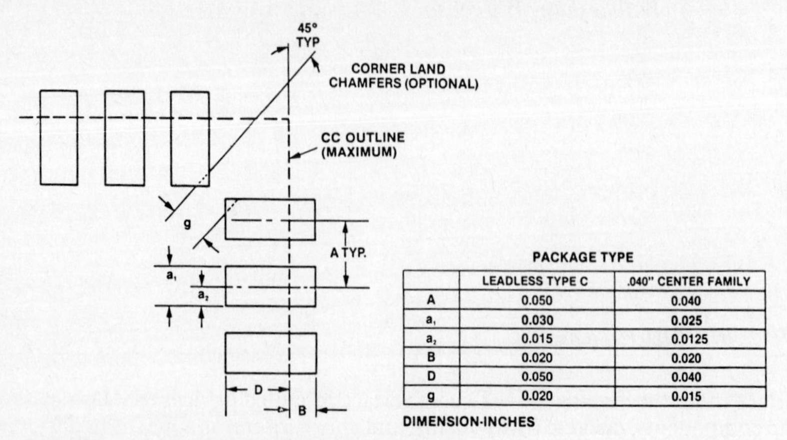

	LEADLESS TYPE C	.040" CENTER FAMILY
A	0.050	0.040
a_1	0.030	0.025
a_2	0.015	0.0125
B	0.020	0.020
D	0.050	0.040
g	0.020	0.015

FIG. 27.1 Example of low-density land pattern geometry. *(Reprinted with permission from IPC-CM-78, November 1983, p. 19, Fig. 5.9.)*

In relatively low density applications, it is advantageous to avoid routing conductors between mounting lands, since the absolute width of the lands can then be maximized without reducing the conductor-to-conductor spacing. The use of the wide lands shown will simplify the assembly process and increase yield of acceptable solder joints by easing the requirements on chip carrier positioning.

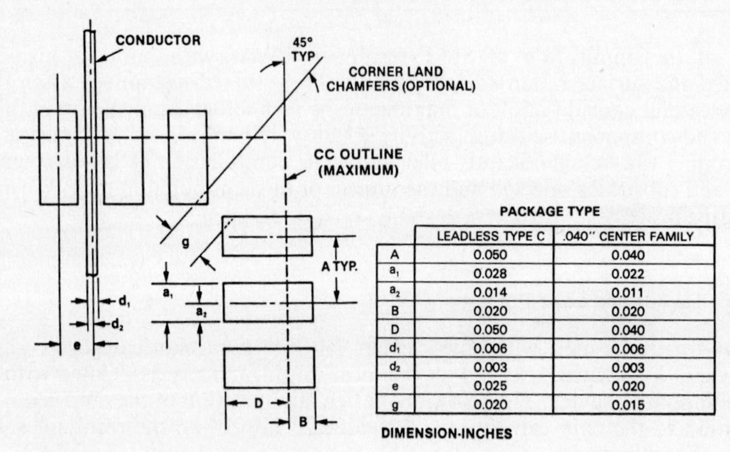

	LEADLESS TYPE C	.040" CENTER FAMILY
A	0.050	0.040
a_1	0.028	0.022
a_2	0.014	0.011
B	0.020	0.020
D	0.050	0.040
d_1	0.006	0.006
d_2	0.003	0.003
e	0.025	0.020
g	0.020	0.015

FIG. 27.2 Example of high-density land pattern geometry. *(Reprinted with permission from IPC-CM-78, November 1983, p. 20, Fig. 5.11.)*

27.2.1.2 High-Density Designs. High-density applications include the added feature of a routed conductor between adjacent lands (see Fig. 27.2).

In addition, the use of these high-density land patterns is usually associated with stringent-tolerance conductor routing. The penalties associated with these design options include reduced conductor spacing, tighter assembly tolerance, and the requirement for a solder mask to help avoid shorting of conductors.

27.2.1.3 Via Effects. Vias are used for interconnecting chip carriers to layers other than the mounting layer. Wicking of solder from component pads into vias during soldering should be minimized to avoid poor component solder joints. A neckdown in the land pattern or fanout as illustrated in Fig. 27.3 can serve to minimize the effect of the via on the solder flow. Placement of solder mask can also be used for the same effect. Vias placed in this manner can also be used as locations for test probing during final assembly testing.

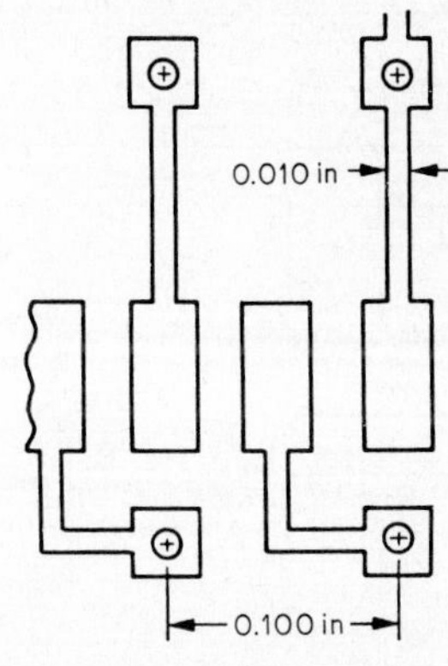

FIG. 27.3 Land pattern, via relationship to limit solder wicking. *(Reprinted with permission from IPC-CM-78, November 1983, modified Fig. 5.7a.)*

27.2.1.4 Layout of Chip Components. Layout of components of this type for wave soldering is critical to provide for good solder penetration to all desired sites. In general, component orientation perpendicular to the direction of progression of the assembly through the wave results in higher yields of good solder joints. Spacing of chip components in a staggered pattern rather than end to end also aids solder quality.

27.3 CLASSES OF ASSEMBLIES

The introduction of SMT has been an *evolutionary* rather than a *revolutionary* process. Mixed-technology assemblies [combinations of through-hole mounted components (THMC) and surface mounted components (SMC)] comprise a high percentage of the products utilizing SMT.

These mixed-technology assemblies have been classified in many different ways, leading to some confusion. Table 27.1 illustrates one method of classification of PWAs. This classification system is used by the Institute for Interconnecting and Packaging Electronic Circuits (IPC) and is recommended. Typical soldering techniques have been added by the author.

The assembly and soldering processes for each class may be somewhat different and vary throughout the industry. However, Fig. 27.4 depicts some accepted assembly and solder process sequences used for the various assembly classes.

27.4 COMPONENTS

The construction and physical properties of components selected should be considered when choosing soldering methods and parameters.

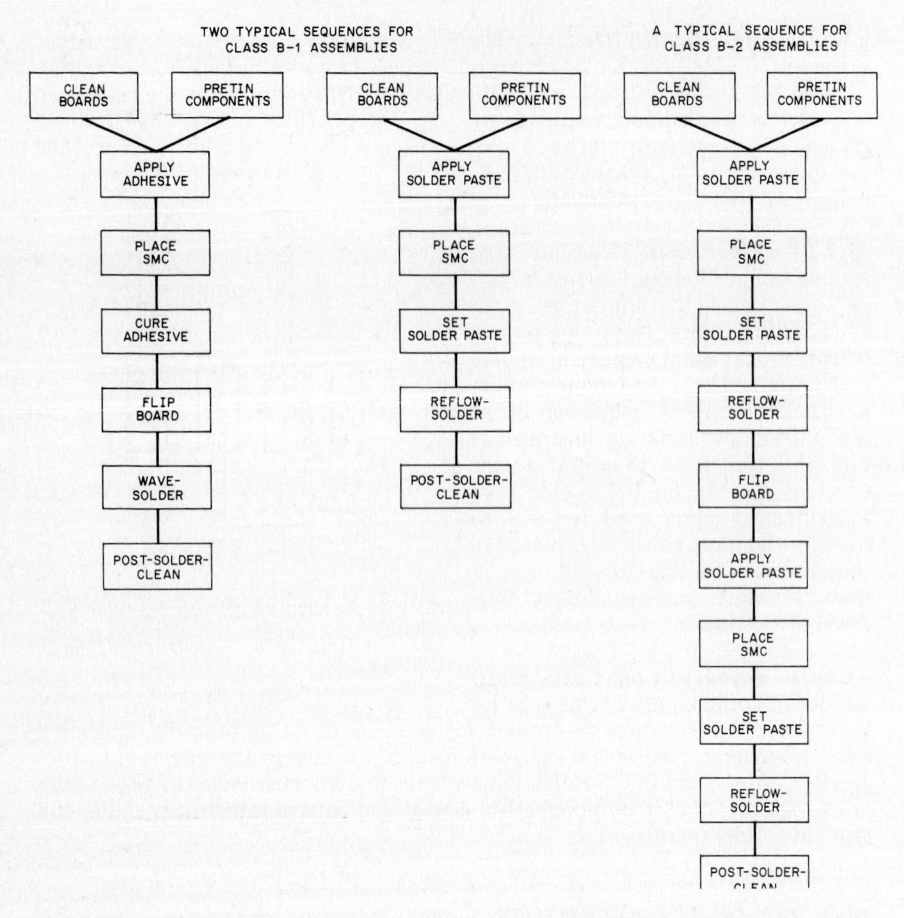

FIG. 27.4 Typical assembly and solder process sequences. (See Table 27.1 for asembly class definition.)

27.4.1 Chip Components

Passive components available in either rectangular or cylindrical (metal electrode face bonding, MELF) packages are provided with terminations as shown in Fig. 27.5. Five-face or three-face terminations with a solder-plated or solder-dipped finish provide for solder fillet geometries with increased mechanical and electrical reliability.

A copper or nickel barrier layer under the solder-dipped or -plated finish is preferred to protect the internal contact against leaching or degradation resulting from intermetallic formation during exposure to soldering materials and temperatures.

Thermal shock damage to chip capacitors can occur without adequate preheat and controlled cooling during soldering operations.

27.4.2 Discrete Components

Discrete components such as transistors and diodes are available for SMT in small outline (SO) packages. These plastic-packaged devices typically have rec-

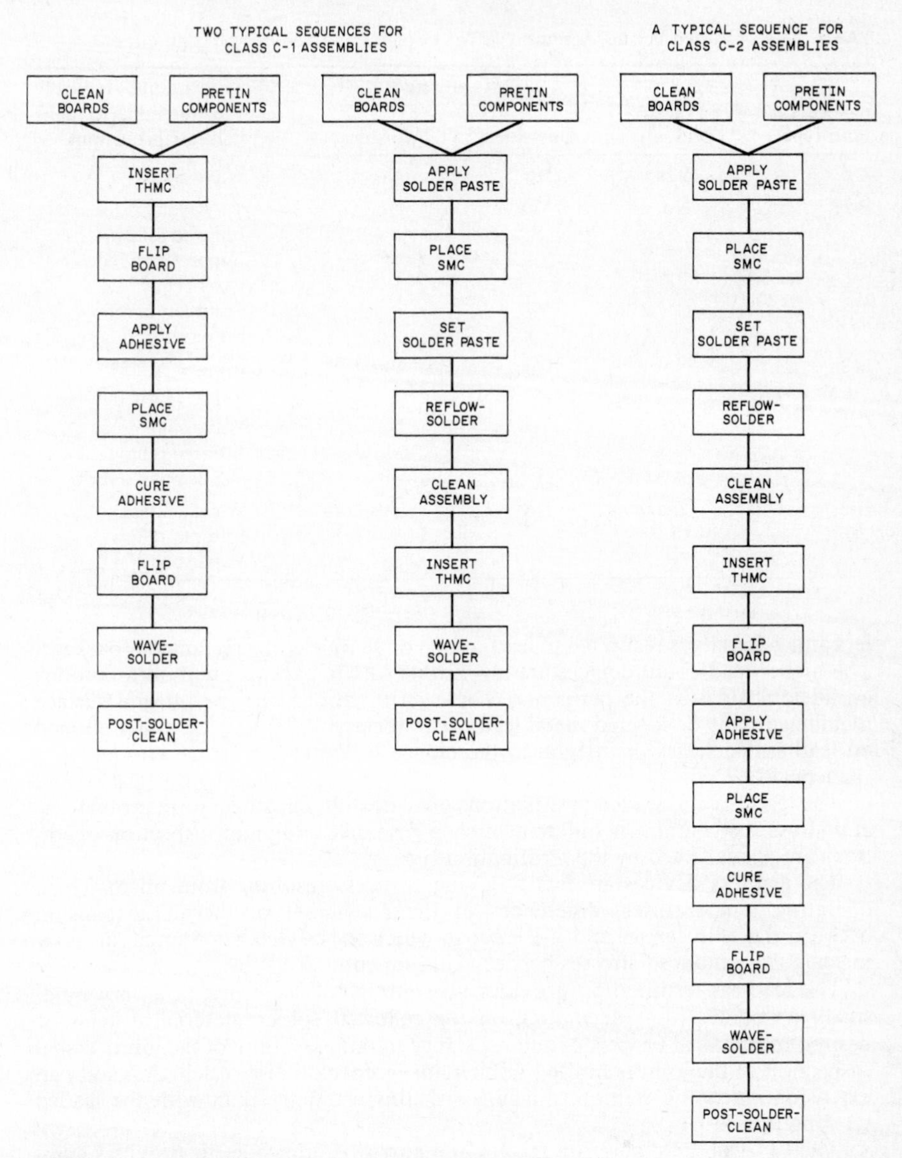

FIG. 27.4 (*continued*)

tangular leads formed into a "gullwing" shape exiting from the middle of the body. This configuration provides a flat surface to contact the pad area on the printed wiring board (PWB). These low-pin-count devices are successfully soldered with wave soldering as well as the reflow processes.

27.4.3 Chip Carriers

These active devices are packaged in either ceramic or plastic bodies.

The small-outline integrated circuit (SOIC) has gullwing leads on 50-mil cen-

TABLE 27.1 Printed Wiring Assembly (PWA) Classes and Soldering Techniques

PWA class and type	THMC present?	SMC present? One side	SMC present? Two sides	Solder techniques typically used for medium- to high-volume throughput
A	Yes	No	No	Wave solder
B-1	No	Yes	No	Wave solder Infrared reflow Vapor phase reflow Laser reflow Conduction reflow Convection reflow
B-2	No	No	Yes	Infrared reflow Vapor phase reflow
C-1	Yes	Yes	No	Wave solder Infrared reflow Vapor phase reflow
C-2	Yes	No	Yes	Wave solder Infrared reflow Vapor phase reflow

ters and generally is restricted to lead counts of 28 or less in a rectangular package. The higher-lead-count devices are low-profile square packages with terminations on all four sides of the perimeter. These terminations may be integral surface metallized pads or formed metal leads. The formed leads may be either clipped on leadless packages or integral to the device in the case of plastic leaded chip carriers (PLCCs).

The leadless and leaded terminations provide different solder joint geometries, and stress distribution is different in each case. Cleaning and inspection operations are also affected by the termination type.

The leadless devices are less forgiving of stress resulting from mismatch in operating temperatures, differences in the coefficient of thermal expansion between the chip carrier and the PWB to which the device is mounted, or other mechanically induced stresses between component and PWB.

The leadless termination provides no compliancy; it results in a very rigid, small lap solder joint depending on the reflowed solder material system for desired mechanical properties and resistance to fatigue failure of the joint. Visual inspection of the joint is limited to fillet appearance on any castellation and pad extension. Cleaning is more difficult with this geometry than with the leaded termination.

The "J" lead configuration results in a narrow solder fillet. It provides compliancy which can compensate for some degree of mechanical flexing and mismatch in expansion between the component package and the substrate. Visual inspection of solder joints is somewhat easier with this geometry.

Cleaning operations are also aided by this geometry, since it can provide clearance between the bottom of the component and the substrate. Excess solder fillets on this geometry stiffen the lead and reduce the compliancy advantage.

27.4.4 Other Components

Other PWA mounted components such as sockets and connectors are also available in surface mount configurations. These may have leads in a gullwing or butt

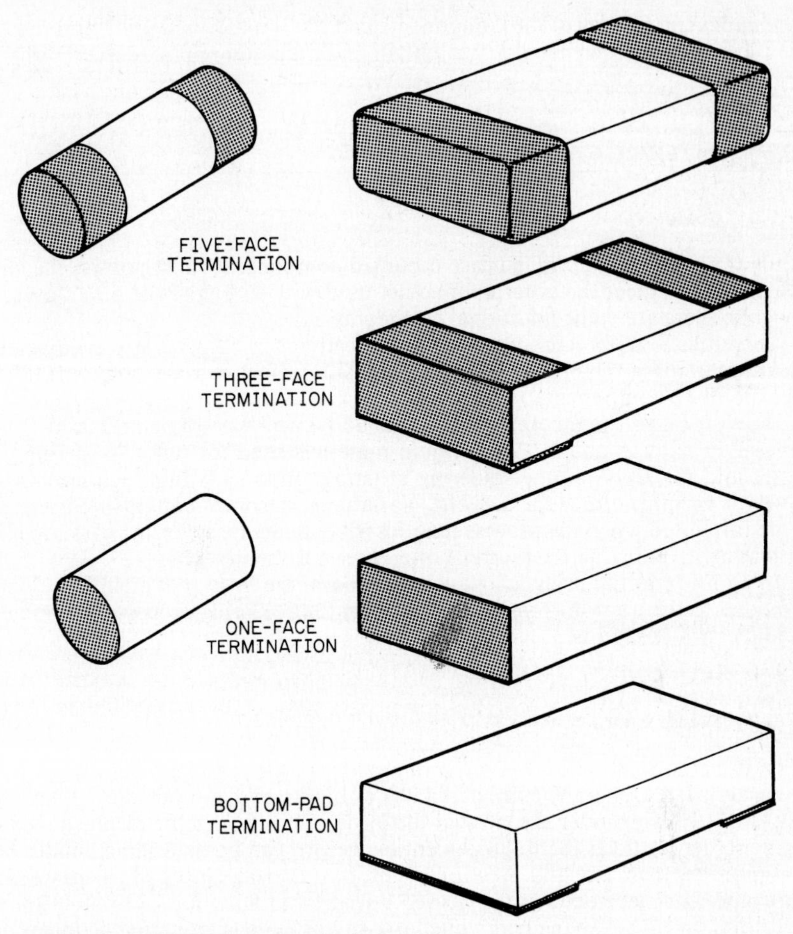

FIG. 27.5 Chip component termination options.

joint configuration, sometimes created by reforming the through-hole leads. Effects of mating and withdrawal forces are a major concern with components of this type.

27.4.5 Substrates

Much attention has been given in the literature to the impact of the relation of the coefficient of thermal expansion (TCE) of PWA substrates and components mounted on the substrate in producing PWAs with reliable solder joints.

Various approaches have been utilized to affect the mismatch or alter the substrate to reduce the mismatch. In general, chip carriers with more than 28 terminations require close attention to TCE matching between component and substrate.

The ability to achieve coplanar component lead–to–solder pad contact—in order to achieve an acceptable solder joint and provide compliancy and strength to maintain the integrity of the solder joint—becomes more critical as the lead count of surface mounted components increases. Consideration of physical and mechanical properties of the components and PWB material and management of the thermal stresses is necessary to obtain and maintain reliable solder joints.

27.5 MATERIAL SYSTEMS

27.5.1 Solder

While the solder used with surface mounted components must possess the characteristics and meet the criteria for solder used with through-hole mounted components, there are some additional considerations.

The eutectic 63 percent tin–37 percent lead and a 60 percent tin–40 percent lead alloy remain the most common compositions for the wave soldering of surface mounted components.

In addition to these compositions, the 62 percent tin–36 percent lead–2 percent silver alloy is one of the common materials used for reflow-soldering processes and for high-volume soldering of chip components in which leaching of silver or metallization is a concern. Variations in solder composition are also sometimes used when the process requires more than one soldering pass. An alloy with a higher reflow temperature is often used for the first pass.

In addition to the "classic" wave-solder bath, the form in which the solder is provided to the joint may vary depending on the soldering process used and end product considerations.

27.5.2 Solder Paste

Paste is a common method of providing solder in the reflow processes.

Solder pastes generally consist of a flux or binder and a solder alloy. The rheology of the flux or binder is such that the paste viscosity and the dispersion of the heavy solder particles (85 to 88 percent by weight) can be maintained under normal room temperature and pressure conditions. During curing of the paste prior to component placement (often carried out at 60 to 80°C for 15 to 30 min), the flux flows out to cover the pad while viscosity of the paste retains the metal particles in place. If the percentage of metal content falls below the indicated range, there appears to be more tendency for some metallic particles to be carried out between pads, increasing the possibility of bridges.[2]

The flux or binder may contain pure rosin (or synthetic resins), activators, solvents, lubricants, and thickeners. The activator removes surface oxides and tarnishes during the soldering process. The solvents are used to adjust the solder paste to a proper viscosity. Lubricants are used to reduce the sticky nature of the rosin, enhancing paste-deposition characteristics.

Since the solvents and lubricants do not provide adequate "body" to support the metal powder, thickeners are added to give the solder paste the proper texture. Powder particle shape and size also influence deposition characteristics significantly. Well-rounded particles flow more readily through screen or stencil openings and give better pattern definition, while irregularly shaped particles and residual large particles clog the deposition equipment.

Increased oxidation present on extra-fine particles will result in gradual reaction of the oxidized particle with the flux to cause flocculation (formation of clumps) of the solder paste.

Short shelf life for the solder paste and subsequent formation of solder balls during the reflow-soldering operation are experienced when "fines" occur in a paste mix. Spherical particles from 325- to 200-mesh size are preferred, with "fines" (less than 10 or 20 μm in diameter) excluded.

Characterization and specification of significant parameters of solder paste are important to repeatable high-quality results. Specification of the parameters should be sufficient to meet process variables and end product requirements. A solder paste specification is available from the IPC as IPC-SP-819, *General Requirements for Solder Paste for Electronic Application.* Generally, samples and vendor specifications are also available from material suppliers.

Pastes should be stored in closed containers in a desiccated environment to eliminate or minimize the rate of oxidation. Manufacturers' shelf life instructions should be strictly followed. Paste oxidation is a prime source of subsequent solder nonwetting and contributes to formation of solder balls.

Solder balls on soldered assemblies are frequently a troublesome occurrence when using solder paste. In a following section on the reflow-soldering process, solder balls are discussed in more detail.

27.5.2.1 Solder Paste Application. Pastes can be applied to substrates, using three basic methods: screen printing, stenciling, and dispensing. Pastes are generally stirred by hand before use. Power mixers can deform or flake the metal particles and may cause excessive entrapment of air in the paste.

27.5.2.1.1 Screen Printing. To ease the passage of solder paste through the screen for a complete printed image, the mean particle size of the solder paste should be about one-third of the mesh size.

Screen mesh size has significant impact on the thickness of the paste deposit; the finer the mesh, the thinner the deposit. Mesh sizes of 80 to 105 are commonly used. These provide reflowed deposits of 0.002 to 0.005 in with the 105-mesh screen and the more common 0.005 to 0.010 in with 80 mesh.

The screen, generally made from stainless steel wire, should be positioned approximately 0.013 to 0.063 in above the substrate and be in a plane that is parallel to the substrate. The substrate should be located in exact registration with the screen image and held in place with a reliable tooling pin setup, double-sided adhesive tape, or other reliable means. One of the major drawbacks of screen printing is the inability to position the screen accurately with the maximum open area registered over the pads to achieve uniform paste disposition at each location desired.

The solder paste is drawn across and through the screen by a squeegee which should have a sharp rubber edge. The squeegee should be adjusted so that it contacts the screen at an angle of approximately 60 to 80°. For long screen life and good detail definition, pressure on the squeegee should be minimized. The screen and screen printer must be cleaned in accordance with good practice and manufacturing procedures. Solvents such as methylethylketone, trichloroethane, and acetone are used as cleaners.

27.5.2.1.2 Stencil Printing. Patterned metal-foil stencils may be used to enhance both the solder volume deposited and its uniformity. Stencils prevent the concerns associated with exact registration of the deposition patterns with a screen mesh, and the etched open-pattern area is maximized and uniform at each deposition site. Stencils are used on conventional screen-printing equipment, with the equipment adjusted for contact printing. Squeegee pressure should be sufficient to provide a clean wiping action of the metal foil without "dishing out" paste from the etched pattern.

Stencil printing has been successfully used for depositing solder paste for chip carrier assembly with deposits up to 0.015 in thick and with center-to-center land spacings as small as 0.020 in.

The fabrication of metal-foil stencils by chemical etching becomes increasingly more difficult for opening aspect ratios (stencil opening width to stencil thickness) less than 2. In addition, for aspect ratios less than 2, paste begins to cling to the sidewalls of the stencil, and irregular paste deposits may result.

If a stencil is used, the openings in it should be at least the same as the outlines for the surface attachment lands. Since solder paste tends to pull toward wettable lands, the stencil openings may be designed to be larger than the lands in order to provide additional solder volume; however, spacing between deposits must be sufficient to prevent solder bridges.

Metal stencils generally have a longer serviceable life than mesh screen. The screens become less resilient and may sag, making control of deposits more difficult.

27.5.2.1.3 Pressure Dispensing. In the dispensing process, paste from a supply cartridge is pneumatically or mechanically forced through small-diameter tubes or orifices to cause small discrete deposits of paste to be placed at interconnection sites.

Single-point dispensing is best suited for repair and low-production volume applications and in mixed technology, where surface-mounted chip carriers are added after the wave-soldered through-hole mounted components. Multipoint dispensers are used for the local deposition of paste at a complete chip carrier site, whereas single-site dispensers are best suited to the local repair of individual joints.

27.5.2.2 Solder Paste Set. This step of the processing flow is intended to allow some pastes to become tacky prior to the placement of the chip carrier packages. The time required for this paste set varies depending on the type of paste being used. While minimum paste set times may be 30 min or less, some assembly operations have allowed pastes to set for several hours. However, chip carriers should be placed before the paste loses its tackiness.

The utilization of a paste-set operation minimizes the potential for smearing the solder paste during the assembly operation.

27.5.3 Solder Preforms

Solder preforms are used in applications requiring large solder volumes or precise solder-volume control. In some applications, spherical solder preforms are used to provide a space between the chip carrier package and substrate that is sufficient for cleaning or the subsequent introduction of conformal coating. The spherical preforms are mechanically arrayed over the metallized terminals of the chip carrier and subsequently reflowed. After reflow, the chip carrier is inspected for solderability. Finally, the solder-coated chip carriers are attached to the substrate with another fluxing operation and a reflow cycle.

27.5.4 Other Methods of Solder Placement

Heavier deposits of tin-lead alloy (0.006 to 0.008 in total thickness) are sometimes applied to the PWB lands to provide a source of solder as well as space the component body off the board. This additional material is applied by plating or paste application and subsequent reflow prior to component mounting.

In special cases, cast pillars of tin-lead alloy as much as 0.060 in in height are provided on components to serve the same purpose.[3]

The spacing enhances flux removal by cleaning agents, aids in joint inspection and in application of conformal coating, and can provide additional joint compliance with some stress conditions.

27.5.5 Flux

Flux is incorporated in the solder paste used in reflow processes. In all SMT soldering processes, flux considerations are in general the same as described in preceding chapters dealing with general PWA soldering. Special issues related to SMT are covered in the following section on soldering processes and equipment.

Specific concerns relative to the effect of fluxes on surface mounted assemblies center on criticality of the ability to remove any corrosive residue after soldering because of the reduced spacing between SM components and the substrate surface. Use of fluxes more aggressive than R or RMA may require specialized cleaning processes to assure removal of potentially corrosive residues. IPC-SF-818, *General Requirements for Electronic Solder Fluxes,* provides general classifications and test methods for fluxes. Cleaning requirements are discussed later in this chapter.

27.5.6. Adhesives

Adhesives or staking compounds are used to provide temporary retention of surface mounted components to substrates during processing and handling prior to the soldering operation. The most common need arises when a wave-soldering process is used with components placed on the surface of the substrate to be passed through the wave. In reflow-soldering processes, solder pastes are used to provide the source of solder and also retain the components in position on the substrate surface.

27.5.6.1 Properties. Desirable characteristics of adhesives for use with surface mount technology are listed below:

1. Long shelf life
2. Long potlife or working life
3. Lack of stringing during application
4. Fast and complete curing—UV- or temperature-activated
5. Cure temperature compatible with components and substrate
6. Resistance to fluxes, soldering temperatures, cleaning agents
7. Ability to withstand repeated exposure to soldering temperature
8. Absence of outgassing at soldering temperatures
9. Low slump properties
10. Gap filling to accommodate components held off substrate by terminations
11. Strength after cure low enough to be compatible with repair processes
12. Color providing an aid to inspection
13. Dielectric properties compatible with circuit operation

27.5.6.2 Application. The three most commonly used methods of adhesive placement are screen printing, pressure dispensing, and pin transfer.

27.5.6.2.1 Screen Printing. With this technique, a squeegee forces adhesive through screen openings onto desired sites of the substrate, depositing material on the complete pattern in one stroke. Its characteristics are as follows:

1. Generally used in low-volume application
2. Requires unpopulated substrate

3. May not provide sufficient gap fill
4. Requires screen control, as in paste dispensing
5. Can apply many dots at a time
6. Adhesive open to contamination prior to application

27.5.6.2.2 Pressure Dispensing. In pressure dispensing, air or mechanical pressure dispenses adhesive through a hollow needle onto desired sites one at a time. Its characteristics are as follows:

1. Most common method
2. Can be used on populated substrate
3. Controlled, variable dispensed quantity
4. Adaptable to manual or machine-controlled process
5. Applies adhesive to one site at a time
6. Flexibility in application pattern and dot size
7. Adhesive not exposed until application

27.5.6.2.3 Pin Transfer. The pin-transfer process utilizes small-diameter pins which are immersed in a reservoir of adhesive, then positioned over desired sites very close to the substrate surface to transfer adhesive, using viscosity of adhesive to effect the transfer. The transfer can be one site or multiple sites at a time, depending on fixtures used. Its characteristics are as follows:

1. High volume
2. Fixed dot size
3. Can apply many dots at a time on unpopulated substrate
4. Expensive fixturing for multiple dots at one time
5. Adhesive open to contamination prior to application
6. Works best with flat surface
7. Uniform adhesive viscosity critical to control stringing
8. Fixed pattern when multiple site transfer done simultaneously

27.5.6.3 Curing. The curing process should meet the following criteria:

1. Take place at a temperature below that which would damage components used or cause warping of the substrate.
2. Proceed at a speed compatible with production rates required.
3. Provide cured adhesive strength sufficient to hold components in place during subsequent assembly and handling prior to soldering.

Convection oven curing and ultraviolet (UV) cure processes are the principal processes used for curing these adhesives. Oven cure tends to be longer, while UV curing—though faster—generally requires subsequent convection oven or infrared (IR) oven curing, since the portion of the adhesive under the component is shielded from the UV energy. Some applications utilize UV adhesives off center and depend on initiation of the cure process by exposure of a portion of the adhesive. The polymerization begins in the exposed area and progresses to the material under the component.

27.5.6.4 Available Adhesives. One- and two-part adhesive materials are available from several sources. Single-component adhesives require no mixing, have generally shorter potlife, and may exhibit undesirable outgassing as a result of solvent incorporated in the material. Two-component systems have longer shelf life but require mixing with accurate metering to assure uniformity during processing. The mixing may generate air bubbles which cause process problems in dispensing.

Adhesive systems used for SMT include epoxies, acrylics, cyanoacrylates, and urethanes. Each of these systems has advantages and disadvantages.

Conductive adhesives present possibilities as a mounting material, providing electrical contact without subjecting the assembly to high-temperature processes. At the time of this writing, these materials were not yet widely used. Advantages include elimination of soldering and the need for fluxes as well as a postsoldering defluxing operation. The metal content provides improved ductility of the joints. Disadvantages include lower conductivity than 100 percent metal joints and material cost.

27.6 SOLDERING PROCESSES AND EQUIPMENT

SMT solder joints must serve for mechanical attachment of components to the substrate as well as provide electrical continuity. Processes and equipment utilized range from those required for mass soldering of many joints at a time in a continuous flow to those required to generate one joint at a time. This section will deal with several methods in this spectrum which are being applied to providing interconnection of SMDs to substrates generally referred to as "printed wiring boards," creating a PWA. Table 27.2 provides a matrix of several of these soldering processes and pertinent characteristics.

27.6.1 Wave Soldering

This mass-soldering process can be applied to soldering SMDs when the devices are mounted on a surface of the PWB which can contact the solder wave. PWAs containing THM as well as discrete passive SM chips, SOICs, and SOTs can be wave-soldered with the THM components mounted on one side and SMD on the other.

The wave-solder process has the advantage of being a developed technology which is reasonably well understood. It provides the opportunity to solder THM and SM components in one operation with an unlimited source of solder. This requires the component termination and PWB pad to control the solder joint volume.

TABLE 27.2 Surface Mount Solder Process Comparison

Process	Potential production volumes	Capital cost	Operating cost	Heat-sensitive-component stress	Process control
Wave solder	High	Medium	Medium	High	Good
Infrared reflow	High	Medium	Medium	Medium	Good
Vapor phase reflow	High	High	High	High	Excellent
Laser reflow	Low	High	High	Low	Good
Belt conduction	Medium	Low	Low	Medium	Good
Belt convection	Medium	Low	Medium	High	Good

The use of other techniques with solder preforms and with screened or stencilled solder paste provides control of solder volumes through the deposition process.

In wave soldering, the SMDs are mounted with an adhesive to the bottom side of the board, the THM components inserted, and the unsoldered PWA presented to the wave-solder equipment.

Pitfalls of this process include the following:

1. Layout of SMD in patterns which "shadow" the terminations from the wave, causing skips or excessive opportunity for shorts.

2. High-solid-content fluxes, resulting in excessive outgassing. Fluxes with lower solids and higher activity appear to be advantageous.

3. Low preheat temperature (requiring lower soldering speeds), longer time in the wave, and resultant possibility of leaching, or loss of material from the chip

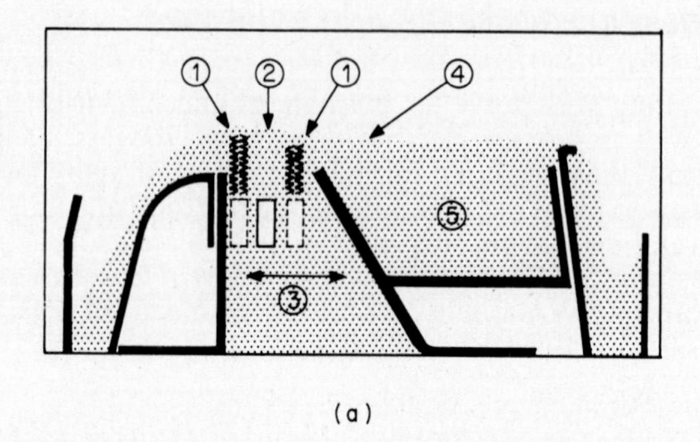

(a)

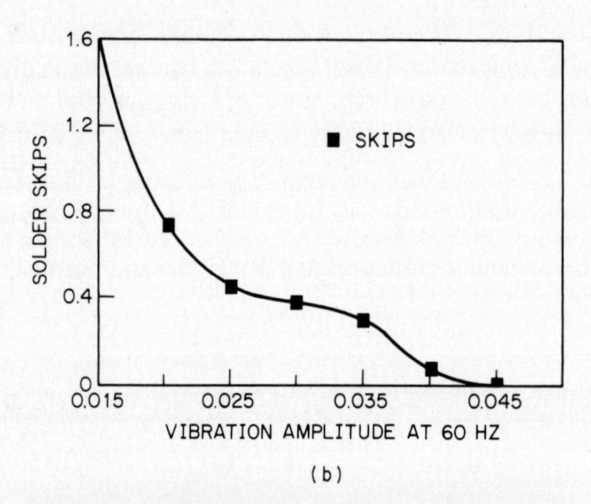

(b)

FIG. 27.6 Proprietary single oscillating wave. (*a*) Cross section of proprietary single oscillating wave. Shown are (1) vibration zones, (2) dead zone, (3) amplitude, (4) smooth laminar exit section, (5) solder. (*b*) Impact of oscillation amplitude on the occurrence of solder skips. *(Adapted from Surface Mount Technology, June 1986, pp. 8 and 9, and reprinted with permission.)*

contacts by solution in the wave. Reducing the solder-bath temperature to 245 to 250°C (478 to 482°F) is advantageous in overcoming leaching.

Low preheat can also result in thermal shock as the PWA enters the wave, resulting in component damage, particularly in the case of chip capacitors.

Equipment manufacturers have developed machines with wave dynamics which aid in solving the solder skip problem. This equipment provides a turbulent wave form which successfully introduces solder to all the joint areas and then a second smooth wave which serves to remove excess solder, thereby eliminating shorts or bridges and improving fillet shape.

Other equipment is available with a single oscillating wave, created by transducer-induced low-frequency vibrations with controlled amplitude. This oscillating wave is intended to prevent skips and bridges by breaking up gas pockets, thereby enhancing solder coverage. Figure 27.6 illustrates a schematic of this technique and results achieved.[4]

Another approach employs hot-air-knife technology as an effective method in removing or preventing solder shorts on high-density designs.[5]

The air knife directs hot air under pressure against the solder side of the board as the assembly leaves the wave. While the solder is still molten, the air flow removes excess solder and aids in shaping the remaining joint to a preferred profile with low contact angles (see Fig. 27.7).

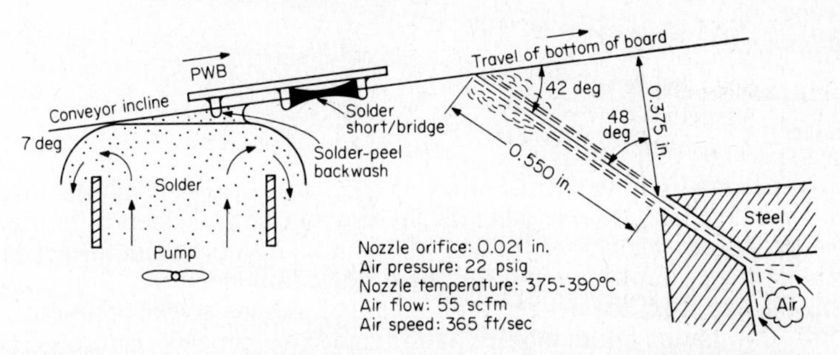

FIG. 27.7 Action of proprietary hot-air knife in removing shorts and bridges. *(Adapted from Electronic Packaging and Production, February 1984, and reprinted with permission.)*

A number of manufacturers of wave-soldering equipment market equipment which incorporates microprocessor control of critical wave-soldering parameters. These systems are advantageous in the soldering of PWAs with either SM or THM components. Parameters typically monitored and controlled with sensors and feedback loops include the following:

1. Conveyor width
2. Conveyor speed
3. Flux density
4. Preheat temperature
5. Solder-pot temperature
6. Wave height

This computer-controlled equipment provides accurate, fast, and reproducible process parameter control.

Coupled with wave-solder equipment capable of soldering assemblies with SM components, the production of quality products becomes dependent primarily on design, component solderability, equipment setup, maintenance, and understanding of the metallurgical processes.

27.6.2 Reflow Soldering

This soldering process, used in the hybrid industry for some time, has become the most widely used process for soldering PWAs populated with large numbers of multileaded SMDs.

In this process, the solder source consists of preforms, paste, or plated-up depositions of a tin-lead alloy on the substrate to which the components are to be soldered. Flux required is provided as a component of the solder paste deposit or added separately. Components are placed on the substrate in contact with the solder source and generally held there by the viscosity and tackiness of the paste or through use of adhesive in the case of preforms or plated-up lands.

Several techniques are used to activate the flux and bring the solder to its melting point to effect the metallurgical bond and solder fillet. All of these techniques use conduction, convection, radiation, or a combination of these heat sources.

The reflow techniques most commonly used are as follows:

1. Infrared
2. Vapor phase
3. Hot platen
4. Hot-gas furnace
5. Laser

Each has similarities as well as unique advantages or disadvantages that are in part dependent on characteristics of the product being soldered. In the use of reflow processes, the surface tension of the molten solder is important in maintaining position, allowing self-centering of the components and limiting wicking of solder away from the desired joint area. Pad design, solder paste-flux composition, termination solderability, and soldering process controls must be addressed to maintain consistent and predictable forces.

With reflow processes, a condition termed "tombstoning" is sometimes experienced with discrete passive chip components (see Fig. 27.8). This upending of the component with one termination off the board appears to result from a difference in the surface tension of the solder at the two terminations, holding one end fast and allowing the other end to raise off the substrate pad. The difference in surface tension can be attributed to any of the following: rate of approach to the liquidus temperature and joint temperature, wettability of the terminations, presence of condensed vapor phase fluids in the molten solder, outgassing of volatiles from fluxes, and difference in solder alloy composition.

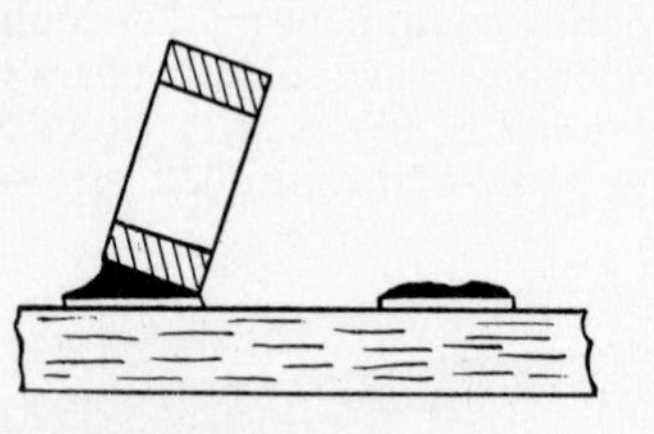

FIG. 27.8. Tombstoning; solder defect observed with chip components.

27.6.2.1 ***Infrared (IR) Reflow Soldering.*** Nonfocused IR sources providing energy with intensity and wavelengths compatible with the materials of the PWAs are a common source of IR energy for soldering.

In this indirect use of IR energy, emitter panels diffuse energy in the middle to far IR range to the product within an oven enclosure generally equipped with a transport belt and multiple zones. The emitter panels provide IR energy by secondary emission, in which energy from an IR element heats a flat ceramic panel which emits IR energy across its full surface. Source life of the panel emitters is up to 8000 h whereas lamp-type emitters have a life of less than 5000 h[6] and can progressively degrade with accumulation of residues on the bulbs.

This energy source provides for more even heating and reduced color selectivity than that found with focused sources. Lower operating temperature, high efficiency, and heating beneath the surface of some materials minimize cost and temperature stresses.

The low heating rates achievable with this technique provide some advantage in reduced spattering from solder paste solvent evaporation. This can reduce or eliminate the need for precuring solder paste other than in one of the first zones of the IR oven. Setup of the specific IR energy profile through the oven is critical and needs to be determined for each assembly design.

Differences in layout resulting in shielding of some joint locations and variations in absorption properties of different materials can result in large differences in the heating rates resulting in hot spots and material damage or unsoldered joints. Use of an inert atmosphere such as nitrogen in the IR oven is recommended to aid in eliminating oxidation of substrate and component materials degradation of fluxes and of solderability properties.

Time at temperature must be controlled to obtain a quality solder joint and also to prevent degradation of the assembly. This parameter can be controlled by speed of travel and IR energy profile through the furnace and with some recent equipment by varying the distance between the emitter and the substrate.[7] An example of a typical IR reflow temperature profile is shown in Fig. 27.9. The total time is controlled by thermal mass of the PWA, prior paste set time, energy profile, and belt speed.

Tungsten filament and Nichrome quartz lamps are sources of near-range, high-temperature focused IR energy.

Drawbacks of using these sources include hot spots with different-colored com-

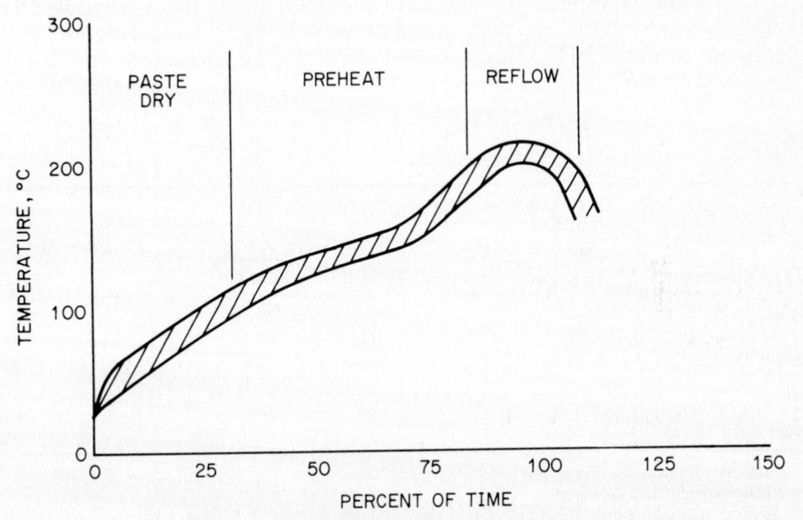

FIG. 27.9 IR reflow process temperature profile.

ponents as a result of differential absorption of the IR in the near-range wavelengths; shadowing of joints by large components; board and component damage as a result of improper adjustment; and gradual degradation and clouding of the source, requiring profile adjustment or early lamp replacement. An advantage is the ability to heat very local areas in cases in which only a few SMCs are located on an assembly. The balance of the assembly can be shielded and the focused source applied to the specific location. Because of the rapid heating rate, solder paste spatter may result in occurrence of more solder balls than with nonfocused IR. The temperature-time profile may also be varied with this equipment, with good results reported.

27.6.2.2 Vapor Phase Reflow Soldering (VPS). This process is an application of condensation heating to PWA soldering. A vaporized fluid condenses on a PWA immersed in the vapors over the boiling liquid. The latent heat of the vapor is given up to the PWA, raising the temperature of the entire workpiece to the temperature of the vapor.

27.6.2.2.1 Materials and Equipment. Synthetic, inert fluids are available with boiling points in the range providing sufficient heat to reflow-solder alloys used for SMD assembly. Some of these fluids have a narrow boiling range, providing a stable soldering temperature, while others possess a wider range over which the boiling temperature may shift with contamination and loss of constituents. Properties of some typical primary vapor phase soldering fluids in use are illustrated in Table 27.3.[8]

In vertical batch-process equipment, a second fluid with a lower boiling point is typically used as a vapor blanket over the soldering fluids. In Fig. 27.10[9] *(a)* illustrates a unit of this type heated with conventional immersion heaters, while *(b)* depicts a vertical unit with an alternate heating system. The secondary vapor blanket reduces loss of the primary vapor, which is more expensive, typically $500/gal versus $15/gal (30:1).

Replacing conventional heaters immersed in a liquid reservoir with a heated thermal mass in contact with the vapor as shown can reduce the fluid requirements and aid in maintaining a constant vapor generation rate. This thermal mass can aid in avoiding vapor collapse under heavy workloads when heaters might be unable to maintain rapid and sufficient heat input to reboil the condensate.[10] Horizontal in-line systems are also available. A schematic representation of this equipment in Fig. 27.11[11] illustrates the use of cooled surfaces to condense the

TABLE 27.3 Properties of Primary Fluids

	Montedison		3M			Air products,
Fluid distributor	Galden LS	Galden HS	Flutech	Fluorinert FC-70	Fluorinert FC-71	Multifluor APF 215
Boiling point, °C	221–227	270–276	215	215	253	215
Latent heat, cal/9m	15	15	16*	15	16	14*
Liquid density, kg/L	1.8	1.8	2.0	1.9	1.9	2.0
Evidence of toxic gas and HF acid	—	—	No†	Yes	Yes	No†

*Estimated value.

†Information from supplier.

Source: Donald Ford, *Electonic Packaging and Production,* February 1985, p. 100. (Used by permission.)

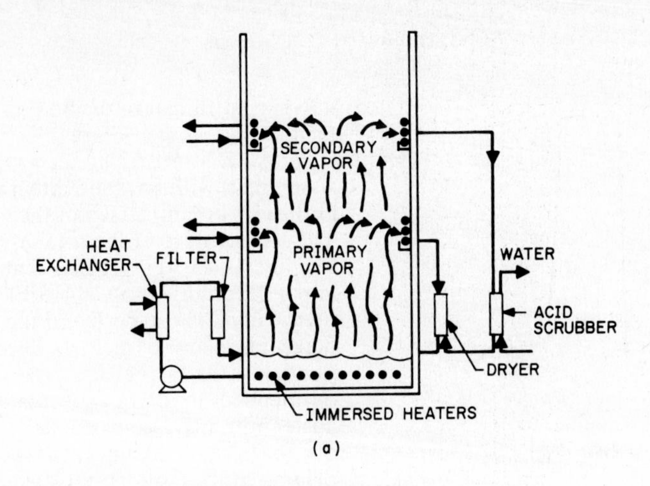

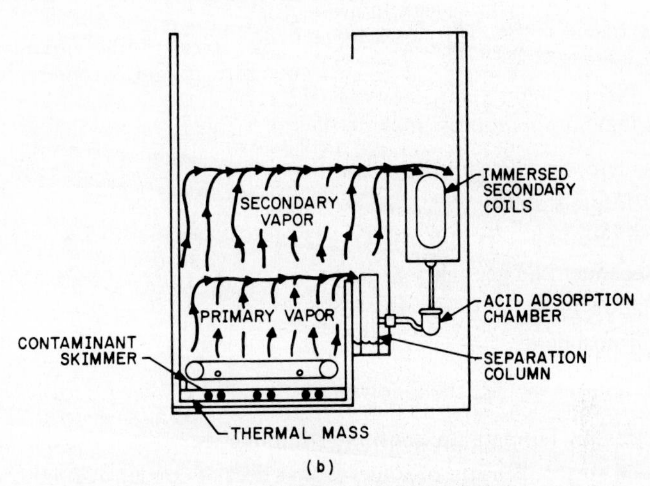

FIG. 27.10 Vertical, dual fluid batch vapor phase system with (*a*) conventional immersion heaters and (*b*) heated thermal mass as heat source. *(Reprinted with permission from Electronic Packaging and Production, December 1984, p. 87, Figs. 5 and 6.)*

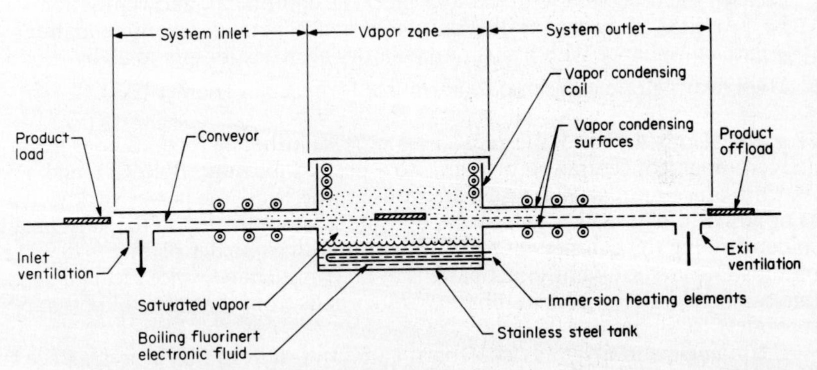

FIG. 27.11 In-line single-vapor heating system schematic. *(Reprinted with permission from Electronic Packaging and Production, November 1982, p. 63, Fig. 1.)*

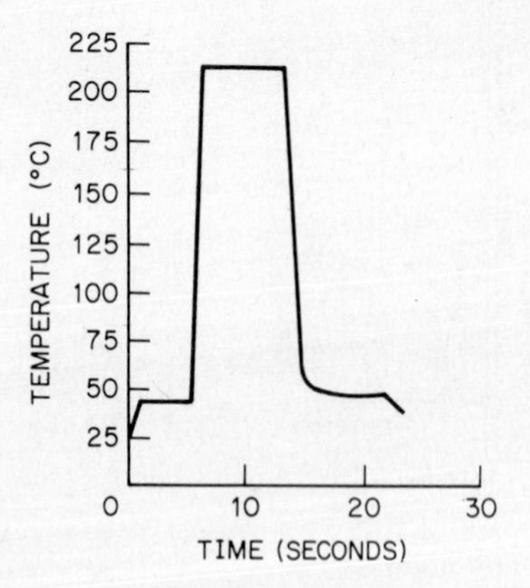

FIG. 27.12 Typical in-line vapor phase reflow temperature profile. *(Reprinted with permission from Electri-onics, July 1985, p. 31, Fig. 3.)*

vapor, returning it to the boiling liquid tank.

27.6.2.2.2 Soldering Process. Assemblies with SMD on either one or both sides are immersed in the vapors. The temperature of the total assembly quickly (10 to 30 s) reaches that of the vapor [typically 215°C (419°F) vapor used with 63:37 solder], and the solder deposit is reflowed (melted). Because of rapid heating, a prebake cycle is recommended to drive solvents out of paste, minimizing the ocurrence of solder balls and aiding in preventing delamination of multilayer substrates.

A typical in-line vapor phase temperature profile is represented in Fig. 27.12.[12]

Advantages of this process include the following:

1. Precise temperature control
2. Temperature stability
3. Uniform heating
4. Not dependent on geometry or size
5. No shadowing
6. Inert atmosphere

Disadvantages include the following:

1. Overstressing temperature-sensitive components.
2. Reduced surface tension of solder from effect of condensing fluid, which can cause movement, misalignment, tombstoning, and loss of components.
3. High operating costs.
4. High maintenance cost; by-products of the operation can include acidic and toxic materials (hydrofluoric acid and perfluoroisobutylene). A neutralizing cycle in the process is required to protect the equipment, and ventilation must be provided in accord with equipment manufacturer recommendations to assure compliance with limits of regulatory agencies for any toxic by-product.
5. High heating rate can cause or aggravate formation of solder balls.

27.6.2.3 Laser Reflow Soldering. This process utilizes a focused laser beam to deliver a high concentration of energy to a desired location. Both CO_2 and YAG (yttrium-aluminum-garnet) lasers are in use to provide low-power steady energy as opposed to high-power pulsed energy. Solder is introduced to the joint area by solder-dipping the component terminations and thick solder plating (~0.002 in) on substrate lands. Use of solder paste is not recommended, since the high heating rate can produce excessive numbers of solder balls from the paste. Flux is applied separately to the joint.

Beam angle and assembly coplanarity are important to this process, since solder surface tension will prevent fillet formation if large clearances exist between component terminations and substrate pads across the surface of larger multi-

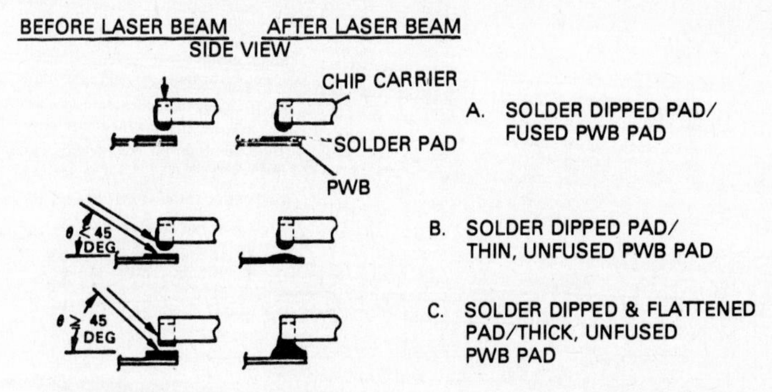

FIG. 27.13 Effects of laser beam angle and solder coating configurations. *(Reprinted from E. Lish, Application of Laser Microsoldering to Printed Wiring Assemblies, IPC TP 538, April 1985, Fig. 4.)*

leaded devices. One approach utilizes a 45° angle beam, "flattening" of the solder deposited on the component terminations by reflowing on a hot plate and increased solder thickness on the substrate pad (see Fig. 27.13[13]).

Advantages of this process include the following:

1. Precise, selective application of heat
2. Ability to solder temperature-sensitive components, since heating is highly localized
3. Low-stress joints with minimum formation of intermetallic compound
4. Noncontact with fluids or oxidizing atmospheres
5. Applicable to individual joints at otherwise inaccessible locations or bonded to large heat sinks

Disadvantages include the following:

1. Requires accurate device placement
2. High initial cost
3. Forms only one joint at a time; low throughput rate

27.6.2.4 ***Hot Platen Reflow Soldering.*** This process uses transfer of heat by conduction, generally from a resistance-heated surface. The assembly is placed on that surface in static direct contact or moved across via a very thin transfer belt (see Fig. 27.14[14]). The transfer belt process temperature profile is controlled by varying the temperature of the platens and the speed of the PWA across the platens.

Advantages include the following:

1. Low capital and operating costs.
2. Highly flexible; temperature profile and dwell time variations allow for paste cure and cool-down.
3. Moderate to high throughput.
4. Good for process development; reflow point can be observed and process-tuned.
5. Can be used with temperature-sensitive components.

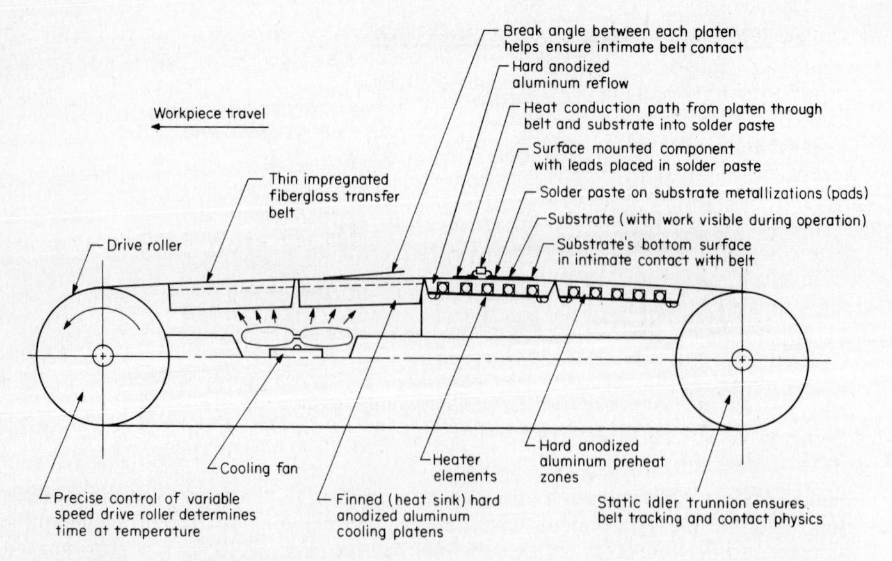

FIG. 27.14 Hot platen linear conduction soldering system. *(Reprinted with permission from Electri-onics, August 1985, p. 28, Fig. 4.)*

The disadvantages stem from the fact that heat conduction is through the substrate. Thus the following are required:

1. Flat substrate
2. A substrate that is a satisfactory conductor
3. No warp, twist, or degradation of substrate at reflow temperature
4. Oxidation of solder unless provided with inert-gas hood
5. Limitation to smaller sizes because of warp
6. Limitation, in general, to single-sided assemblies without use of specialized tooling

27.6.2.5 Contact (Hot-Bar) Reflow Soldering. This process technology uses pressure applied by resistance-heated elements to formed leads of leaded chip carriers or leaded components with the leads formed to be coplanar to the substrate.

Contact soldering must provide controlled pressure and heat to the component-substrate termination area. Applicability is selective, requiring gullwing or flatpack-type construction and accessibility of the lead-substrate contact surface.

The process is advantageous for short runs of temperature-sensitive packages with low-density assemblies.

27.6.2.6 Hot-Gas (Convection) Reflow Soldering. Convection heating from circulating hot gases is used in both high-volume oven configurations and in single-component soldering through the use of directed hot-gas streams.

In the oven process, component-substrate assemblies incorporating deposited solder in the form of paste, preforms, or plated substrate lands are transported on a belt through a muffled oven containing a heated inert atmosphere.

This process provides a highly stable thermal environment for reflow with the temperature profile for individual assemblies controlled by varying the speed through the oven. Good atmosphere control and high throughput are attainable with well-designed equipment. Slow startup and stabilization at desired temperature do not provide high flexibility for this process, and damage can occur to temperature-sensitive components.

Hand-held or mounted hot-gas-stream sources provide a tool for selective placement or removal of individual components. They are used primarily in repair and rework applications.

In this process, inert gas (nitrogen or argon) is superheated by passing over electric heater coils and then directed through a nozzle or nozzles to the location where reflow of solder is desired. Some systems work on one joint at a time; others reflow several joints simultaneously. Control of flow time, gas temperature, and volume are critical to successful application of the process. Gas temperatures can be used in the range of 150 to 700°C, and nozzle diameters less than 0.1 in are used to control gas flow.[15]

Because of the high temperature of the gas and the high heat content of the impinging stream, accurate control of time is necessary to prevent damage to the substrate. Adjacent components subject to damage by high temperatures must be shielded from the gas flow. In the case of components subject to damage by discharge of static charge, the air ionization resulting from the gas flow must be protected against by good practices in grounding equipment, material, and personnel.

Further discussion of this process can be found in a later section, "Repair and Rework of SMD Assemblies."

27.6.2.7 ***Hand-Soldering Reflow.*** For surface mount technology, the use of hand-held soldering irons to join components to substrates has been limited to very low production needs and repair of previously soldered assemblies. State-of-the-art equipment provides special tip configurations as well as accurate tip-temperature control, both of which are necessary for use with SMDs.

Since the predominant use of this technique is in the removal and replacement of SMDs during repair or rework of assemblies, further discussion has been deferred to a later section.

27.7 REPAIR AND REWORK

Design changes, unsatisfactory initial solder joints, and component failures detected in test all may require removal and replacement of SMDs just as is the case with through-hole mounted components. Densely populated assemblies, small joints, high-pin-count devices, varying thermal masses of the assemblies, and double-sided mounting present conditions complicating the operation for surface mount technology.

There are different areas of particular concern, including those associated with through-hole technology:

1. Pad or trace lifting or damage
2. Component damage
3. Substrate material degradation
4. Contamination or damage to nearby joints
5. High density, limiting access to joints and lack of space for rerouting signal traces

27.7.1 Equipment

Repair and rework are most generally performed with either the classical hand-held soldering iron conductive reflow method or the hot-gas-stream convection heating method.

Though use of hand-held irons is difficult because of density configuration and thermal damage considerations, there are custom-designed tips for SMDs and small tip irons with tip-temperature controls which are applicable in many cases. Equipment for the use of hot-gas convection ranges from hand-held devices to systems incorporating parameter controls, accurate placement, and magnification of the workpiece.

27.7.2 Process

The process requires preparation, reflow, removal, cleaning, and replacement. Recleaning and recoating may also be necessary if called for by specific end use.

Preparation should include removal of adjacent heat sinks or hardware wherever possible to aid in accessing the area to be repaired, minimizing thermal mass and obstruction to solder iron or hot-gas flow. Removal of any conformal coating and cleaning with a suitable solvent should be followed by application of a flux to aid in obtaining a clean, reflowed surface after component removal.

Reflow of the joints desired should be performed quickly (2 to 3 s), with the heat applied to the desired joint or joints. Control of the time, temperature, and distribution of the heat applied is critical in preventing damage to components and substrates. In the case of ceramic or other substrates subject to thermal cracking, preheat of the area to be reworked to a temperature within 50°C of the reflow temperature is recommended. This preheat will also serve to shorten the time at reflow temperature for each joint repaired, reducing gas usage and increasing operator productivity.

Removal of the device can be accomplished with tweezers or a vacuum pickup. These devices can be positioned and controlled manually or by machine.

Replacement of the component removed involves following adequate cleaning, pretinning, and fluxing of the component terminations. If replacement immediately after prior device removal is possible, the solder surface remaining on an uncontaminated substrate should be suitable for use. If this is not the case, cleaning and addition of solder paste or flux to the pads may be required.

Accurate placement of the component being added is less critical with leadless components, which will tend to self-position themselves as the solder reflows. Leaded chip carriers need to be precisely positioned. Attaining reflow temperature for attachment is achieved using the same techniques used for removal.

The reworked assembly should be inspected for solder joint acceptability, cleaned, and recoated as required.

27.8 POSTSOLDER CLEANING

Use of SMT and fine-line technology have simultaneously increased the difficulty of effectively cleaning PWAs and the necessity of doing so in order to carry out bed-of-nails testing as well as to assure product reliability in service.

Removing residues left after soldering requires the cleaning media to penetrate very small clearances between the underside of the device and the surface of the substrate as well as between components and circuitry. Clearances under components can vary from 0.025 in for PLCCs to less than 0.010 in for SOICs and less than 0.005 in for passive monolithic chip components.

The close spacing can make the impact of remaining residue more critical because of increased susceptibility to current leakage through bridging residue and the reduced distance for electromigration.

The source of residues include handling as well as flux residues. The flux residues include ionics such as flux activators and nonionic rosin residue. In SMT, the residues are in some cases more resistant to removal because of exposure to the soldering heat source for a longer period of time in reflow processes.

A wide range of defluxing media are available. Systems using either (1) hot deionized water with detergent or (2) stabilized chlorinated or fluorinated solvents are most generally in use. A dual process utilizing solvent cleaning followed by deionized water rinsing has gained acceptance by suppliers of military hardware as well as by many organizations producing soldered assemblies for commercial use and controlling their process to meet the ionic cleanliness specifications of MIL-P-28809.

The choice of defluxing agent is highly dependent on the flux system used, since the nature of the residue depends on the flux chemistry. Industry and military organizations have conducted extensive studies in the attempt to identify the most effective defluxing materials and processes. Some data indicate that solvent defluxers can penetrate spacings less than 0.006 in, while aqueous detergent cleaning is inefficient under 0.010 in.[16] Other data indicate little difference, if any, between solvent and aqueous defluxing following wave soldering of chip components and SOICs.[17]

Solvent defluxing equipment currently available makes use of vapor, spray, and immersion processes or a combination of all three. The spray provides some impingement forces which may help penetration but also may cause electrostatic discharge damage to sensitive components. Ultrasonics are used with some solvent defluxing equipment to aid in penetrating the small clearances.

Military specifications (as of this writing) do not permit the use of ultrasonics for cleaning PWAs, and commercial suppliers also express concern for potential damage to fine wire and wire bonds in SMDs as a result of unpredictable resonant frequencies.

Some aqueous defluxing equipment available for use with the highly active water-soluble rosin or organic acid fluxes makes use of air knives to "sweep" the water off the assembly and from under components as an aid in ensuring that the radiant or convective drying system does not fix the contaminant in place on the assembly, permitting reaction at a later time.

As with through-hole mounted assemblies, selection of equipment and processes for defluxing PWAs with surface mounted assemblies requires consideration of the parameters of the total material system and soldering process in use as well as environmental and safety regulations affecting handling, use, and disposal of residue-contaminated waste defluxing materials.

27.9 SOLDER JOINT REQUIREMENTS AND EVALUATION

27.9.1 Requirements

With surface mount technology, the integrity of the solder joints achieved between components and the substrate takes on a more significant role in mechanical attachment in addition to electrical continuity. At the same time, the

Dimensional criteria, mm and [in]
(Minimum values apply)

Feature	Dim.	Class I	Class II	Class III
Max. side overhang	A	0.5 (W or P)	0.25 (W or P)	0.25 (W or P) or 0.5 [0.02]
Min. end overlap	L	NR*	0.005†	0.005†
Min. end joint width	B	0.5 (W or P)	0.5 (W or P)	(W or P) − A
Min. joint height	M	F‡	0.25H or 0.5 [0.02]	0.25H or 0.5 [0.02]
Side fillet	S	NR*	NR*	NR*
Max thickness	X	F‡	F‡	0.5 [0.02]

* Not required.
† Across length of fillet B.
‡ Properly wetted fillet evident.

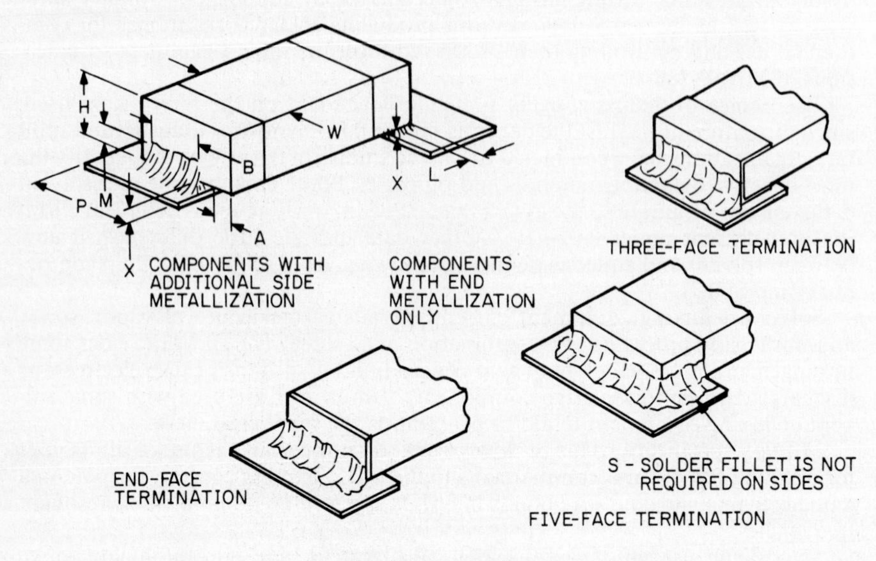

Acceptance criteria

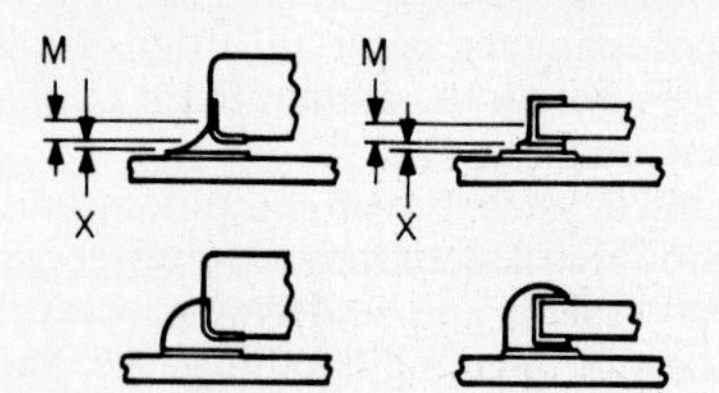

Minimum:
Solder fillet flows up end more than 25% or 0.020 in, whichever is less; for Class II and III product.

Maximum:
Solder covers entire pad and end cap but *does not* overhang either. Also, there is *no* evidence of dewetting at pad or end cap metallization.

FIG. 27.15 Chip components: end, 3, and 5 face terminations; solder joint requirements. *(Reprinted with permission from IPC. Adapted from IPC-S-815B for presentation by L. Hymes, "Surface Mount Workmanship Requirements," IPC annual meeting, April 1986.)*

size of the component termination is generally smaller than that encountered in through-hole technology; the pad-to-pad as well as pad-to-trace spacing is less; the number of joints per assembly is greater; and the joints are often on both side of the assembly as well as under components out of sight.

Attention to generation and preservation of good solderability characteristics of component terminations and substrate pads as well as control of the accuracy of component placement and soldering process parameters is particularly essen-

Dimensional criteria, mm and [in]

Feature	Dim.	Class I	Class II	Class III
Min. overlap and joint width	O	0.5(W or P)	0.5(W or P)	0.75(W or P)
Max. overhang	E	U*	U*	O
Max. thickness	X	U*	U*	0.5 [0.02]

*Unspecified parameter.

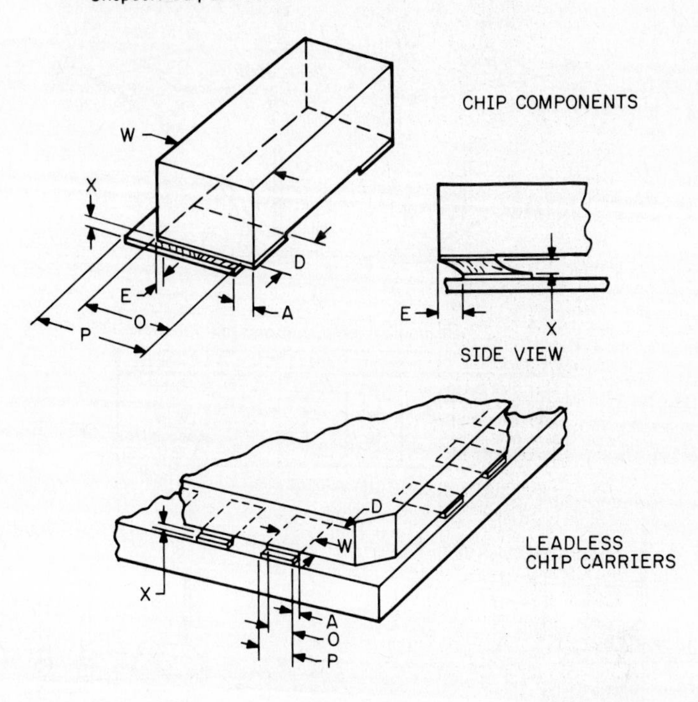

Acceptance criteria

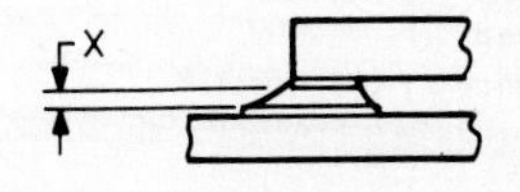

Minimum:
Solder fillet flows to top of lead.

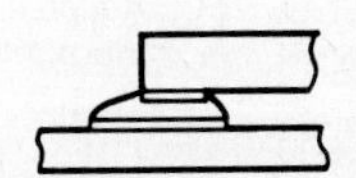

Maximum:
Solder covers entire pad but *does not* overhang. Solder *does not* bulge at the component body. *No* evidence of dewetting at the pad.

FIG. 27.16 Bottom-only terminations; solder joint requirements for chip components and leadless chip carriers. *(Reprinted with permission from IPC. Adapted from IPC-S-815B for presentation by L. Hymes, "Surface Mount Workmanship Requirements," IPC annual meeting, April 1986.)*

tial to achieving high-quality, reliable solder joints when using surface mount technology.

As with through-hole mounted technology, the PWA design and layout has a significant impact on the ability to consistently produce high-quality solder joints. That is, there are process and machine-controllable defects and those which occur

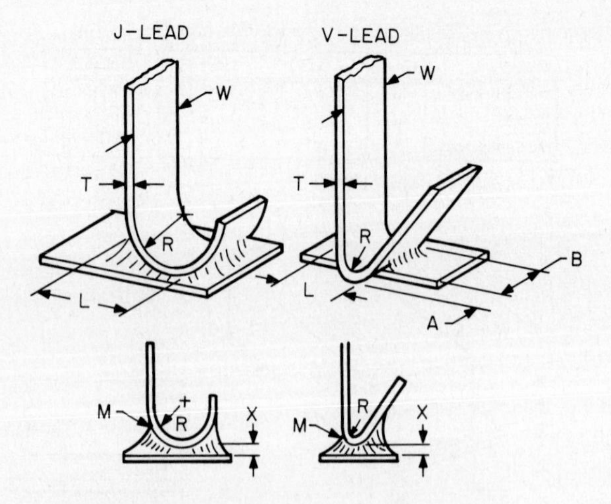

Dimensional criteria, mm and [in]

Feature	Dim.	Class I	Class II	Class III
Max. side overhang	A	0.5W	0.5W	0.5W
Min. side joint length	L	F*	1.5W	2W
Min. joint width	B	F*	0.5W	(W−A)
Min. heel fillet	M	F*	F*	Midpoint†
Max. thickness	X	F*	F*	0.75 [0.03]

*F = Properly wetted fillet evident.

† = Midpoint of bend as shown.

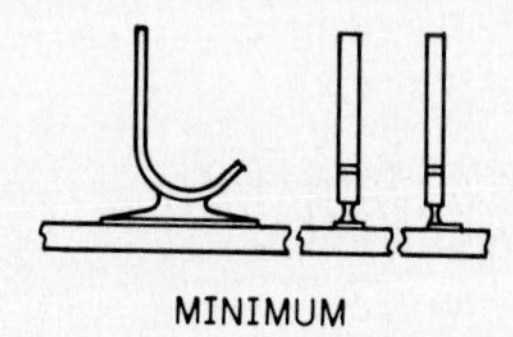

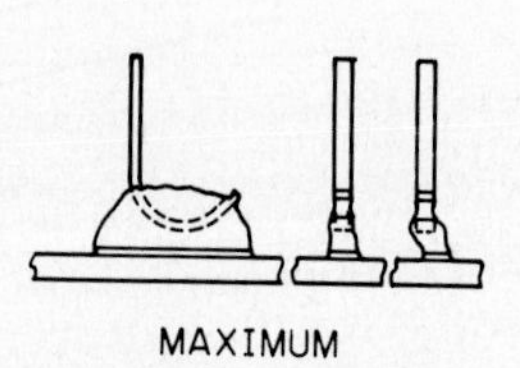

Acceptance criteria

Class III: Wettable sides; side fillets required.

Classes I & II: Where sides not wettable by design, e.g., leads sheared from preplated stock; side fillets not required; design for inspection of wetting to wettable surfaces.

FIG. 27.17 J and V leads typical of plastic leaded chip carriers; solder joint requirements. *(Reprinted with permission from IPC. Adapted from IPC-S-815B for presentation by L. Hymes, "Surface Mount Workmanship Requirements," IPC annual meeting, April 1986.)*

because of design philosophies and rules (or lack thereof), independent of the manufacturing process and equipment. Knowledge of process parameters and capabilities is necessary for design and layout personnel, and close interaction with process engineers in review of manufacturability and testability characteristics of designs is essential. The design affects not only joint formation but also capability to inspect and test the assembly.

The use of statistical process control (SPC) techniques, always important to good manufacturing practice, becomes increasingly critical with SMT for many of the reasons mentioned earlier in this section. The capabilities of humans as well as machine inspection and testing techniques become more limiting with this technology than heretofore experienced.

There are many diverse opinions as to what constitutes an acceptable SMD solder joint in light of the relatively small area and the short-term as well as long-term cyclical, mechanical, and electrical stresses imposed by different classes of service and reliability needs. IPC-S-815B, *General Requirements for Soldering Electronic Interconnections,* encompasses some widely accepted acceptability requirements arrived at through a consensus of producers and users. Figures 27.15 through 27.19 are based on this document and illustrate requirements for several typical SMT solder joints.[18] The classes referred to in the figures are defined by IPC as the following:

Class I: Consumer products

Class II: General industrial

Class III: High reliability

Cosmetic evaluation is never an adequate technique for solder joint acceptance. In particular, comparison of the characteristics of through-hole mounted joints formed with the wave-solder process to surface mounted device joints

Dimensional criteria, mm and [in]
(minimum values apply)

Feature	Dim.	Class I	Class II	Class III
Max. side overhang	A	0.5W	0.5W	0.5W
Min. fillet extension	E	F*	0.5(M+X) or P	0.5(M+X) or P
Min. fillet height	M	F*	0.25(H)	0.25(H)
Max. thickness	X	F*	F*	0.75[0.03]

*Properly wetted fillet evident.

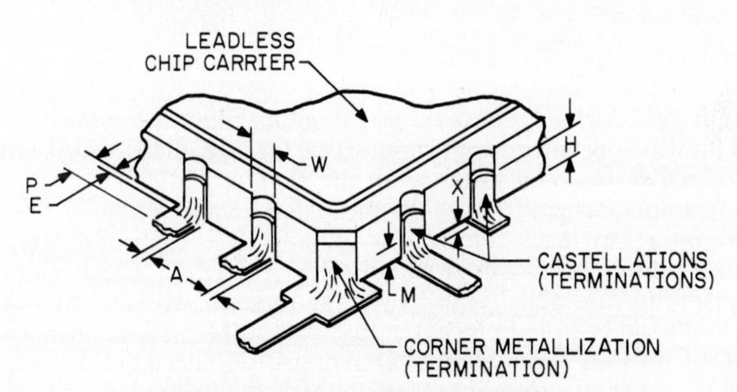

FILLET REQUIRED IF LAND IS PRESENT.

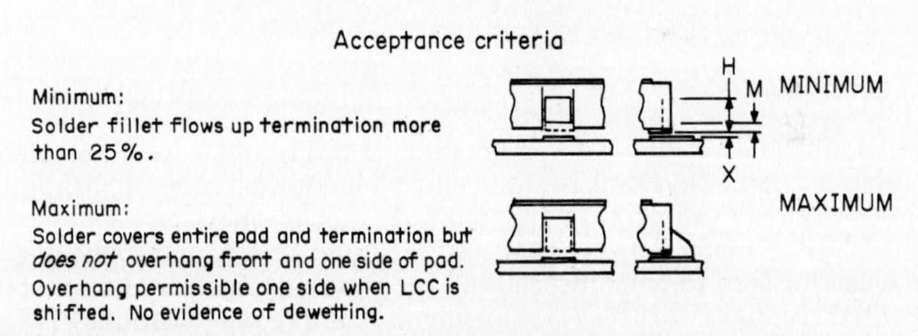

FIG 27.18 Castellated termination typical of ceramic leaded chip carriers; solder joint requirements. *(Reprinted with permission from IPC. Adapted from IPC-S-815B for presentation by L. Hymes, "Surface Mount Workmanship Requirements," IPC annual meeting, April 1986.)*

Dimensional criteria, mm and [in]
(Both sides of lead)

Feature	Dim.	Class I	Class II	Class III
Min. height	H	0.25 [0.01]	0.5 [0.02]	0.5 [0.02]
Min. width	B	0.5 W	0.75 W	W

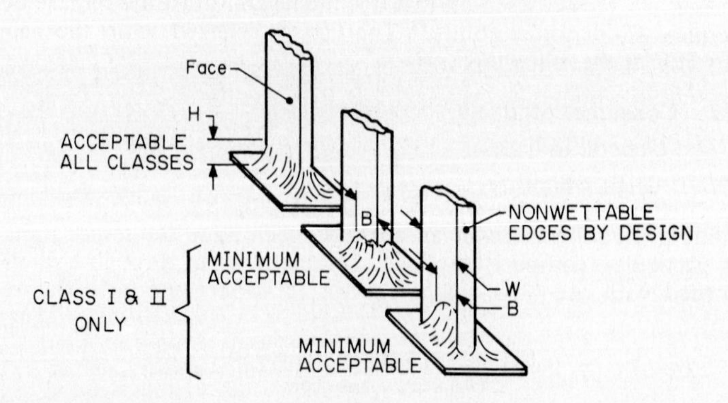

FILLETS MUST EXIST ON BOTH FACES OF LEAD

FIG. 27.19 Butt joints; solder joint requirements. *(Reprinted with permission from IPC. Adapted from IPC-S-815B for presentation by L. Hymes, "Surface Mount Workmanship Requirements," IPC annual meeting, April 1986.)*

formed with the reflow process should not be relied upon as a highly reliable inspection tool.

While smooth, well-formed, concave, bright solder fillets are typical of good wave-soldered joints using eutectic or near-eutectic tin-lead alloys, this is not necessarily true for reflow-soldered joints, with the same or different composition alloys. Typically, joints formed from solder paste in the vapor phase or IR reflow process will exhibit a gray luster, thought by some to be a result of cooling rate, off eutectic composition, joint geometry, and solder thickness. These conditions do not necessarily indicate poor joints according to some sources.[19] To others, these are conditions indicative of process or material problems when using eutectic or near-eutectic tin-lead solder alloys: "Dull or gritty solder joints are not normal, and are always an indication that something in the process or components to be soldered is out of control."[20]

27.9.2 Evaluation

The need for reliable methods for evaluation of SMT solder joint acceptability has accelerated the development and introduction of new applications of machine-controlled evaluation as opposed to test techniques.

Three of these methods—optical-based machine vision, x-ray imaging, and laser-induced infrared thermal inspection—are receiving significant development efforts with three major thrusts: (1) implementing improved techniques, (2) eliminating operator errors, and (3) achieving increased rates of evaluation.

27.9.2.1 Optical-Based Machine Vision. Techniques utilize diffuse light, precise high-speed positioning, and pattern-recognition algorithms with a microprocessor-controlled system to determine solder joint acceptability. With most equipment, it is necessary to have the equipment "learn" what is acceptable as well as joint location from a good sample board. This can take up to 4 h per assembly type. Evaluation of individual joints for solder-related defects takes place at rates in the range of 6 joints per second.[21]

27.9.2.2 X-Ray Imaging. This technique applies x-ray or fluoroscopic technology to evaluation of the integrity of the joints. This technique is currently best adapted to SMT rather than THM joints. The soldered assembly is subjected to a field of radiation which passes through the components, joints, and substrate. The solder alloy is more opaque than the balance of the assembly. Because of the contrast, voids, fractures, solder balls, lack of bond, etc., are observable much as in medical x-ray images. Joints under components or on the bottom side of the board are generally as accessible to the x-ray and open to unobstructed viewing as are those on top. The x-ray technique allows interrogation into the joint, which is not possible with optical methods. The joint image can be processed or otherwise compared with acceptance criteria either preprogrammed or learned from a good sample.

27.9.2.3 Laser-Induced Infrared Thermal Inspection. This technique utilizes a metered pulse of laser beam energy to produce thermal change (warming) in the solder joint.

Time-temperature relationships (affected by surface properties and internal features of the joint) are measured by an IR detector, amplified, digitized, and plotted as a curve depicting thermal emission rise and fall. With an exposure of 30-ms duration used to measure heating rate and multiple 5-ms readings during cooldown, up to 10 joints per second can be interrogated by one such system.

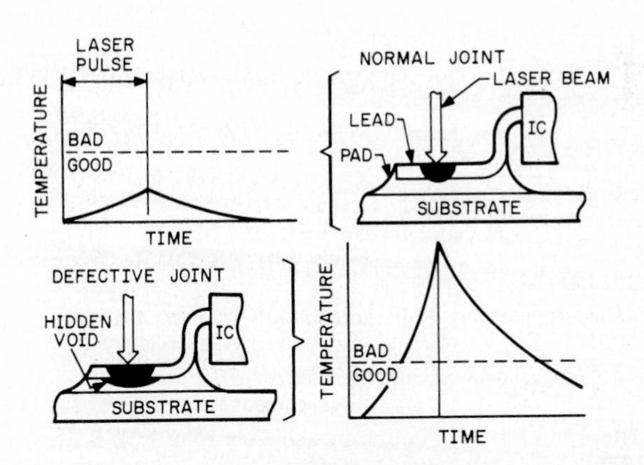

FIG. 27.20 Laser thermal inspection indications with normal and defective lap joints. *(Reprinted with permission from Electri-onics, October 1985, p. 45, Fig. 3.)*

Comparison of the "thermal signature" of the joint with results previously learned from good samples of solder joint types allows accept-reject decisions to be made by computer or by human operators. In general, defective joints absorb more energy and reach higher temperatures. Figure 27.20[22] illustrates indications given by normal and defective lap joints.

REFERENCES

1. IPC-CM-78, *Guidelines for Surface Mounting and Interconnecting Chip Carriers,* November 1983.
2. David Schoenthaler, "Soldering Surface Mounted Chip Carriers to Printed Circuits," *The Western Electric Engineer,* vol. 27, no. 1, 1983, pp. 73–79.
3. John R. Fisher, *Cast Lead Process and Surface Mount Reliability,* IPC-TP-536, April 1985.
4. Craig Biggs, "An Overview of Single Wave Soldering for SMT Applications," *Surface Mount Technology,* June 1986, pp. 8–10.
5. Leo Lambert, "Air Knives Have an Edge on Solder Defect Control," *Electronic Packaging and Production,* February 1984.
6. Stephen Dow, *The Use of Radiant Infrared in Soldering Surface Mounted Devices to Printed Circuit Boards,* Vitronics Corp. technical report, December, pp. 3, 6.
7. James Smorto, "IR Solder Reflow Update," *Circuits Manufacturing,* February 1986, pp. 49–50.
8. Donald Ford, "Vapor Phase Soldering," *Circuits Manufacturing,* February 1985, p. 100, Table 2.
9. Ronald Pound, "Smaller PCB Geometrics Shape Soldering Systems Trends," *Electronic Packaging and Production,* December 1984, pp. 84–87, Figs. 5 and 6.
10. Ibid., p. 87.
11. H. J. Peck, "In-line Vapor Phase Soldering: An Economic Reality," *Electronic Packaging and Production,* November 1982, pp. 62–66, Fig. 1.
12. Michael Stein, "How to Choose Solder Pastes for Surface Mounting," *Electri-onics,* July 1985, pp. 28–32, Fig. 3.
13. Earl Lish, "Application of Laser Microsoldering to Printed Wiring Assemblies," presented at IPC annual meeting, April 1985.
14. James Holloman, "Soldering for Present and Future Surface Mount Applications—Part 2," *Electri-onics,* August 1985, pp. 26–30, Fig. 28.
15. J. Werth and C. Bassman, "Hot Gas Desoldering Cuts Hybrid and PCB Repair Costs," *Electri-onics,* December 1983, pp. 35–37.
16. "Solvents Flush Solder Flux from Tight Spaces," *Electronic Packaging and Production,* August 1984, pp. 58–63.
17. M. F. Comerford, "Cleaning Printed Wiring Assemblies: The Effect of Surface Mounted Components," *Electri-onics,* November 1984.
18. L. Hymes, "Surface Mount Workmanship Requirements," presented at IPC annual meeting, April 1986.
19. J. Keller, "Re-defining Solder Joint Acceptability," *Electronic Packaging and Production,* February 1986, pp. 101–103.
20. J. Mather and J. Hagge, personal letter to *Electronic Packaging and Production* in rebuttal to comments attributed to them in Ref. 18 above, April 1986.
21. G. L. Schoenbaum, "Process Monitoring and Control of PCB Soldering Utilizing an Automated Machine Vision System," presented at IPC annual meeting, April 1986.
22. Alan C. Traub, "Overview of Laser/Thermal System Used to Detect Faulty Solder Joints," *Electri-onics,* October 1985, pp. 43–47.

P · A · R · T · 6

QUALITY AND RELIABILITY

CHAPTER 28
QUALITY ASSURANCE

George A. Smith
Vice President, Trace Laboratories, Linthicum, Maryland

28.1 INTRODUCTION

This chapter discusses quality assurance considerations applicable to printed circuits. The boards, or flexible circuits, may consist of two or more layers of conductor patterns on insulating material and be interconnected by a continuous plated-through-hole metallic connection. For purposes of the chapter, the term "boards" will be used hereafter when referring to finished printed wiring boards.

28.2 QUALITY CLASSES

Multilayer printed board assembly has been separated into classifications according to intended use. Toward this end three general classes of assemblies have been established by the Institute for Interconnecting and Packaging Electronic Circuits (IPC). It should also be recognized that there may be overlap of equipment between classes. The user has the responsibility to determine the class into which this product belongs. The three classes are as follows:

- *Class 1—Consumer Products:* Includes TV sets, toys, entertainment electronics, and noncritical consumer or industrial control devices. Boards in this class are suitable for applications where cosmetic defects are not important and the only requirement is the functioning of the completed circuit. These boards represent a low-cost alternative because of the lack of required inspection and testing.
- *Class 2—General Industrial:* Includes computers, communication equipment, sophisticated business machines, instruments, and certain noncritical military applications. Boards in this class are suitable for use where circuit design, process yield, and specification conformance requirements allow localized areas to be defective. These boards have minimal inspection, testing, and application requirements.
- *Class 3—High Reliability:* Includes equipment whose continued performance is critical, such as life-support equipment. Boards in this class are suitable for applications where high levels of assurance are required. The major difference is improved service and a higher level of inspection, testing, and certification.

These classifications have been created so that the performance of multilayer products may be tested according to the requirements of any one of the three classes. The use of one class for a specific attribute does not mean that all other attributes must meet that same class. Selection should be based on minimum need. However, crossover between classes requires complete definition of test requirements in the procurement specification.

28.3 QUALITY CONFORMANCE TEST CIRCUITRY

A series of test coupons, which are referred to as "quality conformance test circuitry," was designed to permit testing without destroying boards (Fig. 28.1).

In Fig. 28.1 the following is suggested: (1) For economy and ease of artwork layout, the location of each coupon of the quality conformance test circuitry is optional. All test coupons illustrated must appear on each panel. (2) The number of layers must be identical with the number of layers in the boards derived from the panel. Layers shown are for illustration purposes only. (3) Etched letters on the coupons are for identification purposes only. (4) Length of coupons D, E, and G is dependent upon the number of layers in the panel. (5) All pads should be 0.070 ± 0.005 in in diameter or 0.070 ± 0.005 in square. Holes in pads should be the diameter of the smallest hole in the associated board. (6) All conductor lines should be 0.020 ± 0.003 in wide unless otherwise specified. (7) All tolerances must meet the requirement of the imposed specification.

The quality conformance test circuitry also provides the feature of a perpetual preproduction sample that is representative of each multilayer board made throughout the project. The design allows for either placing all coupons in a strip configuration, or separating and placing them at various locations on the panel. Strategically locating the individual coupons on the panels can be a definite advantage to manufacturers, because they can then observe in-process steps more accurately. For example, they may (1) observe the drilling done by each drill bit on panels being processed by multiple drilling, (2) examine the plating thickness at various time intervals of the plating process, (3) compare the plating thickness at many locations on the panel, and (4) compare the plating thickness at any location of all panels in the plating tanks. If a process should go out of control, the manufacturer can immediately shift to 100 percent testing until the process is corrected, without needlessly destroying end-product boards during the screening operation. Obviously, accurate identification, storage, and retrieval of coupons is essential. In-process testing early in production is a self-governing evaluation system.

28.4 TESTING

28.4.1 Sampling and Testing

After the capability of testing representative coupons has been incorporated, the next step is to develop, with an eye toward minimizing costs, a realistic sampling technique for testing. Certainly any testing is expensive, but not testing at all is ultimately more expensive. Sampling need not be risky, especially if we keep in mind that all tests can be performed at a later time simply by retrieving the desired coupons for lot screening or for diagnostic work.

Preproduction samples must pass all requirements of the specification in order

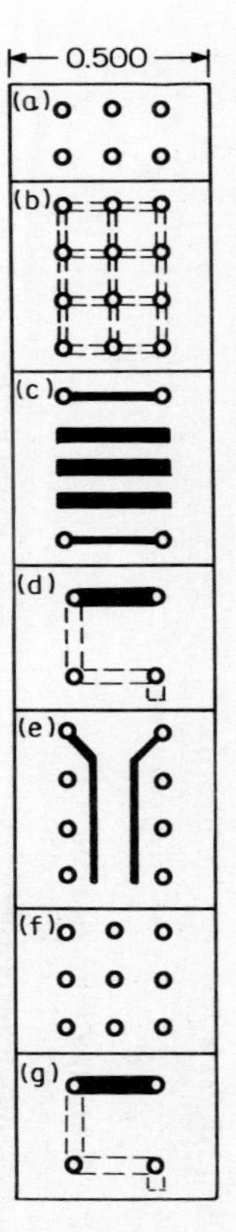

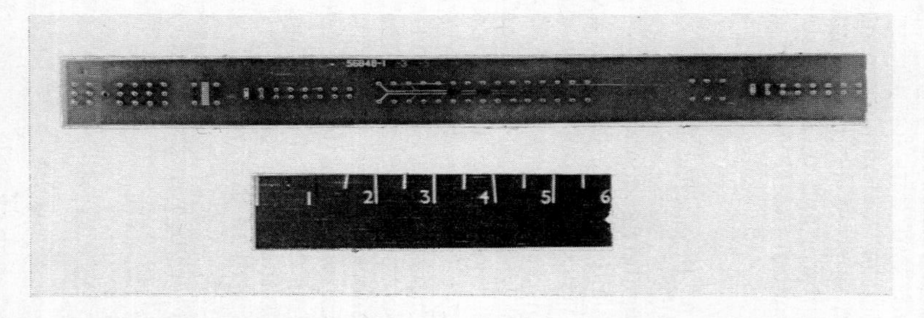

FIG. 28.1 Quality conformance test circuitry.

for manufacturers to prove their ability to fabricate boards of acceptable quality. Caution must be used in evaluating preproduction boards, because they are often made in a laboratory hand-line operation. It is very difficult and unusual for a manufacturer to arrange all the special provisions necessary to process preproduction samples in the true production mode. Examples of special provisions are (1) preparation of numerical control tape for drilling, (2) fabrication of racks and other fixtures that would normally be used in production, (3) rearranging plating anodes, board orientation, agitation, etc., and (4) interrupting other production work to process a few sample boards.

28.4.2 Preproduction Testing

Preproduction testing includes visual and dimensional examination, microsectioning, and tests for plating adhesion, terminal pull, warp and twist, water absorption, solderability, plated-through-hole structure, thermal shock, thermal stress, interconnection resistance, moisture and insulation resistance, current breakdown, and dielectric strength. In special applications, fungus resistance, mechanical shock, vibration, and outgassing characteristics also are tested for.

28.4.3 Production Testing

During the day-to-day production phase, testing is held to a minimum provided the quality is maintained at a sufficiently high level. Under normal conditions, only visual and dimensional examination, microsectioning in one plane, plating adhesion and moisture and insulation resistance tests on the coupon, and circuitry tests on the boards are required. In an effective sampling procedure, destructive testing is performed on the coupons every four weeks or after every 5000th board is produced, whichever occurs first. The testing includes visual and dimensional examination, microsectioning in three principal planes, and tests for terminal pull, water absorption, solderability, plated-through-hole structure, thermal shock, thermal stress, interconnection resistance, current breakdown, current-carrying capacity, and dielectric strength. If failures occur during those tests, retesting of associated coupons is performed. Repeating the failure causes temporary shutdown, recall of test coupons, and a fact-finding study, although not necessarily in that order. Normal production is resumed only after the cause is detected, corrective action is taken, and a complete screening is accomplished. However, tightened inspection sampling should be in effect for a prescribed period of time to assure that corrective action was indeed effective and permanent.

Heavy emphasis is therefore placed on reliable plated-through holes. Hence, periodic microsectioning of the x, y, and z planes and periodic removal of all glass cloth and epoxy resin for viewing the outer structure of plated-through holes (plated-through-hole structure test) was developed. Appropriate test methods are discussed later in this chapter. During production, the periodic test coupons mentioned above are sent to the customer regardless of whether boards were made in-house or by a subcontractor.

28.4.4 Final Testing

For multilayer boards, only the two most critical requirements must be tested 100 percent: (1) microsectioning of one test coupon from each production panel and

(2) electrical circuitry testing of the finished boards. It is not economically feasible to wait for the final assembly test before testing for bare board electrical performance. To have hundreds of dollars worth of components on defective boards and not find electrical failures until later is extremely costly. It is estimated that the bare (unpopulated) boards represent 7 percent of the cost of finished electronic equipment but that assembled boards represent 85 percent. There obviously is a great cost significance to be considered. The testing cost for low-priced radio and TV boards is practically nil because there are very few requirements, but the testing cost of high-reliability multilayer boards can be approximately 3 percent of the board cost. This is still negligible when one considers the total cost of sophisticated equipment.

Since the function of the multilayer board is to provide reliable electrical interconnection, mechanical integrity of the plated-through hole is of paramount importance.

28.4.5 Tested Characteristics

The characteristics which may be required on various levels of quality boards are shown in Table 28.1. The letters "LR" designate commercial boards that have limited requirements. The numbers "1," "2," and "3" designate the IPC's three levels of quality. "MIL" designates military specifications, and "Hi-rel" indicates high-reliability requirements (outer-space requirements).

TABLE 28.1 Characteristics Which May Be Required on Various Levels of Quality

Requirement	LR	1	2	3	MIL	Hi-rel
Visual and dimensional characteristics	x	x	x	x	x	x
Etchback						x
Plating adhesion		x	x	x	x	x
Microsectioning					x	x
Terminal pull		x	x	x	x	x
Copper strike						x
Warp and twist			x	x	x	x
Traceability						x
Water absorption						x
Copper pyrophosphate						x
Solderability	x			x		x
PTH structure					x	x
Fungus resistance						x
Mechanical shock						x
Vibration						x
Thermal shock					x	x
Thermal stress					x	x
Outgassing						x
Interconnection resistance				x	x	x
Insulation resistance		x	x	x	x	x
Dielectric strength		x	x	x	x	x
Hot-oil resistance		x	x	x	x	
Moisture resistance				x	x	x
Current-carrying capacity	x		x	x		x
Internal shorts	x	x	x	x		x
Circuitry electrical test (100%)					x	x
Flammability			x	x	x	x

28.4.6 High-Reliability Concerns

One of the most serious problems can be administrative delay in obtaining test coupons, especially when subcontractors are involved, for large volumes of boards may be manufactured in the interim. Coupons should be tested promptly to help in the assessment of the quality of production. Another problem is that if manufacturers incorrectly drill holes, plate a quantity of the boards, and then store the boards until the assembly people order the particular part number, they do not know that they are producing a bad product up to the plating stage. Because of these problems, it is advisable, especially in high-reliability programs, to expedite things by having a coupon from each board or panel sent directly from the production floor to the customer's test department, bypassing all engineering, production, and administrative channels. Appropriate follow-up can come later. One finished board should be required periodically for float-solder testing, because the coupons have very little surface area for that test, and delaminations and outgassing are important properties.

28.5 MICROSECTIONING REQUIREMENTS

This section describes multilayer printed wiring board plated-through-hole microsection requirements.

Emphasis is placed on end-product acceptance (but bear in mind that microsectioning is also effective for in-process checks and as a failure-analysis tool for the finished assembly).

The military services rely heavily on microsectioning techniques in order to verify that the product they are purchasing does in fact meet the end-product requirements.

Some simple ground rules for a good analysis are as follows:

1. Make the evaluation within the specified magnification limits (both minimum and maximum) of the specification.
2. If magnification limits are not stated, arrange a vendor-customer agreement before production begins.
3. Common magnification is 100 ± 5× for visual inspection and dimensional measurements and 200 ± 10× for referee testing.
4. Do not use metallographic techniques for measurements under 0.00003 in.
5. Look at the actual specimen rather than making critical decisions based solely on photomicrographs.
6. Examine and/or photograph each side of the barrel independently.

Before the actual evaluations, carefully inspect the microsection to be sure it is of sufficient quality to perform a valid examination, especially if the microsection has been made by someone else.

When inspecting a cross section, check the smoothness of the surface of the mount and determine if all the plated-through holes are completely filled with molding compound. If not, the plating metals may have burred over and the thickness determinations may be incorrect. Also examine the mount under the microscope to ensure that the specimen has been polished to the exact center line for accurate barrel measurements and that there are no scratches, particularly in the areas of plating interfacing or where dimensional measurements will be taken. If the mount has been made some time previously, check the chemical etching to determine if the metals are too dark because of staining to afford an accurate evaluation. Repolishing and re-etching may be necessary. Also, check to ensure that the specimen has not been overetched, giving false impressions of grain boundaries. Carefully examine all laminate, metal-foil, and plating areas. Underetched specimens could have copper smear, which masks defects.

28.5.1 Specific Requirements

For acceptability standards see Chap. 29, "Acceptability of Fabricated Circuits." Listed below, however, are some important examples of microsection inspection (see Fig. 28.2):

1. Nailheading up to a 50 percent increase in foil thickness is allowed. Nailheading indicates that there has been degradation in the drilling operation. Look carefully for resin smear.

2. Etchback (removal of glass and resin) measurements are 0.0002 to 0.003 in. Shadowing is allowed on one side of each inner conductor, and wicking may extend an additional 0.003 in maximum.

3. No resin smear is allowed. In order to examine for resin smear, evaluate the specimen before chemical etching to assure against false impression of resin smear.

4. Resin recession is a series of voids along the interface between the copper-plated barrel and the laminate. Resin recession may total up to 40 percent maximum of the total dielectric thickness on *each* side of the barrel as seen in the microsection. (See Sec. 28.6.)

5. Three plating voids are allowed in any specific plated-through hole. The total length of voids cannot exceed 10 percent of the board's total thickness. Voids also are not permitted at any copper-plating–foil interface, nor on both sides of a barrel in the same plane. Plating voids may be completely disallowed in the next revision of MIL-P-55110 and MIL-P-50884.

6. Plating nodules are acceptable if they do not reduce the inside diameter of the barrel below minimum requirements. A word of caution, however: nodules can be a serious problem where automatic insertion is employed.

7. Cracks in the copper-plated barrel and copper foil are not allowed. Foil cracks usually develop at about 0.0003 in out from the end of the conductor foil (whether or not etchback is employed).

8. Complete bonding of all surfaces (no voids or delamination) is required. Also, delamination of copper foil from the laminate is not allowed.

9. Plating contamination is usually in the form of dark spots due to foreign matter deposited into the copper plating and is not allowed.

10. No particles of unetched metal, where etching is intended, are allowed. Examine for unetched foil on every layer across the microsection.

11. Complete solderability of all surfaces is required. The presence of tin-lead diffusion into the copper, forming a ternary compound and complete wetting, must be evident.

28.5.2 Additional Evaluations

In addition to the plated-through-hole evaluations, the specimen must also be examined for:

1. Laminate thicknesses (and lay-up) between metal-foil conductors for thickness compliance (0.0035 in minimum when cured, and two plies minimum).

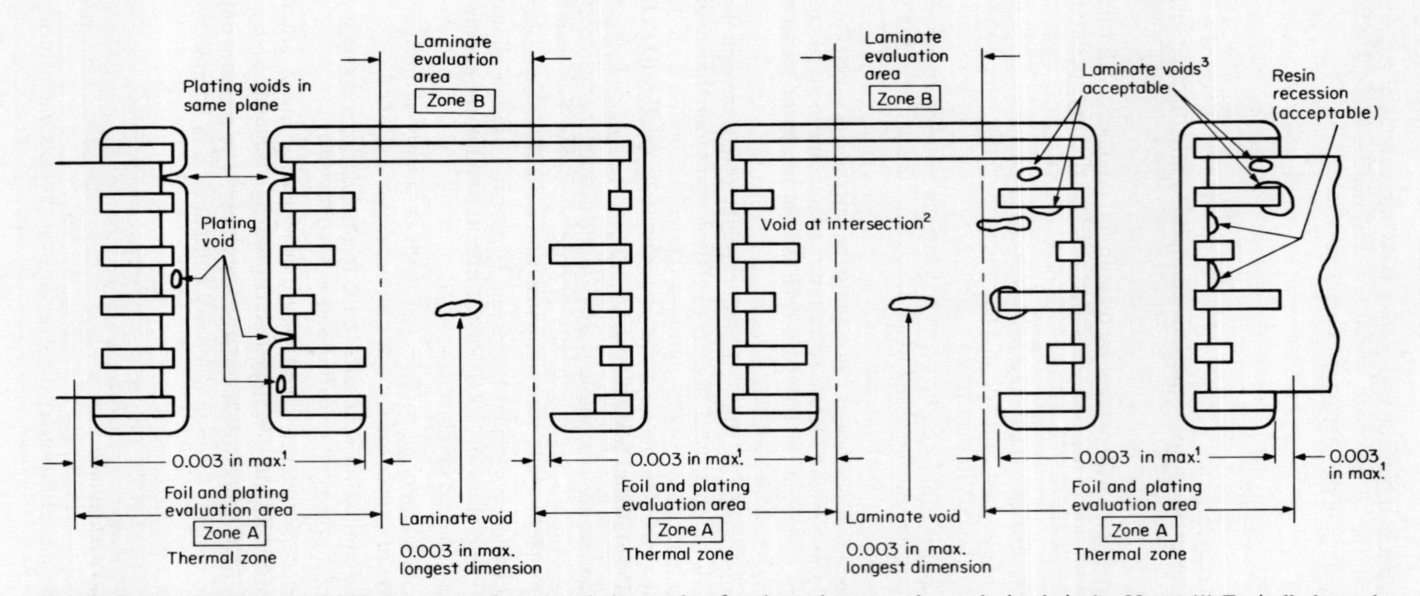

FIG. 28.2 Typical plated-through-hole cross section (three-hole sample, after thermal stress and rework simulation). Notes: (1) Typically beyond land edge most radially extended. (2) Void at intersection of zone A and zone B; laminate voids greater than 0.003 in (0.08 mm) in length which extend into the laminate evaluation area are rejectable. (3) Laminate voids are not evaluated in zone A; laminate voids greater than 0.003 in that extend into zone B are rejectable.

2. Measles, haloing, or crazing, a separation of the resin from the glass cloth (IPC-A-600C Class 3).

3. Laminate voids, which should be 0.003 in maximum in any direction, providing the required minimum spacing remains between conductors in the lateral plane.

4. Inclusions, including inadvertently unetched metal foil, which are not acceptable.

5. Proper lay-up of metal foils and compliance with the thickness requirements.

6. Registration (0.014 in maximum registration), which is a measure of the relative location of the terminal pads and, contrary to popular belief, has nothing to do with the hole sizes or locations. Check registration in both the *x* and *y* planes by preparing two separate vertical microsections.

7. Measurement of annular rings. The external annular ring is measured from the inside of the copper-plated hole to the outer edge of the pad on the surface of the board. The annular ring on internal layers is measured from the edge of the *drilled* hole to the outer edge of the internal pad. Typical annular ring requirements (from MIL-P-55110C) are (*a*) external pad: unsupported holes, 0.015 in minimum; supported holes, 0.005 in (may be 0.004 in on the opposite side due to drill splay); (*b*) internal pad: plated-through holes, type 3, 0.002 in minimum.

8. Pad "breakout," which is not allowed.

9. Undercutting at the edges (of conductors *and* pads), which cannot exceed the total thickness of the metal foil. In the case of external conductors, the plated metal is included in the calculation.

10. Overhang. There is no overhang allowed when fused tin-lead is specified. Where other plating metals are specified, the overhang may not exceed 0.001 in. In any case, metal slivers are not allowed.

28.6 THE THERMAL ZONES

When the first government multilayer board specification was written with the help of the Electrodeposition Section of the National Bureau of Standards, the thermal requirements were levied to evaluate the integrity of plating and copper foil only. Unfortunately, during the combining of later military specifications, the evaluation for laminate integrity was added to the thermal tests. The situation was further aggravated because bromides were added to material for flame retardancy, thereby even further lowering the epoxy-resin system's physical and chemical resistance to high-temperature thermal exposures. The thermal zone concept is to correct the problem (see Figs. 28.2 and 28.3).

28.6.1 Solutions

The following solutions have been accepted by the military and industry and are now incorporated into the appropriate specifications:

1. In the "as received" condition, all areas of each microsection must meet all plating, foil, and laminate requirements presently in MIL-P-55110C and the "rigid" portions of MIL-P-50884 and IPC-ML-950.

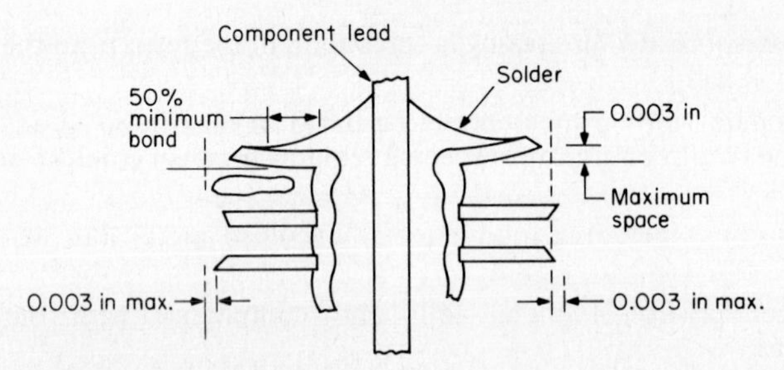

FIG. 28.3 Lifted lands (pads).

2. After testing for thermal stress or bond strength, specimens must meet all laminate requirements presently in MIL-P-55110C, MIL-P-50884, and IPC-ML-950 in the zone B areas.

28.6.2 Justifications

The principle justifications for the approval of these changes are as follows: (1) many successful years of procurement history in government; (2) IPC round-robin test reports, especially Round Robin V; (3) history of no failures when waivers, materials review board (MRB) authorizations, and engineering change proposals were granted; and (4) corroborating reports of thermal behavior of epoxy resin.

28.6.3 Impact

This revision to all appropriate specifications avoids rejection of a good product solely because the epoxy-resin systems are not capable of meeting the structural tests for plating ahd foil.

With the correction of this problem, millions of dollars per year in savings are predicted, especially on military and other government programs for which we as taxpayers must pay.

In addition, the undue administrative expenses for repeatedly processing waivers, MRB actions, and engineering change proposals will be eliminated, thereby further reducing costs.

None of these changes has adversely effected board quality or reliability.

28.7 TESTING

28.7.1 Production and Inspection Survey Checklist

To ensure that all process, production, inspection, and test stations are checked, a board and inspection survey checklist is suggested. It ensures that during inspection audits, none of the many steps are inadvertently overlooked. Perhaps even more important, the checklist immediately shows up any significant materials or process changes.

28.7.2 Testing Conditions

Three main elements must be considered in connection with obtaining data required at room conditions: (1) the test setup for supporting and operating the equipment, (2) the instrumentation for measuring the data from the tests, and (3) the instrumentation for measuring the room conditions.

Considerable emphasis is placed on thorough consideration for the factors involved in obtaining the data, which serve as standard, comparative, environmental data when a test report is presented properly. In most cases, the step consists of a simple recording of the barometric pressure, relative humidity, and temperature. Since ambient room conditions are relative and not absolute, the temperature, pressure, humidity, vibration, etc., both of the room and of the area immediately adjacent to the test setup should be considered. The conditions to which the specimen is to be subjected and the type of test being performed will vary, but at least temperature, pressure, and humidity are recorded in nearly all cases.

28.7.3 Calibration and Inspection Interval

Consideration must be given to the instrumentation necessary for measuring the required data. Attention must be given to (1) the required accuracy, (2) the effects of the means of measurement on the quantity being measured, and (3) the method of calibrating the instrumentation. Judgment must be employed in correlating the accuracy requirements of the final test data with the individual accuracies obtainable from the instruments employed.

If a choice is permitted by the setup, the most direct means of measurement is usually most reliable. For example, the manometer is preferable to the face-reading pressure gauge. Of course, the choice of instrument depends not only on the accuracy requirements but also on the specific conditions under which the test is conducted.

Depending on the accuracy requirements, the standards against which the instrument calibrations are made must be considered. The probable accuracy of the laboratory standards, the frequency with which the standards are checked against primary standards, the techniques of calibration, and the environmental conditions under which the calibration is performed must be assessed against the determined accuracy requirement of the test data.

28.8 QUALITY CONTROL PROGRAM

28.8.1 Controls

28.8.1.1 Incoming Material Control. Quality control samples all raw materials received for production and inspects materials for compliance with drawings, specifications, and all other required contractual criteria. Material testing which requires specialized equipment, chemical analysis, and/or tests which cannot be performed by the inspector should be forwarded to the laboratories for testing. No material should be used until the inspector certifies that it is in compliance with all contract requirements or is accepted on a waiver. A material history file should be maintained on all items received.

28.8.1.2 In-Process Control. Random sampling of material in process is performed by the inspector at established inspection stations located at critical points within the production flow. That type of inspection will alert production personnel if a process goes out of control and requires remedial action. It will also reveal nonsalvageable rejects soon after they occur and thereby preclude further processing of such items.

At the end of each shift of plating on either automatic plating line at least one board is selected at random for determination of plating thickness. The determination is performed by the microscopic method. A plating adhesion test also is performed by the inspector at least three times during each shift of operation.

28.8.1.3 ***Tooling, Equipment, and Solution Control.*** Production and inspection tooling and equipment are inspected at specific intervals to determine wear, mechanical alignment, operation, and electrical calibration. Records are kept to indicate when replacement, repair, calibration, or realignment will be required.

28.8.1.4 ***Electrical Inspection Control.*** Each board receives an electrical test for shorts, current-carrying capacity, circuit continuity, and circuit verification. The electrical test equipment is checked at least once each shift to ensure that it is operating according to requirements. The tests are performed by the inspector.

28.8.1.5 ***Complete Product Control.*** The inspector selects at least 1 board from each 300 completely packaged boards; inspects for completeness of packaging, identification, and overall product quality; and performs a functional electrical test. A permanent record is maintained on all boards shipped. The record includes the serial numbers, the quantity shipped, and the name of the person who originated shipment.

28.8.2 Sampling Plans

The sampling plan to be used varies considerably with board type. In deciding which plan is best suited for a particular category, five factors are considered: (1) allowable delay time for in-process lot quantity, (2) required frequency of examination per lot quantity, (3) the complexity of the operations to be controlled, (4) the size of the lot to be sampled, and (5) the defect occurrence level of the operations through which the lot quantity has passed. In the following examples, the sampling plan used in each control category and the method used to perform the quality control check are outlined.

28.8.2.1 ***Incoming Material Sampling.*** Acceptability of incoming material is determined by use of sampling plan K as outlined in FED-STD-105D, *Gen. Insp. Lvl. II Norm.*, 1 percent acceptable quality level (AQL) in 2 to 3000 pieces.

As an example, this sampling plan chosen for incoming material requires that 125 pieces be drawn at random from 3000 for evaluation. If 3 or more are found to have major defects, the lot is rejected and the tightened sampling plan as outlined in FED-STD-105D is used on all subsequent lots until quality returns to the acceptable level.

28.8.2.2 ***Sampling of the Product in Process.*** The complexity of the manufacturing process increases the possibility of defect occurrence. Therefore, frequent lot sampling is done to reduce buildup of material at any point and reveal any out-of-control production process which should have corrective action.

One may find it necessary to deviate from FED-STD-105D in establishing sample size per lot quantity, because the complexity of manufacturing processes dictates tighter sampling procedures. A sample of each product lot quantity is checked to determine the presence of major and minor defects. The sample size to be checked at each station varies proportionally to the complexity of the operations the boards have passed through since their last quality control check. The defects sampled for at each station are those that conceivably could result from out-of-control operations that the boards have passed through since their last quality control check.

Sampling is performed at those designated stations. The check sheet shows which defects will be inspected for and reported on at each quality control check station. When a series of similar major or minor defects at any given sampling station has been determined, the quality control representative will determine the operation number(s) of the out-of-control process(es) and will notify the production supervisors immediately, to preclude the occurrence of additional defects.

28.8.2.3 Tooling, Equipment, and Solution Control. Inspection records are maintained on all production tooling, inspection gauges, and electrical test equipment. It is the responsibility of quality control to maintain such records. The records should indicate when replacement, repairs, calibration, or realignment will be required.

28.8.3 Sequential Testing Groups

Typical Group A and B testing requirements are given as follows:

Group A: Visual and dimensional examination, microsectioning, and tests for bow and twist, plating thickness, plating adhesion, thermal stress, solderability, and circuitry.

Group B: Group B inspection is performed on production boards or on the quality conformance test circuitry area that has passed Group A inspection. Two boards and related quality conformance coupons should be selected from lots which have passed Group A inspection. The frequency should be every 4 weeks, or after every 5000th board is produced, whichever occurs first. In addition, the manufacturer must, every 4 weeks, submit at least one sample of a quality conformance test circuitry area to the customer for testing.

Characteristics to be tested include cleanliness, bond strength, rework simulation, interconnection resistance, moisture and insulation resistance, and dielectric withstanding voltage.

28.9 TRACEABILITY

Because of the possibility of latent defects, it is essential that the fabricator establish and maintain a traceability program for all boards used in high-reliability programs. The traceability program must begin at the lamination panel level and continue to reflect the exact disposition of each board from that panel, including the serial number of the equipment in which the board is located. A record of rejected boards should be kept, and the reason for rejection should be identified. The program may be set up in such a manner that the data can be sorted and compiled by electronic data processing.

28.10 DEFECTIVE CHARACTERISTICS AND CAUSES

The following list of frequently encountered deficiencies, together with the most likely causes, will be useful to the in-process or final product inspector or tester:

28.10.1 Annular Ring (Failure to Meet Minimum Requirements)

Design
Drilling accuracy
Improperly registered layers
Unstable artwork
Lay-up pinning
Overetching
Worn tooling plates and registration pins
Improper reduction of artwork

28.10.2 Delamination

Aged prepreg
Silicone release agent
Low temperature and pressure during lamination
Moisture
Inadequate baking
Uncontrolled lay-up area (temperature and humidity)
Handling during lay-up
Improper heat transfer through stack
Improperly cleaned layers
Improperly vented ovens
Excessive antistaining agent on copper
Temperature control during assembly

28.10.3 Dewetting

Contaminated surfaces
Contaminated reflow media
Porous copper that may result in volatile migration
Inadequate volume of fusing media resulting in temperature variation
Improper fluxing
Oxides
Improper angle of product into fusing oil
Improper heat during reflow

28.10.4 Inadequate Etchback

Incorrect agitation
Improper lamination and curing
Nailheading
Hardened epoxy smear
Unbalanced and/or depleted bath
Improper bath temperature
Improper time exposure
Presence of Novalac in epoxy resin
Incompatibility of materials

28.10.5 Laminate Voids

Improperly cured resin
Improper flow of resin
Border barriers
Too little prepreg to allow fill
Poor material selection
Improper temperature and/or pressure

28.10.6 Mealing

Insufficient removal of fluxes
Improper cleaning
Moisture entrapment
Hygroscopic coating

28.10.7 Measling

Moisture
Pressure
Internal stresses
Volan in lieu of silane gas treatment
Circuit design geometry

28.10.8 Misregistration

Dimensional stability of base materials
Registration of artwork
Care in punching layers and artwork
Unparallel laminating plates
Unstable artwork
Worn or deformed holes in artwork

28.10.9 Plating Adhesion

Improperly cleaned surface
Resist residue
Embedded pumice on surface
Fingerprints
Silicones
Out-of-control electroless or underplating process

28.10.10 Plating Cracks

Brittle copper
Excessive brightness
Excessive levelers
Resin content contributing to z expansion
Deletion of nonfunctional pads
Contaminated copper bath
Improper current density
Design
Excessive etchback resulting in elbowing during plating

28.10.11 Plating, Peeling

Improperly cleaned surface
Embedded pumice
High-stress plating
Careless handling during processing
Plating at improper current density

28.10.12 Plating Voids

High current density
Inadequacies in electroless-copper process
Depleted electroless bath
Particles of photoresist not removed

Improper anode displacement
Storing boards with no copper strike following electroless deposition
Improper agitation (cavitation)
Contamination in bath
Unbalanced plating solution

28.10.13 Resin Starvation

Improper mix of resin
Unparallel laminating plates
Weave count of fabric
Resin content
Excessive pressure during lamination

28.10.14 Slivers

Excessive overhang
Overetching
Brittle overplate

28.10.15 Solderability

Oxidation
Photoresist residue
Porous copper

28.10.16 Weave Exposure

Inadequate butter coat
Excessive pressure
Overexposure to electroless tin
Careless workmanship practices
Fabric size too large for end-product thickness

APPENDIX A
IMPORTANT QUALITY ASSURANCE DEFINITIONS

CLASSIFICATION OF DEFECTS

A defect is any aspect of the board that does not conform to specified requirements. Defects will normally be classified according to degree of seriousness: critical, major, or minor.

1. *Critical defect:* A critical defect is a defect that judgment and experience indicate is likely to result in hazardous or unsafe conditions for individuals using, maintaining, or depending upon the product; or a defect that judgment and experience indicate is likely to prevent performance of the tactical function of a major end item such as a ship, aircraft, tank, missile, or space vehicle.

2. *Major defect:* A major defect is a defect, not a critical one, that is likely to result in failure or to reduce materially the usability of the unit or board for its intended purpose.

3. *Minor defect:* A minor defect is a defect that is not likely to reduce materially the usability of the unit or product for its intended purpose, or is a departure from established standards having little bearing on the effective use or operation of the unit.

DEFECTIVES

A defective is a unit or product which contains one or more defects. Defectives will usually be classified as follows: critical, major, or minor.

1. *Critical defective:* A critical defective contains one or more critical defects and may also contain major and/or minor defects.

2. *Major defective:* A major defective contains one or more major defects, perhaps some minor defects, but no critical defect.

3. *Minor defective:* A minor defective contains one or more minor defects but no critical or major defect.

DEFECTS PER HUNDRED UNITS

$$\text{Percent defective} = \frac{\text{number of defectives}}{\text{number of units inspected}} \times 100$$

The number of defects per hundred units of any given quantity of units of product is 100 times the number of defects contained therein (one or more defects being possible in any unit of product) divided by the total number of units of product, i.e.:

$$\text{Defects per hundred units} = \frac{\text{number of defects}}{\text{number of units inspected}} \times 100$$

ACCEPTABLE QUALITY LEVEL (AQL)

The AQL is the maximum percent defective (or the maximum number of defects per hundred units) that, for purposes of sampling inspection, can be considered satisfactory as a process average.

When some specific value of AQL is designated by the consumer for a certain defect or group of defects, it means that the consumer's acceptance sampling plan will accept the great majority of the lots or batches that the supplier submits, provided that the process average level of percent defective (or defects per hundred units) in these lots or batches is no greater than the designated value of AQL. Thus the AQL is a designated value of percent defective (or defects per hundred units) that the consumer indicates will be accepted most of the time by the acceptance sampling procedure to be used.

LOT OR BATCH

The term "lot" or "batch" means "inspection lot" or "inspection batch," i.e., a collection of units of a product from which a sample is to be drawn and inspected to determine conformance with the acceptability criteria, and which may differ from a collection of units designed as a lot or batch for other purposes (e.g., production, shipment, etc.). The boards shall be assembled into identifiable lots, sublots, batches, or in such other manner as may be prescribed.

APPENDIX B

REFERENCE SPECIFICATIONS

COMMERCIAL SPECIFICATIONS

IPC-T-50B	Terms and Definitions
IPC-CF-150E	Copper Foil for Printed Wiring
IPC-A-600D	Acceptability Guidelines
IPC-TM-650	Test Methods
IPC-R-700	Guidelines for Modifications and Repair, Printed Wiring
IPC-S-801	Edge Dip Solderability Test
IPC-S-815B	General Requirements for Soldering, Printed Wiring Assemblies
IPC-ML-950C	Performance Specification for Multilayer Printed Wiring

GOVERNMENT SPECIFICATIONS

MIL-Y-1140	Yarn, Cord, Sleeving, Cloth, and Tape-Glass
MIL-T-10727	Tin Plating, Electrodeposited or Hit-Dipped
MIL-L-13808	Lead Plating, Electrodeposited
MIL-P-13949	Plastic Sheet, Laminated, Metal Clad for Printed Wiring
MIL-F-14256	Flux, Soldering, Liquid (Rosin Base)
MIL-C-14550	Copper Plating, Electrodeposited
MIL-P-18177	Plastic Sheet, Laminated, Thermosetting, Glass Fiber Base, Epoxy-Resin
MIL-C-28809	Printed Wiring Assemblies
MIL-G-45204	Gold Plating, Electrodeposited
MIL-S-45743	Soldering, Manual Type (M1)
MIL-I-46058	Insulation Compound, Electrical (for Coating Printed Circuit Assemblies)
MIL-R-46085	Rhodium Plating, Electrodeposited
MIL-P-46843	Printed Circuit Assemblies, Design and Production of (M1)
MIL-S-46844	Solder Bath Soldering of Printed Wiring Assemblies, Automatic Machine Type
MIL-P-50884	Rigid/Rigid-Flex Printed Wiring
MIL-P-55110	Printed Wiring Boards
MIL-P-81728	Tin-Lead, Electrodeposited

MIL-STD-130	Identification, Marking of U.S. Military Property
MIL-STD-202	Test Methods for Electronic and Electrical Components
MIL-STD-275	Printed Wiring for Electronic Equipment
MIL-STD-454	Standard General Requirements for Electronic Equipment
MIL-STD-810	Environmental Test Methods
MIL-STD-883	Test Methods and Procedures for Microelectronics
QQ-N-290	Nickel Plating, Electrodeposited
QQ-S-571	Solder, Tin Alloy, Lead-Tin Alloy, and Lead Alloy
A-A-113	Tape, Pressure Sensitive Film, Office Use
FED-STD-151	Metals: Test Methods
FED-STD-406	Plastics: Test Methods

VIDEOTAPES

The following videotapes are useful educational tools and should be viewed periodically:

IPC-VT-01F	Polyimide Multilayers
IPC-VT-016	Product Assurance Considerations on Printed Wiring
IPC-VT-021	Automation of Rigid Printed Board Production
IPC-VT-023	Voltage Clearance Recommendations for Printed Boards
IPC-VT-026	Mechanical Testing of In-House Copper Plating
IPC-VT-028	Solderability Defect Analysis
IPC-ID-203	Basic Microsectioning Techniques
IPC-ID-204	Microsection Evaluation

CHAPTER 29

ACCEPTABILITY OF FABRICATED CIRCUITS

A. D. Andrade
Sandia National Laboratories, Livermore, California

29.1 INSPECTION OPERATIONS

Quality assurance operations can be performed by the purchaser or contracted out, partially or totally, to an independent company. Present printed wiring board fabrication technology is capable of conductor widths and spacing less than 0.008 in (0.20 mm), and they have been reported as low as 0.003 in (0.07 mm). Visual inspection of surface patterns on pattern densities less than 0.008 in (0.20 mm) is questionable. Many companies are replacing surface visual inspection of high-density patterns with automated electrical or image-scanning methods. Wherever visual inspection is performed, however, there must be access to a facility equipped with the required types of mechanical gauges, as well as the necessary equipment for microsectioning, chemical analysis, dimensional measuring, and electrical and environmental testing.

Quality assurance requirements are usually met by one of the following methods:

1. Inspection data submitted by the fabricator are reviewed for compliance to design requirements.
2. Inspection data submitted by the fabricator are reviewed, and a sample lot inspection is performed.
3. A complete inspection, including destructive testing, is performed to test for compliance with all design requirements.

Where and how the inspections are performed is a question of economics and availability of experienced personnel. Prior to making a decision, management should review all aspects of the question, such as equipment cost and maintenance, work volume, time to and from quality assurance facility location, and personnel availability.

29.2 USE OF TEST PATTERNS

The use of test patterns located outside the finished pattern contour area for monitoring all processes and also for microsectioning is a controversial subject. Some persons state that test patterns located outside the contour area usually have thicker plating thicknesses than the actual printed board pattern and, therefore, are not representative of the actual pattern. Others debate that the cost savings of not destroying actual circuits overshadow any slight differences in plating thicknesses. Test patterns are useful mechanisms for performing destructive tests for such characteristics as peel strength, flammability, and solder shock as an indication of the integrity of the finished printed boards.

The use of test patterns for the acceptance of printed boards is therefore an individual question that must be resolved for each design requirement and the economic advantage of various inspection operations.

29.3 DETERMINATION OF ACCEPTABILITY

The question of acceptability must be answered for individual printed board design functions. "Functionality" should be the ultimate criterion for acceptance. A printed board which will see extreme environmental conditions should not be inspected according to the same requirements applied to a printed board for an inexpensive radio. Some of the characteristics normally inspected come under the heading of workmanship. In the majority of designs those characteristics are cosmetic; they pertain to how the printed board looks and not how it performs. In most cases, the characteristics are inspected to establish a confidence level for integrity. The inspection results should be used to establish a fabrication quality level and not to scrap parts that will meet the functional criteria. Scrapping functional parts has a considerable impact on unit cost. However, the same workmanship characteristics can have an effect on the function of some designs, and the effects of the different characteristics will be discussed later in this chapter. The determination of what to do with printed boards that do not meet specified acceptance requirements but are functional should be made by a materials review board. The levels of acceptance have to be established by each company; they are dependent on the functional criteria to which the printed board will be subjected.

The Institute for Interconnecting and Packaging Electronic Circuits (IPC) established recommended guidelines for acceptance categorized as Class 1, Class 2, and Class 3. The classes are defined as follows:

Class 1—Consumer Products: Includes TV sets, toys, entertainment electronics, and noncritical consumer or industrial control devices.

Class 2—General Industrial: Includes computers, telecommunications equipment, sophisticated business machines, instruments, and certain noncritical military applications.

Class 3—High Reliability: Includes equipment whose continued performance is critical, such as life-support equipment.

Each group is then subdivided into three categories: preferred, minimum acceptable, and rejectable. The acceptance guidelines in this chapter, however, utilize only two categories: preferred and nonpreferred. The illustrations in the nonpreferred category are typical examples of rejectable conditions. They will guide individuals in establishing minimum acceptance levels applicable to their functional requirements.

29.4 THE MATERIALS REVIEW BOARD

The materials review board (MRB) is usually composed of one or more representatives from the departments of quality, production, and design. The purpose of the MRB is to effect positive corrective action to eliminate the cause of recurring discrepancies and prevent occurrence of similar discrepancies. The board's responsibilities include:

1. Reviewing questionable printed boards or materials to determine compliance or noncompliance with quality and design requirements
2. Reviewing discrepant boards for effects on design functionality
3. Authorizing repair or rework of nonconforming materials, when appropriate
4. Establishing responsibility and/or identifying causes for nonconformance
5. Authorizing the scrapping of excessive quantities of materials

29.5 VISUAL INSPECTION

Visual inspection is the inspection of characteristics which can be seen in detail with the unaided eye. Magnification, approximately 10×, is advantageous for viewing questionable characteristics after their initial location with the unaided eye.

The viewing of printed board microsections for defects is the exception. Defects such as resin smear and plated-through-hole quality require magnifications of 50 to 500×, depending on the characteristic. Visual inspection criteria are difficult to define verbally, because individuals interpret words differently.

An effective way to define visual inspection criteria is to use line illustrations or photographs. IPC utilitized this method in the publication *Acceptability of Printed Boards Manual* to "visually standardize the many individual interpretations to specifications on printed boards." Another method is the use of an audio-visual aid projector. Slides of line illustrations and photographs depicting preferred and nonpreferred criteria for visual inspection, along with the related description, in audio, provide a reproducible inspection method.

Of the different types of inspection, surface visual inspection costs the least. Visual inspection by the unaided eye is usually performed on 100 percent of the printed boards or on a sample taken following an established sampling plan. MIL-STD-105, *Sampling Procedure and Tables for Inspection by Attributes,* is frequently used for this purpose. Inspection for visual defects according to sampling plan is done on the premise that the boards were 100 percent visually screened during the fabrication process. Defects which usually can be detected by visual inspection may be divided into three groups: surface defects, base material defects, and other defects.

29.5.1. Surface Defects

Surface defects include dents, pits, scratches, surface roughness, voids, pinholes, inclusions, and markings. Dents, pits, scratches, and surface roughness usually fall in the class of workmanship. The defects, when minor, are normally considered to be cosmetic, and they usually have little or no effect on functionality. However, they can be detrimental to function in the edge-board contact area (Fig. 29.1). Voids in conductors, lands, and plated-through holes can be detrimental to function depending on the degree of defect. Pinholes and inclusions are in the same category. Voids or pinholes, either of which reduce the effective conductor width, reduce current-carrying capacity and can affect other electrical design characteristics, i.e., inductance, impedance, etc. Voids in the hole-wall plating area result in reduced conductivity, increased circuit resistance, and voids in plated-through-hole solder fillets (Fig. 29.2).

Voids in lands are also detrimental to solderability. Pinholes or voids can undermine the top metal plate (Figs. 29.3 and 29.4). The degree of undermining depends on when during the fabrication process the defect occurred.

The defects in this group are defined as follows:

1. *Dent:* A smooth depression in the conductive foil which does not significantly decrease foil thickness.

2. *Pit:* A depression in the conductive layer that does not penetrate entirely through it.

3. *Scratch:* Slight surface mark or cut.

4. *Surface roughness:* Surface that is not smooth or level, having bumps, projections, etc.

5. *Void:* The absence of substances in a localized area.

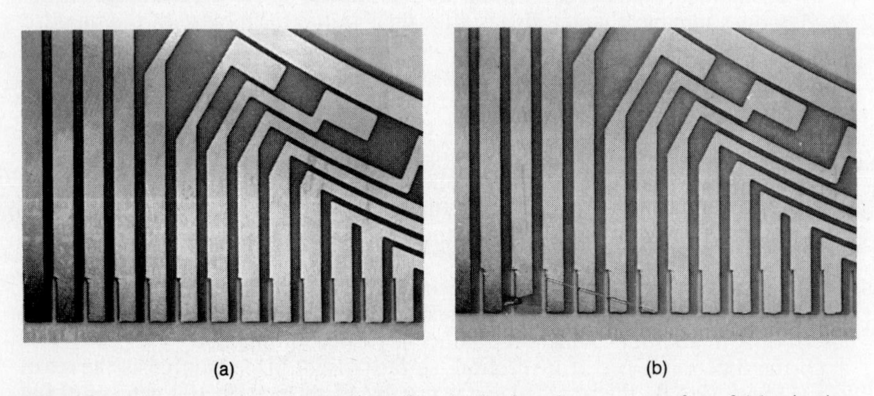

FIG. 29.1 Edge-board contact area. (*a*) Preferred: edge-board contact area free of delamination, pits, pinholes, dents, nodules, and scratches; (*b*) nonpreferred: (1) scratch depth at the contact area exceeds the microinch surface-roughness requirements; (2) delamination of one of the edge-board contacts. *(IPC.)*

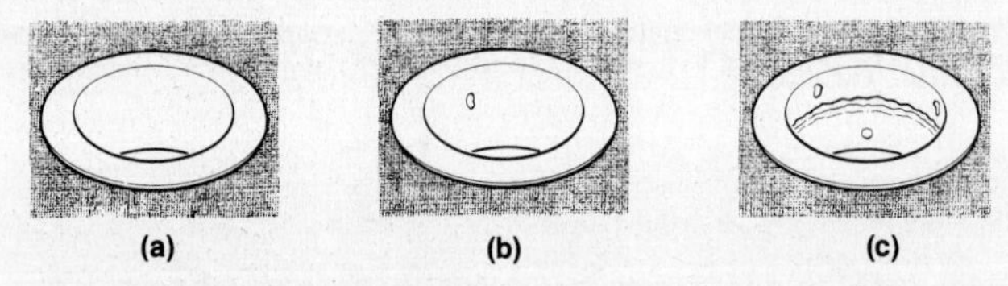

FIG. 29.2 Voids in hole. (*a*) Preferred: no voids in hole; (*b*) Acceptable: no more than three voids in the hole; total void area does not exceed 10 percent of the hole-wall area; (*c*) nonpreferred: voids exceed 10 percent of the hole-wall plating area; circumferential void present. *(IPC.)*

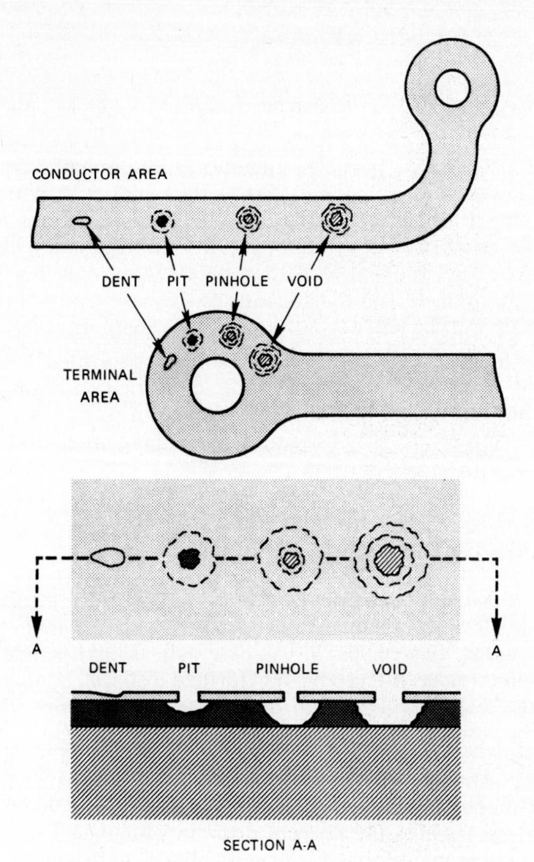

FIG. 29.3 Pits, dents, pinholes, and voids. *(Sandia Laboratories.)*

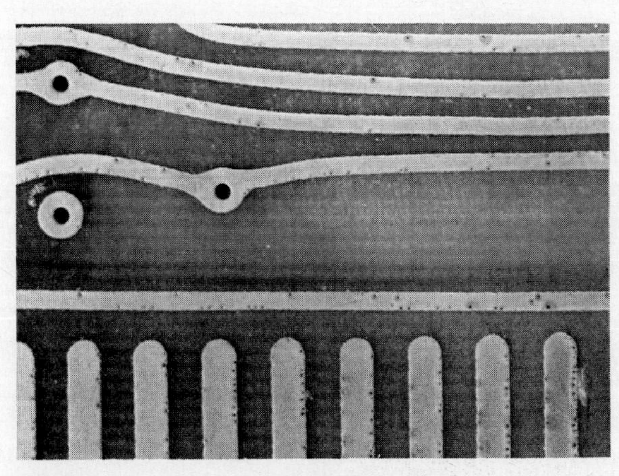

FIG. 29.4 Severe pitting and pinholing. *(IPC.)*

6. *Pinhole:* A small hole occurring as an imperfection which penetrates entirely through a layer of material.

7. *Inclusion:* A foreign particle, metallic or nonmetallic, in a conductive layer, plating, or base material. Inclusions in the conductive pattern, depending on degree and material, can affect plating adhesion. Inclusions in a conductor greater than 0.005 in (0.127 mm) in their greatest dimension are usually cause for rejection, but inclusions less than 0.0001 in (0.00254 mm) in their largest dimension are usually allowed. Metallic inclusions in the base material reduce the electrical insulation properties and are not normally acceptable.

8. *Markings (legend):* A method of identifying printed boards with part number, revision letter, manufacturer's code, etc. The condition is usually considered minor; but if there are different revisions to the same part number, markings that are missing or partially obscured could have an effect on functionality (Fig. 29.5).

29.5.2 Base Material Defects

Visual inspection is also used to detect the following base material defects: measling, crazing, blistering, delamination, weave texture, weave exposure, fiber exposure, and haloing (Figs. 29.6, 29.7). These defects have been a source of controversy as to what is acceptable. The IPC formed a special committee in 1971 to consider base material defects and to define them better with illustrations and photographs. The conditions are defined and discussed here.

1. *Measling:* An *internal* condition occurring in laminated base material in which the glass fibers are separated from the resin at the weave intersection. This condition manifests itself in the form of discrete white spots or "crosses" below the surface of the base material and is usually related to thermally induced stress. A report compiled by the IPC titled *Measles in Printed Wiring Boards* was released in November 1973. The report stated that "measles may be objectionable cosmetically, but their effect on functional characteristics of finished products are, at worst, minimal and in most cases insignificant."

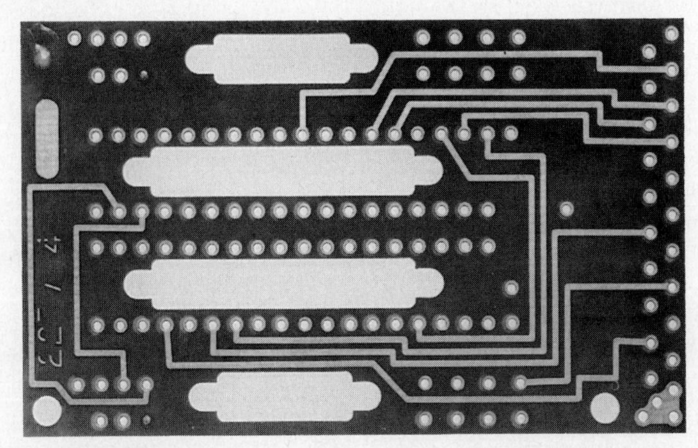

FIG. 29.5 Part numbers partially obscured and missing. *(Sandia Laboratories.)*

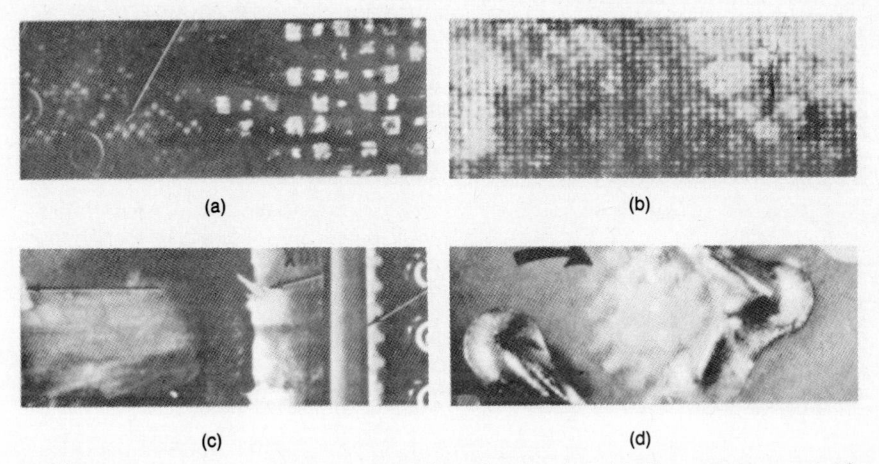

FIG. 29.6 Base material defects. (*a*) Measling; (*b*) blistering; (*c*) fiber exposure; and (*d*) crazing. *(IPC.)*

2. *Crazing:* An internal condition occurring in laminated base material in which the glass fibers are separated from the resin at the weave intersections. This condition manifests itself in the form of "connected" white spots or "crosses" below the surface and the base material and is usually related to "mechanically" induced stress.

3. *Blistering:* A localized swelling and separation between any of the layers of a laminated base material, or between base material and conductive foil. (It is a form of delamination.)

4. *Delamination:* A separation between plies within the base material, or between the base material and the conductive foil, or both.

Blistering and delamination are considered to be major defects. Whenever a separation of any part of the board occurs, a reduction in insulation properties and adhesion occurs. The separation area could also house entrapped processing solutions that contribute to corrosion and other detrimental effects in certain environments (Fig. 29.7). There also is the possibility that the delamination or blister area will increase to the point of complete board separation, normally during assembly soldering operation.

Last, but not least, is the question of solderability in plated-through holes. Entrapped moisture, when subjected to soldering temperatures, has been known to create steam that blows holes through the plated sidewalls of the plated-through holes and creates large voids in the solder fillet.

5. *Weave texture:* A surface condition of base material in which a wave pattern of glass cloth is apparent, although the unbroken fibers of the woven cloth are completely covered with resin.

6. *Weave exposure:* A surface condition of base material in which the unbroken fibers of woven glass cloth are not completely covered by resin. Weave texture and weave exposure differ in the degree of defect (Fig. 29.7). A condition of weave texture after the board has been completely fabricated is considered a minor defect. However, if the condition materializes during processing, a judgment must be made concerning the possible attack of subsequent processing chemicals. Weave texture, usually caused by the lack of sufficient resin, can become weave

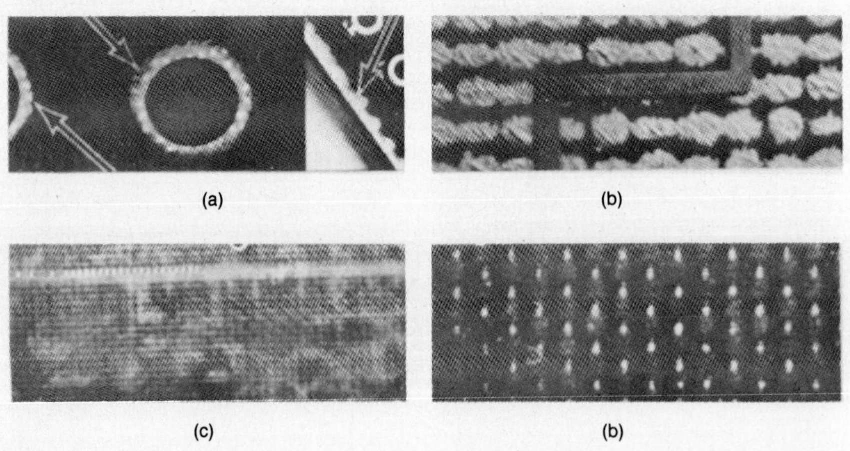

FIG. 29.7 Base material defects. (*a*) Haloing; (*b*) weave exposure; (*c*) delamination; and (*d*) weave texture. *(IPC.)*

exposure if processing chemicals do attack the thin resin layer. Weave exposure is considered a major defect. The exposed bundles of glass fibers allow wicking of moisture and entrapment of processing chemical residues.

7. *Fiber exposure:* A condition in which reinforcing fibers within the base material are exposed in machined, abraded, or chemically attacked areas. (See also "weave exposure.")

8. *Haloing:* Mechanically induced fracturing or delaminating on or below the surface of the base material; it is usually exhibited by a light area around holes, other machined areas, or both.

29.5.3 Resin Smear

Resin smear, or resin transferred from the base material onto the surface or edge of the conductive pattern, is normally caused by drilling. Excessive heat generated during drilling softens the resin in holes and smears it over the exposed internal copper areas. The condition creates an insulator between the internal land and subsequent plated-through holes, and the result is "open circuits." The defect is removed by chemical cleaning.

Inspection for resin smear is performed by viewing vertical and horizontal microsections of plated-through holes. Chemical cleaning is the chemical process used in the manufacture of multilayer boards. Its purpose is to remove only resin from conductive surfaces exposed in the inside of the holes (Fig. 29.8). Sulfuric or chromic acid is used in the chemical cleaning process; subsequent inspection indicates whether it has fulfilled its purpose.

29.5.4 Layer-to-Layer Registration, X-Ray Method

The x-ray method provides a nondestructive way to inspect layer-to-layer registration of internal layers of multilayer boards. It utilizes an x-ray machine and, usually, Polaroid film. The multilayer board is x-rayed in a horizontal position. The x-ray photos are then examined for hole breakout of the internal lands. The lack of an annular ring denotes severe misregistration (Fig. 29.9).

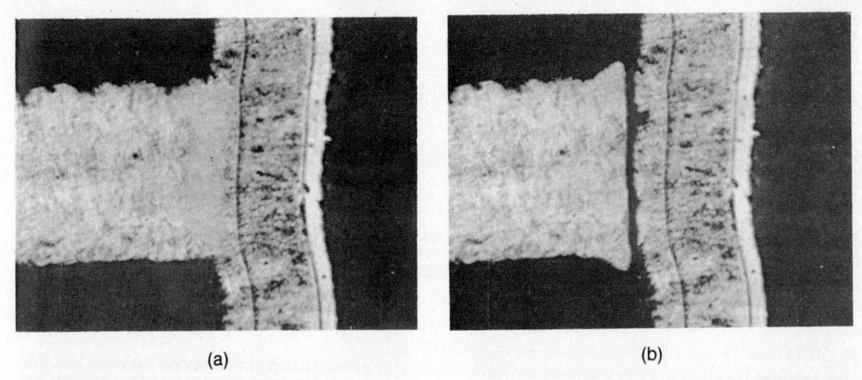

FIG. 29.8 Resin smear. (*a*) Preferred: no evidence of resin smear between layer and plating in the hole; (*b*) nonpreferred: evidence of resin residue or resin between internal layer and plating in the hole. *(IPC.)*

29.5.5 Plated-Through Holes: Roughness and Nodulation

Roughness is an irregularity in the sidewall of a hole; nodulation is a small knot or an irregular, rounded lump. Roughness and/or nodulation creates one or more of the following conditions:

1. Reduced hole diameter
2. Impaired lead insertion
3. Impaired solder flow through the hole
4. Voids in solder fillet

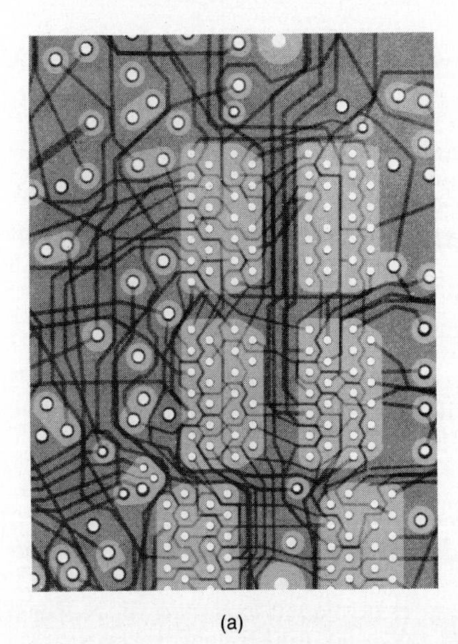

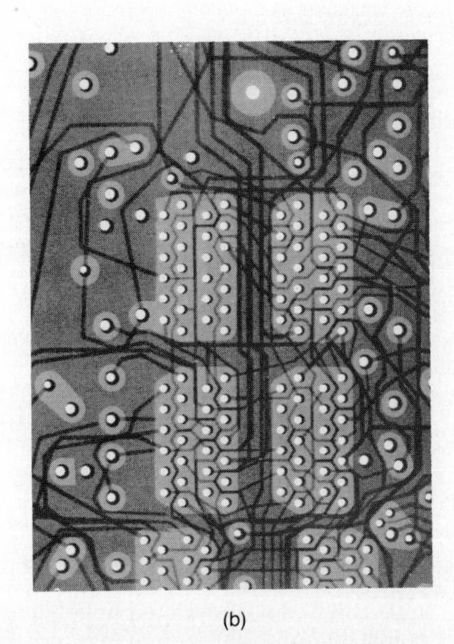

FIG. 29.9 Layer-to-layer registration, x-ray method. (*a*) Preferred: all layers accurately registered; (*b*) nonpreferred: extreme misregistration; insufficient measurable annular ring exists on a segment of circumference. *(IPC.)*

5. Possible entrapment of contaminants
6. Highly stressed areas in the plating

Although roughness and nodulation are not desirable, they are allowable in small amounts. Specifications have a tendency to use simple, generalized statements such as "good uniform plating practice" when defining acceptability criteria. Such statements require that a judgment be made by the inspector as to what is acceptable. The use of visual aids allows judgment to be made by different inspectors within a close degree of consistency. Figure 29.10 illustrates the acceptance criteria recommended by the IPC.

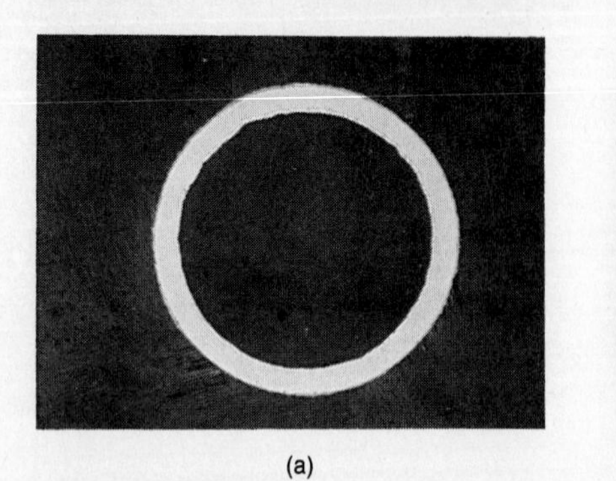

(a)

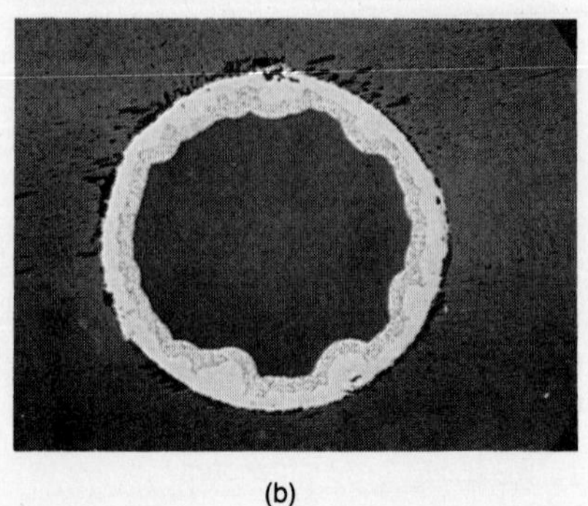

(b)

FIG. 29.10 Plated-through holes: roughness or nodulation. (*a*) Preferred: (1) plating is smooth and uniform throughout the hole; (2) there is no evidence of roughness or nodulation; (*b*) nonpreferred: (1) roughness or nodulation reduces plating thickness below minimum requirements; (2) roughness or nodulation reduces finished-hole size below minimum requirements; (3) excessive roughness or nodulation permits outgassing of the hole when it is solder-dipped. *(IPC.)*

29.5.6 Eyelets

Metallic tubes, the ends of which can be bent outward and over to fasten them in place, are called eyelets. Eyelets are used to provide electrical connections with mechanical strength on printed boards. Acceptability of eyeleted printed boards is based on eyelet installation. Eyeleted boards should be inspected for the following conditions:

1. Form of the flange and/or roll should be set in a uniform spread and be concentric to the hole.
2. Splits in the flange or roll should be permissible provided they do not enter the barrel and provided they allow proper wicking of solder through the eyelet and around the setting.
3. Eyelets should be set sufficiently tight so that they cannot move.
4. Eyelets should be inspected for improper installation, deformations, etc.
5. A sample lot of eyeleted holes should be microsectioned to inspect for proper installation. IPC acceptance criteria for roll- and funnel-flange eyelets are shown in Figs. 29.11 and 29.12.

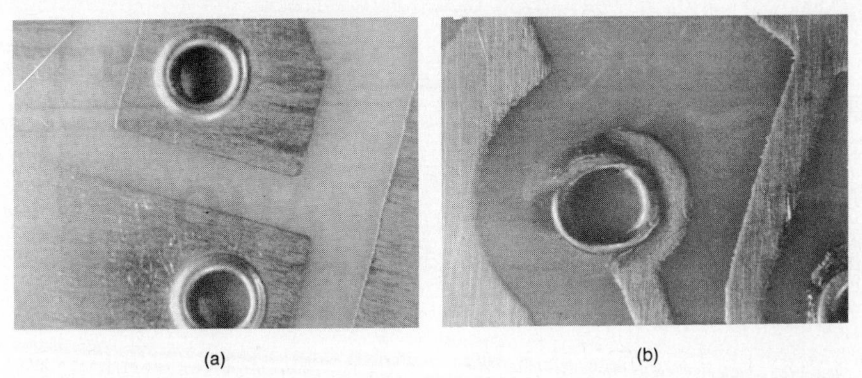

FIG. 29.11 Roll-flange eyelets. (*a*) Preferred: (1) eyelet set uniformly to be concentric with hole; (2) strain or stress marks caused by rollover kept to a minimum; (*b*) nonpreferred: (1) eyelet flange uneven or crushed; (2) splits entering the barrel. *(IPC.)*

29.5.7 Base Material Edge Roughness

Base material edge roughness occurs on the printed board edge, cutouts, and non-plated-through-hole edges (Fig. 29.13). It is classified as a workmanship condition, and it is created by dull cutting tools that cause a tearing action instead of a clean cutting action.

29.5.8 Solder Mask

Solder mask is a coating material used to mask off or to protect selected areas of a pattern from the action of solder. Solder mask is also used as a conformal coating. Attributes inspected are registration, wrinkles, and delamination. Misregistration at land areas reduces or prevents adequate solder fillet formation. Minimum land area to obtain adequate solder fillets should be the minimum acceptance criterion. Wrinkles and delamination provide sites for moisture absorption and contaminates.

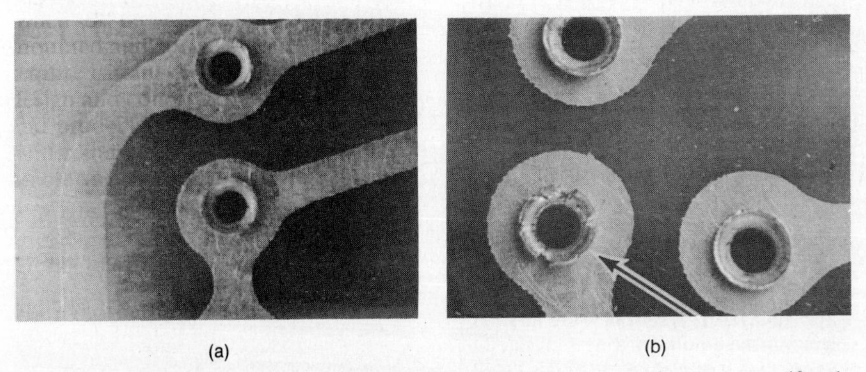

FIG. 29.12 Funnel-flange eyelets. (*a*) Preferred: (1) funnel-flange eyelet (funnelet) set uniformly and concentric to hole; (2) strain or stress marks caused by setting kept to a minimum; (*b*) nonpreferred: (1) funnelet periphery uneven or jagged; (2) splits enter into barrel. *(IPC.)*

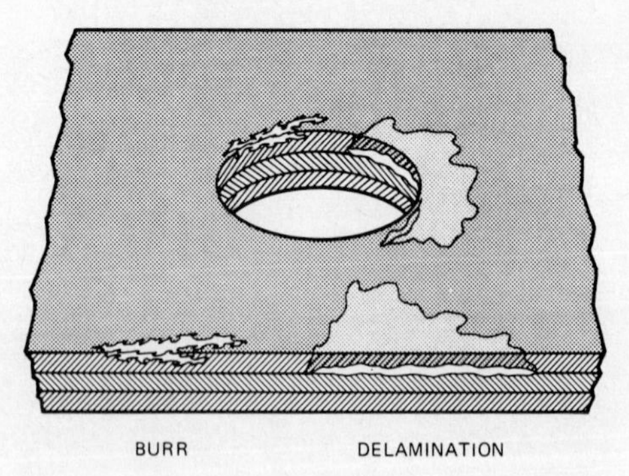

FIG. 29.13 Base material edge roughness. *(Sandia Laboratories.)*

29.5.9 Summary: Visual Inspection

The board attributes subjected to nondestructive and destructive visual inspection are listed in Table 29.1.

29.6 DIMENSIONAL INSPECTION

Dimensional inspection is the measurement of the printed board attributes such that dimensional values are necessary to determine compliance with functional requirements. The methods of inspection vary, but the basic inspection equipment consists of gauges and measuring microscopes. More sophisticated equip-

TABLE 29.1 Visual Inspection Chart

Subject	Nondestructive	Destructive
Dents	X	
Pits	X	
Scratches	X	
Voids	X	X
Pinholes	X	
Inclusions	X	
Surface roughness	X	
Markings	X	
Measling	X	
Crazing	X	
Blistering	X	
Delamination	X	
Weave texture	X	
Weave exposure	X	
Haloing	X	
Chemical cleaning		X
Layer-to-layer registration, x-ray method	X	
Plated-through-hole roughness and nodulation	X	
Eyelets	X	X
Base material edge roughness	X	X
Solder mask	X	

ment is available, including comparators, numerical control measuring equipment, coordinate measuring systems, microohm meters, and beta backscatter gauges.

Dimensional inspection is usually performed on a sampling plan basis. One such plan, termed acceptable quality level (AQL), is specified as the maximum percent of defects which, for the purpose of sampling, can be considered statistically satisfactory for a given product. The following attributes fall in the category of dimensional inspection.

29.6.1 Annular Ring

That portion of conductive material completely surrounding a hole is called the annular ring (Fig. 29.14). The primary purpose of the annular ring is that of a flange surrounding a hole; it provides an area for the attachment of electronic component leads or wires. Annular ring width of 0.010 in (0.254 mm) is a standard requirement, but some specifications have allowed rings as small as 0.005 in (0.127 mm). Figure 29.15 shows holes neatly centered in the land, holes on the extreme edge of the land, and holes breaking the edge. Holes that extend beyond the land are generally not acceptable.

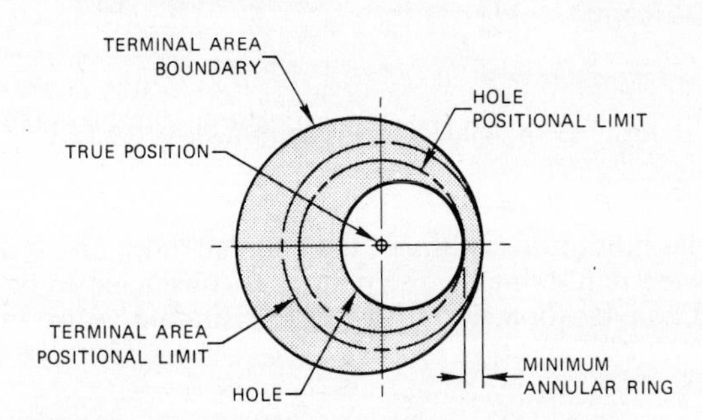

FIG. 29.14 Annular ring. *(Sandia Laboratories.)*

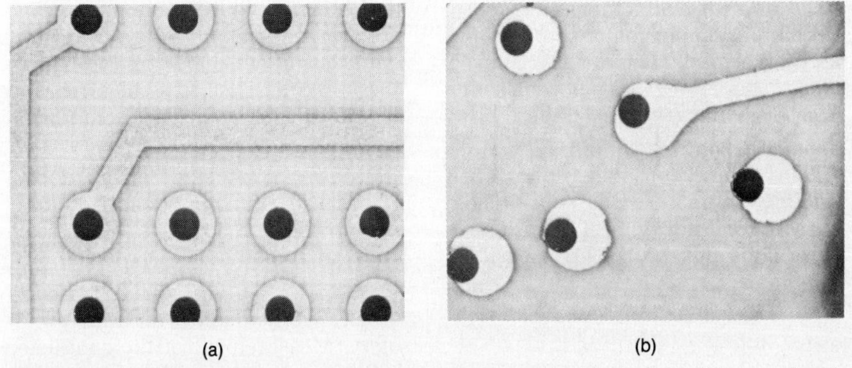

FIG. 29.15 Land registration. (*a*) Preferred: holes neatly centered in the land; (*b*) nonpreferred: holes not centered in the land. *(IPC.)*

29.6.2 Pattern-to-Hole Registration

The annular ring can also be used for determining registration between the pattern and the holes (Fig. 29.16). Some designers dimension a land to each datum on the master drawing. By verifying the dimensions on the printed boards and then

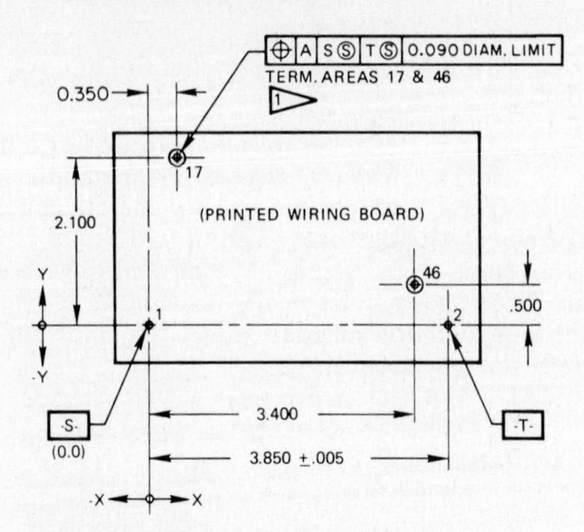

FIG. 29.16 Pattern-to-hole registration. *(Sandia Laboratories.)*

verifying that the minimum widths of the annular rings, on all other lands, are within the drawing requirements, the pattern is considered to be in registration with the drilled hole location. Front-to-back registration is also inspected in this manner.

29.6.3 Conductor Width

The observable width of a conductor at any point chosen at random on the printed board is normally viewed vertically from above unless otherwise specified. Although referred to as the minimum conductor width, the measurement is actually performed at the widest point on the conductor (Fig. 29.17). The conductor width affects the electrical characteristics of the conductor. A decrease in conductor width decreases the current-carrying capacity and increases electrical resistance (Fig. 29.18). Most graphs of current-carrying capacity take the fabrication process into consideration and adjust the current-carrying capacity to allow for a margin of safety. Measurement of the conductor width is always nondestructive and easily performed (see Fig. 29.19, dimension C).

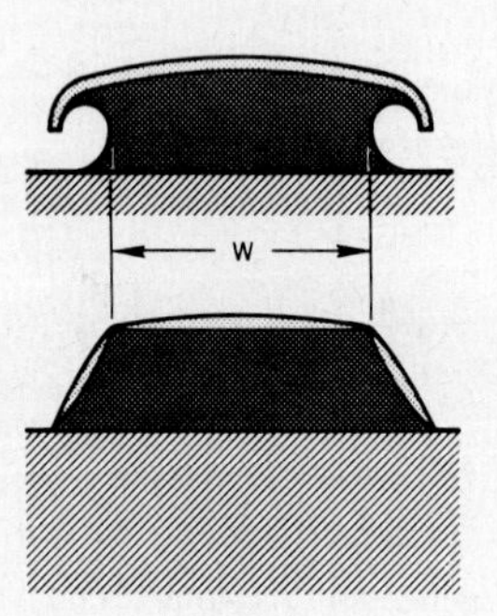

FIG. 29.17 Conductor width. *(Sandia Laboratories.)*

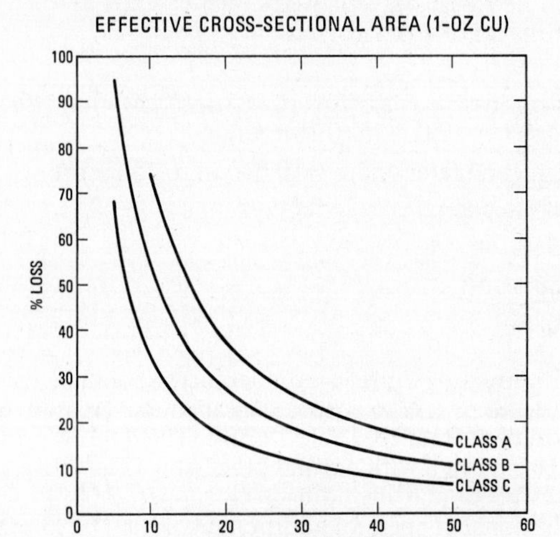

FIG. 29.18 Cross-sectional area reduction. Depicts the percentage loss in cross-sectional area that can be expected with the IPC conductor width measurement Class A, Class B, and Class C tolerances for 1-oz copper, 0.0014 in (0.036 mm). *(Sandia Laboratories.)*

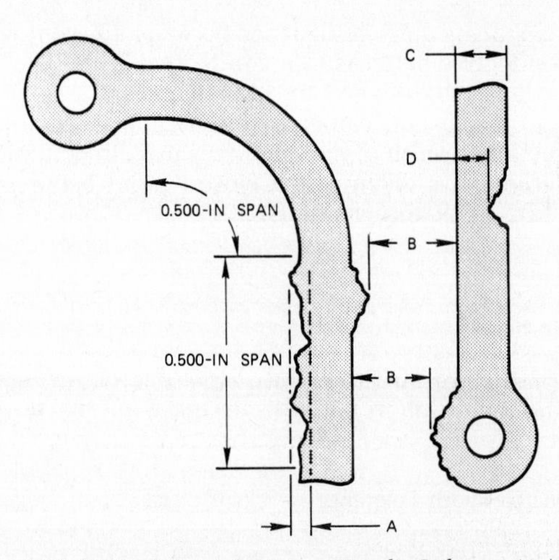

FIG. 29.19 Edge definition. *(Sandia Laboratories.)*

The difference between the actual minimum conductor width and the measured conductor width, which is measured vertically from above, can have an effect on the current-carrying capacity, inductance, and impedance. The difference can be significant on narrow conductors, 0.015 in (0.38 mm) or less, especially if fabricated by the panel-plating process.

As an example:

Copper-cladding thickness = 1 oz, 0.0014 in (0.036 mm)

Copper electroplate thickness = 1 oz

Etch factor = 1 to 1

Conductor width loss = Class A: −0.008 in (0.20 mm), Class B: −0.006 in (0.15 mm), Class C: −0.004 in (0.10 mm)

Nominal design conductor width = 0.010 in (0.254 mm)

Conductor cross-sectional area reduction:

Class A = 96%

Class B = 86%

Class C = 68%

IPC Class B minimum or C preferred tolerances should be considered for conductor widths 0.015 in or less to ensure that adequate cross-sectional area exists for design electrical performance. Table 29.2 (see IPC-D-320) presents process tolerances that can be expected with normal processing. The tolerances are based on copper thicknesses up to and including 1-oz copper, 0.0014 in (0.04 mm). For each ounce of additional copper, an additional 0.001-in (0.03-mm) variation per conductor edge can be expected. Therefore, check the design drawing notes and associated specifications to determine the minimum conductor width required. If not specified, verify the minimum conductor width required with the purchaser.

29.6.4 Conductor Spacing

Conductor spacing is defined as the distance between adjacent edges (not centerline) of isolated conductive patterns to a conductive layer. The spacing between conductors and/or lands is designed to allow sufficient insulation between circuits. A reduction in the spacing can cause electrical leakage or affect the capacitance. The cross-sectional width of conductors is usually nonuniform. Therefore, the spacing measurement is taken at the closest point between the conductors and/or lands (see Fig. 29.19, dimension B).

29.6.5 Edge Definition

The fidelity of reproduction of the pattern edge relative to the original master pattern is called edge definition. It falls into the cosmetic effects category and does

TABLE 29.2 Conductor Width Tolerance

Feature	Class A	Class B	Class C
Without plating	+0.004 in −0.006 in (+0.10 mm) (−0.15 mm)	+0.002 in −0.004 in (+0.05 mm) (−0.10 mm)	+0.001 in −0.002 in (+0.03 mm) (−0.05 mm)
With plating	+0.008 in −0.006 in (+0.20 mm) (−0.15 mm)	+0.004 in −0.004 in (+0.10 mm) (−0.10 mm)	+0.003 in −0.003 in (+0.08 mm) (−0.08 mm)

Source: IPC.

not normally affect functionality. It can, however, have an effect on high-voltage circuits: a corona discharge may be caused at the irregular conductor edges.

Measurement of edge definition is performed by measuring the distance from the crest to the trough (see Fig. 29.19, dimension A). A popular specification is 0.005 in (0.127 mm) from crest to trough. Isolated indentations which do not reduce the conductor width by more than 20 percent are usually allowed (see Fig. 29.19, dimension D). Also allowed are isolated projections which do not reduce the conductor spacing below specification requirements. As stated above, projections can produce corona discharge in high-voltage circuits. Figure 29.20 illustrates isolated indentation and projection and edge definition.

29.6.6 Hole Specifications

Hole size is the diameter of the finished plated-through or unplated hole. A plated-through hole is a hole in which electrical connection is made between internal or external conductive patterns, or both, by the deposition of metal on the wall of the hole. An unplated hole, termed an "unsupported hole," is a hole containing no conductive material or any other type of reinforcement. Measurement of hole size is performed to verify that the hole meets minimum and maximum drawing requirements. The size requirement is usually associated with a fit requirement of a component lead, mounting hardware, etc., plus adequate clearance for solder. Two basic methods are used to measure hole size: (1) drill blank plug or suitable gauge (Fig. 29.21) and (2) optical. The latter method, utilized when soft coatings over the copper are used, prevents deformation of the soft coatings within the hole. When Kwik-Chek or drill blank plug gauges are used, the inspector should

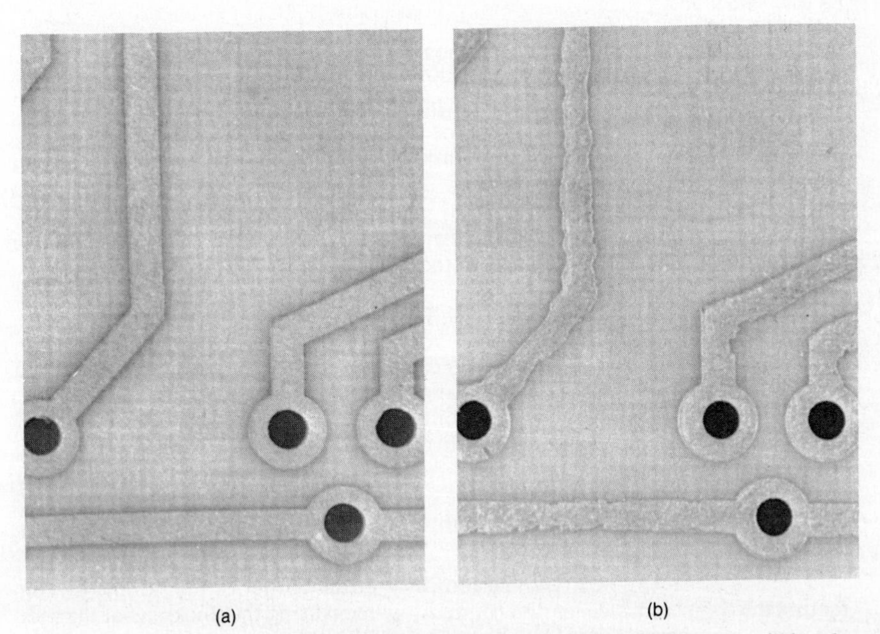

FIG. 29.20 Edge definition. (*a*) Preferred: conductor edges are smooth and even within tolerance; (*b*) nonpreferred: conductor edges are poorly defined and outside tolerance. *(IPC.)*

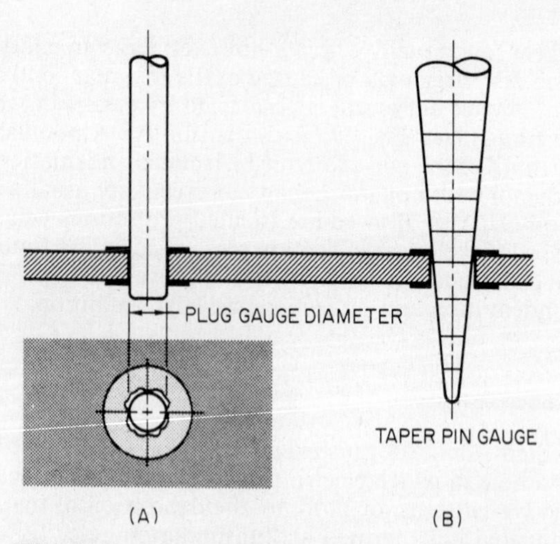

FIG. 29.21 Hole-measuring methods. (*a*) Drill blank plug gauge method; (*b*) Kwik-Chek or taper-pin method. *(Sandia Laboratories.)*

acquire a soft touch to prevent damage to the hole. The Kwik-Chek or drill blank plug gauges should be cleaned, prior to use, to prevent solderability degradation.

Plating nodules are sometimes present in the hole and restrict the penetration of the gauges. Forcing the gauges into the holes causes the nodules to be dislodged, and that results in voids in the plated-through-hole sidewall.

29.6.7 Bow and Twist

Bow is a deviation from flatness of a board characterized by a roughly cylindrical or spherical curvature such that, if the board is rectangular, its four corners are in the same plane. Twist is a deformation parallel to a diagonal of a rectangular sheet such that one of the corners is not in the plane containing the other three corners. Bow and twist, on a printed board, are inspected when the conditions impair function. Bow and/or twist on a board is detrimental when the board must fit in card guides or in packaging configurations in which space is limited. Two methods are recommended for measuring the degree of bow and/or twist: the indicator height gauge method (Fig. 29.22) and the feeler gauge method (Fig. 29.23).

29.6.7.1 Procedure No. 1 (Bow). See Fig. 29.24.

1. Place the sample to be measured on the datum surface with the convex surface of the sample facing upwards. For each edge apply sufficient pressure on both corners of the sample to ensure contact with the surface. Take a reading with a dial indicator at the maximum vertical displacement of this edge, denoted as R_1. Repeat this procedure until all four edges of the sample have been measured. It may be necessary to turn the sample over to accomplish this. Identify the edge with the greatest deviation from datum. This is the edge to be measured in steps 2 and 3.

2. Take a reading with the dial indicator at the corner of the sample by contacting the datum surface, or determine R_2 by measuring the thickness of the sample with a micrometer (denoted R_2 in Fig. 29.25).

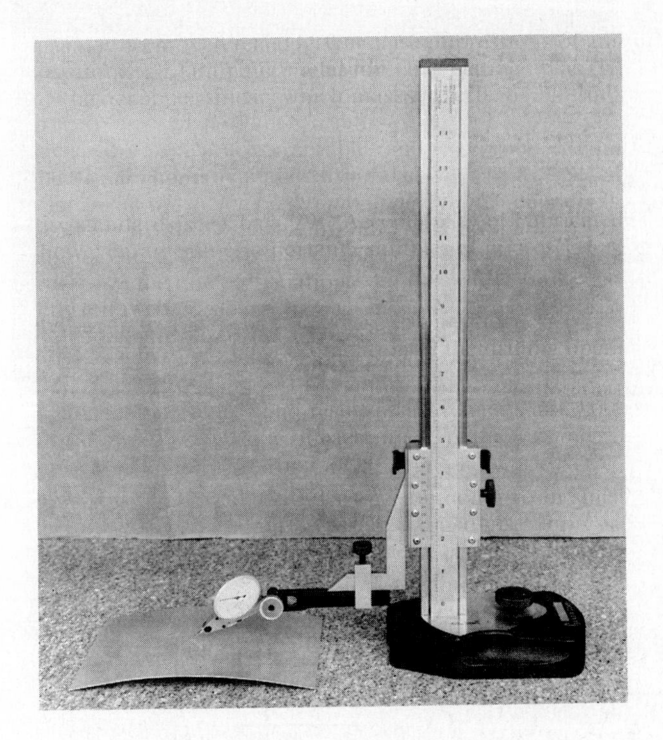

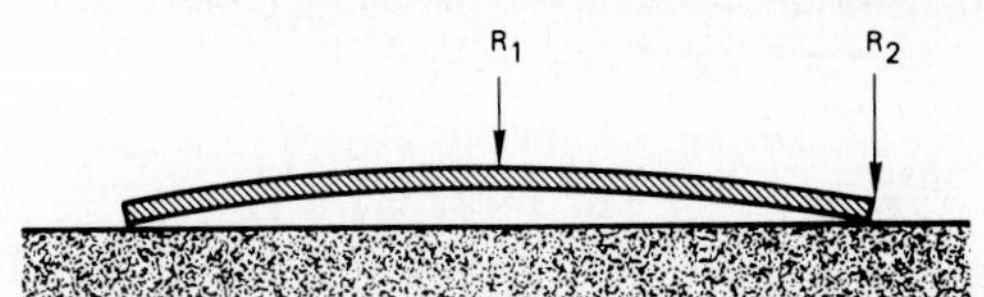

FIG. 29.22 Indicator height gauge method. *(Sandia Laboratories.)*

3. Apply sufficient pressure so that the entire edge contacts the datum surface. Measure the length of the edge and denote as L.

4. Calculate bow for this edge as follows:

$$\text{Percent bow} = \frac{R_1 - R_2}{L} \times 100$$

The result of this calculation is the percent of bow.

Repeat the procedure for the other three edges, and record the largest value of percent of bow for the sample.

The percentage of allowable bow and twist is dependent on the board thickness. See Table 29.3 for thickness codes. These thickness codes are used in Tables 29.4 and 29.5 to determine the percentage of acceptable bow and twist for Classes A, B, and C of paper-base materials and glass-base materials, respectively.

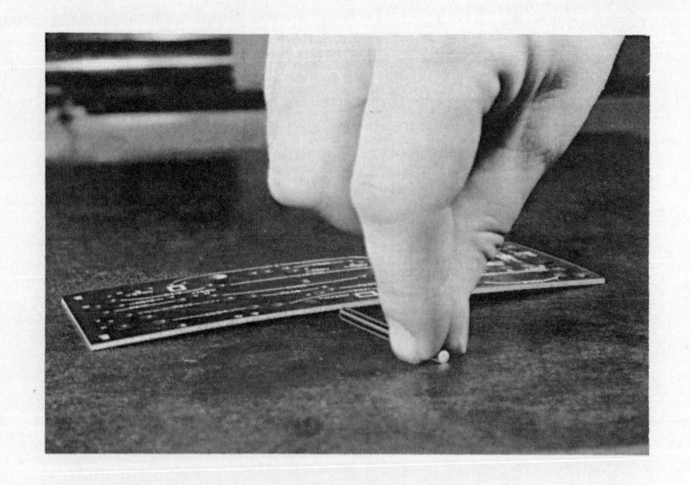

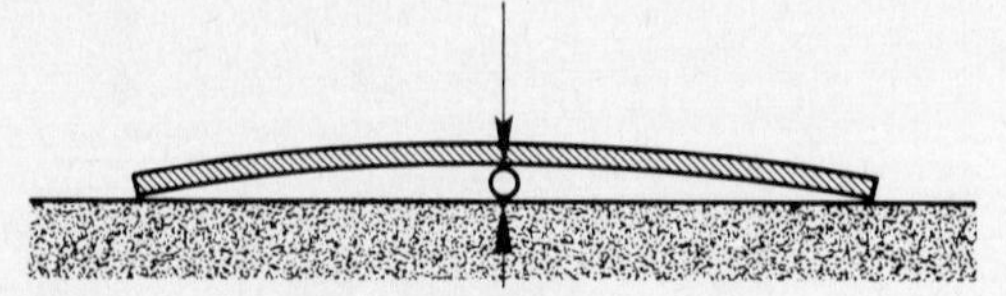

FIG. 29.23 Feeler gauge method. *(IPC.)*

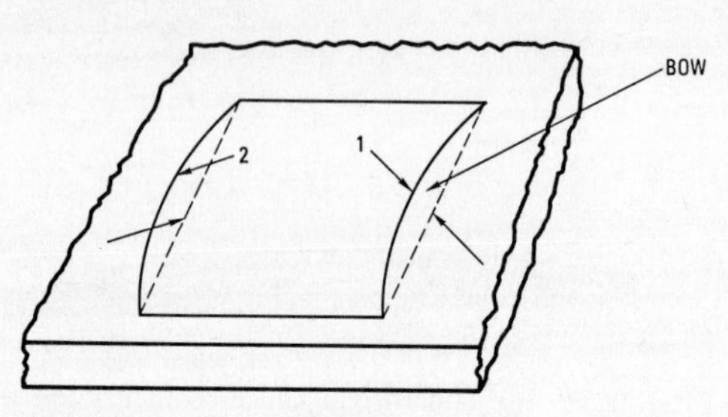

FIG. 29.24 Bow. *(IPC.)*

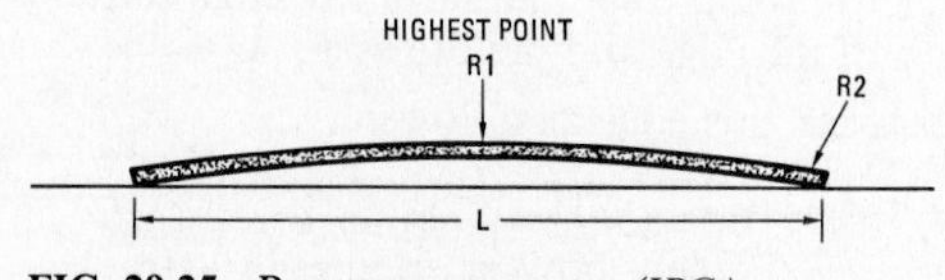

FIG. 29.25 Bow measurement. *(IPC.)*

29.6.7.2 Procedure No. 2 (Twist). See Fig. 29.26.

1. Place the sample to be measured on the datum surface with any three corners of the sample touching the surface. Apply sufficient pressure to ensure that three corners are in contact with the datum surface. Take a measurement from the datum surface to the lifted corner, and record the reading. Repeat this pro-

TABLE 29.3 Board Thickness Codes

Thickness code	Nominal board thickness
T1	0.2 mm (0.008 in), 0.5 mm (0.020 in), 0.7 mm (0.028 in), 0.8 mm (0.030 in), 1.0 mm (0.039 in)
T2	1.2 mm (0.047 in), 1.5 mm (0.059 in), 1.6 mm (0.063 in)
T3	2.0 mm (0.079 in), 2.4 mm (0.094 in)
T4	3.2 mm (0.125 in) and above

TABLE 29.4 Bow and Twist Tolerance, Paper-Base Material

Pattern	Thickness code	Class A	Class B	Class C
1-sided	T1	No requirement	2.5%	1.5%
	T2	2.5%	2.0%	1.0%
	T3	2.0%	1.2%	0.8%
	T4	1.5%	0.8%	0.6%
2-sided	T1	No requirement	2.0%	1.0%
	T2	2.0%	1.5%	0.8%
	T3	1.5%	1.0%	0.6%
	T4	1.0%	0.7%	0.7%

Source: IPC.

TABLE 29.5 Bow and Twist Tolerance, Glass-Base Material

Pattern	Thickness code	Class A	Class B	Class C
1-sided	T1	2.5%	2.0%	1.5%
	T2	2.0%	1.5%	1.0%
	T3	1.5%	1.0%	0.8%
	T4	0.8%	0.6%	0.6%
2-sided	T1	2.0%	1.5%	1.5%
	T2	1.5%	1.0%	0.9%
	T3	1.0%	0.7%	0.6%
	T4	0.6%	0.5%	0.5%
Multi	All categories	3.0%	2.0%	1.0%

Source: IPC.

cedure until all four corners of the sample have been measured. It may be necessary to turn the sample over to accomplish this. Identify the corner with the greatest deviation from datum. This is the corner to be measured in steps 2 and 3.

2. Place the sample to be measured on the datum surface with three corners touching the surface, and insert suitable shims under the raised corner so that it

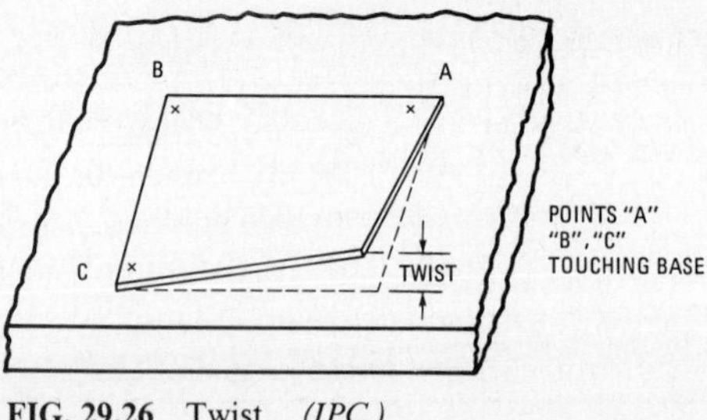

FIG. 29.26 Twist. *(IPC.)*

is just supported. When the correct shim thickness is used, the three corners will be in contact with the datum surface without applying pressure to any corner.

3. Without exerting any pressure on the sample, take a reading with the dial indicator at the maximum vertical displacement, denoted R_1 in Fig. 29.27, and

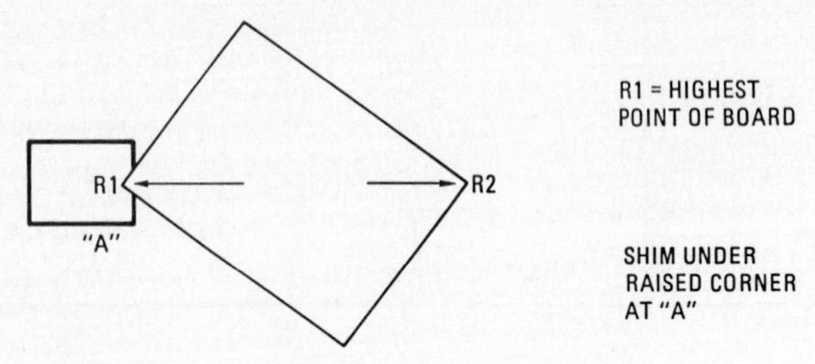

FIG. 29.27 Twist measurement with shims. *(IPC.)*

record the reading. Without disturbing the sample, take a reading with the dial indicator on the top surface of the sample at the edge contacting the datum surface, or determine R_2 by measuring the thickness of the sample with a micrometer. *Note:* For fabricated boards, both readings must be made on base material.

4. Measure the diagonal of the sample (for rectangular boards), and record the reading. For nonrectangular boards measure from the corners exhibiting displacement diagonally to the point on the opposite end of the board.

5. Calculate as follows:

a. Deduct R_2 reading from R_1 reading and divide the result by 2.

$$\text{Deviation} = \frac{R_1 - R_2}{2}$$

b. Divide the measured deviation (step 1) by the recorded length and multiply by 100. The result of this calculation is the percent of twist.

$$\text{Percent twist} = \frac{R_1 - R_2}{L} \times 100$$

The percentage of allowable bow and twist is dependent on the board thickness. See Table 29.3 for thickness codes. These thickness codes are used in Tables 29.4 and 29.5 to determine the percentage of acceptable bow and twist for Classes A, B, and C of paper-base materials and glass-base materials, respectively.

The two procedures described above are used to determine a percentage of allowable bow and twist, which may be applied to either metric or customary dimensions; it is a percentage per unit of measure. Thus 2 percent allowable bow and twist translates to 0.02 × inches or millimeters.

29.6.8 Conductor Pattern Integrity

Several methods are used to determine conductor pattern integrity, i.e., the quality or state of being complete. Some methods include the use of comparison equipment and overlays. The use of positive or negative overlays made from the master pattern is an inexpensive and effective method. By overlaying the film on the finished printed board, differences are easily detected. Overlays can also be used to determine if conductor widths are within tolerance and annular rings and contours are within drawing requirements.

29.6.9 Contour Dimensions

Inspection of contour dimensions verifies that the outside border dimensions are within the drawing requirements. Contour dimension requirements can be considered fit requirements. Both undersized and oversized printed boards can affect functionality, depending upon the degree of requirement violation. Measuring methods vary from the use of a ruler or calipers to sophisticated numerical control equipment. The sophistication of the method is naturally dependent on the required dimensions and tolerances.

29.6.10 Plating Thickness

29.6.10.1 Nondestructive Methods. The plating process is used in the fabrication of numerous printed board designs to produce plated-through holes and conductive pattern circuitry. When the plating process is utilized, a plating thickness requirement is usually specified on the drawing or in the accompanying specifications. Verification of the plating thicknesses on the pattern and in the holes is usually required to ensure circuit functionality. Two of the presently popular ways to measure plating thicknesses nondestructively are the beta backscatter and microohm methods.

1. *Beta backscatter method:* (See ASTM manual B-567-72.) This method utilizes an energy source (a radioisotope) and a detector (a Geiger-Müller tube). It functions on the beta radiation backscatter principle. It allows thickness measurements to be made within a short period of time, and it can be utilized on metallic, nonmetallic, magnetic, and nonmagnetic materials. The beta backscatter method can be used to measure plating thicknesses on the circuitry and in the plated-through holes during or after the board fabrication process and prior to or after the etching operation. It is useful only in measuring the top metal thickness, and it reads the thickness as an average.

2. *Microohm method:* (See IPC-TM-650, 2.2.13.1.) The microohm method functions as a four-wire resistance bridge circuit that imposes a constant current across the area under measure and measures the voltage drop across the same area. The answer is given in microohms. In plated-through-hole measurements, it treats the hole as a cylindrical resistor. The method is effective only after the circuitry has been etched. When two different metals are used on top of one

another, only the lowest-resistance metal thickness is detected. Like the beta backscatter method, the microohm method reads the thickness only as an average.

29.6.10.2 Destructive, Microsection Method. The verification of plating thicknesses on the surface and in the holes by the microsection method usually requires three measurements at three different locations on each plated-through-hole sidewall (Fig. 29.28). The results are reported either individually or as an average.

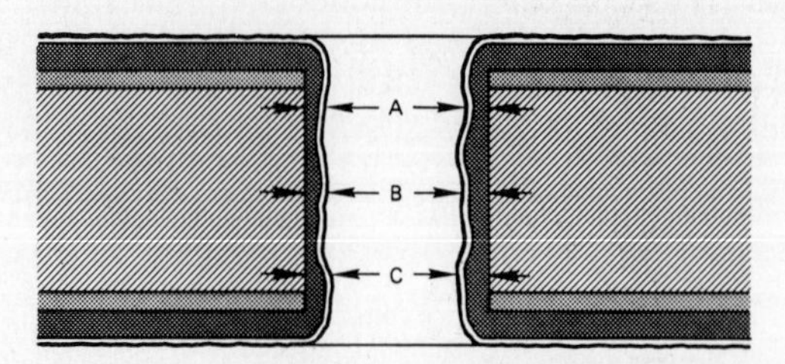

FIG. 29.28 Verification of plating thickness in holes by use of vertical cross sections. *(Sandia Laboratories.)*

Specifications also define the minimum plating thickness allowed in the hole, either as the minimum thickness or as an average minimum thickness. Care must be exercised to select locations free from voids; they are usually 0.010 in (0.254 mm) apart. Examples of plating thickness are shown in Fig. 29.29. Typical plating thickness requirements are as follows:

Copper	0.001 in (0.0254 mm) minimum
Nickel	0.0005 to 0.0010 in (0.0127 to 0.0254 mm)
Gold	0.000050 to 0.000100 in (0.00127 to 0.00254 mm)
Tin-lead	0.0003 in (0.00762 mm) minimum

(a)

(b)

FIG. 29.29 Examples of plating thickness. (*a*) Preferred: plating is uniform and meets thickness requirements; (*b*) nonpreferred: plating thickness less than requirements. *(Sandia Laboratories.)*

Rhodium	0.000005 to 0.000020 in (0.000127 to 0.000508 mm)

Polishing vertical plated-through-hole cross sections to the mean of the hole is critical. If the hole is polished less than or greater than the mean of the hole, a plating thickness error is introduced. Horizontal plated-through-hole cross sections are recommended as references (Fig. 29.30). Although horizontal plated-through-hole cross sections are usually more accurate for plating thickness measurements, they do not allow adequate inspection of other attributes such as voids, plating uniformity, adhesion, etchback, and nodules. Surface plating thickness measurements are taken on vertical cross sections of conductor areas. Microsection mounts are usually etched with an appropriate etchant to show grain boundaries between the copper-clad foil and copper plating. Copper plating surface measurements usually exclude the copper foil thickness unless otherwise specified.

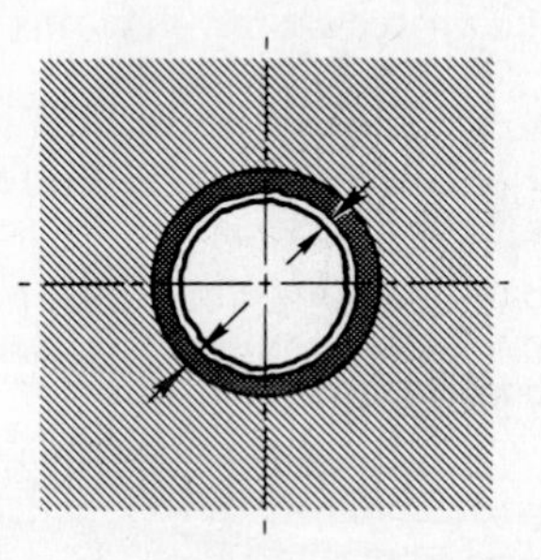

FIG. 29.30 Verification of plating thickness in holes by use of horizontal cross sections. *(Sandia Laboratories.)*

29.6.10.3 Method of Preparing Microsections. (Institute for Interconnecting and Packaging Electronic Circuits, test method 2.1.1.) Cut from the printed board or test coupon a specimen for vertical evaluation that contains at least three of the smallest holes adjacent to edge. Cut additional specimens from the board or test coupon for horizontal evaluation.

The test equipment and apparatus necessary for preparation of the specimen is as follows:

Glass plate, 5 × 7 in (127 × 177.8 mm)
Aluminum rings, $1\frac{1}{4}$ in ID (31.75 mm)
Silicone release agent
Room-temperature-curing potting material
Wooden spatulas
Plastic cups (at least 6-oz cups)
Saw or shear
Engraver
240-grit abrasive
Double-coated tape
Metallographic polishing tables
240-, 320-, 400-, and 600-grit disks
Number 2 liquid alumina polish
Polishing cloths
Chemical etchants

For metallographic evaluation the following equipment is necessary:

Microscope and camera accessories

Filar eyepieces or graduated reticle
Engraver
Photographic film or Polaroid film
Filter lens

The procedure for preparing the specimen is as follows:

1. Clean glass plate and aluminum rings and dry thoroughly.
2. Apply strip of double-coated tape to plate to support specimen. Apply thin film of release agent to glass plate and ring; then place ring on plate.
3. Sand the long edge of the perpendicular specimen until the edges of the conductor pads appear and specimen will stand on edge on a flat surface. Use 240-grit abrasive.
4. When secondary plating thickness is being measured, overplate specimen with a harder electroplated metal. Specimens may be overplated as per ASTM method E3-58T.
5. Measure inside diameter of plated-through holes prior to encapsulation.
6. Stand specimen on edge on double-coated tape in aluminum ring with the plated-through-hole edge down. For parallel specimens, delete tape and lay specimen flat on glass plate inside ring.
7. Mix potting material and pour to one side of the specimen until it flows through the holes. Support the specimen in the vertical position if necessary. Continue pouring until the ring is full. Avoid entrapment of air.
8. Allow specimen to cure at laboratory temperature. Accelerated curing at a higher temperature following manufacturer's instructions is permissible, provided cracking or deformation does not occur.
9. Identify specimen promptly by engraving.
10. If more than one specimen is potted in one ring, the specimens should be spaced apart to facilitate filling the holes. Specimens shuld be identified with a strip of paper marked with traceability information and molded into the mounting.

The grinding and polishing procedure is as follows:

1. Rough-grind the face of the specimen to the approximate center of the plated-through holes by using 240-, 320-, 400-, and 600-grit disks, in that order.
2. Flush away all residue by using tap water at room temperature. Wash hands repeatedly to avoid carrying over coarse grids.
3. Rotate specimen 360° about the axis of the wheel and opposite the direction of rotation of the wheel. Keep the face of the specimen flat on the wheel.
4. Micropolish by using a nylon disk and alumina polish number 2 until the specimen is smooth and free of sanding marks and a clear sharp image of plating lines is evident.
5. Rinse specimen thoroughly, and while it is still wet, chemically etch by using a cotton swab to highlight plating boundaries.
6. Lightly rub twice across the specimen, and immediately rinse in distilled water. Repeat if necessary.
7. Rinse thoroughly after etching to eliminate carryover of acids to the microscope.
8. Dry specimen prior to viewing through the microscope.

To make a microscope examination, proceed as follows:

1. Place specimen on the microscope stage, and adjust to get specimen centered with the eyepiece.
2. Focus and adjust lighting for best viewing; then scan specimen.
3. With filar micrometer or graduated reticle make three thickness measurements of each plating on both walls of the plated-through hole.

The following is the procedure for photomicrographing (composite photographs):

1. Locate specimen and mount camera to microscope tube. Insert film pack.
2. Set lighting and exposure time to prearranged settings.
3. Focus on one part of one wall and push camera plunger.
4. Immediately remove negative from camera and develop.
5. Take sufficient photographs to illustrate the entire length of the wall.
6. Repeat steps 1 to 5 and photograph the opposite wall.
7. If defective junctions, voids, etc., appear, enlarged photographs may be made. Enlargements should be referenced to the photographs.

29.6.11 Undercut (after Fabrication)

Undercut is the distance on one edge of a conductor measured parallel to the board surface from the outer edge of the conductor, excluding overplating and coatings, to the maximum point of indentation on the same edge. Measurement of the degree of undercut present is not usually required, because undercut is included in the minimum conductor-width measurement. If, however, the undercut measurement is required separately, it is performed by first measuring the minimum conductor width and then subtracting that dimension from the design width of the conductor and dividing by 2 (Fig. 29.31). For example:

Design width of conductor = (a)
Minimum conductor width = (b)
Undercut = (c)

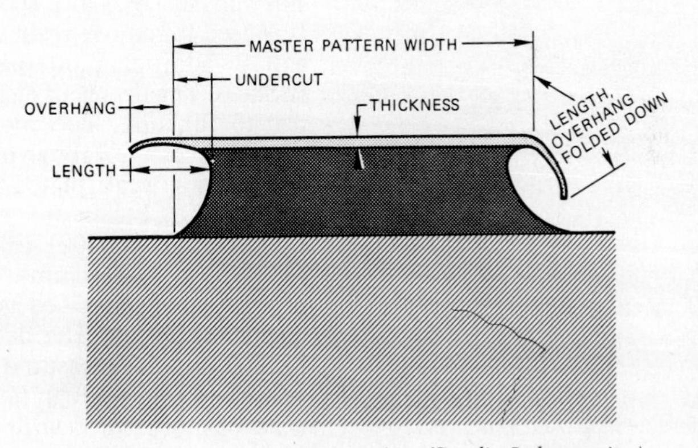

FIG. 29.31 Undercut and outgrowth. *(Sandia Laboratories.)*

Then

$$\frac{a - b}{2} = c$$

29.6.12 Outgrowth

The increase in conductor width at one side of a conductor, caused by plating buildup, over that delineated on the production master is called outgrowth. The measurement of the degree of outgrowth is required in some specifications. Excessive outgrowth can eventually lead to "slivering," a thin metallic piece that breaks off of the conductive pattern.

Outgrowth length dimensions of $1\frac{1}{2}$ times, or less, the outgrowth thickness, at the point where the outgrowth material extends beyond the main conductor configuration, is usually allowed. The degree of outgrowth is measured on a conductor cross section (Fig. 29.31). An example of excessive outgrowth is depicted in Fig. 29.32. A method of determining if the outgrowth is prone to slivering utilizes an ultrasonic cleaner; the total printed board is suspended in water in an ultrasonic cleaner for 1 to 2 min. If the board is prone to slivering, metallic slivers will be evident along pattern edges (Fig. 29.33).

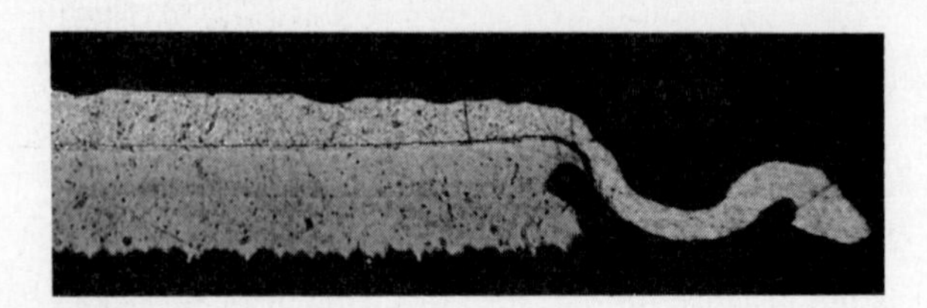

FIG. 29.32 Outgrowth extension is approximately 10 times the thickness. *(Sandia Laboratories.)*

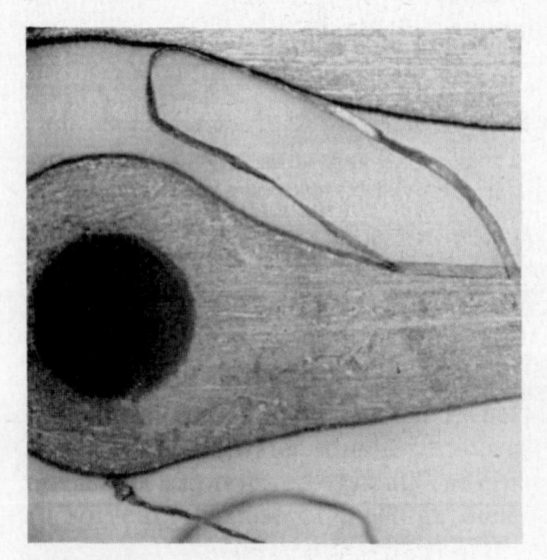

FIG. 29.33 Metallic slivers present after ultrasonic test. *(Sandia Laboratories.)*

29.6.13 Etchback

Etchback is a process for the controlled removal of nonmetallic materials from sidewalls of holes to a specified depth. It is used to remove resin smear and to expose additional internal conductor surfaces. The degree of etchback is critical to function. Too much etchback creates excessively rough hole sidewalls and causes weak plated-through-hole structures. Etchback requirements range from 0.0002 to 0.0030 in (0.00508 to 0.0762 mm). The requirement is usually specified as a minimum and a maximum. The degree of etchback is usually measured by utilizing vertical plated-through-hole cross sections of multilayer boards (Fig. 29.34).

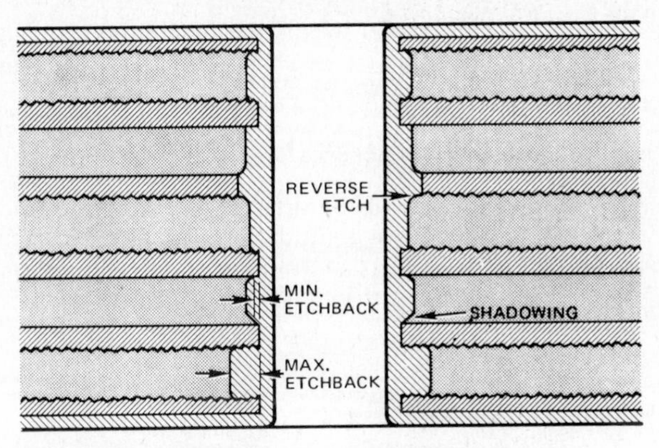

FIG. 29.34 Etchback configurations. *(IPC.)*

Typical etchback acceptance criteria recommended by the IPC are shown in Fig. 29.35.

29.6.14 Layer-to-Layer Registration

Layer-to-layer registration is the degree of conformity of the position of a pattern, or portion thereof, with its intended position or with that of any other conductor layer of a board. It is required to ensure electrical connection between the plated-through holes and the internal layers. Misregistration of internal layers increases electrical resistance and decreases conductivity. Severe misregistration creates an open-circuit condition: complete loss of continuity. Two popular methods of measuring layer-to-layer registration are the x-ray method described in Sec. 29.5.4. and the microsection method, which is similar to the plating thickness inspection method.

The microsection method consists of measuring each internal land area on a vertical plated-through-hole cross section and determining the centerline. The

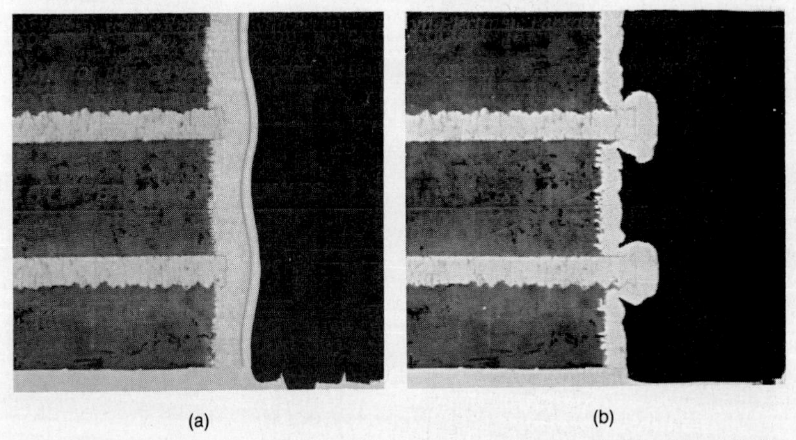

FIG. 29.35 Etchback. (*a*) Preferred: uniform etchback of base laminate; uniform plating in the plated-through hole; (*b*) nonpreferred: nonuniform and excessive etchback of base laminate results in unacceptable nonuniform plating in the hole. *(IPC.)*

maximum variation between centerlines is the maximum misregistration (Fig. 29.36).

29.6.15 Flush Conductor, Printed Boards

A flush conductor is a conductor whose outer surface is in the same plane as the surface of the insulating material adjacent to the conductor. Flush conductors are primarily used in rotating switches, commutators, and potentiometers. Common

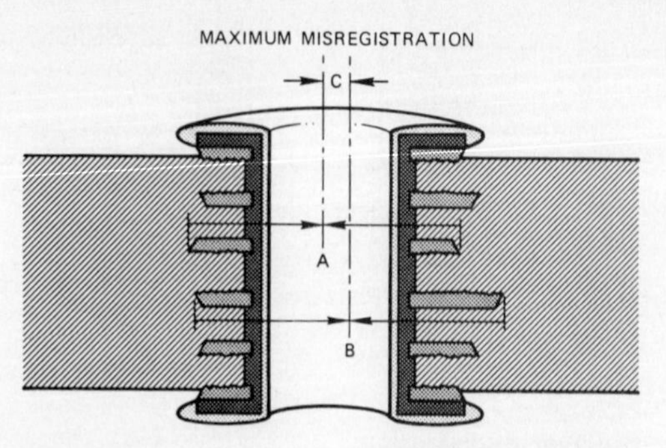

FIG. 29.36 Layer-to-layer registration microsection method. *(IPC.)*

to those uses are brush or wiper and conductor pattern combinations. Making the conductors flush in the above applications reduces wiper vibrations, wear, and intermittent or noisy signals. The height of the step allowed between pattern and base material is dependent on the relative wiper speed, the materials used, and the degree of signal error and electrical noise tolerable in the circuit. A commonly accepted height allowance is as follows:

Up to 50 r/min	80 to 200 μin (2.032×10^{-4} to 5.08×10^{-3} mm)
51 to 125 r/min	Better than 50 μin (1.27×10^{-4} mm)
126 to 500 r/min	Better than 30 μin (0.762×10^{-4} mm)

Inspection of the degree of flushness is performed with height gauges or measuring microscopes (Fig. 29.37).

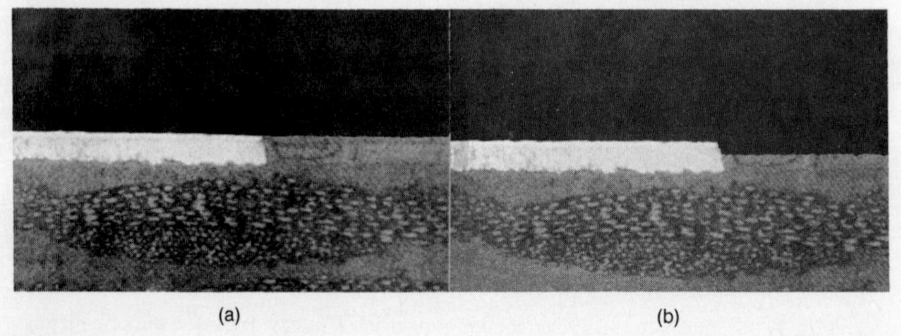

FIG. 29.37 Flush printed circuits. (*a*) Preferred: conductor is flush to the board surface; (*b*) non-preferred: conductor is not flush to the board surface and does not meet the specified tolerance. *(IPC.)*

29.6.16 Summary: Dimensional Inspection

The dimensional board attributes subjected to nondestructive and destructive inspection are listed in Table 29.6.

TABLE 29.6 Dimensional Inspection Chart

Subject	Nondestructive	Destructive
Annular ring	X	
Pattern-to-hole registration	X	
Conductor width	X	X
Conductor spacing	X	
Edge definition	X	
Hole size	X	
Bow and twist	X	
Conductor pattern integrity	X	
Contour dimensions	X	
Plating thickness	X	X
Undercut		X
Outgrowth		X
Etchback		X
Layer-to-layer registration		X
Flush circuits	X	X

29.7 MECHANICAL INSPECTION

Mechanical inspection is applied to characteristics which can be verified with a qualitative physical test. Mechanical inspection methods can be either nondestructive or destructive. The following sections describe methods that are used to verify printed board fabrication integrity.

29.7.1 Plating Adhesion

A common method of inspecting for plating adhesion is the tape test, which is described in detail in IPC-TM-650, 2.4.10, and MIL-P-55110. The basic test method is shown in Fig. 29.38.

Another method of determining plating adhesion is applied during the preparation and viewing of the microsection mounts. If adhesion is poor, the layers of plating will separate during preparation of the microsection specimen or will indicate lack of adhesion at plating boundaries during viewing of the microsection mount.

Figure 29.39 indicates lack of adhesion at the different plating boundaries, i.e., copper foil to copper plate and copper plate to tin-lead.

29.7.2 Solderability

Solderability inspection measures the ability of the printed pattern to be wetted by solder for the joining of components to the board. It involves the use of three terms: (1) wetting, (2) dewetting, and (3) nonwetting. The terms are defined as follows:

1. *Wetting:* The formation of a relatively uniform, smooth, unbroken, and

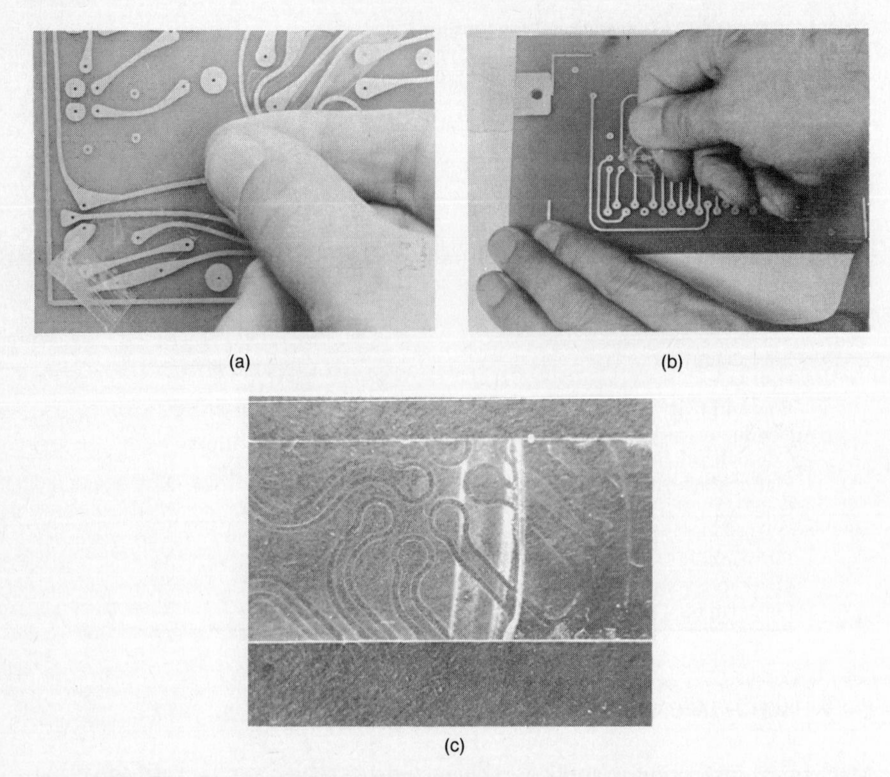

FIG. 29.38 The tape test. (*a*) Step 1: Place transparent cellophane tape across the circuits to be tested, and press the tape onto the circuits. Eliminate all air bubbles with finger. (*b*) Step 2: Lift tape on one end enough to get a grip. Pull tape off the printed board at approximately 90° to the board. Use a rapid pull. (*c*) Step 3: A clear tape is the preferred test result. *(IPC.)*

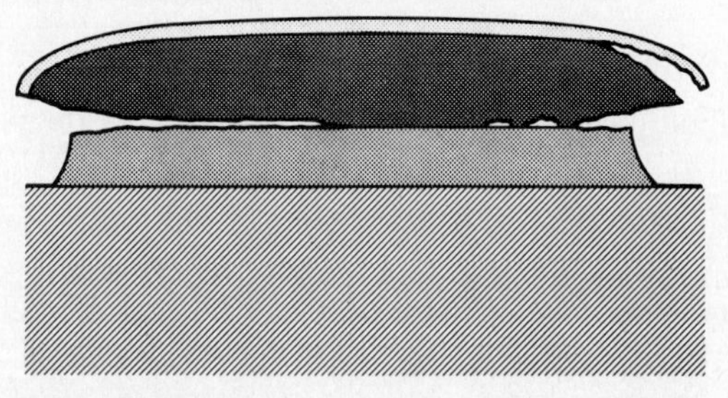

FIG. 29.39 Plated metal adhesion. *(Sandia Laboratories.)*

adherent film of solder to a basis material, conductors, and land areas to form an adherent bond.

2. *Dewetting:* A condition which results when molten solder has coated a surface and recessed, leaving irregularly shaped mounds of solder separated by areas covered with a thin solder film: basis metal is not exposed. Most of the solder balls up at random locations on the surface.

3. *Nonwetting:* A condition whereby a surface has contacted molten solder, but the solder has not adhered to all of the surface; basis metal remains exposed.

Many methods for inspecting printed boards for solderability both quantitatively and qualitatively have been established. However, the most meaningful method is that which will be used in the assembly soldering operation—hand soldering, wave soldering, drag soldering, etc. The IPC has adopted the criteria illustrated in Fig. 29.40 for printed board solderability acceptance.

29.7.3 Alloy Composition

Two popular alloys used in printed board manufacture are tin-lead and tin-nickel. The composition of the tin-nickel bath is usually 65 percent tin and 35 percent nickel. The composition of the deposit, during electroplating, remains nearly constant despite fluctuations in bath composition and operation conditions. However, the composition of the tin-lead deposits can vary with bath composition and operation conditions, and that will influence the melting temperature of the alloy and affect solderability. Specifications usually require the tin content to be between 50 and 70 percent.

Methods available for analyzing the alloy composition on the plated printed board include wet analysis, atomic absorption, and the beta backscatter. The beta backscatter method is relatively new and is gaining in popularity because of the ease in obtaining the alloy composition nondestructively.

29.7.4. Thermal Stress (Solder Float Test)

Temperature-induced strain or straining force that is exerted upon the printed board and tends to stress or deform board shape can be a serious problem during

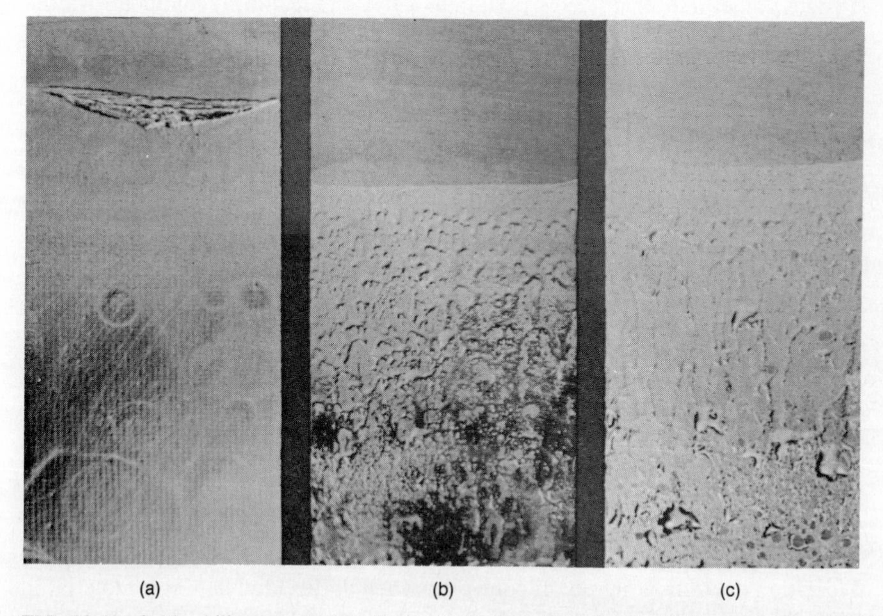

FIG. 29.40 Solderability acceptance criteria. (*a*) Preferred: good solderability; smooth, bright, good wetted copper; and good solder coverage; (*b*) acceptable: some dewetting is visible at left surface; (*c*) nonpreferred: surface shows severe dewetting and nonwetting. *(IPC.)*

the soldering operation. Thermal stress inspection is performed to predict the behavior of the printed boards after soldering. Plated-through-hole degradation, separation of platings or conductors, or laminate delamination is detectable by the thermal stress test.

The printed board specimen is (1) conditioned at 250 to 300°F for a period of 2 h minimum to reduce moisture; (2) placed in a dessicator, on a ceramic plate, to cool; (3) fluxed with type RMA flux; (4) floated in a solder bath (63 ± 5 percent tin maintained at 550 ± 10°F) for 10 s; and (5) after stressing, placed on an insulator to cool. Visual inspection for defects is followed by cross-sectioning the plated-through holes and inspecting them with magnification for integrity. Test patterns also are used for thermal stress inspection. (See Fig. 29.41 for a test pattern used in this and other connections.)

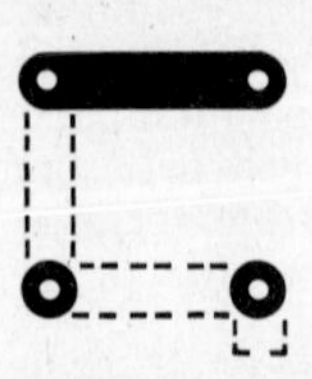

FIG. 29.41 Continuity, thermal stress, moisture resistance test pattern. *(IPC.)*

29.7.5 Peel Strength

Peel strength is the force per unit width required to peel the conductor or foil from the base material. It is usually associated with the acceptance testing of copper-clad laminate material upon receipt, but it is sometimes utilized to test the adhesion of the conductors to the finished board.

Peel strength tests of conductors are usually performed after the specimens have been dip- or flow-soldered. Peel strength values in MIL-P-13949, after solder dip, also are used for printed board conductor peel strength tests. However, the values must be adjusted for the particular conductor width, because the peel strength values in MIL-P-13949 pertain to a 1.00-in (25.4-mm) conductor width. As an example:

$$\text{Conductor width} = 0.050 \text{ in (1.27 mm)}$$

$$\text{Peel strength value in MIL-P-13949} = 10 \text{ lb}$$

$$\text{Conductor width in MIL-P-13949} = 1.00 \text{ in (25.4 mm)}$$

$$\frac{0.050 \times 10.0}{1.00} = 0.5 \text{ lb}$$

A calculation with width in milimeters will, of course, yield the same results.

Peel strength tests of conductors are a good method of assuring that the conductor-to-laminate adhesion is sufficient to withstand assembly soldering operations. Actual printed boards or test patterns can be utilized for the test. Figure 29.42 is an example of a typical peel strength test pattern.

29.7.6 Bond Strength (Terminal Pull)

The bond strength test is a test of the plating adhesion to the laminate in the hole.

Poor adhesion in the holes can occur when the drilling or electroless copper deposition operations are out of control or when undercured laminate material is used. Whatever the cause, poor plating adhesion in the hole affects functionality.

When lack of plating adhesion to the laminate in the hole is detectable during microsection analysis, a terminal pull test should be performed to substantiate the condition (see Fig. 29.43).

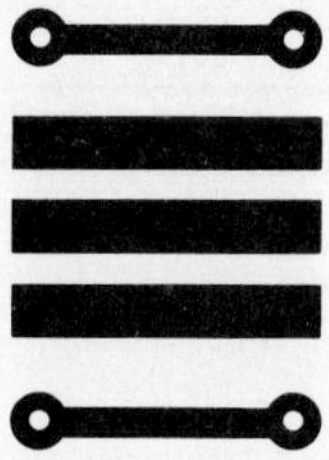

FIG. 29.42 Peel strength test pattern. *(IPC.)*

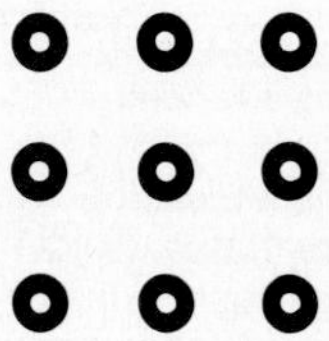

FIG. 29.43 Typical terminal pull test coupon. *(IPC.)*

29.7.7 Cleanliness (Resistivity of Solvent Extract)

The cleanliness test is used to verify the state of cleanliness. Printed boards are fabricated using wet chemistry and mechanical techniques. Some of the chemical baths have metallic salts or are corrosive in nature. These conditions could affect the functionality of the boards, i.e., reduce electrical insulation resistance, corrode the metal pattern, etc. Another common problem is electromigration of metal between conductors. This condition is usually associated with low operating voltage, 10 V or less, and requires that three components be present: (1) moisture, (2) metallic contaminate, and (3) voltage.

The cleanliness test measures the resistivity of the wash solution that the board has been washed with. This test (see MIL-P-55110) can be performed by open setup or with commercially available equipment.

29.7.7.1 Open Setup Procedure. *Note:* All laboratory ware must be scrupulously clean before use.

1. Position a conveniently sized funnel over an electrolytic beaker.

2. Suspend the printed board within the funnel.

3. Prepare a wash solution of 75 percent by volume of ACS reagent grade isopropyl alcohol and 25 percent by volume of distilled water. *Note:* This wash solution must have a resistivity equal to or greater than $6 \times 10^6\ \Omega \cdot \text{cm}$.

4. Direct a fine stream of solution, from a wash bottle, onto both sides of the printed board until 100 mL of the wash solution is collected for each 10 in^2 of board surface (including both sides of the board). Wash activity time should be a minimum of 1 min. *Note:* The initial wash solution must be included in the sample to be measured for resistivity.

5. Measure the resistivity of the collected wash solution with a conductivity bridge or some other instrument of equivalent range and accuracy.

29.7.7.2 Alternative Test Procedure. Commercial equipment is available to perform the cleanliness testing. See MIL-P-55110 for source and equipment listing.

29.7.8 Summary: Mechanical Inspection

The mechanical board attributes subjected to nondestructive and destructive inspection are listed in Table 29.7.

TABLE 29.7 Mechanical Inspection Chart

Subject	Nondestructive	Destructive
Plating adhesion	X	X
Solderability		X
Alloy composition	X	X
Thermal stress (solder float test)		X
Peel strength, printed wiring conductor		X
Bond strength (terminal pull)		X
Cleanliness	X	

29.8 ELECTRICAL INSPECTION

Electrical inspection is performed to verify electrical circuit integrity after processing and also to ensure that the electrical characteristics of the processed board meet design intent.

Electrical inspection methods are both destructive and nondestructive. Nondestructive tests are usually performed on the actual printed boards. Destructive tests are performed either on printed circuit boards utilized for destructive acceptance testing or on test patterns fabricated outside the border of the board.

Two popular nondestructive electrical tests are the tests for insulation resistance and continuity. They are usually performed on 100 percent of complex printed boards, especially multilayer ones. Care should be taken to prevent arcing as the probes approach the printed board pattern when the insulation resistance test is performed. An easily activated switch in series with one of the probes allows the probes to make contact with the pattern prior to current flow. To prevent probe impressions on soft metal coatings, apply the same coating on the probe tips.

29.8.1 Continuity

Continuity tests are performed basically on multilayer circuits to verify that the printed circuit pattern is continuous. They can be performed with an inexpensive multimeter or with more elaborate equipment such as a constant-current power supply and a voltmeter if currents greater than 50 mA are required.

In the printed circuits industry the continuity test is performed in one of two ways: (1) as a go/no-go test to verify that the pattern is continuous or (2) to verify that the pattern is continuous, as well as to verify the integrity of the measured pattern. The latter result is reported in electrical resistance values (ohms). The preferred method is to perform the continuity test on all printed boards submitted for acceptance. This is especially recommended for multilayer circuitry, the internal patterns and interconnections of which cannot be inspected visually after fabrication.

Test coupons are sometimes used for the continuity test. See Fig. 29.41 for a pattern used on a typical coupon.

29.8.2 Insulation Resistance

The purpose of the insulation resistance test, as stated in MIL-STD-202, is to measure the resistance offered by the insulation members of a printed board to an impressed direct voltage that tends to produce a leakage of current through or on the surface of those members. Low insulation resistances can disturb the operation of circuits intended to be isolated, by permitting the flow of large leakage currents and the formation of feedback loops. The test also reveals the presence of contaminants from processing residues.

In printed circuitry the test is performed either between conductors on the same layer or between two different layers. The test is also utilized before and after thermal shock and temperature cycling tests. Test voltages of 100 and 500 V dc and minimum insulation resistances of 500 MΩ are popular. The insulation resistance test is performed on either the actual printed board or a test coupon fabricated on the same panel as the board (Fig. 29.44).

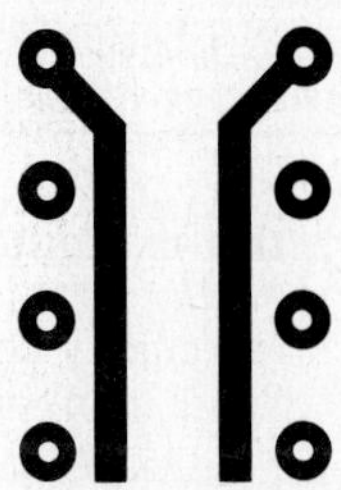

FIG. 29.44 Insulation resistance, dielectric withstanding voltage, and moisture resistance test patterns. *(IPC.)*

When special conditions such as isolation, low atmospheric pressure, humidity, and immersion in water are required, they should be specified in the test method instruction.

29.8.3 Current Breakdown, Plated-Through Holes

The current breakdown test is used to determine if sufficient plating is present within the plated-through hole to withstand a relatively high current potential. The time and current selected determine if this test is destructive or nondestructive. IPC-TM-650, 2.5.3, recommends a current of 10 A for 30 s. The test is performed as follows:

1. Place a load resistor, of predetermined value, across the negative and positive terminals of a current-regulated power supply.
2. Adjust the supply for a current of 10 A or any other desired value.
3. Remove one end of the resistor from the positive supply terminal.
4. Connect the desired plated-through hole between the disconnected end of the resistor and the positive supply terminal.
5. Perform the test for the desired time. The test is performed either on an actual printed board or on a test pattern (Fig. 29.45).

FIG. 29.45 Typical current-breakdown test pattern. *(IPC.)*

29.8.4 Dielectric Withstanding Voltage

The test for dielectric withstanding voltage is used to verify that the component part can operate safely at its rated voltage and withstand momentary overpotentials due to switching, surges, and similar phenomena. The test also serves to

TABLE 29.8 Electrical Inspection Chart

Subject	Nondestructive	Destructive
Continuity	X	
Insulation resistance	X	
Current breakdown, plated-through holes	X	X
Dielectric withstanding voltage	X	X

determine whether insulating materials and spacings in the component part are adequate.

The test is thoroughly defined in IPC-ML-950 (multilayer PWBs) and MIL-STD-202 (electronic components method 301). One of three different test voltages—500, 1000, and 5000 V—is usually specified. The test is performed on either an actual board or a test coupon. Voltage is applied between mutually insulated portions of the specimen or between insulated portions and ground. The voltage is increased at a uniform rate until the specified value is reached. The voltage is held for 30 s at the specified value and then reduced at a uniform rate.

Visual examination of the part is performed during the test for evidence of flashover or breakdown between contacts. See Fig. 29.44 for a typical test pattern. The test can be either destructive or nondestructive depending on the degree of overpotential used.

29.8.5 Summary: Electrical Inspection

The electrical board attributes subjected to nondestructive and destructive inspection are listed in Table 29.8.

29.9 ENVIRONMENTAL INSPECTION

Environmental inspection consists of performing specific tests to ensure that the printed board will function under the influence of climatic and/or mechanical forces to which it will be subjected during use. Environmental tests are performed on preproduction printed boards to verify design adequacy. Specific tests are sometimes specified as part of the printed board acceptance procedure to expose a prospective failure situation. Specific environmental tests, as part of the acceptance procedure, are prevalent in high-reliability programs.

Note: It is recommended that a component assembly soldering simulation, consisting of a two-soldering-cycle preconditioning, be performed prior to performing environmental testing. Test information will then relate to actual conditions that the printed wiring board will be subjected to during use. Environmental tests are performed on either actual printed boards or coupon test patterns. This section briefly reviews some of the popular environmental tests. Specific details on performing the tests are referenced in Sec. 29.11.

29.9.1 Thermal Shock

The thermal shock test is particularly efficient in identifying (1) printed board designs with areas of high mechanical stress and (2) the resistance of the printed board to exposure of high and low temperature extremes.

A thermal shock is induced on a printed board by exposure to severe, and rapid, differences in temperature. The test is usually performed by transferring the board rapidly from one temperature extreme (e.g., 125°C) to another (e.g., −65°C), usually within 2 min. Thermal shock effects on the board include cracking of plating in the holes and delamination. See Fig. 29.46 for a typical thermal shock test pattern (see MIL-P-55110 for requirements).

Note: Continuous electrical monitoring during thermal shock cycling will display intermittent electrical circuits. Intermittent circuits usually occur at temperature extremes.

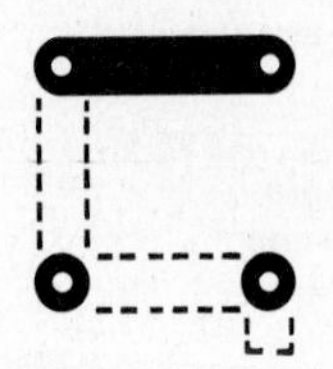

FIG. 29.46 Typical thermal shock and moisture resistance test pattern. *(IPC.)*

29.9.2 Moisture and Insulation Resistance

The moisture resistance test is an accelerated method of testing the printed board for deteriorative effects of high humidity and heat conditions typical of tropical environments. The test conditions are usually a relative humidity of 90 to 98 percent at a temperature of 25 to 65°C. After the required test cycles are completed, the board is subjected to insulation resistance testing. Test specimens should exhibit no blistering, measling, warp, or delamination after the moisture resistance test. See Fig. 29.44 or 29.46 for typical test patterns.

29.10 SUMMARY

To select the printed board attributes to be inspected in an individual program, proceed as follows:

1. Review the environment in which the printed board will operate and the life expectancy of the board.
2. Review the electrical and mechanical parameters associated with functionality.
3. Consider the total assembly unit cost and the importance of the printed board before determining if a quality assurance program is economical.
4. Consider both functionality and economics in quality assurance program selection. Functionality should prevail, however.
5. Design the quality assurance program for at least a 90 percent confidence level by using sampling plans when they are appropriate.
6. Select for inspection the attributes which will satisfy requirements 1 and 2.
7. Select test methods which verify the functionality and integrity of the printed board.

Test methods may require modification or creation to satisfy the quality assurance requirements.

29.11 TEST SPECIFICATIONS AND METHODS RELATED TO PRINTED BOARDS

Attributes	Method
Annular ring	IPC-TM-650, 2.2.1
Base material edge roughness	IPC-TM-650, 2.1.5
Blistering	IPC-TM-650, 2.1.5
Bow and twist	IPC-TM-650, 2.4.22
Chemical cleaning	IPC-TM-650, 2.2.5 and 2.1.1
Conductor spacing	IPC-TM-650, 2.2.2
Conductor width	IPC-TM-650, 2.2.2
Crazing	IPC-TM-650, 2.1.5
Current breakdown, plated-through hole	IPC-TM-650, 2.5.3
Dents	IPC-TM-650, 2.1.5
Dielectric withstanding voltage	IPC-TM-650, 2.5.7; MIL-STD-202, 301
Edge definition	IPC-TM-650, 2.2.3
Etchback	IPC-TM-650, 2.2.5
Eyelets	IPC-TM-650, 2.1.5
Haloing	IPC-TM-650, 2.1.5
Hole size	IPC-TM-650, 2.2.6
Inclusions	IPC-TM-650, 2.1.5
Insulation resistance	IPC-TM-650, 2.5.9, 2.5.10, 2.5.11; MIL-STD-202, 302
Layer-to-layer registration, destructive	IPC-TM-650, 2.2.11
Markings	IPC-TM-650, 2.1.5
Measling	IPC-TM-650, 2.1.5
Microsections, methods of preparing	IPC-TM-650, 2.1.1
Moisture resistance	IPC-TM-650, 2.6.3; MIL-STD-202, 106C
Overhang	IPC-TM-650, 2.2.9
Pinholes	IPC-TM-650, 2.1.5
Pits	IPC-TM-650, 2.1.5
Plating adhesion	IPC-TM-650, 2.4.10
Plating thickness, nondestructive	IPC-TM-650, 2.2.13.1; ASTM-B-567-72
Scratches	IPC-TM-650, 2.1.5
Solderability	IPC-TM-650, 2.4.12 and 2.4.14
Surface roughness	IPC-TM-650, 2.1.5
Bond strength (terminal pull)	IPC-TM-650, 2.4.20 and 2.4.21
Thermal shock	IPC-TM-650, 2.6.7; MIL-STD-202, 107C
Thermal stress	IPC-TM-650, 2.6.8
Voids	IPC-TM-650, 2.1.5
Weave exposure	IPC-TM-650, 2.1.5
Weave texture	IPC-TM-650, 2.1.5

29.12 GENERAL SPECIFICATIONS RELATED TO PRINTED BOARDS

The following printed board specifications are some of those used throughout the industry. The specifications usually cover both processing and acceptance requirements.

IPC specifications	
IPC-FC-240B	Specifications for Flexible Printed Wiring
IPC-FC-250	Specifications for Double Sided Flexible Wiring with Interconnections
IPC-D-300	Printed Circuit Board Dimensions and Tolerances
IPC-D-320	Printed Wiring Board Rigid, Single and Two Sided End Product Specifications
IPC-TC-500	Copper Plated Through Hole Two Sided Boards, Rigid
IPC-TC-510	Procedure for Installing and Inspecting Clinched Wire Type Interfacial Connections in Rigid Printed Circuit Boards
IPC-TC-550	Procedure for the Design, Assembly, and Testing of Fused-in-Place Interfacial Connections in Rigid Printed Circuit Boards
IPC-A-600	Acceptability of Printed Wiring Boards
IPC-TM-650	Test Methods Manual
IPC-ML-910	Design and End Product Specification for Rigid Multilayer Printed Wiring Boards
IPC-ML-950	Performance Specifications for Multilayer Printed Wiring Boards
IPC-ML-990	Performance Specification for Flexible Multilayer
Military specifications	
MIL-Q-9858	Quality Assurance
MIL-P-55110	Printed Wiring
MIL-P-55640	Multilayer Printed Wiring
MIL-P-50884	Flexible Printed Wiring
MIL-STD-105	Sampling Procedure and Tables for Inspection by Attributes
MIL-STD-202	Test Methods for Electronic and Electrical Component Parts
MIL-STD-275	Printed Wiring for Electronic Equipment
MIL-STD-429	Printed Circuit Terms and Definitions
MIL-STD-810	Environmental Test Methods
MIL-STD-1495	Multilayer Printed Wiring Boards for Electronic Equipment
Other publications	
ASTM-B-567-72	Measurement of Coating Thickness by the Beta Backscatter Principle
ASTM-A-226-58	Standard Method of Test for Local Thickness of Electrodeposited Coatings
EIA-RS-326	Solderability of Printed Wiring Boards
IEC No. 326	Performance Specification for Single and Double Sided Printed Wiring Boards
NSA No. 68.8	Multilayer Printed Wiring
UL 796	Printed Wiring Boards

REFERENCES

1. C. R. Draper, *The Production of Printed Circuits and Electronics Assemblies,* Robert Draper Ltd., London, 1969.
2. C. A. Harper, *Handbook of Electronic Packaging,* McGraw-Hill Book Company, New York, 1969.
3. F. W. Kear, "The Design and Manufacture of Printed Circuits," part 14, *PCM-PCE Magazine,* February 1970.

4. IPC-A-600, *Acceptability of Printed Wiring Boards,* Institute for Interconnecting and Packaging Electronic Circuits.
5. IPC-Design Guide, Sec. 9, "Quality Assurance," Institute for Interconnecting and Packaging Electronic Circuits.
6. IPC-TM-650, *Test Methods Manual,* Institute for Interconnecting and Packaging Electronic Circuits.
7. MIL-STD-202, *Test Methods for Electronic and Electrical Component Parts.*
8. J. A. Scarlett, *Printed Circuit Boards for Microelectronics,* Van Nostrand Reinhold Company, New York, 1970.
9. H. R. Shemilt, "Inspection of Printed Circuits," *Electronics Manufacturer Magazine,* vol. 15, 1971.

CHAPTER 30
RELIABILITY

Thomas A. Yager, Ph.D.
Hewlett-Packard Company, Palo Alto, California

30.1 INTRODUCTION

Fabrication of a printed wiring board (PWB) to perform reliably throughout its intended life is becoming an ever-increasing requirement. As board complexity increases, along with the cost of the finished printed wire assembly (PWA), the demand for reliable interconnection and isolation increases. This chapter reviews topics which are essential for understanding PWB reliability. The first is plated-through-hole (PTH) failure due to fracture. PTHs are the Achilles' heel of the PWB, since they consist of thermally incompatible materials and are the basis of multilayer interconnection. The second major cause of PWB failure is the loss of electrical insulation, termed insulation-resistance (IR) failure. Although this hazard can influence all electrical circuits, it is more critical in circuits where high isolation is essential to the performance of the circuit.*

30.2 PLATED-THROUGH-HOLE RELIABILITY

30.2.1 The PTH Failure Theory

Plated-through holes form the basis for two-sided and multilayer interconnection on conventional PWBs. A hole is drilled through the PWB, passing through copper connection pads (lands) on the outer and inner layers and leaving exposed copper for interconnection. After the hole is cleaned up and prepared, copper and/or other metals are deposited in the through-hole, electrically connecting the copper connection pads on all layers. In the final product, the circuit may function properly, but if a PTH is fabricated improperly, or the physical properties of the copper in the hole do not meet certain requirements, a reliability hazard may result. The intention of this section is to illustrate this problem and to describe what measures can be taken to avoid open circuits or intermittent circuits resulting from PTH failure.

An idealized diagram of the cross section of a PTH in a four-layer PWB is shown in Fig. 30.1. Interconnections on the two outer layers and two inner layers

*Throughout this chapter, statistical analysis of failure rates is used. References 1, 2, and 3 at the end of the chapter may be useful for additional understanding and interpretation of the data.

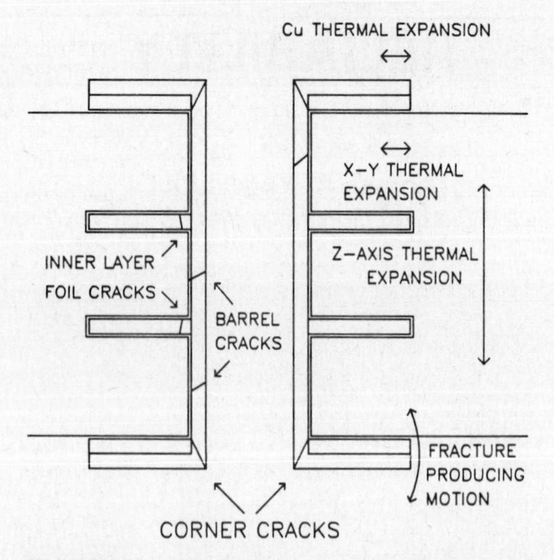

FIG. 30.1 Idealized diagram of a four-layer PTH cross section.

are shown. Because of the thermal expansion mismatch between the copper and the FR-4 epoxy-glass composite, reliability problems can be encountered during thermal stressing of the circuit board. These thermal stresses include wave soldering, rework, or thermal cycling during use or storage.

The problem of differential thermal expansion derives from the fact that thermal expansion of copper is approximately 17×10^{-6}°C from -65 to 260°C, and the thermal expansion of epoxy-glass is 75×10^{-6}°C below 110°C (the glass transition temperature, T_g) and 400×10^{-6}°C above T_g. T_g usually spans a 20 to 30°C temperature range and varies with the type of material and the extent of cure. These relationships are shown graphically in Fig. 30.2. Here the epoxy-glass-free

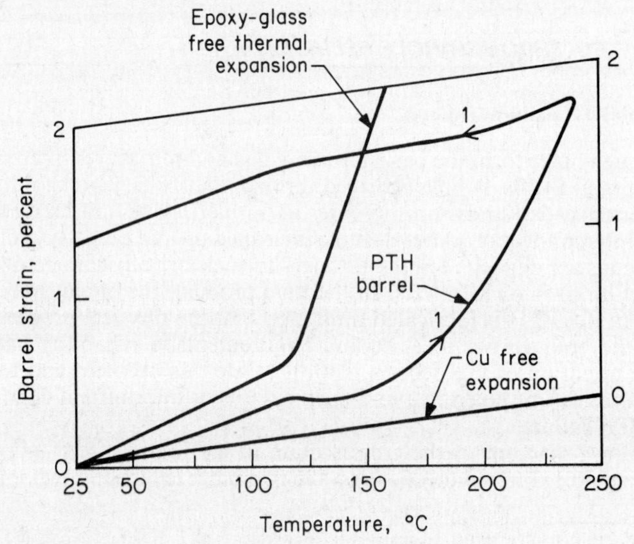

FIG. 30.2 Strain vs. temperature for epoxy-glass PTH barrel and copper.[4]

thermal expansion, the copper-free expansion, and the composite PTH expansion are plotted as strain vs. temperature. The behavior of the PTH at low temperatures is the combined expansion of both the copper and the epoxy-glass. At higher temperatures, the dip in the curve is the result of softening of the epoxy-glass following T_g. Near soldering temperatures, the expansion is the combined thermal expansion of the epoxy-glass and plastic deformation of the copper. A maximum deformation of 1.7 percent has been noted.[5] In the case shown in Fig. 30.2, the cooling portion of the curve is characteristic of plastic deformation and some debonding of the epoxy-glass from the copper barrel. A 1.3 percent residual axial strain occurred only for the initial thermal stress, and all subsequent thermal cycles produced much smaller strains.

Stress and deformation during thermal excursions are the basis for most PTH reliability problems. As illustrated in Fig. 30.1, fractures such as corner cracks, inner-layer cracks, or barrel cracks may result. The mechanisms of cracking have been analyzed both theoretically and experimentally.[6] Experimental values for the stress-strain of plated copper, the thermal expansion of plated copper, the Young's modulus of the laminate, the thermal expansion of the laminate, and the total strain energy of plated copper have been measured. Figure 30.3 shows test results for stress and elongation vs. temperature for acid sulfate copper and pyrophosphate copper. The number of thermal cycles to failure was calculated by dividing the total strain energy required to cause fracture by the strain energy accumulated in one thermal cycle. Figure 30.4 shows a comparison of experimental data for pyrophosphate copper to the calculated curve for the number of thermal cycles to failure vs. temperature. The parameters used for these calculations are:

U_0 = the total strain energy required to cause fracture, J/cm^3

b = the hole radius, mm

c = the distance from the hole center to the free end, mm (or the radius of deformation)

hc = the plated thickness, mm

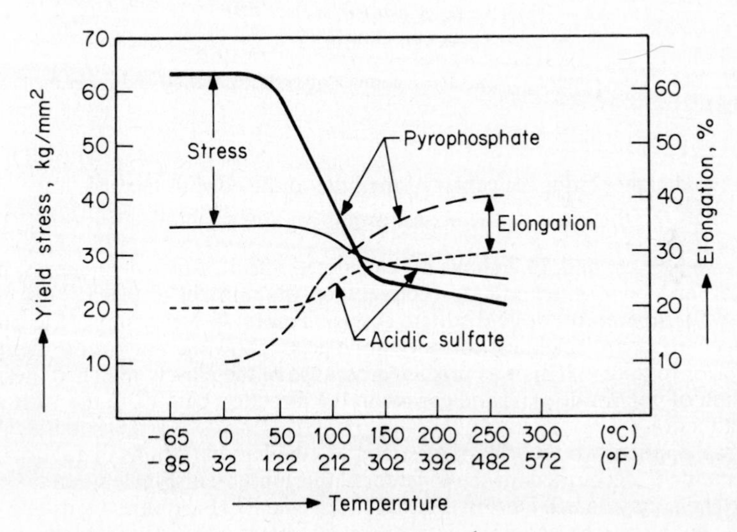

FIG. 30.3 Stress and elongation vs. temperature.[6]

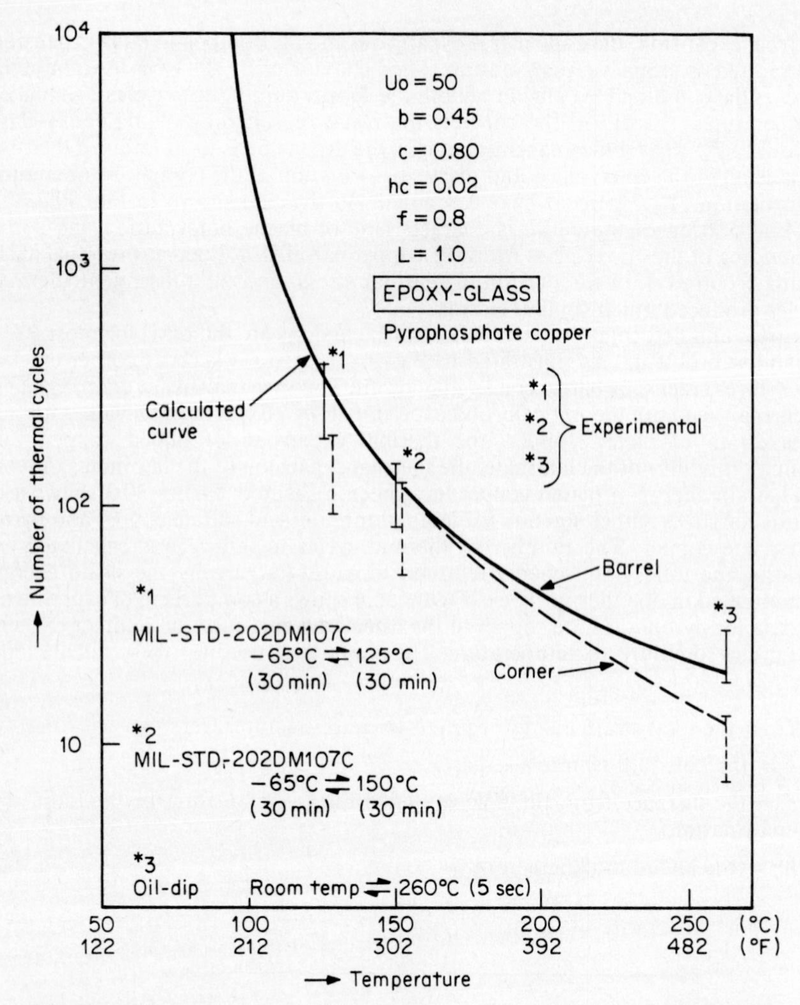

FIG. 30.4 Temperature vs. calculated number of cycles to failure and experimental values.[6]

f = the distance from the center of the hole to the edge of the pad, mm

L = half the thickness of the board, mm

Figures 30.5, 30.6, and 30.7 show the calculated effect of plated thickness, hole diameter, and board thickness, respectively, on the number of thermal cycles to failure vs. temperature for acid sulfate copper. Figure 30.7 also shows the calculated effect of board thickness for polyimide-glass, showing many more thermal cycles prior to failure. This was predicted because of the closely matched thermal expansion of polyimide-glass and copper in the direction parallel to the PTH axis (the *z* direction).

Torres[7] applied two theories of elasticity to evaluate the strains in a PTH. The "force method" accounted for the differential thermal expansion between the copper and the epoxy board. During thermal shock, the PTH acquires a tensile stress and the epoxy board acquires a compressive stress. The "pressure method"

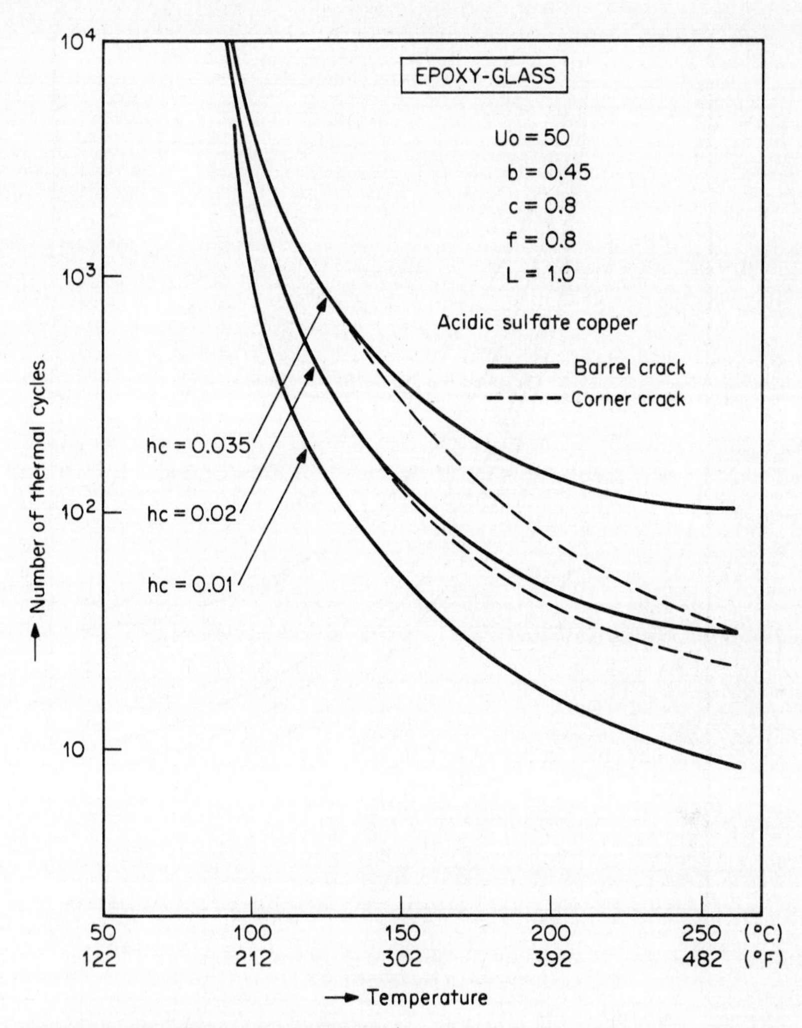

FIG. 30.5 The effect of plated thickness.[6]

extended these calculations to include the effect of the pad on the copper barrel strain. He concluded from this analysis that higher strains would result from:

1. An increase in laminate thickness
2. A decrease in hole diameter
3. A decrease in plating thickness
4. An increase in pad diameter

The maximum tensile stress and the maximum equivalent stress of the copper in the PTH region was analyzed theoretically using finite element analysis by Zakraysek et al.[8] Values for the modulus and thermal expansion for Kevlar*-epoxy

*Kevlar is a trademark of E. I. du Pont de Nemours & Company.

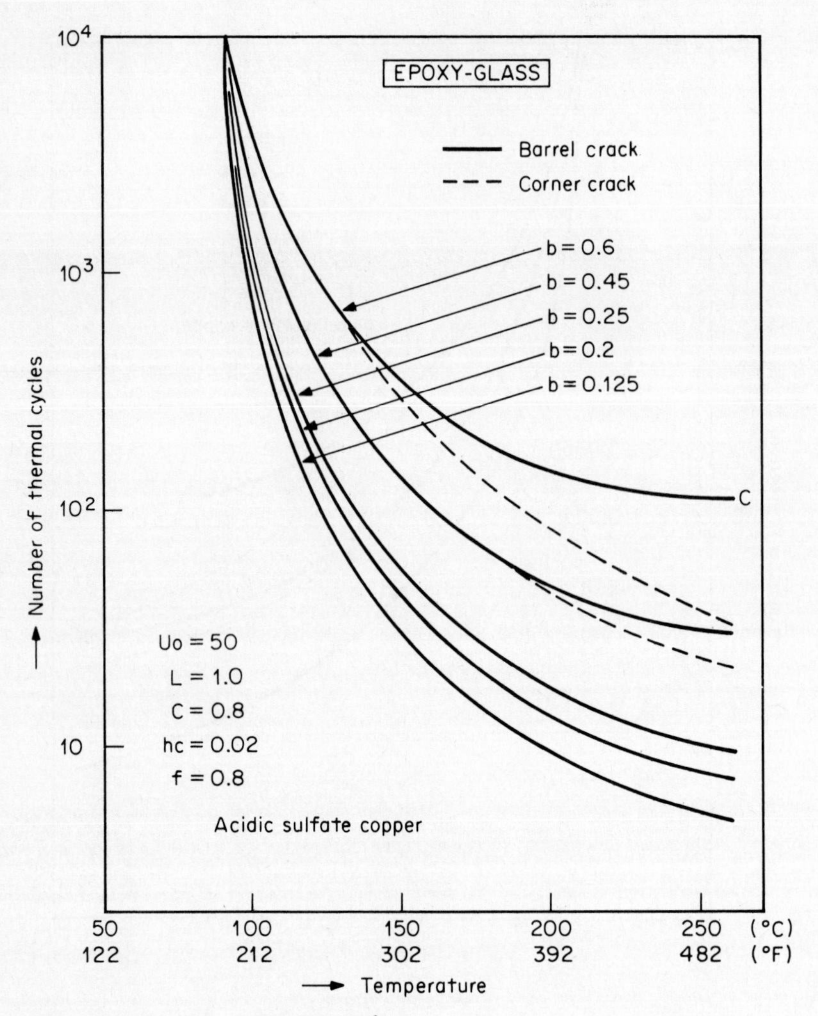

FIG. 30.6 The effect of hole diameter.[6]

composite laminate, with electrodeposited copper, were used for the calculation. Three modes of failure were investigated, involving the interaction between the copper and the laminate with displacement: (1) in the *z* direction, (2) in the lateral direction, and (3) as a result of bending moments. These calculations suggested that corner cracks would be predominant for brittle copper, while barrel cracks would be predominant for ductile copper.

30.2.2 PTH Reliability Testing

Plated-through-hole reliability testing should simulate the thermal excursions of a PTH throughout its life. The first thermal shock is wave soldering. Typically, the board is preheated and exposed to the hot solder wave at temperatures exceeding 250°C for a couple of seconds. If rework is necessary, the PWB may be sent

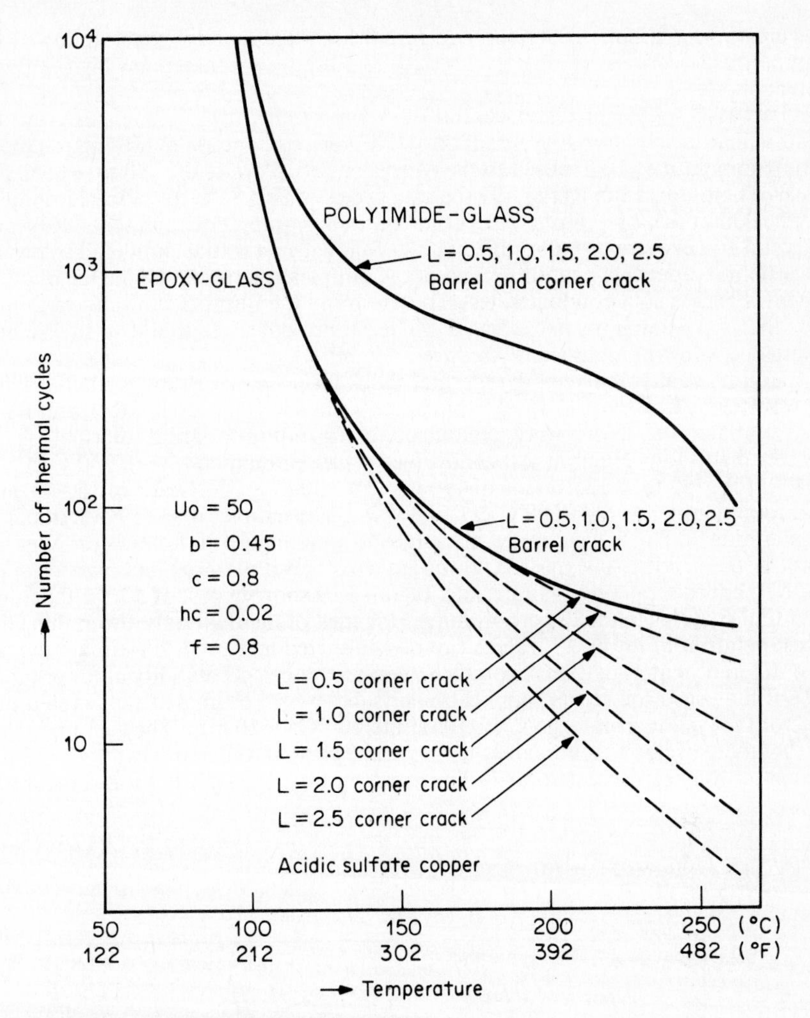

FIG. 30.7 The effect of board thickness.[6]

back through this procedure several times. An even more severe thermal excursion is the hand rework procedure. During hand rework, a PTH is heated to temperatures far exceeding that of wave soldering, while a component lead is pulled from it until freed.

Failure resulting from thermal shock was investigated by Wild,[9] who measured the number of thermal cycles (−62 to +125°C) to failure. The assumption was the formation of an initial fracture followed by crack propagation at lower temperature cycles. He found that unsoldered PTHs exceeded 1300 cycles, soldered PTHs failed at less than 500 cycles, PTHs with copper pins failed at less that 350 cycles, PTHs with nickel pins failed at less than 275 cycles, PTHs which were hand-solder-repaired 10 times failed at less than 150 cycles, and PTHs exposed to three solder-sucking repair procedures failed at less than 75 cycles. Oien[4] suggested that large-amplitude thermal cycles below the T_g will generally stimulate barrel cracking, while large-amplitude cycles above T_g tend to stimulate interface or debonding failures. Therefore, an initial high-temperature shock followed by numerous lower-temperature cycles may be more severe than a few high-temperature shocks.

Farkass[10] suggested that a central office environment may be exposed to the equivalent of 10 thermal cycles of 0 to 85°C per year, and an outdoor equipment environment may be exposed to the equivalent of 25 cycles of −40 to +85°C per year. Therefore, a life test of 400 thermal cycles of 0 to 85°C for office equipment and 1000 cycles of −40 to +85°C for outdoor equipment represents 40 years. Several PWBs were tested in this manner following an initial simulated soldering transient. Under these conditions all of the samples greatly exceed the life requirement. Farkass also conducted tests to determine the number of cycles to failure for thermal excursions exceeding T_g. The failure data was found to fit Weibull statistics, showing significantly reduced reliability.

Honma et al.[11] analyzed the thermal shock resistance of PTHs in double-sided PWBs formed by the CC-4* additive process on XXXPC, FR-2, FR-3, and FR-4 PWB materials. Tests were conducted using liquid-to-liquid thermal shock according to four different test conditions: (1) room temperature to 260°C (ΔT of 240°), (2) −65 to +125°C (ΔT of 190°C), (3) −30 to +100°C (ΔT of 130°C), and (4) room temperature to 100°C (ΔT of 80°C). Eight samples of each PWB material containing 50 PTHs were tested at each condition. Figure 30.8 shows the Weibull plot of the results. As expected, lower grades of PWB material and higher values of ΔT show an increased failure rate. Figure 30.9 shows a plot of ΔT vs. the number of test cycles to 0.025 percent failure for each of the materials. From this plot, extrapolation to lower ΔT values can be conducted and used to estimate the life for different materials. This analysis helped prove the acceptability of lower-grade PWB materials for certain applications. Isobe et al.[12] estimated the accelerating factor for cycling from −65°C for 10 min to 100°C for 10 min. They assumed that 2.7 of these cycles are equivalent to one year in actual use. Samples surviving 100 cycles of this test were assumed to be capable of high reliability for more than 30 years.

*CC-4 is a registered trademark of Kollmorgen Corporation.

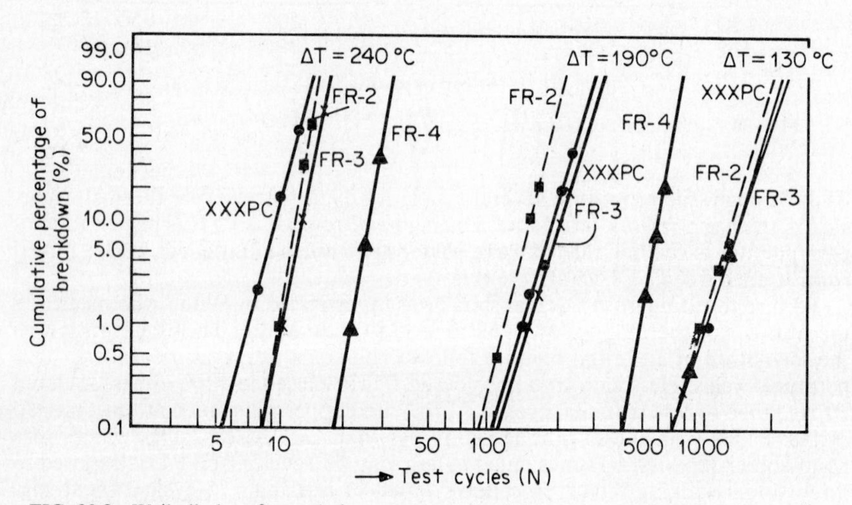

FIG. 30.8 Weibull plot of cumulative percentage breakdowns vs. number of test cycles.[11]

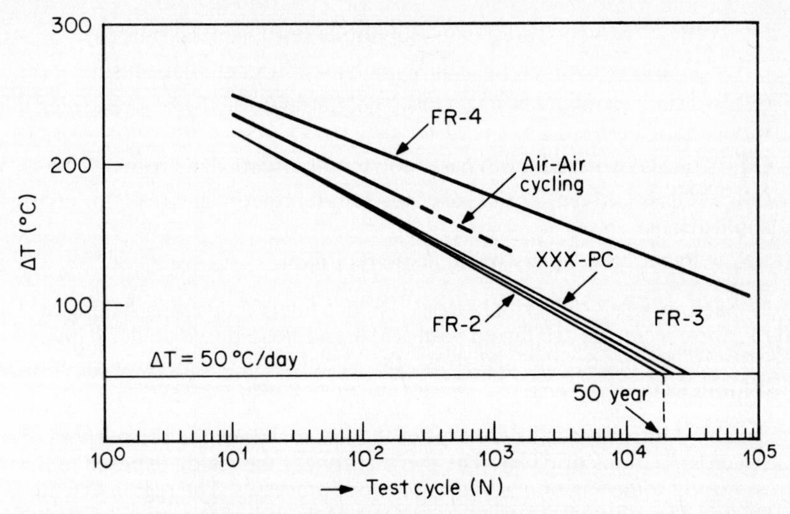

FIG. 30.9 ΔT vs. number of test cycles.[11] The FR-4 curve represents results for FR-4 material tested by liquid-to-liquid cycling. The air-air cycling curve represents results for all paper-base materials tested by air-air cycling. The bottom curves represent results for paper-base materials tested by liquid-to-liquid cycling.

30.2.3 Test Specifications

The military specification which applies to testing the thermal shock of PTHs in printed wiring boards is MIL-P-55110 (Military Specification for Printed Wiring Boards). The test conditions are the following:

1. Section 4.8.6, "Thermal Shock." Depending on laminate type, the specimens are tested using the following conditions:

Low temperature, °C	Time, min	High temperature, °C	Time, min	Type
−65	15	105	15	PX
−65	15	125	15	GE, GF
−65	15	150	15	GB, GH, GP, GT, GX, GR, FEP
−65	15	204	15	GI

The samples are then electrically tested for continuity and shorts. The military designations for laminates are as follows:

FEP = Fluorocarbon unfilled

GB = Glass (woven-fabric) base, epoxy resin, heat-resistant

GE = Glass (woven-fabric) base, epoxy resin, flame-retardant

GH = Glass (woven-fabric) base, epoxy resin, heat-resistant and flame-retardant

GI = Glass (woven-fabric) base, polyimide resin, general purpose

GP = Glass (nonwoven-fiber) base, polytetrafluoroethylene resin

GR = Glass (nonwoven-fiber) base, polytetrafluoroethylene resin, for microwave application

GT = Glass (woven-fabric) base, polytetrafluoroethylene resin

GX = Glass (woven-fabric) base, polytetrafluoroethylene resin for microwave applications

PX = Paper base, epoxy resin, flame-retardant

2. Section 4.8.7, "Thermal stress" (solder float). Following baking at 250 to 300°F, the specimens are fluxed with RMA and floated in a solder bath at 550°F for 10 s. Following the thermal stress, the samples are cross-sectioned and examined for defects.

Smith[13,14] introduced the concept of the thermal zone for the purpose of evaluating cross sections of PTHs. The thermal zone is the region exposed to the most stress during soldering or desoldering-rework processes. The region extends 0.003 in beyond the ends of the farthest pad on each end of the microsection. Smith suggested that defects that are produced as a result of thermal stress are acceptable only within the thermal zone. Additionally, within the thermal zone, some copper cracks are acceptable so long as they do not interfere with functional circuitry.

30.2.4 Test Coupon

Several types of test coupons exist for testing the reliability of PTHs. One coupon in use is shown in Fig. 30.10. This coupon contains 3000 PTHs interconnected in series on four layers. Three different PTH sizes were incorporated (0.052-, 0.028-, and 0.018-in drill size) on a 0.1-in grid. Of these, one third had a pad width of 0.005 in, one third a pad width of 0.01 in, and one third a pad width of 0.015 in. Additionally, the drill pattern was specified to determine the effect of the number of drill hits on reliability. This test coupon has proved to be useful to evaluate several process parameters. It is often rare that PTHs have identical plating distributions, wall roughness, epoxy smear, plating ductility, etc., from day to day. Efforts should be made to incorporate this test coupon as a process control.

Electrical testing of PTH resistance is typically conducted using four point resistance measurements. When several PTHs are connected in series, small resistance changes of one PTH are often undetected. Rudy[15] studied the four point resistance measurements on single PTHs to detect the onset of cracking. Using this technique, he was able to evaluate flaws in PTHs nondestructively prior to catastrophic failure.

30.3 FABRICATION PROCESSES THAT INFLUENCE PTH RELIABILITY

Several studies[16] have been conducted to relate processes to PTH reliability. These studies suggest the following corrective actions:

1. Use nonfunctional pads wherever possible.
2. Use 2-oz copper on inner layers.
3. Use ductile copper.

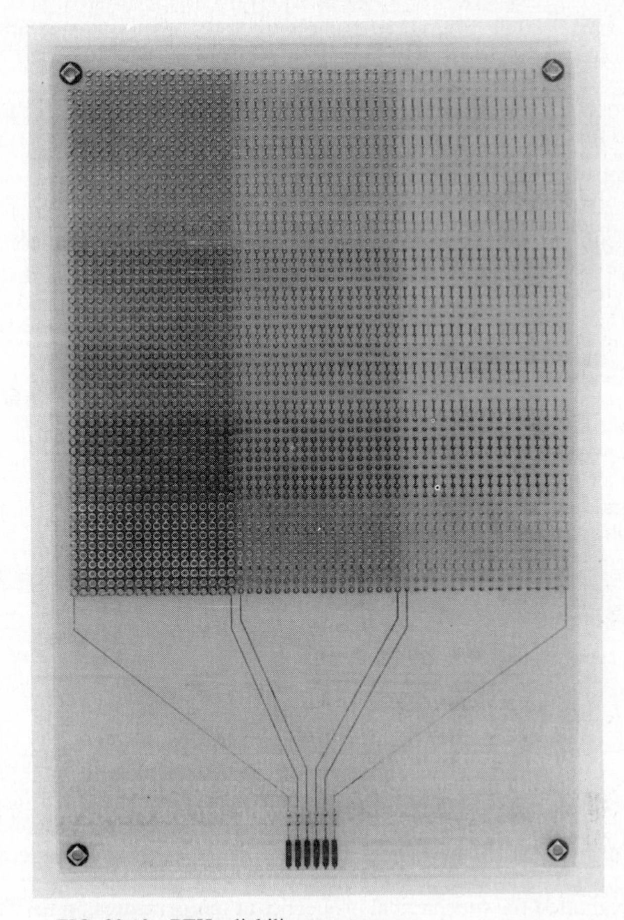

FIG. 30.10 PTH reliability test coupon.

4. Eliminate the use of small holes.
5. Minimize etchback.
6. Bake laminates prior to processing.
7. Cure after lamination.
8. Bake after drill and/or etchback.
9. Increase plating thickness.

30.3.1 Electroless Deposition

Although etchback should be minimized, drill smear should be sufficiently removed to ensure a good metallurgical contact between the inner layers and the PTH barrel. Bartlett et al.[17] investigated the influence of etching and accelerator treatment steps of the electroless process used to provide the metallurgical bond between the original copper foil of the inner layers and electroplated copper in the PTH barrel. The initial electroless deposit must satisfy three specific requirements:

1. The copper must uniformly cover the dielectric materials, both epoxy and glass, in the hole.
2. The deposit must be sufficiently thick to withstand subsequent cleaning and plating operations.
3. The electroless copper must adequately bond to the inner layers so that the interface between the original copper and the plated copper does not crack or break under stresses induced by soldering operations.

Using the ultimate tensile strength test, they found that:

1. The copper foil–plated copper bond strength can exceed the ultimate tensile strength of the copper foil.
2. Fracture occurred between the copper foil and the catalyst.
3. Short residence times in the copper etchant prior to electroless plating lead to a deterioration in the bond strength.
4. The bond strength is reduced if the accelerator step is not used in the electroless plating process.

30.3.2 Electroplated Copper

Broache and Poch[18] found that copper deposited from a cyanide bath was stronger and more deductile than other commercially available baths such as fluoroborate or acid baths. The drawback of the cyanide bath was its chemical attack on epoxy and resist materials. To overcome this problem, an initial envelope of 0.0002 to 0.0003 in of pyrophosphate copper followed by 0.001 in of cyanide copper produced the most crack-resistant PTH.

Fox[19] evaluated the mechanical properties at temperatures up to 250°C of electroplated copper for PWB applications and wrought, annealed, electrolytic test pitch copper. Figure 30.11 shows the stress-strain curves for these materials at room temperature. The electroplated copper had lower ductility than wrought, annealed copper, but the ductility was believed to be adequate for PWB applications. The electroplated copper was found to strain-harden more rapidly that wrought annealed copper and had a higher tensile strength. Mayer and Barbieri[20] evaluated the influence of three types of additives on the thermal shock resistance of electrodeposits of acid copper. The additives were as follows:

1. A carrying agent, to guide the other two components to create the equiaxed structure. (Striations occurred with insufficient carrying agent.)
2. A leveling agent, to smooth over surface imperfections. (Without the leveling agent, imperfections are reproduced in the deposit.)
3. A ductility-promoting agent, which functions to produce the equiaxed grain structure.

Additive levels below a certain level make the bath more susceptible to impurity effects. Iron contamination of 100 mg/L without the recommended concentration

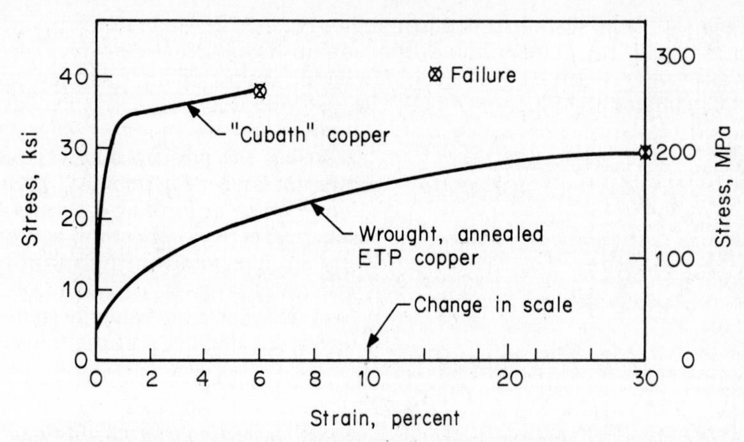

FIG. 30.11 Stress-strain curves for plated and wrought, annealed, electrolytic test pitch (ETP) copper. *(A. Fox, Journal of Testing and Evaluation, 1981, pp. 74–84.)*

of ductility-promoting agents will produce a columnar grain structure at the hole corners. Similarly, organic contaminants such as photoresist can produce laminar deposits. The influence of photoresist and plating resists on copper ductility was indicated by Isobe et al.[12] to be one justification for using panel-plating processes. Panel-plating processes do not use plating-resist materials and therefore reduce plating-bath contamination. Sherlin and Bjelland[21] found that higher concentrations of additives in pyrophosphate copper and acid copper reduce the recrystallization temperature of the copper deposit. Baking of the deposit above the recrystallization temperature caused a decrease in the ultimate tensile strength, hardness, and electrical resistivity and an increase in elongation.

30.3.3 Copper Deposit Ductility

Paul[22] found the major cause of inner-layer foil cracks to be due to the larger variation in ductility of foils obtained from various vendors. The ductility variation was correlated to the microstructure observed in metallographic cross sections. Ductility effects were found to influence PTH reliability more than plating thickness and depth of etchback. A minimum of 8 percent elongation is required for 1-oz foil to eliminate the problem.

Ogden and Tench[23] evaluated the influence of additive concentration, annealing treatments, current density, pH, ammonia concentration, and contaminants on the elongation and tensile strength of pyrophosphate copper electrodeposits, as well as the influence of additive concentration, chloride level, current density, and copper sulfate concentration on sulfate copper electrodeposits. Pilot lines were monitored to optimize the electroplating conditions. Techniques such as cyclic voltametric stripping (CVS) analysis and rotating cylinder (RC) tensile test were used to control the bath composition and control deposit properties.

Copper foil ductility up to 300°C has been measured by Zakraysek et al.[8] using the hot-bulge test method. A sample 8 cm in diameter is caused to bulge under gas pressure, at varying test temperatures, until rupture occurs. The bulge height and burst pressure can be used to quantitatively evaluate the ductility of the foil. Foils plated in a Hull cell can be easily evaluted for room temperature ductility using the 180° bend test. This is illustrated in Fig. 30.12, where the sample panel is bent flat with the bend running from low to high current density. Fractures occur at current densities producing nonductile copper. This technique can be used to evaluate the influence of bath chemistry on ductility, and it can also be used as a bath monitor. Zakraysek[24] describes the hot-cup ductility test to monitor copper-foil ductility. The technique involves forming 6.3-mm hemispherical cups in copper foil at progressively increasing temperatures. A visual examination is used to observe fracture, which indicates the temperature of the ductile-to-brittle transition.

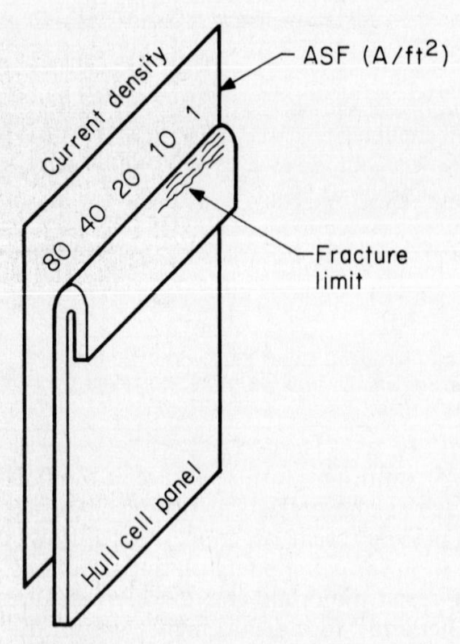

FIG. 30.12 Ladwig panel ductility test.[8]

30.4 INSULATION RESISTANCE RELIABILITY

The primary function of a printed wiring board is to provide either low impedance or high impedence where needed, thus forming the basis for electrical interconnection. Reliability problems are encountered when there is a degradation of either (opens or shorts). For the case of PTH thermal shock resistance, the low impedance of the circuit needs to be retained. The fracture which may form often produces an abrupt change from low to high impedance. For the case of insulation resistance failure, the change in impedance is from high impedance to low impedance, and it often (but not always) occurs slowly over a long period of time. Insulation resistance deterioration is particularly harmful for the case of analog measurement circuits. If these circuits are used to measure low-voltage, high-impedance sources, changes in circuit impedance can result in the deterioration of the instrument's performance. Medical products are another area of concern. When sensors are attached to a patient, any deterioration of insulation resistance has the potential of electrical shock. In this review, only the effects of direct current or low-frequency alternating current will be discussed.

30.4.1 Leakage Paths

Unwanted current paths on PWBs (often termed "current leakage") can occur either on the surface or through the bulk of the board. Leakage in the bulk can occur through the polymer material at glass-to-polymer or polymer-to-polymer interfaces. The current leakage (I) between two conductor traces is made up of several contributers: I (trace-to-trace) $=$ I (material) $+$ I (contamination) $+$ I (moisture) $+$ I (metal migration), where I (trace-to-trace) $=$ applied voltage/R (trace-to-trace).

R (material) is a function of the intrinsic resistivity of the materials used. The extent of cure of the polymer, the porosity, the temperature, the humidity, or the amount of free ions may have a large influence on R (material).

R (contamination) is due to leakage paths formed from conductive contaminants. The contamination may be metallic, such as metal slivers or solder bridges; or the contamination may be residues from PWB processing, soldering, fingerprints, or dust. Ionic residues become activated in the presence of moisture, and because of their hygroscopic nature, they will absorb moisture from the atmosphere, which may lead to the production of a conductive solution at a sufficiently high humidity.

R (moisture), as discussed, is strongly influenced by contamination. Leakage may be across the surface, in the bulk, or at interfaces.

R (material) is strongly influenced by the absorption of moisture into the polymer or into pores. It is reasonable to assume that 90 percent of insulation resistance problems can be blamed on moisture. Figure 30.13 shows a plot of surface resistivity vs. relative humidity at 25 and 80°C,[25] illustrating the strong influence of humidity. For example, at 25°C a relative humidity (RH) of 30 percent results in a surface resistivity of 10^{15}, while at 90 percent RH a surface resistivity of less than 10^{11} results. Thus in typical operating conditions a variation of four orders of magnitude in surface resistivity may occur.

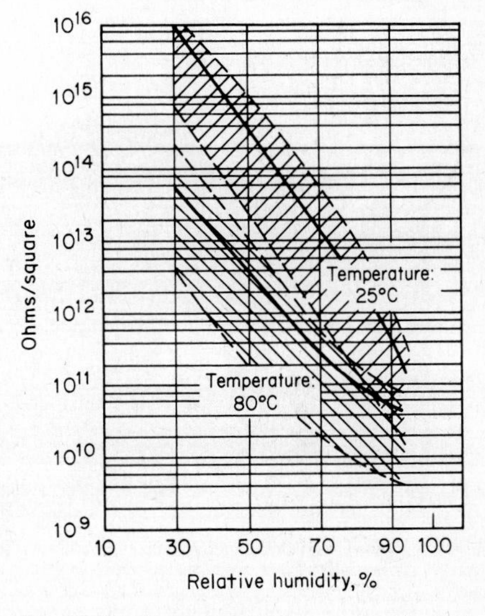

FIG. 30.13 Surface resistivity as a function of relative humidity at 25 and 80°C.[25]

R (metal migration) is typically an aging effect and may lead to a short circuit between conductors. One form of metal migration is known as whisker growth. This phenomenon has plagued tin platings for many years,[26] although it is known to influence other metal deposits.[27] Whiskers are typically single-crystal metal growths a few microns in diameter which may extend several millimeters. Their growth has been correlated to environmental conditions, the stress of the metal deposit, surface roughness, and coating thickness. Ways to minimize this effect include using tin alloys, thicker coatings, or polymer coatings over the metal deposit, or reducing the stress by melting or reflowing the tin.

Another form of metal migration is the result of corrosion. Corrosive byproducts, such as chlorides or sulfides formed in industrial environments, tend to migrate across both metallic and insulating surfaces. Figure 30.14 shows the migration of corrosion products across the surface of FR-4,[28] bridging two conductors. Although this material is somewhat insulating, it may absorb moisture to form leakage paths.

A third form of metal migration is the result of electromigraton. Under a dc potential this process involves the electrolytic transfer of metal ions formed on the anode to the cathode. A metal dendrite forms on the cathode where the metal ion is reduced. Eventually, the dendrite will bridge the anode and cathode, causing deterioration of the insulation resistance. This process occurs on surfaces, in pores in materials, or at interfaces. Figure 30.15 is a transmitted light photograph through an FR-4 PWB that has failed in the field because of insulation resistance deterioration. A dendrite has formed at the interface of a UV-cured screened sol-

FIG. 30.14 Migration of corrosion products across the surface of FR-4.[28]

der mask and the FR-4 surface, briding the cathode at the left with the anode at the right. Figure 30.16 is a transmitted light micrograph through a multilayer PWB showing metal migration occurring within the FR-4 material. This board was exposed to 85°C, 85 percent RH, and 100 V dc bias for 1000 h. The bend of the migration at right angles suggests the migration along the glass fibers.

30.4.2 Dendrite Growth on Printed Wiring Boards

A model for dendrite growth within FR-4 material was proposed by Lando et al.[29] They described two steps: (1) degradation of the resin-glass bond and (2) electrochemical reaction. Degradation at the resin-glass interface provided a path along which electrodeposition could occur. This was attributed to poor glass treatment,

FIG. 30.15 Transmitted light micrograph through a PWB that has failed in the field. Dendritic growth has formed at the interface of a UV-cured screened solder mask and the FR-4 surface.

hydrolysis of the silane glass finish, or the release of mechanical stress. It was proposed that the absorption of moisture produced an electrochemical cell with the following possible electrode reactions:

At the anode:

$$Cu \rightarrow Cu^{n+} + ne^-$$
$$H_2O \rightarrow \tfrac{1}{2}O_2 + 2H^+ + 2e^-$$

At the cathode:

$$H_2O + e^- \rightarrow \tfrac{1}{2}H_2 + 2OH^-$$

where the majority of the leakage was attributed to the electrolysis of water. These equations also indicate that a pH gradient would form from the acidic anode to the basic cathode. Copper metal is dissolved at the anode and migrates to the cathode, where it is no longer soluble. The dendrite that forms follows the pH

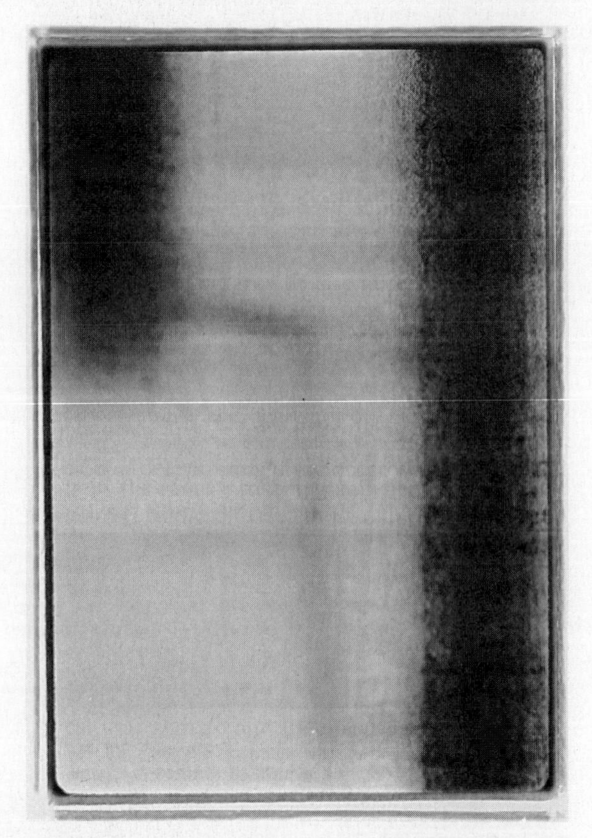

FIG. 30.16 Transmitted light micrograph through a multilayer PWB showing metal migration along glass fibers.

concentration gradient. A failure model was developed based on the existence of a debonding region at the resin-glass interface, which provides a path for electrical connection between conductors.

30.4.3 Time to Failure

Mechanisms accelerating insulation resistance failures of FR-4-based flexible PWBs coated with a photocuring polyblend acrylate-thioether resin (solder mask) were investigated by Boddy et al.[30] A plot of log (insulation resistance) vs. time at 85°C, 80 percent RH, and 78 V is shown in Fig. 30.17. For most test samples, failure was associated with an abrupt drop in resistance. Other samples were not as abrupt, but deviation from the main sample population was taken as failure. Failure prediction was conducted by fitting the data for time-to-failure to log-normal statistics, and it is shown in Fig. 30.18. From the log-normal plot, the median time to failure represented the 50th percentile. This type of analysis was conducted at different humidities, voltages, and temperatures to obtain the following equations of mean time to failure (t_{FM}):

$$t_{FM}(85°C,\ 78\ V) = 600\ e^{-0.068\ RH}\ (days)$$
$$t_{FM}\ (65°C,\ 80\%\ RH) = 60\ e^{-V/80}\ (days)$$

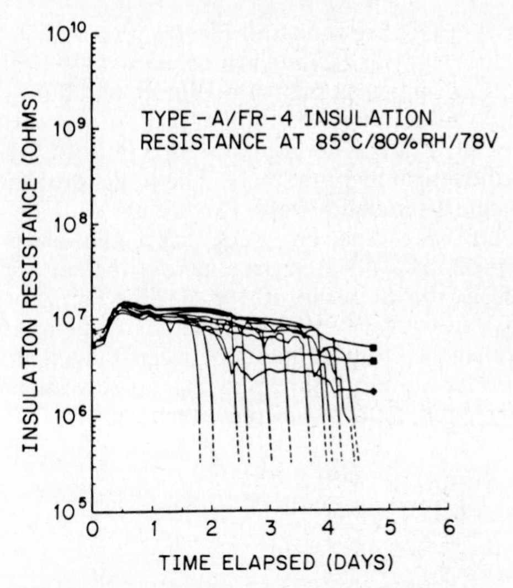

FIG. 30.17 A log plot of insulation resistance vs. time for FR-4 PWBs.[30]

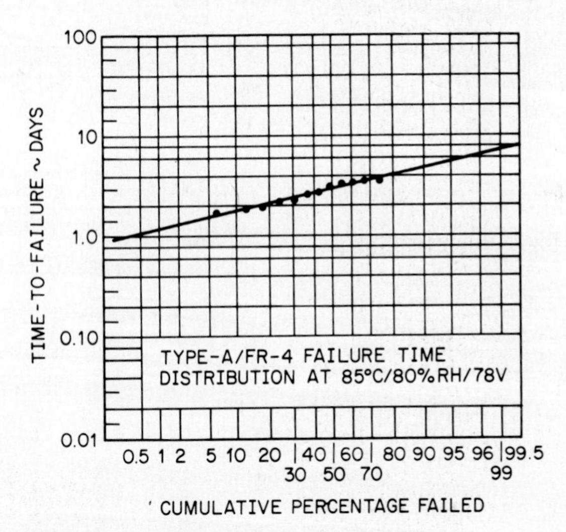

FIG. 30.18 A log-normal plot of time to failure vs. cumulative percent failed.[30]

30.4.4 Insulation Resistance Failure Mechanisms

The failure mechanism was found to follow temperature-activated Arrhenius curves with an activation energy of 0.41 eV above and 1.41 eV below 75°C (80 percent RH). Extrapolation of this failure data to 25 to 35°C and 40 to 60 percent RH suggested that these boards had a medial life measured in tens of years for voltages below 100 V in a typical air-conditioned equipment environment.

Boddy et al.[30] also analyzed the failure modes in the samples they tested. Diagrams of these failure modes are shown in Fig. 30.19(*a–e*). The through-substrate short shown in Fig. 30.19*a* was believed to be caused by local separation of the epoxy and glass reinforcement. At high temperature and humidity, contact of the glass fibers with the conductors made the PWB more susceptible to this type of failure. A subsurface substrate short is represented in Fig. 30.19*b*, where conductive material fills voids around glass fibers. The voids propagate along the fiber bundles and finally short the conductors. Cover-coat or solder-mask failures are represented by Fig. 30.19*c–e*. If the cover coat has a high sensitivity, the dendrites grow from the cathode toward the surface, across the surface, and through the coating to the anode, as shown in Fig. 30.19*c*. At high temperature and humidity, this may eventually lead to charring of the cover coat and failure as shown in Fig. 30.19*d*. The entrapment of contaminants between the substrate and the cover coat leads to the interfacial failure shown in Fig. 30.20*c*. This is characterized by delamination, blisters, and dendrites at the interface.

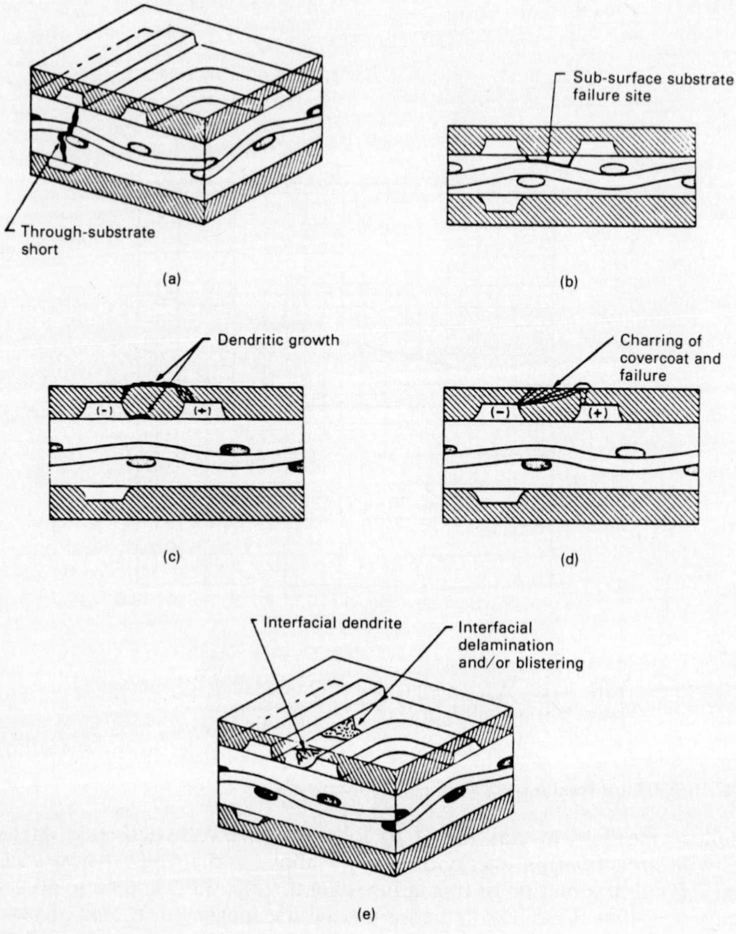

FIG. 30.19 Insulation resistance failure modes in PWBs.[30]

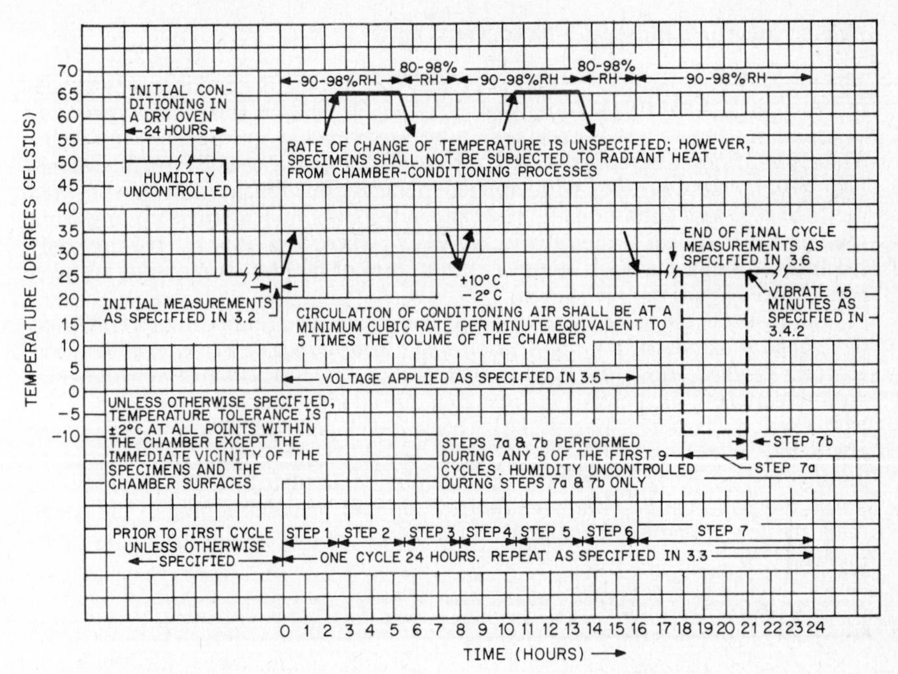

FIG. 30.20 MIL-STD-202, Method 106, Preconditioning and Thermal Cycling.

30.4.5 Contamination Effects

The deterioration of the electrical performance of a printed circuit may be due to the presence of contaminating conductive solids or solutions. Conductive solids include metal slivers, dust, or dendrites. As indicated above, dendrites often grow as the result of current passing through a conductive solution containing ionic contaminants. These contaminants may originate from residues following fabrication or assembly processes. At low humidities, ionic contaminants are inactive and do not strongly influence the electrical performance. At higher humidities, moisture is absorbed resulting in the formation of conductive solutions. Polymers are nonhermetic and permit water vapor to permeate. Condensation within a polymeric material can occur to form conductive paths if a contaminant is present within a polymer, laminated within a PWB, or is present at the interface of materials.

Correlation between contamination content of the PWB and leakage current has been attempted with little success. This is attributed to the distribution of the contaminant, the degree to which the ionic solution is capable of bridging conductors, and the secondary effect of metal migration. Der Marderosian and Murphy[31] found a humidity threshold for surface contaminants below which no migration occurs. The relative humidity thresholds for KCl, NaCl, and KI were found to be 65, 52, and 44 percent, respectively. As the amount of contaminant or the bias voltage was reduced, the time to form dendrites was reduced. At low humidities, Ulbricht[25] found low contamination levels. At humidities approaching 100 percent, the difference became more pronounced, but the effect of humidity was more significant than contamination.

30.4.6 Insulation Resistance Reliability Testing

The military specification of the test procedure for moisture and insulation resistance is MIL-P-55110 *(Military Specification for Printed-Wiring Boards)*. The moisture resistance test is specified to be performed in accordance with MIL-STD-202 *(Military Standard Test Methods for Electronic and Electrical Component Parts)*, method 106, with applied polarization voltage (100 V dc), and method 302, test condition A. The preconditioning and thermal cycling from method 106 is shown graphically in Fig. 30.20, where MIL-P-55110 requires only the first six steps, cycled 10 times. A minimum of 500 MΩ is required between conductors.

The Institute for Interconnecting and Packaging Electronic Circuits (IPC) has developed a standard in *Qualification and Performance of Permanent Polymer Coating (Solder Mask) for Printed Boards* (see IPC-SM-840A). Environmental requirements of this standard include moisture and insulation resistance and electromigration. Two test boards have been developed specifically to test these attributes, as shown in Fig. 30.21. The IPC-B-25 test board "B" is used to qualify the process, while the "Y" coupon is designed to be incorporated on production boards. Following preparation and preconditioning of the specimen, the test coupons are exposed to the following test conditions and requirements:

Class	Temperature	Humidity	Time	Bias voltage	Minimum requirement
1	35°C	90+% RH static	4 days	100 V dc	100 MΩ
2	50°C	90+% RH static	7 days	100 V dc	100 MΩ
3	25–65°C	90+% RH cycling	7 days	100 V dc	500 MΩ

The "Y" coupon uses the parallel trace pattern. Every 24 h the pattern is electrified at 500 V dc for 1 min, followed by the resistance measurement at 500 V dc. For Class 3 the measurements are always made at 65°C.

A test for the resistance to electromigration of polymer solder mask is also contained in IPC-SM-840A. Following preparation, either the IPC-B-25 test board "B" or the "Y" coupon is exposed for 7 days to test conditions of 85°C and 90 percent RH and biased at 10 V dc with a limiting current of 1 mA. The current is recorded before and after chamber conditioning to note any change. The samples are also microscopically examined for electromigration or other causes of failure.

The material parameters used to specify insulation resistance are bulk (ohm-centimeters) or surface (ohms per square) resistivity.[32] In theory, once these parameters are known, the trace-to-trace, layer-to-layer, hole-to-hole, or signal line–to–signal line resistance can be calculated. This is complicated by the complex interaction of surface resistance, bulk resistance, and the nonisotropic or homogeneous nature of printed circuit materials. For comparative purposes, the data collected from the IPC-B-25 test board or the "Y" coupon can be converted from measured resistance (in ohms) to surface resistivity (in ohms per square) by multiplying the measured resistance by the square count of the pattern. The square count is determined geometrically by measuring the total length of the parallel traces between the anode and cathode and dividing by the separation distance. Values for these patterns are given below:

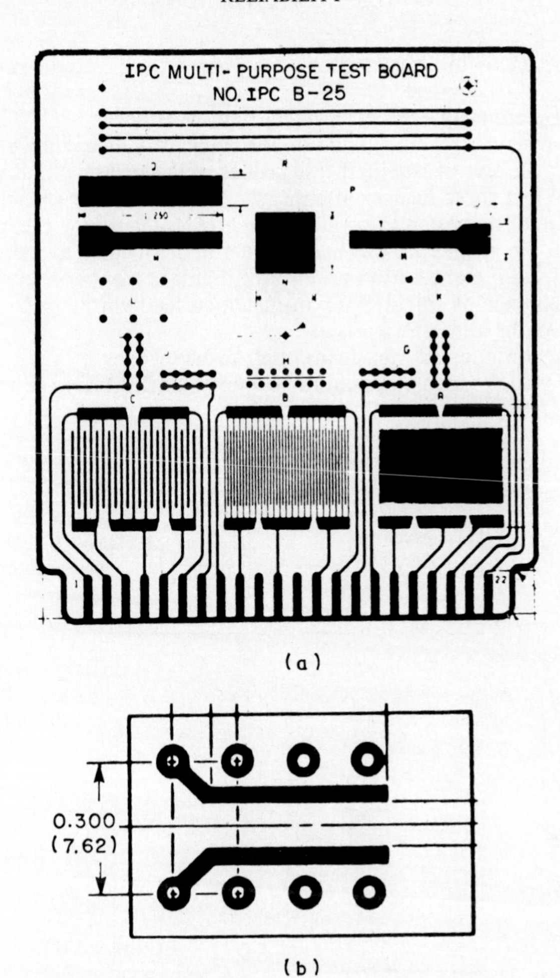

FIG. 30.21 (*a*) the IPC-B-25 test board; (*b*) the "Y" coupon for insulation resistance testing. *(IPC-SM-840 A, Institute for Interconnecting and Packaging Electronic Circuits, Evanston, Ill., 1983.)*

Specimen	Spacing width, in	Pattern length	Number of squares
IPC-B-25 pattern A and D	0.0065	50	7692
IPC-B-25 pattern B and E	0.0125	25	2000
IPC-B-25 pattern C and F	0.025	12.5	500
"Y" coupon	0.025	1.0	40

Special precautions are needed to make accurate measurements of insulation resistance. See ASTM D 257–78, *Standard Test Methods for DC Resistance or Conductance of Insulating Materials.* Measurements of resistance below 10^{12} are very difficult and require careful shielding. Measurements of resistance below 10^{12}

can be conducted in most laboratory environments if certain precautions are taken.

The simplest technique for measuring high resistances on PWBs is with the use of a high-resistance meter (such as the Hewlett-Packard 4329A, or equivalent). Through the use of shielded test probes, resistances as high as $2 \times 10^{16}\ \Omega$ are possible. When these measurements are conducted, stray capacitance or electromagnetic interference can introduce large errors. Therefore, the best results are obtained when the probes are not hand-held and are located away from electrical equipment. Measurement within an environmental chamber requires additional precautions. The use of shielded teflon-insulated lead wire is necessary to avoid faulty leakage paths and interference.

A second technique for measuring high resistances is by measuring current. This technique can be implemented in two ways, as illustrated in Fig. 30.22. Figure 30.22*a* uses a constant-voltage power supply (of voltage V_P), a current-limiting resistor (of resistance R_1), and a current meter capable of measuring picoamps. The resistance of the test sample (R_s) can be calculated by:

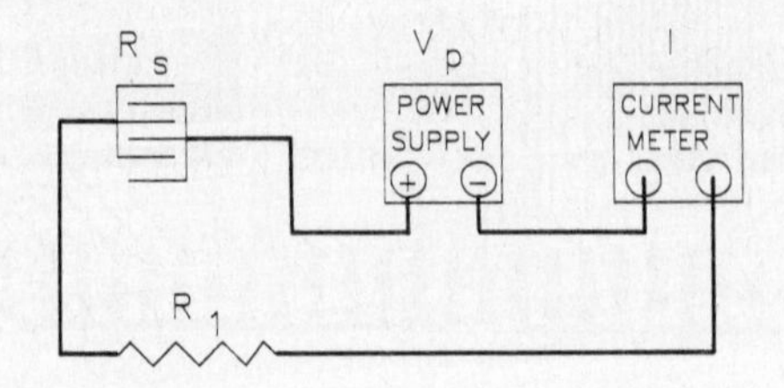

a) HIGH RESISTANCE MEASUREMENTS USING A CURRENT METER

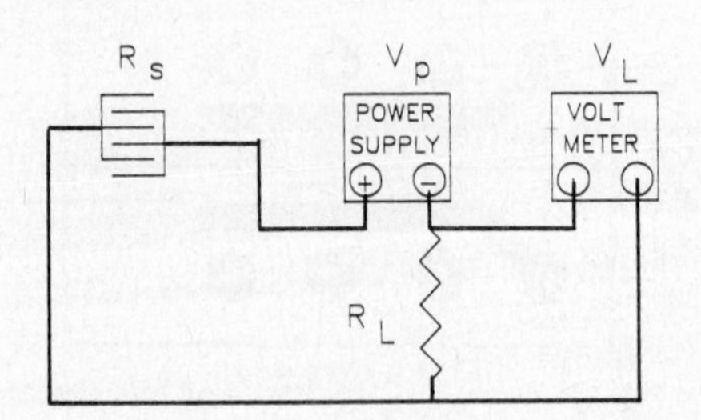

b) HIGH RESISTANCE MEASUREMENTS USING A VOLT METER

FIG. 30.22 High-resistance measurements by measuring current.

$$R_s = V_P/I$$

where I is the measured current. This equation is valid so long as the $R_1 \gg R_s$. For additional accuracy and elimination of errors due to thermal emf's generated at junctions of dissimilar metals, the current should be measured several times with the voltage in the forward bias and then remeasured several times with the voltage in the reverse bias. The corrected value for resistance is calculated by:

$$R_s = 2\ V_p/(\bar{I}_F - \bar{I}_R)$$

where $\bar{I}_F$ and $\bar{I}_R$ are the average measured currents with the voltage in forward and reverse bias, respectively. Care must be taken to allow sufficient time for the measured current to stabilize when reversing polarity.

A second way of measuring current is illustrated in Fig. 30.22*b*. This technique uses a voltmeter to measure the voltage across a resistor in series with the test pattern. The corrected value of resistance is calculated by:

$$R_s = (\overline{V}_F - \overline{V}_R - \overline{V}_{LF} + \overline{V}_{LR})R_L/(\overline{V}_{LF} - \overline{V}_{LR})$$

where $\overline{V}_F$ and $\overline{V}_R$ are the average power supply voltages in the forward and reverse direction, $\overline{V}_{LF}$ and $\overline{V}_{LR}$ are the average measured voltage across the current-limiting resistor in the forward and reverse direction, and R_L is the resistance value of the current-limiting resistor.

The current-limiting resistor should be carefully chosen to allow the measured voltages to be within the range of the voltmeter. If a current-limiting resistor of 1 MΩ, a voltmeter capable of measuring 1 μV, and a supply of voltage of 100 V are used, a resistance of 10^{14} Ω is possible, and 10^{12} Ω is routine. Sufficient time must be allowed for the signal to stabilize following reversal of the bias, and the voltmeter must have an input impedance much greater than the current-limiting resistor.

Several other measurement schemes are available for high-resistance measurements, as documented in ASTM D 257-78. All of these techniques described above have been used successfully in our laboratory and permit the system to be automated with the use of relays or scanning systems. Care must be taken to eliminate leakage paths or other sources of error throughout the electronic system.

30.4.7 Processes That Influence Insulation Resistance Reliability

Five precautions should be taken to reduce the possibility of insulation resistance failures on PWBs.

1. Cleanliness in processing. This precaution is the most critical. Any ionic or hygroscopic residues left on the board during processing are destined to cause problems.
2. Careful use of metals that tend to form whiskers or migrate.
3. Use of protective coatings over exposed metal.
4. Use of the final product in an environmentally controlled area. Of particular concern is the presence of corrosive pollutants or dust, and high temperature and humidity.
5. Careful design of PWBs to reduce the proximity of oppositely biased metal traces.

Electronic hardware is presented in *Blue Ribbon Committee Report,* IPC-TR-476.[33] The literature, corrective actions, and suggestions on design, material, or process considerations are reviewed in this document. The following are some of the considerations presented:

Design:

- Minimum of two plies in all layers
- Separate ground and voltage planes
- Separate ground and voltage connection pins

- Design for heaters in equipment (keeps boards dry)
- Establishment of realistic environmental test requirements
- Increase in component lead spacings

Materials:

- Solder masking
- Conformal coating
- Package protection
- Selection of board materials
- Selection of plating metal
- Adhesive insulator under components
- Elimination of component spacers
- Selection of solder
- Selection of fluxes
- Avoidance of tin plating
- Elimination of glycol-activated fluxes

Processes:

- Cleaning of components
- Improvement in adhesion of glass to epoxy
- Sealing of board edges
- Use of proper bake cycles
- Ultrasonic cleaning of boards
- Vacuum baking
- Smoothing of edges
- Stress relief of boards
- Elimination of plating overhang
- Fusing of boards
- Dehumidification during storage
- Increase in plating thickness
- Glove handling
- Reduction of external pressures
- Alloying of tin-plating metal
- Reverse polarity plating
- Ultrasonic agitation during plating
- Reduction of hydrogen during plating

REFERENCES

1. C. Lipson and N. J. Sheth, *Statistical Design and Analysis of Engineering Experiments,* McGraw-Hill, New York, 1973.

2. W. H. Beyer (ed.), *Handbook of Tables for Probability and Statistics,* 2d ed., CRC Press, Boca Raton, Fla., 1979.
3. R. S. Burington and D. C. May, Jr., *Handbook of Probability and Statistics with Tables,* 2d ed., McGraw-Hill, New York, 1970.
4. M. A. Oien, "Methods for Evaluating Plated-Through-Hole Reliability," *14th Annual Proceedings of IEEE Reliability Physics,* Las Vegas, Nev., April 20–22, 1976.
5. M. A. Oien, "A Simple Model for the Thermo-Mechanical Deformations of Plated-Through-Holes in Multilayer Printed Wiring Boards," *14th Annual Proceedings of IEEE Reliability Physics,* Las Vegas, Nev., April 20–22, 1976.
6. K. Kurosawa, Y. Takeda, K. Takagi, and H. Kawamata, *Investigation of Reliability Behavior of Plated-Through-Hole Multilayer Printed Wiring Boards,* IPC-TP-385, IPC, Evanston, Ill., 1981.
7. L. A. Torres, *Thermally Induced Strain in Plated-Through-Holes,* IPC-TP-510, IPC, Evanston, Ill., 1984.
8. L. Zakraysek, R. Clark, and H. Ladwig, "Microcracking in Electrolytic Copper," *Proceedings of Printed Circuit World Convention III,* Washington, D.C., May 22–25, 1984.
9. R. N. Wild, *Thermal Characterization of Multilayer Interconnection Boards,* IPC-TR-470, IPC, Evanston, Ill., 1974.
10. I. Farkass, "Results of Reliability Tests with Rigid Printed Wiring Boards," *Insulation/Circuits,* vol. 25, no. 8, July 1979, pp. 20–23.
11. M. Honma, T. Komatsu, and H. Nakahara, *An Evaluation Method of PTH Reliabliity,* IPC-TP-466, IPC, Evanston, Ill., April 1983.
12. M. Isobe, H. Kobuna, K. Susuki, T. Sato, and S. Noguchi, "Application and Reliability of Copper Plated-Through-Hole Printed Wiring Boards," *NEC Research and Development,* vol. 55, October 1979, pp. 37–47.
13. G. Smith, "The Thermal Zone Concept," *Fabrication,* December 1983.
14. G. Smith, "A Look into Future Amendments and Revisions of Commercial Mil Specs," *Printed Circuit Fabrication,* November 1983.
15. D. A. Rudy, "The Detection of Barrel Cracks in Plated-Through-Holes Using Four Point Resistance Measurements," *14th Annual Proceedings of IEEE Reliability Physics,* Las Vegas, Nev., April 20–22, 1976.
16. G. Smith, B. W. Zimmerman, and J. T. Strohner, "Multilayer Board Plated-Through-Hole Barrel and Foil Cracks, Causes and Suggested Remedies," *Printed Circuit Fabrication,* December 1980.
17. C. J. Bartlett, R. D. Rust, and R. J. Rhodes, "Electroless Copper Plating of Multilayer Printed Circuit Boards," *Plating and Surface Finishing,* July 1978, pp. 36–41.
18. E. W. Broache and J. A. Poch, "Elimination of Fractures in Plated-Through-Hole Printed Circuit Boards by the Use of Ductile Plating," *IEEE Transactions on Parts, Materials and Packaging,* PMP-2 (4), December 1966, pp. 107–109.
19. A. Fox, "Mechnical Properties at Elevated Temperature of CuBath Electroplated Copper for Multilayer Boards," *Journal of Testing and Evaluation,* vol. 4, no. 1, January 1976, pp. 78–84.
20. L. Mayer and S. Barbieri, "Characteristics of Acid Copper Sulfate Deposits for Printed Wiring Board Applications," *Plating and Surface Finishing,* March 1981, pp. 46–49.
21. D. E. Sherlin and L. K. Bjelland, "Improve Electrodeposited Copper with Organic Additives and Baking," *Insulation/Circuits,* September 1978, pp. 27–32.
22. G. T. Paul, "Cracked Innerlayer Foil in High Density Multilayer Printed Wiring Boards," *Proceedings of Printed Circuit Fabrication West Coast Technical Seminar,* San Jose, Aug. 29–31, 1983.
23. C. Ogden and D. Tench, "Manufacturing Technology for Printed Wiring Board (PWB) Electrodeposition Process Control," Technical Report AFWAL-TR-81-4027, Wright-Patterson Air Force Base, Ohio, May 1981.

24. L. Zakraysek, "Embrittlement in Electro-Deposited Copper," *Circuits Manufacturing,* December 1983, pp. 38–40.
25. H. Ulbricht, "Surface Resistivity in Printed Circuits," *Metal Finishing,* vol. 81, no. 9, September 1983, pp. 43–47.
26. S. C. Britton, "Spontaneous Growth of Whiskers on Tin Coatings: 20 Years of Observation," *Transactions of Institute of Metal Finishing,* vol. 52, 1974, p. 95.
27. C. W. Jennings, "Filament Formation on Printed Wiring Boards," *IPC Technical Review,* February 1976, pp. 25–32.
28. D. W. Rice, "Corrosion in the Electronics Industry," Corrosion/85, paper no. 323, copyright 1985, National Association of Corrosion Engineers, Houston.
29. D. J. Lando, J. P. Mitchell, and T. L. Welsher, "Conductive Anodic Filaments in Reinforced Polymeric Dielectrics: Formation and Prevention," *17th Annual Proceedings of IEEE Reliability Physics Symposium,* San Francisco, April 24–26, 1979, pp. 51–63.
30. P. J. Boddy, R. H. Delaney, J. N. Lahti, E. F. Landry, and R. C. Restrick, "Accelerated Life Testing of Flexible Printed Circuits, Part I and II," *14th Annual Proceedings of IEEE Reliability Physics,* Las Vegas, Nev., April 20–22, 1976.
31. A. Der Marderosian and C. Murphy, "Humidity Threshold Variations for Dendrite Growth on Hybrid Substrates," *IPC Technical Review,* July 1977, pp. 11–20.
32. A. G. Osborne, "Surface Resistance," *Printed Circuit Fabrication*, June 1984, pp. 27–30.
33. "How to Avoid Metallic Growth Problems on Electronic Hardware," *Blue Ribbon Committee Report,* IPC-TR-476, IPC, Evanston, Ill., September 1977.

P · A · R · T · 7

MULTILAYER CIRCUITS

CHAPTER 31

DESIGNING FOR MULTILAYER

Jerome J. Mikschl
Unisys Corporation, San Diego, California

31.1 INITIAL DESIGN CONSIDERATIONS

Multilayer printed wiring boards are designed and used because of one or two requirements: (1) the high density of the electronic components resulting in the need for several layers of etched interconnection and/or (2) the need for electrical characteristics of the board not obtained by one- or two-sided printed wiring boards. Although there are other high-performance interconnect methods available, etched/plated-through-hole multilayer printed wiring boards are the major technology used by the electronics industry. They have proven reliability, are available from a broad base of qualified manufacturers, and can be designed to meet most cost and performance requirements of various electronic systems.

The design of these boards has become complicated. As the functional circuitry has become more complex, circuit packages have more input-output pins in smaller areas. In addition, circuit speeds have increased dramatically demanding minimum distances between packages, while power requirements have also increased significantly, demanding more critical thermal management within the board. These requirements have made circuit lines smaller, spaces between conductors smaller, and component and via holes smaller. At the same time, board sizes have generally become larger, and the demand for power and electrical requirements has become greater, requiring more and/or heavier copper, ground, and voltage levels. Finally, the competition has become stronger, and cost and performance goals have become more ambitious.

The multilayer board must be designed as part of the overall system. The cost of the board alone should not be considered in isolation but must be looked at along with the cost of the components, connectors, backplanes, and other elements of the system. It is the cost of the total system together with the performance of that system which must be evaluated. The board itself might appear complex and expensive, but if it can contribute to fewer boards, fewer connectors, a smaller backplane, and higher system performance, it could be the optimum design. All elements which add cost to the board (number of holes, hole sizes, pad sizes, line widths and spaces, number of layers, copper thickness, dielectric material, board size, and board technology) must be considered. As the technology gets more complex, manufacturing yields go down, costs go up, and fewer vendors are available. This chapter will try to supply the multilayer printed wiring board designer with aids to help select design features which will give a cost-effective, functional, and producible multilayer printed wiring board.

31.2 MULTILAYER BOARD TYPES

31.2.1 Conventional Plated-Through Hole

The conventional plated-through hole represents the most common and least expensive layer-to-layer interconnection technique. Inner layers of copper-clad laminate are first etched with signal lines or power-ground patterns. These are then laminated together under heat and pressure with dielectric adhesive (prepreg) with either a layer of unetched copper laminate or copper foil on the external surfaces. The panel is then drilled, plated, and etched as a conventional double-sided plated-through-hole board. In the conventional process, all holes are drilled through the panel whether they are used as component holes or as via holes.

Figure 31.1 is an example of a typical six-layer through-hole board. Note that the through-hole via takes up valuable real estate on *all* layers, regardless of the number of layers the hole is connecting.

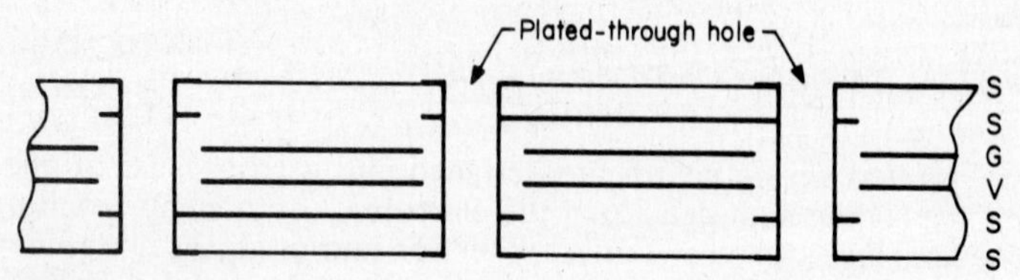

FIG. 31.1 Conventional multilayer board. S = signal plane, G = ground plane, and V = voltage plane.

31.2.2 Buried Via

A buried via can be described as a plated-through hole connecting two or more layers of a multilayer board, buried inside the board structure, but not coming to the external surface of the board. Figure 31.2 shows a cross section of this multilayer construction. Note in this example that two via holes can be stacked up over one another in the *x-y* direction, whereas in a conventional multilayer, only one could be used in this location. Also note that if an area or location can be saved from being used as a through-hole via, that area can then be used for signal lines. The buried via construction is therefore used when signal trace routing is very dense, requiring more via sites connecting signal layers and more channels for signal traces than a conventional multilayer can offer. This construction also allows smaller via holes (less than 0.008 in in diameter) because it is drilled and plated in a thin piece of laminate, resulting in an aspect ratio of 1:5 or higher. When smaller holes are used, smaller pads (or lands) can be used, which then allows even more room for signal lines.

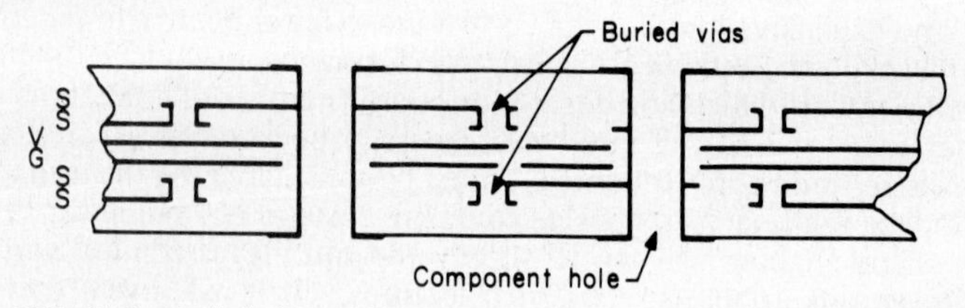

FIG. 31.2 Buried via multilayer board.

These routing-density advantages will result in a more costly board because of the added process steps. For example, the buried via layers are first made just as conventional double-sided boards. The buried via layers, together with additional voltage, ground, and surface layers, are then laminated together using prepreg. The component through-holes are drilled, and the board is completed just as a conventional multilayer. If two buried via levels are used, a total of three drilling and three plating operations are required, contrasted to a conventional multilayer board, which requires only one drilling and one plating operation regardless of the number of layers or holes. As a result, the cost goes up for buried via multilayer compared with conventional multilayer of the same area; however, the total number of interconnections is also increased sharply.

The decision regarding when to use buried vias is influenced by many factors, dependent upon the capabilities and available design tools of the designer. The ability to route the signal lines of the board in either buried or through-hole technology must first be available. If, for a particular dense board, these technologies can be quantified in terms of the number of routed or unrouted signal lines, decision elements can be defined. The number of layers and overall board construction features, such as lines and spaces and the number, size, and types of vias, can then be determined. From this information, a board cost study can be made comparing both technologies, and a decision (conventional versus buried via) can be made.

31.2.3 Blind Via

The blind via hole is a plated-through hole connecting the surface layer to one or more layers of a multilayer board which does not go through the entire board thickness. These holes can be used on both sides of a multilayer board and can be used in conjunction with via and component holes which go through the board. Figure 31.3 is an example of such technology. As in the buried via construction, blind vias can be stacked on top of one another, and they can be made smaller, leaving more room for signal lines. There are several manufacturing processes used to make blind vias, and all are more complex than the buried via board process.

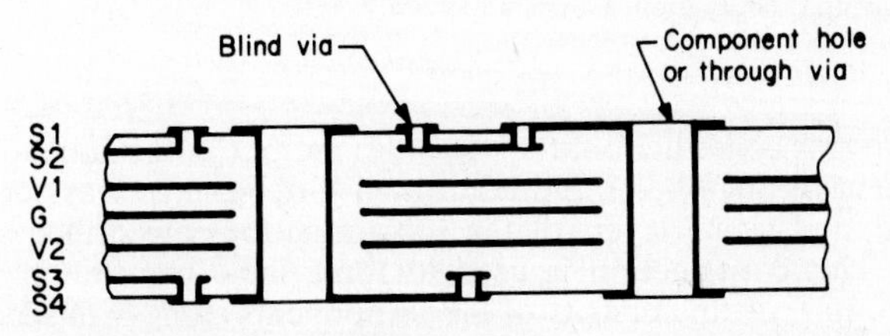

Fig. 31.3 Blind via multilayer board.

One method used is to etch all internal layers and laminate as a conventional multilayer, but without external copper layers. The through-holes are mechanically drilled, and the blind vias are drilled either by using a mechanical drill just through the internal copper layer or by laser drilling just down to the copper pad of the internal layer. Since these blind holes are small and do not go through the entire board, subsequent hole cleaning and plating operations become more difficult. This is due to the restricted flow of chemicals in the holes during these operations. However, plasma etching for hole cleaning and copper additive plating rather than electroplating methods have been used to achieve plating into these holes.

A second method used to make blind vias is to drill the via holes in a piece of thin laminate with copper on both sides. The internal layer is imaged with the desired signal or power pattern, leaving the external layer solid copper. The panel is then copper-electroplated and etched to make the connections between the two layers. These panels, together with additional internal conductor layers, are laminated, drilled, plated, and etched as a conventional multilayer. This method has the advantage of not using more complex manufacturing processes in drilling and plating; however, very dense external circuit patterns are more difficult to achieve because of the additional plating operation on the board surface.

A major use of blind via technology is with surface mounted components and connectors. Surface mount does not require large component holes; only small via holes are needed to connect the external surface to internal layers. This is best achieved by small blind vias which take up little space and allow the designer to fully utilize the size- and weight-reduction characteristics of surface mount technology for very dense and thick multilayer boards.

31.2.4 Sequential Lamination

Sequential lamination is a technology which takes two or more plated-through multilayer boards and laminates them together to produce one many-layered board. Many varieties of this construction can be made depending upon the desired end-product design. Figure 31.4 is an example of this construction.

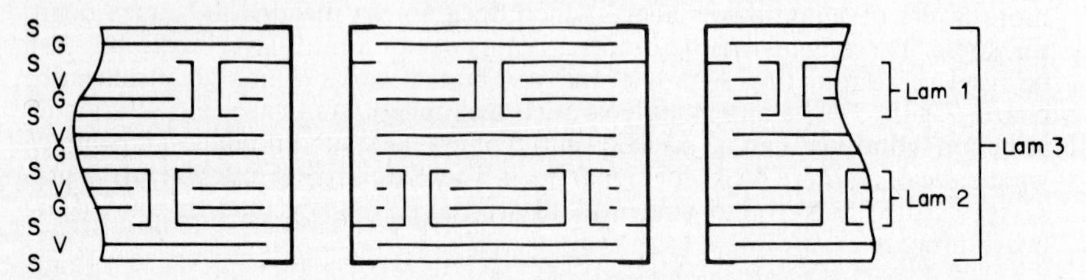

Fig. 31.4 Sequential lamination. Lam 1 and Lam 2: internal layers laminated separately first; Lam 3: conventional multilayer process.

Usually in this construction, the via holes are put in the sublaminations, and the component through-holes and additional through-hole vias connecting voltages, ground, and signal layers of the sublaminations are drilled after the final lamination. This construction is used in some large mainframe computers in which the complexity and density of the components along with specific electrical requirements are such that other, less complex, interconnect designs cannot be used.

31.3 MATERIALS

Almost without exception, multilayer boards are made using glass cloth coated or impregnated with a resin as the dielectric material. The glass cloth gives mechanical strength to the board, acts as a vehicle to carry or contain the resin, and, because of the controlled thickness of the glass cloth, allows the manufacturer to build the multilayer board to a total thickness with some reasonable tolerance. This section will therefore consider glass-resin as the dielectric material and copper as the conductive material.

31.3.1 Multilayer PWB Trends

Finer lines
Increased density
Smaller holes
Closely spaced holes
Larger boards
Higher speeds
Surface mount
Higher power dissipation

A list of multilayer trends published at any time in the past would have looked identical. These trends will continue. Manufacturing processes must continue to meet these requirements, and materials and properties must continually be improved to meet these demands. There are many continuing technical programs within the industry which will allow more sophisticated multilayer board designs with high quality and reliability levels:

- Low-loss and low-dielectric-constant materials as speeds become faster
- Dimensional stability as features become smaller and boards bigger
- Better coupling agents to enhance glass-to-resin adhesion
- Improved thermal properties as power becomes higher
- Improved copper adhesion as etched features become smaller and surface mounted components become more common
- Resistance to copper electromigration as line and hole spacings become smaller
- Lower *z*-axis expansion as board thickness increases

31.3.2 Resin

Epoxy is the most common resin used in multilayer boards and produces boards that meet the demands of most commercial and military applications. There is a broad base of manufacturers who support the industry with epoxy technology. This material is relatively easy to machine, laminate, and process chemically, and it exhibits sufficient dimensional stability for most applications. Epoxies are cost-effective for most board designs. However, where higher temperature resistance is required, other resins are needed.

Epoxy-glass has a continuous operating temperature rating of approximately 130°C. High-temperature resins used are modified epoxies, Triazine, and polyimides. These resins are applied to glass cloth similar to epoxy and are used by the multilayer manufacturer as laminate or as prepreg. They have the added advantages of being more chemical- and solvent-resistant, and they have a lower *z*-axis expansion at the elevated temperatures. The polyimides exhibit excellent thermal stability up to 200°C. This higher temperature resistance may become an important property during the repair of large, expensive boards. Repeated soldering and desoldering will degrade any multilayer board to some degree. For this reason alone, one of the high-temperature resins might be a wise selection. The disadvantages of these materials are cost, requirement of different or more critical processing operations and controls, and availability from a smaller vendor base.

31.3.3 Glass Cloth

As laminate thickness increases, generally heavier glass cloth is used. Heavy glass cloth carries a smaller percentage of epoxy resin than thinner glass weaves. As the percentage of glass in a laminate goes up, the dielectric constant of the material goes up. In a laminate of a specific thickness, it is cheaper to use fewer thick pieces of prepreg rather than thinner sheets. Laminates with a higher glass content have a lower thermal coefficient of expansion and greater dimensional stability. Glass, however, is hard to machine, and drill bit wear is greater with more glass. The use of thin glass prepreg, having a higher resin content, is beneficial when laminating together etched voltage and ground layers which have many etched isolations to fill.

A variety of glass fabrics are available. The weaves, ranging in thickness from 0.001 to 0.007 in, are used in multilayer boards (see Table 31.1). Generally, as the thickness increases, the size of the individual glass filaments becomes larger.

TABLE 31.1 Multilayer Glass Fabric Styles

Fabric no.	Thickness, in
104	0.0010
106	0.0015
108	0.0020
112	0.0030
116	0.0048
7628	0.0068

31.3.4 Prepreg

The selection of a prepreg for the finished multilayer board depends upon the electrical requirements for dielectric constant and thickness, overall board thickness, required resin fill between etched layers, ease of handling, and cost. The lighter materials require more layers per desired thickness, need additional and more careful handling, and cost more for a finished thickness.

Table 31.2 gives approximate properties for various epoxy prepregs used in

TABLE 31.2 Epoxy Multilayer Prepregs

Prepreg type	Pressed thickness, in	Resin content, %	Dielectric constant
104	0.0016	76	4.3
106	0.0023	73	4.3
108	0.0025	63	4.4
112	0.0038	58	4.6
116	0.0046	52	4.7
7628	0.0066	41	5.0

multilayer boards. These properties in the finished board will vary slightly with the manufacture of the prepreg and with the processes of the manufacturers of the multilayer board.

31.3.5 Copper Foil

Most copper foil used in multilayers is made by an electroplating process in which the copper is plated from a solution onto a polished drum and then peeled off as a foil. The side against the drum is smooth; the other side is relatively rough and makes a better bond with the prepreg resin systems. To further enhance this bond, the copper foil is plated with proprietary metal compositions on one or both sides. This electroplating process is used for copper foils up to approximately 14 oz/ft^2 (0.019 in thick) for multilayer boards. Below 0.5 oz/ft^2 (commonly called "half-ounce"), the copper is plated onto a carrier, often copper or aluminum foil. It is then laminated to the prepreg in this form, and the carrier metal is subsequently etched or peeled off. This process allows easy handling of thin copper foil during lamination and also eliminates resin bleeding through any pinholes in the copper to the surface.

The considerations in the copper thickness selection are current-carrying capacity and the need for precision tight tolerance etched features. Heavy copper is difficult to etch and more difficult to fill with resin during lamination. When very fine lines (0.005 in and less) or lines with tight width tolerances are required ($\pm$0.0005 in or less), copper thicknesses of 0.5 oz and less are commonly used.

For special applications, where thin laminate is required and only one piece of prepreg can be used between two foil sheets, resin-coated copper foil is used. By coating and curing the resin on the copper foil before lamination, the glass cloth is kept away from the copper. This prevents copper migration from one sheet of copper to another through one bundle of glass fibers which possibly was not completely saturated with resin during prepreg coating or lamination operations (see Fig. 31.5). This illustration shows what can happen without this cured resin coating on the copper.

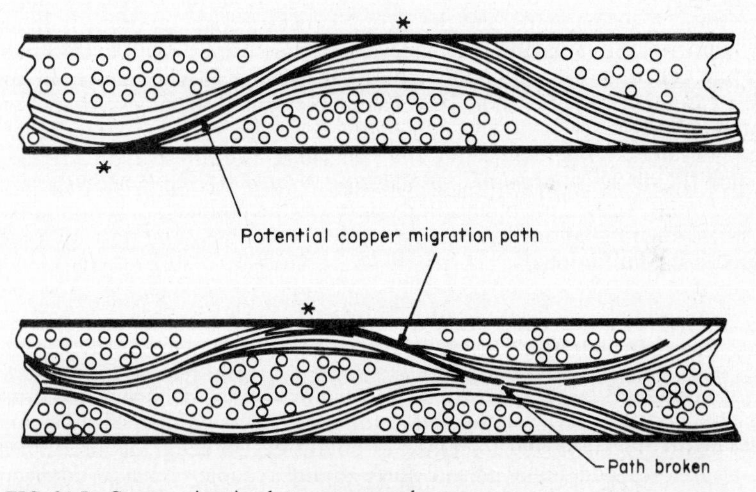

FIG. 31.5 Copper migration between copper layers.

31.4 MECHANICAL DESIGN FEATURES

31.4.1 Board Size

Determining optimum size depends on many factors. Aside from maximum size limitations determined by a system cabinet, the needs for large multilayer boards are many, and the problems associated with large boards are just as numerous.

Large boards minimize connectors, keep the backplane smaller, allow a higher input-output connection count per board, allow a shorter circuit path between many components giving higher speeds, and can reduce the number of active components in some systems. Ideally, the system designer would prefer to have the entire system on one large board, since he or she can then achieve many of the above advantages. In some designs, this has been accomplished successfully. Examples of this can be seen in some computer terminals and personal computers, where nearly all functions are contained on one board.

Problems associated with large boards from a design standpoint, however, can outweigh the many possible advantages. These include difficult routing of signal lines on a board, possibly requiring more signal levels or finer lines and spaces, and difficult thermal management in air-cooled systems because many heat-producing components are placed in close proximity, one heating up the next. In addition, alignment of connectors and other hardware becomes more critical; power requirements per board are higher, requiring heavier or more copper voltage and ground planes; critical vibration frequencies are more easily reached; and large board costs can actually be relatively higher because of lower yields in the multilayer board manufacturing process. Furthermore, testing of both the bare board and the functionally assembled board is more difficult and complex.

The difficult task before the designer is to weigh these many advantages and disadvantages and select the proper size for the particular application. During this process, he or she must know the limitations and capabilities of the board manufacturer. For example, the industry has generally been trying to establish 18 × 24 in as the standard panel size which will go through the various fabrication processes most efficiently. From this panel size, the designer must subtract a border area which is used for material handling, tooling holes, plating thieves, and test coupons and as an aid to dimensional stability through various process steps. The size of this border will vary with the complexity of the board, the number and location of test coupons, and the unique process capability of the individual board fabricator. Some boards are made successfully using a 0.50- to 0.75-in border, while other, more complex, boards may require 2 in or more. In some extreme cases, a 4-in border may be considered essential to achieve overall yield and quality levels.

The Institute for Interconnecting and Packaging Electronic Circuits (IPC) has published IPC-D-322, *Guidelines for Selecting Printed Wiring Board Sizes Using Standard Panel Sizes.* This publication considers the laminator's common panel sizes, printed wiring fabrication equipment sizes and limitations, test equipment, and processing limitations.

31.4.2 Board Thickness

A major consideration in selecting board thickness is the connector to be used. One-piece, or "edge-card," connectors using gold fingers on the board demand a specific board thickness and tolerance to obtain design goals for insertion and withdrawal force and for a normal force giving reliable electrical connection between the two. If a two-piece through-hole connector is used, critical board thickness is not required, and the board can then be designed to optimize electrical properties (impedance), the number of signal layers, and the number and thickness of power planes. Other factors in the selection of board thickness and tolerance are aspect ratio of hole diameter to thickness for quality drilling and plating; component lead length requirements for automatic insertion, clinching, solderability, and inspection; card guides; weight; and vibration.

Finished board thicknesses and tolerances should be called out on the drawing only in areas of the board where they are important to the board performance, or if it is critical for an assembly operation. Tight thickness tolerances on a complex board with many layers are difficult to maintain. Therefore, yields can usually be increased and, at the same time, manufacturing controls relaxed by using wider tolerances. As a result, the finished board cost will be less. A reasonable tolerance on the board thickness is 10 percent of the overall thickness.

After the designer knows the thickness and number of copper layers required, and critical dielectric thicknesses if required, a knowledge of the manufacturing process of the board vendor and the preferred materials (laminates and prepregs) is very important in establishing and specifying overall board thickness and tolerances.

31.4.3 Board Lay-up

The layering of a multilayer board should be kept symmetrical, and it is best to have an even number of copper layers. These requirements will help the manufacturer to keep the finished board flat. Constructions which are asymmetrical are more prone to warp, which subsequently causes many problems in board assembly, soldering, and testing, as well as in the system cabinet. A particular multilayer board can be made from laminate clad with copper on one or two sides, bare laminate, copper foil, and prepreg, and in many different combinations. Unless it is important for electrical or other reasons, do not specify on the drawing how the board should be constructed or which of the above materials should be used. Keep the lay-up symmetrical, and call out critical dielectric thicknesses only when necessary. Copper thickness and the density of the copper pattern on the layers should also be kept symmetrical. The use of laminate and prepreg should be symmetrical, and the selection of glass cloth for these materials should also be considered so that the finished multilayer board will be as symmetrical as possible. This will give the best opportunity to get a flat board that will remain flat during the soldering operations.

The minimum dielectric thickness in the finished board should be 0.0035 in, regardless of whether this thickness is made from a piece of laminate or from prepreg in lamination. Additionally, higher-performance boards require that this thickness be made from two pieces of glass cloth. This further minimizes the possibility of metal migration along glass fibers between copper layers if the glass cloth is not completely wetted with epoxy resin when the prepreg was made. Figure 31.5 shows how this can happen under the right conditions and how two layers of glass will prevent this from happening.

Two pieces of thin prepreg are more expensive than one thicker piece. For some applications in which product cost is important and reliability requirements are not high, one piece of glass cloth is allowed and can be very satisfactory.

31.4.4 Inner-Layer Copper

The selection of copper thicknesses in a multilayer board must first be made considering the current-carrying capacity requirements and impedance requirements, if any. However, the designer must also be aware of the manufacturing and reliability problems associated with copper thicknesses in the board. For signal layers, 1-oz copper (one ounce of copper foil per square foot of surface area, or 0.001374 in thick) is most commonly used, and the military specifications require 1 oz minimum. Two-oz (0.00275-in) copper for signal layers is not used unless required for higher currents. Heavy copper requires more etching, and it is more difficult to hold line width and spacing tolerances with the heavier copper. The use of 0.5-oz (0.0007-in) copper for internal signal layers, however, has applications. As copper thickness goes down and dielectric thickness remains constant, impedance goes up. In some complex, dense boards which require tight impedance control, and where board thickness is a major concern, 0.5-oz copper is being used. A further advantage is that finer lines and spaces with tighter tolerances can be etched on 0.5-oz copper.

The disadvantage is the higher tendency to crack insider the multilayer board close to the plated-through-hole and also its smaller interconnect area where it meets the plated-through hole. However, with advanced hole cleaning, etchback, and plating processes, this potential failure mechanism seems to be of less concern.

Electrical designers like a lot of copper in their voltage and ground planes. However, anything heavier than 2-oz copper requires special processing by the board manufacturer. Etching heavy copper results in less control of the desired pattern. In lamination, it is mandatory that the etched signal hole isolation patterns in the voltage and ground planes be totally filled with epoxy resin from the prepreg. With heavy copper, this isolation is large and sometimes must be prefilled with epoxy before lamination. If voids are left in this area, after drilling and plating, there is a good possibility for shorting from the plated-through hole to the power plane (see Fig. 31.6). Poor plating may also result because the various cleaning, plating, and rinsing solutions cannot get down into these small voids.

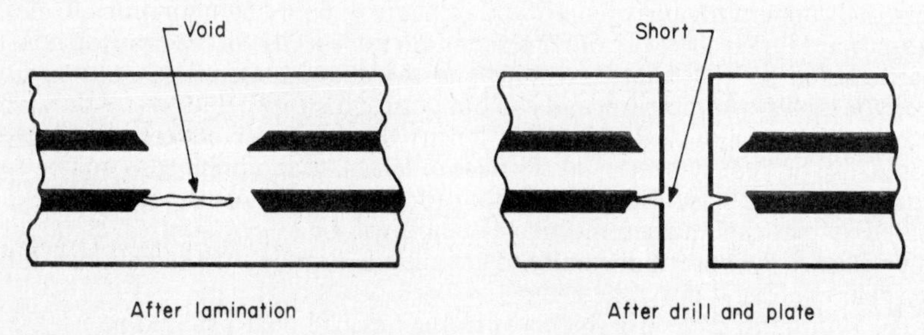

FIG. 31.6 Lamination voids in heavy copper plane area.

If this heavy section of copper isolation is totally filled with epoxy in lamination or some other operation, the area is "epoxy-rich," with no glass fibers from prepreg or laminate in the area. The overall adhesion of the electroless copper and subsequent electroplated copper to the drilled hole wall is significantly enhanced by the presence of the glass fibers protruding into the plating. Therefore, in these "epoxy-rich" areas, adhesion is the poorest, and in subsequent solder reflow or component soldering operations, hole-wall pullaway becomes a serious problem and reason for rejection (see Fig. 31.7).

31.4.5 Holes

Holes in multilayer boards are used to mount electronic components and other hardware, to interconnect internal and external layers to one another, and to facilitate fixturing for test and assembly.

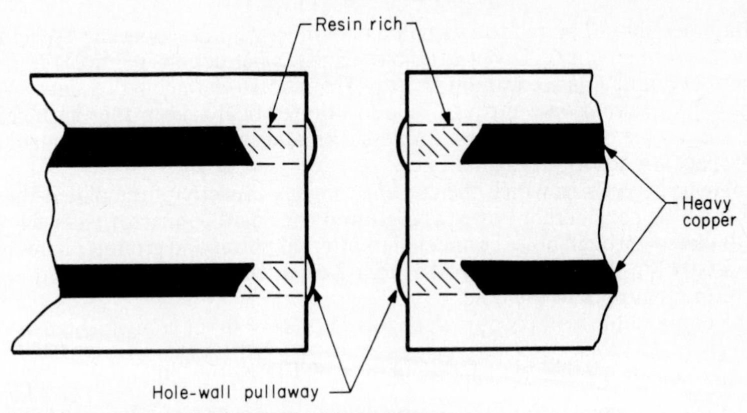

FIG. 31.7 Hole-wall pullaway in resin-rich area.

Component holes and via holes must be plated through. Others can be unsupported (nonplated). The designer would like the component and via holes to be kept as small as possible with small associated land areas, which will then allow more area to route signal lines. The board manufacturer would like to see the holes reasonably large to assure quality and cost-effective drilling and plating. The automatic assembly machine likes the holes large. The wave-soldering machine likes them just the right size to assure proper filleting. The board designer must recognize these needs and the importance of these requirements as well as the necessary trade-offs.

For component holes, military standards require the plated-through-hole diameter to be no more than 0.028 in and not less than 0.010 in from the nominal component lead diameter or diagonal. This assures a volume between the lead and the plated-through-hole wall that is sufficient for good soldering. Working within this range, it is still best to keep the number of hole sizes to a minimum. This allows maximum production efficiency and lessens the possibility of error during production planning. For automatic component insertion equipment, the hole size should be calculated as follows:

$$\text{Hole diameter} = \text{effective lead diameter} + \text{hole location tolerance} + 0.008 \text{ in}$$

Example:

$$\text{DIP lead } 10 \times 20 \rightarrow \sqrt{10^2 + 20^2} = 0.023 \text{ in}$$

$$\text{Hole location} \pm 0.003 \text{ in (total of 0.006 in)}$$

Thus,

$$\text{Hole diameter} = 0.023 + 0.006 + 0.008 = 0.037 \text{ in}$$

31.4.6 Aspect Ratio

For the board manufacturing processes, an aspect ratio of 3:1 is considered standard. The aspect ratio is the thickness of the board compared with the diameter of the drilled hole. The considerations are in drilling, smear removal or etchback, and plating. Aspect ratios of 5:1 are not unusual, and there are some boards being made with an 8:1 aspect ratio or greater. These boards require special manufacturing processes and controls. Know the capability of your board manufacturer when designing these special features.

Via holes should be made as small as possible, while keeping the aspect ratio within a producible range. With the buried via construction, via holes less than 0.008 in in diameter are common in laminate of 0.016-in thickness or less.* When designing a conventional through-hole multilayer board, keep the via holes the same size as the component holes unless a smaller via hole and associated land allow space for additional signal lines.

There are boards in which the current-carrying capacity of the plated-through holes must be considered. Power and ground edge-card connector pads generally lead to plated-through holes connected to internal power and ground planes. Consider a 0.040-in-diameter hole plated to a 2-oz internal power plane with 0.001-in-minimum copper in the hole.

The copper hole wall is equivalent to 125×10^{-6} in^2 of copper:

$$0.040 \text{ in} \times 0.001 \text{ in} \times \pi = 125 \times 10^{-6} \text{ in}^2$$

At the interconnection to the power plane, there is 352×10^{-6} in^2 of copper.

$$0.040 \text{ in} \times 0.0028 \text{ in}\dagger \times \pi = 352 \times 10^{-6} \text{ in}^2$$

The copper plating of the hole wall is the smallest conductor in the path, and for purposes of calculation in current-carrying capacity, consider it an internal conductor of a multilayer. Referring to the current-carrying capacity, curves of Military Standard 275, a 125×10^{-6} in^2 conductor will carry approximately 2 A with a 10°C rise above ambient (see Fig. 31.8).

31.5 ETCHED DESIGN FEATURES

31.5.1 Line Width and Spacing

Selection of the signal line width and tolerance is dependent upon the following:

Electrical requirements
Current-carrying requirements
Interconnection density
Manufacturing process and yields

The first three requirements are discussed elsewhere in this chapter. In general, if the electrical tolerance requirements are small (±10 percent for impedance as an example) and if there is more than one signal line between holes on 0.100-in centers, which equates to approximately 0.008-in lines and spaces, it is recommended that lines be kept on the internal layers of the multilayer. The two external layers are often kept for pads only unless board construction is solder mask over bare copper (SMOBC). Table 31.3 is an example typical of a board manufacturer who is building quality boards and wants to hold line width tolerances to a minimum. These relationships will vary with materials, etching and plating methods, and manufacturer, but they do point out design alternatives which will give different results. Note that the print-and-etch process which is used for internal layers of multilayers gives not only a finished line close to the artwork but also

*Holes down to a diameter of 0.004 in are in production, but they require special equipment and are usually drilled one board in a stack, putting great strain on available drilling capacity.

†2-oz copper thickness = 0.0028 in.

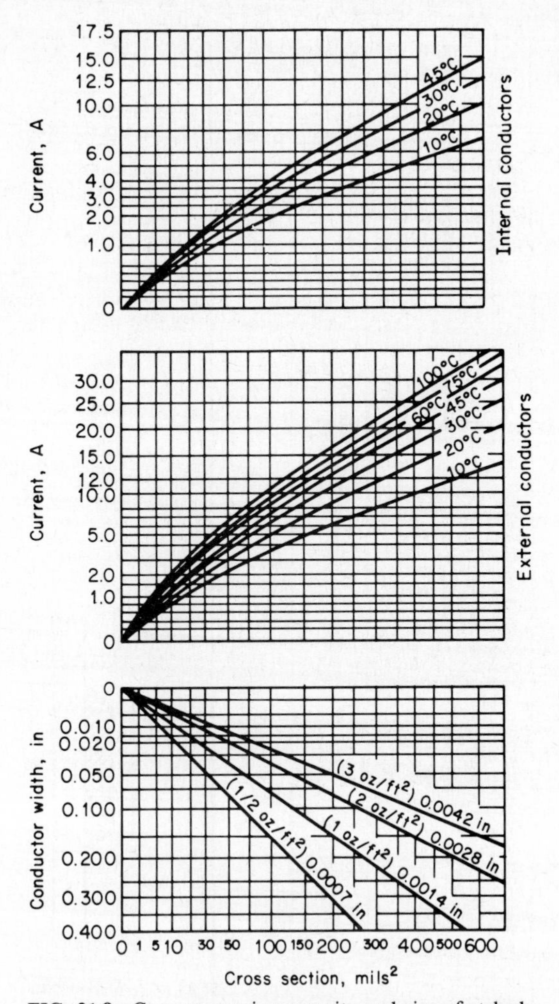

FIG. 31.8 Current-carrying capacity and size of etched conductors on multilayer boards. (1) The design chart has been prepared as an aid in estimating temperature rises (above ambient) vs. current for various cross-sectional areas of etched copper conductors. It is assumed that for normal design, conditions prevail where the conductor surface area is relatively small compared with the adjacent free panel area. The curves as presented include a nominal 10 percent derating (on a current basis) to allow for normal variations in etching techniques, copper thickness, conductor width estimates, and cross-sectional area. (2) Additional derating of 15 percent (currentwise) is suggested under the following conditions: (*a*) for panel thickness of $\frac{1}{32}$ in or less and (*b*) for conductor thickness of 0.0042 in (3 oz/ft^2) or thicker. (3) For general use, the permissible temperature rise is defined as the difference between the maximum safe operating temperature of the laminate and the maximum ambient temperature in the location where the panel will be used. (4) For single conductor applications, the chart may be used directly for determining conductor widths, conductor thickness, cross-sectional area, and current-carrying capacity for various temperature rises.

TABLE 31.3 Etched Line Cross Section

		Typical relationship of artwork to etched feature					
Copper thickness	Process	Artwork	Nominal final line at B	Nominal Δ, A/W to B	Line width range	Line width tolerance	Δ, A to B
½ oz Cu	Print and etch	10.0	10.0	0	9.3–10.7	±0.7	1.0
1 oz Cu	Print and etch	10.0	10.0	0	9.0–11.0	±1.0	1.5
2 oz Cu	Print and etch	10.0	10.0	0	8.0–11.0	±1.5	2.5
½ oz Cu	Pattern-plate and etch	10.0	8.5	−1.5	7.0–10.0	±1.5	1.0
1 oz Cu	Pattern-plate and etch	10.0	8.0	−2.0	5.5–10.5	±2.5	1.5

Note: Dimensions are in 10^{-3} in.

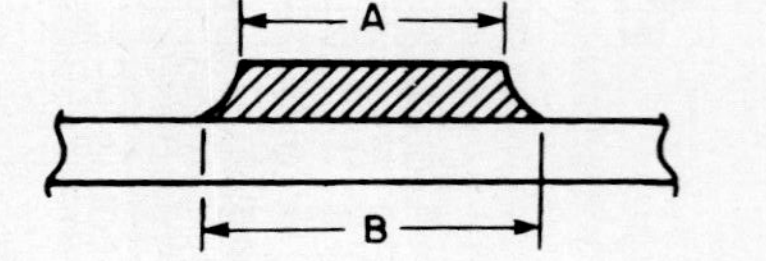

a line with the smallest tolerance on the width. As base copper becomes thicker, and when plating is necessary (as when lines are on the surface), line width control becomes more difficult. Note also that the line width is different top to bottom, depending upon the base copper and type of etchant used. In applications where this difference is critical to the electrical performance of the board, make sure this is clearly defined on the drawing.

If all fine lines, fine spaces, and critical etched features are removed from the external surfaces, yields will usually be higher, handling problems will be kept to a minimum, and soldering of the final assembly will be simplified. By using this "pads only" process on the outside surfaces, a solder mask may not be necessary in some designs.

It is generally easier for manufacturers to make small narrow lines than it is for them to hold narrow spaces. Therefore, designers must make this decision, and if there is a choice, they should take the narrower line. "Narrower" and "smaller" are relative terms and do not mean too much unless one becomes specific. When using a dry-film photoresist process, lines under 0.008 in fall into this category. There are boards being made in volume for commercial applications with line widths of 0.005 in, and one will see claims down to 0.002 in. These line widths can be achieved with special processes but generally have higher yield losses.

The silk screen resist method generally produces lines wider than the photoresist process. There are quality manufacturers, however, who routinely screen boards with lines down to 0.008 in.

31.5.2 Land Areas (Pads)

Whenever a conductor enters a plated-through hole, there will be a land area, or pad, of larger diameter than the drilled hole. This is a requirement both in military standards and in IPC standards. There are boards designed and made without pads; however, there are concerns about cost and reliability. Therefore, the through-hole connection used most often has a pad, and the designer is advised to consider all trade-offs before committing to a padless hole. The extra space may be too expensive.

31.5.2.1 Size. The size of the land for a plated-through hole is of great concern to the designer, to the manufacturer, and to the reliability function. The designer likes to keep it as small as possible so more space is available to run lines between holes. The manufacturer would like it large so the drilled hole can have a large target and yields can be high. Reliability likes the pads large so there is assurance that the minimum land area requirements are met and signal integrity will be high under a variety of stress conditions.

To determine the size of the pad for a plated-through hole, the following items must be considered:

Drilled hole diameter

Minimum annular ring requirement

Etchback requirements

Fabrication allowances

The drilled hole size is calculated from the desired plated-through-hole size for a particular component. Calculation for this finished hole size is explained above in Sec. 31.4.4. To determine drilled hole size, add 0.006 to 0.008 in to the minimum finished board size. This allows for copper plating, reflowed solder, and other processing variables.

The minimum annular ring requirement will vary with the customer and the program. For a consumer-grade product, the IPC allows some breakout of the drilled hole to the pad. For computer-grade boards, a 0.002-in annular ring is required. Military Standard 275 requires a 0.005-in external ring and a minimum of 0.002 in for internal lands. Some commercial applications require a 0.002-in minimum land where the line meets and 0.001 in elsewhere for both internal and external (see Fig. 31.9). The requirements of the finished board must be known before an intelligent selection of pad size can be made.

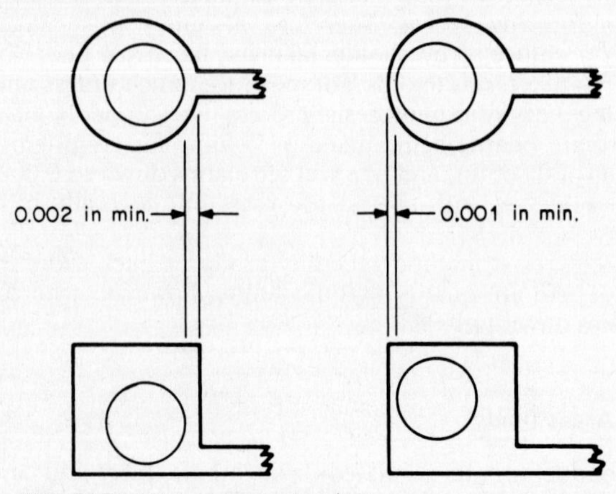

FIG. 31.9 Minimum annular ring.

When etchback is required by the contract or customer, an allowance is made for the maximum allowable etchback. This is usually 0.001 to 0.003 in.

Finally, an allowance is made for fabrication tolerances depending upon the size of the panel. This allowance takes into account the dimensional stability of the material, artwork tolerances, movement during lamination, drilling accuracy, etc. Military Standard 275 suggests the allowances in Table 31.4 for three classes of multilayer boards based upon the degree of difficulty to produce. The IPC suggests allowances about 0.002 in less.

If the above information is known, the recommended land size can be determined by using the following formula:

$$\text{Land size} = \text{drilled hole} + 2 \times \text{the minimum annular ring} + \text{allowance for etchback} + \text{the allowance for fabrication}$$

TABLE 31.4 Standard Fabrication Allowances

Greatest board-panel dimension	Preferred, in	Standard, in	Reduced producibility, in
Up to 12 in	0.028	0.020	0.012
More than 12 in	0.034	0.024	0.016

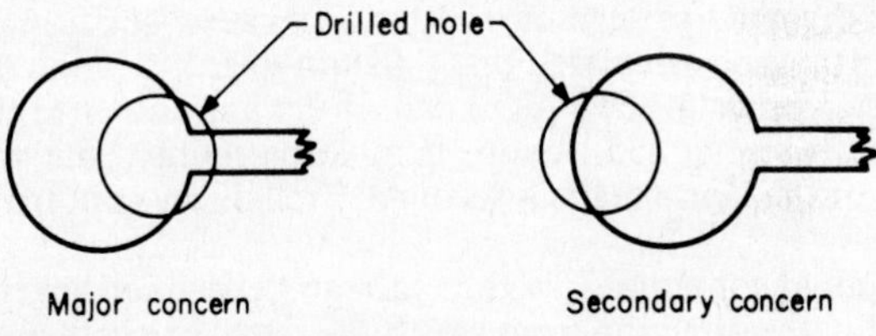

FIG. 31.10 Broken-out lands.

31.5.2.2 Land Shapes. Round lands are easiest to use from a design standpoint. The signal line is routed into the center of the circle without difficulty. There are other land shapes which allow for more misregistration between the etched pattern and the drill, do not require additional space, and still give the same degree of reliability. When a signal line is isolated from its land area, reliability is of major concern. Of secondary concern is the drilled hole being tangent to the land (see Fig. 31.10).

Consider a 0.055-in round pad attached to a 0.010-in line and 0.035-in drilled hole. Then consider a 0.055-in square pad with a 0.010-in line attached to the corner and drilled with a 0.035-in drill. Figure 31.11 shows that a misregistration

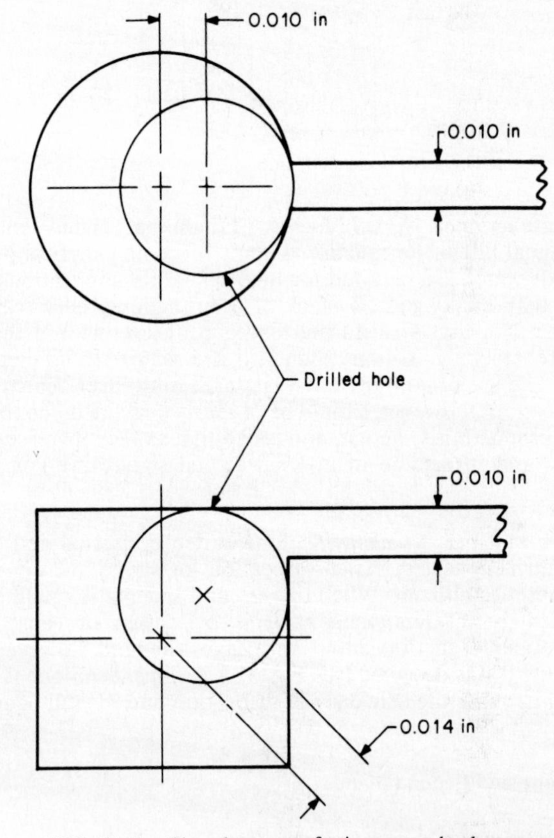

FIG. 31.11 The advantage of using square lands.

between drill and etched pattern of 0.010 in will cause isolation of the line from the round hole. It can be calculated that 0.014-in misregistration is required if the line comes to the corner of the square pad. If the line comes to the center of the square pad, nothing is gained. Again, it must be pointed out that tangency or slight breakout of the land area is a secondary reliability concern compared with line isolation.

A second method sometimes used to gain an "effective" larger land area is to enlarge the land in the direction from which the signal line comes, using a smaller-diameter pad offset from the center of the larger land (see Fig. 31.12). By offsetting a 0.045-in pad 0.020 in from the center of the larger 0.055-in pad, the "effective" land is enlarged by 0.017 in.

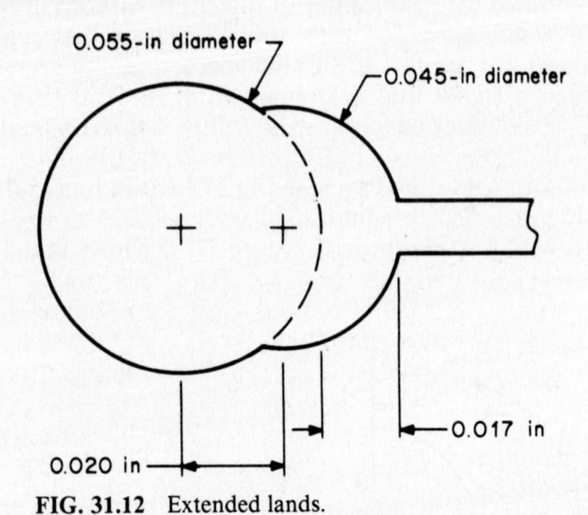

FIG. 31.12 Extended lands.

31.5.2.3 Nonfunctional Signal Lands. Functional signal lands are those attached to signal lines. Nonfunctional lands serve no electrical purpose but are used on boards which are required for high-reliability applications. Since copper expands less than epoxy-glass, a stack of nonfunctional lands can stabilize the multilayer board in that area during temperature cycling and therefore reduce potential barrel cracking. Military Standard 275 requires nonfunctional lands on signal layers. The use of nonfunctional lands on multilayer boards used for commercial applications, however, should be properly evaluated, looking at the trade-offs of board producibility, design, and reliability, as they increase internal shorting problems, and inspection of internal etched signal layers is more effective without them.

31.5.2.4 Line-to-Land Attachment. Unless dense internal and external signal layers are electrically tested, visual inspection for shorts and opens is necessary although sometimes difficult. When line-to-land spacing is small, it is difficult to determine visually if an attached line is by design or caused by spurious (unwanted) copper. If the line enters and leaves the land, however, there is little doubt as to how it was designed (see Fig. 31.13). This technique is easy to implement and can increase the efficiency of inspection and overall board yields.

31.5.3 Voltage and Ground Planes

31.5.3.1 Voltage and Ground Isolation Patterns. The copper planes in a multilayer board are used to bring power to the components and control electrical

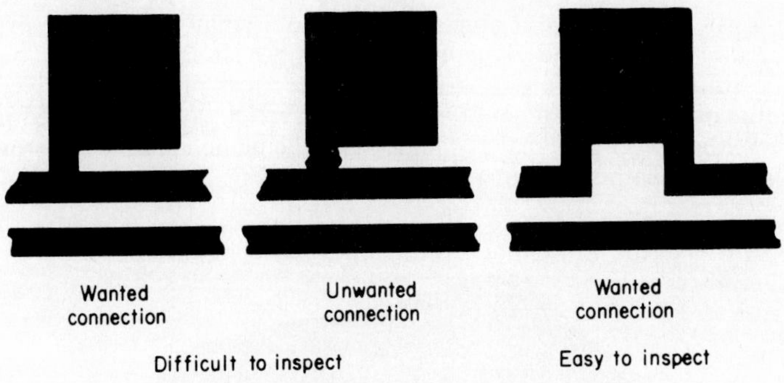

FIG. 31.13 Line attachment to land.

characteristics such as impedance and capacitance. For boards where the density of the components and via holes is not great, there is usually enough open area left on the conductive copper plane to carry sufficient current. Additionally, if the board is not dense, the remaining copper area is sufficient to act as a reference plane above or below a signal line to control the impedance. However, if the board is dense, much of the copper plane may be reduced by the necessary etched isolations for component and via signal holes. On some boards when large isolations are used, little is left of the copper plane to do its job. If they are too small, misregistrations between etched pattern and drilled plated-through hole will cause shorts.

To determine the size of these isolations, use a method similar to that used to determine land size. Again, four items are considered:

Drilled hole size

Minimum conductor spacing requirements

Etchback, if used

Fabrication allowances (see Table 31.4)

Assume that a 0.045-in hole is drilled in an 18-in board. Minimum conduction spacing requirement is dependent upon the voltages involved. For 0 to 15 V, Military Standard 275 requires a 0.005-in minimum space between conductors. Etchback is usually called out as 0.0002 to 0.0003 in. To calculate the isolation diameter, use the following equation:

$$\text{Isolation} = \text{drilled hole} + 2 \times \text{required conductor spacing} + \text{etchback allowance} + \text{the fabrication allowance}$$

In the above example, the isolation diameter is

$$0.043 + 0.010 + 0.0003 + 0.016 = 0.0693 \text{ in}$$

However, as a safety factor both for reliability and maximum manufacturing yield, try to keep the isolation as large as possible while still maintaining the required copper areas for electrical purposes.

31.5.3.2 Voltage and Ground Connections. For a good electrical connection, it is necessary to drill only into the solid copper power plane to provide the interconnection with subsequent plating. In cases in which soldering is not necessary, such as when using pressed fit pins, no pattern on the plane is needed. However, when wave-soldering the components to the board, all holes must act the same thermally, whether they are a hole connected to a small signal trace or a hole connected to a large ground plane which acts as a heat sink. If these connected ground or voltage holes are drilled into a solid copper plane, unreliable "cold" solder joints result. To overcome this, a thermal relief pattern is etched into the plane where the power hole makes contact with the plane, making a narrow, thermally conductive path simulating a signal line (see Fig. 31.14).

FIG. 31.14 Thermal relief lands.

The internal diameter of this feature can be calculated using the same equation for calculating the size of a land area on a single layer (see Sec. 31.5.2). The external diameter should be approximately 0.020 in larger and should be wide enough to facilitate economical application of the pattern (screening or photoresist) and etching. The width of the webs connecting the land to the plane should be 0.010 in or more.

31.5.3.3 Nonfunctional Lands. The advantages and disadvantages of using nonfunctional lands on power layers are similar to those outlined for nonfunctional lands on signal layers. There are, however, additional specific considerations, which are listed below.

When nonfunctional lands are used in power planes, there is the added possibility of shorting to the plane across the small isolations because of problems in imaging or etching. Visual inspection for this unwanted copper is difficult, very time-consuming, and never 100 percent accurate. However, if the internal power and ground layers are electrically tested with a "bed-of-nails" tester, a "nail" contact centered in a ground isolation without a land will not screen out an etching defect (unwanted copper) that later, after drilling and plating, could become a short. If the nonfunctional land is present, the nail will make contact with the land, to the unwanted copper, and on to the power plane, and it will show up in the tester as a fault. If internal ground planes of multilayer boards are to be electrically tested, it is best to use the nonfunctional lands.

An additional consideration is the necessity of filling the many isolations of a ground plane during lamination with resin from the prepreg. If voids are present in this area, shorting of the plane to the plated-through hole can occur (see Fig. 31.15). Filling these voids becomes more difficult with heavyweight copper. If a nonfunctional land is used and voids are present, the drilled hole will be within the land, and shorting will not be possible in subsequent plating operations. The voids will still be present but will not cause shorts (see Fig. 31.15).

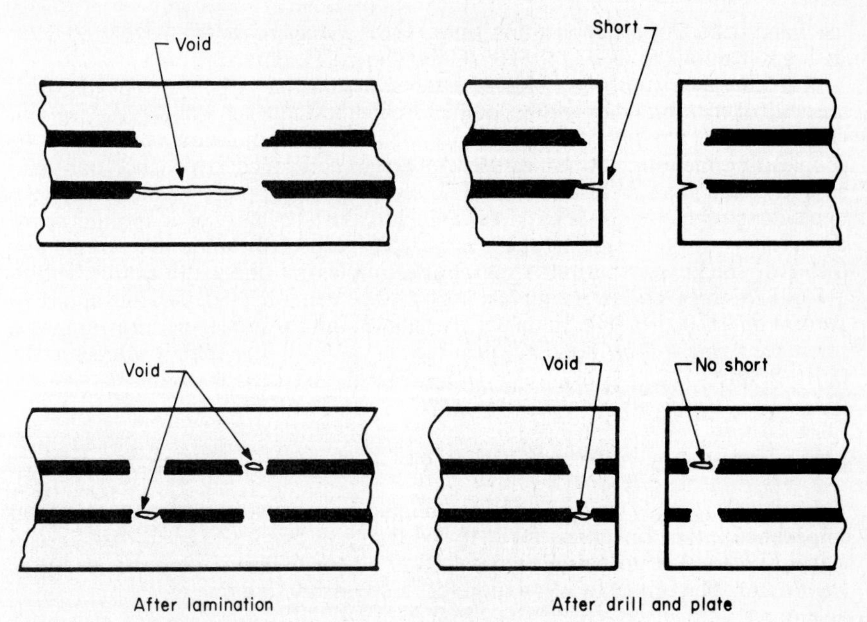

FIG. 31.15 Use of nonfunctional lands on power planes.

31.6 ELECTRICAL DESIGN FEATURES

As circuit-switching speeds become faster, the integrity of signals becomes a greater concern to the designer. A multilayer board in a high-performance, high-speed system must be more than copper lines connecting one circuit to another and a few copper planes to conduct power and ground levels. The multilayer board has become a necessary and critical electronic component of the system, with controlled impedance characteristics. As frequencies increase and signal rise times decrease, signal reflections and line lengths become critical, and the board must be designed to accommodate these effects.

Features which must be specified on the drawing or specifications, and which affect these electrical requirements, are as follows: the dielectric constant of the laminate and prepreg, conductor line width, spacing between conductors on one layer, dielectric thickness between layers, and thickness of the copper conductors. These items, together with the layering sequence of conductors in the multilayer board and the method or sequence in which the signal nets are connected, are all critical in high-speed applications.

31.6.1 Impedance Control

"Impedance" may be defined as the apparent opposition in a circuit to the flow of alternating current. It comprises inductive reactance, capacitive reactance, resistance, and conductance. In general, when the electrical "length" of the signal path exceeds one-third the signal rise or fall time, the controlled impedance interconnect should be used. Electrical length is a time unit that is equal to the physical length of the line multiplied by the propagation delay. Selection of the impedance value and allowable tolerance is dependent upon the family of circuitry being used and other circuit application rules. These values range from 50 to 80 Ω for ECL applications to 80 to 110 Ω for CMOS and TTL devices.

The characteristic impedance of a transmission line depends upon the relationship of the conductor width, conductor thickness, dielectric thickness between conductor and ground-voltage reference plane, and the dielectric constant of the propagating medium. Formulas used to design controlled impedance transmission lines are given below. It should be recognized that these variables are very dependent upon one another. Test vehicles must be built of a specific board design to verify the validity of these variables, especially when line widths and dielectric thicknesses become small. For example, if a line width can be held to ±0.001 in, this 0.001 in is equivalent to ±10 percent on a 0.010-in line and ±20 percent on a 0.005-in line. In the following formulas for impedance, note that this tolerance swing will be more significant for the 0.005-in line than it will be for the 0.010-in line.

31.6.2 Microstrip

The microstrip transmission lines are a popular means used to provide controlled impedance interconnections for high-speed digital circuits. The microstrip is a signal line exposed to air and referenced to a near voltage on ground level (see Fig. 31.16). The impedance equation for this transmission line is

$$Z_o = \frac{87}{\sqrt{E_r + 1.41}} \ln \frac{5.98h}{0.8W + t}$$

where Z_o = ohms
E_r = dielectric constant
h = distance between signal line and reference plane, in
W = width of the line, in
t = thickness of the line, in

The above equation is accurate to ±5 percent when the ratio of W to h is 0.6 or less. When W to h is between 0.6 and 2.0, the equation is accurate to ±20 percent.

The width of the line should be measured at its surface closest to the reference plane. Remember that this line is made by an etching process, and the line width can vary significantly depending upon where it is measured. This will be dependent upon the copper thickness, the etchant used, and the manufacturing process.

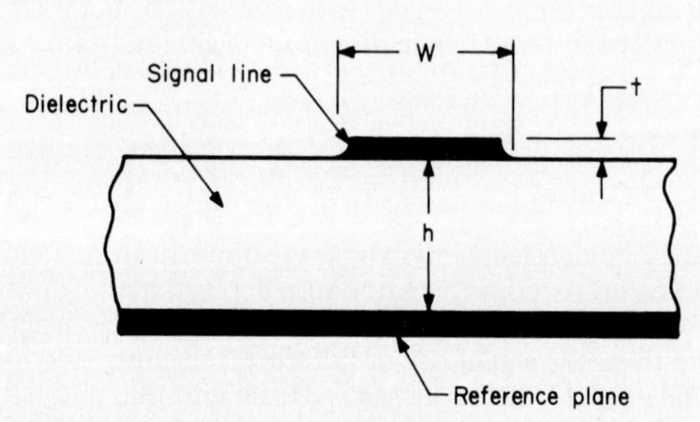

FIG. 31.16 Microstrip transmission line.

The signal propagation delay of the line may be calculated by using $t_{Pd} = 1.017\sqrt{0.475\,E_r + 0.67}$ ns/ft. Note that the propagation delay depends only on the dielectric constant and is not a function of the line width, thickness, or spacing. For FR-4 boards with a dielectric constant of 4.6, the propagation delay for a microstrip line is 1.72 ns/ft.

31.6.3 Embedded Microstrip

The embedded microstrip is similar to the microstrip, but the signal line is covered by a dielectric material—solder mask, thin laminate, or conformal coating (see Fig. 31.17). If the line is embedded with a material having the same dielectric constant of the multilayer base material and is at least 0.008 to 0.010 in thick, the effect of the surrounding air will have little effect on the effective dielectric constant. Embedding the microstrip will protect the signal lines in handling and from the environment and will help prevent shorts in soldering. However, if the line is fully embedded with dielectric, the characteristic impedance is lowered about 20 percent, and the propagation delay is increased by approximately 20 percent.

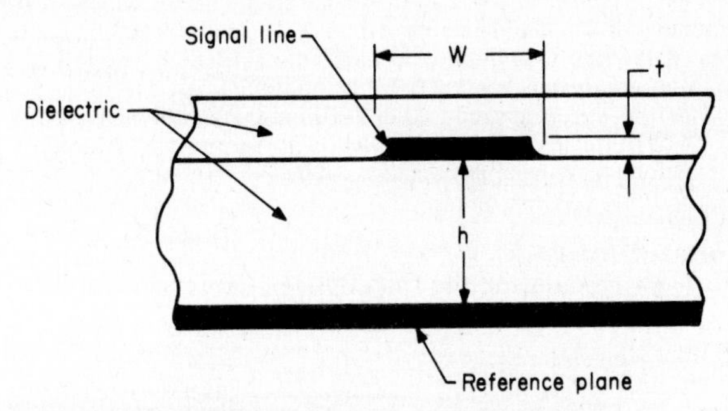

FIG. 31.17 Embedded microstrip transmission line.

The propagation delay for a fully embedded microstrip can be calculated by $t_{Pd} = 1.017\sqrt{E_r}$ ns/ft. Again, the delay is only dependent upon the dielectric constant and not the line geometry. For an FR-4 board with a dielectric constant of 4.6, the propagation of a fully embedded microstrip is 2.2 ns/ft.

Figure 31.18 shows a series of impedance curves for various line widths and dielectric spacings for the microstrip transmission line. The notes give approximate impedance changes for small changes in the dielectric constant, for reduced copper line thickness, and for embedding the microstrip in dielectric material such as a solder mask.

31.6.4 Stripline

The stripline is a transmission line embedded in a dielectric material and sandwiched between two reference planes, i.e., ground or power. This construction is used widely when the signal lines of the multilayer are not on the surface layers. Figure 31.19 is an example of such a balanced stripline where the line is equally spaced between the two reference planes. This configuration significantly reduces the cross-talk effect when compared with microstrip lines of similar impedances.

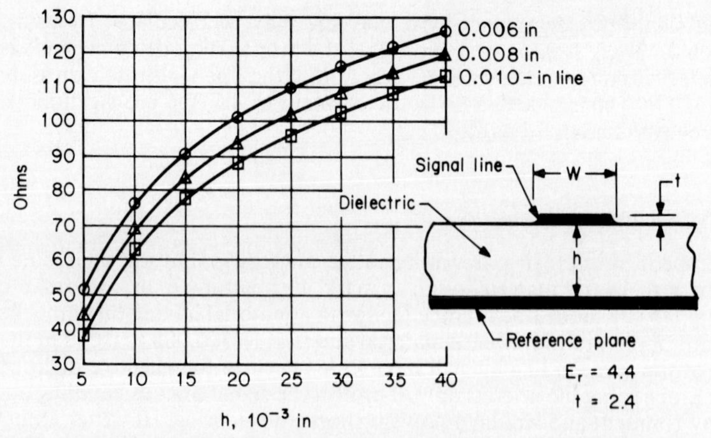

FIG 31.18 Characteristic impedance of microstrip lines. In the 50-Ω area, if E_r is increased by 0.2, Z_o will decrease by approximately 0.9 Ω. In the 90-Ω area, if E_r is increased by 0.2, Z_o will decrease by approximately 1.5 Ω. In the 50-Ω area, if the line is *embedded* in 0.002 in of 4.4 dielectric, the Z_o will be reduced by approximately 4.5 Ω; if 0.001 in approximately 2 Ω; and if 0.003 in, 5.8 Ω. In the 90-Ω area, if the line is *embedded* in 0.002 in of 4.4 dielectric the Z_o will be reduced by approximately 8.3 Ω; if 0.001 in, approximately 5 Ω. In the 50-Ω area, if the *copper line thickness* is reduced to 0.0017 in the Z_o will increase by approximately 1.4 Ω. In the 90-Ω area, if the *copper line thickness* is reduced to 0.0017 in, the Z_o will increase by approximately 2.4 Ω.

The equation for calculating the impedance of this transmission line is given by Eq. 31.1:

$$Z_o = \frac{60}{\sqrt{E_r}} \ln \frac{4h}{0.67\,\pi W\left(0.8 + \dfrac{t}{W}\right)}\ \Omega \qquad (31.1)$$

where E_r = dielectric constant
h = distance between reference plane, in
W = width of the line, in
t = thickness of the line, in

Equation 31.1 is a very close approximation of what can be expected in an actual board and is good to approximately ±5 percent. Test boards should be made to prove out the design before it is committed to volume production.

The propagation delay for this transmission line can be calculated by using the formula $t_{\text{Pd}} = 1.017\sqrt{E_r}$ ns/ft.

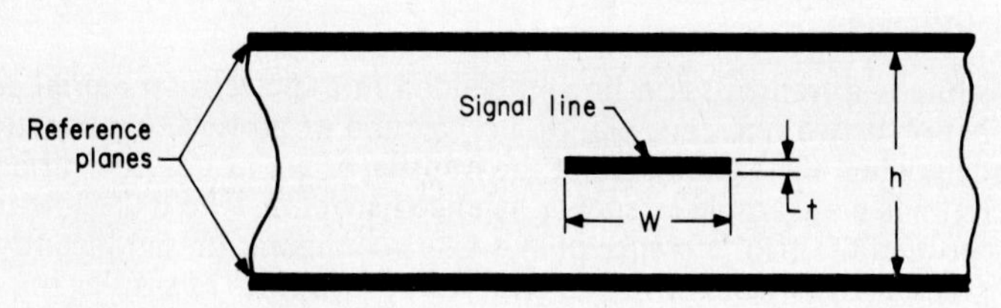

FIG. 31.19 Balanced stripline.

Note that this is the same equation used for calculating delay for the embedded microstrip, since the signal line is completely surrounded by dielectric material without the influence of air.

A variation of the balanced stripline is the dual stripline (see Fig. 31.20). Usually, one signal layer is run perpendicular to the other to accommodate routing of the signal lines, and cross talk is reduced significantly if the lines are not stacked on top of each other and parallel.

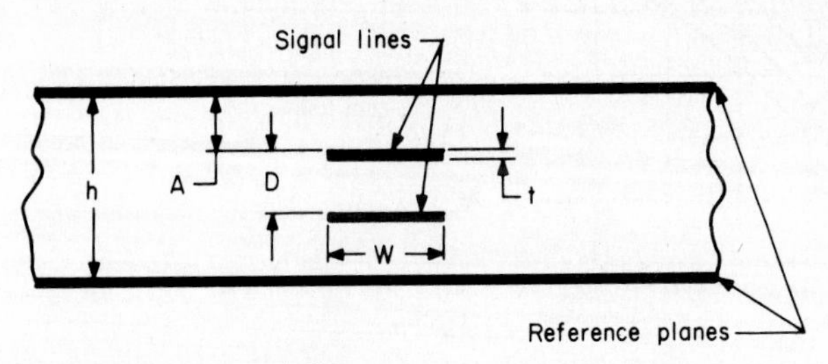

FIG. 31.20 Dual stripline.

Close approximations of characteristic impedance of the dual stripline can be made using the following formula:

$$Z_o = \frac{2\,\Gamma_1\,\Gamma_2}{\Gamma_1 + \Gamma_2}$$

$$\text{where } \Gamma_1 = \frac{60}{\sqrt{E_r}} \ln \frac{8A}{0.67\pi W\left(0.8 + \dfrac{t}{W}\right)}$$

$$\Gamma_2 = \frac{60}{\sqrt{E_r}} \ln \frac{8(A + D)}{0.67\pi W\left(0.8 + \dfrac{t}{W}\right)} \tag{31.2}$$

$h = 2A + D$ (see Fig. 31.20)*

Note that this equation, with slight modification, can be used for a single stripline not equally centered between two reference planes. In this case, A is still the distance from the center of the line to the nearest reference plane, but D would become the distance from the center of the line to the farthest plane.

Character impedance curves are shown in Fig. 31.21 for various line widths and dielectric spacings. The notes give approximations of impedance changes as the line is off-centered, as the line thickness is decreased, and if the dielectric constant is increased.

31.6.5 Capacitance

"Capacitance" can be defined as the ability of a capacitor to store electrical energy. This capacitance tends to minimize voltage transients between voltage

*J. Belanger, Teradyne Connection Systems, Inc.

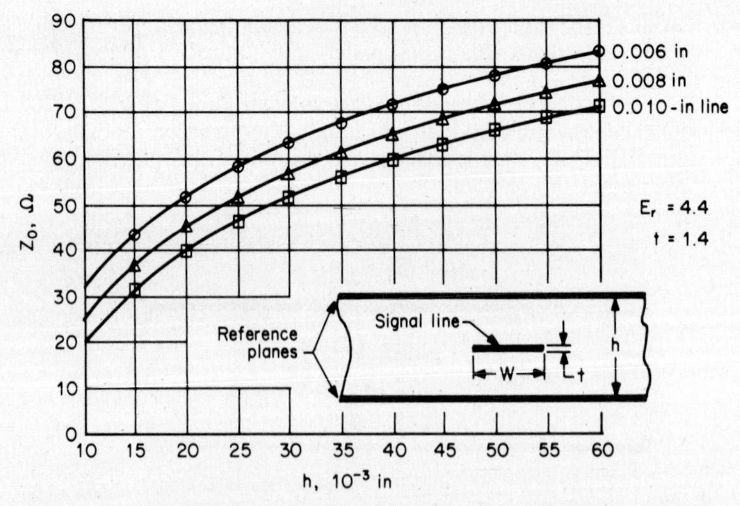

FIG. 31.21 Characteristic impedance of striplines. In the 50-Ω area, moving the line 0.002 in *off center* will decrease the impedance by about 0.7 Ω; 0.004 in, about 4.2 Ω. In the 80-Ω area, moving the line 0.010 in *off center* will decrease Z_o by about 4 Ω. In the 50-Ω area, if the *copper thickness* is decreased by 0.0007 in, Z_o will decrease by 4.5 Ω. In the 80-Ω area, if the *copper thickness* is decreased by 0.0007 in, Z_o will increase by about 3 Ω. In the 50-Ω area, a decrease in E_r by 0.2 will increase Z_o by 0.7 Ω; by 0.4, about 2 Ω . In the 80-Ω area, a decrease in E_r by 0.2 will increase Z_o by 2 Ω ; by 0.4, about 3.5 Ω.

and ground planes in multilayer boards. This parallel-plate capacitance between two planes can be calculated as follows:

$$\text{Cpp} = \frac{0.2249 E_r A}{D} \tag{31.3}$$

where Cpp = capacitance, pF
E_r = dielectric constant of the insulating media between the plates
A = common area between the plates, in^2
D = distance between the plates, in

31.6.6 Dielectric Constant

In the equations in the preceding paragraphs, note that the dielectric constant plays a major role in the determination of impedance, propagation delay, and capacitance. As the dielectric constant gets smaller, the delay time gets smaller. In critical high-speed systems in which system performance and speed are of prime concern, lower-dielectric-constant materials such as Teflon-glass can be used which have a dielectric constant of about 2.5, and they can increase the line propagation speed by approximately 25 percent compared with epoxy-glass. However, the line delay is only approximately 25 percent of the total delay time of a typical ECL circuit, and the net result in total circuit delay will be 6 or 7 percent. The cost of these materials is 2 to 4 times that of epoxy-glass, and they present new and difficult processing problems not associated with conventional epoxy-glass multilayer manufacturing.

For these reasons, most multilayer boards are made with epoxy-glass materials. The epoxy resin has a dielectric constant of approximately 3.45 and glass approximately 6.2. Depending upon the percentage of each of these materials, the dielectric constant of the laminate will range from approximately 5.3 to 4.2.

Figure 31.22 plots the dielectric constant of epoxy-glass against the percent of resin content. It can be seen that if the electrical characteristics of the multilayer board are critical, the dielectric constant of the laminate and prepreg can be carefully controlled by the proper selection of resin content of the materials.

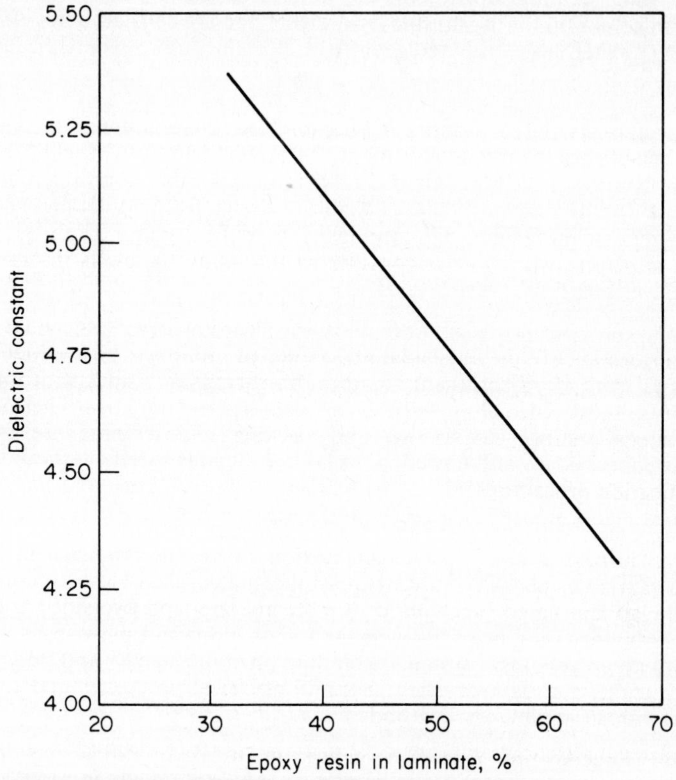

FIG. 31.22 Dielectric constant vs. percent of resin content.

It is recommended that the board manufacturer work closely with the laminate and prepreg manufacturer to establish these critical relationships. Since the thickness and weight of glass cloth can be very closely held, and since the ratio of glass to epoxy resin is predictable, once the glass cloth construction is established for a particular laminate, thickness measurements can become a reasonable means to determine and control dielectric constant.

31.7 DESIGN STANDARDS AND PERFORMANCE SPECIFICATIONS

There are two major sources for specifications and standards for the multilayer board industry. The U.S. Department of Defense has been a driving force in developing and establishing specifications primarily for high-performance boards.

The Institute for Interconnecting and Packaging Electronic Circuits (IPC) is the industry organization devoted primarily to the printed wiring board business; it is concerned with all facets of the industry. The IPC membership consists of PWB users, PWB fabricators, manufacturers of materials and equipment used by the industry, and various government agencies that have a direct interest in the use of printed wiring boards. The joint efforts of these various groups within the IPC have resulted in many standards and specifications which are used widely by the industry. Some of these documents are described below. They are a source of valuable information for the multilayer designer and are constantly revised as the technology, materials, and requirements become more sophisticated and complex.

Military Standard 275, *Printed Wiring for Electronic Equipment:* This standard establishes design requirements for rigid printed wiring boards—the design features of the bare board itself as well as for the mounting of the components. In this standard, there is a table of design feature tolerances classified by degrees of manufacturing complexity or difficulty (see Table 31.5). This guide is helpful when one is considering the many trade-offs in designing a multilayer board.

Military Specification P-55110: This specification covers the qualification and performance of the finished double-sided or multilayer board. Mechanical, chemical, and electrical quality feature dimensions or values with tolerances or limits are specified along with the appropriate test methods. Thermal stress and shock testing, cleanliness, and packing requirements are specified. Required test board and test coupons are detailed for specific tests and vendor qualification procedures.

Military Specification P-13949, *Plastic Sheet, Laminated, Metal Clad for Printed Wiring Boards:* This specification covers the requirements of fully cured laminate, partially cured resin-impregnated glass cloth (prepreg), and metal-clad laminated sheets used to fabricate one- and two-sided and multilayer printed wiring boards. It includes glass cloth and paper-base materials and all acceptable resin systems including phenolic, epoxy, and polyimide. It also includes mechanical, chemical, and electrical properties and tolerances along with acceptable test methods.

Military Specification Y-1140, *Specification for Glass Cloth:* This document includes many glass cloth types, including glass cloth made from "E" glass filament, which is used in multilayer laminate and prepreg. Details of filament, yarn size and construction, and cloth weave are specified.

IPC-D-949, *Design Standard for Rigid Multilayer Printed Boards:* This document covers the design requirements for rigid multiwire boards based upon industry manufacturing capabilities. Three design complexities are considered based upon degrees of precision required to make the board in terms of tooling, artwork, materials, and processing. This is a companion document to IPC-ML-950, which is the performance specification for the finished multilayer boards. Additionally, design aids are detailed for effective board assembly and test.

IPC-ML-950, *Performance Specifications for Rigid Multilayer Printed Boards:* Three classes of boards are recognized in this specification based upon end usage. Class 1 is consumer products (i.e., TV sets, games, and noncritical controls). Class 2 is general industrial (computers, office machines, medical equipment, etc.). Class 3 is high-reliability (life support, aircraft, military equipment, and equipment for which downtime cannot be tolerated). This document details mechanical and electrical properties of the finished

TABLE 31.5 Composite Board Design Guide
Re: MIL-STD-275

	Preferred	Standard	Reduced producibility
Number of conductor layers (maximum*)	6	12	20
Thickness of total board (maximum), in	0.100 (2.54)	0.150 (3.81)	0.200 (5.08)
Board thickness tolerance	±10% of above nominal or 0.007 (0.18), whichever is greater		
Thickness of dielectric (minimum)	0.008 (0.20)	0.006 (0.15)	0.004 (0.09)
Minimum conductor width:			
Internal	0.015 (0.38)	0.010 (0.25)	0.008 (0.20)
External	0.020 (0.51)	0.015 (0.38)	0.008 (0.20)
Conductor width tolerance:			
Unplated 2 oz/ft^2	+0.004 (0.10) −0.006 (0.15)	+0.002 (0.05) −0.005 (0.13)	+0.001 (0.025) −0.003 (0.08)
Unplated 1 oz/ft^2	+0.002 (0.05) −0.003 (0.08)	+0.001 (0.025) −0.002 (0.05)	+0.001 (0.025) −0.001 (0.025)
Protective plated (metallic etch resist over 2 oz/ft* copper)	+0.008 (0.20) −0.006 (0.15)	+0.004 (0.10) −0.004 (0.10)	+0.002 (0.05) −0.002 (0.05)
Minimum conductor spacing	0.020 (0.51)	0.010 (0.25)	0.005 (0.13)
Annular ring plated-through hole (minimum):			
Internal	0.008 (0.20)	0.005 (0.13)	0.002 (0.05)
External	0.010 (0.25)	0.008 (0.20)	0.005 (0.13)
Feature location tolerance [master pattern, material movement, and registration (rtp)]:			
Longest board dimension 12 in or less	0.008 (0.20)	0.007 (0.18)	0.006 (0.15)
Longest board dimension over 12 in	0.0010 (0.25)	0.009 (0.23)	0.008 (0.20)
Master pattern accuracy (rtp):			
Longest board dimension 12 in or less	0.004 (0.10)	0.003 (0.08)	0.002 (0.05)
Longest board dimension over 12 in	0.005 (0.13)	0.004 (0.10)	0.003 (0.08)
Feature size tolerance	±0.003 (0.08)	±0.002 (0.05)	±0.001 (0.025)
Board thickness to plated-hole diameter (maximum)	3:1	4:1	5:1
Hole location tolerance (rtp):			
Longest board dimension 12 in or less	0.005 (0.13)	0.003 (0.08)	0.002 (0.05)†

TABLE 31.5 Composite Board Design Guide (*Continued*)

	Preferred	Standard	Reduced producibility
Longest board dimension over 12 in	0.007 (0.18)	0.005 (0.13)	0.003 (0.08)†
Unplated-hole-diameter tolerance (unilateral):			
Up to 0.032 (0.81)	0.004 (0.10)	0.003 (0.08)	0.002 (0.05)
0.033 (0.84)–0.063 (1.61)	0.006 (0.15)	0.004 (0.10)	0.002 (0.05)
0.064 (1.63)–0.188 (4.77)	0.008 (0.20)	0.006 (0.15)	0.004 (0.10)
Plated-hole-diameter tolerance (unilateral)—for minimum hole diameter–maximum board thickness ratios greater than 1:4 add 0.004 (0.01):			
0.015 (0.38)–0.030 (0.76)	0.008 (0.20)	0.005 (0.13)	0.004 (0.10)
0.031 (0.79)–0.061 (1.56)	0.010 (0.25)	0.006 (0.15)	0.004 (0.10)
0.062 (1.59)–0.186 (4.75)	0.012 (0.31)	0.008 (0.20)	0.006 (0.15)
Conductor to edge of board (minimum):			
Internal layer	0.100 (2.54)	0.050 (1.27)	0.025 (0.64)
External layer	0.100 (2.54)	0.100 (2.54)	0.100 (2.54)

*The number of conductor layers should be the optimum for the required board function and good producibility.

†To be used only in extreme situations warranted by the application.

Note: All dimensions and tolerances are in inches unless otherwise specified; data in parentheses is expressed in millimeters.

board and tolerances or defect levels based upon the end product usage. Vendor qualification procedures, quality conformance tests, and frequency of testing are detailed.

IPC-D-300G, *Printed Board Dimensions and Tolerances:* This document provides rules, principles, and methods for establishing dimensions and tolerances used to define end product requirements. Again, this standard provides for three classes of tolerances to reflect progressive increases in complexity and sophistication of tooling material and processes.

IPC-D-325, *End Product Documentation for Printed Wiring Boards:* This standard defines the end product documentation requirements for boards.

IPC-CF-150, *Copper Foil for Printed Wiring Applications:* Various copper foils, treatments, weights, and specifications are detailed in this specification. Test methods, defects, and limits are detailed.

IPC-L-110, *Prepreg—B-Stage E Glass Cloth for Multilayer Printed Wiring Boards:* This specification calls out properties, limits, and test procedures for the various prepreg materials used for multilayer boards.

CHAPTER 32
MULTILAYER MATERIALS

Paul C. Marx, Ph.D.
Fortin Industries, Inc., Sylmar, California

32.1 INTRODUCTION

Three types of materials are used to manufacture conventional, rigid multilayer boards (MLBs): thin laminates, typically less than 0.030 in thick; prepreg (semicured) for bonding circuitized laminates and foil lamination; and copper foil, coated and uncoated, for foil lamination. The modular concept is an integral part of MLB manufacture. The above materials are made from an even more basic set of raw materials: a resin system, the reinforcing fabric, and treated metal foil. It is these raw materials together with special processing techniques that ultimately determine MLB performance. Increasingly, performance requirements have been raised as circuit densities and signal velocities have increased along with concerns about heat management, impedance control, chemical and thermal resistance, and *x*-, *y*-, and *z*-axis expansion.

Not all the approaches and materials used to meet these concerns can be covered in this chapter. However, the raw materials will be examined, followed by a description of laminate and prepreg testing and test evaluations.*

Properties and tests not discussed in this chapter are common to both thin and rigid materials and are covered in Chap. 6.

32.2 RAW MATERIALS

32.2.1 The Resin System

Three resin types are widely used to make thin laminates and bonding grade prepreg: conventional flame-resistant epoxies; modified, high-performance epoxies; and polyimides.

The structure of the standard FR-4 epoxy is shown in Fig. 32.1. This is a difunctional epoxy with the average number of repeating units n equal to 2.5. Every other bisphenol A group contains four aromatic bromines. It is the relative weakness of the aromatic carbon-bromine bond that contributes to the formation

*The assistance of Dwight Naseth and Richard LoDolce is gratefully acknowledged.

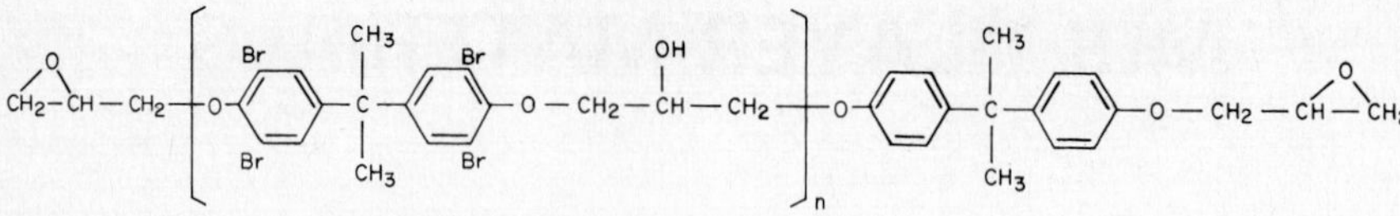

FIG. 32.1 Brominated difunctional epoxy.

of flame-retardant gases during combustion. If the bromines are replaced with hydrogens, the structure shown would resemble a typical G-10 epoxy resin.

Brominated bisphenol A epoxies are the workhorse laminating resin in the printed wiring board (PWB) industry. They have excellent adhesion to copper and other metals, exhibit low shrinkage during cure, have good chemical and moisture resistance, and are easy to process and B-stage. With a dicyandiamide hardener their glass transition temperature (T_g) lies in the range of 120 to 130°C. Their weakness is a high-percent expansion when heated to solder temperatures.

This weakness is addressed by modified epoxies, which are blends of the difunctional bisphenol A type with multifunctional epoxies, BT resin, triazine, and even polyimide. These modifiers increase the cross-link density, which in turn raises T_g and improves chemical and thermal stress resistance. As a disadvantage, there is some loss in the ease of processing, increased brittleness and drill wear, and generally higher material costs. With some modifiers, dicyandiamide can be eliminated as the hardener. This is an additional advantage (see Sec. 32.4.11).

BT resin is supplied by Mitsubishi Gas Company, a cyanate ester triazine by Interez, a tetrafunctional epoxy by Shell Chemical and Ciba Geigy, and a tris epoxy novolac by Dow Chemical. The structure of a tetrafunctional epoxy is shown in Fig. 32.2.

Polyimides represent the upper end of the scale, with T_g values typically greater than 240°C. Property improvements parallel the modified epoxies. Additionally, they have a high continuous operating temperature, 170°C (FR-4 epoxy: 125°C); high copper-to-resin bond strength at soldering temperatures (making rework reliable); a low degree of drill smear; and less than 1 percent expansion from room temperature to 250°C. Disadvantages include high cost, a lower flammability rating (UL-94V1), higher moisture absorption, and a special post-

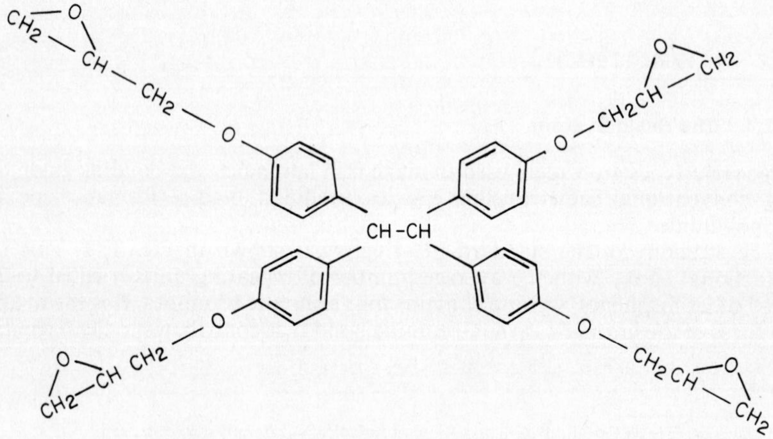

FIG. 32.2 Tetrafunctional epoxy.

bake cycle (4 h at 230°C). Also, the common varnish solvent, dimethylformamide, is a high-boiling solvent difficult to remove in the treater oven. Retained solvent in the prepreg can lead to later processing and performance problems. The structure of bismaleimide, the most widely used polyimide monomer, is shown in Fig. 32.3.

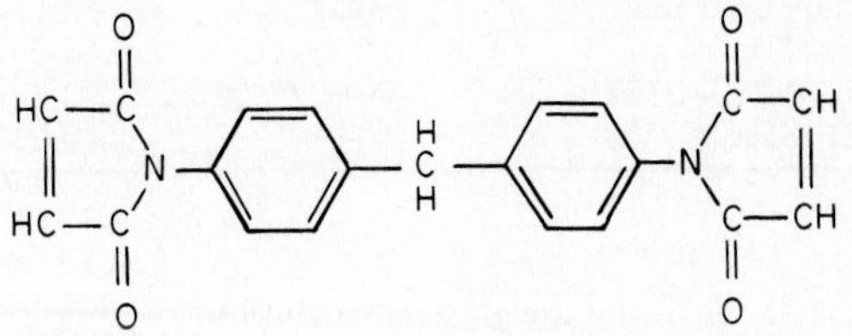

FIG. 32.3 Bismaleimide monomer.

32.2.2 Reinforcement Materials

The reinforcement material in widest use for making laminates and bonding-grade prepreg is E (electrical) glass, a borosilicate type with good resistance to water, fair resistance to alkali, and poor resistance to acid. E-glass fibers are formed from a melt and then combined to form strands that are woven into the cloth fabric used as the reinforcement material. Glass cloth carries a style designation (three- or four-digit number) which determines its nominal weight, thickness, and thread count. Glass styles used in bonding-grade prepreg are listed in Table 32.1, along with prepreg resin content and cured-thickness ranges. The glass

TABLE 32.1 Fabric and Prepreg Properties

Fabric	Style	Nominal thickness, mils	Thread count, per in	Average weight, oz/yd^2	Typical resin content range, %	Prepreg cured thickness, mils
E*	104	1.0†	60 × 52‡,†	0.56§,†	70–80¶	1.2–2.2¶
E	106	1.3	56 × 56	0.74	70–80	1.5–2.5
E	108	2.3	60 × 47	1.40	55–70	2.2–3.0
E	1080	2.1	60 × 47	1.40	55–70	2.2–3.0
E	112	3.6	40 × 39	2.09	50–65	3.0–4.0
E	2112	3.1	40 × 39	2.06	50–65	3.0–4.0
E	113	3.3	60 × 64	2.40	50–65	3.0–4.0
E	2113	3.0	60 × 56	2.30	50–65	3.0–4.0
E	2313	3.1	60 × 64	2.40	50–65	3.0–4.0
E	116	3.9	60 × 58	3.11	45–60	4.0–5.0
E	2116	3.5	60 × 58	3.04	45–60	4.0–5.0
E	1675	4.0	40 × 32	2.86	45–60	4.0–5.0
E	7628	6.9	44 × 32	6.03	35–50	6.2–7.5
Q	503	5.0	50 × 40	3.30	40–60	4.8–5.6
Q	507	4.0	27 × 25	2.00	40–60	2.8–3.6
Q	525	3.0	50 × 50	2.00	40–60	2.8–3.6
A	108	2.0	60 × 60	0.80	55–70	2.0–2.6
A	120	4.0	34 × 34	1.70	50–65	4.0–4.8
A	177	2.0	70 × 70	0.93	50–65	2.0–2.6

*E = E glass; Q = quartz; A = aramid. (Du Pont style designation.)
†Taken from IPC-EG-140.
‡The warp count is given first.
§To convert to $g/38.5\ in^2$ (scaled flow ply), multiply by 0.843.
¶For reference only, not design.

TABLE 32.2 Reinforcement Materials

Property	E glass	Aramid	Quartz
Specific gravity	2.54	1.44	2.20
Tensile strengh (psi at 25°C)	500,000	400,000	500,000
Tensile modulus (psi at 25°C)	10,500,000	18,000,000	700,000
Coefficient of thermal expansion (longitudinal, ppm/°C)	5.0	−2.1	0.54
Dielectric constant (1 MHz at 25°C)	6.0	3.8	3.8
Dissipation factor (1 MHz at 25°C)	0.002	0.024	0.0001

weaver, after thermally cleaning the glass fabric, applies a coupling agent or finish to the surface to enhance adhesion to the resin. The coupling agent, normally a silane, is bifunctional. One functional group reacts with the glass surface to form very stable siloxy bonds. The other, an organofunctional group (e.g., epoxy), is exposed and chosen to have maximum reactivity with the resin system under conditions of cure. Specifications for E-glass fabric, including defects, are found in IPC-EG-140.

Quartz (high-purity silica) and aramid (aromatic polyamide polymer) are used in special applications where their lower longitudinal expansion coefficients and dielectric constant are an advantage over E glass. Both materials have processing problems, notably drilling and machining.

Reinforcement properties are listed in Table 32.2. Because of its wider use, E glass will be the reinforcement emphasized in the discussion which follows.

32.2.3 Metal Foil

The principal metal foil used to manufacture multilayer boards is electrodeposited (ED) copper. Increased circuit densities leading to finer circuit lines and thicker MLBs have made it necessary for the foil supplier to consider three new requirements:

1. Improved elongation properties at elevated temperatures
2. Reduced treatment profiles
3. Special bonding treatments for modified epoxies and polyimides

Foil developed with a minimum 2 percent elongation value at 180°C [high-temperature-elongation (HTE) foil] has proved to be effective in controlling foil cracking under thermal stress and shock. This foil is specified as Class 3 by IPC-CF-150. The foil-cracking problem is illustrated in Fig. 32.4. Here the percent of expansion of copper and the base material are plotted with foil elongation as a function of the temperature. Potential foil-cracking problems occur when the percent of expansion exceeds the foil-elongation value at a known processing temperature, e.g., a soldering temperature.

Foil suppliers also have reduced the matte surface treatment profile (see Fig. 32.5). High profiles are undesirable. They reduce the effective dielectric spacing, cause erratic characteristic impedance values, and lead to undercut problems during etching because of the longer etch time required.

Special bonding treatments have been developed for the modified epoxies and polyimide resins. These were needed to resist bond degradation through oxidation at the high postbake temperature conditions sometimes required.

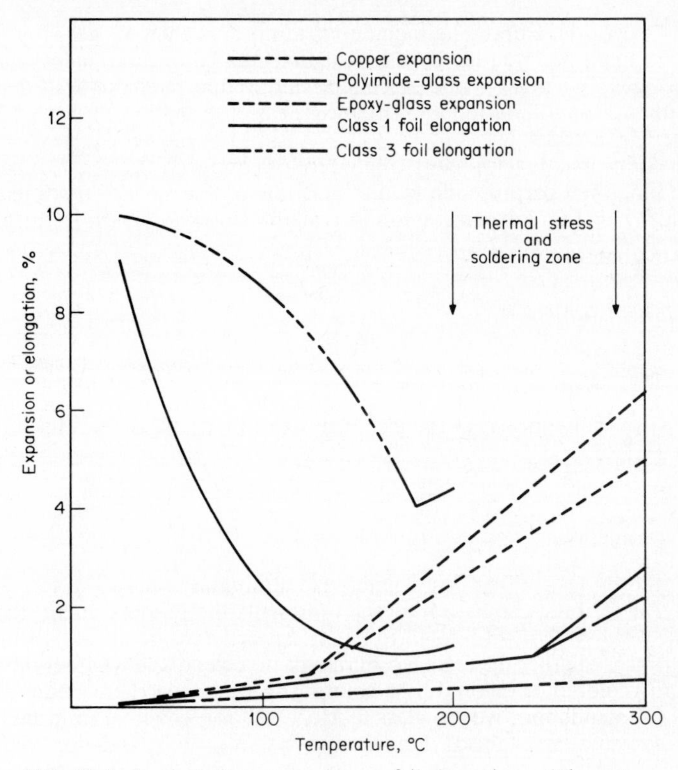

FIG. 32.4 Laminate base expansion vs. foil elongation. *(Elongation data provided by R. Steiner, Gould, Inc.)*

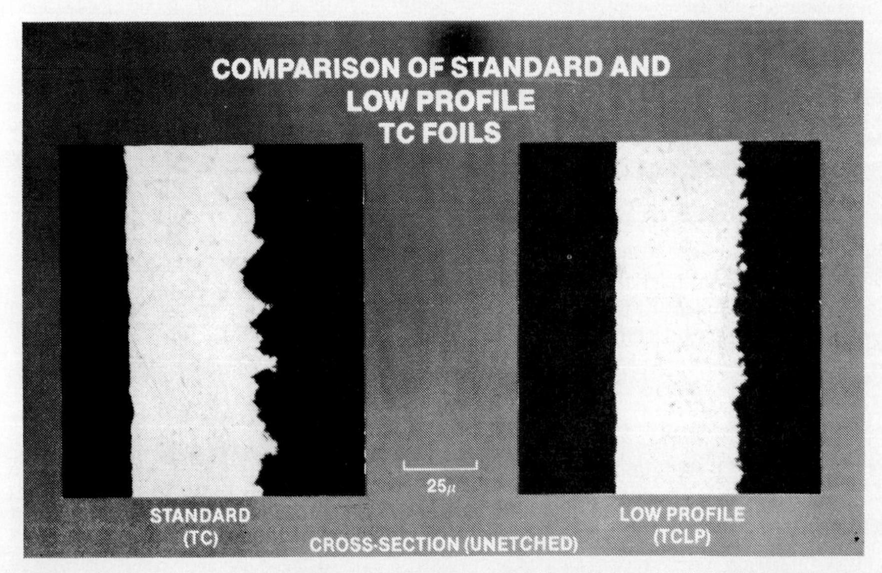

FIG. 32.5 Comparison of standard and low-profile TC foils. *(Courtesy of Gould, Inc.)*

Other recent developments include ultrathin (less than 12 μm) foil for high-resolution, fine-line circuits; double-treat foil (treated on both sides) that eliminates the need for a black-oxide processing step; and resin-coated foils that protect the treatment from oxidation and enhance the peel strength. The latter foil is covered by IPC-CF-148.*

In an additional development, treated rolled copper (IPC Class 7) is being used where its high-temperature elongation and low-profile surface treatment are an advantage. This more expensive foil is available down to 12 μm in thickness.

32.3 THIN LAMINATES

32.3.1 Testing

Testing of special concern in the use of thin laminates is covered in Secs. 32.3.4 through 32.3.13.

32.3.2 Designation

The method of designating thin copper-clad laminates is shown in Fig. 32.6. The first letter describing the base material represents the reinforcement; the second letter, the resin system; and the third letter, "special consideration," including the resin T_g range, whether the resin is a pure type or a blend and whether the material is naturally colored or colored by a coloring agent or opacifier. Under reinforcements, "G" stands for "woven glass fabric," "A" for woven aramid fabric," and "Q" for woven quartz fabric." Under resin systems, "F" stands for "flame-resistant epoxy," "E" for "non-flame-resistant epoxy," and "I" for "polyimide." The "special considerations" designator is described in Table 32.3.

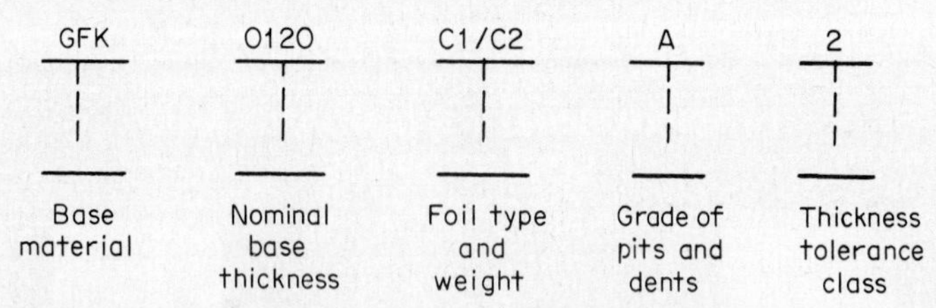

FIG. 32.6 Thin laminate designation. *(MIL-P-13949.)*

The nominal base thickness is given in ten-thousandths of an inch (0120 represents a thickness of 0.012 in).

The copper-foil cladding is described in the third set. Here the indication is electrodeposited copper, drum side out (C), with a foil weight of 1 oz/ft² on side 1 (commonly referred to simply as "1-oz copper") and 2 oz/ft² on side 2 ("2-ox copper").

The grade of pits and dents is given a letter designation, and the class of thickness tolerance is given a number designation.

The NEMA designation for MIL-type GF base material is FR-4; for type GE it is G-10.

*A standard developed and distributed by the Institute for Interconnecting and Packaging Electronic Circuits (IPC).

TABLE 32.3 "Special Considerations" Designators: Prepeg and Base Material

Special prepreg designator	Flow test*	Special base material designator	Resin system†	Color‡	Blend§	T_g range, °C¶
R	R	N/P	Any	N/P	Y/N	—
D	S	N/P	Any	N/P	Y/N	—
N	R	K	F	N	N	110–150
A	R	L	I	N	N	>250
G	R	G	F	N	N	150–200
H	S	G	F	N	N	150–200
T	R	T	F	N	Y	170–220
C	S	T	F	N	Y	170–220
J	R	J	I	N	Y	200–260
L	S	J	I	N	Y	200–260
E	S	K	F	N	N	110–150
K	S	L	I	N	N	>250
P	R	M	F	P	N	110–150
M	S	M	F	P	N	110–150

*S = scaled flow test; R = percent flow test.
†F = flame-resistant epoxy; I = polyimide.
‡N = natural color; P = coloring agent or opacifier added.
§Y = blend of two or more resin types; N = unblended.
¶Glass transition temperature range of cured resin.

32.3.3 Thin Laminate Selection Criteria

The user of thin laminates should consider the following general rules when ordering material:

1. Be exact when stating requirements and specifications for dimensions, tolerances, metal cladding, and expected board performance. Do not overspecify.
2. Use the supplier's standard material and board constructions whenever possible.
3. Remember that the supplier buys copper foil to a weight standard and that you, the user, may have a thickness standard.
4. Do not sacrifice board performance for ease of processing.
5. Make sure that board performance, controlled by the resin system, reinforcement, and copper-foil class and treatment, satisfies all production processing steps, including thermal stressing, plating, and the use of stripping solvents, developers, fluxes, and flux removers.
6. First consider the performance needed for service use and board processing; then consider cost and processibility.
7. Govern testing by the extremes to which laminate materials will be subjected during processing and service.

32.3.4 Surface Standards: Metal Cladding

Surface standards for metal-clad thin laminates are covered by MIL-P-13949G and IPC-L-108A. In the category of pits and dents, several grades are specified on the basis of a per-square-foot point count. The specification does not apply to copper foil treated on both sides.

Metal-clad wrinkles visible to normal vision are not allowed. Scratches deeper than 5 percent of the nominal foil thickness (1 oz = 70 μin) and over 4 in long are not allowed within the working area of panels or sheets by the military specification. The "working area" is defined as the area measured 1 in inward from all four edges. The IPC specification allows no scratches deeper than 140 μin. These surface defects prevent the tight conformance and adhesion of dry-film and liquid resists to the metal surface. This can cause shorts and opens in etched circuit lines after further processing.

32.3.5 Surface Standards: Base Materials

After the copper cladding is removed from a test specimen by etching (method 2.3.6 or 2.3.7 in IPC-TM-650), the base material is examined with an optical aid providing a minimum of 4× magnification. Both IPC-L-108A and MIL-P-13949G allow not more than one piece of residual metal per 5 ft^2 of surface. The piece may not have a diameter greater than 0.005 in.

Other surface and subsurface imperfections are allowed by the military specification provided that they are nonconductive, that the reinforcement fiber is not exposed, that they do not propagate during thermal stress testing, that the longest dimensions of voids are less than 0.003 in, that the longest dimensions of opaque foreign inclusions are less than 0.020 in, and that such inclusions occur no more than twice per square foot of specimen area.

32.3.6 Base Material Thickness

Both IPC-L-108A and MIL-P-13949G set nominal thickness tolerances for the dielectric base material (see Table 32.4). Generally, the thickness is determined by mechanical measurement of the base material after the cladding is removed by etching. However, the military specification allows for a new tolerance class where the thickness is determined by microsection. The thickness is taken to be

TABLE 32.4 Thin Laminate Thickness Tolerances

	Base laminate tolerance, ± mils	
Base thickness, mils	Class 1	Class 2
1.0 to 4.5	1.0	0.75
4.6 to 6.0	1.5	1.0
6.1 to 12.0	2.0	1.5
12.1 to 20.0	2.5	2.0
20.1 to 31.0	3.0	2.5

Source: IPC-L-108A.

the closest point between metal claddings (Fig. 32.7). The microsection test can be important in determining the characteristic impedance of the board and also ensuring against clad-to-clad shorting in very thin laminates. The military specification provides for a test area measured 1 in inward from all four test panel edges and further specifies a minimum base thickness of 0.0035 in for double-clad laminates (copper on two sides).

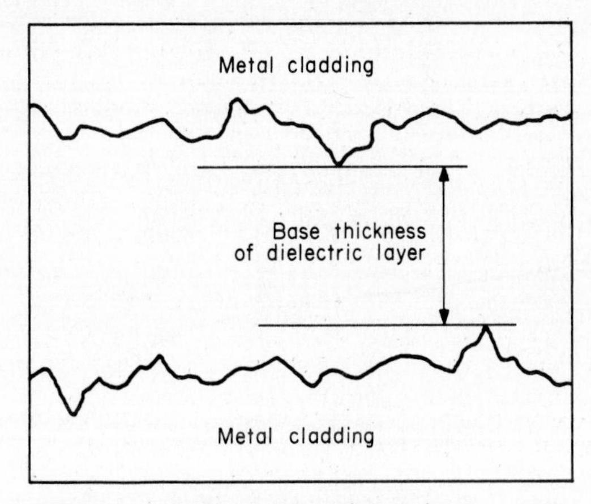

FIG. 32.7 Base thickness by microsection.

32.3.7 Peel Strength

Peel strength is a measure of the adhesion between the copper cladding and the resin. It is reported as pounds per inch of test strip width. Thin laminates should be bonded to a rigid surface to get a true reading. A reference test method is 2.4.8, in IPC-TM-650.

Inadequate adhesion can result in the movement or swimming of circuit lines during the melt phase of the lamination cycle. This can be a problem with the increasingly finer circuit lines being used. Finer lines will have a proportionately lower effective bond strength. Two developments help this situation. New high-performance resins have better peel-strength retention at elevated temperatures. And low-profile foils have reduced the undercut problem, which gives more strength for each nominal line width.

32.3.8 Glass Transition Temperature

The glass transition is a second-order property change in a polymer where the rate of expansion or rate of heat absorption changes over a narrow temperature interval. (A first-order transition is a phase change, such as melting.) Certain materials, e.g., natural rubber, seem to pass from a glasslike to a rubbery state as this interval is exceeded. The midpoint of the interval is taken as the glass transition temperature (T_g). The IPC provides two test methods for measuring the T_g: method 2.4.24 in IPC-TM-650 uses thermomechanical analysis (TMA), and method 2.4.25 uses differential scanning calorimetry (DSC). The results generally compare within a few degrees centigrade.

The glass transition is an important factor in choosing a multilayer resin system because it sets the temperature where the board's *z*-axis expansion rate can increase a factor of 4 to 6 times over the rate below the T_g.

Higher degrees of cross-linking between polymeric chains increase the T_g of thermoset resins such as epoxies and polyimides. Multifunctional epoxies added to the difunctional base resin are effective in improving the T_g because they have more functional "tentacles" to react with and cross-link adjacent polymeric chains. Higher T_g's usually mean improved moisture and solvent resistance and better high-temperature peel strengths. They also cause increased brittleness and

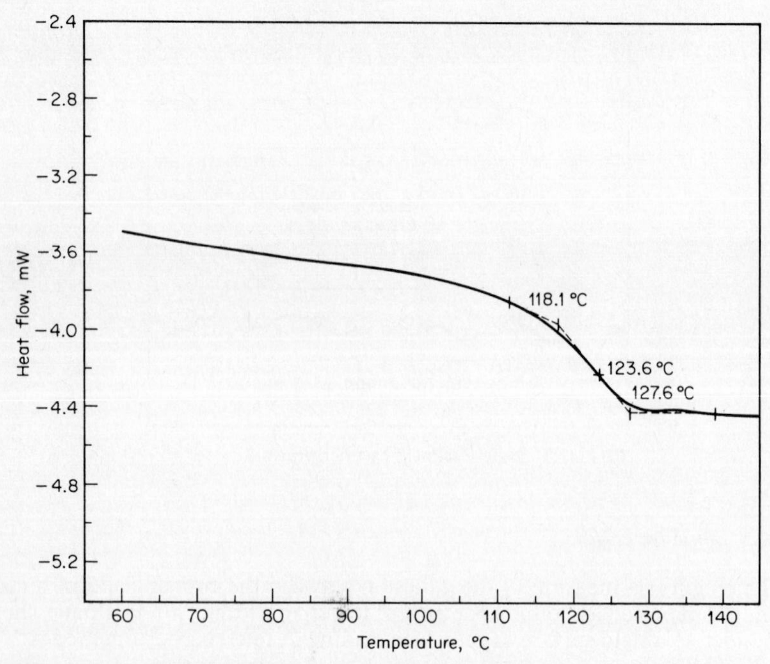

FIG. 32.8 Epoxy-glass laminate T_g determination by DSC.

a tendency to fracture on drilling. They do not necessarily ensure improved high-temperature thermal stability. Epoxies, e.g., are limited here by the weak carbon-bromine bond in the brominated base resin. In general, however, a higher T_g means a higher-performance resin system.

Figure 32.8 is a DSC determination of the T_g of an epoxy-glass laminate, where the heating rate is 20°C/min. Heat flow in milliwatts is plotted against the temperature. Two baseline plateaus are shown. Their relative shift represents a difference in the rate of heat absorption, a second-order transition. Tangents to both baselines are drawn and the mid- or inflection-point temperature between them chosen as the T_g.

Absorbed moisture, acting as a plasticizer, can lower the measured T_g. Single-sided laminates, in particular, should be desiccated before DSC testing.

32.3.9 Coefficient of Thermal Expansion

The coefficient of thermal expansion of the base material along the *z* axis (out-of-plane) is conveniently measured by TMA. A typical TMA plot is shown in Fig. 32.9. A break in the expansion rate occurs at the T_g, creating two expansion coefficients, one below and one above this temperature. The T_g is chosen as the point of intersection of the tangents to the different expansion rate curves. The higher rate can be 4 to 6 times the lower rate.

It is important to be aware of board dimensional changes because of the stresses they can place on the copper foil and PTH copper.

A convenient way to present expansion data is to report it as the percent of expansion from room temperature to 250°C.

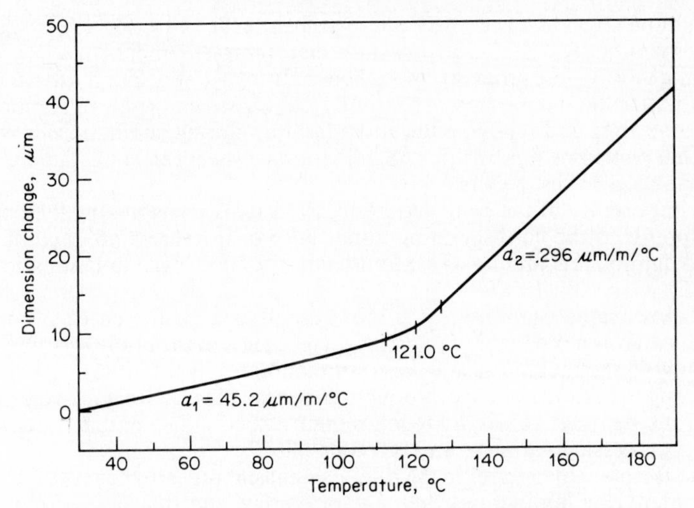

FIG. 32.9 TMA measurement of *z*-axis epoxy-glass expansion coefficient.

32.3.10 Thermal Stress

Thin laminates are subjected to a thermal stress test that can be described as an elevated temperature test for copper-to-resin adhesion, glass-to-resin adhesion, and an excessive level of trapped volatiles (double-clad specimen).

In MIL-P-13949G, etched and unetched specimens (2 × 2 in) of single- and double-sided laminates are tested, after fluxing, on the surface of a solder bath maintained at 287 ± 6°C for 10 s. Intimate contact is important. To prevent curling, a cardboard backing may be used. The etched specimens are baked for 6 h at 135 ± 14°C and allowed to cool to room temperature in a desiccator before testing.

Any sign of surface blistering, interlaminar blistering, delamination, or crazing is cause for rejection. Excessive measling is also not allowed.

32.3.11 Dimensional Stability

Dimensional stability is the resistance of a thin laminate to planar dimensional changes, either shrinkage or expansion, through processing. It is much more critical in multilayer than in standard materials.

The number value is inches of movement per inch of board length in the x (warp or machine) and y (fill) directions of the base reinforcement fabric.

Method 2.4.39 in IPC-TM-650 is the referenced test method. The test specimen is 12 in (machine) by 11 in, and it is prepared by locating 4 points on a 10-in grid by drilling or scribing. The distance between pairs of x and y directional points is measured to within 0.0005 in and recorded.

With the location points protected by small pieces of tape, the copper cladding is removed by cupric chloride spray etching. The specimen then is vertically racked and baked at 105 ± 5°C for 4 h within 4 h of the etching step. After baking, the specimen is placed in a desiccator for a 1-h stabilization period; the distance between pairs of points is remeasured within 5 min of panel exposure to ambient conditions.

A thermal stress test (vertical-rack oven baking at 150°C for 2 h) is also conducted, followed by desiccation and point-pair remeasurement.

During an IPC test program, more board shrinkage was found after the stress test than after the baking step. Apparently the higher temperature is more effective in annealing and relieving the stress built up during lamination. It was also found that panel position within the full laminate sheet can contribute to significant differences in test results.

The importance of the bake and desiccation steps in stabilizing the board for testing points to the corresponding importance of in-process production baking prior to lamination and to the minimization of the bake-to-lamination time interval.

The dimensional stability test is more useful as a quality check of incoming materials than as a predictor of registration problems in the multilayer board. One problem with the method as a process simulation test is the almost total removal of the copper cladding (etched production boards have at least 20 percent cladding). Consistent test values, lot to lot, remain a good indicator that the raw materials and processing variables are well controlled.

Some factors that determine the dimensional stability are the resin system and resin content, the lamination cycle, the properties and thickness of the copper, and the glass construction. In general, dimensional stability is improved by lower composite resin contents, lower lamination pressures, thinner foils with lower treatment profiles, and heavier glass fabrics.

For predictable performance, the warp direction in thin laminates always should be consistently aligned with the same direction in the bonding plies of prepreg. This requires proper identification on the cladding surface.

32.3.12 Electrical Testing

The electrical properties of thin laminate base materials have grown in importance as circuit densities have increased, and more attention is now paid to the board's characteristic impedance.

Properties that require testing include:

1. The volume and surface resistivity, which are a measure of current leakage through and across the surface of the base material
2. The dielectric constant (permittivity), which measures the contribution of the base material to circuit capacitance, characteristic impedance, signal velocity, and the power-loss factor
3. The dissipation factor (loss tangent), which measures the power dissipation
4. The dielectric strength, which measures the ability of the base material to withstand a high-voltage electric stress
5. The arc resistance

The dielectric constant and dissipation factor, in particular, are dependent on test temperature, frequency, and humidity conditions. The military test conditions are 23°C, 1 MHz, and 50 percent relative humidity. The recommended procedure is the two-fluid method, where the results are independent of the specimen's physical dimensions. Care must be taken to prevent invisible levels of corrosion from forming on cell contacts. This should be checked periodically with a reference specimen.

The dielectric constant of the base material, a composite, will vary with the resin and absorbed moisture content. This is shown in Fig. 32.10 for FR-4 epoxy

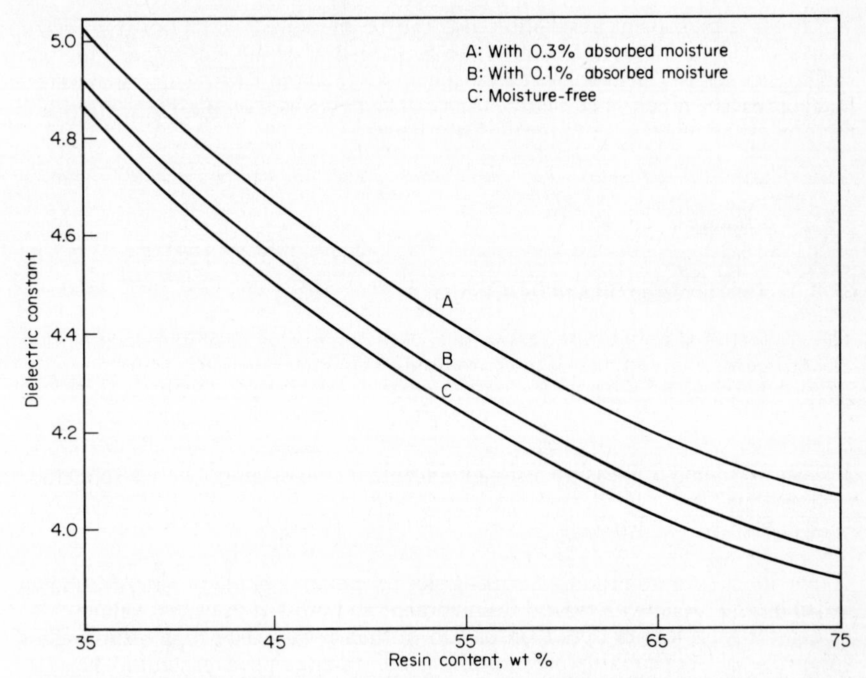

FIG. 32.10 Dielectric constant vs. resin content. *(Data supplied by T. Newton, Norplex-Oak, Inc.)*

laminates. The large effect of water comes from its high dielectric constant, 79.3 at 23°C.

32.3.13 Chemical Resistance

During processing, the base material of a thin laminate is exposed to a variety of solvents and aqueous solutions. All will be absorbed to some degree. The weight or volume absorbed will depend on the solvent, the resin system, the processing temperature, the exposed surface area, and the exposure time. It is noted that the surface area of a clad (1-oz copper) laminate that has been etched can be approximately 5 times that of an unclad laminate.

Subsequent baking can remove nearly all the absorbed solvent and leave the resin largely intact. However, solvent penetration to and wicking along the glass fiber bundles can be more serious if even traces of solvent or moisture at this level are not expelled by baking. Excessively thin levels of resin over the weave crossover points (knuckles) can accentuate this problem.

Solvent penetration into the resin immediately below the circuit lines can irreversibly weaken the copper bond even with a postbake. This problem has become more prevalent with increased circuit densities and finer lines. A reference test on bond strength retention is method 2.4.8 in IPC-TM-650.

Undercured or overbaked resin (in air) can lead to a relatively high solvent sensitivity. Curing tests by T_g measurement and carefully controlled oven temperature and bake times are important in preventing damaging solvent attack.

Solvent or water absorption can be reported as the weight absorbed with the specimen area, thickness, and test temperature and time given.

Moisture or solvent absorption also can be measured by thermogravimetric analysis (TGA). The test specimen can be heated after solvent immersion, or it can be exposed to a high-solvent atmosphere while in the TGA. Isothermal weight loss curves as a function of heating time can be developed on the TGA, providing the process engineer with optimum bake conditions.

32.4 PREPREG

32.4.1 Characterization and Designation

The properties of prepreg are very critical to the manufacture of high-quality multilayer boards. Three types of tests are used to characterize these properties:

1. Tests designed to measure the quantity of resin coated onto the reinforced fabric.
2. Tests designed to measure resin advancement or B-staging. These also include flow tests.
3. A test of the volatile content.

The most common reinforcement styles for bonding-grade prepreg are shown in Table 32.1 along with typical resin content and pressed-thickness values.

Lighter glass fabrics (less than 0.0025 in thick), with their high resin-to-glass volume ratio, are primarily used for filling (etched circuit encapsulation). Heavier fabrics are used to build up dielectric thickness and reduce dimensional shifts. A heavy fabric, such as style 7628, can be used between two outer plies of any of the four light fabrics (104, 106, 108, 1080). This combines the dimensional stability of 7628 with the circuit-filling capacity of the lighter glass styles.

The prepreg type designation in MIL-P-13949G is shown in Fig. 32.11. "PC" identifies the material as a preimpregnated B-staged reinforcement.

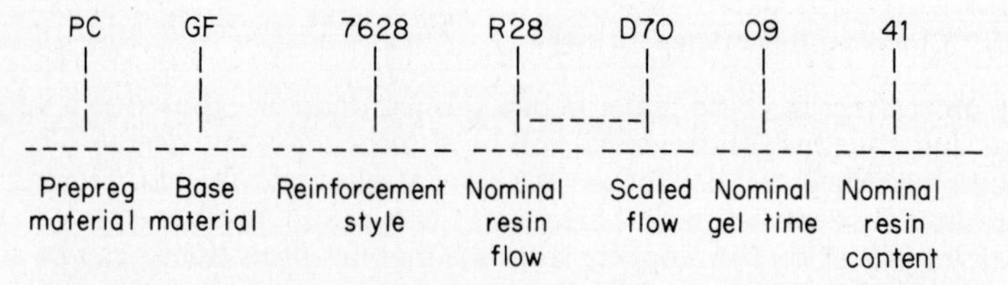

FIG. 32.11 Prepreg type designation. The letters "R" and "D" are special designators. *(MIL-P-13949.)*

The second two letters identify the base material: the first letter represents the reinforcement and the second letter the resin type.

The next three- or four-digit number identifies the reinforcement style. In the example, "7628" represents reinforcement style 7628.

This is followed by a "special considerations" letter designator (see Table 32.3), which specifies the flow test and special base material requirements. The next two-digit number is either the nominal percent flow or the scaled flow per ply pressed-thickness value. In the example, "28" represents 28 wt % flow (32.4.6) and 70, a 0.007-in pressed thickness.

The next two digits represent the gel time in tens of seconds. In the example, "09" stands for 90 nominal seconds. The last two digits represent the nominal weight percent of resin. The example shows 41 percent.

32.4.2 Prepreg Selection Criteria

Some general rules can guide the user in selecting prepreg:

1. Be specific in stating requirements, specifications, and tolerances.

2. Use the supplier's standard grade of prepreg whenever possible.

3. Use at least two plies of prepreg for bonding etched laminates.

4. If more than two plies are needed to meet a dielectric spacing requirement, minimize the total number used.

5. After ensuring that there is sufficient resin to fill or encapsulate the circuitry, minimize the volume percent of resin in those plies used for building the dielectric spacing. This will help reduce *z*-axis expansion and dimensional shifts or misregistration in the multilayer board.

6. Wherever possible, use those reinforcement styles which offer the greatest permeability to resin flow through the fabric. This will help distribute the resin used for filling and will also minimize voiding.

7. Choose a resin system that will provide adequate *z*-axis expansion control for the particular multilayer thickness you are trying to build.

8. Choose a resin system that will perform well, when cured, under all known board processing and service conditions.

32.4.3 Resin Content

Resin content is a measure of the weight percent of resin coated onto the reinforcement fabric. Each reinforcement type and style seems to have an optimum range of resin content, where the prepreg has a smooth, resin-rich surface. Lighter fabrics characteristically have higher resin contents than heavier fabrics.

The reference measurement method is 2.3.16 in IPC-TM-650. A 4- × 4-in die-cut specimen is weighed to the nearest 0.001 g and placed in a previously fired and weighed crucible. The crucible and specimen are then placed in a muffle furnace maintained at 524 to 593°C for 15 min or until all the resin has been burned off. The specimen should have a pure white appearance and show no evidence of glass fusion. The crucible is removed from the furnace and allowed to cool to room temperature in a desiccator. The glass residue and crucible are weighed to the nearest 0.001 g and the resin content calculated by dividing the loss in weight by the original specimen weight and multiplying by 100.

Aramid-reinforced prepreg, which is totally combustible, must be tested differently. The weight of a 4- × 4-in die-cut prepreg specimen is weighed to the nearest 0.001 g. A 4- × 4-in die-cut sample of aramid cloth from the same roll of fabric used to make the prepreg is also weighed to the nearest 0.001 g. This weight is subtracted from the prepreg weight, and the result is divided by the prepreg weight and multiplied by 100 to give the resin content.

32.4.4 Total Weight

Prepreg weight is more reliable than resin content in controlling the dielectric thickness where there are variations in fabric weight. Resin content is a weight ratio. Holding it constant when the fabric weight increases requires a corresponding increase in resin weight. The two increases are additive in causing a thickness increase. But when the total weight is held constant, an increase in fabric weight requires a corresponding decrease in resin weight. The net effect is subtractive, with the thickness change proportional to the difference in densities of resin and fabric. Where the densities are equal, weight control is perfect thickness control.

Nominal resin-content values remain excellent reference points for setting target values for total weight. By experience, we know that these are usable resin-to-glass weight values.

32.4.5 Volatile Content

The control of residual solvent or moisture in prepreg materials is essential in making a sound multilayer board. Both residual solvent and moisture can increase resin flow, affect the rate of cure, and at higher concentrations cause voiding, blistering, and even interlaminar delamination. The volatile content can be measured using method 2.3.19 in IPC-TM-650. The following test is used to determine the amount of residual solvent and moisture present.

Three 4- × 4-in specimens are bias-cut, with a hole punched in one corner of each, and individually weighed to the nearest 0.001 g. A wire hook coated with a release agent is used to suspend each sample in a circulating air oven. For epoxy materials, the oven is set at 163 ± 2°C; for polyimides, it is set at 225 ± 2°C. After 15 min for epoxy material and 30 min for polyimide, the samples are removed and placed in a desiccator for cooling to room temperature. The volatile content is determined by dividing the weight loss by the original specimen weight and multiplying the result by 100. The reported volatile content is an average of the three specimen values.

Test temperature, more than time, is a major factor in controlling the volatile content value. For each resin system there will be an optimum test temperature, where a temperature increase will result in no further change in the volatile value. The standard temperatures, however, provide a reasonably good measure of volatile content and a constant test base for comparing values between laboratories.

The rate of moisture absorption by standard epoxy prepreg is a function of time, relative humidity, and prepreg resin content. Absorption rates, normalized to a fixed resin weight, are shown in Fig. 32.12.

Recommended volatile contents are 0.5 wt % or less for epoxy prepreg and 1.5 wt % or less for polyimide prepreg.

32.4.6 Resin Flow

The most direct measurement of prepreg performance during lamination is a flow test. Gel time and heat of cure only indirectly predict flow.

Two flow tests are now widely used to characterize prepreg: a high-pressure test (IPC-TM-650, method 2.3.17) and a low-pressure test, commonly called the scaled flow test (IPC-TM-650, method 2.4.38).

In the high-pressure test, 4 plies of prepreg are bias-cut 4 ± 0.05 in on a side. The sample is weighed to within 0.005 g and then placed between release sheets and caul plates. The package is inserted into a press preheated to 171 ± 3°C and immediately subjected to 200 ± 10 psi for 10 min. A pneumatic press capable of continuously maintaining 200 psi is the preferred press type. The package is removed from the press and allowed to cool to room temperature. A circular section with a diameter of 3.192 ± 0.010 in is cut from the center of the sample using a steel rule or punch die. This circular section is weighed to the nearest 0.005 g and the weight percent of resin flow calculated as follows:

$$\text{Resin flow} = \frac{(\text{sample weight} - 2 \times \text{circle weight}) \times 100}{\text{sample weight}}$$

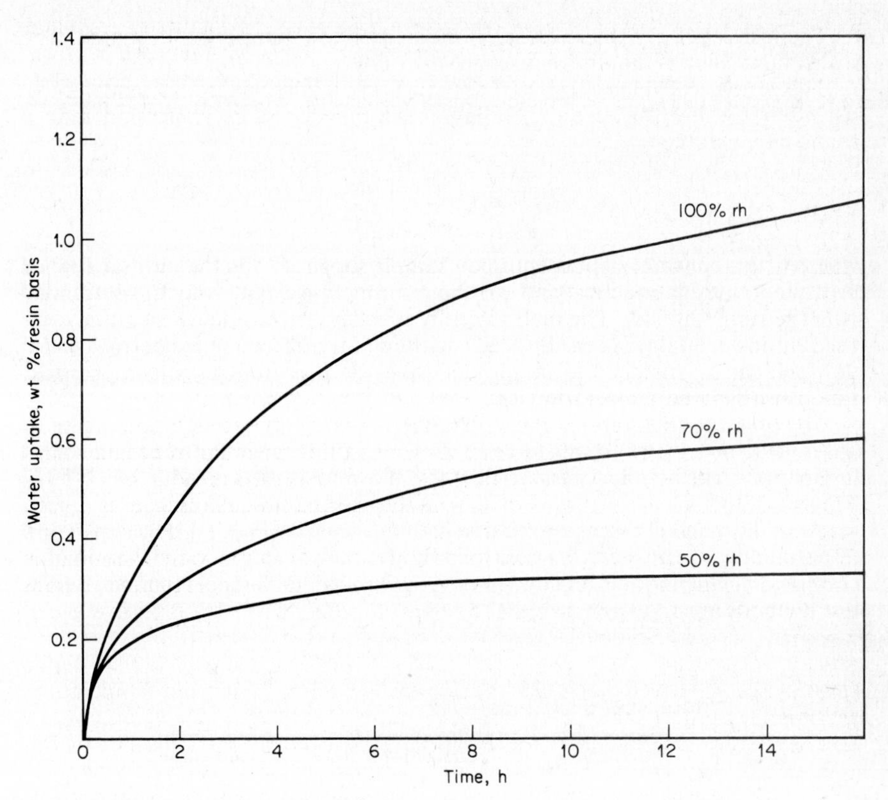

FIG. 32.12 Prepreg moisture absorption.

In the scaled flow test, the prepreg sample consists of 10 to 20 plies cut 5.5 ± 0.05 in by 7.0 ± 0.05 in. The 7-in dimension is cut parallel to the warp direction of the glass cloth. For glass thicknesses up to 0.0025 in (see Table 32.1), 18 or 20 plies are used. For thicker fabrics, 10 plies are used.

Cut specimens are placed in a desiccator for 24 h prior to testing. Absorbed moisture can affect the scaled flow test value; it acts as a plasticizer. After conditioning, the prepreg stack is weighed to the nearest 0.01 g and placed between two tedlar release sheets, with corner tape holding the package to the lower sheet.

No caul plates are used in this test, but a press plate 4.50 ± 0.01 in by 6.00 ± 0.01 in (0.125 to 0.250 in thick) is centered on the press platen and brought to a temperature of 150 ± 2°C. Other temperatures can be used where the prepreg user and supplier agree. The prepreg stack is centered over the press plate, and a ram force of 840 pounds (31 psi) ± 5 percent is applied for 10 min. The stack is then removed from the press and allowed to cool. Then, with the release sheets peeled off, the stack thickness is measured with a micrometer at three equally spaced points along one diagonal to the nearest 0.0001 in.

The three measurements are averaged to give the final cured thickness. Final thickness per ply can be calculated by dividing the stack thickness by the number of plies. The initial stack thickness H_0, is determined by calculation, using the initial stack weight W_0 and the stack glass weight W_g, measured after the resin is burned off. The relationship is:

$$H_0 = 1.17W_0 - 0.55W_g \quad \text{mils}$$

The initial per ply thickness is H_0 divided by the number of plies used. The scaled flow value is the thickness change per ply.

This test was developed by Bell Laboratories* and derived from a model showing that resin flow during any lamination is a function of an expression defined as the flow parameter *F*:

$$F = C\left(\frac{P}{A}\right)\int \frac{dt}{\eta(t)}$$

where C is a constant dependent upon sample shape, P/A is the ratio of applied pressure to lamination area, and η is the resin-melt viscosity, which is integrated over the time t of flow. The melt viscosity is controlled by the temperature cycle used during the lamination. By trial, the flow integral for the isothermal scaled flow test can be matched to the production-cycle flow integral, which normally involves a ramped temperature rise.

The pressure/area ratio is the scaling factor which gives this test its name. It has a value of 1.16 psi in this test and remains in this range for most large-panel laminations. In the high-pressure flow test, however, it has a value of 12.5 psi. This value, which is an order of magnitude greater, frequently causes "glass-stopped" flow and prevents correlation with full-scale lamination. The scaled flow test generally is resin-viscosity-controlled and is the key to production lamination correlation, which also is mostly viscosity-controlled. Differences in the two flow-test methods are shown in Table 32.5.

TABLE 32.5 Flow Tests

Condition	High pressure	Scaled flow
Applied pressure, psi	200	31
Lamination area, in^2	16	27
Pressure/area ratio	12.5	1.16
Test temperature, °C	171	150
Plies prepreg	4	10–20

As a true flow test related to resin advancement, the scaled flow value in mils of movement per ply can be converted to weight percent flow simply by using the resin density as a conversion factor. Scaled flow expressed as a final thickness in mils per ply is simply a cured thickness, largely dependent upon the initial prepreg resin content or weight per ply. It is not a measure of flow or prepreg advancement unless the initial weight also is controlled.

A minimum amount of flow is required during multilayer lamination to encapsulate and bond the etched circuitry and ensure void-free laminates. This is also known as the resin needed for fill. It can be specified as a minimum scaled flow value requirement.

Standard IPC scaled flow testing by the prepreg supplier is done to ensure a consistent product. It is not intended to guarantee a particular multilayer dielectric thickness, which is determined only by both the properties of the prepreg and the user's press cycle.

*D. Bloechle, *Journal of Elastomers and Plastics,* vol. 10, 1978, p. 375.

32.4.7 Gel Time

The gel-time measurement is widely used to determine prepreg reactivity or advancement. It is a test that is not easy to repeat between different operators because of small differences in technique. Testing is conducted on the bottom platen of a lab press or on an electrically heated hot plate. The surface temperature can be measured by using a thermal insulator to press a precalibrated thermocouple against the plate or platen. It should be adjusted to 171 ± 0.5°C. Dry resin powder for the test is extracted from the prepreg by crushing or folding, and then the collected resin is poured through a 60-mesh wire sieve. The required sample size is 200 ± 20 mg. A clock is started as soon as the resin powder is placed on the hot plate. After the powder melts, a tapered wooden stick is used to stir or work the melt, using a motion describing small circles with diameter of approximately $\frac{1}{2}$ in. It is important that the stirring be uniform and continuous until the resin stiffens and finally separates from the heated surface. The clock is then stopped and the gel time noted.

The gel-time test is an isothermal test measuring the length of time the resin remains flowable at 171°C. The flowable stage will be extended in time during MLB lamination in direct proportion to the rate of the press temperature rise (see Sec. 32.4.9).

Basically, the gel time of the prepreg must be sufficiently long to ensure good encapsulation of the circuitry and allow volatiles and air to escape, preventing internal voids. An excessively long prepreg gel time for a particular press cycle can cause heavy flow and lead to resin starvation and a weakened interlaminar bond. The reference method is given in section 2.3.18 in IPC-TM-650.

32.4.8 Heat of Cure

Another test of prepreg reactivity or advancement is the heat of cure or cure exotherm, conveniently measured by means of a differential scanning calorimeter (DSC). The same resin powder used in the gel-time test can be used as the DSC test specimen, but here it is limited to 4 to 7 mg per test and requires a microbalance for exact weighing. The heat of cure is expressed in joules or calories per gram of prepreg resin. The test has a repeatability of approximately 5 percent. Comparison of a prepreg's heat of cure with that of uncured (only dried) reference prepreg gives the material's degree of cure or advancement. A heat-of-cure "use range" can be established by correlating this value with actual lamination performance experience.

Figure 32.13 shows a typical DSC determination of cure exotherm on epoxy resin. The area under the bell curve is directly proportional to the heat of cure. The heating rate, 20°C/min, is an important test parameter, which must be specified when reporting heat-of-cure data.

32.4.9 Parallel Plate Rheometry

A thermomechanical analyzer (TMA) fitted with a parallel plate rheometer (PPR) attachment* can be used to measure resin melting and viscosity as a function of temperature or time. The test specimen is the resin powder used in the gel-time

*Du Pont Instruments.

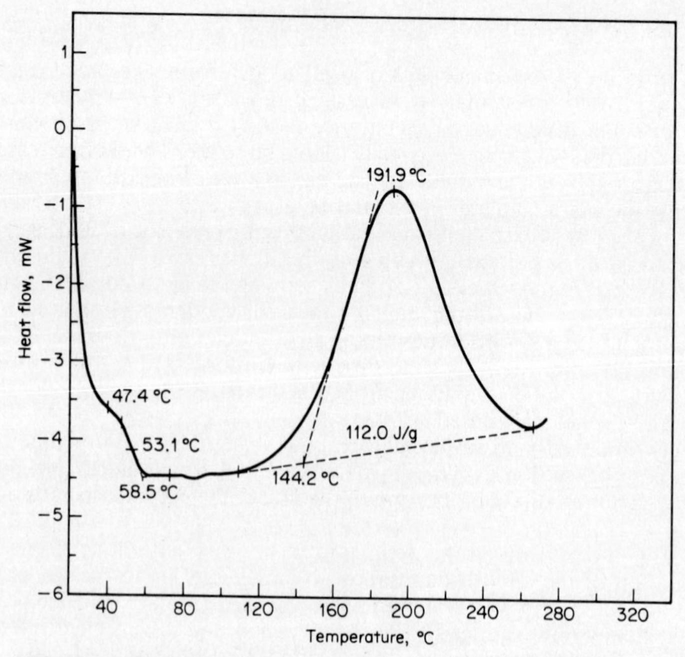

FIG. 32.13 Heat of cure of epoxy prepreg by DSC.

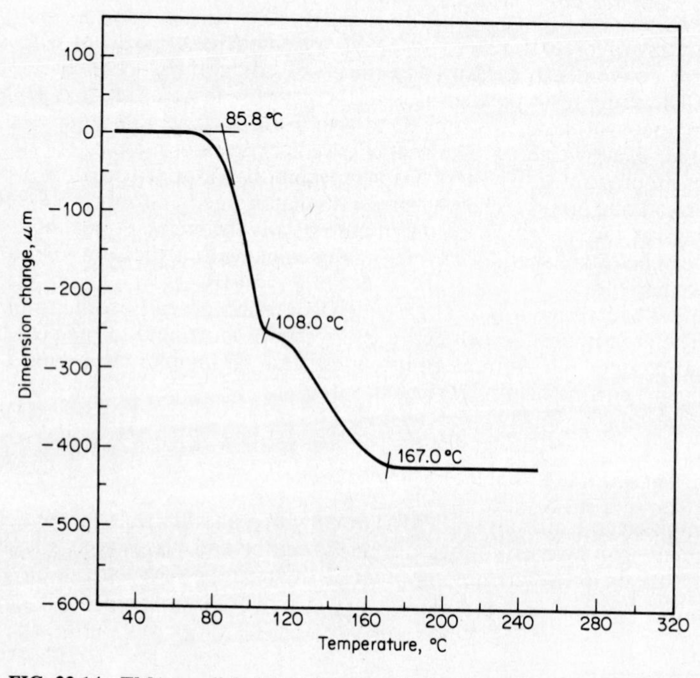

FIG. 32.14 TMA parallel plate rheometry of epoxy prepreg.

test. It is compressed into a wafer, inserted between parallel steel or quartz plates, and subjected to a selected load.

Figure 32.14 shows a TMA/PPR analysis of FR-4 epoxy B-stage resin. The deflection at 86°C is the onset of powder melting. The melting range extends to 108°C. The changing slope between 108 and 167°C is related to a changing melt viscosity. The flattened curve at 167°C represents infinite viscosity or gelation. For the heating rate used, 10°C per minute, this corresponds to a gel time of 486 s. As the heating rate is increased, gel times are shortened and gel temperatures increased. This test is another indication of prepreg advancement and performance in a temperature-ramped press.

32.4.10 Electrical Testing

The electrical properties of cured prepreg are as important as those of the base material of thin laminates for overall multilayer board performance. The military specification MIL-P-13949G requires the testing of three properties: dielectric strength, dielectric constant, and dissipation factor. Test specimens are prepared from two plies of prepreg made into an unclad laminate. The test methods and specifications are the same as those used for thin laminates.

32.4.11 Dicyandiamide

"Dicy" is used as a hardener in epoxy and modified epoxy-resin systems. It is a white crystalline solid initially dissolved in the resin varnish. As the solvents are driven off in the treater oven, dicy crystals can appear in the prepreg unless treater conditions are well controlled. This could create several problems. First, dicy is hygroscopic, and this can cause a moisture-absorption problem. Second, the concentration of dicy in the form of crystals could allow an excess to appear in one area and a deficiency (causing undercuring) to appear in another. It is important that the dicy remain evenly dispersed and not be allowed to form crystals.

To find possible dicy crystals, the prepreg is examined under 50× magnification using polarized light. A second polarizing filter, turned 90° to the first, is placed in the light path to create a dark background for viewing. The dicy crystals cause scattering of the polarized light, which passes through the second filter and makes them visible. To the inexperienced operator, prepreg crazing or fracturing may look like dicy crystals. Therefore, it is recommended that operators use reference specimens of prepreg containing known concentrations and patterns of dicy crystals.

32.4.12 Storage of Prepreg

The advancement or curing of prepreg continues from the time of manufacture until it is used for lamination. As with all catalyzed resins, the rate of advancement will be a function of the storage temperature. The degree of advancement will change with storage time. The shelf or storage life can be defined as the time prepreg can be safely stored at a specified temperature and remain suitable for use, meeting all specified requirements.

The military specification for shelf life is not less than 6 months after receipt of shipment when stored at a maximum temperature of 4.5°C (40°F) and not less than 3 months when stored at a relative humidity between 30 and 50 percent and a maximum temperature of 21°C (70°F).

Prepreg should be tested within 10 days of receipt, and after long storage times, it should be retested immediately prior to use.

After testing, prepreg should be returned to its original packaging and resealed until used. Moisture-barrier packaging is recommended to prevent or slow the ingress of moisture or other contaminants during storage. Prepreg stored in a cold environment should be stabilized without opening the moisture-barrier packaging a minimum of 4 h prior to processing.

If temperature indicators are used in the prepreg package, they should be checked on receipt of the prepreg to ensure that the material was not exposed to excessive temperatures during transit.

CHAPTER 33
FABRICATING MULTILAYER CIRCUITS

Allan G. Osborne, Ph.D.
Osborne Technical Associates, Seattle, Washington

33.1 INTRODUCTION

Multilayer printed circuits were developed in the 1960s as a result of electronic circuits becoming more complex and more densely packed. "Real estate" or board space, became progressively dear, and two-sided boards were inadequate. Military hardware, avionics systems, and some advanced computers were the first to adopt multilayer boards (MLBs). The Minuteman III computer used a combination of MLBs and some of the earliest integrated circuits (ICs). As the IC advanced to large-scale integration (LSI) and on to very large scale integration (VLSI), the requirements for more dense circuitry rose almost exponentially, although ICs were not the sole driving force for more and more space. Data from the IPC (Institute for Interconnecting and Packaging Electronic Circuits) show that multilayer circuit board production grew from less than 10 percent of the total circuit board production ($100 million total MLB value) in 1964 to 46 percent ($2 billion total MLB value) in 1984.

Some early MLBs were fabricated by laminating plated-through-hole two-sided boards to form four-, six- and eight-layer MLBs with holes that did not go all the way through the board. These feed-throughs or "vias" were called "buried" if they did not terminate at the surface and "blind" if they did. Components were attached to these boards by soldering to surface pads or "lands." The driving force for this technique was *necessity* because of the absence of holes through the boards or "normal" through-hole mounting. This bit of history is pertinent because surface mount technology (SMT) has become a major force in influencing the component industry and the design and production of circuit boards, mostly MLBs. "Real estate" and cost are driving the SMT revolution (surface mount components are smaller and will be cheaper than their through-hole counterparts), but the need for unbroken surface space coupled with the reduced need for through-holes has made the major impact on the resurrection of the blind and buried via.

Essentially all MLBs, from the outset up to 1983, were fabricated with through-holes for mounting components; large numbers will continue to be so produced for many more years. However, as the trend toward SMT continues, some of these will evolve into through-hole vias (feed-throughs) that, since they will not require a soldered-in lead, will be much smaller in diameter (e.g., 0.020 versus

0.045 in), yielding more board space for component mounting. The process described here, therefore, can be expected to provide the basis for multilayer production of boards required by SMT as well as traditional needs.

33.2 MULTILAYER BOARD PROCESS GENERAL DESCRIPTION

The typical multilayer circuit board is made up of successive layers of copper, C-stage (fully cured) epoxy-glass dielectric (the same as is used in two-sided boards), and B-stage (prepreg) epoxy-glass that has been dried and partially cured, as bonding material—"glue"—to hold the board together. The stack of layers is heated under pressure until the prepreg is cured. Materials other than glass-reinforced epoxy are also used, but to a considerably lesser degree: polyimide-glass is used for high-temperature applications, and Teflon-glass is used for high-frequency applications. Kevlar, quartz fiber, or even graphite fiber is infrequently used in place of fiberglass for certain special applications.

The inner layers are formed by printing and etching circuits and planes on copper-clad dielectric materials (drilled and plated through in the case of buried vias). Dielectric thickness for these inner layers will depend upon the number of layers and overall board thickness specifications. Common thin-core dielectrics are 0.031 in, 0.016 in, 0.012 in, 0.010 in, 0.008 in, and 0.005 in, which are referred to as "*x*-core"; e.g., material with a dielectric thickness of 0.005 in is "5-core." Copper foil cladding is typically 2 oz/ft^2, 1 oz/ft^2, $\frac{1}{2}$ oz/ft^2, with 1 oz/ft^2 being the one most often used. Copper foil that weighs 1 oz/ft^2 is normally 0.0014 in thick.

Interconnection between layers is accomplished by drilling holes that penetrate features that are to be connected electrically. Since drilling produces heat and heat melts epoxy, the drilled-through edges of the inner layers of copper may be "smeared" with epoxy, producing a dielectric layer that will prevent electrical continuity to the inner layers during subsequent plating steps. Therefore, after drilling and deburring, a smear-removal step is necessary. After smear removal, electroless and electroplated copper are deposited on the board and in the holes in the same manner as in two-sided, through-hole plated boards. The remainder of the steps are identical to those in the two-sided, through-hole process.

33.3 FABRICATION PROCESS STEPS

The process chart in Fig. 33.1 and the ensuing process descriptions provide an outline of a good basic process. Only minor variations in these processes will be encountered among MLB manufacturers.

This chapter and the next are intended to show the basics of MLB manufacture. However, it cannot be a complete recipe book, nor can it describe all the details of the entire multilayer technology, using either exotic or standard materials, with more than 8 or 10 layers. Most such MLBs are for highly specific applications, and while they are based on this process, they require more detail than is possible here. However, the process described is appropriate for the greatest part of multilayer production and is the foundation for almost all the rest.

1. *Inspect raw material:* Check C-stage (fully cured clad or unclad material) for dimensional stability, copper peel strength, and embedded foreign particles.

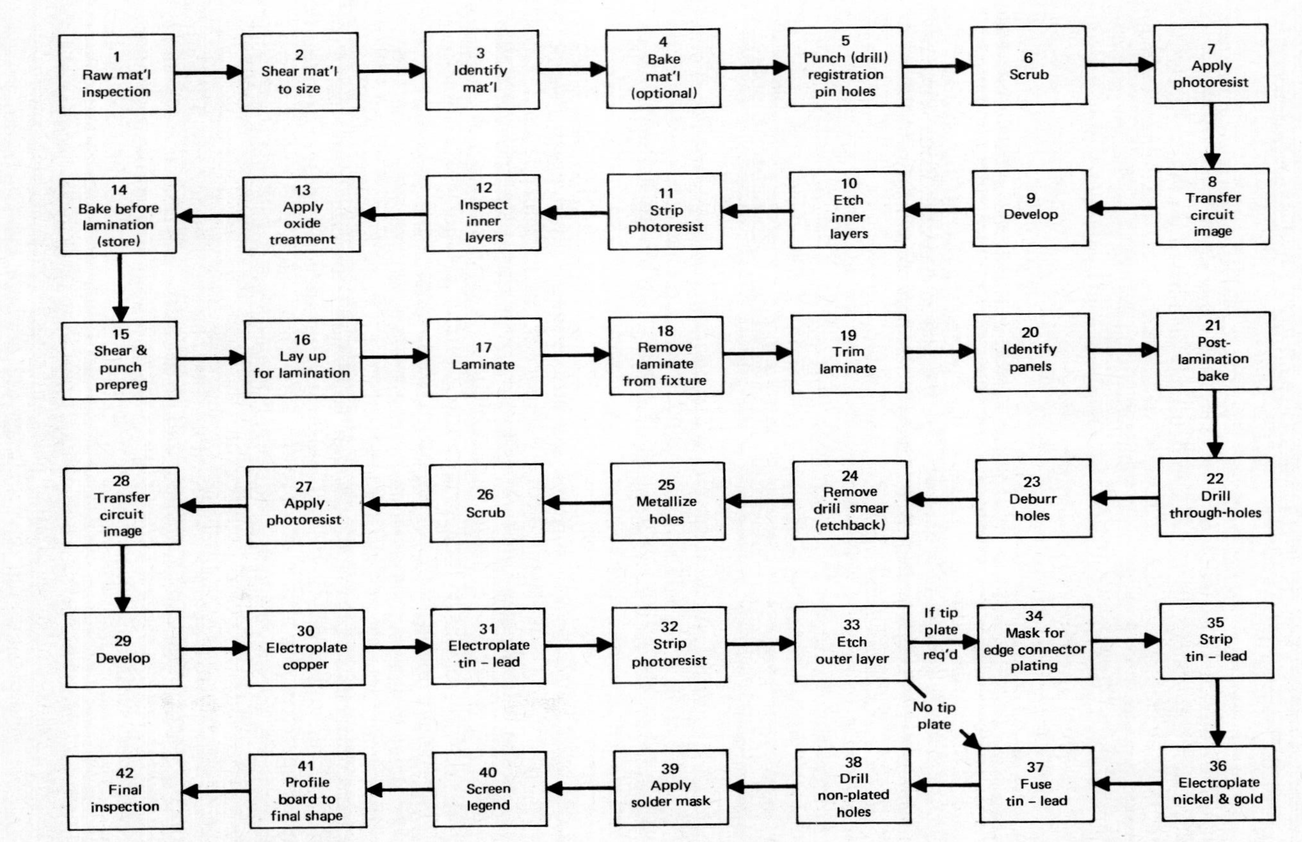

FIG. 33.1 Process flowchart.

Most tests need only be run on a sampling (AQL) basis unless there is a military specification or customer requirement otherwise. Check prepreg (B-stage) material for gel time, resin flow, resin content, epoxy type, and glass weave type—in approximate order of importance. (See Chap. 32.)

2. *Shear material into panel sizes:* Most companies purchase precut panels in a limited number of standard sizes. The IPC recommends 18 × 24 in as the basic standard panel size. Panel size is usually also based on these criteria: (*a*) minimum trim scrap from standard raw material sheet size, (*b*) number of circuit images per panel (step and repeat), (*c*) press platen size, (*d*) raw material dimensional stability, and (*e*) production volume of a given circuit board.

3. *Identify material:* Once sheared, sawed, or routed, raw material is difficult to identify for copper-clad and dielectric thicknesses. Storage in marked bins presents material control difficulties once it is issued to the line. A stamped code along a panel edge is recommended. The marking should include an arrow to indicate the "strong" thread or machine direction (warp thread) of weaving. It is important that all material be laid up so that all layers have the same weave orientation. "Cross-laid" lay-ups tend to bow or warp and twist badly.

4. *Bake material (optional):* Most manufacturers of raw material give their material a postcure in the press in "sheet" (3 × 4 ft) form. Many fabricators use a further bake or postcure after cutting to panel size (either by the supplier or in-house) in the belief that better dimensional stability is achieved. Baking temperature should be above the second-order glass transition temperature of epoxy, or a minimum of 275 °F.

5. *Punch (drill) registration-pin holes:* Registration systems use holes, orthogonal slots, or hole-and-slot combinations. These features must be accurately located, since they directly affect circuit-to-circuit registration between MLB layers.

6. *Scrub material:* Material is mechanically or chemically cleaned to remove oxides, oxide conversion coatings, fingerprints, or other soils from panel surfaces so that photoresist will adhere to panels. Pumice is not recommended as a scrubbing medium. No matter how thoroughly a panel is rinsed after pumice scrubbing, some pumice will remain embedded in the copper cladding. When dry, pumice can easily be accidently transported around a manufacturing area and contaminate plating baths.

7. *Apply photoresist.* The photoresists that are most often used are dry-film, positive-acting (which polymerize and become insoluble when exposed), and solvent-, semiaqueous-, or aqueous-developing. Photoresist is applied to a panel so that a circuit image may be transferred to it. This and the exposure operation must be performed in a humidity-, temperature-, light- and dust-controlled environment. Particle-caused, etched-out circuit features not detected by inner-layer inspection will be "buried" after pressing. (See Chap. 11.)

8. *Transfer circuit image:* A phototool (silver halide or diazo copy of 1:1 artwork) is placed in an exposure frame and given a time exposure to a UV light source which polymerizes the photoresist under the clear areas of the phototool. Polymerized areas are resistant to developing, etching, and plating solutions, whereas the unexposed areas of resist may easily dissolve by appropriate developing solutions. Accurate, repeatable image transfer to the panel is accomplished while using pins to align registration holes and slots in the panel and in the phototool. The phototool should carry an image identification that will be transferred to the panel border.

Direct-imaging systems with a computer-controlled raster or boustrophedonic laser beam that "writes" the image directly with no intervening phototool are also available.

9. *Develop circuit image:* Photoresist-coated and exposed panels are developed by washing away unexposed photoresist with the appropriate solvent or aqueous system. This exposes bare copper that may then be etched to form the inner-layer circuit patterns.

10. *Etch inner patterns:* Conventional MLBs are made with "print-and-etch" inner layers as described in steps 8, 9, and 10. "Buried or "blind" vias (Fig. 33.2) or feed-throughs (holes that carry current but have no parts soldered into them) are formed by fabricating two-sided plated-through holes (PTHs) with thin inner-layer material, filling the holes with epoxy resin, and then laminating the PTH layers together as if they were regular print-and-etch inner layers.

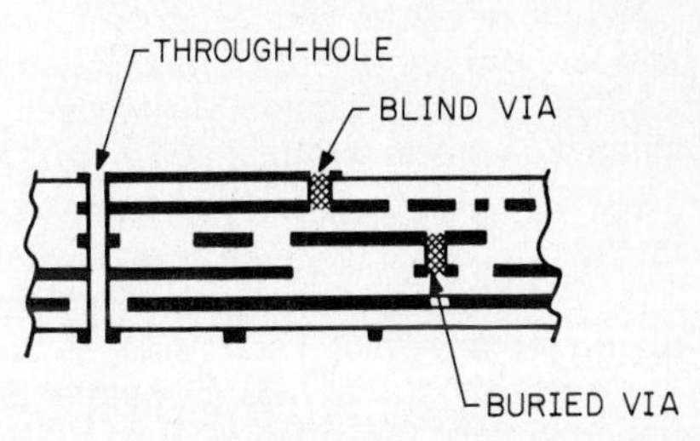

FIG. 33.2 Buried and blind vias.

All exposed copper is etched from the panel. *Note:* A ring of copper should be left around all registration holes for reinforcement, particularly on very thin laminates. A "broken" frame (Fig. 33.3) should be left around the outer edge of all panels to prevent too rapid a flow of resin while pressing but to still allow air bubbles and some resin to escape.

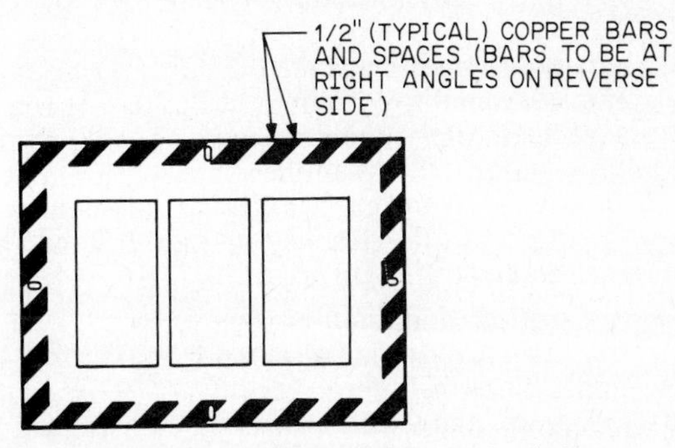

FIG. 33.3 Copper "frame" and registration holes.

11. *Strip exposed photoresist:* Photoresist has served its purpose and must be removed using the appropriate solvent or aqueous caustic solution.

12. *Inspect inner layers:* Inspection should be made for opens or line-width reduction (neither one recommended for repair) and shorts or other unwanted copper. These may be removed; unetched copper in open areas may frequently be ignored. Painstaking inner-layer inspection is economically important, since a scrapped inner layer is much less expensive than a laminated MLB; the more layers to the finished board, the higher the value added.

As inner layers become more complex and larger numbers of layers render boards more expensive, traditional visual inspection becomes more and more inadequate. Often 18- × 24-in panels can take from 30 to 60 min to inspect, visually. Even so, flaw escape rates may reach 40 to 50 percent in some cases. This is devastating to the yield of even a six-layer board. Automatic optical inspection (AOI) systems are commercially available; while expensive, they provide the only viable alternative to the human eye. Electrical continuity testers for inner layers have the severe limitation that they do not detect design-rule violations: near-shorts and near-opens. State-of-the-art AOI electronic vision systems are capable of inspecting 18- × 24-in panels with 0.010-in line circuitry in less than 30 s with a flaw escape rate of 0.5% while 0.003-in line circuitry is said to require 1 min or more.[1]

13. *Apply oxide treatment:* Epoxy-to-epoxy adhesion is not difficult to attain in multilayer lamination. Copper-to-epoxy bonds are more difficult to form and require that the copper have some form of pretreatment, the most common of which is an oxide layer. High-temperature black oxides are giving way to lower-temperature red or brown oxides (the latter two are required for polyimide).

Prior to oxiding, panel surfaces may be mechanically scrubbed and are usually given a chemical cleaning and a light etch.

14. *Bake before lamination:* Water, solvents, and other volatiles absorbed by the inner layers in previous processes must be removed to prevent voids, blisters, and delamination that can occur during the lamination process or in subsequent high-temperature processes such as tin-lead fusing and wave soldering.

Panels should be baked at ambient pressure at 175 to 200°F for 2 to 4 h or under reduced pressure at 100 to 125°F for 1 to 3 h. While awaiting lay-up, panels should be stored in a desiccating cabinet up to 24 h, after which they should be rebaked. Handling should be only by operators wearing clean protective gloves or by automated equipment.

15. *Shear prepreg (B-stage) and punch registration-pin holes:* As with inner-layer material, prepreg may be purchased cut to size—punched, if so specified. Since prepreg is not registered to the inner-layer images, the holes should be oversize.

Prepreg should be stored under controlled conditions (50°F and 50 percent RH). It should be removed from refrigeration in sealed bags to remain sealed until room temperature is reached. After punching, prepreg should be stored in a desiccating cabinet at approximately 70°F at ambient pressure or under vacuum. Prepreg must be handled carefully to prevent breakage and particulate contamination (glass particles and epoxy dust) of the clean lay-up area. It should be handled only by operators wearing clean protective gloves.

16. *Lay-up for lamination:* Lamination is performed by using fixtures with registration pins to maintain layer-to-layer spatial relationships. A typical lay-up (Fig. 33.4) consists of the following:

- ***a.*** Thermal insulation material to control rate of temperature rise.
- ***b.*** Bottom lamination fixture or "caul plate."
- ***c.*** Sheet of release material such as Teflon-glass cloth or Tedlar film.
- ***d.*** Bottom circuit panel: (1) unetched bottom copper-clad, dielectric, and etched inner layer of clad copper, or, more likely, (2) unetched bottom copper-clad, dielectric, and no inner copper (single-clad capping method).
- ***e.*** Required number of sheets of prepreg.
- ***f.*** Inner circuit panel (double-clad with copper, etched both sides).
- ***g.*** Required sheets of prepreg.
- ***h.*** More inner layers and sheets of prepreg to make up numbers of layers called out on drawing.
- ***i.*** Top circuit panel (see *d*).
- ***j.*** Sheet of release material.
- ***k.*** Top lamination fixture. Boards may be multiply stacked in the fixture during lamination. (Separated by release material with or without "ferrotype plates.")
- ***l.*** Thermal insulation material.

17. *Laminate lay-up:* Application of heat and pressure typically (350°F and 150 to 350 psi pressure; presses are usually of 100 to 150 tons capacity) converts

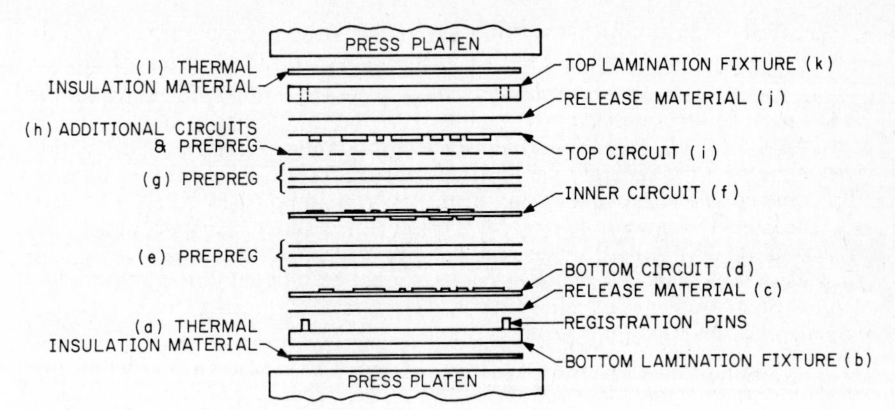

FIG. 33.4 Typical multilayer board (MLB) lamination process lay-up.

prepreg between etched circuits to fully cured (C-state) epoxy. During the process, the prepreg resin liquifies for a few minutes, flows to fill voids and spaces in the etched circuitry, replaces the air, and finally cures to bond the layers together. Vacuum lamination has become an alternative to the above pressure and temperature process. Equipment may be either a vacuum "bag" for the lay-up, enclosed in a pressure chamber in which an isostatic pressure of heated carbon dioxide is used in place of a press, or a standard hydraulic press adapted to accommodate a vacuum chamber surrounding the platens. Vacuum lamination will be covered in more detail in Chap. 34.

18. *Remove laminate from fixture:* After the stack is cooled to near room temperature in the press or in the cooling portion of a hot-transfer press, registration pins are pushed out carefully (e.g., with an arbor press), and the top lamination fixture and succeeding MLBs are carefully removed, with care taken not to scratch or mar the surfaces of either the lamination fixtures or the copper-clad MLBs.

19. *Trim laminate:* Shear prepreg flash (typically $\frac{1}{4}$ to $\frac{1}{2}$ in) from the perimeter of panels, and drill any plugged tooling holes. Abnormal flash can indicate lamination problems and should be noted. For example, an inch of flash could indicate too much flow and possible resin starvation in the finished board. Bubbles in the flash can indicate excess moisture in the prepreg or inner layers that can cause possible blistering of the MLB in the tin-lead fusing or wave-soldering operations.

20. *Identify panels:* Internal identification is not visible until after etching. Some type of identification (part number, work order) should be marked on each panel.

21. *Post-lamination bake:* Bake panels under weights (e.g., a steel caul plate) at a minimum of 275°F for at least 2 h at temperature to stress-relieve and postcure them.

22. *Drill holes for "plating through":* Holes to be plated through are drilled just as for two-sided boards, except that feed rate (chip load) and drill speeds are reduced. MLBs may be stacked for drilling within the limitations of drill size. Table 33.1, in Sec. 33.5, gives typical ranges for drilling parameters. An entry sheet, usually of solid aluminum, and a backup sheet, of thin aluminum-clad composite material with epoxy binder, are included in each drilling stack to reduce copper burring and drill temperature. (See Chap. 10.)

23. *Deburr holes:* Drilled panels are run through a deburring machine to remove drilling burrs. In-hole debris is best removed with pressurized water.

24. *Remove drill smear (etchback):* Heat from the drilling process causes the epoxy to melt. The melted epoxy smears over the inner rings of copper formed by drilling through internal pads. This dielectric coating, if not removed, will prevent complete formation of internal circuit connections to produce opens or resistive connections. Removal of drill smear may be accomplished using liquid honing, chemical treatments, or plasma. Etchback, if required, removes more hole-wall material than simple smear removal and must include a process to remove glass as well as epoxy. Registration holes should be plugged during etchback to protect them from undue enlargement. The chemical smear-removal–etchback process includes a neutralization step and a bake-out to drive out any retained solutions.

25. *Metallize holes (electroless copper):* This process consists of a series of chemical cleaning steps, rinses, and a sensitizing catalyst bath. The sensitizer activates the exposed epoxy surfaces in the drilled holes to initiate deposition and enhance adhesion of electroless copper, which is deposited to a thickness of 60 to 100 μin. The practice of applying thinner electroless copper and an electrolytic copper strike has virtually disappeared from use.

26. *Scrub for photoresist application:* See step 6.

27. *Apply photoresist:* See step 7.

28. *Transfer circuit image:* See step 8. Since the circuit area is to be plated upon, positive phototools (areas to be plated, circuit traces, pads, etc., will be dark) will be used, assuming negative-acting resist (polymerizes on exposure).

29. *Develop outer circuit images:* See step 9. Development of the photoresist causes the removal of unexposed resist in all the areas to be electroplated. All other areas are covered with exposed resist.

30. *Electroplate copper:* (See Chap. 12.) This process consists of a series of cleaning, rinsing, and microetching steps followed by the electroplating bath, usually "acid" (copper sulfate) electrolyte with appropriate brighteners and levelers. Anodes are of purified copper, while the panel to be plated serves as the cathode workpiece. Current densities, calculated on the basis of exposed copper, are most commonly 25 to 50 A/ft^2 but high-speed proprietary coppers with much higher current densities are becoming popular. Target copper thickness is usually 0.001 in minimum inside the plated hole.

31. *Electroplate tin-lead:* (See Chap. 12.) Tin-lead is the most common overplate used on PWBs. Following copper electroplating, panels are processed directly into the tin-lead line without unracking or drying. Tin-lead is electroplated much the same way as copper. The anodes are tin-lead alloy, and the electrolyte is usually a tin and lead fluoborate solution with appropriate additives. Plating thickness is usually in the range of 0.0003 to 0.0005 in but may be thicker for special applications.

32. *Strip photoresist:* See step 11. In this step, photoresist is stripped from the panel with caustic solution, leaving exposed the copper to be etched.

33. *Etch copper:* Exposed copper must be removed to leave only the tin-lead-plated outer MLB surface circuit pattern.

34. *Mask circuit board for edge connection plating:* Chemical-resistant plating tape applied with pressure rollers is commonly used to provide masking of circuitry above the edge connector fingers to prevent wicking of stripping and plating solutions along the circuit traces (Fig. 33.5).

35. *Strip tin-lead:* Tin-lead plating is removed from the edge connector pattern to expose bare copper.

36. *Electroplate nickel and gold:* Low-stress nickel is electroplated onto the edge connector fingers, followed by hard (200 Knoop) gold electroplating. Currently, nickel plating thickness is usually 200 μin, and gold 30 to 50. Plating tape is removed mechanically, and adhesive with solvent, if necessary.

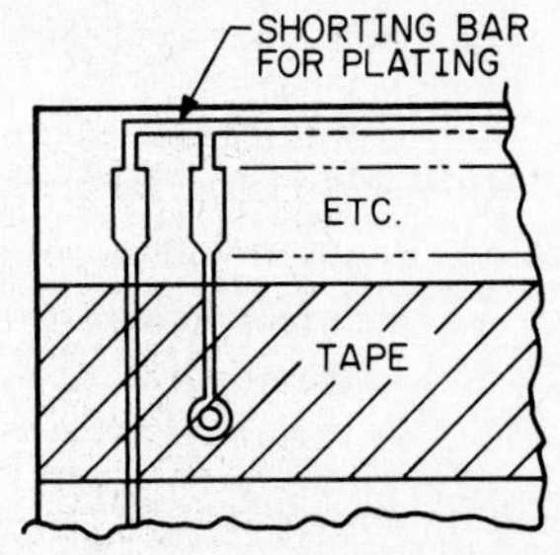

FIG. 33.5 Edge connector, tape-masked for stripping.

37. *Fuse (reflow) tin-lead:* Fusing of tin-lead causes the porous deposit of discrete areas of tin and lead to become a homogeneous solder alloy less susceptible to oxidation than the as-plated material. Fusing is usually carried out in a conveyorized infrared fuser with an in-line washer. Some fabricators still use hot oil (synthetic).

38. *Drill nonplated holes:* Many MLB designs have holes in which plating is undesirable. The alternative is to plug these holes during plating.

39. *Solder-mask panels:* Most MLBs require solder mask, either screened or dry-film. Component identification and other markings are either etched in or screened.

40. *Screen legend (component identification):* Component identification and other markings such as diode clocking are normally screened in white or yellow ink over solder mask.

41. *Profile board to final outline:* While template routing is often used for profiling MLBs, CNC (computer numerical control) routers are now commonplace. The key to good routing, whether manual or computerized, is the use of *sharp* tools. This also applies to bevelers.

42. *Final inspection:* Final inspection should include (*a*) visual defect inspection, (*b*) dimensional inspection, (*c*) bare-board electrical continuity and leakage test, and (*d*) coupon cross section for plating thickness, hole quality, and other features that may be required (military, special customer requirements).

33.4 REGISTRATION AND LAMINATION

33.4.1 Circuit Registration System

Circuit registration is crucial to the entire process in any MLB fabrication facility. Each layer of an MLB must be located as accurately as possible in relation to all other layers. Otherwise, during drilling, a mislocated feature may be improperly attached to a plated-through hole, causing an internal short and a scrap MLB. Accuracy of the registration system itself does not include the effects of photoresist exposure, material dimensional stability, etching, copper thickness, drilling accuracy, or other processing variables, but all of these and others, such as temperature-humidity instability of phototools, have a measurable effect on the alignment of the final product.

Registration system accuracy is the tolerance variable between alignment pins and holes on equipment using or generating those features.

It is important that the registration system be carefully considered for the following reasons:

1. It is additive to processing variables and overall fabrication accuracy.
2. It can be handled with accuracy and consistency from machine to machine.
3. If it is not considered properly at an early stage of setting up a facility, it becomes increasingly difficult to change because of the many specialty tools and fixtures that are generated.

While three-hole systems have been prevalent for many years, the most common and most accurate method used is the "orthogonal slot" system or some variation thereof. There are four slots, with those of the x axis being at right angles (orthogonal) to those of the y axis (Fig. 33.6). One advantage of this system is that since movement is from the center of the panel, errors in expansion or contraction of phototools are spread over an entire panel instead of being accumulated at one edge.

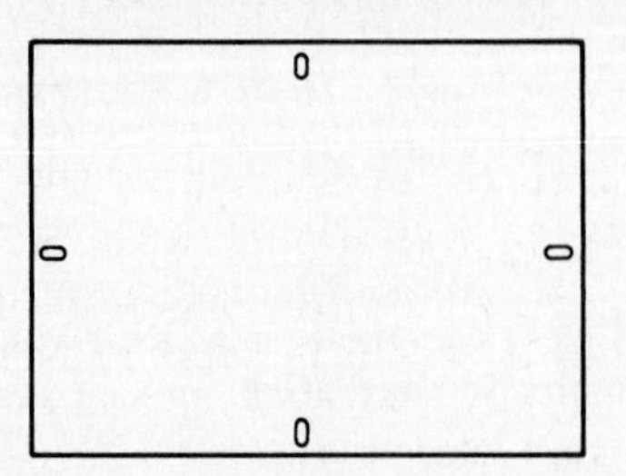

FIG. 33.6 Orthogonal slot registration system.

Overall registration is not usually expressed as "plus or minus some amount" from front to back or from outer to inner layers; instead, final acceptance is based on the MLB's conformance to design rules and annular ring specifications, such as the requirement that the hole be completely surrounded by the annular ring without any breakouts.

33.4.2 Lamination

Lamination is the first MLB process step in which process errors cause the scrapping of more than an individual layer. Special materials and process control will pay off in reduced scrap. Whatever the material to be used, rigid cleanliness precautions and environmental controls are mandatory.

1. *Environmental precautions:* Lamination bonding material (prepreg or B-stage) should be stored in a chamber with controlled temperature (50°F) and humidity (50 percent RH maximum) in sealed bags. Prior to use, bags of prepreg should be allowed to come to room temperature before being opened. This prevents condensation of airborne moisture during the equilibration process. It is recommended that prepreg be purchased precut and even prepunched. If it is to be sheared or punched, the packages must be equilibrated in the shearing or punching area. Both prepreg and C-stage (cured) material should be handled only by operators wearing clean gloves. The lay-up area should have a filtered, positive-pressure air supply and controlled temperature and relative humidity (72°F and 50 percent RH maximum).

2. *Laminate preparation:* Epoxy-to-epoxy bonds are readily attainable with clean inner-layer laminates and prepreg that has not exceeded its shelf life. Bonding to inner-layer copper is more difficult, particularly where circuitry is closely spaced or where large ground planes are present. (Such planes, where electrically feasible, should be cross-hatched or perforated.) Copper surface treatment has moved from scrubbing and abrading to chemical etching (e.g., nitric-acetic acid) to growing "black-oxide" and finally brown-oxide coatings. Originally designed for copper and brass as a base for painting these metals, oxiding has become the treatment of choice for adhesion enhancement of inner-layer copper. Brown oxides, sometimes described as red oxides, are applied at lower temperatures than black oxides (170°F versus 200°F) and are gaining widespread use both for energy conservation and for the fact that brown oxide is mandatory for adequate adhesion in polymide MLB.

The typical black-red-brown oxide solution is a mixture of sodium chlorite as the oxidizer with sodium hydroxide and trisodium phosphate as stabilizers, dissolved in water. Most solutions will also include a surfactant (wetting agent). Other additives and congeners of apochryphal value will be found in proprietary preparations. A red oxide may also be produced by depositing a heavy black oxide and partially removing it with dilute acid.[2]

After the application of an oxide, inner layers are dried and baked and must then be kept in a desiccator for not more than 8 h before lamination (rebake if longer) and must be handled by operators wearing clean lint-free gloves.

3. *Material testing:* It is important to check incoming B-stage prepreg lots or to check lots that may have exceeded shelf life or optimum environmental conditions during storage. Commonly tested for the former are resin content, resin flow, gel time, and volatiles content; for the latter, gel time. While these parameters are still important, *scaled flow*[3] has replaced percent of resin flow as the accepted test to allow a fabricator to evaluate a prepreg sample and set press conditions for its use.

The older resin flow test uses a 4- × 4-in sample at 200 psi laminating pressure (most prepregs are single-stage and have no "kiss" cycle). It is pointed out in Ref. 3 that an important factor in the modern resin flow model, P/A (the ratio of pressure to the area of the sample, having units of psi/in^2 or lb/in^2/in^2), is tenfold higher in the resin flow test sample than in a typical lamination using, say, a 12- × 18-in panel. In some cases in the resin flow test, flow may be "glass-stopped" because the pressure plates are resting on the glass fibers. Scaled flow prevents this and other problems and gives pressed-thickness values for a given resin content. Prepreg should be purchased with a requirement for percent of resin and press thickness as well as minimum gel time or open flow percent.

33.4.3 Lamination Fixtures

Lamination fixtures (caul plates; see Fig. 33.7) are most commonly made of steel, since steel approximates the thermal expansion coefficient of copper-glass-epoxy

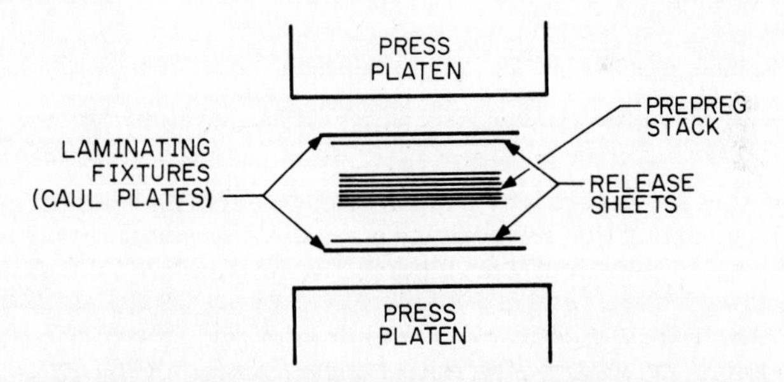

FIG. 33.7 Typical lamination press lay-up for scaled flow to check prepreg.

resin composite materials below the glass transition temperature (250 to 260°F). The lamination fixture must contain at least two registration pins to align the MLB circuit laminates. MLBs of many thin dielectric layers (eight layers or more) will require multiple-pin fixtures. Such fixtures may contain (orthogonal) slots or pins (holes) spaced every 3 to 5 in around the perimeter of the fixture. Laminate inner layers must have the same pattern registered to the circuitry. Foil or single-clad "caps" need not be registered and may even have oversize holes.

33.5 DRILLING MULTILAYER HOLES FOR PLATING

Drilling holes in the copper-clad epoxy-glass material primarily used for MLB fabrication is an area that contains as many identifiable variables as any other process step, if not more. Drill machine technology has progressed rapidly, and state-of-the-art equipment is fast and accurate. What is usually referred to as the "drill machine" or just "drill" is actually a complex, relatively expensive, highly precise *x-y* table equipped with multiple (usually four) drill spindle motors that change tools automatically, upon command. The secret to good drilling is the drill bit itself. Fragile, often no larger than a sewing machine needle, its configuration, sharpness, rate of rotation (speed), and rate of penetration (feed, chip load) are the keys to hole quality.

For a detailed description of the drilling process, refer to Chap. 10.

Drilling machines in general use can produce 200 to 300 holes per minute using table travel speeds of over 400 in/min and spindle speeds of 80,000 r/min. Hole production rate is also dependent on feed rate (chip load) and programming (optimized drill path). In production, hole location accuracy is 0.002 in or better.

Computer control, numerical control tapes or disks, and combinations of the two drill table control modes offer a wide range of flexibility and automation to the hole-drilling operation. Drill data from computer-aided design (CAD) may be put onto floppy disks at either the CAD site or the drilling site.

TABLE 33.1 Comparison of Two-Sided and MLB Drilling Conditions. *Industry usage.*

	Two-sided	Multilayer
Surface feet per minute	600–700	0.500–600
Chip load	0.003–0.005	0.002–0.004
Number of repoints	3–5	0–2
Number of hits	3000–5000	1000–2000
Stack height	3–4	1-3
Entry material	Al-clad phenolic.	0.010-in aluminum
Backup material	Al-clad phenolic.	

Drilling holes for subsequent plating through entails a carefully orchestrated combination of parameters, whether two-sided PTH boards or MLBs are being drilled. Table 33.1 illustrates the differences and similarities between the two types of boards.

Backup material is normally 0.003 aluminim-clad, epoxy-bound (nonglass) cellulosic composite. Entry material is most commonly 0.010-in aluminum. These materials reduce entry and exit burrs and help to cool the drill before and after entry.

As we have previously mentioned, MLBs are susceptible to "drill smear" from melted epoxy caused by dull drills and improper feeds and speeds. While smear cannot be eliminated, it can be reduced. For MLBs, compared with two-sided boards, we must use slower drill speed, lower chip load, restrict the number of "hits" (number of times the drill penetrates the material before it is repointed or discarded), repoint fewer times, and use fewer boards per stack. Stack height also depends on board thickness and on drill size. It is recommended that boards be drilled individually (one-high) when drills with diameters of less than 0.015 in are used (no. 79 and up: the smaller the drill, the larger the number). Also, boards should be stacked two-high for drills 0.015 in to 0.020 in (nos. 78, 77, and 76) and no more than three-high for all drills larger than 0.020 in. Besides being easily broken, drills smaller than no. 80 (0.0135 in) are restricted to one-high stacking because of their short (0.180-in) flute length.

33.5.1 Common Drilling Problems

1. *Epoxy smear:* Drilling generates, on the drill-face surfaces, local temperatures which can melt epoxy resin. Good drilling practices minimize the effect through a satisfactory combination of feed rate, drill speed, and drill sharpness. Poor drilling through a different combination of the same parameters may leave varying amounts of a thin epoxy coating across the face of each internal pad penetrated (Fig. 33.8). This epoxy is an insulator that, if not removed, will insulate hole plating from internal MLB circuitry. The result will be an open or high-resistance connection and cause for a scrap board. Even with mechanical parameters in good control, epoxy smear can be caused by lower-melting, partially cured prepreg in the board bond line. Proper laminating conditions, careful monitoring of prepreg cure by thermal analysis techniques, and a postbake following lamination provide insurance against this cause of epoxy smear.

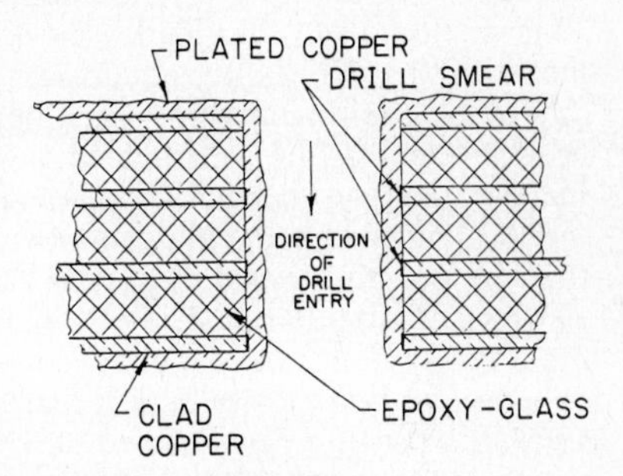

FIG. 33.8 Drill smear.

2. *Burrs:* Copper burrs can occur on the top and bottom of the MLB. They are usually caused by (*a*) dull drills, (*b*) improper backup and entry materials, (*c*) improperly adjusted pressure foot on the drill machine, or (*d*) incorrect speed and feed combination on the drill.

3. *Hole quality or appearance:* "Hole quality" refers to the straightness and smoothness of the walls of drilled holes. It is a function of drill feed, speed, and sharpness as the drill bit does the work of penetrating, cutting, and removing material from the hole.

4. *Nailheads:* "Nailheading" describes an effect sometimes observed on internal copper layers, so named for its appearance in cross section (Fig. 33.9). The copper can be widened to as much as twice normal thickness at the hole wall. Dull drills or improper feed rate and speed are the contributing factors.

All of these problem areas are so interactive that there is some controversy as to exactly how some of the problems are caused and what conditions are optimum.

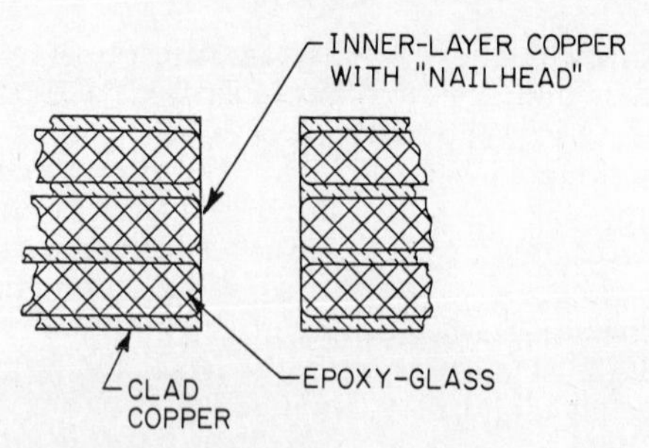

FIG. 33.9 Nailheading.

33.5.2 Causes and Mechanisms of Drilling Problems

1. *Drill feeds and speeds:* Probably the most common causes of drilling problems are chipped or dull drills or improperly set drill feed rate and speed.

Drilling is a material cutting and removal process. Feed rate is the rate at which the drill penetrates. It can be defined in inches per minute or, as is more usual, mils (thousandths of an inch) of drill penetration per revolution of the drill. This is known as "chip load." We want rapid penetration to minimize drill contact with the material to reduce heat and wear caused by friction, but penetration that is too rapid results in punching, producing bad exit burrs and internal nailheads or extrusion, which also produces nailheads. Copper is a soft, poorly machinable material that does not readily form chips but rather smears (galls) at the tool cutting edge and generates heat from the friction. Epoxy is more brittle than copper and forms chips, but the heat from drilling friction causes melting, as we have already noted. The softening of the epoxy causes more drag and thus more friction and compounds the problems. Glass is even more brittle and highly abrasive. Its main contribution to drilling problems is drill wear—uneven wear—since it is not homogeneously interspersed with the copper-clad epoxy-glass composite. Slowing the penetration rate of the drill allows more material contact time, more friction, more heat, and consequently more epoxy smear and more drill wear.

Drill speeds of 80,000 r/min are readily available on drill machines currently used. The higher speeds generally contribute to straighter, smoother hole walls that yield fewer pockets to entrap gas bubbles and tend to produce fewer plating voids.

One dependent parameter not yet discussed is "surface feet per minute," the velocity of a point on the periphery of a drill. This parameter, which must be controlled, requires that drill speed must be reduced as drill diameter gets larger.

A delicate balance of feed rate (chip load) and drill speed of rotation is required for each drill size (or group of closely sized drills).

2. *Drill configuration:* Most drills, smaller than 0.0125 in, are made with 130° points. Those larger than 0.125 in are produced with 165° points.

3. *Drill sharpness and sharpening:* Drill sharpening (repointing) is routinely performed to extend the life of solid carbide drills. New or repointed drills must have straight, uniform cutting edges without nicks and with a fine surface finish. It is of utmost importance that highly trained technicians be employed and high-quality equipment be purchased for an in-house repointing operation. Meticulous microscopic inspection of all drills is a key to quality drilling, whether drills are repointed in-house or by an outside service organization. Handling is of equal importance, from the delivery of new or repointed drills to their placement in the tool holder of the drill machine. Carbide drills are *fragile!* Most PWB fabricators store their drill points upright in a base plate drilled with a matrix of holes. Even handling drills in individual vials can cause nicks and chips.

4. *Drill life:* Drill life is most commonly defined in terms of numbers of holes drilled, or "hits." In Table 33.1, we note that fabricators use a range of 1000 to 2000 hits for multilayers before repointing and maximum of two repoints (after two repointings, drills may still be used for two-sided boards).

5. *Drill spindle pressure feet:* During drilling, the laminate stack, entry, and backup material must be clamped securely to the drill table close to where the hole is being drilled. That function is performed by the parts of the drill machine called the "pressure feet." Improper action of these parts can cause severe burrs and drill breakage or MLB delamination.

33.5.3 Deburring

Following drilling, holes are mechanically deburred on top and bottom surfaces. Most deburring is performed in machines containing rotating, end-to-end oscillating, composite abrasive "brushes." Such machines operate wet, and debris is removed from the circuit board by high-pressure water and air knives. To prevent discharge of copper particles in plant waste, water deburring systems must be equipped with filters; many are closed systems, recycling the water.

33.5.4 Hole Cleaning

Cleaning holes is a preparatory step to metallizing the internal surfaces of a drilled hole. The internal surfaces can retain numerous particles and imperfections plus epoxy smear, on internal copper surfaces, that must be removed to provide a good plating base for adequate adhesion of electroless-copper plating to the epoxy, glass, and copper areas of the hole. Debris and a thin layer of the hole wall may be removed chemically, mechanically, or with plasma (or by a combination of the processes).

Two levels of hole cleaning are recognized: (1) smear removal and (2) etchback. Smear removal involves removing less epoxy than in etchback, which also requires glass removal (Fig. 33.10).

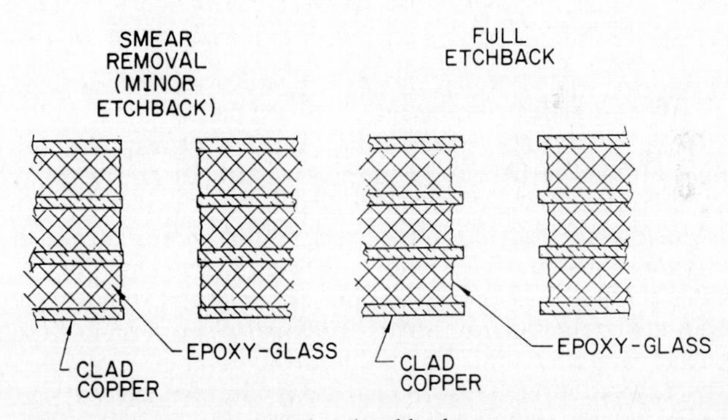

FIG. 33.10 Smear removal and etchback.

33.5.4.1 Mechanical Hole Cleaning. Mechanical cleaning is preferred by some MLB fabricators because it is simple and no strong or difficultly disposable chemicals are used in the process. It normally consists of conveyorized liquid honing equipment, in which a slurry of water and abrasive is forced by compressed air onto the surface of a drilled panel and through the holes. Nozzles are designed to oscillate with variable speed (cycle) control. The slurry blast is followed in line by a high-pressure water rinse and air knife drying section. With good burr control in the drilling process, liquid honing can be used for both deburring and hole cleaning. Etchback is accomplished by longer exposure to the abrasive slurry. Both epoxy resin and glass fibers are removed simultaneously.

33.5.4.2 Chemical Hole Cleaning. Chemical hole cleaning is often preferred because it is simpler and requires less investment.

33.5.4.2.1 Smear Removal. Concentrated sulfuric acid (must be 92 percent minimum) is used as solvent for removal of epoxy debris and from 0 to 0.0003 in of hole wall as well as the thin layer of epoxy on the inner-layer copper (drilling smear). Merely quenching the concentrated sulfuric acid with water precipitates an unwanted gelatinous residue. This residue can be removed with caustic permanganate solution.[4] The permanganate oxidizer also etches the inner-hole epoxy surface for better electroless-copper adhesion. Chromic acid may also be used for smear removal, but it is falling into disfavor because of waste disposal problems.

33.5.4.2.2 Etchback. Concentrated sulfuric acid is used just as in smear removal, except that a longer application is required. Any reasonable amount of material can be removed. While military specifications have required etchback and allowed up to 0.003 in, it is suggested that 0.001 in is more than adequate and that excessive etchback can cause copper cracking in the holes (Fig. 33.11) where plating drastically changes direction.

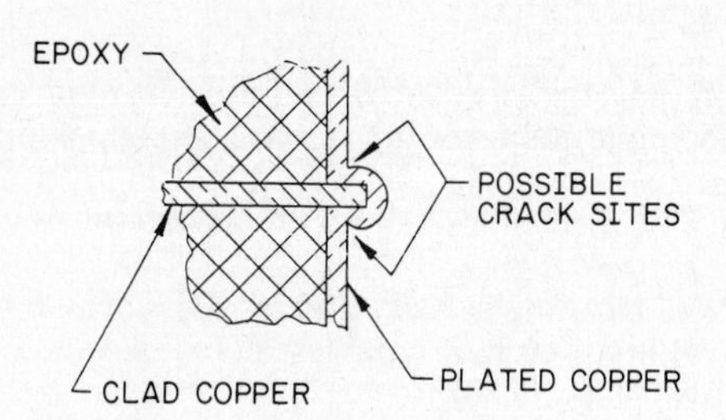

FIG. 33.11 Plated excessive etchback.

The major difference between etchback and smear removal is that in etchback, glass-fiber bundles project from the hole wall a distance equal to the amount of etchback. Plating over these projections causes protrusions similar to those formed from plating over debris. Therefore, it is important to remove the glass projections. Early glass etchants were concentrated hydrofluoric acid or fluosulfonic acid. Most fabricators now use ammonium bifluoride and hydrochloric acid. Some use the glass etch even with smear removal to etch the tips of the glass fibers for better electroless-copper adhesion. The majority of the MLB fabricators purchase their etchback/smear removal "chemistry" from one of the half dozen major proprietary suppliers; the remainder formulate their own solutions.

No bake cycle is recommended. In fact, it is recommended that panels be taken wet from the smear removal/etchback process into the electroless-copper line.

33.5.4.3 Electrical (Plasma). Exposure of drilled boards to a low-pressure carbon tetrafluoride–oxygen atmosphere backfill in a vacuum chamber with an RF voltage applied will clean holes, removing debris and smear and producing etchback depending on duration of exposure (both epoxy and glass are removed). Plasma treatment leaves an ash that must be removed before application of electroless copper. The major advantages of the process are that it is "dry" (no chemical solutions) and there is no effluent, except a small amount of "scrubbable" gas. The major disadvantages are that the plasma is not completely homogeneous throughout the chamber, and very close control is required of process conditions.

33.6 ELECTROLESS-COPPER PLATING

Electroless-copper plating is a process whereby sites on the surface of a nonconductive material are sensitized so that copper ions will deposit and form a continuous metal coating and an electroplating process may be used to build up enough copper thickness to conduct current at required voltages through the holes of the MLB. Sensitizing is accomplished with a palladium catalyst solution that is available from all major chemical suppliers to the printed circuit industry, as is the electroless-copper bath itself.

33.7 ELECTROPLATING

To plate to a given thickness, it is important to know plating deposition rates at various plating current densities. These rates are measurable and are frequently available in the literature of proprietary preparations. The surface and the in-hole plating area also must be carefully calculated to make it possible to select the appropriate current density and plating time necessary to achieve the desired thickness of copper.

33.7.1 Outer Circuit Preparation

At this point the MLB is totally covered with a thin layer of copper. The next step is to remove the electroless film from the clad copper, usually by wet, rotating mechanical brushes. This process will also prepare the copper surface for adhesion to photoresist. The MLB fabricator has three options for increasing the copper thickness by electroplating at this point:

1. *Panel plating:* The entire panel is immersed in an electrolytic plating bath and plated to a thickness of 0.001 in minimum in the holes. A pattern of tin-lead is plated on the surface of the copper, and the final circuit is etched using the tin-lead as the etch resist. This method causes severe line-width reduction (undercutting)—shown in Fig. 33.12—in comparison with copper pattern plating.

2. *Pattern plating:*

a. *Copper strike:* A thin layer of electroplated copper is panel-plated over the electroless-copper layer, after which photoresist is exposed to a positive (black-line) phototool, developed, and pattern-plated to a total minimum thickness of 0.001 in in the holes.

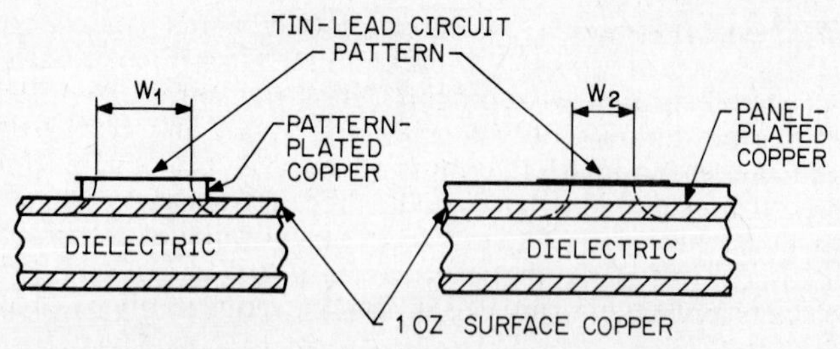

FIG. 33.12 Reduction of circuit lines: pattern- versus panel-plated copper process.

b. *No copper strike:* Modern electroless-copper baths are capable of rapidly depositing 60 to 100 μin of void-free copper that can withstand the cleaning and microetch steps that precede electroplating. Elimination of the strike saves time and copper. Most fabricators use this process. Whether or not a strike is used, after the copper has been plated, a thin layer of tin-lead (0.0003 to 0.0005 in up to 0.0015 in in special cases) is plated on the copper pattern. As in the case of panel plating above, the tin-lead serves as the etch resist, later enhances the solderability of the holes and pads, and protects the circuit traces from attack from ambient conditions.

33.8 INSPECTION OF MULTILAYER BOARDS

Inspection requirements for MLBs vary greatly with customers' needs and specifications. Military specifications are rigid and formal. Commercial customers usually have much more flexibility than military customers, while consumer electronics may have even more relaxed requirements. On the other hand, computer OEMs and test and measurement OEMs may have requirements equal to or exceeding those of the military. Environmental extremes are more frequently encountered by military electronics than in the mostly benign laboratory conditions of test and measurement instrument usage.

Military MLB inspection requirements are defined in MIL-P-55110 and its revisions. Individual company requirements may be based on IPC or other industry specifications, or a combination thereof. No matter what differences exist between customers, there is one common performance requirement for MLBs: The circuitry must have continuity for conducting the electrical signals for which it was designed; that is, it must be a "good" board.

Excluding environmental testing, inspection is commonly performed in four categories:

1. In-process inspections
2. Completed MLB visual and dimensional inspection
3. Coupon sample evaluation
4. Completed MLB electrical inspection

The first two of these inspections are routinely performed by the fabricator as part of the operation cost. The latter two are usually negotiated items depending on the fabricator's capability and the buyer's requirements.

33.8.1 In-Process Inspection

In-process inspection is essential to MLB production. Since errors are not usually reworkable and those that are must be reworked in sequence, inspection points should be established as a way to minimize scrap loss and as an indicator of process control. Typical inspection points include the following:

1. Phototool damage and accuracy check.
2. Visual check of photoresist lamination for wrinkles and foreign particles.
3. Visual scan after photoresist exposure and development for areas of poor image transfer or deposits of photoresist debris or "strings" on the pattern.
4. Visual or AOI inspection of etched inner layers.
5. Check of drilled holes for debris after hole cleaning.
6. Check for complete coverage in-hole by electroless copper.
7. Check of plating resist as in steps 2 and 3.
8. Inspection after outer circuit etching for lamination defects.

33.8.2 Completed MLB Visual Inspection

The completed MLB is inspected for imperfections on the outer circuit pattern, such as reduced line width, etched-out spots in the tin-lead plating, peeled plating, inclusions in the laminate, lamination flaws, shorted circuit features, and spots of copper remaining on the MLB surface. Inspection may be performed using low-power (5 to 10×) magnifiers or AOI vision systems. Rework is rarely possible at this point, but unwanted copper may be carefully removed, and copper spots having no functional impact may be left untouched.

1. *Completed MLB dimensional inspection:* Multilayer boards are inspected for outline dimension, hole size, thickness, and flatness (bow and twist). The latter two are most important if the MLB must fit into card edge guides or edge connectors.

2. *Coupon sample evaluation:* Coupons, when used, are test patterns located singly or in multiples on MLB panels. They are used to inspect plating thickness, dielectric thickness, plating quality, drilling quality, layer-to-layer registration, epoxy smear, barrel cracking, and thermal shock resistance (solder float) of the MLB. These characteristics are observed by the process of cross-sectioning the sample. The coupon or a portion of it is mounted in clear plastic. After rough-sawing the plastic-mounted coupon near the areas to be inspected, the cut face is sanded to expose the centerlines of one or more holes, and the face of the sample is progressively polished with successively smaller grit to remove scratches. Examination with a high-power microscope (200 to 800×) allows measurement of plating thickness, allows observation for defects listed above, and allows, after appropriate etching, analysis of the metal grain structure.

33.8.3 Completed MLB Electrical Inspection

The circuit board is usually the most expensive component in an electronic sub-assembly, and testing experience has shown that reject rate is proportional to the number of test points (and thus proportional to the cost of the board. It is therefore imperative that MLBs be bare board–tested. Most electronic components are tested after receipt to tight AQLs or are pretested by the supplier where today's emphasis is on a low "parts per million" defect level of delivered components. If the components are good and the MLB is defective, low turn-on rates occur, resulting in higher factory cost of product.

Continuity and leakage testers (bare board testers) in the 1960s used cross-bar switching and could be "programmed" for as low as 1 Ω to as high as 1 MΩ at voltages as high as 500 V. These have often been replaced by MOSFET-based testers. While they are much faster (500 to 5000 contacts/min) than cross-bar testers, they can only be used to test for opens above the 1000-Ω range. Care must be taken in choosing a bare board tester. The consensus of the independent board-testing industry is that values of 5 Ω for opens and 500K Ω for shorts will pick up more than 99 percent of all such defects. These ranges cannot be achieved with high-leakage MOSFETs.

Initial capital investment is only part of the expense of starting and maintaining a bare board test operation. Unique handling fixtures consisting of a matrix of spring-loaded electrical contacts in a pattern that matches the MLB hole pattern are also required, although some large "universal" fixtures with contacts on a grid pattern are available. For organizations that cannot justify the capital expenditure for a bare board tester, there are several contract testing houses available.

33.8.4 Process Control

Inspection of MLBs is an essential part of production, but no amount of inspection can improve a product. Product improvement and high quality are the results of process control. Process control differs from quality control in that it looks for potential causes rather than the effects of process problems. Process causes can range from fading light intensity on an exposure machine to impurity levels in plating baths, bath temperatures, acid concentration in an etchback system, and the resistivity of the deionized water supply. These are typical examples of the many process steps that contain observable and measurable characteristics. Understanding the limits of each process step that must be maintained to produce consistently good results from that process step is the first step toward establishing process control. Suppliers are one source of information; they know the strong points of their own products and often the weak points of their competitors' products. Supplemented by first-hand product experience, that is one approach to establishing process step limits.

Second, process characteristics must be measured or monitored routinely to observe their relations to established limits. Appropriate action must be taken to restore a process as it exceeds a limit condition, even though product quality may not suffer immediately. "Bleed and feed" chemical replenishment systems equipped with microprocessors and automatic chemical analysis units are readily available to assist in mainataining control of wet processes.

Next, on a long-term time base, process step characteristics should be plotted to show trends not observable in simple routine measurements (statistical process control). As an adjunct, scrap loss records on a similar time base for each process step can assist in general troubleshooting or help define areas in which process or equipment improvement can reduce operating costs.

Finally, charts are useful only if they are used. They should be displayed in the work area, near the processes involved, to stimulate interest and action among operators, supervisors, inspectors, and managers.

REFERENCES

1. H. Gilutz, "AOI Assists Process Control of Fine PC Board Production," *Electronic Packaging and Production,* September 1985, p. 68. V. Bunze, "AOI: Paving the Way to Higher Profits," *Printed Circuit Fabrication,* November 1985, p. 45.
2. A. G. Osborne, "An Alternate Route to Red Oxide for Inner Layers," *Printed Circuit Fabrication,* August 1984, p. 28.
3. C. J. Bartlett, D. P. Bloechle, and W. A. Mazeika, "Use of Scaled Flow Testing for B-Stage Prepreg," IPC 22d Annual Meeting, Boston, Mass., April 1979.
4. M. J. Barmuta, "Smear Removal for Consistent Hole Wall Reliability," *Printed Circuit Fabrication,* April 1984, p. 50.

CHAPTER 34
LAMINATING MULTILAYER CIRCUITS

Allan G. Osborne, Ph.D.
Osborne Technical Associates, Seattle, Washington

34.1 INTRODUCTION

This chapter deals with the materials, processes, and equipment involved in laminating multilayer boards. The lamination process marks the difference between two-sided and multilayer boards (MLBs). It comprises a number of individual process steps and related equipment. The process steps and equipment involved explain the increased cost normally experienced when a circuit design goes from a one- or two-sided to a multilayer board with three or more layers. In return for this investment, the designer has, at his or her disposal, an extremely versatile design tool. It is obvious that the fabricator must be familiar with the laminating processes, but it is also important that the designer understand the basic capabilities and limitations of these processes.

34.2 MATERIALS

The materials most commonly used for MLBs are G-10 and FR-4 (fire-retardant) grades of epoxy resins supported on woven glass fabric, unclad or clad on one or both sides with electrolytic sheet copper. Special-purpose materials, such as polyimide for high temperature or Teflon for high frequency, are also laminated to make MLBs.

34.2.1 Epoxy Resin

34.2.1.1 Definition. In a broad sense, the term "epoxy" refers to a chemical group with an oxygen bonded to two carbon atoms to form a three-member ring, ethylene oxide (I) being the simplest example (Fig. 34.1).[1] However, the epoxy group must in turn be bonded to a larger molecule to be able to form epoxy polymers. The basis for almost all printed circuit epoxies is the adduct (IV) of epichlorohydrin (II) and bisphenol A (III).

The adduct is polymerized into a linear multiple-epoxy molecule which can be "cured" by cross-linking.

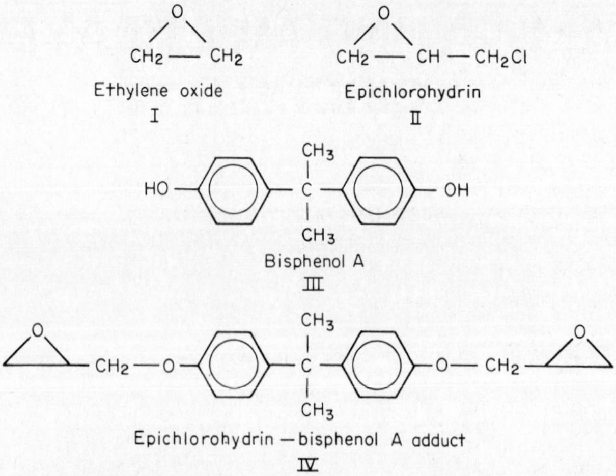

FIG. 34.1 Epoxy structures.

*34.2.1.2 **Characteristics.*** The low-viscosity and easy-cure properties of epoxy resin used for MLBs simplify the manufacture of the basic raw materials. The epoxy resin curing agent system can be modified readily to provide special properties such as room-temperature latency, and the base resin can be modified for high-temperature resistance and flame retardancy. Room-temperature latency is of particular interest, since that feature is used to formulate a dry-handling, semicured epoxy adhesive resin used to bond circuit layers together in MLBs. The same resin system with heavier glass fabric is used to make the cured laminate (C-stage) material used in fabricating multilayer boards as well as one- and two-sided boards. Availability and ease of handling of the semicured material, called "B-stage" or "prepreg," has contributed to the ease of manufacture of MLBs and to the greatly increased use in their development in the 1960s.

Epoxy resins as a material category possess excellent electrical, mechanical, and chemical properties. The general cured-state properties vary little. Even though the exact chemical composition is usually proprietary and not available from resin suppliers, there is little real variation in the composition of curing agents, stabilizing agents, and flame retardants. The MLB fabricator, however, should not mix prepreg from more than one source, since the curing times and temperatures vary from supplier to supplier.

34.2.2 Glass Cloth

Woven glass cloth is used as the support vehicle for epoxy resins in the manufacture of printed board raw material for structural strength. Continuous rolls of glass cloth fabric are drawn through vats of liquid resin, sizing rollers, and curing ovens to produce dry, semicured rolls of epoxy-resin-impregnated glass-base material.

The glass cloth is of interest to the MLB fabricator in the lamination process, primarily for the reason of its impact on material dimensional stability of the individual circuit layers within an MLB. One glass cloth characteristic which has an influence on dimensional stability is weave style. Glass cloth is available in a wide variety of weaves, although there is a continuing trend toward standardization by the suppliers of raw material (copper-clad glass-epoxy), as well as the glass suppliers. Regardless of weave type, the thinner the glass (the thinner the inner-layer material), the less dimensional stability it has.

Glass cloth is surface-treated with silane-organic coupling agents which promote glass-to-resin bonding by providing chemical bonds, both to the inorganic glass (silane part) and to the epoxy resin (organic part). This intimate bonding reduces wicking of moisture or chemicals up the glass fibers and contributes to the material strength through transfer of stresses to the high-strength glass fibers.

34.2.3 B-Stage Material (Prepreg)

34.2.3.1 Description. Semicured glass-cloth-reinforced epoxy resin is referred to either as "B-stage" or as "prepreg." It is most common to find producers of cured laminate (C-stage) calling it "B-stage," while MLB fabricators call it "prepreg."

In the semicured state, the epoxy resin has been "B-staged," that is, not fully polymerized, or cross-linked; therefore, although it may be handled easily, it may also be returned to the liquid state by application of heat and pressure. That feature makes prepreg a convenient bonding material for MLBs.

Technically, prepreg is available in all the glass weave styles; but for bonding layers in multilayer boards, the thinner lightweight fabrics are most commonly used because they carry a higher resin-to-glass ratio. That characteristic helps provide a void-free laminated bond, since more resin is available to fill circuit patterns where copper has been removed.

34.2.3.2 Prepreg Resin Properties. Prepreg is specified by type designation, which identifies resin material, glass style, resin flow, gel time, and resin content. A typical designation from MIL-P-13949 is as follows:

PC = prepreg material

GE = resin material (general-purpose)

1528 = glass style

28 = nominal resin flow, 28%

090 = nominal gel time, 90 s

55 = nominal resin content, 55%

34.2.4 G-10 and FR-4 Copper-Clad Raw Material (Laminate)

Single- and two-sided printed wiring boards are fabricated from a variety of paper-base phenolic materials as well as glass-filled epoxy. Multilayer boards, on the other hand, are made principally from epoxy-glass. Phenolics do not lend themselves to the multilayer lamination process.

G-10 and FR-4 epoxy materials are available in many forms:

1. Copper-clad one side (1*/0, 2/0, etc.)
2. Copper-clad both sides (1/2, 1/2, 2/2, etc.)
3. No copper cladding (0/0)

*Copper foil weight in ounces per square foot; "1/0" means 1 oz of copper on one side and none on the other, "1/1" means 1 oz of copper on each side, or a two-sided board.

4. Varied dielectric material thickness (0.005, 0.010, 0.016, etc.) in inches
5. Varied glass weave styles within the material (108, 1080, 116, 7632, etc.)

G-10 or FR-4 fully cured material (C-stage) will soften during the heat and pressure of lamination but will not liquefy again. Excessive lamination temperatures will turn the laminate brown or char it.

34.2.5 Miscellaneous Multilayer Board Materials

In addition to the predominantly used G-10 or FR-4 material, a number of other materials are used in the MLB industry. Some of them and their advantages and disadvantages are discussed in the next paragraphs.

Polyimide is a high-temperature material utilizing a glass cloth base, and it is available as prepreg and as copper-clad C-stage material. Maximum service temperature is of the order of 500°F, compared with about 275°F for G-10 and FR-4. The high service temperature and glass transition temperature (Fig. 34.2) extend

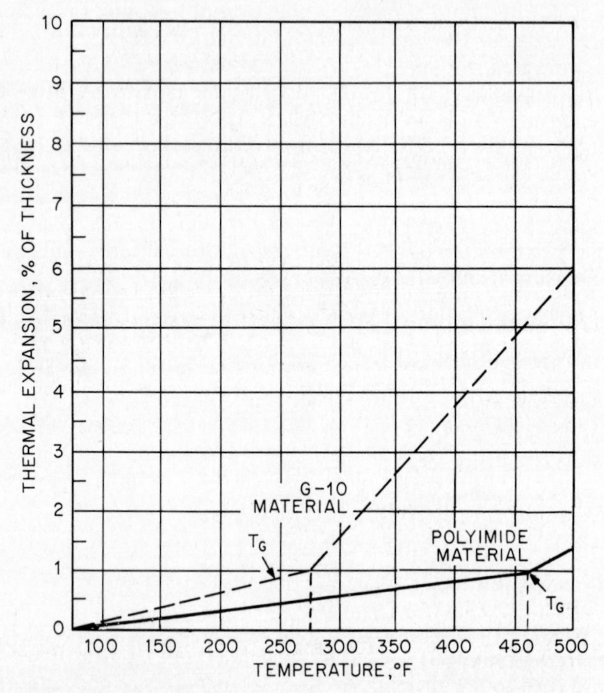

FIG. 34.2 Comparison of typical glass transition temperatures T_g for G-10 versus polyimide.

polyimide into MLB applications, where high-temperature thermal excursions would eventually create excessive *z*-axis expansion, which would fracture plated-through-hole barrels in regular G-10 and FR-4 epoxy. Electrical characteristics of polyimide are similar to those of the epoxy materials, but polyimide has improved chemical resistance and somewhat less flexural strength. Polyimide is also more dimensionally stable than epoxy. Major disadvantages are the following:

1. *Cost:* Material cost is 3 to 6 times that of epoxy.
2. *Copper adhesion:* Typical peel strengths are approximately half those of copper-clad epoxy.
3. *Lamination temperature:* Cure temperature is about 425°F versus 350°F for epoxy. Longer postcure is required.

*Glass-base Teflon,** with either random fiber or woven glass cloth base, has excellent low-loss electrical characteristics, which make it useful in high-frequency applications. Teflon is also a high-temperature material, but its main use is for microwave and strip-line applications. Teflon is much more difficult to process than epoxy.

1. *Teflon-to-glass filament adhesion:* Teflon does not penetrate the glass bundles well or adhere to the glass filaments. Therefore, during processing through chemical solutions, the glass fiber bundles wick solution, which is retained or released at other processing steps.

2. *Teflon chemical resistance:* The carbon-fluorine bonds of Teflon are resistant to most chemical agents. The Teflon surface is highly hydrophobic (water-repellant), and mechanical abrasion alone does not improve its wettability. Metallic sodium is one of the few reagents that will attack Teflon, and two proprietary preparations, Tetra-Etch and Bond-Aid, which are principally sodium-naphthalene compounds that act as if they are free sodium, will etch the surface, allowing it to be plated.

Flexible materials—one- and two-sided copper-clad flexible polyimide materials (chemically different from rigid polyimide)—are used to fabricate flexible circuits for many special applications, including replacement of discrete wiring harnesses and contoured circuits in telephones, cameras, and automobiles.

Flexible materials are amenable to high-volume processing, since they can be processed from rolls or reels in a continuous "web" and may be stamped or punched rather than drilled for through-hole fabrication.

34.3 LAMINATION FIXTURES AND TOOLING

Successful multilayer circuit board lamination technique really requires only two things to happen simultaneously:

1. Layer-to-layer registration of circuit pads—which are eventually interconnected by drilling and electroplating—must be maintained.
2. Void-free laminate bond lines must resist thermal shock. Uniform and predictable thickness of finished laminate is desirable as well.

Layer-to-layer circuit registration is maintained by the use of tooling pins in the material outside the circuit area.

*Registered trade name of E. I. du Pont de Nemours & Company.

34.3.1 Registration Pin Systems

While two- and three-pin systems have been in use since the early days of multilayers and are still being used for four- and six-layer boards, the most accurate system in use is the four-hole "orthogonal slot" system (Fig. 34.3*a*). However, at

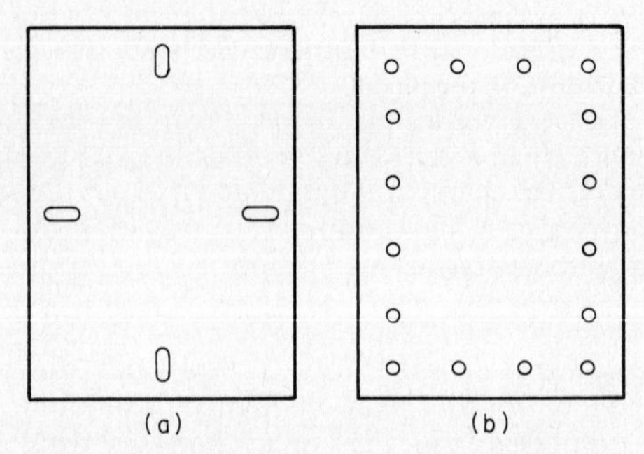

FIG. 34.3 Two multilayer registration pin systems. (*a*) Orthogonal slots; (*b*) multiple holes around periphery.

more than eight layers or for thin-core material (less than 0.010 in), a large number of registration holes around the periphery of the panel (Fig. 34.3*b*) greatly aid the material dimensional stability problem.

Whatever pin registration system is used, it becomes the common denominator for all process steps, from initial shearing to size to photoimaging to drilling to profiling the final board.

34.3.2 Lamination Fixtures

Ideally, lamination fixtures (often referred to as "caul plates") should have an expansion rate which exactly matches that of the laminate throughout the lamination cycle, thus minimizing induced thermal stresses from coefficient of thermal expansion (CTE) mismatch between the metal fixture and the laminate layup. However, since laminates are a composite of copper, glass, and epoxy, and since the lamination temperature exceeds the glass transition temperature of the material (at which point the CTE increases radically), the laminate is an unknown entity for which the thermal expansion may only be estimated.

Glass-base epoxy laminates do approximate the thermal expansion rates of steel reasonably well up to the glass transition temperature (approximately 265°F); steel, therefore, provides a compatible fixture material with a minimum of induced thermal stresses throughout the lamination cycle.

For vertical stability and laminate registration, lamination plates should be thicker than the registration pin diameter. Surfaces of lamination plates should be examined for scratches which can cause dents and broken traces on outer laminate circuits during subsequent process steps. To protect press platens, registration pin length should be less than the combined thickness of the top and bottom lamination fixture, plus the thinnest laminate fabricated (Fig. 34.4); that is, T must always be greater than L.

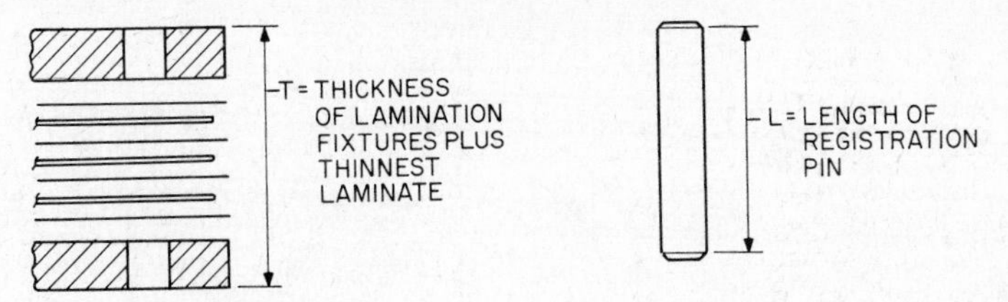

FIG. 34.4 Laminate buildup versus registration pin length.

34.3.3 Lamination Presses

Lamination presses for the production of multilayer circuit boards are principally hydraulic for applying ram pressure, operated in ambient atmospheric pressure, and heated by either electricity or steam, the latter being more frequently used for production of base laminate (C-stage). Many fabricators use two presses in series: one for the heat-up, flow of the prepreg, and initial cure; a second one for "hot transfer," in which the remainder of the cure takes place.

The use of vacuum is rapidly increasing and takes one of three forms:[2]

1. *Vacuum hydraulic lamination (VHL):* lay-up enclosed in a vacuum bag
2. *VHL:* platens and lay-up surrounded by a vacuum chamber
3. *Vacuum autoclave lamination (VAL):* lay-up enclosed in a vacuum bag within a pressure chamber heated by isostatic gas (usually carbon dioxide)

Most presses have multiple platens (multiple openings or "daylights"). This is discussed in more detail later in this chapter.

34.3.4 Lamination Press Auxiliary Equipment and Procedures

1. *Thermocouple:* A small thermocouple embedded in the laminate bond line within each press opening provides a necessary backup to the press temperature indicators for process control.

2. *Temperature recorder:* Thermocouples are connected to a data logger or recording electronic thermometer. The press may be manually controlled to operate at the desired laminate temperature, or the electronics can be set up as closed-loop to control temperature automatically.

3. *Pressure transducer:* Lamination press hydraulic systems and pressure gauges can malfunction on occasion. The most direct method of detecting such problems is to use a pressure load cell directly in the press opening to calibrate load versus pressure or force gauge readings.

4. *Platen flatness check:* Press platens should be parallel and flat within 0.004 in to produce consistent-quality MLBs. A simple way to check that is to start with a cold press and lay a spiral of solder (approximately 0.06 in in diameter) across the face of the press platen in each press opening. Solder spirals should be approximately 3 to 4 in apart, as shown in Fig. 34.5. Straight pieces of solder work equally well. Apply enough load to the press to produce approximately 0.02 to 0.03 in of deflection on the solder, and then, with a micrometer, prepare a cross-the-corners profile of measurements on the solder. If measurements vary more than 0.005 in, it is likely that a corresponding or even increased variation

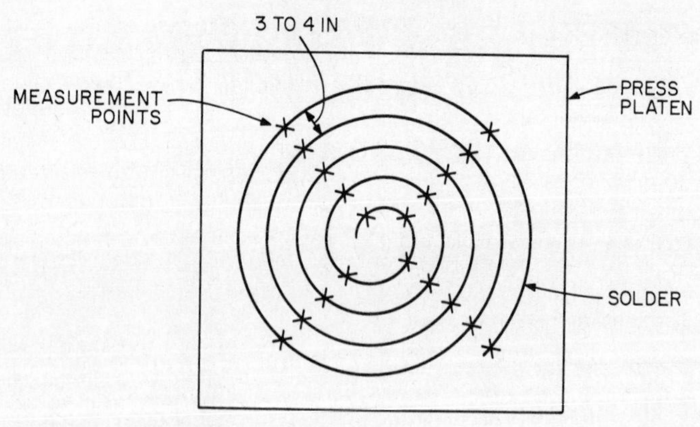

FIG. 34.5 Solder spiral.

will be noticed on laminates produced in the press. Press manufacturers can usually correct such problems by adjustments or rework, although extensively warped platens must sometimes be replaced.

5. *Platen heating uniformity check:* Nonuniform heat distribution (greater than ±3°F variation of surface temperature) on press platen surfaces can be a problem on both steam and electric presses. Greater temperature variation can cause voids in the laminate bond line. In general, a 2-in outside border zone of any platen has a greater heat loss than the central portion of the platen and should not be used for lamination. Two methods of checking platen temperature uniformity are to prepare a grid pattern (2 to 3 in) of thermocouples or temperature-indicating paint dots. Either pattern should be on a piece of nonmetallic material approximately $\frac{1}{4}$ in thick. Cover the thermocouples or paint dots with a sheet of release material and the appropriate thermal insulation material (Fig. 34.6). This lay-up is loaded into a preheated (340°F) press for 4 min and then removed. Thermocouples should be connected to a data logger or recorder to read temperature just prior to removal of the lay-up from the press.

The purpose of the test is to leave the lay-up in the press long enough to get a temperature indication resulting from a point on the adjacent platen, with as little influence as possible from the opposite platen and the surrounding area of the adjacent platen. With the lay-up and temperatures described, monitored temperatures should be in the 300 to 310°F range. Temperatures on a press platen surface area should be within 6°F. Both platens of a press opening must be checked. The checkout procedure is recommended annually, or when lamination problems become prevalent.

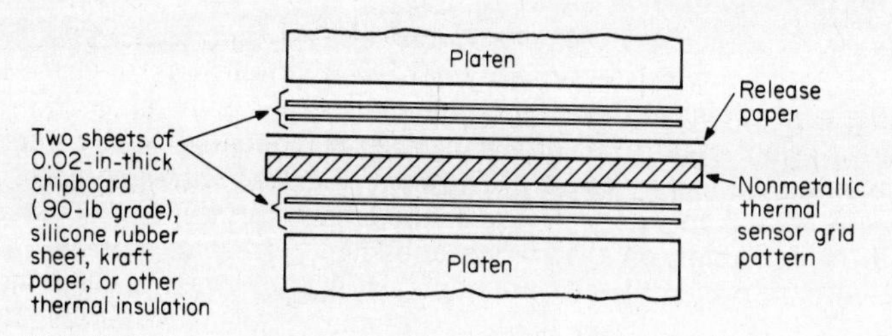

FIG. 34.6 Temperature distribution buildup.

6. *Press-loading equipment:* Depending upon the size and number of press openings, a variety of carts, conveyors, and elevator lifts can be used to transport lamination fixtures and ensure accurate placement of the fixtures in the press openings.

7. *Registration pin removal:* Following lamination, registration pins are surrounded by resin from the laminate. It is important that pins and mating holes in fixtures be coated with spray or liquid mold-release material prior to being used. Removal of the pins is then readily accomplished with an air- or hydraulic-assisted press. A hand-operated arbor press also may be used, but hammer and drift pin-removal methods are not recommended because of the possibility of marring lamination fixture features.

8. *Lamination fixture maintenance:* Lamination fixtures should be routinely protected from resin squeeze-out from the laminate package. Protection is secured by the use of spray or liquid mold release applied to registration pins and the insertion of a sheet of release material between each half of the lamination fixture and the laminate. Also, between use cycles, the fixtures should be inspected for deposits of resin on either surface which will cause localized high-pressure points or depressions in the surface of the laminate. Excess epoxy must be removed.

Lamination is the first MLB process step which requires a complete restart because of a process problem. Also, since several laminates are usually processed in a given press load, the importance of consistent lamination process, skilled operators, and good equipment cannot be overemphasized.

34.4 LAMINATE PREPARATION

34.4.1 Prepreg Characteristics

To maintain a consistent lamination process, the MLB fabricator must be concerned with at least five specific characteristics. (See Chap. 6 for a detailed discussion of multilayer materials and parametric testing.)

1. *Glass weave type:* This and residual resin determine cured inner-layer dielectric thickness.

2. *Resin flow:* (*a*) Too high a flow may result in a resin-starved bond line.* (*b*) Too low a flow may result in too thick a bond line or voids in the bond line.

3. *Resin content:* (*a*) Too low a resin content may result in a resin-starved bond line. (*b*) Too high a resin content may result in too thick a bond line.

4. *Gel time:* "Gel time" and "tack time" are commonly used interchangeably. Both terms are meant to indicate the point when a prepreg resin, upon being heated, has become liquid and is *starting* to "gel" or cross-link into a cured thermoset condition. Measurement consistency is a function of the test technique and is highly operator-dependent. A more reproducible, quantitative result may be achieved by parallel-plate rheometry, but this is not in widespread use.

*"Bond line" is actually "bonding volume," the space between the inner layers that is filled with prepreg resin and fiberglass, but "bond line" is part of normal usage, since it is usually graphically represented in a cross section.

5. *Volatile content:* Volatiles consisting of retained solvents or moisture in prepreg must be controlled at less than 1 percent by weight to prevent formation of gas bubbles and voids in laminate bond lines. These characteristics and the procedures for measuring them have been defined in an attempt to allow consistent testing by prepreg suppliers and users. Standard test procedures are available from the Society of Plastics Industry, Inc.; NEMA; IPC; and MIL-P-13949, which has been revised to include both MIL-G-55635 and MIL-P-55617. Despite the standards, inconsistent results between user and suppliers can still occur owing to operator dependency, sampling techniques, and test equipment difference not specified in the standards.

However "standard" a test may be, it may not be representative of the lamination process as it is performed on the production line. Therefore, in addition to standard testing, it is good shop practice to make a representative size laminate of several pieces of each new batch of prepreg to observe the way it responds to the normal press cycle. For example, resin starvation or voids (for which there is no direct standard test) in the sample will give clues to adjustments in the press cycle or a warning that standard tests should be repeated.

34.4.2 Prepreg Storage and Handling Prior to Lamination

Prepreg is very susceptible to moisture, solvent, airborne soil, and oil particle contamination. Any of these foreign materials are likely to produce spots of delamination within the bond line. In addition, time and temperature gradually advance the cure of the semicured material.

To protect and maintain material usefulness, the following shop practices and conditions should be observed:

1. Material should be packaged by the supplier in airtight containers (sealed plastic bags). Preferably, it should be cut to standard shop sizes.
2. Material should be stored in a cool, dry environment (less than 50°F and 50 percent RH). Unopened bags must be brought to room temperature before opening to prevent water vapor from condensing on the prepreg.
3. Material should be handled only by operators wearing clean cotton gloves.
4. Material should be punched and cut to size in a positive-pressure room.
5. Material should be placed in a vacuum chamber to remove volatiles prior to use.

Observation of these practices will resolve many lamination process problems.

34.4.3 Circuit Laminate Preparation Prior to Lamination

Circuit laminates are the circuit patterns etched in copper-clad to C-stage glass-filled epoxy material. Laminates may have circuitry on one or both sides. They are frequently referred to as "inner layers." After printing and etching of the circuit pattern, and prior to lamination with prepreg, the laminates should undergo some combination of the following preparatory steps:

1. *Chemical cleaning:* Dilute (10 percent) sulfuric acid dip followed by water rinse and forced-air drying. If boards will not be stacked and are going directly to scrubbing, they need not be dried.

2. *Scrubbing, rinsing, and drying:* Scrubbing with a rotating, composite-material brush followed by water rinse and forced-air drying.

3. *Oxide treatment:* Except for material that is already pretreated (double-treated), inner layers must be treated with a copper-oxidizing preparation. This preparation is principally a mixture of sodium chlorite, sodium hydroxide, and trisodium phosphate and may be purchased as a proprietary solution or formulated by the fabricator. Oxiding solutions produce black, brown, or red oxides, depending on solution composition and temperature. Most fabricators have a favorite formulation. (See Chap. 33 for a more detailed discussion of "oxide" processes.)

4. *Bake cycle:* One-hour (minimum, at temperature) bake at 250°F to remove moisture during oxide treatment. Store in a dessicator for no more than 8 h without rebake before lay-up.

34.5 LAMINATION

34.5.1 Laminate Lay-up Procedures

Laminate lay-up should be performed in a controlled-environment room for best results. Desirable conditions include:

Positive air pressure

Source of filtered (5-μm) air

Daily cleaning of the room (damp-mopping of floor)

Clean protective gloves and lint-free lab coats and hats for operators

Controlled temperature and humidity (70 ± 5°F, 50 percent RH, max.)

Precleaned lamination fixtures with nonsilicone mold release applied to registration pins prior to entering lay-up room

Entry vestibule with tack mat and lockers for protective clothing

A typical laminate lay-up, as it would be in the press ready for lamination, is shown in Fig. 34.7.

To increase press output, multiple multilayer stacks (one MLB or panel of

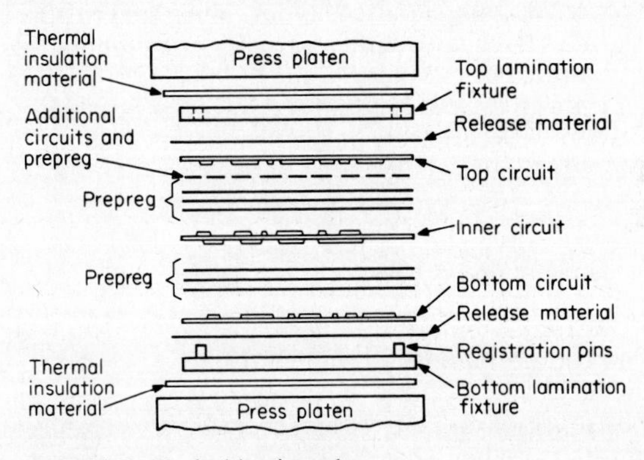

FIG. 34.7 Typical laminate lay-up.

MLBs) may be placed one on top of another, preferably with 0.020- to 0.030-in steel "ferrotype" plates in between to form a "book" several MLBs thick. The size of the book will be determined by the process requirements for rate of heat transfer.

The amount of prepreg in the bond lines of each MLB type will vary because of the amount of etched circuit area to be filled and the copper thickness used. As a benchmark, for 108-glass prepreg, about 0.0025 in thick before pressing, a bond line should contain a minimum of two sheets of prepreg plus one more sheet for every 2 oz of copper circuitry in the bond line. Side-by-side loading of two books in a single press opening requires that the pressed-book thicknesses be nominally equal.

Bond line requirements are as follows:

1. No delamination after thermal shock (10-s, 550°F solder float, 2- × 3-in specimen)
2. No bubbles or voids in the bond line
3. No resin starvation (while glass fabric pattern in prepreg is visible) in the bond line
4. No dirt or foreign particles in the bond line
5. *Good layer-to-layer circuit registration* to allow drilled holes to stay within internal circuit pads (with a minimum 0.002-in annular ring)

34.5.2 Press Cycles

Early prepregs required two-stage pressure cycles: the first, approximately 10 psi ("kiss" cycle); the second, full pressure of 150 to 300 psi. With current prepregs, fewer than 10 percent of multilayer fabricators use the two-stage system.

Steam presses use a cold press start; electric presses, a hot-start cycle. To better understand the reason, see the typical platen heat rate in electric and steam presses (Fig. 34.8) and the prepreg resin viscosity curve (Fig. 34.9). Electric presses can take twice as long to heat as steam presses. Prepreg resin is cured by the combination of temperature and time, or thermal energy, to which it is exposed. Viscosity of the resin is also a function of the absorbed thermal energy. Figure 34.9 shows that during a fast press heat-up cycle, there will be a longer period of time T_1 when the resin has a useful viscosity for fabricating multilayer boards. A slower heat-up cycle such as on the electric press presents a shorter period of time T_2 when the resin has an acceptable viscosity for fabricating MLBs. Therefore, electric presses are usually preheated either partially or to full temperature for MLB lamination use. To prevent excessive resin runout from a full-temperature preheated press, the heat-up cycle can be slowed by inserting a sheet or two of 0.020-in-thick cardboard, or several sheets of kraft paper, between the press platens and the lamination fixture.

34.5.3 Vacuum Lamination Process

The purpose of vacuum lamination is to remove air, moisture, and residual solvents which can produce bubbles in the bond line. At the same time, a further yield improvement can be realized by using lower pressure and reducing the amount of misregistration from movement of the layers in respect to one another. Lower pressures are possible, since less resin flow is necessary to remove gas bubbles.

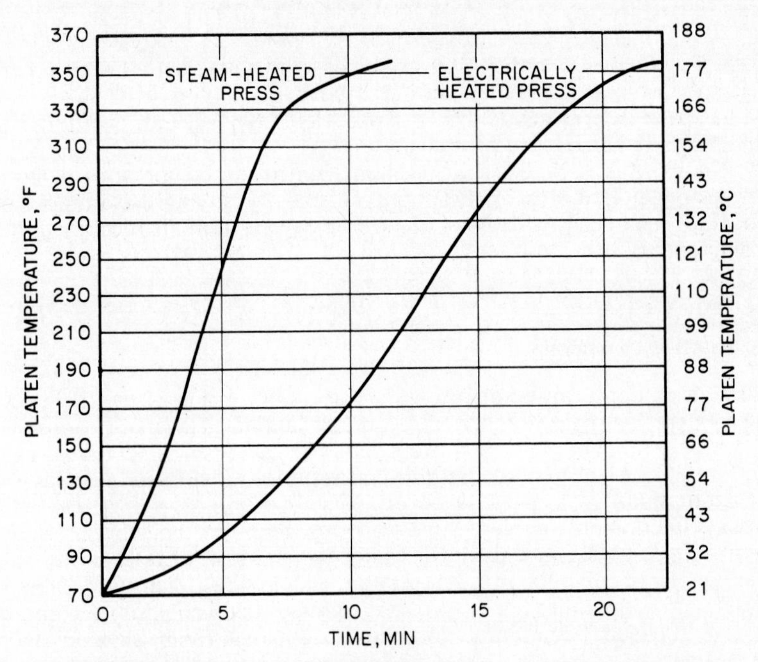

FIG. 34.8 Typical heat-up curves for steam-heated and electrically heated presses.

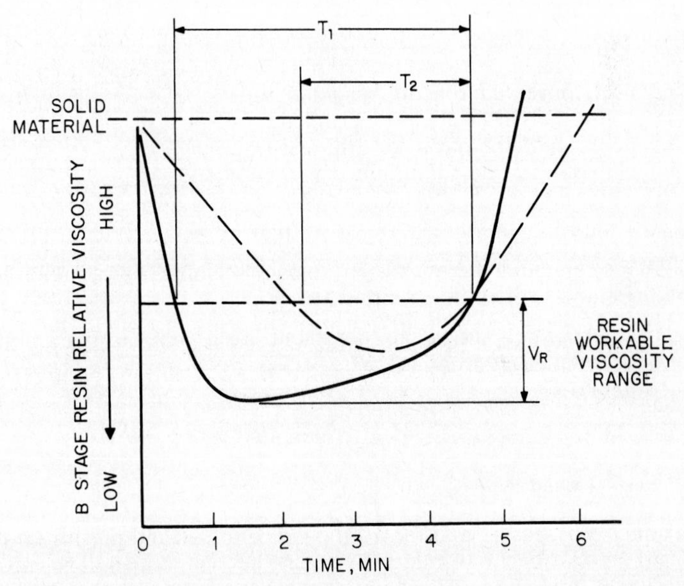

FIG. 34.9 Typical high-flow prepreg resin viscosity curves during lamination. Solid line indicates viscosity curve for fast-heat-up press. Broken line indicates viscosity curve for slow-heat-up press.

Because of better overall removal of air and moisture, there is also an increase in final performance characteristics of the MLB.

It has been determined that a 95 percent or better vacuum is required. This translates to 38 mmHg or 50 millibars.

The consensus is that despite vacuum lamination equipment's being more costly than standard presses, the improvement in yield and performance of the product for high-density MLBs of 10 or more layers[3] more than compensates for the difference.

34.5.4 Mass Lamination

"Mass lamination" can be defined as a partial fabrication of multilayer circuit panels by the laminate supplier. This process laminates layers of copper foil to the top and bottom of pre-etched two-sided inner-layer panels. Registration is not a concern during lamination of four layers, as the location of tooling holes is done after lamination.

"Mass lamination" employs large-scale production methods by massing multiple images of PWBs on a single very large panel. Most MLB fabricators manufacture *panels* of circuit boards, but they are seldom larger than 18 × 24 in. Mass laminators, on the other hand, have large presses that can easily accommodate 36- × 48-in panels. After lamination, these large panels can be sheared to a more manageable 18- × 24-in or 12- × 18-in size and returned to the customer, the MLB fabricator. Registration for drilling is accomplished by penetrating the outer copper layer with an end mill ("spot-facing") to pick up a fiducial mark that has been etched on the next inner layer.

In the beginning, only four-layer boards were fabricated, since the two inner layers (on double-clad laminate) were easily registered after laminating by the spot-facing technique—the outer layers were formed from copper foil and prepreg and did not need to be registered during lamination. By the mid-1980s, however, a survey showed that fully two-thirds of mass-laminated panels were six layers and more.[4] With more than one double-sided inner layer, a pinning system must be used to assure that all layers are registered. The spot-facing technique is still used to expose the fiducial marks for drill registration.

The major reason for mass lamination is economics:

1. The larger the panel laminated, the lower the cost per individual MLB.
2. Small shops can enter the multilayer business before they can afford a lamination press and registration equipment.
3. Multilayer fabrication shops can "farm out" peak loads to mass laminators and avoid capital expenditure for extra presses which would be idle during low production periods.

34.5.5 Thermal Analysis

Thermal analyzers measure and record the temperatures and magnitudes of various transitions that occur in materials when they are heated. The most commonly measured transition for epoxies and other circuit board materials is the second-order glass transition temperature T_g, the point at which, along with changes in heat capacity and viscosity, the CTE increases markedly.[5] The T_g may be detected using differential scanning calorimetry (DSC), thermomechanical analysis (TMA), or dynamic mechanical analysis (DMA), the former two being most frequently used.

T_g in and of itself is of little interest to the laminator, but as a measure of the degree of advancement of cure of prepreg (Fig. 34.10) and as a measure of final cure of the finished laminate (Fig. 34.11), it is a valuable adjunct to process control. Since the analytical procedure involves heating the laminate sample, it undergoes further curing, and since there is variation in T_g from resin to resin, the rule of thumb for indication of proper cure is that if the second T_g run on the same sample shows an increase of less than 4°C from the previous run, the sample is from a fully cured laminate. Representative industry standard measurements methods are test methods 2.4.24 and 2.4.25 in IPC-L-108.

34.6 LAMINATION PROCESS TROUBLESHOOTING

Lamination process problems are extremely costly to the MLB fabricator because they are not always obvious at the completion of the process; it is not until final circuit etch that lamination problems are evident. To prevent losses that can amount to thousands of dollars in a very short period of time, some manufacturers routinely (each day, shift, or lot) etch the outer copper from a board which has just been laminated to check for laminate quality by both visual test and thermal shock resistance. If production volume warrants, that practice saves money by reducing scrap cost; e.g., a bad lot can be rejected before drilling and other value-added processes are performed. Furthermore, earlier detection of problems resulting in the scrapping of a lot or run can save schedule time for a remake.

When problems do occur, many causes and effects are interactive and overlapping, making analysis difficult. Some lamination problems, potential causes, corrective actions, and analytical procedures are described in the following section.

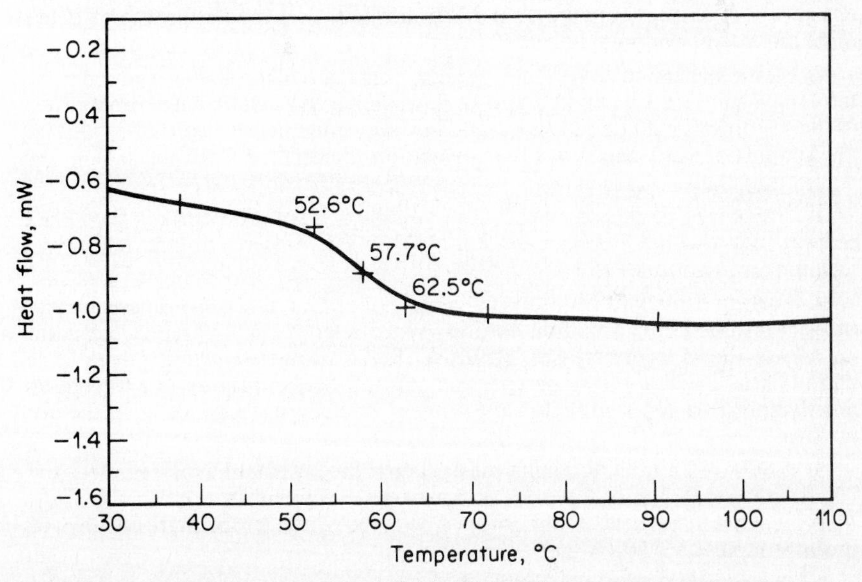

FIG. 34.10 Prepreg analysis by differential scanning calorimetry (DSC). Sample temperature was raised 10° per minute.

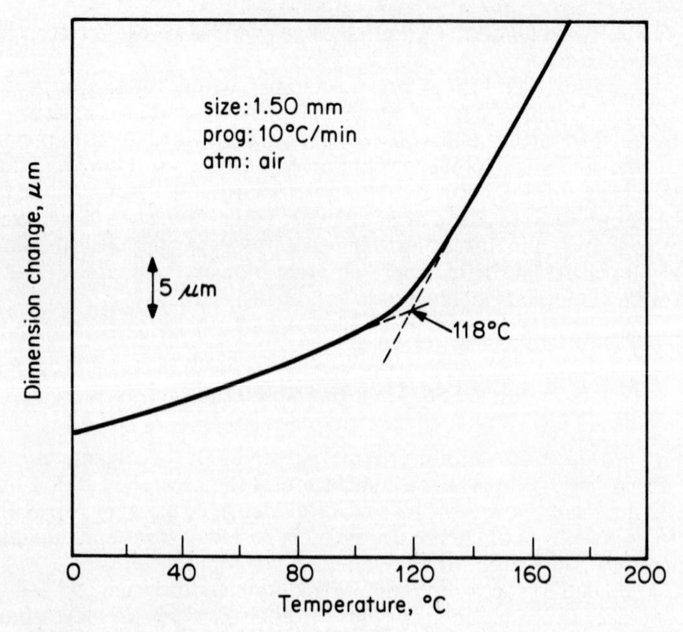

FIG. 34.11 Determining glass transition temperature T_g using thermomechanical analysis (TMA).

34.6.1 Lamination Problems—Causes and Corrective Actions

1. Resin starvation (bond line appears white, glass weave is visible) occurs because of too few plies of prepreg in the laminate. *Corrective action:* If not obvious, section the laminate or perform burnout in a muffle furnace to determine the amount of prepreg actually used.

2. Resin starvation may occur because prepreg has too low a resin content, too long a gel time, or too high a resin flow. *Corrective action:* Any combination of the conditions may be accommodated by adjusting the heat-up rate by varying the amount of kraft paper or other insulation material between the platen and laminating fixture.

3. Voids may be caused by excessive volatile content. *Corrective action:* Place prepreg in a vacuum chamber for 2 h at 27 in Hg, or perform lamination in a vacuum lamination press.

4. Surface dents and resin on the surface of laminate may be caused by particles of resin left on lamination fixtures, by particles of dust or prepreg collecting on release sheets, or by creases in release sheets. *Corrective action:* Clean lamination fixtures with a plastic scraper and solvent. Perform lay-up in a filtered-air positive-pressure room to reduce the amount of airborne particulates in the lay-up area.

5. Voids and a thick laminate may occur if the press temperature is too low. *Corrective action:* Verify the press temperature during lamination by inserting a thermocouple in the edge of the lay-up. Check the platen temperature profile with thermocouples and data logger.

6. Placement of two or more lamination fixtures side by side that are more than 0.008 in different in thickness can cause "wedging" or circuit shift. *Corrective action:* When placing lamination fixtures side by side, be sure that the "book" in each fixture is made up of the same number of MLB lay-ups of the same part number.

7. Insufficient prepreg plies can cause voids or resin starvation. *Corrective action:* Make certain that at least two plies of 108 glass prepreg are used, adding third or fourth plies if the inner layers are 2-oz copper.

8. Board warp and twist may be caused by nonsymmetrical lay-ups—or designs. It can also be caused by cross-laying the laminate weaves. *Corrective action:* Make sure that design rules mandate that all multilayers containing ground and supply planes be symmetrically placed about the *z* axis of the MLB and that partial planes be avoided. When laying up, make sure that the strong direction (machine direction or warp thread) of the laminate material is laid in the same direction. Require all material to be marked in the strong direction with an arrow by the vendor. Some fabricators insist that the prepreg also be laid with its warp thread in the same direction as the inner-layer laminate.

34.6.2 Resin Flow

Most fabricators rely on the resin flow test as specified in MIL-P-23949 to measure the percent of resin flow. A growing number of manufacturers are using the technique of "scaled flow,"[6] briefly mentioned in the previous chapter, as a better overall method for determining resin flow. With the ratio of lamination pressure to panel area in psi/in^2 (lb/in^2/in^2) higher by a factor of 10 in the standard resin flow test sample than in a typical lamination of a 12- × 18-in panel, neither a true measure of resin flow nor a true measure of pressed thickness may be obtained using the pressure called out in traditional percent of resin flow tests.

34.7 MLB RAW MATERIAL SPECIFICATIONS

34.7.1 Military

MIL-P-13949	Plastic sheet, laminated, metal-clad (for printed wiring boards), general specification

34.7.2 Commercial

IPC-L-108*	Specification for thin laminates, metal-clad primarily for high-temperature multilayer printed boards
IPC-L-109*	Specification for glass cloth, resin-preimpregnated, (B-stage) for high-temperature multilayer printed boards
IPC-L-110†	Preimpregnated, B-stage epoxy-glass cloth for multilayer printed circuit boards

*Under major revision and retitling to include both high-temperature and general-purpose materials.
†To be included in -108, -109 revisions.

IPC-L-130*	Specification for thin laminates, metal-clad, primarily for general-purpose multilayer printed boards

REFERENCES

1. Henry Lee and Kris Neville, *Handbook of Epoxy Resins,* McGraw-Hill Book Company, New York, 1967, Chap. 1.
2. Paul Marx, Bob Forcier, and Dwight Naseth, "An Overview of Vacuum Systems for Multilayer PWB Fabrication," *Electri-Onics,* June 1986.
3. G. Muller, "Quality Improvement by Vacuum Laminating of Complex Multilayer Boards," *Circuit World,* vol. 12, no. 1, 1985.
4. Mike Brody, "Mass Lamination: An Overview," *Printed Circuit Fabrication,* June 1985.
5. W. J. Sachina, P. S. Gill, and K. F. Baker, "Thermal Analysis for the PC Shop," *Circuits Manufacturing,* August 1985.
6. C. J. Bartlett, D. P. Bloechle, and W. A. Mazeika, "Use of Scaled Flow Testing for B-Stage Prepreg," *Proceedings of the IPC 22nd Annual Meeting,* Boston, Mass., April 1979.

*To be included in -108, -109 revisions.

P · A · R · T · 8

FLEXIBLE CIRCUITS

CHAPTER 35

FLEXIBLE CIRCUIT DESIGN, MATERIALS, AND FABRICATION

Steve Gurley

35.1 INTRODUCTION

Flexible circuits are a unique type of interconnection system (see Fig. 35.1 for examples). Although many of the manufacturing processes for flexible circuits are similar to those for rigid boards, it is necessary to modify these processes to take into consideration such special factors as handling of thin films and foils. Specifications relating to adhesive systems are also important, since each type has different processing and electromechanical characteristics which must be tailored to the application. This chapter provides a general overview of the subject. However, for some detailed process description it relies on material discussed in preceding, specialized chapters. By relying on this body of detailed information on the general techniques, it is fairly straightforward to relate to the differences in materials and processes presented here for an understanding of flexible circuits as a separate technology.

The Institute for Interconnecting and Packaging Electronic Circuits (IPC) publishes an industry guideline to technical terms for the printed circuit industry. In the document ANSI/IPC T-50 C, Rev. C., "flexible circuits" are defined as "a patterned arrangement of printed circuits and components utilizing flexible base materials with or without flexible cover layers." This definition may be accurate, but it leaves a lot to be desired relative to the actual use of flexible circuits and the proper choices of materials necessary for each and every application.

To understand flexible circuits, one must be familiar with metal foils, plastic films, and adhesive systems and must understand how all three are selected and used.

35.2 FILMS

One of the primary considerations in designing a flexible circuit is the selection of the proper base and cover film. (The cover film insulates the etched pattern of the circuit much as solder mask or resist does in rigid boards.) Although there are many substrates which could conceivably be used, there are only two kinds of film

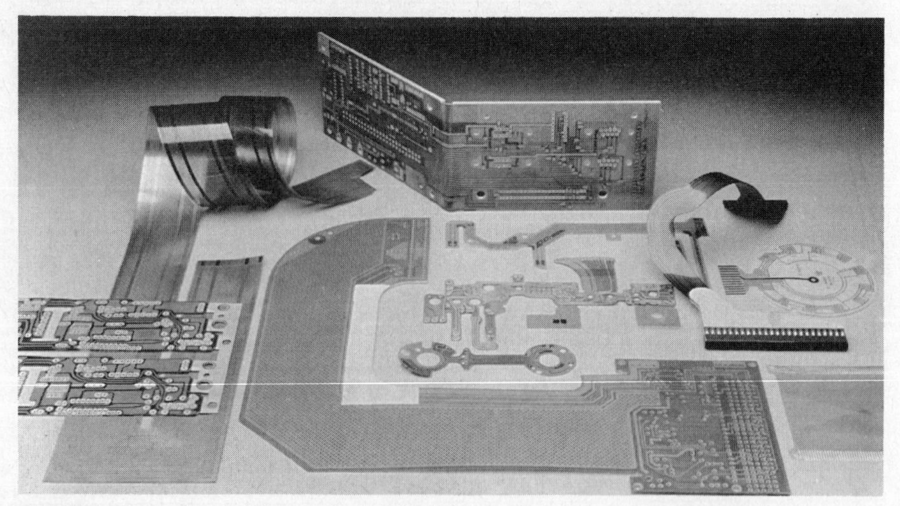

FIG. 35.1 Examples of different kinds of flexible circuits. *(Courtesy of Rogers Corporation, Rogers, Connecticut.)*

in general use for flexible circuits. These films are polyimide and polyester. Films such as Teflon and Ultem are also used, but they appear only in limited and very specialized applications.

35.2.1 Polyimide (Kapton)

The first choice of film in most circuit applications is polyimide film. Polyimide film can withstand the temperatures required in soldering operations. It has no known organic solvent, and it cannot be fused. This film is also used in wire insulation and transformer insulation, and as insulation in motors.

Polyimide films are offered in standard thicknesses of 0.0005, 0.001, 0.002, 0.003, and 0.005 in. Table 35.1 shows the various properties of this film and should be referred to when the operating and environmental requirements of the flexible circuit are established. The trade name for the product used to generate this table is Kapton, which is a trademark of E. I. du Pont de Nemours & Company.

There are some very specialized versions of Kapton film, but these are available for use in applications requiring very special properties. Examples of these materials are designated as follows:

1. *Kapton XT:* Improved thermal conductivity for better heat dissipation as a dielectric insulator, or for higher speeds in thermal-transfer printers
2. *Kapton XC:* Film with conductive fillers providing a range of electrical conductivities for specific uses

While Kapton XT film is readily available, the Kapton XC products were still in experimental use only at the time of this writing, and the manufacturer of the film should be contacted for availability. Kapton XT properties are compared with standard HN film in Table 35.2.

TABLE 35.1 Typical Properties of Kapton Type H Film—25 μm (1 mil)

PHYSICAL	Typical Values 78K (−195°C)	296K (23°C)	473K (200°C)	Test Method
Ultimate (MD) Tensile Strength, MPa (psi)	241 (35,000)	172 (25,000)	117 (17,000)	ASTM D-882-64T
Ultimate (MD) Elongation	2%	70%	90%	ASTM D-882-64T
Tensile Modulus, GPa (MD) (psi)	3.5 (510,000)	3.0 (430,000)	1.86 (260,000)	ASTM D-882-64T
Tear Strength — Propagating (Elmendorf), g	—	8	—	ASTM D-1922-61T
Tear Strength — Initial (Graves), g(g/mil)	—	510 (510)	—	ASTM D-1004-61

MD — Machine Direction

THERMAL	Typical Values	Test Condition	Test Method
Zero Strength Temperature	1088K (815°C)	.14MPa (20 psi) load for 5 seconds	Du Pont Hot Bar Test
Coefficient of Linear Expansion	2.0×10^{-5}m./m./K (2.0×10^{-5}in./in./°C)	259 to 311K (−14°C to 38°C)	ASTM D-696-44
Flammability	94 VTM-O		UL-94 (1-24-80)
Limiting Oxygen Index	100H-38		ASTM D-2863-74

ELECTRICAL	Typical Value	Test Condition	Test Method
Dielectric Strength 25 μm (1 mil)	276 v/μm (7,000 v/mil)	60 hertz 1/4" electrodes	ASTM D-149-61
Dielectric constant 25 μm (1 mil)	3.5	1 kilohertz	ASTM D-150-59T
Dissipation Factor 25 μm (1 mil)	.0025	1 kilohertz	ASTM D-150-59T

CHEMICAL		
Chemical resistance	Excellent (except for strong bases)	
Moisture Absorption 25 μm (1 mil)	1.3% Type H 2.9% Type H & V	50% Relative Humidity at 296K (23°C) Immersion for 24 hours at 296K (23°C)

(Du Pont, Polymer Products Division.)

35.2.2 Polyester (Mylar)

Polyester film represents a good value for use in many flexible circuit applications. Technically, it is known as "polyethelyne terephthalate," which is a polymer formed by the condensation reaction of ethylene glycol and terephthalic acid. One of the largest producers of this film is E. I. du Pont de Nemours & Company, which sells the product under the trade name of Mylar.

Polyester film is low in cost: about $\frac{1}{20}$ the cost of polyimide film. It contains no plasticizers and therefore does not become brittle under normal conditions. It is very resistant to solvents and other chemicals and has a high tensile strength (25,000 psi) and a good dielectric strength (7.5 kV/10^{-3} in for 0.001-in film).

The service temperature for this film ranges from 70 to 150°C. The low temperature resistance of the film is a drawback when the finished circuit must be exposed to soldering temperatures over 230°C, but with careful engineering of the product, even this problem can be circumvented by designing a circuit with a heavy 0.005-in base. It should have large solder pads and wide traces, and 2-oz*

*The expression "2-oz copper foil" refers to the standard technique of defining the thickness of copper laminated to an insulating substrate. The weight of copper per square foot of laminate area describes the thickness: for example, 2 oz/ft^2 equates to a thickness of 0.00275 in, and 1 oz defines a thickness of 0.001375 in.

TABLE 35.2 "Kapton" XT Properties

Description			
"Kapton" polyimide film with alumina incorporated in the polymer to increase thermal conductivity			
Major feature			
Up to twice the heat-dissipating capacity of regular "Kapton" type H film			
Intended uses			
Magnet wire and motor coils Semiconductor insulators Thermal printer transfer substrate Metal/XT composites for circuit substrates Any application where more rapid heat dissipation improves performance			
		Typical values	
Property	Units	200H	200XT
Tenacity, 23°C	kpsi	34	20
Elongation, 23°C	%	90	30
Shrinkage, 400°C	%	1	1
Moisture absorption, 23°C	%	3	5
Dielectric strength, 23°C	V/mil	5900	4000
Volume resistance, 200°C	Ω·cm	10^{14}	10^{14}
Dielectric constant, 23°C	—	3.4	3.4
Dissipation factor, 23°C*	—	0.0025	0.0024
Thermal conductivity, 23°C	W/M·K	0.155	0.24

*At 1 kHz, 50% RH.

Note: "Kapton" type XT is a thermally conductive "Kapton" with up to twice the heat-dissipating capacity of "Kapton" type H film, which is an all-purpose, all-polyimide film that has been used successfully in application temperatures from −269°C to as high as 400°C. The high thermal conductivity of "Kapton" type XT is achieved by incorporating alumina in type H film.

Recent customer development programs have demonstrated the benefits of a heat-conducting, dielectrically insulating, thin film in several applications such as motor wire insulation, semiconductor insulating pads, electronic circuit substrates made from metal/XT composite, and thermal printers.

copper foil should be specified. This will result in a circuit which can be carefully hand-soldered or even wave-soldered using an appropriate mask or jig to keep the heat away from all parts of the circuit except the portions being soldered. Polyester is widely used in automotive and communications circuitry and is most cost-effective in very large applications. See Table 35.3 for typical values of polyester (Mylar) film.

35.2.3 Aramid Material (Nomex)

Although the two films previously discussed are the most common ones in use, there are other base insulation materials which have attractive properties. Some of these include Nomex and Dacron-epoxy. Nomex is Du Pont's random-fiber aramid material. Nomex is a high-temperature paper which withstands soldering temperatures very well. Its main drawback in flexible circuit applications is that it is very hygroscopic and absorbs processing chemicals, which must be carefully removed from one wet process point to the next. Nomex has fairly low initiation

TABLE 35.3 Typical Property Values of Mylar* Polyester Film Type El for Printed Circuits

THERMAL	
Melting Point	Approximately 250°C (480°F) (Fisher Johns Method).
Coefficient of Thermal Expansion	1.7×10^{-5} in/in/°C (30°C-50°C, ASTM D 696-44 modified).
Service Temperature	−70°C to 150°C (−100°F to 300°F). (Soldering of circuits on "Mylar" can be done at solder temperatures up to about 275°C (530°F).
Strain Relief	1.5% (30 minutes at 150°C [300°F]**
PHYSICAL	
Tensile Strength	25,000 psi (25°C, ASTM D 882-64T Method A).
Elongation	120% (25°C, ASTM D 882-64T Method A).
Tensile Modulus	550,000 psi (25°C, ASTM D 882-64T Method A).
Density	1.395 (25°C, ASTM D 1505-63T modified).
CHEMICAL	
Moisture Absorption	Less than 0.8% (ASTM D 570-63; 24 hour immersion at 23°C).
ELECTRICAL	
Dielectric Strength (1-mil)	7500 volts/mil (25°C, 60 Hz, ASTM D 149-64).
Dielectric Strength (1-mil)	5000 volts/mil (150°C, 60 Hz, ASTM D 149-64).
Dielectric Constant	3.0-3.7 (25°C-150°C, 60 Hz-1 MHz, ASTM D 150-65T).
Dissipation Factor	0.005 (25°C, 1 KHz, ASTM D 150-65T).
Volume Resistivity	10^{18} ohm-cm (25°C, ASTM D 257-66). 10^{13} ohm-cm (150°C, ASTM D 257-66).

*Reg. U.S. Pat. Off.
**Typical strain relief for experimental low shrink film is 1.0% at 150°C (300°F) for 30 minutes.

(Du Pont, Industrial Films Division, Polymer Products Department.)

and propagation tear strengths. The material has a fairly low dielectric constant—about half that of Kapton. See Table 35.4 and Fig. 35.2 for typical properties of Nomex film.

35.2.4 Polyester-Epoxy (BEND/flex)

Dacron-epoxy base insulation material is available in 0.005-, 0.0085-, 0.015-, 0.020-, and 0.030-in thicknesses. The material is manufactured using a nonwoven mat of Dacron polyester and glass fibers which is saturated with a B-stage epoxy. This saturated mat is then combined with copper foil and laminated into single- and double-clad material without using a separate, adhesive system. One of the more interesting forms of this material is in a product called BEND/flex, which looks similar to rigid printed circuit materials but which can be bent into three-dimensional shapes without heating. The material will hold a set. This unique bendability is due to a low and broad glass transition temperature of the fully cured material. The general materials properties are shown in Table 35.5.

TABLE 35.4 Typical Properties of Nomex Aramid Paper Type 410

Typical Physical Properties* of NOMEX® Aramid Paper Type 410
(MD = machine direction, XD = cross direction of paper)

Nominal Thickness	mils		**2**	**3**	**5**	**7**	**10**	**12**	**15**	**20**	**24**	**29**	**30**
	(mm)		(0.05)	(0.08)	(0.13)	(0.18)	(0.25)	(0.30)	(0.38)	(0.51)	(0.61)	(0.74)	(0.76)
Basis Weight	oz/yd²		1.2	1.9	3.4	5.1	7.3	9.1	11	16	20	25	25
	(g/m²)		(40)	(63)	(120)	(170)	(250)	(310)	(370)	(540)	(680)	(850)	(850)
Tensile Strength	lb/in	MD	22	40	83	130	180	230	280	350	460	540	510
(ASTM** D-828-60)		XD	11	19	39	64	90	110	150	230	310	370	350
	(N/cm)	MD	(39)	(70)	(150)	(230)	(320)	(400)	(490)	(610)	(810)	(950)	(890)
		XD	(19)	(33)	(68)	(110)	(160)	(190)	(260)	(400)	(540)	(650)	(610)
Elongation	%	MD	9	12	16	20	21	21	21	21	19	17	20
		XD	7	9	13	16	18	18	18	18	14	13	16
Finch Edge Tear	lb	MD	21	41	85	130	180	200	220	210	170	200	270
(ASTM D-827-47)		XD	9	16	33	58	71	86	89	100	79	86	130
	(N)	MD	(93)	(180)	(380)	(580)	(800)	(890)	(980)	(930)	(760)	(890)	(1200)
		XD	(40)	(71)	(150)	(260)	(320)	(380)	(400)	(450)	(350)	(380)	(580)
Elmendorf Tear	g	MD	84	120	220	340	560	690	890	1200	—	—	—
(ASTM D-689)		XD	160	250	580	750	1000	1200	1500	1900	—	—	—
	(N)	MD	(0.8)	(1.2)	(2.2)	(3.3)	(5.5)	(6.8)	(8.7)	(12)	—	—	—
		XD	(1.6)	(2.5)	(5.7)	(7.4)	(9.8)	(12)	(15)	(19)	—	—	—
Shrinkage at 300°C	%	MD	1.7	1.1	0.6	0.5	0.5	0.3	0.3	0.3	0.3	0.3	0.3
		XD	1.3	0.9	0.5	0.4	0.4	0.3	0.3	0.3	0.3	0.3	0.3
Limiting Oxygen Index (ASTM D-2863-70)			0.24 – 0.28										

*Not to be used for product specification purposes.
**American Society for Testing and Materials, Philadelphia, PA

Typical Electrical Properties* of NOMEX® Aramid Paper Type 410
(Measured at 23°C, 50% R.H.)

Nominal Thickness	mils	**2**	**3**	**5**	**7**	**10**	**12**	**15**	**20**	**24**	**29**	**30**
	(mm)	(0.05)	(0.08)	(0.13)	(0.18)	(0.25)	(0.30)	(0.38)	(0.51)	(0.61)	(0.74)	(0.76)
Dielectric Strength AC Rapid Rise (ASTM D-149**)	V/mil	500	610	730	900	880	880	860	780	750	700	660
	(kV/mm)	(20)	(24)	(29)	(35)	(35)	(35)	(34)	(31)	(30)	(28)	(26)
Dielectric Constant (ASTM D-150***)	60 Hz	1.6	1.6	2.4	2.7	2.7	2.9	3.2	3.4	3.7	3.7	3.7
Dissipation Factor (ASTM D-150***)	60 Hz	0.004	0.005	0.006	0.006	0.006	0.007	0.007	0.007	0.007	0.007	0.007

*Not to be used for product specification purposes.
**2-in (51-mm) diameter electrodes
***1-in (25-mm) diameter electrodes under 20 lb/in² (140 kPa) pressure

(Courtesy of E. I. du Pont de Nemours & Company, Wilmington, Delaware.)

35.3 FOILS

Metal foils used to create the circuit patterns are usually made of copper. Copper foil is measured in ounces per square foot (see above), a carryover from the building trades industry. Copper foil is available in weights of ½, 1, and 2 oz, usually desired for most flexible circuit applications. It is also available in heavier weights if desired for high current applications in which a larger cross section of material in the conductors is necessary.

There are two fundamental differences in the kinds of copper foils used in flexible printed circuitry. One product is produced by electrolytic deposition, and the other is produced by cold rolling.

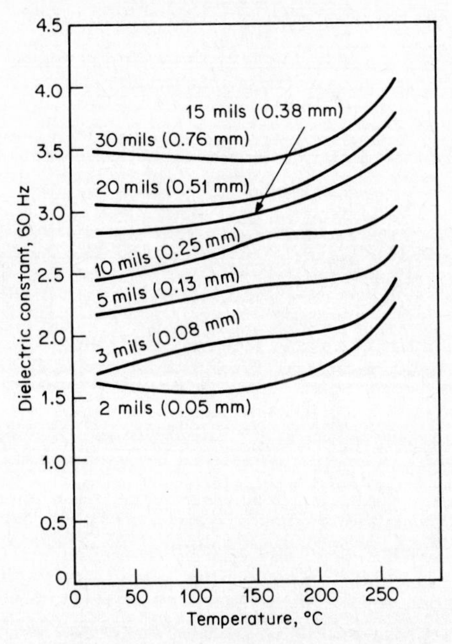

FIG. 35.2 Dielectric constant vs. temperature for Nomex aramid paper type 410. *(Courtesy of E. I. du Pont de Nemours & Company, Wilmington, Delaware.)*

35.3.1 Electrodeposited Copper

Electrodeposited (ED) copper is made by electroplating copper onto a stainless steel drum. The longer this plating action continues, the thicker the copper foil becomes. After the material is coated onto the drum, it is removed in coil form. The "drum side" of the material has a very smooth, shiny finish, whereas the outside, or dull side, of the material has a tooth which provides a very good surface for adhesion to take place with the adhesive. The grain structure of ED foil is very vertical in nature, and although this gives excellent bonds to various film bases, it is less ductile than the rolled, annealed product that is used primarily in dynamic, or moving applications. Typical ED foil properties are shown in Table 35.6.

35.3.2 Rolled Annealed Copper

Rolled annealed copper is manufactured by melting cathode copper, which is produced electrolytically, and then forming this copper into large ingots. This direct chill casting method allows for controlled solidification, which provides continuous purity monitoring and grain size selection and also eliminates existing defects, such as voids which would influence the quality of the foil when it is rolled into its final form.

The copper ingots are large and weigh several tons. They are hot-rolled to an intermediate gauge and then milled on all surfaces to ensure that there are no

TABLE 35.5 BEND/flex 2400 Properties

Construction: 1 ounce copper (307 gr/m²) 1 or 2 sides
Insulation Thickness (ASTM D-374A): .020" (0.15mm)*

	TEST METHOD	UNITS	REQUIRED VALUE
PHYSICAL PROPERTIES			
Dimensional Stability	IPC-TM-650, Section 2.2.4		
On Etching MD		Max, Percent	±0.10
On Etching TD			±0.10
On Heat Aging MD		Max, Percent	±0.15
On Heat Aging TD			±0.15
Peel Strength	IPC-TM-650, Section 2.4.9, Method A	Min, lb_f/in., (KN/m)	8, (1.4)
Initiation Tear Strength	IPC-TM-650, Section 2.4.16		
MD		Min, lb_f, (N)	18, (80)
TD		Min, lb_f, (N)	18, (80)
Propagation Tear Strength	IPC-TM-650, Section 2.4.17.1		
MD		Min, lb_f, (N)	1.2, (5.3)
TD		Min, lb_f, (N)	1.3, (5.3)
ELECTRICAL PROPERTIES			
Insulation Resistance	IPC-TM-650, Section 2.5.9	Min, Megohm	1×10^5
Dielectric Strength	ASTM-D-149	Min, Volts/mil	300
CHEMICAL PROPERTIES			
Flammability	IPC-TM-650, Section 2.3.8	Min, Percent O_2	28
U.L. 94 Flame Class	U.L. 94	—	94V-0
Solder Float	IPC-TM-650, Section 2.4.13, Method A		Pass
ENVIRONMENTAL PROPERTIES			
Moisture Absorption	IPC-TM-650, Section 2.6.2	Max, Percent	1.0

*Required values for .015" and .030" are available on request.

Note

BEND/flex™ material is easily processed using standard hardboard processing techniques and chemicals. It can be easily plated with tin or solder using standard roller coating or hot air leveling processes.

When bending circuits into shape, it must have a minimum inside bend radius of 10 times the material thickness, and must be bent approximately 50% beyond the desired angle of set.

The circuit will then relax to its required angle of bend, or it can be manually positioned with little return force.

The information presented is, to the best of our knowledge, accurate. HOWEVER, values obtained are subject to handling and process variations. Rogers Corporation makes no warranties with respect to such data, and assumes no responsibility for performance characteristics resulting from conditions which differ from those used in laboratory tests. This data should not be used as specification limits.

(© 1985, Rogers Corporation, Rogers, Connecticut.)

defects. After this milling operation, the copper is cold-rolled and annealed to specifications before being processed in a specially designed rolling mill called a Sendzimir mill.

Rolled copper is very flexible and should be used in dynamic applications requiring constant flexing. Figures 35.3 and 35.4 show the rolled copper foil's horizontal grain structure compared with the very vertical grain structure of electrodeposited foil. See Table 35.7 for typical specifications and properties of rolled copper. The horizontal grain allows the rolled copper, after being properly processed into a flexible circuit, to flex hundreds of millions of times with no conductor failures.

There are almost as many kinds of copper alloys as there are birds in the air.

TABLE 35.6 Mechanical Properties of Electrodeposited Copper

CLASS	DESCRIPTION
1	Standard electrodeposited
2	High ductility electrodeposited (flexible applications with minor strain)
3	High temperature elongation electrodeposited (see special products section) (high reliability multilayer applications where severe stresses are imposed)
4	Annealed electrodeposited (high reliability multilayer and flexible applications where severe strains are imposed without high tensile requirements)

Minimum Values

Electrodeposited Copper Class	Copper Weight oz.	At room temperature 23° C			At elevated temperature 180° C		
		Tensile		% Elongation (2.0" GL)	Tensile		% Elongation (2.0" GL)
		lbs/in²	kgs/mm²	CHS 2"/M	lbs/in²	kgs/mm²	CHS 0.05"/M
1	½	15,000	10.55	2.0	Not applicable		
	1	30,000	21.09	3.0			
	2+	30,000	21.09	3.0			
2	½	15,000	10.55	5.0	Not applicable		
	1	30,000	21.09	10.0			
	2+	30,000	21.09	15.0			
3	½	15,000	10.55	2.0	–	–	–
	1	30,000	21.09	15.0	20,000	14.09	2.0
	2+	30,000	21.09	22.0	25,000	17.62	3.0
4	1	20,000	14.09	10.0	15,000	10.55	4.0

Foil Specifications

Thickness by Weight		Thickness by Gauges	
oz/ft²	gr/m²	Nom. Inch	Nom. Millimeter
⅛	44	—	0.005
¼	80	—	0.009
⅜	106	—	0.012
½	153	0.0007	0.018
1	305	0.0013	0.035
2	610	0.0028	0.071
3	915	0.0042	0.106
4	1221	0.0056	0.143
5	1526	0.0070	0.178
6	1831	0.0084	0.213
7	2136	0.0098	0.246
10	3052	0.0140	0.353
14	4272	0.0196	0.492

(Courtesy of Gould, Inc., Eastlake, Ohio.)

FIG. 35.3 Rolled copper grain structure. *(Courtesy of Olin Brass, Olin Corporation, Waterbury, Connecticut.)*

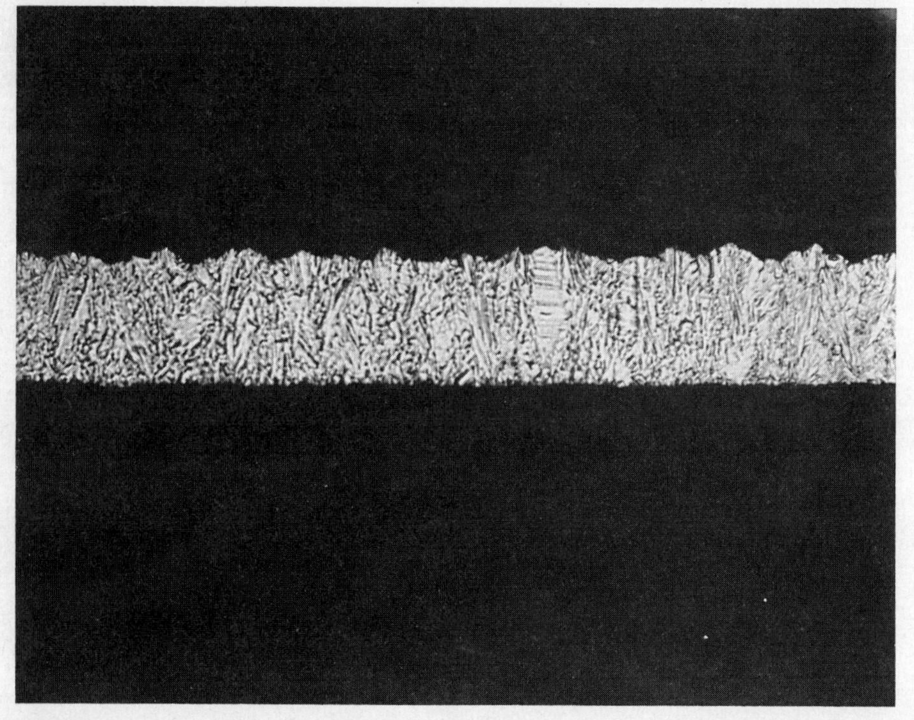

FIG. 35.4 Electrodeposited copper grain structure. *(Courtesy of Gould, Inc., Eastlake, Ohio.)*

TABLE 35.7 Specifications and Physical Properties of Rolled Copper

COMPOSITION LIMITS:	Copper plus Silver 99.9% Min Oxygen .03% Max	
APPLICABLE SPECIFICATIONS:	ASTM B-451 Copper 110 MIL-F-55561 IPC-CF-150	
PHYSICAL PROPERTIES:	English Units	Metric Units
Melting Point	1981° F	1083° C
Density	.322 lbs/in³	8.91 gm/cu cm
Coefficient of Thermal Expansion	.0000098/°F (68-572° F)	.0000177/°C (20-300° C)
Thermal Conductivity (Annealed)	226 Btu-ft/sq ft-hr-°F @ 68° F	.934 cal-cm/sq cm-sec-°C @ 20° C
Electrical Resistivity (Annealed)	10.3 ohm circ mils/ft @ 68° F	1.71 microhm-cm @ 20° C (volume) .152361 ohm-g/m² (weight)
Electrical Conductivity (Annealed)	101% I.A.C.S.① @ 68° F	.586 megohm/cm @ 20° C
Thermal Capacity (Specific Heat)	.092 Btu/lb/°F @ 68° F	.092 cal/gm/°C @ 20° C
Modulus of Elasticity (Tension)	17,000,000 psi	12,000 kg/sq mm
Modulus of Rigidity	6,400,000 psi	4,500 kg/sq mm

① International Annealed Copper Standard.

(Courtesy of Olin Corporation, Sommers Brass Division, Waterbury, Connecticut.)

Flexible circuits, however, use only a few types. Some of the types are designated in hardness values. The temper of rolled copper can vary from $\frac{1}{8}$ hard to $\frac{3}{4}$ hard, (see Table 35.8), most commonly specified. These are also available as rolled and annealed copper (see Table 35.8). There is also a trademarked type available from the Thin Strip Brass Group, Sommers Division of Olin Corporation, called LTA copper. It is an alloy called "110" and is fairly hard when received by the flexible circuit materials laminator. It anneals at low temperatures which are the same as processing temperatures for laminations. This gives the laminator a copper which is easily handled without stretching or wrinkling and yet anneals into a very soft foil after it is combined with the film (see Table 35.9 for typical properties).

One other important feature to specify when using rolled annealed copper is a treatment which enhances the bond. The horizontal grain structure of the rolled copper is fairly smooth. An additional process which is similar to an electrolytic flash of copper on the surface enhances the bond without deteriorating the superior flexing characteristics of the rolled annealed copper.

Finally, Table 35.10 compares wrought annealed copper properties with those of electrodeposited flex-grade copper.

35.4 ADHESIVES

The third component of most flexible circuit materials is the adhesive system. There are many brand names available, but in the final analysis there are three major types of systems: polyesters, epoxies, and acrylics. Each system and the dozens of modifications which exist offer properties suitable for a variety of applications.

TABLE 35.8 Mechanical Property Data for Wrought Copper

ANNEALED AND AS ROLLED TEMPERS

Temper Name	Min. Gauge (Inches)	Max. Width (Inches)	Tensile Strength ksi (Min)	Elongation % in 2" Min.
Annealed	.0007 .0014 .0028 or greater	24 28 28	15 20 25	5 10 20
As Rolled	.0007 .0014 .0028 or greater	24 28 28	50 50 50	1/2 1/2 1

ROLLED TEMPERS

Temper Name	Min. Gauge (Inches)	Max. Width (Inches)	Tensile Strength ksi	Elongation % in 2" Min.
1/8 Hard	.0095	12	32/40	17
1/4 Hard	.0057	12	34/42	13
1/2 Hard	.003	12	37/50	5
3/4 Hard	.002	12	40/55	3

(Courtesy of Olin Brass, Waterbury, Connecticut.)

35.4.1 Epoxy Adhesives

Epoxy systems include modified epoxies known as phenolic butyrals and nitrile phenolics. These systems are widely used and are generally lower in cost than acrylics but higher in cost than polyesters. Epoxy has good high-temperature resistance and remains in good condition in all approved soldering systems. It also has very long-term stability at elevated temperatures in environmental conditions up to 250°F. (See Table 35.11 for typical properties of epoxy adhesives.)

35.4.2 Polyester Adhesives

Polyesters are the lowest-cost adhesives used and the only adhesives which can be used properly with polyester films for base laminate and polyester cover film. The major drawback of this system is low heat resistance, which may not be a drawback at all if the application for the circuit does not require soldering, as in many automotive instrument cluster applications. (See Table 35.12 for typical properties of polyester adhesives.)

35.4.3 Acrylic Adhesives

Acrylic systems are most often used when the completed circuits are used in high-temperature soldering applications. They have the best resistance to short-term, high-temperature exposure. (See Table 35.13 for typical properties of acrylic adhesives.)

TABLE 35.9 Specific Properties of LTA Copper

PROPERTIES AS SUPPLIED
1/2 OR 1 OUNCE

Temper	Tensile Strength (psi)	Elongation % in 2"
As Rolled	50,000 min	1/2 min

PROPERTIES FOLLOWING TYPICAL ADHESIVE CURE CYCLE
FOR 1/2 OUNCE FOIL

Cure Temperature And Time (temp/minutes)	Nominal Tensile Strength (psi)	Nominal Elongation % in 2"	D_f ①
350° F/15	15,000	5	25
350° F/45	15,500	7	32
350° F/180	17,000	7	34

PROPERTIES FOLLOWING TYPICAL ADHESIVE CURE CYCLE
FOR 1 OUNCE FOIL

Cure Temperature And Time (temp/minutes)	Nominal Tensile Strength (psi)	Nominal Elongation % in 2"	D_f ①
350° F/15	21,000	12	52
350° F/45	19,800	11	56
350° F/180	20,600	11	58

① Parallel (MD) & Transverse (XMD) Average

(Courtesy of Olin Corporation, Sommers Brass Division, Waterbury, Connecticut.)

TABLE 35.10 General Comparison of Sommers Wrought Copper with Electrodeposited Flex Grade

	WROUGHT-ANNEALED			ELECTRODEPOSITED FLEX GRADE		
Gauge (Weight)	.0007" (1/2 oz)	.0014" (1 oz)	.0028" (2 oz)	.0007" (1/2 oz)	.0014" (1 oz)	.0028" (2 oz)
Tensile Strength (psi) (NOMINAL)	23,000	25,000	34,100	39,000	40,000	44,000
Yield Strength (psi) 0.2% Offset (NOMINAL)	NA	11,100	13,600	NA	13,700	20,700
Elongation % in 2 Inches (NOMINAL)	5	17	35	3	12	18
Mullens Bulge Height	0.175"	0.292"	0.363"	0.140"	0.197"	0.267"
Electrical Resistivity (Weight: ohms-g/m²)	.152361			<.16359	<.15940	NA

(Courtesy of Olin Corporation, Sommers Brass Division, Waterbury, Connecticut.)

TABLE 35.11 Flexible Adhesive Bonding Films—Epoxy

MATERIAL TYPE: Unsupported Epoxy Adhesive
MATERIAL DESIGNATION: 000L

PROPERTY TO BE TESTED AND TEST METHOD	CLASS 1	CLASS 2	CLASS 3
1. Peel strength, min., lb./in.—width IPC-TM-650, method 2.4.9			
As received .... Method A Method B	6.0*	7.0*	8.0*
After solder float .... Method C Method D	5.0*	6.0*	7.0*
After temp. cycling .... Method E Method F	6.0*	6.0*	8.0*
2. Low temperature flexibility IPC-TM-650, method 2.6.18	N/A	NA	PASS (5 cycles)
3. Flow, max. (squeeze out mil/mil of adhesive) IPC-TM-650, method 2.3.17.1	5	5	5
4. Volatile content (max. percent) IPC-TM-650, method 2.3.37	4.0	3.0	2.0
5. Flammability, min. percent O_2 IPC-TM-650, method 2.3.8	15	18	20
6. Solder float Method A IPC-TM-650, method 2.4.13 Method B	N/A N/A	PASS N/A	N/A PASS
7. Chemical resistance, percentage IPC-TM-650, method 2.3.2 Method A	70	75	80
8. Dielectric constant, max. (at 1 MHz) IPC-TM-650, method 2.5.5.3	N/A	4.0	4.0
9. Dissipation factor, max. (at 1 MHz) IPC-TM-650, method 2.5.5.3	N/A	0.06	0.06
10. Vol. resistivity (damp heat), min. megohms-cm IPC-TM-650, method 2.5.17	N/A	10^6	10^6
11. Surface resistance (damp heat), min., megohms IPC-TM-650, method 2.5.17	N/A	10^3	10^4
12. Dielectric strength, min. volts/mil ASTM-D-149	500	500	500
13. Insulation resistance, min., megohms IPC-TM-650, method 2.6.3.2 at ambient	10^2	10^3	10^4
14. Moisture and insulation resistance, min., megohms IPC-TM-650, method 2.6.3.2	N/A	10^2	10^3
15. Moisture absorption, max., percent IPC-TM-650, method 2.6.2	N/A	4	4
16. Fungus resistance IPC-TM-650, method 2.6.1	N/A	N/A	NON-NUTRIENT

N/A = NOT APPLICABLE *REPRESENTS VALUES FOR PEEL WITH BONDING TO TREATED SIDE OF COPPER. VALUES ARE HALVED WHEN BONDING TO UNTREATED COPPER SURFACES.

(Courtesy of IPC, Lincolnwood, Illinois, IPC-FC-233A/2, February 1986.)

35.5 FLEXIBLE CIRCUIT DESIGN

Flexible circuits are much different from their rigid board cousins in material composition, handling requirements, processing requirements, design rules, and interconnection technology.

TABLE 35.12 Flexible Adhesive Bonding Films—Polyester

MATERIAL TYPE: Unsupported Polyester Adhesive
MATERIAL DESIGNATION: 000N

PROPERTY TO BE TESTED AND TEST METHOD	CLASS 1	CLASS 2	CLASS 3
1. Peel strength, min., lb./in.—width IPC-TM-650, method 2.4.9 As received .... Method A Method B	N/A	3.0	5.0
After solder float .... Method C Method D	N/A	N/A	N/A
After temp. cycling .... Method E Method F	3.0	3.0	5.0
2. Low temperature flexibility IPC-TM-650, method 2.6.18	N/A	NA	PASS (5 cycles)
3. Flow, max. (squeeze out mil/mil of adhesive) IPC-TM-650, method 2.3.17.1	10	10	10
4. Volatile content (max. percent) IPC-TM-650, method 2.3.37	2.0	1.5	1.5
5. Flammability, min. percent O_2 IPC-TM-650, method 2.3.8	15	18	20
6. Solder float IPC-TM-650, method 2.4.13 Method A	N/A	N/A	N/A
Method B	N/A	N/A	N/A
7. Chemical resistance, percentage IPC-TM-650, method 2.3.2 Method A	70*	80*	90*
8. Dielectric constant, max. (at 1 MHz) IPC-TM-650, method 2.5.5.3	N/A	4.6	4.6
9. Dissipation factor, max. (at 1 MHz) IPC-TM-650, method 2.5.5.3	N/A	0.13	0.13
10. Vol. resistivity (damp heat), min. megohms-cm IPC-TM-650, method 2.5.17	N/A	10^6	10^6
11. Surface resistance (damp heat), min., megohms IPC-TM-650, method 2.5.17	N/A	10^4	10^5
12. Dielectric strength, min. volts/mil ASTM-D-149	1000	1000	1000
13. Insulation resistance, min., megohms IPC-TM-650, method 2.6.3.2 at ambient	10^3	10^4	10^5
14. Moisture and insulation resistance, min., megohms IPC-TM-650, method 2.6.3.2	N/A	10^3	10^4
15. Moisture absorption, max., percent IPC-TM-650, method 2.6.2	N/A	2.0	2.0
16. Fungus resistance IPC-TM-650, method 2.6.1	N/A	N/A	NON-NUTRIENT

N/A = NOT APPLICABLE *EXCEPT CHLORINATED SOLVENTS & KETONES

(Courtesy of IPC, Lincolnwood, Illinois, IPC-FC-233A/3, February 1986.)

35.5.1 Basic Rules

Several rules are almost universally true. (1) The material is less dimensionally stable than rigid material, so usually artwork must be developed to allow for material shrinkage during processing. (2) Retrofits almost never work. The design must be started from scratch. (3) All designs must be thought of in terms of a

TABLE 35.13 Flexible Adhesive Bonding Films—Acrylic

MATERIAL TYPE: Unsupported Acrylic Adhesive
MATERIAL DESIGNATION: 000M

PROPERTY TO BE TESTED AND TEST METHOD	CLASS 1	CLASS 2	CLASS 3
1. Peel strength, min., lb./in.—width IPC-TM-650, method 2.4.9 As received .... Method A Method B After solder float .... Method C Method D After temp. cycling .... Method E Method F	 6.0* 5.0* 6.0*	 8.0* 7.0* 7.0*	 8.0* 7.0* 8.0*
2. Low temperature flexibility IPC-TM-650, method 2.6.18	N/A	NA	PASS (5 cycles)
3. Flow, max. (squeeze out mil/mil of adhesive) IPC-TM-650, method 2.3.17.1	5	5	5
4. Volatile content (max. percent) IPC-TM-650, method 2.3.37	4.0	3.0	2.0
5. Flammability, min. percent O_2 IPC-TM-650, method 2.3.8	15	15	15
6. Solder float Method A IPC-TM-650, method 2.4.13 Method B	N/A N/A	PASS N/A	N/A PASS
7. Chemical resistance, percentage IPC-TM-650, method 2.3.2 Method A	70	75	80
8. Dielectric constant, max. (at 1 MHz) IPC-TM-650, method 2.5.5.3	N/A	4.0	4.0
9. Dissipation factor, max. (at 1 MHz) IPC-TM-650, method 2.5.5.3	N/A	0.05	0.05
10. Vol. resistivity (damp heat), min. megohms-cm ** IPC-TM-650, method 2.5.17	N/A	10^6	10^6
11. Surface resistance (damp heat), min., megohms ** IPC-TM-650, method 2.5.17	N/A	10^4	10^5
12. Dielectric strength, min. volts/mil ASTM-D-149	1000	1000	1000
13. Insulation resistance, min., megohms IPC-TM-650, method 2.6.3.2 at ambient	10^2	10^3	10^4
14. Moisture and insulation resistance, min., megohms IPC-TM-650, method 2.6.3.2	N/A	10^2	10^3
15. Moisture absorption, max., percent IPC-TM-650, method 2.6.2	N/A	6.0	6.0
16. Fungus resistance IPC-TM-650, method 2.6.1	N/A	N/A	NON-NUTRIENT

N/A = NOT APPLICABLE
*REPRESENTS VALUES FOR PEEL WITH BONDING TO TREATED SIDE OF COPPER. VALUES ARE HALVED WHEN BONDING TO UNTREATED COPPER SURFACES.
**DATA INDICATED IS FOR 50% RELATIVE HUMIDITY AND 73 ± 2°F. DATA IS BEING DEVELOPED FOR DAMP HEAT.

(Courtesy of IPC, Lincolnwood, Illinois, IPC-FC-233A/1, February 1986.)

three-dimensional form, since the purpose of flexible circuits is to interconnect on multiplanar fields.

35.5.2 Applications

The two primary applications are static and dynamic. In static applications, the circuits are usually flexed once or bent into position, and they remain in that position for the life of the product. Dynamic applications must be specified differently to provide for maximum flex life, while using a combination of materials. Table 35.14 shows the different choices available by application. Figure 35.5 shows a typical static application.

Most dynamic, flexing applications are designed with the copper in a neutral axis, with 1 oz of copper foil encapsulated on both sides with 0.001-in Kapton

TABLE 35.14 Materials and Applications

Application	ED copper	Rolled copper	Polyimide film	Polyester film	Composites
Static	Best	Good	Good	Good	Good
Dynamic	Poor	Best	Best	Good	Poor
Instrument	Best	Good	Best	Good	Good
Automotive	Best	Good	Poor	Best	Poor
Telecommunication	Best	Good	Good	Best	Good

FIG. 35.5 Typical static circuit application. *(Courtesy of Rogers Corporation, Rogers, Connecticut.)*

film. Flexible circuits fabricated in this way have been known to operate for over 500 million cycles with no conductor failures. Loop diameters in this kind of flexing application are always successful with $\frac{1}{2}$-in diameters, and even diameters as small as $\frac{1}{4}$-in have a high level of flex life. Figure 35.6 shows a typical dynamic application.

35.5.3 Solderability Applications

If there is any question as to whether to specify polyester or polyimide in a flex circuit, the answer usually lies with solderability. If either hand soldering or wave soldering is a requirement in an application, then polyimide film must be speci-

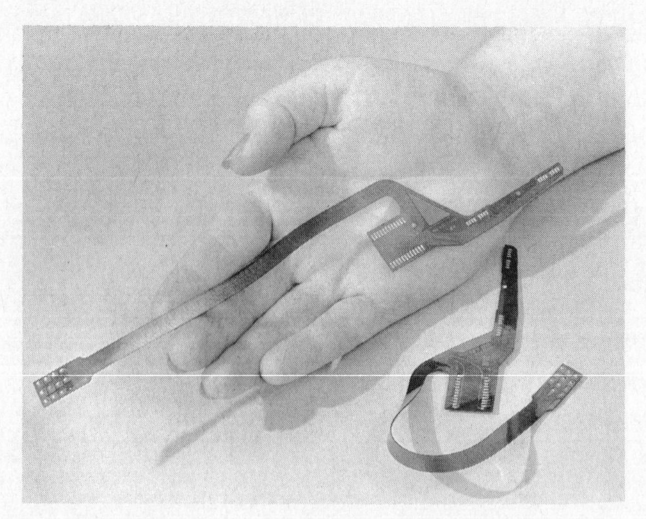

FIG. 35.6 Typical disc drive flexible circuit. *(Courtesy of Rogers Corporation, Rogers, Connecticut.)*

fied to provide a good heat-resistant material which can be processed with few problems, such as delamination or distortion of the plastic film. The use of polyester will save a few cents per circuit in material cost, but the expense reduction does not justify the means except in very high volume applications (several hundred thousand per month).

35.5.4 Design for Processing

The best flexible circuit designs follow some rules which are unique to flexible circuits and not rigid printed circuits. One should always design traces by minimizing spaces. The more copper which is left on the circuit, the more stable it will be during process. It will also be easier to hold dimensional tolerances. All traces which cross a bend should cross at 90° to the bend. Tooling should be hard and sharp and designed so that the outline has no sharp inside corners where a tear could begin. This is especially true for dynamic circuit design. If components are to be mounted on the flexible circuit, then some reinforcement or rigidizing should be added to the circuit. This enforcement is commonly added before final drilling or punching, and the adhesive used is usually the same as or similar to that which is used in the circuit construction. If the circuit is to be bent against the rigid member in assembly, a strain relief, or at least a chamfer, on the rigid member should be employed to reduce the likelihood of trace breakage caused by sharp bending of the circuit against the reinforcing member (see Fig. 35.7).

35.5.5 Plating Considerations

Plating considerations should be made on the basis of a cost/need analysis. It would be simple to gold-plate everything and let it go at that; however, cost is important, so solder plating or tin plating is most commonly used. Roll-to-roll

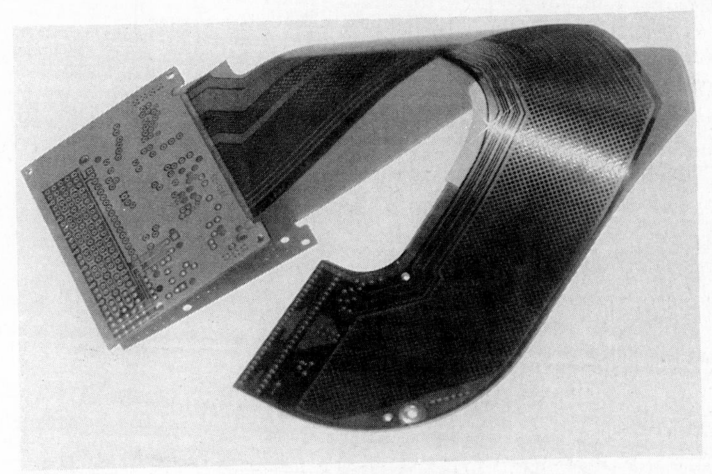

FIG. 35.7 Typical flexible circuit showing a rigidized portion and strain relief techniques. *(Courtesy of Rogers Corporation, Rogers, Connecticut.)*

processing is impractical for all but the highest-volume applications, so panel plating is used in most cases. There are many methods of solder plating, and they vary between electrolytic pattern plating, where the plating becomes an etch resist, to molten roll soldering, which is usually applied to the exposed circuit pads only after a cover film is applied.

35.5.6 Imaging

Imaging a flex circuit laminate with etch resist can be done either with the classical silk screen method or with the more popular photoresist method discussed elsewhere in this book.

35.5.7 Electrical Requirements

The ability of the circuit to do its job in a system depends on the design as well as on good engineering practices with respect to copper foil thickness and trace width.

Cross-sectional analysis is made simpler by using Fig. 35.8. Other electrical characteristics are important and are listed in Table 35.15.

35.5.8 Specifications and Standards

Industry standards which will assist the designer are listed below:

General Specifications

MIL-STD-429	Printed wiring and printed circuits terms and definitions
IPC-T-50	Terms and definitions

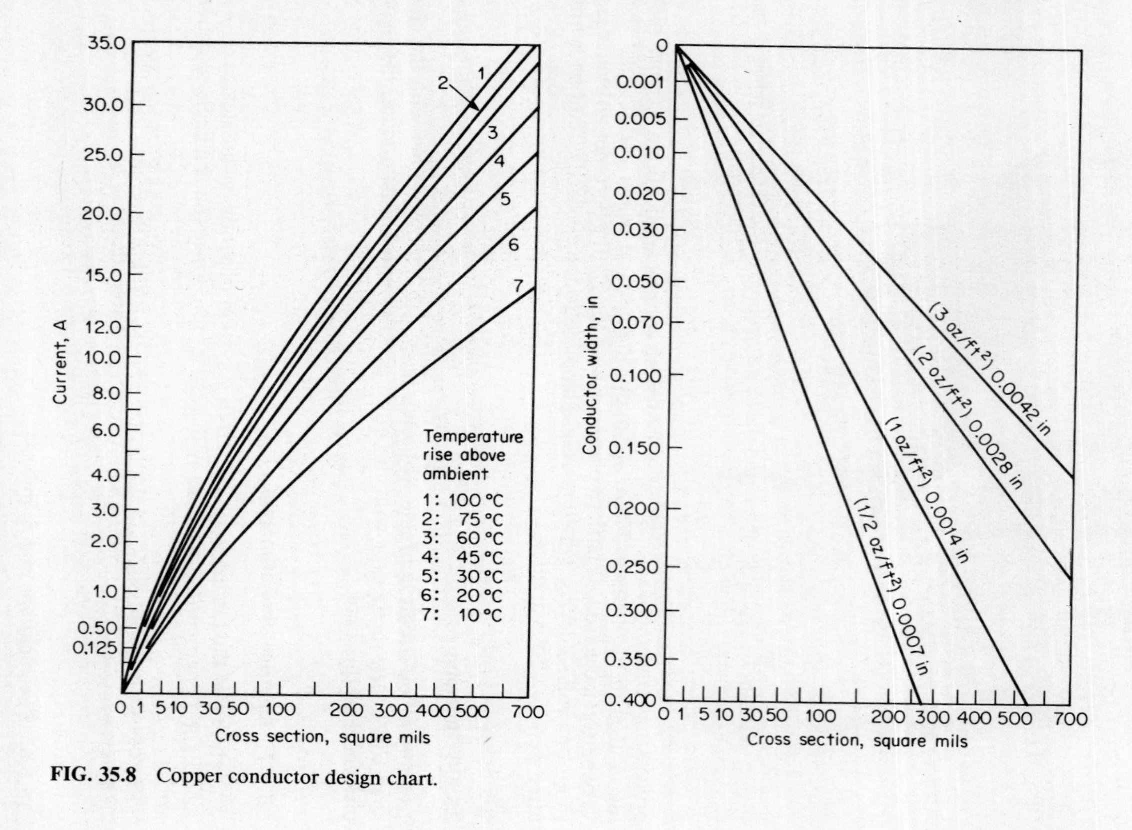

FIG. 35.8 Copper conductor design chart.

TABLE 35.15 Electrical Characteristics of Materials

Material, generic name	Sample thickness, in	Dielectric strength, V	Dielectric constant, 1 kHz	Dissipation factor, 1 kHz
Polyester	0.001	7000	3.2	0.005
Polyimide	0.001	7000	3.5	0.003
Aramid	0.004	600	2.0	0.007
Dacron/epoxy	0.004	3100	3.2	0.015

Material Specifications

MIL-M-55627	Materials for flexible printed wiring
IPC-FC-231	Flexible bare dielectrics for use in flexible printed wiring
IPC-FC-232	Specifications for adhesive-coated dielectrics films for use as cover sheets for flexible printed wiring
IPC-FC-241	Metal-clad flexible dielectrics for use in fabrication of flexible printed wiring

General Circuitry Specifications

MIL-P-55110	Printed wiring boards
MIL-P-50884	Printed wiring, flexible
IPC-FC-240	Specification for single-sided flexible printed wiring
IPC-FC-250	Specification for double-sided flexible wiring with interconnections

Testing Specifications

MIL-STD-202	Test method for electronic and electrical component parts
MIL-STD-454	Standard general requirements for electronic equipment
MIL-STD-810	Environmental test methods

35.6 PRODUCTION PROCESSES

Production processes are detailed in other parts of this book and in other references.* A brief review, however, is in order here to show the differences in processing between rigid printed circuits and flexible circuits.

*See Steve Gurley, *Flexible Circuits, Design and Application,* Marcel Dekker, New York, 1984.

35.6.1 Artwork Master

Imaging is much the same, with the exception that the artwork master must be checked to make sure it includes allowances for manufacturing tolerances present in flexible materials. To be absolutely the best, master art should be on glass, since it allows the least amount of dimensional change for a tightly toleranced circuit.

35.6.2 Tooling

Tooling is one of three basic types: hard-hard, hard-soft, and soft. The hard-hard tool provides the most accurate compound blanking and piercing, with tolerances as close as 0.003 to 0.005 in, depending on the feature. The soft tool, on the other hand, is usually a steel rule die with tolerances up to 5 times looser, or from 0.015 to 0.025 in. The circuit design usually dictates what kind of tooling is needed. The cost has a wide range, from as little as $300 to $500 for steel rule dies to $15,000 to $20,000 for hard-hard steel compound dies.

35.6.3 Indexing

Indexing, or sprocketing material with process guide holes, is used in many flexible circuit processes, especially in the large roll-to-roll production lines (see Fig. 35.9). The object is to have several very accurately punched holes, which will allow the manufacturer to align the tooling for through holes and outlines with the artwork and image placing processes.

FIG. 35.9 Etching flexible circuits in a continuous roll-to-roll etcher. *(Courtesy of Sheldahl Inc., Northfield, Minnesota.)*

35.6.4 Etching

Etching is done in standard spray etchers with a conveyer belt that pulls the circuitry through. Some very lightweight flexible circuitry panels require that they be taped to G-10 leader boards so they can be pulled through the etcher without being lost in the etchant spray turbulence. Many standard etchants are used, ranging from alkaline types to chlorides (cupric and ferric).

35.6.5 Lamination

After the circuit is etched, the resist is removed, and the surface is cleaned in preparation for the top layer of insulation (cover layer).

The next step in actual production is the lamination of cover film to the top of the circuit. The cover film is an adhesive (B-stage) coated film that is usually of the same type and thickness as the base film.

The cover film is prepunched with holes for process and for solder pads. This material is then heat-tacked to the base circuit using pins and the punched holes for alignment. The resulting assembly is then put into a platen press, and the B-stage adhesive is cured under pressure of around 350 psi at a temperature of about 350°F for 1 h or so and cooled under pressure for another 20 min.

The circuit then goes into a cleaning operation prior to solder application. The parts are coated with solder, cleaned again, and then moved to the blank and pierce or drilling area, where holes are put in the exposed pads and, finally, where the outline or cutline of the circuit is created.

35.6.6 Process Flowcharts

Figures 35.10, 35.11, and 35.12 show different flexible circuit process charts.

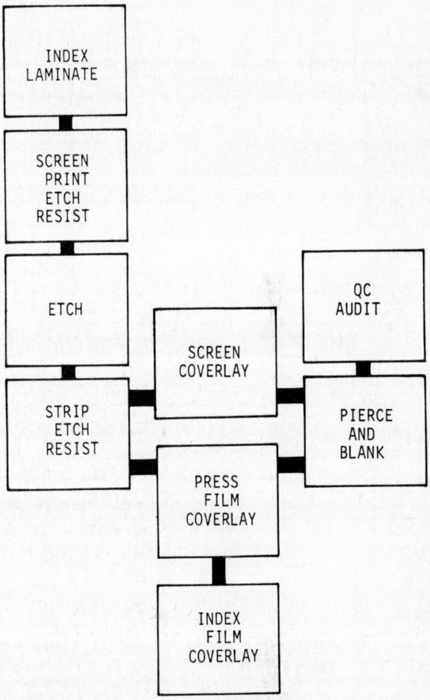

FIG. 35.10 Print and etch with coverlay.

35.7 RIGID-FLEX

Rigid-flex is a combination of rigid printed circuits and flexible printed circuits. The various combinations of layers are all laminated together. After lamination, the through-holes are drilled and then plated so that they connect the various layers together electrically. This kind of interconnect circuitry can be made with as many as 20 layers of material in panel sizes up to 16 × 20 in.

This circuitry is used when cumbersome wiring needs to be replaced with a simpler system to decrease weight, reduce assembly time, and improve reliability. Rigid-flex assemblies eliminate the need for jumpers and mother-daughter board combinations, and they reduce wiring errors while increasing package density.

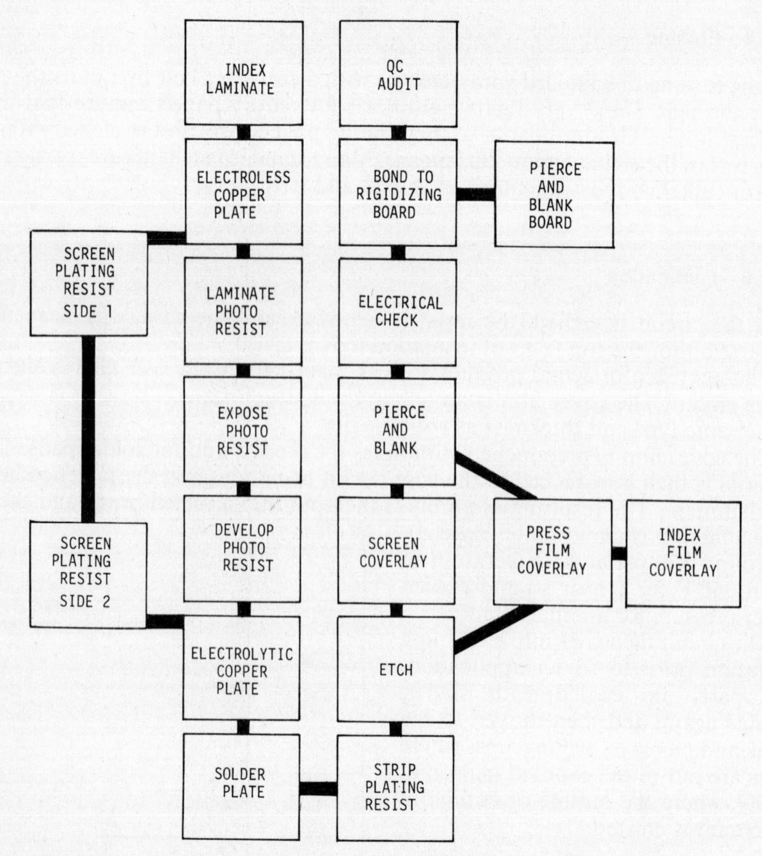

FIG. 35.11 Two-sided pattern-plated through-hole circuit with coverlay.

35.7.1 Process Overview

Figure 35.13 is a process chart showing the "merger" of flexible circuit technology and rigid board technology into "rigid-flex" technology. The two different technologies are brought together at the lamination step of the process.

35.7.2 Materials

The selection of materials is very important and must be done very carefully, since all the component parts must withstand processing steps without damage. As discussed with flexible circuits, dimensional changes occur in the x and y axes of the product, but in rigid-flex, because of the thickness, a third, vertical stress comes into play called "z-axis expansion." This distortion of materials can cause copper hole barrel cracking when the materials are exposed to final lamination and soldering temperatures. Tables 35.16 and 35.17 show some of the most popular material choices for rigid-flex assemblies.

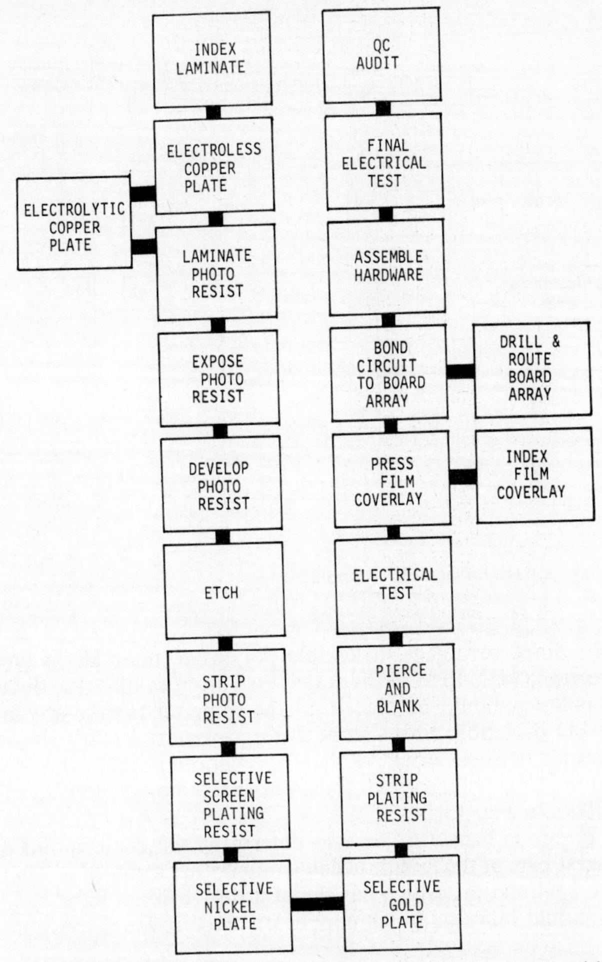

FIG. 35.12 Two-sided panel-plated through-hole circuit with selective nickel-gold plating, film coverlay, rigid board array, and hardware.

35.7.3 Process Concerns

The four most difficult areas of process in rigid-flex products are lamination, drilling to meet the center of all successive pads in all the layers, removing the adhesive deposited into the holes after drilling (smear removal), and the subsequent plating of the holes previously described after the adhesive smear is removed. These problem areas will cause anomalies such as adhesive smear from drilling, delamination, copper embrittlement, plated-through-hole cracking, and misregistration of the various layers. There is a multitude of reasons for these problems. Many can be overcome with careful processing.

Selection of materials is first. Polyimide materials are used in the flexible portions. The polyimide should be as stable as possible; Du Pont's type VN Kapton

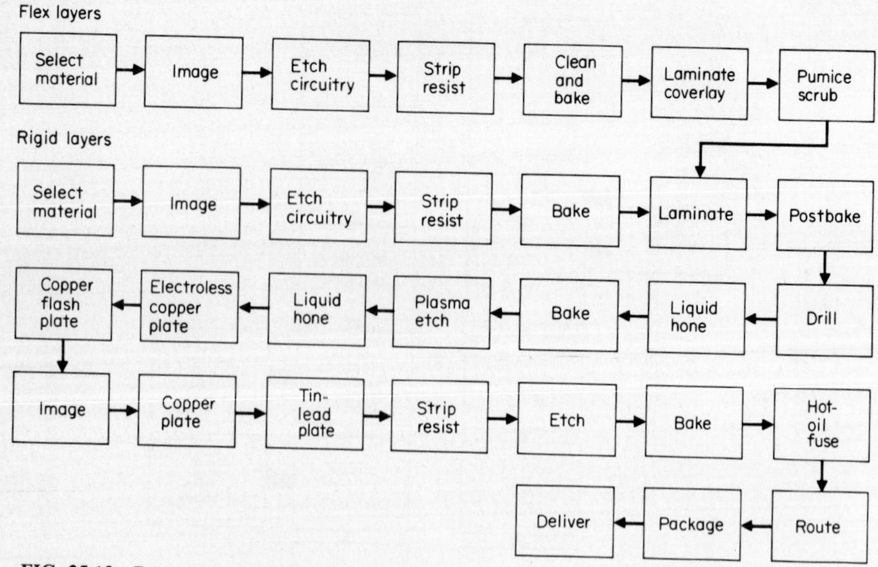

FIG. 35.13 Process steps for rigid-flex manufacturing. *(Courtesy of IPC.)*

is very stable. Since shrinkage of flexible polyimide materials is greater in the machine direction (MD) than in the transverse or cross-machine direction (TD), the artwork pattern should be placed on the material to take advantage of the more stable TD direction. Sheet-manufactured material is usually more stable than roll-manufactured material. All materials should be aligned so that the grain and stress orientation of the copper foils, polyimide films, and cover films are all aligned as shown in Fig. 35.14.

After the design is firm, it is easy to determine the center point of a circuit. Once the longest part of the circuit is determined, that dimension can be oriented in the transverse direction, which has the best dimensional stability. In addition, the designer should leave as much copper on the circuits as possible, including

TABLE 35.16 Rigid-Flex Material Combinations

Flexible section		
Copper-clad	Adhesive system	Cover layer
Polyimide	Modified epoxy	Polyimide
Polyimide	Acrylic	Polyimide

Rigid section	
Copper-clad	B-stage/bond plies
Epoxy-glass	Epoxy-glass B-stage
Epoxy-glass	Modified epoxy adhesive
Epoxy-glass	Acrylic adhesive
Polyimide-glass	Polyimide-glass B-stage
Polyimide-glass	Modified epoxy adhesive
Polyimide-glass	Acrylic adhesive

TABLE 35.17 Rigid-Flex System Combination

Copper-clad rigid material	Copper-clad flexible material	B-stage
Epoxy-glass	Polyimide Modified epoxy	Epoxy-glass
Epoxy-glass	Polyimide-acrylic	Epoxy-glass
Epoxy-glass	Polyimide Modified epoxy	Modified epoxy Cast film
Epoxy-glass	Polyimide-acrylic	Acrylic
Polyimide-glass	Polyimide-acrylic	Acrylic—cast film
Epoxy-glass	Polyimide Modified epoxy	Acrylic
Polyimide-glass	Polyimide Modified epoxy	Acrylic—cast film
Polyimide-glass	Polyimide-acrylic	Acrylic-glass
Polyimide-glass	Polyimide epoxy	Epoxy-glass
Polyimide-glass	Polyimide-acrylic	Modified epoxy Cast film

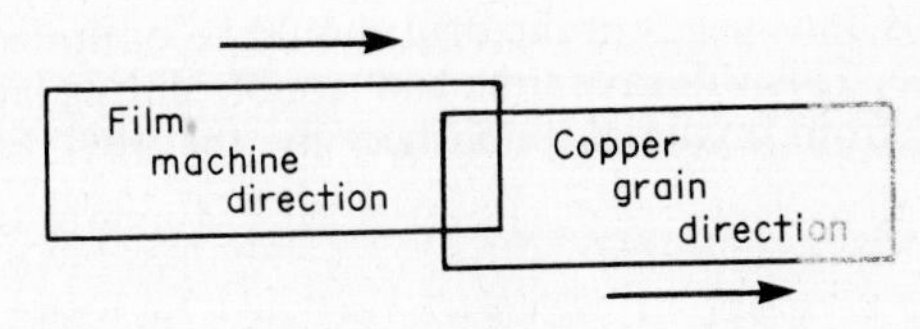

FIG. 35.14 Grain orientations.

the borders. The copper left on the circuit after etching improves stability of the material.

35.7.4 Rigid-Flex Lamination

When the inner-layer material is being prepared, the adhesive thickness should be called out to provide good fill-flow characteristics in the spaces between the copper traces. Usually 1-oz copper is combined with 0.001 in of adhesive, 2 oz with 0.002 in, 3 oz with 0.003 in, etc.

After all the component parts of the rigid-flex have been put together, the job of combining the flexible circuits to the rigid circuits is at hand. This is the most difficult part of the job, since most of the component parts are made of dissimilar materials.

Although many companies use standard platen presses for this job, the trend is toward vacuum lamination. There are as many steps in this final assembly portion of the job as there are in all the other steps combined. There are also a great number of process parameters and specifications for these steps. They differ widely because of the selection and processing characteristics of the materials. Only one type will be outlined here. Suppliers of flexible and rigid materials can give valuable technical assistance when asked.

The steps for one process are as follows (see also Chap. 34, "Laminating Multilayer Circuits"):

1. Bake all materials to remove moisture.

2. Arrange the stackup of layers. (Pins can be used to keep layers aligned.)
3. Sandwich the lay-up between caul plates and laminating materials, which can include Teflon or Tedlar and glass-reinforced Teflon. Silicone rubber and kraft paper are also used.
4. Place in a vacuum press and draw down 28 to 30 in Hg.*
5. Bump the package a few times (add pressure and release several times) to help remove air trapped in the stackup.
6. Put under full pressure of 325 to 350 psi and a temperature of 350 to 375°F for 2 h, with a thermocouple at the load; make sure that the stackup is at temperature for at least 1 h.
7. Cool the load while still under pressure for at least 20 min.
8. Remove the material from the press and immediately put in a postcure oven for 2 h at 250°F. Keep the pin tooling in place with its fixture to keep the package as stable as possible.

35.7.5 Drilling

The next step is drilling the laminated package. Because many different materials will be drilled, feeds and speeds of the drill should be optimized for the cleanest drilled hole possible. This will save time and trouble during later steps.

Table 35.18 is from material published by the IPC Further Rigid-Flex Workshop.

35.7.6 Hole Cleaning

After the drilling process, the holes through the stacked materials must be cleaned so that the subsequent copper-plating process will yield a hole with no cracks or voids in the barrel.

Since most rigid-flex circuits are made with acrylic adhesive systems, the most popular etchback or hole-cleaning process in use is the plasma system. There are

*This applies when vacuum lamination is used.

TABLE 35.18 Recommended Drilling Parameters for Rigid-Flex Printed Wiring Boards

Diameter range, in	Drill size range		Feed rate, in/min	r/min × 10³
	No.	mm		
0.0135–0.0180	80–77	0.35–0.45	200	75
0.0197–0.0240	76–73	0.50–0.60	195	66
0.0250–0.0280	72–70	0.65–0.70	160	54
0.0292–0.0350	69–65	0.75–0.85	135	44
0.0354–0.0400	64–60	0.90–1.00	110	38
0.0410–0.0472	59–56	1.05–1.20	95	33
0.0492–0.0595	55–53	1.25–1.50	75	26
0.0610–0.0736	52–49	1.55–1.85	65	21
0.0748–0.0890	48–43	1.90–2.25	55	17
0.0906–0.1100	42–35	2.30–2.75	45	14
0.1102–0.1250	34–1/8″	2.80–3.10	35	12

other methods in use, such as chromic sulfuric for epoxy-based systems. For the epoxy system, sulfuric acid relieves the excess epoxy resin, while the chromic portion enhances the action of the sulfuric. Fluorosulfonic acid is used to dissolve the exposed glass fibers left in the rigid portion of the stackup. The actual process is more complex than described; information concerning the actual processes can be obtained from Chap. 12 of this book and from the chemical manufacturers.

The plasma etchback system used with acrylic-based materials works with ionized gasses (see Table 35.19). The gasses are generated by a radio-frequency current which ionizes CF_4 (Freon) mixed with oxygen. It does a good job of desmearing the flexible circuit portion of the assembly but will not do anything to glass fibers that might be in the holes of the rigid portion. Proper drill speeds and feeds and *sharp* drills will minimize problems with the cleanliness of the holes. When only a few holes need to be desmeared, a mask–heat sink system is designed to expose the hole only. This can reduce the etchback time as well as keep the rigid-flex cool, since the mask protects the other polyimide surfaces and acts as a heat sink to protect the rest of the assembly during processing.

TABLE 35.19 Typical Plasma Etch Parameters

RF power	1.0–1.5 kW
Pressure	0.4 torr
Gas	70% O_2 and 30% CF_4
Time	12–15 min

After the plasma treatment, an organic residue is left in the holes. This is difficult to remove with ordinary cleaning, but it can be removed with an alkaline cleaner at 140°C for 2 to 3 min in an ultrasonic cleaner.

35.7.7 Plating and Etching

Now that the hole is drilled and cleaned, the holes are copper-plated with electroless copper. Etch-resist placement is next to provide the pattern for the top and/or bottom surfaces. The final surface and hole electrolytic plating is done, and then the panels are solder-plated. The resist is then removed, allowing the solder plating to become the "new resist." Etching then removes the unwanted copper. The solder plating can be fused with hot oil. The panels are then cut (routed) into the final size and shape.

35.7.8 Final Cover Layers

The final cover layers on the top and bottom of the finished rigid-flex assembly are applied. This process is very similar to that in a single-sided flexible circuit operation in that a prepunched or drilled cover-film panel complete with position registration holes is tacked down in a clean environment to the etched circuit. The resulting finished circuit is then laminated so the cover film is bonded and cured to the top and bottom etched circuits.

Some circuit manufacturers prefer not to deal with solder-plated circuitry, especially if it is to have a film or liquid cover coat on it. The problem with this kind of system is that when the final circuit is sent through a solder wave or IR reflow system, the solder on the traces melts and causes tunneling, which is a form of delamination. This means that the entire assembly is not a completely "bonded" homogeneous unit. The alternative approach is to use a positive etch resist, etch the copper, cover-film the circuit, and then solder-plate the pads only, prior to drilling out the final pad configuration.

35.7.9 Multilayer Concerns

The best rule for rigid-flex assemblies is to limit the number of layers as much as possible. The greater the number of individual layers, the more difficult it is to hold tolerances. There are increasing difficulties of dealing with larger coefficients of expansion and then resulting low yields, as layers increase in number. There is an exponentially increasing cost per square inch of circuitry as layers increase in number.

The very best way to start the package design is with a "paper doll" layout of the circuit (see Fig. 35.15). In this way it is possible to anticipate what areas of difficulty there will be in fitting the finished circuit into its proper place in the electronic system chassis. Thicker circuit stackups also have problems forming around bends or corners. If this is a problem in the final design, make sure that a "bookbinder" approach is used in the design.

Design the circuit layers at the *outside* of the bend so they are slightly longer than the circuit layers of the *inside* of the bend. These layers must be kept separate

FIG. 35.15 An eight-layer rigid-flex board showing layers prior to lamination. *(Courtesy of Sheldahl Inc. Certel/Cerpac.)*

and not bonded together unless the designer wants a permanent "bend" laminated into the assembly. Dimensional stability can be improved by leaving a copper pattern similar to a cross-hatch pattern etched into the copper. This will prevent excess shrinkage of the layers after the etching process.

There are many sources for design and processing information for flexible circuits. The best sources are always the producers of flexible circuits and/or the suppliers of process materials and chemicals. Using this information in combination with supplier information and some carefully designed experiments to prove process integrity, it is possible to fabricate dependable flexible and multilayer circuits to meet almost any need.

GLOSSARY

ACCELERATOR: A chemical that is used to speed up a reaction or cure, as cobalt naphthenate is used to accelerate the reaction of certain polyester resins. It is often used along with a catalyst, hardener, or curing agent. The term "accelerator" is often used interchangeably with the term "promoter."

ACCURACY: The ability to place the hole at the targeted location.

ADDITIVE PROCESS: A process for obtaining conductive patterns by the selective deposition of conductive material on an unclad base material.

ADHESIVE: Broadly, any substance used in promoting and maintaining a bond between two materials.

AGING: The change in properties of a material with time under specific conditions.

AMBIENT TEMPERATURE: The temperature of the cooling medium, such as gas or liquid, which comes into contact with the heated parts of an apparatus (or the normal temperature of the surrounding environment).

ANNULAR RING: The circular strip of conductive material that completely surrounds a hole.

ARC RESISTANCE: The time required for an arc to establish a conductive path in a material.

ARTWORK MASTER: An accurately scaled configuration used to produce the production master.

BACKUP MATERIAL: A material placed on the bottom of a laminate stack in which the drill terminates its drilling stroke.

BASE MATERIAL: The insulating material upon which the printed wiring pattern may be formed.

BASE MATERIAL THICKNESS: The thickness of the base material excluding metal foil cladding or material deposited on the surface.

BLISTERING: Localized swelling and separation between any of the layers of the base laminate or between the laminate and the metal cladding.

BONDING LAYER: An adhesive layer used in bonding other discrete layers during lamination.

BOND STRENGTH: The force per unit area required to separate two adjacent layers by a force perpendicular to the board surface; usually refers to the interface between copper and base material.

BOW: A laminate defect in which deviation from planarity results in a smooth arc.

B STAGE: An intermediate stage in the curing of a thermosetting resin. In it a resin can be heated and caused to flow, thereby allowing final curing in the desired shape.

B-STAGE LOT: The product from a single mix of B-stage ingredients.

B-STAGE RESIN: A resin in an intermediate stage of a thermosetting reaction. The material softens when heated and swells when in contact with certain liquids, but it may not entirely fuse or dissolve.

BURR: A ridge left on the outside copper surfaces after drilling.

CAPACITANCE: The property of a system of conductors and dielectrics which permits the storage of electricity when potential difference exists between the conductors.

CAPACITIVE COUPLING: The electrical interaction between two conductors caused by the capacitance between the conductors.

CARBIDE: Tungsten carbide, formula WC. The hard, refractory material forming the drill bits used in PWB drillings.

CATALYST: A chemical that causes or speeds up the cure of a resin but does not become a chemical part of the final product.

CERAMIC LEADED CHIP CARRIER (CLCC): A chip carrier made from ceramic (usually a 90–96% alumina or beryllia base) and with compliant leads for terminations.

CHIP CARRIER (CC): An integrated circuit package, usually square, with a chip cavity in the center; its connections are usually on all four sides. (See *leaded chip carrier* and *leadless chip carrier.*)

CHIP LOAD (CL): The movement of the drill downward per revolution; usually given in mils (thousandths of an inch) per revolution.

CHLORINATED HYDROCARBON: An organic compound having chlorine atoms in its chemical structure. Trichloroethylene, methyl chloroform, and methylene chloride are chlorinated hydrocarbons.

CIRCUIT: The interconnection of a number of electrical devices in one or more closed paths to perform a desired electrical or electronic function.

CLAD: A condition of the base material, to which a relatively thin layer or sheet of metal foil (cladding) has been bonded on one or both of its sides. The result is called a metal-clad base material.

CNC: Computer numerically controlled. Refers to a machine with a computer which stores the numerical information about location, drill size, and machine parameters, regulating the machine to carry out that information.

COAT: To cover with a finishing, protecting, or enclosing layer of any compound.

COLD FLOW: The continuing dimensional change that follows initial instantaneous deformation in a nonrigid material under static load. Also called creep.

COLLIMATION: The degree of parallelism of light rays from a given source. A light source with good collimation produces parallel light rays, whereas a poor light source produces divergent, nonparallel light rays.

COMPONENT HOLE: A hole used for the attachment and electrical connection of a component termination, including pin or wire, to the printed board.

COMPONENT SIDE: The side of the printed board on which most of the components will be mounted.

COMPOUND: A combination of elements in a stable molecular arrangement.

CONDUCTIVE FOIL: The conductive material that covers one side or both sides of the base material and is intended for forming the conductive pattern.

CONDUCTIVE PATTERN: The configuration or design of the electrically conductive material on the base material.

CONDUCTOR LAYER 1: The first layer having a conductive pattern, of a multilayer board, on or adjacent to the component side of the board.

CONDUCTOR SPACING: The distance between adjacent edges (not centerline to centerline) of conductors on a single layer of a printed board.

CONDUCTOR THICKNESS: The thickness of the copper conductor exclusive of coatings or other metals.

CONDUCTOR WIDTH: The width of the conductor viewed from vertically above, i.e., perpendicularly to the printed board.

CONFORMAL COATING: An insulating protective coating which conforms to the configuration of the object coated and is applied on the completed printed board assembly.

CONNECTOR AREA: The portion of the printed board that is used for providing external (input-output) electrical connections.

CONTACT BONDING ADHESIVE: An adhesive (particularly of the nonvulcanizing natural rubber type) that bonds to itself on contact, although solvent evaporation has left it dry to the touch.

COPOLYMER: See *polymer.*

CORE MATERIAL: The fully cured inner-layer segments, with circuiting on one or both sides, that form the multilayer circuit.

CORNER MARKS: The marks at the corners of printed board artwork, the inside edges of which usually locate the borders and establish the contour of the board.

COUPON: One of the patterns of the quality conformance test circuitry area. (See *test coupon.*)

CRAZING: A base material condition in which connected white spots or crosses appear on or below the surface of the base material. They are due to the separation of fibers in the glass cloth and connecting weave intersections.

CROSS-LINKING: The forming of chemical links between reactive atoms in the molecular chain of a plastic. It is cross-linking in the thermosetting resins that makes the resins infusible.

CROSS TALK: Undesirable electrical interference caused by the coupling of energy between signal paths.

CRYSTALLINE MELTING POINT: The temperature at which the crystalline structure in a material is broken down.

CTE: Coefficient of thermal expansion. The measure of the amount a material changes in any axis per degree of temperature change.

CURE: To change the physical properties of a material (usually from a liquid to a solid) by chemical reaction or by the action of heat and catalysts, alone or in combination, with or without pressure.

CURING AGENT: See *hardener.*

CURING TEMPERATURE: The temperature at which a material is subjected to curing.

CURING TIME: In the molding of thermosetting plastics, the time in which the material is properly cured.

CURRENT-CARRYING CAPACITY: Maximum current which can be carried continuously without causing objectionable degradation of electrical or mechanical properties of the printed board.

DATUM REFERENCE: A defined point, line, or plane used to locate the pattern or layer of a printed board for manufacturing and/or inspection purposes.

DEBRIS: A mechanically bonded deposit of copper to substrate hole surfaces.

DEBRIS PACK: Debris deposited in cavities or voids in the resin.

DEFINITION: The fidelity of reproduction of the printed board conductive pattern relative to the production master.

DELAMINATION: A separation between any of the layers of the base laminate or between the laminate and the metal cladding originating from or extending to the edges of a hole or edge of the board.

DIELECTRIC CONSTANT: The property of a dielectric which determines the electrostatic energy stored per unit volume for a unit potential gradient.

DIELECTRIC LOSS: Electric energy transformed into heat in a dielectric subjected to a changing electric field.

DIELECTRIC LOSS ANGLE: The difference between 90° and the dielectric phase angle. Also called the dielectric phase difference.

DIELECTRIC LOSS FACTOR: The product of dielectric constant and the tangent of dielectric loss angle for a material.

DIELECTRIC PHASE ANGLE: The angular difference in phase between the sinusoidal alternating potential difference applied to a dielectric and the component of the resulting alternating current having the same period as the potential difference.

DIELECTRIC POWER FACTOR: The cosine of the dielectric phase angle (or sine of the dielectric loss angle).

DIELECTRIC STRENGTH: The voltage that an insulating material can withstand before breakdown occurs, usually expressed as a voltage gradient (such as volts per mil).

DIMENSIONAL STABILITY: Freedom from distortion by such factors as temperature changes, humidity changes, age, handling, and stress.

DISSIPATION FACTOR: The tangent of the loss angle of the insulating material. Also called loss tangent or approximate power factor.

DRILL FACET: The surface formed by the primary and secondary relief angles of a drill tip.

DRILL WANDER: The sum of accuracy and precision deviations from the targeted location of the hole.

DUMMY: A cathode with a large area used in a low-current-density plating operation for the removal of metallic impurities from solution. The process is called "dummying."

DWELL POINT: The bottom of the drilling stroke before the drill bit ascends.

EDGE-BOARD CONTACTS: A series of contacts printed on or near an edge of a printed board and intended for mating with a one-part edge connector.

EDX: Energy dispersive x-ray fluorescent spectrometer.

ELASTOMER: A material which at room temperature stretches under low stress to at least twice its length but snaps back to its original length upon release of the stress. Rubber is a natural elastomer.

ELECTRIC STRENGTH: The maximum potential gradient that a material can withstand without rupture. It is a function of the thickness of the material and the method and conditions of test. Also called dielectric strength or disruptive gradient.

ELECTROLESS PLATING: The controlled autocatalytic reduction of a metal ion on certain catalytic surfaces.

ELEMENT: A substance composed entirely of atoms of the same atomic number, e.g., aluminum or copper.

EMULSION SIDE: The side of the film or glass on which the photographic image is present.
ENTRY MATERIAL: A material placed on top of a laminate stack.
EPOXY SMEAR: Epoxy resin which has been deposited on edges of copper in holes during drilling either as a uniform coating or as scattered patches. It is undesirable because it can electrically isolate the conductive layers from the plated-through-hole interconnections.
ETCHBACK: The controlled removal of all the components of the base material by a chemical process acting on the sidewalls of plated-through holes to expose additional internal conductor areas.
ETCH FACTOR: The ratio of the depth of etch to lateral etch.
EXOTHERM: A characteristic curve which shows heat of reaction of a resin during cure (temperature) vs. time. The peak exotherm is the maximum temperature on the curve.
EXOTHERMIC REACTION: A chemical reaction in which heat is given off.
FIBER EXPOSURE: A condition in which glass cloth fibers are exposed on machined or abraded areas.
FILLER: A material, usually inert, added to a plastic to reduce cost or modify physical properties.
FILM ADHESIVE: A thin layer of dried adhesive. Also, a class of adhesives provided in dry-film form with or without reinforcing fabric and cured by heat and pressure.
FLEXURAL MODULUS: The ratio, within the elastic limit, of stress to corresponding strain. It is calculated by drawing a tangent to the steepest initial straight-line portion of the load-deformation curve and using the equation $E_B = L^3m/4bd^3$, where E_B is the modulus, L is the span (in inches), m is the slope of the tangent, b is the width of beam tested, and d is the depth of the beam.
FLEXURAL STRENGTH: The strength of a material subjected to bending. It is expressed as the tensile stress of the outermost fibers of a bent test sample at the instant of failure.
FLUOROCARBON: An organic compound having fluorine atoms in its chemical structure, an inclusion that usually lends stability to plastics. Teflon* is a fluorocarbon.
GEL: The soft, rubbery mass that is formed as a thermosetting resin goes from a fluid to an infusible solid. It is an intermediate state in a curing reaction, and a stage in which the resin is mechanically very weak.
GEL POINT: The point at which gelation begins.
GLASS TRANSITION POINT: The temperature at which a material loses properties and becomes a semiliquid.
GLASS TRANSITION TEMPERATURE: The temperature at which epoxy, for example, softens and begins to expand independently of the glass fabric expansion rate.
GLUE-LINE THICKNESS: Thickness of the fully dried adhesive layer.
GRID: An orthogonal network of two sets of parallel lines for positioning features on a printed board.
GROUND PLANE: A conducting surface used as a common reference point for circuit returns, shielding, or heat sinking.
GULL WING LEAD: A surface mounted device lead which flares outward from the device body.
HALOING: A light area around holes or other machined areas on or below the surface of the base laminate.
HARDENER: A chemical added to a thermosetting resin for the purpose of causing curing or hardening. A hardener, such as an amine or acid anhydride for an epoxy resin, is a part of the chemical reaction and a part of the chemical composition of the cured resin. The terms "hardener" and "curing agent" are used interchangeably.
HEAT-DISTORTION POINT: The temperature at which a standard test bar (ASTM D 648) deflects 0.010 in under a stated load of either 66 or 264 psi.
HEAT SEALING: A method of joining plastic films by simultaneous application of heat and pressure to areas in contact. The heat may be supplied conductively or dielectrically.
HOLE PULL STRENGTH: The force, in pounds, necessary to rupture a plated-through hole or its surface terminal pads when loaded or pulled in the direction of the axis of the hole. The pull is usually applied to a wire soldered in the hole, and the rate of pull is given in inches per minute.
HOOK: A geometric drill bit defect of the cutting edges.
HOT-MELT ADHESIVE: A thermoplastic adhesive compound, usually solid at room temperature, which is heated to fluid state for application.

*Trademark of E. I. du Pont de Nemours & Company.

HYDROCARBON: An organic compound containing only carbon and hydrogen atoms in its chemical structure.

HYDROLYSIS: The chemical decomposition of a substance involving the addition of water.

HYGROSCOPIC: Tending to absorb moisture.

I-LEAD: A surface mounted device lead which is formed such that the end of the lead contacts the board land pattern at a 90° angle. Also called a butt joint.

IMPREGNATE: To force resin into every interstice of a part, as of a cloth for laminating.

INHIBITOR: A chemical that is added to a resin to slow down the curing reaction and is normally added to prolong the storage life of a thermosetting resin.

INORGANIC CHEMICALS: Chemicals whose molecular structures are based on other than carbon atoms.

INSULATION RESISTANCE: The electrical resistance of the insulating material between any pair of contacts, conductors, or grounding devices in various combinations.

INTERNAL LAYER: A conductive pattern contained entirely within a multilayer board.

IPC: Institute for Interconnecting and Packaging Electronic Circuits. A leading printed wiring industry association that develops and distributes standards, as well as other information of value to printed wiring designers, users, suppliers, and fabricators.

IR: Infrared heating for solder-reflow operation.

J-LEAD: A surface mounted device lead which is formed into a "J" pattern folding under the device body.

JUMPER: An electrical connection between two points on a printed board added after the printed wiring is fabricated.

LAMINATE: The plastic material, usually reinforced by glass or paper, that supports the copper cladding from which circuit traces are created.

LAMINATE VOID: Absence of epoxy resin in any cross-sectional area which should normally contain epoxy resin.

LAND: See *terminal area.*

LANDLESS HOLE: A plated-through hole without a terminal area.

LASER PHOTOPLOTTER (laser photogenerator, or LPG): A device that exposes photosensitive material, usually a silver halide or diazo material, subsequently used as the master for creating the circuit image in production.

LAYBACK: A geometric drill bit defect of the cutting edges.

LAYER-TO-LAYER SPACING: The thickness of dielectric material between adjacent layers of conductive circuitry.

LAY-UP: The process of registering and stacking layers of a multilayer board in preparation for the laminating cycle.

LCCC: Leadless ceramic chip carrier.

LEADED CHIP CARRIER: A chip carrier (either plastic or ceramic) with compliant leads for terminations.

LEADLESS CHIP CARRIER: A chip carrier (usually ceramic) with integral metallized terminations and no compliant external leads.

LEGEND: A format of lettering or symbols on the printed board, e.g., part number, component locations, or patterns.

LOOSE FIBERS: Supporting fibers in the substrate of the laminate which are not held in place by surrounding resin.

MAJOR WEAVE DIRECTION: The continuous-length direction of a roll of woven glass fabric.

MARGIN RELIEF: The area of a drill bit next to the cutting edge is removed so that it does not rub against the hole as the drill revolves.

MASTER DRAWING: A document that shows the dimensional limits or grid locations applicable to any or all parts of a printed wiring or printed circuit base. It includes the arrangement of conductive or nonconductive patterns or elements; size, type, and location of holes; and any other information necessary to characterize the complete fabricated product.

MEASLING: Discrete white spots or crosses below the surface of the base laminate that reflect a separation of fibers in the glass cloth at the weave intersection.

MICROSTRIP: A type of transmission line configuration which consists of a conductor over a parallel ground plane separated by a dielectric.

MINOR WEAVE DIRECTION: The width direction of a roll of woven glass fabric.

MIXED ASSEMBLY: A printed wiring assembly that combines through-hole components and surface mounted components on the same board.

MODULUS OF ELASTICITY: The ratio of stress to strain in a material that is elastically deformed.

MOISTURE RESISTANCE: The ability of a material not to absorb moisture either from air or when immersed in water.
MOUNTING HOLE: A hole used for the mechanical mounting of a printed board or for the mechanical attachment of components to a printed board.
MULTILAYER BOARD: A product consisting of layers of electrical conductors separated from each other by insulating supports and fabricated into a solid mass. Interlayer connections are used to establish continuity between various conductor patterns.
MULTIPLE-IMAGE PRODUCTION MASTER: A production master used to produce two or more products simultaneously.
NAILHEADING: A flared condition of internal conductors.
NC: Numerically controlled. Usually refers to a machine tool, in this case a drilling machine. The most basic type is one in which a mechanical guide locates the positions of the holes. NC machines are usually controlled by punched tape.
NEMA STANDARDS: Property values adopted as standard by the National Electrical Manufacturers Association.
NOBLE ELEMENTS: Elements that either do not oxidize or oxidize with difficulty; examples are gold and platinum.
OILCANNING: The movement of entry material in the *z* direction during drilling in concert with the movement of the pressure foot.
ORGANIC: Composed of matter originating in plant or animal life or composed of chemicals of hydrocarbon origin, either natural or synthetic.
PAD: See *terminal area.*
PADS ONLY: A multilayer construction with all circuit traces on inner layers and the component terminal area only on the surface of the board. This construction adds two layers but may avoid the need for a subsequent solder resist, and since inner layers usually are easier to form, this construction may lead to higher overall yields.
PH: A measure of the acid or alkaline condition of a solution. A pH of 7 is neutral (distilled water); pH values below 7 represent increasing acidity as they go toward 0; and pH values above 7 represent increasing alkalinity as they go toward the maximum value of 14.
PHOTOGRAPHIC REDUCTION DIMENSION: The dimension (e.g., line or distance between two specified points) on the artwork master to indicate the extent to which the artwork master is to be photographically reduced. The value of the dimension refers to the 1:1 scale and must be specified.
PHOTOMASTER: An accurately scaled copy of the artwork master used in the photofabrication cycle to facilitate photoprocessing steps.
PHOTOPOLYMER: A polymer that changes characteristics when exposed to light of a given frequency.
PINHOLES: Small imperfections which penetrate entirely through the conductor.
PITS: Small imperfections which do not penetrate entirely through the printed circuit.
PLASTICIZER: Material added to resins to make them softer and more flexible when cured.
PLASTIC LEADED CHIP CARRIER (PLCC): A chip carrier packaged in plastic, usually terminating in compliant leads (originally "J" style) on all four sides.
PLATED-THROUGH HOLE: A hole in which electrical connection is made between printed wiring board layers with conductive patterns by the deposition of metal on the wall of the hole. (See *PTH.*)
PLATING VOID: The area of absence of a specific metal from a specific cross-sectional area: (1) When the plated-through hole is viewed as cross-sectioned through the vertical plane, it is a product of the average thickness of the plated metal times the thickness of the board itself as measured from the outermost surfaces of the base copper on external layers. (2) When the plated-through hole is viewed as cross-sectioned through the horizontal plane (annular method), it is the difference between the area of the hole and the area of the outside diameter of the through-hole plating.
PLOWING: Furrows in the hole wall due to drilling.
POLYMER: A high-molecular-weight compound made up of repeated small chemical units. For practical purposes, a polymer is a plastic. The small chemical unit is called a mer, and when the polymer or mer is cross-linked between different chemical units (e.g., styrene-polyester), the polymer is called a copolymer. A monomer is any single chemical from which the mer or polymer or copolymer is formed.
POLYMERIZE: To unite chemically two or more monomers or polymers of the same kind to form a molecule with higher molecular weight.

POTLIFE: The time during which a liquid resin remains workable as a liquid after catalysts, curing agents, promoters, etc., are added. It is roughly equivalent to gel time.

POWER FACTOR: The cosine of the angle between the applied voltage and the resulting current.

PRECISION: The ability to repeatedly place the hole at any location.

PREPRODUCTION TEST BOARD: A test board (as detailed in IPC-ML-950) the purpose of which is to determine whether, prior to the production of finished boards, the contractor has the capability of producing a multilayer board satisfactorily.

PRESS PLATEN: The flat heated surface of the lamination press used to transmit heat and pressure to lamination fixtures and into the lay-up.

PRESSURE FOOT: The tubelike device on the drilling machine that descends to the top surface of the stack, holding it firmly down, before the drill descends through the center of the pressure foot. The vacuum system of the drilling machine separates through the pressure foot to remove chips and dust formed in drilling.

PRINTED WIRING ASSEMBLY DRAWING: A document that shows the printed wiring base, the separately manufactured components which are to be added to the base, and any other information necessary to describe the joining of the parts to perform a specific function.

PRINTED WIRING LAYOUT: A sketch that depicts the printed wiring substrate, the physical size and location of electronic and mechanical components, and the routing of conductors that interconnect the electronic parts in sufficient detail to allow for the preparation of documentation and artowrk.

PRODUCTION MASTER: A 1:1 scale pattern used to produce one or more printed wiring or printed circuit products within the accuracy specified on the master drawing.

PROMOTER: A chemical, itself a feeble catalyst, that greatly increases the activity of a given catalyst.

PTH: Plated-through hole. Also refers to the technology that uses the plated-through hole as its foundation.

QUADPACK: Generic term for surface mount technology packages with leads on all four sides. Commonly used to describe chip carrier–like devices with gull wing leads.

QUALITY CONFORMANCE CIRCUITRY AREA: A test board made as an integral part of the multilayer printed board panel on which electrical and environmental tests may be made for evaluation without destroying the basic board.

RAW MATERIAL PANEL SIZE: A standard panel size related to machine capacities, raw material sheet sizes, final product size, and other factors.

REFRACTIVE INDEX: The ratio of the velocity of light in a vacuum to the velocity in a substance. Also, the ratio of the sine of the angle of incidence to the sine of the angle of refraction.

REGISTER MARK: A mark used to establish the relative position of one or more printed wiring patterns, or portions thereof, with respect to desired locations on the opposite side of the board.

REGISTRATION: The relative position of one or more printed wiring patterns, or portions thereof, with respect to desired locations on a printed wiring base or to another pattern on the opposite side of the base.

RELATIVE HUMIDITY: The ratio of the quantity of water vapor present in the air to the quantity which would saturate the air at the given temperature.

REPAIR: The correction of a printed wiring defect after the completion of board fabrication to render the board as functionally good as a perfect board.

RESIN: High-molecular-weight organic material with no sharp melting point. For current purposes, the terms "resin," "polymer," and "plastic" can be used interchangeably.

RESIST: A protective coating (ink, paint, metallic plating, etc.) used to shield desired portions of the printed conductive pattern from the action of etchant, solder, or plating.

RESISTIVITY: The ability of a material to resist passage of electric current through its bulk or on a surface.

RIFLING: Spiral groove or ridge in the substrate due to drilling.

ROCKWELL HARDNESS NUMBER: A number derived from the net increase in depth of an impression as the load on a penetrator is increased from a fixed minimum load to a higher load and then returned to minimum load.

ROUGHNESS: Irregular, coarse, uneven hole wall on copper or substrate due to drilling.

SCHEMATIC DIAGRAM: A drawing which shows, by means of graphic symbols, the electrical interconnections and functions of a specific circuit arrangement.

SEM: Scanning electron microscope.

SHADOWING: Etchback to maximum limit without removal of dielectric material from conductors.

SHORE HARDNESS: A procedure for determining the indentation hardness of a material by means of a durometer.

SINGLE-IMAGE PRODUCTION MASTER: A production master used to produce individual products.

SINGLE-IN-LINE PACKAGE (SIP): Component package system with one line of connectors, usually spaced 0.100 in apart.

SMC: Surface mounted component. Component with terminations designed for mounting flush to printed wiring board.

SMD: Surface mounted device. Any component or hardware element designed to be mounted to a printed wiring board without penetrating the board.

SMEAR: Fused deposit left on copper or substrate from excessive drilling heat.

SMOBC: Solder mask over bare copper. A method of fabricating a printed wiring board which results in the final metallization being copper with no other protective metal; but the non-soldered areas are coated by a solder resist, exposing only the component terminal areas. This eliminates tin-lead under the solder mask.

SMT: Surface mount technology. Defines the entire body of processes and components which create printed wiring assemblies without components with leads that pierce the board.

SOIC: Small-outline integrated circuit. A plastic package resembling a small dual-in-line package (DIP) with gull wing leads on two sides for surface mounting.

SOJ: SOIC package with J-leads rather than gull wing leads.

SOT: Small outline transistor. A package for surface-mounting transistors.

SPECIFIC HEAT: The ratio of the thermal capacity of a material to that of water at 15°C.

SPINDLE RUNOUT: The measure of the wobble present as the drilling machine spindle rotates 360°.

STORAGE LIFE: The period of time during which a liquid resin or adhesive can be stored and remain suitable for use. Also called shelf life.

STRAIN: The deformation resulting from a stress. It is measured by the ratio of the change to the total value of the dimension in which the change occurred.

STRESS: The force producing or tending to produce deformation in a body. It is measured by the force applied per unit area.

SUBSTRATE: A material on whose surface an adhesive substance is spread for bonding or coating. Also, any material which provides a supporting surface for other materials used to support printed wiring patterns.

SURFACE RESISTIVITY: The resistance of a material between two opposite sides of a unit square of its surface. It may vary widely with the conditions of measurement.

SURFACE SPEED: The linear velocity of a point on the circumference of a drill. Given in units of surface feet per minute—sfm.

TERMINAL AREA: A portion of a conductive pattern usually, but not exclusively, used for the connection and/or attachment of components.

TEST COUPON: A sample or test pattern usually made as an integral part of the printed board, on which electrical, environmental, and microsectioning tests may be made to evaluate board design or process control without destroying the basic board.

TETRA-ETCH*: A nonpyrophoric (will not ignite when exposed to moisture) proprietary etchant.

TETROFUNCTIONAL: Describes an epoxy system for laminates that has four cross-linked bonds rather than two and results in a higher glass transition temperature, or T_g.

T_g: Glass transition temperature. The temperature at which laminate mechanical properties change significantly.

THERMAL CONDUCTIVITY: The ability of a material to conduct heat; the physical constant for the quantity of heat that passes through a unit cube of a material in a unit of time when the difference in temperatures of two faces is 1°C.

THERMOPLASTIC: A classification of resin that can be readily softened and resoftened by repeated heating.

THERMOSETTING: A classification of resin which cures by chemical reaction when heated and, when cured, cannot be resoftened by heating.

*Trademark of W. L. Gore and Associates, Inc.

THIEF: An auxiliary cathode so placed as to divert to itself some current from portions of the work which would otherwise receive too high a current density.

THIXOTROPIC: Said of materials that are gel-like at rest but fluid when agitated.

THROUGH-HOLE TECHNOLOGY: Traditional printed wiring fabrication where components are mounted in holes that pierce the board.

THROWING POWER: The improvement of the coating (usually metal) distribution ratio over the primary current distribution ratio on an electrode (usually a cathode). Of a solution, a measure of the degree of uniformity with which metal is deposited on an irregularly shaped cathode. The term may also be used for anodic processes for which the definition is analogous.

TWIST: A laminate defect in which deviation from planarity results in a twisted arc.

UNDERCUT: The reduction of the cross section of a metal foil conductor caused by the etchant removing metal from under the edge of the resist.

VAPOR PHASE: The solder-reflow process that uses a vaporized solvent as the source for heating the solder beyond its melting point, creating the component-to-board solder joint.

VIA: A metallized connecting hole that provides a conductive path from one layer in a printed wiring board to another. (1) *Buried via*—connects one inner layer to another inner layer without penetrating the surface. (2) *Blind via*—connects the surface layer of a printed wiring board to an internal layer without going all the way through the other surface layer.

VOID: A cavity left in the substrate.

VOLUME RESISTIVITY: The electrical resistance between opposite faces of a 1-cm cube of insulating material, commonly expressed in ohm-centimeters. The recommended test is ASTM D 257 51T. Also called the specific insulation.

VULCANIZATION: A chemical reaction in which the physical properties of an elastomer are changed by causing the elastomer to react with sulfur or some other cross-linking agent.

WATER ABSORPTION: The ratio of the weight of water absorbed by a material to the weight of the dry material.

WEAVE EXPOSURE: A condition in which the unbroken woven glass cloth is not uniformly covered by resin.

WEAVE TEXTURE: A surface condition in which the unbroken fibers are completely covered with resin but exhibit the definite weave pattern of the glass cloth.

WETTING: Ability to adhere to a surface immediately upon contact.

WICKING: Migration of copper salts into the glass fibers of the insulating material.

WORKING LIFE: The period of time during which a liquid resin or adhesive, after mixing with catalyst, solvent, or other compounding ingredients, remains usable. (See *potlife.*)

INDEX

ABOUT THE EDITOR

Clyde F. Coombs, Jr., is currently marketing manager for the Printed Circuit Board Division of Hewlett-Packard Company. This division is responsible for the supply of printed circuits for all of Hewlett-Packard's internal needs. He developed and installed the company's original production processes for printed circuits. This experience led to the organization and publication of the first edition of *Printed Circuits Handbook* in 1967 and the two subsequent editions.

In addition, Mr. Coombs developed and managed the Hewlett-Packard core memory operation and served as manager of corporate process engineering. He has also served as a functional manager in manufacturing, quality assurance, and marketing.

He was the organizer and six-term president of the California Circuits Association, an influential industry association that has grown with the electronics business in that state.

Mr. Coombs was named Production Engineer of the Year for 1967 by *Electronic Packaging and Production* magazine, and he has been a frequent contributor to publications on the subject of printed circuits.